Photosynthetic Picoplankton

Edited by
Trevor Platt and William K. W. Li

Department of Fisheries and Oceans
Marine Ecology Laboratory
Bedford Institute of Oceanography
Dartmouth, Nova Scotia B2Y 4A2

DEPARTMENT OF FISHERIES AND OCEANS
Ottawa 1986

The *Canadian Bulletins of Fisheries and Aquatic Sciences* are designed to interpret current knowledge in scientific fields pertinent to Canadian fisheries and aquatic environments.

The *Canadian Journal of Fisheries and Aquatic Sciences* is published in annual volumes of monthly issues. *Canadian Special Publications of Fisheries and Aquatic Sciences* are issued periodically. These series are available from authorized bookstore agents and other bookstores, or your may send your prepaid order to the Canadian Government Publishing Centre, Supply and Services Canada, Ottawa, Ont. K1A OS9. Make cheques or money orders payable in Canadian funds to the Receiver General for Canada.

Editorial Office: Department of Fisheries and Oceans
Communications Directorate
Information and Publications Branch
200 Kent Street
Ottawa, Ontario, Canada K1A 0E6

Director General: Dixi Lambert
Director: Johanna Reinhart, M.Sc.
Editorial and Publishing Services: Gerald J. Neville
Typesetter: Aubut & Associates Ltd., Ottawa, Ontario
Printer: Friesen Printers, Altona, Manitoba
Cover Design: André, Gordon and Laundreth Inc., Ottawa, Ontario

Correct citation for this publication:

PLATT, T., AND W. K. W. LI. [ED.] 1986. Photosynthetic picoplankton. Can. Bull. Fish. Aquat. Sci. 214: 583 p.

Contents

Abstract

PLATT, T., AND W. K. W. LI. [ED.] 1986. Photosynthetic picoplankton. Can. Bull. Fish. Aquat. Sci. 214: 583 p.

This book is a collection of 16 essays with the aim of summarising the present state of knowledge about photosynthetic picoplankton in the ocean. Both prokaryotic and eukaryotic picoplankton fall within its scope. It is based on the lectures given in an Advanced Study Institute (San Miniato, Italy, October 1985) at which both field (oceanography) and laboratory (biochemical, physiological) specialists pooled their knowledge on the photosynthetic picoplankton. The general subject areas dealt with include: physiological and ecological limitations of extreme smallness in cell size; the distribution, abundance and taxonomic status of prokaryotic and eukaryotic picoplankton; the metabolism of cyanobacteria; the relationship between nutrition and culture of prokaryotic and eukaryotic picoplantkon; the physiology of picoplankton in the field and in the laboratory; discrimination between autotrophic and heterotrophic picoplankton; the optical properties and pigments of picoplankton with implications for remote sensing; and the role of picoplankton in the pelagic ecosystem.

Résumé

PLATT, T., AND W. K. W. LI. [ED.] 1986. Photosynthetic picoplankton. Can. Bull. Fish. Aquat. Sci. 214: 583 p.

Ce livre est un recueil de 16 études dont le but est de résumer l'état actuel des connaissances sur le picoplancton photosynthétique dans l'océan. Tant le picoplancton prokaryotique qu'eukaryotique tombent dans cette catégorie. Cet ouvrage s'inspire de cours donnés à un institut d'études supérieures (San Miniato, Italie, octobre 1985) dans le cadre desquels les spécialistes, œuvrant tant en laboratoire (biochimie, physiologie) que sur le terrain (océanographie), ont pu partager leurs connaissances sur le picoplancton photosynthétique. Les sujets abordés concernaient, entre autres, les limitations physiologiques et écologiques dues à la taille extrêmement petite des cellules; la répartition, l'abondance et le statut taxonomique du picoplancton prokaryotique et eukaryotique; le métabolisme des cyanobactéries; les rapports qui existent entre la nutrition et la culture du picoplancton prokaryotique et eukaryotique; la physiologie du picoplancton sur le terrain et en laboratoire; la discrimination entre le picoplancton autotrophique et hétérotrophique; les propriétés optiques et les pigments du picoplancton qui peuvent servir aux applications en télédétection; et le rôle du picoplancton dans l'écosystème pélagique.

Foreword

This Bulletin is a compilation of the invited contributions to an Advanced Study Institute, sponsored by NATO (Scientific Affairs Division) and held in San Miniato, Italy in October 1985. The goals were the exchange of ideas and results between field ecologists and laboratory physiologists working on photosynthetic picoplankton, and the consolidation of the present state of knowledge about these organisms. Where data were lacking, lecturers were invited to speculate about the properties of picoplankton on the basis of experience with other algal groups, a device that led to many plans for new research.

The meeting was planned by an Organising Committee that included Prof. Noel Carr, Dr. Robert L. Guillard, Dr. Luigi Lazzara, Dr. William K. W. Li and Dr. Giuseppe Magazzú.

The accommodation and meeting facilities were subsidised by the Cassa di Risparmio di San Miniato, to whom all the participants are indebted. We also received important help and advice from Prof. F. Faranda and Dr. L. Guglielmo (Messina); Prof. M. Innamorati (Florence); Mr. G. Neville (Ottawa); Dr. C. Sinclair (Brussels); and Dr. J. Watson (Ottawa). I am most grateful to all of these people, and particularly to the principal lecturers, for the contributions. It is a pleasure to thank also my secretary, Mrs. M. Landry, for her invaluable help, as always, at every step of the way.

TREVOR PLATT
Director of the
Advanced Study Institute

Physiological Consequences of Extremely Small Size for Autotrophic Organisms in the Sea

JOHN A. RAVEN

Department of Biological Sciences, University of Dundee, Dundee DD1 4HN, Scotland

Ruder heads stand amazed at these prodigeous pieces of nature, whales, elephants, dromedaries, and camels; these, I confess, are the colossus and majestick pieces of her hand; but in these narrow engines there is more curious mathematicks; and the civility of these little citizens more easily sets forth the wisdom of their Maker.

Sir Thomas Browne, 1642

Introduction

The marine photosynthetic picoplankton have only recently achieved prominance as potentially very important primary producers in the world's oceans (Johnson and Sieburth 1979, 1982; Waterbury et al. 1979), although hints as to their significance have a lengthy history (see Johnson and Sieburth 1982).

The purpose of this article is to consider some of the physiological consequences of the extremely small size of these organisms for their life in the ocean. The analysis starts with some considerations of how miniaturization of a phototroph influences its ability to transform resources when these resources are supplied at an optimal rate, i.e. how smallness can constrain the value of μ_{max}. We deal in this first Section, then, with the influence which "non-scalable" cell components might have on the content in the biomass of catalysts whose activity constrains the rate at which the biomass can increase itself.

The next point to be considered brings us rather closer to reality of existence for picoplankter under natural conditions, i.e. the response of the picoplankton cell to a supply to the cell surface of an essential resource at a rate inadequate to support μ_{max}. Here it is possible that both the *rate* of resource transformation and the *efficiency* of resource use in producing new cells is biologically (selectively) significant. The discussion here involves an analysis of the effects which smaller cell size have on the manipulation of resources arriving at the cell surface.

We next consider the ways in which small size alters the rate at which resources arrive at the cell surface: the main theme here is a comparison of unstirred layer effects for picoplankton cells and for large cells and the effects this difference may have on resource supply in a given, resource-depleted environment.

Finally, the conclusions which can be drawn from the discussions are enumerated, with particular emphasis on the likely lower limit for cell size in phototrophs, and on the possible selective advantages of small size and the means by which these possible advantages may be investigated.

The Effect of Small Size on the Maximum Rate of Resource Transformation in Phototrophs

PROLOGUE

A number of investigations have attempted to produce *mechanistic* models of the rate of growth or micro-organisms under conditions of optimal resource supply (i.e., modelling μ_{max} (log$_e$ 2/minimum generation time) and under conditions of growth-limiting rates of resource supply Shuter 1979; Churchward et al. 1982; Ehrenberg

1

and Kurland 1984; Raven 1986a). Shuter (1979) and Raven (1986a) specifically address the growth of phototrophs. Shuter (1979) compartments the biomass of a phytoplankton organism into (1) structural materials, (2) storage materials and the catalytic apparatus, i.e. the (3) photosynthetic apparatus which supplies photosynthate to (4) the remaining parts ("heterotrophic apparatus") of the synthetic apparatus. Raven (1986a) adopts (without explicitly acknowledging Shuter (1979)!) a similar formalism; he expresses μ_{max} in terms of equation (1):

(1)	μ_{max}	=	B.R.F.
where	μ_{max}	=	maximum specific growth rate: units of mol C assimilated (mol C in cells)$^{-1} \cdot$ s^{-1} or, in brief, s^{-1};
	B	=	concentration of the catalyst in the biomass: units of mol C in the catalyst (mol C in cell)$^{-1}$;
	R	=	maximum specific reaction rate of the catalyst: units of mol C transformed by the catalyst (mol C in the catalyst)$^{-1} \cdot$ s^{-1}:
and	F	=	Fraction of the maximum specific reaction rate of the catalyst which is required to account for the measured μ_{max}.

Although the definition of R is in terms of mol C transformed, for generality we must also include catalysis which involves non-C-containing substrates such as photons for light harvesting pigments, and nitrate for nitrate transport catalysts and nitrate reductase. Furthermore, we must include substrates which contain C but where C is not transformed by the catalyst (e.g. the C in light-harvesting pigments, whose excited states act as substrates for excitation energy transfer within the pigment bed, and for photochemistry; and the C in the ATP which is hydrolyzed in kinase reactions). In these cases, the *equivalent* maximum specific reaction rate of the catalyst in terms of mol C handled (mol C in the catalyst)$^{-1} \cdot$ s^{-1} can be arrived at *via* the ratio of the amount of resource (e.g. mol photons, mol nitrate) handled by the catalyst to mol carbon needed in cell synthesis. Accordingly, for photons, the appropriate ratio would be the photon requirement for growth (mol photons absorbed per mol C assimilated from CO_2 into cell material); for nitrate, the mol nitrogen: mol carbon ratio in the biomass. For the photosynthetic carboxylase RUBISCO (ribulose bisphosphate carboxylase-oxygenase), application of equation (1) requires that only the C in CO_2, but not that in the co-substrate RuBP, be counted in terms of the catalysis by this enzyme which contributes to net C assimilation into cell material. Furthermore cognisance must be taken of the possibility that some of the C fixed by the carboxylase activity of RUBISCO may be lost as CO_2 ("dark" respiration and photorespiration) or as organic C (extracellular organic C): see Raven (1986a).

A further important point about the use of equation (1) involves the definition of F. F can have values between 0 and 1. A value near 1 means that the catalyst under consideration is close to being growth-limiting; any decrease in its concentration in the biomass (B) could not be compensated by a compensatory increase in F, thus maintaining (at constant R) the same value of μ_{max}. A high F value for a given catalyst does not, of course, preclude high F values for other catalysts in that cell. Inspection of equation (1) shows that for a given set of values of R for the totality of catalysts in a cell, μ_{max} will be maximum when F values are uniformly high, since this reduces (*via* term B) the total number of mols of C needed to produce a cell. However, there may well be very important regulatory functions for low F values, and low R values, for certain of the catalysts in any metabolic sequence (Heinrich et al. 1977; Raven 1986a). Procedures by which F values can be determined for particular catalysts are discussed by Kacser and Burns (1973), Heinrich et al. (1977), Tager et al. (1983), Raven (1981, 1984a, 1986a), and Fasham and Platt (1983).

What relevance do these considerations, and the model of Shuter (1979), have for limitations on the magnitude of μ_{max} in picoplankton organisms relative to the value in comparable nanoplanktonic and larger microphototrophs? The fact that the majority of growth processes in O_2 – evolvers are catalysed by homologous catalysts, and that the sequences in which they operate are also common to all of these organisms, means that relatively little effect on μ_{max} would be expected as a result of changes in R or F (equation (1)) between picoplanktonic and larger phototrophs.

However, there are several important reactions in planktophytes which are catalysed by *analogous*, rather than *homologous*, catalysts; occurrence of many of these different catalysts is important in defining Divisions and classes of planktophytes (Raven 1984a, 1986b). Examples are the light-harvesting pigment–protein complexes; enzymes of synthesis and degradation of low M_r carbohydrates and alditols ("compatible solutes", etc) and of storage polysaccharides; and redox catalysts (soluble c – type cytochrome or plastocyanin) of electron transfer from the (reduced) cytochrome b-f-non-haem-iron complex to (oxidised) P_{700} in photosynthesis. Some examples of the distribution of these catalysts among important taxa of planktophytes are given in Table 1. In the case of these alternative catalysts there can be substantial differences among Divisions and Classes of planktophytes with respect to the value of R (equation (1)). This is most dramatically illustrated by the light-harvesting pigment–protein complexes where R must be computed in terms of the molar specific absorption coefficient of the chromophore over a particular wavelength range and the mol C contained in the complex per mol chrmophore in the complex. This yields a value in m^2 (mol chromophore)$^{-1}$ (mol C in complex (mol chromophore)$^{-1}$)$^{-1}$ or m^2 (mol C in complex)$^{-1}$. In the absence of "package" or "scattering" effects, the interaction of such a complex with a photon flux density in the specified wavelength range of *a certain number of mol photon • m^{-2} • s^{-1}* yields a photon absorption rate in (mol photon • m^{-2} • s^{-1})$\times$(m^2 (mol C in complex)$^{-1}$) or mol photon absorbed (mol C in complex)$^{-1}$ • s^{-1}. If the photon efficiency of C assimilation from CO_2 into cell C is known (mol C assimilated (mol photon absorbed)$^{-1}$), then the number of mol C assimilated per s per mol C in chromophore can be computed from (mol C assimilated/mol photon absorbed)$^{-1}$)$\times$(mol photon absorbed/mol C in complex)$^{-1}$ • s^{-1}). Application of these methods leads to R values for 400–600 or 400–500 nm light (deep ocean light depleted in larger wavelengths) which are substantially lower for phycocyanin than for phycoerythrin, which in turn is smaller than the values for Chlorophyta and Dinophyta (see Raven 1984a, b). Granted that light saturation of growth of planktophytes saturates at, at most, a photon flux density incident on individual cells of a few hundred mol photon • m^{-2} • s^{-1}, we note that the observed μ_{max} value requires that B for light-harvesting pigment-protein complexes amounts to 0.03–0.1 or so even for "sun" cells of a constitution (e.g. small size) which permits F to be near 1 (little "package effect"). Furthermore, the B value to achieve a given μ_{max} at a given incident photon flux density (400–500 nm) is predicted to be highest for phycocyanin-containing cells; lower for phycoerthrin-containing cells; and lowest for chlorophyll(s) b or chlorophyll c_2 ($\pm c_1$) -containing cells (Raven 1984a, b) as a result of the different R values. Since the B values for these complexes are substantial, we would predict that, *other things being equal*, cells in which B for light-harvesting complexes is higher, μ_{max} must be lower. The rationale here is that the higher B value for the light-harvesting complexes in phycocyanin-containing cells means lower B values for other catalysts which (*ex hypothesis*) have the same R and F values as in other planktophyte cells which have lower B values for their light-harvesting complexes.

The other catalysts mentioned in Table 1 have lower B values. While comparisons of F, R, and B values among the analogous enzymes catalysing synthesis and degradation of low M_r carbohydrates and alditols, and of storage polysaccharides, many data are available for cytochrome c and plastocyanin. It would appear that F is high for both catalysts, and that B and R are essentially equal for the two catalysts (see Raven 1984a). Thus swapping cytochrome c for the analogous plastocyanin would

TABLE 1. Distribution of analogous catalysts among some major taxa of planktophytes.

| Pigment Groups | 'Phycobiliphytes' | 'Chlorophytes' | | 'Chromophytes' | | | |
| Divisions/ classes Characteristics | CYANOBACTERIA | CHLOROPHYTA | | | | | |
		CHLOROPHYCEAE	MICROMONADOPHYCEAE (PRASINOPHYCEAE)	BACILLARIOPHYTA	DINOPHYTA	PRYMNESIOPHYTA	EUSTIGMATOPHYTA
Light-harvesting pigment protein complexes: nature of chromophores	phycocyanobilin phyoerythrobilin phycourobilin	chlorophyll (s) *a* chlorophyll (s) *b* lutein	chlorophyll (s) *a* chlorophyll (s) *b* ± Mg-2, 4-divinyl phaeoporphyrin a_5 monomethyl ester. ± lutein ± prasinoxanthin ± siphonoxanthin ± siphonein	chlorophyll (s) *a* chlorophyll c_1 chlorophyll c_2 fucoxanthin	chlorophyll (s) *a* chlorophyll c_2 Peridinin	chlorophyll (s) *a* chlorophyll c_1 chlorophyll c_2 fucoxanthin	chlorophyll (s) *a* violoxanthin
kg of complex per mol chromophore	16 (phycocyanobilin in C-phycocyanin) 5.7 (phycoerythrobilin, ± phycourobilin, in C-phycoerythrin).	≤ 4.8 (chlorophyll $a + b$ + carotenoids)	?	?	3.5–7.0 (chlorophyll *a* + peridinin)	?	?
Low M_r carbohydrates cyclitols, and alditols synthesised/degraded ("compatible solutes")	sucrose; trehalose; glycosylglycerol	sucrose glycerol sorbitol	Mannitol (glycerol?)	[Glucose — not a 'compatible solute']	Glycerol [Glucose — not a 'compatible solute']	Cyclohexane tetrol [Glucose — not a 'compatible solute']	Mannitol
Storage polysaccharides synthesised/degraded	α 1,4 glucan α 1,6 glucan	α 1,4 glucan α 1,6 glucan	α 1, 4 glucan α 1,6 glucan	β 1,3 glucan	α 1,4 glucan	β 1,3 glucan	β 1,3 glucan (?)
Nature of redox catalysts mediating oxidation of reduced cyt b-f-Fe_{nh}	Cytochrome c and plastocyanin	Cytochrome c and plastocyanin	(Cytochrome c?)	Cytochrome c	(Cytochrome c?)	(Cytochrome c?)	(Cytochrome c?)

4

Volume regulation mechanism	Peptidoglycan cell wall around protplast with osmolarity in excess of that in the medium	Non-flagellate organisms/stages: polysaccharide wall round protoplast with osmolarity in excess of that in the medium. Flagellate organisms/stages: (marine): isosmolar with environment, peptidoglycan "wall" (if present) not functional in volume regulation (freshwater): cells hyper-osmolar to environment; volume regulation by contractile vacuoles; peptidoglycan "wall" (if present) not functional in volume regulation	Non-flagellate organisms (phycoma stages of *Halosphaera*, etc.): polysaccharide wall round protoplast with osmolarity in excess of that in medium. Flagellate organisms/stages: as for Chlorophyceae, but *never* having a "cell wall"/theca (*Tetraselmis* = *Platymonas*, = *Prasinocladus*) is now in the class Pleurastrophyceae!)	Vegetative cells: ± silicified wall of polysaccharide round protoplast with osmolarity in excess of that in the medium. Flagellate/amoeboid reproductive stages: as for Prasinophyceae	Cysts: turgor-resisting walls? Vegetative cells usually flagellate, no "wall" outside plasmalemma; isosmolar with medium (marine) or hyperosmolar with contractile vacuoles (pusules) in freshwater	Cysts: turgor resisting wall? Vegetative cells often flagellate usually lacking turgor-resistant cell wall; volume regulation as for Chlorophyceae (though wall probably not peptidoglycan)	Vegetative cells often flagellate; some coccoid with a cell wall
Mechanisms which increase cell density relative to medium or otherwise promote sinking	Large fraction of cell occupied by 'storage' polysaccharide	Flagella can → downward movement	Flagella can → downward movement	Silicification	Calcification, rare, occurs in *Thoracosphaera*, flagella can → downward movement	Calcification, common flagella can → downward movement	Flagella can → downward movement
Mechanisms which decrease cell density relative to medium or otherwise decrease the rate of sinking, or cause upward movement	Large fraction of cell occupied by expanded gas vacuoles; can be + ve bouyant	Flagella can → upward movement	Flagella can → upward movement	Vacuoles containing 'light' ions (works best in seawater!)	Vacuoles containing 'light' ions (works best in seawater). Flagella can → upward movement	Flagella can → upward movement	Flagella can → upward movement

Data compiled from Bold and Wynne (1985); Raven (1984a, 1986a); Reed and Stewart (1983); Reed et al. (1984a); Reed et al. (1984b); Warr et al. (1985); Hellebust (1976); Brown and Hellebust (1980a,b); Kremer (1980).

not be expected to have a significant effect on μ_{max} (other things being equal) in a planktophyte. However, under Fe- or Cu- limiting growth conditions plastocyanin (containing Cu but no Fe) or cytochrome c (containing Fe but not Cu), respectively, might be selectively favoured. For another pair of analogous redox catalysts in photosynthesis, the Fe-containing ferredoxin and the metal-free flavodoxin, insufficient chemotaxonomy data are available to permit their inclusion in Table 1; it is, however, clear that flavodoxin has a significantly lower R value than does ferredoxin (see Raven 1984a, 1986a).

Despite the intrinsic interest of these analagous catalysts with markedly different R values in different major taxa, we must not try to make too much of these differences as far as the picoplankton is concerned. The reason is that the picoplankton contains well-characterised representatives of the "phycobiliphytes" (Cyanobacteria such as species of *Synechococcus*) and the "Chlorophytes" (Micromonadophyceae = Prasinophyceae such as species of *Micromonas* and *Mantoniella*; species of *Chlorella* and *Nannochloris* in the Chlorophyceae): Johnson and Sieburth (1979, 1982), Waterbury et al. (1979), Manton (1959), Manton and Parke (1960). For the organisms close to extremes of the range of R values for light-harvesting pigment protein complexes (low R in the phycocyanins dominant in many Cyanobacteria; high R in the chlorophyll $a + b$ complexes of Chlorophyta) which have substantial B values, it is clear that both picoplankton organisms and larger planktophytes are represented at these extremes. Accordingly, the predicted effect of a large B value (both relative *and* absolute) in phycocyanin-dminated "phycobiliphytes" in reducing the fraction of the biomass available to accomodate all other catalysts (and structures), and hence reducing the μ_{max}, should occur in both picoplankton organisms and larger planktophytes. Similarly, the predicted effect of a smaller B value (absolutely and also relative to the phycocyanin-dominated Cyanobacteria) in increasing the fraction of the biomass available to accommodate all other catalysts (and structures), and hence increasing the μ_{max} should occur in both large planktophytes and in picoplankton cells of the "chlorophyte" pigment group.

The occurrence of "chromophyte" planktophytes less than 2 μm diameter is less well authenticated. While chlorophyll(s) c is found in particles from the sea which are less than ~1 μm in diameter, the carotenoids typically found in "chromophytes" (e.g. fucoxanthin and peridinin) were not detected (Yentsch 1983). Furthermore, the smallest characterised "chromophyte' cells such as *Imantonia rotunda* (Prymmesiophyta: Green and Pienaar 1977) and *Nannochloropsis* (Eustigmatophyta: Turner and Gowen 1984) have radii of ~1.5 μm, i.e. rather larger than the maximum for picoplankton (1 μm radius). The chromophytes, therefore, are not well established to be picoplankters (cf. Murphy and Haugen 1985) and should not feature in our inter-pigment group comparisons of the predicted effects of varying R and B values for light-harvesting complexes on μ_{max} values for picoplankton organisms.

An implicit assumption made here is that the catalysts in question are completely "scaleable" over the size ranges involved. That this assumption is valid will be tested for cytochrome c/plastocyanin, molecules which are present at $\frac{1}{30} - \frac{1}{100}$ the quantity of the individual molecular species of the light-harvesting pigment-protein complexes, and at *approximately* the same quantity as that of reaction centre one and two core complexes, cytochrome b–f complexes, and CF_0–CF_1 ATP synthetase.

We deal with 5 and 0.5 μm radius cells with 0.25 pg dry weight per μm^3 cell volume, i.e. 131 pg dw for the 5 μm radius cell and 0.131 pg dw for the 0.5 μm radius cell. If chlorophyll is 0.01 for the dry weight, and the molar ratio of chlorophyll to cytochrome c_{554} or plastocyanin is 500 (see Raven 1984a) then the 5 μm radius cells has 2.9×10^{-18} mol cytochrome c_{554} or plastocyanin per cell, or 1 745 000 molecules per cell. The 0.5 μm radius cell has 1745 molecules per cell. 1745 molecules per cell is by no means an unacceptably small number of molecules of an individual catalyst to have in each cell. If we turn to the respiratory analogue of cytochrome c_{554} (or plastocyanin), i.e. the mitochondrial cytochrome c of eukaryotes (the cyanobacteria use the same soluble redox catalyst in respiration and in photosynthesis), we find that the number of catalyst molecules per cell is only about $\frac{1}{10}$ of the number of cytochrome C_{554} or plastocyanin (table 6.2 of Raven 1984a), or 174 500 molecules per 5 μm radius cell and 175 molecules per 0.5 μm radius cell. The lowest concentration of inner mitochondrial membrane redox components is found for the NADH-Q oxide-reductase, which is present at about 0.25 the concentration of the cytochrome b–c, or cytochrome a–a_3 complexes, or the concentration of cytochrome c (Raven 1984a). Even for these "scarce" complexes the 5 μm radius cell has about 43 625 molecules per cell, and the 0.5 μm radius cell has 44 molecules per cell. Thus there is no obvious difficulty in diminishing cell size from 5 to 0.5 μm radius in terms of maintaining reasonably numbers of copies of individual copies of each catalyst in the cell *without* increasing the relative numbers of the "rarer" catalysts, at least in relation to the catalysts that have been considered heretofore.

Srere (1967) notes that, for a eukaryotic cell with 200 kg protein $\cdot$ (m^3 cell volume)$^{-1}$, and 1000 different proteins with a mean M_r of 10^5, the "average" protein is present at a concentration of 2 mmol $\cdot$ m^{-3}. Table 2 of Srere (1967) quotes concentrations of enzymes from various mammalian tissues which are in the range of 0.3–45 mmol $\cdot$ (m^3 cell water)$^{-1}$. Assuming that $\frac{1}{3}$ of the intraplasmalemma volume is "cytoplasm" and $\frac{2}{3}$ vacuole (Berlin et al., 1982; Smart and Trewavas 1983; Hajibagheri, et al. 1984), then the values for catalyst concentrations in higher plant leaf cells (mmol catalyst per m^3 of intraplasmalemma volume) in tables I, II, III, V, and VI of Hewitt (1983) are in the range 0.006 mmol $\cdot$ (m^3 "cytoplasm")$^{-1}$ for xanthine dehydrogenase to 24 mmol $\cdot$ (m^3 "cytoplasm")$^{-1}$ for plastocyanin. If 0.60 of the "cytoplasm" volume of the leaves of C_3 higher plants is chloroplast (Berlin, et al. 1982; Smart and Trewavas 1983; Hajibagheri, et al., 1984), then the concentration of active sites of RUBISCO (8 per 500 000 M_r), given as 3.4 mol $\cdot$ (m^3 chloroplast)$^{-1}$ by Ashton (1982), is 1.8–2.5 mol $\cdot$ (m^3 "cytoplasm")$^{-1}$.

Clearly, if there are 1 000 enzymes in the cytosol of an individual cell, some must be present at less than Srere's 2 mmol $\cdot$ m^{-3} if RUBISCO (to name but one) is present at 1.8–2.5 mol $\cdot$ m^{-3}. In the context of our two "model" cells, the 5 μm radius cell (volume 524×10^{-18} m^3) will have $524.18^{-18} \times 2.10^{-3} = 1.048 \times 10^{-18}$ mol, or 628 800 molecules of that protein in each cell; the 0.5 μm radius cell (volume 0.524×10^{-18} m^3) will have 629 molecules per cell of that particular protein. For the lowest enzyme concentration mentioned by Hewitt (1983), i.e. 0.006 mmol $\cdot$ m^{-3} for xanthine dehydrogenase, there would be 1 886 molecules per cell for a 5 μm radius cell, and 2 molecules per cell for a 0.5 μm radius cell. We are here at the lower limit of reduction of cell size without altering the ratio of enzymes, i.e. any further decrease in cell size while maintaining the diversity of enzymes would imply that B for the enzymes with the lowest B value would have to increase in order to keep any cell complement of that enzyme, with the corresponding necessity to reduce the content of (B value of) more abundant proteins, with corresponding effects on μ_{max} as some catalyst(s) with low R, and high F, values which is rate-limiting for growth and has a reduced B value (equation (1)).

Finaly we deal with a non-protein catalyst, the Co-containing co-enzyme (vitamin) B_{12}. For chemostat cultures of the 200 μm^3 – volume Prymnesiophyte *Monochrysis* (= *Pavlova*) *lutheri* (Green, 1975), Droop (1968) showed that an intracellular vitamin B_{12} concentration of $3.8 \cdot 10^{-5}$ mol $\cdot$ m^{-3} was needed for μ_{max} for a 0.5 μm radius picoplankton cell this would correspond to about 12 molecules per cell, so that this cofactor is present in rather larger amounts than the least abundant enzyme considered above.

NON-SCALEABLE CELL CONSTITUENTS AND PROPERTIES

1. *DNA Content*

Clearly there is a minimum DNA content for a cell if it is to code for a given number of RNAs (tRNA, rRNA and mRNA) and, via the diversity of mRNAs, of proteins. The tremendous range of C-DNA values in eukaryotes (Cavalier-Smith 1978; Moore 1984) poses a number of paradoxes, regarded by Moore (1984) as aspects of the "C-value paradox": (1) Morphologically similar organisms can have very different C-DNA values (2) C-DNA content of major groups correlates poorly with organismal complexity and (3) genome coding capacity and number of proteins actually synthesised do not match. Partial explanations enumerated by Moore include the innacuracy of measurements of C-DNA contents which may exaggerate differences within groups; the observation that *minimal* C-DNA content of a group agrees with organismal complexity of that group; and the likelihood that there are similar amounts of coding DNA in organism of similar complexity, with the 5–10–fold excees of DNA over that required for the observed number of proteins in *Drosophila*, and the 25–50–fold excess of DNA in *Mus*, being non-protein-coding DNA. It is, nevertheless, clear that the statement mode at the beginning of this paragraph must be correct. The smallest reported genome size for an O_2-evolving phototroph (Table 2) is that of Cuhel and Waterbury (1984) for *Synechococcus* WH 7803 at 126.10^9 Daltons per cell ($2.1.10^{-15}$ g DNA per cell); this is sufficient to code for several hundred polypeptides of M_r 10KDa–100kDa in addition to the regulatory apparatus, rRNA and tRNA. If we take this as the minimum genome size for an O_2-evolving phototroph, and one which will cope with the coding requirements of any size of prokaryotic O_2-evolving unicell, then the 2.1×10^{-5} g DNA of a single complete genome is 1.6×10^{-3} of the wet mass of a cell 5 μm in radius, but 0.016 of the mass of a cell of 0.5 μm in radius. The largest recorded cyanobacterial genome size ($14.3.10^{-15}$ g) then represents 8.5×10^{-5} of the dry mass of a 5 μm radius cell, or 0.085 of the dry mass of a cell 0.5 μm in radius. In terms of the B values for catalysts, the decrement due to the presence of DNA is negligible in the 5 μm radius cells, but not in the 0.5 μm radius cells.

TABLE 2. DNA content of cyanobacterial and algal cells with (for comparison) that of certain other cells.

Organism	pg DNA $\cdot$ cell^{-1}	GDa DNA $\cdot$ cell^{-1}	References
PROKARYOTIC PHOTOTROPHS			
Cyanobacteria			
Range of cyanobacteria in pure culture pre-1979	0.0027–0.0143	1.6–8.6	Herdman et al. (1979)
Synechococcus WH 7803	0.0021	1.26	Cuhel and Waterbury (1984)
Chloroxybacteria			
Prochloron sp.	0.0060	3.59	Herdman (1981)
EUKARYOTIC PHOTOTROPHS			
Range of eukaryotic algae	0.044–200	24–120 000	Cavalier Smith (1978)
Chlorophyta			
Chlorophyceae			
Nanochlorum eucaryotum	0.0605	36.3	Wilhelm, et al. (1982)
Chlorella fusca	0.10–0.40	60–240	McCullough and John (1972)
Chlamydomonas reinhardtii	0.12–0.20	72–120	Witeway and Lee (1977)
Micromonodophyceae			
Pedinomonas minor	0.10	60	Ricketts (1966)
Bacillariophyta			
Navicula pelliculosa	0.10	60	Holm-Hansen (1969)
Prymnesiophyta			
Monochrysis lutheri	0.10	60	Holm-Hansen (1969)
Range in Bryophyta, Tracheophyta	0.64–310	384–186 000	Cavalier Smith (1978)

For the eukaryotes, the smallest reported C-DNA content (including not only nuclear DNA — probably the haploid quantity — but also the organelle DNAs — probably more than one copy of each) in eukaryotic phototrophic cells is 41 fg DNA per cell. This corresponds to only $2.44.10^{-4}$ of the mass of a 5 μm radius cell, but to 0.24 of the mass of a 0.5 μm radius cell. It is important to note that the smallest known CDNA content of eukaryotic chemo-organotrophs is found in certain fungi with as little as 5.10^{-15} g DNA per cell. This is only one-eight of the C-DNA content of the alga with the smallest C-DNA content; the corresponding ratio for chemo-organotrophic eubacteria (i.e. excluding molicutes) relative to cyanobacteria is $\sim \frac{1}{2}$ (Table 2). It could be argued that a similar ratio pertains for eukaryotes if the smallest examples of photolithotrophs and chemoorganotrophs are compared; the smallest algal cells examined for their C-DNA content *Nanochlorum eucarotum* have a mass of some 0.45 pg (DNA $\sim$ 0.12 of cell mass); a 0.5 μm radius algal cell, with a mass of 0.17 pg, would only have 0.01 pg of DNA if it obeyed the "twice the smallest fungal DNA content" suggestion. More work is needed!

To conclude this brief discussion of DNA contents of O_2-evolving phototrophic prokaryotic and eukaryotic unicells of varying sizes, it would seem that the *minimal* DNA content for the prokaryotes account for 0.016 of the mass of a 0.5 μm cell and $1.6.10^{-3}$ of the mass of a 5 μm cell; for eukaryotes, likely values are 0.059 for a 0.5 μm cells and $5.9.10^{-3}$ of the mass of a 5 μm cell. The minimal fraction of the cell mass taken up by DNA is negligible in the 5 μm cells but has a significant potential effect on B values, and hence μ_{max} for a 0.5 μm eukaryotic cell and a smaller effect for a 0.5 μm prokaryotic cell.

2. Cell Walls

The inclusion of cell walls here needs some explanation, since Raven (1982, 1984a) has recently re-emphasised the Laplace relationship for cell walls. In brief, the argument relates to a series of congruent cells of different sizes; for simplicity we only deal here with spherical organisms. We assume a cell turgor pressure (defined from $\Delta\psi_w = \Delta\psi_p + \Delta\psi_\pi$, where $\Delta\psi_w$ = water potential of the cell relative to the medium; $\Delta\psi_p$ = pressure potential of the cell relative to the medium and $\Delta\psi_\pi$ = osmotic potential of the cell relative to the medium, when $\Delta\psi_w = 0$ (all $\Delta\psi$ terms are measured in *Pa*), and wall mechanical properties, which are independent of cell radius $r(m)$. The stress in the cell wall engendered by this $\Delta\psi_p$ is given by equation (2):

$$(2) \qquad \Delta\psi_p = \frac{2\sigma\, t}{r}$$

where σ is the tangential stress in the wall (*Pa*) and t is the wall thickness (*m*). For a constant $\Delta\psi_p$, maintenance of a constant σ requires that t is a constant fraction of r (granted our assumption of constant mechanical properties of cell walls whereby the stress is dealt with, and a constant "safety factor"). Accordingly, with our assumptions, t/r is constant. In practice t/r is small ($\sim$ 0.01); denoting t/r as x (i.e. $t = xr$), we see that while the volume contained by the cell wall is $\frac{4}{3} \pi r^3$, the volume of the cell wall is $4\pi r^3 x$ (i.e. $4 \pi r^2\, t$). Accordingly, the ratio of cell wall volume to protoplast volume is $\frac{4\,\pi\, r^3 x}{4/3\, r^3} = 32x$. For our example of $x = 0.01$, the cell wall volume is 0.03 of the protoplast volume *independent of the size of the* (spherical) *cell*. Accordingly, granted our assumptions, the cell wall volume is a constant fraction of the protoplast volume (or, indeed, the (protoplast + cell) volume).

With these preliminaries, we can ask if there are any grounds for believing that the cell wall is "non scaleable"; specifically, are there constraints on the extent to which the wall thickness t can be reduced (keeping t/r constant) as r is reduced to 0.5 μm without altering its properties such that a "scaled" value of t is no longer capable of resisting the same value of σ with a constant safety factor? Pirie (1973) suggests that cellulose microfibrils cannot be reduced to a diameter of less than 10 nm; furthermore, their mechanical (turgor-resisting) function requires at least two, non-parallel layers of cellulose microfibrils around the cells. For a picoplankton cell with

TABLE 3. Fraction of (protoplast + wall) volume of ± spherical chlorococcalean cells (Chlorophyta: Chlorophyceae) occupied by the cell wall.

Organism	Radius of (protoplast + wall) μm	Volume of (protoplast + wall) μm^3	Fraction of (protoplast + wall) volume occupied by wall	References
Nannochloris sp. (UTEX 1999)	1.06	5	0.048	Brown and Elfman (1983)
Nannochloris maculatus (CCAP 251/3)	1.06	5	0.054	Brown and Elfman (1983)
Nannochloris sp. (UTEX 2055)	1.12	6	0.035	Brown and Elfman (1983)
Chlorella nana (marine)	1.125	6	0.159	Thinh and Griffiths (1985)
Nannochloris bacillaris (UWASH 20-2-2)	1.19	7	0.058	Brown and Elfman (1983)
Chlorella minutissima (fw)	1.45	13	0.120	Dempsey et al. (1980)
Nannochloris atomus	2.0	33	0.109	Brown and Elfman (1983)
Chlorella emersonii (cultured in 1 mol NaCl $\cdot$ m^{-3})	3.52	183	0.072	Munns et al. (1983)
(cultured in 335 mol NaCl $\cdot$ m^{-3})	4.26	324	0.078	
Hydrodictyon africanum (freshwater)	10^3	4.19.10^9	0.014	Raven (1984), chapter 8

$r = 0.5 \ \mu m$ and $t = 0.02 \ r$ or $0.010 \ \mu m$ ($= 10$ nm), we can see that cellulose is not "scaleable" down to the dimensions of a picoplankton cell wall if Pirie's suggestion is correct. However, Brown and Elfman (1983) found a wall thickness of only 13 nm for a cell with a radius of $1.13 \ \mu m$ in the Chlorococcalean *Nannochloris* sp. UTEX 2055, casting doubt on Pirie's suggestion on minimal wall thickness. While we note (Table 1) that the cell wall of cyanobacteria is composed of peptidoglycan, and that the wall of Chlorophycean picoplankters probably do not contain cellulose microfibrills as their major tension-resisting elements (Raven 1986), it is possible that there *is* a lower limit on the thickness of cell walls not described by the equation $\Delta\psi_p = \frac{2 \ \sigma \ t}{r}$. A doubling of the fraction of the total cell volume occupied by the wall, i.e. from $t = 0.01r$ to $t = 0.02 \ r$, as the cell size is decreased means that the wall is 0.06 of the dry weight in larger cell (e.g. *Chorella ellipsoidea*: Takeda and Hirokawa, 1978; see Table 3) and 0.12 of the dry weight of the smaller cell. If this is the required increase in the fraction wall volume in going from a 5 μm radius cell to a 0.5 μ radius cell (due to a requirement for the cell wall to have a certain minimum thickness), it implies a significant decrease in the fraction of cell volume (or dry matter) which is available for catalytic processes which can contribute (via equation (1) to growth rate. When other "non-catalytic" components are considered (e.g. DNA; storage material; cytoskeleton), the decrement in cell catalytic capacity due to an increase in wall fraction is increased from 3.1% (a decrease from 0.97 to 0.94 based on total cell C) to perhaps 4.5% (a decrease from 0.67 to 0.64 based on "catalytic C" assuming DNA, cytoskeleton and storage material amount to 0.30 of the cell volume or organic weight). Further progress in this analysis requires more data on the comparative mechanical properties as a fraction of cell radius. The data assembled in Table 3 do not appear to indicate a significantly greater fraction of wall plus protoplast volume occupied by walls in the smallest Chlorococcalean alga listed (i.e. a greater wall thickness as a fraction of cell radius) over the cell radius range $1.0–4.25 \ \mu m$, although the fraction is significantly smaller for the much larger ($10^3 \ \mu m$ radius) cell of *Hydrodictyon*. We note that the assumed constraints of 'scaleability' of the microfibrils — based turgor — resisting wall may not apply (in the picoplankton size range) to prokaryotes with their peptidoglycan cell walls equivalent to a "bag-shaped macromolecule".

The other mechanism whereby marine planktophytes manage their osmotic/turgor/volume regulation problem (Cram 1976; Reed 1984; Raven 1984a) and the only one open to flagellate cells, does not rely on turgid cells with cell walls. These (effectively) wall-less cells (e.g. the marine *Micromonas*, *Mantoniella*, *Dunaliella* and *Imantonia*) are probably isosmotic with their seawater environment. The mechanism whereby cell volume is regulated is probably based on the "Double Doman Equilibrium" in which the osmotic contribution to the cells by (almost!) impermeant organic metabolites (including compatible solutes) is balanced by reduced permeability to some major extra-cellular solute(s) (usually Na^+), together with a mechanism for the active efflux of that solute(s). This "exclusion plus extrusion" mechanism (Raven and Smith 1982) is considered further under (6) below when "Leakage and Maintenance processes" are considered.

Finally, we can briefly consider the possible role of a cell membrane in a "naked" cell in resisting cell turgor. Weak as the cell membrane is in pressure resistance (Gruen and Wolfe 1982), the membrane *is* an essentially non-scaleable entity (thickness indenpedent of cell radius) and application of a Laplace-like relationship to the plasmalemma of a wall-less cell suggests that the $\Delta\psi_w$ needed to burst the cell is 0.8 kPa for a 5 μm radius cell and 8 kPa for a 0.5 μm radius cell, corresponding to an excess osmolarity in the cell of 0.377 osmol $\cdot$ m^{-3} for a 5 μm radius cell and 3.77 osmol $\cdot$ m^{-3} for a 0.5 μm radius cell. This is a negligible turgor difference, even for a 0.5 μm radius cell, in terms of the seawater osmolarity of ~ 1 k osmol $\cdot$ m^{-2}. Parenthetically, we see that, in the context of a freshwater cell, the minimum excess of osmolarity in the cell over that outside is at least 15 osmol $\cdot$ m^{-3} (Raven 1982,

1984a) so that the plasmalemma could not contain safely the minimum turgor pressure of a (as yet hypothetical) freshwater picoplankton cell lacking contractile vacuoles or their functional equivalent, even if it had the lowest possible intracellular osmolarity.

3. Flagella

If we assume (see below; and Purcell 1977) that the main function of flagella in flagellate planktophytes is to move the cell relative to the surrounding water body in order to gain access to a better ratio of resource availability (photons, N, P, etc.), then we can compare a 5 μm radius with a 0.5 μm radius cell in terms of the ratio of flagellar volume to cell body volume which would be needed *to achieve a certain velocity of cell movement*. Our chosen velocity is 50 μm $\cdot$ s^{-1} which can be achieved by representatives, with a range of cell sizes, of many classes of phytoflagellates, and can be comfortably exceeded by many dinophytes (see Manton and Parke 1960; Roberts 1981; Raven and Richardson 1984). The analysis of the cost of synthesizing the flagellar apparatus necessary to achieve such velocities follows that employed by Raven and Beardall (1981a and Raven and Richardson (1984). We note that while such movement relative to the bathing medium is *not* required to attain μ_{max} because the resource supplies are optimal, the motile apparatus is *retained* under conditions yielding μ_{max} in flagellates.

The minimum power to move a 0.5 μm radius (1) cell at a velocity (v) 50 μm $\cdot$ s^{-1} through a medium of dynamic viscosity (η) 9.0 10^{-4} kg $\cdot$ m^{-1} $\cdot$ s^{-1} is 21.2 10^{-18} W $\cdot$ cell^{-1}, while for a 5 μm radius cell it is 212 10^{-18} W $\cdot$ cell^{-1}, using the relationship (equation (3)):

$$(3) \qquad P = 6 \, \pi v^2 \, r\eta$$

Assuming that the effective *in vivo* ΔG of ATP hydrolysis is 55 kj $\cdot$ mol^{-1}, and that there are 600 dynein ATPase molecules per μm length of flagellum, each ATPase molecule hydrolysing 10 ATP molecules to ADP and P$_i$ per second, the energy transduction rate is 550 10^{-18} W (μm length of flagellum)$^{-1}$. If the efficiency of conversion of the chemical energy of ATP hydrolysis into mechanical work in overcoming the resistance to movement at 50 μm $\cdot$ s^{-1} is 10% (a rather high value, according to Purcell 1977 and Roberts 1981), we find that the 21.2 10^{-18} W per 0.5 μm radius cell needs a flagellar length of $\frac{21.2 \ 10^{-18} \times 10}{550 \ 10^{-18}}$ or 0.385 μm, while the 212 10^{-18} W per 5 μm radius cell needs a minimum flagellar length of $\frac{212.10^{-18} \times 10}{550.10^{-18}}$ or 3.85 μm. For a 0.1 μm radius flagellar axis, the volume of the 0.385 μm long flagellum is 0.012 μm^3, while that of the 3.85 μm long flagellum is 0.121 μm^3. The volume of the cell body of a 0.5 μm radius cell is 0.524 μm^3, so that the ratio flagellar volume: cell body volume is 0.012/0.524 or 0.023 in the picoplankton cell, while for the 5 μm radius cell (cell body volume of 5.24 μm^3) the ratio is 0.00023. This rather naive analysis of the minimum volume of flagellum for the two sizes of cell in order to permit movement at 50 μm $\cdot$ s^{-1} shows that it is a much more significant fraction of the volume of the cell body in the case of the 0.5 μm radius than of the 5 μm radius cell. The fraction of total cell volume (and, probably, organic mass: Raven and Richardson 1984) occupied by the flagellum is, for the 0.5 μm radius cell, very similar to the computed minimum fraction of cell organic mass which is occupied by DNA; for both flagellum mass and DNA mass the minimum fraction of cell organic mass is much lower for the 5 μm radius cells. When we take into account the contribution of other "non-catalytic" material (see (2) above) to the organic mass of the cell assumed to be 30%, the decrement of the "catalytic biomass' due to the presence of the flagellum in a 0.5 μm radius cell is, then, from 0.70 to 0.677 of the total biomass while, for a 5 μm radius cell, the decrement is from 0.70 to 0.69977; the effect is clearly much greater in the smaller cell. Both of these decrements are increased if the basal bodies, and associated structures, are considered in the accounting of the biomass of the flagellar apparatus (Raven 1982; Raven and Richardson 1984).

What data are available as to the achieved ratio of flagellar volume to cell body volume is eukaryotic planktophytes of varying sizes? The smallest well characterised flagellate planktophyte is *Micromonas pusilla* (formerly *Chromulina pusilla*, before its "Chlorophyte" rather than "Chromophyte" affinities were clear) in the Micromonadophyceae (formerly Prasinophyceae): Manton 1959; Manton and Parke 1960; Moestrup 1979; Norris 1980; Okometo and Witman 1981; Mattox and Stewart 1984). The cells of *Micromonas pusilla* are comma-shaped, with the blunt end facing *forwards* during swimming, which can occur at $> 50 \mu m \cdot s^{-1}$; the cell body is 1–3 μm long and 0.7–1.0 μm wide so that the cell volume is about 1.5 μm^3. The single flagellum has a basal portion with a typical "9 + 2" microtubular structure, and a slender hair-point which only certainly contains the "central 2" microtubules; both parts of the flagellum, as well as the cell body, have a surface (plasma) membrane which lacks obvious surface embellishments such as scales. The hair-part is some 3 μm long, and the flagellum proper is $\sim 1 \mu m$ long. Manton (1959) claims a diameter of 0.14 μm for the flagellum proper; this would yield a flagellar volume of 0.015 μm^3, giving a ratio of flagellar volume: cell body volume ratio of 0.015/1.5 or 0.01; if the flagellar diameter is 0.2 μm (the "normal" value), than the flagellar volume: cell body volume ratio is 0.031/1.5 or 0.02. This ratio (0.01–0.02) is rather lower than our computed ratio for a 0.5 μm radius cell moving at the "*Micromonas*-like" velocity of 50 $\mu m \cdot s^{-1}$, but is substantially higher than our computation for a 5 μm-radius cell. This is reasonable in view of the "effective spherical radius" of a 1.5 μm^3 *Micromonas pusilla* cell of $\sim 0.75 \mu m$. Manton (1959) was of the opinion that the "hair-tip" of the flagellum moved passively in response to the active movement of the flagellum proper, so that it would not contribute to the length of flagellum involved in ATP consumption related to flagellar movement.

A small, *truly* "chromophyte" phytoflagellate is *Imantonia rotunda* (Prymnesiophyta: Green and Pienaar 1977) which has spherical cells with a diameter of 2–4 μm, and 2 equal flagellar, each 4.5 μm long. A mean diameter of the cell body of 3 μm implies a volume of 14.14 μm^3, while the 9 μm total length of flagellar (radius 0.1 μm) has a volume of 0.28 μm^3, i.e. a ratio of flagella volume: cell body volume of 0.28/14.14 or a ~ 0.02, a similar value to that computed for *Micromonas pusilla*, and is again intermediate between the values suggested earlier for a 0.5 μm radius cell and a 5 μm radius cell, but much closer to the former value.

Moving to even large phytoflagellates, such as *Chlamydomonas*, with a mean volume of the cell body (for a cell cycle generating 4 autospores) of 120 μm^3 (Table 4) and a total flagellar length of perhaps 20 μm and a flagellar volume of 0.58 μm^3, the flagellar volume: cell body volume ratio is 0.58/120 or 0.0046. This is smaller than the *Micromonas* or *Imantonia* values, but is much higher than minimal computed value for a 5 μm radius cell. Further discussion of the fraction of cell material devoted to the flagella in planktophytes may be found in Raven and Richardson (1984).

The consideration of flagella here has been limited to the eukaryotic organelles, i.e. to what Margulis (1981) terms "undulipodia" in order to distinguish them from the much smaller prokaryotic flagella. No well substantiated case of the occurrence of flagella in O_2-evolving phototrophic picoplanktons has yet been reported, although a number of non O_2-evolving prokaryotic phototrophs do have such flagella (Trüper and Pfennig 1978). Since the non-O_2-evolving phototrophs do not function photosynthetically (or, indeed, in some cases, such as the Chlorobioceae, at all) in the presence of O_2, we do not consider here the cost of producing bacterial flagella necessary to give a certain velocity of movement of variously sized cells. The recently reported swimming motility of some cynobacteria (strains of marine *Synechococcus*) does not appear to involve prokaryotic flagella (Waterbury et al. 1985); the cost of producing the ill-defined motility mechanism is unkown.

TABLE 4. Specific growth rates of micro-organisms at 20°C. Where the original data were given for growth temperatures other than 20°C, the data have been recalculated assuming a Q_{10} of 2 (see Raven 1986a) or by interpolation.

| Organism | Growth source of | | | Growth Temperature/°C | $\mu_{20°C}/s^{-1}$ | Cell volume/μm^3 | References |
	Energy	C	N				
PROKARYOTES							
Cyanobacteria							
Synechococcus leopolensis (= *Anacystis*	Light	CO_2	inorganic	40	24.10^{-6}	~ 3.0	Kratz and Myers (1955a,b)
nidulans)	Light	CO_2	inorganic	25	10.10^{-6}	~ 3.0	
strain UTEX 625	Light	CO_2	inorganic	31	11.10^{-6}		Turpin et al. (1984)
strain TEX 625	Light	CO_2	inorganic	40	14.10^{-6}	~ 1.0	Utkillen (1982)
Synechococcus sp. strain 53	Light	CO_2	inorganic	44	15.10^{-6}		Peary and Castenholz (1964)
(thermophilic)	Light	CO_2	inorganic	30	8.10^{-6}		
Synechococcus sp. (strain PC_2 = WH 7803, PE — dominant, marine)	Light	CO_2	inorganic	20	14.10^{-6}		Morris and Glover (1981); Glover and Morris (1981)
	Light	CO_2	inorganic	"room temperature"	13.10^{-6}	2.4 (assuming C = ⅛ fresh weight)	Barlow and Alberte (1985)
Synechococcus sp. (strain syn = WH 5701; PC — dominant, marine)	Light	CO_2	inorganic	20	17.10^{-6}		Morris and Glover (1981); Glover and Morris (1981)
Synechococcus linearis Cuture 2–7, FW Institute Collection	Light	CO_2	inorganic	"room temperature"	17.10^{-6}	~ 8.0 (assuming C = ⅛ fresh weight)	Healey (1985)
Anabaena variabilis	Light	inorganic	inorganic	25	9.10^{-6}	25	Reed et al. (1981) Kratz and Myers (1955a,b)
Anabaena cylindrica	Light	inorganic	inorganic	25	13.10^{-6}		Fogg (1949); Krauss (1958)
Anabaena flos-aquae	Light	inorganic	inorganic	25	7.10^{-6}		Elder and Parker (1984)
Gloeobacter violacea (PC 7421)	Light	inorganic	inorganic	25	$1.9.10^{-6}$	3.5	Rippka et al. (1974)
Synechococcus sp. (PC 7001)	Light	inorganic	inorganic	25	$3.9.10^{-6}$	0.5	Rippka et al. (1974)

Molicutes							
Spiroplasma sp.	organic	organic	organic	20	106.10^{-6}	0.05–0.10	Whitcomb (1984)
Eubacteria							
(other than							
(Cyanobacteria)							
Escherichia coli	organic	organic	organic	37	$137–283.10^{-6}$	assuming	Ingraham (1958);
	organic	organic	organic	20	68.10^{-6}	cells are	Altman and Dittmer (1962)
	organic	organic	organic	37	110.10^{-6}	4 μm $\times$	Chohji et al.
	glucose	glucose	inorganic	37	68.10^{-6}	1 μm	(1976)
Lactobacillus casei	organic	organic	organic	25	214.10^{-6}		Altman and Dittmer (1962)
							Pradel and Clement-
Rhodopseudomonas	Light	organic	inorganic	21	50.10^{-6}	0.8	Metral (1977); Trüper
sphaeroides	Light	organic	inorganic	30	50.10^{-6}	0.8	and Pfennig (1978)
							Göbel (1978);
Rhodopseudomonas	Light	organic	inorganic	30	46.10^{-6}	1.2	Trüper and Pfennig
capsulata		(lactate)					(1978)
Chlorobium limicola		$CO_2 +$					Broch-Due
forma sp.	Light	organic	inorganic	26	42.10^{-6}	0.5	et al. (1978);
thiosulphotophilum		(acetate)					Trüper and Pfennig (1978)
							Lippert and Pfennig (1969)
Chlorobium	Light	CO_2	inorganic	25	$15–27.10^{-6}$	0.5	Kelley (1974)
thiosulphotophilum							Trüper and Pfennig (1978)

EUKARYOTES:

Algae

Bacillariophyta

"small"	Light	inorganic	inorganic	various	30.10^{-6}	30	Banse (1982)
"Large"	Light	inorganic	inorganic	various	10.10^{-6}	10^6	Banse (1982)

Chlorophyta:

Chlorophyceae

Choricystis coccoides	Light	inorganic	inorganic	5	$7.2.10^{-6}$	0.5	Vincent (1962)
							Brown (1982, 1985);
Nannochloris bacillaris	Light	inorganic	inorganic	20	$6.4.10^{-6}$	(0.9 –) 3–7	Brown and Elfman (1983);
strain UWASH 20-2-2					$8.8.10^{-6}$		Brown and Hellebust (1980a,b)
Nannochlorum eucarotum							
(marine)	Light	inorganic	inorganic	25	$2.5.10^{-6}$	1.8	Wilhelm et al. (1982);
Friedmannia sp. (fw)	Light	inorganic	inorganic	5	$9.6.10^{-6}$	4.4	Vincent (1962)
						9	
Coelastrum microporum	Light	inorganic	inorganic	15–27°C	13.10^{-6}	(assuming C =	Schlesing et al. (1981)
Texas strain 281)						⅛ of fresh weight)	

Species							Reference
Monoraphidum contortum	Light	inorganic	inorganic	5	$3.8.10^{-6}$	18.2	Vincent (1982)
Chlorella nana (marine)	Light	inorganic	inorganic	25	12.10^{-6}	6.0	Thinh and Griffiths (1985)
Chlorella pyrenoidosa (Emerson strain)	Light	inorganic	inorganic	25–26	18.10^{-6}	≥ 30	Sorokin (1959) Myers and Graham (1961)
Chlorella pyrenoidosa (TX 7-11-05)	Light	inorganic	inorganic	39	20.10^{-6}	≥ 30	Sorokin (1959)
Chlorella vulgaris	Light	inorganic	inorganic	25 / 15–27	17.10^{-6} / 8.10^{-6}	≥ 30 100 (assuming C = ⅛ of fresh weight)	Schlesinger et al. (1981)
Chlamydomonas pulsatilla (rock pools)	Light	inorganic	inorganic	20	12.10^{-6} (3% SW) 6.10^{-6} (100% SW)	$\geq 2,600$ $\geq 3,200$	Hellebust and Le Gresley (1985)
Chlamydomonas reinhardtii	Light	inorganic	inorganic	25	21.10^{-6}	≥ 63	Sorokin and Krauss (1958)
"*Stichococcus* sp." (fw)	Light	inorganic	inorganic	38	12.10^{-6}	24–55	Sorokin and Nishino (1983) Sorokin (1975)
Dunaliella sp. (marine)	Light	inorganic	inorganic	28	13.10^{-6}	60	Owens and Seliger (1975)
Dunaliella tertiolecta (marine)	Light	inorganic	inorganic	15	$13–14$ 10^{-6}	232 (asuming C = ⅛ of fresh weight) 800 (assuming C = ⅛ of fresh weight)	Falkowski and Owens (1980) Borowitzka and Brown (1974) Schlesinger et al. (1981)
Tetraedron bitridens (fw)	Light	inorganic	inorganic	15–27	6.10^{-6}		
Hydrodictyon africanum (fw)	Light	inorganic	inorganic	15	$1.5.10^{-6}$	10^7	Raven and Glidewell (1975)
Prototheca zopfii (fw)	organic	organic	organic	25	44.10^{-6}	1436 (mean size of cells in multiple fission cycle)	Poynton (1973) Poynton and Branton (1972)
	organic	organic	inorganic	25	44.10^{-6}		Lloyd and Turner (1968)

Polytomella uvella (fw)	organic	organic	organic	25	31.10^{-6}		Altman and Dittmer (1962)
Dinophyta							
"Small"	Light	inorganic	inorganic	various	8.10^{-6}	400	Banse (1982)
"Large"	Light	inorganic	inorganic	various	$3.7.10^{-6}$	5.10^4	Banse (1982)
"Small ciliates"	organic	organic	organic	various	39.10^{-6}	10^3	Banse (1982)
"Large ciliates"	organic	organic	organic	various	$4.6.10^{-6}$	5.10^5	Banse (1982)
Tetrahymea geleii	organic	organic	organic	24	$46-88.10^{-6}$		Altman and Dittmer (1962)
Tetrahymea pyriformis	organic	organic	organic	29	38.10^{-6}		Altman and Dittmer (1972)
Naegleria gruberi	organic	organic	organic	30	48.10^{-6}		Altman and Dittmer (1972)
Saccharomyces cerevisae	organic	organic	organic	20–30	$60-96.10^{-6}$	70	Fenchel (1974) Boehlke and Frierson (1975); Waldron and Lacroute (1975);
	glucose	glucose	inorganic	30	45.10^{-6}	70	
Achlya bisexualis	organic	organic	organic	24	170.10^{-6}		Griffin et al. (1974)
	glucose	glucose	glutamate	24	77.10^{-6}		

4. Mineral (and Organic) Phases Related to Regulation of Cell-Density Relative to That of the Medium

In addition to motility induced by flagella activity, planktophytes also exhibit motility related to decreased density of cells (*upward* movement relative to the medium) or increased density of cells (*downward* movement relative to the medium) (Table 1). As with flagella-induced movement, so with "passive" movement relative to the medium: while the movement is, apparently related to attaining optimal resource availabilities by moving the cells through the medium from places with less optimal resource availabilities to places with more optimal resource availabilities, the capacity for "passive" movement may not be lost under conditions which permit μ_{max} to be achieved.

Raven (1984a, chapter 9) discusses the mineral phases which can alter cell density, and thus determine the upward or downward movement of the cells relative to the medium. These mineral phases can be divided into three categories. First there are solids (SiO_2, $CaCO_3$, $BaSO_4$), which are denser than other cell components, and denser than the surroundings, and thus lead to downward movement, second there are aqueous (salt) solutions in vacuoles which, by emphasizing "heavy" ions, such as Ca^{2+}, Mg^{2+}, K^+ and SO_4^{2-}, can increase the overall cell density and thus lead to downward movement, or, by emphasizing "light" ions, such as NH_4^+, H^+, and Na^+, can decrease the overall density of the cell and, in extreme cases, but only in seawater, lead to upward movement. Finally, gas phases (mainly N_2, O_2, CO_2 and H_2O), contained in prokaryotic gas vacuoles, can greatly reduce the overall cell density, and lead to upward movement in fresh — or sea — water. Polysaccharide deposits are important in increasing cell density, while lipid deposits may be important in decreasing cell density.

The influence which these "density-increasing" or "density-decreasing" changes in cell composition in have on the rate of vertical movement relative to the medium in spherical cells or radius 0.5, 5 and 50 μm is shown in Table 5. Taking first the cells *without* specific density-altering mineral (or organic) phases, line 1 of Table 5 gives the sinking rate of our three sizes of cells computed from equation (4):

$$(4) \qquad v \; = \; \frac{2r^2 \cdot g \cdot (p^1 - p)}{9\eta}$$

where
- v = velocity of downward movement/m $\cdot$ s^{-1}
- r = radius of spherical cell/m
- g = acceleration due to gravity/m $\cdot$ s^{-2}
- $\varrho\,'$ = density of cells/kg $\cdot$ m^{-3}
- ϱ = density of medium/kg $\cdot$ m^{-3}
- η = dynamic viscosity of the medium/kg $\cdot$ m$^{-1} \cdot$ s^{-1}.

Values in the terms of equation (4) are g = 9.807 m $\cdot$ s^{-1}, η = 9.10^{-4} kg $\cdot$ m$^{-1} \cdot$ s^{-1}, ϱ = 1000 kg $\cdot$ m^{-3} (freshwater) or 1026 kg $\cdot$ m^{-3} (seawater), $\varrho\,'$ = 1050 kg $\cdot$ m^{-3} (cells with no mineral phases in freshwater) or 1075 kg $\cdot$ m^{-3} (cells with no mineral phases in seawater); justification of these values may be found in Smayda (1971). Walsby and Reynolds (1980), and Reynolds (1984). These values were used to obtain the sinking rates shown in line 1 of Table 5. We see that the predicted sinking rate for a 50 μm radius cell is 300 μm $\cdot$ s^{-1} or 26 m $\cdot$ d^{-1}, i.e. the cell could sink through a stratified 50 m euphotic zone in less than 2 d. The 5 μm radius cell is predicted to sink at 3 μm $\cdot$ s^{-1} or 0.26 m $\cdot$ d^{-1}, so that the cell would take more than 6 mo to sink through a stratified 50 m euphotic zone. The 0.5 μm radius cell is predicted to sink at only 0.03 μm $\cdot$ s^{-1} or 2.6 mm $\cdot$ d^{-1}, and would accordingly take more than 50 yr to sink through a stratified 50 m euphotic zone! While the 50 μm radius cell sinks a distance equal to its own diameter in 0.33 s, the 5 μm radius cell takes 3.3 s, and the 0.5 μm radius cell takes 33 s.

18

These computed sinking rates for the different cell sizes must, in terms of the assumed selective significance of movement of cells relative to the surrounding medium in moving the cells to a region of superior resource supply be viewed in relation to the spatial variability of resource supply. Raven and Richardson (1984) quote gradients of $[NO_3^-]$ in stratified water bodies of up to 1.5 mmol $\cdot$ m^{-3} $\cdot$ m^{-1}, with higher concentrations in deeper water. Even this high value of a nutrient gradient only permits a 0.5 μm radius cell to increase its bulk phase $[NO_3^-]$ by 3.9 μmol $\cdot$ m^{-3} $\cdot$ d^{-1} by sinking at our computed rate of 0.03 μm $\cdot$ s^{-1}. Against an assumed initial $[NO_3^-]$ of 0.5 mmol $\cdot$ m^{-3}, the sinking rate of our 0.5 μm radius cell only allows it to increase its bulk phase $[NO_3^-]$ by 0.0079, or 0.79%. In terms of the likely kinetic values for NO_3^- influx by these cells, i.e. a linear increase in NO_3^- influx with increase in bulk phase $[NO_3^-]$ in the range of 0–1 mmol $\cdot$ m^{-3}, we see that even sinking through a relatively steep gradient of $[NO_3^-]$ only increases the NO_3^- availability to the cells by less than 1% per day. The 5 μm radius cell could, by contrast, increase by 78% (almost double) its NO_3^- availability in 1 day by sinking 0.26 m from an initial concentration of 0.5 mmol $\cdot$ m^{-3} through a gradient in which $[NO_3^-]$ increases downward by 1.5 mmol $\cdot$ m^{-3} $\cdot$ m^{-1}.

The other side of the resource supply to a cell sinking in a stratified environment is the decrease in light supply with depth, granted a constant diel variation in surface photon flux density (I_o, mol photon $\cdot$ m^{-2} $\cdot$ s^{-1}) and a constant attenuation coefficient (k, m^{-1}). Raven and Richardson (1984) used a value of 0.12 m^{-1} for k as applying to the environment with the (NO_3^-) gradient of 1.5 mmol $\cdot$ m^{-3} $\cdot$ m^{-1}. Assuming a mean photon flux density of 10 μmol photon $\cdot$ m^{-2} $\cdot$ s^{-1} (I_o^1) during the photoperiod at the initial station of our "model" planktophyte cells, we can compute the photon flux density (I_d) d metres below the initial station from Beer's law (Equation (5)).

$$(5) \qquad I_d = I_o^1 e^{-kd}$$

One day of sinking for the 0.5 μm radius cell moves it through a distance d of 2.6 10^{-3}m, i.e. to a station with a mean P.F.D. of 9.997 rather than 10 μmol photon $\cdot$ m^{-2} $\cdot$ s^{-1} at the initial station. The availability of light for growth and maintenance is only decreased, in this "light limited" range of P.F.D.'s for growth, by 0.03% per day. For a 5 μm radius cell sinking 0.26 m, i.e. to a station with a mean PFD of 9.69 rather than 10 μmol photon $\cdot$ m^{-2} $\cdot$ s^{-1} at the initial station, so that the decrement of (limiting) PFD due to a days worth of sinking is 3%.

This analysis of the role of cell size in regulating the downward movement of planktophytes relative to the bulk medium shows that a 0.5 μm cell moves through a distance corresponding to a very small change in resource supply in a stratified water body as a result of one days sinking, and that larger planktophytes corresponding make much more progress in terms of the change in resource availability corresponding to 1 days' sinking. Accordingly, *granted our assumptions as to cell density relative to the density of the bulk medium*, the sinking rate of picoplankton cells is very small relative to the spatial variation in resource availability, at least over periods of days or even weeks.

The utility of sinking of picoplankton cells in terms of movement to environments of different resource supply (more nutrient solutes per m^3 of solution which might alleviate nutrient deficiency; lower photon flux density relative to an initial high value which might decrease the incidence of photoinhibition of photosynthesis) is shown to be of even *less* significance than the earlier analysis might indicate if we consider *temporal variations* in resource availability to the volume element of medium which contains the cells. Even if the medium is *completely unstirred* (solutes only move by diffusion in the volume element; no vertical mass movement of medium to alter the position in the PFD gradient of the volume element), resource availability can be changed with time by variations in mineralisation rate in neighbouring volume elements and by seasonal (and stochastic) changes in PFD at the depth of the volume element. Any stirring of the medium which in extreme cases lead to vertical movements of our volume element at velocities of up to 1 m $\cdot$ s^{-1} in well-mixed water bodies, can clearly completely overwhelm the sinking rates of planktophytes and, especially, that of picoplankton organisms by $\sim 10^7$ − fold!).

It would, then, seem that the rate of sinking attained by a picoplankton cell of "normal" density is *probably* negligible with respect to moving organisms to more favourable resource-supply conditions. Can changes in overall cell density (at constant cell size) so increase sinking (replace some of the organic material by dense materials) as to give significant increases in resource supply in a stratified environment? The densest components commonly found in planktophytes are SiO_2 ($\varrho = 2.17$–2.66) and $CaCO_3$ ($\varrho = 2.71$–2.93) (Table 1); $BaSO_4$ is substantially denser (see Raven et al. 1986), but is relatively rare in planktophytes (it occurs in some non-picoplanktonic Prymnesiophyta: Fresnel et al. 1979) possibly as a result of the relative unavailability of Ba in the environment (Bowen 1979; cf. Brook 1981). Table 5 (row 2) shows the predicted sinking rates for cells of various sizes with their volume half made up of the organic components assumed for the cells in row 1, of Table 5, and half of silica.

TABLE 5. Computed rates of vertical movement of spherical cells of various sizes, with a range of density differences (ϱ_{cell}–ϱ_{medium}), using Stokes Law.

	Sinking rate, m $\cdot$ s^{-1} (m $\cdot$ d^{-1}) for cells of radius:		
(ϱ_{cell}–ϱ_{medium})	0.5 μm	5 μm	50 μm
$+50$ kg $\cdot$ m^{-3}[a]	3.10^{-8} ($2.6.10^{-3}$)	3.10^{-6} (0.26)	3.10^{-4} (26)
$+820$ kg $\cdot$ m^{-3}[b]	5.10^{-7} ($4.3.10^{-2}$)	3.10^{-5} (4.3)	5.10^{-3} (430)
-420 kg $\cdot$ m^{-3}[c]	-3.10^{-7} ($-2.6.10^{-2}$)	-3.10^{-5} (-2.6)	-3.10^{-3} (-260)

[a]The density difference of $+50$ kg $\cdot$ m^{-3} assumes a cell (no pure mineral phases) with a density of 1050 kg $\cdot$ m^{-3} in freshwater (1000 kg $\cdot$ m^{-3}) or with a density of 1075 kg $\cdot$ m^{-3} in seawater (1025 kg $\cdot$ m^{-3}).
[b]The density difference of $+820$ kg $\cdot$ m^{-3} assumes a cell with half the volume having a density of 2600 kg $\cdot$ m^{-3} (SiO_2), while the remainder has a density as described under (a) above (actual density difference $+ 825$ kg $\cdot$ m^{-3} for freshwater cells; $+812.5$ kg $\cdot$ m^{-3} for seawater cells).
[c]The density difference of -480 kg $\cdot$ m^{-3} assumes a cell with half the volume having a density of 1 km $\cdot$ m^{-3} (lumen of gas vacuoles), while the remainder has a density as described under (a) above (actual density difference -475 kg $\cdot$ m^{-3} for freshwater, -487.5 kg $\cdot$ m^{-3} for seawater).

Even with this very large fraction of density-increasing substances in the cell volume, we see that the predicted sinking velocity of a 0.5 μm radius cell is only increased from the 0.03 μm $\cdot$ s^{-1} of the non-silicified cell to 0.5 μm $\cdot$ s^{-1}; this is still less than the sinking velocity of an "organic" 5 μm radius cell (3 μm $\cdot$ s^{-1}). Our hypothetical "high silica" picoplankton cell would still only sink 43 mm $\cdot$ d^{-1}, and would take over 3 yr to sink through a stratified 50 m deep euphotic zone. While the comments made earlier as to the quantitative inadequacy of the sinking rate of an "organic" 0.5 μm radius cell to materially improve its resource supply in a resource-limited natural environment must be somewhat muted for the "half SiO_2" 0.5 μm radius cell, we must bear in mind what costs, in terms of potential μ_{max}, are likely to be incurred by such a large non-catalytic investment of cell volume. If 0.3 of the cell volume of our "organic" cell (ignoring compatible solutes) is devoted to "non-catalytic" fixed costs (DNA, cytoskeleton, storage compounds), and the same fraction is occupied by these components in the "high silica" cell, then the "high silica" cell (which already has 0.5 of its volume pre-empted by SiO_2) only has 0.2 of its volume available for "catalysts of growth" while the "organic" cell has 0.7 of its volume occupied by "catalysts of growth". Application of equation (1), assuming 125 kg C $\cdot$ (m^3 "organic" cell)$^{-1}$ and 62.5 kg C $\cdot$ (m^3 "high silica" cell)$^{-1}$, suggests a ratio of μ_{max} (high silica): μ_{max} (organic) of $^2/_7$ or 0.286. Any allocation of C to form a template for SiO_2 deposition or in any other role not found in the "organic" cell, would *decrease* this ratio.

Similar general conclusions relate to the occupation of half the cell volume of a 0.5 μm cell with gas-filled vacuoles, an option only, apparently, open to prokaryotic picoplankton organisms. Such extensive gas vacuolation would render the cells substantially less dense than the medium (row 3 of Table 5). However, in the case of our "model" 0.5 μm radius cell, the velocity of upward movement relative to the surrounding water is only 0.3 μm $\cdot$ s^{-1}. Assuming the 'usual' inverse gradients of nutrient concentration and PFD (nutrient concentration lowest near the surface where the PFD is highest) the assumed "function" of upward movement is to move the cell to a higher (but not photoinhibitory) mean PFD, thus giving access to a higher mean PFD to energize growth and maintenance processes. With the attenuation coefficient of 0.12 m^{-1} assumed earlier, and a mean PFD in the light period of 1 μmol photon $\cdot$ m^{-2} $\cdot$ s^{-1} at the initial position of the cell, the upward movement of 26 mm $\cdot$ d^{-1} would only increase the mean PFD to 1.003 μmol photon $\cdot$ m^{-2} $\cdot$ s^{-1} in a day (i.e. by 0.3%). Accordingly, we can conclude that even the massive dilution of other cell components attendant on devoting half of the cell volume to the gas phase in the gas vesicles, which reduces the potential μ_{max} to not more than 0.286 of that of a cell of the same volume which lacks the gas phase (see above) can only increase the mean PFD available to the cell in an otherwise constant environment by 0.3% per day. This seems to be a very small benefit resulting from a substantial cost.

This analysis of the velocities of vertical movement which can be attained by picoplankter cells, even granted very substantial allocations of cell volume to phases which increase or decrease cell densities, suggests that the achieved velocities are small relative to likely gradients of resources, the changeability of resource supply conditions and potential cell growth rates. Even if the formation of the density-changing phases were repressible under conditions of high resource availability thus reducing the penalty in terms of μ_{max} attendant on reducing the per cell quantity of catalysts resulting from the presence of these phases, it would appear that their occurrence could only yield a marginal effect on the resource availability to picoplankter cells. Accordingly, it is not surprising that such density-changing phases are generally absent from picoplankton cells. Among the eukaryotes, the picoplanktonic *Nannochloris/Chlorella* — type of Chrlorophyceae have no obvious density-altering deposits; such deposits are rare in the Chlorococcales, with some larger-celled exception are *Gloeotaenium* ($CaCO_3$: Deviprasad and Chowdary 1981) and a few silicified species (Round 1981). While many motile members of the Prasinophyceae (Micromonadophyceae) have superficial scales, these are absent from the smallest reported member of this class (*Micromonas pusilla*), and in any case lack the extensive mineralisation common in the scales of flagellate members of the Chrysophyceae (SiO_2) and Prymnesiophyceae ($CaCO_3$); Manton (1959); Manton and Parke (1960); Norris (1980); Romanovicz (1981). SiO_2 probably occurs in the walls of the non-flagellate phycoma stage (up to 500 μm diameter) of *Halasphaera* (Norris 1980). In the cyanobacteria, neither specific density-increasing inorganic phases nor gas vacuoles are common in picoplankton cells (Walsby and Reynolds 1980; Walsby 1981).

5. Membranes

The thickness of "fluid mosaic" membranes, based as they are on bi-layers of (mainly) diglycerides with C_{16} and C_{18} fatty acids, does not permit of much variation. As long as membranes can be considered as contributing a fixed number of molecules of some catalyst contained in or on them to unit volume of cell they are as "scaleable" as are soluble catalysts. This applies to membranes such as the thylakoid membranes of eukaryotic plastid and of cyanobacteria, and the inner membrane of eukaryotics mitochondria. Data reviewed by Raven (1984a, Chapter 3 and 4) does not reveal any detectable phydogenetic trends in the density of occupation of these membranes by their major catalysts such as chlorophyll in the case of the thylakoids, consistent with the scalability notion; more recent data (Utkilen et al. 1983; Rosen and Lowe 1984) support this view. With a constant value for mol chlorophyll per m² of thylakoid membrane, a given quantity of chlorophyll per unit volume of cell is achieved with a constant *area* of thylakoid membrane thickness, and thus a constant *volume* of thylakoid membrane per unit *volume* of cell.

Non-scaleability of membrane area (and hence membrane volume) per unit cell volume occurs when the volume enclosed by a given membrane system is a constant fraction of the cell volume, and the shape of the contained volume is constant at different cell sizes. In the case of spherical membrane — bounded compartments, the volume enclosed is proportional to r^3, while the area of the enclosing membrane is proportional to r^2, and the volume of the enclosing membrane is proportional to $10^{-8}r^2$ if r is measured in m and the membrane is 10^{-8} m thick. For spherical cells, with spherical organelles this non-scaleability applies to the outer membrane of the (gram-negative) prokaryotic cyanobacteria, the plasmalemma of cyanobacteria and of eukaryotes; and the membranes round the spherical organelles whose volume is a constant fraction of the cell volume, e.g. the nuclear envelope and the membranes round single — membrane — bounded "vacuoles" (*P* phases); the outer membrane of mitochondria, and the envelope (and, where appropriate, plastid e.r.: see Table 1) membranes of chloroplasts are less firmly placed here because of their usually non-spherical shape. We note that non-scaleability is exacerbated for concentric membrane

TABLE 6. Fraction of volume of prokaryote and eukaryotic cells occupied by membranes which are not "scaleable". Membranes assumed to be 10 nm thick.

	Cell radius = 0.5 μm	Cell radius = 5 μm
Prokaryotic outer membrane	0.06	0.006
Prokaryotic cytoplasmic membrane, with wall thickness of 10 nm in 0.5 μm radius cell, 100 nm in 5 μm radius cell	0.055	0.0057
Total of outer membrane + cytoplasmic membrane of prokaryote	0.115	0.0117
Eukaryotic plasmalemma (for wall-less cell)	0.06	0.006
Eukaryotic outer envelope membrane of plastid in a Chlorophyte, assuming one spherical plastid occupying half the cell volume	0.038	0.0038
As above, but inner envelop membrane separated from outer envelope membrane by 10 nm	0.037	0.0037
Total envelope membrane + intermembrane space of a Chlorophyte	0.105	0.01145
Total envelope membrane + inter-membrane space of a Dinophyte (3 × 10 nm thick membranes in series, with 10 nm "N" phase between each), if one spherical plastid occupies half cell volume	0.166	0.0191
Total envelope membrane + intermembrane space of a diatom or Prymnesiophyte (4 × 10 nm thick membranes in series, with 10 nm "N" phase between each), if one spherical plastid occupies half cell volume	0.221	0.0258

systems (e.g. inner and outer plastid envelope membranes) which have a constant (cell size-independent) separation of the membranes. We note that this constraint does *not* apply to thylakoids and inner mitochondrial membranes where the appropriate, constant, ratio of membrane area (and volume) to stroma (or matrix) volume is maintained as the volume of stroma (or matrix) decreases by changes in the number and size of thylakoids and the degree of invagination of the inner mitochondrial membrane.

Quantitation of the fraction of the cell volume occupied by membranes in prokaryotes and eukaryotes of various sizes has been carried out for a number of membranes in Table 6. For the prokaryotes, the sum of the "gram-negative" outer membrane *plus* the cytoplasmic membrane (plasmalemma) occupies only 1.17% of the cell volume of a 5 μm radius cell, but 11.5% of the cell volume of a 0.5 μm radius cell. If we assume that all of the necessary functions of the outer-membrane and the plasmalemma can be prosecuted, under optimal resource supply conditions, by the quantity of membrane found in the 5 μm radius cell, it would appear that the 0.5 μm radius cell is over-provided with outer membrane and plasmalemma materials (e.g. barriers to, and catalysts of, transport). Accordingly, if the 5 μm radius cell can achieve a certain μ_{max}, the 0.5 μm radius cell has more membrane material in the outer membrane and plasmalemma than is necessary to provide a balanced ratio of outer membrane- and plasmalemma-associated barriers plus catalysts to other cell components in achieving the same μ_{max}. The problem is exacerbated if these two membranes fulfill in the 0.5 μm radius cell only those functions which they carry out in the 5 μm radius cell if we consider the quantity of C in membranes relative to that in the cell as a whole. We have assumed earlier that it is one eighth of the cell fresh weight; C is unlikely to be less than one half of membrane fresh weight, so that, even without allowing for the greater mean density of membranes (Bowles et al. 1979) than of whole cells quoted in Table 5, an increased volume fraction of membrane in cells means an increased content of C per unit cell volume. This in turn means that more c has to be processed *via* photosynthesis and downstream reactions to make unit volume of cell material, with a corresponding penalty in terms of a decreased potential μ_{max} (see equation (1)), since essential catalysts have been displaced from cell volume by the additional, and *ex hypothesis* superfluous, membrane material.

A possible "use" for the apparent excess of membrane area in the plasmalemma of a 0.5 μm radius photolithotrophic cell (relative to a 5 μm radius cell with similar metabolism) would be to transfer some, or all, of the energy transduction processes which occur in and around the intracellular thylakoid membranes to the plasmalemma.

If we take $30.10^{-6} \cdot s^{-1}$ as the highest μ_{max} value attainable for cyanobacteria (or eukaryotic microalgae) at 20°C in continuous light (Table 1), then we can compute a *minimum* membrane area to provide reductant and ATP for photolithotrophic growth by assuming that growth equivalent to 1 mol CO_2 converted to 1 mol cell C needs 2 mol NADPH and 5 mol ATP, so that ATP is needed at $1.5 \ 10^{-4}$ mol ATP (mol cell C)$^{-1} \cdot s^{-1}$ (Raven 1984a, b). If the maximum H^+ recirculation through the thylakoid membrane is 1 μmol H^+ (m^2 thylakoid area)$^{-1} \cdot s^{-1}$ at 20°C (i.e. 1 μmol H^+ (m^2 thylakoid area)$^{-1} \cdot s^{-1}$ moved into the intrathylakoid space by photoredox reactions and out again through the ATP synthetase: Raven 1984a), and 1 mol ATP is generated per 3 mol H^+ recirculated (Raven 1984a, b) then 450 m^2 thylakoid area is needed per mol cell C to attain the μ_{max} of $3.10^{-5} \cdot s^{-1}$. If one eighth of the cell fresh weight is C, and approximating 1000 kg $\cdot$ m^{-3} as the cell density, 4.68 μm^2 of thylakoid area is needed per μm^3 cell volume to achieve the specified μ^{max}. A 5 μm radius cell only has 0.6 μm^2 plasmalemma area per μm^3 cell volume, so that its need for internalized thylakoids to achieve a μ_{max} of $3.10^{-5} \cdot s^{-1}$ is clear. A 0.5 μm radius cell, by contrast, has 6 μm^2 of plasmalemma area per μm^3 of cell volume, so that the need for internalized thylakoids to provide the membrane area necessary for photoredox and photophosphorylation reactions is less clear. The 0.6 μm^2 of plasmalemma per μm^3 of cell volume in a 5 μm radius cell is clearly adequate for the influx of nutrient solutes and the efflux of excreta (O_2, H^+/OH^-: see Raven 1984a). The assumption of the need for as little as 5 mol ATP per mol CO_2 converted to cell C above implies a high exogenous CO_2 concentration which *represses* the energy-requiring CO_2 accumulating mechanism (Raven and Lucas 1985) and *suppresses* the energy-requiring RuBPo activity (with subsequent PCOC activity or glycollate excretion) of RUBISCO; there is no well-defined upper limit on the CO_2 influx by diffusion across the plasmalemma. The assumptions made earlier about cell C content shows that a 5 μm radius cell has a CO_2 influx of 0.52 μmol CO_2 $\cdot$ (m^2 plasmalemma)$^{-1}$ $\cdot$ s^{-1} during growth at 3.10^{-5} $\cdot$ s^{-1}. A P_{CO_2} pf $\cdot$ 3.10^{-3} m $\cdot$ s^{-1} (Raven 1984a) means that the excess of CO_2 concentration at the outer surface of the plasmalemma over at the inner surface during steady-state photosynthesis is only some 0.17 mmol CO_2 $\cdot$ m^{-3} in order to provide the driving force for diffusive CO_2 entry: this is very small relative to the intracellular CO_2 concentration required to saturate RuBPc and suppress RuBPo in cyanobacteria (several hundred mmol CO_2 $\cdot$ m^{-3}; Raven 1984a). Similar considerations apply for O_2 exit during photosynthesis (Raven 1977; Samish 1975). For mediated fluxes, the largest *net* mediated fluxes needed for microalgal growth with CO_2 as C source relate to N entry and H^+ fluxes required for pH regulation. Even if the mol N: mol C ratio in the biomass is as low as 0.2, the required N(NH_4^+ or NO_3^-) influxes in a 5 μm radius cell are only 0.104 μmol $\cdot$ m^2 plasmalemma)$^{-1}$ $\cdot$ s^{-1}, with net H^+ efflux (NH_4^+) or influx (NO_3^-) of a similar magnitude; these fluxes are not excessive in terms of portermediated primary active transport at up to 1 μmol $\cdot$ m^{-2} $\cdot$ s^{-1} at 20°C, and of secondary active transport and mediated uniport with even higher capacities (Raven 1984a). Accordingly, there is no obvious difficulty with nutrient influx/excreta efflux through the 0.6 μm^2 of plasmalemma per μm^3 cell volume for a 5 μm radius cell growing at 3.10^{-5} $\cdot$ s^{-1}. For a 0.5 μm radius cell also growing at 3.10^{-5} $\cdot$ s^{-1}, with a similar C and N content per μm^3 cell volume to the 5 μm radius cell, the 6 μm^2 of plasmalemma area per μm^3 cell volume represents a substantial overprovision of membrane area for these necessary transplasmalemma fluxes. If we assume that 0.6 μm^2 of plasmalemma area per μm^3 cell volume (as in the 5 μm radius cell) is adequate for CO_2, N source, P source, etc., influx, O_2 efflux and pH-regulating H^+ fluxes, then (6.0–0.6) μm^2 or 5.4 μm^2 per μm^3 cell volume of the plasmalemma area of the 0.5 μm radius cell is potentially available for photoredox and ATP synthesis reactions. The computations performed earlier suggest that 4.68 μm^2 of thylakoid area is needed per μm^3 cell volume for photoredox and ATP synthesis reactions during growth at 3.10^{-5} $\cdot$ s^{-1}, so that a 0.5 μm radius cell could, in theory, accommodate all of these thylakoid-associated processes in its plasmalemma without displacing essential nutrient transport and excretion processes. There is a precedent for a thylakoid-less photolithtrophic cyanobacterium in the form of *Gloeobacter violaceae* (Rippka et al. 1974; Stanier and Cohen-Bazire 1977) which has all of its photosynthetic pigments, and other photoredox and photophosphorylation activities, associated with the plasmalemma. By contrast, the available data from cell fractionation studies on *Anacystis nidulans* and *Synechocystis* sp. (Omata and Murata 1983, 1984a,b) suggest that essentially all of the photosynthetic catalysts which are membrane-associated are in the thylakoid membranes, although the plasmalemma of this group of cyanobacteria clearly contains redox components and, perhaps, a vanadate-insensitive ATPase which constitutes a dark respiratory chain which might function in oxidative phosphorylation (Omata and Murata 1984b; Craig et al. 1984; Peschek et al 1984; Scherer and Böger 1984). *Gloeobacter violacea* has a very low μ_{max} (20°C) of $1.9.10^{-6}$ $\cdot$ s^{-1} (Table 4), although the growth conditions may not have been optimized for *Gloeobacter* or the other stains of cyanobacteria investigated by Rippka, et al. (1974). A possible problem for cyanobacteria with photoredox reactions located in their plasmalemma relates to the pH-dependence of the H_2O dehydrogenase reac-

24

tions of photosynthesis which occurs on the "P" side of photosynthetic membranes; i.e. on the face of the thylakoid which faces the intrathylakoid space and on the periplasmic (cell-wall-facing) side of the *Gloeobacter* plasmalemma (Raven 1984a). The pH optimum for this reaction is 6.0 or below, which corresponds to the intra-thylakoid pH in illuminated photosynthetic O_2-evolvers, (Raven and Smith 1981; Delrieu et al., 1985), but is lower than the likely periplasmic pH of most cyanobacteria (Brock 1973; Coleman and Colman 1981). Internalization of photoredox reactions enables the N and P sides of the photosynthetic (thylakoid) membrane to each be at their optimal pH values for the reactions which occur during photosynthesis, with most of the $\Delta\bar{\mu}$ H^+ across the membrane being as ΔpH rather than $\Delta\psi$ (see Raven and Smith 1981). These might, of course, be other regulatory problems related to the occurrence of photoredox and photophosphorylation as well as nutrient-transport, excretion and respiratory processes in the plasmalemma even if the area of membrane available were adequate, in a small cell, for sufficient catalytic activity of these processes to yield a high rate of growth.

Turning now to the eukaryotes, Table 6 shows (as is inevitable!) that the fraction of cell volume occupied by the plasmalemma in a 0.5 μm radius cell is substantially greater than in a 5 μm radius cell in a quantitatively very similar way to that in pro-karyotes. Similar arguments may be used here as were used for the prokaryotes with respect to the possible overprovision of plasmalemma area in 0.5 μm radius cells as far as transport processes related to growth at μ_{max} in a resource-rich environment is concerned. However, the eukaryotes clearly do not have photosynthetic redox and ATP-generation reactions in their plasmalemma, so the 0.5 μm radius eukaryotes cannot "take up the slack" in this way. The photoredox and photosynthetic ATP-generating processes of eukaryotes are *always* in thylakoids which are *always* sequestered, along with the stroma, behind two (Chlorophyta, some Dinophyta) or three (most Dinophyta) chloroplast envelope membranes while plastids of the Bacillariophyta and Prymnesiophyta have two envelope membranes and two chloroplast endoplasmic reticulum membranes (Gibbs 1981; Whatley and Whatley 1981; Cavalier-Smith 1982; Ford 1984; Durand and Berkaloff 1985). Table 6 shows that the two envelope membranes of a Chlorophyte, together with the intermembrane space, occupies 1.145% of the volume of a 5 μm radius cell but 11.5% of the volume of a 0.5 μm radius cell, assuming that a single spherical plastid occupies half of the cell volume. Table 6.4B of Raven (1984a) shows that the plastid volume of *Chlorella* spp, and *Chlamydomonas reinhardtii* (Chlorophyta: Chlorophyceae) is 0.33–0.42 of the total cell volume, i.e. rather less than the value assumed here; however, the single plastid of cells of these algae is much more non-spherical in shape than is the cell itself, and it is likely (from semi-quantitative inferences from cell reconstructions based on serial section) that the area of *each* of the two membranes of the plastid envelope is similar to that of the plasmalemma (see Raven 1980a, b). Accordingly, the fraction of the cell volume occupied by plastid envelope quoted in Table 6 is not likely to be widely in error. In terms of transport requirements, the 0.5 μm and the 5 μm radius cells seem to be well provided with plastid envelope area. For a 5 μm radius cell growing photolithotrophically at 3.10^{-5} $\cdot$ s^{-1}, we have seen (above) that the mean net CO_2 influx at the plasmalemma is 0.52 μmol $\cdot$ m^{-2} $\cdot$ s^{-1}. For a spherical plastid occupying half the cell volume the surface area of the envelope is 198 μm^2 relative to the 314 μm^2 of the plasmalemma, so the CO_2 influx at the plastid envelope is 0.82 μmol $\cdot$ m^{-2} $\cdot$ s^{-1}; this flux is *probably* by passive diffusion (Marcus, et al., 1984), and thus poses no problem, granted reasonable values for the P_{CO_2} (see above). If the reduced product of photosynthesis is all exported from the plastid as a C_3 compound (an approximation to the multifarious fluxes across that membrane) *via* porters in the inner envelope membrane the required flux is 0.27 μmol $\cdot$ m^{-2} $\cdot$ s^{-1}; this is not an excessive flux (see Raven 1980a, b, 1984a). For the 0.5 μm radius cell growing at the same rate the fluxes are 82 μmol CO_2 influx $\cdot$ m^{-2} $\cdot$ s^{-1} and 27 μmol C_3 efflux $\cdot$ m^{-2} $\cdot$ s^{-1}; it would seem that there is a substantial overprovision of plastid envelope area. Accordingly the volume fraction of the envelope is greater than transport considerations dictate.

Turning briefly to the non-Chlorophycean algae, Prasinophycean (Micromonodophycean) algae are probably similar to their Chlorophycean relatives since they have but a single plastid in each cell, with a similar fraction of cell volume probably occupied by the plastid (note that the *Platymonas* entry in table 6.4B of Raven (1984a) is really for a member of the Pleurastrophyceae, *Tetraselmis* = *Prasinocladus* = *Platymonas*: Raven 1986b).

There are no known Dinophyte examples of picoplankters, so that the even greater fraction of cell volume occupied by envelope membranes in the majority of Dinophytes (3 membranes) than in Chlorophytes (2 membranes) for otherwise similar cells is not a consideration for picoplankters. While Dinophytes do not store their starch in their plastids, and hence may have smaller volume fractions of plastids than do the Chlorophytes (cf. Sicko-Good, Stoermer and Ladewiski 1977) they almost invariably have more than one plastid per cell (Dodge 1983; cf. Inoyye and Pienaar 1983) which, for a given plastid shape and total plastid volume increases the area and thus volume of the plastid envelope.

Somewhat similar considerations apply to the Bacillariophyta and Prymnesiophyta as have just been mentioned for the Dinophyta, although the smallest Prymnesiophytes are *almost* of picoplankter size (Green and Pienaar 1977). As for the Dinophyta, the storage polysaccharide is extraplastidic with a rather smaller plastid volume fraction than in Chlorophytes even if vacuole volume in diatoms is allowed for (Sicko-Good, Stoermert and Ladewski 1977; Sicko-Good 1982; Rosen and Lowe 1984); furthermore, there are generally two (Prymnesiophyta) or at least two (Bacillariophyta) plastids per cell (cf. Murphy and Haugen 1985). Table 6 shows that the 0.5 μm radius Prymnesiophyte–Bacillariophyte cell would have a very substantial fraction of its cell volume (22% or so) occupied by plastid envelope if a single plastid occupied half of the cell volume and all forms of the membranes (two plastid envelope membranes; two plastid endoplasmic reticulum membranes) around the plastid, and the three aqueous phase between them, are included in the computation. It is a moot point as to whether the plastid endoplasmic reticulum should be included in their computations, since it is possible that the "plastid" endoplasmic reticulum fulfills at least some of the functions in the cell which would otherwise be carried out by "non-plastid" endoplasmic reticulum. At all events, the volume of plastid-investing membranes in these three "Chromophyte" Divisions of planktophytes *could* impose a constraint on the capacity of these three Divisions to be picoplankters, especially if vegetative cells are constrained to have two or mor plastids.

Before attempting to analyse quantitatively the membrane–volume-associated constraints on the miniaturization of eukaryotic phototrophic cells we will look at the constraints related to DNA content of the nucleus and the fraction of the cell occupied by the nucleus and the nuclear envelope. Table 6.4B of Raven (1984a) and Table 6 of this paper quote minimal vacuolar volumes as a fraction of cell volume of 0.08 for *Chlorella fusca* var. *vacuolata*, with a minimal cell volume (for "autospores" with the C-DNA content of 0.1 pg DNA) of 33 μm³, while for *Chlamydomonas reirhardtii* the nucleus occupies at least 0.044 of the cell volume which is at least 63 μm³ for autospores with C-DNA content of 0.12 pg DNA. If we temporarily ignore plastid and mitochondrial DNA (i.e. assume that all of the DNA is nuclear), these values yield DNA per μm³ of nucleus of 0.038–0.043 pg per μm³. If we use the "possible" lowest value of C-DNA per photolithotrophic eukaryotic cell (see (9) above) of 0.01 pg DNA, the computed nuclear volume (assuming volume is proportional to DNA content) is 0.0233–0.0263 μm³, or 0.044–0.050 of the cell volume (i.e. rather less than the mean of *Chlorella* and *Chlamydomonas*). If we take the thickness of the nuclear envelope (2 membranes + the space between them) as 30 nm, then the volume of the nuclear envelopes

0.010–0.011 μm^3 or 0.019–0.021 of the cell volume. For *Chlorella* and *Chlamydomonas*, the computed volume of the nuclear envelope is 0.0046–0.0074 of the cell volume. Even with a ratio of nuclear volume in *Chlorella* or *Chlamydomonas* (autospore radius 2.0–2.5 μm) to that in our hypothetical 0.5 μm picoplankter (which somewhat exceeds the ratio of the cell volumes) we see that the non-scaleability of membrane thickness means that the fraction of the cell volume taken up by the nuclear envelope is still substantially higher in the picoplankter than in the larger algae. For a 5 μm radius cell with the same sized nucleus as *Chlorella* or *Chlamydomonas* the volume fraction of nuclear envelope is 0.001 (Table 7).

This sort of analysis of the plasmalemma, plastid envelope and nuclear envelope volume as a fraction of cell volume of eukaryotes as a function of cell size can also be performed for the outer mitochondrial membrane and lysosome/microbody membranes. In the case of the mitochondria, a 0.5 μm radius cell with a volume of 0.524 μm^3 would have a mitochondrial volume of only 0.0157 μm^3 if mitochondria occupied 0.03 of the cell volume (see table 4B of Raven, 1984a). If the radius of a single cylindrical mitochondrion with hemispherical ends is 0.1 μm (Atkinson et al. 1974), then the computed area of outer membrane for a 10 μm thick membrane yields an outer membrane volume of 0.0036 μm^3 which is 0.007 of the total cell volume (Table 7). For a 5 μm radius cell, with the same volume fraction (0.03) of mito-chondria, the 15.7 μm^3 of mitochondria would, as a cylinder of radius 0.10 μm (cf. Blank et al. 1980; Schotz, et al. 1972; Atkinson et al. 1974), the outer mitochondrial membrane occupies 0.006 of the cell volume (see Table 7).

For a 0.5 μm radius cell, a single spherical vacuole, microbody or lysosome occupying 0.05 of the cell volume would have a membrane volume of 0.008 of the cell volume; 3 spherical organelles occupying in total 0.05 of the cell volume would have a membrane fractional volume of 0.011; we assume a value of 0.01 (Table 7). For *Chlorella*, Atkinson, Gunning, John and McCullough (1972) find a vacuole mem-brane area equal to 0.35 that of the plasmalema although 'vacuoles' only occupy 0.08–0.13 of the cell volume. If a vacuole fractional volume of 0.05 corresponds to a membrane area in a 5 μm radius cell which is 0.16 (scaled from 0.35 for a volume fraction of 0.11) that of the plasmalemma, then the volume fraction of vacuole mem-brane is 0.001 (Table 7).

TABLE 7. Fraction of cell volume for Chlorophyte eukaryotic algae with spherical cells of radius 5 μm or 0.5 μm occupied by various "non-scaleable" membrane systems, and the sum of these volume fractions. Values from Table 6, and text of paper.

	Volume fraction of cell of radius	
	0.5 μm	5 μm
Plasmalemma	0.060	0.006
Plastid envelope	0.105	0.011
Nuclear envelope	0.020	0.001
Outer mitochondrial membrane	0.007	0.006
Microbody + lysosome + 'vacuole' membranes	0.010	0.001
Total	0.202	0.025

The total fractional volumes of these "incompletely scaleable" membrane systems (Table 7) amounts to only 0.025 of the 5 μm radius cell, but 0.202 of the 0.5 μm radius cell. Since the larger eukaryotic photolithotrophic cells have at least as high a μ_{max} as do the smaller eukaryotic photolithotrophic cells (Table 4), it would seem that the area, catalytic activity, and barrier function of the plasmalemma, plastid and nuclear envelopes, outer mitochondrial membrane, and membranes of lysosomes, microbodies and vacuoles which correspond to a total volume which is 0.025 of the cell volume, are adequate to permit the eukaryotic photolithotroph to grow rapidly. Accordingly, the additional (0.202–0.025) or 0.177 of the cell volume which is occupied by these 'incompletely scaleable' membrane components can be construed as not strictly necessary to achieve a high growth rate but represents a necessary concomittant of "scaling down" the cells. Even if it is allowed that the membrane thickness (and intermembrane space) of 10 nm used in the computations in Tables 6 and 7 is an overestimate and the value is reduced to 7 nm, we are still dealing with 0.141 of the cell volume occupied by "incompletely scaleable" membrane systems in a 0.5 μm radius cell as against 0.017 in a 5 μm radius cell, a difference of 0.124. We have already seen that an increased volume fraction of membrane in a cell increases the C content per unit volume, so that doubling the cell volume prior to cell division in a binary fission cycle involves more C assimilation in a cell with relatively more membrane; if much of the "extra" membrane in the 0.5 μm radius eukaryotic is superfluous to growth at μ_{max} in a high-resource environment, then the fraction of total cell C, and the absolute amount of C per unit cell volume, which is devoted to catalysts (equation (1)) is decreased, with a corresponding decrement in μ_{max}. An increase in the fraction of the cell which is occupied by membranes must decrease the *absolute* quantity of cell C which is available for other structures and catalysts by 0.177. Assuming, as we did earlier, that 0.7 of the C in a basic (ignoring compatible solutes) 5 μm radius cell is in "non-catalytic" cytoskeleton, DNA and storage material, then a volume fraction of 0.177 of "non-catalytic" membrane material in a 0.5 μm radius cell which was absent from a 5 μm radius cell means a reduction in "catalytic" C from 87.5 kg • m^{-3} in a 5 μm radius cell to 65.4 kg • m^{-3} in a 0.5 μm radius cell, assuming a 'basic cell" value of C per m^3 cell volume of 125 kg • m^{-3}. Accordingly, the 'catalytic capacity' *for increasing cell volume* in the 0.5 μm radius cell is only 65.4/87.5 or 0.75 of its value in the 5 μm radius cell. Furthermore, a given increment in cell volume of the 0.5 μm radius cell involves more C assimilation, as membranes have more C/volume than the average for the cell.

Replacing 0.177 of the cell volume with a 'mean cell' C content of 125 kg C • m^{-3} by a similar volume of membrane (+ intermembrane space) with 333 kg C • m^{-3} increases the C which must be assimilated in order to make unit volume of cell material to 1.29 times that in a cell with 125 kg C • m^{-2} so that 65.4 units of "catalytic C" per unit volume in a 0.5 μm radius cell has to assimilate 1.29 times more C per cell doubling than does 87.5 units of catalytic C per unit volume in a 5 μm radius cell, giving a ratio of μ_{max} values of 0.75/1.29 or 0.58. The conclusion that the 0.5 μm radius eukaryote has a potential μ_{max} only 0.58 that of 5 μm radius eukaryote can be questioned quantitatively in several ways, e.g. in relation to the implicit assumption that the catalytic capacity of the 'catalytic C' (mol C assimilated/mol catalytic C. second) is independent of the membrane content of the cell being synthesized, but it does serve to show that an increased volume fraction membrane of membranes in small eukaryotes *could* decrease their μ_{max}.

Finally, we compare prokaryotes with eukaryotes with respect to "membrane scaleability" and its effect on μ_{max}. Table 6 shows that "non-scaleable" membranes occupy 0.0117 of the volume of a 5 μm radius prokaryote and 0.115 of the volume of a 0.5 μm radius prokaryote; Table 7 shows that "incompletely scaleable" membranes occupy 0.025 of the volume of a 5 μm radius eukaryote and 0.202 of the volume of a 0.5 μm radius eukaryote. Assuming no transfer of function from intracellular membranes of a 5 μm radius prokaryote to the plasmalemma of a 0.5 μm radius prokaryote then the "superfluous" membranes amount to 0.177 of the volume of the 0.5 μm radius eukaryote and 0.103 of the volume of a 0.5 μm radius prokaryote. Applying a similar approach to the analysis of the effect of the "superfluous" membrane on μ_{max} of small prokaryotes, to that conducted above for small prokaryotes, (but assuming that the C content of membranes is 500 kg • m^{-3} since no intermembrane spaces are involved) we see that the catalytic C per unit cell volume in the 0.5 μm radius is reduced to $\frac{74.625 \text{ kg C} \cdot \text{m}^{-3}}{87.5 \text{ kg C} \cdot \text{m}^{-3}}$ or 0.85 of its value in the 5 μm radius cell, while the C assimilation per unit volume increment is increased to $\frac{(0.103 \times 500) + (0.897 \times 125)}{125}$ or 1.31-fold that in the 0.5 μm radius cell, so that the predicted μ_{max} of the 0.5 μm radius cell is 0.85/1.31 or 0.65 that of the 5 μm-radius cell. Thus, the computed constraint on μ_{max} is rather smaller for the prokaryotic than for the eukaryotic picoplankter. However, these values of 0.65 for the prokaryote and 0.58 (eukaryote) will *under*-estimate differences in μ_{max} for 0.5 μm radius *versus* 5 μm radius cells between prokaryotes and eukaryotes to the extent that the prokaryotic (cyanobacterial) plasmalemma can probably carry out functions in small cells which occur at intracellular membranes in larger cells, while there are a number of reasons for believing that "superfluous" membrane volume in the eukaryotes (Table 7) has been underestimated (see above). We can conclude that "non-scaleable" or "incompletely scaleable" membranes can contribute to a decreased μ_{max} for picoplankters relative to larger cells, and that this effect is more significant for eukaryotes than for prokaryotes.

6. *"Leakage" and Maintenance Processes*

The processes considered under this heading differ from those considered in Sections (1)–(5) above in that "leakage" and related processes do not involve a stoichiometric production of cell components during growth. Rather, these processes involve a loss of materials which had been previously synthesized (e.g. compatible solutes) or accumulated by an energy-requiring process (e.g. inorganic C species; ammonium; nitrate; phosphate; chlrodie; potassium) by the cell, or the entry of "unwelcome" substances into the cell (e.g. H^+, Na^+, Ca^{2+}). In all of these cases the trans-membrane flux is thermodynamically downhill, and occurs *via* "lipid solution" flux, in some cases supplemented by specific mediated transport mechanisms (uniporters). Homoiostasis of the individual solutes concerned, as well as of turgor (for walled cells) and volume (for non-walled cells) demands that the intracellular concentrations of the solutes be maintained by energy-dependent synthesis (in the case of leaked compatible solutes) or by energy-dependent influx (for accumulated solutes which have been lost) or efflux (for excluded solutes which have crept in).

Since, by definition, μ_{max} implies that the cells are not energy-limited, it might be asked why these "leakage" processes which are energy-demanding are considered in the context of constraints on μ_{max}. The reason is that the synthesis (replacing "leaked" compatible solutes) and the mediated, energy-dependent transport (of "leaked" solutes) requires an addition of catalysts of synthesis or transport to the cell's complement over and above what would be needed in a "leak-proof" cell growing at the same rate. This diverts carbon from some other function or demands additional C, in either case potentially restricting μ_{max}. Furthermore, provision of the substrates for synthesis of compatible solutes, or for energizing solute fluxes, demands additional C fluxes (or their equivalent: see above); which again has implications for the balance of catalysts within a certain C budget in the cell, with thus for μ_{max}.

The leakage rate of a solute per unit of cell volume is directly proportional to the area of plasmalemma per unit cell volume if the permeability of the plasmalemma to that solute, and the driving force ($J \cdot mol^{-1}$) across the membrane for that solute are held constant. We note that, in a given environment, the intracellular (cytosol) concentrations of these solutes which permit the maximum specific growth rate to be achieved are likely to be essentially independent of cell size.

An example is the concentration of compatible solutes; this is related to extracellular osmolarity, and to whether the cell has a functional cell wall: cells with walls which can resist turgor tend to have higher intracellular osmolarities (and, at high external osmolarities, correspondingly higher intracellular concentrations of compatible solutes) than do otherwise similar cells which lack turgor-resisting walls. There is no obvious reason why the concentration of the solutes should be a function of cell size.

A second example concerns the "inorganic carbon accumulation mechanism", where the intracellular free CO_2 concentration which is required for the likely function of this mechanism (saturating the RuBPc function of RUBISCO while very substantially inhibiting the RuBPo activity: see Raven 1984a), is, for constant kinetics of RUBISCO and constant intracellular concentrations of RUBISCO, RuBP, Mg^{2+} and H^+, independent of cell size. If the intracellular carbonic anhydrase: RUBISCO activity ratio is also constant, then the $CO_2 \rightleftharpoons HCO_3^-$ reaction will be at a constant displacement from equilibrium during photosynthesis, so that the mean total inorganic C is also constant. We shall consider this system in more detail later under photon-limited and carbon-limited growth, since the conditions for μ_{max}, if they include high $[CO_2]$ in the medium, would repress the "inorganic carbon accumulation mechanism": diffusive entry of CO_2 could suffice. Similar arguments apply to "leaky" mediated influx of NH_4^+ (Raven 1980, 1984a; Kleiner 1985a,b; c.f. Boussiba et al. 1984). A final suite of leaks which *will* be considered here are those concerned with the inorganic solutes whose leakage across the plasmalemma effectively short-circuits energy-dependent influx mechanisms for solutes (e.g. NO_3^-, $H_2 PO_4^-$) whose influx is mediated and energy dependent even under μ_{max} conditions (cf. inorganic C, ammonium), or which are normally partially excluded, and whose leakage into the cell could cause problems for acid-base balance (H^+ influx), Ca^{2+} balance (Ca^{2+} influx) or volume regulation and enzyme activation (Na^+ influx): see Raven (1984a); Raven and Smith (1982); Smith and Raven (1979).

It is important to note that the electrical component of the driving force on ions (ψ_{co}) is not intrinsically dependent on the cell size (see Walker 1976). Consequently, with the cell-size independence of cytosol ionic concentrations and a constant extracellular environment, there should be little or no effect of cell size on the electrochemical driving force on the ions at the plasmalemma.

Having made these general points, we can now turn to some specific cases. Most of the discussion centres on the compatible solutes, which are tabulated for the main groups of organisms (Cyanobacteria, Chlorophyceae and Micromonadophyceae) under discussion in Table 8. It will be seen that quite a range of solutes are employed, with sucrose, glucosylglycerol and quarternary ammonium compounds in the Cyanobacteria, sucrose, proline, glycerol, sorbitol and tertiary sulphonium compounds in the Chlorophyceae, and mannitol (and glycerol and tertiary sulphonium compounds) in the Micromonadophyceae. Ignoring any differences in the degree of "compatibility" of these "compatible solutes" (Warr et al. 1985), we can see two conflicting effects on μ_{max} related to the use of different compatible solutes. The obvious, cell-size independent effect is the additional C which must be assimilated in producing unit volume of cell as the compatible solutes used have more C (and N or S, since equation (1) can express N or S use in "C equivalents") per osmol. Table 9 gives values for C content per m^3 cell volume for a "basic" freshwater cell with 125 kg C per m^3 cell volume and for various marine cells with 500 mol $\cdot m^{-3}$ ($= 500$ osmol $\cdot m^{-3}$ if the activity coefficient $= 1.0$) of compatible solutes with different numbers of C atoms per molecule. The use of a C_{12} compound such as sucrose increase the C content per m^3 cell by more than half relative to a "basic" freshwater cell. We would predict, using the sorts of arguments based in equation (1), that have been used earlier in relation to the occurrence of more membrane material per unit cell volume, that a penalty in terms of μ_{max} would be incurred by increasing the quantity of "non-catalytic" cell C required to double cell volume.

The size-dependent effect relates to the extent to which the compatible solutes leak out of the cells. The sort of analysis of the relationship between molecular structure and permeability coefficient through lipid bilayers which Stein (1967) performed suggests that $P_{\text{compatible solute}}$ should decrease substantially in the order glycerol > mannitol $\cong$ sorbitol, $\cong$ glycine betaine $\cong$ dimethylsulphonium propionate >

30

glucosylglycerol > sucrose. Data from work on lipid bilayers shows that $P_{glycerol}$ is about 5.10^{-8} m $\cdot$ s^{-1} while $P_{mannitol}$ is $\sim 10^{-12}$ m $\cdot$ s^{-1} (table 3.5 of Raven 1984a). Accordingly the leakage rate for glycerol should be $\sim 5.10^4$ times that for mannitol, with a corresponding loss of $2.5.10^4$ times as much C per unit time for glycerol leakage as from mannitol leakage, other things being equal. Table 9 gives values for C leakage rates per unit C per second for 5 μm and 0.5 μm radius cells based on a concentration difference of 500 mol $\cdot$ $^{-3}$ glycerol *or* mannitol across the plasmalemma, and the "lipid bilayer" values for permability coefficients (values with superscript "a"). With glycerol as the compatible solutes the computed specific C effluxes in compatible solutes are very high relative to the empirical upper limit (Table 4) of 30.10^{-6} mol C (mol cell C)$^{-1}$ $\cdot$ s^{-1} for μ_{max} of planktophyte cells. For 5 μm radius cells, the ratio specific C efflux: $\mu_{max} = 126$, while for a 0.5 μm radius cells this ratio is 1256. Clearly the cells could not support such a high leakage rate. For mannitol the ratio of specific C efflux: $\mu_{max} = 0.0046$ for a 5 μm radius cell and 0.0446 for a 0.5 μm radius cell. Thus, even for the picoplankton cell, C efflux in mannitol is only 0.0446 of the C flux into components retained in the cell, and the volume-independent flux of C into mannitol retained (at 500 mol $\cdot$ m^{-3}) in the cell ($6.79.10^{-6}$ mol C (mol cell C)$^{-1}$ $\cdot$ s^{-1}) is 5.0 times the flux of C into effluxed mannitol; for the 5 μm radius cell this ratio is 50. For the picoplankton cell, the efflux of C in mannitol of 0.0446 times the C flux to components retained in the cell would, if it demanded an increment of 0.0446 in C flux per unit C incorporated into cell material without changing the cellular "catalytic C", decrease μ_{max} by 0.0446 to 0.9554 of its value in a cell producing mannitol as its compatible solute but which does not leak. Mannitol leakage thus extols a levy of about 4.5% in terms of growth rate in a 0.5 μm radius cell but only 0.45% in a 5 μm radius cell. Much larger decrements can be computed if the $P_{mannitol}$ of 10^{-10} m $\cdot$ s^{-1} for *Daucus* plasmalemma (superscript "c" in Table 9) are used; however, mediated fluxes of mannitol may be involved here, so these values will not be discussed further.

TABLE 8. Compatible solutes of some major taxa which contribute to the picoplankton as well as to the nanoplankton.

Taxon	Compatible solutes	References
Cyanobacteria:		
Synechococcus (three strains from freshwater habitats)	Sucrose	Reed et al. (1984a)
Synechococcus (three strains from marine habitats)	Glucosylglycerol	Reed et al. (1984a)
Synechococcus DUN 52 (from hypersaline habitat)	Quarternary ammonium compounds	Reed et al. (1984b)
Rivularia atra	Trehalose	Reed and Stewart (1983)
Chlorophyceae:	Sucrose, sorbitol, glycerol, Tertiary sulphonium compounds	Raven (1986b)
Nannochloris bacillaris		Brown ad Hellebust (1980a,b)
UWASH 20-2-2 (marine)	Sucrose, proline	Brown (1982, 1985)
Chlorella emersonii	Sucrose, proline	Setter and Greenway (1979)
Dunaliella spp.	Glycerol	Ben-Amotz and Avron (1983); Hellebust (1976); Wegman (1979)
Chlamydomonas pulsatilla	Glycerol	Hellebust and Le Gresley (1985)
Micromonadophyceae	Mannitol (Glycerol?) Tertiary sulphonium compounds	Raven (1986b)

Returning to the glycerol data, it is clear (Table 8) that glycerol *is* used as a compatible solute by planktophyte cells which have effective radii (assuming spherical cells) of 2.4–9 μm, i.e. spanning the radius of 5 μm for which the "lipid bilayer" permeability values predicted a C efflux in glycerol which is 126 times the C flux into material retained in the cell for a growth rate of 30.10^{-6} $\cdot$ s^{-1}. It would appear that some molecular trick in the plasmalemma of these cells (species of *Dunaliella, Chlamydomonas* and *Asteromonas*) which greatly reduces $P_{glycerol}$ (to values of $\leq 5.10^{-13}$ m $\cdot$ s^{-1}, Table 9, superscript "b", which in turn means that the ratio

TABLE 9. Effects of producing different compatible solutes on the C content per m^3 of cell volume, and on specific C leakage rates (mol C in compatible solutes leaked $\cdot$ (mol cell C)$^{-1} \cdot$ s^{-1}) for 0.5 μm and 5 μm radius cells.

Compatible solute employed	C content (mol) per m^3 cell volume of 5 μm radius *or* 0.5 μm radius cell	Specific C efflux in compatible solute $\cdot$ (mol C (mol cell C)$^{-1} \cdot$ s^{-1} for two cell sizes with different assumed permeability coefficients	
		5 μm radius cell	0.5 μm radius cell
None (freshwater)	10417 (125 kg C $\cdot$ m^{-3})	0	0
500 mol $\cdot$ (m^3 cell volume)$^{-1}$ glycerol (C$_3$)	11917	[a]3.77.10^{-3} (P$_{glycerol}$ = 5.10^{-8} m $\cdot$ s^{-1}) [b]3.77.10^{-8} (P$_{glycerol}$ = 5.10^{-13} m $\cdot$ s^{-1})	[a]3.77.10^{-2} (P$_{glycerol}$ = 5.10^{-8} m $\cdot$ s^{-1}) [b]3.77.10^{-7} (P$_{glycerol}$ = 5.10^{-13} m $\cdot$ s^{-1})
500 mol $\cdot$ (m^3 cell volume)$^{-1}$ mannitol or sorbitol (C$_6$)	13417	[a]1.34.10^{-7} (P$_{mannitol}$ = P$_{sorbitol}$ = 10^{-12} m $\cdot$ s^{-1}) [c]1.34.10^{-5} (P$_{mannitol}$ = P$_{sorbitol}$ = 10^{-10} m $\cdot$ s^{-1})	[a]1.34.10^{-6} (P$_{mannitol}$ = P$_{sorbitol}$ = 10^{-12} m $\cdot$ s^{-1}) [c]1.34.10^{-4} (P$_{mannitol}$ = P$_{sorbitol}$ = 10^{-10} m $\cdot$ s^{-1})
500 mol $\cdot$ (m^3 cell volume)$^{-1}$ glucosylglycerol (C$_9$)	14917	[d]__	[d]__
500 mol $\cdot$ (m^3 cell volume)$^{-1}$ sucrose (C$_{12}$)	16417	[d]__	[d]__

[a] = permeability values from work on lipid bilayers (Raven 1984a).
[b] = permeability values from work on plasmalemma of *Dunaliella* cells (Raven 1984a).
[c] = permeability values from work on plasmalemma of *Daucas* cells (Cram 1984).
[d] = permeability values not available.

of C efflux in glycerol to C flux into material retained in the cell at μ_{max} = $30.10^{-6} \cdot s^{-1}$ is 0.00126 for the 5 μm radius cell and 0.0126 for the 0.5 μm radius cell. For the 0.5 μm radius cell, the C efflux equal to 0.0126 of the C flux in growth would, be demanding an increment of 0.0126 in C flux per unit C incorporated into all material without changing the cellular "catalytic C", decreases μ_{max} by about 0.0126 to 0.9874 of its value in a cell producing glycerol as its compatible solute but which does not leak. Glycerol leakage thus involves a levy of about 1.2% in terms of μ_{max} for a 0.5 μm radius cell, but only 0.12% for a 5 μm radius cell.

The discussion thus far just deals with the effects on μ_{max} of the C flux to the compatible solute which is synthesized to replace the quantity leaked from the cell. Another effect on μ_{max} relates to the fraction of cell C which is used to produce the additional catalysts involved in synthesizing the compatible solute which replaces the "leaked" compound. However, it is not likely that this "extra" catalytic C has to be synthesized at more than 0.1 of the rate at which compatible solute is leaking out of the cells.

These computations suggest that the decrement of μ_{max} due to replacing leaked compatible solutes is only up to a few per cent relative to a "leak-proof" 0.5 μm radius cell, and an order of magnitude less for 5 μm radius cells, for growth at $30.10^{-6} \cdot s^{-1}$.

We next turn to the leakage of ions, an occurrence which could, if not countered by an active flux of the ions involved in the opposite direction, compromise a number of very important cell functions even under non-growing conditions. For a cell (like that analysed in Table 10) in which Cl^- is close to electrochemical equilibrium, efflux of K^+ (and Cl^-) for a walled cell would (like efflux of compatible solutes) diminish turgor, while influx of Na^+ (and Cl^-) would upset volume regulation in a wall-less cell (see Raven and Smith 1982). Exchange of (accumulated) K^+ for (excluded) Na^+ would have very substantial effects on enzyme activity (Raven 1984a). Efflux of $H_2PO_4^-$ would not only diminish the cells content of this potentially growth-limiting resource, but would also have dire consequences, as cytosol $H_2PO_4^-$ decreased, for the maintenance of the *in vivo* free energy of hydrolysis of ATP without an excessively high [ATP] : [ADP] ratio. H^+ passive influx could, if not balanced by active H^+ efflux, upset cellular acid-base balance and hence the ratio of activities of different catalysts, while unchecked Ca^{2+} influx could inhibit enzymes and upset any 'messenger' role of cytosolic Ca^{2+}.

Table 10 shows, granted "reasonable" values for cytosol ion concentrations for microphytes grown in seawater, and for ψ_{co}, the transplasmalemma ion fluxes which would occur *if the only mechanism for ion flux across the plasmalemma were by "lipid solution"*, since the permeability value used in the computation of fluxes are those for lipid bilayers without added "porter proteins", or are derived from such measurements (see chapters 3 and 6 of Raven 1984a). These fluxes are, accordingly, the *maximum* net fluxes of the specified ions across the plasmalemma, granted maintenance of ψ_{co} by other ion fluxes on the quoted external and cytosol ion concentrations. These data will be used later in discussing maintenance energy requirements for cells of different radii. For the moment, they will be used in relation to their effects on growth rate due to any additional catalysts of active transport, or of energy supply to the porters, which may be needed in a "leaky" growing cell relative to a "non-leaky" cell, and of the C flux (or its equivalent) which powers these active transport processes.

During growth there must clearly be a net influx of all externally derived nutrients, so that the net influxes (Table 10) of Na^+, Ca^{2+}, and Mg^{2+} are in the 'growth' direction, while the net effluxes of K^+, Cl^- and $H_2PO_4^-$ *oppose* the required net influx during growth. Thus, for K^+, Cl^- and $H_2PO_4^-$ (as well as SO_4^{2-} and, when it is the N source, NO_3^-), it is clear that "lipid solution" efflux implies additional active influx at the plasmalemma to maintain the growth rate relative to what would

TABLE 10. Leakage fluxes and energy dissipation during "lipid-solution" fluxes of H^+, K^+, Na^+, Cl^-, $H_2PO_4^-$, Ca^{2+}, Mg^{2+} across the plasmalemma of a 5 μm radius and of a 0.5 μm radius cell in seawater.

	H^+	K^+	Na^+	Cl^-	$H_2PO_4^-$	Ca^{2+}	Mg^{2+}	Total
1) Assumed cytosol concentration/mol $\cdot$ m^{-3} (cf. Raven 1980a, 1984a)	3.10^{-5}	200	10	50	1	10^{-4}	1	262
2) Assumed external concentration/ mol $\cdot$ m^{-3}	10^{-5}	10	450	500	0.001	10	50	1020
3) Electrochemical potential of ion in cytosol relative to that in medium/kJ $\cdot$ mol^{-1} (assuming $\psi_{co} = -0.06$ V); equation 3.4 of Raven 1984a)	-3.34	$+1.38$	-15.17	$+0.31$	$+22.71$	-34.22	-21.33	—
4) Lipid-solution permeability coefficient/m $\cdot$ s^{-1} (from Table 3.5 of Raven 1984a, assuming $P_{Ca}{}^{2+}$ and $P_{Mg}{}^{2+} = 0.5 \times P_{Na}{}^+$ and $P_K{}^+$)	10^{-5}	2.10^{-12}	2.10^{-12}	2.10^{-12}	2.10^{-12}	10^{-12}	10^{-12}	—
5) Lipid-solution flux from medium to cytosol/ pmol $\cdot$ m^{-2} $\cdot$ s^{-1} (from rows 1, 2 and 4, using equation in table 3.5 of Raven 1984a)	$+196$	-48	$+2390$	-32	-5	$+49$	$+245$	—

6) Energy dissipation in leakage fluxes/ W cell^{-1}, in a 5 μm radius cell	$2.05.10^{-16}$	$0.21.10^{-16}$	$113.8.10^{-16}$	$0.031.10^{-16}$	$0.36.10^{-16}$	$5.26.10^{-16}$	$16.4.10^{-16}$	$138.1.10^{-16}$
7) Energy dissipation in leakage fluxes/ W cell^{-1} in a 0.5 μm radius cell	$2.05.10^{-18}$	$0.21.10^{-18}$	$113.8.10^{-18}$	$0.031.10^{-18}$	$0.36.10^{-18}$	$5.26.10^{-18}$	$16.4.10^{-18}$	$138.1.10^{-18}$
8) Energy used in pumping leaked ions if 2 mol charge moved per mol ATP, $\Delta G_{ATP} = 55$ kJ $\cdot$ mol^{-1}, in a 5 μm radius cell/ W $\cdot$ cell^{-1}	$1.69.10^{-15}$	$0.41.10^{-15}$	$20.64.10^{-15}$	$0.28.10^{-15}$	$0.04.10^{-15}$	$0.85.10^{-15}$	$4.23.10^{-15}$	$28.1.10^{-15}$
9) Energy used in pumping leaked ions if 2 mol charge moved per mol ATP, $\Delta G_{ATP} = 55$ kJ $\cdot$ mol^{-1}, in a 0.5 μm radius cell/ W $\cdot$ cell^{-1}	$1.69.10^{-17}$	$0.41.10^{-17}$	$20.64.10^{-17}$	$0.28.10^{-17}$	$0.04.0^{-11}$	$0.85.10^{-17}$	$4.23.10^{-17}$	$28.1.10^{-17}$

be needed in a "leak-proof" cell, with a likely need for additional active transporters for these ions.

For the cations H^+, Na^+, Ca^{2+} and Mg^{2+} the situation is less clear; is the lipid solution influx in excess of that needed for growth? For H^+ the lipid solution influx amounts to $6.15.10^{-20}$ mol $H^+ \cdot$ cell$^{-1} \cdot$ s^{-1} for a 5 μm radius cell, and $6.15.10^{-22}$ mol $H^+ \cdot$ cell$^{-1} \cdot$ s^{-1} for a 0.5 μm radius cell. H^+ is a nutrient for cells growing with NO_3^- as N source and CO_2 as C source; for a C:N ratio of 5, a 5 μm radius cell growing at $30.10^{-6} \cdot$ s^{-1} will assimilate NO_3^- at $1.96.10^{-17}$ mol N $\cdot$ cell$^{-1} \cdot$ s^{-1}, and (see Raven and Smith 1976; Raven 1980a, 1984a) produce OH^- at about 0.7 the rate of NO_3^- assimilation, or $1.37.10^{-17}$ mol N $\cdot$ cell$^{-1} \cdot$ s^{-1}. Accordingly, the H^+ influx by lipid solution in the 5 μm radius cell only provides about 0.005 of the H^+ needed for cell acid-base homoiostosis; for a 0.5 μm radius cell, the value is 0.05. If NH_3 is the N source and CO_2 is the C source, a C:N ratio and growth rate as described above, there is a net H^+ excess of some $5.85.10^{-18}$ mol $H^+ \cdot$ cell$^{-1} \cdot$ s^{-1} for a 5 μm radius cell, or $5.85.10^{-21}$ mol $H^+ \cdot$ cell$^{-1} \cdot$ s^{-1} for a 0.5 μm radius cell. Accordingly, lipid solution H^+ influx increases the need for H^+ net active extrusion by about 0.01 in a 5 μm radius cell and 0.1 in a 0.5 μm radius cell.

For Na^+ the lipid solution Na^+ influx is $7.5.10^{-19}$ mol $\cdot$ cell$^{-1} \cdot$ s^{-1} and $7.5.10^{-21}$ mol $\cdot$ cell$^{-1} \cdot$ s^{-1} for 5 μm radius and 0.5 μm radius cells, respectively. If the mean intracellular Na^+ is equal to the quoted cytosol value (10 mol $\cdot$ m^{-3}), the required Na^+ influx for growth at $30.10^{-6} \cdot$ s^{-1} is $1.56.10^{-19}$ mol $\cdot$ cell$^{-1} \cdot$ s^{-1} and $1.57.10^{-22}$ mol $\cdot$ cell$^{-1} \cdot$ s^{-1} for 5 μm and 0.5 μm radius cells, respectively. The lipid solution Na^+ influx is accordingly 4.8 times the required flux for a 5 μm cell and 48 times for a 0.5 μm radius cell. This means that active Na^+ efflux is needed in cells of both sizes even during rapid growth.

For Ca^{2+}, analogous calculations show a computed influx of $1.54.10^{-20}$ mol $Ca^{2+} \cdot$ cell$^{-1} \cdot$ s^{-1} in a 5 μm radius cell and $1.54.10^{-22}$ mol $Ca^{2+} \cdot$ cell$^{-1} \cdot$ s^{-1} for a 0.5 μm radius cell. These fluxes would give, in the absence of active Ca^{2+} efflux, a steady-state Ca^{2+} of $2.94.10^{-5}$ mol $Ca^{2+} \cdot$ m^{-3} for a 5 μm radius cell and $2.94.10^{-4}$ mol $Ca^{2+} \cdot$ m^{-3} for a 0.5 μm radius cell. While these values are similar to the *free* Ca^{2+} concentration of 10^{-4} mol $\cdot$ m^{-3} quoted for cytosol, the mean total $[Ca^{2+}]$ in the cell in usually at least two orders of magnitude higher than the free Ca^{2+} concentration (accommodated by a total of 10 mol $\cdot$ m^{-3} of a Ca chelator such as citrate with a pK_{Ca} of 4: see Raven 1986c). Accordingly, some additional Ca^{2+} influx is needed, and there is no obvious requirement for active Ca^{2+} efflux in a 5 μm or a 0.5 μm radius cell growing at $30.10^{-6} \cdot$ s^{-1}.

Mg^{2+} influx by lipid solution is $7.7.10^{-20}$ mol Mg $\cdot$ cell$^{-1} \cdot$ s^{-1} for a 5 μm radius cell, and $7.7.10^{-22}$ mol Mg $\cdot$ cell$^{-1} \cdot$ s^{-1} for a 0.5 μm radius cell. In addition to the free Mg^{2+} in the cells of 1 mol $\cdot$ m^{-3}, there is at least another 1 mol $\cdot$ m^{-3} chelated to phosphorylated compounds and several mol $\cdot$ m^{-3} in chlorophyll(s) (see Table 12)). A total of 5 mol Mg $\cdot$ m^{-3} would require an influx, for growth at $30.10^{-6} \cdot$ s^{-1}, of $7.9.10^{-20}$ mol Mg $\cdot$ cell$^{-1} \cdot$ s^{-1} for a 5 μm radius cell, and $7.9.10^{-23}$ mol Mg $\cdot$ cell$^{-1} \cdot$ s^{-1} for a 0.5 μm radius cell. Lipid solution Mg influx is about equal to the cell's need for a 5 μm radius cell, but about 10 times in excess of the needs of a 0.5 μm radius cell, but about 10 times in excess of the needs of a 0.5 μm radius cell, so active Mg efflux during growth would be needed for the smaller cell.

If we sum the minimum requirement for *additional* active ion fluxes at the plasmalemma during growth at $30.10^{-6} \cdot$ s^{-1} on NH_3 as N source and CO_2 as C source resulting from the occurrence of leak fluxes, these come to $9.71.10^{-19}$ mol charge $\cdot$ cell $^{-1} \cdot$ s^{-1} for a 5 μm radius cell and $12.5.10^{-21}$ mol charge $\cdot$ cell$^{-1} \cdot$ s^{-1} for a 0.5 μm radius cell. With a stoichiometry of 1 mol ATP used per 2 mol charge actively transported and a mol ATP synthesized per mol CO_2 produced in oxidative phosphorylation of 6, these active fluxes involve an additional

C flux of $8.09.10^{-20}$ mol C $\cdot$ cell^{-1} $\cdot$ s^{-1} for a 5 μm radius cell and $1.04.10^{-21}$ mol C $\cdot$ cell^{-1} $\cdot$ s^{-1} for a 0.5 μm radius cell, or $1.48.10^{-8}$ mol C $\cdot$ (mol cell C)$^{-1}$ $\cdot$ s^{-1} for a 5 μm radius cell and $1.91.10^{-7}$ mol C $\cdot$ (mol cell C)$^{-1}$ $\cdot$ s^{-1} for a 0.5 μm radius cell. These C fluxes represent increases of 0.005 and 0.0064, respectively, relative to those in a "leak-proof" cell. It is unlikely that C fluxes required for active transport of "leaked" ions not accounted for in Table 10, or C fluxes into additional catalysts of C transformation and of active transport required by these "leak-correcting" fluxes, would double the C fluxes quoted above, so upper limits on additional C fluxes associated with "leak-correcting" are ≤ 0.001 and ≤ 0.01, respectively, for a 5 μm radius cell and a 0.5 μm radius cell.

It would, then, seem that leakage of ions has rather less effect on potential μ_{max} of cells growing at 30.10^{-6} $\cdot$ s^{-1} than does leakage of compatible solutes, and that the two together only amount to an additional decrement of growth rate of a 0.5 μm radius cell relative to a 5 μm radius cell of about 0.01.

7. Conclusions as to Possible Restrictions on μ_{max} of Picoplanktors due to "Non-Scaleable" Properties: Comparisons with Measured Values of μ_{max}

Table 11 gives some values, based on subsections (1)–(6) above, for the fraction of cell C involved in certain "non-scaleable" properties of a 0.5 μm radius cell relative to that in a 5 μm radius cell. For all of the fractions *except* the leakage flux the numbers quoted are independent of growth rate. Furthermore, we note that turgor-resisting cell walls and flagella are assumed, for the purposes of summing the various values for eukaryotes, to be mutually exclusive. The volume regulation mechanisms of wall-less cells are subsumed under the heading of "leakage". We do not consider, in this salty article, the volume regulation in wall-less freshwater cells using contractile vacuoles, which probably requires a greater power consumption per unit volume to operate the contractile vacuole mechanism, as well as requiring a larger fraction of cell biomass to be devoted to the contractile vacuole structure, in small cells relative to large cells with a similar shape and driving force for, and permeability to, water at the plasmalemma (Raven 1982).

The reported sum of the increments of cell C in the "non-scaleable" cell components in the 0.5 μm radius cell relative to the 5 μm radius cell would substantially decrease the fraction of the cell C which is devoted to potentially rate-limiting catalysts

TABLE 11. Fraction of cell C involved in non-scaleable properties in a 0.5 μm radius cell in excess of that in a 5 μm radius cell.

	Fraction of cell C involved in specified function in a 0.5 μm radius cell in excess of that in a 5 μm radius cell, based on a cell with no compatible solute or, in brackets, a cell with 500 mol $\cdot$ m^{-3} of mannitol	
	Prokaryotes	Eukaryotes
DNA content	0.016 (0.013)	0.059 (0.047)
Cell wall	0? (0)	$\leq 0.06^{a}$ (0.05)
Flagella	not applicable	0.023[b] (0.018)
Membranes	0.103 (0.091)	0.117 (0.093)
Leakage of compatible solutes, etc. in a cell growing at 30.10^{-6} $\cdot$ s^{-1}	0.010 (0.008)	0.0 (0.00)
Total	0.129 (0.112)	$\leq 0.24^{a}$ (≤ 0.198) 0.209[b] (0.166)

[a] For non-flagellate, walled cells only.
[b] For flagellate, non-walled cells only.

(equation (1)). If the non-catalyst C is 0.3 of the total cell C in the 5 μm radius cells (prokaryote or eukaryote), then the 0.5 μm radius prokaryote has effectively increased this to 0.424, and the 0.5 μm radius eukaryote to 0.504 for flagellate non-walled cells and, perhaps up to 0.541 for non-flagellate walled cells. We would, accordingly, predict a decrement of μ_{max} in the ratio $\frac{1-0.424}{1-0.3}$ or 0.823 in the prokaryote, and 0.656–0.708 in the eukaryotes, when the 0.5 μm radius cell is compared with the 5 μm radius cell.

Table 4 shows some of the μ_{max} values available in the literature for various sizes of marine and freshwater prokaryotic and eukaryotic planktophytes. Since Hoogenhout and Amesz (1964) produced their compilation, various authorities (e.g. Eppley and Strickland 1969; Droop 1974), have counselled against such tabulations of μ_{max} values in view of variations in the techniques involved in obtaining the data reported, and the varying rigour of the precautions taken to establish that μ_{max} was *really* obtained. Furthermore, the data in Table 4 have been massaged by assuming a Q_{10} of 2 to normalize the data to 20°C if the original measurements were made neither at 20°C nor a range spanning 20°C from which the μ_{max} at 20°C can be obtained by interpolation. Finally, it must be acknowledged that the μ_{max} values quoted for such "laboratory weeds" as the Chlorophyceans *Chlorella* spp. and *Chlamydomonas* spp., and the Cyanobacterium *Anacystis nidulans* (= *Synechococcus leopolensis*) may have been grown under more nearly optimal culture conditions than were less frequently cultured organisms. These caveats notwithstanding, it is felt that the compilation in Table 4 is of use in determining if there are any *obvious* variations of μ_{max} with cell size over the range which our theoretical analysis has covered.

For the Cyanobacteria, the highest μ_{max} values are found for various strains of *Synechococcus* with cell volumes of 0.5–8.0 μm³ (volume of spherical cell, 1 μm radius = 4.19 μm³; 0.5 μm radius = 0.524 μm³); although the smallest strain tested (PC 70001) had the lowest μ_{max} it is not clear that it was grown at light saturation. The highest reported μ_{max} values were lower for species of *Anabaena* (larger cells in filaments) and for *Gloeobacter violacea* (small cells, but lacking intracytoplasmic membranes) than for the fastest-growing *Synechococcus* strains. It is accordingly difficult to find evidence in Table 4 to support the view that cell volumes of ~ 1 μm³ incur a penalty in terms of μ_{max} such as is suggested from Table 10 relative to cell volumes of tens or hundreds of μm³. However, it is important to note that the fastest-growing *Synechococcus* strains have rather lower μ_{max} values than do the fastest-growing diatoms, and that *Anabaena* sp. (whose cell volumes more closely accord with those of the smallest diatoms) only attain μ_{max} values half as great as those of the diatoms. It would seem that the larger cyanobacteria do not exhibit μ_{max} values as high as would be expected from their cell organisation characteristics, thus rendering difficult a comparison of their μ_{max} with that of smaller-celled dyanobacteria. Before leaving the prokaryotes, it is worth pointing out that the fastest-growing photolithotrophic but non-O_2-evolving prokaryote (*Chlorobium thiosulphotophilum*) does not grow significantly faster than the fastest-growing O_2-evolving cyanobacterium. Faster growth rates can be found in the order: obligate photolithotroph with organic supplement (*Chlorobium limicola* form sp. *thiosulphotophilum*); photoorganotrophically growing Rhodospirillaceae; chemoorganotrophs growing on minimal medium; chemoorganotrophs growing on complex media. Such behaviour can be at worst rationalised and at best predicted (Raven 1986a).

Among the eukaryotes a somewhat clearer picture emerges. The highest photolithotrophic μ_{max} values are found for Chlorophyceans (*Chlorella* sp., *Chlamydomonas* sp.) with cell volumes of at least 30 μm³ (μ_{max} 20°C $\cong$ 20.10^{-6} · s^{-1}) and "small diatoms" with cell volumes also of at least 30 μm³ (μ_{max} 20°C $\cong$ 30.10^{-6} · s^{-1}). In both the Chlorophyceae and the Bacillariophyceae we find that the fastest-growing larger cells (volumes of many hundreds of μm³ or more) have lower specific growth rates than do the fastest-growing cells with cell volumes immediately after cell division of less than 100 μm^3. This inverse relationship between specific growth rate and cell size is *not* predicted from the sorts of analysis carried out earlier in this paper. However, the inverse relationship is in accord with some other possible restrictions on μ_{max} which apply to cells with much smaller surface areas per unit volume of which three are now listed. The self-shading/package effects which imply that more pigment/unit volume is needed to obtain the same photon absorption rate per unit cell volume in full daylight (Raven 1984a, b). A lack of membrane (plasmalemma) area to accommodate all of the porters needed to transport nutrients *in* and end-products *out* and maintain a high solute flux per unit volume even with porters working at their highest rate and with optimal resource input (Raven 1980a, 1984a). An increased DNA content (in some proportion with cell size) lead to longer minimal cell cycle times (Cavalier-Smith 1978) although replicon size probably does not increase with DNA content: Francis et al. (1985); Waterborg and Shall (1985).

Turning to the smaller sizes of eukaryotic photolithotrophs (0.5–50 μm^3 cell volume), the data in Table 4 deal solely with members of the order Chlorococcales of the class Chlorophyceae. While *Chlorella* spp. with cell volumes $\geq 30 \mu m^3$ can grow up to $20.10^{-6} \cdot s^{-1}$, *Chlorella nana* and other strains with volumes less than 20 μm^3 grow at $\leq 13.10^{-6} \cdot s^{-1}$. Thus, for the Chlorococcales, the predictions of lower μ_{max} as cell size decreases in the range 100 μm^3 down to 0.5 μm^3 appear to be broadly in accord with reality.

As with the prokaryotes, Table 4 shows the eukaryotic chemoorganotrophs have greater μ_{max} values for a given cell size than do eukaryotic photolithotrophs (cf. rationalisations/predictions of Raven 1986a).

The Effect of Small Size on the Rate and Efficiency of Net Transformation of Resources During Resource-Limited Growth, and on Maintenance Energy Requirements Relative to Energy Availability

PROLOGUE

Resource limited growth is probably closer to ecological reality for the average picoplankton than is growth at or near μ_{max} such as was considered in the last Section. In this Section we consider mainly the effects on the use of limited resources which are intrinsic to the picoplankton cell, treating the (limited) supply of resources to the cell surface as outwith the cell's control. In the next Section we see what constraints the small size of cells have on the supply of resources to the cell surface.

In this Section we deal with the dependence on cell size of the capacity for nutrient influx at the plasmalemma, and of energy (photon) trapping, under conditions of restricted supply of solutes and/or photons to the cell surface, and on the requirements for nutrients and energy for growth, and of energy for maintenance.

NUTRIENT SOLUTE INFLUX AT THE PLASMALEMMA

Again comparing a 5 μm radius spherical cell with a 0.5 μm radius spherical cell, the plasmalemma area per unit volume of cytoplasm is 0.6 μm^2 plasmalemma per μm^3 cell volume for the 5 μm radius cell and 6.0 $\mu m^2 \cdot \mu m^{-3}$ for the 0.5 μm radius cells. For a diffusive entry of a nutrient solute such as CO_2, with a constant activity of assimilating enzyme (RuBPc activity of RUBISCO) per unit cell volume, a constant (limited) potential for CO_2 supply to the cell surface (mol $CO_2 \cdot (m^2$ plasmalemma$)^{-1} \cdot s^{-1}$) and a constant CO_2 permeability, the CO_2 influx and fixation rate per unit cell surface is substantially increased. The same argument applies to O_2 efflux, with further advantages for CO_2 fixation in terms of the $[CO_2]/[O_2]$ ratio in the cell and the RuBPo ratio (Raven 1977; Samish 1975).

For a nutrient solute whose influx is mediated, the area of plasmalemma available for the insertion of porters is ten times higher, per unit volume of cytoplasm, in the 0.5 μm radius cell than in the 5 μm radius cell. For porters associated with the entry of potentially growth-limiting this factor could be even higher, since the occupancy of the membrane by "constitutive" porters, performing volume-dependent functions (e.g. efflux of H^+ generated in growth-related processes at a rate proportional to the volume of the cell for a given cell composition and specific growth rate. These arguments apply to de-repressible porters as well as to constitutive ones. Examples of repressible porters whose occurrence depends on nutrient supply are the NO_3^- porter which may be absent when some other N source is available, the NH_4^+ porter, which may be absent at high NH_4^+ concentrations where, in seawater, the NH_3 concentration could be sufficient to provide a diffusive entry of NH_3 adequate to support growth; and the "inorganic C porter" which is, apparently, absent at high external CO_2 concentrations: see Raven (1984a). The content (per unit area of plasmalemma) of porters related to the recouping of leaked solutes should be independent on cell size, since leakage is area-dependent.

Accordingly, we see that the smaller cells have an advantage in terms of the acquisition of dissolved which arrive at the cell surface at a cell area-related rate.

MINIMUM REQUIREMENT OF NUTRIENT SOLUTES IN A CELL

We have previously discussed the per-cell requirement of nutrients which are only required in very small quantities per μm^3 of cell (of the order of molecules) in the context of μ_{max}, in relation to a possible need to increase the per μm^3 requirement for very small cells in order to maintain more than one molecule per cell of the essential nutrient. While no such nutrient was located in that context, it behoves us to consider the nutrient-limitation case, since it is here, when the cell is approaching its "subsistence quota" rather than being on the verge of "luxury accumulation", that we would expect to find the need for a higher-than-expected quote in very small cells.

The most comprehensive set of data is to be found for Vitamin B_{12} (Cyanocobolamin), the only known Co-requiring component of planktophyte cells. Many planktophytes cannot synthesize the organic part of the molecule, and so require an exogenous source of Vitamin B_{12} rather than of just Co. A remarkable consistency is found for the subsistence quota of Vitamin B_{12} expressed as molecules per μm^3 of cell volume; data provided, and compiled by Droop (1957), Guillard and Cassie (1963) and Bradbeer (1971) shows values of 2–18.4 molecules of Vitamin B_{12} per μm^3 cell volume as the subsistence quota of a range of eukaryotic microalgae from 5 Divisions whose cell volumes range from 12 μm^3 (*Stichococcus* sp.) to 2665 μm^3 (*Thalossiosira fluviatilis*). Extrapolation of the lowest value (2 molecules μm^{-3}) to our 0.5 μm radius picoplankter (volume 0.524 μm^3) gives us a subsistence quota of 1 molecule of Vitamin B_{12} per cell! It would appear that the subsistence quota of Vitamin B_{12} for picoplankton (molecules μm^{-3}) must be higher than that for larger cells in order to ensure that each cell has > 1 molecule of vitamin B_{12}. This in turn means that the biomass of picoplankton which could be produced from a fixed quantity of Vitamin B_{12} (as the limiting resource) is less than that of larger cells although, of course, more picoplankton cells could be produced.

PHOTON ABSORPTION

This topic is dealt with elsewhere in this volume by J.T.O. Kirk, so only a brief outline will be given here, based in part on earlier work by Kirk (1975a,b, 1976, 1983) and Raven (1982, 1984a,b, 1986a), Raven and Richardson (1984), and Osborne and Raven (1986). Size-dependent scattering effects (see Kirk 1983; Osborne and Raven 1986) are not dealt with here.

An important initial point to establish is the extent to which the increase with decreasing cell size in the fraction of the cell volume which is occupied by "non-scaleable" components (Table 10), thus reducing the fraction of the cell volume available to house catalysts, including the photosynthetic chromophores together with apoproteins and related membranes. This constraint applies at the lower end of the volume range (0.524–524 μm^3) which we are considering; at the other end of the range, vacuolation may pre-empt an increasing fraction of the cell volume, thus restricting the maximum pigment content per unit cell volume.

Table 11 presents some comparisons of pigment (chromophore) content per unit cell volume for picoplankters and related, larger oganisms. Before commenting on the data presented, it is worth pointing out some restrictions on their validity. Firstly, comparisons of chromophore content per unit volume should only be made *within* a "pigment group" ("Chlorophyte", "Chromophyte" or "Phycobiliphyte"), since the quantity of apoprotein per mol chromophore, and hence the maximum chromophore content per unit cell volume, varies substantially between these groups (Raven 1984a,b). The "Chlorophytes" and "Chromophytes" generally have lower values of g aproprotein per mol chromophore, and hence a larger potential to accommodate chromophores in unit volume of cell, than do "Phycobiliphytes"; it may be significant that the phycoerythrin-dominated phycobiliphytes, such as many of the picoplanktonic cyanobacteria, are less disadvantaged in this respect than are the phycocyanin-dominated phycobiliphytes (Raven 1984a,b; Alberte et al. 1984). Secondly, to determine the maximum extent to which biomass can be occupied by pigments in cells of various sizes, we need to compare the most "shade-adapted" (genotypically *and* phenotypically) examples of the cell sizes and pigment groups under consideration. Finally, interpretation of the data in terms of absorption of photons in the natural environment involves a number of considerations (package effect, scattering, specific absorption coefficient of pigments in the wavelength range to which they are exposed); these factors will be briefly mentioned later.

The data in Table 12 deals with Cyanobacteria among the Phycobiliphytes and, among the "Chlorophytes", the Chlorophyceae (Chlorococcales and Volvocales). Neither the Cyanobacteria nor the Chlorophyceae show pronounced trends of maximum chromophore content per m^3 cell volume with cell size, provided the precision of the data is taken into account. In addition to the lack of data for some strains (e.g. *Chlorella nana*) growing at less than saturating photon flux densities, there are problems with the use of scaling factors to obtain cell volumes, and internal inconsistencies in some of the data, presumably as a result of systematic errors in measurement. Thus, for the Cyanobacteria, 4-fold different chromophore contents came from computing cell volume from cell C (assuming 125 kg C $\cdot$ m^{-3}) rather than from electron micrographs of *Synechococcus* WH 7803. For the Chlorophyceae, the pigment/cell, and cell volume of *Monoraphidium contortum*, together with the cell C deduced by Vincent (1982) from a scaling factor relating cell C to cell volume in other algae, gives a (chlorophyll plus associated protein) of the light-harvesting complex value of 618 g per kg dry weight, even with a *low* estimate of the apoprotein: chlorophyll ratio (Raven 1984a)! For *Dunaliella tertiolecta*, the quoted cell volume and cell C for cells grown at 20 μmol photon $\cdot$ m^{-2} $\cdot$ s^{-1} implies 370 kg $\cdot$ m^{-3} or (if C = 0.5 of the dry mass) 740 kg dry mass per m^3 cell volume (Falkowski and Owens 1980). Finally, the *Nanochlorum eucarotum* data (Wilhelm and Wild 1982) showing 24.9 mg chlorophylls $a + b$ per g dry weight and 35.34 mg chlorophylls $a + b$ per mL packed cell volume implies, even without void spaces in the packed cells, 1419 kg dry mass per m^3 cell volume! These observations are not to be taken as criticisms of the workers concerned, but are meant to show how different approaches can give discrepant results. Accordingly, even without discussing problems of pigment analysis, we must conclude that the precision of the data to hand does not let us draw firm conclusions as to the occurrence of trends in chromophore concentration in cells of cyanobacteria over the volume range considered (0.524–25

TABLE 12. Chromophore contents per unit cell volume in shade-adapted "phycobiliphytes" and "chlorophytes" of different cell volumes.

Organism	Cell volume/μm^3	Chromophore concentration mol chromophore (m^3 cell volume)$^{-1}$	Reference
'Phycobiliphytes' *Synechococcus* WH 7803 (marine, high-phycoerythrin seratin)	2.35 (from C cell^{-1}, assuming 125 kg C m^{-3})	Chlorophyll a = 1.04 (grown at 60 μmol photon $\cdot$ m^{-2} $\cdot$ s^{-1}) Carotenoids = 1.58 (grown at ?)	Cuhel and Waterbury (1984) Guillard (1985); cell size from Cuhel and Waterbury (1984)
	0.524 (from Figure 5 of Alberte, Wood, Kursar and Guillard, 1984)	Chlorophyll a = 4.08 Phycoerythrin = 13.00 (grown at 25 μmol photon $\cdot$ m^{-2} $\cdot$ s^{-1}; assuming 6.10^3 g phycoerythrin per mol chromophore: Alberte, Wood, Kursar and Guillard, 1984)	Barlow and Alberte (1985)
		Chlorophyll a = 4.60 Phycoerythrin = 15.11 (grown at 10 μmol photon $\cdot$ m^{-2} $\cdot$ s^{-1})	Barlow and Alberte (1985)
Synechococcus (freshwater, high phyco-cyanin) strain	8.0	Chlorophyll a = 7.60 Phycoerythrin = 3.13 Phycocyanin = 2.08 (grown at 10 μmol photon $\cdot$ m^{-2} $\cdot$ s^{-1}; assuming 6.10^3 g phycoerythrin per mol chromophore, 12.10^3 g phycocyanin per mol chromophore)	Healey (1985)
Synechococcus leopolensis (= *Anacystis nidulans*) UTEX 625 TX20	0.9	Chlorophyll a = 25.0 (grown at 30 μmol photon $\cdot$ m^{-12} $\cdot$ s^{-1} — limiting) Chlorophyll a = 7.75 Phycocyanin = 4.55 (grown in near-saturating light?)	Utkilen (1982) Utkilen et al. (1983) Myers et al. (1978)

		Chlorophyll a = 9.4 (grown in limiting photon flux densities)	Myers and Kratz (1985)
Anabaena variabilis (freshwater)	25	Chlorophyll a = 11 mol $\cdot$ m^{-3} Phycobilins = 5.7 mol $\cdot$ m^{-3} (grown under limiting light of 8–12 μmol photon $\cdot$ m^{-2} $\cdot$ s^{-1})	Kawamura et al. (1979); Reed et al. (1981)
'Chlorophytes' *Chlorella nana* (marine)	6.0	Chlorophyll a = 10.7 Chlorophyll b = 3.0 (grown at 126 μmol photon $\cdot$ m^{-2} $\cdot$ s^{-1} saturating)	Thinh and Griffiths (1985)
Monoraphidium contortum (freshwater)	18.2	Chlorophyll a = 71.4 Chlorophyll b = 45.8 (grown at 2.5 μmol photon $\cdot$ m^{-2} $\cdot$ s^{-1}; growth-limiting)	Vincent (1982)
Chlorella pyrenoidosa (Emerson strain) (freshwater)	$\geq$ 30	Chlorophyll a = 6.7 Chlorophyll b = 1.2 (grown at a limiting photon flux density; from chlorophyll/packed cell volume, assuming cells = 7/10 of pellet)	Myers and Graham (1983)
Dunaliella tertiolecta	84	Chlorophyll a = 37 Chlorophyll b = 16 (grown at 20 μmol photon $\cdot$ m^{-2} $\cdot$ s^{-1} — limiting)	Falkowski and Owens (1980)
Nanochlorum eucaryotum (marine)	1.33	Chlorophyll a = 27.0 Chlorophyll b = 11.5 (grown at a limiting photon flux density)	Wilhelm and Wild (1982)

μm^3) or in cells of Chlorophyceae over the range 1.33–84 μm^3. If we take half of the dry weight as an absolute upper limit on the fraction occupied by chromophores *plus* apoproteints in a 5 μm radius cell, then the maximum expected chromophore concentrations are ~ 31 mol • m^{-3} in "Chlorophytes" (1 mol chromophore per 4 kg light-harvesting complex), ~ 21 mol • m^{-3} in phycoerythrin-dominated "Phycobiliphytes" (1 mol chromophore per 6 kg light-harvesting complex) and ~ 13 mol • m^{-3} in phycocyanin-dominated "Phycobiliphytes" (1 mol chromophore per 10 kg light-harvesting comples): Raven (1984a,b). Data in Table 10, showing the extent to which the "non-scaleable" functions reduce the fraction of cell C available for catalysts, means that maintaining the fraction of "catalytic C" in light-harvesting chromophores + proteins at 0.71 (0.5/0.7) means a reduction to ~ 21 mol • m^{-3} for a 0.5 μm radius "Chlorophyte", and ~ 17 and ~ 11 mol • m^{-3} respectively for 0.7 μm radius "phycoerythrin" and "phycocyanin" Phycobiliphytes.

Turning to the effect of cell size on the efficiency with which chromophores can harvest light in a given light field, we note that the number of photons absorbed by cells in a given vector light field are directly proportional to their volumes if they are (1) spherical with (2) the same chromophores, (3) homogeneously distributed at (4) the same concentration (mol • m^{-3}) which is (5) not high enough to give a significant "package effect" for the largest cells considered (Kirk 1975a, b 1976, 1983, Raven 1984a,b). While the (theoretical) package effect is not very significant at even the highest likely chromophore concentrations in a 0.5 μm radius cell, it is likely to be significant for a 5 μm radius cell (Raven 1984b) *assuming homogeneous pigment distribution*. For the "real world", with non-homogeneous chromophore distribution, there are a number of reasons for expecting the picoplankton-sized cell to have a *less* homogeneous distribution of pigments, e.g. the limited range of positions available for thylakoids in small cyanobacteria (Johnson and Sieburth 1979) and of thylakoids in plastids, and of the plastids themselves, in eukaryotes such as *Nannochloris* or *Micromonas*. This could lead to a greater (relative to the "homogeneous" model) "package" effect in smaller cells relative to larger cells with a more nearly uniform (on the cell scale) pigment distribution. For large, vacuolate cells, another sort of inhomogeneity obtrudes (see Papageourgiou 1971).

An index of these various effects of cell size has been sought in the extent to which the *in vivo* specific absorption coefficient for chlorophyll *a* is decreased below its *in vitro* (solution) value. Figure 26 of Harris (1978, see also Reynolds 1984) plots the *minimum* self-shading coefficient, $\epsilon_{s\,min}$, in m^2 • mg^{-1} at 530–575 nm, as a function of cell volume, and finds quite a good fit with the predictions of Kirk (1975b) for cell volumes of 100–100 000 μm^3 (equivalent to radius of spherical cells of 2.9–29 μm), but with a markedly lower-than-predicted values of $\epsilon_{s\,min}$ for cells of 10–100 μm^3 volume (equivalent to radii of spherical cells of 1.3–2.9 μm).

Much of the data on which Harris' analysis was based came from field observations, with (generally) unknown chromophore/unit cell volume values, and with representatives of different pigment groups. Data obtained for laboratory cultures (Osborne, Geider and Raven, unpublished) for various cyanobacteria, Chlorophyceans and diatoms suggest that there is *no* significant *increase* in apparent self-shading as cell volumes decline over the range 100 μm^3 to 10 μm^3. For the Chlorophyceae the size range covered is ~ 7 μm^3 (*Nannochloris atomus*) to ~ 200 μm^3 (*Dunaliella tertiolecta*). For the 677 nm absorption peak of chlorophyll *a* the ratio of *in vivo* to *in vitro* specific absoprtion coefficient is ~ 0.7 for *Dunaliella tertiolecta* with 7 kg chlorophyll *a* • m^{-3}; ~ 0.5 for *Chlamydomonas rhienhardii* (volume ~ 150 μm^3) with 28 g chlorophyll *a* m^{-3}; ~ 0.85 for "high light" *Nannochloris atomus* with 16 g chlorophyll *a* m^{-3}, and ~ 0.72 for "low light" *Nannochloris atomus* with 65 kg chlorophyll *a* m^{-3}. We see that the departures from the trend of increasing similarity of *in vivo* and *in vitro* specific absorption coefficients can be readily interpreted in terms of variations in chromophore per m^3 of cells.

The sum of the data on chromophore concentration per m^3 of cell volume, and on "self-shading" (package effect), as a function of cell size, suggests that the potential for photon absorption per unit volume of cell is greater for a picoplankton cell exceeds that of a larger planktophyte cell. The *rationale* here is that the decreased "package effect" in the smaller cells more than offsets any decrease in the fraction of the biomass available for occupation by chromophore–protein complexes, or increased heterogeneity (at the scale of the cell) in chromophore distribution. The efficacy of individual chromophore molecules in photon absorption from a given light field is probably even more markedly increased in a picoplankton relative to a large cell than is absorption per unit cell volume.

An important aspect of the relatively low absorptance of picoplankton cells (despite the efficiency with which individual pigment molecules are used) means that gradients of photon flux density within the cell are small relative to those occurring in larger cells with the same chromophore concentration and in the same light field. The simplest case is that of a spherical cell in a scalar light field, or rotating at $\geq$ 0.1 Hz in a vector light field; here the lowest mean photon flux density is at the centre of the cell or, more precisely, at the centre of the pigmented area. Under these conditions computations in Raven (1984b) suggest that the photon flux density at the centre of a 0.5 μm radius cell averages as much as 0.9 of that at the periphery, while in a 5 μm radius cell it might only be 0.6. This means that there is much less likelihood of a *radial* sun-shade "adaptation" in the smaller than in the larger cells such as occurs in the abaxial–adexial direction through a horizontally disposed higher plant leaf (Terashima and Inoue 1984, 1985).

EFFICIENCY OF USE OF ABSORBED PHOTONS IN PHOTOSYNTHESIS

Most determinants of the photon (quantum) efficiency of photosynthesis (mol CO_2 fixed or mol O_2 evolved per mol photon absorbed) would not be expected to be size dependent. Assuming that the basic mechanisms of photoredox systems are universal in O_2-evolvers, and that the "leakage" (H$^+$ flux by lipid-solution across the photosynthetic membrane), and "slippage" (short-circuiting of the O_2 evolution "S states") reactions which lead to proportionately larger reductions in photon efficiency at low rates of photon absorption (Raven and Beardall 1981b, 1982; Richardson et al. 1983) are independent of cell size. The decrement of efficiency due to "slippage" reactions, related as they are to individual water dehydrogenase complexes, is probably related to the size of the photosynthetic unit, in this case, defined as mol chromophore per mol water dehydrogenase complex, with less slippage associated with a large unit (Raven and Beardall 1982). In view of the relatively large number of reaction centres which are found in a single *Synechococcus* cell of 2.35 μm^3 volume, i.e. 6900 photoreaction 1 centres and 6600 photoreaction 2 centres, it would appear that even a 0.524 μm^3 volume cell (with perhaps 1500 of each kind of photoreaction centre per cell) has no cell-size-related problems in manipulating the size of photosynthetic units (cf. Wilhelm and Wild 1982). For leakage, the determinant is the area of membrane associated with the transformation of a given, limiting number of photons per second, granted a constant permeability to H$^+$ and a constant $\Delta\bar{\mu}_{H+}$ across the membrane (Raven and Beardall 1982). Since the thylakoid area is "scaleable" (see above), it would appear that there is no *cell-size-related* constraint on the area of leaking membrane associated with unit capacity for photon absorption. This is true, provided that we deal with comparisons within a "pigment group", since it is possible that "Phycobiliphytes" are more frugal of membrane area per unit capacity for photon absorption than are "Chlorophytes" or "Chromophytes" (Richardson et al. 1983). Furthermore, it would not apply to organisms of various sizes which, like *Gloeobacter*, have all of their photoredox reactions in the plasmalemma whose area is not independent of cell size in spherical cells (see above). H^+ (and other solute) leakage across the plasmalemma *is* an energetic penalty which is greater in small than in large cells, but these are *not* costs to be directly debited to the photon requirement of *gross* photosynthesis unless, as for CO_2 leakage which short-circuits the "inorganic carbon accumulation mechanism", it is directly related to substrate acquisition for photosynthesis and is not an energy cost for cells in the dark (see below).

This brief analysis suggests that there is no size-dependence of photon efficiency of ATP and NADPH photoproduction. In the case of the efficiency of use of ATP and NADPH in "gross" fixation of CO_2 (defined as net CO_2 fixation plus dark CO_2 evolution), there are two important ways in which the "classical" stoichiometry of the use of 2 mol NADPH and 3 mol ATP in the fixation of 1 mol CO_2 could be increased during photosynthesis at "natural" CO_2 and O_2 concentrations. If there is a diffusive entry of CO_2 *from* ~ 10 mmol $CO_2 \cdot m^{-3}$ and a diffusive efflux of O_2 *to* ~ 230 mmol $O_2 \cdot m^{-3}$, then the *intracellular* steady-state $[CO_2]$ is less than 10 mmol $CO_2 \cdot m^{-3}$ while $[O_2]$ is greater than 230 mmol $O_2 \cdot m^{-3}$, with a similar *absolute de*crement for CO_2 as there is an *in*crement for O_2 (Samish 1975; Raven 1977). Since this $[CO_2]$ of $\leq$ 10 mmol $\cdot m^{-3}$ at a $[O_2]$ of $\geq$ 230 mmol $\cdot m^{-3}$ would not saturate the carboxylase activity, or completely suppress the oxygenase activity, of RUBISCO of either cyanobacteria or of eukaryotic microalgae of the Chlorophyta (Jordan and Ogren 1981, 1983; Raven 1984a), diffusive gas exchange implies the occurrence of phosphoglycolate synthesis and hence of either the very energy-expensive excretion of glycollate or, much more commonly, the less energy-expensive operation of the PCOC or the tartronate semialdehyde pathway (Raven and Beardall 1981a; Raven 1984). The enzymology of this process is independent of cell volume, although there may be some advantage for small cells in that, granted identical steady-state $[CO_2]$ and $[O_2]$ at the cell surface and identical intracellular enzyme activities and cofactor supply rate *per unit cell volume*, the large surface area per unit volume and small internal diffusion distances in a picoplankton-size cell would mean higher steady-state $[CO_2]$ and lower steady-state $[O_2]$ within the cell, with a consequent diversion of cofactors from P-gylcolate synthesis and the PCOC to the PCRC, with an increased CO_2 fixation rate per unit cell volume. As against this potential advantage we must bear in mind the possible disadvantage of the increased loss from small cells of the more lipid-soluble intermediates of the PCOC such as glycolate, glyoxylate, (?) hydrox-ypyruvate, and (?) glycerate, or of the tartronate semialdehyde pathway, such as glycolate, glyoxylate or (?) tartronic semialdehyde.

The alternative mode of coping with oxygenase activity of RUBISCO at low external $[CO_2] : [O_2]$ is the operation of an inorganic carbon accumulation mechanism (Raven 1984a; Beardall and Entwistle 1984). By maintaining a higher steady-state intracellular CO_2 than could be provided by diffusive entry of CO_2, the active influx at the plasmalemma of some inorganic carbon species causes a higher ratio of carboxylase to oxygenase activity of RUBISCO in the cell. The saving of ATP and NADPH per unit CO_2 fixed in the cell which the presence of the high CO_2 in the cells permits must be set against the energy (ATP) cost of active inorganic C influx at the plasmalemma, including the extra pumping which must occur to recoup losses of CO_2 by leakage through the plasmalemma. The leakage problem would, with other things being equal, be ten-fold greater on a unit volume basis in a 0.5 μm radius cell as in a 5 μm radius cell.

The balance of energetic advantage between diffusive CO_2 entry and the inorganic carbon accumulation mechanism in microalgae and cyanobacteria has been discussed, for a specific cell size in each case, by Spalding and Portis (1985) for *Chlamydomonas*, by Badger et al. (1985) for *Synechococcus*, and in a more general context by Raven and Lucas (1985) and Raven (1980a, 1984a). Clearly, any energetic advantage would shift away from the "inorganic carbon accumulation mechanism" as the cell size was decreased from 5 μm radius to 0.5 μm radius if other parameters stayed constant. Using low values for P_{CO_2} (but values which are derived with measurements on the organisms concerned) it would seem that the energetic advantage could still be with the inorganic carbon accumulation mechanism even with cells as small as the *Synechococcus* used by Badger et al. (1985). It is clear that the use of "normal" P_{CO_2} values derived from experiments on lipid bilayers would lead to far greater CO_2 leakages than are observed, or are compatible with observed photon efficiences of photosynthesis by plants operating the "inorganic carbon accumulation mechanism". We seem to be dealing here with some modification of the plasmalemma which substantially reduces its P_{CO_2}, although not perhaps by quite as large a factor as found for $P_{glycerol}$ in *Dunaliella* and related Chlorophyceans. However, unlike the situation with $P_{glycerol}$, it is possible that the same organism can produce "high P_{CO_2}" and "low P_{CO_2}" variants of its plasmalemma, with the former operative when the "inorganic carbon accumulation mechanism" is repressed at high external $[CO_2]$ values, and the latter when the "inorganic carbon accumulation mechanism" is operative at low external $[CO_2]$ (see Raven 1980a, 1984a). The molecular basis for any such effect is unclear.

If we now turn to the *evidence* as to the occurrence of the "inorganic carbon accumulation mechanism" rather than "RuBPo + PCOC (or tartronate semialdehyde pathway)" in various sizes of planktophyte organisms, we can fairly rapidly dismiss the Chlorophyta. No data appear to be available for the Micromonodophyceae. For the Chlorophyceae, the two orders of interest to us in terms of picoplankton are possibly the Volvocales and certainly the Chlorococcales. *Direct* evidence of the occurrence of this mechanism, i.e. a higher mean intracellular inorganic C concentration than can be accounted for in terms of diffusive CO_2 entry, taking into account the mean intracellular pH value and the external pH and inorganic carbon concentration, has been found (see Raven 1984a, 1985) for *Chlamydomonas reinhardtii* (freshwater) and *Dunaliella tertiolecota* (marine) in the Volvocales, and various relatively large-celled strains of freshwater *Chlorella* species in the Chlorococcales. More of these cells has a cell volume of less than 30 μm^3 (equivalent to a radius for a spherical cell of 1.92 μm); no data seem to be available for the picoplankters *Nannachloris* spp. or *Chlorella nan* (marine) or such freshwater picoplankters as *Choricystis* spp.

As for the Cyanobacteria, the presence of an inorganic carbon accumulation mechanism has been demonstrated by the direct measurement of intracellular inorganic C accumulation (see Raven 1984c, 1985; Shelp and Canvin 1984) for *Anabaena variabilis* (cell volume $\leq 25 \ \mu m^3$), *Coccochloris pentocystis* (cell volume $\leq 14 \ \mu m^3$), *Synechococcus leopolenis* (= *Anacystis nidulans*; cell volume 0.7–3.0 μm^3) and *Synechococcus* sp. (cell volume ~ 2.4 μm^3). We thus see that the capacity for "inorganic carbon accumulation" is manifested in Cyanobacteria of "picoplankton" size, as well as in larger-celled organisms. In these organisms the properties of photosynthetic gas exchange are consistent with a suppression of RuBP oxygenase and an active accumulation of inorganic C, e.g. low O_2-sensitive CO_2 compensation concentration, small O_2 inhibition (21 kPa cf. 1 kPa) of CO_2 fixation from air-equilibrium CO_2 levels; and high affinity for external CO_2.

All of the Cyanobacteria mentioned above have phycocyanin as their dominant phycobilin. What of the picoplanktonic cyanobacteria with their phycobilin complement dominated by phycoerythrin, and which are believed to be especially significant as components of the oceanic phototrophic plankton? The only data available are for *Synechococcus* strains WH 7803 and WH 8018 (Glover and Morris 1981; Morris and Glover 1981), which, like the cyanobacteria tested, have CO_2 fixation kinetics consistent with the operation of the PCRC with no obligatory preliminary $C_3 + C_1$ carboxylation/decarboxylation step. These two strains show substantial inhibition of photosynthetic gas exchange at seawater levels of inorganic carbon by 21 kPa O_2 relative to 1 kPa O_2, a feature which is suggestive of a substantial *in vivo* activity of RuBPo relative to that of RuBPc. While complete kinetics were not presented, it would seem that these two strains had *in vitro* RuBPo: RuBPc activities in air-equilibrated solutions which are even higher than the high values found for other cyanobacteria (Morris and Glover 1981; Glover and Morris 1981; Badger 1980; Badger and Andrews 1982; Jordan and Ogren 1981, 1983). The possibility that these marine phycoerythrin-rich strains of cyanobacteria lack an inorganic carbon accumulation mechanism even when they are grown at "air-equilibrium" CO_2 and O_2 levels clearly demands further investigation.

For the picoplankton-size cyanobacteria which *do* exhibit the inorganic carbon accumulation mechanism, we can legitimately ask how even a small CO_2 leak influences their photosynthetic efficiency at low photon flux densities. A picoplankton cell with a radius of 0.5 μm has a projected area of 0.785 μm^2, so, if its absorptance is 0.1, and a vector light field of 10 μmol photon $\cdot$ m^{-2} $\cdot$ s^{-1} is incident on it, its rate of photon absorption is $7.85.10^{-19}$ mol photon $\cdot$ cell^{-1} $\cdot$ s^{-1}. If it accumulates CO_2 to a steady-state concentration of 100 mmol $\cdot$ m^{-3} in excess of that in the medium, a P_{CO_2} of 10^{-6} m $\cdot$ s^{-1} gives an efflux per cell (area of 3.14 μm^2) of $3.14.10^{-19}$ mol CO_2 $\cdot$ cell^{-1} $\cdot$ s^{-1}. Maintaining the CO_2 gradient, assuming 1 mol ATP is used per inorganic C pumped, and the photon efficiency of ATP synthesis by cyclic photophosphorylation is 0.67 mol ATP/mol photon absorbed (Raven 1984a, b), uses $4.71.10^{-19}$ mol photon $\cdot$ cell^{-1} $\cdot$ s^{-1}, leaving $3.14.10^{-19}$ mol photon $\cdot$ cell^{-1} $\cdot$ s^{-1} to energize net inorganic C pumping for net CO_2 fixation; net CO_2 fixation; growth based on photosynthate and exogenous nutrients; and maintenance. Even if the growth and maintenance components of photon use are ignored, we see that CO_2 leakage has reduced the energy available for photosynthesis to less than half of the original amount. By contrast, for a 5 μm radius cell, the photon absorption rate (for an absorptance of 0.8, assuming the same chromophore per unit volume of cell but a greater package effect: see Raven 1984b) is $6.27.10^{-16}$ mol photon $\cdot$ cell^{-1} $\cdot$ s^{-1}. The photon requirement for CO_2 reaccumulation is $4.71.10^{-17}$ mol photon $\cdot$ cell^{-1} $\cdot$ s^{-1}, i.e. only involving a decrease in photon availability for photosynthesis (plus growth and maintenance) of 7.5%. A reduction of P_{CO_2} of almost 10-fold in the picoplankton cell would be needed to reduce the energy cost of maintaining the intracellular CO_2 concentration at 100 mmol $\cdot$ m^{-3} in excess of that in the medium to the same fraction as occurs in the 5 μm radius cell. Since our assumed P_{CO2} is already only ~ 10^{-3} that of lipid bilayer membranes, we can see that the potential decrement of photon efficiency of photosynthesis is very substantial. Raven and Lucas (1985) note that the O_2 exchange *versus* photon flux density relationship (Badger and Andrews 1982) for "low CO_2"-grown *Synechococcus* differ from those of cells grown in "high CO_2" in a manner consistent with a substantial energy input for maintaining intracellular the CO_2 pool in the face of substantial CO_2 leakage, other possibilities have not been ruled out.

It would appear that CO_2 leakage from the CO_2 accumulative mechanism would impose a substantial decrement of photon efficiency (yield) on microphytes, and that this decrement would be substantially (~ 10-fold greater for a 0.5 μm radius cell than for a 5 μm radius cell. However, it would not be manifest in measurements of the photon efficiency of photosynthesis which involve the quotient photon absorption rate $(R_2 - R_1)/(A_2 - A_1)$, where R_2 = rate of net photosynthesis, resulting from photon absorption rate A_2, R_1 = rate of net photosynthesis resulting from photon absorption rate A_1, where $A_2 > A_1$, and $A_2 < <$ light needed for saturation, and $A_1 > $ that needed to eliminate "sigmoidal" or "Kok" effects.

Be all that as it may, the measured photon yields for gross O_2 evolution in a variety of eukaryotic microalgae does not seem to show any significant relationship to cell size (25–8000 μm^3) except for a possible decrease in an organism (*Prorocentrum micons*: Dinophyta) with a very large cell volume (~ 8000 μm^3): Emerson (1958), Kok (1960), Senger (1982), Welschmeyer and Lorenzen (1981), Falkowski et al. (1985), Geider, Osborne and Raven (1986); cf. Pirt (1983), Raven (1984).

Unfortunately, no measurements seem to be available for eukaryotic picoplankters (cell volume $\leq$ 4.19 μm^3).

Effect of Cell Size on the Efficiency of Photon Use in Cell Growth

We deal briefly here with the effect of cell size on the photon use efficiency of growth, defined here as the conversion of photosynthate plus exogenous solutes into cell material. (We deal with maintenance activities (volume-related homoiostasis of turgor of walled cells, of volume of wall-less cells, of individual intracellular "leaky" solutes; and area-related energization of flagellar movement) in the next section.)

Growth-related processes *per se* are mainly size-independent. The biochemical stoichiometries of the individual reactions are presumably *independent* of cell size; the only size-dependence would relate to the differences in cell composition due to, *inter alia*, the non-scaleability of certain cell structures such as the plasmalemma. Since unit mass of lipids and proteins are relatively energy-expensive to synthesise (compared with unit mass of carbohydrates) from photosynthate (see Raven 1982, 1984a), we might expect that extra membrane area per unit volume in small cells increases the energy demand for synthesis per unit cell volume. As against *this* scale effect, it could well be that the greater efficiency of photon absorption, in a given light field, of a given pigment-protein complex in a small as opposed to a larger cell, less of these components may need to be synthesized per unit cell volume produced. Since pigment-protein complexes are relatively energy-expensive to make (Raven 1984b), this might decrease the energy input needed to synthesise unit volume of cells.

Cell Size Dependence Photon Use in Maintenance Processes

We have just seen that, while the protein turnover 'aspect of maintenance costs' is essentially a volume-dependent cost, there are a number of maintenance costs which are functions of cell surface area. These non-growth-rate-dependent processes which are cell-area-dependent are "transport" phenomena in the broad sense, and include the leakage phenomena which compromise volume regulation in wall-less cells, turgor regulation in walled cells and the maintenance of the intracellular concentration of specific solutes; and the propulsion of the cell at a given velocity.

We have already considered the magnitude of some of the fluxes related to leakage in the context of additional carbon fluxes during growth which are required in "leaky" as opposed to "leak-proof" cells both to provide extra catalysts to "resynthesize" leaked compatible solutes and to actively transport leaked inorganic ions, and to provide the substrates for synthesis and active transport (subsection (6) above; Tables 9 and 10). The data assembled in these Tables can be used to compute the energy costs of maintenance processes in the dark for cells of 5 μm and 0.5 μm radius.

For cells using glycerol as their compatible solutes, the "low" estimate of P$_{glycerol}$ yields specific C loss rates in glycerol leakage of $3.77.10^{-8}$ mol C $\bullet$ (mol cell C)$^{-1}$ $\bullet$ s^{-1} from the 5 μm radius cell and $3.77.10^{-7}$ mol C $\bullet$ (mol cell C)$^{-1}$ $\bullet$ s^{-1} for the 0.5 μm radius cell. The energy loss to the cell is greater than this, since the conversion of starch to glycerol requires an input of 0.5 mol ATP and 1 mol NADPH per mol glycerol synthesized (Wegman 1979; Kaplan et al. 1980), and a further 0.208 mol C from starch must be oxidized to CO_2 in order to provide the cofactors needed to produce 1 mol C in glycerol from 1 mol C in starch, increasing the specific C loss rate to $4.55.10^{-8}$ $\bullet$ s^{-1} for the 5 μm radius cell and $4.55.10^{-7}$ $\bullet$ s^{-1} for the 0.5 μm radius cell. The ATP requirement for active ion transport making good losses due to ion leakage (see rows (8) and (9) of Table 10) involve, with 6 mol ATP produced per mol C oxidized in oxidative phosphorylation, a specific C use rate of $1.73.10^{-8}$ $\bullet$ s^{-1} for a 5μm radius cell. The sum of the costs of synthesising replacements for glycerol molecules which have leaked out, plus the cost of actively transporting outwards the ions that have leaked in, and of actively transporting outwards the ions that have leaked in, and of actively transporting inwards replacements for ions which have leaked out, is thus equivalent to a specific C loss rate of $6.28.10^{-8} \cdot s^{-1}$ for the 5 μm radius cell and $6.28.10^{-7} \cdot s^{-1}$ for the 0.5 μm radius cell.

If we consider the photon flux density which is needed to support this specific C loss rate, we posit (see figure 1 of Raven 1984b), a 5 μm radius Chlorophyte with 20 mol chromophore $\cdot$ (m^3 cell volume)$^{-1}$ and an absorptance of 0.85, and a 0.5 μm radius Chlorophyte with the same chromophore concentration but, with less self shading, an absorptance of 0.25. If the photon requirement for the fixation of one mol exogenous CO_2 into starch is 8 mol photons absorbed (mol ATP/mol photon absorbed $= 0.5$ in non-cyclic photophosphorylation; mol NADPH/mol photon absorbed $= 0.25$ in non-cyclic photophosphorylation; 3 mol ATP and two mol NADPH used in fixing CO_2 into sugar phosphate via the PCRC with negligible RuBPo activity; one mol ATP per mol C fixed used in powering the "inorganic C accumulation mechanism" which suppresses RuBPo (5/6 mol ATP) and in starch synthesis (1/6 mol ATP), then the production of enough starch in a 12 h photoperiod to support the specific C loss rates derived earlier over a 12 h light–dark cycle requires the absorption of $8.04.10^{-8}$ mol photon $\cdot$ cell$^{-1} \cdot s^{-1}$ for a 5 μm radius cell and $8.04.10^{-9}$ mol photon $\cdot$ cell$^{-1} \cdot s^{-1}$ for a 0.5 μm radius cell. For a vector light field and the absorptances quotes above, the incident photon flux density required is 0.047 μmol photon $\cdot$ m$^{-2} \cdot s^{-1}$ for the 5 μm radius cell and 0.120 μmol photon $\cdot$ m$^{-2} \cdot s^{-1}$ for the 0.5 μm radius cell. These values represent the *maintenance* photon flux densities over the 12 h photoperiod which would keep the cells in C balance over the 24 h cycle if there were *no other* maintenance requirements (e.g. protein resynthesis, recouping leakage at internal membranes).

We have already seen that intracellular membranes are intrinsically more scaleable than is the plasmalemma, i.e. the area of the intracellular membranes can be *directly* proportional to cell volume provided the volumes they enclose are not necessarily spherical and/or their number can change (cf. Bont 1975, 1985). Thus, while leakage through such membranes is a "maintenance" cost, it need not vary with cell size, i.e. its magnitude on a unit volume basis need not vary with cell size within "Chlorophytes" or within "Phycobiliphytes".

Two other important maintenance-type energy costs involve leakage across membranes. These are H^+ leakage across photosynthetic thylakoids, and CO_2 leakage across the plasmalemma during the functioning of the "inorganic carbon accumulation mechanism" (see above). Both of these leaks are essentially independent of the rate of photosynthesis (limited by light or by exogenous inorganic carbon supply) and so would be expected to have a greater relative effect on photosynthetic rate when the latter was already low provided that the driving forces for H^+ (across the thylakoid membrane) and CO_2 (across the plasmalemma) do not decline in direct proportion to resource (photons, exogenous inorganic C) availability. In both cases the restriction on the rate of photosynthesis acts to reduce substrate supply for other maintenance processes; this could particularly disadvantage the smaller cell when energy was limiting and the CO_2 leakage (related to plasmalemma area) restricted substrate supply to deal with other, area-related plasmalemma leakage processes. In the case of the H^+ leakage across thylakoid membranes, the relatively lower absorptance of smaller cells may mean that relatively fewer thylakoids are subjected to photon flux densities low enough to cause severe leakage-related inhibition of ATP synthesis in a fluctuating light field (cf. comments above about possible radial variation in the size of system two photosynthetic units, and hence slippage-related losses, in large unicells). It is also possible that the lower potential absorptance of "Phycobiliphytes" than of "Chlorophytes" of the same cell size, due to differences in the mass of light-harvesting complexes per mol of chromophore, could be in part offset by a smaller area of thylakoid membrane per mol chromophore, and hence less potential for leakage per mol of chromophore, due to the peripheral rather than the integral nature of the light-harvesting pigment–protein complexes in phycobiliphytes (Raven and Beardall 1981b, 1982; Richardson et al. 1983; Raven 1984b). Direct evidence (mol chromophore • (m^2 thylakoid area)$^{-1}$) is equivocal (Raven 1984a).

Our computations here suggest that the apparent energy cost of making good the depradations of solute leakage at the plasmalemma of marine Chlorophytes of 0.5–5.0 μm radius involves an incident photon flux density for 12 h per 24 h of about 0.1 μmol photon • m^{-2} • s^{-1}. The 0.5 μm radius cell needs 2–3 times the photon flux density to offset plasmalemma leakage ($\sim$ 0.12 μmol photon • m^{-2} • s^{-1}) than does the 5 μm radius cell ($\sim$ 0.47 μmol photon • m^{-2} • s^{-1}), indicating that decreased self-shading of chromophores in the smaller cells, increasing the efficiency of individual chromophore molecules in photon absorption does not completely offset the high plasmalemma area per unit cell volume. These computed incident photon flux densities for combatting leakage at the plasmalemma could be revised upwards in relation to interactions with other "leakage" factors (CO_2 at the plasmalemma short-circuiting the "inorganic carbon accumulation mechanism"; H^+ at the thylakoid reducing the photon efficiency of ATP synthesis) at these low photon flux densities; and in relation to mediated passive uniport by specific passive ion iniporters and by "slippage" of action ion cotransporters and uniporters. An example of the latter suite of possibilities relates to our assumption of equal plasmalemma permeabilities to K^+ and Na^+ which contrasts with the invariably greater permeability to K^+ than to Na^+ in "real" plasmalemmas (Raven 1976). We also note that, with the same permeabilities as assumed for the marine cells, comparable freshwater cells only dissipate 0.1–0.01 of the power in plasmalemma leakage as the marine cells do, due to much lower fluxes resulting from lower external ion concentrations and internal solute concentrations (note that lines (6), (7) and (8) of table 7.4 of Raven 1984a, are too low by a factor of 10^6, although the conclusions are correct!).

Turning now to another important maintenance energy-consumer, i.e. cell motion caused by flagellar activity, we have seen earlier that movement of a 5 μm radius cell at 50 μm • s^{-1} involves an energy input rate of at least 212.10^{-18} W • cell^{-1}, while to move a 0.5 μm radius cell at 50 μm • s^{-1} at least $21.2.10^{-18}$ W • cell^{-1} is needed. Assuming, as we did earlier, that the efficiency of conversion of the chemical energy of ATP hydrolysis into movement of the cell is 10%, then the ATP consumption rate for motility is, assuming the *in vivo* free energy of ATP hydrolysis is 55 kJ • mol^{-1}, $3.85.10^{-20}$ mol ATP • cell^{-1} • s^{-1} for the 5 μm radius cell and $0.385.10^{-20}$ mol ATP • cell^{-1} • s^{-1} for the 0.5 μm radius cell. With 6 mol ATP produced per mol C in starch oxidized to CO_2, and $6.24.10^{-12}$ mol C • cell^{-1} for the 5 μm cell and $6.24.10^{-15}$ mol C • cell^{-1} for the 0.5 μm cell (assuming a "glycerol cell" from Table 9), the energy consumption rate for motility corresponds to specific C loss rates of $1.03.10^{-9}$ • s^{-1} for the 5 μm cell and $1.03.10^{-7}$ • s^{-1} for the 0.5 μm radius cell. With similar assumptions to those used earlier (for computing photon costs of combatting leakage) with respect to absorptance of cells, the cost in absorbed photons of producing C in starch, and the vector nature of the light field, the *minimum* mean incident photon flux density for 12 h per day to power motility for 24 h per day is 0.00155 μmol photon • m^{-2} • s^{-1} for the 5 μm radius cell and 0.0528 μmol photon • m^{-2} • s^{-1} for the 0.5 μm radius cell. These photon requirements must be added to other maintenance requirements for photons (e.g. in dealing with protein breakdown and solute leakage) in these cells.

To conclude and summaries this discussion of the size-dependence of the energy demands for maintenance processes, and of the energy supply to these processes, we note that there is a hierarchy of size-dependences in these various processes. The per cell content of chromophores can be approximated to a constant concentration per unit cell volume, i.e. is directly proportional to r^3; the photon absorption rate per chromophore molecule in a given light field decreases with cell size due to the "package effect"; and the stoichiometry of conversion of the energy of absorbed photons into energy usable for maintenance processes is size-independent. Energy requirements for protein resynthesis are probably also a function of cell volume, i.e. proportional to r^3 on a per cell basis. Energy requirements for active solute transport which combats leakage of ions, and for synthesising replacements for leaked organic solutes, are a function of plasmalemma area, i.e. proportional to r^2 on a per cell basis, while energy requirements for flagellar motility at a fixed velocity are a function of cell radius, i.e. proportional to r on a per cell basis. At a given, limiting photon flux density, therefore, it appears that the "package effect" could *increase* the ratio energy available for maintenance: energy required for protein resynthesis in small cells relative to its value in larger cells, but that the ratio of energy available:energy required is lower in the smaller cells for maintenance processes related to "leakage", and is even lower for motility at a given velocity in smaller cells.

CONCLUSIONS

We attempt to summarize the effects of cell size on growth and maintenance processes under restricted resource supply conditions by listing various attributes of resource acquisition and use which we have considered, together with the likely influence of cell size on these processes.

1) Nutrient solute influx at the plasmalemma (mediated or diffusive) — smaller cells can transport more solute per unit cell volume from a given surface concentration of solute.

2) Nutrient solute requirement — smaller cells may need more of certain solute (required in low amounts), per unit cell volume for survival if the "scaled" quantity needed by small cells falls to the order of a molecule or two per cell.

3) Photon absorption — the "package effect" favours photon absorption by small cells: for a given chromophore concentration, a small cell has a higher photon absorption rate per unit cell volume. However, there may be constraints on the chromophore content per unit cell volume in small cells due to non-scaleability of other essential components causing a reduction in the fraction of the cell volume which is available for occupation by chromophores.

4) Efficiency of photon use in photosynthesis — the main factor here which may disadvantage small cells is cell-area-dependent CO_2 leakage from cells with an "inorganic carbon accumulation mechanism". By contrast, small cells with diffusive CO_2 entry (and, at "natural" CO_2 and O_2 concentrations, have RuBPo and PCOC activity) may be at an advantage relative to larger cells (see (1) above).

5) Efficiency of photon use in growth, using photosynthate and exogenous nutrients — no very clear advantages are seen for small or large cells.

6) Balance of photon availability from a given light field and energy requirement for maintenance processes — on a cell volume basis, a small cell with a given chromophore concentration per unit volume in a given light field can achieve a higher ratio of energy absorbed to energy needed to protein maintenance (due to the package effect), but has a lower energy absorbed:energy required ratio for combatting leakage at the plasmalemma, and an even lower ratio for motility using flagella, than does a larger cell.

The Effect of Small Size on the Supply of Resources to the Cell Surface

PREAMBLE

The discussion up to this point has taken the resource supply to the cell surface as fixed (*either* sufficient to support μ_{max}, *or* limiting for growth), and has attempted to analyse the effect of cell size on the way in which the resources are handled. We now attempt to put the picoplankter back into its environment and see how its ability to influence the resource supply to the cell surface compares with that of larger cells.

PHOTON SUPPLY TO THE CELL SURFACE

We have already seen that the daily distance which non-flagellate picoplankters can move vertically in a stratified water body is extremely limited in relation to vertical gradients of photon flux density (granted a constant surface photon flux density) and to likely temporal variations in the surface photon flux density. Larger non-flagellate planktophytes which can alter their mean density move fast enough to achieve substantial changes in their light environment in a stratified water body, usually (in the case of cyanobacteria) in relation to the relative supply of photons and of nutrient solutes (Walsby and Reynolds 1980).

Flagellate picoplankters can swim at velocities which enable them to achieve meaningful changes in their incident photon flux density in a day's vertical movement in a stratified water body. *Micromonas pusilla* can swim at 50 μm • s^{-1}, and can accordingly move up to 4.32 m in 24 h. If this movement is vertically upwards in a water body with an attenuation coefficient of 0.12 m^{-1}, equation (5) tells us that the cell can increase its mean incident photon flux density to 1.68 times the original value in one day's swimming. This could well be a result worth achieving in terms of increased specific growth rate if the initial photon flux density was limiting; conversely, downward swimming could take a cell from a region of *high*, photo-inhibitory photon flux density to one with a lower, but still growth-saturating photon flux density (see Raven and Richardson 1984). However, we have also seen that the capital cost (fraction of cell devoted to the structure) and the running cost (power used per unit cell volume, and per unit photon absorbed by a cell in a given light field with a certain chromophore concentration) of flagellar synthesis and operation to enable a picoplankter to swim at 50 μm • s^{-1} is greater than that for a larger cell. With these larger costs of motility, it is likely that larger (fitness) benefits must accrue to a picoplankter than to larger cells to make the occurrence of flagella selectively advantageous.

NUTRIENT SOLUTE SUPPLY TO THE CELL SURFACE

Since the classical study of Munk and Riley (1952) a number of workers have attempted to quantify the rate of supply of nutrient solutes from the bulk phase to the surface of a nutrient-consuming cell, and the effect of motion of the cell relative to the bulk medium on that rate of supply (e.g. Pasciak and Gavis 1974, 1975; Gavis and Ferguson 1975; Gavis 1976; Purcell 1977; Berg and Purcell 1977; Lehman 1978; Vogel 1981; Mierle 1985a,b,c). The brief discussion here is intended to present some of the major conclusions from these publications insofar as they apply to picoplankters (and, for comparison, to somewhat larger planktophytes).

We start with the existence of unstirred layers (boundary layers), around all solid objects in fluids. The notion of the unstirred layer requires that there is essentially no component of mass movement of medium at right angles to the surface of a cell within the boundary layer; accordingly, exchange of material across the boundary layer (between the surface of the solid and the bulk medium) is by molecular diffusion (we ignore the small effects of net water uptake during cell growth). It is convenient to speak of the *effective thickness* of a boundary layer if diffusion is being considered; this is the thickness of a hypothetical layer with *no* radial movement of solution, distinct from the mixed bulk medium; in practice the distinction is less marked.

The size of the cells we are considering (i.e. 0.5–5 μm radius) together with the properties of water, and the maximum velocity at which the cells can move relative to the bulk medium determines the minimum unstirred layer thickness. Such motion can occur as a result of turbulent mixing in the bulk medium (Gavis 1976), by density differences between cells and the medium (equation (4); Table 5); or by swimming, using flagella, at up to hundreds of μm • s^{-1} in Dinophytes, and rather more slowly in other phytoflagellates (Roberts 1981; Raven and Richardson 1984). These various motions have little impact on unstirred layer thicknesses (Gavis 1976; Purcell 1977; Vogel 1981).

For spherical cells, the effective thickness of the unstirred layer is directly proportional to the radius of the cell for the limiting case of no motion of the cell relative to the bulk medium (Langmuir 1918; Gavis 1976; Purcell 1977). The steady-state flux to the cell surface is given (Gavis 1976; Purcell 1977) by equation (7):

$$(7) \qquad J = \frac{3(C_b - C_s)\, D}{r^2}$$

where
J = steady state flux to the cell surface of the specified solute/ mol $\cdot$ m^{-3} $\cdot$ s^{-1}
r = cell radius/m
C_b = bulk phase concentration of the specific solute in the steady-state/ mol $\cdot$ m^{-3}
C_s = cell surface concentration of the specified solute in the steady state/ mol $\cdot$ m^{-3}
D = diffusion coefficient of the specified solute/mol $\cdot$ m^{-3}

In the limiting case where C_s = O, J becomes the maximum flux to the surface of the cell which is effectively a "perfect sink" for the solute. Raven (1984a) terms this the "potential flux", although he uses units of area rather than volume of cell in expressing the flux.

With phosphate as the specified solute, C_b = 10^{-4} mol $\cdot$ m^{-3}, C_s = O, and D = 10^{-9} m^2 $\cdot$ s^{-1}, the "potential flux" is 1.2 mol $\cdot$ m^{-3} $\cdot$ s^{-1} for a 0.5 μm radius cell and 12 mmol $\cdot$ m^{-3} $\cdot$ s^{-1} for a 5 μm radius cell. For a cell with 125 kg C $\cdot$ (m^3 cell volume)$^{-1}$ and the "Redfield ratio" of 106 mol C:1 mol P in cell material, the intracellular P concentration is 98 mol $\cdot$ m^{-3}. The "potential fluxes" could, then, support specific growth rates (computed from the rate of increase of cell P) of 11.2/98 or 0.0122 s^{-1} for the 0.5 μm radius cell and 0.012/98 or 1.2.10^{-4} $\cdot$ s^{-1} for the 5 μm radius cell. The actual μ_{max} is unlikely to exceed 30.10^{-6} $\cdot$ s^{-1} at 20°C (Table 4), so the potential flux is 407 times the actual flux for the 0.5 μm radius cell and 4.07 times the actual flux for the 5 μm radius cell. Another way of expressing this imbalance is to compute C_s from equation (7), using J (computed from μ_{max} = 30.10^{-6} $\cdot$ s^{-1} and cellular P concentration = 98 mol $\cdot$ m^{-3}) = 2.94.10^{-3} mol P $\cdot$ (m^3 cell volume)$^{-1}$ $\cdot$ s^{-1} and previously quoted values of C_b and D, yielding C_s = 9.9755.10^{-5} mol $\cdot$ m^{-3} for the 0.5 μm radius cell and C_s = 7.75.10^{-5} mol $\cdot$ m^{-3} for the 5 μm radius cell. These two approaches agree in indicating that the fraction of the overall limitation (cf. Jones 1973) on the rate of P acquisition which resides in diffusion through the unstirred layer is 0.00246 for the 0.5 μm radius cell and 0.246 for the 5 μm radius cell. Thus, while the unstirred layer is a significant (but not the major) factor restricting phosphate uptake at high growth rates in 5 μm radius cells, it is not an important limitation in 0.5 μm radius cells. It is important to note that, if growth rate is restricted by some other factor (endogenous or exogenous) and the cell phosphorus concentration is maintained at 98 mol $\cdot$ m^{-3}, the reduced phosphate uptake rate means that the unstirred layer makes a smaller contribution to the limitation of phosphate influx. Conversely, if phosphorus-deficient cells are recovering from that deficiency in 10^{-4} mol $\cdot$ m^{-3} exogenous phosphate with a phosphate influx in excess of the rate found in steady-state growth at 30.10^{-6} $\cdot$ s^{-1} (see Raven 1980a), then the 5 μm radius cell can *only* increase its influx to 4.07 times the steady-state rate even if the plasmalemma phosphate transport system acts as a "perfect sink" for phosphate; no such restriction is found for 0.5 μm radius cells.

Mierle (1985a,b,c) has presented data on phosphate transport by *Synechococcus leopolensis* UTEX 625 which bear upon the role of unstirred layers in limiting phosphate influx in picoplankters. Mierle's *Synechococcus* cells has an effective spherical radius of ~ 1 μm (based on the quoted value for cell surface area per unit

54

dry weight of 126 cm^2 per mg : Mierle 1985a). The relationship between phosphate influx and phosphate concentration were on the "Blackman" rather than the "Michaelis–Menten" type, i.e. had a much more rapid transition to saturation than does a rectangular hyperbola (see Smith and Walker 1980). This is consistent with diffusive limitation dominating the ascending portion of the rate/concentration curve and phosphate porter activity defining the phosphate-saturated rate of uptake. This is how Mierle (1985a,b,c) interpreted the data, and used a model with diffusion through the unstirred layer in series with a porter showing Michaelis-Menten kinetics to compute estimates of the diffusion resistance of the unstirred layer, and the K_m (half-saturation value) and J_{max} (phosphate saturated rate) for the porter. The values obtained were 2 990 s • m^{-1} for the diffusion resistance of the unstirred layer, $1.57.10^{-5}$ mol • m^{-3} for K_m, and 10.7 nmol • (m^2 cell surface)$^{-1}$ • s^{-1} for J_{max}.

The diffusion resistance of the unstirred layer (2 990 s • m^{-1}) was suggested by Mierle (1985c) to be consistent with theoretical predictions for a cell of the size and shape (between prolate spheroid and a cylinder) of the *Synechococcus* cells. However, his analysis did not take into account the resistance to diffusion imposed by the cell wall and, particularly, by the gram-negative outer membrane of these cells (Nikaido and Nikae 1979; Nikaido and Rosenberg 1981; Koch and Wang 1982; Raven 1984a, Benz and Bohme 1985). The permeability coefficient of the outer membrane of *Escherichia coli* to glucose, for example, is only $2.0–3.7.10^{-5}$ m • s^{-1}, equivalent to a resistance of $2.7–5.0.10^5$ s • m^{-1}. The effective radius of the protein pores ("porin") which permit the diffusion of low M_r solutes across the membrane is rather larger in the cyanobacterium *Anabaena cylindrica* (0.9 nm: Benz and Bohme 1985) than in *Escherichia coli* (0.6 nm: Nikoido and Rosenberg 1981); the cyanobacterial porin is slightly (2×) selective for cations (K$^+$) over anions (Cl$^-$): Benz and Bohme (1985). While the permeability of a porin-containing membrane for phosphate is probably in excess of that for glucose, the diffusion resistance computed by Mierle (1985c) probably has a component attributable to the outer membrane. Another possible cause of overestimation of the diffusion resistance of the unstirred layers in Mierle's (1985a) analysis of the phosphate influx: phosphate concentration relationship comes from the possibility that the assumption of a single Michaelis–Menten relationship for phosphate influx as a function of C_s (equation (7)) is in error. A similar influx *vs.* concentration curve could come about from limitation of porter activity by phosphate concentration for the lower phosphate concentrations and limitation by a co-substrate for the porter (ATP, if it is a primary active transport mechanism; H$^+$ or Na$^+$ if it is a cotransport mechanism) at higher phosphate concentrations. Failure to recognize this possibility led to an over-estimate of diffusion resistances for the photosynthesis of C$_3$ land plants (see Farquhar and Caemmerer 1982; MacFarlane and Raven 1985). Such a phenomenon in the case of phosphate transport by *Synechococcus leopolensis* would not necessarily be inconsistent with the observed lower activation energy for the "low phosphate concentration" than for the "high phosphate concentration" phosphate influx (Mierle 1985a).

Caution is clearly needed in interpreting solute influx *vs.* solute concentration relationships. It is likely that the data of Mierle (1985a) reflect an unstirred layer thickness of $\sim$ 1 μm (cf. equation 4a of Mierle 1985c) if the cells are approximated to 1 μm radius spheres, that the outer membrane imposes a significant barrier to phosphate uptake at low external concentrations, and that the computed K_m and J_{max} values are approximately correct. The computed J_{max} value of 10.7 nmol • (m^2 cell surface)$^{-1}$ • s^{-1} is equivalent to a cell-volume based flux of 0.0321 mol P • (m^3 cell volume)$^{-1}$ • s^{-1} which, with an assumed P concentration of 98 mol P • (m^3 cell volume)$^{-1}$ could support a growth rate of $3.28.10^{-4}$ • s^{-1}, or scaled from 25°C to 20°C with the assumption of a Q$_{10}$ of 2, $2.32.10^{-4}$ • s^{-1}. This is an order of magnitude greater than μ_{max} of *S. leopolensis* at 20°C ($\leq 0.24.10^{-4}$ • s^{-1}: Table 4), and the computed total extracellular diffusion resistance of Mierle (1985a), i.e. $\sim$ 3.10^3 s • m^{-1}, would only require a ($C_b - C_s$) value (equation (7)) of $2.3.10^{-6}$

mol $\cdot$ m^{-3} to account for the μ_{max} value of phosphate influx of 0.78 nmol $\cdot$ m^{-2} $\cdot$ s^{-1} at 20°C.

Accordingly, for the *growth* of *Synechococcus leopolensis*, phosphate diffusion from the bulk phase to the plasmalemma need not be a major constraint even at low external phosphate concentrations; *short-term uptake* experiments may show diffusion to the plasmalemma to be more important if the influx is higher than that associated with steady-state growth. In other words, the phosphate transport system in the plasmalemma of *S. leopolensis does not* achieve diffusive limitation (Koch 1971; Koch and Wang 1982); this is also true of wild-type *Escherichia coli*, and of chemorganotrophic eubacteria from oligotrophic freshwater habitats, but less so of a chemostat-selected strain of *E. coli* (Koch and Wang 1982). Raven (1980a) notes that approaching this limit involves an increased quantity of porter per m^2 of membrane area, and/or an increase in the ratio of specific reaction rate at substrate saturation: K_m for substrate of individual porter molecules. The costs for the first option are more readily quantified than are those of the second (cf. Raven and Richardson 1984).

The occurrence of the gram-negative outer membrane in cyanobacteria probably offsets some of the advantages which picoplanktonic strains enjoy relative to their large-celled relatives in terms of reduced unstirred layer resistance to solute uptake per unit of cell volume. We have seen from equation (7) that the volume-based "potential flux" from a given bulk phase concentration is inversely proportional to the square of cell radius. We can define the 'potential flux' through a gram-negative outer membrane in terms of the volume of the cell which is summarised using equation (8):

$$(8) \qquad J_{vol} = \frac{3\,C_o\,D'}{r\,l}$$

where J_{vol} = solute flux per unit volume of cell/mol $\cdot$ m^{-3} $\cdot$ s^{-1}

$\quad\;\; C_o$ = solute concentration at the outer surface of the plasmalemma (the concentration at the inner surface is assumed to be zero, the cell acting as "perfect sink"

$\quad\;\; D'$ = Diffusion coefficient for the solute in water, corrected for the fraction of the outer membrane surface occupied by pores/m^2 $\cdot$ s^{-1}

$\quad\;\; l$ = length of diffusion path (= thickness of outer membrane/m

$\quad\;\; r$ = cell radius/m (the factor $\frac{3}{r}$ converts an area-based flux to a volume-based (flux)

If the potential flux through the outer membrane greatly exceeds that through the unstirred layer, then the potential flux to the cell surface is inversely proportional to r^2 (equation (7)); if the reverse is the case, the potential flux is inversely proportional to r (assuming that D' and l in equation (8) do not vary). Equation (8) also describes the potential "lipid-solution" flux through the plasmalemma on a volume basis provided D' is now the diffusion coefficient in the membrane substance corrected for the distribution ratio for the solute between water and the membrane substance. Mediated transport on a volume basis through the plasmalemma is also (granted a fixed aerial density of porters) proportional to the reciprocal of r. Thus, more generally, the potential for membrane-limited fluxes (on a unit volume basis) into planktophytes are inversely proportional to the cell radius, while limitation by unstirred layers (the cell surface approximating to a 'perfect sink') is inversely proportional to the square of the cell radius.

The gram-negative outer membrane can thus be construed as a liability in relation to the capacity for uptake (on a cell volume basis) of nutrients present at low concentrations in the bulk medium, albeit a liability which is relatively less significant for smaller cells. What might be the "use" of the outer membrane in cyanobacteria? A function in coliforms seems to be in resisting digestion in the gut (Nikaido and Nakae 1979) and, to the extent that picoplankters are ingested by herbivores, a similar

56

function might be imputed here, enhancing the viability of cells voided from the herbivore.

Another possible function relates to leakage of a solute, such as CO_2, which thereby short-circuit the active accumulation of CO_2 by the "inorganic carbon accumulation mechanism". We have seen that, other things being equal, the larger surface area per unit cell volume in picoplankters exacerbates this problem; we can now see that, with the effective maximum thickness of the unstirred layer around a spherical cell being proportional to the cell radius, the problem is further exacerbated. Using equations (7) and (8) to predict "potential leakage", it is clear that "potential leakage" per unit cell volume is inversely proportional to r if leakage is determined by the plasmalemma, or the outer membrane, but is inversely proportional to r^2 if leakage is determined by the unstirred layers. Thus the presence of an outer membrane, by leading to periplasmic accumulation of leaked inorganic C *might* help to increase net influx of inorganic C.

These various potential influences of the presence of the gram-negative outer membrane *may* have their analogues in certain Chlorophyte microalgae, in that the outer layer of the cell wall of some strains of *Chlorella* (and, probably, *Nannochloris*) contains a sporopallerin-like substance which prevents the passage of molecules of the size of NADH and larger, and probably impedes the flux of smaller molecules (see chapter 9 of Raven 1984a, and references in Table 3). However, not all walled eukaryotic picoplankters have this layer, and many eukaryotic picoplankters lack a cell wall.

Overall, it would appear that picoplankton cells have an advantage over larger cells (0.5 μm radius compared to 5 μm radius) with respect to their capacity to acquire dissolved nutrients from low bulk phase concentrations of solutes bearing those nutrients when acquisition is expressed as mol solute $\cdot$ (m^3 cell volume)$^{-1} \cdot$ s^{-1}. Important factors here are the small effect of achievable motion of 0.5–5 radius cells relative to their medium on the thickness of the unstirred layer around the cells, and the smaller maximum thickness of unstirred layers in the smaller cells (0.5 μm in the 0.5 μm radius cell; 5 μm in the 5 μm radius cell). These considerations show that the potential flux of nutrients to the cell surface (i.e. the flux of a solute in mol $\cdot$ (m^3 cell volume)$^{-1} \cdot$ s^{-1} from a given concentration in the bulk phase to a cell surface which acts as a perfect sink for the solute) is 100 times higher for a picoplankton cell (0.5 μm radius) than in a 5 μm radius cell (equation (7)). The solute flux across the plasmalemma is, again on a cell volume basis, 10 times higher in the 0.5 μm than in the 5 μm radius cell, other factors being constant; this applies to non-mediated passive fluxes and to mediated active and passive fluxes. Potential leakage (per unit cell volume) across the plasmalemma is, other factors being constant, 10 times higher in 0.5 μm than in 5 μm radius cells. Finally, the resistance to diffusion imposed by the gram-negative outer membrane, and to analogous cell wall layers in eukaryotes, can help to restrict leakage (by favouring re-accumulation) but acts to restrict net mediated uptake of most exogenous solutes.

With these advantages, it would seem that picoplankters could have a significant advantage in terms of nutrient acquisition rate per unit cell volume, relative to larger cells in the same, limiting, bulk phase concentration. We reiterate that the very limited motion of picoplankters resulting from density differences between cells and the medium, as well as the faster movement found in flagellate picoplankters, does not significantly enhance solute acquistion from a bulk phase of constant solute concentration. The substantial migrations which flagellates can perform are, however, important in relation to moving the cells into parcels of water with higher solute concentrations, as well as being important in relation to the mean photon flux density incident on the cell (see Koch 1971; Purcell 1977; Berg and Purcell 1977; Vogel 1981). In the context of cost-benefit analyses of flagellar motility in energy-limited environments, we note that the minimum energy input rate for motility increases as the square of the velocity of movement for a cell of given size (and shape).

Dealing first with interception of photons, the area available for interception of a strictly vector light field is πr^2, while the area available for interception of a totally diffuse (scalar) is $4\pi r^2$, in a spherical cell. The absorptance (fraction of incident photons that are absorbed by the cell) is, for a given chromophore at a certain concentration (mol chromophore • $(m^3$ cell volume)$^{-1}$) directly proportional to the cell radius (provided that the chromophore concentration in the largest cell considered is insufficient to give a significant package effect). Thus, under these constraints, the photon absorption rate per unit cell volume in a given light field is *independent* of cell radius. In a stratified water body the "optimal" photon flux density (that at which the growth rate is maximal) can be reached by motile cells, albeit at a minimum energy cost per unit cell volume which is in direct inverse proportion to the cell radius for flagellar motility at a given velocity. Smaller cells are, then, at an advantage in terms of a smaller package effect than occurs in larger cells at a given chromophore concentration but may be at a disadvantage in terms of a lower maximum chromophore concentration (because more of the volume of a small cell may be occupied by non-scaleable components) and of a greater power requirement per unit cell volume for locomotion at a given velocity.

For solute acquisition, the small cell has a clear advantage in terms of the potential solute flux from a bulk phase of given concentration and the surface of a cell which acts as a perfect sink for the solute, i.e. maintains a zero solute concentration at the outer surface of the plasmalemma. This "potential flux" per unit cell volume is directly proportional to the square root of the cell radius, so a ten-fold decrease in radius increase 10^2 times the potential flux per unit of cell volume. This advantage may be partly offset by a capacity for more rapid movement to a parcel of water with higher bulk-phase nutrient concentrations by the larger cells.

Thus, for spherical cells of 0.5–5 μm radius (Table 5) a given density difference relative to the environment lead to "passive" movement relative to the immediate environment at a velocity which is in direct proportion to the square of the cell radius in a spherical cell, while the power requirement per unit cell volume for movement at a given velocity is directly proportional to the reciprocal of the cell radius. We reiterate here that the movements mentioned here only significantly enhance nutrient acquisition by the cells if they move the cells to a parcel of water with a higher bulk phase concentration of the nutrient; the movement of the cells at an achieveable velocity through a medium of constant nutrient concentration does not significantly enhance nutrient uptake.

Finally, it is worth noting that the *decreased* cell size commonly found in a given genotype of picoplankter when it is resource-limited (photons *or* nutrient solutes) tends to *increase* the capacity to acquire these resources on the basis of mol photon (or mol solute) absorbed • $(m^3$ cell volume)$^{-1}$ • s^{-1} from a low-resource environment (see Raven 1986d).

Conclusions and Prospects

THE MINIMUM SIZE OF PHOTOLITHOTROPHIC O_2-EVOLVERS

The minimum size of chemo-organotrophs has been discussed by Morowitz (1967) and Pirie (1964, 1973); further relevant papers are Churchward et al. (1982), Maniloff (1983), Birky and Skavaril (1984) and Ehrenberg and Kurland (1984). The most quantitatively detailed of these analyses is that of Morowitz (1967). On the basis of the genome size needed to code for a minimum number of kinds of soluble protein catalysts (assumed to be 100), together with tRNA, rRNA and "structural proteins", the occurrence of three copies of the ribosome and of each tRNA and soluble protein species, and of one species of mRNA, the whole ensemble enclosed in a "unit membrane" plasmalemma, and with three times as much water mass as dry matter. Morowitz (1967), computes a minimum radius for a spherical cell of 62 nm (0.062 μm). Morowitz (1967) also computed minimum cell size for smaller numbers of soluble protein catalysts (45 catalysts means a radius of 52 nm; 50 catalyst requires a radius of 53 nm). However, the protein catalyst requirements for protein synthesis have certainly not *de*creased in number since the publication of Morowitz (1967).

Furthermore, Morowitz (1967) did not include membrane transporters in his computations of minimum genome size; these would be needed for substrate uptake even if the environment were such that pH and volume regulation did not require membrane porters (see Raven and Smith 1982), since minimizing the number of intracellular protein-catalysed transformations of low M_r compounds in cell growth demands that the medium shall supply a wide range of organic substrates and that the membrane shall transport them. While the volume occupied by these porters has already been accounted for in terms of plasmalemma volume in the computations of Morowitz (1967), the presence of such porter protein requires a larger minimum genome size. Accordingly, we opt for the larger estimate of Morowitz (1967), i.e. a radius of 62 nm as a reasonable minimum theoretical size for a chemo-organotroph.

Morowitz (1967) points out that this "smallest theoretical cell" has a volume $\sim$ $\frac{1}{10}$ that of the smallest characterized cell (a mollicute or mycoplasma) i.e. 0.001 μm^3 rather than 0.01 μm^3. Morowtiz (1967) suggests that his computations may underestimate the minimum size due to the need for more synthetic apparatus (more steps in syntheses) in relation to minimizing "mistakes" in synthesis, and because more catalysts must be coded for (even if they are not always present) to cope with environment fluctuations. To these two we can add another reason, again related to error-avoidance and controlability; this relates to the need to have kinetic heterogeneity in catalysts involved in metabolic sequences, which in turn means that a rigid "three copies of each catalyst" must involve very large inequalities of catalytic capacity per cell for different reactions, and hence a low potential μ_{max} (see above, and equation (1)). Rectifying the situation by having different numbers of copies of different catalysts must clearly involve having more than three copies per cell of catalysts of low specific reaction rate rather than less than three copies per cell of catalysts of high specific reaction rate. Certainly a chemo-organotroph of 0.05–0.10 μm^3 volume can have a high specific growth rate (Table 4), and thus must substantially overcome the restrictions on μ_{max} which were just mentioned. We note that there are few data on the efficiency of resource use in growth by very small cells ($\sim$ 0.01–0.05 μm^3).

Raven (1986a), and discussion earlier in the present paper, suggest that *minimum genome size*, and number of kinds of catalyst per cell, and hence minimum cell volume and mass, increase in the order: chemo-organotrophs on complex media; chemo-organotrophs on simple media; lithotrophs. Accordingly, we would expect the smallest photolithotrophs to have two or three times the volume of the smallest chemo-organotroph growing on a very complex medium. This implies a volume of $\sim$ 0.05–0.10 μm^3, or a spherical cell with a radius of 0.23–0.29 μm. Thus some of the smaller picoplankters may well be approaching the lower size limit.

THE MINIMUM SIZE OF MEMBERS OF PARTICULAR MAJOR TAXA OF O$_2$-EVOLVING PHOTOLITHOTROPHS

The minimum size of a photolithotrophic cyanobacterium is probably less than that of the smallest photolithotrophic eukaryotes due to the presence of more "non-scaleable" components in the latter (Tables 5 and 6). Simiarly we might expect to find a smaller minimum cell size for "Chlorophyte" eukaryotes than for "Chromophyte" eukaryotes in view of the greater fraction of the cell volume taken up by "non-scaleable" plastid envelopes. This argument does not, however, explain the absence of picoplanktonic (eukaryotic) Rhodophyta, since they, like the

"Chlorophytes" (excluding the Euglenophyta and Chlorarachniophyta; Raven 1985b), have only 2 plastid envelope membranes. While a number of "Chromophyte" Divisions (Prymnesiophyta, Chrysophyta, Cryptophyta) have representatives in the picoplankter size range (Murphy and Haugen 1985; Yentsch 1983) the diatoms (Bacillariophyta) and dinoflagellates (Dinophyta) are absent. Is the diploid nature of diatoms in part responsible (Lewis 1985)? We note that the mechanism of vegetative cell division in a diatom involves a decrease in cell volume of perhaps 16-fold between successive sexual events which restore vegetative cell sizes to its upper limit for that taxon (Werner 1971; Maske 1982) with a further 4-fold decrease in cell volume when a diploid vegetative cell produces 4 gametes. Such a (16 × 4) on 64-fold decrease in cell volume between the largest vegetative cell and the gametes requires a 2.5-fold decrease in the radius of a spherical entity, so that a "good" picoplanktonic gamete with a radius of 0.5 μm would have been produced from an equally picoplanktonic vegetative cell of 0.8 μm radius, but would give rise (after sexual fusion and valuation of the zygote to a non-picoplanktonic vegetative cell of radius 1.25 μm! Similar arguments apply to the number of autospores which Chlorococcoid Chlorophyte picoplankters (see Brown and Elfman 1983; Raven 1986d), can produce if the whole cell cycle is to stay within the radius bounds of the picoplankter.

In this context it is of interest that the propagules and gametes of benthic eukaryotic macroalgae approach, but do not enter, the picoplankter size range (Grubb 1925; Margruder 1984; Manton 1964), and that the functional size of some of them (especially in the Rhodophyta) is increased by being contained in slime strands (Boney 1981; Fetter and Neushul 1981; cf. Searles 1980). While it is not likely that picoplankton-sized disseminules and gametes would have insurmountable problems in effecting "benthic-planktonic-benthic" transitions (Neushul 1972; Silvester and Sleight 1985); such picoplankton-sized propagules do not seem to exist, thus avoiding further problems in the identification of picoplankton-sized eukaryotic pigmented cells! Maybe the cDNA content of macroalgae is too great to be packaged in a picoplankton-sized cell (cf. Table 2).

We conclude this brief discussion of the minimum size of different major taxa of photolithotrophs by enquiring as to the likely size of the earliest eukaryotic photolithotrophs in relation to the minimum size of extant eukaryotic photolithotrophs. We follow here the reasoning of Stewart and Mattox (1980) that the earliest eukarotes are phagocytotic chemoorganotrophs, exploiting the "predator niche" and able to ingest and digest prokaryotic phototrophs and saprotrophic chemoorganotrophs. The arguments of Stewart and Mattox (1980) suggest that such cells would have had to be large and complex enough to find, capture and ingest prey organisms, and were naked and were motile (probably flagellate). Endosymbiotic origin of plastids could, by making some eukaryotes photolithotrophic, release them from some of the constraints on cell size, although many eukaryotic microphototrophs retain flagellar motility, and/or are wall-less (although often not benefit of structures outside the plasmalemma). We thus regard the smallest eukaryotic picoplankters as "derived".

SMALL SIZE AS A CONSTRAINT ON μ_{MAX}?

The lengthy discussion in the second section of this paper identifies a number of incompletely scaleable components of phototrophic cells which might, by reducing the volume available for rate-limiting catalysts in a fixed total cell volume, reduce μ_{max}. The sum of such components could lead to a reduction in μ_{max} of the fastest-growing 0.5 μm radius cell to ~ 0.8–0.9 of the μ_{max} of the fastest growing 5 μm radius cell, with a larger predicted decrement in eukaryotes than in prokaryotes. Comparison with experimental values of μ_{max} (normalised to 20°C) for various photolithotrophic microorganisms suggests that such a decrement is more consistent with the eukaryotic than with the prokaryotic data. However, compilations of μ_{max}

60

values of the kind shown in Table 4 are subject to a number of criticisms (Eppley and Strickland 1968; Droop 1974), so we should regard the predicted decrement of μ_{max} as "not proven" in terms of available data.

It is not clear how close to the conditions for achieving μ_{max} a picoplankton cell *in situ* is likely to come in the oligotrophic ocean (cf. Goldman et al. 1984; Laws et al. 1984), and, consequently, what the potential contribution to fitness might acrue from a high μ_{max} value *per se*. Exploitation of episodic increases in nutrient availability may well depend on factors other than attaining a very high μ_{max} during the nutrient pulse (Rivkin and Swift 1985).

SMALL SIZE AS AN AID TO RESOURCE ACQUISITION IN LOW-LIGHT AND/OR LOW-NUTRIENT ENVIRONMENTS?

0.5 μm radius cells are probably capable of a higher photon absorption rate per unit cell volume in a given light field than are otherwise similar cells 5 μm in diameter: the decrease package effect in the smaller cells outweights restrictions on the fraction of the cell available to house chromophores as a result of "incompletely scaleable" components. Similarly, the intrinsically thinner unstirred layers around the smaller cells, and the greater area of plasmalemma per unit cell volume, make the potential for nutrient absorption from a low bulk phase concentration per unit of cell volume much higher in the smaller cell; the higher velocity of movement of larger cells relative to the bulk medium for a given density difference between cells and the medium is *not* a significant factor in reducing unstirred layer thickness and thus enhancing nutrient uptake. Flagellar motility is only useful in resource acquisition in terms of movement from one resource supply regime to another in a stratified water body; it does not intrinsically enhance nutrient acquisition in a uniform environment.

SMALL SIZE AS A CONSTRAINT ON RESOURCE RETENTION?

The larger surface area per unit volume of a spherical 0.5 μm radius cell than of a spherical 5 μm radius cell can increase resource (energy, C, N, P, etc.) loss per unit cell volume. Leakage of solutes down free energy gradients is, for fixed plamalemma properties, faster per unit volume in the small than the larger cell. For solutes directly accumulated from the medium (NO_3^-, NH_4^+, phosphate, inorganic C) this faster leakage means that more energy must be used per mol of *net* solute acquisition by the smaller cell. There is also an energetic penalty of leaking solutes which are the product of intracellular metabolism; compatible solutes must be replaced, as must NH_4^+ which is lost as NH_3 after active influx and reduction of NO_3^-. In the case of these solutes which were not directly acquired from the medium there is an additional energetic constraint related to small size, in that the possibilities of reacquisition of the solutes by the cell are diminished in the smaller cells due to more rapid dissipation to the bulk medium (essentially free of those solutes) due to the thinner unstirred layers. To these energetic penalties of smaller cells may be added, for flagellates, the larger power requirement per unit cell volume for movement at a given velocity. It would seem that the superiority of the smaller cells in acquiring nutrients from low-nutrient environments is tempered at low photon flux densities by the additional energy costs, per unit cell volume, of nutrient retention, and (for motile cells) of swimming to parcels of water with higher nutrient availability. These energetic penalties of the smaller cells are manifest (to a smaller extent) during growth at higher nutrient concentrations (see Foy, Gibson and Smith 1976) especially with respect to leakage of 'compatible solutes' in marine organisms. More data are needed on actual leakage rates to quantify these apparent energetic penalties on small cells. This is especially necessary in view of the apparently *lower* specific maintenance loss rates in smaller than in larger-celled photolithotrophs in the 0.5–200 μm^3 cell volume (Van Gemerden 1980), although we must note that not enough data are available to compare the specific maintenance loss rates for cells of widely different volumes (including picoplankton volumes of $\leq 4 \ \mu m^3$) some major taxa (classes, orders, families).

Cell Size in Relation to Resource Storage

The capacity to store resources is important for survival, and (probably) in maximizing growth, in conditions of temporally variable resource supply (Cohen and Parnas 1976; Parnas and Cohen 1976; Rivkin and Swift 1985). Are there differences between a 0.5 μm radius cell and a 5 μm radius cell with respect to storage capacity per unit volume such as could change the number of cell doublings which could occur in a low-P environment at the expense of P stored during a "P pulse", or the length of time that an obligate photolithotroph could survive in the dark on its energy (e.g. polysaccharide) stores? Raven (1984a, chapter 8) considers some of the constraints on storage by aquatic photolithotrophs. For cells in the size range 0.5–5 μm radius, we note that the problem of "incomplete scaleability" of certain cell components may restrict the capacity for storage per unit cell volume in the 0.5 μm radius cell relative to the larger cells. There is evidence consistent with a decreased capacity to store organic C relative to maintenance requirements for organic C in smaller cyanobacterial cells (Foy et al. 1976; Laws 1975; Reynolds 1984), probably due to both a decreased value of mol stored organic C $\cdot$ (mol cell C)$^{-1}$ and an increase in mol C used in maintenance $\cdot$ (mol cell C)$^{-1} \cdot$ s^{-1} in the smaller cells.

What is the Ecological and Evolutionary Significance of the Size-Related Phenomena Discussed in this Paper?

The significance of cell and colony size in the ecology of phytoplankton has been the subject of much discussion (e.g. Banse 1976, 1982; Fogg 1975; Kirk 1983; Laws 1975; Lewis 1976; Malone, 1980; Reynolds 1984; Sournia 1982), albeit mainly using examples from cell sizes in excess of that of picoplankter. However, even when the costs and benefits (see Raven 1984a,b; Raven and Richardson 1984) of the various size-dependent traits of planktophytes have been more rigorously and quantitatively analysed than has been possible in this paper, we shall still not be able to conclude much as to the contribution of these traits to the fitness of the organisms in a given environment until analysis at a different conceptual and experimental level has been carried out (see Osmond et al. 1980).

Acknowledgements

Work in my laboratory related to the topics addressed in this paper has been generously supported by S.E.R.C. and N.E.R.C. Many past and present colleagues have contributed to my thinking on the functioning of planktophytes; special thanks in this regard are due to (alphabetically!) Drs John Beardall, Richard Geider, Sheila Glidewell, Jeff MacFarlane, Bruce Osborne and Kath Richardson.

References

Alberte, R. S., A. M. Wood, T. A. Kursar, and R. R. L. Guillard. 1984. Novel phycoery-thrins in marine *Synechococcus* ssp. Characterization and evolutionary and ecological implications. Plant Physiol. 75: 732–739.

Altman, P. L., and D. S. Dittmer. [ed.]. 1962. Growth. Federation of American Societies for Experimental Biology, Washington D.C.

———. 1972. Biological data book, Volume 1. Federation of American Societies for Experimental Biology, Washington D.C.

Andrews, T. J., and K. M. Abel. 1981. Kinetics and subunit interactions of RUBISCO from the cyanobacterium, *Synechococcus* sp. J. Biol. Chem. 256: 8445–8451.

Ashton, A. R. 1982. A role of ribulose bis-phosphate carboxylase as a metabolite buffer.

F.E.B.S. Letters 145: 1–6.

ATKINSON, A.W., JR., P. C. L. JOHN, AND B. E. S. GUNNING. 1974. The growth and division of the single mitochondrion and other organelles during the cell cycle of *Chlorella*, studied by quantitative stereology and three dimensional reconstruction. Protoplasma 81: 77–110.

ATKINSON, A. W., JR., B. E. S. GUNNING, P. C. L. JOHN, AND W. MCCULLOUGH. 1972. Dual mechanisms of ion absorption. Science 176: 694–695.

BADGER, M. R. 1980. Kinetic properties of ribulose 1,5 bisphosphate carboxylase/oxygenase from *Anabaena variabilis*. Arch. Biochem. Biophys. 201: 247–254.

BADGER, M. R., AND T. J. ANDREWS. 1982. Photosynthesis and inorganic carbon usage by the marine cyanobacterium, *Synechococcus* sp. Plant Physiol. 70: 517–523.

BADGER, M. R., M. BASSETT, AND H. N. COMINS. 1985. A model for HCO_3^- accumulation and photosynthesis in the cyanobacterium *Synechococcus* sp. Theoretical predictions and experimental observations. Plant Physiol. 77: 465–471.

BANSE, K. 1976. Rates of growth, respiration and photosynthesis of unicellular algae as related to cell size — a review, J. Phycol. 12: 135–140.

———— 1982. Cell volumes, maximal growth rates of unicellular algae and ciliates, and the role of ciliates in the marine pelagial. Limnol. Oceanogr. 27: 1059–1071.

BARLOW, R. G., AND R. S. ALBERTE. 1985. Photosynthetic characteristics of phycoerythrin-containing marine *Synechococcus* sp. I. Response to growth photon flux density. Mar. Biol. 86: 63–74.

BEARDALL, J., AND L. ENTWISTLE. 1984. Evidence for a CO_2 concentrating mechanism in *Botrydiopsis* (Tribophyceae). Phycologia 23: 511–513.

BEN-AMOTZ, A., AND M. AVRON. 1983. Accumulation of metabolites by halo-tolerant algae and its industrial potential. Annu. Rev. Micro-biol. 37: 95–119.

BENZ, R., AND H. BOHME. 1985. Pore formation by an outer membrane protein of the cyanobacterium *Anabaena variabilis*. Biochim. Biophys. Acta 812: 286–292.

BERG, H. C., AND E. M. PURCELL. 1977. Physics of chemoreception. Biophys. J. 20: 193–219.

BERLIN, J., J. E. QUISENBERRY, F., BAILEY, M. WOODWORTH, AND B. L. MCMICHAEL. 1982. Effect of water stress on cotton leaves. I. An electron microscopic study of the palisade cells. Plant Physiol. 70:: 238–243.

BIRKY, C. W., AND R. V. SKAVARIL. 1984. Random partitioning of cytoplasmic organelles at cell division: the effect of organelle and cell volume. J. Theor. Biol. 106: 441–448.

BLANK, R., E. HAUPTMAN, AND C-G. ARNOLD. 1980. Variability of mitochondrial population in *Chlamydomonas reinhardtii*. Planta 150: 236–241.

BOLD, H. C, AND W. J. WYNNE. 1985. Introduction to the algae. Structure and Reproduction. Second Edition. Prentice-Hall, Inc., Englewood Cliffs, NJ.

BONEY, A. D. 1981. Mucilage: the ubiquitous algae attribute. Br. Phycol. J. 16: 115–132.

BONT, W. S. 1978. The diameter of membrane vesicles fit in geometric series. J. Theor. Biol. 74: 361–375.

———— 1985. The geometric series of nuclear volumes. J. Theor. Biol. 112: 505–512.

BOROWITZKA, L. J., AND D. A. BROWN. 1974. The salt-relations of marine and halophilic species of the unicellular alga *Dunaliella*. Arch. Microbiol. 96: 37–52.

BOUSSIBA, S., C. M. RESCH, AND J. GIBSON. 1984. Ammonia uptake and retention in some cyanobacteria. Arch. Microbiol. 138: 287–292.

BOWEN, H. J. M. 1979. Environmental Chemistry of the Elements. Academic Press, London.

BOWLES, D. J., P. H. QUAIL, D. J. MORRE, AND G. C. HARTMANN, 1979. Use of markers in plant-cell fractionation, p. 207–223. *In* E. Reid [ed.] Plant Cell Organelles. Ellis Horwood, London.

BRADBEER, C. 1971. Transport of vitamin B_{12} by *Ochromonas malhamensis*. Arch. Biochem. Biophys. 144: 184–192.

BROCH-DUE, M., J. G. ORMEROD AND B. S. FJERDINGEN. 1978. Effect of light intensity on vesicle formation in *Chlorobium*. Arch. Microbiol. 116: 269–274.

BROCK, T. D. 1973. Lower pH limit for the existence of blue-green algae: evolutionary and ecological implications. Science 179: 480–483.

BROOK, A. J. 1981. The biology of Desmids. Blackwell Scientific Publications, Oxford.

BROWN, L. M. 1982. Photosynthetic and growth responses to salinity in a marine isolate of *Nannochloris bacillaris* (Chlorophyceae). J. Phycol. 18: 483–488.

———— 1985. Stepwise adaptation to salinity in the green alga *Nannochloris bacillaris*. Can. J. Bot. 63: 327–332.

BROWN, L. M., AND B. ELFMAN. 1983. Is autosporulation a feature of *Nannochloris*? Can. J. Bot. 61: 2647–2657.

BROWN, L. M., AND J. A. HELLEBUST. 1980a. Some new taxonomic characteristics applied to *Stichococcus bacillaris* (Chlorophyceae). Can. J. Bot. 58: 1405–1441.

———— 1980b. The contribution of organic solutes to osmotic balance in some green and eustigmatophyte algae. J. Phycol. 16: 265–270.

CAVALIER-SMITH, T. 1978. Nuclear volume control by nucleoskeletal DNA, selection for cell volume and cell growth rate, and the solution of the DNA C-value paradox. J. Cell Sci. 34: 247–278.

———— 1982. The origin of plastids. Biol. J. Linn. Soc. 17: 289–306.

———— 1985. Genetic and epigenetic control of the plant cell cycle, p. 179–197. *In* J. A. Bryant and D. Francis [ed.] The Cell division cycle in plants. Cambridge University Press.

CHOHJI, T., T. SAWADA, AND S. KUNO. 1976. Macromolecule synthesis in *Escherichia coli* BB under various growth conditions. Appl. Environ. Microbiol. 31: 864–869.

CHURCHWARD, G., H. BREMER, AND R. YOUNG. 1982. Macro-molecular composition of bacteria. J. Theor. Biol. 94: 651–670.

COHEN, D., AND H. PARNAS. 1976. An optimal policy for the metabolism of storage materials in unicellular algae. J. Theor. Biol. 56: 1–18.

COLEMAN, J. R., AND B. COLMAN. 1981. Inorganic carbon accumulation and photosynthesis in a blue-green alga as a function of external pH. Plant Physiol. 67: 917–921.

CRAIG, T. A., F. L. CRANE, P. C. MISRA, AND R. BARR. 1984. Transplasma-membrane electron transport in *Anacystis nidulans*. Plant Sci. Letts. 35: 11–17.

CRAM, W. J. 1976. Negative feedback inhibition of transport in cells. The maintenance of turgor, volume and nutrient supply, p. 284–316. *In* U. Lüttge and M. G. Pitman [ed.] Encyclopedia of plant physiology, New Series, Volume 2A. Springer-Verlag, Berlin.

———— 1984. Mannitol transport and suitability as an osmoticum in root cells. Physiol. Plant. 61: 396–404.

CUHEL, R. L., AND J. B. WATERBURY. 1984. Biochemical composition and short term nutrient incorproation patterns in a unicellular marine cyano-bacterium *Synechococcus* (WH 7803). Limnol. Oceanogr. 29: 370–374.

DELRIEU, H-J., S. P. N. HUNG, AND F. DE KOUCHKOVSKY. 1985. pH dependence of the S_2-S_3 transition associated with O_2 evolution in inside-out thylakoids. F.E.B.S. Letters 187: 321–326.

DEMPSEY, G.P., D. LAWRENCE, AND V. CASSIE. 1980. The ultrastructure of *Chlorella minutissima* Fott et Navakova (Chlorophyceae, Chlorococcales). Phycologia 19: 13–19.

DEVIPRASAD, P. V., AND Y. B. K. CHOWDARY. 1981. Effect of metabolic inhibitors on the calcification of a freshwater algal, *Gloeotaenium Ioitlesbergarianum* Harsging. 1. Effects of some photosynthetic and respiratory inhibitors. Ann. Bot. 47: 451–459.

DODGE, J. D. 1983. Dinoflagellates: investigation and phylogenetic speculation. Br. Phycol J. 18: 335–356.

DROOP, M. R. 1957. Vitamin B_{12} in marine ecology. Nature (London) 180: 1041–1042.

———— 1968. Vitamin B_{12} and marine ecology. IV. The kinetics of uptake, growth and inhibition of *Monochrysis lutheri*. J. Mar. Biol. Assoc. U.K. 48: 689–733.

———— 1974. Heterotrophy of carbon, p. 530–539. *In* W. D. P. Steward [ed.] Algae physiology and biochemistry. Blackwell Scientific Publications, Oxford.

DURAND, M., AND C. BERKALOFF. 1985. Pigment composition and chloroplast organisation of *Gambierdiscus toxicus* Adachi and Fukuyo. Phycologia 24: 217–223.

DHRENBERG, M., AND C. G. KURLAND. 1984. Costs of accuracy determined by a maximal growth rate constraint. Quart. Rev. Biophys. 17: 45–82.

ELDER, R. G., AND M. PARKER. 1984. Growth responses of a nitrogen fixer (*Anabaena flosaquae*; Cyanophyceae) to low nitrate. J. Phycol. 20: 298–301.

EMERSON, R. 1958. The quantum yield of photosynthesis. Annu. Rev. Plant Physiol.

9: 1–24.

EPPLEY, R. W., AND J. D. H. STRICKLAND. 1968. Kinetics of marine phytoplankton growth. Adv. Microbiol. Sea. 1: 23–62.

FALKOWSKI, P. G., AND T. G. OWENS. 1980. Light-shade adaptation. Two strategies in marine phytoplankton. Plant Physiol. 66: 592–595.

FALKOWSKI, P. G., Z. DUBINSKI AND K. WYMAN. 1985. Growth-irradiance relationships in phytoplankton. Limnol. Oceanogr. 30: 311–321.

FARQUHAR, G. D., AND S. VON CAEMMERER. 1982. Modelling of photosynthetic responses to environmental conditions, p. 549–587. *In* O. L. Lange, P. S. Nobel, C. B. Osmond, and H. Ziegler [eds.] Encyclopedia of plant physiology (New Series), Volume 12B, Springer Verlag, Berlin.

FASHAM, M. J. R., AND T. PLATT. Photosynthetic response of phytoplankton to light: a physiological model. Proc. Roy. Soc. Lond. B 219: 355–370.

FENCHEL, T. 1974. Intrinsic rate of natural increase: the relationship with body size. Oecologia 14: 317–326.

FETTER, R., AND M. NEUSHUL. 1981. Studies on developing and released spermatia in the red alga, *Tiffaniella snyderae* (Rhodophyta). J. Phycol. 17: 141–159.

FOGG, G. E. 1949. Growth and heterocyst production in *Anabaena cylindrica* Lemm. II. In relation to carbon and nitrogen metabolism. Ann. Bot. 13: 241–259.

———— 1975. Algae cultures and phytoplankton ecology. University of Wisconsin Press, WI.

FORD, T. W. 1984. A comparative ultrastructural study of *Cyanidium caldarium* and the unicellular red alga *Rhodosorus marinus*. Ann. Bot. 53: 285–294.

FOY, R. H., C. E. GIBSON, AND R. V. SMITH. 1976. The influence of daylength, light intensity and temperature on the growth rates of planktonic blue-green algae. Br. Phycol. J. 11: 151–163.

FRANCIS, D., A. B. KIDD, AND M. D. BENNETT. 1985. DNA replication in relation to DNA C values, p. 61–82. *In* J. A. Bryant and D. Francis [ed.] The cell division cycle in plants, Cambridge University Press.

FRESNEL, J., P. GALLE, AND P. GAYRAL. 1979. Resultats de la microanalyse des cristaux vacuolaires chez deux chromophytes unicellulaires marines: *Exanthemachrysis gayraliae, Pavlova* sp. (Prymnesiophyceae, Pavlovacees). Compt. Rend. Acad. Sci. Paris D. 288: 823–825.

GAVIS, J. 1976. Munk and Ridley revisited: nutrient diffusion transport and rates of phytoplankton growth. J. Mar. Res. 34: 161–179.

GAVIS, J. AND J. F. FERUSON. 1975. Kinetics of carbon dioxide uptake by phytoplankton at high pH. Limnol. Oceanogr. 20: 211–221.

GEIDER, R. J., B. A. OSBORNE, AND J. A. RAVEN. 1986. Ligth dependence of growth and photosynthesis in *Phaeodactylum triconutum* (Bacillariophyceae). J. Phycol. 21: 39–48.

GIBBS, S. P. 1981. The chloroplast endoplasmic reticulum: structure, function and evolutionary significance. Int. Rev. Cytol. 72: 49–99.

GLOVER, H. E., AND I. MORRIS. 1981. Photosynthetic characteristics of coccoid marine cyanobacteria. Arch. Microbiol. 129: 42–44.

GÖBEL, F. 1978. Quantum efficiency of growth, p. 907–925. *In* R. K. Clayton and W. R. Sistrom [ed.] The photosynthetic bacteria, Plenum Press, New York, NY.

GOLDMAN, J. C., J. J. McCARTHY, AND D. G. PEAVEY. 1979. Growth rate influence on the chemical composition of phytoplankton in oceanic waters. Nature (London) 279: 210–215.

GREEN, J. C. 1975. The fine-structure and taxonomy of the haptophycean flagellate *Pavlova lutheri* (Droop) Comb. Nov. [= *Monochrysis lutheri* Droop). J. Mar. Biol. Assoc. U.K. 55: 785–793.

GREEN, J. C., AND R. N. PIENAAR. 1977. The taxonomy of the order Isochrysidales (Prymnesiophyceae) with special reference to the general *Isochrysis* Parke, *Dicrateria* Parke and *Imantonia* Reynolds. J. Mar. Biol. Assoc. U.K. 57: 7–17.

GRIFFIN, D. H., W. E. TIMBERLAKE, AND J. C. CHENEY. 1974. Regulation of macromolecular synthesis, colony development and specific growth rate of *Achlya bisexualis* during balanced growth. J. Gen. Microbiol. 80: 381–388.

GRUBB, V. M. 1925. The male organs of the florideae. Bot. J. Linn. Soc. 47: 177–255.

GRUEN, D. W. R., AND J. WOLFE. 1982. Lateral tensions and pressures in membranes and lipid monolayers. Biochim. Biophys. Acta 688: 572–580.

GUILLARD, R. R. L., AND V. CASSIE. 1963. Minimum cyanocobalamin requirements of some marine centric diatoms. Limnol. Oceanogr. 8: 161–165.

GUILLARD, R. R. L., L. S. MURPHY, P. FOSS AND S. LIAAEN-JENSEN. 1985. *Synechococcus* spp. as likely zeaxanthin-dominant ultraplankton in the North Atlantic. Limnol. Oceanogr. 30: 412–414.

HAJIBAGHERI, M. A., J. L. HALL, T. J. FLOWERS. 1984. Stereological analysis of leaf cells of the halophyte *Suaeda maritima* (L) Dunn. J. Exp. Bot. 35: 1547–1557.

HARRIS, G. P. 1978. Photosynthesis, productivity and growth: the physiological ecology of phytoplankton. Ergeb. Limnol. 10: 1–171.

HEALEY, F. P. 1985. Interacting effects of light and nutrients on the growth rate of *Synechococcus linearis* (Cyanophyceae). J. Phycol. 21: 134–146.

HEINRICH, R., S. M. RAPOPORT, AND T. A. RAPOPORT. 1977. Metabolic regulation and mathematical models. Prog. Biophys. Mol. Biol. 32: 1–82.

HELLEBUST, J. A. 1976. Osmoregulation. Annu. Rev. Plant Physiol. 27: 485–505.

HELLEBUST, J. A., AND S. M. L. LE GRESLEY. 1985. Growth characteristics of the marine rock pool flagellate *Chlamydomonas pulsatilla* Wollenweber (Chlorophyta). Phycologia 24: 225–229.

HERDMAN, M. 1981. DNA base composition and genome size of *Prochloron*. Arch. Microbiol. 129: 314–316.

HERDMAN, M., M., JANVIER, R. RIPPKA, AND R. Y. STANIER, 1979. Genome size of cyanobacteria. J. Gen. Microbiol. 111: 73–85.

HEWITT, E. J. 1983. A perspective of mineral nutrition: essential and functional minerals in plants, p. 277–323. *In* D. A. Robb and W. S. Pierpoint [ed.] Metals and micronutrients. Uptake and utilization by plants. Academic Press, London.

HOLM-HANSEN, O. 1969. Algae: amounts of DNA and organic carbon in single cells. Science 163: 87–88.

HOOGENHOUT, H., AND J. AMESZ. 1965. Growth rates of photosynthetic microorganisms in laboratory cultures. Arch. Mikrob. 50: 10–25.

INGRAHAM, J. L. 1958. Growth of psychrophilic bacteria. J. Bacteriol. 76: 75–80.

INOYYE, I., AND R. N. PIENAAR. 1983. Observations on the life history and microanatomy of *Thoracasphaera heimii* (Dinophyceae) with special reference to its systematic position. S. Afr. J. Bot. 2: 63–75.

JOHNSON, P. W. AND J. M. SIEBURTH. 1979. Chroococcoid syanobacteria in the sea: a ubiquitous and diverse phototrophic biomass. Limnol. Oceanogr. 24: 928–935.

1982. *In situ* morphology and occurrence of eucaryotic phototrophs of bacterial size in the picoplankton of estuarine and oceanic waters. J. Phycol. 18: 318–327.

JONES, H. G. 1973. Limiting factors in photosynthesis. New Phytol. 72: 1089–1094.

JORDAN, D. B., AND W. L. OGREN. 1981. Species variation in the specificity of ribulose bisphosphate carboxylase-oxygenase. Nature 291: 513–515.

1983. Species variation in the specificity of ribulose bisphosphate carboxylase-oxygenase. Arch. Biochem. Biophys. 227: 425–433.

KACSER, H., AND J. A. BURNS. 1973. The control of flux. Symp. Soc. Exp. Biol. 27: 65–104.

KAPLAN, A., U. SCHREIBER, AND M. AVRON. 1980. Salt-induced metabolic changes in *Dunaliella salina*. Plant Physiol. 65: 810–813.

KAWAMURA M., M. MIMURO, AND Y. FUJITA. 1979. Quantitative relationship between two reaction centres in the photosynthetic system of blue-green algae. Plant Cell Physiol. 20: 697–705.

KELLY, D. P. 1974. Growth and metabolism of the obligate photolithotroph *Chlorobium thiosulphotophilum* in the presence of added organic nutrients. Arch. Microbiol. 100: 163–178.

KIRK, J. T. O. 1975a. A theoretical analysis of the contribution of algal cells to the attenuation of light within natural waters. I. General treatment of suspensions of pigmented cells. New Phytol. 75: 11–20.

1975B. A theoretical analysis of the contribution of algal cells to the attenuation of light within natural waters. II. Spherical cells. New Phytol. 75: 21–36.

1976. A theoretical analysis of the contribution of algal cells to the attenuation of light in natural waters. II. Cylindrical and spheroidal cells. New Phytol. 77: 341–358.

_____ 1983. Light and photosynthesis in aquatic ecosystems. Cambridge University Press.

KLEINER, D. 1985a. Energy expenditure for cyclic retention of NH_3/NH_4^+ during N_2 fixation by *Klebsiella pneumoniae*. F.E.B.S. Letters 187: 237–239.

_____ 1985b. Bacterial ammonium transport. F.E.M.S. Microbiol. Rev. 32: 87–100.

KOCH, A. L. 1971. The adaptive responses of *Escherichia coli* to a feast and famine existence. Adv. Microb. Physiol. 6: 147–217.

KOCH, A. L. AND C. H. WANG. 1982. How close to the theoretical diffusion limit do bacterial uptake systems function? Arch. Microbiol. 131: 36–42.

KOK, B. 1960. The efficiency of photosynthesis p. 566–633. *In* W. Ruhland [ed.] Encyclopedia of plant physiology, Volume V/1. Springer Verglag, Berlin.

KRATZ, W. A. AND J. MYERS. 1955a. Nutrition and growth of several blue-green algae. Am. J. Bot. 42: 282–287.

_____ 1955b. Photosynthesis and respiration of three blue-green algae. Plant Physiol. 30: 275–280.

KRAUSS, R. W. 1958. Pysiology of the freshwater algae. Annu. Rev. Plant Physiol. 9: 207–244.

KREMER, B. P. 1980. Taxonomic implications of algal photoassimilate patterns. Br. Phycol. J. 15: 399–409.

LANGMUIR, I. 1918. Evaporation of small spheres. Phys. Rev. 12: 368–370.

LAWS, E. A. 1975. The importance of respiration losses in controlling the size distribution of marine phytoplankton. Ecology 56: 419–426.

LAWS, E.A., D. G. REDALJE, L. W. HAAS, P. K. BIENFANG, R. W. EPPLEY, W. G. HARRISON, D. H. KARL, AND J. MARRA. 1984. High phytoplankton growth and production rates in obligotrophic Hawaiian coastal waters. Limnol. Oceanogr. 29: 1161–1169.

MANTON, I., AND M. PARKE 1960. Further observations on small green flagellates with special reference to possible relatives of *Chromulina pusilla* Butcher. J. Mar. Biol. Assoc. U.K. 40: 275–298.

MARCUS, Y., M. VOLOKITA, AND A. KAPLAN. 1984. The location of the transporting system for inorganic carbon and the nature of the form translocated in *Chlamydomonas reinhardtii*. J. Exp. Bot. 35: 1136–1144.

MARGULIS, L. 1981. Symbiosis in cell evolution. W. H. Freeman and Co., San Francisco, CA.

MASKE, H. 1982. Ammonium-limited continuous cultures of *Skeletonema costatum* in steady and transitional state: experimental results and model simulations. J. mar. Biol. Assoc. U.K. 62: 919–943.

MATTOX, K. R., AND K. D. STEWART. 1984. Classification of the green algae: a concept based on comparative cytology, p. 29–72. *In* D. E. G. Irvine and D. M. John [Ed.] Systematics of the green algae. Academic press, London.

MIERLE, G. 1985a. Kinetics of phosphate transport by *Synechococcus leopoliensis* (Cyanophyta):

_____ evidence for diffusion limitation of phosphate uptake. J. Phycol. 21: 177–181.

_____ 1985b. A method for estimating the diffusion resistance of the unstirred layer in microorganisms. Biochim. Biophys. Acta 812: 827–834.

_____ 1985c. The effect of cell size and shape on the resistance of unstirred layers to solute diffusion. Biochim. Biophys. Acta 812: 835–839.

MOESTRUP, Ø 1979. Identification by electron microscopy of marine nanoplankton from New Zealand, including the description of four new species. N. Z. J. Bot. 17: 61–95.

MOORE, G. P. 1984. The C-value paradox. BioScience 34: 425–429.

MOROWITZ, H. J. 1967. Biological self-replicating systems. Prog. Theor. Biol. 1: 35–58.

MORRIS, I., AND H. E. GLOVER. 1981. Physiology of photosynthesis by marine coccoid cyanobacteria — some ecological implications. Limnol. Oceanogr. 26: 957–961.

MUNK, W. H. AND G. A. RILEY. 1952. Absorption of nutrients by aquatic plants. J. Mar. Res. 11: 215–240.

MUNNS, R., H. GREENWAY, K. L. SETTER, AND J. KUO. 1983. Turgor pressure, volumetric elastic modulus, osmotic volume and ultrastructure of *Chlorella emersonii* grown at high and low external NaCl. J. Exp. Bot. 34: 144–155.

MURPHY, L. S. AND E. M. HAUGEN. 1985. The distribution and abundance of photoautotrophic ultraplankton in the North Atlantic. Limnol. Oceanogr. 30: 47–58.

MYERS, J., AND J-R. GRAHAM. 1961. On the mass culture of algae. III. Light diffusers: high *vs* low temperature Chlorellas. Plant Physiol. 36: 342–346.

_____ 1983. On the ratio of photosynthetic reaction centres RC2/RC1 in *Chlorella*. Plant Physiol. 71: 440–442.

MYERS, J., AND W. A. KRATZ. 1956. Relations between pigment content and photosynthetic characteristics in a blue-green alga. J. Gen. Physiol. 39: 11–22.

MYERS, J., J-R. GRAHAM, AND R. T. WANG. 1978. On spectral control of pigmentation in *Anacystis nidulans*. J. Phycol. 14: 513–518.

NEUSHUL, M. 1972. Functional interpretation of benthic algal morphology, p. 47–73. *In* I. A. Abbott and M. Kurogi [Ed.] Contributions to the systematics of benthic marine algae of the North Pacific. Japanese Society of Phycologists, Kobe.

NIKAIDO, H., AND T. NAKAE. 1979. The outer membrane of gram-negative bacteria. Adv. Microb. Physiol. 20: 163–250.

LEHMAN, J. T. 1978. Enhanced transport of inorganic C into algal cells and its implications for the biological fixation of C. J. Phycol. 14: 33–42.

LEWIS, W. M. 1976. Surface volume ratio: implications for phytoplankton morphology. Science 192: 885–887.

LEWIS, W. M., JR. 1985. Nutrient scarcity as an evolutionary cause of haploidy. Am. Nat. 125: 692–701.

Lippert, K. D., and N. Pfennig. 1969. De Verwetung von molekularen Wasser-staff durch *Chlorobium thiosulphatophilum*. Arch. Mikr. 65: 29–47.

Lloyd, D., and G. Turner. 1968. The cell wall of *Prototheca zopfii*. J. Gen. Microbiol. 50: 421–427.

McCullough, W., and P. C. L. John. 1972. Temporal control of the *de novo* synthesis of isocitrate lyase during the cell cycle of the eucaryote *Chlorella pyrenoidosa*. Biochim. Biophys. Acta 296: 287–296.

MacFarlane, J. J., and J. A. Raven. 1985. External and internal CO_2 transport in *Lemanea*: interactions with the kinetics of ribulose bisphosphate carboxylase. J. Exp. Bot. 36: 610–622.

Magruder, W. H. 1984. Specialized appendages on spermatia from the red alga *Agloethamnion neglectum* (Ceramiales, Ceramiaceae) specifically bind with trichogynes. J. Phycol. 20: 436–440.

Malone, T. C. 1980. Algal size, p. 433–463. *In* I. Morris [ed.] The physiological ecology of phytoplankton. Blackwell Scientific Publications, Oxford.

Maniloff, J. 1983. Evolution of wall-less prokaryotes. Annu. Rev. Microbiol. 37: 477–499.

Manton, I. 1959. Electron microscopic observations on a very small flagellate: the problem of *Chromulina pusilla* Butcher. J. Mar. Biol. Assoc. U.K. 38: 319–333.

———— 1964. A contribution towards understanding of "the primitive fucoid". New Phytol. 63: 244–254.

Nikaido, H., and E. Y. Rosenberg. 1981. Effect of solute size on diffusion rates through the transmembrane pores of the outer membrane of *Escherichia coli*. J. Gen. Physiol. 77: 121–135.

Norris, R. E. 1980. Prasinophytes, p. 85–145. *In* E. R. Cox [ed.] Phytoflagellates. Elsevier, New York, NY.

Okomoto, C. K., and G. B. Witman. 1981. Functionally significant central-pair rotation in a primitive eukaryotic flagellum. Nature 290: 708–710.

Omata, T., and N. Murata. 1983. Isolation and characterization of the cytoplasmic membranes from the blue-green alga (Cyanobacterium) *Anacystis nidulans*. Plant Cell Physiol. 24: 1101–1112.

———— 1984a. Isolation and characterization of three types of membranes from the Cyanobacterium (blue-green alga) *Synechocystis* PCC 6714. Arch. Microbiol. 139: 113–116.

———— 1984b. Cytochromes and prenylquinones of cytoplasmic and thylakoid membranes from the Cyanobacterium (blue-green alga) *Anacystis nidulans*. Biochim. Biophys. Acta 766: 395–402.

Osborne, B. A., and J. A. Raven. 1986. Light absorption by plants and its implications for photosynthesis. Biol. Rev. 61: 1–61.

Osmond, C. B., O. Björkman, and D. J. Anderson. 1980. Physiological processes in plant ecology: toward a synthesis with *Atriplex*. Springer Verlag, Berlin.

Owens, O. van H., and H. H. Seliger. 1975. Effect of light intensity on dinoflagellate photosynthesis, p. 355–364. *In* F. J. Vernberg [ed.] Physiological Adaptation to the Environment, Intext Educational Publishers, NY.

Papageorgiou, G. 1971. Absorption of light by non-refractive spherical shells. J. Theoret. Biol. 30: 249–254.

Parnas, H., and D. Cohen. 1976. The optimal strategy for the metabolism of reserve materials in micro-organisms, J. Theor. Biol. 56: 19–55.

Pasciak, W. J., and G. Gavis. 1974. Transport limitation of nutrient uptake in phytoplankton. Limnol. Oceanogr. 19: 881–888.

———— 1975. Transport limited nutrient uptake rates in *Ditylum brightwellii*. Limnol. Oceanogr. 20: 604–617.

Peary, J., and R. W. Castenholz. 1964. Temperature strains of a thermophilic blue-green alga. Nature (London) 202: 720–721.

Peschek, G. A., W. M. Nitscham, and M. Springer-Lederer. 1985. Respiration linked proton extrusion from the unicellular Cyanobacterium *Anacystis nidulans*. Naturwiss 72: 273–274.

Pirie, N. W. 1964. The size of small organisms. Proc. Roy. Soc. London B 160: 149–166.

———— 1973. "On being the right size". Annu. Rev. Microbiol. 27: 119–132.

Pirt, S. J. 1983. Maximum photosynthetic efficiency: a problem to be resolved. Biotechnol. Bioeng. 25: 1915–1922.

Poynton, R. O. 1973. Effect of growth rate on the macromolecular compositon of *Prototheca zopfii*, a colourless alga which divides by multiple fission, J. Bacteriol. 113: 203–211.

Poynton, R. O., and D. Branton. 1972. Control of daughter-cell number variation in multiple fission: genetic *versus* environmental determinants in *Prototheca*. Proc. Nat. Acad. Sci. Wash. 64: 2346–2350.

Pradel, J., and J. Clemant-Metral, 1977. Cell division and photosynthetic apparatus construction in the cell envelope deficient mutant of *Rhodopseudomonas spheroides*. J. Gen. Microbiol. 102: 365–374.

Purcell, E. M. 1977. Life at low Reynolds Number. Am. J. Phys. 45: 3–11.

Raven, J. A. 1976. Transport in algal cells, p. 129–188. *In* U. Lüttge and M. G. Pitman [ed.] Encyclopedia of plant physiology (New Series), Volume 2A. Springer Verlag, Berlin.

———— 1977. Ribulose bisphosphate carboxylase activity in terrestrial plants: significance of O_2 and CO_2 diffusion. Curr. Adv. Plant Sci. 9: 579–590.

———— 1980a. Nutrient transport in micro-algae. Adv. Microb. Physiol. 21: 47–226.

———— 1980b. Chloroplasts of eukaryotic microorganisms. Symp. Soc. Gen. Microbiol. 30: 181–205.

———— 1981. Introduction to metabolic control, p. 3–27. *In* D. A. Rose and D. A. Charles-Edwards [ed.] Mathematics and plant physiology.

1982. The energetics of freshwater algae: energy requirements for biosynthesis and volume regulation. New Phytol. 92: 1–20.

1984a. Energetics and transport in aquatic plants. A. R. Liss and Co., New York, NY.

1984b. A cost-benefit analysis of photon absorption by photosynthetic unicells. New Phytol. 98: 593–625.

1985. The CO_2 concentrating mechanism, p. 67–82. In W. J. Lucas and J. A. Berry [ed.] Inorganic carbon uptake by photosynthetic organisms. American Society of Plant Physiologists, Rockville, Maryland.

1986a. Limits to growth. In M. A. Borowitzka and L. J. Borowitzka [ed.] Microalgal biotechnology. Cambridge University Press. (in press).

1986b. Biochemistry, biophysics and physiology of chlorophyll b-containing algae: implications for taxonomy and phylogeny. Progr. Phycol. Res. 4. (In press)

1986c. Long distance transport of calcium, 241–250. In A. Trewavas [ed.] Molecular and cellular aspects of Calcium in plant development. Plenum Press, New York. (In press)

1986d. Plasticity in algae. Symp. Soc. Exp. Biol. 40: 347–372.

RAVEN, J. A., AND J. BEARDALL. 1981a. Respiration and photorespiration, p. 55–82. In T. Platt [ed.] Physiological bases of phytoplankton ecology. Can. Bull. Fish. Aquat. Sci. 210.

1981b. The intrinsic permeability of biological membranes to H^+: significance for low rates of energy transormation. F.E.M.S. Microbiol. Letters 10: 1–5.

1982. The lower limit of photon fluence rate for phototrophic growth: the significance of 'slippage' reactions. Plant Cell Environ. 5: 117–124.

RAVEN, J. A., AND S. M. GLIDEWELL. 1975. Photosynthesis, respiration and growth in the shade alga Hydrodictyon africanum. Photosynthetica 9: 361–371.

RAVEN, J. A. AND W. J. LUCAS. 1985. The energetics of carbon acquisition, p. 305–324. In W. J. Lucas and J. A. Berry [ed.] Inorganic carbon uptake by aquatic photosynthetic organisms. American Society of Plant Physiologists, Rockville, Maryland.

RAVEN, J. A. AND K. RICHARDSON. 1984. Dinophyte flagella: a cost-benefit analysis. new Phytol. 98: 259–276.

RAVEN, J. A, AND F. A. SMITH. 1976. Nitrogen assimilation and transport in vascular land plants in relation to intracellular pH regulation. New Phytol. 76: 415–431.

1981. H^+ transport in the evolution of phytosynthesis. BioSystems 14: 95–111.

1982. Solute transport at the plasmalemma and the early evolution of cells. BioSystems 15: 13–26.

RAVEN, J. A., F. A. SMITH, AND N. A. WALKER. 1986. Biomineralisation in the Charophyceae, p. 125–139. In R. Riding and B. E. S. Leadbetter [ed.] Symposium on Biomineralisation in Lower Plants and Animals. Systematics Association Series Number 30. Oxford University Press.

REED, R. H. 1984. Use and abuse of osmoterminology. Plant Cell Environ. 7: 165–170.

REED, R. H., AND W. D. P. STEWART. 1983. Physiological responses of Rivularia atra to salinity: osmotic adjustment in hyposaline media. New Phytol. 95: 595–603.

REED, R. H., P. ROWELL, AND W. D. P. STEWART, 1981. Characterization of the transport of potassium ions in the cyanobacterium Anabaena variabilis. Eur. J. Biochem. 116: 323–330.

REED, R. H., J. A. CHUDEK, R. FOSTER, AND W. D. P. STEWART. 1984a. Osmotic adjustment in cyanobacteria from hypersaline environments. Arch. Microbiol. 138: 333–337.

REED, R. H., D. L. RICHARDSON, S. R. C. WARR, AND W. D. P. STEWART, 1984b. Carbohydrate accumulation and osmotic stress in cyanobacteria. J. Gen. Microbiol. 130: 1–4.

REYNOLDS, C. S. 1984. The ecology of freshwater phytoplankton. Cambridge University Press.

RICHARDSON, K., J. BEARDALL, AND J. A. RAVEN. 1983. Adaptation of unicellular alage to irradiance: an analysis of strategies. New Phytol. 93: 157–191.

RICKETTS, T. R. 1966. On the chemical composition of some unicellular algae. Phytochemistry 5: 67–76.

RIPPKA, R., J. WATERBURY, AND G. COHEN-BAZIRE. 1974. A cyanobacterium which lacks thylakoids. Arch. Microbiol. 100: 419–436.

RIVKIN, R. B., AND E. SWIFT. 1985. Phosphorus metabolism of oceanic dinoflagellates: phosphate uptake, chemical composition and growth of Pyrocystis noctiluca. Mar. Biol. 88: 189–198.

ROBERTS, A. M. 1981. Hydrodynamics of protozoan swimming, p. 5–66. In M. Lewardowsky and S. H. Hutner [ed.] Biochemistry and physiology of Protozoa, 2nd Edition, Volume 4. Academic Press, New York, NY.

ROMANOVICZ, D. K. 1981. Scale formation in flagellates, p. 27–62. In O. Kiermayer [ed.] Cytomorphogenesis in plants, Springer Verlag, Wien.

ROSEN, B. H., AND R. L. LOWE. 1984. Physiological and ultrastructural responses of Cyclotella meneghiniana (Bacillariophyta) to light intensity and nutrient limitation. J. Phycol. 20: 173–183.

ROUND, F. E. 1981. Morphology and phyletic relationships of the silicified algae and the archetypal diatom — Monophyly and Polyphyly, p. 97–128. In T. L. Simpson and B. E. Volcani [Ed.] Silicon and siliceous structures in biological systems, Springer Verlag, New York, NY.

SAMISH, Y. B. 1975. Oxygen build-up in photosynthesising leaves and canopies is small. Photosynthetica 9: 372–375.

SCHERER, S., AND P. BÖGER. 1984. Vanadate-sensitivie proton efflux by filamentous cyanobacteria, F.E.M.S. Microbiol. Letters 22: 215–218.

SCHLESINGER, D. A., L. A. MOLOT, AND B. J.

SHUTER. 1981. Specific growth rates of freshwater algae in relation to cell size and light intensity, J. Fish. Res. Board Can. 36: 1052–1058.

SCHOTZ, F., H. BATHELT, C-G. ARNOLD, AND O. SCHIMMER. 1972. Die Architetkur un organization der *Chlamydomonas* — Zelle. Ergebnisse der Elektronmikroscopie von Serierschnitten und der daraus resultierenden driedimensionalen Rekonstruktion. Protoplasma 75: 229–254.

SEARLES, R. B. 1980. The strategy of the red algae life history. Am. Nat. 115: 113–120.

SENGER, H. 1982. Efficiency of incident light energy utilization and quantum requirement for microalgae, p. 55–58. *In* O. R. Zaborsky [ed.] C. R. C. Handbook of Biosolar Resources, C. R. C. Press, Boca Raton, Florida.

SETTER, T. L., AND H. GREENWAY. 1979. Growth and osmoregulation of *Chlorella emersonii* in NaCl and neutral solutes. Austr. J. Plant Physiol. 6: 47–60 (see also p. 569–570).

SHELDON, R. W. 1984. Phytoplankton growth rates in the tropical ocean. Limnol. Oceanogr. 29: 1342–1346.

SHELP, B. J., AND D. T. CANVIN. 1984. Evidence for bicarbonate accumulation by *Anacystis nidulans*. Can. J. Bot. 62: 1398–1403.

SHUTER, B. 1979. A model of physiological adaptation in unicellular algae. J. Theor. Biol. 78: 519–552.

SICKO-GOAD, L. 1982. A morphometric analysis of algal responses to low dose, short-term heavy metal exposure. Protoplasma 110: 75–86.

SICKO-GOAD, L., E. F. STOERMER, AND B. G. LADEWSKI. 1977. A morphometric method for correcting phytoplankton cell volume estimates. Protoplasma 93: 1–6.

SILVESTER, N. R., AND M. A. SLEIGH. 1985. The forces on microorganisms at surfaces in flowing water. Freshwater Biol. 15: 433–448.

SMART, C. C., AND A. J. TREWAVAS. 1983. Abscisic-acid-induced turion formation in *Spirodela polyrrhiza* L. II. Ultrastructure of the turion: a stereological analysis. Plant Cell Environ. 6: 515–522.

SMAYDA, T. J. 1971. The suspension and sinking of phytoplankton in the sea. Oceanogr. Mar. Biol. Annu. Rev. 8: 353–414.

SMITH, F. A., AND J. A. RAVEN. 1979. Intracellular pH and its regulation. Annu. Rev. Plant Physiol. 30: 289–311.

SMITH, F. A., AND N. A. WALKER. 1980. Photosynthesis by aquatic plants: effects of unstirred layers in relation to assimilation of CO_2 and HCO_3^- and to carbon isotope discrimination. New Phytol. 86: 245–269.

SOROKIN, C. 1959. Tabular comparative data for the low and high temperature strains of *Chlorella*. Nature (London) 184: 613–614.

——— 1975. Morphological adaptation in unicellular algae, p. 333–353. *In* F. J. Vernberg [ed.] Physiological adaptation to the environment, Intext Education Publishers, NY.

SOROKIN, C, AND R. W. KRAUSS. 1958. The effects of light intensity on the growth rates of green algae. Plant Physiol. 33: 109–113.

SOROKIN, C., AND S. NISHINO. 1973. Transformation of rod-shaped algal cells into spherical cells. Am. J. Bot. 60: 907–914.

SOURNIA, A. 1982. Form and function in marine phytoplankton. Biol. Rev. 57: 347–394.

SPALDING, M. H., AND A. J. PORTIS JR. 1985. A model of carbon dioxide assimilation in *Chlamydomonas reinhardii*. Planta 164: 308–320.

SRERE, P. A. 1967. Enzyme concentrations in tissues. Science 158: 936.

STANIER, R. Y. AND G. COHEN-BAZIRE. 1977. Phototrophic prokaryotes: the cyanobacteria. Annu. Rev. Microbiol. 31: 225–274.

STEIN, W. D. 1967. The movement of molecules across cell membranes. Academic Press, New York, NY.

STEWART, K. D. AND K. R. MATTOX. 1980. Phylogeny of phytoflagellates, p. 433–462. *In* E. R. Cox [ed.] Phytoflagellates. Elsevier, New York, NY.

TAGER, J. M., R. J. A. WANDERS, A. K. GROEN, W. KUNZ, R. BOHNENSACK, U. KÜSTER, G. LETKO, G. BÖHME, J. DUSZYNSKI, AND L. WOJTCZAK. 1983. Control of mitochondrial respiration. F.E.B.S. Letters 15: 1–9.

TADEDA, H., AND T. HIROKAWA. 1978. Studies on the cell wall of *Chlorella*. I. Quantitative changes in cell wall polysaccharides during the cell cycle of *Chlorella ellipsoidea*. Plant Cell Physiol. 19: 591–598.

TERASHIMA, I., AND Y. INOUE. 1984. Comparative photosynthetic properties of palisode tissue chloroplasts and spongy tissue chloroplasts of *Camellia japonica* L.: functional adjustment of the photosynthetic apparatus to light environment within a leaf. Plant Cell Physiol. 25: 555–564.

——— 1985. Vertical gradient in photosynthetic properties of spinach chloroplasts dependent on intra-leaf light environment. Plant Cell Physiol. 26: 781–785.

THINH, L-V., AND D. J. GRIFFITHS. 1985. A small marine *Chlorella* from the waters of a coral reef. Bot. Mar. 28: 41–46.

TRÜPER, H. G., AND N. PFENNIG. 1978. Taxonomy of the Rhodosprillales, p. 19–27. *In* R. K. Clayton and W. J. Sistrom [ed.] The photosynthetic bacteria. Plenum Press, New York, NY.

TURNER, M. F., AND R. J. GOWEN. 1984. Some aspects of the nutrition and taxonomy of fourteen small green and yellow-green algae. Bot. Mar. 27: 249–255.

TURPIN, D. H., A. G. MILLER, AND D. T. CANVIN. 1984. Carboxysome content of *Synechococcus leopoliensis* (Cyanophyta) in response to inorganic carbon. J. Phycol. 20: 249–253.

UTKILEN, H. C. 1982. Magnesium-limited growth of the cyanobacterium *Anacystis nidulans*. J. Gen. Microbiol. 128: 1849–1862.

UTKILEN, H. C., T. BRISIED, AND B. ERIKSSON. 1983. Variation in photonsynthetic membrane and pigment content with ligth intensity for *Anacystis nidulans* grown in continuous culture. J. Gen. Microbiol. 129: 1717–1720.

VAN GEMERDEN, H. 1980. Survival of *Chromatium vinosum* at low light intensities. Arch. Micro.

125: 115–121.

VINCENT, W. F. 1982. Autecology of an ultraplanktonic shade alga in Lake Tahoe. J. Pycol. 18: 226–232.

VOGEL, S. 1981. Life in moving fluids: The physical biology of flow. Willard Grant Press, Boston, MA.

WALDRON, C., AND F. LACROUTE. 1975. Effect of growth rate on the amounts of ribosomal and trasfer ribonucleic acids in yeast, J. Bacteriol. 122: 855–865.

WALKER, N. A. 1976. Membrane Transport: theoretical background, p. 26–52. *In* U. Lüttge and M. G. Pitman [ed.] Encyclopedia of plant physiology, New Series, Volume 2A. Springer Verlag, Berlin.

WALSBY, A. E. 1981. Cyanobacteria: planktonic gas-vacuolate forms, p. 224–235. *In* M. P. Starr, M. Stolp, H. G. Trüper, and H. G. Shlegel [ed.] The Prokaryotes, Volume 1. Springer, Berlin.

WALSBY, A. E., AND C. S. REYNOLDS. 1980. Sinking and floating, p. 371–412. *In* I Morris [ed.] The physiological ecology of phytoplankton. Blackwell Scientific Publications, Oxford.

WARR, S. R. C., R. H. REED, AND W. D. P. STEWART. 1985. Carbohydrate accumulation in osmotically stressed cyanobacteria (blue-green algae): interactions with temperature and salinity. New Phytol. 100: 285–292.

WATERBORG, J. H., AND S. SHALL. 1985. The organisation of replicons, p. 15–35. *In* J. A. Bryant and D. Francis [ed.] The cell division cycle in plants, Cambridge University Press.

WATERBURY, J. B., S. W. WATSON, R. R. L. GUILLARD, AND L. E. BRAND. 1979. Widespread occurrence of a unicellular, marine, planktonic, cyanobacterium. Nature (London) 227: 293–294.

WATERBURY, J. B., J. M. WILLEY, D. G. FRANKS, F. W. VALOIS, AND S. W. WATSON. 1983. A cyanobacterium capable of motility. Science 230: 74–76.

WEGMAN, K. 1979. Biochemische Anpassung von *Dunaliella* an weschselnde Salinitat und Temperature. Ber. Dtsch. Bot. Ges. 92: 43–54.

WELSCHMEYER, N. A., AND C. J. LORENZEN. 1981. Chlorophyll-specific photosynthesis and quantum efficiency at subsaturating light intensities. J. phycol. 17: 283–293.

WERNER, D. 1971. Der Entwicklungscyclus mit sexual phase beider marinen Diatomee *Coscinodiscus asteromphalus*. II. Oberflachenabhangige Differenzierung wahrend der vegetativen Zellverklierung. Arch. Mikr. 80: 115–133.

WHATLEY, J. M., AND F. R. WHATLEY. 1981. Chloroplast evolution. New Phytol. 87: 233–247.

WHITCOMB, R. F. 1980. The genus *Spiroplasma*. Annu. Rev. Microbiol. 34: 677–709.

WILHELM, C., AND A. WILD. 1982. Growth and photosynthesis of *Nanochlorum eucaryotum*, a new and extremely small eucaryotic green alga. Z. Naturforsch. 37C: 115–119.

WHILHEM, C., G. ELSENBEIS, A. WILD, AND R. ZAHN. 1982. *Nanochlorum eucarotum*: a very reduced coccoid species of marine Chlorophyceae. Z. Naturforsch. 37C: 107–114.

WITEWAY, M. S., AND R. W. LEE. 1977. Chloroplast DNA content increases with nuclear ploidy in *Chlamydomonas*. Mol. Gen. Genet. 157: 11–15.

YENTSCH, C. S. 1983. A note on the fluorescence characteristics of particles that pass through glass-fibre filters. Limnol. Oceanogr. 28: 597–599.

Biological and Ecological Characterization of the Marine Unicellular Cyanobacterium *Synechococcus*

JOHN B. WATERBURY, STANLEY W. WATSON, FREDERICA W. VALOIS, AND DIANA G. FRANKS

Woods Hole Oceanographic Institution, Woods Hole, Massachusetts 02543, USA

Introduction

The cyanobacteria (blue-green algae) are one of the most morphologically and developmentally diverse groups of procaryotes. They range from simple unicellular forms that reproduce by binary fission to complex filamentous organisms that are capable of true branching and the differentiation of a variety of highly specialized cell types (Carr and Whitton 1982; Rippka et al. 1981a).

In contrast to their morphological and developmental diversity cyanobacteria are nutritionally conservative. Their dominant mode of nutrition is oxygenic plant type photosynthesis. Some are also capable of anoxygenic bacterial-type photosynthesis and some have a limited heterotrophic capacity. Despite this apparent metabolic uniformity, their ecological diversity is remarkable. They occur both as free-living organisms in a wide variety of habitats including terrestrial, fresh water and marine, and as symbionts in association with both plants and animals. See the monograph of Carr and Whitton (1982) for a review of recent advances in their biology.

Cyanobacteria have an ancient marine history which can be traced back almost three billion years in the fossil record (Brock 1973). During the Precambrian, they were probably the dominant photosynthetic organisms in the world's oceans, being especially prevalent in the intertidal zone and in shallow tropical seas. Today they are less conspicuous but are still an important component of marine habitats.

Marine cyanobacteria are found in all the world's oceans but are the most abundant in temperate and tropical regions. Their taxonomic diversity does not differ markedly from that of cyanobacteria found in terrestrial and freshwater habitats. However, particular marine habitats (e.g., intertidal and subtidal zones, coral reefs, salt marshes and the open ocean) often contain a characteristic diversity of forms which may differ markedly from season to season and between different geographical locations (Fogg 1973, 1982; Fogg et al. 1973; Whitton and Potts 1982).

In contrast to freshwater habitats (Whitton 1973; Gibson and Smith 1982), where over 100 species of planktonic cyanobacteria have been described and approximately 20 are capable of forming dense water blooms, the open ocean harbors a restricted number of planktonic forms. The two best known of these are members of the genera *Trichodesmium* and *Synechococcus*. *Trichodesmium* spp. are oscillatorian cyanobacteria that often form dense blooms covering vast areas of ocean in the tropical and subtropical seas. These filamentous organisms are of considerable interest to biological oceanographers because of their reputed ability to fix nitrogen (Fogg 1973; Fogg et al. 1973; Carpenter 1973).

Synechococcus spp. are small unicellular forms that occur abundantly in surface waters of the temperate and tropical oceans. Since their discovery (Waterbury et al. 1979; Johnson and Sieburth 1979) they have received considerable attention from the oceanographic community (Whitton and Carr 1982) resulting in an abundant literature that has recently been reviewed by Glover (1985). This chapter will describe aspects of the biology and ecology of this important group of "picoplankton".

SECTION I

The Organism

The name *Synechococcus* (sensu Rippka et al. 1979) has been applied loosely to the group of small unicellular cyanobacteria observed in and isolated from marine waters. Rippka et al. (1979) placed in the genus *Synechococcus* all the small unicellular cyanobacteria with ovoid to cylindrical cells that reproduce by binary transverse fission in a single plane and that lack sheaths. It was realized that the strains then in culture, even though morphologically similar, showed considerable genetic heterogeneity that would eventually warrant splitting the group into several genera.

The marine isolates of *Synechococcus* are also a heterogenous assemblage that fall into two quite distinct subgroups. One group, exemplified by strains WH 5701 and WH 8101, lacks phycoerythrin and does not have elevated salt requirements for growth. Strains of this type have been isolated from coastal waters within the continental shelf margin but have never been observed in or cultured from the open ocean. We interpret strains of this type as being halotolerant, possibly representing terrestrial forms that have invaded the marine environment. The possession of phycocyanin as their primary light harvesting pigment places strains of this type at a distinct spectral disadvantage in seawater (Wood 1985) and probably accounts for their inability to compete successfully in the open ocean.

The second group of marine *Synechococcus* contains phycoerythrin as its primary light harvesting pigment and has elevated salt requirements for growth. Strains of this type occur abundantly within the euphotic zone and have been isolated from both coastal and oceanic waters in the world's temperate and tropical oceans.

It is this second group, which we will refer to as "marine *Synechococcus*", that will be the subject of this monograph.

Collection and Isolation

Isolation of *Synechococcus* is facilitated by the fact that these cyanobacteria can be counted easily and specifically and can be separated from other phytoplankters by differential filtration.

Precautions should be taken to avoid contamination in collecting and storing water samples from which *Synechococcus* is to be isolated. In practice, water collected with clean Nisken bottles and stored in plastic containers has worked well. Isolation procedures should be initiated immediately following water collection to avoid loss of viability or overgrowth by other microorganisms. Direct microscopic counts of *Synechococcus* abundance provides baseline information on morphology and the extent to which samples can be diluted during the enrichment process.

The isolation of *Synechococcus* can be enhanced by physically separating *Synechococcus* cells from other microorganisms either by serial dilution in the enrichment medium or by differentially filtering the original sample through 10 μm and 1 μm Nuclepore filters. In practice, many of the isolates of *Synechococcus* in the Woods Hole Collection have appeared as contaminants during single cell isolations for a variety of eucaryotic algae.

Two enrichment media have been used for *Synechococcus*, SNAX and F/40 + 10 μM ammonium chloride (Guillard 1975) (Table 1). Enrichment media were prepared by adding sterile stock solutions of nutrients to seawater that had been autoclaved separately in teflon containers. Successful enrichments were made in a variety of containers including glass and plastic flasks, bottles and tubes. Enrichments were incubated in low light (10–20 μE $\cdot$ m^{-2} $\cdot$ s^{-1} measured with a Licor 2π sensor) at temperatures ranging from 10° to 25°C. *Synechococcus* appeared as a small red to orangish pellet in the bottom of the culture vessel following incubation 2–4 wk.

Purification of *Synechococcus* strains was accomplished by streaking of solid media, by serial dilution or a combination of these two techniques. Strains of marine *Synechococcus* do not grow vigorously on solid media, consequently satisfactory results (i.e., well isolated colonies derived from a single cell) were obtained by rigorous care in the preparation of media and a combination of patience and perseverance in monitoring the growth of plate cultures.

Solid media were prepared using Difco Bacto-Agar that had been further purified using the following protocol: ¼ lb of agar was washed by stirring with 3 L of double distilled water in a 4-L beaker. After 30 min of stirring the agar was allowed to settle, the wash water was siphoned off and the agar was filtered onto Whatman F4 filter paper in a Buckner funnel. This procedure was repeated once more or until the filtrate was clear. The agar was then washed with 3 L of 95% ethanol followed by a final 3-L wash with analytical grade acetone. The agar was then dried at 50°C in glass baking dishes for 2–3 days and stored in a tightly covered container. Solid media prepared with the purified agar at a final concentration of 0.6% were sufficiently stable for streaking.

To prepare 40 agar plates from 1 liter of medium, the three following solutions were prepared and autoclaved separately: 1). 750 mL of filtered seawater in a teflon bottle, 2). 6.0 g super clean agar in 200 mL double distilled water in a 2-L glass flask

TABLE 1. Composition of media.

Ingredient	SN	SNAX	F medium Guillard (1975)
		Amount Litre^{-1}	
Double distilled water	250 mL	250 mL	F/40 is F medium
Filtered seawater	750 mL	750 mL	diluted 1/40
EDTA (disodium salt)	15 μM	1.5 μM	
NaNO$_3$	9.0 mM	1.0 mM	
NH$_4$Cl	—	100 μM	10 μM
K$_2$HPO$_4$	90 μM	9.0 μM	
Na$_2$CO$_3$ • H$_2$O	100 μM	—	
Cyano-Trace Metals*	1.0 mL	0.1 mL	

For SN: Mix double distilled water, filtered seawater and minerals together; dispense into glassware and sterilize. Add Vitamin B$_{12}$**, 1 drop/50 mL media.

For SNAX: Add NH$_4$Cl (100 μM) and Va vitamin mix (Davis and Guillard 1958) (1 drop/50 mL medium) from sterile stocks (after autoclaving).

**Cyano-Trace Metals:*

Compound	g Litre^{-1}	
ZnSO$_4$ • 7H$_2$O	0.222	Dissolve each compound
MnCl$_2$ • 4H$_2$O	1.400	separately, then add
Co(NO$_3$)$_2$ • 6H$_2$O	0.025	together and bring to
Na$_2$MoO$_4$ • 2H$_2$O	0.390	1 litre.
Citric Acid • H$_2$O	6.250	
Ferric Ammonium Citrate	6.00	

***Vitamin B$_{12}$ Stock Solution:*
0.1 mg in 100 mL of double distilled water. Sterilize by autoclaving and keep refrigerated.

and 3). the mineral salts for 1 L of medium (Table 1), in 50 mL double distilled water in a 125 mL glass flask. After autoclaving, the seawater and minerals were added to the agar flask. Vitamins (Va, Table 1) and sterile sodium sulfite (2 mM final concentration) were added aseptically to the hot agar solution which was then cooled to 50° before the plates were poured. The surface of agar plates should be dry prior to streaking. Following inoculation the plates were stored upside down in clear plastic vegetable crispers to minimize evaporation and contamination by fungi. Best results were obtained by incubation in continuous light at intensities between 20 and 40 μE $\cdot$ m^{-2} $\cdot$ s^{-1} at 25°C. Colonies appeared in 2-4 wk and were removed from the agar surface with drawn Pasteur pipettes, inoculated into liquid media and allowed to grow up between successive streakings.

Purity of cultures was verified by microscopic examination of senescent cultures and by inoculation of purity media (e.g., a seawater mineral medium containing 0.2% glucose and 0.02% casamino acids).

Marine *Synechococcus* strains are best maintained in liquid cultures (50 mL of medium SN (Table 1) in 125 mL glass flasks) incubated at 20-25°C in either continuous light or a light–dark cycle (e.g., 14 h L, 10 h D) at intensities between 20 and 40 μE $\cdot$ m^{-2} $\cdot$ s^{-1}. Transfer of stock cultures at three week intervals using a heavy inoculum was sufficient for most strains. However some strains, particularly those containing phycoerythrin high in urobilin content, proved to be more sensitive and required more frequent transfers.

Long-term preservation of marine *Synechococcus* strains was accomplished in liquid nitrogen. Cultures in the exponential phase of growth were gassed with nitrogen to drive off oxygen, dimethylsulfoxide was added to a final concentration of 10%, the culture dispensed into plastic cryogenic ampules and placed in the vapor phase of a cryogenic chamber for freezing and storage. Frozen strains were reactivated by thawing the ampules at 37°C in a water bath and then serially diluting the culture in liquid medium SN and incubating them under the conditions described for culture maintenance.

A more detailed discussion of the isolation, purification, growth and maintenance of cyanobacteria can be found in Chapters 8 and 9 of the *Prokaryotes* (Rippka et al 1981b; Waterbury and Stanier 1981). Specific problems associated with the isolation and growth of picoplankton are discussed in a chapter in this volume by L.E. Brand.

The sources and histories of the strains of marine *Synechococcus* in the Woods Hole Culture Collection, including the halotolerant phycocyanin containing strains, are given in Table 2.

Properties of Marine *Synechococcus*

Morphology: In natural water samples examined by epifluorescence microscopy, marine *Synechococcus* cells are predominantly coccoid in shape and range in size from 0.6 to 1.6 μm in diameter. Within this size range a tendency toward increased cell size with depth in the euphotic zone has often been observed (Murphy and Haugen 1985). In culture the marine strains of *Synechococcus* are small coccoid to rod-shaped cells 0.6–0.8 $\times$ 0.6–1.6 μm that divide by binary transverse fission in a single plane (Fig. 1). The differences in cell size and shape between individual strains are well within the range of variability caused by differing growth conditions. For example, strains maintained in continuous illumination have a tendency toward more rod-shaped cells whereas the cells of the same strain maintained in a light–dark cycle (14-hL–10-hD) are more coccoid and have cell volumes about 30% less than cells grown in continuous light as determined by carbon cell^{-1} (Cuhel and Waterbury, unpublished data). Cultures grown in a light–dark cycle more closely resemble natural populations of *Synechococcus* in both size and shape and should be used in physiological experiments where various parameters are equated to carbon cell^{-1},

TABLE 2. Woods Hole culture list (Cyanobacteria).

Woods Hole strain designation	Location	Sampling date	Old strain designation	Cruise no.	Isolated by:
WH 5701	Long Island Sound	1957	Syn		R.R.L. Guillard
WH 6501	8°44′N, 50°50′W	1965	48-Syn		R.R.L. Guillard
WH 7801	33°44.9′N, 67°29.8′W	7/1/78	DA-3	Oceanus 48	L. Brand
WH 7802	" "	"	DB-3	"	"
WH 7803	" "	"	DC-2	"	"
WH 7804	33°44.8′N, 67°30′W	6/30/78	CF-3	"	"
WH 7805	" "	"	CF-4 (L1602)[a]	"	"
WH 7806	38°19.5′N, 69°34.5′W	6/28/78	A-5	"	"
WH 8001	19°45′N, 92°25′W	4/15/80	636	Researcher	L. Brand
WH 8002	" "	"	696	"	"
WH 8003	" "	4/17/80	1011 (L1601)	"	"
WH 8004	" "	4/16/80	830	"	"
WH 8005	" "	4/15/80	748	"	"
WH 8006	" "	4/16/80	864	"	"
WH 8007	" "	"	838 BG	"	L. Provosoli
WH 8008	" "	"	L 1606	"	L. Brand
WH 8010	38°40.7′N, 69°19′W	6/26/80	25 m	Oceanus 82	J. Waterbury
WH 8009	" "	"	50 m	"	"
WH 8020	" "	"	50 m	"	"
WH 8011	Northern Sargasso Sea	7/80	I	Oceanus 83	L. Brand
WH 8012	34°N, 65°W	7/7/80	Orange-L	"	"
WH 8013	"	"	Blue-4	"	"
WH 8015	Woods Hole	6/80	5B		F. Valois
WH 8016	"	6/80	1E (L1603)		"
WH 8017	"	6/80	GR		"
WH 8018	"	6/80	C-7 (L1604)		"
WH 8101	"	1981	WH 1μ filtrate		"
WH 8102	22°29.7′N, 65°36′W	3/15/81	2×10^{-2}	Oceanus 92	J. Waterbury
WH 8103	28°30′N, 67°23.5′W	3/17/81	5×10^{-1}	"	"
WH 8104	31°59′N, 68°18.86′W	3/18/81	7×10^{-2}	"	"
WH 8105	38°20.85′N, 69°38.94′W	3/19/81	10×10^{-2}	"	"
WH 8106	40°0.6.83′N, 70°26.33′W	3/20/81	11×10^{-2}	"	"

TABLE 2. (cont.) Woods Hole culture list (Cyanobacteria).

WH 8107	39°28.6′N, 70°27.72′W	6/81	2 Red A-8	Oceanus 100	L. Brand
WH 8108	" "	"	1 Red D-6	"	"
WH 8109	" "	"	1 Red C-4	"	"
WH 8110	Sydney Harbor, Aust.	8/27/81	Sydney Harbor Red		J. Waterbury
WH 8111	36°N, 66°W	10/81	6–2	Oceanus 105	F. Valois
WH 8112	"	"	6–20	"	"
WH 8113	"	"	6–60	"	"
WH 8201	20°44.5′N, 109°.04′9′W	5/6/82	Vent Site Filter	21°N Cruise New Horizon	J. Waterbury
WH 8202	29°15.6′N, 85°54.2′W	2/12/80	L 1162		L. Brand
WH 8203	St. Georges Harbor, Bermuda	10/82	B.D.A. #1		M. Wood
WH 8205	Bermuda	10/82	Left sort		"
WH 8406	30°N, 88°W	12/84			R. Olson

[a]L numbers (e.g., L1601) denote Larry Brand strain designations

chlorophyll cell^{-1}, protein cell^{-1}, etc., especially if the results of such experiments will be compared to the behavior of *Synechococcus* in nature.

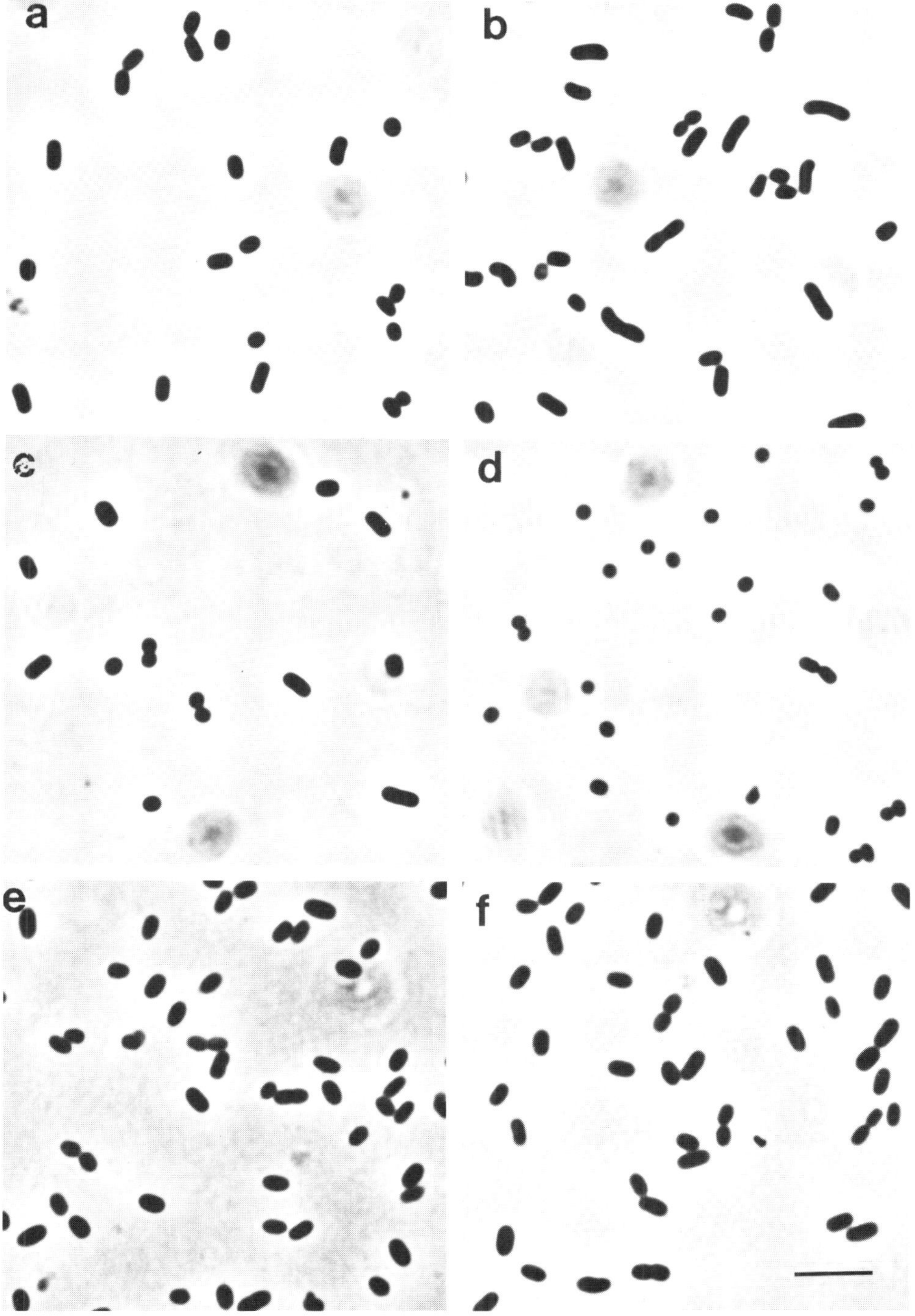

FIG. 1. Phase contrast light micrographs of six axenic strains of *Synechococcus* grown at 25°C in continuous light (30μE $\cdot$ m^{-2} $\cdot$ s^{-1}); (a) WH 7803, (b) WH 8001; (c) WH 8105; (d) WH 8012; (e) WH 8011; (f) WH 8112. Bar equals 5μm.

Ultrastructurally the marine strains of *Synechococcus* possess the procaryotic features common to all cyanobacteria. The cell envelope is made up of the cytoplasmic membrane and a gram negative cell wall consisting of a peptidoglycan layer and an

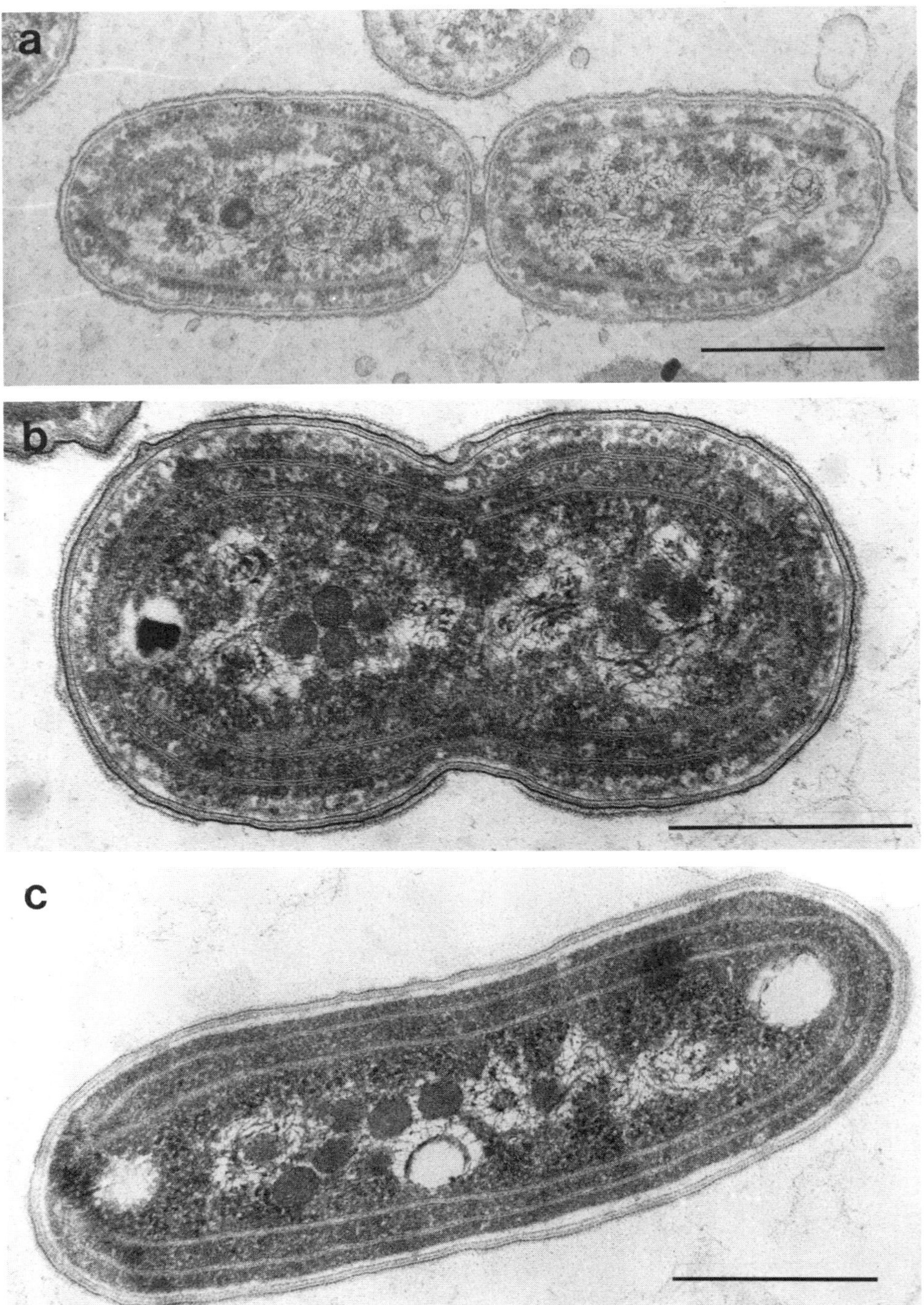

FIG. 2. Electron micrographs of thin sections of three strains of *Synechococcus* showing their typical synechococcoid ultrastructure. (a) WH8112, (b) Strain WH7802, (c) Strain WH7804. Fixation and preparation for TEM were conducted as described by Waterbury and Stanier (1978). Bar equals 0.5μm.

outer membrane (Fig. 2). Some strains possess external structures such as the highly structured outer envelope of strains WH 7803 and WH 7802 (Kursar et al. 1981; Fig. 2) and external projections (spinae) observed by Perkins et al. (1981).

The location and spacing between the photosynthetic thylakoids is a characteristic feature of members of the genus *Synechococcus*. They are located peripherally just internal to the cytoplasmic membrane and are separated by a distance of 40–50 nm that corresponds to the space necessary to house the phycobilisomes that are attached to the outer surface of the thylakoid membranes.

Polyphosphate inclusions, glycogen granules and carboxysomes are frequently observed in transmission electron micrographs of thin sections of the marine strains but other cytoplasmic inclusions present in some other cyanobacteria are noticeably absent including the storage products cyanophycin and poly-β-hydroxybutyrate.

Motility: Two forms of motility occur among cyanobacteria. Many filamentous and some unicellular cyanobacteria are capable of gliding motility when in contact with solid surfaces (Castenholz 1982). Recently it has been shown that some strains of marine *Synechococcus* are capable of a novel form of swimming motility at speeds of 5–30 μm $\cdot$ s^{-1} (Waterbury et al. 1985). Their behavior, when examined by light microscopy, possesses many of the features of bacterial flagellar motility. More than half of the cells in exponentially growing cultures are actively motile. Long rod-shaped cells swim in relatively straight paths, whereas more coccoid cells loop and spiral about. Individual cells occasionally rotate end over end at 3–5 revolutions $\cdot$ s^{-1} and more rarely cells will attach to the slide or coverslip surface and pivot about one pole at 0.5–1 revolution $\cdot$ s^{-1}. However, attached cells have never been observed to glide along a surface. Translocation is accompanied by cell rotation and occurs at speeds varying from 5–30 μm $\cdot$ s^{-1}. Variability in cell speed is associated with the age and condition of individual cultures and not with variation in light intensity. Surprisingly, the motile cells lack bacterial flagella or other visible external organelles of motility. A plausible mechanism responsible for their motility is presently lacking.

To date, all motile strains have been isolated from the open ocean, none have been found in coastal waters. They may be one of the dominant forms of marine *Synechococcus* found in the open ocean as evidenced by the frequency of their isolation, 9 of 40 strains, and the fact that actively motile cells have been observed in natural *Synechococcus* populations concentrated by filtration and examined in wet mounted preparations by epifluorescence microscopy.

The ecological advantage of motility in marine *Synechococcus* has not been determined but several properties of the strains indicate that motility is not associated with response to light (Waterbury et al. 1985). The motile cells lack classical photokinetic and photophobic behavior. In addition, a *Synechococcus* cell swimming constantly in a straight line at 25 μm $\cdot$ s^{-1} could cover only 2 m in 24 h, a distance of almost no consequence with respect to light quantity or quality in the open ocean. Finally, because of their size, *Synechococcus* cells behave like colloidal particles in seawater resulting in their movement being dominated by the physical mixing processes in the water column. It seems more likely that motility enables these cyanobacteria to respond chemotactically to nutrient-enriched micropatches or microaggregates, which are currently believed to be important to nutrient cycling within the euphotic zone (Goldman 1984).

Pigments: The strains in the marine *Synechoccocus* cluster, like all other cyanobacteria, contain chlorophyll *a* as their primary photosynthetic pigment and phycobiliproteins as accessory light harvesting pigments (Waterbury et al. 1979). All cyanobacteria contain at least three different types of phycobiliproteins, listed in their order of relative abundance within the cell: phycocyanin, allophycocyanin and allophycocyanin B. Many, but not all cyanobacteria also contain a fourth phycobiliprotein which is a phycoerythrin (Stanier and Cohen-Bazire 1977). The relative abundance of the major phycobiliproteins (i.e., phycocyanin and phycoerythrin) is responsible for the color of cyanobacterial cells. For example, cyanobacteria which contain no phycoerythrin will appear blue-green (hence the common name blue-green algae). Cyanobacteria that contain phycoerythrin range in color from olive-green to reddish-orange depending on the ratio of phycocyanin to phycoerythrin. All the open ocean isolates of *Synechococcus* are reddish-orange due to a predominance of phycoerythrin.

In the cyanobacteria that contain phycoerythrin two modes of synthesis control the relative amounts of phycoerythrin and phycocyanin. In some strains, the relative rates of synthesis of these two phycobiliproteins are strongly affected by light quality. These strains display a process known as chromatic adaptation (Tandeau de Marsac 1977). Cyanobacteria of this type preferentially synthesize phycocyanin when grown in red light and phycoerythrin when grown in green light. However, many phycoerythrin containing cyanobacteria including marine *Synechococcus* do not exhibit chromatic adaptation. In cyanobacteria of this type the relative rates of synthesis of phycoerythrin and phycocyanin are constant for any given strain and unaffected by light quality.

Phycoerythrins are present in most red algae, many cyanobacteria and some cryptomonads. They owe their intense visible absorption properties to chromophores that are open-chain tetrapyrroles covalently linked to their apoproteins. The spectral properties of the phycoerythrins depend on several factors including the properties of the attached phycobilins, chromophore-chromophore interactions and chromophore-protein interactions. Several spectrally different phycoerythrins are present in strains of marine *Synechococcus* (Kursar et al. 1981; Ong et al. 1984; Alberte et al. 1984; Table 3). They are the result of the presence and proportions of the two chromophores, phycoerythrobilin (PEB) and phycourobilin (PUB). The PEB groups give rise to the absorption peaks between 545–560 nm and the PUB groups to the peaks between 490–500 nm. The diversity of phycoerythrins within the marine cluster of *Synechococcus* strains is greater than in all of the other groups of cyanobacteria and encompasses almost the entire diversity of phycoerythrins found within the red algae (Rhodophyta) (Glazer et al. 1982).

Major ionic growth requirements: Physiologically, cyanobacteria isolated from marine habitats can be divided into two categories based on their major ionic requirements for growth. Some are halotolerant and grow equally well on a medium with either a seawater or freshwater base. Others have obligate requirements for elevated concentrations of sodium, magnesium, calcium and chloride that preclude their growth in freshwater media even when supplemented with 3% NaCl (Waterbury 1976; Stanier and Cohen-Bazire 1977).

All the strains in the open-ocean cluster of *Synechococcus* so far tested belong to the latter group indicating that they are intrinsically marine, whereas the marine isolates of *Synechococcus* that lack phycoerythrin (e.g., WH 5701, WH 8101) are all halotolerant forms.

Heterotrophic growth: Most cyanobacteria are photoautotrophs, but many do have the ability to grow photoheterotrophically in the light and some are capable of facultative chemoheterotrophic growth in the dark, albeit at rates considerably slower than growth in the light (Rippka 1972; Rippka et al. 1979).

Strains of marine *Synechococcus* thus far tested all appear to be obligate photoautotrophs (i.e., organic compounds cannot serve as sole carbon sources) (Table 3). This does not preclude the possibility that they are able to assimilate a variety of organic compounds into some but not all cell constituents. For example, we have shown that strain WH 7803 can assimilate acetate and adenine but not thymidine (Cuhel and Waterbury 1984).

Nitrogen sources: All our strains of *Synechococcus* can utilize nitrate and ammonia as their sole nitrogen source for growth and approximately one half the strains tested can utilize urea as the sole nitrogen source (Table 3). Eighteen axenic strains have been rigorously analyzed for their ability to induce nitrogenase synthesis under strictly anaerobic conditions using the technique of Rippka and Waterbury (1977). They have

TABLE 3. Properties of the isolates of marine *Synechococcus*.

Strain no.	Axenic	Mean DNA base composition mol % G+C	Phycobiliproteins — Phycoerythrin (PE) Present (+) Absent (−)	Phycobiliproteins — A495*/A545	Phycobiliproteins — Chromatic adaptation (+) or (−)	Obligately** marine (+) or (−)	Nitrogen sources — Urea (+) or (−)	Nitrogen sources — Synthesis of nitrogenase (+) or (−)	Photohetero-trophy (+) or (−)	Swimming motility (+) or (−)
WH 7805	+	59.7	+	No PUB***	−	+	+	−	−	−
WH 8010	+	58.6	+	"	−	+		−	−	−
WH 8018	+	57.6	+	"	−	+		−	−	−
WH 8110	+	56.6	+	"	−	+		−	−	−
WH 8006	+	57.6	+	"	−	+		−	−	−
WH 8008	+	55.8	+	"	−	+		−	−	−
WH 8009	+	57.5	+	"	−	+	+	−		−
WH 8017	+	54.5	+	0.42	−					−
WH 8105	+	54.9	+	0.59	−	+		−	−	−
WH 8015	+	55.3	+	0.44	−	+		−	−	−
WH 8016	+	55.5	+	0.40	−					−
WH 8001	+	56.4	+	0.39	−	+	−			−
WH 7802	+	57.9	+	0.40	−		−	−	−	−
WH 8003	+	58.5	+	0.40	−	+	+	−	−	−
WH 8005	+	58.6	+	0.44	−	+	+	−	−	−
WH 7801	+	59.7	+	0.38	−					−
WH 8002	+	60.0	+	0.48	−	+	+	−	−	−
WH 8004	+	60.6	+	0.43	−	+	−	−	−	−
WH 7803	+	61.3	+	0.39	−		−	−	−	−
WH 6501	+	62.3	+	0.43	−	+	+	−	−	−
WH 8012	+	62.4	+	0.40	−		−	−	−	−
WH 8106	+		+	0.44	−					−
WH 7806	−		+	0.38	−		−			−
WH 7804	−		+	0.37	−					−
WH 8104	−		+	0.42	−					−
WH 8111	−		+	0.47	−					−
WH 8203	−		+	0.42	−					−
WH 8205	−		+	0.48	−					−
WH 8020	−		+	0.78	−	+				−
WH 8108	−		+	0.76	−					−

TABLE 3. (cont.) Properties of the isolates of marine *Synechococcus*.

WH 8109	+		+	0.89	−					−
WH 8201	−		+	0.94	−					−
WH 8011	+	59.3	+	0.84	−	+	−	−	−	+
WH 8406	−		+	0.84	−					+
WH 8103	+	58.9	+	2.40	−	+	−	−	−	+
WH 8102	+	60.4	+	2.06	−			−	−	−
WH 8013	−		+	1.84	−					−
WH 8107	−		+	1.71	−					−
WH 8112	+	59.8	+	Variable****	−	+	+	−	−	+
WH 8113	+	60.5	+	"	−	+	−	−	−	+
Properties of other marine isolates of Synechococcus										
PCC 7002*****	+	49.1	−		−	−		−	+	−
PCC 73109	+	49.0	−		−	−		−	+	−
PCC 7003	+	49.4	−		−	+		−	+	−
PCC 7001	+	69.5	−		−	−		−	−	−
WH 5701	+	65.8	−		−	−		−	−	−
WH 8202	+		−		−	−				−
WH 8101	+	63.9	−		−	−	+		−	−
WH 8007	+	62.5	−		−	+		−	−	−
PCC 7335	+	47.4	+	No PUB***	+	+		+	+	−

*$\frac{A_{495}}{A_{545}}$ is the ratio of absorbance at 495 nm and 545 nm that is related to the bilin chromophore content (phycourobilin and phycoerythrobilin, respectively) of the phycoerythrins.

**Obligately marine strains have elevated growth requirements for sodium chloride and magnesium and calcium ions.

***These strains contain C-phycoerythrin that lacks phycourobilin.

****These strains have phycoerythrins whose chromophore ratios vary with light intensity.

*****PCC = Pasteur Institute Culture Collection (see Rippka et al. 1979) all been negative and it is likely that the ability to fix nitrogen will not be found in this group of cyanobacteria.

Temperature range permitting growth: Field observations in Woods Hole Harbor indicate that the spring *Synechococcus* bloom begins when the water temperature reaches 6°C and that growth is severely inhibited in the winter when the water temperature falls below 5°C. Most of our cultures grow optimally and are maintained between 20–25°C and fail to grow at temperatures in excess of 30°C.

Growth factors: The vast majority of all the cyanobacteria now in axenic culture can grow photoautotrophically on simple mineral media without added growth factors. In the few cases where strains have an obligate growth factor requirement, it is invariably fulfilled by Vitamin B_{12} (Rippka et al. 1979). We routinely add a vitamin mixture (Davis and Guillard 1958) to all our enrichment media and Vitamin B_{12} to maintenance medium SN. To date, an absolute requirement for Vitamin B_{12} has not been demonstrated for any of the strains now in culture.

DNA base ratio: DNA base composition has become a fundamental taxonomic character in systematic bacteriology. The DNA base composition (i.e., the mol % G + C content) is constant for a given organism. Closely related bacteria have similar DNA base ratios. However, bacteria with similar base ratios are not necessarily closely related because the mol % G + C values do not take into acount the linear arrangement of the nucleotides within the DNA. Thus, strictly speaking, DNA base ratios provide an index of dissimilarity.

To date we have determined the base ratios of 27 strains in the marine cluster of *Synechococcus*. Their base ratios have fallen between 54.9 and 62.4 mol % G + C (Table 3).

Taxonomic Position

Differentiation between small unicellular cyanobacteria has been problematic in the botanical system for many years because these organisms lack sufficient morphological and phenotypical characters for the adequate description of species and even genera. As an interim measure Rippka et al. (1979) placed in the genus *Synechococcus* all the small unicellular cyanobacteria with cylindrical to ovoid cells that reproduced by binary transverse fission in a single plane and that lacked sheaths. The name *Synechococcus* has been applied to unicellular cyanobacteria observed in the field and to all the marine isolates now in culture using the broad definition of Rippka et al. (1979).

At the time of the 1979 monograph, the loosely defined *Synechococcus* group contained 28 axenic strains primarily of terrestrial and freshwater origin. It was realized that even though the strains were morphologically similar they showed considerable genetic heterogeneity as reflected by the span of the mean DNA base composition within the group (39 to 71 mol % G + C). It was in fact suggested that this group might subsequently warrant subdivision into three groups corresponding to the DNA base compositional subgroups with spans of 39–43, 47–56, and 66–71 mol % G + C. This subdivision has now been formally proposed by Rippka and Cohen-Bazire (1983) but not formally accepted by the International Committee on Bacterial Nomenclature. Rippka and Cohen-Baziere have proposed the genus designation *Cyanobacterium* for the low G + C cluster (i.e., 39–43 mol % G + C) *Synechococcus* for the middle G + C cluster (i.e., 47–56 mol % G + C), and *Cyanobium* for the high G + C cluster (i.e., 66–71 mol % G + C). The fact that the DNA base ratios of the open ocean isolates fall between and completely span the gap between two of Rippka's DNA base ratio clusters (Rippka et al. 1979; Rippka and Cohen-Bazire 1983), in effect eliminates the basis on which Rippka and Cohen-Bazire defined the boundaries of their proposed genera *Synechococcus* and *Cyanobium* (sensu Rippka and Cohen-Bazire 1983). We are thus left with approximately 60 strains (both freshwater and marine) forming a cluster with DNA base ratios that span from 47 to 71 mol % G + C with no obvious break-off points. This genetically heterogeneous cluster almost certainly encompasses three and maybe more genera, but final resolution of generic boundaries will have to await further taxonomic studies using DNA–DNA hybridization and 16S rRNA sequence analysis. It is likely that the marine phycoerythrin containing cluster of isolates will represent a discrete generic unit. However, until generic boundaries are firmly established it is best to continue to refer to the marine isolates as marine *Synechococcus*.

Summary

A large collection of clonal isolates of *Synechococcus* have been purified from the open ocean by repeated streaking of single colonies on solid media. These strains are housed in the culture collection at the Woods Hole Oceanographic Institution and representative strains for distribution to the research community are available from the marine algal collection at the Bigelow Laboratory for Marine Science, Boothbay Harbor, Maine, USA. Research to date indicates that this group of cyanobacteria share many features, but that individual strains or small groups of strains possess properties that distinguish them from one another. Genetic analysis at the level of DNA base ratios has not revealed clear-cut strain clusters. However, the total spread in DNA base ratios from 54 to 63 mol % G + C is sufficient to indicate that the marine cluster of *Synechococcus* represents a number of species.

SECTION II

Standing Stock

The natural abundance of *Synechococcus* results from a complex interaction of a number of physical and biological factors. Properties such as light quantity and quality, water temperature, mixing and nutrient availability are factors affecting growth whereas grazing is primarily responsible for the removal of individuals from the population.

Counting Technique

Ecological studies are greatly facilitated by the fact that marine *Synechococcus* can be counted directly, specifically, and easily as a result of the unique combination of phycoerythrin autofluorescence and cell size. Phycoerythrin fluoresces orange, and is easily distinguished from the red fluorescence of chlorophyll contained in other phytoplankters. Phycoerythrin is not unique to *Synechococcus*. Many other cyanobacteria, including the marine planktonic forms *Trichodesmium* and *Synechocystis*, possess it as well as almost all red algae (Rhodophyta) and many cryptomonads (Cryptophyta). In principle, members of these other phycoerythrin containing organisms might interfere when attempting to count *Synechococcus*, but in practice differences in size, form and chloroplast morphology permit the unambiguous identification and enumeration of marine *Synechococcus*.

Samples to be counted should be processed immediately following collection; phycoerythrin fluorescence does not preserve well in liquid samples stored in the cold or fixed with biological preservatives. The sample size to be filtered varies with the in situ concentration of *Synechococcus*. Typically between 5 and 50 mL are filtered to give approximately 30 cells per field in the occular grid. Filtration of larger volumes can lead to *Synechococcus* cells being obscured by detritus or larger cells.

Samples are filtered onto 0.4 μm Nuclepore filters (25 mm ϕ) using a Millipore glass chimney and base in which the cross-sectional area of the chimney has been carefully calibrated. It is critical that the sample is evenly distributed over the filter surface. This is accomplished by using a backing filter (Whatman GF/F) between

84

the membrane filter and the filter base, and by acid-washing the filter base and changing the backing filter frequently. The evenness of distribution can be monitored during counting by comparing the number of counts per field between fields. If the counts per field vary considerably, it is necessary to make a new preparation. Before placing the membrane filter on the base, the backing filter is flooded with a sterile 3% NaCl solution. The membrane filter is then floated on the salt solution and a vacuum applied to the system to pull the membrane evenly onto the backing filter before clamping the chimney in place. Following filtration of the sample under low vacuum (100–125 mm of mercury), the membrane filter is lifted off the backing filter while still under vacuum and blotted on the backing filter to remove any remaining liquid. The filter is then placed sample side up on top of a drop of immersion oil (Cargille type A) on a microscope slide. An additional drop of oil is placed on the sample surface of the membrane filter and a cover slip is placed on top. The cover slip is gently pressed into place and the slide inverted on a paper towel to blot away excess oil and to flatten the preparation. Excessive pressure when blotting the preparation can be damaging. The preparations can be counted immediately or wrapped in aluminum foil and frozen to be counted later. In the case of shipboard counting, it is possible to accurately enumerate the overall population of *Synechococcus*, but vibrations make it impossible to accurately count the percent of dividing cells. These counts are best performed on frozen prepared samples upon returning to the laboratory. Frozen samples have been counted successfully following storage for 3–4 wk.

Preparations are counted using a Zeiss standard microscope equipped with Neofluar objectives and an epifluorescence illumination system containing either a 100 W halogen or a 50 W mercury lamp. The Zeiss acridine orange filter set 48.77.09 is a good overall set in which phycoerythrin fluoresces orange and chlorophyll fluoresces red. This filter set can be used with the halogen light source and is excellent for routine counting of laboratory cultures and experiments. Zeiss filter set 48.77.15, used with a mercury lamp, produces the most intense phycoerythrin fluorescence and is used for critical counts of field material. However, this filter set is not good for chlorophyll fluorescence. Zeiss filter set 48.77.05 works well for chlorophyll fluorescence and has been used to monitor chlorophyll containing phytoplankters in field material.

Calculations

The multiplication factor equating the area of the ocular grid to the cross-sectional area of the filter chimney was calculated by carefully measuring the inside diameter of the chimney (approx. 16 mm in a Millipore 25 mm filter holder) to get its cross-sectional area and measuring the size of the ocular grid in the microscope with a stage micrometer to get its area (approx. 100×100 μm with the Zeiss, using a $100 \times$ oil objective). The factor of approximately 2×10^4 is calculated by dividing the cross sectioned area of the filter chimney by the area of the ocular grid.

The number of *Synechococcus* cells mL^{-1} is calculated from:

$$\frac{\text{cells field}^{-1*}}{\text{mL sample filtered}} \times (2 \times 10^4)$$

*1 field = cells within the 100×100 μm ocular grid.

The counting technique described above provides a rapid and accurate way of monitoring in situ populations of marine *Synechococcus*. It cannot, however, discriminate between species or clones of *Synechococcus*. Once species boundaries have been established it may be possible to identify characters that can be used to identify species in the field. Preliminary studies using immunology (Campbell et al. 1983), light microscopy (Wood et al. 1985) and flow cytometry (Wood et al. 1985 and Olson et al. 1985) already show promise of providing a means to differentiate *Synechococcus* populations.

Occurrence of Marine *Synechococcus*

Since the discovery of marine *Synechococcus* in 1977 many investigators using epifluorescence microscopy have observed them over a wide geographic area at concentrations ranging from a few cells mL^{-1} to nearly 10^6 cells mL^{-1} in seawater collected within the euphotic zone.

We have found *Synechococcus* cells in virtually every sample we have collected in the euphotic zone with the exception of samples collected in Antarctica. No *Synechococcus* cells were found during a sampling program conducted in McMurdo Sound on the Ross Ice Shelf between November 1980 and February 1981 even though it was possible to concentrate several liters of seawater on a 0.2 μm filter (25 mmΦ).

Vertical Distribution and Abundance

Numerous vertical profiles have been taken to document the distribution and concentration of *Synechococcus* within the water column (Waterbury et al. 1979; Krempin and Sullivan 1981; Johnson and Sieburth 1979, 1982; Murphy and Haugen 1985; Davis et al, 1985; El Hag and Fogg 1986). Characteristic features of that distribution are shown in the six profiles illustrated in Fig. 3.

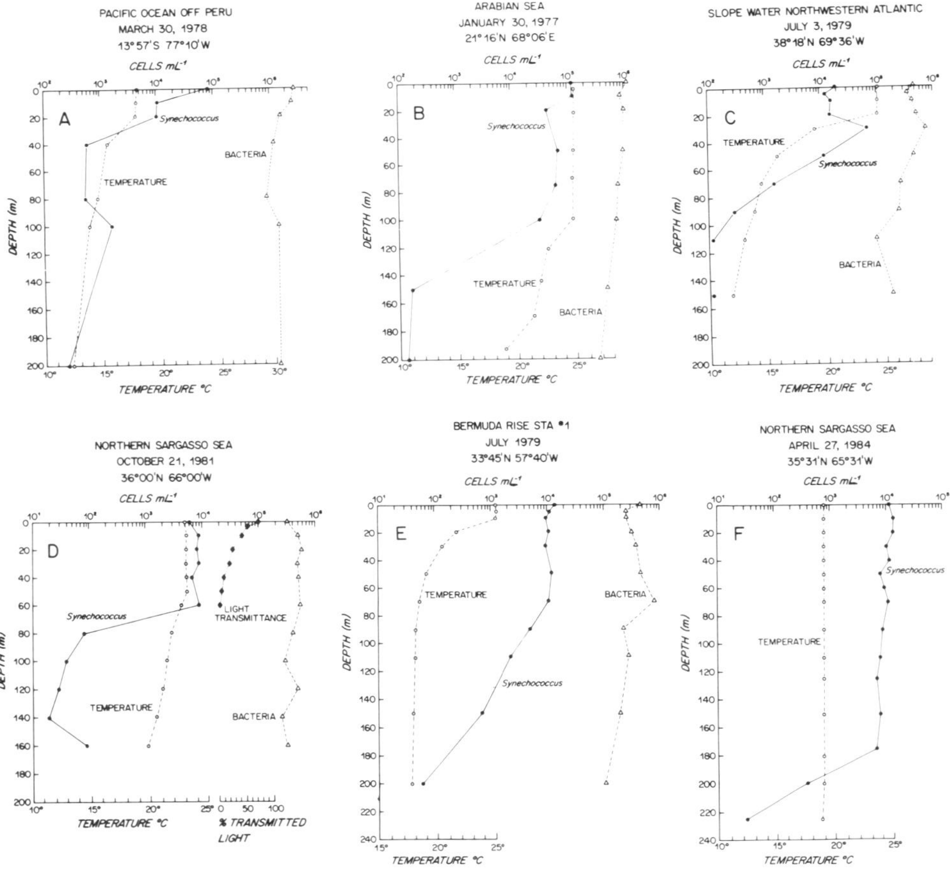

FIG. 3. Vertical profiles of *Synechococcus* abundance from various locations in the world's oceans. Bacteria were counted by the method of Watson et al., (1977). Temperature profiles were determined by XBT and the percent transmitted light was determined using a Licor light meter equipped with a 2π sensor.

86

Profile A was taken at a 3400 m station off the coast of Peru (13°57′S, 77°10′W) during an intense algal bloom. The surface chlorophyll concentration was 16 μg • L^{-1} and the NO$_3^-$ concentration ranged from 8 μgA at the surface to 20 μgA at 20 m where the oxygen concentration also dropped dramatically. Under these stratified nutrient rich conditions the *Synechococcus* concentration was high at the surface (9 × 10^4 cells • mL^{-1}) and dropped off quickly as the photosynthetically available light became rapidly attenuated within the first 20 m as a result of the high algal biomass. A second sharp decline in *Synechococcus* concentration occurred at the boundary layer between 20 and 40 m.

Profile B was taken at a 375 m station off the coast of Pakistan (21°16′N, 68°06′E) where the water column was stratified. The surface layer was mixed to a depth of 100 m at which point there was a break in oxygen concentration and salinity. Nitrate ranged from 1.5 μgA at the surface to 2.4 μgA at 100 m and then increased rapidly to 8 μgA at 120 m and 13 μgA at 140 m. As in profile A the maximum concentration of Synechococcus was high (1.5 × 10^5 cells • mL^{-1}) and occurred at the surface. It remained high (in the mid 10^4 cells • mL^{-1}) throughout the mixed layer and then dropped dramatically below the boundary layer at 100 m.

Profile C was taken at Station DOS-2 (38°18′N, 69°36′W) in slope water north of the Gulf Stream during July when the water column was stratified. The 24.5°C mixed layer occurred to a depth of only 20 m. The *Synechococcus* concentration was nearly constant within the mixed layer (1.25 to 2.0 × 10^4 cells • mL^{-1}). Just below the base of the mixed layer at 30 m there was a pronounced subsurface maximum (7.5 to 10^4 cells • mL^{-1}) in the *Synechococcus* abundance and then a rapid decline in cell concentration to 100 cells • mL^{-1} at 110 m.

Profile D was taken in the northern Sargasso Sea (36°00′N, 66°00′W) in October when the water column was still stratified. The temperature of the mixed layer was 23°C and had a depth of 50 m. The *Synechococcus* concentration was relatively constant throughout the mixed layer and below to a depth of 60 m which corresponds to the one percent light level. *Synechococcus* then dropped off rapidly to below 100 cells • mL^{-1} by 80 m.

Profile E was taken on the Bermuda Rise (33°45′N, 57°40′W) during June when only the upper 10 m was mixed. The *Synechococcus* concentration (10^4 cells • mL^{-1}) was constant to a depth of 70 m, well below the bottom of the mixed layer, and then decreased gradually to less than 100 cells • mL^{-1} by 200 m.

Profile F was taken in the northern Sargasso Sea (35°31′N, 65°31′W), at approximately the same location as Profile D, during late April before the water column had begun to stratify as indicated by a constant temperature through the upper water column of 18.8°C. Here the *Synechococcus* concentration remained relatively constant at 10^4 cells • mL^{-1} to a depth of 175 m and then dropped off rapidly.

The profiles in Fig. 3 demonstrate that a number of factors contribute to the vertical distribution of *Synechococcus*. One of the most important is the depth to which photosynthetically active light penetrates the water column with one percent transmittance, the approximate lower limit for this group of cyanobacteria. In oligotrophic waters the one percent light level occurs at approximately 70 m (see Profile D) but may be considerably shallower in more nutrient rich areas where the biomass (particularly algal biomass) attenuates the photosynthetically available light (see Profile A). In situations where the *Synechococcus* concentration is high below the one percent light level, mixing is most likely to be the mechanism responsible (see Profiles B and F). In Profile F one might have expected *Synechococcus* to be mixed uniformly throughout the 19°C mixed layer. The fact that the concentration drops dramatically at 175 m suggests that wind driven mixing may have been responsible for this vertical pattern. Boundary layers between water masses can also effect the vertical distribution of *Synechococcus*. In Profiles A and B the *Synechococcus* concentration drops dramatically below the boundary layers at 30 m in Profile A and 100 m in Profile B. Usually the maximum concentration of *Synechococcus* occurs at or near the surface. Occasionally a distinct subsurface maximum is evident (see Profile C) just below the mixed layer at a depth where the light level exceeds one percent of the transmitted irradiance, suggesting that the population making up the subsurface maximum may benefit from nutrients supplied from below the mixed layer.

Horizontal Distribution and Abundance

Reported values indicate that the surface concentration of *Synechococcus* can range from a few cells • mL^{-1} to as many as 10^6 cells • mL^{-1}. Figure 4 is an example of how this range in *Synechococcus* concentration manifests itself on a transect from the southern tip of South America (53°S latitude) to Woods Hole, Massachusetts (42°N latitude). Figure 4 incorporates surface counts from three transects collected during March 1981, March 1984, February 1985, and a few points for comparative purposes from a cruise in July 1982. Together they provide a snap shot of the range in *Synechococcus* concentrations encountered at one time of year, although during different years, on a roughly north–south transect covering 5500 nautical miles.

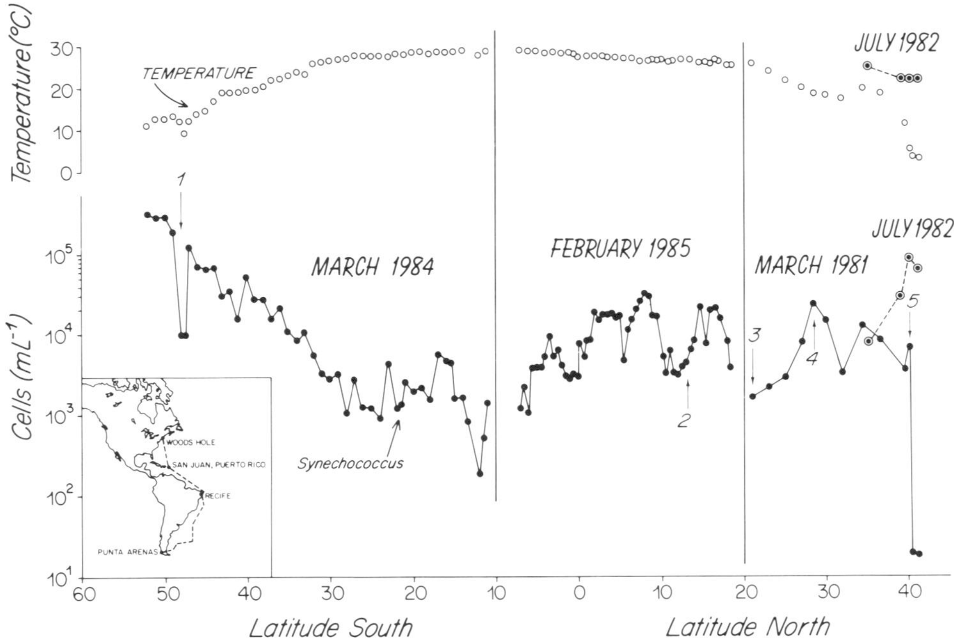

Fig. 4. Horizontal distribution of *Synechococcus* cells in surface waters and surface temperatures from latitude 53°S to 42°N. Each point in the March 1984 and February 1985 transects represents an average of four measurements from counts made every 2 h. Each point in the March 1981 and July 1982 transects represents a single determination.

During the transect between Punta Arenas, Chile and Recife, Brazil in March 1984, the *Synechococcus* concentration decreased gradually from a high of 3×10^5 cells • mL^{-1} at the tip of South America where the water temperature was 10°C to 1×10^3 cells • mL^{-1} off Recife where the water temperature was 28°C. In addition to the more general trend seen in the Punta Arenas–Recife transect, dramatic changes in the *Synechococcus* concentration, such as the event labeled (1) in Fig. 4, are also evident. This sharp drop in *Synechococcus* concentration, accompanied by a drop in water temperature, is an example of how a change in water mass can be reflected by a dramatic change in *Synechococcus* concentration.

The trend of decreasing cell concentration with increasing water temperature that occurred during the Punta Arenas–Recife transect was opposite to that reported by

Murphy and Haugen (1985). In their transect, made from 37°N latitude to 63°N latitude during April and May 1982, the cell concentration gradually decreased from 10^5 cells $\cdot$ mL^{-1} when the water temperature was 18°C to 10^3 cells $\cdot$ mL^{-1} when the water temperature was 5°C. Seasonal differences may provide an explanation for these opposing trends. Although the two transects were made about the same time of year, the Murphy and Haugen transect was made in the Northern Hemisphere springtime whereas the Punta Arenas-Recife transect was made in the Southern Hemisphere fall. These opposing trends may reflect the effects of nutrient availability and water temperature on growth and grazing rates. During the spring season when nutrients were available, the *Synechococcus* bloom occurred as the temperature rose, resulting in the trend seen by Murphy and Haugen. At this time *Synechococcus* counts were lowest at the poleward end of the transect where seasonal warming had progressed the least and was limiting the growth rate of *Synechococcus*. The inverse relationship between *Synechococcus* concentration and temperature seen in the fall, may have resulted when the warm water at the equatorial end of the transect became stratified and nutrient-depleted as the season progressed. This limited the growth of *Synechococcus*, even though the water temperature was near their growth optimum. The high concentration of *Synechococcus* at the poleward end of the fall transect reflected the maximal effects of seasonal heating of the water and nutrient availability.

On the transect from San Juan, Puerto Rico to Recife, Brazil during February 1985 the *Synechococcus* concentration varied between 3.5 $\times$ 10^4 cells $\cdot$ mL^{-1} and 1 $\times$ 10^3 cells $\cdot$ mL^{-1} while the temperature ranged from 26°C to 28°C. The *Synechococcus* concentration was 5 $\times$ 10^3 cells $\cdot$ mL^{-1} upon leaving San Juan, Puerto Rico and rose to 2 $\times$ 10^4 cells $\cdot$ mL^{-1} while passing down the west side of the Leeward Islands in the Caribbean, then dropped again at point (2) in Fig. 4 where the transect passed Barbados. The *Synechococcus* concentration then rose to 3.5 $\times$ 10^4 cells $\cdot$ mL^{-1} from which it declined gradually to 10^3 cells $\cdot$ mL^{-1} as the transect passed along the northern coast of Brazil to Recife.

During the transect from San Juan, Puerto Rico to Woods Hole, Massachusetts in March 1981, the *Synechococcus* concentration showed considerable variation as the temperature decreased from a high of 26°C north of Puerto Rico to a low of 3°C on the continental shelf south of Cape Cod, Massachusetts. In the 26°C stratified and nutrient-depleted water north of Puerto Rico at point (3) in Fig. 4, the *Synechococcus* concentration was 1.7 $\times$ 10^3 cells $\cdot$ mL^{-1}. Coming north, the *Synechococcus* concentration increased steadily, reaching a high of 2.6 $\times$ 10^4 cells $\cdot$ mL^{-1} at point (4) in Fig. 4 when the surface water temperature reached 19°C, indicating that a *Synechococcus* bloom occurred at the front where stratification was occurring. The *Synechococcus* concentration then fluctuated between 5 $\times$ 10^3 and 1 $\times$ 10^4 cells $\cdot$ mL^{-1} until reaching the shelf break south of Woods Hole at point (5) in Fig. 4. There the *Synechococcus* concentration dropped from 6.5 $\times$ 10^3 where the water was 5°C to a few cells mL^{-1} when the temperature dropped to 3.5°C within 20 nautical miles. In contrast, the northern Sargasso Sea concentration of *Synechococcus* was similar in July 1982 when the water temperature was 24°C as it was in March, but was considerably higher in the slope water north of the Gulf Stream and on the continental shelf where the water was 21°C.

Summary

Extension of the existing data indicates that the following global patterns of *Synechococcus* distribution and abundance will emerge. Marine *Synechococcus* should be present within the euphotic zone in virtually all marine waters at concentrations ranging from a few to as many as 10^6 cells $\cdot$ mL^{-1}. A notable exception seems to be the absence of *Synechococcus* from those regions of the polar seas where the water temperature remains below 5°C throughout the year. In oligotrophic tropical waters *Synechococcus* is present throughout the year at concentrations ranging from 10^3 to the low 10^4 cells $\cdot$ mL^{-1}, but may reach higher concentrations in nutrient rich coastal waters. *Synechococcus* concentrations in temperate waters are strongly influenced by seasonal changes in water temperature (see Discussion in section IV). In offshore temperate waters the *Synechococcus* concentration ranges from 10^3 to 5×10^5 cells $\cdot$ mL^{-1}. In temperate coastal waters where the water temperature drops below 5°C during winter months the range of *Synechococcus* concentration is even greater and varies from a few cells $\cdot$ mL^{-1} in the winter to as many as 5×10^5 cells $\cdot$ mL^{-1} in the summer.

SECTION III

Contribution of *Synechococcus* to Primary Production

It has become apparent that photosynthetic "picoplankters" (i.e., phytoplankton smaller than 2 or 3 μm) play a major role in oceanic primary production (Gieskes et al. 1979; Waterbury et al. 1980; Bienfang and Takahashi 1983; Joint and Pomroy 1983; Li et al. 1983; Platt et al. 1983; Douglas 1984; Glover et al. 1985; Putt and Prézelin 1985; Iturriaga and Mitchell 1986). However, quantification of their role has been complicated by the realization that the traditional ^{14}C methodology of Steemann-Nielsen (1952) for the measurement of primary production may be fraught with potential errors. With this in mind we have attempted to develop techniques based on the incorporation of ^{14}C sodium bicarbonate to specifically measure the contribution of *Synechococcus* to primary productivity. Two inherent properties of *Synechococcus* have aided this process: it can be counted specifically and accurately, and it can be separatd from virtually all other phytoplankters by differential filtration due to its small size (0.6 $\times$ 1.4 μm).

Differential Filtration

Differential filtration has been used by many investigators to examine processes attributable to different size classes of organisms. In interpreting data from size fractionated experiments it is critical to recognize that size fractionation does not effect a quantitative separation of size classes. Invariably a certain percentage of each size class gets trapped on the filters that have pore sizes larger than the particular size class of interest. In the case of *Synechococcus*, typically 20–50% of the in situ population is lost in the preparation of a 1 μm filtrate by differential filtration. The specific contribution of *Synechococcus* to primary production using a fractionated population can be quantified because *Synechococcus* abundance in the 1 μm fraction can be monitored by direct counts, enabling the results to be normalized to the total population in the original sample.

Post-Incubation and Pre-Incubation Fractionation

Both post-incubation fractionation (i.e., size fractionation of the natural population following incubation with ^{14}C sodium bicarbonate) and pre-incubation fractionation (i.e., size fractionation of the natural population before incubation with ^{14}C sodium bicarbonate) have been used to assess the contribution of *Synechococcus* to primary productivity. In our experience post-incubation fractionation has the serious potential disadvantage of greatly overestimating the amount of ^{14}C-bicarbonate assimilated by *Synechococcus*. In some post-incubation experiments when the carbon fixed by *Synechococcus* in the 1 μm filtrate was normalized to the in situ abundance of *Synechococcus* in the unfiltered seawater, results were obtained that indicated that *Synechococcus* was responsible for more than 100% and as much as 175% of the total primary productivity. Our interpretation of these experiments is that some eucaryotic phytoplankters were disrupted during the post-incubation filtration process, even under a vacuum of 125 mm mercury. Some of the resulting eucaryotic cell fragments were caught on the small pore size filters and incorrectly attributed to carbon fixed by *Synechococcus*. Post-incubation fractionation experiments do not consistently yield abnormally high values for the contribution of *Synechococcus* to primary productivity. In many experiments, notably in inshore waters, the results of post-incubation and pre-incubation experiments yield comparable results. Pre-incubation fractionation has been used routinely in this study. In addition to avoiding the serious potential problem described above, pre-incubation fractionation of *Synechococcus* into a 1 μm filtrate has the advantage of both separating *Synechococcus* from other phytoplankters and from its grazers.

Fractionation Procedure

Each depth sampled required 3.5 L of 1 μm filtrate for the ^{14}C incubation and an additional 2 L if chlorophyll was measured. The procedure for fractionating seawater is shown at the top of Fig. 5. The seawater was first gravity filtered through 30 μm Nitex screening, then filtered through 10 μm Nuclepore filters (47 mm ϕ) using Millipore glass chimneys and bases, 2-L glass filter flasks and a vacuum of 125 mm of mercury. To avoid excessive cell breakage and clogging, the 10 μm filters were changed after filtering 250 mL in coastal waters and after filtering 500 mL in the Sargasso Sea. The 10 μm filtrate was pooled in polyethylene containers and then filtered through 1 μm Nuclepore filters (47 mm ϕ) that were changed after filtering 250 mL using a vacuum of 125 mm of mercury. The 1 μm filtrate was pooled in a polyethylene container and subsampled for *Synechococcus* and bacterial counts.

Potential Problems Arising from Filter Fractionation of *Synechococcus*

Several potential problems may affect the validity of data obtained from fractionation experiments designed to assess the role of *Synechococcus* as a primary producer. In weighing the advantages and disadvantages of post- and pre-incubation experiments it has been argued that post-incubation fractionation has the distinct advantage of not disturbing the sample prior to incubation with ^{14}C-bicarbonate thus eliminating the risk of physiologically damaging *Synechococcus* cells. In our own experience several lines of evidence indicate that *Synechococcus* is not significantly impaired physiologically by pre-incubation filtration. In experiments where both post- and pre-incubation fractionation have been used the results were often comparable, and where they were not, the results from post-incubation fractionation experiments were usually unreasonably high. Secondly, in pre-incubation fractionation experiments growth rates calculated from the amount of carbon fixed $\cdot$ cell^{-1} $\cdot$ h^{-1} were in good agreement with in situ growth rates calculated from direct cell counts made at frequent intervals from the same body of water used for the ^{14}C incorporation studies. Finally, *Synechococcus* cells are extremely resistant to breakage. Rupture of laboratory cultures by shearing in a French Pressure Cell requires the application of 20000 psi.

A second potential problem concerns the specificity of the fractionation procedure. Comparisons have been made between filtrates prepared using 2 and 1 μm Nuclepore filters. The use of 2 μm filters had the advantage of allowing a greater percentage of the total *Synechococcus* population to pass into the filtrate, but had the insurmountable disadvantage of also allowing many small eucaryotic phytoplankters to pass as well, making it impossible to calculate the amount of ^{14}C-bicarbonate fixed by *Synechococcus* alone.

Conversely, it might be argued that fractionation through 1 μm filters, which typically resulted in a loss of 20–50% of the in situ population, may be artificially fractionating the *Synechococcus* population. This problem was minimized by collecting samples at dawn when the percentage of dividing cells was lowest (Fig. 16, 17, 20–22) and the cells within the natural population were the most uniform in size.

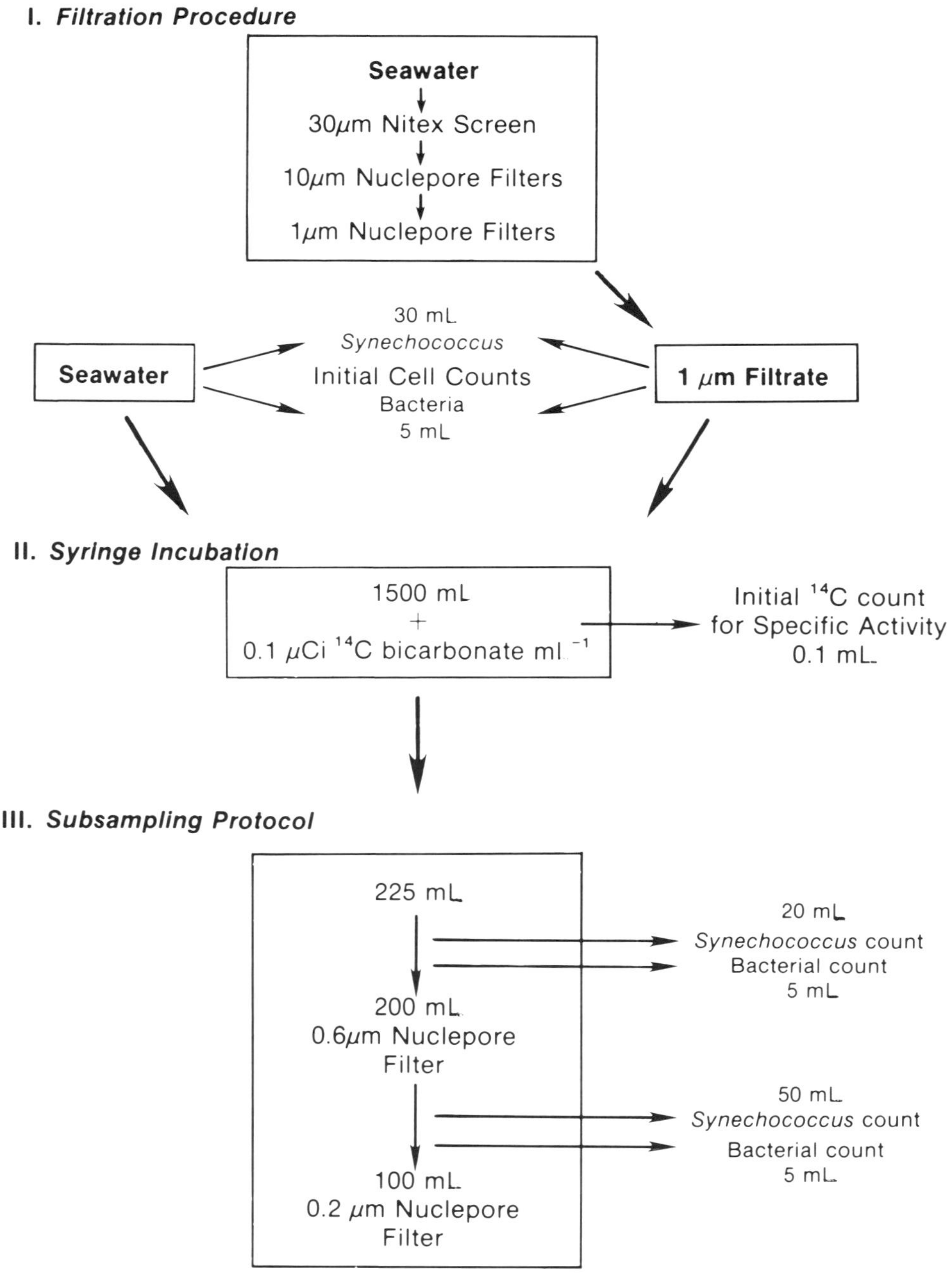

FIG. 5. Experimental Flow Sheet for ^{14}C incorporation studies.

Presence of Eucaryotic Phytoplankters in the 1 μm Filtrate

The use of 1 μm Nuclepore filters had the advantage of producing a filtrate that contained a significant portion of the in situ *Synechococcus* population (usually between 50 and 80%) and excluded virtually all other phytoplankters. The 1 μm filtrates were routinely monitored for contamination by very small eucaryotic phytoplankters using direct observation of chlorophyll autofluorescence by epifluorescence microcsopy with a Zeiss 48.77.05 filter set. An additional check of the 1 μm filtrate can be provided by measuring the amount of chlorophyll per *Synechococcus* cell (Cuhel and Waterbury 1984). Values of 1–2 fg chlorophyll per cell were typical for natural populations of *Synechococcus*. Higher values would have indicated the presence of either very small eucaryotic phytoplankters or cell fragments from larger cells broken during filteration. Haugen and Murphy (1985) have reported instances where significant numbers of small eucaryotic phytoplankton pass into the 1 μm filtrate. In instances where this occurs the 1 μm filtrate cannot be used to assess the contribution of *Synechococcus*.

Bacteria in the 1 μm Filtrate

In addition to *Synechococcus* the 1 μm filtrate contained approximately 90% of the in situ bacterial assemblage, typically in the range of 10^6 cells $\bullet$ mL^{-1}. In theory such large numbers of bacteria might be responsible for a significant amount of $^{14}CO_2$ fixation in the 1 μm filtrate through both auto- and heterotrophic processes. In practice the contribution of bacteria other than *Synechococcus* to carbon fixation in the 1 μm filtrate was monitored through the use of a dark control and by an additional fractionation step that effectively separated *Synechococcus* from the bacterial component.

In cyanobacteria, CO_2 fixation is a tightly regulated process, that only occurs in the light (Stanier and Cohen-Bazire 1977; Smith 1982) (Fig. 18). Consequently dark CO_2 fixation, obtained from a dark control for each experimental condition, was not attributable to *Synechococcus* and was subtracted from the corresponding values of light CO_2 fixation. The amount of dark CO_2 fixation in whole seawater normally ranged from 1 to 5% of the light values. In the pre-fractionated 1 μm filtrate the amounts of dark CO_2 fixation were somewhat higher, ranging from 1 to 10% of the light values. On several occasions in the Sargasso Sea the values of dark CO_2 fixation in the 1 μm filtrate were extremely high, ranging from 20 to 70% of the light values. It is unlikely that either *Synechococcus* or the bacterial component contributed to these high values. In the absence of an explanation for this phenomenon we have continued to subtract the high dark values from our calculations.

A second check on the contribution of the bacterial assemblage to CO_2 fixation in the 1 μm filtrate was achieved by an additional filter fractionation step (Fig. 5). Subsamples of the 1 μm filtrate were filtered through 0.6 μm Nuclepore filters that effectively captured the *Synechococcus* cells, but allowed between 80–85% of the bacteria to pass through into the filtrate. This bacterial component was subsequently captured on a 0.2 μm Nuclepore filter and monitored for $^{14}CO_2$ incorporation. $^{14}CO_2$ fixation by this bacterial fraction was consistently low, ranging from 3 to 5% of the carbon fixation found in the 1 μm filtrate. It was even low in those instances, described above, where the dark values in the 1 μm filtrate were exceptionally high, leading to the conclusion that the bacteria were not responsible for those high dark values.

Incubation of Samples

The choice of techniques and conditions of incubation used to quantify primary production by *Synechococcus* is of critical importance. It is necessary to minimize the introduction of errors leading to measurements that deviate dramatically from what is actually occurring in situ. One of the most serious potential problems arises from sample containment necessitated by the use of radioisotopes. Sample containment changes the natural assemblage from an essentially open to a closed system initiating the possibility that processes such as nutrient limitation, chemical toxicity, light effects and changes in species composition may alter natural processes in the closed system.

Our initial experiments were incubated in an in situ buoy system that had the advantage of reproducing in situ conditions of temperature, light quantity and light quality, but had the distinct disadvantage of providing a single endpoint determination making it impossible to determine time course rate measurements of $^{14}CO_2$ incorporation. When using endpoint determinations, the assumption is made, often

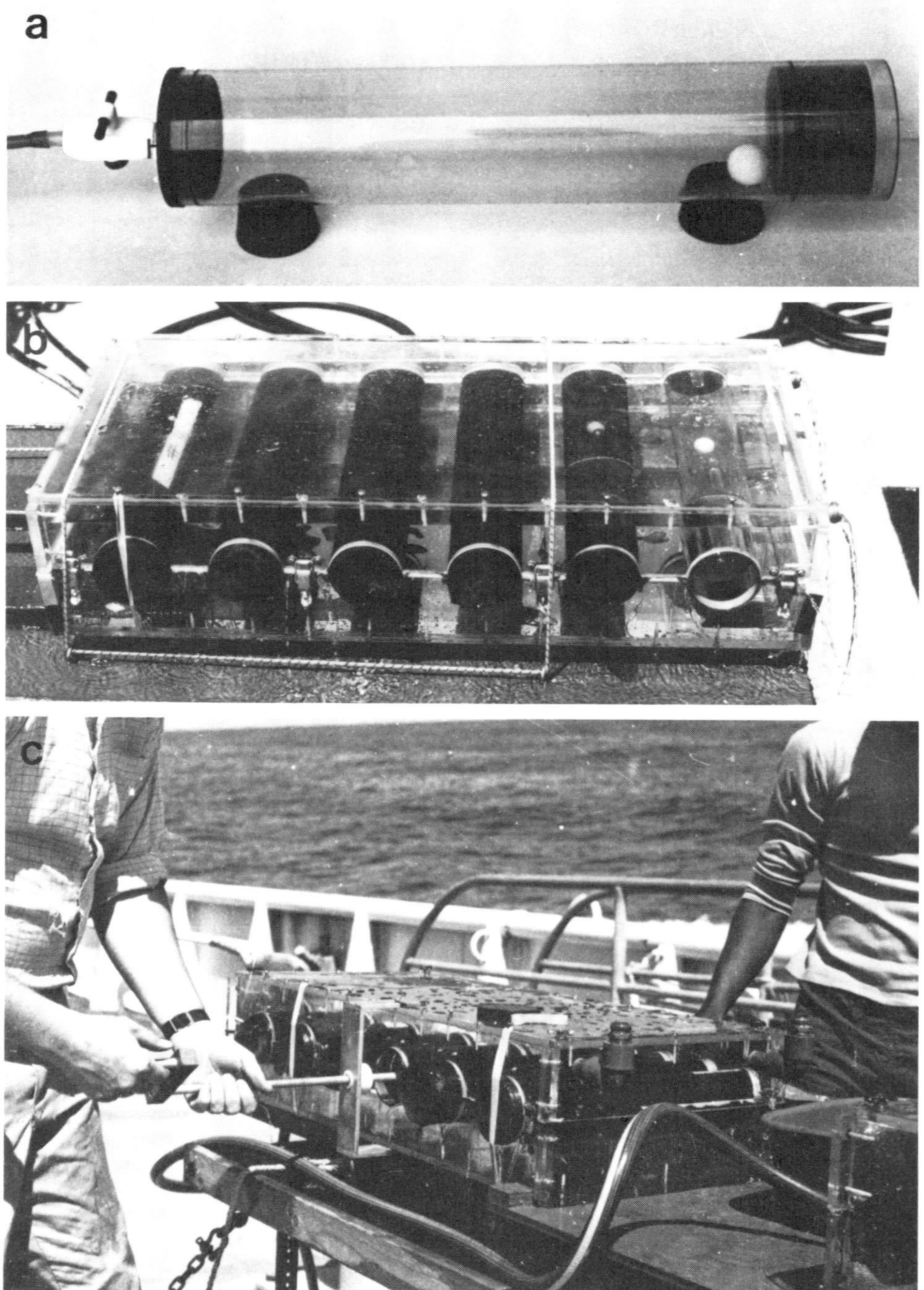

FIG. 6. Syringe Incubation System. (A) Individual syringe showing: polycarbonate endcap with teflon valve, filling port and silicon O-ring seal; the movable piston with silicon O-ring seal; and the teflon stirring ball. (B) Incubation chamber holding six syringes screened with neutral density filters to simulate in situ light intensity. (C) Incubation system showing procedure for removing subsamples.

94

incorrectly, that rate processes have occurred linearly over the course of incubation, which in our experiments ranged from 4 to 12 h.

Subsequent experiments have been incubated in a simulated in situ system conceptually modeled after the Sampler Incubation Device (SID) described by Taylor et al. (1983). It consists of a series of polycarbonate syringes enclosed in a water jacket (Fig. 6). The syringes (Fig. 6a) have a volume of 1600 mL and are made of 3-inch diameter polycarbonate tubing, with polycarbonate end-caps and pistons sealed with silicone "O" rings. A teflon valve and teflon filling port are attached to the end-cap. Gentle stirring is accomplished in each syringe by a 1″ teflon ball that slowly traverses the syringe as the ship rolls. A water jacket filled with running seawater holds six syringes (Fig. 6b,c). Both ends of each syringe are accessible during the course of incubation. A crank can be attached to one end to drive the piston for obtaining subsamples through the teflon valve. Individual syringes can be wrapped with neutral density filters (combinations of plastic window screen and mylar film) to simulate in situ light intensities as well as being completely covered to act as dark controls. Two of these systems containing a total of 12 syringes were used. Typically each depth sampled required four syringes (a seawater, light and dark, and a 1 μm filtrate, light and dark) so that a total of three depths could be run per experiment.

The principal advantage of this syringe incubation system is that it is designed so that time course rate measurements can be made conveniently. The trade off for accessibility of frequent subsamples is the loss of in situ incubation. In situ conditions of light intensity and water temperature can be reasonably simulated in the syringe deck incubation system, but to date we have not attempted to simulate in situ conditions of light quality.

Subsampling Protocol

The time course measurements were typically made over a period of five to six hours centering around noon. Subsamples were taken at 30 min intervals for the first 2 h and subsequently at hourly intervals. The subsampling protocol is shown in Fig. 5.

Following the completion of the subsampling procedures the 47 mm ϕ Nuclepore filters, with the ^{14}C labeled subsamples, were placed directly in scintillation vials which were treated with concentrated hydrochloric acid fumes in a closed plastic container for 1 h to drive off non-cellular $^{14}CO_2$. Washing of filters to remove non-cellular labeled material with seawater or salt solution should be avoided because of the risk of cell rupture and consequent loss of fixed carbon (Goldman and Dennett 1985).

Productivity Experiments

Three experiments, one in the northern Sargasso Sea (Fig. 7–9) and two in Woods Hole (Fig. 10–12), have been chosen to illustrate the contribution of *Synechococcus* to primary production.

The site in the northern Sargasso Sea (35°30′N, 65°30′W) was occupied between 24 April and 3 May 1984. The water temperature was 18.8°C and was isothermal to a depth of 250 m. The *Synechococcus* concentration varied by almost an order of magnitude during a diel cycle and was changing from day to day as the result of bloom conditions (Fig. 20). The *Synechococcus* concentration was relatively uniform to a depth of 180 m, probably as a result of wind-driven mixing (Fig. 3F). The weather was overcast and windy with surface light readings ranging from 500 to 1200 μE • m^{-2} • s^{-1}, measured with a Licor light meter equipped with a 2π sensor. This meter and sensor were also used to determine the light levels within the water column.

Three depths were measured on each of four successive days (28 April–1 May). The data from each of 10 depths are shown in Fig. 7–9. Each data point represents the amount of ^{14}C fixed into cell material in the light minus its dark control.

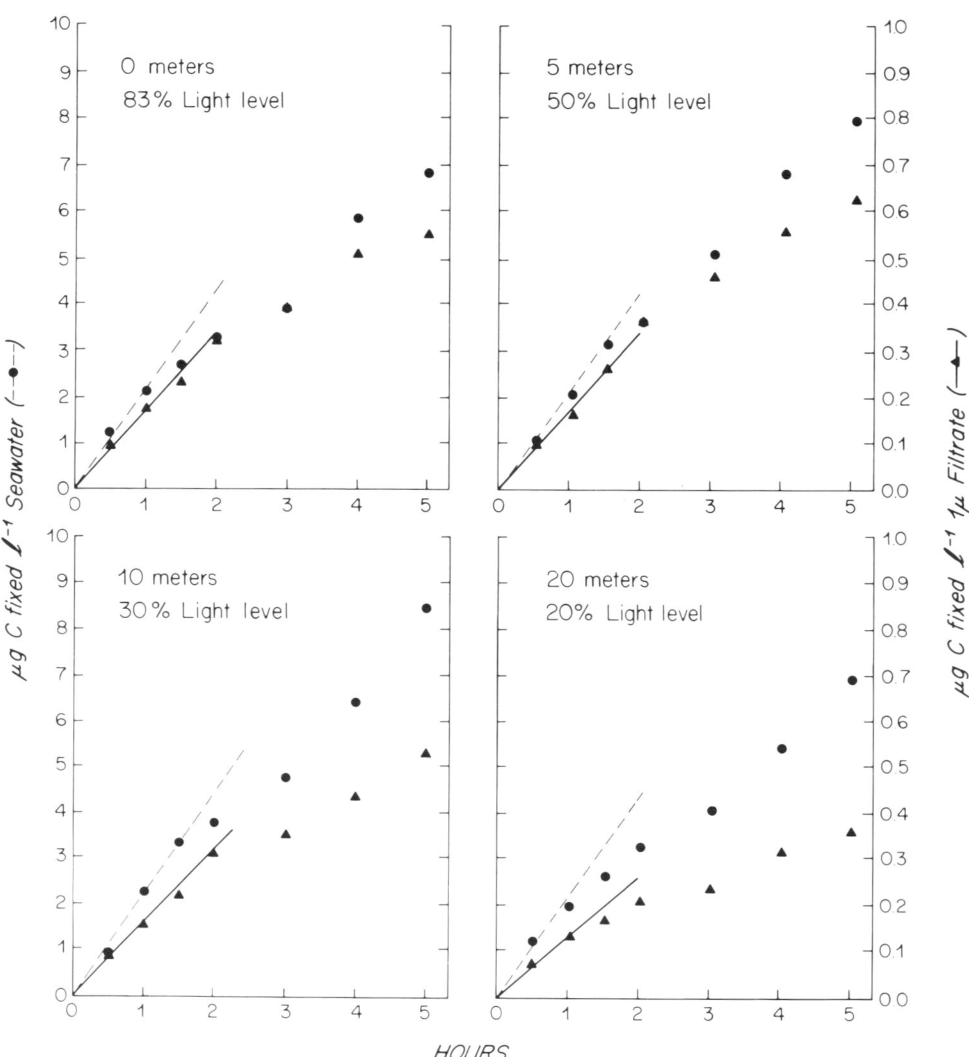

FIG. 7–9. Time course rate measurements of ^{14}C sodium bicarbonate incorporation into cell material in the northern Sargasso Sea (35°N, 65°W) in April 1984. Each point represents the amount of ^{14}C sodium bicarbonate incorporated in the light minus its dark control. The solid and dashed lines indicate the initial rates of incorporation; (– – –) whole seawater, (———) the 1 µm filtrate.

A striking feature of this data set was the nonlinear incorporation of ^{14}C sodium bicarbonate at every depth for both whole seawater and the 1 µm filtrate. In our experience this was unusual since most often incorporation of ^{14}C sodium bicarbonate is linear over the 5 or 6 h of incubation. Nonlinear assimilation of ^{14}C sodium bicarbonate has been encountered in surface samples, where it can usually be attributed to photoinhibition, and in samples from the bottom of the euphotic zone, where it can usually be attributed to light limitation. The reason for the nonlinear incorporation in this series of Sargasso Sea experiments was not evident, but might have resulted from nutrient depletion or contamination associated with sample containment of the natural assemblages. In those instances when the assimilation of ^{14}C sodium bicar-

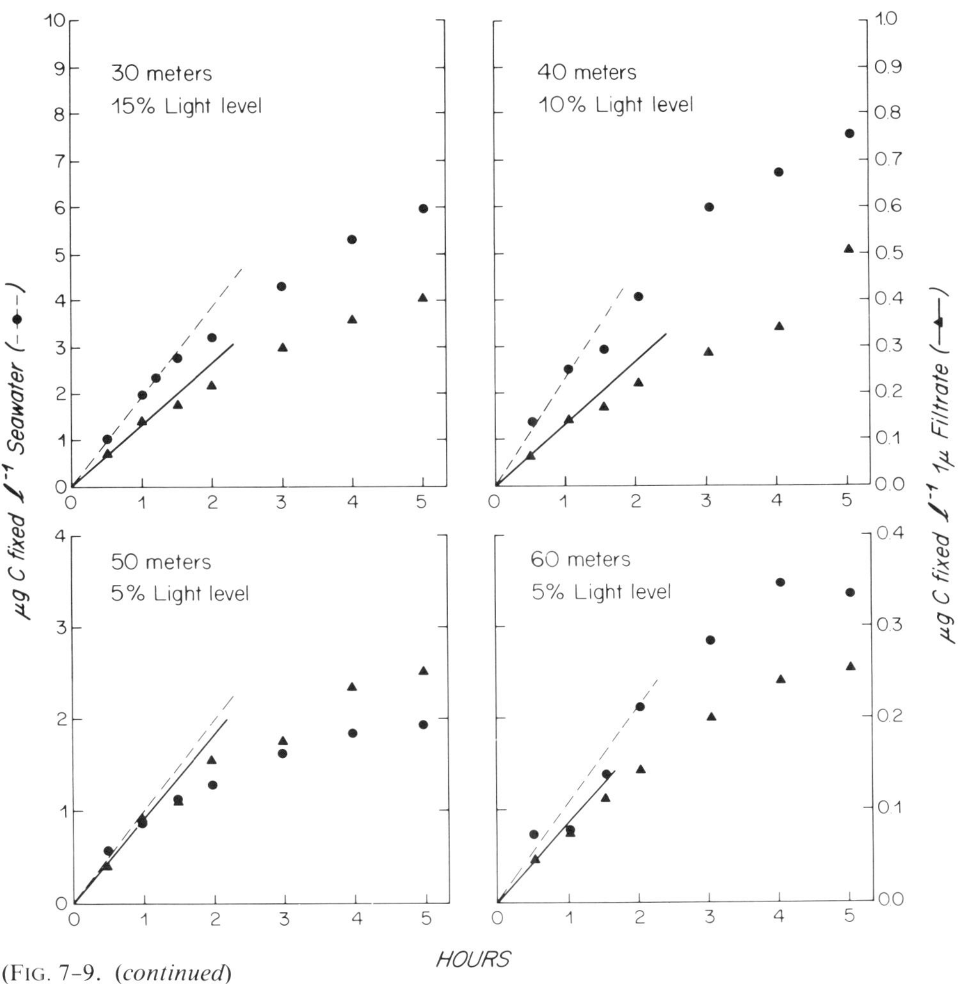

(FIG. 7-9. (*continued*)

bonate has been nonlinear, the rates of ^{14}C incorporation have been determined from the initial slopes (the dashed and solid lines in Fig. 7-9).

A summary of the data from the April 1984 Sargasso Sea experiments is shown at the bottom of Fig. 9. The amount of carbon fixed • L^{-1} • h^{-1} in the 1 μm filtrates has been adjusted to reflect the in situ concentration of *Synechococcus* at each depth by multiplying the fg C fixed • *Synechococcus* cell^{-1} • h^{-1} in the 1 μm filtrate by the concentration of *Synechococcus* in the original seawater sample.

The *Synechococcus* population at this site was changing dynamically as a result of rapid growth and grazing as evidenced by the growth rates determined by direct cell counts (Fig. 20) and the ^{14}C incorporation rates • cell^{-1} • h^{-1} (Fig. 9). A very striking feature of this data set was the high rates of carbon fixed • *Synechococcus* cell^{-1} • h^{-1} throughout the upper 80 m of the water column, 40 fg C • cell^{-1} • h^{-1} at the surface to 16 fg C • cell^{-1} • h^{-1} at 80 m. *Synechococcus* was responsible for 11% of the total primary productivity, calculated by integration from 0 to 80 m. Their contribution was highest near the surface (18%) and at 80 m near the bottom of the euphotic zone (20%).

The contribution of *Synechococcus* to primary productivity in coastal waters is shown in two experiments conducted in Woods Hole waters on 25 May 1983 (Fig.

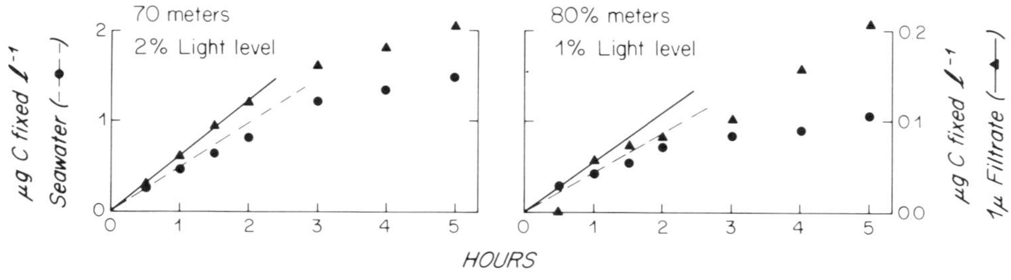

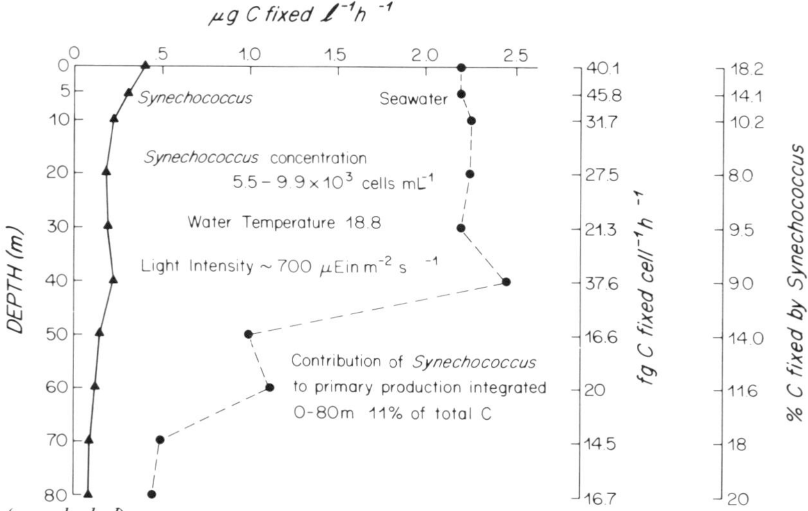

Fig. 7-9. (*concluded*)

Fig. 9. A summary of data from the April 1984 Sargasso Sea experiment. For each depth the contribution of *Synechococcus* (—▲▲—) was determined by multiplying the fg carbon fixed *Synechococcus* cell⁻¹ · h⁻¹ by the in situ concentration of *Synechococcus*.

10, 12) and on 25 July 1984 (Fig. 11, 12). The Woods Hole experiments were each conducted from a single water sample that was screened to reflect the light levels at various depths within the euphotic zone. As a result only two dark controls were needed (a seawater and a 1 μm filtrate) so that five simulated depths could be run with the remaining 10 syringes.

The experiment conducted on 25 May 1983 occurred during the spring bloom when the water temperature was 13°C, the *Synechococcus* concentration had reached 4.15 × 10⁴ cells · mL⁻¹ and the surface light intensity ranged from 1300 to 1500 μE · m⁻² · s⁻¹. The ¹⁴C incorporation rates were linear for both seawater and the 1 μm filtrate at each of the five light levels (Fig. 10).

The summary of the data from the 25 May 1983 experiment is shown at the top of Fig. 12. Photoinhibition was evident in both the whole seawater and in the 1 μm filtrate in the samples screened to reflect a depth of 0.25 m. The rates of carbon incorporation · *Synechococcus* cell⁻¹ · h⁻¹ were the same in the samples screened to the 50, 30, and 15% light levels, again indicating that *Synechococcus* does well throughout the euphotic zone. *Synechococcus* was responsible for 24% of the primary productivity integrated from 0.25 to 10 m and its relative contribution was evenly distributed with depth.

The second Woods Hole experiment was conducted on 25 July 1984 when the water temperature was 22°C, the *Synechococcus* concentration was 4.5 × 10⁴ cells · mL⁻¹

98

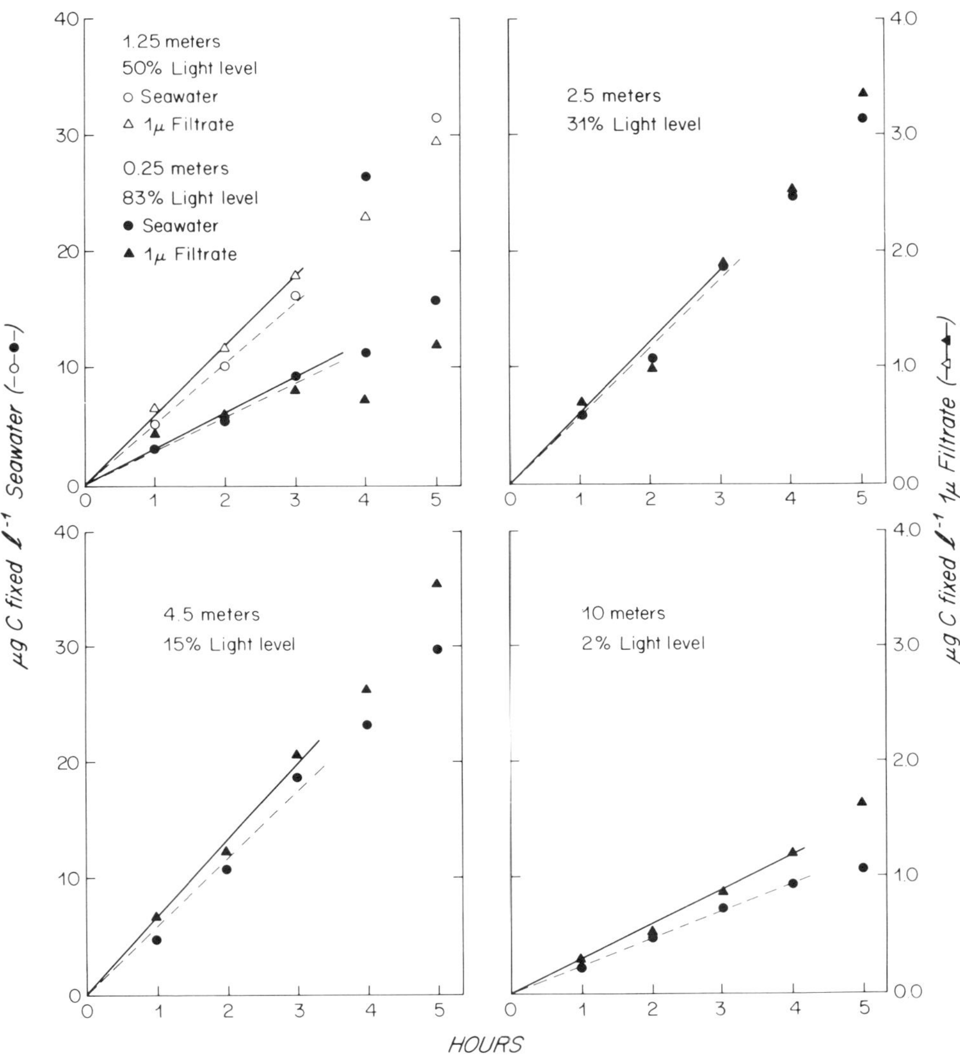

FIG. 10. Time course rate measurements of ^{14}C sodium bicarbonate incorporation into cell material determined in Woods Hole waters on 25 May 1983. Each point represents the amount of ^{14}C sodium bicarbonate incorporated in the light minus its dark control. The solid and dashed lines indicate the initial rates of incorporation; (– – –) whole seawater, (————) 1 μm filtrate.

and the surface light intensity was 1700 μE $\cdot$ m^{-2} $\cdot$ s^{-1}. The ^{14}C incorporation rates were linear for both the whole seawater and the 1 μm filtrate for each of the five light levels except that the whole seawater screened to the 1% light level increased non-linearly (Fig. 11).

The summary of the data for the 25 July 1984 Woods Hole experiment is shown at the bottom of Fig. 12. Photoinhibition was again evident in both the whole seawater and the 1 μm filtrate in the unscreened surface samples. The rates of carbon incorporation $\cdot$ *Synechococcus* cell^{-1} $\cdot$ h^{-1} were highest at the surface (40–50 fg C $\cdot$ cell^{-1} $\cdot$ h^{-1}), dropped to 28 fg C $\cdot$ cell^{-1} $\cdot$ h^{-1} in the samples screened to the 20% and 10% light levels and then dropped dramatically to 5 fg C $\cdot$ cell^{-1} $\cdot$ h^{-1} in the sample screened to the 1% light level.

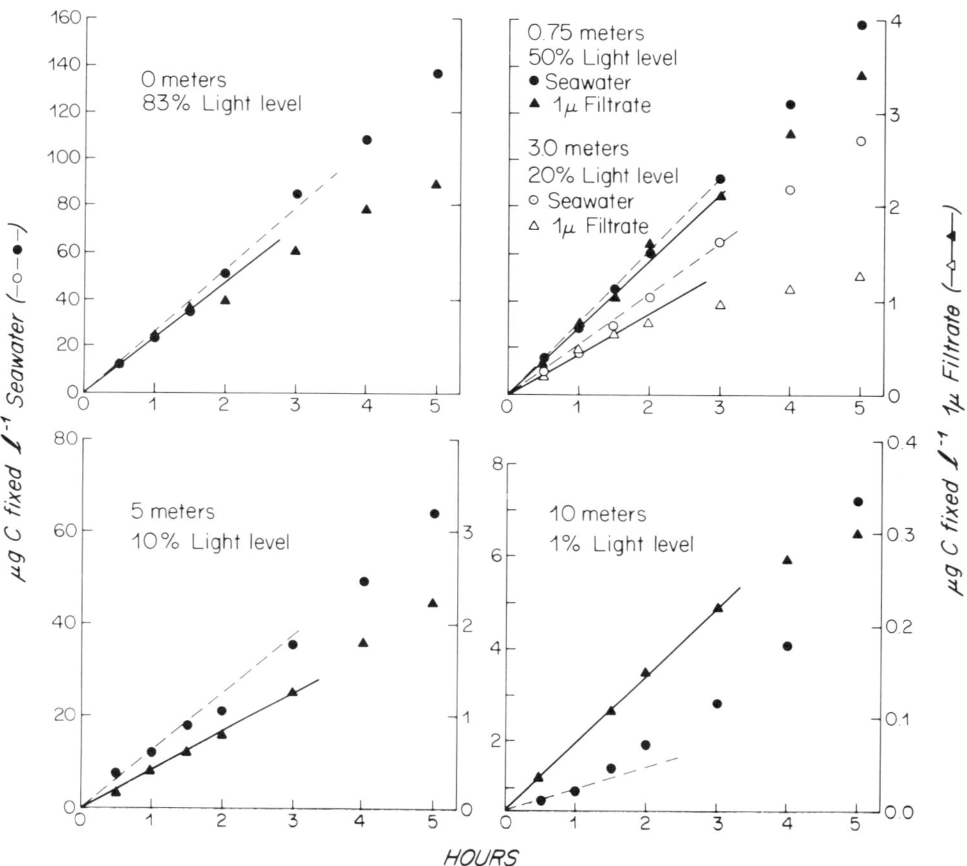

FIG. 11. Same as Fig. 10 except date is 25 July 1984.

Photoinhibition

Photoinhibition in near surface samples, such as the surface whole seawater and 1 μm filtrates in the two Woods Hole experiments (Fig. 10–12), has been encountered frequently in both near shore and open ocean waters. In our opinion this phenomenon results from sample containment of the natural assemblages at high light intensities for artificially long periods of time (i.e., the 5–6 h of incubation). Under natural conditions, individual phytoplankters probably move about sufficiently within the water column, either passively by mixing or actively by motility in the case of some of the larger eucaryotic phytoplankters, so that they are not subjected to high light intensities for periods long enough to induce photoinhibition. This is certainly the case in Woods Hole where there is strong tidal mixing and probably at many open ocean sites as the result of wind-driven mixing. However, photoinhibition almost certainly occurs at times in oligotrophic waters when a shallow mixed layer is present and subjected to high solar radiation.

Summary

With respect to primary productivity one of the most striking features of *Synechococcus* was the significance of their contribution throughout the euphotic zone. The rates of carbon fixed • *Synechococcus* cell⁻¹ • h⁻¹ were highest near the

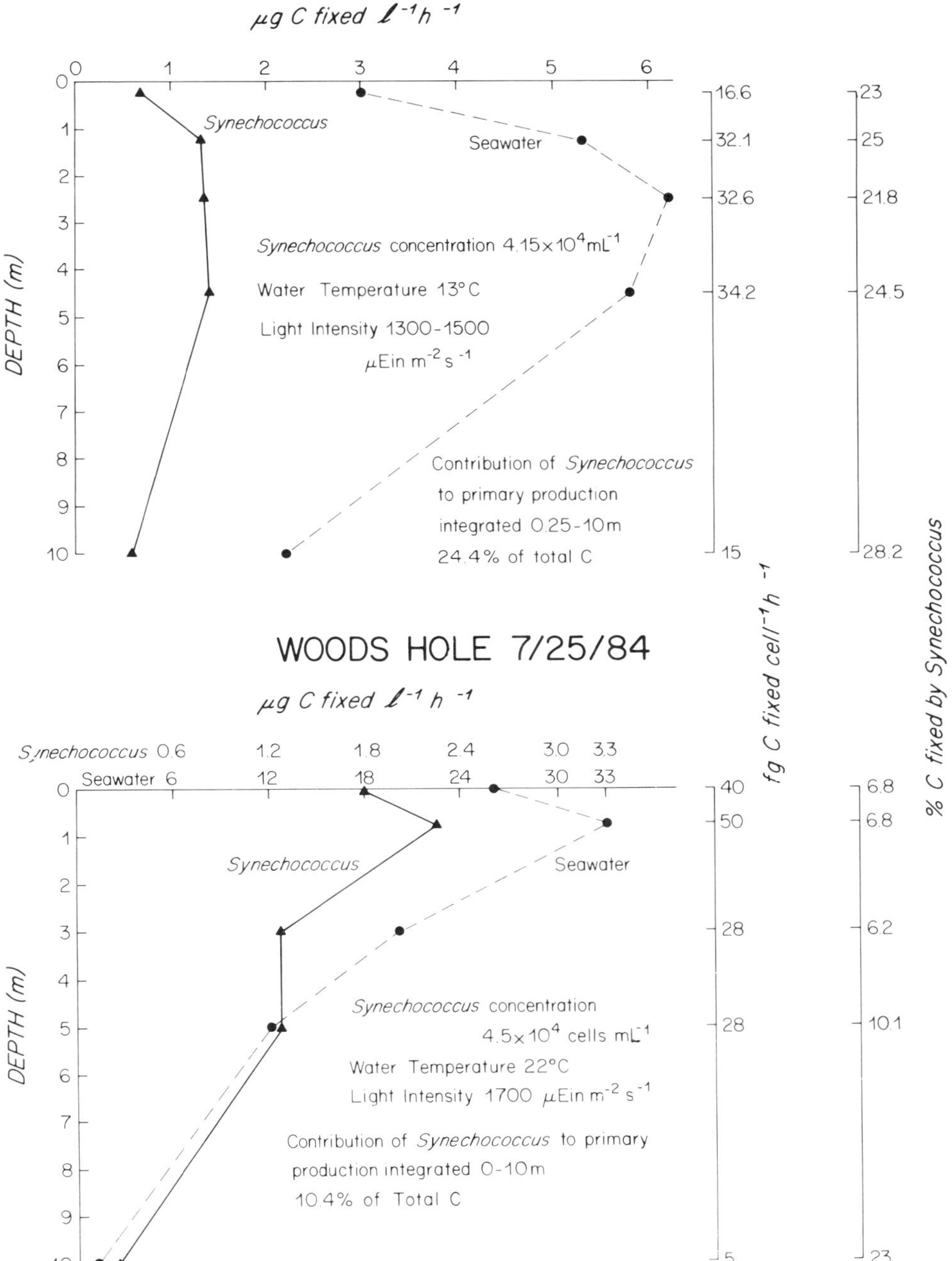

Fig. 12. Summaries of the 25 May 1983 and 25 July 1984 Woods Hole time course experiments. The contribution of *Synechococcus* (▲▲) was determined by multiplying the fg carbon fixed • *Synechococcus* cell⁻¹ • h⁻¹ by the in situ abundance of *Synechococcus*.

surface and dropped off slowly toward the bottom of the euphotic zone. This is in contrast to the hypothesis of Glover et al. (1985) that the contribution of *Synechococcus* in surface waters may be negligible due to photoinhibition and photorespiration. However, as emphasized by Glover et al. (1985) the relative contribution of

Synechococcus to primary productivity often increases toward the bottom of the euphotic zone because *Synechococcus* is better adapted than the eucaryotic phytoplankters to carry out photosynthesis at low light intensities and at the light quality present below the 10% light level.

Another feature of the productivity data in the three experiments presented here is the strong correlation between the rates of carbon fixed • *Synechococcus* cell^{-1} • h^{-1} and water temperature. As would be predicted from the known effect of temperature on the rates of chemical reactions and microbial growth, the rates of carbon fixed • *Synechococcus* cell^{-1} • h^{-1} were highest during the July 1984 Woods Hole experiment when the water temperature was 22°C (Fig. 12), slightly lower at the northern Sargasso Sea site in April 1984 when the water temperature was 19°C (Fig. 9) and lowest during the May 1983 Woods Hole experiment when the water temperature was 13°C (Fig. 12). These data indicate that at these sites and times temperature played an important role in the dynamics of *Synechococcus* growth and suggests that nutrient limitation if operative was of secondary importance.

In direct contrast, the rates of carbon fixed • *Synechococcus* cell^{-1} • h^{-1} were dramatically lower (3–7 fg C • cell^{-1} • h^{-1}) in experiments conducted at the northern Sargasso Sea site (35°N, 65°W) in July 1982 when the water temperature was 26°C. This temperature was probably not yet high enough to adversely affect growth. It is much more likely in this instance that nutrient limitation, rather than water temperature, was of primary importance in regulating *Synechococcus* growth.

The rates of carbon fixed • *Synechococcus* cell^{-1} • h^{-1} at the Sargasso Sea site during the spring bloom (Fig. 9) and in Woods Hole during the summer (Fig. 12) were comparable, indicating that in the absence of severe nutrient limitation *Synechococcus* is equally successful in coastal and open ocean waters. Differences in the relative contribution of *Synechococcus* to total primary productivity usually reflect how well the eucaryotic phytoplankters are doing at a particular site and time rather than reflecting the relative success of *Synechococcus* between sites and times. This is particularly evident if one compares the data from the two Woods Hole experiments in Fig. 12. The rates of carbon fixed • *Synechococcus* cell^{-1} • h^{-1} did not vary dramatically between these two experiments whereas the amount of carbon fixed • litre^{-1} • h^{-1} by the entire phytoplankton assemblage increased five-fold between the May and July experiments.

From the data we have gathered to date the following patterns of the contribution of *Synechococcus* to marine primary productivity emerge when the values are integrated throughout the euphotic zone to the depth of the 1% light level. During three cruises to the northern Sargasso Sea site (35°N, 65°W), two during the spring when the water temperature was 19°C and one during the summer when the surface water temperature was 26°C, *Synechococcus* was responsible for 10–25% of the total primary productivity. In waters north of the Gulf Stream and in slope water south of Cape Cod, visited during the three cruises described above, the contribution of *Synechococcus* to total primary productivity ranged between 5–15%. In the coastal waters near Woods Hole, documented on numerous occasions between May and November, the contribution of *Synechococcus* has typically ranged from 5 to 10% of the total primary production, but on occasion has been as high as 23% (Fig. 10,12).

It is now recognized that the contribution of picoplankters to primary productivity in the oceans is high. Values in the literature range according to the size fractions examined. Putt and Prézelin (1985) reported values of 75% for the <5 µm fraction; Takahashi and Bienfang (1983) and Glover et al. (1985) reported values from 35 to 80% for the <3 µm fraction. Reported values in the <1 µm fraction range from 20 to 30% in Joint and Pomroy (1983); 20 to 80% in Li et al. (1983); 12–65% in Douglas (1984) and 65% in the <1.5 µm fraction in Iturriaga and Mitchell (1986).

The values reported by these investigators are, in general, considerably higher than the values reported here for the contribution of *Synechococcus* to oceanic primary productivity. Some of the discrepancy can be related to the size classes examined.

The <5 and <3 μm fractions may often be numerically dominated by *Synechococcus* but will also contain appreciable numbers of small eucaryotic phytoplankters that may be responsible for a considerable portion of the primary productivity associated with these size classes.

The higher values reported in some of the studies using <1 μm fractions and <1.5 μm fractions may be attributable to several factors. In some studies, the values reported are for individual depths rather than integrated values throughout the euphotic zone. It is well documented that the relative contribution of *Synechococcus* to primary productivity often increases with depth within euphotic zone (Li et al. 1983; Platt et al. 1983; Glover et al. 1985) so that values as high as 50% or 60% for the contribution of *Synechococcus* to primary productivity at the 1% light level can be achieved. In other cases the discrepancies may reflect differences in methodology or may represent instances where *Synechococcus* contributed more to total primary productivity than at the sites we have examined.

We believe that with the methodologies described here, the specific contribution of *Synechococcus* to total primary productivity can be accurately assessed. We would encourage other investigators who wish to measure the contribution of *Synechococcus* to primary productivity to adopt the following procedures:

1) Choose a size fraction that contains only *Synechococcus* and report how it was prepared.
2) Monitor the abundance of *Synechococcus* and other phytoplankters by direct cell counts using epifluorescence microscopy.
3) Make time course rate measurements over short time periods.
4) Report values in fg carbon fixed *Synechococcus* $\cdot$ cell^{-1} $\cdot$ h^{-1}

SECTION IV

Patterns of *Synechococcus* Abundance

Synechococcus abundance varies not only spatially as shown in Fig. 3 and 4, but also varies temporally over a variety of time scales resulting in characteristic patterns of abundance ranging from annual to diurnal cycles.

Annual Cycle of *Synechococcus*

The annual cycle of *Synechococcus* abundance has been followed in Woods Hole Harbor since 1978 and is shown in Fig. 13. The *Synechococcus* concentration ranges from tens of cells $\cdot$ mL^{-1} in the winter to 10^5 cells $\cdot$ mL^{-1} during peak periods in the summer. The onset of the spring bloom occurs at the beginning of April when the water temperature reaches 6°C. It is seeded by the over-wintering population of *Synechococcus*, that ranges in concentration from 10 to 100 cells $\cdot$ mL^{-1}. The bloom period continues through April, May and peaks at 5×10^4 to 10^5 cells $\cdot$ mL^{-1} between the middle and end of June when the water temperature is 18°C. The population drops following the peak at the end of the bloom period and then varies between 10^4 and 10^5 cells $\cdot$ mL^{-1} from July through November. There has also been a small but predictable bloom of *Synechococcus*, except in 1982, during the middle of November when the water temperature falls to 11°C. The *Synechococcus* abundance begins to drop off rapidly during December when the water temperature falls below 6°C. Thus 6°C signals both the onset of the spring bloom and the decline of *Synechococcus* in the fall. The population remains low from January through the beginning of April when the water temperature again rises above 6°C signalling the onset of the spring bloom.

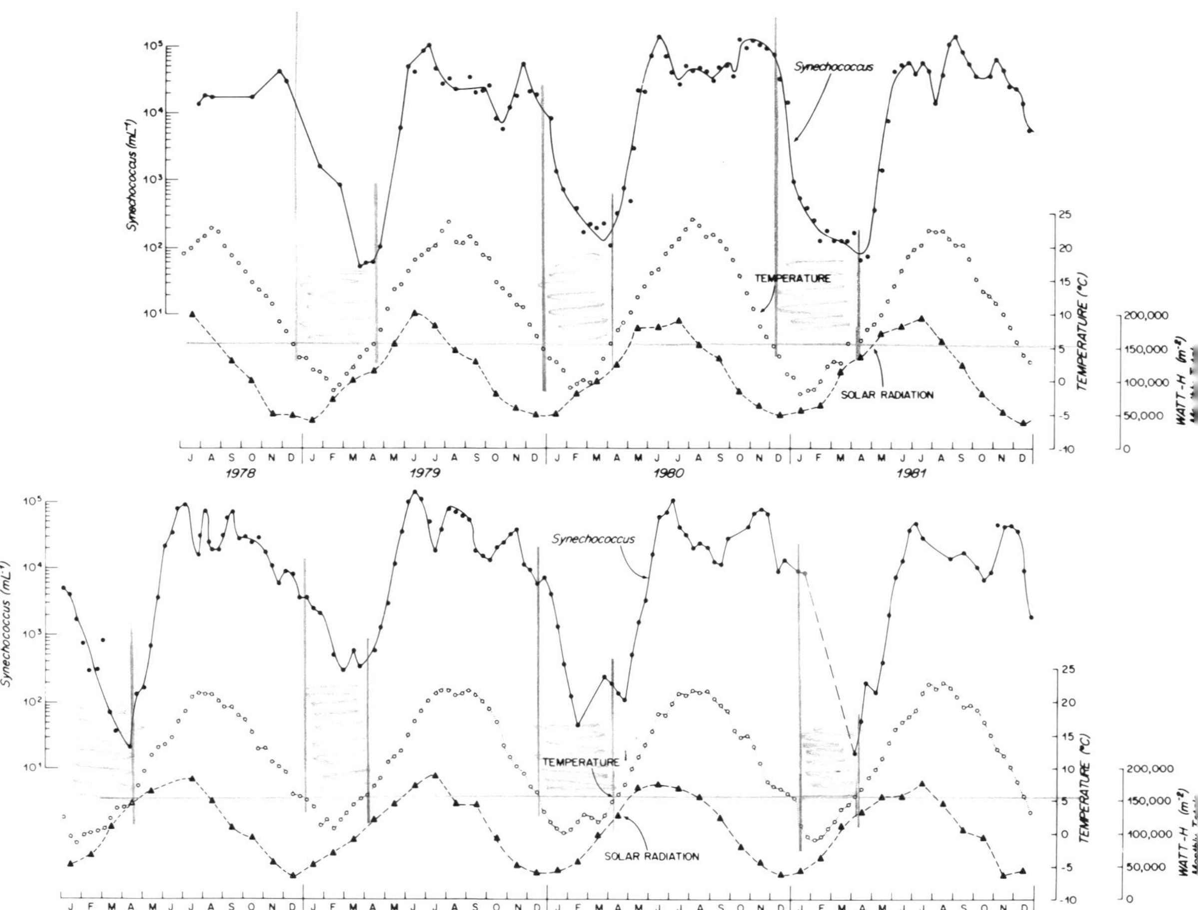

FIG. 13. Annual cycles of *Synechococcus* abundance in Woods Hole Harbor, 1978–85. *Synechococcus* abundance, determined from surface samples, is compared to surface water temperature and to monthly totals of solar radiation. Throughout much of each annual cycle the individual points are weekly averages from several counts.

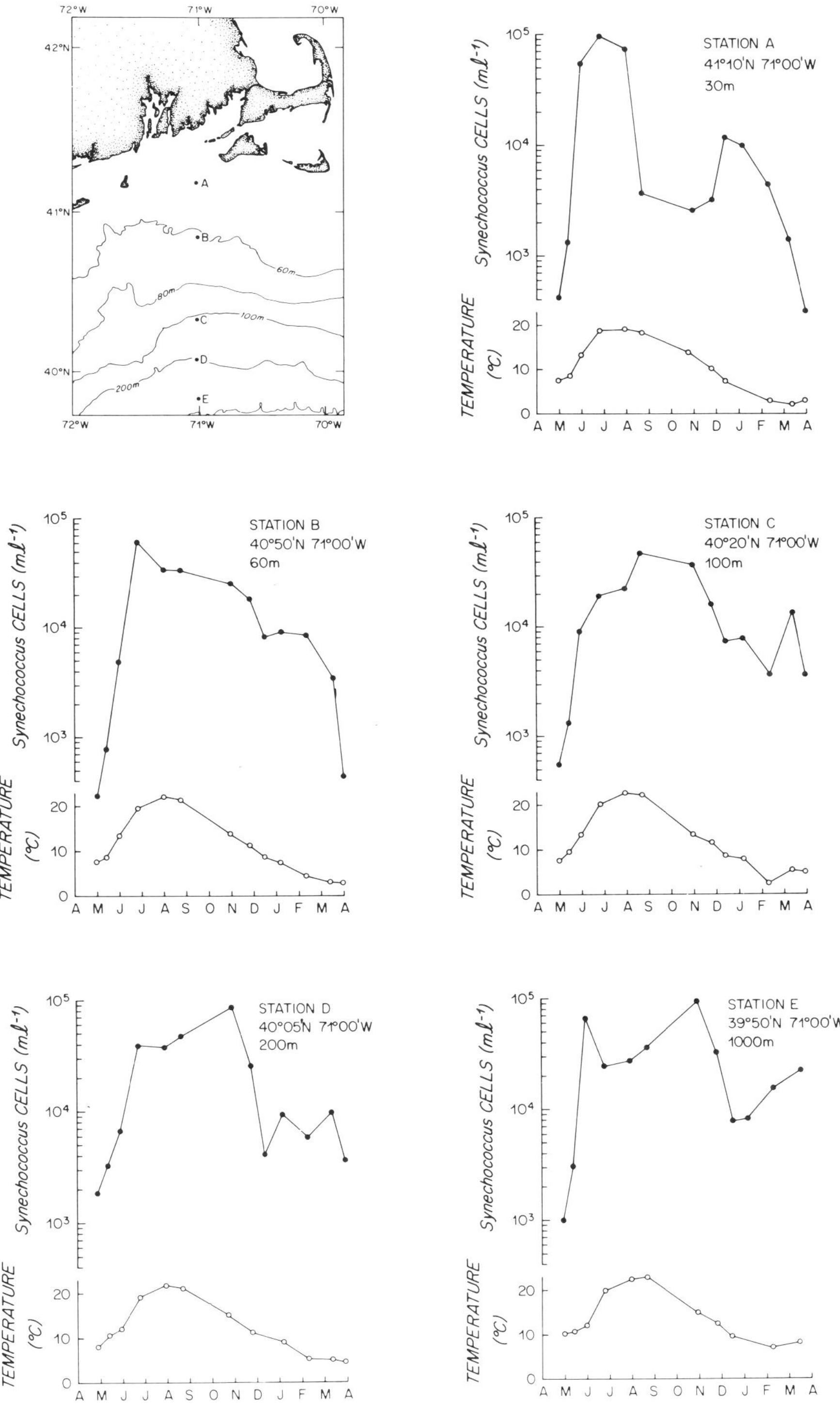

FIG. 14. Annual cycles of *Synechococcus* abundance and water temperature in surface waters at five stations on the continental shelf and slope south of Cape Cod, Massachusetts, 1981–82.

105

The annual cycle of *Synechococcus* abundance was also recorded at a series of stations in shelf and slope water south of Cape Cod, Massachusetts from April 1981 to March 1982 (Fig. 14). The general features of the three shelf stations (A–C) were similar to those of the annual cycle documented for Woods Hole Harbor (Fig. 13). A spring bloom in May and June was followed by a peak in *Synechococcus* abundance ranging from 5×10^4 to 10^5 cells $\cdot$ mL^{-1} during July and August coincident with the annual maximum water temperature. The annual minimum concentration of *Synechococcus* of several hundred cells $\cdot$ mL^{-1} occurred during the month of April at the three shelf stations.

The spring bloom of *Synechococcus* also occurred in May and June at the shelf break station D and the slope station E. The maximum abundance of *Synechococcus* 10^5 cells $\cdot$ mL^{-1} at the two offshore stations was similar to the inshore stations, but it occurred in November, two to three months after the annual water temperature maximum. This was coincident with the fall bloom that has been present in all but one year in Woods Hole Harbor (Fig. 13). The annual minimum cell concentration at the two offshore stations did not fall below 10^3 cells $\cdot$ mL^{-1}, probably as a result of a moderation of the minimum temperature in the annual cycle of water temperature.

There is a remarkable degree of repeatability from year to year in the annual cycle of *Synechococcus* abundance documented in Woods Hole Harbor. This is especially evident for the threshold of 6°C that both initiates the spring bloom and fall decline in *Synechococcus* abundance and in the timing of the peaks of abundance at the end of the spring bloom in late June and the end of the growth season in late November. However, there are both differences in temporal patterns and ranges in cell concentration evident in the annual cycles of *Synechococcus* at different geographic locations (Krempin and Sullivan 1981; El Hag and Fogg 1986) and in different hydrographic regions (e.g., shelf and slope water, Fig. 14). Both the high degree of repeatability from year to year at particular sites and the differences observed between geographic locations and water masses in the annual cycles of *Synechococcus* abundance seem to be closely correlated with the annual cycles of water temperature (El Hag and Fogg 1986).

Even though certain major features of the annual cycle of *Synechococcus* abundance in Woods Hole show a remarkable degree of repetition, there are other periods during the annual cycle which show considerable variability from year to year, for example, the year to year variations in the spring bloom and the variability in *Synechococcus* concentration that occurs during the summer months.

The dynamics of the spring bloom of *Synechococcus* in Woods Hole Harbor is shown in more detail in Fig. 15. A striking and repeatable feature of the bloom is the 3–8 week period during which the *Synechococcus* concentration increases exponentially; although the onset, duration and rate of increase in cell number varies from year to year (Table 4). The onset of the period of exponential cell increase varies from 16 April to 25 May with the average on 10 May. Its duration varies from 18 to 56 days with an average of 35 days and the net population doubling time varies from 3 to 6 days with an average of 4.3 days.

A major component of the variability in the spring bloom in Woods Hole can almost certainly be attributed to year to year differences in weather patterns as shown by differences in the monthly totals of solar radiation and the resulting annual variation in water temperataure (Fig. 13 and 15).

Short-Term Variation in *Synechococcus* Abundance

Patterns of *Synechococcus* abundance may also vary dramatically over periods of days as a result of short-term weather patterns. The effects of an individual storm event on the spring bloom in 1981 can be seen in the insert of Fig. 15. The *Synechococcus* population had been increasing exponentially with a net doubling time of 3.25 days. Following a storm on 24 and 25 May the net doubling time decreased to

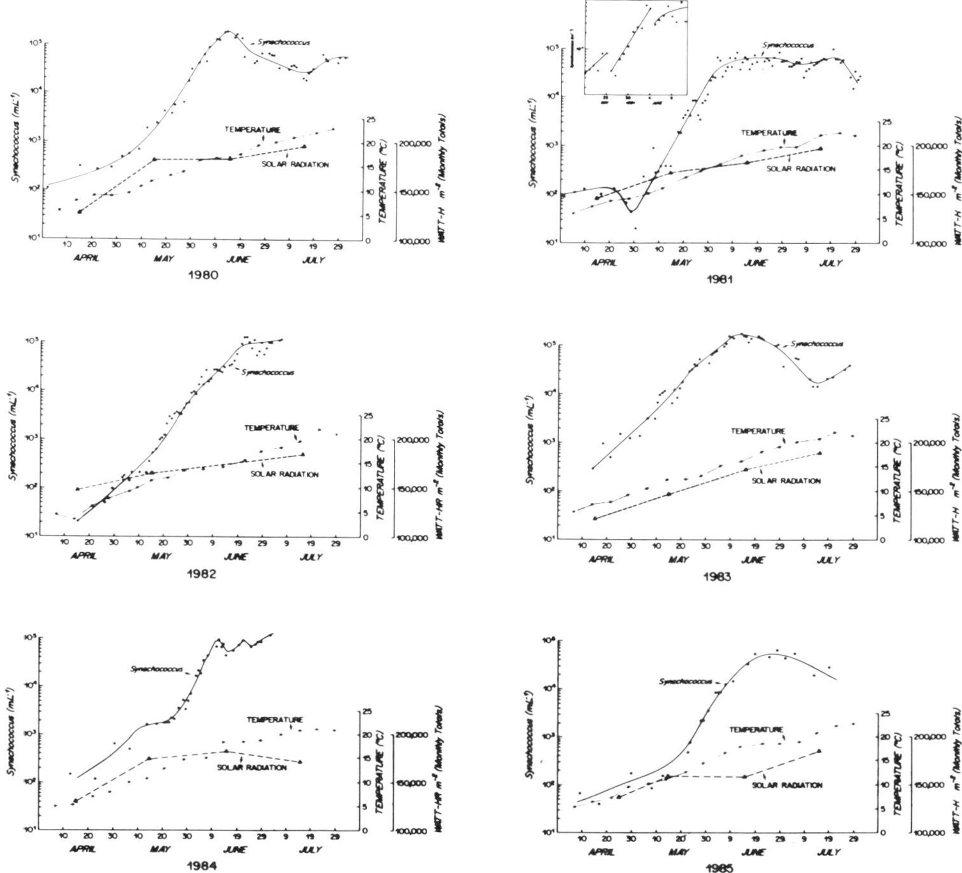

FIG. 15. *Synechococcus* abundance during spring blooms in Woods Hole waters, 1980–85. Variation in onset, duration and the rate of exponential cell increase are compared with water temperature and solar radiation. Insert in 1981 shows the effect of a severe storm on the rate of population increase.

TABLE 4. Spring blooms (Woods Hole Harbor).

Year	Onset of exponential phase	Duration of exponential phase (days)	Net doubling time (days)
1980	10 May	35	4.5
1981	30 April	36	3.25
1982	8 May	43	5.0
1983	16 April	56	6.0
1984	25 May	18	3.0
1985	20 May	30	4.25
Average:	10 May	35	4.3

2 days for a period of 10 days probably as a result of an influx of nutrients and/or a perturbation of the *Synechococcus* grazers.

In another example, *Synechococcus* abundance was followed during the spring bloom of 1982 in Buzzards Bay, Massachusetts (Fig. 16). Surface samples were collected several times daily at Stony Beach and were monitored for *Synechococcus* abundance and the percentage of dividing cells within the population. The bloom had a net doubling time of 3.5 days from 29 May–8 June which is typical of the doubling times seen in Woods Hole (Fig. 15 and Table 4). Between 1–2 June the population decreased sharply and then increased exponentially for two days (2–3 June) with a net doubling time of 1.4 days. A similar pattern followed between 4–8 June, the populations dropped sharply on 4–5 June then increased exponentially for three days with a net doubling time of 1.2 days. In both cases the sharp declines in the *Synechococcus* population occurred during periods of moderate rainfall. The two periods of increased net growth took place following the rainfall and occurred in the first case (2 and 3 June) during sunny weather and in the second case (5–8 June) during very overcast days, suggesting that these increases in net growth of the *Synechococcus* were not associated with the amount of solar radiation reaching the *Synechococcus* population. It seems more likely that the short term increases in the net growth rate might have occurred as the result of nutrient input associated with the rainfall, possibly as the result of nitrogen-enriched acid rain as shown recently by Paerl (1985).

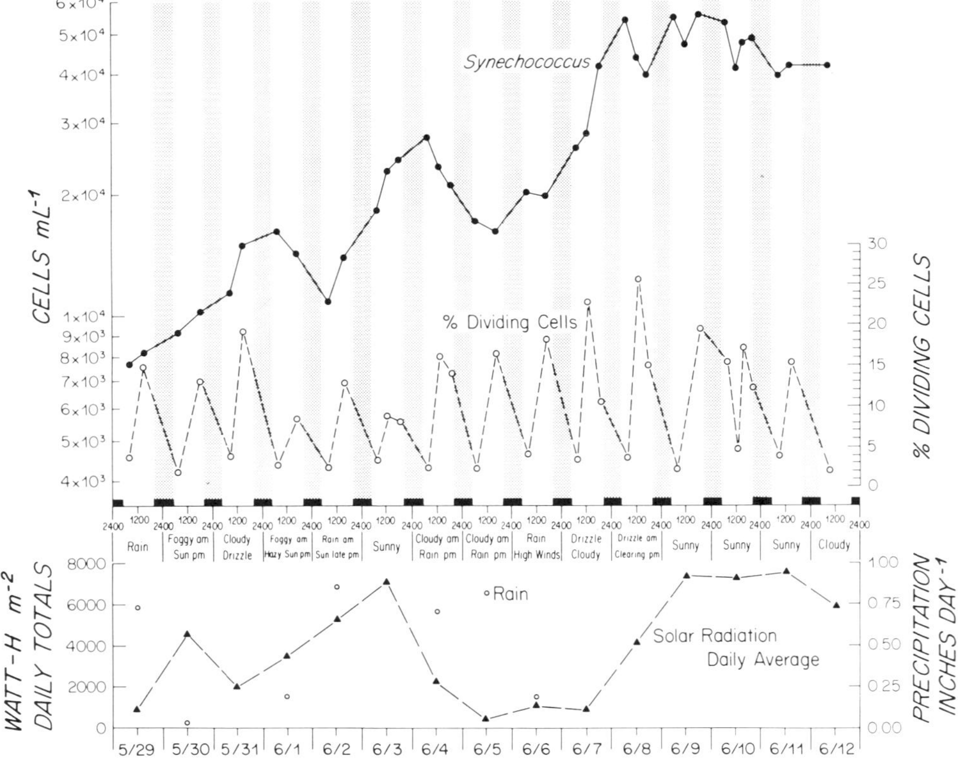

FIG. 16. *Synechococcus* abundance and percentage of dividing cells in surface water of Buzzards Bay, Woods Hole, Massachusetts taken at frequent intervals between 29 May–12 June 1982. Cell counts and percentage of dividing cells are compared to weather conditions including daily totals of solar radiation and rainfall.

Figure 17 is an example of a situation in which the percentage of cloud cover affects the population dynamics of *Synechococcus*. This is in contrast to the situation in Fig. 16 where there did not appear to be a correlation between changes in the amount of solar radiation and variations in *Synechococcus* growth rates. Figure 17 was documented from surface samples collected at frequent intervals in Vineyard Sound between 14–17 September 1981. The first day (14 September) and the last day (17 September) of this series were sunny with 42 and 44% cloud cover, respectively. The two intervening days (15 and 16 September) were very overcast with 85 and 93% cloud cover, respectively. The percentage of dividing *Synechococcus* cells in the population has a characteristic diurnal pattern (discussed in detail in a subsequent section) which typically shows a peak in the late afternoon. The magnitude of this peak was a good relative indicator of *Synechococcus* growth rate (refer to the Chapter by S. W. Chisholm in this volume). On the first sunny day the peak of dividing cells reached 29% and on the ensuing progressively cloudy days the percentage of dividing cells at the peak dropped to 23% and 12% and then increased to 32% on the final sunny day. *Synechococcus* abundance decreased concomitantly with the decline in the percentage of dividing cells over the first three days. Unfortunately the experiment was terminated before it could be determined if the population would have increased following the increase in solar radiation on the fourth day, but it seems clear that the percentage of cloud cover had a dramatic effect on *Synechococcus* growth rate as evidenced by the change in the percent dividing cells.

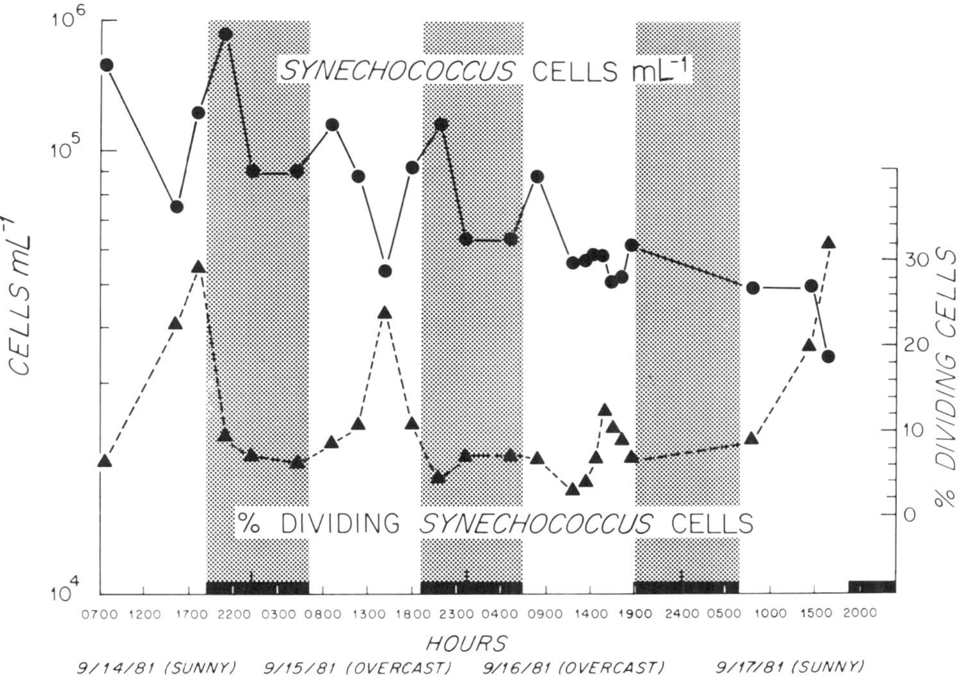

FIG. 17. Diurnal pattern of *Synechococcus* abundance and percent dividing cells in surface water collected in Woods Hole Harbor between 14 and 17 September 1981. The decrease of both the whole population and the percent dividing cells is compared to the daily percent cloud cover. The percentages of cloud cover and the daily total solar radiation (watt-h m²) for each consecutive day were: 42%, 4787; 85%, 1290; 93%, 592; 44%, 4497.

The Diurnal Cycle of *Synechococcus*

The diurnal cell cycle of *Synechococcus* has been characterized with a pure culture (Strain WH 7803) in a 14-h light/10-h dark cycle (Fig. 18). It is provided as a basis for comparison with the features of the diurnal cell cycles of *Synechococcus* observed in natural waters.

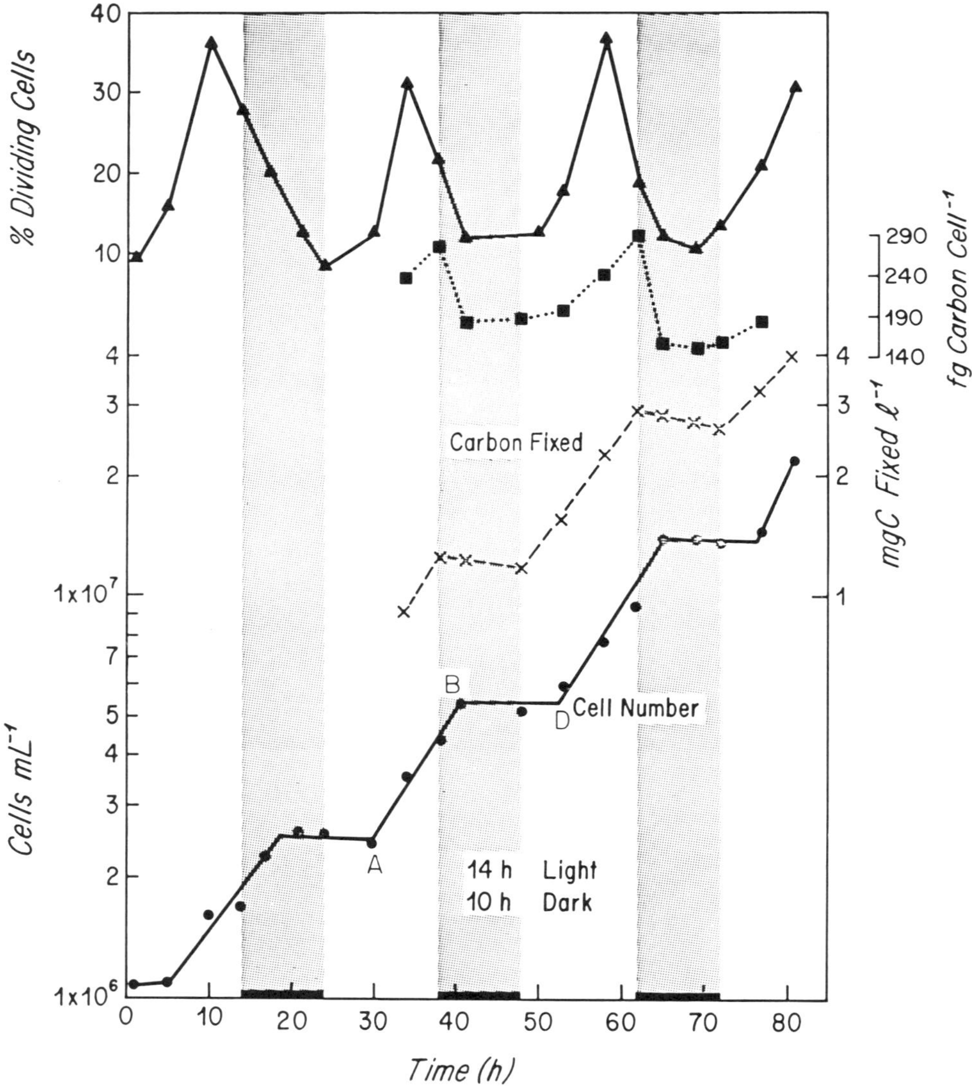

FIG. 18. Diurnal cycle of *Synechococcus* (Strain WH 7803) grown in a 14-h light (40 μE • m^{-2} • s^{-1})/10-h dark cycle at 25°C. Cell counts and percent dividing cells were monitored by direct counts using epifluorescence microscopy. Carbon fixation was determined by ^{14}C sodium bicarbonate assimilation. The radioisotope was added when the cell count was 5×10^5 cells • mL^{-1} so that the cells were equilibrium labeled by hour 35. Carbon • cell^{-1} was determined by CHN analysis on a Perkin Elmer elemental analyzer. Lettered points A, B and D are explained in the text.

The major feature of the diel cycle of *Synechococcus* was its highly repeatable pattern of discontinuous growth. In general, cell number increased exponentially during the light period and remained constant during the dark period. More precisely, cell number began to increase exponentially 4 hours after the beginning of the light period at point A. Exponential growth continued throughout the remainder of the light period and for the initial 3 hours of the dark period (point B). Increase in cell number then ceased and did not resume until 4 hours after the onset of the light period at point D, which marked the beginning of the next diel cycle.

The percentage of dividing cells also showed a characteristic diel cycle. A dividing cell is defined here as a cell in the course of division where median constriction, caused by cross wall formation, is evident, but where the two resulting daughter cells have not yet separated. The percentage of dividing cells was lowest (approximately 10%) at the beginning of the light period. It then increased linearly during the light period, reaching a maximum of 30–35% dividing cells 4 hours before the end of the light period. The point of maximum cell division coincided with the midpoint in the exponential growth phase of the cell cycle.

As previously mentioned, CO_2 fixation in cyanobacteria is a tightly regulated process that occurs in the light but not in the dark (Stanier and Cohen-Bazire 1977; Smith 1982). Thus, in contrast to the increase in cell number, the incorporation of ^{14}C sodium bicarbonate into cell material occurred only during the light period and, in this instance, increased exponentially in parallel with the increase in cell number because the cells were uniformly labeled. The isotope was added when the cell density was 5×10^5 cells mL^{-1}. Dark respiration in *Synechococcus* was small accounting for the loss of approximately 5% of the carbon fixed during each light period (Fig. 18).

Changes in the amount of carbon per cell during the diel cycle were measured using a Perkin Elmer CHN analyzer. The values range from 150–290 fg C • cell^{-1} with the lowest values occurring late in the dark period and early in the light period when the percentage of dividing cells and the intracellular glycogen pool were at a minimum. Correspondingly the highest values occurred at the end of the light period when the percentage of dividing cells and the intracellular glycogen pool were at a maximum.

Growth of *Synechococcus* is expressed here as doublings of cell number • day^{-1}. As a result of the discontinuous diel growth cycle, the exponential increase in cell number occurs during a portion of the light and early dark periods (Fig. 18 between points A and B) whereas no increase in cell number occurs during most of the dark and early light periods (Fig. 18 between points B and D). The diel growth rate (i.e., growth over a 24-h cycle) is calculated here as the observed growth rate, expressed by line AB, multiplied by the fraction of the diel cycle over which it occurs. In the case of the pure culture shown in Fig. 18, the diel growth rate of 1.23 doublings • day^{-1} was calculated by multiplying the observed rate between points A and B of 2.28 doublings • day^{-1} by 13/24, which corresponds to the 13 h of the 24-h cycle over which rate AB was expressed. The same diel growth rate may be obtained by calculating the doublings • day^{-1} expressed by a line drawn between points A and D which corresponds to one 24-h cycle.

In describing Fig. 17 we indicated that the diel cycle in the percentage of dividing cells can be used as a relative indicator of *Synechococcus* growth rate. When *Synechococcus* is actively growing there is a marked difference between the percentage of dividing cells found at the beginning of the light period (the daily minimum) and in the latter half of the light period (the daily maximum). This is particularly evident for the pure culture (Fig. 18) and in the natural populations shown in Fig. 16, 17, 20–22. Figure 19 illustrates the difference between the percentage of dividing cells in the Woods Hole *Synechococcus* population during the morning and afternoon over an annual cycle. During the periods when *Synechococcus* was actively growing there was a marked difference between the percent dividing cells present in the morning and afternoon. During the fall when growth in the population was decreasing, and in the spring at the beginning of the bloom the difference between the percentage of dividing cells present in the morning and afternoon decreased and remained relatively constant at 10–15% dividing cells throughout the diel cycle. Dividing cells were rarely observed during the winter months when the water temperature fell below

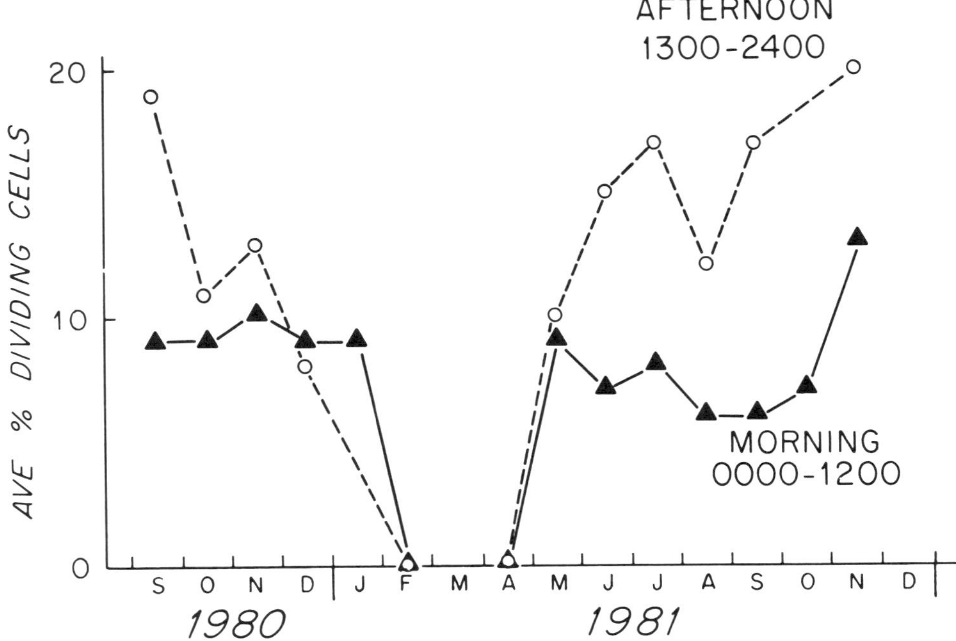

FIG. 19. The annual distribution in 1980-81 of the percent dividing cells in the *Synechococcus* population counted during the morning and afternoon from Woods Hole Harbor surface samples.

6°C. Thus, monitoring the percent dividing cells present throughout the diel cycle in natural populations of *Synechococcus* can provide a good qualitative indicator of growth. A marked difference in the percentage of dividing cells between morning and afternoon indicates that the population is actively growing. If the percentage of dividing cells remains relatively constant at 10-15% throughout the day, it is a good indication that the population is senescent.

The in situ diurnal cell cycle of *Synechococcus* has been examined at three sites (Fig. 20-22). The individual cycles were generated from surface samples collected at frequent intervals over several diel cycles and were documented by direct cell counts using epifluorescence microscopy.

Figure 20 shows the diurnal cycle of *Synechococcus* recorded between 29 April and 1 May 1984 at a site in the northern Sargasso Sea (35°N, 65°W). The in situ diurnal cycle of *Synechococcus* possessed many of the characteristic features of the diurnal cycle shown by the pure culture illustrated in Fig. 18. Like the pure culture, cell numbers did not begin to increase until 5 hours after dawn, growth was then exponential for the remainder of the light period and for 2 hours following dusk. The most striking departure from the pure culture cycle is the sharp decline in the in situ *Synechococcus* abundance during the remainder of the night and for the first 5 hours of the light period. This decline in abundance can almost certainly be attributed to grazing. As in the pure culture, the percentage of dividing cells was lowest at dawn (about 2-3%), and increased linearly during the day to a maximum of 12-15% 3 hours before dusk at a point coincident with the mid-point in the exponential growth phase between points A and B.

We made two assumptions in order to calculate the in situ diel growth rate of *Synechococcus* from cell numbers at this site. First, the in situ population possessed a discontinuous growth pattern, similar to that displayed by the pure culture where no growth occurred between points B and D. As a result, the rate expressed by a

112

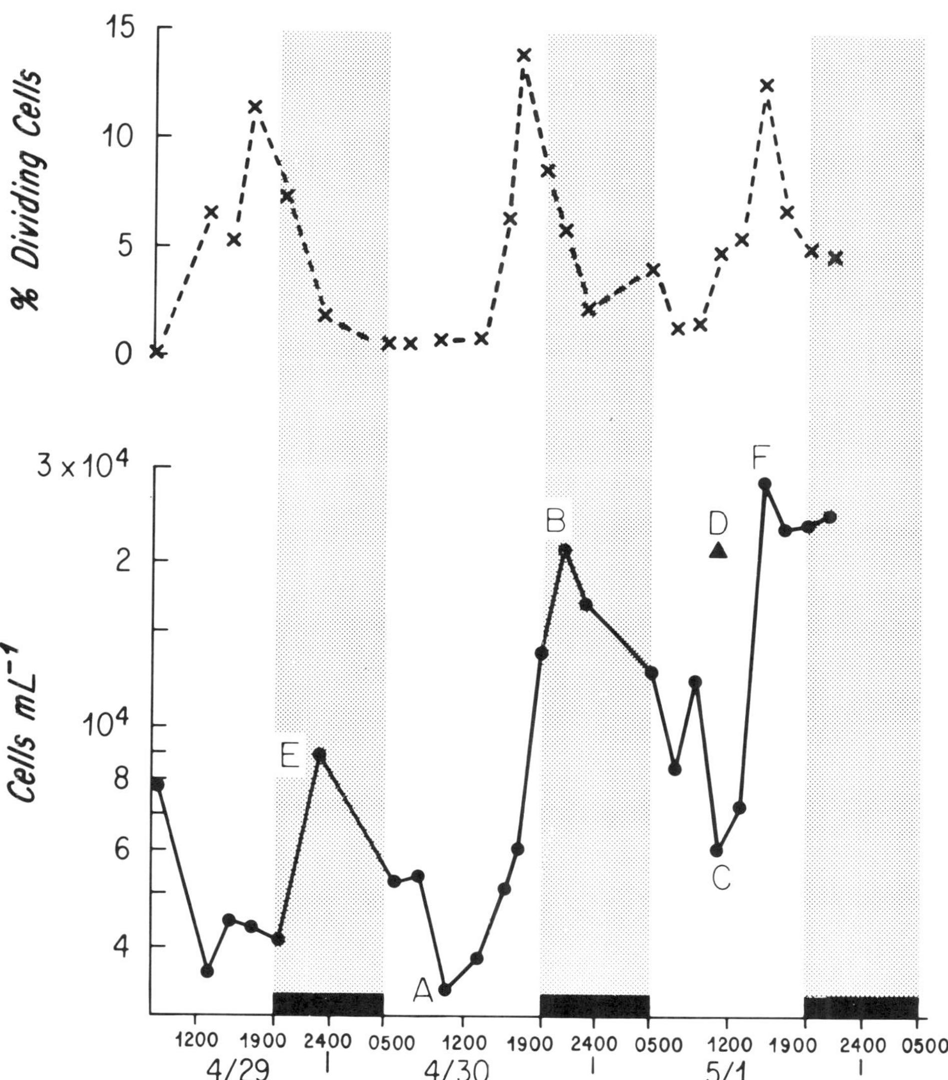

FIG. 20. Diurnal cell cycle and percent dividing cells determined by direct cell counts from surface samples collected at frequent intervals from 29 April–1 May 1984 at a site in the northern Sargasso Sea (35°N, 65°W). Lettered points A–F are explained in the text.

line drawn between B and C equals the observed grazing rate. The second assumption, that grazing at this site was discontinuous and occurred only during the latter portion of the dark period and for the first part of the light period, is less obvious. This assumption was made because to have assumed continuous grazing pressure would have resulted in growth rates that we believe would be unrealistically high for a marine *Synechococcus*. We are unaware of examples in the literature where an organism as small as *Synechococcus* was grazed discontinuously. It will be of considerable interest to examine the feeding patterns of the *Synechococcus* grazers, especially small flagellates, to determine if their feeding patterns show periodicity.

Synechococcus at this Sargasso site achieved a diel growth rate of 2.9 doublings • day^{-1} assuming that grazing was discontinuous. This rate was obtained by multiplying the observed growth rate between points A and B of 6.4 doublings • day^{-1} by 11/24 corresponding to the 11-h of the 24-h cycle over which rate AB was expressed. Similarly the grazing rate of -1.7 doublings • day^{-1} was calculated by multiplying rate BC (-3.2 doublings • day^{-1}) by 13/24. The diel growth and grazing rates of 2.9 and -1.7 doublings • day^{-1}, respectively, resulted in a calculated net population increase of 1.2 doublings • day^{-1} and an observed net population increase of 1.1 doublings • day^{-1} expressed by a line from E to F.

By comparison, if it were assumed that grazing was continuous, the observed growth rate AB would equal the actual growth rate plus the grazing rate. The resulting actual growth rate of 9.6 doublings • day^{-1} would be determined by subtracting the observed grazing rate (-3.2) from the observed growth rate (6.4). This would result in a diel growth rate of (9.6 × 11/24) 4.4 doublings • day^{-1}, a rate that we feel is unrealistically high.

The in situ diel growth rate of *Synechococcus* has also been estimated from ^{14}C sodium bicarbonate incorporation studies conducted at the same time and site where the diurnal cell cycle was examined (Fig. 7–9). The specific carbon uptake rates (μ') were estimated from the following expression:

$$\mu' = \frac{1}{\Delta t} \ln \left[\frac{C \text{ cell}^{-1} + \Delta C^{14}\text{cell}}{C \text{ cell}^{-1}} \right]$$

where C = carbon, ΔC is the ^{14}C fixed • cell^{-1} during time Δt determined from the syringe incubated time course measurements of ^{14}C sodium bicarbonate incorporation discussed in the previous section. In calculating μ' we have used Δt in hours, resulting in μ' having units of h^{-1} rather than day^{-1} as used by Iturriaga and Mitchell (1986). The time interval (Δt) must be shorter than the doubling time of the organism to avoid introducing a considerable error into the calculation of the carbon specific growth rate (μ').

It is assumed for these calculations that the rate of carbon fixed • cell^{-1} • h^{-1} was constant throughout the light period. This assumption almost certainly introduces some error, but seems reasonable because marine *Synechococcus* both saturates for photosynthesis at low irradiance and does well over nearly the entire range of irradiances encountered within the euphotic zone (see Fig. 9 and 12). The weighted mean value of 210 fg of carbon per *Synechococcus* cell was determined from the data shown in Fig. 18. The ^{14}C estimated diel growth rates were calculated from the carbon specific uptake rates using the following conversion:

$$\text{doublings day}^{-1} = \frac{\mu' \bullet \text{ hours of daylight}}{0.693}$$

A ^{14}C estimated diel growth rate of 2.8 doublings • day^{-1} was calculated for the northern Sargasso Sea site. This calculation was made using a weighted mean value of 31 fg C fixed • *Synechococcus* cell^{-1} • h^{-1} (Fig. 9) and 14 h for the duration of the light period (Fig. 20). The weighted mean was determined by integrating the values of ^{14}C fixed • cell^{-1} • h^{-1} from the surface to 40 m (Fig. 9). This value was used rather than the surface value of 40 fg C • cell^{-1} • h^{-1} because we felt it more closely represented the carbon fixed by *Synechococcus* cells that were moving within the euphotic zone as a result of wind driven mixing. The value of 2.8 doublings • day^{-1} calculated from the ^{14}C data agrees well with the doubling time of 2.9 doublings • day^{-1} from the in situ cell counts. However, it must be emphasized that small variations in the values determined for the amount of carbon per *Synechococcus* cell and the rates of ^{14}C incorporation • cell^{-1} • h^{-1} will change the resulting growth rate significantly.

114

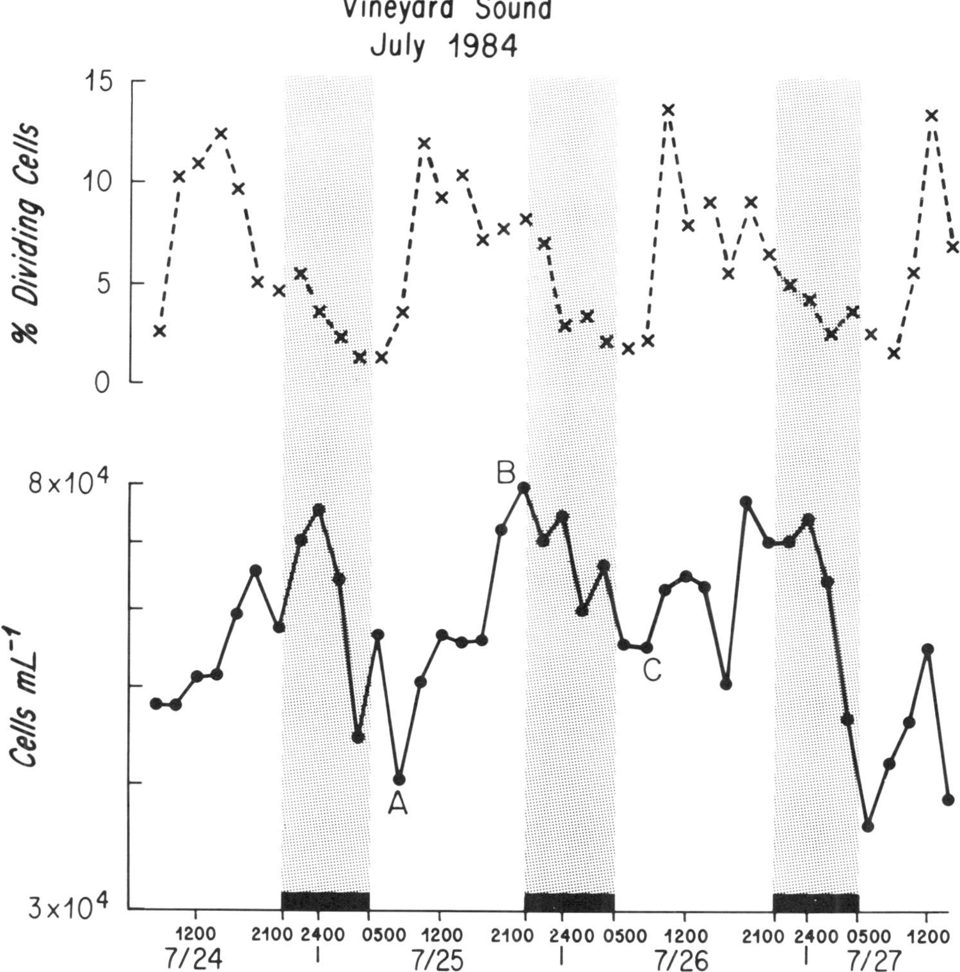

FIG. 21. Diurnal cell cycle and percent dividing cells determined by direct cell counts from surface samples collected at frequent intervals from 24–27 July 1984 in Vineyard Sound, Woods Hole, Massachusetts. Lettered points A–C are explained in the text.

The in situ diurnal cell cycle of *Synechococcus* has also been documented in Vineyard Sound, Massachusetts on two occasions, 24–27 July 1984 (Fig. 21) and 3–6 September 1985 (Fig. 22). On both occasions the major features of the diel cell cycle were similar to those documented in the pure culture (Fig. 18) and at the site in the northern Sargasso Sea (Fig. 20). The regions of cell growth (AB) and grazing (BC) are clearly discernible even though there is more noise in the data probably due to considerable tidal mixing.

The diel pattern of dividing cells also showed similarities with both the pure culture (Fig. 18) and the site in the Sargasso Sea (Fig. 20) but differed in one important respect. The peak of dividing cells was shifted toward the middle of the light period and no longer coincided with the midpoint in the exponential growth phase of the cycle between points A and B. The shape of the diel cycle of dividing cells gave the impression that its peak had been shifted to earlier in the light period. The possibility exists that some grazers may have been preferentially cropping the larger dividing cells causing the perturbation in the diel cycle of dividing cells. Qualitatively, it has been observed in Woods Hole waters that a wide variety of protozoans and small

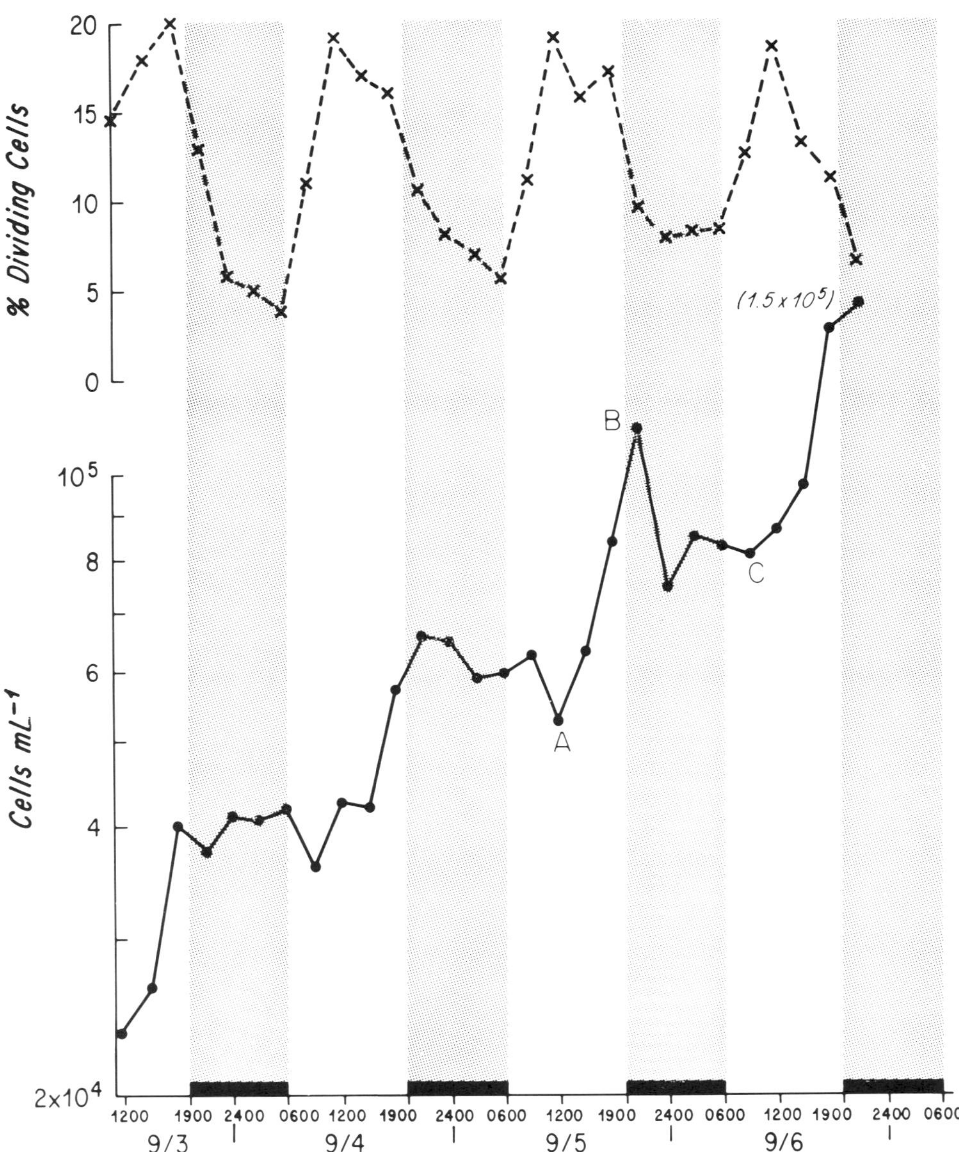

FIG. 22. Diurnal cell cycle and percent dividing cells determined by direct cell counts from surface samples collected at frequent intervals from 3–6 September 1985 in Vineyard Sound, Woods Hole, Massachusetts. Lettered points A–C are explained in the text.

invertebrates are active grazers of *Synechococcus*. This has led us to assume that grazing was probably continuous at the times when the Vineyard Sound cycles were documented. The assumption that grazing was discontinuous at the Sargasso Sea site and continuous in Vineyard Sound seems reasonable considering the diversity of *Synechococcus* grazers in coastal waters. Further experiments will be necessary to establish whether grazing of *Synechococcus* is continuous, discontinuous, or variable.

116

The following calculations of the in situ diel growth rates of *Synechococcus* at the Vineyard Sound sites were made with the assumption that grazing was continuous throughout the diel cycle. In July 1984, the observed growth rate (AB) of 2.0 doublings • day^{-1} consisted of the actual growth rate plus the grazing rate (Fig. 21). The resulting actual growth rate of 3.0 doublings • day^{-1} was determined by subtracting the grazing rate (BC) of -1.0 doublings • day^{-1} from the observed growth rate (AB) of 2.0 doublings day. This resulted in a diel growth rate for *Synechococcus* of (3.0 × 12/24) 1.5 doublings • day^{-1}.

A ^{14}C estimated diel growth rate of 2.7 doublings • day^{-1} by *Synechococcus* in Vineyard Sound on 25 July 1984 was calculated using a weighted mean value of 26 fg C • *Synechococcus* cell^{-1} • h^{-1} calculated by integrating the ^{14}C fixed • cell^{-1} • h^{-1} from 0–10 m (Fig. 12) and 16 h for the duration of the light period (Fig. 21). We felt that integrating the fg C • cell^{-1} • h^{-1} more closely represented the carbon fixed by *Synechococcus* cells that were being moved rapidly throughout the euphotic zone as a result of the vigorous tidal mixing that occurs in Vineyard Sound. The diel growth rate of 2.7 doublings • day^{-1} estimated from the ^{14}C incorporation data is considerably higher than the diel growth rate of 1.5 doublings • day^{-1} calculated from the direct cell counts. One possible explanation for this discrepancy is that the weighted mean value of 26 fg ^{14}C fixed • *Synechococcus* cell^{-1} • h^{-1} is an overestimate. *Synechococcus* in Vineyard Sound may spend an appreciable amount of time at very low light intensities as the result of strong tidal mixing and the shallow euphotic zone.

During the diel cycle of *Synechococcus* documented during September 1985 (Fig. 22), the observed growth rate (AB) of 3.2 doublings • day^{-1} consisted of the actual growth rate plus the grazing rate. The actual growth rate of 4.1 doublings • day^{-1} was determined by subtracting the grazing rate (BC) of -0.9 doublings • day^{-1} from the observed rate of 3.2 doublings • day^{-1}. This resulted in a diel growth rate for *Synechococcus* of (4.1 × 11/24) 1.9 doublings • day^{-1}. ^{14}C incorporation studies were not conducted during the period when the September diel cycle was documented.

Summary

Synechococcus abundance has been shown to vary temporally over a variety of time scales resulting in patterns of abundance ranging from annual to diurnal cycles. The annual cycle of *Synechococcus* in Woods Hole has major features that are repeatable from year to year and closely correlated with the annual cycle of water temperature. Other periods in the annual cycle varying from weeks to days show year-to-year variation attributable to differences in weather patterns. These range from variations in the dynamics of the spring bloom to changes in the net population dynamics over periods of days caused by storms, changes in the percentage of cloud cover and rainfall.

Synechococcus has also been shown to have a dramatic diel cycle of cell abundance characterized by a repeatable pattern of discontinuous growth. It was possible to determine both growth and grazing rates from natural populations by direct cell counts made from samples collected in situ at frequent intervals throughout the diel cycle.

Growth and grazing rates calculated from direct cell counts have shown that *Synechococcus* was capable of rapid growth in both coastal and offshore waters during certain periods of the year and that *Synechococcus* was being actively grazed at rates comparable to their growth rates. The rates measured here by cell counts have the advantage of directly and specifically measuring growth and grazing rates of *Synechococcus* on unmanipulated in situ populations. The only uncertainty in these measurements is the duration of grazing during the diel cycle.

Growth rates for picoplankton and *Synechococcus* have been estimated from increases in chlorophyll concentration (Bienfang andd Takahashi 1983), from $^{14}CO_2$

incorporation (Douglas 1984; Iturriaga and Mitchell 1986) and by direct cell counts in diffusion chambers (Landry et al. 1984). Grazing losses of *Synechococcus* have been calculated by the dilution technique in diffusion chambers (Landry et al. 1984) and by the incorporation of radioisotope from labeled *Synechococcus* cells (Iturriaga and Mitchell 1986). These techniques suffer from one or more of the following deficiencies: they are not specific for *Synechococcus*, they require confinement and/or manipulation of the natural population, they require conversion factors that are not determinable for natural populations. For example, growth rates estimated from the incorporation of ^{14}C sodium bicarbonate are unreliable because small changes in the parameters used in their estimation (i.e., the amount of carbon *Synechococcus* cell^{-1} and the rate of ^{14}C fixed *Synechococcus* cell$^{-1} \cdot h^{-1}$) result in considerable changes in growth rate.

Conclusion

The analysis of the ecological role of *Synechococcus* in the oceans has been possible as a result of a unique set of properties: *Synechococcus* can be cultured and studied physiologically in the laboratory, it can be counted directly and specifically in situ, and it can be physically separated from other microorganisms in natural samples. Using these properties, studies have been conducted that have shown that *Synechococcus* is a major component of the picoplankton that, as a result of high in situ growth rates, is capable of contributing significantly to primary productivity in both coastal waters and the open ocean.

Many details of their biology and ecology remain to be studied. In particular, the grazing of *Synechococcus*, by what now appears to be a diverse group of microorganisms, is one of a major ecological problems yet to be resolved.

Acknowledgements

We thank R. R. L. Guillard, L. E. Brand, and R. J. Olson for providing enrichments containing *Synechococcus*, P. H. Wiebe for providing samples from his Slope Water Time Series (NSF OCE-80-19055), R. E. Payne for supplying us with the solar radiation and meteorological data, Hans Jannasch for sampling in Antarctica and numerous students who helped on cruises. This work was supported by grants from NSF (OCE-) 78-19595, 80-19057, 82-14889, and 84-16960. Woods Hole Contribution Number 6250.

References

ALBERTE, R. S., M. WOOD, T. A. KURSAR, AND R. R. L. GUILLARD. 1984. Novel phycoerythrins in marine *Synechococcus* spp.: Characterization and evolutionary and ecological implications. Plant Physiol. 75: 732–739.

BIENFANG, P. K., AND M. TAKAHASHI. 1983. Ultraplankton growth rates in a subtropical ecosystem. Mar. Biol. 76: 213–218.

BROCK, T. D. 1973. Evolutionary and ecological aspects of the cyanophytes, p. 487–500. *In* N. G. Carr and B. A. Whitton [ed] The biology of blue-green algae. Blackwell, Oxford.

CAMPBELL, L., E. J. CARPENTER, AND U. I. IACONO. 1983. Identification and enumeration of marine chroococcoid cyanobacteria by immunofluorescence. Appl. Environ. Microbiol. 46: 553–559.

CARPENTER, E. J. 1973. Nitrogen fixation by *Oscillatoria* (*Trichodesmium*) thiebautii in the southwestern Sargasso Sea. Deep Sea Res. 20: 285–288.

CARR, N. G., AND B. A. WHITTON. 1982. The biology of cyanobacteria. University of California Press, Berkeley and Los Angeles. 688 p.

CASTENHOLZ, R. W. 1982. Motility and taxes, p. 413–439. *In* N. G. Carr and B. A. Whitton [ed] The biology of cyanobacteria. University of California Press, Berkeley and Los Angeles.

CUHEL, R. L., AND J. B. WATERBURY. 1984. Biochemical composition and short-term nutrient incorporation patterns in a unicellular marine cyanobacterium, *Synechococcus* sp. Limnol. Oceanogr. 29: 370–374.

DAVIS, P. G., D. A. CARON, P. W. JOHNSON, AND J. McN. SIEBURTH. 1985. Phototrophic and apochlorotic components of picoplankton and nanoplankton in the North Atlantic: geographic, vertical, seasonal and diel distributions. Mar. Ecol. 21: 15–26.

DAVIS, H. C., AND R. R. L. GUILLARD. 1958. Relative value of ten genera of micro-organisms as food for oyster and clam larvae. U.S.F.W.S. Fish Bull. 136 (Vol. 58): 293–304.

DOUGLAS, D. J. 1984. Microautoradiography-based enumeration of photosynthetic picoplankton with estimates of carbon-specific growth rates. Mar. Ecol. 14: 223–228.

EL HAG, A. G. D., AND G. E. FOGG. 1986. The distribution of coccoid blue-green algae (Cyanobacteria) in the Menai Straits and the Irish Sea. Br. Phycol. J. 21: 45–54.

FOGG, G. E. 1973. Physiology and ecology of marine blue-green algae, p. 368–378. *In* N. G. Carr and B. A. Whitton [ed] The biology of blue-green algae. Blackwell, Oxford.

1982. Marine plankton, p. 491–513. *In* N. G. Carr and B. A. Whitton [ed] The biology of cyanobacteria. University of California Press, Berkeley and Los Angeles.

FOGG, G. E., W. D. P. STEWART, P. FAY, AND A. E. WALSBY. 1973. The blue-green algae. Academic Press, NY. 495 p.

GIBSON, C. E., AND R. V. SMITH. 1982. Freshwater plankton, p. 463–489. *In* N. G. Carr and B. H. Whitton [ed] The biology of cyanobacteria. University of California Press, Berkeley and Los Angeles.

GIESKES, W. W. C., G. W. KRAAY, AND M. A. BAARS. 1979. Current ^{14}C methods for measuring primary production: gross underestimates in oceanic waters. Neth. J. Sea Res. 13: 58–78.

GLAZER, A. N., J. A. WEST, AND C. CHAN. 1982. Phycoerythrins as chemotaxonomic markers in red algae: a survey. Syst. Ecol. 10: 203–215.

GLOVER, H. E. 1985. The physiology and ecology of the marine cyanobacterial genus *Synechococcus*. Adv. Aquat. Microbiol. 3: 49–107.

GLOVER, H. E., D. A. PHINNEY, AND C. S. YENTSCH. 1985. Photosynthetic characteristics of picoplankton compared with those of larger phytoplankton populations in various water masses in the Gulf of Maine. Biol. Oceanogr. 3: 223–248.

GOLDMAN, J. C 1984. Conceptual role for micro-aggregates in pelagic waters. Bull. Mar. Sci. 35: 462–476.

GOLDMAN, J. C., AND M. R. DENNETT. 1985. Susceptibility of some marine phytoplankton species to cell breakage during filtration and post filtration rinsing. J. Exper. Mar. Biol. Ecol. 86: 47–58.

GUILLARD, R. R. L. 1975. Culture of phytoplankton for feeding marine invertebrates, p. 29–60. *In* W. L. Smith and M. H. Chanley [ed.] Culture of marine invertebrate animals. Plenum Publishing, N.Y.

ITURRIAGA, R., AND B. G. MITCHELL. 1986. Chroococcoid cyanobacteria: a significant component in the food web dynamics of the open ocean. Mar. Ecol. 28: 291–297.

JOHNSON, P. W., AND J. McN. SIEBURTH. 1979. Chroococcoid cyanobacteria in the sea: a ubiquitous and diverse phototrophic biomass. Limnol. Oceanogr. 24: 928–935.

JOHNSON, P. W., AND J. McN. SIEBURTH. 1982. In situ morphology and occurrence of eucaryotic phototrophs of bacterial size in the picoplankton of estuarine and oceanic waters. J. Phycol. 8: 318–327.

JOINT, I. R., AND A. J. POMROY. 1983. Production of picoplankton and small nanoplankton in the Celtic Sea. Mar. Biol. 77: 19–27.

KREMPIN, D. W., AND C. W. SULLIVAN. 1981. The seasonal abundance, vertical distribution, and relative microbial biomass of chroococcoid cyanobacteria at a station in southern California coastal waters. Can. J. Microbiol. 27: 1341–1344.

KURSAR, T. A., H. SWIFT, AND R. S. ALBERTE. 1981. Morphology of a novel cyanobacterium and characterizations of light-harvesting complexes from it: Implications for phycobiliprotein evolution. Proc. Nat. Acad. Sci. U.S.A. 78: 1–5.

LANDRY, M. R., L. W. HAAS, AND V. L. FAGERNESS. 1984. Dynamics of microbial plankton communities: experiments in Kaneohe Bay, Hawaii. Mar. Ecol. 16: 127–133.

LI, W. K. W., D. V. SUBBA RAO, W. G. HARRISON, J. C. SMITH, J. J. CULLEN, B. IRWIN, AND T. PLATT. 1983. Autotrophic picoplankton in the tropical ocean. Science, 219: 292–295.

MURPHY, L. S., AND E. HAUGEN. 1985. The distribution and abundance of phototrophic picoplankton in the North Atlantic. Limnol. Oceanogr. 30: 47–58.

OLSON, R., D. VAULOT, AND S. W. CHISHOLM. 1985. Marine phytoplankton distributions measured using shipboard flow cytometry. Deep-sea Res. 32: 1273–1280.

ONG, L. J., A. N. GLAZER, AND J. B. WATERBURY. 1984. An unusual phycoerythrin from a marine cyanobacterium. Science 224: 80–83.

PAERL, H. W. 1985. Enhancement of marine primary production by nitrogen-enriched acid rain. Nature 316: 747–749.

PERKINS, F. O., L. W. HAAS, D. E. PHILLIPS, AND K. L. WEBB. 1981. Ultrastructure of a marine *Synechococcus* possessing spinae. Can. J. Microbiol. 27: 318–329.

PLATT, T., D. V. SUBBA RAO, AND B. IRWIN. 1983. Photosynthesis of picoplankton in the oligotrophic ocean. Nature, Lond. 301: 702–704.

PUTT, M., AND B. B. PRÉZELIN. 1985. Observations of diel patterns of photosynthesis in cyanobacteria and nanoplankton in the Santa Barbara Channel during 'El Nino'. J. Plank. Res. 7: 779–790.

RIPPKA, R. 1972. Photoheterotrophy and chemoheterotrophy among unicellular blue-green algae. Arch. Mikrobiol. 87: 93–98.

RIPPKA, R., AND G. COHEN-BAZIRE. 1983. The Cyanobacteriales: A legitimate order based on the type strain *Cyanobacterium stanieri*? Ann. Microbiol. (Inst. Pasteur) 134B: 21–36.

RIPPKA, R., J. DERUELLES, J. B. WATERBURY, M. HERDMAN, AND R. Y. STANIER. 1979. Generic assignments, strain histories and properties of pure cultures of cyanobacteria. Soc. Gen. Microbiol. 111: 1–61.

RIPPKA, R., AND J. B. WATERBURY. 1977. The synthesis of nitrogenase by non-heterocystous cyanobacteria. FEMS Microbiol. Let. 2: 83–86.

RIPPKA, R., J. B. WATERBURY, AND R. Y. STANIER. 1981a. Provisional generic assignments for Cyanobacteria in pure culture, p. 247–256. *In* M. P. Starr, H. Stolp, H. G. Trüper, A. Balows, and H. G. Schlegel [ed.] The prokaryotes. Springer-Verlag, Berlin, Heidelberg.

——— 1981b. Isolation and purification of cyanobacteria: Some general principals, p. 212–220. *In* M. P. Starr, H. Stolp, H. G. Trüper, A. Balows, and H. G. Schlegel [ed.] The prokaryotes. Springer-Verlag, Berlin, Heidelberg.

SMITH, A. J. 1982. Modes of cyanobacterial carbon metabolism, p. 47–85. *In* N. G. Carr and B. A. Whitton [ed.] The biology of cyanobacteria. University of California Press, Berkeley and Los Angeles.

STANIER, R. Y., AND G. COHEN-BAZIRE. 1977. Phototrophic prokaryotes: The cyanobacteria. Ann. Rev. Microbiol. 31: 255–274.

STEEMANN-NIELSEN, E. 1952. The use of radioactive carbon (C^{14}) for measuring organic production in the sea. J. Cons., Cons. Perm. Int. Explor. Mer 18: 117–140.

TAKAHASHI, M., AND P. K. BIENFANG. 1983. Size structure of phytoplankton biomass and photosynthesis in subtropical Hawaiian waters. Mar. Biol. 76: 203–211.

TANDEAU DE MARSAC, N. 1977. Occurrence and nature of chromatic adaptation in cyanobacteria. J. Bacteriol. 130: 82–91.

TAYLOR, C. D., J. J. MOLONGOSKI, AND S. E. LOHRENZ. 1983. Instrumentation for the measurement of phytoplankton production. Limnol. Oceanogr. 28: 781–787.

WATERBURY, J. B. 1976. Purification and properties of some fresh water and marine cyanobacteria belonging to the orders Chamaesiphonales and Pleurocapsales. Thesis, Univ. of California, Berkeley, CA.

WATERBURY, J. B., AND R. Y. STANIER. 1978. Patterns of growth and development in Pleurocapsalean Cyanobacteria. Microbiol. Rev. 42: 2–44.

——— 1981. Isolation and growth of Cyanobacteria from marine and hypersaline environments, p. 221–223. *In* M. P. Starr, H. Stolp, H. G. Trüper, A. Balows and H. G. Schlegel [ed.] The prokaryotes. Springer-Verlag Berlin, Heidelberg.

WATERBURY, J. B., S. W. WATSON, R. R. L. GUILLARD, AND L. E. BRAND. 1979. Widespread occurrence of a unicellular, marine, planktonic, cyanobacterium. Nature 277: 293–294.

WATERBURY, J. B., S. W. WATSON, AND F. W. VALOIS. 1980. Preliminary assessment of the importance of *Synechococcus* spp. as oceanic primary producers, p. 516–517. *In* P. G. Falkowski [ed.] Primary production in the sea. Plenum Press, New York and London.

WATERBURY, J. B., J. M. WILLEY, D. G. FRANKS, F. W. VALOIS, AND S. W. WATSON. 1985. A cyanobacterium capable of swimming motility. Science 30: 74–76.

WATSON, S. W., T. J. NOVITSKY, H. L. QUINBY, AND F. W. VALOIS. 1977. Determination of bacterial number and biomass in the marine environment. Appl. Environ. Microbiol. 33: 940–946.

WHITTON, B. A. 1973. Freshwater plankton, p. 353–367. *In* N. G. Carr and B. A. Whitton [ed] The biology of blue-green algae. Univ. of California Press, Berkeley, CA.

WHITTON, B. A., AND N. G. CARR. 1982. Cyanobacteria: current perspectives, p. 1–8. *In* N. G. Carr and B. H. Whitton [ed] The biology of cyanobacteria. Univ. of California Press, Berkeley, CA.

WHITTON, B. A., AND M. POTTS. 1982. Marine Littoral, p. 515–542. *In* N. G. Carr and B. H. Whitton [ed] The biology of cyanobacteria. Univ. of California Press, Berkeley, CA.

WOOD, A. M. 1985. Photosynthetic apparatus of marine ultraphytoplankton adapted to natural light fields. Nature (London) 316: 253–254.

WOOD, A. M., P. K. HORAN, K. MUIRHEAD, D. A. PHINNEY, C. M. YENTSCH, AND J. B. WATERBURY. 1985. Discrimination between types of pigments in marine *Synechococcus* spp. by scanning spectroscopy, epifluorescence microscopy, and flow cytometry. Limnol. Oceanogr. 30: 1303–1315.

A Survey of the Smallest Eucaryotic Organisms of the Marine Phytoplankton

HELGE ABILDHAUGE THOMSEN

*Institut for Sporeplanter, University of Copenhagen,
Ø. Farimagsgade 2 D, DK-1353 Copenhagen K, Denmark*

Introduction

The presence of very small, often fragile, algae passing through the ordinary plankton nets was thoroughly demonstrated very early in this century by Lohmann (1908, 1911). A detailed description of the internal and external ultrastructure of these numerous small phytoplankton organisms was, however, delayed for approximately 40 years. One of the major triggers to stimulate this type of research in the early fifties was of course the introduction of the electron microscopical techniques (Manton and Clarke 1950 et seq.). From 1950 until 1970 these studies were mostly based on cultured material. Most prominent — and successful — among the scientists trying to grow marine flagellates in culture was Dr. Mary Parke from Plymouth. For two decades she provided most of the cultures that were subsequently investigated with the electron microscope by Professor Manton and co-workers from the University of Leeds. The first to make use of electron microscopical techniques on freshly collected marine samples, to obtain field data also on the most fragile flagellates, was Leadbeater (1972a,b) studying shadowcast whole mounts prepared from Norwegian coastal water samples. The more sturdy groups of planktonalgae, the diatoms and the coccolithophorids, had already been studied for many years using EM-techniques on fixed net- and watersamples (Halldall and Markali 1955). In 1971 freshly collected seawater samples were for the first time processed directly for sectioning (Manton and Leadbeater 1974). The studies evolving from these pioneer works mostly concentrated on working out ultrastructural, biogeographical and taxonomical details of the nanoplankton organisms (i.e. those that range from 2 to 20 μm in length (Sieburth et al. 1978)). Although some of the organisms studied in previous years (e.g. *Micromonas pusilla*) were smaller than 2 μm, the elegant paper by Johnson and Sieburth from 1982, presenting ultrastructural details from thin sections on a variety of minute cells collected from the North Atlantic, in many respects makes up the formal starting point of ultrastructural research on eucaryotic picoplankton (i.e. organisms that are smaller than 2 μm (Sieburth et al. 1978)).

The present paper aims at summarizing some of the present day information on minute eucaryotic, marine, phytoplankton. As will appear from Table 1 the number of algal classes comprising genuine eucaryotic picoplankton forms is very limited. This paper, however, not only covers these forms, but also concentrates on genera that comprise species smaller than 5 μm (ultraplankton sensu Jørgensen (1966)). It is impossible here to elaborate in any length on morphological details of the individual species. To compensate for this, the most recent comprehensive publication dealing with the genus, or species, in question, has been cited. For a more general description of the individual algal classes one should consult some of the recent textbooks on phycology (e.g. Bold and Wynne 1985; Christensen 1980; Lee 1980; van den Hoek 1978) and also more specialized books and papers such as Dodge (1973), Cox (1980), Sieburth (1979), Lee et al. (1985), and Moestrup (1982).

Table 2 (kindly prepared by Dr. Moestrup, University of Copenhagen) furthermore summarizes some of the major ultrastructural features of the algal cells. This table hopefully will be of some guidance when undertaking preliminary identification of marine plankton prepared for electron microscopy.

TABLE 1. Algal classes and occurrence in marine plankton.

	Present in marine plankton	Includes specimens < 5 μm	Includes specimens from 0 to 2 μm
Rhodophyta			
Bangiophyceae	+		
(Rhodophyceae/Red algae)		(spermatia)	
Cryptophyta			
Cryptophyceae	+	+	
(Recoiling algae)			
Heterocontophyta			
Chrysophyceae	+	+	
(Golden-Brown algae)			
Fucophyceae	(swarmers)	(swarmers)	
(Phaeophyceae/Brown algae)			
Diatomophyceae	+	+	+
(Bacillariophyceae/Diatoms)			
Tribophyceae			
(Xanthophyceae/Yellow-green algae)			
Raphidophyceae	+		
Eustigmatophyceae	+	+	(+)
Haptophyta			
Prymnesiophyceae	+	+	
(Haptophyceae)			
Euglenophyta			
Euglenophyceae	+		
Chlorophyta			
Loxophyceae	+	+	+
Prasinophyceae	+	+	+
Chlorophyceae	+	+	(+)
Charophyceae			
Dinophyta			
Dinophyceae	+	(+)	
(Peridinians)			

While writing the first draft of this paper it became obvious that it was impossible to deal with all the algal groups according to a general scheme. The reasons for this are plentiful and diverse. Within the Cryptophyceae the classical principles used to separate genera have been more or less rejected, and the search for new and more reliable criteria has been intensive in recent years. This fact makes it — for the moment — meaningless to summarize in great detail the present day knowledge of this group of algae. The diatoms, also the small ones, are so numerous that only a few selected genera have been highlighted in the text. The opposite extreme is represented by the Eustigmatophyceae and the Prasinophyceae. In both cases all marine genera yet described have been mentioned, and some details on individual species is also included.

Another important piece of information when classifying autotrophic organisms (in addition to ultrastructural details), is the pigment composition of the alga in question. For a recent summary of algal pigmentation see Jeffrey (1980).

Survey of Ultraplanktonic Phytoplankton

BANGIOPHYCEAE (RED ALGAE)

The vast majority of the red algae are macroscopic plants forming multicellular thalli. The very few known unicellular red algae are all allocated to the order Porphyridiales, comprising *Rhodella maculata* Evans (1970), a species which occurs

TABLE 2. Fine structural features of the algae, visible in thin sections for electron microscopy.

	Chromosomes visible in the interphase nucleus	Nucleus connected to chloroplast via outer nuclear membrane	No. of membranes surrounding chloroplast	No. of thylakoids in chloroplast lamellae	Girdle lamella in chloroplast	Eyespot part of the chloroplast; closely appressed flagellar swelling (P)	Trichocysts (T) present in flagellated cells
Bangiophyceae (Red algae)			2	1, with phycobilisomes	+		
Cryptophyceae		+	4	2, mostly swollen	−	Yes	T
Chrysophyceae s.l.		+(−)	4	3	+	Yes, P	(discobolocysts)
Fucophyceae (Brown algae)		+	4	3	+	Yes, P	
Diatomophyceae (Diatoms)		+	4	3	+		
Tribophyceae		+(−)	4	3	+(−)	Yes, P	
Raphidophyceae			4	3	±	(mucocysts)	
Eustigmatophyceae		+(marine) −(freshwater)	4	3	−	No, P	
Prymnesiophyceae (Haptophyceae)		+	4	3	−	Yes, (Pavlovales)	
Euglenophyceae	+		3	ca. 3	−	No, (P) (mucocysts)	
"Loxophyceae" (Micromonas, Pedinomonas)			2	irregular	−	Yes	
Prasinophyceae (sensu Moestrup 1982)			2	irregular	−	Yes	(T)
Chlorophyceae (Green algae s.l.)			2	irregular, sometimes grana	−	Yes	
Dinophyceae	+		2 or 3	ca. 3	−	Rarely	T
Chlorarachniophyceae			4	irregular	−	No	T

	Tubular hairs on front (F) or all flagella	Flagellar transition region with short helix (TH) or starshaped body (SB)	Special cell wall features apart from scales	Body scales on motile cells	No. of flagella visible also in LM	Flagellar scales	Mitochondrial cristae
Bangophyceae (Red algae)					0	0	flat
Cryptophyceae	+		mostly very thin plates, sometimes on both sides of cell membrane	+, seen best in freeze etching/ fracturing	2	+, seen best in freeze etching/ fracturing	flat
Chrysophyceae s. l.	F	TH (rarely lacking)		+(−)	1–2	(+)	tubular
Fucophyceae (Brown algae)	F				2		tubular
Diatomophyceae (Diatoms)	F		silica, probably homologous with scales		1 (sperms)	tubular	
Tribophyceae	F	TH			2		tubular
Raphidophyceae	F				2		tubular
Eustigmatophyceae	F	TH			1–2		tubular
Prymnesiophyceae (Haptophyceae)				+	(1)–2	(Pavlovales)	tubular
Euglenophyceae			thin plates beneath the membrane		2		flat circular plates
"Loxophyceae" (Micromonas, Pedinomonas)		SB			1		flat

Prasinophyceae (sensu Moestrup 1982)	− except possibly Mesostigma	SB	fused scales in Tetraselmis and Scherfelia	+	1,2,4,8	+	flat
Chlorophyceae (Green algae s.l.)		SB			2,4	(+)	flat
Dinophyceae			plates in vesicles beneath the cell membrane	(+)	2	Oxyrrhis	tubular
Chlorarachniophyceae					1		tubular

regularly in the plankton of the Oslo Fjord (Paasche and Throndsen 1970). The spherical *R. maculata* cells are 7-24 μm in diameter. There is a central, very conspicuous pyrenoid, an eccentrically located nucleus, and a single parietal, much-branched pink chloroplast (Evans 1970).

The non-motile male gamete (spermatium) of a red alga is a small pale or colourless cell (diam. 2-5 μm (Dixon and Irvine 1977)). Under certain circumstances (a favourable combination of time and locality) a mass release of such male gametes, or monospores (asexual reproductive cells) may be expected to contribute to the pico/ultraplankton biomass. During spermatial formation a progressive degeneration of the photosynthetic organelle takes place, to the extent that mature spermatia of *Corallina*, *Furcellaria* and *Nemalion* (Peel and Duckett 1975) are completely devoid of plastids. This is mentioned here to stress the fact that while red algal gametes may be of some significance when evaluating the pico/ultraplankton biomass of a certain area, they hardly make any contribution to primary productivity.

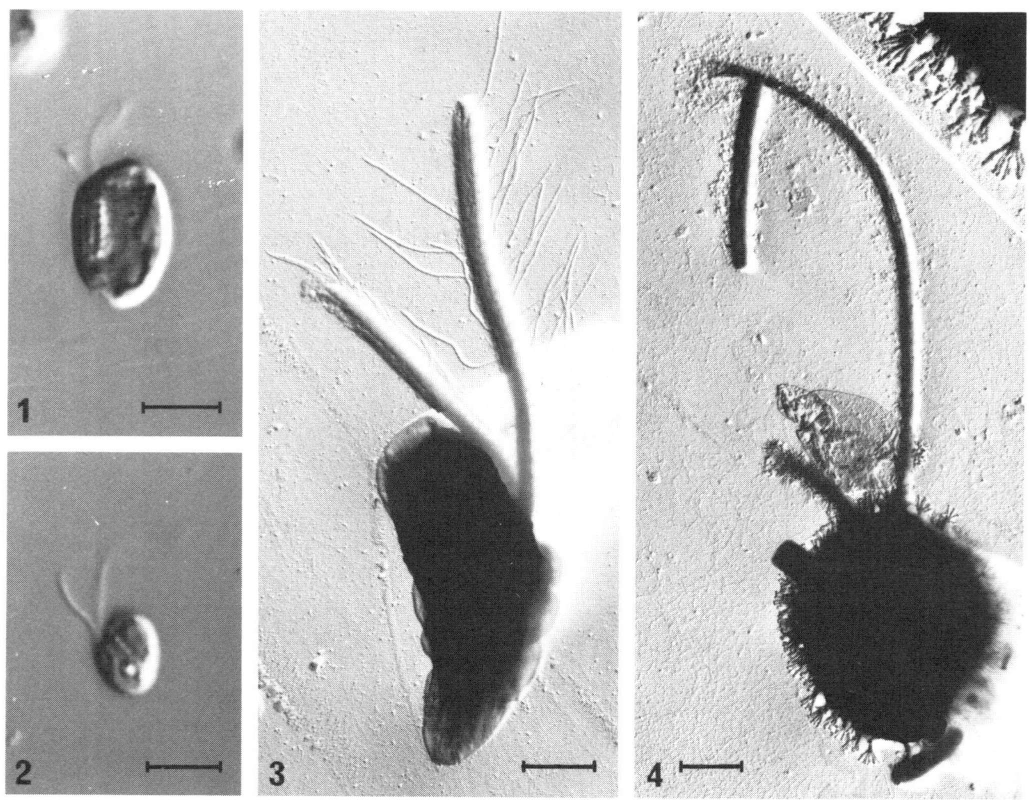

FIG. 1-3. Cryptophyceae. FIG. 4. Chrysophyceae.
FIG. 1. *Chroomonas* sp. (LM, interference contrast optics); material from the Baltic Sea (Gulf of Bothnia); × 2 000. FIG. 2. *Hemiselmis anomala* aff. (LM, interference contrast optics); material from the Baltic Sea (Gulf of Bothnia); × 2 000. FIG. 3. Shadowcast whole mount (TEM) of a cryptomonad flagellate; material from Isefjorden (Denmark); micrograph T2841, × 10 000. FIG. 4. *Sphaleromantis marina*; shadowcast whole mount (TEM) showing the typical chrysophycean heterokont flagellation; material from the Danish Waddensea; micrograph T 3381, × 7 500; inset shows high magnification (× 20 000) of body-scales.
Scale bars: Fig. 1,2 = 5 μm; Fig. 3,4 = 1 μm.

CRYPTOPHYCEAE

This algae class is well represented in coastal marine habitats. Butcher (1967), in his classical account of the cryptophycean flagellates of the British coastal waters, separated genera mostly on the basis of variation in the depression–furrow–gullet system. Not only are these details difficult to analyze in the light microscope, but it has also been shown that characteristics such as the number and arrangement of the trichocysts, the position of flagellar insertion, the length of the vestibular region etc., do vary within the very same taxon (Meyer and Pienaar 1984) rendering the search for new criteria relevant. Much effort at the moment is actually being directed towards establishing such new sets of characteristics to separate the pigmented cryptophycean genera (Santore 1977, 1982a,b, 1985a,b; Hill and Wetherbee 1985; Brett and Wetherbee 1985; Wetherbee 1985). There is much hope that details of periplast structure, once worked out in full detail, will form a solid basis for generic separation.

The following list of taxa is included here to provide references to some of the smallest cryptophycean taxa yet described (nomenclature according to Butcher (1967)). Light and electron microscopical illustrations of brackish water cryptophycean flagellates are shown in Fig. 1–3.

Hillea Schiller
    H. fusiformis (Schiller)Schiller 1925 — (5–10 × 4–5 μm)
    H. marina Butcher 1952 — (2 × 2.5 μm)
Hemiselmis Parke
    H. rufescens Parke 1949 — (4–8.5 × 3.5–5 × 2–3 μm)
    H. virescens Droop 1955 — (4.5–7 × 2.5–3 μm)
Chroomonas Hansgirg
    C. diplococca Butcher 1967 — (4–6 × 3.5–5 μm)
    C. africana Meyer and Pienaar 1984 — (6–8 × 4–5 μm)
Cryptomonas Ehrenberg
    C. rhynchophora (Conrad) Butcher 1967 — (6–9 × 4–5 × 4–7 μm)

CHRYSOPHYCEAE

The Chrysophyceae (in particular members of the Ochromonadales) is a very important group of organisms when fresh water ponds and lakes are considered. In the marine and brackish water environment chloroplast containing Chrysophyceae are, on the contrary, infrequently reported. This applies in particular to the Ochromonadales, whereas members of the somewhat atypical chrysophycean orders Pedinellales and Dictyochales (Silicoflagellates) may occasionally contribute significantly to the biomass and productivity of coastal areas.

OCHROMONADALES

Marine species smaller than approximately 5 μm are mainly found within the following genera (here each exemplified by one species): *Ochromonas minima* Throndsen (1969), *Sphaleromantis marina* Pienaar (1976) (Fig. 4), *Polylepidomonas vacuolata* (Thomsen) Preisig and Hibberd (Preisig and Hibberd 1983) (Fig. 8), *Syncrypta glomifera* Clarke and Pennick (1975) (Fig. 5–7), and *Chromulina pleiades* Parke (1949).

It is a typical feature of several members of the Ochromonadales that the cell surface is covered by scales, mostly silicified structures of a highly diverse species-specific architecture. Three of the taxa listed above carry such scales, and the appearance of these structures from electron microscopical whole mounts are shown in Fig. 4,7,8. In many cases it is impossible to detect the scales from light microscopy (LM) alone. It is furthermore quite common that even when looking at cells carrying larger scales

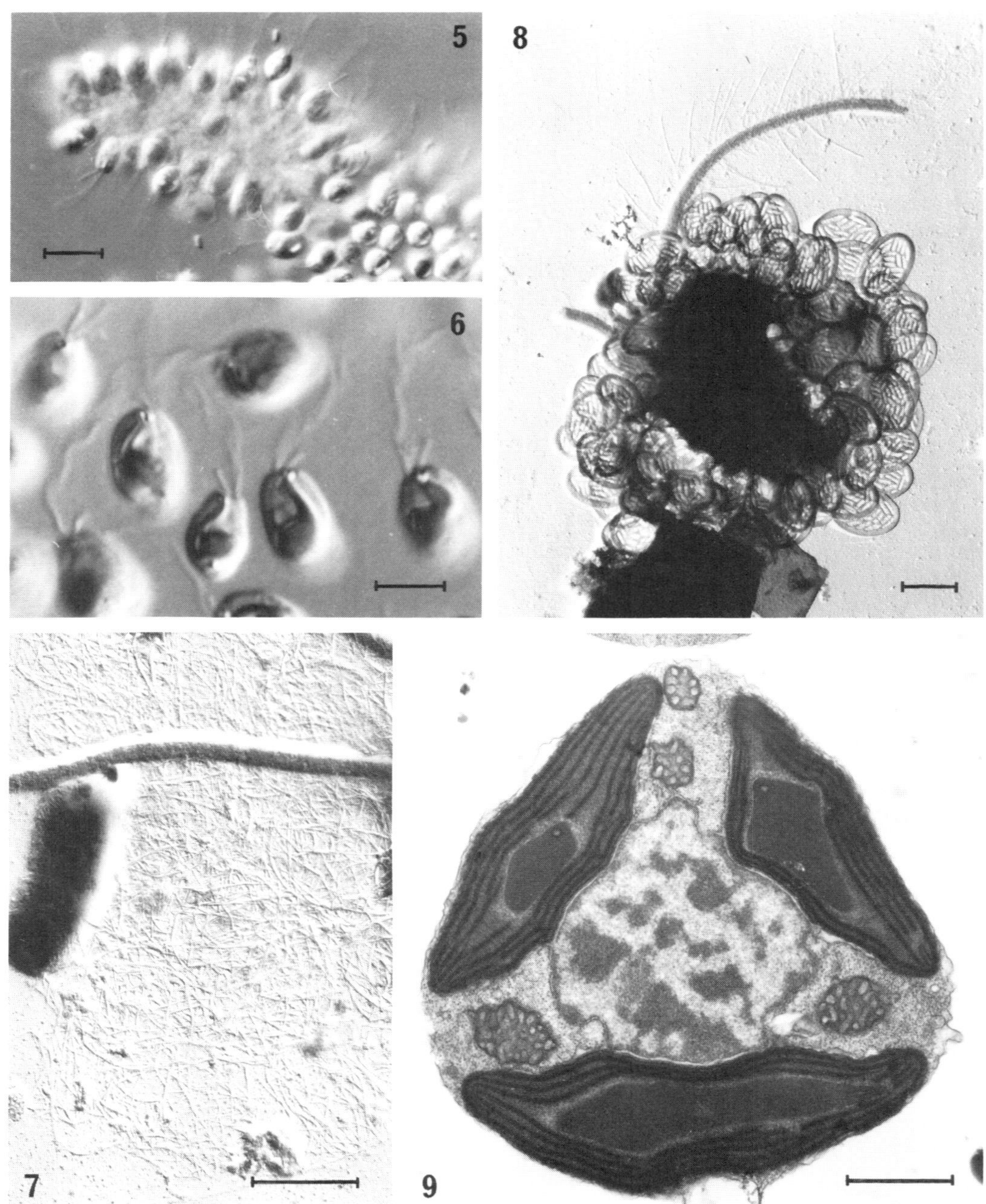

FIG. 5–9. Chrysophyceae.
FIG. 5,6. *Syncrypta glomifera*, colony and selected cells; (LM, interference contrast optics); material from the Baltic Sea (Gulf of Bothnia); × 800 (Fig.5), × 2 000 (Fig. 6). FIG. 7. Shadowcast whole mount (TEM) of *Syncrypta glomifera* scales, also showing part of the hairy flagellum and a bacterial cell. Material from the Baltic Sea (Gulf of Bothnia); micrograph T 4839, × 15 000. FIG. 8. *Polylepidomonas vacuolata*, shadowcast whole mount (TEM) showing body scales and flagellation; material from freshwater (Denmark); micrograph T 150, × 7 500. FIG. 9. *Pedinella tricostata*, transverse section of cell (TEM) showing the three chloroplasts, a central nucleus and profiles of mitochondria; material from Isefjorden (Denmark); micrograph 20452, × 15 000.
Scale bars: Fig. 5 = 10 μm; Fig. 6 = 5 μm, Fig. 7–9 = 1 μm.

128

visible in the LM, not enough details can be resolved to ensure a proper species identification. In both cases it is evident that a prerequisite to identification (species level) will be at least an electron microscopical examination of whole mounts.

None of the species mentioned above, and no other ultraplanktonic member of the Ochromonadales recorded from a marine environment, have been frequently reported. This may of course be due to the very obvious problems of identification, but is, in my opinion, more likely to reflect the fact that ultraplanktonic members of the Ochromonadales do not, either quantitatively or qualitatively, form an important part of the marine plankton. The only possible exception is perhaps that in the most dilute brackish waters (salinity range from 0 to 5 °/oo S), species such as *Polylepidomonas vacuolata*, and both solitary and colonial forms of *Syncrypta glomifera* turn up regularly in both wild material and cultured material (Thomsen, unpublished data).

CHRYSOSPHAERALES

Pelagococcus subviridis Norris (Lewin et al. 1977) is one of the very few examples (if not the only one) of a truly oceanic chrysophyte. *Pelagococcus subviridis* is a pale green coccoid alga, 2.5–5.5 μm in cell diameter, with an organelle fine structure and pigment composition that closely resembles that of members of the Chrysophyceae (Lewin et al. 1977). This alga appeared in enrichment cultures based on several water samples collected over a wide stretch of the North Pacific Ocean.

PEDINELLALES

Members of this order, characterized by radial symmetry, one emergent flagellum and mostly six chloroplasts per cell, occur regularly in brackish and marine environments. Most of the species from the genera *Pedinella* Wyssotzki, *Pseudopedinella* N. Carter and *Apedinella* Throndsen, are, however, larger than 5 μm in cell diameter (for a recent review-like paper see Zimmermann et al. 1984). Only *Pedinella tricostata* Rouchijajnen (1966) is smaller than 5 μm. Contrary to all other known autotrophic members of the Pedinellales, this species is characterized by having three chloroplasts only (Fig. 9), a fact that of course allows *P. tricostata* to attain smaller overall cell dimensions (type material: 4–5.5 μm). *Pedinella tricostata* was described from the Black Sea (Rouchijajnen 1966) and has later been reported from Finnish coastal waters (Zimmermann et al. 1984) and Japanese coastal waters (Throndsen 1984). In February 1985 a bloom (approx. 75×10^6 cells per litre) of *P. tricostata* developed underneath the ice in the Danish Isefjord (Thomsen 1985). The yellowish water samples clearly accentuated the fact that members of this order, contrary to the Ochromonadales, may occasionally contribute significantly to the ultraplanktonic biomass and primary productivity of coastal waters.

DICHTYOCHALES (SILICOFLAGELLATES)

Members of this order may be very common in the marine plankton. *Distephanus Stöhr* and *Dictyocha* Ehrenberg cells encountered are, however, typically in the size range from 10 to 30 μm, which together with the characteristic siliceous skeleton of these forms should prevent these cells from interfering with picoplankton size-fractionating procedures. It should be mentioned, however, that atypical non-skeleton forming silicoflagellate cells, first reported by van Valkenburg and Norris (1970) from cultured material of *Dictyocha fibula* Ehr., may also occur in nature. A dense bloom (up to 15×10^6 cells per litre) of a non-skeleton forming *Distephanus speculum* (Ehr.) Haeckel population thus took place in Danish waters in May 1983 (Thomsen and Moestrup 1984, Moestrup, Thomsen and Pedersen, unpublished data). Not only are these cells without skeletons much more variable in outer dimensions, and flexible in shape, but the population also comprised amoeboid forms that were capable of attaining any body shape and therefore also most likely capable of bypassing even some very finely perforated filters.

DIATOMOPHYCEAE (DIATOMS)

It has for decades been a well-known fact that the diatoms make up one of the most important phytoplankton groups in all types of saltwater. It is perhaps less well-known that a considerable number of very small diatom species have been described in detail in literature from the very recent years, and that these minute forms have in some cases been shown to be ecologically very important organisms (Guillard and Ryther 1962). Because of the number of taxa described, the presentation here has to be limited to a few selected genera and species.

Within the Eupodiscales, the centric diatoms, the most important genus to be mentioned here is *Thalassiosira* Cleve, a genus which in addition to a large number of "netplankton" forms, comprises also some very small nanoplanktonic forms (Hasle 1983; Table 2: *T. profunda* (Hendey) Hasle, *T. pseudonana* Hasle and Heimdal, *T. proschkina* Makarova, *T. malo* Takano). The above list of species comprises those that are known to have a valve diameter that may be as small as 2 μm. Figure 10 shows a shadowcast whole mount preparation of *T. pseudonana* Hasle and Heimdal (1970) (diam.: 3.5 μm). Detailed information on morphology, taxonomy and distribution of the species of *Thalassiosira* can be found in a series of papers by Hasle, Fryxell, and Takano (see by way of example Johansen and Fryxell 1985; Hasle 1983; Takano 1981).

Also the genus *Minidiscus* Hasle (1973), another centric diatom genus, comprises at least one very small species, *M. trioculatus* (Taylor)Hasle (Hallegraeff 1984a, Australian material, diam.: 1.7–3.8 μm). *Minidiscus trioculatus* (Fig. 11,12 valve diam.: 3.3 μm/c. 2μm) is a cosmopolitan species (Hasle 1976).

Chaetoceros galvestonensis Coller and Murphy (1962) is yet another example of a centric diatom with very modest cell dimensions (1.5–3 μm).

In a recent paper by Hasle et al. (1983) presenting morphological details of members of a new centric diatom family, the Cymatosiraceae, some very minute plankton forms have been carefully examined. *Minutoccellus polymorphus* (Hargraves and Guillard) Hasle, von Stosch and Syvertsen (basionym: *Bellerochea polymorpha* Hargraves and Guillard (1974)) is a very elegant species, the larger cells carrying some very conspicuous pili (Fig. 14, arrows). *Minutoellus polymorphus* is, however, a morphologically extremely variable species. In cultured clonal material (Hasle et al. 1983) as well as in wild material (Hargraves and Guillard 1974) and Fig. 13,14 (Thailand material) cells are seen to cover a morphological continuum from minute (2.2–3.6 μm (Hargraves and Guillard 1974)), subspherical cells lacking the pili (Fig. 13), to very much larger cells like the one shown in Fig. 14 (up to 14 μm (Hargraves and Guillard 1974)) thus contributing to both the nano- and ultra/picoplanktonic stock). According to Hasle et al. (1983) *M. polymorphus* is a cosmopolitan species.

FIG. 10–14. Bacillariophyceae (diatoms). FIG. 15 Fucophyceae. (See facing page) FIG. 10. *Thalassiosira pseudonana*, shadowcast whole mount (TEM); material from Finland; micrograph T 2674, × 15 000. FIG. 11,12. *Minidiscus trioculatus*, shadowcast whole mounts (TEM) showing valve with four strutted tubuli (Fig. 11) and a folded over specimen with only one strutted tubulus (Fig. 12); material from Phuket (Thailand) (Fig. 11) and from Friday Harbour (USA) (Fig. 12); micrographs T4157/T5162, × 15 000. FIG. 13,14. *Minutocellus polymorphus*, shadowcast whole mounts (TEM) showing (Fig. 14) a typical valve with pili (arrows) and spinulose area (arrowheads), and (Fig. 13) a small cell (process valve); material from Phuket (Thailand); micrographs T3996/T4157, × 15 000/× 10 000. FIG. 15. *Sargassum decipiens*, shadowcast whole mount (TEM) of swarmer, showing the typical heterokont flagellation, with the hairy flagellum in front; the proboscis is pointed out; material from New Zealand (micrograph M 4630, (courtesy Ø. Moestrup) × 7 500.
Scale bars = 1 μm.

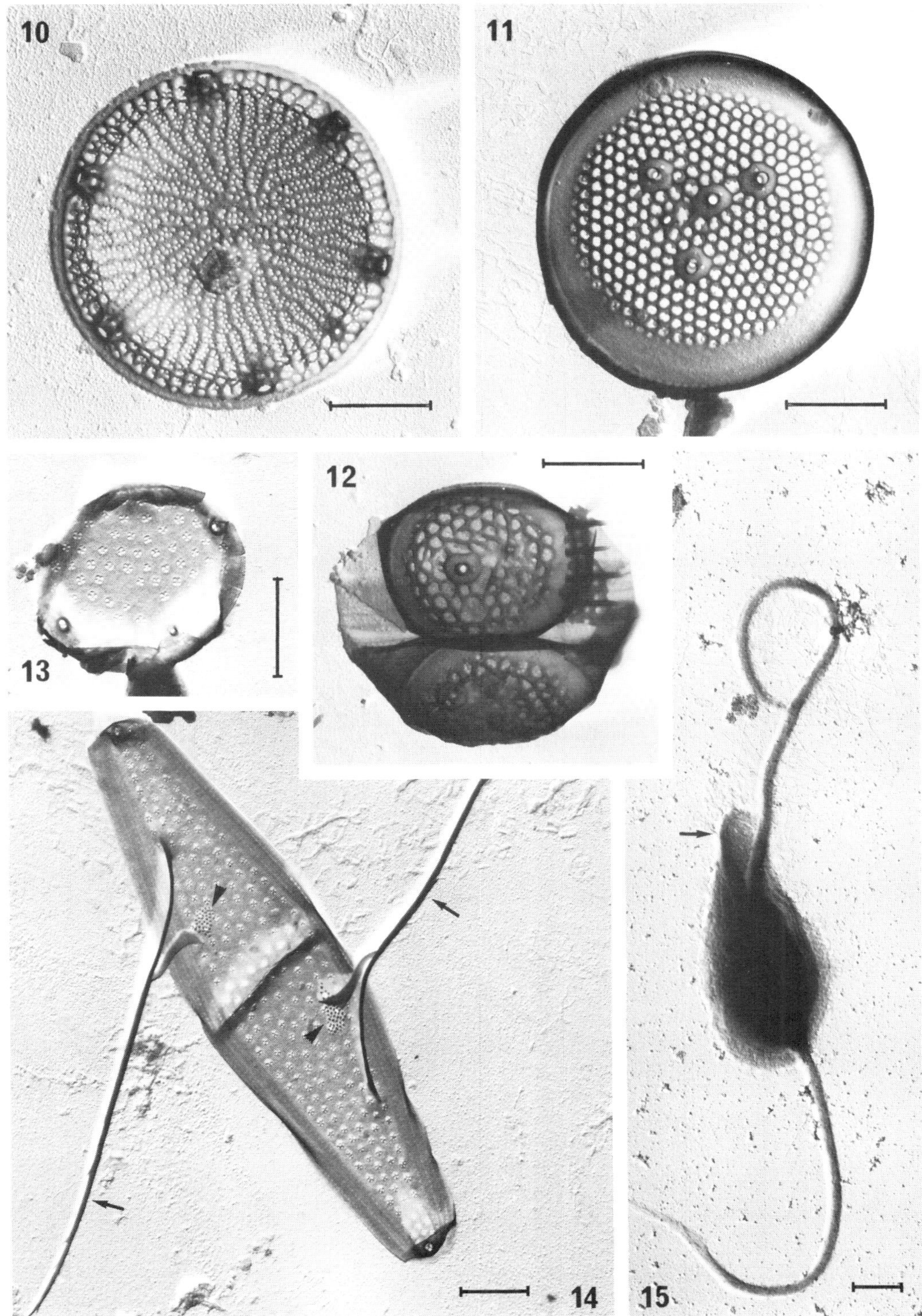

Arcocellulus cornucervis Hasle, von Stosch and Syvertsen (1983) is another planktonic member of the Cymatosiraceae likewise showing an extreme morphological and dimensional variability (Hasle et al. 1983, apical axis: 1.2–13 μm, transapical axis: 0.7–1.5 μm). This species is so far especially found in cooler waters of both hemispheres (Hasle et al. 1983).

Extubocellulus spinifer (Hargraves and Guillard) Hasle, von Stosch and Syvertsen (1983) (basionym: *Bellerochea spinifera* Hargraves and Guillard (1974)) is the last example from the Cymatosiraceae to be mentioned here. Characteristic dimensions of this taxon is according to Hargraves and Guillard (1974) 1.9–7.0 μm in cell length.

Considering the fact that the vegetative centric diatom cell may, as documented above, attain almost picoplanktonic dimensions, it is obvious that when these specimens produce male gametes in preparation of sexual reproduction, these swarmers (one hairy flagellum) must certainly be genuine picoplanktonic forms. It is likely that diatom spermatozoids may occasionally contribute significantly to the picoplankton eucaryotic biomass (e.g. during diatom blooms in coastal areas). For information on ultrastructural details of diatom spermatozoids see Manton and von Stosch (1966) and Heath and Darley (1972).

Certainly there are also among the Bacillariales, the pennate diatoms, some examples of taxa that regarding overall cell dimensions approach the picoplankton size class. Such examples are given by Hargraves and Guillard (1974, *Fragilaria rotundissima* Hargraves and Guillard (2.0–5.5 μm), *F. pinnata* Ehr. (2.8–22.5 μm); *Synedra fragilarioides* Hargraves and Guillard (2.8–5.3 μm)).

The taxa referred to above have been selected because they all (when considering the largest dimension of the cell) at least occasionally approach the 2 μm size limit chosen to distinguish picoplankton from other plankton size-classes. It must be emphasized, however, that when filtering a water sample in order to differentiate between different size-classes of plankton organisms it is more often the minimum dimension (e.g. the diameter of a cylinder) that actually determines to which size-class a certain algal cell is actually referred. Many pennate diatoms are very long, slender forms (diameter <2 μm) that because of flow-characteristics may easily pass through a 2 μm filter (e.g. species of *Thalassiothrix* Cleve and Grunow and *Nitzschia* Hassall). Similarly one might expect also some of the colony-forming centric diatoms (e.g. species of *Leptocylindrus* Cleve and *Skeletonema* Greville) to pass through a 2 μm filter.

Fucophyceae (Brown algae)

The swarmers of the benthic brown algae are indeed very small cells that may in certain localities contribute to the eucaryotic pico/ultraplankton biomass. The chloroplast thylakoids of brown algal spermatozoids are usually reduced to an extent that this algal group should be neglected when primary productivity of the different size classes is considered. In *Fucus* (Manton and Clarke 1956) the thylakoids are thus reduced to an extreme extent and the whole organelle is filled with globules and functions as an enormous eyespot.

Brown algal swarmers show typical heterokont flagellation (Fig. 15, *Sargassum decipiens* (R. Brown) J. Ag.). In some cases the hairy flagellum carries highly characteristic spines (Moestrup 1982). Other characteristic features include the lateral insertion of the flagella and the very conspicuous anterior proboscis (Fig. 15, arrow).

Dimensions of some brown algal swarmers are given in Table 3. Dimensions based on electron micrographs of whole mounts are given as 1.5 times the measured values, due to the shrinkage of the cells following this preparational procedure (factor determined from Manton and Clarke (1951) showing LM and EM photographs of *Fucus serratus* spermatozoids at the same magnification; Manton and Clarke 1951, fig. 1c and fig. 5).

TABLE 3. Dimensions of brown algal swarmers (μm).

	Dimensions	References
Fucus serratus L.	6 × 2	Manton and Clarke (1951)
Ascophyllum nodosum (L.)le Jol.	10 × 5	Manton and Clarke (1953)
Pelvetia canaliculata (L.)Dcne & Thur.	5 × 3	—
Himanthalia lorea (L.)S.F. Gray	6 × 5	—
Dictyota dichotoma (Huds)Lamour.	5 × 3	—
Hormosira banksii (Turn.)Dcne	5-6 × 2	Osborn (1948)
Sargassum decipiens (R.Brown) J. Ag.	7 × 4	Fig. 15
Chorda tomentosum Lyngb.	5-10×3-6	Maier (1984)

TRIBOPHYCEAE (XANTHOPHYCEAE)

Members of this algal class occur predominantly in freshwater habitats. A limited number of unicellular taxa are listed in the checklist of British marine algae (Parke and Dixon 1976). The knowledge of the majority of these very rarely reported forms, is, however, quite limited (even the systematic position is questionable in some cases). In general it appears that the conclusion to be drawn from looking through published lists of marine plankton, is that the Tribophyceae is very unlikely to be of any importance in the brackish and marine environments. There has to the best of my knowledge not been any single paper based on modern techniques (e.g. electron microscopy or chromatography) documenting the presence of any tribophycean taxon from a saline environment.

EUSTIGMATOPHYCEAE

The algal class Eustigmatophyceae was segregated from the Tribophyceae (Xanthophyceae) on the basis of differences in ultrastructure and pigment composition (Hibberd and Leedale 1970, 1971, 1972; Whittle 1976; Whittle and Casselton 1975). For recent reviews of this algal class, see Hibberd (1980, 1981).

All presently known eustigmatophycean taxa are included in the Eustigmatales, which is further divided into four families (Hibberd 1981). One of these, the Monodopsidaceae, contains the euryhaline genus *Nanochloropsis* Hibberd (1981) with two species.

Nanochloropsis salina Hibberd (1981) (synonym: *Monallantus salina* Bourrelly (1958), nomen nudum) comprises cylindrical, straight or slightly reniform cells (1.5—1.7 × 3-4 μm). The species was originally found in large numbers in supralittoral pools.

Nanochloropsis oculata (Droop) Hibberd (1981) (Basionym: *Nannochloris oculata* Droop (1955)) is a genuine marine species. Each cell is subspherical, 2-4 μm in diameter, and greenish coloured.

The number of published findings of *Nanochloropsis* species is yet rather limited. It seems, however, likely that eustigmatophytes will be found to be more prevalent in nature, in particular in eutrophicated coastal areas, once the group has become more widely recognized. Several laboratories at present keep *Nanochloropsis*-like cultures grown from their own local waters.

For electron microscopical details of *Nanochloropsis* see Antia et al. (1975).

PRYMNESIOPHYCEAE (HAPTOPHYCEAE)

Most Prymnesiophyceae occur in salt water, and members of the group may be both qualitatively and quantitatively dominant in coastal waters as well as oceanic regions.

The majority of the Prymnesiophyceae carry species-specific organic scales on the cell-body. In the majority of the species described, a morphologically distinct $CaCO_3$ deposit is found on top of the outermost organic scales, forming the so-called coccoliths. In most genera the two naked flagella are more or less equal in length. Between them there is usually another thread-like organelle, the haptonema, which differs structurally from the flagella.

Almost without exception the identification to the species level requires at least the examination of whole mounts prepared for the transmission or the scanning electron microscopes. Some large-spined species of *Chrysochromulina*, and also some of the coccolithophorids, may be unambiguously identified from light microscopy alone.

The taxonomy of the group is somewhat uncertain for the moment. Christensen (1980) referred all taxa to one single order, Prymnesiales, whereas Parke and Dixon (1976) made use of four orders (Isochrysidales, Coccosphaerales, Prymnesiales, Pavlovales). Members of the Pavlovales certainly do differ in many respects from the majority of the Prymnesiophyceae and should be kept as a separate order. It is, however, most likely that the orders Isochrysidales, Coccosphaerales and Prymnesiales cannot in the future be kept apart. The main difference between members of the Isochrysidales and the Prymnesiales relates to the haptonema which is either absent or rudimentary in the former order (Green and Pienaar 1977). The Coccosphaerales is distinguished from the Prymnesiales by the presence of calcification on top of the organic scales, a system that obviously separates closely related forms (*Wigwamma arctica* Manton, Sutherland and Oates (1977) and *Chrysochromulina pyramidosa* Thomsen (1977)). In this paper a two-order system will be adopted.

PRYMNESIALES

A. A haptonema is either lacking or very much reduced; organic scales may be present.

Marine representatives (<5 μm) fitting the description above are *Isochrysis galbana* Parke (1949), *Dicrateria inornata* Parke (1949), and *Imantonia rotunda* Reynolds (1974). These species were examined ultrastructurally by Green and Pienaar (1977). Figure 16 shows the minute flagellate *Imantonia rotunda* from Danish coastal waters. The characteristic bicycle-wheel-like scales (Fig. 17) of the species has resulted in the recognition of this taxon from a number of widely separated localities (Friday Harbour area, Washington, USA (Green and Pienaar 1977), Spitsbergen (Reynolds 1974), Australia (Hallegraeff 1983)).

B. A haptonema (visible in the LM) is present; organic scales present.

The genus *Chrysochromulina* Lackey comprises 46 species (43 marine). All taxa, except *C. brachycylindra* Hällfors and Thomsen (1985) and *C. apheles* Moestrup and Thomsen (1986), are listed in Estep et al. (1984).

The species of *Chrysochromulina* are distinguished on the basis of the shape and size of the cell (4–25 μm), the length of the haptonema (from a few microns to more than 150 μm) and, above all by differences in morphology of the organic scales covering the cell-body. *Chrysochromulina* cells from shadowcast whole mounts are shown in Fig. 18–22. The species selected are among the smallest yet described (*C. apheles* Moestrup and Thomsen (1986) (Fig. 18,19), *C. minor* Parke and Manton (Parke et al. 1955) (Fig. 20), *C. pyramidosa* Thomsen (1977) (Fig. 21,22)).

Members of this genus are found in all marine environments (Thomsen 1982; Leadbeater 1972a; Hallegraeff 1983; Estep et al. 1984; Manton and Oates 1983). Recent results from the Baltic Sea indicate that a considerable number of species are able to form dense populations in even the most dilute brackish water (Hällfors 1981; Thomsen, unpublished data). Species of *Chrysochromulina* occasionally outgrow

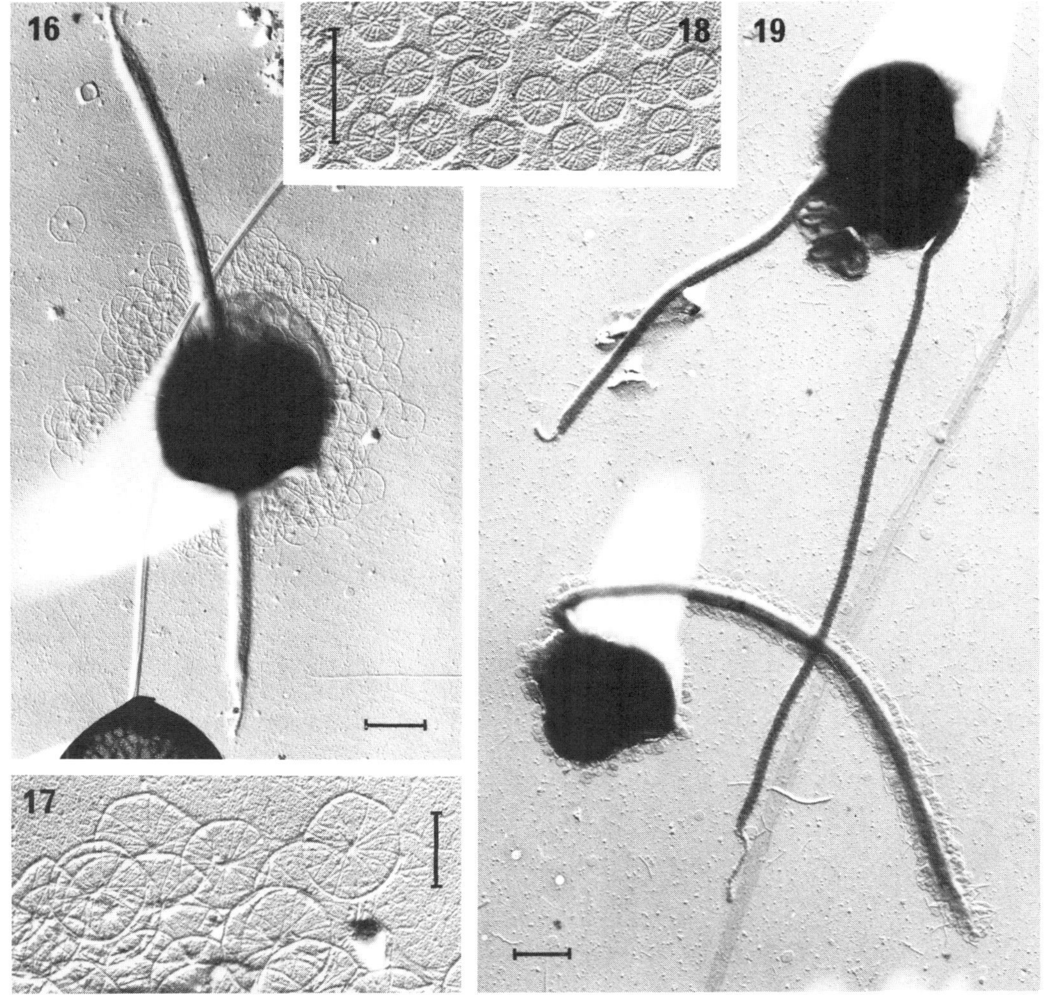

FIG. 16–19. Prymnesiophyceae (Haptophyceae).
FIG. 16,17. *Imantonia rotunda*, shadowcast whole mounts (TEM) showing complete cell with two flagella (Fig. 16) and body-scales (Fig. 17); material from Isefjorden (Denmark); micrograph T 3810, × 7 500/× 20 000. FIG. 18,19. Shadowcast whole mounts (TEM) of *Chrysochromulina apheles*; scales of two size classes are shown in Fig. 18; a complete cell with tightly curled-up haptonema appears at the top of Fig. 19; the second organism present is *Mantoniella squamata* (Prasinophyceae); material from Isefjorden (Denmark); micrographs T 5268/5275, × 30 000/×7 500.
Scale bars: Fig. 16,19 = 1 μm, Fig. 17,18 = 0.5 μm.

almost all other phytoplankton organisms (e.g. *C. birgeri* G. Hällfors and Niemi (1974)).

The genera *Prymnesium* Conrad with seven species (see Chang and Ryan 1985) and *Platychrysis* Geitler with four species (see Gayral and Fresnel 1983) are indeed very difficult to distinguish. Neither cell morphology and dimensions (ca. 6–20 μm), nor the scaly coverings offer details that render a separation between the two genera possible. The only difference is that while species of *Platychrysis* appear to be perfectly adapted to a neustonic lifestyle (i.e. cells very often transform into non-flagellated

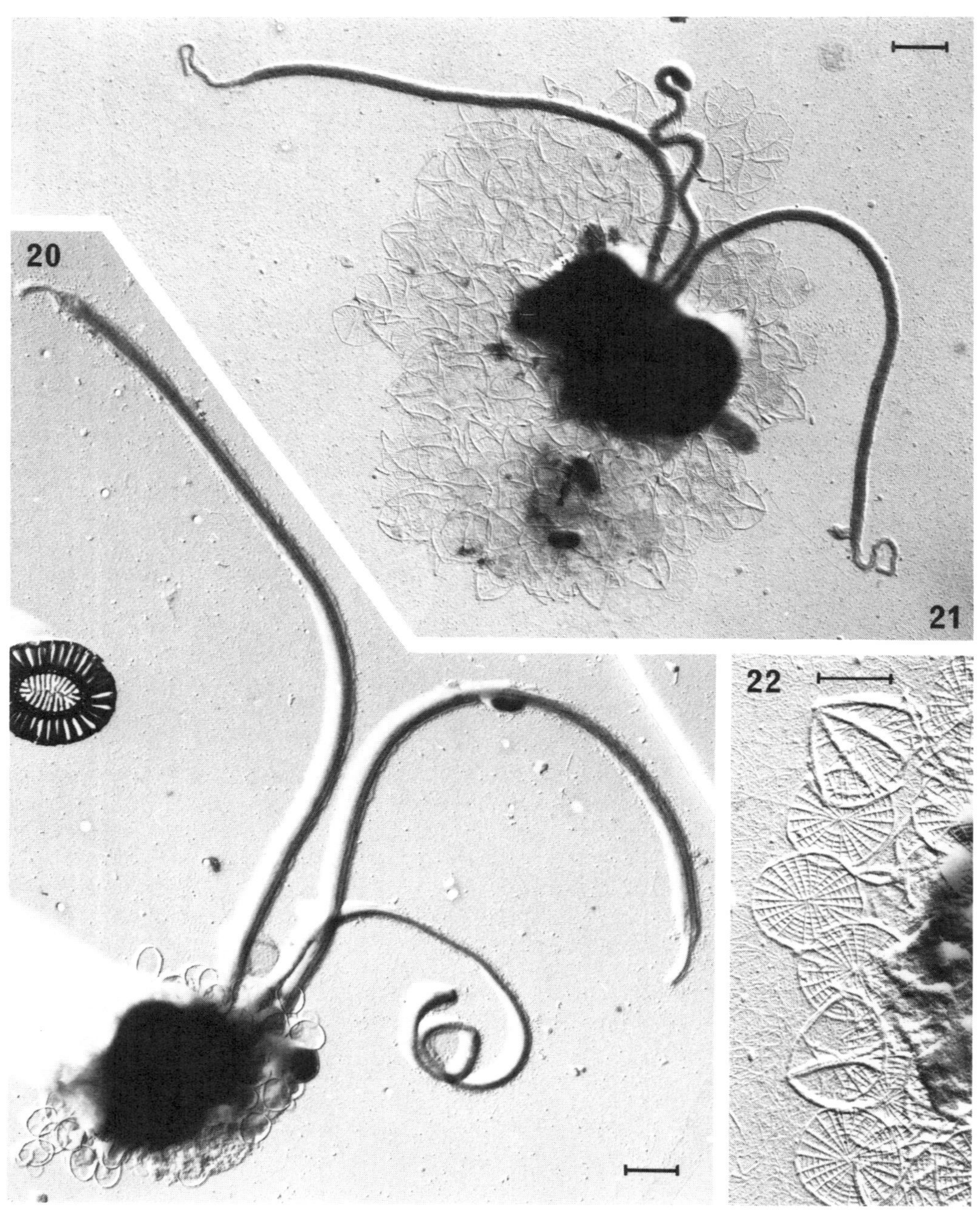

FIG. 20–22. Prymnesiophyceae (Haptophyceae).
FIG. 20. *Chrysochromulina minor*, shadowcast whole mount (TEM) of a complete specimen; material from Isefjorden (Denmark); micrograph T 5450, × 7 500.
FIG. 21,22. *Chrysochromulina pyramidosa*, shadowcast whole mounts (TEM) of a whole cell (Fig. 21) and scales (Fig. 22); material from Asnæs and Isefjorden (Denmark); micrographs T 1075/18662, × 7 500/ × 20 000.
Scale bars: Fig. 20,21 = 1 μm, Fig. 22 = 0.5 μm.

stages), species of *Prymnesium* appear to be monads always. The separation of both genera from *Chrysochromulina* also warrents careful consideration in the future.

There are, however, no difficulties in distinguishing between the individual species, provided that the cells can be examined by electron microscopy, in order to resolve the morphological details of the scales.

The recognition of several taxa within each genus is a rather recent event (*Prymnesium*: Green et al. 1982; Billard 1983; Chang and Ryan 1985; *Platychrysis*: Gayral and Fresnel 1983). There is every reason to believe that most of the species described will eventually turn up in most coastal areas. It is interesting that Estep et al. (1984) report *Platychrysis pienaarii* Gayral and Fresnel as well as *Prymnesium patellifera* Green and Hibberd from North Atlantic oceanic localities. *Prymnesium parvum* N. Carter (1937) (Fig. 23,24) can form very dense (and ofter toxic) blooms in brackish water areas.

In the present context it seems relevant to refer to a recent finding by Throndsen (1983) of a very small *Prymnesium* sp. from Japanese coastal waters (2.5–3 μm in diameter). No electron microscopical data on this species is yet available.

The life cycle of the cosmopolitan species *Phaeocystis pouchetii* (Hariot) Lagerheim comprises two characteristic phases, large floating gelatinous colonies, and minute swarmers (3–8 μm) with a typical prymnesiophycean morphology (Fig. 25). The swarmers are easily recognized from electron microscopy of whole mounts (and also from LM when examining "air-mounted" cells) due to the production and release of thread-likle structures showing highly characteristic pentagonal configurations (Fig. 26). For details on ultrastructure and geographical distribution see Parke et al. (1971).

A somewhat larger species of *Phaeocystis*, *P. scrobiculata* Moestrup (1979), has been described from the sea off New Zealand and later recorded from Australia (Hallegraeff 1983) and North Atlantic localities (Estep et al. 1984). It differs from *P. pouchetii* in scale morphology and arrangement of the thread-like material (compare Fig. 26 and Fig. 27).

C. Haptonema present or absent; coccoliths present.

This prymnesiophycean group, comprising the coccolith-bearing forms, is morphologically a very heterogenous group. The coccolithophorids are mainly, but not exclusively marine, and occur at all temperatures, becoming one of the dominant groups of phytoplankton in tropical seas. A limited number of species tolerates and survives in brackish water habitats (e.g. the Baltic Sea) forming a stable plankton element that can be detected in successive growth seasons (Thomsen, unpublished data). Most coccolithophorids are of nanoplanktonic size (5–20 μm). *Gephyrocapsa oceanica* Kamptner (Fig. 30) and *Emiliana huxleyi* (Lohmann) Hay and Mohler (Fig. 29) are examples of widely distributed species that frequently are not any bigger than 5 μm in cell diameter.

The arctic and subarctic waters, previously considered almost devoid of coccolithophorids, have quite recently been shown to contain a considerable number of very small, hitherto unknown coccolithophorids, some of which are as small as 2 μm in cell diameter (Fig. 28: *Trigonaspis minutissima* Thomsen (1980)). A complete list of references, and a summary of the known distribution of these very small cold-water coccolithophorids is found in Thomsen (1981a). Note that *Balaniger balticus* Thomsen and Oates (1978), one of the species mentioned by Thomsen (1981a), is not autotrophic. Recent light microscopical examination of this species, which appears regularly in samples from the Baltic Sea, has shown without doubt that no functional chloroplasts are present inside the cell, and that *B. balticus* is thus the first known heterotrophic member of this algal class (Thomsen 1981b).

For recent, very comprehensive entries to the literature on coccolithophorids see Okada and McIntyre (1977) Heimdal and Gaarder (1980,1981), and Hallegraeff

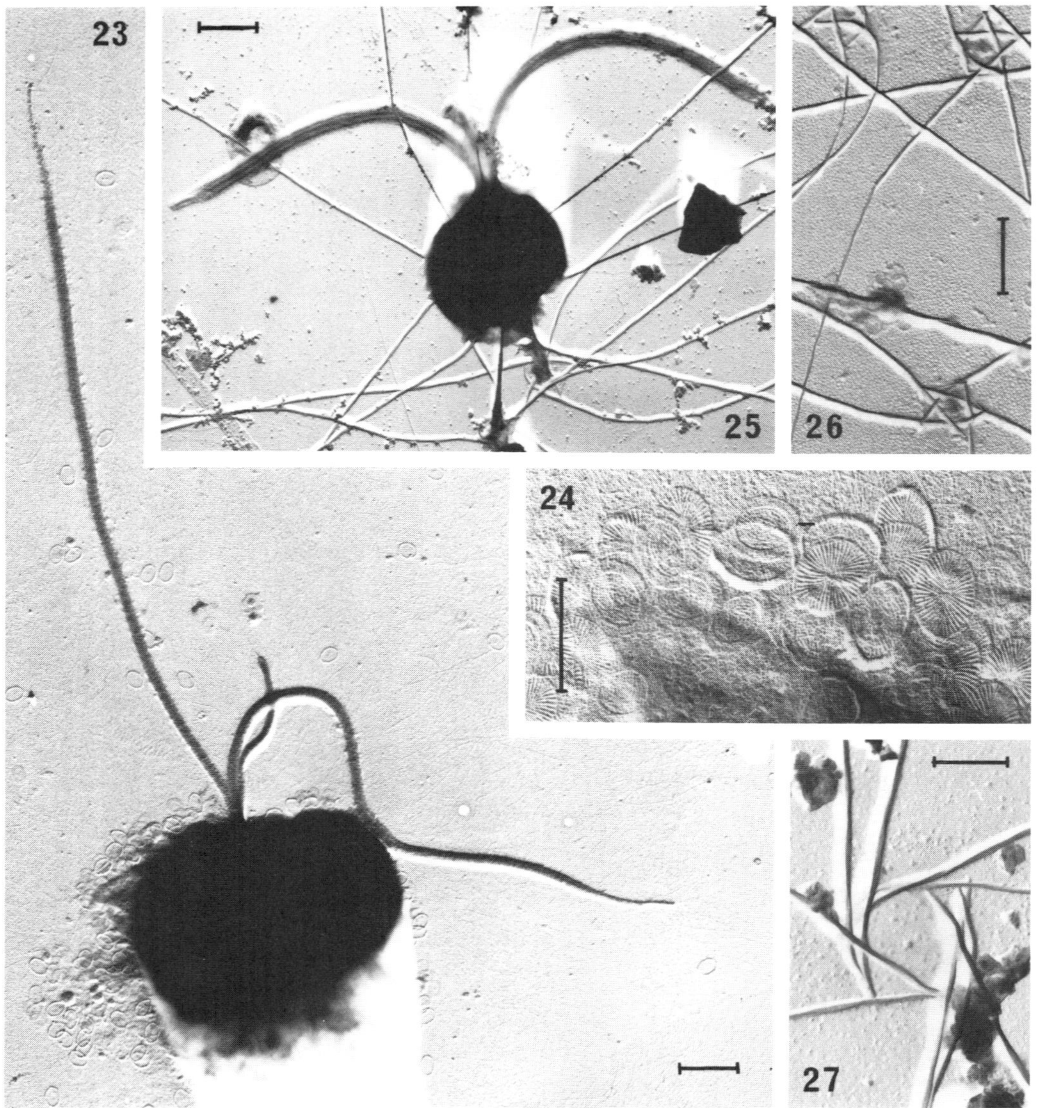

FIG. 23–27. Prymnesiophyceae (Haptophyceae).
FIG. 23,24. *Prymnesium parvum*, shadowcast whole mounts (TEM) of a complete specimen (Fig. 23) and scales (Fig. 24); material from Randers Fjord (Denmark); micrographs T5565/5566, × 7 500/x 30 000. FIG. 25. Shadowcast whole mount (TEM) of *Phaeocystis pouchetii*; material from Læsø Rende (Denmark); micrograph T1085, × 7 500. FIG. 26. Thread-like material from *Phaeocystis pouchetii* lying freely in the preparation; note the pentagonal structures; material from Phuket (Thailand); micrograph T 4573, × 10 000. FIG. 27. Thread-like material from *Phaeocystis scrobiculata* (TEM); notice that nine rays form part of this structure; material from Phuket (Thailand); micrograph T 4574, × 10 000.
Scale bars: Fig. 23,25–27 = 1 µm, Fig. 24 = 0.5 µm.

(1984b). These papers also provide an excellent selection of electron micrographs of numerous coccolithophorid taxa.

PAVLOVALES

This order, which comprises three genera — *Pavlova* Butcher (ten species), *Diacronema* Prauser (one species) and *Exanthemachrysis* Lepailleur (one species) —

138

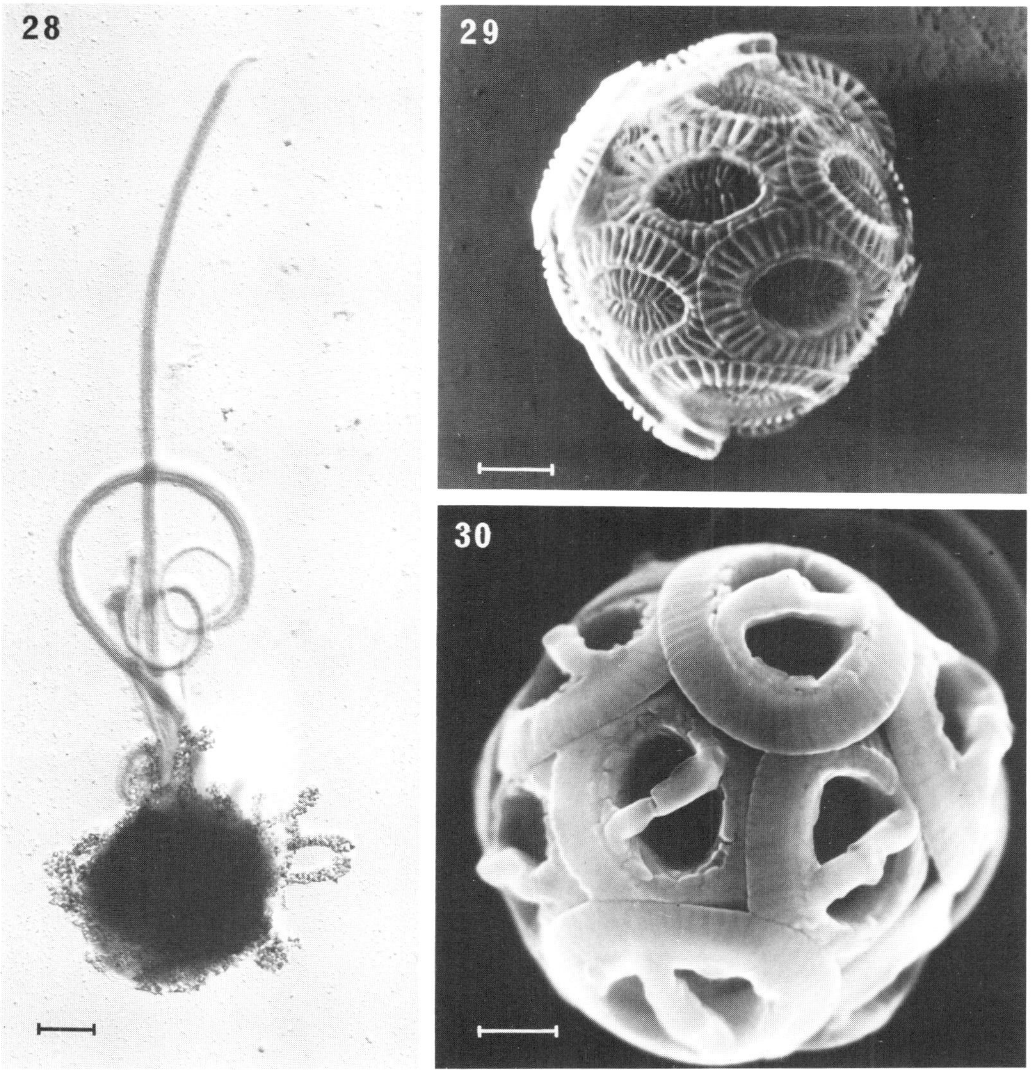

FIG. 28-30. Prymnesiophyceae (Haptophyceae).
FIG. 28. *Trigonaspis minutissima*, shadowcast complete specimen (TEM); material from Godhavn (Greenland); micrograph T2782, × 7 500. FIG. 29. *Emiliana huxleyi* (SEM); material from Isefjorden (Denmark); micrograph b0702, × 10 000. FIG. 30. *Gephyrocapsa oceanica* (SEM); material from Phuket (Thailand); micrograph b2702, × 10 000. Scale bars = 1 μm.

was surveyed in great detail by Green (1980). This paper includes a key to the genera and species of the Pavlovales based on the characters of the motile cells.

In general the motile cell of members of the Pavlovales is elongated and with cell dimensions indicating that these cells, although ultra/nanoplanktonic in size, may cause trouble when filtering to obtain separated picoplankton samples (e.g. *Pavlova gyrans* Butcher (1952) (4–10 × 3–6 × 2–2.5 μm) and *Diacronema vlkianum* Prauser (1958) (3.5–7.5 × 4–5 × 1.5–3 μm)).

Members of the Pavlovales may be found in oceanic, coastal, brackish and fresh water habitats.

139

Euglenophyceae

This algal class does not contain a single species that fits into the 5 μm limit used in this paper (maximum dimension). Members of this class, species of the genus *Eutreptiella* da Cunha in particular, occasionally form very dense blooms in coastal waters. A characteristic of species of *Eutreptiella* is the quite considerable form plasticity that these cells show. In certain highly metabolic species (e.g. *E. gymanstica* Throndsen (1969); cell length 17–38 μm) the cell appears to be able to pass through perforations that are of the same size as the individual chloroplast, or matching the diameter of the nucleus (5–7 μm; Throndsen 1973a). It is thus virtually impossible, by means of filtration, to size fractionate a water sample properly if it contains some of these highly metabolic forms.

Loxophyceae

There has recently been much controversy regarding the class-level grouping of the chlorophyll *a* and *b* containing green flagellates. Several suggestions have been

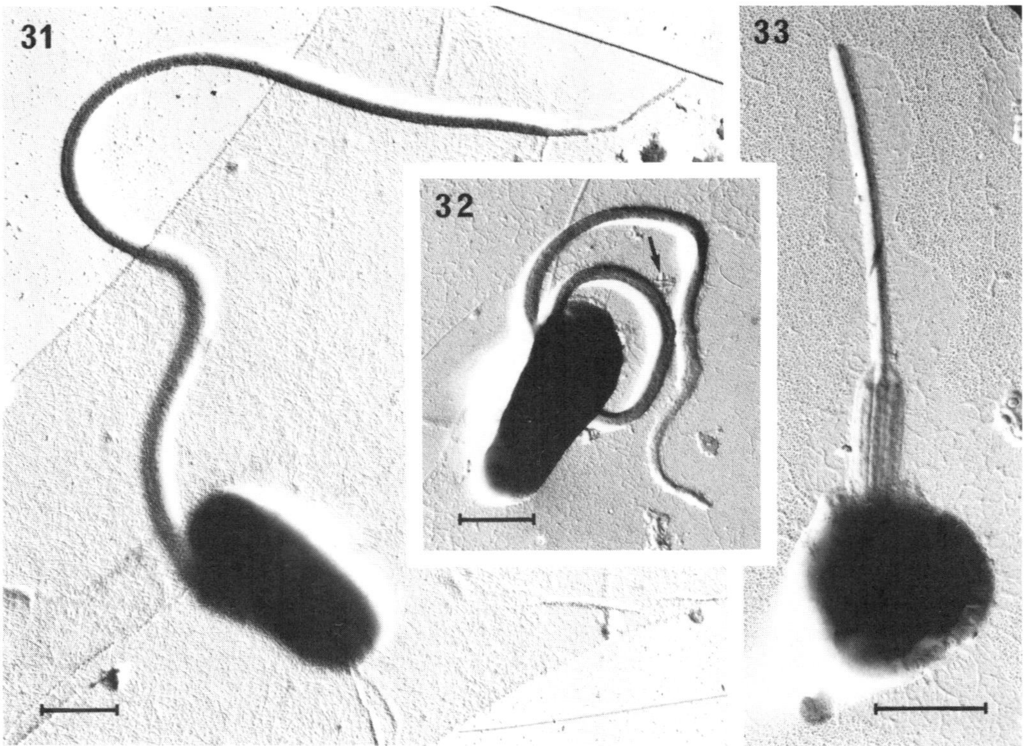

Fig. 31–33. Loxophyceae.
Fig. 31. *Pedinomonas mikron*, shadowcast cell (TEM) overlying a *Dinobryon balticum* (Chrysophyceae) lorica; material from Isefjorden (Denmark); micrograph T3693, × 10 000.
Fig. 32. Shadowcast whole cell (TEM) of *Pedinomonas mikron* from Finnish coastal waters; the scales (arrow) originate from the dinoflagellate *Heterocapsa triquetra*; micrograph T2875, × 10 000. Fig. 33. *Micromonas pusilla* shadowcast whole cell (TEM); the peripheral flagellar tubules stand out clearly in the lower part of the flagellum; material from Godhavn (Greenland); micrograph T 2556, × 15 000.
Scale bars = 1 μm.

140

put forward, but none of these has yet received general approval. The choice made here of retaining the algal class Loxophyceae (Christensen 1962), to accommodate the genera *Pedinomonas* and *Micromonas*, should not be considered a contribution to the ongoing phylogenetic discussion, but be looked upon merely as a practical step, to avoid further confusion while awaiting more ultrastructural and biochemical information on some of the chlorophyll *a* and *b* genera (including those mentioned above). For a recent attempt to classify green algae on the basis of comparative cytology see Mattox and Stewart (1984).

Pedinomonas Korschikoff

Small laterally compressed cells with one flagellum. No scales present on either cell body or flagellum. Three marine species, two of which are known from symbiotic associations only: *P. symbiotica* Cachon and Caram (1979) (4 × 2.5 μm; cells living in the jelly of the radiolarian *Thallassolampa magarodes*) and *P. noctilucae* (Subrahmanyan) Sweeney (Sweeney 1976) (2 × 5μm; symbionts in the dinoflagellate *Noctiluca miliaris*).

Pedinomonas mikron Throndsen (1969) is the only free-living marine species yet described (cell body: 1.5–2.5 μm; flagellum 7–12 μm). This species was described on the basis of light microscopy alone (material from Norwegian coastal waters). It is likely that the shadowcast cells shown in Fig. 31,32 from Finland (Fig. 32) and Denmark (Fig. 31) should be referred to *P. mikron*. *Pedinomonas*-like cells are occasionally very common in brackish water samples from the Baltic Sea (Thomsen, unpublished data).

Pedinomonas mikron has furthermore been recorded from Japanese coastal waters by its author (Throndsen 1983) and from North Atlantic localities (Estep et al. 1984).

Micromonas Manton and Parke

Monotypic genus comprising only the cosmopolitan species *Micromonas pusilla* (Butcher) Manton and Parke (1960) (basionym: *Chromulina pusilla* Butcher (1952)). Cells measure: 1.0–3.0 × 0.75–1.0 μm. The distal 3 μm of the ca. 4 μm long flagellum consists of a two-stranded hairpoint (Fig. 33). The well-known 9 + 2 flagellar substructure is only present in the proximal ca. 1 μm part of the flagellum (Manton 1959; Manton and Parke 1960). No scales present.

Due to the very small size of this genuine picoplanktonic flagellate, a light microscopical identification is obviously difficult, swimming behaviour (described in detail by Manton and Parke (1960)) being the only usable criterion. Because of the very unusual flagellum, this taxon (despite the absence of species-specific scales) is usually easily recognized in electron microscopical whole mount preparations.

Micromonas pusilla has a world wide distribution, occurring both in coastal and oceanic samples (see Throndsen 1976). Cell numbers commonly exceed 10^6 cells per litre. *Micromonas pusilla* has been shown to occur further down in the sea than flagellates in general, occurring well below the euphotic zone (Manton and Parke 1960; Throndsen 1973b, 1983).

PRASINOPHYCEAE

The circumscription of this algal class is in accordance with Moestrup (1982). The grouping into three orders was suggested by Moestrup (1984).

Mamiellales

(comprising monads whose flagella are scale-covered, but lack an inner layer of square or diamond shaped scales).

141

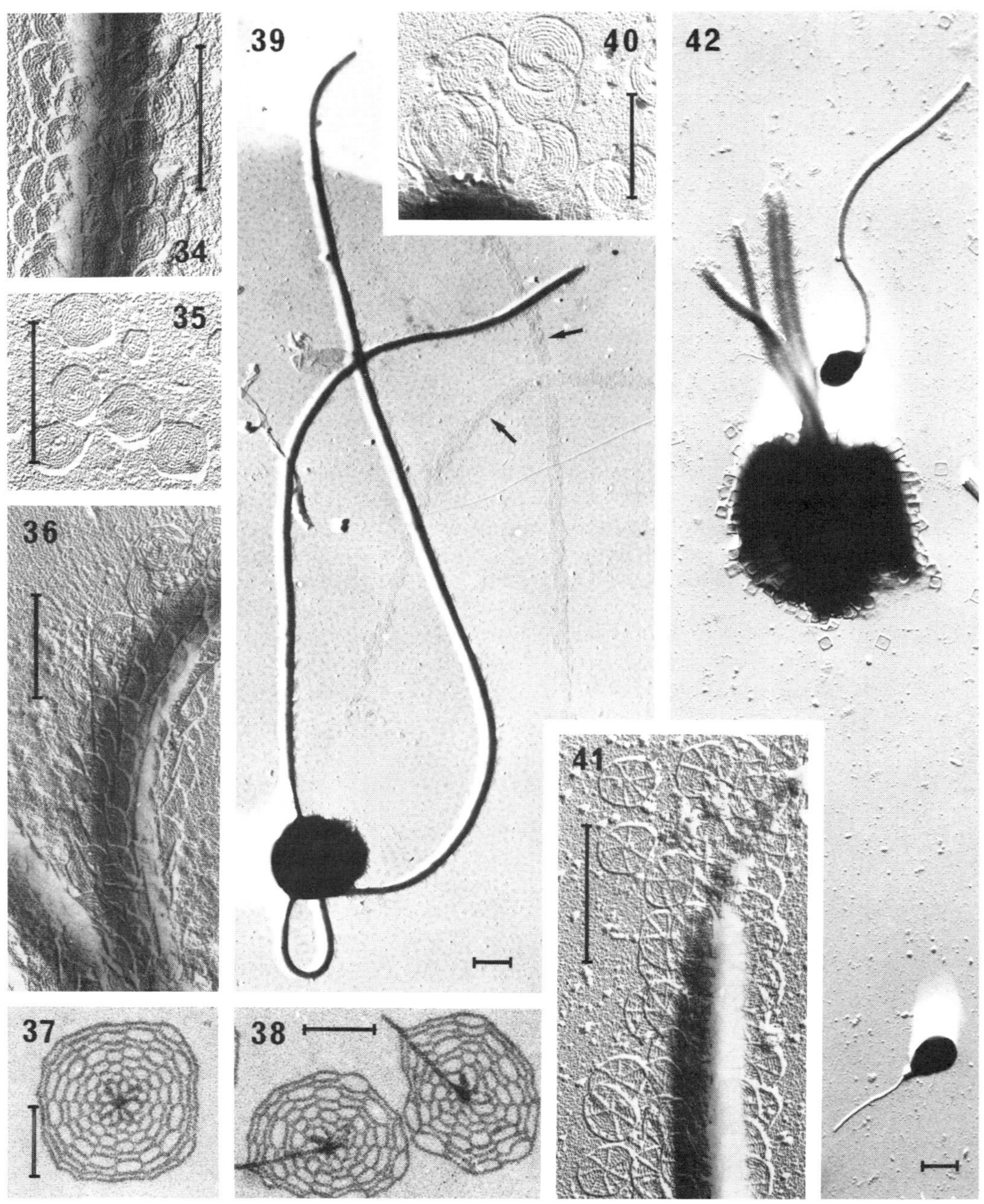

FIG. 34–42. Prasinophyceae.
FIG. 34,35. *Mantoniella squamata*, shadowcast whole mounts (TEM) of flagellar scales (Fig. 34) and body scales (Fig. 35); material from Isefjorden (Denmark); micrograph H178, × 40 000. FIG. 36,37,38. *Mamiella gilva*; flagellar scales (shadowcast preparation, TEM) are shown in Fig. 36., whereas Fig. 37,38 are negative stained preparations (TEM) showing a body scale (Fig. 37), and flagellar scales (with spines) (Fig. 38); material from (Fig. 36) Isefjorden (Denmark), and (Fig. 37,38) type-culture No. 150/1 (Culture Collection of Algae and Protozoa, Cambridge); micrographs T 5103, × 30 000; M 7316 (courtesy Ø. Moestrup), × 100 000. FIG. 39,40. *Dolichomastix nummulifera*, shadowcast whole mounts (TEM) of a complete cell (Fig. 39) to show the very long flagella, and (Fig. 40) to show morphology of body scales; arrows (Fig. 39) point to detached rows of flagellar scales; material from Isefjorden (Denmark);

142

Mamiella Moestrup

Monotypic genus: *M. gilva* (Parke and Rayns) Moestrup (basionym: *Nephroselmis gilva* Parke and Rayns (1964)). The cell carries two long flagella that emerge from a small anterior depression. Typical cell dimensions are: 4–6.5 × 3–4.5 × 3–4 μm (Parke and Rayns 1964). The cell body is covered by two differently sized types of "spider's web" scales (Fig. 37) and on the flagellar surfaces (Fig. 36,38) eight longitudinal rows of "spider's web" scales, each furnished with a short anteriorly directed spine, appear together with flagellar hairs (hair-scales). A detailed light microscopical description of this taxon was given by Parke and Rayns (1964). For electron microscopical details see Moestrup (1984).

Mamiella gilva is a widely distributed species, known from northern and southern hemisphere temperate seas, as well as from the tropics (Moestrup 1984).

A possible second species of *Mamiella* was described from the Mediterranean Sea by Leadbeater (1974) as "*Nephroselmis* sp." (Moestrup 1984).

Mantoniella Desikachary

Monotypic genus: *M. squamata* (Manton and Parke) Desikachary (1972) (basionym: *Micromonas squamata* (Butcher) Manton and Parke (1960)).

Biflagellate cells 3–5 μm diam. (Manton and Parke 1960), with "spider's web" scales completely covering the cell body and the flagella (Fig. 34,35). Those covering the body are slightly bigger than those on the flagella. Otherwise the two types appear to be morphologically indistinguishable. The long flagellum (Fig. 19) also carries typical prasinophycean hairs. One flagellum is 2.5–4 times the length of the cell (Fig. 19), whereas the second flagellum is very much reduced in size and not be observed in the light microscope (up to 1 μm in length) (Barlow and Cattolico 1980).

Mantoniella squamata is distinguished from *Mamiella gilva* by the type of flagellation and also by the absence of anteriorly directed spines on the flagellar scales (compare Fig. 34 and Fig. 36). Manton and Parke (1960) provided a very detailed light microscopical description of this taxon. Electron microscopical details have been accounted for by Manton and Parke (1960) and Barlow and Cattolico (1980).

Mantoniella squamata is a cosmopolitan species commonly found in samples from temperate seas, but also present in arctic waters (Thomsen 1982), tropical seas (Thailand coastal waters, Thomsen, unpublished data) and brackish water (Finnish coastal waters, Thomsen, unpublished data).

Dolichomastix Manton

Three species, all marine: *D. nummulifera* Manton (1977), *D. lepidota* Manton (1977), *D. eurylepidea* Manton (1977).

Each cell carries two very conspicuous flagella (Fig. 39), in length approaching ten times the cell diameter (2.5–5 μm). Body scales and flagellar scales are morphologically very similar, but those covering the flagellar surfaces are significantly smaller than the plate-scales on the body. Flagellar hairs are also present. *D. nummulifera* is characterized by having plate- scales with a surface pattern of conspicuous concentric groves within a thickened margin (Fig. 40). Scales of the other

---

(Fig. 34–42 — *Continued*)
micrographs T 3472/3596, × 5 000/× 30 000. Fig. 41. *Dolichomastix lepidota* (TEM); scales scattered around tip of flagellum; material from Isefjorden (Denmark); micrograph T 3775, × 40 000. Fig. 42. Multispecimen shadowcast micrograph (TEM) showing a large *Pyramimonas* sp. (with two types of box scales), *Pedinomonas mikron* (above) and *Micromonas pusilla* (below); material from Isefjorden (Denmark); micrograph T 3569, × 5 000.
Scale bars: Fig. 34–36, 40,41 = 0.5 μm; Fig. 37,38 = 0.1 μm; Fig. 39,42 = 1 μm.

two species described are constructed from more or less closely woven concentric
threads crossed by a number of radiating ribs (Fig. 41).

No light microscopical information is yet available for any of the *Dolichomastix*
species. From electron microscopy only details of the outer morphology has yet been
accounted for (Manton 1977).

Apart from the findings listed by Manton (1977) reporting all three species from
the Cape Town area, and *D. nummulifera* additionally from three arctic localities,
the only subsequent published finding is Thomsen (1982; *D. nummulifera* from West
Greenland). *Dolichomastix nummulifera* is, however, also capable of living in tropical
seas (Thailand coastal waters; Thomsen, unpublished data) and both *D. nummulifera*
and *D. lepidota* occur regularly in Danish coastal waters (Thomsen, unpublished data).
Based on these additional findings it appears that this genus also has a cosmopolitan
distribution.

Scaly prasinophyte

The tiny (0.5 × 1.0 µm), flagella-less, scaly cells illustrated from thin sections of
embedded material, originating from a number of North Atlantic localities (Johnson
and Sieburth 1982), most likely represent a species that is closely related to the genera
of the order Mamiellales (Moestrup 1984). Cell numbers varied from 10^6–10^7 cells
L^{-1} (Johnson and Sieburth 1982). This yet undescribed taxon has also been found
in water samples from the Pacific Ocean (Silver et al. 1986).

PYRAMIMONADALES

(monadoid forms possess a layer of square or diamond-shaped scales on the flagella
and often also on the cell body).

Pyramimonas Schmarda

A very large genus comprising more than 70 species occurring in a wide range of
habitats. The cells are very often of an inverse pyramidal shape (2–20 µm in length),
and furnished with four or occasionally eight flagella arising form an anterior
depression (Fig. 42). The flagella are covered by three different types of organic scales:
an innermost layer of small diamond-shaped scales, and external layers of limuloid
scales (Fig. 44) and flagellar hairs (Fig. 43). The cell body is typically covered by
three layers of scales: small square scales (Fig. 43), box-scales (Fig. 42), and crown
scales (Fig. 45). Additional scale-layers may be present, e.g. "footprint-scales"
(McFadden et al. 1982).

Most species of *Pyramimonas* are described from light microscopy alone. In many
cases the amount of information included in the species description is insufficient
to prove a possible conspecificity between material examined by modern techniques
and previously examined material. To date electron microscopical details of 21 species
is available. For a complete list of references and an up-to-date review of this genus
see McFadden et al. (1986) and Inouye et al. (1985). Butcher (1959) provided light
microscopical descriptions of several taxa.

Pyramimonas virginica Pennick (1977) (Fig. 43) is the smallest species of the genus
yet encountered (type material: 2.7–3.5 × 1.9–2.4 µm). Whereas the flagella are
furnished with the typical three tiers of scales, the cell body is covered by only two
layers of scales (an innermost layer of small, square scales (Fig. 43, arrow) and an
outermost layer of somewhat atypical *Pyramimonas* scales that are basket-like and
with a six-fold symmetry (Fig. 45)). *Pyramimonas virginica* is furnished with
trichocysts (Fig. 43). This species is known from the Atlantic coast of the USA (Pennick 1977), West Greenland (Thomsen 1982), and Australia (McFadden et al. 1986).
It is furthermore found in samples from New Zealand (Moestrup, personal com-

munication), and Thailand coastal waters (Thomsen, unpublished data), indicating a cosmopolitan distribution.

Tetraselmis Stein

Synonyms (according to Norris et al. 1980): *Prasinocladus* Kuckuck; *Platymonas* G.S. West; *Aulacochlamys* Margalef.

Tetraselmis is also a very large genus comprising approximately 35 species (Ettl 1983). Flagellated cells of *Tetraselmis* are characterized by having four flagella of

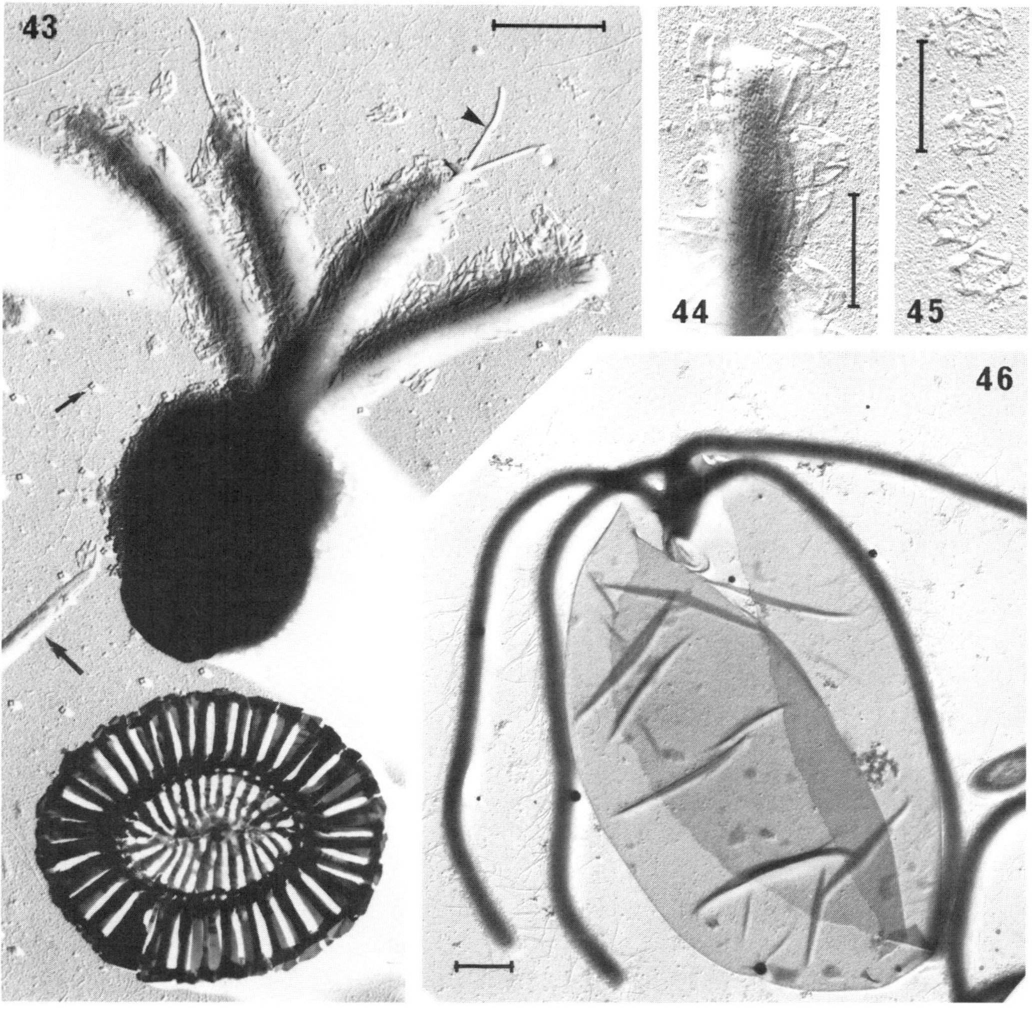

FIG. 43–46. Prasinophyceae.

FIG. 43–45. *Pyramimonas virginica*, shadowcast whole mounts (TEM) of a complete cell (Fig. 43), flagellar scales (Fig. 44) and hexagonal outer layer body scales (Fig. 45). The inner-most, square body scales are present in Fig. 43 (arrow), as are also flagellar hairscales (arrowhead) and discharged trichocyst (big arrow). The coccolith (Fig. 42) comes from *Emiliana huxleyi* (cf. Fig. 29). Material (Fig. 43) from Isefjorden (Denmark), and (Fig. 44,45) from Godhavn (Greenland); micrographs T5470/1765/1957, × 15 000/× 30 000. FIG. 46. *Platymonas* sp. shadowcast whole mount (TEM) of empty scale theca and four flagella; material from Gulf of Aquaba (Israel); micrograph T1307 × 7 500.
Scale bars: Fig. 43,46 = 1 μm; Fig. 44,45 = 0.5 μm.

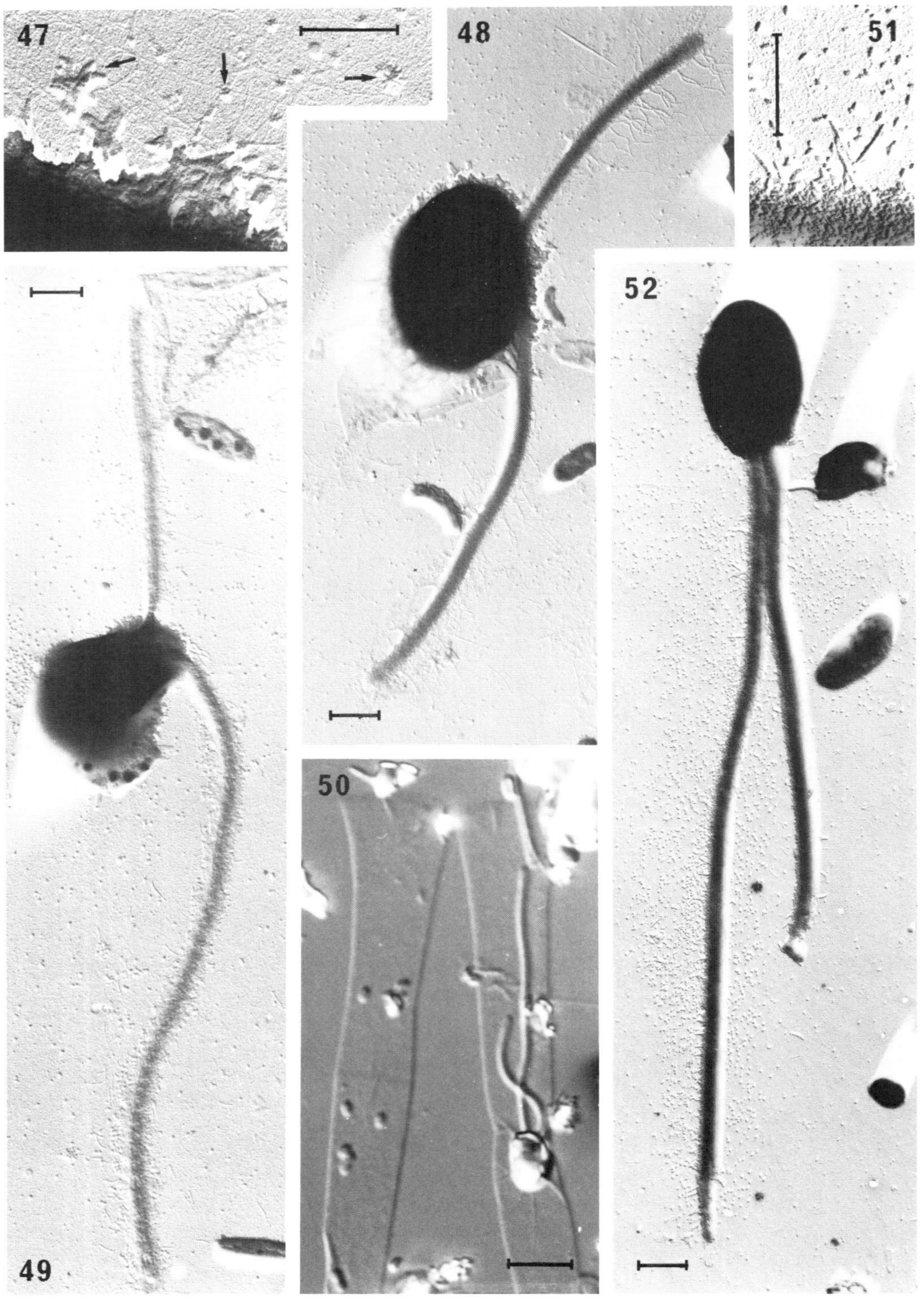

FIG. 47–52. Prasinophyceae.
FIG. 47,48. *Nephroselmis minuta*, shadowcast whole mounts (TEM) of different types of body scales (Fig. 47, arrows) and a complete cell (Fig. 48) (notice the hairscales on the front flagellum); material from Hov Vig (Denmark); micrographs T2952/2955, × 30 000/× 7 500.

146

equal length that arise from an apical, trough-shaped cell depression. Very often the motile cells are somewhat compressed. The flagella are covered by loosely attached flagellar hairs (hair-scales) (Fig. 46) and two layers of minute scales covering the entire flagellar membrane. The outermost layer consists of rod-shaped scales very similar to those of *Pseudoscourfieldia* (Fig. 51), whereas the innermost layer consists of scales which are square or pentagonal in face view, thus being very similar to those found in species of *Pyramimonas*. The cell body is closely enveloped by a theca formed by the coalescence of minute stellate scales (Fig. 46; shadowcast empty theca of *Tetraselmis* sp.).

Most species of *Tetraselmis* are rather big (i.e. cell length 10–20 μm), but at least one species, *T. inconspicua* Butcher (1959) is known to be as small as 4.5 μm.

A revision of the genus *Tetraselmis* is currently being undertaken using ultrastructural details to delimit species and subgenera (Norris et al. 1980; Hori et al. 1982,1983). A light microscopical study of several marine species was published by Butcher (1959).

The genus *Tetraselmis* has a worlwide distribution. Species of *Tetraselmis* always turn up in enriched coastal water samples.

Nephroselmis Stein

Synonyms (according to Ettl 1983): *Bipedinomonas* N. Carter; *Heteromastix* Korschikoff.

This genus comprises for the moment four marine species (viz. *N. rotunda* (N. Carter) Fott (Manton et al. 1965); *N. pyriformis* (N. Carter) Ettl (Moestrup 1983); *N. minuta* (N. Carter) Butcher; *N. astigmatica* Inoye and Pienaar (Inoye and Pienaar 1984)). The references given are the most recent electron microscopical treatments of the taxon in question.

Species of *Nephroselmis* are characterized by a flattened cell body with two unequal flagella. The surface of the flagella is covered by at least two layers of scales; minute rod-like scales, and rounded plates. Hair-scales are also present. The cell surface is coated by two to four (e.g. *N. astigmatica*) tiers of scales. In *N. pyriformis* (Fig. 49,50), which is the most commonly reported species, the inner layer consists of minute square plates arranged in rows. Each scale has a small raised rim and a central spine. The outer layer consists of stellate scales in size corresponding to the squares.

Nephroselmis astigmatica is a relatively large species (type material: 9–13.5 × 5–11.5 μm) so far known only from South Africa and Japan (Inoye and Pienaar 1984). *Nephroselmis rotunda* and *N. pyriformis* are rather similar in size (4–8 μm), whereas *N. minuta* is slightly smaller (type material: 3 × 3.5 × 1.5 μm; N. Carter 1937). The latter species has recently been established in culture from Baltic Sea brackish water samples, and is currently being examined from electron microscopical thin sections (G. and S. Hällfors and Moestrup, personal communication). Based on their preliminary results it is evident that the cell shown in Fig. 47,48, originating from Danish coastal water, represents *N. minuta* (differing form the other species particularly as regards the length of flagella relative to the cell body (compare with *N. pyriformis* (Fig. 49,50)).

(FIG. 47–52 — *Continued.*)
FIG. 49,50. *Nephroselmis pyriformis*, shadowcast whole cell (Fig. 49) (TEM); material from Isefjorden (Denmark); micrograph T2503, × 7 500. Fig. 50. LM, interference contrast optics, showing a cell that has taken over a *Dinobryon balticum* lorica; material from Godhavn (Greenland), × 2 000. FIG. 51,52. *Pseudoscourfieldia marina*, shadowcast whole mounts (TEM) showing flagellar scales (Fig. 51) and a complete cell (Fig. 52) surrounded by bacteria; material from Isefjorden (Denmark); micrographs 00266/00265, × 30 000/× 7 500. Scale bars: Fig. 47,51 = 0.5 μm; Fig. 48,49,52 = 1 μm; Fig. 50 = 5 μm.

Whereas *N. rotunda* and *N. minuta* are so far only known from a very limited number of localities, *N. pyriformis* has been shown to have a worldwide distribution (Moestrup 1984), yet listed under very many different names (for a list of synonyms see Moestrup 1984).

Although the species of *Nephroselmis* are indeed very small organisms it is possible with some degree of certainty to distinguish light microscopically between living cells. Cell symmetry, flagellar lengths, swimming behaviour and the flagellar resting positions form the most important features to look for (for further details see Manton et al. (1965) and Moestrup (1984)).

Pseudoscourfieldia Manton

Monotypic genus: *P. marina* (Throndsen) Manton (basionym: *Scourfieldia marina* Throndsen (1969)). *Pseudoscourfieldia marina* is a very small species (type material 3–3.5 × 2.5–3 × 1.5 μm) originally described on the basis of light microscopy and electron microscopy of shadowcast whole mounts (Throndsen 1969). An electron microscopical reinvestigation of this taxon was undertaken by Manton (1975). It is evident that *P. marina* (Fig. 51,52) is in many respects similar to *Nephroselmis pyriformis* (Fig. 49). The scaly coverings of the two taxa are thus basically similar although minor morphological details distinguish them, e.g. the stellate outer layer body scales. A useful distinguishing characteristic is, however, the insertion of the flagella. In *P. marina*, which swims with the two flagella directed backwards, the flagella are almost parallel (Fig. 52), whereas in *N. pyriformis* the two flagella are not only used in a very much different way, but also diverge at an angle of approximately 90 degrees (Fig. 49).

Pseudoscourfieldia marina has been found in water samples from Norway (Throndsen 1969), South Africa and Denmark (Manton 1975), North Atlantic localities (Estep et al. 1984).

HALOSPHAERALES

(life cycle includes a phycoma stage, otherwise like the proceeding order).

The cosmopolitan genus *Pterosperma* Pouchet should be mentioned here, inasmuch as the monadoid life cycle stages of this genus vary in size from 4 to 10 μm (Parke et al. 1978). Flagellated cells of *Pterosperma* are highly conspicuous due to the four very long flagella (3–10 times body diameter in length) attached to a rather small cell body (Fig. 53). All flagellated cells are scale covered. It is, however, not possible to distinguish between the different species of *Pterosperma* on the basis of scale morphology alone (Fig. 54). Examination of morphological details of the phycoma stage is a must in this context. For details of cell morphology, taxonomy, distribution etc. see Parke et al. (1978).

CHLOROPHYCEAE

The amount of information on marine planktonic green algae, other than members of the Euglenophyceae, Loxophyceae and Prasinophyceae, is surprisingly sparce. This is most probably due to difficulties in carrying out species identification of both the monadoid and the coccoid members of this group, rather than implying unimportance of these algae in the marine ecosystem. There is reason to believe that green algae once recognized may be shown to contribute significantly to the pico/ultraplankton biomass and productivity of eutrophicated coastal areas in particular (see Guillard et al. 1975). See also Joint and Pipe (1984) and Takahashi and Hori (1984), who illustrated *Chlorella*-like cells from thin sections of embedded picoplankton samples from the Celtic Sea and the South China Sea respectively.

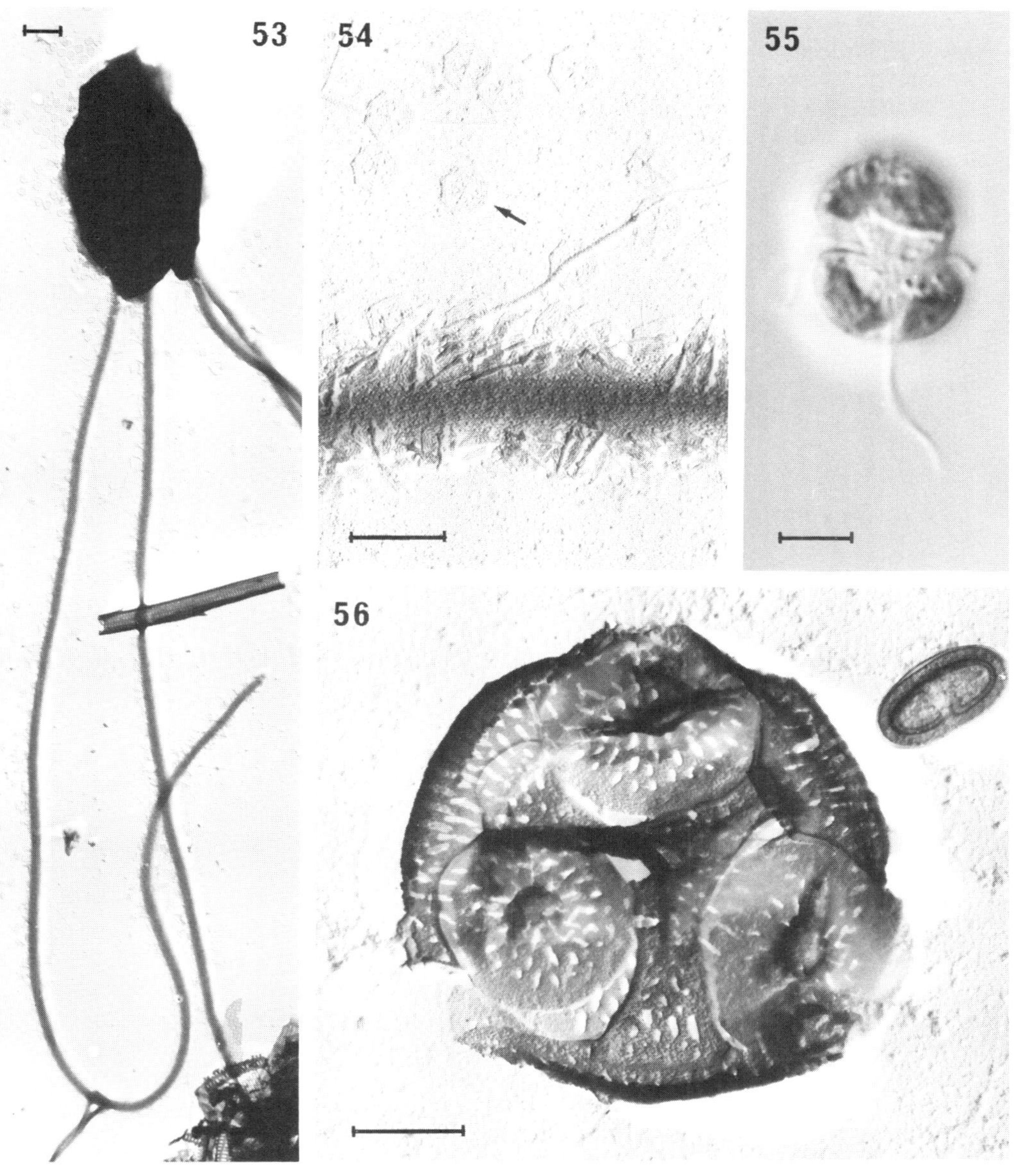

FIG. 53,54. Prasinophyceae. FIG. 55. Dinophyceae. FIG. 56. Incertae sedis.
FIG. 53,54. *Pterosperma* sp., shadowcast whole mounts (TEM) of flagellated stage (Fig. 53) with four long flagella, and (Fig. 54) body scales (arrow) and flagellar scales (bottom of figure); material from Isefjorden (Denmark); micrographs T2925/709, × 5 000/× 25 000. FIG. 55. *Gymnodinium simplex* aff. (LM, interference contrast optics); material from the Baltic Sea (Gulf of Bothnia), × 2 000. FIG. 56. Siliceous cyst (TEM) showing characteristic circular plates separated by triradiate structure; material from the Baltic Sea (Gulf of Bothnia); micrograph T 4962, × 15 000.
Scale bars: Fig. 53,56 = 1 μm; Fig. 54 = 0.5 μm; Fig. 55 = 5 μm.

The following list of taxa (far from complete) is included to provide references to some of the small planktonic green algae recorded from marine habitats. *Dunaliella* and *Chlamydomonas* are monadoid forms (2 equal flagella) whereas all the others are coccoid forms.

Dunaliella Teodorescu
  D. primolecta Butcher (1959) (5–7 μm)
Chlamydomonas Ehrenberg
  C. coccoides Butcher (1959) (4.5–5 μm)
  C. concordia Green in Green et al. (1978) (6–10 × 2.5–8.5 μm)
Chlorella Beijerinck
  C. marina Butcher (1952) (4–6 × 7–10 μm)
  C. ovalis Butcher (1952) (3–5 × 5–10 μm)
  C. salina Butcher (1952) (4–7 μm)
  C. stigmatophora Butcher (1952) (4–6 μm)
  C. sp. Johnson and Sieburth (1982) (2–3 μm)
Halochlorococcum Dangeard
  H. saccatum Guillard et al. (1975) 2.5–60 μm)
Mychonastes Simpson and van Valkenburhg
  M. ruminatus Simpson and van Valkenburg (1978) (4–10 μm)
    (2 × 10⁷ cells L⁻¹; Chesapeake Bay, USA)
Nanochloris Naumann
  N. atomus Butcher (1952) (3 μm)
  N. maculatus Butcher (1952) (3 μm)
Stichococcus Naegeli
  S. cylindricus Butcher (1952) (2 × 3–4 μm)

DINOPHYCEAE

Looking through the textbooks on dinoflagellates (e.g. Dodge 1982) in a search for organisms smaller than 5 μm, is a very little profitable task. There is hardly any autotrophic dinoflagellate smaller than 10 μm. Even a species such as *Katodinium rotundatum* (Lohm.)Loeblich, which is very small when compared with dinoflagellates in general, rarely measures less than 10 μm in length (Dodge 1982; length 8–17 μm, width 6–12 μm).

Only *Gymnodinium simplex* (Lohm.)Kof. and Swezy (basionym: *Protodinium simplex* Lohm. (1908)) deserves to be mentioned specifically here. According to Dodge (1974), examining the ultrastructure of this taxon, on the basis of cultured material from the Plymouth Sound, typical cell dimensions are 7–9 × 6–7.5 μm. Lebour (1925) was likewise able to examine cultures obtained from Plymouth Sound samples. These organisms were, however, considerably smaller (1.7–7 μm long) than either those studied by Lohmann (1908) or Dodge (1974). The smallest cells examined by Lebour were apparently lacking both grooves and flagella. Dodge (1974) lists some reported occurrences of *G. simplex* indicating a worldwide distribution. The cell shown in Fig. 55 (12 × 10 μm) is most likely *G. simplex*, corresponding to e.g. the material examined by Dodge (1974) as regards cell-shape and number, position of chloroplasts and dimensions.

Conrad and Kufferath (1954) illustrate and describe (without too many details) a considerable number of small dinoflagellates from the Belgian coast. Some of these forms are no longer than 10 μm.

It is very difficult to avoid reflecting on the reasons why dinoflagellates rarely produce any small specimens. Dinoflagellates certainly differ ultrastructurally in many respects from the other groups of organisms (fairly large nucleus with permanently condensed chromosomes, theca, pusule etc.). In general dinoflagellates appear to have a more complex cell ultrastructure than most other unicellular phytoplankton-organisms, and also the individual dinoflagellate organelles appear to be less easily reduced in size–facts which together may account for the apparent lack of pico/ultraplanktonic forms within this class.

150

Siliceous cysts

Siliceous cysts, not yet classified, have recently been described from the Gulf of Alaska (Booth et al. 1980,1981; size range: 2.5–5.5 μm) and from the Weddell Sea (Silver et al. 1980: size range: 2.5–15 μm). These siliceous forms, each furnished with a cell wall made of eight siliceous plates, have been shown to be very abundant, sometimes the most numerous organisms (Booth et al. 1980; 7 $\times$ 10^5 cells L^{-1}). It is unknown whether or not these cysts contain a chloroplast. It is likewise unknown how these siliceous forms relate to other plankton organisms. The possibility that these cysts form part of the life cycle of the heterotrophic choanoflagellates was put forward by Booth et al. (1980) and Silver et al. (1980). As already stressed by these authors the evidence for this hypothesis is circumstantial. The cyst shown in Fig. 56 occurred in a brackish water sample from the Baltic Sea, indicating that these cysts have a considerably wider distribution than might be expected from previous reports, which presented results from oceanic localities only.

Archaeomonadaceae

Until recently, these siliceous forms (probably related to the Chrysophyceae) have been thought to be extinct. Following the recognition of these cells in recent samples also (Mitchell and Silver 1982; Weddell Sea) archeomonad cysts have been reported by Buck and Garrison (1983; Weddell Sea; max. concentration 7 $\times$ 10^4 cells L^{-1}) and Booth et al. (1982; Gulf of Alaska; max. concentration 5.4 $\times$ 10^4 cells L^{-1}).

Identification Techniques and Problems

When working with organisms smaller than ca. 5 μm, light microscopy (even when performed on living cells) is not always sufficient to ascertain a reliable species identification (for details on how to optimize light microscopy on naked flagellates see Throndsen (1980)). To obtain adequate information on features such as flagellation, mineralization of scales, thecae, periplasts or frustules, some kind of electron microscopical technique is needed in addition to light microscopy. The basic techniques are as follows (the first reference given is a technical description of the method; the second reference is a selected paper presenting results based on the technique in question):

Transmission electron microscopy (TEM)
Shadowcast whole mounts
  (Moestrup and Thomsen 1980; Leadbeater 1972a,b)
Negative staining (Fig. 37,38)
  (Gantt 1980; Moestrup 1984)
Carbon replica technique
  (Gantt 1980; Okada and McIntyre 1977)
Freeze-etch techniques
  (Staehelin 1980; McFadden et al. 1986)
Embedding/sectioning
  (Reimann et al. 1980; Johnson and Sieburth 1982)
  See also e.g. Reymond and Pickett-Heaps (1983) presenting a routine flat embedding technique, which allows the selection and precise orientation of single cells by light microscopy for subsequent sectioning.

Scanning electron microscopy (SEM) (Fig. 29,30)
Whole mounts, incl. critical point drying
  (Paerl and Shimp 1973; Booth et al. 1982)

In most cases large numbers of cells will be needed to prepare a detailed description of a certain species. This means that some kind of concentration of the organisms present will be needed in most cases. The generally used methods for this purpose include: filtering (Li 1986), centrifugation (Throndsen 1978a) or culturing (Brand 1986; Throndsen 1978b).

Methods Used to Estimate Abundance of Nanoplanktonic Cells

A number of different methods have recently been used to obtain data on the abundance of the smallest organisms of the sea. None of these methods has proved to be universally applicable, inasmuch as all have been shown to bias the results obtained somewhat. There follows a short listing of the major techniques used (including references to recent works in which the method in question is either described or employed. Excluded from the list of techniques are those routines that provide only information on cell numbers, according to size-classes, omitting qualitative information.

1. Counts of living cells using e.g. a haemacytometer chamber
   (Guillard 1978; Booth et al. 1982)
2. Counts of cells using inverted light microscopy on preserved water samples (Utermöhl-technique)
   (Hasle 1978; Throndsen 1973b; Booth et al. 1982; Buck and Garrison 1983)
3. Counts of cells using SEM (critical point dried samples)
   (Paerl and Shimp 1973; Booth et al. 1982)
4. Estimates of cell abundancies obtained from the serial dilution culture technique.
   (Throndsen 1969, 1976, 1978b,c, 1983).

A whole list of advantages and disadvantages may be added to each of the enumeration techniques listed. When working on living material the cells are generally in perfect shape, but identification is hampered by the limited resolution of the light microscope, and also by the limited ability of many fragile flagellates to survive for any length of time under these conditions. Also this technique may be difficult to use at sea. Any known fixative seriously changes the morphology of one or another of the groups of organisms present in the sample, thus making light microscopical recognition even more difficult. Certain groups of organisms are completely destroyed following fixation (see Booth et al. 1982). There is thus no doubt that attemps to quantify the smallest organisms following the Utermöhl procedure must be accepted with some reservation. The dilution culture technique has certainly provided much knowledge concerning the abundance and distribution of small flagellates in particular (see Throndsen 1978c). The technique, however, probably underestimates the cell densities in the original sample, and at the same time also favours certain groups of organisms (depending on the culture medium used). There is no doubt, however, that this technique should be considered seriously, at least when working in coastal waters. The enumeration of the smallest forms using a SEM-based technique is very promising (see Booth et al. 1982). One of the only disadvantages of this method is the inability to distinguish photosynthetic from non-photosynthetic organisms, and likewise the difficulties relating to distinguishing living organisms from certain detritus particles. These problems could be minimized by examining filters using a fluorescence microscope at a very early preparational stage. On the basis of chlorophyll auto-fluorescence it should be possible to prepare a diagram with the coordinates of the photosynthetic forms. Likewise the addition of proflavin to the sample (to stain the DNA areas) renders possible, by means of fluorescence microscopy the preparation of a map showing the "living" organisms on the filter. By using some of the modern picture analysis systems, these procedures might even be automatized, and if furthermore the coordinates of the photosynthetic/non-photosynthetic forms can be fed directly into the SEM unit and located automatically, the whole enumeration and identification procedure might become very routine.

Acknowledgements

I wish to thank Ø. Moestrup for providing Table 2 of this paper, and also for supplying some micrographs reproduced here as Fig. 15,37,38. Technical assistance was carefully provided by Lene Christiansen and Lisbeth Haukrogh. Fundings from the Danish Natural Science Research Council (providing microscopes, travel grants etc.) have been of importance to the present work.

References

ANTIA, N. J., T. BISALPUTRA, J. Y. CHENG, AND J. P. KALLEY. 1975. Pigment and cytological evidence for reclassification of *Nannochloris oculata* and *Monallantus salina* in the Eustigmatophyceae. J. Phycol. 11: 339–343.

BARLOW, S. B., AND R. A. CATTOLICO. 1980. Fine structure of the scale-covered flagellate *Mantoniella squamata* (Manton et Parke) Desikachary. Br. Phycol. J. 15: 321–333.

BILLARD, C. 1983. *Prymnesium zebrinum* sp.nov. et *P. annuliferum* sp.nov., deux nouvelles espèces apparantées à *P. parvum* Carter (Prymnesiophyceae). Phycologia 22: 141–151.

BOLD, H. C., AND M. J. WYNNE. 1985. Introduction to the algae. 2nd ed., Prentice Hall, Inc. 720 p.

BOOTH, B. C., J. LEWIN, AND R. E. NORRIS. 1980. Siliceous nanoplankton I. Newly discovered cysts from the Gulf of Alaska. Mar. Biol. 58: 205–209.

1981. Silicified cysts in North Pacific nanoplankton. Biol. Oceanogr. 1: 57–80.

1982. Nanoplankton species predominant in the subarctic Pacific in May and June 1978. Deep-Sea Res. 29: 185–200.

BOURRELLY, P. 1958. Note systématique sur quelques algues microscopiques des cuvettes supralittorales de la région de Dinard. Bull. Lab. Mar. Dinard 43: 111–118.

BRAND, L. E. 1986. Nutrition and culture of autotrophic ultraplankton and picoplankton, p. 205–233. *In* T. Platt andd W. K. W. Li [ed.] Photosynthetic picoplankton. Can. Bull. Fish. Aquat. Sci. 214.

BRETT, S. J., AND R. WETHERBEE. 1985. A comparative study of the periplast in two species of *Cryptomonas*. Second Int. Phycological Congress, Copenhagen (abstract).

BUCK, K. R., AND D. L. GARRISON. 1983. Protists from the ice-edge region of the Weddell Sea. Deep-Sea Res. 30: 1261–1277.

BUTCHER, R. W. 1952. Contributions to our knowledge of the smaller marine algae. J. Mar. Biol. Assoc. U.K. 31: 175–190.

1959. An introductory account of the smaller algae of British coastal waters. Part 1: Introduction and Chlorophyceae. Fish. Invest. London, Ser. IV, p. 1–74.

1967. An introductory account of the smaller algae of British coastal waters. Part IV: Cryptophyceae. Fish. Invest., London, Ser. IV, p. 1–54.

CACHON, M., AND B. CARAM. 1979. A symbiotic alga, *Pedinomonas symbiotica* sp.nov. (Prasinophyceae), in the radiolarian *Thalassolampe margarodes*. Phycologia 18: 177–184.

CARTER, N. 1937. New or interesting algae from brackish water. Arch. Protistenk. 90: 1–68.

CHANG, F. H., AND K. G. RYAN. 1985. *Prymnesium calathiferum* sp. nov. (Prymnesiophyceae), a new species isolated from Northland, New Zealand. Phycologia 24: 191–198.

CHRISTENSEN, T. 1962. Alger. *In* T. W. Böcher, M. LANGE, and T. Sørensen [ed.] Botanik, II, No. 2 Munksgaard, Copenhagen. 178 p.

1980. Algae. A taxonomic survey. Fasc. 1. AiO Tryk as, Odense. 216 p.

CLARKE, K. J., AND N. C. PENNICK. 1975. *Syncrypta glomifera* sp.nov., a marine member of the Chrysophyceae bearing a new form of scale. Br. Phycol. J. 10: 363–370.

COLLIER, A., AND A. MURPHY. 1962. Very small diatoms: Preliminary notes and description of *Chaetoceros galvestonensis*. Science 136: 780–781.

CONRAD, W., AND H. KUFFERATH. 1954. Recherches sur les eaux saumâtre des environs de Lilloo. II. Partie descriptive. Mém. Inst. r. Sci. Nat. Belg. 127: 1–346.

COX, E. R. [ed.]. 1980. Phytoflagellates. Elsevier/North-Holland, New York, NY. 473 p.

DESIKACHARY, T. V. 1972. Notes on Volvocales. I. Curr. Sci. 41: 445–447.

DIXON, P. S., AND L. M. IRVINE. 1977. Seaweeds of the British Isles. Vol. 1 Rhodophyta. Part 1 Introduction, Nemaliales, Gigartinales. British Museum (Natural History), London. 252 p.

DODGE, J. D. 1973. The fine structure of algal cells. Academic Press, London. 261 p.

1974. A redescription of the dinoflagellate *Gymnodinium simplex* with the aid of electron microscopy. J. Mar. Biol. Assoc. U.K. 54: 171–177.

1982. Marine dinoflagellates of the British Isles. London: Her Majesty's Stationery Office. 303 p.

DROOP, M. R. 1955. Some new supra-littoral Protista. J. Mar. Biol. Assoc. U.K. 34: 233–245.

ESTEP, K. W., P. G. DAVIS, P. E. HARGRAVES, AND J. M. SIEBURTH. 1984. Chloroplast containing microflagellates in natural populations of North Atlantic nanoplankton, their identification and distribution; including a description of five new species of *Chrysochromulina* (Prymnesiophyceae). Protistologica 20: 613–634.

ETTL, H. 1983. Süsswasserflora von Mitteleuropa.

Bd. 9: Chlorophyta I. Phytomonadina. Gustav Fischer Verlag, Stuttgart. 807 p.

EVANS, L. V. 1970. Electron microscopical observations on a new red algal unicell, *Rhodella maculata* gen.nov., sp.nov. Br. Phycol. J. 5: 1–13.

GANTT, E. 1980. Replica production and negative staining, p. 377–383. *In* E. Gantt [ed.] Handbook of phycological methods; development & cytological methods, Cambridge University Press.

GAYRAL, P., AND J. FRESNEL. 1983. *Platychrysis pienaarii* sp.nov. et *P. simplex* sp.nov. (Prymnesiophyceae): description et ultrastructure. Phycologia 22: 29–45.

GREEN, J. C. 1980. The fine structure of *Pavlova pinguis* Green and a preliminary survey of the order Pavlovales (Prymnesiophyceae). Br. Phycol. J. 15: 151–191.

GREEN, J. C., AND R. N. PIENAAR. 1977. The taxonomy of the order Isochrysidales (Prymnesiophyceae) with special reference to the genera *Isochrysis* Parke, *Dicrateria* Parke and *Imantonia* Reynolds. J. Mar. Biol. Assoc. U.K. 57: 7–17.

GREEN, J. C., D. J. HIBBERD, AND R. N. PIENAAR. 1982. The taxonomy of *Prymnesium* (Prymnesiophyceae) including a description of a new cosmopolitan species, *P. patellifera* sp.nov., and further observations on *P. parvum* N. Carter. Br. Phycol. J. 17: 363–382.

GREEN, J. C., D. NEUVILLE, AND P. DASTE. 1978. *Chlamydomonas concordia* sp.nov. (Chlorophyceae) from oyster ponds on the Île d'Oléron, France. J. Mar. Biol. Assoc. U.K. 58: 503–509.

GUILLARD, R. R. L. 1978. Counting slides, p. 182–189. *In* A. Sournia [ed.] Phytoplankton manual, UNESCO Monograph.

GUILLARD, R. R. L., AND J. H. RYTHER. 1962. Studies on marine planktonic diatoms. I. *Cyclotella nana* Hustedt and *Detonula confervacea* (Cleve)Gran. Can. J. Microbiol. 8: 229–239.

GUILLARD, R. R. L., H. C. BOLD, AND F. J. MACENTEE. 1975. Four new unicellular chlorophycean algae from mixohaline habitats. Phycologia 14: 13–24.

HALLDAL, P., AND J. MARKALI. 1955. Electron microscope studies on coccolithophorids from the Norwegian Sea, the Gulf Stream and the Mediterranean. Avh. Norske VidenskAkad. I. Mat. Nat. Kl. 1955, 1: 1–30.

HALLEGRAEFF, G. M. 1983. Scale-bearing and loricate nanoplankton from the East Australian Current. Bot. Mar. 26: 493–515.

1984a. Species of the diatom genus *Thalassiosira* in Australian waters. Bot. Mar. 27: 495–513.

1984b. Coccolithophorids (calcareous nanoplankton) from Australian waters. Bot. Mar. 27: 229–247.

HÄLLFORS, G., AND Å. NIEMI. 1974. A *Chrysochromulina* (Haptophyceae) bloom under the ice in the Tvärminne archipelago, southern coast of Finland. Mem. Soc. Fauna Flora Fennica 50: 89–104.

HÄLLFORS, S. 1981. The algal genus *Chrysochromulina* Lackey (Prymnesiophyceae) in the plankton of the archipelago of southern Finland, and a preliminary division of the genus into sections. M.Sc. thesis, Dep. of Bot., Univ. of Helsinki. 76 p. (in Finnish).

HÄLLFORS, S., AND H. A. THOMSEN. 1985. *Chrysochromulina brachycylindra* sp.nov. (Prymnesiophyceae) from Finnish coastal waters. Nord. J. Bot. 5: 499–504.

HARGRAVES, P. E., AND R. R. L. GUILLARD. 1974. Structural and physiological observations on some small marine diatoms. Phycologia 13: 163–172.

HASLE, G. R. 1973. Thalassiosiraceae, a new diatom family. Norw. J. Bot. 20: 61–69.

1976. The biogeography of some marine planktonic diatoms. Deep-Sea Res. 23: 319–338.

1978. Using the inverted microscope, p. 191–196. *In* A. Sournia [ed.] Phytoplankton Manual, UNESCO Monograph.

1983. The marine, planktonic diatoms *Thalassiosira oceanica* sp.nov. and *T. partheneia*. J. Phycol. 19: 220–229.

HASLE, G. R., AND B. R. HEIMDAL. 1970. Some species of the centric diatom genus *Thalassiosira* studied in the light and electron microscopes. Beih. Nova Hedwigia 31: 545–581.

HASLE, G. R., H. A. VON STOSCH, AND E. E. SYVERTSEN. 1983. Cymatosiraceae, a new diatom family. Bacillaria 6: 9–156.

HEATH, I. B., AND W. M. DARLEY. 1972. Observations on the ultrastructure of the male gametes of *Biddulphia levis* Ehr. J. Phycol. 8: 51–59.

HEIMDAL, B. R., AND K. R. GAARDER. 1980. Coccolithophorids from the northern part of the eastern central Atlantic I. Holococcolithophorids. "Meteor" Forsch.-Ergebnisse, Reihe D, 32: 1–14.

1981. Coccolithophorids from the northern part of the eastern central Atlantic II. Heterococcolithophorids. "Meteor" Forsch.-Ergebnisse, Reihe D, 33: 37–69.

HIBBERD, D. J. 1980. Eustigmatophytes, p. 319–334. *In* E. R. Cox [ed.] Phytoflagellates, Elsevier/North Holland, New York, NY.

1981. Notes on the taxonomy and nomenclature of the algal classes Eustigmatophyceae and Tribophyceae (synonym Xanthophyceae). Bot. J. Linn. Soc. 82: 93–119.

HIBBERD, D. J., AND G. F. LEEDALE. 1970. Eustigmatophyceae — a new algal class with unique organization of the motile cell. Nature, London 225: 758–760.

1971. A new algal class — the Eustigmatophyceae. Taxon 20: 523–525.

1972. Observations on the cytology and ultrastructure of the new algal class, Eustigmatophyceae. Ann. Bot. 36: 49–71.

HILL, D. R. A., AND R. WETHERBEE. 1985. Observations on two morphologically distinct forms of a single species of cryptomonad flagellate. Sec. Int. Phycological Congress, Copenhagen (abstract).

HOEK, C. van den. 1978. Algen. Einführung in die Phykologie. Georg Thieme Verlag, Stuttgart. 481 p.

HORI, T., R. E. NORRIS, AND M. CHIHARA. 1982. Studies on the ultrastructure and taxonomy of the genus *Tetraselmis* (Prasinophyceae) I. Subgenus *Tetraselmis*. Bot. Mag. Tokyo 95: 49–61.

——— 1983. Studies on the ultrastructure and taxonomy of the genus *Tetraselmis* (Prasinophyceae) II. Subgenus *Prasinocladia*. Bot. Mag. Tokyo 96: 385–392.

INOUYE, I., AND R. E. PIENAAR. 1984. Light and electron microscope observations on *Nephroselmis astigmatica* sp.nov. (Prasinophyceae). Nord. J. Bot. 4: 409–423.

INOUYE, I., T. HORI, AND M. CHIHARA. 1985. Ultrastructural characters of *Pyramimonas* (Prasinophyceae) and their possible relevance in taxonomy, p. 314–327. *In* H. Hara [ed.] Origin and evolution of diversity in plants and plant communities. Academia Scientific Book Inc., Tokyo.

JEFFREY, S. W. 1980. Algal pigment systems, p. 33–58. *In* P. G. Falkowski [ed.] Primary productivity of the sea. Plenum Publ. Corp., New York, NY.

JOHANSEN, J. R., AND G. A. FRYXELL. 1985. The genus *Thalassiosira* (Bacillariophyceae): studies on species occurring south of the Antarctic Convergence Zone. Phycologia 24: 155–179.

JOHNSON, P. W., AND J. McN. SIEBURTH. 1982. In-situ morphology and occurrence of eucaryotic phototrophs of bacterial size in the picoplankton of estuarine and oceanic waters. J. Phycol. 18: 318–327.

JOINT, I. R., AND R. K. PIPE. 1984. An electron microscope study of a natural population of picoplankton from the Celtic Sea. Mar. Ecol. Prog. Ser. 20: 113–118.

JØRGENSEN, C. B. 1966. Biology of suspension feeding. Pergamon Press, Oxford. 357 p.

LACKEY, J. B. 1939. Notes on plankton flagellates from the Scioto Riwer. Lloydia 2: 128–143.

LEADBEATER, B. S. C. 1972a. Fine structural observations on six new species of *Chrysochromulina* (Haptophyceae) from Norway with preliminary observations on scale production in *C. microcylindra* sp.nov. Sarsia 49: 65–80.

——— 1972b. Identification, by means of electron microscopy, of flagellate nanoplankton from the coast of Norway. Sarsia 49: 107–124.

——— 1974. Ultrastructural observations on nanoplankton collected from the coast of Jugoslavia and the Bay of Algiers. J. Mar. Biol. Assoc. U.K. 54: 179–196.

LEBOUR, M. V. 1925. The dinoflagellates of northern seas. Plymouth, 250 p.

LEE, R. E. 1980. Phycology. Cambridge Univ. Press. 478 p.

LEE, J., S. H. HUTNER, AND E. C. BOVEE [ed.] 1985. An illustrated guide to the Protozoa. Society of Protozoologists. Allen Press, Lawrence, KS. 629 p.

LEWIN, J., R. E. NORRIS, S. W. JEFFREY, AND B. E. PEARSON. 1977. An aberrant chrysophycean alga *Pelagococcus subviridis* gen. nov. et sp.nov. from the North Pacific Ocean. J. Phycol. 13: 259–266.

LI, W. K. W. 1986. Experimental approaches to field measurements: methods and interpretation, p. 251–286. *In* T. Platt and W. K. W. Li [ed.] Photosynthetic picoplankton. Can. Bull. Fish. Aquat. Sci. 214.

LOHMANN, H. 1908. Untersuchungen zur Feststellung des vollständigen Gehaltes des Meeres an Plankton. Wiss. Meeresuntersuch. Abt. Kiel, N.F. 10: 129–370.

——— 1911. Über das Nannoplankton und die Zentrifugierung kleinster Wasserproben zur Gewinnung desselben in lebendem Zustande. Int. Rev. Gesamten Hydrobiol. Hydrogr. 4: 1–38.

MAIER, I. 1984. Culture studies of *Chorda tomentosa* (Phaeophyta, Laminariales). Br. Phycol. J. 19: 95–106.

MANTON, I. 1959. Electron microscopical observations on a very small flagellate: the problem of *Chromulina pusilla* Butcher. J. Mar. Biol. Assoc. U.K. 38: 319–333.

——— 1975. Observations on the microanatomy of *Scourfieldia marina* Throndsen and *Scourfieldia caeca* (Korsch.) Belcher et Swale. Arch. Protistenk. 117: 358–368.

——— 1977. *Dolichomastix* (Prasinophyceae) from arctic Canada, Alaska and South Africa: a new genus of flagellates with scaly flagella. Phycologia 16: 427–438.

MANTON, I., AND B. CLARKE. 1950. Electron microscope observations on the spermatozoid of *Fucus*. Nature 166: 973–974.

——— 1951. An electron microscope study of the spermatozoid of *Fucus serratus*. Ann. Bot. 15: 461–471.

——— 1956. Observations with the electron microscope on the internal structure of the spermatozoid of *Fucus*. J. exp. Bot. 7: 416–432.

MANTON, I., AND B. S. C. LEADBEATER. 1974. Fine-structural observations on six species of *Chrysochromulina* from wild Danish marine nanoplankton, including a description of *C. campanulifera* sp.nov. and a preliminary summary of the nanoplankton as a whole. Biol. Skr. Dan. Vid. Selsk. 20,5: 1–26.

MANTON, I., AND K. OATES. 1983. Nanoplankton from the Galapagos Islands: *Chrysochromulina vexillifera* sp.nov. (Haptophyceae = Prymnesiophyceae), a species with semivestigial body spines. Bot. Mar. 26: 517–525.

MANTON, I., AND M. PARKE. 1960. Further observations on small green flagellates with special reference to possible relatives of *Chromulina pusilla* Butcher. J. Mar. Biol. Assoc. U.K. 39: 275–298.

MANTON, I., AND H. A. VON STOSCH. 1966. Observations on the fine structure of the male gamete of the marine centric diatom *Lithodesmium undulatum*. J. R. Microsc. Soc. 85: 119–134.

MANTON, I., B. CLARKE, AND D. GREENWOOD. 1953. Further observations with the electron microscope on spermatozoids in the brown algae. J. Exp. Bot. 4: 319–329.

MANTON, I., D. G. RAYNS, AND H. ETTL. 1965. Further observations on green flagellates with scaly flagella: the genus *Heteromastix* Korshikov. J. Mar. Biol. Assoc. U.K. 45: 241–255.

Manton, I., J. Sutherland, and K. Oates. 1977. Arctic coccolithophorids: *Wigwamma arctica* gen. et sp.nov. from Greenland and arctic Canada, *W. annulifera* sp. nov. from South Africa and S. Alaska and *Calciarcus alaskensis* gen. et sp.nov. from S. Alaska. Proc. R. Soc. Lond. B. 198: 145-168.

Mattox, K. R., and K. D. Stewart. 1984. Classification of the green algae: A concept based on comparative cytology, p. 29-72. *In* D. E. G. Irvine and D. M. John [ed.] Systematics of the green algae, Systematics Association Special Vol. No. 27, Academic Press, London.

McFadden, G., D. R. A. Hill, and R. Wetherbee. 1986. A study of the genus *Pyramimonas* (Prasinophyceae, Chlorophyta) from south-eastern Australia. Nord. J. Bot. 6: 209-234.

McFadden, G. I., Ø. Moestrup, and R. Wetherbee. 1982. *Pyramimonas gelidicola* sp.nov. (Prasinophyceae), a new species isolated from Antarctic sea ice. Phycologia 21: 103-111.

Meyer, S. R., and R. N. Pienaar. 1984. The microanatomy of *Chroomonas africana* sp.nov. (Cryptophyceae). S. Afr. J. Bot. 3: 306-319.

Mitchell, J. G., and M. W. Silver. 1982. Modern archaeomonads indicate sea-ice environments. Nature, London 296: 437-439.

Moestrup, Ø. 1979. Identification by electron microscopy of marine nanoplankton from New Zealand, including the description of four new species. N.Z. J. Bot. 17: 61-95.

—— 1982. Flagellar structure in algae: a review, with new observations particularly on the Chrysophyceae, Phaeophyceae (Fucophyceae), Euglenophyceae, and *Reckertia*. Phycologia 21: 427-528.

—— 1983. Further studies on *Nephroselmis* and its allies (Prasinophyceae). I. The question of the genus *Bipedinomonas*. Nord. J. Bot. 3: 609-627.

—— 1984. Further studies on *Nephroselmis* and its allies (Prasinophyceae). II. *Mamiella* gen.nov., Mamiellaceae fam. nov., Mamiellales ord.nov. Nord J. Bot. 4: 109-121.

Moestrup, Ø., and H. A. Thomsen. 1980. Preparation of shadowcast whole mounts, p. 385-390. *In* E. Gantt [ed.] Handbook of phycological methods; developmental & cytological methods, Cambridge University Press.

—— 1986. Ultrastructure and reconstruction of the flagellar apparatus in *Chrysochromulina apheles* sp.nov. (Prymnesiophyceae = Haptophyceae). Can. J. Bot. 64: 593-610.

Norris, R. E., T. Hori, and M. Chihara. 1980. Revision of the genus *Tetraselmis* (class Prasinophyceae). Bot. Mag. Tokyo 93: 317-339.

Okada, H., and A. McIntyre. 1977. Modern coccolithophores of the Pacific and North Atlantic oceans. Micropaleontology 23: 1-55.

Osborn, J. E. M. 1948. The structure and life history of *Hormosira banksii* (Turner)Decaisne. Trans. R. Soc. N.Z. 77: 47-71.

Paasche, E., and J. Throndsen. 1970. *Rhodella maculata* Evans (Rhodophyceae, Porphyridiales) isolated from the plankton of the Oslo Fjord. Nytt Mag. Bot. 17: 209-212.

Paerl, H. W., and S. L. Shimp. 1973. Preparation of filtered plankton and detritus for study with scanning electron microscopy. Limnol. Oceanogr. 18: 802-805.

Parke, M. 1949. Studies on marine flagellates. J. Mar. Biol. Assoc. U.K. 28: 255-286.

Parke, M., and P. S. Dixon. 1976. Check-list of British marine algae — third revision. J. Mar. Biol. Assoc. U.K. 56: 527-594.

Parke, M., and D. G. Rayns. 1964. Studies on marine flagellates VII. *Nephroselmis gilva* sp.nov. and some allied forms. J. Mar. Biol. Assoc. U.K. 44: 209-217.

Parke, M., G. T. Boalch, R. Jowett, and D. S. Harbour. 1978. The genus *Pterosperma* (Prasinophyceae): Species with a single equatorial ala. J. Mar. Biol. Assoc. U.K. 58: 239-276.

Parke, M., J. C. Green, and I. Manton. 1974. Observations on the fine structure of zoids of the genus *Phaeocystis* (Haptophyceae). J. Mar. Biol. Assoc. U.K. 51: 927-941.

Parke, M., I. Manton, and B. Clarke. 1955. Studies on marine flagellates II. Three new species of *Chrysochromulina*. J. Mar. Biol. Assoc. U.K. 34: 579-609.

Peel, M.C., and J. C. Duckett. 1975. Studies of spermatogenesis in the Rhodophyta, p. 1-13. *In* J. G. Duckett and P. A. Racey [ed.] The biology of the male gamete. Biol. J. Linn. Soc. 7. Suppl. No. 1.

Pennick, N. C. 1977. Studies on the external morphology of *Pyramimonas* 4. *Pyramimonas virginica* sp.nov. Arch. Protistenk. 119: 239-246.

Pienaar, R. N. 1976. The microanatomy of *Sphaleromantis marina* sp.nov. (Chrysophyceae). Br. Phycol. J. 11: 83-92.

Prauser, H. 1958. *Diacronema alkianum*, eine neue Chrysomonade. Arch. Protistenk. 103: 117-128.

Preisig, H. R., and D. J. Hibberd. 1983. Ultrastructure and taxonomy of *Paraphysomonas* (Chrysophyceae) and related genera 3. Nord. J. Bot. 3: 695-723.

Reimann, B. E. F., E. L. Duke, and G. F. Floyd. 1980. Fixation, embedding, sectioning, and staining of algae for electron microscopy, p. 285-303. *In* E. Gantt [ed.] Handbook of phycological methods; developmental & cytological methods, Cambridge University Press.

Reymond, O. L., and J. D. Pickett-Heaps. 1983. A routine flat embedding method for electron microscopy of microorganisms allowing selection and precisely oriented sectioning of single cells by light microscopy. J. Microsc. 130: 79-84.

Reynolds, N. 1974. *Imantonia rotunda* gen. et sp.nov., a new member of the Haptophyceae. Br. Phycol. J. 9: 429-34.

Rouchijajnen, M. I. 1966. Duae species novae e chrysophytis mobilibus maris nigri. Nov. Syst. Plant Non Vasc. 1966: 10-15.

Santore, U. J. 1977. Scanning electron microscopy and comparative micromorphology of the

156

periplast of *Hemiselmis rufescens, Chroomonas* sp., *Chroomonas salina* and members of the genus *Cryptomonas* (Cryptophyceae). Br. Phycol. J. 12: 255–270.

1982a. Comparative ultrastructure of two members of the Cryptophyceae assigned to the genus *Chroomonas* — with comments on their taxonomy. Arch. Protistenk. 125: 5–29.

1982b. The ultrastructure of *Hemiselmis brunnescens* and *Hemiselmis virescens* with additional observations on *Hemiselmis rufescens* and comments on the Hemiselmidaceae as a natural group of the Cryptophyceae. Br. Phycol. J 17: 81–99.

1985a. A cytological survey of the genus *Cryptomonas* (Cryptophyceae) with comments on its taxonomy. Arch. Protistenk. 130: 1–52.

1985b. Species concepts in the Cryptophyceae. Sec. Int. Phycological Congress, Copenhagen (abstract).

SCHILLER, J. 1925. Die planktonischen Vegetationen des adriatischen Meeres. B. Chrysomonadina, Heterokontae, Cryptomonadina, Eugleninae, Volvocales. I. Systematischer Teil. Arch. Protistenk. 63: 59–123.

SIEBURTH, J. McN. 1979. Sea microbes. Oxford University Press. New York, NY. 491 p.

SIEBURTH, J. McN., V. SMETACEK, AND J. LENZ. 1978. Pelagic ecosystem structure: Heterotrophic components of the plankton and their relationship to plankton size-fractions. Limnol. Oceanogr. 23: 1256–1263.

SILVER, M. W., M. M. GOWING, AND P. J. DAVOLL. 1986 The association of photosynthetic picoplankton and ultraplankton with pelagic detritus through the water column (0–200 m), p. 311–341. *In* T. Platt and W. K. W. Li [ed.] Photosynthetic picoplankton. Can. Bull. Fish. Aquat. Sci. 214.

SILVER, M. W., J. G. MITCHELL, AND D. L. RINGO. 1980. Siliceous nanoplankton. II. Newly discovered cysts and abundant choanoflagellates from the Weddell Sea, Antarctica. Mar. Biol. 58: 211–217.

SIMPSON, P. D., AND S. D. VAN VALKENBURG. 1978. The ultrastructure of *Mychonastes ruminatus* gen. et sp.nov., a new member of the Chlorophyceae isolated from brackish water. Br. Phycol. J. 13: 117–130.

STAEHELIN, L. A. 1980. Freeze-fracture and freeze-etch techniques, p. 355–365. *In* E. Gantt [ed.] Handbook of phycological methods, developmental & cytological techniques, Cambridge University Press.

SWEENEY, B. M. 1976. *Pedinomonas noctilucae* (Prasinophyceae), the flagellate symbiotic in *Noctiluca* (Dinophyceae) in southeast Asia. J. Phycol. 12: 460–464.

TAKAHASHI, M., AND T. HORI. 1984. Abundance of picophytoplankton in the subsurface chlorophyll maximum layer in subtropical and tropical waters. Mar. Biol. 79: 177–186.

TAKANO, H. 1981. New and rare diatoms from Japanese marine waters VI. Three new species in Thalassiosiraceae. Bull. Tokai Reg. Fish. Res. Lab. 105: 31–43.

THOMSEN, H. A. 1977. *Chrysochromulina pyramidosa* sp.nov. (Prymnesiophyceae) from Danish coastal waters. Bot. Notiser 130: 147–153.

1980. Two species of *Trigonaspis* gen.nov. (Prymnesiophyceae) from West Greenland. Phycologia 19: 218–229.

1981a. Identification by electron microscopy of nanoplanktonic coccolithophorids (Prymnesiophyceae) from West Greenland, including the description of *Papposphaera sarion* sp.nov. Br. Phycol. J. 16: 77–94.

1981b. Species composition of nanoplankton in brackish water. XIII Int. Bot. Congress, Sydney (abstract).

1982. Planktonic choanoflagellates from Disko Bugt, West Greenland, with a survey of the marine nanoplankton of the area. Meddr Grønland, Biosci. 8: 1–35.

1985. Ice-related occurrence of *Pedinella tricostata* (Chrysophyceae) — an ultrastructural investigation of the flagellated stage and the cyst. Sec. Int. Phycological Congress, Copenhagen (abstract).

THOMSEN, H. A., AND Ø. MOESTRUP. 1984. Is *Distephanus speculum* a fishkiller? A report on an unusual algal bloom from Danish coastal waters. Int. Symp. on Marine Plankton, Shimizu (abstract).

THOMSEN, H. A., AND K. OATES. 1978. *Balaniger balticus* gen. et sp.nov. (Prymnesiophyceae) from Danish coastal waters. J. Mar. Biol. Assoc. U.K. 58: 773–779.

THRONDSEN, J. 1969. Flagellates of Norwegian coastal waters. Nytt Mag. Bot. 16: 161–216.

1973a. Fine structure of *Eutreptiella gymnastica* (Euglenophyceae). Norw. J. Bot. 20: 271–280.

1973b. Phytoplankton occurrence and distribution on stations sampled during the SCOR WG 15 cruise to the Caribbean Sea, Pacific Ocean and Sargasso Sea in May 1970, based on direct cell counts. Data Report SCOR Discoverer Expedition May 1970, SIO Ref. 73–16, Univ. California, D 1–28.

1976. Occurrence and productivity of small marine flagellates. Norw. J. Bot. 23: 269–293.

1978a. Centrifugation, p. 98–103, *In* A. Sournia [ed.] Phytoplankton manual, UNESCO Monograph.

1978b. The dilution-culture methods, p. 218–224. *In* A. Sournia [ed.] Phytoplankton manual, UNESCO monograph.

1978c. Productivity and abundance of ultra- and nanoplankton in Oslofjorden. Sarsia 63: 273–284.

1980. Bestemmelse av marine nakne flagellater. (Identification of marine naked flagellates.) Blyttia 38: 189–207 (In Norwegian with English summary).

1983. Ultra- and nanoplankton flagellates from coastal waters of southern Honshu and Kyushu, Japan (including some results from the western part of the Kuroshio off Honshu). Working Party on Taxonomy in the Akashiwo Mondai Kenkyukai. Gakujutsu Tosha Printing Co., Tokyo. 62 p.

VALKENBURG, S. VAN, AND R. E. NORRIS. 1970. The growth and morphology of the silicoflagellate *Dictyocha fibula* Ehrenberg in culture. J. Phycol. 6: 48–54.

WETHERBEE, R. 1985. Structure and assembly of cell surface components in the Cryptophyceae. Sec. Int. Phycological Congress, Copenhagen (abstract).

WHITTLE, S. J. 1976. The major chloroplast pigments of *Chlorobotrys regularis* (West) Bohlin (Eustigmatophyceae) and *Ophiocytium majus* Naegeli (Xanthophyceae). Br. Phycol. J. 11: 111–114.

WHITTLE, S. J., AND P. J. CASSELTON. 1975. The chloroplast pigments of the algal classes Eustigmatophyceae and Xanthophyceae. I. Eustigmatophyceae. Br. Phycol. J. 10: 179–191.

ZIMMERMANN, B., Ø. MOESTRUP, AND G. HÄLLFORS. 1984. Chrysophyte or heliozoon: Ultrastructural studies on a cultured species of *Pseudopedinella* (Pedinellales ord.nov.), with comments on species taxonomy. Protistologica 20: 591–612.

Cyanobacteria: Their Biology in Relation to the Oceanic Picoplankton

N. G. Carr and M. Wyman

Department of Biological Sciences, University of Warwick, Coventry, U.K. CV4 7AL

Range, Form and Structure

The cyanobacteria are the most diverse and widespread of the photosynthetic prokaryotes and constitute one of the largest subgroups of Gram-negative bacteria. Unlike purple and green photosynthetic bacteria, they have a photosynthetic apparatus similar in structure and function to that of the eukaryotic (strictly speaking, the rhodophytan) chloroplast. Their capacity to perform oxygenic photosynthesis together with certain features of their ecology, for many years led to the treatment of cyanobacteria as a class or division of algae. The fundamental differences in the cellular organization of cyanobacteria and eukaryotes were recognized by Stanier and van Niel (1962) but even a decade later these organisms continued to be known by the common name of blue-green algae (Carr and Whitton 1973; Fogg et al. 1973). In more recent times, however, the true taxonomic position of cyanobacteria has been firmly established and a preliminary classification of genera based upon the principles of the Bacteriological Code rather than the Botanical Code has been published (Rippka et al. 1979).

PHYSICO-CHEMICAL FACTORS AFFECTING GROWTH AND DISTRIBUTION

Cyanobacteria have a global distribution and in certain habitats they are often ubiquitous components of the microbial flora. Apart from the availability of light and nutrients (which are considered elsewhere in this review) there remain a number of other physico-chemical factors which have profound influence on the distribution of these organisms and the species composition of particular ecosystems.

Although representatives of a few genera have been isolated from acid *Sphagnum* peat bogs (Rippka et al. 1979), cyanobacteria are most frequently associated with neutral to alkaline environments. Indeed certain species are extremely alkalotolerant, e.g. *Spirulina* which occurs in equatorial soda lakes. Whereas many cyanobacteria are mesophiles, temperature optima for growth may be as high as 50–60°C for the thermophilic species of hot springs (Castenholz 1968). At the other extreme, active populations of cyanobacteria may exist under permanent ice cover in the Arctic and Antarctic at temperatures close to 0°C (Fogg and Stewart 1968; Fogg et al. 1973). Many cyanobacteria isolated from the frozen soils of the polar regions display a remarkable resistance to desiccation although this property is most developed amongst the soil cyanobacteria of tropical and arid desert regions. While only active during the wet season, viability may be maintained in the desiccated state for extended periods of time. Most terrestrial species, however, show a distinct preference for habitats of high relative humidity and occur most frequently in shaded regions, in waterlogged soils, by the sides of streams or at altitude (Fogg et al. 1973).

Cyanobacteria are particularly abundant in intertidal and estuarine areas although they differ considerably in their tolerance of changes in salinity. A number of freshwater forms are able to withstand relatively high concentrations of sodium chloride and it appears that many of those 'marine' cyanobacteria isolated from coastal environments are halotolerant rather than halophilic. Stanier and Cohen-Bazire (1977) defined marine cyanobacteria as those species which demonstrated an obligate requirement for elevated concentrations of the ions Na^+, Cl^-, Mg^{2+} and Ca^{2+}. Cyanobacteria are also frequent colonizers of euhaline environments such as salt works and salt marshes and some species have been shown to be capable of growth at combined salt concentrations as high as 3–4 M (Mackay et al. 1984; Reed et al. 1984).

For many years cyanobacteria were considered to be obligate aerobic photoautotrophs but an increasing number are now known to be capable of utilizing organic carbon for growth in the absence of light. Although most cyanobacteria are free-living, a number are epiphytic or occur as endophytes in highly organized symbioses and it is in this latter group that the capacity for heterotrophic growth appears to be most developed. Just as some cyanobacteria may assimilate carbon in the dark, a few may switch between oxygenic (plant-like) and anoxygenic (bacterial-like) photosynthesis utilizing sulphide rather than water as an electron donor (Cohen et al. 1975; Padan 1979). The discovery that certain cyanobacteria are facultative anoxygenic photoautotrophs has helped to explain the common ocurrence of these organisms in anaerobic sediments and sulphide-rich ecosystems.

FORM AND STRUCTURE

Rippka et al. (1979) in their revision of the classification of the cyanobacteria recognized five principal subgroups and 22 separate genera (Table 1). Generic assignments were based on the differences in form and structure of pure cultures and the mode of reproduction. Characters such as colony formation, the consistency of the sheath material and 'false-branching' which were long used by phycologists to distinguish genera and even species in the field were excluded, for these taxonomic features were either lost or were highly variable in culture.

Unicellular cyanobacteria were divided between two of the major subgroups (Section I, Section II; see Table 1). Section I comprised those rods or cocci formerly placed in the family Chroococcaceae and includes *Synechococcus* and *Synechocystis* amongst six form genera recognised. These two genera which divide by binary fission were separated from other members of Section I by their mode of reproduction and also by the absence of a sheath. Whereas *Synechococcus* divides in one plane only, division may be in two or three planes in *Synechocystis*.

The pleurocapslean cyanobacteria of Section II were placed in five form genera and one assemblage, the pleurocapsa group. Reproduction may be by multiple fission only (*Dermocarpa, Xenococcus*) or also by binary fission. Multiple fission of the vegetative cell results in the production of small reproductive cells (baeocytes) which may be capable of gliding motilty following their release through the ruptured vegetative cell wall. The baeocytes of the genera *Xenococcus* and *Chroococcidropsis*, however are immotile.

The filamentous cyanobacteria were placed in three separate subgroups (Table 1). Members of Section III produce filaments (trichomes) composed entirely of vegetative cells and divide in one plane only. The trichome may be straight or helical (*Spirulina*), is motile (except in certain members of the LPP assemblage) and may have pronounced constrictions between the cells (*Pseudanabaena*, some of the LPP group). The trichomes of *Spirulina* and *Oscillatoria* are enclosed in a thin sheath or like *Pseudanabaena* and certain strains of the LPP group they are not ensheathed.

The genera of Sections IV and V all form heterocysts in media free of combined nitrogen and some may also produce akinetes. Six form genera were recognized in Section IV (Table 1) including three which produce hormogonia (*Nostoc, Scytonema, Calothrix*). Heterocysts may be produced in a terminal position only (*Cylindrospermum, Calothrix*) or, in addition, in an intercalary position in mature trichomes. *Calothrix* produces trichomes which taper from base to apex and in this genus, *Cylindrospermum*, and certain strains of *Anabaena* and *Nodularia*, akinetes are formed by differentation of the vegetative cell adjacent to the heterocyst. Akinetes may be either adjacent to or at some distance from the heterocyst in the remaining strains of Section IV. Rippka et al. (1979) recognised only two genera of heterocystous cyanobacteria which divide in more than one plane (Section V; see Table 1); *Fischerella*

TABLE 1. Characteristics of the major subgroups of cyanobacteria (after Rippka et al. 1979).

Section	Definition of Subgroup	Form Genera
I	Unicellular or colonial cyanobacteria reproducing by binary fission or its variant, budding.	*Synechococcus* *Gloeothece* *Gleobacter* *Synechocystis* *Gloeocapsa* *Chamaesiphon*
II	Unicellular or colonial cyanobacteria reproducing by multiple fission or by both multiple fission and binary fission	*Dermocarpa* *Xenococcus* *Dermocarpella* *Myxosarcina* *Chroococlidiopsis* *Pleurocapsa group*
III	Filamentous cyanobacteria composed only of vegetative cells reproducing by trichome fragmentation or by the formation of hormogonia	*Spirulina* *Oscillatoria* *Pseudanabaena* LPP (*Lyngbya, Phormidium Plectonema*) group
IV	Filamentous heterocyst-forming cyanobacteria that divide in one plane only and reproduce by trichome fragmentation, by the formation of hormogonia or by the germination of akinetes	*Anabaena* *Nodularia* *Cylindrospermum* *Nostoc* *Scytonema* *Calothrix*
V	Filamentous heterocyst-forming cyanobacteria that divide in more than one plane reproducing by trichome fragmentation, by hormogonium formation or by the germination of akinettes.	*Fischerella* *Chlorogloeopsis*

and *Chlorogloeopsis*. In the former genera lateral uniseriate branches are produced by mature trichomes. Heterocysts in the primary trichome (which may be partly multiseriate) are predominantly terminal or lateral whereas in *Chlorogloeopsis* the heterocysts are exclusively terminal. This genus does not produce lateral branches.

ULTRASTRUCTURE

The advent of electron microscopy and the development of ultrastructural analysis of microbial cells was the driving force that lead to the recognition of cyanobacteria as prokaryotes which share the same basic cellular organisation as all other bacteria (see Stanier and van Niel 1962). There had previously been perceptive comments that related the blue-green algae to bacteria (Stanier and van Niel 1941; see Stanier 1982). In contrast to the eukaryotic microalgae, the cyanobacteria do not possess membrane bound sub-cellular organelles; have no discrete membrane bound nucleus; have a wall structure based upon a peptidoglycan layer; possess ribosomes of sizes around 70S, rather than 80S. It is not intended here to give an exhaustive account of our current knowledge of cyanobacterial fine structure; this has been well reviewed in its various aspects elsewhere (Lang 1968; Whitton et al. 1971; Drews 1973; Lang and Whitton 1973; Gorlecki and Drews 1982; Drews and Weckesser 1982; Cohen-Bazire and Bryant 1982).

There are, however, aspects of recent cyanobacterial fine structure study which do command at least a brief mention in the present context. Cyanobacteria, like other Gram-negative bacteria, may be surrounded by layers of sheath material the appearance of which is often influenced by the growth conditions and by the procedures used in analysis (see Stanier and Cohen-Bazire 1977; Drews and Weckesser 1982). Such external layers may be fibrous and closely attached to the outer membrane of the cell, or they may be more diffuse and to an extent dissolve in the culture medium. These outer layers have various suggested functions, including those of dessication protection and involvement in gliding motility. Many bacterial cells in natural, especially low nutrient, ecosystems have an extracellular capsule termed a glycocalyx, whose functions may be involved in adhesion. Balkwell and Stevens (1980) have described the network of small fibrils that surround the marine cyanobacterium *Agmenellum quadruplicatum* as a glycocalyx and showed that it was stable to nutritional and environmental variations. Gugliemi and Cohen-Bazire (1982) have described in detail the perforations that occur in the peptidoglycan wall of a range of cyanobacteria — both filamentous and unicellular species — and discuss their work in relation to early observations of such pores. These small holes are usually between

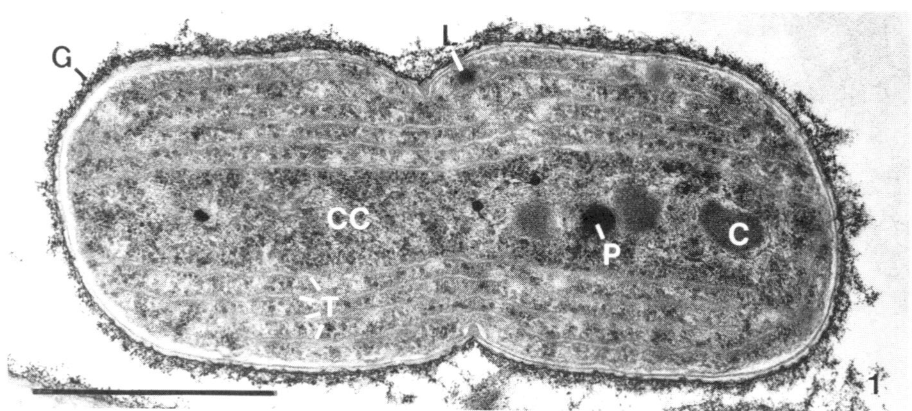

FIG. 1. EM of a longitudinal thin section of *A. quadruplicatum*. Bar, 1.0 m. C, carboxysome; CC, central cytoplasmic region; G, glycocalyx; L, lipid body; P, polyphosphate body; T, photosynthetic thylakoid membrane (from Nierzwicki-Bauer et al. 1983, with permission).

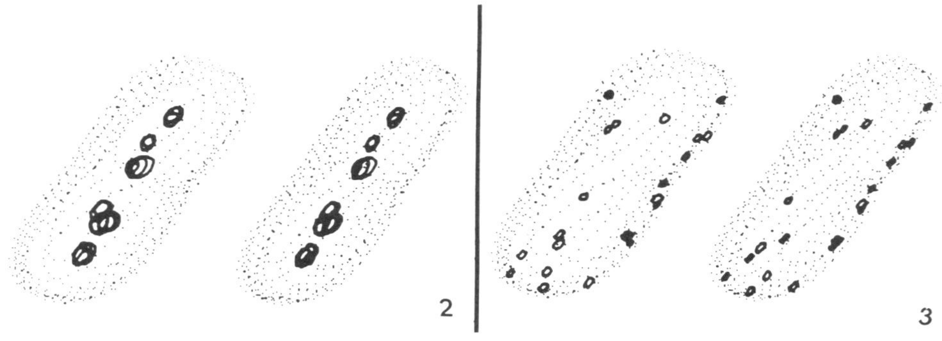

FIGURES 2–3

FIG. 2. A stereo pair of computer-aided reconstructions showing carboxysomes within a typical cell (from Nierzwicki-Bauer et al. 1983, with permission); FIG. 3. As Fig. 2, showing lipid bodies within the cell (from Nierzwicki-Bauer et al. 1983, with permission).

162

10–16 nm in diamter. Their function has been postulated to be concerned in extrusion of slime material in movement, positioning of 'breakage points' in the filament, or with the presence of fibrils outside the cell.

Recently, the application of modern techniques has enabled a three-dimensional picture of the fine structure of *Agmenellum quadruplicatum*, a marine cyanobacterium, to be obtained (Nierzwicki-Bauer et al. 1983). A longitudinal thin-

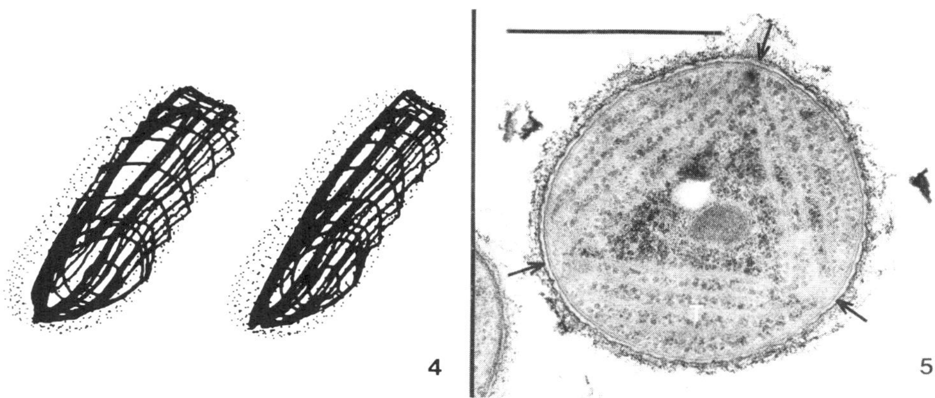

FIGURES 4–5

FIG. 4. As Fig. 2, showing the innermost thylakoid membrane. The thylakoid entirely surrounds the central cytoplasmic region (from Nierzwicki-Bauer et al. 1983, with permission); FIG. 5. EM of a thin cross section and through the cylindrical part of *A. quadruplicatum*. Bar, 1.0 m. Thylakoids join together at three loci (from Nierzwicki-Bauer et al. 1983, with permission).

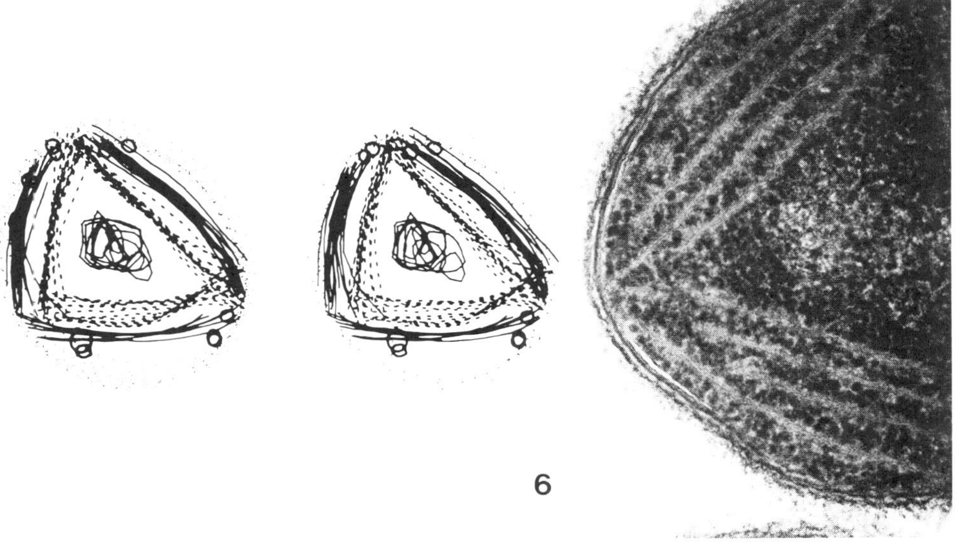

FIGURES 6–7

FIG. 6. As Fig. 2, illustrating two complete thylakoid membranes in cross section (solid and dashed lines) surrounding carboxysomes, with lipid bodies external to the thylakoids (from Nierzwicki-Bauer et al. 1983, with permission); FIG. 7. A high voltage EM of a cross section through the longitudinal portion of *A. quadruplicatum*. Bar, 0.25 m. The thylakoids appear to converge at the cytoplasmic membrane (from Nierzwicki-Bauer et al. 1983, with permission).

163

section of *A. quadruplicatum* is shown in Fig. 1, and the principle structural features indicated. A Gram-negative cell envelope with an extracellular glycocalyx typical of other unicellular cyanobacteria was clearly indicated, as was the peripheral arrangement of several layers of thylakoid membranes. The application of high-voltage electron microscopy allowed relatively thick sections to be examined and consequently fewer of these were required to get a complete picture of the cyanobacterial cell. Computer programs were designed to collect information from several sections and present the data in a three-dimensional form and stereo pairs of the same image were used to enhance the visualisation of specific intracellular features (see Fig. 2, 3, and 4). The arrangements of the thylakoids were particularly clarified by the studies of Nierzwicki-Bauer and her colleagues. An electron micrograph of a cross-section of the cylindrical part of *A. quadruplicatum* illustrates this (Fig. 5); several thylakoids appeared to join together to attach at several peripheral points around the cell. Figure 6 shows a computer-aided reconstruction of this, whilst a thicker section (Fig. 7) of a cross-section showed clearly the tendency of thylakoid membranes to converge at or near the inner surface of the cytoplasmic membrane. In the original publication of Nierzwicki-Bauer et al. (1983), to which we are indebted for these photographs, the computer-graphic reconstructions were presented in six colours, which gave vivid and beautiful visualisations of this cyanobacterial cell.

RELATIONSHIP TO OTHER ORGANISMS

The cyanobacteria comprise the largest section of that group of prokaryotes that can use light as a source of metabolic energy. The relationship of cyanobacterial photosynthesis to that of anoxygenic bacterial photosynthesis is discussed later. The respective pigment compositions of cyanobacteria and photosynthetic bacteria indicate that they usually occupy separate, but sometimes closely associated ecological niches.

It is to chloroplasts of algae and higher plants that cyanobacteria have been most compared. The view propounded around the turn of the century that chloroplasts were evolutionarly derived from cyanobacteria — the endosymbiotic hypothesis of Mereschkowsky — has now gained a high degree of acceptance and the reader is refered to one or two of the numerous reviews on this topic (Margulis 1970; Frederick 1981; Doolittle 1982) for guidance into the literature. What does emerge is the distinct possibility of a polyphyletic origin of chloroplasts based upon more than one endosymbiotic event (Stanier 1970). The discovery (see Lewin 1980) of *Prochloron* which appears to be an obligately symbiotic associate of lower marine animals of the *Didemnum* group provided a possible candidate for the immediate chloroplast precursor. *Prochloron* contains no phycobiliproteins but does contain chlorophyll b, as do green plant chloroplasts. In all other major respects *Prochloron* appears to be a unicellular cyanobacterium. In spite of strenuous efforts *Prochloron* has not been grown free of its symbiotic host. Therefore, the discovery (Burger-Wiersma et al. 1986) of a filamentous prochloron-like organism which is readily cultivated axenically in the laboratory will aid considerably in the understanding of the relationship between *Prochloron*, cyanobacteria and chloroplasts.

Growth Rate: Range and Relationship to Cell Volume

Under optimal conditions of light, temperature and nutrient availability the maximum relative growth rates of cyanobacteria show considerable variation between species and even between individual isolates of the same species. Nevertheless, there is a general correlation between maximum division rate and morphological complexity; unicellular species tending to have more rapid growth rates than filamentous cyanobacteria. Irrespective of morphological type, those cyanobacteria capable of growth at elevated temperatures frequently have more rapid generation times than mesophilic species.

Kratz and Myers (1955) determined that the unicellular cyanobacterium *Anacystis nidulans* (*Synechococcus* sp.) has a mean generation time of 1.7 hours (h) at its optimum temperature for growth, 41°C; the fastest doubling time recorded for a cyanobacterium. At a similar growth temperature (39°C), Stevens et al. (1981) measured a doubling time of 3.6 h for *Agmenellum quadruplicatum* (strain PR6). In the two picoplankton *Synechococcus* strains 'Syn' and DC2, Morris and Glover (1981) reported maximum doubling times of 11 and 13 h, respectively, at a growth temperature of 20°C. At a similar temperature (25°C), Kratz and Myers calculated that the generation time of *Anacystis nidulans* was 8.3 h.

Although capable of growth at high temperature, the generation times of individual species of thermophilic *Oscillatoria* species vary considerably. The doubling time of *O. terebriformis* was approximately 5 h at temperatures of 45–55°C whereas the maximum generation time of *O. princeps* at 40°C was considerably longer at about 48 h (Castenholz 1968). Among mesophilic *Oscillatoria* species, Foy (1980) found a doubling time of 8.8 h for *O. limnetica* FBA L100 at 20°C. Growth rate varied between 0.97–1.39 divisions per day in five strains of *O. redekei* whereas the doubling time of *O. rubescens* and *O. agardhii* var. *isothrix* was approximately 45 h. The growth rate of *Trichodesmium* has only been calculated for field populations as this marine cyanobacterium has not been cultured successfully to date. Estimates of this organisms doubling time have varied between 7 to 43 d depending on the method of growth rate determination (Margue et al. 1977). On the basis of carbon fixation rates, however, McCarthy and Carpenter (1979) calculated a generation time of 180 d for *Oscillatioria* (*Trichodesmium*) *thiebautii* populations in the central North Atlantic Ocean.

A similar variation in the specific growth rates of heterocystous cyanobacteria has been noted. Foy (1980) determined that the doubling time of three strains of *Aphanizomenon flos-aquae* were approximately 21.5 h at 20°C whereas the generation times of five strains of *Anabaena flos-aquae* varied between 16.5 and 26 h. Shaffer et al. (1978) reported a maximum relative growth rate of 1.16 divisions per day for *Anabaena variabilis* grown at 24°C. Stacey et al. (1977) isolated a marine *Anabaena* sp. (strain CA) capable of rapid growth at a temperature of 42–45°C. Maximum relative growth rate of this strain was dependant on nitrogen source and was equivalent to a doubling time of 3.6 h in the presence of NH_4Cl or approximately 4.2 h with either $NaNO_3$ or N_2 as the sole source of nitrogen.

The relationship between growth rate and the cell volume of cyanobacteria has been noted by a number of authors (Castenholz 1968, Foy 1980; Foy and Smith 1980). Castenholz (1968) reported that the maximum doubling time of *Oscillatoria terebriformis* (approximately 5 h) which has a cell diameter of 4–6 μm was considerably faster than the generation time of the wider celled (20–30+ μm) *O. princeps* (approximately 48 h). Foy (1980) found a significant correlation between growth rate and the cell surface to volume (S/V) ratio. Strains with high S/V ratios have higher potential nutrient uptake rates (Gibson and Smith 1982). Insufficient nutrient uptake sites in those cyanobacteria with low S/V ratios, however, may explain only in part the differences in growth rate between these and the smaller celled cyanobacteria (Foy 1980). In a comparison of the growth rates of two *Oscillatoria* species grown in either continuous light or in diel light-dark cycles, Foy and Smith (1980) found that the larger celled *O. agardhii* (cell diameter 3.54 μm) was able to compete successfully with the narrow celled *O. redekei* (cell diameter 2.16 μm) when the light period was 6 h or less. These authors proposed that the greater capacity of *O. agardhii* to store carbohydrate (due to its larger cell volume) during the light period enables this species to continue to provide the energy requirement for growth during the dark period.

In the rather limited number of bacteria that have been studied, defined relationships exist between growth rate, cell size and the content of the principle groups of macromolecules. Maaloe and Kjeldgaard (1966) and Mitchison (1971) provided the beacons that have guided much of our understanding of the ways in which compositional change of DNA, RNA and protein can be related to the cell cycle and to the rate of growth. Such relationships were most profitably studied using cultures in steady-state growth, and cultures that were responding to a nutritional shift that would lead to a new, modified steady state growh rate. Steady-state growth is best defined as when the daughter cells have exactly the same chemical composition and growth rate as the mother cell. Theoretically this can only be achieved in a chemostat because in batch culture the growth of the parental generation will have slightly altered the environment (nutrient availability) available to the daughter generation. In practice an approximation to steady-state growth may be obtained in dilute batch cultures over a period of several generations (Thomas and Carr 1985).

The basic observation that prokaryotic cell size is directly related to growth rate — faster growing cells are bigger — was confirmed in the unicellular *Synechococcus* sp. 6301 (*Anacystis nidulans*) by Mann and Carr (1974) and also in the filamentous *Anabaena cylindrica* (Leach and Carr, unpublished). The expected increase in DNA and RNA content per cell with faster growth rates was also observed in *Synechococcus* sp. 6301 and the Cooper–Helmstetter explanation, that had been developed for *Escherichia coli*, that increase in DNA content resulted from multiple forks of replication in fast-growing cells was applicable. An interesting deviation from the heterotrophic model of DNA and RNA changes was the observation that in the cyanobacteria these two groups of nucleic acids increased in content per cell in a parallel way (Mann and Carr 1974). In contrast, in heterotrophic bacteria RNA content increased at a greater rate than did the DNA content per cell.

The cell cycle of cyanobacteria is discussed fully in the article by Chisholm in this volume, but it is appropriate here to mention briefly some of the modes of cell division and alterations to the cell cycle that are adopted, under specific environmental conditions, by various species of cyanobacteria.

Members of the *Dermocarpa* group will undergo multiple fision, producing large numbers of small cells termed baeocytes (Waterbury and Stanier 1978) which are simultaneously released from the parent cell. Exospores can be formed by budding from the ends of filaments of organisms such as *Chamaesiphon*, although the use of the term 'spore' should not imply characteristics or functions that are associated with bacterial endospores. Akinetes are perennating structures that occur in many of the filamentous cyanobacterial groups and are usually associated with heterocyst development. The arrangement of these two types of modified cell are specified, and varied, with different species. Hormogonia are short filaments of usually small cells and contain no differentiated cells; they are motile and develop from the trichomes of many filamentous cyanobacteria and they would appear to have a defined lifespan and probably serve as a dispersal form. An indication of the range of cell types and division patterns may be found in many earlier works on microalgae and a brief account is given by Fogg et al. (1973). Fuller accounts of the earlier literature on cyanobacterial cell types can be obtained from Fritsch (1945); Desikachary (1959) and Lazaroff (1972).

Cyanobacteria and Light

Photosynthetic Pigments

The organization and structure of the light-harvesting apparatus of cyanobacteria has been characterised extensively during recent years and is described in detail by Bryant elsewhere in this volume.

Unlike eukaryotic algae and higher plants, cyanobacteria possess only one form of chlorophyll, chlorophyll *a*. Thornber (1969) determined that 70% of the chorophyll *a* in *Phormidium luridium* was bound to protein and it is now known that the majority of functional chlorophyll within the cyanobacterial cell is conjugated with specific protein complexes (Ho and Krogman 1982). Chlorophyll a is present *in vivo* in a number of spectroscopically distinct forms owing to differences in the nature and the subcellular environment of the chlorophyll–protein complexes. Approximately 85–95% of the total chlorophyll has been found to be associated with photosystem (PS) I (Mimuro and Fujita 1977; Myers et al. 1977). Manodori et al. (1984) determined that in PSI of *Synechococcus* sp. 6301 140 light-harvesting chlorophyll molecules were associated with the reaction centre P700 chlorophyll–protein and its oxidant X. A similar estimate of the PSI unit size of this organism was reported by Vierling and Alberte (1980) who suggested that variation in irradiance during growth had little effect on the composition of PSI units. A similar conclusiton was reached by Kawamura et al. (1977) who examined the cell chlorophyll: P700 ratio in *Anabaena cylindrica*, *Anabaena variabilis* and *Anacystis nidulans* grown at 1500 or 6000 lux (25 or 100 μE m^{-2} s^{-1}). Wang et al. (1977), however, reported that variation in the light quality experienced during growth resulted in changes in the number of light-harvesting chlorophyll molecules associated with PSI in *A. nidulans*. Although similar in size, the number of PSI units present per cell declined at high irradiance whereas the number PSII reaction centres present per cell was found to be independent of light quantity (Kawamura et al. 1977). The PSII reaction centre (the P680–Q complex) of *Synechococcus* sp. 6301 has been shown to be associated with a small but invariant number (35) of light-harvesting chlorophyll molecules (Manodori et al. 1984). Manodori and Melis (1985) proposed that in this organism two PSII complexes compete for excitation energy from a single phycobilisome (organised aggregates of phycobiliproteins bound to the surface of the thylakoid membrane). In an earlier investigation, however, Khanna et al. (1983) calculated that each phycobilisome was linked with a single PSII reaction centre in wild-type *Anacystis nidulans* and the mutant strain 85Y which was depleted in phycocyanin.

Cyanobacteria produce a number of the carotenoids found in eukaryotic algae and in higher plants, and a few (eg. myxoxanthrophyll, oscillaxanthin) which do not occur in any other phylogenetic group. The structure and distribution of cyanobacterial carotenoids has been reviewed by a number of authors (Goedheer 1969; Hertzberg et al. 1971; Fogg et al. 1973). β-carotene appears to be produced by all cyanobacteria and frequently may be the only carotene present (Healey 1968). Cyanobacteria synthesize a diverse range of xanthrophylls (Table 1) although their occurrence and relative cell concentrations vary between species. Echinenone, however, appears to be present in most cyanobacteria and is a diagnostic feature of these organisms.

In a comparative analysis of the carotenoid complement of nine unicellular cyanobacteria of either freshwater or marine distribution, the carotenoid profile of *Synechococcus* sp. strain DC2 was found to show a number of unique features (Britton, Wyman and Carr, unpublished; Table 2); β-carotene (33%) and zeaxanthin (53%) were the major carotenoids present with smaller amounts of echinenone (1%), β-cryptoxanthin (5%) and several unidentified polar compounds (8%). This picoplanktonic strain differed from other cyanobacteria in lacking the common polar xanthrophylls and glycosidic carotenoids of these organisms and had an overall carotenoid composition similar to that of the red alga *Porphyridium crueutum*. This observation may prove to have evolutionary and phylogenetic implications, for *Synechococcus* sp. strain DC2 is amongst those picoplanktonic cyanobacteria which also synthesize red-algal type phycoerythrins (Alberte et al. 1984).

As in higher plants, the carotenoid complement of cyanobacteria may show considerable variation depending upon environmental conditions. The influence of light quantity on the bulk carotenoid composition of cyanobacteria is similar to that observed in other photosynthetic organisms (Kellar and Paerl 1980; Codd 1981); exposure to high irradiance leads to a stimulation of carotenoid synthesis. Paerl (1984) found that in *Microcystis aeruginosa*, the relative cell concentrations of both β-carotene and the xanthrophylls, zeaxanthin, myxoxanthrophyll and echinenone increased following exposure to ultraviolet radiation. Variation in the quality of visi-

TABLE 2. Carotenoid composition of cyanobacteria.

Carotenoid	Species[a,b]							
	1	2	3	4	5	6	7	8
β-Carotene	+	+	+	+	+	+	+	+
Echinenone	+	+	+	+	+	+	+	+
Cryptoxanthin	+		+	+				+
Zeaxanthin	+	+	+	+			+	+
Canthanxanthin					+	+		
Nostoxanthin							+	+
Oscillaxanthin			+	+	t			
Myxoxanthrophyll		+	+	+	+	+	+	
Caloxanthin							+	+
Aphanizophyll				+				
Unidentified	t	t					t	

[a]Key to species : (1) *Synechococcus* sp. strain DC2, (2) *Microcystis aeruginosa*; (3) L.P.P. sp.; (4) *Aphanizomenon flos-aquae*; (5) *Tolypothrix teunis*; (6) *Fremyella diplosiphon*; (7) *Chlorogloea fritschii*; (8) *Anacystis nidulans*.
[b]Data taken from Paerl 1984 (2); Fiksdahl et al. 1983 (3,4,5,6); Hertzberg et al. 1971 (4); Evans and Britton 1983 (7); Britton, Wyman and Carr, unpublished (1,8).

ble light has also been shown to influence carotenoid composition in a number of cyanobacteria capable of chromatic adaptation (Fiksdahl et al. 1983). In the facultative heterotrophic species *Chlorogloea fritschii*, the carotenoid composition of light and dark grown cells was qualitatively similar although cells were depleted in carotenols when grown in the absence of light (Evans and Britton 1983). Photoheterotrophic cultures grown in the presence of sucrose showed enhanced synthesis of the glycosidic carotenoid myxoxanthrophyll, which Evans and Britton (1983) suggested may be accumulated by those cells of aseriate rather than filamentous morphology.

On the basis of enhancement action spectra, Jones and Myres (1964) determined that photosynthetically active carotenoids donate absorbed light energy to PSI in *Anacystis nidulans*. Goedheer (1969) showed that β-carotene but not xanthrophylls participate in energy transfer to the PSI reaction centre. As is the case in higher plants (Boardman 1970) PSI particles of *Chlorogloea fritschii* were found to be enriched in β-carotene while the majority of xanthrophylls were associated with PSII (Evans and Britton 1983). Although in comparison to PSI less is known of their function in PSII, carotenoids may have a structural role (Ho and Krogman 1982) and may stabilize the attachment of phycobilisomes to the thylakoid membrane (Szalontai and van de Ven 1981). Recently, Laczko (1985) presented evidence that carotenoids may be photosynthetically active in both PSII and PSI of the cyanobacterium *Anabaena cylindrica*.

The principal accessory light-harvesting pigments of cyanobacteria are the phycobiliproteins; intensely coloured chromoproteins which also occur in red algae and cryptomonads. All cyanobacteria synthesize at least three classes of phycobiliproteins; phycocyanin, allophycocyanin, and allophycocyanin B (Table 3). A number of strains produce in addition, phycoerythrin or less commonly, phycoerythrocyanin. The absorption characteristics of cyanobacterial phycobiliproteins depend upon the nature of the chromophore(s) present and interactions between the chromophore and the apoprotein (Table 2). Within the cell, phycobiliproteins are located in supramolecular aggregates, the phycobilisomes, which are distributed on the outer thylakoid surface and transfer absorbed light energy to the reaction centre of PSII. The structure and organization of phycobilisomes has received considerable attention

TABLE 3. Chromophore content of phycobiliproteins.

Biliprotein	α subunit	Chromophore content*		
		β subunit	γ subunit
Allophycocyanin-B	1 PCB	1 PCB	
Allophycocyanin	1 PCB	1 PCB	
C-phycocyanin	1 PCB	2 PCB	
R-phycocyanin	1 PCB	1 PCB, 1 PEB	
Phycoerythrocyanin	1 PXB	2 PCB	
C-phycoerythrin	2 PEB	3 or 4 PEB	
b-phycoerythrin	2 PEB	4 PEB	
B-phycoerythrin	2 PEB	4 PEB	2 PEB, 2 PUB

Phycobilin chromophore abbreviations: phycocyanobilin (PCB); phycoerythrobilin (PEB); phycobiliviolin-type chromophore (PXB); phycourobilin (PUB). After Cohen-Bazire and Bryant (1982).

in recent years and the subject has been reviewed frequently (Gantt 1981; Glazer 1982; Cohen-Bazire and Bryant 1982; see Bryant, this volume).

A number of early studies showed that cell phycobiliprotein concentrations decline in cyanobacteria in response to high irradiance (Myers and Kratz 1955; Halldal 1958; Halldal and French 1958) resulting in an increase in the chlorophyll: phycobiliprotein ratio in many cases (van Liere et al. 1979; Foy and Gibson 1983; Wyman and Fay 1986a) but not in every instance (Kratz and Myers 1955; Carthew 1980). With increasing light both the number of phycobilisomes present per cell and the mean length of the periperal rod substructures (Wyman and Fay 1986a). The phycobilisomes of *Anacystis nidulans*, however, were of a similar size in both white and far-red light-grown cells despite a reduction in the thylakoid area per cell in the latter light regime (Khanna et al. 1983).

The influence of light quality on phycobiliprotein synthesis and phycobilisome structure has been studied extensively in those phycoerythrin-producing cyanobacteria capable of chromatic adaptation. Shifts in light quality result in the preferential synthesis of that phycobiliprotein (phycocyanin in red light, phycoerythrin in green light) which absorbs maximally in the prevailing light quality. Hence, in chromatically adapting species, phycobiliprotein composition is complementary to the light quality experienced during growth. Not all phycoerythrin-producing cyanobacteria modulate phycobiliprotein synthesis in response to light quality, however, and complementary chromatic adaptation has not been reported in a member of the picoplankton. Among those cyanobacteria that do respond to light quality, Tandeau de Marsac (1977) recognised two types of chromatic response. In some, growth in red light was found to suppress the differential rate of phycoerythrin synthesis whereas in others, the synthesis of phycocyanin was also under photocontrol. In these strains the rate of phycocyanin synthesis was stimulated by exposure to red light in response to the decline in the rate of phycoerythrin production. Changes in the differential rate of phycobiliprotein synthesis in chromatic light have been shown to be under transcriptional control (Bennett and Bogorad 1973). Although the existence of photoreversible receptors capable of sensing the quality of the incident light has been proposed (Scheibe 1972; Bogorad 1975) it is still not clear how changes in light quality produce such a rapid and dramatic switch in the transcription of phycobiliprotein genes.

Less is known of the influence of light quality on phycobiliprotein synthesis in those cyanobacteria which either do not adapt chromatically or do not produce phycoerythrin. Gugliemi and Cohen-Bazire (1982), however, have reported that several non-chromatically adapting strains of the genus *Pseudanabaena* produce a spectroscopically distinct form of phycocyanin when grown in red light. In a number of non-chromatically-adapting cyanobacteria, Wyman and Fay (1986b) determined that variation in cell phycobiliprotein content was most closely correlated with differences in growth rate between individual chromatic light regimes rather than with the spectral quality of light provided for growth per se.

MODE AND MECHANISM OF PHOTOSYNTHESIS

All cyanobacteria are photosynthetic in that they are capable of conversion of light to chemical energy. Some of the older, perhaps less cautious literature refers to microorganisms such as *Beggiatoa*, *Leucothrix* and *Caulothrix* as 'colourless' cyanobacteria. There is certainly considerable physical resemblance such un-pigmented filamentous bacteria the *Oscillatoriaceae* family of cyanobacteria. Such descriptions reflect the impossibility of constructing a rational prokaryotic taxomony based upon a limited, usually quite haphazard, readily measurable, set of features. From the view-point of this article the observation of microbial forms comparable to filamentous cyanobacteria in the very non-photosynthetic locale of the deep sea thermal vents (Jannasch 1984) provides the most topical example of possibly related non-photosynthetic organisms.

While all cyanobacteria are genetically capable of photosynthesis, some species are not obligately photosynthetic in that they will grow, usually rather slowly, in the dark by respiration of exogenous sugars. A larger number of species grow photoheterotrophically, that is to say by using light as the principal and essential source of energy but assimilating a proportion of carbon as reduced carbon compounds in place of carbon dioxide. The latter mode of photosynthesis does not require as much reducing capacity as there is less carbon dioxide fixation and consequently, photoproduction of NADPH is less important. The standard procedure for determining whether an organism has photoheterotrophic capacity is to test for growth in the presence of light plus glucose and DCMU, an inhibitor that prevents electron flow between PS II to PSI thus permitting the organism to produce ATP but not reduced pyredine nucleotides from light energy (Rippka 1972).

It is more than likely that the presence of an oxygen-evolving, green plant-like photosynthesis and its associated pigments was a major contributor to the idea that cyanobacteria belonged to the 'algae' with all their connatations of higher plants, rather than they being a type of bacterial photosynthetic microorganism. These two basic types of photosynthesis; the O_2-evolving, as distinct from the anaerobic, sulphide or organic compound utilising, were recognised as separate in the nineteenth century. From a study on the different forms of bacterial photosynthesis and a comparison with that carried out by chloroplasts, van Niel proposed his unifying hypothesis in which all forms of photosynthesis were seen as essentially the same process of light-driven separation of positive and negative charged entities (see van Niel 1941). One can trace the profound effect of the van Niel hypothesis through much of modern photosynthesis research and the considerable present day knowledge of the photosynthetic mechanism accords with the intuitive preception, based upon comparative microbial physiology, which was made by van Niel some 50 years earlier (see Clayton 1980).

There are two main features of the arrangement of the cyanobacterial photosynthetic structures that distinguish them from those of chloroplasts of higher plants (see Ho and Krogmann 1982). The thylakoid membranes of chloroplasts are arranged in 'stacks' in which the two dimensional lipoprotein sheets are adpressed to each other to form the grana of the chloroplast. This arrangement is thought to have physiological significance, and conditions which lead to disruption of the grana are associated with loss of photosynthetic function. Cyanobacteria have photosynthetic membranes which are arranged singly, not in stacks, through much of the cell volume. In some species the photosynthetic membranes occur only in the peripheral parts of the cell, immediately within the cytoplasmic membrane. Of course the prokaryotic nature of cyanobacteria means that there is no subcellular, membrane bound separation of the photosynthetic apparatus as there is in eukaryotes. All the physiological aspects of photosynthesis — CO_2 fixation and O_2 production for example — take place in intimate contact with the rest of cellular metabolism. The other feature of cyanobacterial photosynthesis that differs from that of most chloroplasts (but not the chloroplasts of red algae) is the nature and arrangement of the light-harvesting pigments. The phycobilisomes, which contain several pigmented phycobiliproteins, appear as distinct structures in organised arrays on the outer surface of the photosynthetic membranes (see Glazer 1983 and, Bryant this volume). In green plant chloroplasts, phycobiliproteins are replaced functionally by a chlorophyll *b* containing light harvesting protein complex which is integrated into the photosynthetic membrane of the chloroplast. This difference in the physical arrangement between the light harvesting molecules does, of course, facilitate the considerable alteration in the size of the cyanobacterial phycobilisome that is a prerequisit for its role in nitrogen storage, discussed elsewhere in this Chapter. Apart from these two structural features the other features of photosynthesis are remarkably similar in cyanobacteria and the chloroplasts of plant cells. Indeed, cyanobacteria are being used increasingly to study the basic process of light to chemical energy conversion and their advantages relative to chloroplasts are evident with respect to production, ease of disintergration and biochemical manipulation. This is illustrated by the success in the isolation of functional oxygen-envolving PS II particles from cyanobacteria (Stewart and Bendall 1979 and 1981) and the consequent opportunities afforded for the dissection of the photosynthetic apparatus.

We do not intend to provide here a detailed account of photosynthetic biochemistry (see Foyer 1984), still less to provide an explanation of the physical techniques, such as picosecond time-resolved fluorescence, electron spin resonance studies and Mössbauer spectroscopy which have been applied to this problem. Of the major metabolic processes it is likely that photosynthesis operates over the greatest range of time scales. Thus the initial transfer of energy from one pigment molecule to the next in the light harvesting antennae is measured in terms of less than picoseconds (10^{-13} seconds) and the arrival of energy at a PS reaction centre complex takes picoseconds (10^{-11} s.) Transfer of energy from the modified pigments of the reaction centre to iron quinones takes the process into the microsecond range (10^{-7}–10^{-5} s) while the operation of the electron transfer system of quinones and cytochromes is measured in milliseconds. The subsequent use of the transformed light energy as ATP and NADPH extends into seconds the time scale of the process. The photosynthetic electron transfer chain present in cyanobacteria and derived from studies on several species is shown in Fig. 8. With the exception of the light-harvesting pigments the central features are common to the electron flow pathway thought to operate in chloroplasts, with water acting as reductant leading to the production of O_2 associated with the PS II stage. Photosystem I contains a specialised chlorophyll bound to protein, absorbing at 700 nm, which becomes oxidised by light and donates an electron to the initial acceptor. The subsequent reduction of the oxidised P700 requires a second light stage (PS II), using water as a reductant, and a highly orientated, spacially arranged series of electron carriers, embedded and vectorially arranged in the photosynthetic membrane. The initial electron acceptor (X) has not yet been defined. Mössbauer and ESR spectroscopy indicate that iron-sulphur groups are involved in a temperature-independent manner that would be expected of a photochemical reaction (see Ho and Krogmann 1982; Evans and Heathcote 1983). There is mounting evidence that X is not a single compound and it is not certain that the iron–sulphur groups are in fact the primary electron acceptors from the P700 molecule. These features, and the subsequent electron flow between the reaction centres, reinforce the point that the cyanobacterial and the chloroplast photosynthetic mechanisms

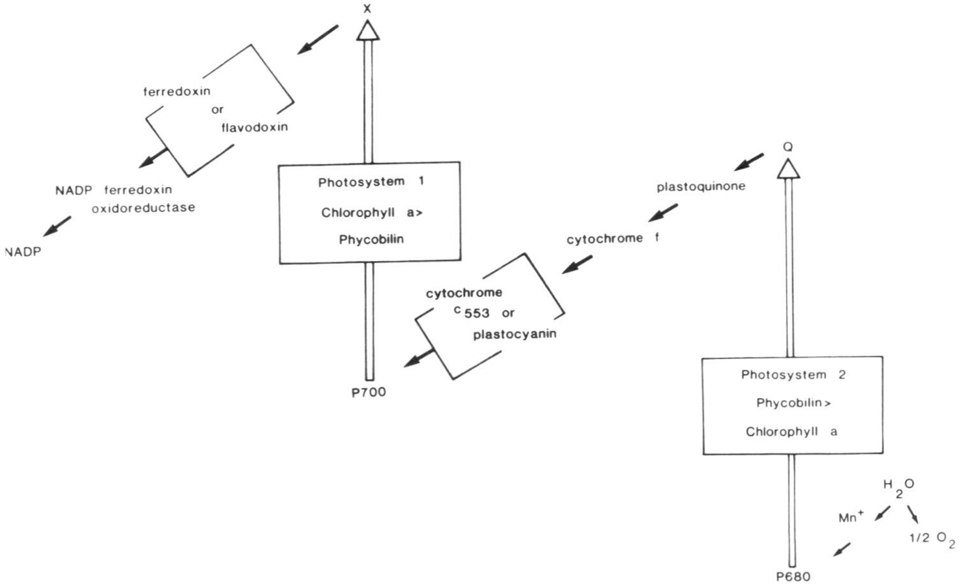

FIG. 8. O_2 evolving photosynthetic electron transfer sequence in cyanobacteria (after Ho and Krogmann 1982).

are essentially of the same nature. Cyanobacteria are increasingly being used for the examination of the very early photochemical processes of light transduction in photosynthesis (Ho and Krogmann 1982).

We have already mentioned the unifying photosynthesis of van Niel which brought into a common framework the basic principles of both green plant and bacterial photosynthesis. Ten years ago this was completely vindicated by the demonstration of what some microbial ecologists had previously suspected: that certain cyanobacteria could utilise sulphide as an electron donor in photosynthesis (Cohen et al. 1975 a, b). This demonstrated at the organism level the unity of the photosynthetic process. Bacterial photosynthesis (see Jones 1982) may be represented as in Fig. 9, in which either an organic molecule, succinate or an inorganic electron source, sulphide or thiosulphate, serve as an electron donor for the production of reduced pyridine nucleotides. The Rhodospirrillaceae (purple non-sulphur photosynthetic bacteria such as *Rhodopseudomonas spheroides*) use organic electron donors, whilst the Chromatiaceae (purple, sulphur bacteria such as *Chromatium* sp.) and *Chlorobacteriaceae* (green, sulphur bacteria such as *Chlorobium* sp.) can use either sulphide or an organic reductant. Depending upon the environmental conditions the cyanobacterium *Oscillatoria limnetica* (and several other species) will switch from using water as electron donor to sulphide (Padan 1979). The S^-/S redox couple permits electron flow to the cytochrome level and under such circumstances *O. limnetica* ceases to use light absorbed by PS II for photosynthesis and converts to the bacterial type of anoxygenic photosynthesis that uses only one photosynthetic reaction centre. Low concentrations (~ 0.1 mM) are inhibitory to oxygenic photoassimilation of carbon dioxide but when higher concentrations of sulphide (~ 3 mM) are available, *Oscillatoria limnetica* adopts a DCMU insensitive photoassimilation of carbon dioxide which is driven only by PS I and uses light of 700 nm wavelength (Cohen et al. 1975b). The anoxygenic mode of photosynthesis appears to require protein synthesis during the necessary two hour induction period, but *O. limnetica* will immediately revert to oxygenic photosynthesis on removal of sulphide. It is clear that the ability,

172

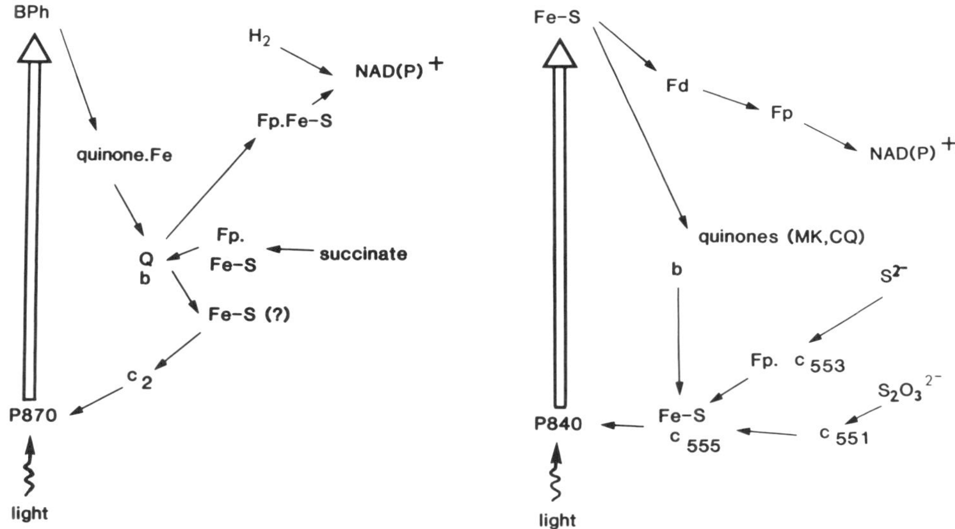

FIG. 9. Light dependent electron transfer in photosynthetic bacteria. *Left* Purple non-sulfur bacteria (e.g. *Rhodospirillaceae*). *Right* Sulphur bacteria (e.g. *Chromatiaceae*) (after Jones 1982).

enjoyed by a significant number of cyanobacteria, to carry out both oxygenic and anoxygenic photosynthesis would be advantagous in environments in which the presence and absence of sulphide alternates. Sediments and stratified water bodies fall into this category. Since reasonably high levels of sulphide are necessary to induce anoxygenic photosynthesis, short term periods of anaerobiosis are unlikely to allow the build up of the necessary sulphide levels. A full discussion of the ecological implications of anoxygenic photosynthesis with an emphasis on its occurrence in the Solar Lake (which is near the coast of the Red Sea) is provided by Padan (Padan 1979; Padan and Cohen 1982). The range of cyanobacteria that are capable of employing sulphide in anoxygenic photosynthesis is quite wide (Garlick et al. 1977). A wide range of sulphide concentrations may be employed, and levels of sulphide optimal for the photosynthesis of some species would prove toxic for others. *Lyngbya* species have only a limited concentration range of sulphide utilisation before toxic levels are reached; this is contrasted with *Oscillatoria limnetica* which has tolerance to high concentrations of sulphide, but has rather low affinity for sulphide and cannot achieve maximum rates of anoxygenic photosynthesis below about 3 mM concentrations.

LIGHT TOLERANCE

Cyanobacteria have a wide geographical distribution and therefore as a group they are subjet to a considerable range of photon flux densities in their natural environment. Richardson et al. (1983) determined that a minimum irradiance of $5~\mu E~m^{-2}~s^{-1}$ was required to support the autotrophic growth of cyanobacteria whereas maximum growth rate was achieved at an average irradiance of $39~\mu E~m^{-2}~s^{-1}$. Their data are biased toward the light requirements of planktonic species, however, and it is apparent that many cyanobacteria thrive in habitats where the incident irradiance is outside of these limites. In Antarctic lakes, benthic species have been shown to survive at photon fluence rates which rarely exceed 1–2 μE $m^{-2}~s^{-1}$ (Fogg et al. 1973). At lower latitudes cyanobacteria have been recorded at the bottom of the euphotic zone in both inland (Jones and IImavirta 1978; Konopka 1980, 1982) and oceanic waters (Waterbury et al. 1979; Takahashi and Hori 1984).

173

Planktonic cyanobacteria may form persistent surface blooms in freshwaters during the summer months, however, where they may be exposed to subsurface light in excess 1300 μE m^{-2} s^{-1} (Paerl et al. 1983; Paerl 1984). A similar tolerance of high irradiance has been noted in a number of thermophilic cyanobacteria growing in hot springs (Castenholz 1968). More or less permanent populations of cyanobacteria occur in certain equatorial lakes where the photon flux density at midday exceeds 2000 μE m^{-2} s^{-1} (Ganf 1974). At the other extreme, a few cyanobacteria capable of heterotrophic carbon assimilation have been isolated from environments subject to prolonged darkness eg. *Nostoc* sp. strain MAC the symbiont of the coralloid roots of the cycad *Macrozamia lutea* (Bowyer and Sherman 1968).

Cyanobacteria have a number of protective mechanisms which enable them to survive exposure to the inimical effects of intense solar radiation. Two principal strategies have been recognised: the avoidance of high irradiance by migration to a more favourable light climate, or the development of structural and physiological responses which enable high irradiance to be tolerated *in situ*.

Fine control of the positioning of cyanobacteria with respect to a light gradient has been described in planktonic (Walsby 1971; Reynolds and Walsby 1975), benthic and terrestrial species (Castenholz 1969, 1982; Whale and Walsby 1984). Many planktonic cyanobacteria control their relative cell density by the light-mediated regulation of gas vacuole content (Walsby 1972). At high irradiance (ie. in surface waters) an increase in cell turgor pressure due to the accumulation of photosynthate (Dinsdale and Walsby 1972) and potassium ions (Allsion and Walsby 1981) has been shown to result in the collapse of the weaker gas vesicles present in the gas vacuoles. Confirming the model proposed by Reynolds and Walsby (1975), Oliver and Walsby (1984) have provided direct evidence that if sufficient gas vesicles are collapsed the cell becomes negatively buoyant and sinks to greater depths in the water column. Prolonged exposure to high irradiance is avoided, thereby, for only at low light levels are sufficient gas vacuoles present to make the cell positively buoyant.

A number of filamentous cyanobacteria capable of gliding motility exhibit phototactic responses. Castenholz (1969) described the clumping behaviour of the normally dispersed filaments within mats of *Oscillatoria terebriformis* exposed to full sunlight. A similar photophobotactic response has been characterised in populations of *Microcoleus chtanoplastes* present in upper littoral mudflats (Whale and Walsby 1984). These workers observed the vertical migration of gliding trichomes toward the mud surface in shaded regions and the avoidance of the surface in areas exposed to direct sunlight. Experimental evidence has shown that the direction of gliding motility (ie. toward or away from a light source) is controlled by differences in electrical potential between each end of the filament generated by changes in the pool of electrons between photosystem I and photosystem II (Hader 1974, 1978, 1979).

Tolerance, rather than avoidance of high irradiance is a strategy common in terrestrial species and in those planktonic cyanobacteria that are of tropical distribution or which form part of the summer phytoplankton bloom at temperate latitudes. Resistance to photoxidation shows little correlation with taxonomic position (Zehnder and Egli 1977) however, and may vary between strains of the same species. Eloff (1978) found that whereas *Microcystis aeruginosa* 7005 was particularly light sensitive, natural populations of this organism dominate surface waters during the summer phytoplankton bloom in many inland lakes and rivers (e.g. Paerl et al. 1983).

In common with other photosynthetic microorganisms, cyanobacteria regulate the rate of synthesis of their photosynthetic pigments and are able to reduce their effective cell light absorption capacity at high irradiance (Myers and Kratz 1955; van Liere et al. 1979; Wyman and Fay 1986a). Cyanobacteria also regulate the efficiency with which absorbed light energy is transferred from light harvesting pigments to the photosynthetic reactions centres. This form of photoprotection has been found to operate in the oceanic *Synechococcus* sp. strain DC2. At high photon flux densities Wyman et al. (1985) found that a substantial fraction of the light energy collected by phycoerythrin was dissipated as autofluorescence whereas at low irradiance this biliprotein was more efficiently coupled to the photosynthetic apparatus. An increase in the *in vivo* autofluorescence yield of phycoerythrin in high light adapted cells was correlated with a decrease in the susceptibility of this strain to photoinhibition at high irradiance (Barlow and Alberte 1985). A reduction in surface photoinhibition has been correlated with a decline in cell chlorophyll and phycobiliprotein concentrations and enhanced carotenoid synthesis in *Microcystis aeruginosa* (Codd 1981; Pearl et al. 1983) and in several *Anabaena* species (Paerl and Kellar 1979; Paerl 1984). The protective role of carotenoids in quenching triplet sensitizers and singlet oxygen produced at high irradiance has been well documented (Krinsky 1979) although the evidence that they perform a similar function in cyanobacteria is largely circumstantial. Resistance to ultraviolet radiation was attributed to the higher carotenoid content of a mutant strain of *Gloeocapsa alpicola* (Buckley and Houghton 1976). These workers found that treatment with diphenylamine (an inhibitor of carotenoid synthesis) reduced the ability of the mutant to resist damage by ultraviolet light. Diphenylamine treatment has also been shown to depress photosynthetic performance at high irradiance in *Microcystis aeruginosa* and *Anabaena oscillarioides* (Paerl 1984). Photoadaptation of *Anabaena variabilis* following a shift-up to high photon flux density was found to require an acclimation period of four days whilst final cell chlorophyll and carotenoid concentrations were established within twenty four hours (Collins and Boylen 1982). This suggests that the enhancement of carotenoid synthesis is only one of the mechanisms which enable cyanobacteria to tolerate high light levels.

Like all other aerobic organisms, cyanobacteria synthesize the enzyme superoxide dismutase (SOD) which catalyses the conversion of superoxide anions (O^{2-}) to hydrogen peroxide and molecular oxygen. SOD has been shown to be induced following transfer of cells of *Anacystis nidulans* from microaerophilic to aerobic culture and to enhance the survival of this organism under photooxidative conditions (Abeliovich et al. 1974). The accumulation of reactive peroxide produced in the dismutase reaction is avoided in other organisms through the action of catalase and this enzyme has been shown to be present in *Microcystis aeruginosa* 7820 (Tytler et al. 1984). However, both SOD and catalase were photoinactivated when low light ($5~\mu E~m^{-2}~s^{-1}$) grown cells were exposed to high irradiance (100 or 1000 $\mu E~m^{-2}~s^{-1}$) and Tytler et al. (1984) question the utility of these enzymes during the early stages of photoinhibition. The susceptibility of SOD to inactivation was found to differ between two isoenzymes isolated from *Plectonema boryanum* (Steinitz et al. (1979). Although the ferri-isoenzyme was particularly sensitive, resistance to photo-oxidation was correlated with an increase in the abundance of the more stable manganese isoenzyme.

A number of the mechanisms present in other bacteria for the repair of DNA following irradiation damage (or treatment with mutagenic agents) exist in cyanobacteria (Herdman 1982). The photoreactivation of chromosomal DNA showed a strong dependence on blue light (425–450 nm) in *Agmenellum quadruplicatum* (Van Baalen and O'Donnell 1972) and on blue (380–480 nm) and near-ultraviolet (355 nm) light in *Plectonema boryanum* (Singh 1975). An increase in sensitivity to ultraviolet light following treatment with agents which inhibit dark excision repair of DNA in other bacteria has been interpreted as evidence for the operation of this process in *Anacystis nidulans* (Asato 1972). Although the activity of enzymes essential for the dark repair of ultraviolet inactivated DNA has been demonstrated in cell-free extracts of *A. nidulans* (Shestakov et al. 1975), Astier et al. (1979) found that inhibition of the repair process with caffeine had little effect on the sensitivity of *Aphanocapsa* 6714 to ultraviolet light.

Light shielding properties have been attributed to certain structural components of the cyanobacterial cell. A number of species produce pigmented sheaths exterior to the cell which may serve to absorb a fraction of the incident light. Exposure to direct sunlight led to the brown colouration of the cell sheath material in mats of

Lyngbya (Birke 1974, quoted by Whitton and Potts 1982). However, other environmental factors (e.g. pH, salinity) may also influence the pigmentation of the sheath (Fogg et al. 1973). Carotenoid pigments present in the cell wall of *Synechocystis* 6714 (Omata and Murata 1984) and therefore these pigments may have a further photoprotective function in screening the underlying photosynthetic lamellae from potentially damaging ultraviolet and visible radiation.

There have been a number of suggestions that the gas vacuoles of planktonic cyanobacteria may act as light shields by back-scattering of the incident light (Waaland et al. 1971; Shear and Walsby 1975; Porter and Jost 1976; Ogawa et al. 1979). Such a role for gas vacuoles appeared to be consistent with the findings of ultrastructural studies. In the cells of *Anabaena flos-aquae* (Shear and Walsby 1975) and *Microcystis aeruginosa* (Barlow et al. 1977) gas valuoles were homogeneously distributed throughout the cytoplasm at low irradiance but were found in a peripheral location exterior to the thylakoids in cells grown in high light. Although gas vesicles have been shown to be extremely effective in scattering light (Walsby 1972) careful photosynthetic measurements performed in dilute culture samples have shown little variation in the photosynthetic performance of *Anabaena flos-aquae* with either intact or collapsed gas vacuoles (Shear and Walsby 1975). Although of doubtful benefit to individual cells in suspension, van Liere and Walsby (1982) suggest that 'sacrificial' light scattering by gas vacuoles present in the peripheral cells of colonial cyanobacteria may protect the underlying cells at high irradiance. The benefit of this may be offset, however, by the reduction in the amount of light penetrating the colony during periods of light limitation.

Respiration

All cyanobacteria that have been carefully examined with respect to respiratory activity show reproducible rates of O_2 uptake in the dark. The fact that these rates are relatively low, Q_{O_2} values of less than 10 being usual, has no doubt contributed to the lack of attention being paid to this aspect of cyanobacterial metabolism. The Q_{O_2} values (μO_2 h^{-1} mg dry wt^{-1}) obtained with cyanobacteria (Kratz and Myers 1955) are comparable to some eukaryotic microalgae but they are considerably less than others (Gibbs 1962). Respiration rates for heterotrophic bacteria show a wide range and they are, of course, environmentally variable; however, Q_{O_2} values around 100 and above are not unusual. In a direct comparison of O_2 uptake per unit biomass it is evident that cyanobacteria make a small contribution to total O_2 consumption relative to heterotrophic prokaryotes. There are, however, many environmental niches in which cyanobacteria predominate over heterotrophs in terms of biomass and their dark respiratory contribution should not be ignored. What is interesting in this context is the apparent inability, albeit of the few species examined, of cyanobacteria to significantly increase their rate of dark oxygen uptake when exogenous respiratory substrates are supplied (Pearce and Carr 1969). To an extent this no doubt is a result of the absence of active-uptake processes, which have been shown to be present to only a limited extent (Beauclerk and Smith 1978). When cyanobacteria were incubated aerobically in the dark, their respiration rate fell and the endogenous reserves became depleted. Addition of glucose at this stage restored the respiratory rate to the original 'unstarved' level, but did not increase it (Pearce and Carr 1969). These observations argue that the rate of entry of exogenous substrates into cyanobacteria is not the limiting factor, and that this should be sought within the respiratory machinary itself.

The feature of cyanobacterial respiration that distinguishes it from that of eukaryotic microalgae and allies it with that of photosynthetic bacteria is the occurrence of light inhibition of oxygen-uptake. Early work with species of *Anabaena* (Brown and Webster 1953) using the heavy isotope of oxygen, O^{18}, measured

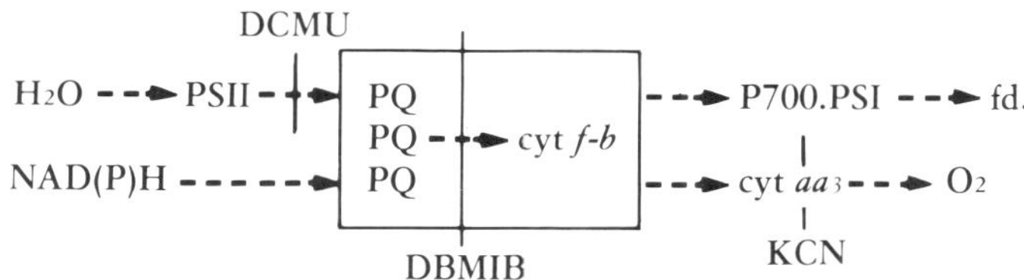

FIG. 10. Common components of the respiratory and photosynthetic electron flow in thylakoid membranes of cyanobacteria. (DCMU, dichlorophenyldimethylurea; DBMIB, dibromothymoquinone.) (after Peschek 1983.)

concurrently the process of O_2 uptake and the production of O_2 from photosynthesis. Photoinhibition of respiration occurred at low oxygen concentrations but was less pronounced at higher O_2 concentrations. Jones and Myers (1963) showed by polargraphic measurements that long wavelength light (700 nm) was most effective in suppressing O_2 uptake and concluded that the electron acceptor from P700 was at least partially a competitor of O_2 as an electron acceptor. In terms of the electron transfer scheme illustrated in Fig. 1 the photoinhibition of respiration can be envisaged in more direct terms. If a component of cyclic photophosphorylation electron flow — possibly a cytochrome or plastoquinone — is also a component of dark, respiratory electron transport then the oxidised P700 could act as a competitor for electrons carried on that component and thus reduce electron flow to cytochrome oxidase. This would in turn cause a reduction in oxygen uptake. The absence of cyanobacterial respiration in the light will influence the extent to which those organisms compete with other microorganisms for the frequently low concentrations of available reduced carbon. Recently Peschek (1983) has produced direct evidence, using membrane fractions of *Anacystis nidulans*, that a cytochrome *f–b* complex was oxidised by either O_2 or by light (Fig. 10). Peschek and Schmetterer (1982) observed the requirement for plastoquinone in the reduction of the cytochrome *f–b* complex. A role for plastoquinone in respiratory electron transport, in addition to its established function in photosynthesis, was indicated by the fact that only plastoquinone, and no ubiquinone, was present in cyanobacteria (Carr and Hallaway 1966).

The physical location of respiratory activity within the cyanobacteria cell is also consistent with the idea that respiration and photosynthesis are interelated and may show common components in their electron transport chains. Bisalputra et al. (1969) and Peschek et al. (1981) showed that the reduction and intracellular deposition of tellurite occured on the thylakoid membranes of cyanobacteria. The latter workers made the additional observation that some tellurite deposition was associated with the plasmamembrane itself, with the implication that this was a possible additional site of respiratory activity.

The *in vitro* formation of ATP by the oxidation of NADPH, and the involvement of several electron transport enzymes and response to inhibitors was shown with *Anabaena variabilis* (Leach and Carr 1971). Bottomley and Stewart (1976) showed that the ATP pool of *Anabaena cylindrica* was maintained with remarkable constancy over several hours provided that the conditions permitted either photophosphorylation or oxidative phosphorylation. On switching to conditions that allowed only substrate level phosphorylation (ie. dark and anaerobic) the ATP pool decreased immediately. Oscillations in the oxygen uptake rate and the NADPH pool were induced over 90 s by short (6-s) light flashes (Metzler 1980). Cytochrome *b* showed similar oscillations in its redox state and information regarding ATP pool behaviour in these types of experiments would be informative.

It is evident that, from the small number of species examined, cyanobacteria carry out dark respiration, and that this process will contribute to the depletion of oxygen in the dark in waters which hold large populations of these organisms. What is less clear is the contribution that cyanobacterial respiratory activity may make to the carbon budget. Clearly respiratory CO_2 is produced in the expected ratio to O_2 uptake. The cyanobacterial respiration rates reported were achieved at the expense of endogenous carbohydrate reserves and were not increased by the supply of exogenous substrates such acetate, glucose or other sugars. This situation still applied after growth of the cyanobacteria in the presence of the potential substrates and when the organisms were incubated with a range of concentrations of substrate, well in excess of those which may be encountered. However, sugars could restore the oxygen uptake rate of respiratory-depleted *Anabaena variabilis* and *Synechococcus* species and ^{14}C-labelled substrates were respired to ^{14}C-carbon dioxide (Pearce and Carr 1969). The picture that emerges from respiratory studies based on a very limited range of cyanobacteria, is that oxygen uptake is a rather constant feature of their physiology and whereas it undoubtably is important as a source of maintenance energy during periods of darkness, is unlikely to be used opportunistically by cyanobacteria to gain benefit from exogeous reduced carbon compounds that may, perhaps intermitantly, become available.

Nutrient Assimilation

INORGANIC CARBON

Photoautotrophy is the dominant mode of carbon nutrition in the majority of cyanobacteria during periods of active growth and therefore considerable interest has been directed toward understanding the mechanism of inorganic carbon (Ci) assimilation in these organisms. In an early investigation, Fogg and Than-Tun (1960) determined that the optimum pCO_2 for the growth of *Anabaena cylindrica* in culture was dependent upon the external temperature (0.1% at 15°C; 0.25% at 25°C and above) and that carbon fixation in *A. cylindrica* was inhibited at a pCO_2 in excess of 0.5%. In the absence of inorganic carbon, cell chlorophyll and phycobiliprotein concentrations declined dramatically and the integrity of the thylakoid membranes was lost within 120 h in *Synechococcus lividus* (Miller and Holt 1977). Although these workers demonstrated that bleaching induced by carbon deprivation was reversible in this organism, prolonged Ci starvation is known to favour the photoxidative death of cyanobacteria (Eloff et al. 1976).

The association of cyanobacteria with neutral to alkaline environments determines that HCO_3^- is the dominant form of available Ci. That the transport of HCO_3^- may act as a substrate for photosynthesis in cyanobacteria was demonstrated in the marine species *Coccochloris peniocystis* (Miller and Colman 1980). In this species, cytoplasmic pH showed little variation (pH 7.6–7.9) with external pH in the range 7.0–10.0 but declined to a pH of 6.6 when the external pH was adjusted to a value of 5.25 (Colman and Coleman 1981). Ribulose bisphosphate carboxylase (Rubisco) activity was optimal at an internal cell pH of 7.5 but Ci-fixation ceased owing to the inactivation of enzyme when the internal pH fell to 6.5. The inability to regulate internal pH in acidic environments led Coleman and Colman (1981) to propose acidification of the cytoplasm and its effect on Rubisco activity as the reason for the poor performance of many cyanobacteria at low pH.

Consideration of the pH gradient generated across the plasmalemma at alkaline pH precludes the possibility that Ci assimilation may occur by simple CO_2 diffusion. Coleman and Colman (1981) concluded that HCO_3^- must be transported in order to account for the measured internal carbon pool in *Coccochloris peniocystis*. Induction of HCO_3^- transporting capacity at low concentrations of external CO_2 was demonstrated in *Anabaena variabilis* (Marcus et al. 1982). Although the affinity of the HCO_3^- transporting mechanism for Ci was similar in cells incubated in either low (0.03%) or high (5%) CO_2, high photosynthetic affinity for Ci in low

178

CO_2-treated cells was associated with a ten-fold increase in the Vmax for HCO_3^- transport. Acclimation to low CO_2 concentrations was found to be sensitive to the inhibitor of protein synthesis, spectinomycin, and was accompanied by changes in the ultrastructure of the cell wall. Examination of the kinetics of CO_2 and HCO_3^- assimilation in *A. variabilis* led Volokita et al. (1984) to conclude that regardless of the source of Ci, HCO_3^- is the carbon species which arrives at the inner membrane surface. A similar conclusion was reached by Badger et al. (1984) in an examination of Ci assimilation by the marine (although not picoplanktonic) *Synechococcus* sp. strain RRIMP IN. Volokita et al. (1984) found that in contrast to HCO_3^- transport, CO_2 assimilation was much more sensitive to the inhibitor of carbonic anhydrase, ethoxyzolamide. Paradoxically, although Badger et al. (1985) were able to demonstrate carbonic anhydrase activity both *in vivo* and *in vitro*, they report that photosynthetic O_2 evolution by *Synechococcus* sp. strain RRIMP IN was insensitive to this sulphonamide. Volokita et al. (1984) have suggested that a carbonic anhydrase-like moiety present in the plasmalemma may enable CO_2 to be assimilated from the external environment prior to its transport into the cell in the form of HCO_3^- via a HCO_3^- porter. The simpler model proposed by Badger et al. (1985), however, assumes that HCO_3^- is the only form of Ci transported. Subsequently HCO_3^- is converted to CO_2 (via carbonic anhydrase), the only form of Ci that can be fixed by Rubisco (Codd and Marsden 1984). Lanaras et al. (1985) found that the carbonic anhydrase of *Chlorogloeopsis fritschii* is localised in a particulate cell fraction other than that recovered with the thylakoids or carboxysomes and suggest the possibility that this enzyme may be associated with either the inner or outer cell membranes.

Organic Carbon

In all the cyanobacteria so far examined carbon dioxide fixation by the Calvin cycle using ATP and NADPH derived from photosynthesis is the principle mode of nutrition. Relatively rare examples demonstrate an increase in growth rate by the supply of reduced carbon compounds. It is likely that this represents a correct, overall picture of the growth of cyanobacteria in many natural environments. However, it is worth recalling that the means employed to isolate the axenic cultures of cyanobacteria on which the above generalisation is based have invariably used autotrophic selection media. It may be that if selection procedures were devised in which heterotrophic growth media were employed a rather different picture would emerge. Such a selection procedure would have to employ some physical means, size selection perhaps, to enrich for cyanobacteria rather than the more rapidly growing heterotrophic bacteria and fungii. The widespread distribution of cyanobacteria in environments which are relatively rich in organic carbon, and sometimes rather limited with regard to irradiance, encourages the belief that the above procedure would be a worthwhile, although technically daunting, approach. Khoja & Whitton (1971) showed that the ability of cyanobacteria to utilise sugars was distributed through several cyanobacterial types.

There have been several studies in which the heterotrophic potential of cyanobacteria have been probed by use of ^{14}C-labelled substrates (see Carr 1973; Smith 1982 for reviews). Whereas it is evident that some isotope was assimilated, with glucose and other sugars proving the most favourable substrates, only a few strains increased their growth rate under such conditions. One approach has been to examine the degree to which cyanobacteria respond to organic substrates in terms of their uptake rate and enzyme adaptation (Carr 1973). The general conclusion reached from these studies being that no increase in metabolism of exogenous carbon compounds was observed in most cyanobacteria, leading to the conclusion that the cyanobacteria studied were unable to adjust their metabolism to possible advantagous environmental changes of this sort (Carr 1973). The comparitive examination of reduced carbon uptake systems present in different cyanobacteria was undertaken by Smith (Beauclerk and

Smith 1978; Smith 1982). In these studies comparison was made between versatile strains (*Nostoc* sp. MAC; Synechocystis 6714) which would grow heterotrophically and several species of cyanobacteria (specialist strains) that would not. It was clear that versatile strains had a relatively effective D-glucose uptake system compared with the specialist strains. Furthermore D-glucose uptake was inhibited by glucose analogues only in versatile strains, indicating the presence of an active uptake system. It will be instructive to apply this type of analysis to a wider range of cyanobacteria, with exogenous carbon utilisations intermediate between the two types above.

The concentration of D-glucose in oceans was estimated to be in the range 10–100 nM (Vaccaro et al. 1968). Using an autoradiographic procedure Saunders (1972) showed that a natural population of planktonic cyanobacteria could take up glucose, and that *Oscillatoria* species were able to accumulate the substrate from an external concentration of 10 nM. The extent to which cyanobacteria remove organic substrates from the environment, and the contribution that this process makes to total carbon assimilation is unknown.

The oxidative dissimilation of sugar molecules, and other metabolites, to carbon dioxide is constrained by the absence in cyanobacteria of a complete tricarboxylic acid cycle (see Carr 1973). In all cyanobacteria so far examined, 2-oxoglutarate dehydrogenase is absent; thus, carbon flow must be directed to glutamate synthesis, rather than to the total oxidation of a 2-carbon unit to carbon dioxide. The formation of malate and succinate (necessary for other biosynthetic purposes) comes from the carboxylation of phosphoenolpyruvate to yield oxalacetate, and the reduction of this molecule by reversible steps of the interrupted TCA cycle. This carboxylation can be considered to be the principal anaplerotic step in cyanobacterial metabolism.

INORGANIC AND ORGANIC NITROGEN

All cyanobacteria are capable of utilizing ammonium- and nitrate-nitrogen for growth (Stanier and Cohen-Bazire 1977; Gibson 1984). Although there are comparatively fewer reports in the literature, growth in the presence of nitrite, hydroxylamine, urea, amino acids and various other organic nitrogen compounds as the sole sources of nitrogen has also been demonstrated in a number of species (Kratz and Myers 1955; Van Baalen 1962; Holm-Hansen 1968; Stanier et al. 1971; Kapp et al. 1975). Whereas nitrogen accounts for approximately 1–3% of the cellular dry weight of eukaryotic phytoplankton (Wheeler 1983), nitrogen represent about 4–9% of the dry weight of the cyanobacterial cell (Fogg et al. 1973). Although the demand for this element in comparatively high, cyanobacteria accumulate substantial macromolecular reserves of nitrogen and therefore the 'essential' cellular nitrogen content may be significantly lower.

Nitrogen deficiency produces dramatic changes in the pigmentation, cellular composition and the ultrastructure of cyanobacteria. Allen and Smith (1969) found that whereas nitrogen-deficient cells of *Anacystis nidulans* contained normal levels of chlorophyll and carotenoids, phycocyanin was totally degraded. Phycocyanin concentrations recovered, however, following the re-introduction of a source of combined nitrogen. Nitrogen starvation elicited the production of proteases specific for phycocyanin degradation in several strains of *Anabaena* (Foulds and Carr 1977; Wood and Haselkora 1980) and was shown to specifically depress the *de novo* synthesis of this biliprotein in *Anacystic nidulans* (Lau et al. 1977). Ownby et al. (1979) showed that amino acids released from protein (principally phycocyanin) turnover during nitrogen starvation were utilized for the synthesis of new cell proteins. It is now accepted that phycobiliproteins constitute a major nitrogen reserve in cyanobacteria and this aspect is discussed elsewhere in this review.

Nitrogen starvation led to the accumulation of carbohydrate in the cells of *Anacystis nidulans* (Allen and Smith 1969); an observation which has been confirmed in *Ana-*

baena cylindrica (de Vasconcelos and Fay 1974) and *Agmenellum quadruplicatum* (Stevens et al. 1981). Mobilization of phycobiliproteins during nitrogen deficiency was followed by the degradation of ribosomes and then the thylakoid membranes in *A. quadruplicatum* (Stevens et al. 1981). However, lipid, polyphosphate bodies, and, significantly, carboxysomes (another potential soure of cell nitrogen) were not degraded. In contrast, nitrogen starvation led to the accumulation of lipid granules and polyphosphate bodies and the degradation of carboxysomes in *Mastigocladus laminosus* (Stevens et al. 1985). Nitrogen-limitation may also influence the gross morphology of cyanobacteria. Heterocyst differentiation is induced by the removal of combined nitrogen in nitrogen-fixing genera. In addition, removal of inorganic nitrogen promoted the production of tapered trichomes and the development of trichome polarity in several *Calothrix* species (Sinclair and Whitton 197). In *Nostoc muscorum* Armstrong et al. (1983) found that the removal of sodium nitrate from the culture medium resulted in the formation of gas-vacuolate hormogonia. Nitrogen deprivation has been shown to promote akinete differentiation in certain cyanobacteria whereas in others nitrogen-limitation may have the opposite effect (see Nichols and Adams 1982).

Kratz and Myers (1955) presented evidence that ammonium was assimilated in preference to nitrate when both nitrogen sources were supplied to cultures of *Anacystis nidulans*. Ammonium ions (or products of their metabolism) have been shown to repress nitrate assimilation in *Synechocystis* sp. and *Anabaena variabilis* (Herrero et al. 1985) and are known to be a potent inhibitor of nitrogenase synthesis (Fogg 1949; see Stewart 1980). Ammonium repression of nitrate assimilation is concentration-dependent and the simultaneous uptake of both ammonium and nitrate by *Oscillatoria agandhii* has been reported by Zevenboom and Mur (1981). The L-isomers of several amino acids inhibited nitrate assimilation by methionine-sulphoximine (MSO)-treated cells of *Anacystis nidulans* (Romero et al. 1985). Although not all of the inhibitory amino acids were equally effective, Romero et al. (1985) found a significant correlation between the inhibition of nitrate assimilation and the ability of these amino acids to act as substrates of α-ketoglutarate-dependent transaminase activity. These authors suggested, therefore, that the ammonium ion promoted inhibition of nitrate uptake involved the participation of organic compounds synthesized following ammonium assimilation. The major pathway of ammonium assimilation in cyanobacteria is the glutamine synthetase-glutamate synthase (GS-GOGAT) pathway (see Stewart 1980). Uptake of ammonium by *Anacystis nidulans* was found to be energy-dependent and was eliminated by transfer of the cells to the dark, anaeraobic conditions, or by the addition of protonophores (Boussiba et al. 1984). Inhibition of GS activity by the addition of MSO strongly inhibited ammonium uptake as well as ammonium metabolism. Other enzymes potentially capable of ammonium assimilation (e.g. glutamate dehydrogenase (GDH), alanine dehydrogenase (ADH)) have been demonstrated in cell-free extracts of cyanobacteria although their activity is generally low or sometimes undetectable (Neilson and Doudoroff 1973; Stewart 1973). However, Meeks et al. (1978) suggested that ammonium assimilation via GDH may be the major route of glutamate formation in *Anacystis nidulans*.

Whereas the pathway of ammonium assimilation by cyanobacteria has been studied extensively (particularly in the context of nitrogen fixation) the nature of the nitrate uptake system remains to be clarified. Meeks et al. (1983) found that the induction of nitrate assimilation in *Anabaena cylindrica* was sensitive to chloramphenical suggesting an obligate requirement for the *de novo* synthesis of at least one component of the nitrate uptake system. Nitrate was required for the induction of nitrate assimilation in *Anabaena variabilis* but not in *Synechocystis* sp. 6714 (Herrero et al. 1985) and these authors proposed that the nitrate regulation of the assimilatory nitrate reductase may be common to all nitrogen-fixing prokaryotes.

The kinetics of inorganic nitrogen assimilation have been examined in a number of cyanobacteria and most extensively in the freshwater planktonic species *Oscillatoria agardhii* (e.g. Zevenboom and Mur 1979, 1981a,b). The characteristics of nitrogen-limited growth and inorganic nitrogen assimilation by cyanobacteria are considered by Zevenboom elsewhere in this volume.

Urea is utilized by most cyanobacteria (Stanier and Cohen-Bazire 1977) although, in general, cyanobacteria grow less well on organic nitrogen than on inorganic sources of this element (Fogg et al. 1973). Some cyanobacteria incorporate amino acids although very few species are capable of growth on this source of nitrogen. Kratz and Myers (1955) reported that *Anabaena variabilis*, *Nostoc muscorum* and *Anacystis nidulans*, although capable of growth on urea, were unable to utilize the mixture of amino acids present in caesein hydrolysate. This may be the result of an inability to regulate the synthesis of the enzymes responsible for the degradation of amino acids once they have been assimilated (Carr 1973) but more likely is a result of the absence of specific uptake systems. Kapp et al. (1975), however, found that the marine species *Agmenellum quadruplicatum* was able to utilize a wide range of amino acids (particularly alanine, asparagine, aspartate, glutamine, histidine and serine). Both urea and casamino acids were shown to serve as nitrogen sources for the growth of *Synechococcus* sp. strain DC2 (Wyman and Carr, unpublished). From the range of organic nitrogen compounds tested, aspartate was shown to support a growth rate comparable to that obtainable in the presence of nitrate under otherwise similar experimental conditions.

The mechanism of amino acid membrane transport has been investigated in detail in the photoautotrophic cyanobacterium *Anacystis nidulans* (Lee-Kaden and Simons 1982). As is the case in other bacteria, different uptake systems were present for the assimilation of neutral and branched-chain amino acids. Assimilation was enhanced by light and was ATP-dependent. Although energy generated through the reactions of photosystem I only were able to support membrane transport, amino acid assimilation was enhanced by the participation of both partial reactions of photosynthesis. The action spectrum for the light-dependent incorporation of arginine by *Oscillatoria rubescens* also showed the participation of both photosystem I and photosystem II (Wyman and Fay, unpublished). Since assimilation of amino acids into protein is essentially dependent upon growth it is not surprising that both PSII and PSI activity are necessary for this process.

Dinitrogen Fixation

Cyanobacteria have for many years been known to fix dinitrogen, and, as with other prokaryotes, the number and range of species known to be capable of carrying out this important process is steadily increasing. This is largely because the means of measuring nitrogenase activity are now more sensitive and because we have a better understanding of the physiological conditions necessary for nitrogen fixation to occur.

Although a variety of nitrogen-fixing cyanobacteria have been isolated from shoreline areas and from saline and upper saline pools, the oceans are noted for the relative absence of nitrogen fixing cyanobacteria (see Fogg 1982). Although the sea is said to be principally nitrogen limited, no major development of nitrogen fixing phototrophs occurs to remedy the situation, this being in contrast to freshwater bodies which, when other nutrients are available, often develop nitrogen-fixing blooms of cyanobacteria.

Nitrogen-fixing cyanobacteria are principally of the filamentous, heterocyst forming groups such as *Anabaena* and *Nostoc* species (see Stewart 1973). It is now accepted that the heterocyst is the site of aerobic nitrogen fixation in these species and considereable information has been assembled on its specialised biochemistry, structure and development (see Adams and Carr 1981; Bothe 1982; Wolk 1982). There are, however, several well documented examples of filamentous forms which do not form heterocysts and which do fix nitrogen. *Plectonema boryanum* was the first such organism to be recognised (Stewart and Lex 1970) and subsequently species of *Lyngbya*, *Oscillatoria* and *Phormidium* were also demonstrated to fix nitrogen under microaerobic conditions. There is very strong circumstantial evidence that the marine filamentous non-heterocystous *Trichodesmium*, creates within its bundles of filaments a relatively microaerobic environment that permits nitrogenase function (Carpenter and Price 1976). This cyanobacterium is rather widespread in its distribution in tropical waters and may be expected to make an important contribution to the nitrogen economy of its habitat (Fogg 1982). Symbiotic associations of heterocyst containing cyanobacteria within the diatom *Rhizosolenia* may also be prevalent in parts of the Pacific (Mague et al. 1974) and these likewise may be expected to have an imput of available nitrogen into certain oceanic areas. However, it has been calculated (Fogg 1978) that the nitrogen budget of the oceans shows a shortfall of a factor of 10^8 tonne per year and, assuming that there is a balance between input and loss, that there must be a major contribution yet to be identified and which could most reasonably come from planktonic cyanobacteria (see Fogg 1982). This possibility clearly should be reviewed in the context of the report by Waterbury (this volume) of the presence of a unicellular cyanobacterium of the *Synechocystis* type in open tropical waters that appears to fix nitrogen. The number of nitrogen-fixing unicellular cyanobacteria is very small. Wyatt and Silvey (1969) were the first to demonstrate nitrogen fixation in *Gloeocapsa* sp. and the isolate that has been most studied is *Gloeothece* sp. PCC 6909 (Gallon et al. 1974 Mullineaux et al. 1980; 1981). When *Gloeothece* sp. was grown under light: dark cycles, nitrogenase activity was only detected in the dark period, when photosynthetically produced oxygen was not present. However, when the organism was grown under continuous light, nitrogen fixation and photosynthesis operated together in the same cell (Gallon et al. 1974), which raised questions as to how the nitrogenase was protected from the photosynthetic oxygen. Recently, data have been presented that suggest that respiration, not photosynthesis, provides ATP for the nitrogenase activity of *Gloeothece* sp. even in the light (Maryan et al. 1986) and this raises the possibility that nitrogenase may be spacially separated from photosynthesis.

The protein structure and enzymology of nitrogenase is remarkably conserved between quite different species. An account of the Mo–Fe protein and Fe protein, and the manner in which electrons are donated and ATP supplied in cyanobacteria is summarised by Bothe (1982). Like all other nitrogen-fixing organisms, the ammonia produced by nitrogenase is assimilated into carbon molecules by the glutamine-synthetase route, as discussed previously in this section.

The subject of phosphate assimilation by cyanobacteria has been reviewed recently by Healey (1982). In comparison to eukaryotic algae (particularly diatoms) cyanobacteria are considered to be relatively poor competitors for phosphate (Tilman et al. 1982) and appear to be favoured only in those waters in which the N:P ratio is low (Smith 1983). However, in mixed phosphate-limited chemostat cultures *Anabaena variabilis* consistently out-competed the green alga *Selenastrum gracile* (Wymer and Thake 1980) while Healey (1982) could find little evidence in the literature that the efficiency of phosphate utilization by cyanobacteria was inferior to that of other phytoplankton. Enrichment with phosphate appears to promote the development of populations of *Anabaena* species over competing *Oscillatoria* species (Tilman et la. 1982) presumably because the shift from a phosphate to a nitrogen limitation favours the growth of N_2-fixing *Anabaena* species.

Inorganic phosphate uptake by cyanobacteria, which is stimulated by the presence of cations (Ca^{2+}, Mg^{2+}), is pH-dependent (optimum at pH 7.5–8.5) and declines markedly in more acidic environments (Healey 1982). This suggests that HPO_4^{2-}

rather than $H_2PO_4^-$ is the major ionic species assimilated. Growth rate showed a dependence upon temperature (below 25°C) in phosphate-limited chemostat cultures of *Oscillatoria agardhii* (Ahlgren 1978). Earlier, Healey (1973) determined a Q_{10} of 2.1 (20–30°C) for phosphate assimilation in *Anabaena variabilis*. Uptake is stimulated by light and is sensitive to uncouplers of both cyclic and non-cyclic photosynthetic electron transport. Simonis et al. (1974) proposed that three phosphate uptake systems were operative in *Anacystis nidulans* which saturated at 10, 50, and 100 μmol phosphate, respectively. Uptake at substrate concentrations in excess of 100 μmol was predominantly passive, however, and was not stimulated following illumination.

Deficiency of inorganic phosphate induces the assimilation of organic phosphate following extracellular hydrolysis by alkaline phosphatase. Optimal activity of this enzyme was recorded at pH 8.0–10.0 and at a temperature of 40–50°C (Healey 1973; Healey and Hendzel 1979). Maximum activity was dependent upon the presence of Ca^{2+} ions in *Anabeana variabilis* particularly at low substrate concentrations (Healey and Hendzel 1979). Both orthophosphate and molybdate competitively inhibit alkaline phosphatase activity (Healey 1973).

Phosphate deficiency results in a number of morphological and ultrastructural changes in cyanobacteria. Abnormalities in cell division lead to the formation of elongated cells in *Plectonema boryanum* and excentricity in septum formation (Jensen and Sicko 1974). Trichome tapering and the formation of apical hairs was promoted by phosphate deficiency in a number of strains of *Rivulariaceae* (Sinclair and Whitton 1977). In a strain of *Calothrix parietina* Livingstone and Whitton (1983) have proposed that the hair cells may be specialised sites of alkaline phosphatase activity.

At the fine structural level, phosphate starvation lead to the loss of polyphosphate bodies (Jensen and Sicko 1974) and, in a number of instances, to the appearance of cyanophycin granules (Stevens and Paone 1981; Lawry and Simon 1982). In *Agmenellum quadruplicatum* 28% of the total cell nitrogen was located in cyanophycin after 32 h of phosphate starvation whereas this polypeptide was not detected in phosphate-sufficient cells (Stevens and Paone 1981). Restoration of the phosphate supply to starved cells resulted in the rapid uptake of phosphate (particularly during the first hour) and led to the mass accumulation of polyphosphate bodies in *Plectonema boryanum* such that total cell P was one order of magnitude greater than in phosphate-sufficient cells (Jensen and Sicko 1974; Sicko-Goad and Jensen 1976; Sicko-Goad et al. 1978). The "polyphosphate overplus" phenomenon appears to be the direct result of an increase in the V_{max} for phosphate uptake during phospate deficiency. Such luxury consumption of phosphate may be of particular value in those environments in which the supply of phosphate is intermittent.

The kinetics of phosphorus-limited growth have been studied most extensively in *Anabaena variabilis* (Healey 1973; Healey and Hendzel 1975) and *Oscillatoria agardhii* (Ahlgren 1977, 1978; Riegman and Mur 1984, 1985; Riegman et al. 1985). Riegman and Mur (1984) found that the affinity of *Oscillatoria agardhii* for phosphate was unaffected by growth rate (μ) although V_{max} decreased with increasing growth rate. The minimum internal concentration permitting growth was 1.3 μg P mg dry weight^{-1} in *O. agardhii* (Riegman and Mur 1984). Half saturation constants for growth were 0.1 μg P L^{-1} and 0.2–1.0 μg P L^{-1} for *A. variabilis and O. agardhii* respectively (Healey and Hendzel 1975; Ahlgren 1978). In contrast to the findings of Healey and Hendzel (1975) (but see Healey 1982) the initial rate of assimilation was dependent on external phosphate concentration in *O. agardhii* (Riegman and Mur 1984). These workers were unable to achieve growth rates near μ_{max} in phosphate-limited chemostats, however, and proposed that short-term feed back by accumulated phosphate influences the activity of the uptake systems. A restricted light climate (low photon fluence rate or short photoperiod) resulted in a reduction in V_{max} and an increase in cell phosphorus concentration in *O. agardhii* (Riegman and Mur 1985) suggesting that the severity of phosphate limitation is reduced under these conditions.

184

Iron is an essential element for the growth of cyanobacteria and in certain instances may limit the growth of these organisms in the natural environment (Morton and Lee 1974; Paerl 1982; Brand et al. 1983). Iron is required for the synthesis of both chlorophyll and phycobiliproteins and forms an essential component of cytochromes, ferredoxin, iron-sulphur proteins and in nitrogen-fixing species, nitrogenase. Although cyanobacteria have a high affinity for iron at low concentrations (Murphy et al. 1976; Armstrong and Van Baalen 1979), it has been suggested that severe iron limitation may favour the growth of green algae over competing cyanobacteria (Morton and Lee 1974).

In batch cultures, Hardie et al. (1983a,b) described a number of sequential changes in the physiology and ultrastructure of the unicellular cyanobacterium *Agmenellum quadruplicatum* following the development of iron limitation. Iron deficiency resulted in the concurrent decline in chlorophyll and phycocyanin content and a progressive vesiculation of the thylakoids. Polysaccharide accumulated throughout the starvation period whereas ribosomes and at a later stage the thylakoids were degraded (Hardie et al. 1983b). The rates of both nitrate and nitrite reductase activity were enhanced during the initial stages of iron starvation but declined thereafter. The observed simultaneous degradation of the photopigments and ribosomes and the accumulation of carbohydrate led Hardie et al. (1983a, b) to suggest that initially iron starvation specifically limited nitrogen metabolism whilst carbon limitation occurred at a later stage. Similar effects of iron-limitation on the growth and pigmentation of *Anacystis nidulans* have been described by Oquist (1971).

The ability of cyanobacteria to produce polypeptides which chelate trace metals was described by Fogg and Westlake (1955) and it is now recognised that like other bacteria, cyanobacteria produce siderochromes which selectively chelate ferric iron (Murphy et al. 1976; Armstrong and Van Baalen 1979). Although Murphy et al. (1976) suggested that the production of siderochromes enabled cyanobacteria to dominate competing algae, the availability of iron may be of particular importance in determining the species composition of mixed phytoplankton populations. Morton and Lee (1974) found that an increase in the concentration of available iron from 0.1 to 1.0 mg per litre resulted in a shift in dominance from green algae to cyanobacteria in mixed batch cultures. The growth of *Synechococcus* sp. strain DC2 was limited at iron concentrations below 10^{-8}M; one order of magnitude higher than that required to maintain the maximal growth rates of a range of other phytoplankton of oceanic distribution (Brand et al. 1983). In bioassays of natural phytoplankton populations, Parl and Smith (1976) found that *Oscillatoria redekei* had a much higher requirement for iron in comparison to the green alga *Selenastrum*. The greater demand of cyanobacteria for iron may in part reflect differences in the cell requirements of these organisms and eukaryotes for the assimilation of nitrogen. Ferredoxin rather than NADH is the electron donor for cyanobacterial nitrate reductase (Losada and Guerrero 1976) and iron is an essential component of nitrogenase which may also utilize ferredoxin as an electron donor (Wolk 1982). In this context it is interesting to note that iron-starvation specifically limits nitrogen metabolism in cyanobacteria (Hardie et al. 1983a,b).

Although nitrogen, phosphorus, carbon, silica and iron are regarded as the nutrients which most frequently limit the growth of phytoplankton, there is evidence that the availability of other nutrients and trace metal elements may also limit growth in the natural environment (Paerl 1982; Brand et al. 1983). Artificial culture media for the growth of cyanobacteria usually include the trace elements manganese, boron, molybdenum, copper, zinc and cobalt (eg. Allen and Arnon 1955; Gorham et al. 1964;

Rippka et al. 1979). In addition, cyanobacteria require sodium, magnesium, calcium and chloride for normal growth; marine species having an obligate requirement for elevated concentrations of these ions (Stanier and Cohen-Bazire 1977; Waterbury and Stanier 1981).

Magnesium is required for the synthesis of chlorophyll and when supplied at the limiting concentration of 5 μm was found to inhibit the formation of cross walls and cell division in *Anacystis nidulans* (Utkitin 1982). Fay (1962, quoted by Fogg et al. 1973) demonstrated a higher calcium requirement for the growth of *Chlorogloea fritschii* on dinitrogen than in the presence of a source of combined nitrogen. There is no evidence, however, that this element is essential for the process of nitrogen fixation. In contrast, molybdenum is required for the synthesis of nitrogenase and is required at elevated concentrations by nitrogen-fixing species (Fogg et al. 1973). Lawry and Simon (1982) found that sulphate deficiency led to the accumulation of cyanophycin and polyphosphate in several cyanobacteria. Deficiency of this element also resulted in a reduction in cell volume and changes in the ultrastructure of the cell wall of *Synechococcus leopoliensis* (Jensen and Rachlin 1984). Cell volume was found to be dependent on the temperature experienced during growth in both magnesium and potassium limited chemostats of *Anacystis nidulaus* (Utkilin 1984). Manganese is essential for the growth of cyanobacteria but may be inhibitory at concentrations close to those required for optimal growth (Gorham et al. 1964). Brand et al. (1983), however, found that although the growth rate of *Synechococcus* sp. strain DC2 was not limited at concentrations as low as 2×10^{-9}M, higher concentrations did not result in growth inhibition.

A number of trace metals have been shown to be toxic to cyanobacteria. The heavy metals mercury and lead were inhibitory to the growth of a range of cyanobacteria at concentrations of 1 and 10 μM, respectively (Lawry and Simon 1982). In the marine environment, copper at concentrations as low as 10^{-10}M has been shown to inhibit photosynthesis in natural populations of *Oscillatoria (Trichodesmium) theibautii* (Rueter et al. 1979) and this element in the form of cupric sulphate has been used to control the growth of freshwater cyanobacteria (Gibson and Smith 1982). Cyanobacteria require copper for growth, however, for this element is essential for components of photosystem I (cytochrome c and plastocyanin). The addition of available copper to natural populations of freshwater cyanobacteria produced a stimulation of both CO_2 and N_2 fixation (Paerl et al. unpublished, quoted by Paerl 1982) apparently through stimulation of photosystem I activity. Whereas freshwater species do not have an obligate vitamin requirement, B_{12} is required by a number of cyanobacteria isolated from marine habitats (van Baalen 1962; Rippka et al. 1979). There are no reports, however, of cyanobacteria demonstrating an obligate requirement for any other vitamin.

Reserves, Storage and Reservoirs

In common with other living organisms, cyanobacteria can accumulate measurable amounts of macromolecular, insoluble materials which are loosely termed 'reserves'. Cyanobacteria are well endowed with ability to accumulate, under appropriate environmental conditions, stores of C, N, P and Fe and perhaps other elements as well; Mo would be a good candidate, especially in nitrogen-fixing species. It is an obvious point, but nevertheless worth making, that the elements listed above are the ones which we know to become growth rate-limiting under many environmental circumstances. Therefore it is the accumulation of these which selection pressure would have make advantageous. Consideration of the natural history of a microorganism can give insights into the evolutionary pressures that have determined its metabolism, just as they may be used to understand the variation in form and structure of higher organisms.

Because many cyanobacteria are much larger than most other prokaryotes the cytochemical localisation of reserve bodies and the observation of these structures in cyanobacteria by light and electron microscopy has a long history (see Lang 1968). Two reviews (Shively 1974; Allen 1984) since that of Lang have described the structure and function of cyanobacterial inclusion bodies, a term which embraces the structures discussed below.

POLYPHOSPHATE BODIES

These structures (sometimes called volulin or metachromatic granules) vary in size (usually less than 0.3 μm) and in appearance in transmission electron microscopy, although they are usually found as spherical, electron-dense granules. The physiology and chemistry of polyphosphate bodies in living organisms generally has been extensively reviewed by Kulaev (1979). Polyphosphate in cyanobacteria certainly has a role as a phosphate reserve; it is lost during phosphate deficiency and reappears on the addition of phosphate to a deficient culture (Livingstone and Whitton 1983). The 'phosphate overplus' phenomena is well established in a range of microorganisms. There is evidence that polyphosphate can serve as an energy reserve in both photosynthetic bacteria and cyanobacteria (Carr and Sandhu 1966). Polyphosphate bodies may bind otherwise toxic metals, such as cadmium (Rachlin et al. 1982) and the sequestering of metals onto polyphosphate bodies, although clearly not unique to cyanobacteria, may be of relevance to the suggested use of microalgae for water purification. There is some evidence that polyphosphate accumulates during sulphur starvation (Lawry and Jensen 1979). Polyphosphate is synthesised by polyphosphate kinase:

$$ATP + (HPO_3)_n \rightarrow ADP + (HPO_3)_{n + 1}$$

Degradation may be accomplished by polyphosphate glucokinase, polyphosphate fructokinase or by polyphosphatases.

CYANOPHYCIN

Cyanophycin is a high molecular weight polymer of equimolar amounts of arginine and aspartate which can be given the chemical name of multi-L-arginyl-poly-[L-aspartic acid] and is often referred to as cyanophycin granule polypeptide (CGP). It was first isolated by Simon (1971) and shown to comprise the "structured granules" (Lang et al. 1972) that had long been observed within cyanobacteria cells (see Lang 1968).

<pre>
        arginine   arginine
        aspartic acid   —   aspartic acid — aspartic acid
                            arginine

                        cyanophycin
</pre>

The widespread distribution of cyanophycin in cyanobacteria and its absence in any other microorganism so far examined, including the colourless sulphur bacteria morphologically similar to cyanobacteria (Lawry and Simon 1982), led to suggestions that it may be used as a cyanobacterial characteristic feature. However, there is a report of its apparent absence from *Anacystis nidulans* (Lawry and Simon 1982) and, pertinently to this article, we have been unable to detect cyanophycin in *Synechococcus* DC2 (WH 7803). The amount of cyanophycin present in cyanobacteria varies considerably, and in exponentially growing organisms is sometimes quite low, less than 0.5% of dry weight. This content is increased towards the end of exponential growth, by phosphate deprivation and, most dramatically, by chloramphenicol addition to a growing culture. The latter procedure inhibits protein synthesis and presumably allows the diversion of non-utilised nitrogen at the amino level of reduction into the formation of arginine and aspartic acid and hence into cyanophycin. Under these circumstances the cyanophycin content may increase to around 7% of dry weight. There is considerable evidence that cyanophycin serves as a reserve of nitrogen (see Allen 1984) and in addition Stanier and Cohen-Bazire (1977) pointed out that it could act as an energy reserve through the breakdown of arginine to ornithine yielding ATP. The amount of ATP realised through this mechanism would be expected to be small compared with the catabolism of glycogen or polyphosphate.

The synthesis of cyanophycin is accomplished by a reaction inserting arginine onto a poly-aspartic primer and requires ATP, KCl, $MgCl_2$, and an SH compound (Simon 1976). This reaction, unaffected by chloramphenicol and not requiring RNA synthesis, confirms the non-ribosomal nature of cyanophycin synthesis. The mobilisation of cyanophycin in *Anabaena* species by an exopeptidase that yields an arg–asp dipeptide has been described (Gupta and Carr 1981). A peptidase that produces arginine and aspartate as individual amino acids has been described in *Aphanocapsa* (Allen et al. 1984). The amino group from aspartic acid may be readily used for the formation of other amino acids by transamination and there are several catabolic enzymes recorded in cyanobacteria that release amino nitrogen from arginine (Hood and Carr 1971). The carbon skeletons of aspartic acid and arginine, after amino group removal, would be readily re-utilised via intermediary metabolism.

These assemblies of globular proteins with linear tetrapyrolle chromophores are fully described in the article by Bryant (this volume). In addition to their light-harvesting function these proteins may act as reserves of nitrogen, which are mobilised when the organism is denied exogenous available nitrogen (Allen and Smith 1969; Boussiba and Richmond 1980). As the phycobiliproteins are mobilised the cyanobacterial cultures change colour, to become a yellow–green in those species in which phycocyanin is the predominant pigment. The loss of phycocyanin is usually not complete, so that photosynthesis may proceed with lowered light harvesting effectiveness (but see below for an exception to this).

Organisms which are largely depleted of phycobiliproteins still have chlorophyll a associated with the photosystem II reaction centre and will carry out photosynthesis (Lemasson et al. 1973). Nitrogen starvation causes coordinate degradation of the two phycocyanin subunits (α and β) and the simultaneous repression of new phycobiliprotein synthesis (Lau et al. 1977). Phycocyanin is not readily degraded by commercial proteolytic enzymes and cyanobacteria contain an enzyme, phycocyaninase which causes the release of trichloroacetic acid soluble material from ^{14}C-phycocyanin protein (Foulds and Carr 1977). Knowledge of the mechanism, and of the intermediate peptides, involved in protein degradation is rudimentary and it would appear that once an initial enzymic attachment is achieved, the subsequent complete breakdown to constituent amino acids is very rapidly achieved. Certainly, when considering the value of phycocyanin as a specific amino-nitrogen store, no useful consequence of its breakdown could be expected until free amino acids are available. Thus all the control processes exerted over protein turnover must be applied at the first stage of degradation.

Since the early days of electron microscopy inclusions with polygonal profiles have been observed in cyanobacteria (Jensen and Bowen 1961), in nitrifying bacteria and in *Thiobacilli* species (see Shively 1974); these were given the general term 'polyhedral bodies'. The important breakthrough in understanding their nature came when Shively et al. (1973) showed that they contained a paracrystalline array of 10 nm particles that were surrounded by a 3.5 nm membrane. The particles were composed of ribulose-1,5-bisphosphate carboxylase (Rubisco) and Shively et al. (1973) gave them the name 'carboxysomes'. Similar paracrystalline membrane enclosed bodies were isolated from cyanobacteria (see Codd and Marsden 1984) and the detection of Rubisco both enzymically and immunologically was shown by Codd and Stewart (1976). Carboxysomes have been found in the vegetative cells of all cyanobacteria examined. They are present in akinetes but not heterocysts of nitrogen fixing cyanobacteria (Stewart and Codd 1975). A comprehensive review of carboxysomes in autotrophic prokaryotes has been prepared recently (Codd and Marsden 1984). There is no unequivocally demonstrated function for carboxysomes. In spite of strenuous efforts there is no evidence that they act as sites of carbon dioxide fixation, although carboxysomes from all species examined possess the Rubisco enzyme. The presence of other polypeptides has been described in some reports but we have no clear evidence of other enzymes present in cyanobacterial carboxysomes. A detailed account of these and other (such as a CO_2-concentrating role) possible functions are evaluated by Codd and Marsden (1984). There are some interesting indications that carboxysomes are involved in partitioning Rubisco into active and inactive forms. Lanarus and Codd (1982) have shown that the proportions of Rubisco in the cytoplasmic pool (enzymically active) were greater than in the carboxysome pool (presumably inactive) during log phase growth of *Chlorogloeopsis fritschii*. This is in accord with other observations that show an increase in carboxysomes during entry into the stationary phase of growth (see Codd and Marsden 1984). *C. fritschii* has been shown not to repress the synthesis of Rubisco under chemoheterotrophic conditions (Joset-Espardellier et al. 1978) so the suggestion (Stewart and Cood 1975) that cyanobacterial carboxysomes act as storage bodies for the Rubisco enzyme may be valuable in understanding the function of these bodies. An organism that is exposed to fluctuating carbon dioxide supply — such as would apply to a bloom of cyanobacteria for example — would clearly benefit from being able to conserve a key enzyme, especially one such as Rubisco that is required in relatively large amounts.

POLYGLUCOSE

Discrete granules located adjacent to thylakoid membranes have been recognized for many years in electron micrographs and termed α-granules. They are now understood, by cytochemical straining and by isolation and analysis, to be composed of polyglucose and are sometimes referred to as glycogen bodies (see Stanier and Cohen-Bazire 1977; Allen 1984). The amount of poly-glucose bodies varies markedly with environmental conditions and with phase of growth. Values as high as 20% of dry weight in exponential culture have been reported (see Allen 1984), although this value does seem high in organisms which usually contain 50–60% of dry weight as protein. Limitation of growth by starvation of N or Fe leads to accumulation of polyglucose whilst light limitation or carbon dioxide starvation leads to its mobilisation indicating its role as a reserve material (Lehmann and Weber 1976). Dark aerobic incubation, with attendant respiratory O_2 uptake, resulted in a decrease in the polyglucose content of *Synechococcus* (Lehmann and Weber 1976), the pentose phosphate pathway being the principal metabolic route employed for polyglucose dissimilation (see Smith 1982). When cyanobacteria were subjected to light-dark cycles of growth the accumulation of polyglucose during the light, and its breakdown during the dark was shown (Gibson 1975; van Liere et al. 1979). Differences in the capacity to store polyglucose in the light have been used to explain the observed relative growth rates of individual species of phytoplankton in natural systems which were found to be at variance with the absolute growth rates that were obtained for the same species under continuous light in the laboratory (Gibson and Smith 1962).

Electron microscopy has yielded an array of other inclusion bodies that are found in some cyanobacteria under certain, sometimes imprecisely defined, conditions (see Allen 1984). There are however two other 'reserve' materials of potential interest. Poly-hydroxybutyrate (PHB) is a well-studied bacterial reserve material (see Dawes and Senior 1973) characterised by its ability to readily generate reduced pyridine nucleotides by virtue of its own high level of reduction. Cyanobacterial PHB was first isolated from *Chlorogloeopsis fritschii* grown in the presence of acetate (Carr 1966); it has since been reported in *Aphanocapsa* (see Allen 1984) and *Spirulina platensis* (Campbell et al. 1982). There is an interesting report that polyesters similar to PHB have been found in *Aphanothece* sp. and isolated from algal mats in Australia, Capon et al. (1983) have found small amounts of a polyester composed primarily of γ-hydroxypentanoate and γ-hydroxybutyrate subunits. The lack of enthusiasm for publishing negative results makes an accurate estimate of PHB distribution in other cyanobacteria species difficult, but it is probable that PHB is not widely distributed in these organisms. In contrast, cyanobacteria appear generally to accumulate iron (Fe) in excess of their immediate needs. The release by cyanobacteria of polypeptides that chelate iron has been known for some years (see Gibson and Smith 1982). Many bacteria, including cyanobacteria (Simpson and Neilands 1976) export siderochromes which are low molecular weight trihydroxymates which will selectively chelate ferric iron and which appear to participate in the transport or iron across the cell membrane. However, ther are indications that the Fe-chelate complex does not pass into phytoplankton and our understanding of the mechanism of iron uptake is still rudimentary (see Huntsman and Sunda 1980). There is evidence derived from Mössbauer spectroscopy that in two species of cyanobacteria iron exists in a paramagnetic form that is associated with the membrane fraction. One of these species of iron is abolished by nutritional limitation and it is proposed that it represents an iron reserve (Evans et al. 1977). Five species of cyanobacteria were shown to concentrate iron up to a factor of 8279 relative to its concentration in the growth medium (Jones et al. 1978) and the same work yielded an average composition of 3626 ppm, much higher that found in heterotrophic bacteria (262 ppm).

Relevance to Picoplankton

It is clear from the above that cyanobacterial physiology has evolved so that C, N, P and Fe may be held as 'reserves' and that exponentially growing cultures expend a significant amount of energy in accumulating these. It is evidence that the constantly changing natural environment makes contingency planning of this nature desirable. We have increasing evidence based upon isotope flow techniques that polyphosphate, polyglucose and cyanophycin in cyanobacteria exist as dynamic reservoirs rather than static reserves. It has been suggested that a proportion, possibly the entirety, of C, N and P after the initial assimilation process, passes first into the appropriate polymeric reservoir before being distributed into the metabolic pools generally (Gupta and Carr 1981; Carr et al. 1982). This process would separate the continual environmental fluctuation in supply from the constant demand for balanced biosynthesis in growing cells. The synthesis of cyanophycin in heterocysts occurs at the final stage of their formation (see Adams and Carr 1981). The very presence of a nitrogen 'reserve' material in a cell specialised for nitrogen fixation and occurring in a culture which is nitrogen limited argues strongly that cyanophycin has a role in addition to that of a storage body. When available nitrogen was added to nitrogen-limited *Aphanocapsa*, cyanophycin synthesis started immediately (Allen and Hutchinson 1980). The enzymes involved in both synthesis and breakdown of cyanophycin (see (b) above) were detected with considerably greater specific activity in heterocyst preparations than in those from vegetative cells (Gupta and Carr 1981), this being consistent with the idea that the heterocyst was the site of considerable turnover of cyanophycin.

The existence, composition and function of these inclusion bodies of cyanobacteria which appear to have a reserve function (and this function should not be considered diminished if, in addition, a role as dynamic reservoir is fully established) has been discussed here at length. There is a reason for this which is directly related to the subject of picoplankton. With the recognition that there was a significant oceanic population of cyanobacteria comprising a part of the picoplankton came the realisation that characteristic features of cyanobacterial metabolism had an input into the understanding of phytoplankton ecophysiology. One of the most relevant of the features was the possession of macromolecular nitrogen reserves in a non-soluble form. Since these have not yet been found in other planktonic microorganisms, their possession by one type of primary producer could have a profound effect on the carbon, as well as the nitrogen, food chain. Accordingly their examination in characteristic species of cyanobacterial picoplankton was desirable. When grown under conditions of nitrogen excess *Synechococcus* strain DC2 has been shown to possess a much higher phycoerythrin to phycocyanin ratio than any other species of cyanobacteria so far examined (Wyman et al. 1985). A significant proportion of the energy absorbed by this phycoerythrin is autofluoresced rather than being passed to a photoreaction centre. When a culture of *Synechococcus* strain DC2 underwent a shift from nitrogen-excess to no exogenously supplied nitrogen, growth continued for an entire cell division period, with concommitant decrease in the phycoerythrin content. It is concluded that in this cyanobacterium the photopigment, phycoerythrin, to an extent becomes specialised so as to act as a nitrogen store with consequent loss of photosynthetic efficiency (Wyman et al. 1985). The other cyanobacterial nitrogen reserve, cyanophycin, presents a differnt picture. So far no cyanophycin has been detected in *Synechococcus* strain DC2 even when grown under light or phosphate limitation or when incubated in the presence of chloramphenicol (Newman, Wyman, and Carr, unpublished), thus appearing to join the very limited number of cyanobacteria which do not possess this polymer (Lawry and Simon 1982).

Extracellular release and interaction with other organisms

EXTRACELLULAR PRODUCTS

The striking feature about cyanobacterial extracellular products is that although they appear to be so widespread and to represent a significant proportion of productivity, we know so little about their composition and function. An early paper by Fogg (1952) provided a detailed account of the excretion of approximately 20% of the total assimilated nitrogen by a culture of *Anabaena cylindrica*. This material appeared mainly as peptides (rather than free amino acids) and was largely unaffected by the growth conditions. The fact that this quantity of nitrogenous material was excreted whilst the organism was fixing atmospheric nitrogen indicates that the process represented a significant energetic cost to the organism and it is difficult to avoid the conclusion that production of extracellular material is important to the growth of *Anabaena cylindrica* in the natural environment. Some way to further analysis of extracellular nitrogenous compounds of *A. cylindrica* was provided by Walsby (1974) who showed the presence of fluorescent peptides with molecular weights in excess of 5000 and which contained large proportions of threonine and serine. Whitton (1965) screened 15 strains of cyanobacteria and showed widespread distribution of non-dialysable nitrogenous material. One polypeptide fraction decreased the toxic effect of polymyxin B against cyanobacteria; polymyxin B is derived from *Bacillus polymyxa* which has been found in association with cyanobacteria (Whitton 1965). Two reviews have discussed the range of products (peptides, carbohydrates, small organic molecules, toxins, etc.) excreted by microalgae generally and include work with cyanobacteria (Fogg 1971; Hellebust 1974). It is clear that substantial amounts of carbon compounds are released and that some of this can be identified as a result of photorespiration in which a molecule of ribulose bisphosphate is cleaved by Rubisco to yield phosphoglycerate and phosphoglycollate, the latter giving rise to excreted glycollate (see Codd and Marsden 1984). The ready assimilation by bacteria of nitrogenous material produced by *Oscillatoria redekei* has been shown by Meffert and Zimmerman-Telschow (1979) and this no doubt reflects the dominant pattern in natural systems.

It is becoming clear that the reasons and advantages of extracellular excretion of metabolic products by cyanobacteria will be diverse with respect to species and habitat. Extracellular polysaccharides will confer protection from desiccation and may be involved in adherance to surfaces. It is known that the transport into *Anabaena* sp. of iron is facilitated by the production of extracellular schizokinen (Lammers and Sanders-Loehr 1982). In a benthic species of *Phormidium* the extracellular release of a highly effective bioflocculant has been described (Fattom and Shilo 1984) and this could have a role in clarification of the water column and hence be beneficial to benthic organisms. There is one well described antibiotic produced by a cyanobacterium that is active against other species of cyanobacteria and several eukaryotic algae, possibly by inhibiting photosynthetic electron transport (Gleason and Paulson 1984). This material is a complex diaryl substituted γ-cylidene γ-butyro lactose with a chlorine substitution in one of the aromatic rings (Mason et al. 1982).

It is in the field of higher animal toxicity that our most detailed knowledge of the epidemiology of cyanobacterial extracellular production occurs. For many years the correlation of sickness and death in domestic animals has been made with the occurrence of cyanobacterial blooms in the drinking water supply (see Carmichael 1981). Cyanobacteria of many species will produce toxins of peptide, phenolic and alkaloid nature, although most of these are of freshwater occurrence. The hepatotoxins from *Microcystis aeruginosa* have been most studied and several have been partially characterized, including a cyclic peptide of around 1000 Daltons molecular weight which contained novel amino acids (see Codd 1984 for a short review). The extensive literature on cyanobacterial toxins and their effects on man and his domestic animals may give an exaggerated impression of the ecological importance of these toxins. The real questions relate to what natural advantage, what competitive edge if any over other aquatic microorganisms do the excretion of these potent chemicals bring to the cyanobacteria? Such questions can be answered only with a much greater understanding of the ways in which microorganisms interact with each other and attempt to control their nutritional microenvironment by chelation and sequestration.

Whereas the majority of cyanobacteria are free-living, a number of species enter into highly specific associations with a diverse range of eukaryotic hosts. Although relatively few genera of each group are involved, symbioses with cyanobacteria have been recorded amongst the algae, fungi, bryophytes, ferns, gymnosperms and flowering plants. For many years the cyanelles of various protists were considered to be endosymbiotic cyanobacteria; however, with knowledge gained through advances in molecular biology these organelles are now treated as chloroplasts and not cyanobionts (Herdman 1982). Invariably symbiotic cyanobacteria fix N_2 and frequently as the result of morphological and physiological modification, become highly specialised for this purpose. Cyanobacteria also enter into more casual associations with other organisms including heterotrophic bacteria as well as eukaryotic partners. Such relationships are probably more widespread than true symbioses but, by their nature, they are less easily characterised.

Cyanobacteria: Other Prokaryotes

In the natural environment epiphytic bacteria are frequently associated with the cells of cyanobacteria. Such associations range from the general occurrence of bacteria adhering to the cyanobacterial sheath material to rather more specific associations with N$_2$-fixing genera (Paerl 1982). Paerl (1976) found that the polar regions of the heterocysts of field populations of *Anabaena circinalis* and *Aphanizomenon flos-aquae* supported a dense microflora of attached bacteria. Subsequently Paerl and Kellar (1978) showed that colonization of axenic cultures of *Anabaena oscillaroides* by lake bacteria produced a significant stimulation of nitrogenase activity. These authors suggested that as a result of bacterial respiration, an O$_2$-depleted microzone was created around the heterocyst thereby providing a more favourable environment for the activity of the O$_2$-sensitive nitrogenase. Inhibition of nitrogenase activity by the addition of NH$_4{}^+$ resulted in a decline in the density of bacteria attached to heterocysts (Paerl and Gallucci 1985) suggesting that the bacteria obtained a supply of fixed nitrogen from this source in the absence of exogenous NH$_4{}^+$. Paerl and Gallucci (1985) proposed that the heterocyst-bacterial association was established in response to the liberation of extracellular products from the heterocyst-vegetative cell junction. In this and an earlier investigation (Gallucci and Paerl 1983), these workers demonstrated positive chemotaxis by pseudomonads toward amino acids and that these bacteria were frequent colonizers of metabolically active heterocysts.

Non-heterocystous cyanobacteria may also be involved in associations with heterotrophic bacteria. Colonies of the marine *Trichodesmium erythraea* harboured numerous attached bacteria particularly in the regions of the trichome bundle thought to be the sites of N$_2$-fixation in this organism (Bryceson and Fay 1981). These authors provided presumptive evidence that nitrogen fixed by *Trichodesmium* was released and assimilated by the associated bacteria and also suggested that O$_2$-depleted microzones may be created in the vicinity of the N$_2$-fixing cells by the action of these bacteria.

CYANOBACTERIA: EUKARYOTES

Endosymbiotic cyanobacteria have been described in some species of the marine diatom *Rhizosolenia* (Mague et al. 1974) and in two freshwater diatoms *Rhopalodia gibba* and *R. gibberula* (Drum and Pankratz 1963). In *Rhizosolenia* the N$_2$-fixing cyanobiont, *Richelia intracellularis* produces short filaments with a single polar heterocyst. In addition to supplying fixed nitrogen to the host, *Richelia* may also provide the bulk of fixed carbon (Fogg *et al.* 1973). Mague et al. (1974) emphasised the ecological importance of N$_2$-fixation by *Richelia* in the North Pacific ocean where *Rhizosolenia* was found to be a significant member of the phytoplankton. Although *Richelia* has been reported occasionally as a free-living cyanobacterium it appears to be most successful as an endophyte. Fogg (1982) has suggested that perhaps free-living *Richelia* is unable to compete for limiting nutrients with other phytoplankton and therefore benefits from the association with *Rhizosolenia* through the superior nutrient uptake capacity of its host.

Approximately 8% of the 17 000 known species of lichens contain cyanobacteria as one or the sole photosynthetic symbiont (Fogg et al. 1973). Species of *Nostoc* are the commonest cyanobionts but lichen associations involving *Stigonema*, *Scytonema*, *Calothrix*, *Dichothrix*, *Gloeocapsa*, *Chroococcus* and *Hyella* have been recorded (Fogg et al. 1973). In bipartite lichens the cyanobiont supplies a source of both fixed carbon and nitrogen to the mycobiont. In tripartite associations, however, the cyanobiont shows a greater specialization for N$_2$-fixation; the phycobiont meeting the carbon requirement of both myco and cyanobiont. Rapid transfer of newly fixed carbon from the cyanobiont to the mycobiont of *Peltigera* species has been demonstrated (Smith et al. 1970). Approximately 40% of the carbon fixed by *Nostoc* was released as glucose and converted by the mycobiont to mannitol prior to assimilation. Millbank (1972) showed that nitrogenase activity in *Peltigera canina* was considerably higher than that measured following isolation of the cyanobiont, *Nostoc*. In an earlier study, Kershaw and Millbank (1970) had shown that the majority of nitrogen fixed by the cyanoboint was subsequently assimilated by the mycobiont. In *Peltigera canina*, fixed nitrogen was shown to be liberated by *Nostoc* in the form of NH_4^+ (Rai et al. 1983) and assimilated by the mycobiont via the glutamate dehydrogenase (GDH) pathway. The cyanobiont also reassimilated a fraction of the liberated NH_4^+ via the glutamine synthetase-glutamate synthase (GS-GOGAT) pathway but at a low rate owing to reduced levels of the assimilatory enzymes (Stewart et al. 1983).

Symbiotic *Nostoc* species occur in specialised cavities on the undersurface of the thallus of a few species of the liverworts *Anthoceros*, *Blasia* and *Cavicularia*. Although *Nostoc* species are the only cyanobacteria involved, the symbiosis is not highly specific. The association has been reconstituted in the laboratory using either the host cyanobiont or cyanobionts isolated from other symbioses as well as with free-living *Nostoc* species (Stewart et al. 1983; Meeks et al. 1985). Symbiotic *Nostoc* species are highly specialised for nitrogen fixation. Heterocyst frequencies of 30–40% have been recorded (in comparison to 3–6% in free-living species) and transfer of fixed carbon from the liverwort to the cyanobiont has been demonstrated (Rogers and Stewart 1977; Stewart and Rogers 1977). Meeks et al (1985) found that 80–90% of the liberated NH_4^+ derived from N_2-fixation by the *Nostoc* cyanobiont was assimilated by *Anthoceros punctatus*. In contrast to the assimilation pathway of lichen mycobionts, however, these workers found that the liverwort assimilated NH_4^+ via the GS-GOGAT pathway.

Each of the six species of the small water fern *Azolla* harbours endophytic cyanobacteria of the genus *Anabaena* (*A. azollae*) in cavities at the base of the dorsal lobes of the aerial leaves. Although both partners of the symbiosis have been shown to be photosynthetically competent, the cyanobacterium made little contribution to carbon fixation in the intact association (Ray et al. 1979). The highest rates of nitrogenase activity were recorded in mature leaves where the heterocyst frequency of *Anabaena* was as high as 30–40%. Following isolation of the *Anabaena* filaments, however, the rate of acetylene reduction (nitrogenase activity) declined rapidly to approximately 60–70% of the activity recorded in the intact association (Ray et al. 1978). NH_4^+ released by the cyanobiont was assimilated by *Azolla coroliniana* via GDH and possibly reassimilated by *Anabaena* via this pathway also (Ray et al. 1978). These authors suggested that although GDH activity is known to be very low in free-living cyanobacteria (Haystead et al. 1973) induction of the GDH assimilation pathway in *Anabaena* may be characteristic of the *Azolla* symbiosis.

A number of cycad species develop nodules on the coralloid roots which harbour cyanobacteria of the genera *Nostoc* or *Anabaena*. The cyanobionts are specific to individual cycad species (Grilli-Caiola 1980) and a number have been isolated in pure culture (Bowyer and Sherman 1968; Rippka et al. 1979). Transfer of ^{15}N fixed by the cyanobiont and released in the form of $^{15}NH_4^+$ to the host tissues has been demonstrated (Halliday and Pate 1976). As the coralloid roots occur at some depth in the soil, it is evident that the cycad meets the fixed carbon requirements of the cyanobiont although the potential for photosynthesis is retained.

Members of the genus *Gunnera* are the only examples of angiosperms which form organized associations with cyanobacteria. Following infection, swellings (nests), which contain intracellular cyanobionts (usually *Nostoc punctiforme*), develop at the bases of the leaf petioles. The cyanobiont shows reduced photosynthetic capacity and liberates fixed nitrogen which is assimilated by the host via specialised filamentous structures within the leaf cavities (Sylvester and Smith 1969; Sylvester and Macnamara 1976).

Concluding Comment

As the assembly of information for this review about cyanobacteria and its relation to oceanic plankton proceeded it became apparent that selection must be ruthless and to an extent, arbitrary. Emphasis has been placed on those aspects of cyanobacterial biology that may be important to the full understanding of the oceanic *Synechococcus* species and to the way in which these organisms may interact with other microorganisms in their environment. Some of the relevant aspects have been treated rather briefly on the grounds that extensive and up to date reviews were available. It is hoped, however, that the diverse nature of cyanobacteria, in terms of environmental requirements, morphology, and cell cycle will have emerged. This contrasts with what appears at this stage to be a rather limited variation in the oceanic cyanobacterial picoplankton. Perhaps the reasons for this contrast will themselves prove illuminating in due course.

References

ABELIOVICH, A., D KELLENBERG, AND M. SHILO. 1974. Effects of photooxidative conditions on levels of superoxide dismutase in *Anacystis nidulans*. Photochem. Photobiol. 19: 379–382.

ADAMS, D. G., AND N. G. CARR. 1981. The developmental biology of heterocyst and akinete formation in cyanobacteria. Crit. Rev. Microbiol. 9: 45–100.

AHLGREN, G. 1977. Growth of *Oscillatoria agardhii* in chemostat culture. 1. Nitrogen and phosphorus requirements. Phykos 29: 209–224.

——— 1978. Growth of *Oscillatoria agardhii* in chemostat culture. 2. Dependence of growth constants on temperature. Mitt. Int. Verein. Theor. Angew. Limnol 21: 88–102.

ALBERTE, R. S., A. M. WOOD, T. A. KURSAR, R. R. L. GUILLARD. 1984. Novel phycoerythrins in marine *Synechococcus* spp. Plant Physiol. 75: 732–739.

ALLEN, M. B., AND D. ARNON. 1955. Studies on nitrogen-fixing blue-green algae. I. Growth & nitrogen fixation by *Anabaena cylindrica*. Plant Physiol. 30: 366–372.

ALLEN, M. M. 1984. Cyanobacterial cell inclusions. Ann. Rev. Microbiol. 38: 1–25.

ALLEN, M. M., AND F. HUTCHISON. 1980. Nitrogen limitation and recovery in the cyanobacterium *Aphanocapsa* 6308. Arch. Microbiol. 128: 1–7.

ALLEN, M. M., R. MORRIS, AND W. ZIMMERMAN. 1984. Cyanophycin granule polypeptide protease in a unicellular cyanobacterium. Arch. Microbiol. 138: 119–123.

ALLEN, M. M., AND A. J. SMITH. 1969. Nitrogen chlorosis in blue-green algae. Arch Mikrobiol. 69: 114–120.

ALLISON, E. M., AND A. E. WALSBY. 1981. The role of potassium in the control of turgor pressure in a gas-vacuolate blue-green alga. J. Exp. Bot. 32 (126): 241–249.

ARMSTRONG, J. E., AND C. VAN BAALEN. 1979. Iron transport in microalgae: the isolation and biological activity of a hydroxymate siderophore from the blue-green alga *Agmenellum quadruplicatum*. J. Gen. Microbiol. 111: 253–262.

ARMSTRONG, R. E., P. K. HAYES, AND A. E. WALSBY. 1983. Gas vacuole formation in hormogonia of *Nostoc muscorum*. J. Gen. Microbiol. 128: 263–270.

ASATO, Y. 1972. Isolation and characterization of ultraviolet light-sensitive mutants of the blue-green alga *Anacystis nidulans*. J. Bact 110: 1058–1064.

ASTIER, C., F. JOSET-ESPARDELLIER, AND J. MEYER. 1979. Conditions for mutagenesis in the cyanobacterium *Aphanocapsa* 6714. Influence of repair phenomena. Arch. Microbiol. 120: 93–96.

BADGER, M. R., M. BASSETT, AND H. N. COMINS. 1985. A model for HCO_3^- accumulation and photosynthesis in the cyanobacterium *Synechococcus* sp. Plant Physiol. 77: 465–471.

BEAUCLERK, A. A. D., AND A. J. SMITH. 1978. Transport of D-glucose and 3-0-methyl-D-glucose in the cyanobacteria *Aphanocapsa* 6714 and *Nostoc* strain MAC. Eur. J. Biochem. 82: 187–197.

BALKWILL, D. L., AND S. E. STEVENS. JR. 1980. Glycocalyx of *Agmenellum quadruplicatum*. Arch. Microbiol. 128: 8–11.

BARLOW, D. J., W. E. SCOTT, AND S. I. FUNKE. 1977. Fine structural changes in response to different light intensities in cultured *Microcystis aeruginosa*. Proc. Electron Microscopy Soc. of South Africa 7: 105–106.

BARLOW, R. G., AND R. S. ALBERT. 1985. Photosynthetic characteristics of phycoerythrin-containing marine *Synechococcus* Spp. I. Responses to growth photon flux density. Mar. Biol. 86: 63–74.

BENNETT, A., AND L. BOGORAD. 1973. Complementary chromatic adaptation in a filamentous blue-green alga. J. Cell Biol. 58: 419–435.

BISALPUTRA, T., D. L. BROWN, AND T. E. WEIER. 1969. Possible respiratory sites in a blue-green alga *Nostoc sphaericum* as demonstrated by potassium tellurite and tetranitro-blue tetrazolium reduction. J. Ultrastruct. Res. 27: 182–197.

BOARDMAN, N. K. 1977. Comparative photosynthesis of sun & shade plants. Ann. Rev. Plant Physiol. 28: 355–377.

BOGORAD, L. 1975. Phycobiliproteins and complementary chromatic adaptation. A. Rev. Plant Physiol. 26: 369–401.

BOTHE, H. 1982. Nitrogen fixation, p. 87–104. *In* N. G. Carr and B. A. Whitton [ed.] The biology of cyanobacteria. Blackwells Scientific Publications, Oxford.

BOTTOMLEY, P. J., AND W. D. P. STEWART. 1976. ATP pools and transients in the blue green alga, *Anabaena cylindrica*. Arch. Microbiol. 108: 249–258.

BOUSSIBA, S., C. M. RESCH, AND J. GIBSON. 1984. Ammonia uptake and retention in some cyanobacteria. Arch. Microbiol. 138: 287–292.

BOUSSIBA, S., AND A. E. RICHMOND. 1980. C-phycocyanin as a storage protein in the blue-green alga *Spirulina platensis*. Arch. Microbiol. 125: 143–147.

BOWYER, J. W., AND V. B. D. SHERMAN. 1968. Production of axenic cultures of soil-borne and endophytic blue-green algae. J. Gen. Microbiol. 54: 299–306.

BRAND, L. E., W. G. SUNDA, AND R. R. L. GUILLARD. 1983. Limitation of marine phytoplankton reproductive rates by zinc, manganese, and iron. Limnol. Oceanogr. 28: 1182–1198.

BROWN, A. H., AND G. C. WEBSTER. 1953. The influence of light on the rate of respiration of the blue-green alga, *Anabaena*, Am. J. Bot. 40: 753–757.

BRYCESON, I., AND P. FAY. 1981. Nitrogen fixation in *Oscillatoria (Trichodesmium) erythraea* in relation to bundle formation and trichome differentiation. Mar. Biol. 61: 159–166.

BUCKLEY, C. E., AND J. A. HOUGHTON. 1976. A study of the effects of near UV radiation on the pigmentation of the blue-green alga *Gloeocapsa alpicola*. Arch. Microbiol. 107: 93–97.

BURGER-WEIRSMA, T., M. VEENHUIS, H. J. KARTHALS, C. C. M. VAN DE WIEL, AND L. R. MUR. 1986. A new prokaryote containing chlorophylls *a* and *b*. Nature 320: 262–264.

CAMPBELL, J. M., S. E. STEVENS JR., D. L. BALKWELL. 1982. Accumulation of poly-β-hydroxybutyrate in *Spirulina platensis*. J. Bacteriol. 149: 361–363.

CAPON, R. J., R. W. DUNLOP, E. L. GHISALBERTI, AND P. R. JEFFERIES. 1983. Poly-3-hydroxyalkanoates from marine and freshwater cyanobacteria. Phytochem. 22: 1181–1184.

CARMICHAEL, W. W. [ed.] 1980. The water environment. Plenum Press, New York, NY. 491 p.

CARPENTER, E. J., AND C. C. PRICE. 1976. Marine *Oscillatoria (Trichodesmium)*: explanation for aerobic nitrogen fixation without heterocysts. Science 191: 12787–1280.

CARR, N. G. 1966. The occurrence of poly-β-hydroxybutyrate in the blue-green alga, *Chlorogloea fritschii*. Biochim. Biophys. Acta 120: 308–310.

——— 1973. Metabolic control and autotrophic physiology, p. 39–65. *In* N. G. Carr and B. A. Whitton [ed.] The biology of blue-green algae. Blackwells Scientific Publications, Oxford.

CARR, N. G., M. GUPTA, J. RYLAH, AND J. McKIE. 1982. The role of cyanobacterial reserve materials as dynamic reservoirs in nutrient assimilation. Abst. IV Int. Symp. Photosynth. Prokaryotes. pB12, Bombannes, France.

CARR, N. G., AND M. HALLAWAY. 1966. Quinones of some blue-green algae, p. 159–162. *In* T. W. Goodwin [ed.] Biochemistry of chloroplasts. Vol. 1. Academic Press, New York, NY.

CARR, N. G., AND G. R. SANDHU. 1966. Endogenous metabolism of polyphosphates in the photosynthetic microorganisms. Biochem. J. 99: 29.

CARR, N. G., AND B. A. WHITTON. [ed] 1973. The biology of blue-green algae. Blackwells Scientific Publications, Oxford.

CARTHEW, R. W. 1980. The thermodynamics of photosynthetic adaptation to photon fluence rate in the blue-green alga *Oscillatoria limnetica*. Photosynthetica 14: 204–212.

CASTENHOLZ, R. W. 1968. The behaviour of *Oscillatoria terebriformis* in hot springs. J. Phycol. 4: 132–139.

——— 1969. Thermophilic blue-green algae and the thermal environment. Bact. Rev. 33: 476–504.

——— 1982. Motility & taxes, p. 413–440. *In* N. G. Carr and B. A. Whitton [ed.] The biology of cyanobacteria. Blackwells Scientific Publ., Oxford.

CLAYTON, R. K. 1980. Photosynthesis: physical mechanisms and chemical patterns. Cambridge University Press, Cambridge. 281 p.

CODD, G. A. 1981. Photoinhibition of photosynthesis and photoinactivation of ribulose bisphosphate carboxylase in algae and cyanobacteria. *In* H. Smith [ed.] Plants and the daylight spectrum. Academic Press, London.

——— 1984. Toxins of freshwater cyanobacteria. Microbiol. Sci. 1: 48–52.

CODD, G. A., AND W. J. N. MARSDEN. The carboxysomes (polyhedral bodies) of autotrophic prokaryotes. Biol. Rev. 59: 389–422.

CODD, G. A., AND W. D. P. STEWART. 1976. Polyhedral bodies and ribulose 1,5 diphosphate carboxylase of the blue-green alga *Anabaena cylindrica*. Planta 130: 323–326.

COHEN, Y., B. B. JORGENSEN, E. PADAN, AND M. SHILO. 1975a. Sulphide dependent anoxygenic photosynthesis in the cyanobacterium *Oscillatoria limnetica*. Nature 257: 489–492.

COHEN, Y., E. PADAN, AND M. SHILO. 1975b. Facultative anoxygenic photosynthesis in the cyanobacterium *Oscillatoria limnetica*. J. Bact. 123: 855–861.

COHEN-BAZIRE, G., AND D. A. BRYANT. 1982. Phycobilisomes: composition and structure. *In* N. G. Carr and B. A. Whitton [ed.] The biology of cyanobacteria. Blackwells Scientific Publications, Oxford.

COLEMAN, J. R., AND B. COLMAN. 1981. Inorganic carbon accumulation and photosynthesis in a blue-green alga as a function of external pH. Plant Physiol. 67: 917–921.

COLLINS, C. D., AND C. W. BOYLEN. 1982. Physiological responses of *Anabaena variabilis (Cyanophyceae)* to instantaneous exposure to various combination of light intensity and temperature. J. Phycol. 18: 206–211.

CUHEL, R. L., AND J. B. WATERBURY. 1984. Biochemical composition and short term nutrient incorporation patterns in a unicellular marine cyanobacterium, *Synechococcus* (WH7803). Limnol. Oceanogr. 29: 370–374.

DAWES, E. A., AND P. J. SENIOR. 1973. The role and regulation of energy reserve polymers in microorganisms. Adv. Microbiol. Physiol. 10: 135–266.

DESIKACHARY, T. V. 1959. *Cyanophyta*. Indian Council of Agricultural Research, New Dehli. 686 p.

DINSDALE, M. T., AND A. E. WALSBY. 1972. The interrelations of cell turgor pressure, gas-vacuolation & buoyancy in a blue-green alga. J. Exp. Bot. 23: 561–570.

DOOLITTLE, W. F. 1982. Molecular evolution, p. 307–331. *In* N. G. Carr and B. A. Whitton [ed.] The biology of cyanobacteria. Blackwells Scientific Publications, Oxford.

DREWS, G. 1973. Fine structure and chemical composition of the cell envelopes, p. 99–116. *In* N. G. Carr and B. A. Whitton [ed.] The biology of the blue-green algae. Blackwells Scientific Publications, Oxford.

DREWS, D., AND J. WECKESSER. 1982. Function, structure and composition of cell walls and external layers. *In* N. G. Carr and B. A. Whitton [ed.] The biology of cyanobacteria. Blackwells Scientific Publications, Oxford.

DRUM, R. W., AND S. PANDRATZ. 1965. Fine structure of an unusual cytoplasmic inclusion in the diatom genus *Rhopalodia*. Protoplasma 60: 141–149.

ELOFF, J. N. 1978. The photooxidation of laboratory cultures of *Microcystis* under low light intensities. Environ. Biogeochem. Geomicrobiol. 1: 95–100.

ELOFF, J. N., Y. STEINITZ, AND M. SHILO. 1976. Photooxidation of cyanobacteria in natural conditions. Appl. Environ. Microbiol. 31: 119–126.

EVANS, E. H., AND G. BRITTON. 1983. Relationship between growth conditions, cell morphology and carotenoid composition of the cyanobacterium *Chlorogloeopsis (Chlorogloea) fritschii*, and carotenoid compositions of photosystem I and photosystem 2 preparations. Arch. Microbiol. 135: 284–286.

EVANS, E. H., N. G. CARR, J. D. RUSH, AND C. E. JOHNSON. 1977. Identification of a non-magnetic iron centre and an iron-storage or transport material in blue-green algal membranes by Mössbauer spectroscopy. Biochem. J. 166: 547–551.

EVANS, M. C. W., AND P. HEATHCOTE. 1983. The early photochemical events in bacterial photosynthesis, p. 35–60. *In* J. G. Ormerod [ed.] The phototrophic bacteria: anaerobic life in the light. Blackwells Scientific Publications Ltd., Oxford.

FATTON, A., AND M. SHILO. 1984. *Phormidium* J-1 bioflocculant: production and activity. Arch. Microbiol. 139: 421–426.

FIKSDAHL, A., P. FOSS, AND S. LIAAEN-JENSEN. 1983. Carotenoids of blue-green algae-II. Carotenoids of chromatically-adapted cyanobacteria. Comp. Biochem. Physiol. 76B: 599–601.

FOGG, G. E. 1949. Growth and heterocyst production in *Anabaena cylindrica* Lemm. II. In relation to carbon and nitrogen metabolism. Ann. Bot. 13: 241–259.

——— 1952. The production of extracellular nitrogenous substances by a blue-green alga. Proc. R. Soc. Lond. 139: 372–397.

——— 1971. Extracellular products of algae in fresh water. Arch. Hydrobiol. 5: 1–25.

——— 1978. Nitrogen fixation in the oceans. *In* U. Granhall [ed.] Environmental role of nitrogen-fixing blue-green algae and asymbiotic bacteria. Ecol. Bull. Stockholm 26: 11–19.

——— 1982. Marine plankton. *In* N. G. Carr and B. A. Whitton [ed.] The biology of cyanobacteria. Blackwells Scientific Publications, Oxford.

FOGG, G. E., AND W. D. P. STEWART. 1968. *In situ* determinations of biological nitrogen fixation in Antarctica. Br. Antarct. Surv. Bull. 15: 39–46.

FOGG, G. E., AND THAN-TUN. 1960. Interrelations of photosynthesis and assimilation of elementary nitrogen in a blue-green algae. Proc. R Soc. B 153: 111–127.

FOGG, G. E., AND D. F. WESTLAKE. 1955. The importance of extracellular products of algae in freshwater. Verh. Int. Verein. Theor. Angew. Limnol. 12: 219–232.

FOGG, G. E., W. D. P. STEWART, P. FAY, AND A. WALSBY. 1973. The blue-green algae. Academic Press, London. 459 p.

FOULDS, I. J., AND N. G. CARR. 1977. A proteolytic enzyme degrading phycocyanin in the cyanobacterium *Anabaena cylindrica*. FEMS Microbiol. Lett. 2: 117–119.

FOY, R. H. 1980. The influence of surface to volume ratio on the growth rates of planktonic blue-green algae. Br. Phycol. J. 15: 279–289.

FOY, R. H., AND C. E. GIBSON. 1982. Photosynthetic characteristics of planktonic blue-green algae. Changes in photosynthetic capacity and pigmentation of *Oscillatoria redikii* Van Goor under high & low light. Br. Phycol. J 17: 183–193.

FOY, R. H., AND R. V. SMITH. 1980. The role of carbohydrate accumulation in the growth of planktonic *Oscillatoria* species. Br. Phycol. J. 15: 139–150.

FOYER, C. H. 1984. Photosynthesis. John Wiley & Son, New York, NY. 219 p.

FREDERICK, J. F. [ED.] 1981. Origins and evolution of eukaryotic intracellular organelles. Annal. N.Y. Acad. Sci. 361: 1–512.

FRITSCH, F. E. 1945. The structure and reproduction of the algae. Vol. II. Cambridge University Press, Cambridge, U.K. 939 p.

GALLON, J. R., T. A. LARUE, AND W. G. W. KURZ. 1974. Photosynthesis and nitrogenase activity in the blue-green alga *Gloeocapsa*. Can. J. Microbiol. 20: 1633–1637.

GALLUCCI, K. K., AND H. W. PAERL. 1983. *Pseudomonas aeruginosa* chemotaxis associated with blooms of N_2-fixing blue-green algae (cyanobacteria). Appl. Environ. Microbiol. 45: 557–562.

GANF, G. G. 1974. Incident solar irradiance and underwater light penetration as factors controlling the chlorophyll a content of a shallow equatorial lake (Lake George, Uganda). J.

Ecol. 62: 593-609.

GANTT, E. 1981. Phycobilisomes. A Rev. Plant. Physiol. 32: 327-347.

GARLICK, S., A. OREN, AND E. PADAN. 1977. Occurrence of facultative anoxygenic photosynthesis among filamentous and unicellular cyanobacteria. J. Bact. 129: 623-629.

GIBBS, M. 1962. Respiration, p. 61-90. *In* R. A. Lewin [ed.] Physiology and biochemistry of algae. Academic Press, New York, NY.

GIBSON, C. E. 1975. A field and laboratory study of oxygen uptake by planktonic blue-green algae. J. Ecol. 63: 867-880.

GIBSON, C. E., AND R. V. SMITH. 1982. Freshwater plankton, p. 463-490. *In* N. G. Carr and B. A. Whitton [ed.] The biology of cyanobacteria. Blackwells Scientific Publications, Oxford.

GIBSON, J. 1984. Nutrient transport by anoxygenic & oxygenic photosynthetic bacteria. Ann. Rev. Microbiol. 38: 135-159.

GLAZER, A. N. 1982. Phycobilisomes: structure and dynamics. A Rev. Microbiol. 36: 173-198.

——— 1983. Comparative biochemistry of photosynthetic light-harvesting systems. A. Rev. Biochem. 52: 125-157.

GLEASON, F. K., AND J. L. PAULSON. 1984. Site of action of the natural algicide, cyanobacterin, in the blue-green alga, *Synechococcus* sp. Arch. Microbiol. 138: 273-277.

GOEDHEER, J. C. 1969. Carotenoids in blue-green and red algae. Progress in Photosynthesis Research II: 811-817.

GOLECKI, J. R., AND G. DREWS. 1982. Supramolecular organization and composition of membranes. p. 125-140. *In* N. G. Carr and B. A. Whitton [ed.] The biology of cyanobacteria. Blackwells Scientific Publications, Oxford.

GORHAM, P. R., J. S. McLACHLAN, U. T. HAMMER, AND W. K. KIM. 1964. Isolation and culture of toxic strains of Anabaena flos-aquae (Lyngb) De Breb. Verh. Internat. Verein. Limnol. 15: 796-804.

GRILLI-CAIOLA, M. 1980. On the phycobionts of the cycad coralloid roots. New Phytol. 85: 537-544.

GUGLIEMI, G., AND G. COHEN-BAZIRE. 1982. Several non-chromatically adapting *Pseudanabaena* strains synthesize constitutively two different molecular types of phycocyanin. Abstr. IV International Symp. Photosynthetic Prokaryotes, pC47, Bombannes.

GUPTA, M., AND N. G. CARR. 1981. Enzyme activities related to cyanophycin metabolism in heterocysts and vegetative cells of *Anabaena* spp. J. Gen. Microbiol. 125: 17-23.

HADER, D-P. 1979. Effect of inhibitors and uncouplers of light-induced potential changes triggering photophobic responses. Arch. Microbiol. 120: 57-60.

——— 1978. Evidence of electrical potential changes in photophobically reacting blue-green algae. Arch. Microbiol. 118: 115-119.

——— 1974. Participation of two photosystems in the photophobotaxis of *Phormidium uncinatum*. Arch. Microbiol. 96: 255-266.

HALLDAL, P. 1958. Pigment formation and growth of blue green alge in crossed gradients of light intensity and temperature. Physiol. Plant 11: 401-420.

HALLDAL, P., AND C. S. FRENCH. 1958. Algal growth in crossed gradients of light intensity and temperature. Plant Physiol. 33: 249-252.

HALLIDAY, J., AND J. S. PATE. 1976. Symbiotic nitrogen fixation by coralloid roots of the cycad *Mairozamia viedlei*: physiological characteristics and ecological significance. Aust. J. Plant. Physiol. 3: 349-358.

HARDIE, L. P., D. L. BALKWILL, AND S. E. STEVENS. 1983a. Effects of iron starvation on the physiology of the cyanobacterium *Agmenellum quadruplicatum*. Appl. Environ. Microbiol. 45: 999-1006.

——— 1983b. Effects of iron starvation on the ultrastructure of the cyanobacterium *Agmenellum quadruplicatum*. Appl. Environ. Microbiol. 45: 1007-1017.

HAYSTEAD, A., M. W. W. DHARMAWARDENE, AND W. D. P. STEWART. 1973. Ammonia assimilation in a nitrogen-fixing blue-green alga. Plant Sci. Lett. 1: 439-445.

HEALEY, F. P. 1968. The carotenoids of four blue-green algae. J. Phycol. 4: 126-129.

——— 1973. Characteristics of phosphorous deficiency in *Anabaena*. J. Phycol. 9: 383-394.

——— 1982. Phosphate, p. 105-124. *In* N. G. Carr and B. A. Whitton [ed.] The biology of cyanobacteria. Blackwells Scientific Publications, Oxford.

HEALEY, F. P., AND L. L. HENDZEL. 1975. Effects of phosphorus deficiency on two algae growing in chemostats. J. Phycol. 11: 303-309.

——— 1979. Fluorometric measurement of alkaline phosphatase activity in algae. Freshwater Biol. 9: 429-439.

HELLEBUST, J. A. 1974. Extracellular products, p. 838-863. *In* W. D. P. Stewart [ed.] Algal physiology and biochemistry. Blackwells Scientific Publications, Oxford.

HERDMAN, M. 1982. Evolution and genetic properties of cyanobacterial genomes, p. 263-305. *In* N. G. Carr and B. A. Whitton [ed.] The biology of cyanobacteria. Blackwells Scientific Publications, Oxford.

HERRERO, A., E. FLORES, AND M. G. GUERRERO. 1985. Regulation of nitrate reductase cellular levels in the cyanobacteria *Anabaena variabilis* and Synechocystis sp. FEMS Microbiol. Lett. 26: 21-25.

HERTZBERG, S., S. LIAAEN-JENSEN, AND H. W. SIEGELMANN. 1971. The carotenoids of the blue-green algae. Phytochem. 10: 3121-3127.

HO, K. K., AND D. W. KROGMANN. 1982. Photosynthesis, p. 191-214. *In* N. G. Carr and B. A. Whitton [ed.] The biology of cyanobacteria. Blackwells Scientific Publications Ltd., Oxford.

HOLM-HANSEN, D. 1968. Ecology, physiology and biochemistry of blue-green algae. A Rev. Microbiol. 22: 47-70.

HOOD, W., AND N. G. CARR. 1971. Apparent lack of control by repression of arginine metabolism in blue-green algae. J. Bact. 107: 365-367.

HUNTSMAN, S. A., AND W. G. SUNDA. 1980. The role of trace metals in regulating phytoplankton growth, p. 285-328. *In* I. Morris [ed.] The physiological ecology of phytoplankton. Blackwells Scientific Publications, Oxford.

JANNASCH, H. W. 1984. Microbes in the oceanic environment, p. 97–122. *In* D. P. Kelly and N. G. Carr [ed.] The Microbe 1984. Vol. II. Cambridge University Press, Cambridge.

JENSEN, T. E., AND C. C. BOWEN. 1961. Organisation of the centroplasm in *Nostoc punctifuromie*. Proc. Iowa Acad. Sci. 68: 89–96.

JENSEN, T. E., AND J. W. RACHLIN. 1974. Effect of varying sulphur deficiency on structural components of a cyanobacterium *Synechococcus leopoliensis*: a morphometric study. Cytobios 41: 35–46.

JENSEN, T. E., AND L. M. SICKO. 1974. Phosphate metabolism in blue-green algae: I. Fine structure of the "polyphosphate overplus" phenomenon in *Plectonema boryanum*. Can. J. Microbiol. 20: 1235–1239.

JONES, L. W., AND J. MYERS. 1964. Enhancement in the blue-green alga, *Anacystis nidulans*. Plant Physiol. 39: 938–946.

JONES, C. W. 1982. Bacterial Respiration and Photosynthesis. Thomas Nelson and Sons, Walton-on-Thames. 106 p.

JONES, G. E., L. MURRAY, AND N. G. CARR. 1978. Trace element composition of five cyanobacteria, p. 967–973. *In* W. E. Krumbein [ed.] Environmental biogeochemistry and geomicrobiology. Ann Arbor Science, Ann Arbor, MI.

JONES, L. W., AND J. MYERS. 1963. A common link between photosynthesis and respiration in a blue-green alga. Nature, Lond. 199: 670–672.

JONES, R. I., AND V. ILMAVIRTA. 1978. Vertical and seasonal variation of phytoplankton photosynthesis in a brown-water lake with winter ice cover. Freshwater Biol. 8: 561–572.

JOSET-ESPARDELLIER, F., C. ASTIER, E. H. EVANS, AND N. G. CARR. 1978. Cyanobacteria grown under photoautotrophic, photoheterotrophic and heterotrophic regimes: sugar metabolism and carbon dioxide fixation. FEMS Microbiol. Lett 4: 261–264.

KAPP, R., S. E. STEVENS, AND J. L. FOX. 1975. A survey of available nitrogen sources for the growth of the blue-green alga, *Agmenellum quadruplicatum*. Arch. Microbiol. 104: 135–138.

KAWAMURA, M., M. MIMURO, AND Y. FUJITA. 1979. Quantitative relationship between two reaction centers in the photosynthetic system of blue-green algae. Plant & Cell Physiol. 20: 697–705.

KELLAR, P. E., AND H. W. PAERL. 1980. Physiological adaptations in response to environmental stress during a N_2-fixing *Anabaena* bloom. Appl. Environ. Microbiol. 40: 587–595.

KERSHAW, K. A., AND J. W. MILLBANK. 1970. Nitrogen metabolism in lichens. II. The partition of cepalodial fixed nitrogen between the mycobiont & phycobionts of *Peltigera aptriosa*. New Phytol. 69: 75–79.

KHANNA, R., J-R. GRAHAM, J. MYERS, AND E. GANTT. 1983. Phycobilisome composition and possible relationship to reaction centers. Arch. Biochem. Biophys. 224: 534–542.

KHOJA, T. M., AND B. A. WHITTON. 1971. Heterotrophic growth of blue-green algae. Arch. Mikrobiol. 79: 280–282.

KONOPKA, A. 1980. Physiological changes wihtin a metalimnetic layer of *Oscillatoria rubescens*. Appl. Environ. Microbiol. 40: 681–684.

1982. Physiological ecology of a metalinnetic *Oscillatoria rubescens* population. Limnol. Oceanogr. 27: 1154–1161.

KRATZ, W. A., AND J. MYERS. 1955a. Photosynthesis and respiration of three blue-green algae. Plant Physiol. 30: 275–280.

1955b. Nutrition and growth of several blue-green algae. Am. J. Bot. 42: 282–287.

KRINSKY, N. I. 1978. Non-photosynthetic functions of carotenoids. Phil. Trans. R. Soc. Lond. B 284: 581–590.

KULAEV, I. S. 1979. The biochemistry of inorganic polyphosphates. John Wiley & Sons, New York, NY. 255 p.

LACZKO, I. 1985. Comparison of photosystem-I and photosystem-II activities of spheroplasts from normal & photobleached *Anabaena cylindrica*. Arch. Microbiol. 141: 112–115.

LAMMERS, P., AND J. SANDERS-LOEHR. 1982. Active transport of ferric schizokinen in *Anabaena* sp. J. Bacteriol. 151: 288–294.

LANARAS, T., AND G. A. CODD. 1982. Variation in ribulose 1,5 bisphospate carboxylase protein levels, activation and sub-cellular distribution during photoautotrophic batch culture of *Chlorogloeopsis fritschii*. Planta 154: 284–288.

LANARAS, T., A. M. HAWTHORNETHWAITE, AND G. A. CODD. 1985. Localization of carbonic anhydrase in the cyanobacterium *Chlorogloeopsis fritschii*. FEMS Microbiol. Lett. 26: 285–288.

LANG, N. J. 1968. The fine structure of blue-green algae. A. Rev. Microbiol. 22: 15–46.

LANG, N. J., R. D. SIMON, AND C. P. WOLK. 1972. Correspondence of cyanophycin granules with structured granules in *Anabaena cylindrica*. Arch. Mikrobiol. 83: 313–320.

LANG, N. J., AND B. A. WHITTON. 1973. Arrangement and structure of thylakoids. *In* N. G. Carr and B. A. Whitton [ed.] The biology of blue-green algae. Blackwells Scientific Publications, Oxford.

LAU, R. H., M. M. MACKENZIE, AND W. F. DOOLITTLE. 1977. Phycocyanin synthesis and degradation in the blue-green bacterium *Anacystis nidulans*. J. Bact. 132: 771–78.

LAWRY, N. H., AND T. E. JENSEN. 1979. Deposition of condensed phosphate as an effect of varying sulfur deficiency in the cyanobacterium. Arch. Microbiol. 120: 1–7.

LAWRY, N. H., AND R. D. SIMON. 1982. The normal and induced occurrence of cyanophycin inclusion bodies in several blue-green algae. J. Phycol. 18: 391–399.

LAZAROFF, N. 1972. Experimental control of nostocacean development. p. 521–544. *In* T. V. Desikachary [ed.] Taxonomy and biology of blue-green algae. University of Madras, India.

LEACH, C. K., AND N. G. CARR. 1971. Electron transport and oxidative phosphorylation in the

199

blue-green alga *Anabaena variabilis*. J. Gen. Microbiol. 64: 55–70.

LEE-KADEN, J., AND W. SIMONIS. 1982. Amino acid uptake and energy coupling dependent on photosynthesis in *Anacystis nidulans*. J. Bact. 151: 229–236.

LEHMANN, M., AND G. WOBER. 1976. Accumulation, mobilization and turnover of glycogen in the blue-green bacterium *Anacystis nidulans*. Arch. Microbiol. 111: 93–97.

LEMASSON, C., N. TANDEAU DE MARSAC, AND G. COHEN-BAZIRE. 1973. Role of allophycocyanin as a light-harvesting pigment in cyanobacteria. Proc Natl. Acad. Sci. 70: 3130–3133.

LEWIN, R. A. 1980. Prochlorophytes, p. 257–266. *In* M. P. Starr, H. Stolp, H. G. Truper, A. Balons, and H. G. Schlegel [ed.] The Prokaryotes. Springer-Verlag, Berlin.

LIVINGSTONE, D., AND B. A. WHITTON. 1983. Influence of phosphorus on morphology of *Calothrix parientina (Cyanophyta)* in culture. Br. Phycol. J. 18: 29–58.

LOSADA, M., AND M. G. GUERRERO. 1979. The photosynthetic reduction of nitrate and its regulation. *In* J. Barber [ed.] Photosynthesis in Relation to Model Systems. Elsevier/North Holland Biomedical.

MAALOE, O., AND N. O. KJELDGAARD. 1966. Control of macromolecular synthesis. W. A. Benjamin, New York, NY.

MACKAY, M. A., R. S. NORTON, AND L. J. BOROWITZKA. 1984. Organic osmoregulatory solutes in cyanobacteria. J. Gen. Microbiol. 130: 2177–2191.

MAGUE, T. A., F. C. MAGUE, AND O. HOLM-HANSEN. 1977. Physiology and chemical composition of nitrogen-fixing phytoplankton in the central North Pacific Ocean. Mar. Biol. 41: 213–227.

MAGUE, T. A., N. M. WEARE, AND O. HOLM-HANSEN. 1974. Nitrogen fixation in the North Pacific Ocean. Mar. Biol. 24: 109–119.

MANN, N., AND N. G. CARR. 1974. Control of macromolecular composition and cell division in the blue-green alga *Anacystis nidulans*. J. Gen. Microbiol. 83: 399–405.

MANODORI, A., AND A. MELIS. 1985. Phycobilisome-photosystem II association in *Synechococcus* 6301 (Cyanophyceae). FEBS Lett. 181: 79–82.

MANODORI, A., M. ALHADEFF, A. N. GLAZER, AND A. MELIS. 1984. Photochemical apparatus organization in *Synechococcus* 6301 (*Anacystis nidulans*). Effect of phycobilisome mutation. Arch. Microbiol. 139: 117–123.

MARCUS, Y. D. ZENVIRTH, E. HAREL, AND A. KAPLAN. 1982. Induction of HCO_3^- transporting capability and high photosynthetic affinity to inorganic carbon by low concentration of CO_2 in *Anabaena variabilis*. Plant Physiol. 69: 1008–1012.

MARGULIS, L. 1970. Origin of eukaryotic cells. Yale University Press, U.S.A. 349 p.

MARYAN, P. S., R. E. EADY, A. E. CHAPLIN, AND J. R. GALLON. 1986. Nitrogen fixation by *Gloeothece* sp. PCC 6909: Respiration and not photosynthesis supports nitrogenase activity in the light. J. Gen. Microbiol. 132: 789–796.

MASON, C. P., K. R. EDWARDS, R. E. CARLSON, J. PIGNATELLO, AND F. K. GLEASON. 1982. Isolation of a chlorine-containing antibiotic from the freshwater cyanobacteria *Scytonema hofunanni*. Science 215: 400–402.

MCCARTHY, J. J., AND E. J. CARPENTER. 1979. *Oscillatoria (Trichodesmium) thiebautii* (Cyanophyta) in the central North Atlantic Ocean. J. Phycol. 15: 75–82.

MEEKS, J. C., C. S. ENDERLIN, C. M. JOSEPH, J. S. CHAPMAN, AND M. W. L. LOLLAR. 1985. Fixation of [^{13}N] N_2 and transfer of fixed nitrogen in the *Anthoceros-Nostoc* symbiotic association. Planta 164: 406–414.

MEEKS, J. C., K. L. WYCOFF, J. S. CHAPMAN, AND C. S. ENDERLIN. 1983. Regulation and expression of nitrate and dinitrogen assimilation by *Anabaena* species. Appl Environ. Microbiol. 45: 1351–1359.

MEFFERT, M-E., AND H. ZIMMERMAN-TELSCHOW. 1979. Net-release of nitrogenous compounds by axenic and bacteria-containing cultures of *Oscillatoria redeki (Cyanophyta)*. Arch. Hydrobiol. 87: 125–138.

METZLER, H. 1980. Oscillations of the NAD(P)H pool size and of the redox state of a cytochrome b during dark respiration of the blue-green alga, *Anacystis nidulans*. Biochim. Biophys. Acta 593: 312–318.

MILLBANK, J. W. 1972. Nitrogen metabolism in lichens. IV. The nitrogenase activity of the *Nostoc* phycobiont in *Peltigera canina*. New Phytol. 68: 721–729.

MILLER, A. G., AND B. COLMAN. 1980. Active transport and accumulation of bicarbonate by a unicellular cyanobacterium. J. Bact. 143: 1253–1259.

MILLER, L. S., AND S. C. HOLT. 1977. Effect of carbon dioxide on pigment and membrane content in *Synechococcus lividus*. Arch. Microbiol. 115: 185–198.

MIMURO, M., AND Y. FUJITA. 1977. Estimation of chlorophyll a distribution in the photosynthetic pigment systems I and II of the blue-green alga *Anabaena variabilis*. Biochim. Biophys. Acta 459–389.

MITCHISON, J. M. 1971. The biology of the cell cycle. Cambridge University Press, Cambridge, U.K. 313 p.

MORRIS, I., AND H. GLOVER. 1981. Physiology of photosynthesis by marine coccoid cyanobacteria — some ecological implications. Limnol. Oceanogr. 26: 957–961.

MORTON, S. D., AND T. H. LEE. 1974. Algal blooms — possible effects of iron. Environ. Sci. Technol. 8: 673–674.

MULLINEAUX, P. M., J. R. GALLON, AND A. E. CHAPLIN. 1981. Acetylene reduction (nitrogen fixation) by cyanobacteria grown under alternating light–dark cycles. FEMS Microbiol. Lett. 10: 245–247.

MURPHY, T. P., D. R. S. LEAN, AND C. NALEWAJKO. 1976. Blue-green algae: their excretion of iron-selective chelators enables them to dominate other algae. Science 192: 900–902.

MYERS, J., AND W. A. KRATZ. 1955. Relations

between pigment content and photosynthetic characteristics in a blue-green algae. J. Gen. Physiol. 39: 11–22.

NIERZWICKI-BAUER, S. A., D. L. BALKWILL, AND S. E. STEVENS JR. 1983. Three-dimensional ultrastructure of a unicellular cyanobacterium. J. Cell Biol. 97: 713–722.

NICHOLS, J., AND D. G. ADAMS. 1982. Akinetes. *In* N. G. Carr and B. A. Whitton [ed.] The biology of cyanobacteria. Blackwells Scientific Publications, Oxford.

OGAWA, T., T. SEKINE, AND S. AIBA. 1979. Reappraisal of the so-called light shielding of gas vacuoles in *Microcystis aeruginosa*. Arch. Microbiol. 122: 57–60.

OLIVER, R. L., AND A. E. WALSBY. 1984. Direct evidence for the role of light-mediated gas vesicle collapse in the buoyancy regulation of *Anabaena flos-aquae* (cyanobacteria). Limnol. Oceanogr. 29: 879–886.

OMATA, T., AND N. MURATA. 1984. Isolation and characterization of three types of membranes from the cyanobacterium (blue-green alga) *Synechocystis* PCC 6714. Arch. Microbiol. 139: 113–116.

OQUIST, G. 1977. Changes in pigment composition and photosynthesis induced by iron-deficiency in the blue-green alga *Anacystis nidulans*. Physiol. Plant 25: 188–191.

OWNBY, J. D., M. SHANNAHAN, AND E. HOOD. 1979. Protein synthesis and degradation in *Anabaena* during nitrogen starvation. J. Gen. Microbiol. 10: 255–261.

PADAN, E., 1979. Facultative anoxygenic photosynthesis in cyanobacteria. A. Rev. Pl. Physiol. 30: 27–40.

PADAN, E., AND Y. COHEN. 1982. Anoxygenic photosynthesis, p. 215–236. *In* N. G. Carr and B. A. Whitton [ed.] The biology of cyanobacteria. Blackwells Scientific Publications, Oxford.

PAERL, H. W. 1976. Specific associations of the blue-green algae *Anabaena* and *Aphanizomenon* with bacteria in freshwater blooms. J. Phycol. 12: 431–435.

1982a. Factors limiting productivity of freshwater ecosystems. Adv. Microb. Ecol. 6: 75–110.

1982b. Interactions with bacteria, p. 411–461. *In* N. G. Carr and B. A. Whitton [ed.] The biology of cyanobacteria. Blackwells Scientific Publications, Oxford.

1984. Cyanobacterial carotenoids: their roles in maintaining optimal photosynthetic production among aquatic bloom forming genera. Oecologia (Berlin) 61: 143–149.

PAERL, H. W., AND K. K. GALLUCCI. 1985. Role of chemotaxis in establishign a specific nitrogen-fixing cyanobacterial-bacterial association. Science 227: 647–649.

PAERL, H. W., AND P. I. KELLAR. 1978. Significance of bacterial-*Anabaena* (*Cyanophyceae*) associations with respect to N$_2$ fixation in freshwater. J. Phycol. 14: 354–260.

1979. Nitrogen-fixing *Anabaena*: physiological adaptations instrumental in maintaining surface blooms. Science 204: 620–622.

PAERL, H. W., J. TUCKER, AND P. T. BLAND. 1983. Carotenoid enhancement and its role in maintaining blue-green algal (*Microcystis aeruginosa*) surface blooms. Limnol. Ocenogr. 20: 847–857.

PARL, M. P., AND R. V. SMITH. 1976. The identification of phosphorus as a growth-limiting nutrient in Lough Neagh using bioassays. Water Res. 10: 1151–1154.

PEARCE, J., AND N. G. CARR. 1969. The incorporation and metabolism of glucose by *Anabaena variabilis* and *Anacystis nidulans*. J. Gen. Microbiol. 49: 301–313.

PESCHEK, G. A. 1983. The cytochrome f-b electron-transport complex. Biochem. J. 210: 269–272.

PESCHEK, G. A., AND G. SCHMETTERER. 1982. Evidence for plastoquinone-cytochrome f/b-563 reductase as a common electron donor to P700 and cytochrome oxidase in cyanobacteria. Biochem. Biophys. Res. Comm. 108: 1188–1195.

PESCHEK, G. A., G. SCHMETTERER, AND N. B. SLEYTOR. 1981. Possible respiration sites in the plasma membrane of *Anacystis nidulans*: ultracytochemical evidende. FEMS Microbiol. Lett. 11: 121–124.

PORTER J. AND M. JOST. 1976. Physiological effects of the presence and absence of gas vacuoles in the blue-green alga, *Microcystis aeruginosa* Kuetz. emend. Elenkin. Arch. Microbiol. 110: 225–231.

RAI, A. N., P. ROWELL, AND W. D. P. STEWART. 1983. Interactions between cyanobacterium and fungus during ^{15}N$_2$-incorporation and metabolism in the lichen *Peltigera canina*. Arch. Microbiol. 134: 136–142.

RAY, T. B., B. C. MAYNE, R. E. TOIA, AND G. A. O. PETERS. 1979. *Azollan Anabaena* relationship. VIII. Photosynthetic characterization of the association and individual partners. Plant Physiol. 64: 791–795.

RAY, T. B., G. A. PETERS, R. E. TOIA, AND B. C. MAYNE. 1978. *Azolla-Anabaena* relationship. VII. distribution of ammonia-assimilating enzymes, protein, and chlorophyll between host and Symbiont. Plant Physiol. 62: 463–467.

RACHLIN, J. W., T. E. JENSEN, M. BAXTER, AND V. JANI. 1982. Utilisation of morphometric analysis in evaluating response of *Plectonema boryanum* (Cyanophyceae) to exposure to light heavy metals, Arch. Environ. Contam. Toxicol. 11: 323–333.

REYNOLDS, C. S., AND A. E. WALSBY. 1975. Water-blooms. Biol. Rev. 50: 437–481.

REED, R. H., J. A. CHUDEK, R. FOSTER, AND W. D. P. STEWART. 1984. Osmotic adjustment in cyanobacteria from hypersaline environments. Arch. Microbiol. 138: 333–337.

RICHARDSON, K., J. BEARDALL, AND J. A. RAVEN. 1983. Adaptation of unicellular algae to irradiance: an analysis of strategies. New Phytol. 93: 157–191.

RIEGMAN, R., M. RUTGERS, AND L. R. MUR. 1985. Effects of photoperiodicity and light irradiance on phosphate-limited *Oscillatoria agardhii* in chemostat cultures. I. Photosynthesis and carbohydrate storage. Arch. Microbiol. 142:

66–71.

RIEGMAN R., AND L. R. MUR. 1984. Regulation of phosphate uptake kinetics in *Oscillatoria agardhii*. Arch. Microbiol. 139: 28–32.

——— 1985. Effects of photoperiodicity and light irradiance on phosphate-limited *Oscillatoria agardhii* in chemostat cultures. II Phosphate uptake and growth. Arch. Microbiol. 142: 72–76.

RIPPKA, R. 1972. Photoheterotrophy and chemoheterotrophy among unicellular blue-green algae. Arch. Mikrobiol. 87: 93–98.

RIPPKA, R., J. DERUELLES, J. B. WATERBURY, M. HERDMAN, AND R. Y. STANIER. 1979. Generic assignments, strain histories and properties of pure culture of cyanobacteria. J. Gen. Microbiol. 111: 1–61.

RODGERS, G. A., AND W. D. P. STEWART. 1977. The cyanophyte–hepatic symbiosis I. Morphology & physiology. New Phytol. 78: 441–458.

ROMERO, J. M., E. FLORES, AND M. G. GUERRERO. 1985. Inhibition of nitrate utilization by amino acids in intact *Anacystis nidulans*. Arch. Microbiol. 142: 1–5.

RUETER, J. G., J. J. McCARTHY, AND E. J. CARPENTER. 1979. The toxic effect of copper on *Oscillatoria (Trichodesmium) thiobautii*. Limnol. Oceanogr. 24: 558–562.

SAUNDERS, G. W. 1972. Potential heterotrophy in a natural population of *Oscillatoria agardhii* var *isothrix* Skuja. Limnol. Oceanogr. 17: 704–711.

SCHEIBE, J. 1972. Photoreversible pigment: occurrence in a blue-green alga. Science 177: 1037–1039.

SHAFFER, P. W., W. LOCKAU, AND C. P. WOLK. 1978. Isolation of mutants of the cyanobacterium *Anabaena variabilis*, impaired in photoautotrophy. Arch. Microbiol. 117: 215–219.

SHEAR, H., AND A. E. WALSBY. 1975. An investigation into the possible light–shielding role of gas vacuoles in a planktonic blue-green alga. Br. Phycol. J. 10: 241–251.

SHESTAKOV, S. V., T. I. POSTNOVA, AND L. G. SHAKHNABATIAN. 1975. *In vitro* repair of UV- and X-irradiated bacteriophage T4 DNA by extract from blue-green alga *Anacystis nidulans*. Mol. Biol. Rep. 2: 89–94.

SHIVELY, J. M. 1974. Inclusion bodies of prokaryotes. A. Rev. Microbiol. 28: 167–187.

SHIVELY, J. M., F. BALL, D. H. BROWN, AND R. E. SAUNDERS. 1973. Functional organelles in prokaryotes: polyhedral inclusions (carboxysomes) of *Thiobacillus neapolitanus*. Science 182: 584–586.

SICKO-GOAD, L., AND T. E. JENSEN. 1976. Phosphate metabolism in blue-green algae. II. Changes in phosphate distribution during starvation and the "polyphosphate overplus" phenomenon in *Plectonema boryanum*. Am. J. Bot. 63: 183–188.

SICKO-GOAD, L., T. E. JENSEN, AND R. P. AYALA. 1978. Phosphate metabolism in blue-green bacterium V. Factors affecting phosphate uptake in *Plectonema boryanum*. Can. J. Microbiol. 24: 105–108.

SILVESTER, W. B., AND P. J. MacNAMARA. 1976. The infection process and ultrastructure of the *Gunnera–Nostoc* symbiosis. New Phytol. 77: 135–141.

SILVESTER, W. B., AND D. R. SMITH. 1969. Nitrogen fixation by *Gunnera–Nostoc* symbiosis. Nature 224: 1231.

SIMON, R. D. 1971. Cyanophycin granules from the blue-green alga *Anabaena cylindrica*: a reserve material consisting of co-polymers of aspartic acid and arginine. Proc. Nat. Acad. Sci. U.S.A. 68: 256–267.

——— 1976. The biosynthesis of multi-L-arginyl-poly (L-aspartic acid) in the filamentous cyanobacterium *Anabaena cylindrica*. Biochim. biophys. Acta 422: 407–418.

SIMPSON, F. B., AND J. B. NEILANDS. 1976. Siderochromes in Cyanophyceae: isolation and characterisation of shizokinen from *Anabaena* sp. J. Phycol. 12: 44–48.

SIMONIS, W., T. BORNEFELD, J. LEE-KADEN, AND K. MAJUMDAR. 1974. Phosphate uptake and photophosphorylation in the blue-green alga *Anacystis nidulans*, p. 220–225. *In* U. Zimmerman and J. Dainty [ed.] Membrane transport in plants. Springer-Verlag, Berlin.

SINCLAIR, C., AND B. A. WHITTON. 1977. Influence of nutrient deficiency on hair formation in the *Rivulariaceae*. Br. Phycol. J. 12: 297–313.

SINGH, P. K. 1975. Photoreactivation of UV-irradiated blue-green algae and algal virus LPP-1. Arch. Microbiol. 103: 297–302.

SMITH, A. J. 1982. Modes of cyanobacterial carbon metabolism, p. 47–85. *In* N. G. Carr and B. A. Whitton [ed.] The biology of cyanobacteria. Blackwells Scientific Publications, Oxford.

SMITH, D. C., L. MUSCATINE, AND D. LEWIS. 1970. Carbohydrate movement from autotrophs to heterotrophs in parasitic and mutualistic symbiosis. Biol. Rev. 44: 17–90.

SMITH, V. H. 1983. Low nitrogen to phosphorus ratios favor dominance by blue-green algae in lake phytoplankton. Science 221: 669–671.

STACEY, G., C. VAN BAALEN, AND F. R. TABITA. 1977. Isolation and characterization of a marine *Anabaena* sp. capable of rapid growth on molecular nitrogen. Arch. Microbiol. 114: 197–201.

STANIER, R. Y. 1982. Foreword, p. IX–X. *In* N. G. Carr and B. A. Whitton [ed.] The biology of cyanobacteria. Blackwells Scientific Publications, Oxford.

STANIER, R. Y., AND G. COHEN-BAZIRE. 1977. Phototrophic prokaryotes: the cyanobacteria. A. Rev. Microbiol. 31: 225–274.

STANIER, R. Y., AND C. B. VAN NIEL. 1962. The concept of a bacterium. Arch. Mikrobiol. 42: 17–35.

STANIER, R. Y., R. KANISAWA, M. MANDEL, AND G. COHEN-BAZIRE. 1971. Purification and properties of unicellular blue-grenn algae (order Chroococcales). Bact. Rev. 35: 171–205.

STEINITZ, Y., Z. MAZOR, AND M. SHILO. 1979. A mutant of the cyanobacterium *Plectonema boryanum* resistant to photooxidation. Plant. Sci. Lett. 16: 327–335.

STEVENS, S. E., D. A. M. PAONE, AND D. L. BALKWILL. 1981. Accumulation of cyanophycin granules as a result of phosphate limitation in *Agmenellun quadruplicatum*. Plant Physiol. 67: 716–719.

STEVENS, S. E., S. A. NIERZWICKI-BAUER, AND D. L. BALKWILL. 1985. Effect of nitrogen starvation on the morphology and ultrastructure of the cyanobacterium *Mastigocladus laminosus*. J. Bact. 161: 1215–1218.

STEWART, W. D. P. 1973. Nitrogen fixation by photosynthetic microorganisms. Ann. Rev. Microbiol. 27: 283–316.

———. 1980. Some aspects of structure and function in N_2-fixing cyanobacteria. Ann. Rev. Microbiol. 34: 497–536.

STEWART, A. C., AND D. S. BENDALL. 1979. Preparation of an active, oxygen evolving photosystem II particle form a blue-green alga. FEBS Lett. 107: 308–312.

———. 1981. Properties of oxygen-evolving photosystem II particles form Phormidium laminosum, a thermophilic blue-green alga. Biochem. J. 194. 877–887.

STEWART, W. D. P., AND G. A. CODD. 1975. Polyhedral bodies (carboxysomes) of nitrogen-fixing blue-green algae. Br. Phycol. J. 10: 273–278.

STEWART, W. D. P., AND M. LEX. 1970. Nitrogenase activity in the blue-green alga *Plectonema boryanum* strain 594. Arch Mikrobiol. 73: 250–260.

STEWART. W. D. P. AND C. A. RODGERS. 1977. The cyanophyte-hepatic symbiosis II. Nitrogen fixation and interchange of nitrogen and carbon. New Phytol. 78: 459–471.

STEWART, W. D. P., P. ROWELL, AND A. N. RAI. 1983. Cyanobacteria–eukaryotic plant symbiosis. Ann. Microbiol. (Inst. Pasteur) 134B: 205–228.

SZALONTAI, B., AND M. VAN DE VEN. 1981. Raman spectroscopic evidence for phycocyanin-carotenoid interaction in *Anacystis nidulans*. FEBS Letts. 131: 155–157.

TAKAHASHI, M., AND T. HORI. 1984. Abundance of picophytoplankton in the subsurface chlorophyll maximum layer in subtropical and tropical waters. Mar. Biol. 79: 177–186.

TANDEAU DE MARSAC, N. 1977. Occurrence and nature of chromatic adaptation in cyanobacteria. J. Bact. 130: 82–91.

TILMAN, D., S. S. KILHAM, AND P. KILHAM. 1982. Phytoplankton community ecology: the role of limiting nutrients. A. Rev. Ecol. Syst. 3: 349–372.

THOMAS, P. H., AND N. G. CARR. 1985. The invariance of macromolecular composition with altered light limited growth rate of *Amphidinium carteri* (*Dinophyceae*). Arch. Microbiol. 142: 81–86.

THORNBER, J. P. 1969. Comparison of a chlorophyll *a*-protein complex isolated from a blue-green alga with chlorophyll-protein complexes obtained from green bacteria and higher plants. Biochim. Biophys. Acta 172: 230–241.

TYTLER, E. M., G. C. WHITELAM, M. F. HIPKINS, AND G. A. CODD. 1984. Photoinactivation of Photosystem II during photoinhibition in the cyanobacterium *Microcystis aeruginosa*. Planta 160: 209–234.

UTKILEN, H. C. 1982. Magnesium-limited growth of the cyanobacterium *Anacystis nidulans*. J. Gen. Microbiol. 128: 1849–1862.

———. 1984. Growth and macromolecular composition of *Anacystis nidulans* at different temperatures in Mg^{2+} -and K^+-limited chemostats. Arch. Microbiol. 138: 244–246.

VACCARRO, R. F., S. E. HICKS, H. W. JANNASCH, AND F. G. CAREY. 1968. The occurrence and role of glucose in sea water. Limnol. Oceanogr. 13: 356–360.

VAN BAALEN, C. 1962. Studies on marine blue-green algae. Botanica Mar. 4: 129–139.

VAN BAALEN, C., AND R. O'DONNELL. 1972. Action spectra for ultraviolet killing and photoreactivation in the blue-green alga *Agmenellum quadruplicatum*. Photochem. Photobiol. 15: 269–274.

VAN LIERE, L., G. J. DE GROOT, AND L. R. MUR. 1979a. Pigment variation with irradiance in *Oscillatoria agardhii* Gomont in nitrogen (nitrate)-limited chemostat cultures. FEMS Microbiol. Lett. 6: 337–340.

VAN LIERE, L., L. R. MUR, C. E. GIBSON, AND M. HERDMAN. 1979b. Growth and physiology of *Oscillatoria agardhii* Gomont cultivated in continuous culture with a light–dark cycle. Arch. Microbiol. 123: 315–318.

VAN LIERE, L., AND A. E. WALSBY. 1982. Interactions of cyanobacteria with light, p. 9–45. *In* N. G. Carr and B. A. Whitton [ed.] The biology of cyanobacteria. Blackwells Scientific Publications, Oxford.

VAN NEIL, C. B. 1941. The bacterial photosyntheses and their importance for the general problem of photosynthesis. Adv. Enzymol. 1: 203–328.

DE VASCONCELOS, L, AND P. FAY. 1974. Nitrogen metabolism and ultrastructure in *Anabaena cylindrica*. I. The effect of nitrogen starvation. Arch. Microbiol. 96: 271–279.

VIERLING, E., AND R. S. ALBERTE. 1980. Functional organization and plasiticity of the photosynthetic unit of the cyanobacterium *Anacystis nidulans*. Physiol. Plant. 50: 93–98.

VOLOKITA, M. ET AL. 1984. Nature of the inorganic carbon species actively taken up by the cyanobacterium *Anabaena variabilis*. Plant Physiol. 76: 599–602.

WAALAND, J. R., S. D. WAALAND, AND D. BRANTON. 1970. Gas vacuoles: light shielding in blue-green algae. J. Cell Biol. 48: 212–215.

WALSBY, A. E. 1971. The pressure relationships of gas vacuoles. Proc. R. Soc. Lond. B 178: 301–326.

———. 1972. Structure and function of gas vacuoles. Bact. Rev. 361: 1–32.

———. 1974. The extracellular products of *Anabaena cylindrica* Lemm. 11. Fluorescent substances containing serine and threonine, and their role in extracellular pigment formation. Br. Phycol. J. 9: 383–391.

WANG, R. T., C. L. R. STEVENS, AND J. MYERS. 1977. Action spectra for photoreactions I & II of photosynthesis in the blue-green alga *Anacystic nidulans*. Photochem. Photobiol. 25: 103–108.

WATERBURY, J. B., AND R. Y. STANIER. 1978. Patterns of growth and development in pleurocapsalean cyanobacteria. Microbiol. Rev. 42: 2–44.

——— 1981. Isolation and growth of cyanobacteria from marine and hypersaline environments. *In* M. P. Starr, H. Stolp, A. Balous, and H. G. Schlegel [ed.] The prokaryotes. Springer-Verlag, Berlin.

WATERBURY, J. B. ET AL. 1979. Widespread occurrence of a unicellular, marine, planktonic, cyanobacterium. Nature 277: 293–294.

WHALE, G. F., AND A. E. WALSBY. 1984. Motility of the cyanobacterium *Microcoleus chtnoplastes* in mud. Br. Phycol 19: 117–123.

WHEELER, P. A. 1983. Phytoplankton nitrogen metabolism. *In* E. Carpenter [ed.] Nitrogen in the marine environment. Academic Press, New York, NY.

WHITTON, B. A. 1965. Extracellular products of blue-green algae. J. Gen. Microbiol. 40: 1–11.

WHITTON, B. A., N. G. CARR, AND I. CRAIG. 1971. A comparison of the fine structure and nucleic acid biochemistry of chloroplasts and blue-green algae. Protoplasma 72: 325–357.

WHITTON, B. A., AND M. POTTS. 1982. Marine littoral, p. 515–542. *In* N. G. Carr and B. A. Whitton [ed.] The biology of cyanobacteria. Blackwells Scientific Publications, Oxford.

WOLK, C. P. 1982. Heterocysts. *In* N. G. Carr and B. A. Whitton [ed.] The biology of cyanobacteria. Blackwells Scientific Publications, Oxford.

WOOD, N. B., AND R. HASELKORN. 1980. Control of phycobiliprotein proteolysis and heterocyst differentiation in *Anabaena*. J. Bact. 141: 1375–1385.

WYATT, J. T., AND J. A. G. SILVEY. 1969. Nitrogen fixation by *Gloeocapsa*. Science 165: 908–909.

WYMAN, M., AND P. FAY. 1986a. Underwater light climate and the growth and pigmentation of planktonic blue-green algae (Cyanobacteria). I. The influence of light quantity. Proc. R. Soc. Lond. B. 227: 367–380.

——— 1986b. Underwater light climate and the growth and pigmentation of planktonic blue-green algae (cyanobacteria). II. The influence of light quality. Proc. R. Soc. Lond. B. 227: 351–393.

WYMAN, M., R. P. F. GREGORY, AND N. G. CARR. 1985. Novel role for phycoerythrin in the marine cyanobacterium *Synechococcus* sp DC2. Science 230: 818–820.

WYMER, P. E. O., AND B. THAKE. 1980. The importance of phosphorus in microalgal growth and species composition in mixed populations: experiments and stimulations. Proc. R. Soc. Lond. B 209: 333–353.

ZEHNDER, A., AND B. EGLI. 1977. Zur resistenz von Blaualgen gegen kurzwellige Strahlung. Schweiz Z. Hydrol. 39: 167–177.

ZEVENBOOM, W., AND L. R. MUR. 1978. N-uptake and pigmentation of N-limited chemostat cultures and natural populations of *Oscillatoria agardhii*. Mitt. Internat. Verein. Limnol. 21: 261–274.

——— 1981a. Ammonium-limited growth and uptake by *Oscillatoria agardhii* in chemostat cultures. Arch. Microbiol. 129: 61–66.

——— 1981b. Simultaneous short-term uptake of nitrate and ammonium by *Oscillatoria agardhii* grown in nitrate- or light-limited continuous culture. J. Gen. Microbiol. 126: 355–363.

Nutrition and Culture of Autotrophic Ultraplankton and Picoplankton

LARRY E. BRAND

*Rosenstiel School of Marine and Atmospheric Science,
University of Miami, 4600 Rickenbacker Causeway, Miami, Florida 33149, USA*

Introduction

Pure laboratory cultures are essential for studying many aspects of the biology of autotrophic ultraplankton and picoplankton. In general, the strategy for isolating phytoplankton species into culture and maintaining them is to recreate their natural environment as well as possible in the laboratory. The inability to culture a particular phytoplankton species is usually attributed to the fact that the laboratory environment is quite different from the natural marine habitat in which the species lives. Despite all the ways phytoplankton culturists may try to recreate the natural environment, there are two ways in which their goals are precisely the opposite. First, for a variety of reasons, an axenic or at least unialgal culture is usually desired. Thus the species in culture is isolated from the rest of the community in which it normally lives. The variety of biotic interactions that can exist among species may be critically important to the health and survival of some phytoplankton. Secondly, high biomass is usually desired in laboratory cultures, making various measurements much easier and reducing the need for excessively large cultures. Laboratory cultures often have cell concentrations a thousand to ten thousand times higher than those found in the natural environment. Such high cell concentrations can have a large effect on the chemical composition of seawater. Autoinhibitory substances that normally remain quite dilute in the natural environment may accumulate to high concentrations in such dense cultures. The production of high biomass also requires that high concentrations of nutrients be added to the culture media. Because the goal of culturing is usually to obtain high cell concentrations of one isolated species, the environment experienced by phytoplankton cells in the ocean cannot be duplicated in the laboratory completely. We must recognize that we are necessarily working with an artificial system and that this may prevent us from successfully isolating some species into culture.

Brief History of Culturing Methods

The history of the culturing of marine phytoplankton is largely one of adding and withdrawing various substances to and from seawater. There has been a general trend toward the use of lower nutrient concentrations and more well-defined and reproducible culture media as the specific nutrients needed become known. The first successful culture of marine phytoplankton was in 1890 by P. Miguel (Provasoli et al. 1957). Nitrate, phosphate and iron were the only nutrients added to natural seawater. The success of these cultures was undoubtedly due to the high concentrations of all the other necessary trace substances being present in the coastal seawater used in the media and to the broad environmental tolerance of the neritic species cultured. In 1934, B. Foyn found soil extracts to be greatly beneficial to the growth of marine phytoplankton (Provasoli et al. 1957). We now know that soil extract performs a number of functions in culture media and it has been replaced, to a large extent, by specific compounds. Soil extract provides various trace elements and vitamins needed for plant growth, metal complexing organic compounds that sequester potentially toxic metals, and organic compounds that keep iron in solution and allow for its photoreduction. In replacing soil extracts, a number of trace elements and vitamins are now usually added to culture media. These include iron, manganese, zinc, cobalt, copper, molybdenum, vitamin B12, thiamine and biotin. Artificial chelators such as ethylenediaminetetraacetic acid (EDTA) are added to keep iron in solution and keep free ionic metal concentrations at nontoxic levels.

Other early efforts went into reducing or eliminating precipitates in the culture media. Autoclaving drives carbon dioxide out of the seawater, causing a shift in the carbonate buffer system. The resulting pH of around 10 causes the precipitation of ferric phosphate and ferric hydroxides. Many culture media contain strong pH buffers such as tris(hydroxymethyl)aminomethane (TRIS), metal chelators such as nitrilotriacetic acid (NTA) or EDTA, and organic phosphate compounds such as glycerophosphate to reduce such precipitates (Provasoli et al. 1957).

Another approach to making culture media is to use totally artificial seawater rather than enriching natural seawater. This is necessary if one does not have access to clean seawater or if the presence of the rather high concentration (around 1 mg/L) of the undefined mixture of organic substances (mostly humic and fulvic acids) found in seawater (Williams 1975; Mopper and Degens 1977) is not desired. One of the most successful artificial medias is ESAW developed by Harrison et al. (1980). Although one would expect such media to be much more chemically defined than natural seawater, in some respects they are not. Because of the large amounts of major salts that must be added, trace contaminants in them can result in higher concentrations of some metals in the artificial seawater than those naturally present in oceanic surface water. The use of ion exchange columns to remove these trace contaminants, as described by Morel et al. (1979), can greatly reduce this problem, but it is a time-consuming technique and unreliable if not done carefully.

In the last decade, chemists have discovered that contamination problems have resulted in large overestimates of the abundance of many trace metals in the ocean (Boyle and Edmond 1975; Bruland et al. 1978, 1979). This has resulted in a reevaluation of the role of trace metals in phytoplankton nutrition and a greater awareness by biologists of trace metal contamination and its potential toxicity (Morel and Morel-Laurens 1983; Carpenter and Lively 1980; Fitzwater et al. 1982). Some (Brand et al. 1981) now autoclave seawater in teflon bottles rather than borosilicate glass flasks to avoid the dissolution of the silica and its associated trace metal contaminants at the high pH generated during the autoclaving process. For the culture of some species, it is preferable to use cleanly-collected oceanic surface water rather than coastal water if possible, as it contains much lower concentrations of many trace metals as well as other anthropogenic pollutants. Bioassays indicate that substantial amounts of trace metals can enter the culture media from autoclave steam (Morel et al. 1979; Brand et al. 1983), so pasteurization of oceanic water in teflon bottles rather than autoclaving is recommended if extremely low trace metal levels are desired.

If we examine some of the most commonly used media for culturing marine phytoplankton (Table 1), we observe no significant differences among them. The major differences are in the concentrations and in the addition of certain compounds that are already present in seawater in excess quantity. At the present time, there is no strong evidence that there is a substantial difference among these media in their ability to culture phytoplankton species. Some other culture media (listed by Guillard and Keller 1984) also have a variety of additional vitamins and organics added to seawater, such as niacin, putrescine, pantothenic acid, riboflavin, pyridoxine, pyridox-amine, para-aminobenzoic acid, choline, inositol, orotic acid, folinic acid, folic acid and adenine, but there is no evidence they are needed.

The first phytoplankton species brought into culture were coastal and estuarine diatoms and other fast growing species (Provasoli et al. 1957). Such cultures were usually established from enrichments, thus there was a strong selection for fast growing "weed" species that can tolerate a wide range of environmental conditions. Enrichment cultures are essentially useless for isolating slow-growing phytoplankton because they are always overgrown by fast growing species. The most successful way of culturing slower growing phytoplankton is by single cell isolation with a micropipet

TABLE 1. Molar additions to natural seawater for standard culture media.

	GPM[a]	IMR[b]	f/2[c]	BWM[d]
NO_3	2×10^{-3}	5×10^{-4}	8.8×10^{-4}	10^{-4}
NH_4			(10^{-4})	10^{-5}
PO_4	2×10^{-4}	5×10^{-5}	3.6×10^{-5}	10^{-5}
Si		5×10^{-4}	10^{-4}	10^{-4}
EDTA	1.3×10^{-4}	1.6×10^{-5}	1.2×10^{-5}	10^{-5}
Fe-EDTA	5×10^{-6}	3.7×10^{-6}	1.17×10^{-5}	10^{-6}
Zn	2.3×19^{-6}	8.7×10^{-8}	8×10^{-8}	10^{-7}
Mn	2×10^{-5}	3.6×10^{-6}	9×10^{-7}	10^{-7}
Co	9×10^{-7}	3.1×10^{-8}	5×10^{-8}	10^{-8}
Cu		2.5×10^{-8}	4×10^{-8}	10^{-9}
Mo		5.4×10^{-7}	3×10^{-8}	
Vit.B12	7.4×10^{-6}	7.4×10^{-9}	3.7×10^{-10}	10^{-8}
Thiamine	3×10^{-5}	6×10^{-7}	3×10^{-7}	10^{-7}
Biotin	4×10^{-9}	4×10^{-9}	2×10^{-9}	10^{-9}
BO_3	5.5×10^{-4}			

[a]Loeblich (1975).
[b]Eppley et al. (1967).
[c]Guillard (1975)
[d]Brand, unpublished data.

immediately at sea without any enrichment. This technique allows one to focus on particular species of interest and record the success rate. It has also allowed the isolation into culture of some of the dominant oceanic species that could not be obtained from enrichments. For autotrophic picoplankton that are too small to see under an isolating microscope and difficult to distinguish from heterotrophic bacteria and flagellates, dilution cultures that result statistically in either one or no cell in each culture vessel can be used. One of the problems with single cell isolations is that the single cell initially placed in an isolation tube is quite susceptible to the preferential partitioning of various toxic substances from a relatively large volume of culture media into the cell's protoplasm. Unlike heavy inoculums, in which large numbers of cells "share" the toxic substances present, a single isolated cell probably tends to absorb the entire amount. This may be the explanation for the lag sometimes seen with small inoculations. Given the much smaller volume of picoplankton, one might expect the single cell isolation of them to be even more difficult and sensitive to trace amounts of toxic substances in the media. A 1 μm cell could end up with a toxic substance a thousand times more concentrated in its protoplasm than a 10 μm cell. The use of clean techniques and the use of teflon bottles in media preparation for single cell isolations has resulted in the successful isolation of more oceanic species than before.

Despite the progress and all the successes up to this time in culturing phytoplankton and the fact that all the different media used today by different people can culture many phytoplankton species, most marine phytoplankton species remain uncultured. Clearly, more work is needed. The types of species we can and cannot culture should give us a clue as to what is wrong with our current techniques.

Biases in Culture Collections

If we examine the types of species that are now in culture and those that no one has been able to culture, we find two general trends — a habitat related one and a phylogenetic one. A very high proportion of estuarine and coastal species can now be cultured using standard techniques. By contrast, very few oceanic species have been brought into culture, and many of these isolations were successful only after the use of clean techniques to reduce the amount of toxic substances in the media. The second trend is a phylogenetic one, with diatoms among the easiest phytoplankton to culture and the dinoflagellates notorious for being some of the most difficult. Although most coastal diatoms and some oceanic diatoms such as *Hemiaulus hauckii* and *Planktoniella sol* are now in culture, many dinoflagellates, even coastal species, are not yet in culture. Even among oceanic species, there is a clear habitat trend in which species can be cultured. This is best seen among the oceanic coccolithophores. *Emiliania huxleyi*, *Gephyrocapsa oceanica*, *Cyclococcolithus leptoporus*, *Umbilicosphaera sibogae*, *U. hulburtiania* and *Crenalithus sessilis* are in culture, but species such as *Calyptrosphaera oblonga*, *Discosphaera tubifera*, *Rhabdosphaera stylifera*, *Umbilicosphaera irregularis* and *U. tenuis* have never been cultured. If we examine the data of Hulburt (1964, 1966, 1967, 1976, 1985), we find that the first group is more abundant in more nutrient rich waters, such as along the edges of the central gyres, in the northern rather than southern Sargasso Sea, and during spring blooms. The second group does not respond to enrichment and is more abundant in the more strongly stratified, nutrient poor waters of the southern Sargasso Sea. This is our challenge of the future — to culture the vast majority of phytoplankton species in the oceanic environment that do not respond to enrichment either naturally in the ocean or in the laboratory.

Phytoplankton groups such as diatoms, coccolithophores and dinoflagellates with relatively large cells, distinctive morphologies and external hard parts can be identified to species at sea with relative ease. This allows one to then determine what fraction of the species in these groups can be cultured using current techniques. We therefore know we cannot culture many species of phytoplankton, especially oceanic dinoflagellates and coccolithophores. With picoplankton and ultraplankton, however, we really have no idea what fraction of the species we have cultured because we cannot identify the species in the field (usually only with great difficulty even in the laboratory). Are the picoplankton and ultraplankton species we presently have in culture a reasonable representation of the ones that live in the ocean? A theoretical consideration of the wide range of types of phytoplankton in the sea and our ability to culture the different types leads one to suspect we have not been very successful at culturing the oceanic picoplankton and ultraplankton.

The argument is based on the premise that the phylogeny of a species is very important in determining its environmental requirements and thus our ability to culture it. To evaluate this influence in a simplistic way, types of phytoplankton and types of habitats have been placed along a hypothetical one-dimensional ecological gradient (Table 2). At one extreme are the fast growing *r*-selected species and the highly variable, nutrient rich environments in which they live. At the other extreme are the *K*-selected species and their oligotrophic, rather predictable environments. Some of the differences listed are well-documented. Some are overgeneralizations, some are highly speculative.

It would appear that the majority of marine cyanobacteria, along with dinoflagellates, would be classified as *K*-selected species, especially those living in oligotrophic waters. They have slower reproduction rates than diatoms (Brand and Guillard 1981) and one suspects they are generally able to grow well at lower nutrient concentrations and have a better competitive ability than diatoms, although this has not yet been adequately tested. Cyanobacteria are a much more important fraction of the phytoplankton community in the warmer, more stratified waters of the subtropical central gyres than in more polar waters or over continental shelves (Murphy and Haugen 1985; Glover 1985). They are also more abundant in the summer than in the winter in temperate regions (J. Waterbury, pers. comm.). In a general way, cyanobacteria have an ecological distribution similar to that of the dinoflagellates as a group and opposite to that of the diatoms, which generally dominate in cold, turbulent, nutrient-rich waters of the polar regions and continental shelves. Diatoms

TABLE 2. Endmembers of an ecological gradient.

Habitats
 eutrophic | oligotrophic
 vertically mixed | stratified
 cold, nutrient rich | warm, nutrient poor
 coastal | oceanic
 polar | tropical
 winter | summer
 early succession | late succession
 water modified little by phytoplankton | water modified substantially by phytoplankton

Organisms
 r-selected | *K*-selected
 rapid growth rate | slow growth rate
 wide environmental tolerance | narrow environmental tolerance
 poor competitive ability | good competitive ability
 highly autotrophic | tendency toward auxotrophy, mixotrophy, photoheterotrophy, amphitrophy
 diatoms | cyanobacteria and dinoflagellates
 evolved only recently | evolved early
 high biomass communities | low biomass communities
 low diversity communities | high diversity communities
 rarely involved in tight symbiotic relationships | often involved in tight symbiotic relationships

Culture Media
 GPM, IMR, f/2, BWM, ESAW | ???

generally initiate seasonal succession in late winter in nutrient rich water that has been below the photic zone for a significant period of time and has been little influenced by the phytoplankton community (Smayda 1980). As seasonal succession proceeds, stratification strengthens, nutrients are depleted, the phytoplankton modify the water by not only taking up inorganic nutrients, but also excreting organic compounds. The diatoms decline in abundance as the cyanobacteria and dinoflagellates become dominant. The cyanobacteria and dinoflagellate remain dominant until stratification starts to break down and inorganic nutrient concentrations in the photic zone increase. It has been hypothesized that the use of organic nutrients becomes more important than the use of inorganic nutrients late in succession as well as in more oceanic waters (Butler et al. 1979; Bonin and Maestrini 1981; Jackson and Williams 1985). If one ignores the benthic pennate diatoms and considers only the plankton, it appears that a higher fraction of cyanobacteria and dinoflagellate species may be capable of utilizing organic matter for energy and nutrients (Wolk 1973; Smith 1982; Droop 1974). Similarly, cyanobacteria and dinoflagellates are involved in the vast majority of close symbiotic relationships that have been found between algae and other organisms (Taylor 1980). Another observation that may be of significance is that the cyanobacteria and dinoflagellates were the earliest groups of phytoplankton to evolve and the diatoms were the most recent group to evolve (Tappan 1980; Lipps 1970). From a macroevolutionary point of view, it appears that new groups of phytoplankton invade the nutrient rich regions of the ocean, forcing older groups into the more oligotrophic habitats. As a result of living in oligotrophic waters, most marine cyanobacteria and dinoflagellates live in plankton communities with low biomass and high species diversity.

Clearly there are many exceptions to the generalizations presented in Table 2 (as well as considerable speculation), but one would expect the environmental requirements of oceanic cyanobacteria to be different from those of coastal diatoms. We should not be too surprised if methods for culturing one group do not work for the other. To date, our culture techniques have been quite successful in culturing the r-selected species from eutrophic habitats, but quite poor at culturing the K-selected species from oligotrophic habitats. This leads one to suspect that there has been little success in the culturing of the primitive oceanic cyanobacteria, as has been the case for oceanic dinoflagellates. One would expect a high diversity of cyanobacteria in the open ocean, yet only a few "types" have been cultured to date. A similar argument can be made with the eukaryotic ultraplankton. As with the oceanic cyanobacteria, the systematics of the tiny eukaryotic algae that populate the open ocean is still quite primitive. Although we know little of their phylogeny, one also suspects only a small fraction of the eukaryotic picoplankton and ultraplankton from the open ocean has been cultured. The picoplankton and ultraplankton that have been cultured from the open ocean are probably not representative dominants from the community, but rather "weed" species that bloom suddenly upon enrichment. This is the result of a biological gradient not shown in Table 2. That is the diversity of ecological strategies of species within a habitat. Within any habitat or water mass, there exists a spectrum of species with strategies spanning the range shown in Table 2. Some species have very high growth rates and respond quickly to occasional nutrient enrichment while slow growing species can dominate by virtue of their competitive abilities. These slow growing species do not respond much to enrichment. Thus oligotrophic waters can be dominated by slow-growing cyanobacteria, ultraplankton and dinoflagellates, but there can still be some fast-growing species present that will respond rapidly to enrichment. It is hypothesized that most of the species that have been isolated from oceanic waters are among this minority of fast-growing species. If so, then the oceanic picoplankton and ultraplankton we presently have in culture are not truly representative of the dominant picoplankton and ultraplankton in the open ocean.

To the extent that our ideas on phytoplankton ecology are based upon the observation of laboratory cultures, our perspective may be strongly modified by this culture-collection bias. The species in culture are not representative of the wide range of species found in the ocean — they represent one end of the spectrum. The question we now face is how to culture phytoplankton species at the other end of the spectrum — the K-selected primitive species that live in oligotrophic waters. A radically different type of culture medium may be needed to grow picoplankton and ultraplankton at this end of the spectrum. The problem with designing new culture media is that one is always working backwards. To culture a species, one must know its environmental requirements. But to determine the environmental requirements of a species, one must study it in culture. We are therefore forced initially to use trial and error methods to obtain into culture species never before cultured.

Several of the trends observed in Table 2 may provide clues to what is wrong with current culturing techniques and perhaps can guide us in making appropriate changes. If the generalizations made above about r- and K-selected phytoplankton species are true, culture media for oceanic picoplankton should be designed for species adapted to very low nutrient concentrations and lower ratios of inorganic to organic nutrients. They are also adapted for living in a community of competing species with high diversity but low biomass. The predictability and persistence of this community makes symbiotic or consortial dependencies more likely. It may also be of significance that the cyanobacteria evolved when the biosphere was still anaerobic (Schopf 1970;

Tappan 1980; Tappan and Loeblich 1973). When they evolved, they had to have adaptations for a very different type of seawater chemistry and redox potential (Williams 1981). The accumulation of oxygen in the biosphere caused a major biotic crisis and cyanobacteria had to evolve a variety of secondary adaptations to deal with this environmental challenge (Chapman and Schopf 1983). This is best seen in the oxygen sensitivity of ribulose bisphosphate carboxylase and nitrogenase and the variety of mechanisms plants use today in an aerobic environment to be able to continue using them (Chapman and Schopf 1983; Raven and Beardall 1981; Kremer 1981; Danks et al. 1983; Fogg 1974; Postgate 1982). Phyletic inertia results in many relict characteristics and often forces groups of organisms with different phylogenetic histories to respond to the same environmental challenge in quite different ways (Gould and Lewontin 1979; Lewontin 1978). Such phylogenetic differences in environmental tolerances and requirements should be kept in mind when attempting to culture organisms that may have relict characteristics that evolved under a totally different set of environmental circumstances. Over half of the evolutionary history of cyanobacteria has been in an anaerobic environment. Such relict characteristics may provide a less than optimum means of functioning in an aerobic world, but the biochemistry of organisms today necessarily reflects the historical development of the biosphere and the transition from an anaerobic environment to an aerobic one (Chapman and Schopf 1983). One might expect these relicts to be observed in some of the mechanisms by which phytoplankton interact with their environment in the ocean or in the laboratory.

Another observation worth noting in attempting to design new culture media is that most oceanic species and most dinoflagellates and cyanobacteria do not attain the very high population densities that coastal diatoms are able to achieve in culture. This may be the result of the depletion of some trace nutrient we are unaware of or the accumulation in the culture of toxic compounds produced by the unnaturally high phytoplankton biomass.

To achieve high biomass levels in culture, high concentrations of nutrients are usually added to culture media. One must ask if these are high enough to be toxic, even though they are needed as nutrients. It is not hard to see how sensitivity to high concentrations of nutrients might evolve. It is quite possible that many oceanic species that experience only very low concentrations of nutrients do not have adequate feed-back controls for handling the very high concentrations of nutrients found in culture media. Good biochemical control depends on substrates being in the linear region of the saturation curve for uptake where concentrations affect the rate. One expects much less control in the plateau region, especially if the substrate is orders of magnitude higher than the half-saturation constant. One might expect genetic drift to lead to oceanic organisms with little feedback control at extremely high substrate concentrations. As illustrated in Fig. 1A, the lack of selection at high concentrations will lead from response C to response A. The range of environmental concentrations considered obviously must include patchiness even on the microscale. How high the concentration of nutrients in zooplankton generated patches can be is a matter of debate at the present time, but the calculations of Jackson (1980) indicate they are not very high for very long. Therefore, it seems plausible that oceanic phytoplankton may not have evolved adequate feedback mechanisms for tolerating extremely high substrate concentrations. If there are energetic and/or nutritional savings in not main-taining the excess feedback mechanisms needed to generate response C in Fig. 1A, then there is positive selection to eliminate the feedback mechanism at high concen-trations and the suite of responses would be as in Fig. 1B. Although one can envision how genetic drift and selection can lead to the loss of feedback control at high nutrient concentrations, there must also be a specific biochemical mechanism leading to the toxicity effects. Some possibilities will be discussed when considering the different nutrients added to culture media.

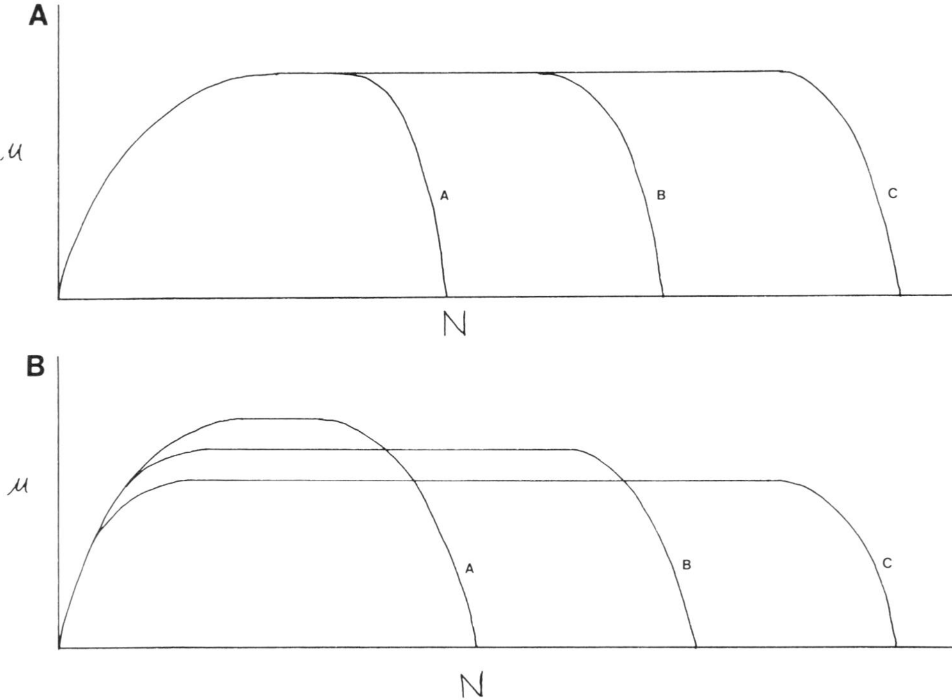

FIG 1. Schematic relationship between growth rate and nutrient concentration showing inhibition at high concentrations. (A) No pleiotropic interactions. (B) Pleiotropic interaction resulting from energetic and nutritional costs of feedback mechanisms.

Inorganic Nutrients

When the cyanobacteria evolved, nitrogen was available in the anaerobic environment in the form of molecular dinitrogen and ammonia. Nitrate and nitrite would not have been present in any significant quantities. It is thought that photodestruction of ammonia and the sedimentary burial of organic nitrogen compounds in the anaerobic environment led to the depletion of the primordial ammonia, leaving only dinitrogen (Schopf 1978; Postgate 1982). This is presumably what led to the evolution of nitrogen fixation. Cyanobacteria are in fact the only plants which are capable of utilizing dinitrogen directly. This mechanism also reflects its history, the nitrogenase being extremely sensitive to molecular oxygen. Even today, the local microenvironment of nitrogenase must be anaerobic for nitrogen fixation to occur (Fogg 1974; Bothe 1982). The ability of plants to take up and reduce nitrate and nitrite probably did not evolve until after oxygen accumulated to a significant degree in the biosphere and chemoautotrophic organisms could produce nitrite and nitrate in the environment (Chapman and Schopf 1983). The mechanisms for taking up and utilizing ammonium and dinitrogen are the primary adaptations cyanobacteria have for obtaining nitrogen. The ability to use nitrate and nitrite are undoubtedly secondary adaptations that evolved 1–2 billion years later. Given that nitrate and nitrite were not important nitrogen sources for the cyanobacteria when they evolved and that they are not very important in the oligotrophic waters of today's ocean because of the rapid recycling of ammonium (McCarthy 1980), many of the cyanobacteria and other picoplankton in the open ocean might depend totally upon only reduced forms of nitrogen. At present there is no experimental evidence for this with laboratory cultures, but then most of the species in culture have been obtained with media containing nitrate, not ammonium, as the nitrogen source. The enrichments used have biased culture collections towards nitrate users. It seems prudent to use small amounts of ammonium in culture media, especially for oceanic species.

It is well known that ammonia can be toxic at high concentrations because of its high permeability through the cell membrane. In multicellular plants, this toxicity is controlled by the rapid incorporation of the incoming ammonia into organic molecules (Givan 1979). Low surface area available for ammonia diffusion and high metabolic rates are advantageous in reducing this toxicity, so multicellular plants are preadapted well for solving this problem. The larger, faster growing coastal phytoplankton that dominate culture collections are probably also reasonably preadapted to tolerate high ammonia concentrations. Slow-growing oceanic picoplankton with high surface to volume ratios, on the other hand, may be particularly sensitive to ammonia toxicity, especially if they are well adapted for taking up the extremely low concentrations found in the open ocean. While many species of phytoplankton already in culture can tolerate 10^{-3} M ammonium and most can tolerate 10^{-4} M, a few species have been shown to not tolerate 10^{-4} M (McLachlan 1973; Antia and Chorney 1968; Abeliovich and Azov 1976). It seems prudent to use 10^{-5} M or less in attempting to culture oceanic picoplankton to avoid any toxic effects. Even 10^{-5} M ammonium can generate easily detectable biomass levels.

High concentrations of nitrate and nitrite are also known to be toxic to some organisms. This appears to be the result of their strong oxidizing characteristics. For example, in animals nitrate and nitrite oxidize ferrous iron in hemoglobin (Emerick 1974). One wonders if excess nitrate in the cell might oxidize the ferrodoxins or other highly reduced compounds in cells. The primitive algae that evolved under anaerobic conditions may be particularly sensitive to such oxidation. Such internal oxidation could occur if oceanic phytoplankton adapted for nitrate concentrations as low as 2×10^{-8} M (Garside 1985) do not have an adequate feedback mechanism for preventing the uptake of excess nitrate in culture media with nitrate concentrations of 10^{-4} to 10^{-3} M. One way to avoid high concentrations of nitrate and ammonium, yet still provide adequate nitrogen would be to use organic nitrogen compounds, as most phytoplankton species have been shown to be able to use them (Fisher and Cowdell 1982; Antia et al. 1975; Antia et al. 1980).

Phosphate concentrations in the ocean were probably not appreciably different in the anaerobic ocean from what they are today (Ochiai 1978). One can conceive of high phosphate concentrations inhibiting the growth of phytoplankton if they competitively inhibit the uptake of arsenate and vanadate and such compounds are nutritional requirements of algae (discussed below). Otherwise, our method of adding phosphate to culture media would not seem to be a problem.

It is thought that silicate was near saturating concentrations until the Jurassic, when diatoms first appeared in great numbers in the fossil record and radiolarians greatly increased in abundance and diversity (Holland 1972; Holland and Schidlowski 1982). There is no indication or reason to believe that the addition of higher concentrations of silicate to culture media would be beneficial. A small amount should always be added, however, because some non-diatoms have been demonstrated to need some silicate (Klaveness and Guillard 1975).

Iron in seawater undoubtedly underwent a dramatic change as the biosphere became aerobic (Lewin 1984). In today's aerobic ocean, iron is in the oxidized form as various ferric hydroxides and thus is rather insoluble in seawater, with total concentrations around $1-6 \times 10^{-10}$ M in the open ocean (Gordon et al. 1982; Symes and Kester 1985). Under the anaerobic conditions that existed when the cyanobacteria evolved, iron was present in the reduced ferrous state and thus much more soluble than today, perhaps as high as 10^{-3} M (Ochiai 1978; Osterberg 1974). Probably at least partially as a result of its high availability at the time that most basic metabolic pathways evolved, iron is the most important metal used in electron transport and other biological redox reactions (Ochiai 1978; Egami 1975; Lewin 1984). The ferrodoxins and other iron-sulfur proteins are the most highly reduced compounds in organisms (Bowen 1966) and probably reflect the highly reduced environment the primitive organisms lived in when the basic electron transport systems were evolving (Yock and Carithers 1979). The transition to an oxidized environment led to a great reduction in the amount of iron available to plants because of the insolubility of the various ferric hydroxides. An extension of Alfred Redfield's concepts (Redfield 1958; Redfield et al. 1963) to trace metals suggests that iron may now be the ultimate geochemically limiting nutrient to phytoplankton in the ocean. A comparison of the elemental composition of plankton with the nutrient concentrations found in deep water indicates that the ratios of nitrogen, phosphorus, zinc and manganese in the deep water are similar to what are needed to produce plankton biomass, but there is proportionally twenty times less iron available in deep water than is apparently needed (Table 3). Such a comparison is oversimplified for a number of reasons (detrital interference, luxury uptake, unrealized optimum nutrient ratios, recycling and shallow upwelling), but it does indicate that not only the chemical form but also the quantity of iron presents a problem to oceanic phytoplankton. The adaptations phytoplankton have for solving it are not known. The exact forms of iron that are available to plants in the photic zone are still not clear (Lewin and Chen 1971; Ahrland 1985; Bruland 1983; Sugimura et al. 1978; Byrne and Kester 1976; Anderson and Morel 1982; Waite and Morel 1984; Wells et al. 1983; Zafiriou and True 1980). Experiments have indicated two different mechanisms by which phytoplankton may obtain iron from seawater. Murphy et al. (1976), Simpson and Neilands (1976), Armstrong and Van Baalen (1979) and Trick et al. (1983a,b) have demonstrated the secretion of iron binding siderophores by phytoplankton. The data of Trick et al. (1983b) indicate that

TABLE 3. Molar ratios of nutrients in plankton and deep water.

Plankton elemental composition (molar ratios)	N	P	Fe	Zn	Mn
Redfield (1958)	16	1			
Collier and Edmond (1984) normalized to P as 1 (collected with 44 micron net)		1	10^{-2}	4×10^{-3}	4×10^{-4}

Dissolved nutrients in deep water	N	P	Fe	Zn	Mn
Redfield (1958) Molar ratio	15	1			
Collier and Edmond (1984)					
ave. Atlantic Conc. (M)		1.4×10^{-6}		1.6×10^{-9}	1.8×10^{-9}
ave. Pacific (M)		2.8×10^{-6}		8.2×10^{-9}	6×10^{-10}
Symes and Kester (1985)					
Atlantic conc. (M)			10^{-9}		
Gordon et al. (1982)					
Pacific conc. (M)			10^{-9}		
ave. conc. for whole ocean (M)		2.5×10^{-6}	10^{-9}	7×10^{-9}	8.5×10^{-10}
Molar ratio (ave. for whole ocean; normalized to P as 1)	15	1	4×10^{-4}	2.8×10^{-3}	3.4×10^{-4}

prokaryotic algae produce much higher concentrations of siderophores in culture than do eukaryotic algae. It is thought that these siderophores not only bind iron to be taken up by the cells, but may also act as sequestering agents to starve competing phytoplankton species of iron (Murphy et al. 1976). Anderson and Morel (1980, 1982) have demonstrated that diatoms utilize ferrous ions generated by photochemical reactions. Thus, there are indications of at least two different mechanisms by which phytoplankton obtain iron from seawater. This could have implications for the techniques needed to culture the different groups of algae. Perhaps the prokaryotic species that produce large amounts of siderophores to obtain iron in dilute seawater starve themselves of iron by producing extremely high concentrations of iron sequestering siderophores in dense cultures. Siderophores that only starve competing species in nature may starve the producing organism as well when the cell and siderophore concentrations in culture are a thousand times higher than in natural waters. Another possibility is that the humic and fulvic substances are extremely important for the proper speciation of iron for it to be available to the cells (Sugimura et al. 1978; Waite and Morel 1984; Finden et al. 1984; Theis and Signer 1974). Is it possible the process of sterilizing the seawater (or using artificial seawater) alters the chemical structure of the organic molecules to the extent that they no longer complex (or photoreduce) the iron properly? Despite a great deal of research, very little is still known about iron in seawater because of its complex chemistry. Much remains to be learned about the chemistry of iron in seawater and the mechanisms by which phytoplankton obtain it. It does appear that phytoplankton do differ in their ability to take up low levels of iron from seawater (Brand et al. 1983), with coastal species more easily limited by low concentrations of iron than oceanic species, and oceanic cyanobacteria more easily limited than oceanic eukaryotic phytoplankton. Excess internal iron is known to complex ATP and result in cellular inhibition (Frausto da Silva 1978; Bowen 1966; May et al. 1978), so we should perhaps also worry about making too much iron available to the cells. Currently, high concentrations of ferric iron are kept in solution in most culture media in the form of a ferric EDTA complex. The usefulness of EDTA may not only be to keep iron in solution, but also photoreduce Fe(III) to Fe(II). Whether weaker chelators more similar to the natural organic compounds that complex iron in seawater would be better for cultures is not known, but their use may be worth trying.

Manganese is another element that decreased dramatically in abundance in seawater as the ocean became aerobic. It is estimated that it was present at around 10^{-5} M in the primitive anaerobic ocean (Ochiai 1978), whereas today it is around 10^{-9} M in oceanic surface waters (Bender et al. 1977; Klinkhammer and Bender 1980; Bruland and Franks 1983; Burton et al. 1983). Like iron, manganese is also a very important trace metal found in many of the basic enzymes that evolved early when the earth was still anaerobic. One might expect the algae that evolved when manganese was much more abundant to have more difficulties obtaining today's low concentrations from seawater. Indeed, the one oceanic cyanobacterium that has been examined for ability to grow in low manganese water is much less able to utilize low concentrations than oceanic diatoms and coccolithophores (Brand et al. 1983). Manganese has a rather unusual vertical profile for a nutrient element, in that the concentrations in the photic zone are higher than in deep water. This is apparently the result of the photochemical production of the more soluble Mn(II) from the highly insoluble Mn(IV) (Sunda et al. 1983). As a result, upwelling, which greatly increases the availability of most nutrients, actually decreases the availability of manganese. It may be of significance that cyanobacteria and dinoflagellates do not proliferate in upwelled waters. The provision of manganese to phytoplankton in culture is thought not to be a problem, as Mn(II) oxidizes to Mn(IV) very slowly and even quite high concentrations are not toxic.

Unlike iron and manganese which decreased in concentration as seawater became aerobic, zinc and copper became more soluble and increased in concentration

(Osterberg 1974; Ochiai 1978). The biological effects of this shift are particularly well demonstrated in superoxide dismutase. All prokaryotes, algae, and mitochondria have iron or manganese in the enzyme, whereas all higher plants and other eukaryotes have copper and zinc in superoxide dismutase (Asada et al. 1977; Lumsden and Hall 1975; Chapman and Schopf 1983; Bannister and Rotilio 1984). This shift reflects the shift in the chemistry of the ocean and the phyletic inertia that results in primitive groups keeping characteristics which are better adapted for an earlier environment. Primitive algae continue to rely upon iron and manganese to a greater extent than more recently evolved groups of organisms, even though the availability of these metals has decreased drastically.

Although the increase in zinc concentration was about three orders of magnitude in deep waters (from 10^{-11} M to 10^{-8} M), it was less in surface waters of the open ocean because of particulate scavenging (surface waters of the open ocean today have around 10^{-11} to 10^{-10} M (Bruland 1980; Bruland and Franks 1983). As a result, the increase in zinc does not appear to have resulted in levels high enough to be toxic to the primitive algae. Only the extremely high levels found in polluted waters are ever toxic to algae (Jensen et al. 1974; Brand, unpublished data). As with manganese, the provision of zinc should not present any problems, but one should be careful not to approach toxic concentrations (Anderson et al. 1978).

Copper increased from 10^{-17} M in anaerobic waters (Ochiai 1978) to 10^{-9} M in aerobic seawater (Bruland and Franks 1983). Apparently as a result of the extreme scarcity of copper in anaerobic waters, copper did not begin to be used by organisms until the earth became aerobic and copper increased in abundance (Osterberg 1974; Williams and Da Sylva 1978; Ochiai 1978; Chapman and Schopf 1983). Even today, anaerobic bacteria do not use copper in their enzymes or electron transport systems (Egami 1975). The primitive cyanobacteria do use copper in plastocyanin and cytochrome oxidase, but it does not appear to be an obligate requirement, as a cytochrome with iron can be substituted for plastocyanin (Wood 1978; Ho and Krogmann 1982) and there appears to be a diversity of terminal cytochromes without copper that can replace the copper containing cytochrome oxidase (Jones 1977). As a result, it appears that most algae either do not need copper or need so little that free ionic copper concentrations of 10^{-19} M are sufficient to sustain maximum growth rates (Brand, unpublished data). It appears that no copper needs to be added to culture media. The increase in copper solubility appears to have resulted in concentrations that can now be toxic at natural levels to the more primitive algae (Brand et al. 1986). Copper is particularly toxic because of the strength of its ligand complexes (Jernelov and Martin 1975; Bowen 1979). Brand et al. (1986) have shown that cyanobacteria are sensitive to the toxic effects of copper at concentrations about ten times lower than those at which diatoms and coccolithophores are affected. Dinoflagellates are intermediate, being less sensitive than the cyanobacteria but more sensitive than the diatoms and coccolithophores. It is hypothesized that the reason for the difference in sensitivity to copper between the primitive algae and more recently-evolved algae is a difference in biochemical structures. Organisms that evolve in anaerobic waters can generate biochemical structures that are not resistant to copper interference (of disulfide bonds, for example), because so little copper is present. The accumulation of oxygen and the resulting increase in copper concentrations would presumably force these species to develop secondary protective devices and detoxification mechanisms. Species that evolved in an aerobic environment with high concentrations of copper present would have evolved biochemical structures that are fundamentally more resistant to copper to begin with and would not be as sensitive to copper toxicity. Whatever the cause of the differential sensitivity to copper among algae, it is clear that cyanobacteria are more susceptible to copper toxicity. Especially when culturing cyanobacteria, one should take care to reduce copper contamination and use chelators to sequester most of the free copper ions. All data to date indicate that it is the free ionic activity of trace metals, not their total concentration, that affects phytoplankton (Sunda and Guillard 1976; Morel and Morel-Laurens 1983; Anderson et al. 1978; Huntsman and Sunda 1980). All these studies, of course, were done with species already in culture. The question remains whether or not some phytoplankton species exist that are affected by the total concentration because of rapid reequilibrations at the cell surface, rather than just the free ions. If so, these species would be exposed to much higher levels of copper. Chelation of culture media would not protect these cells from the toxic effects of the relatively high concentrations of trace metals used in many culture media or from any contaminating trace metals that might be present. Such species would be very difficult to obtain in culture because of their sensitivity to extremely small amounts of trace metals. The question remains as to whether or not this is a reason many species of phytoplankton cannot be cultured. The use of lower concentrations of both trace metals and chelators in culture media may be a fruitful approach.

Molybdenum is particularly important to anaerobic organisms, as it is involved in the anaerobic reduction of nitrate, dinitrogen and carbon dioxide (Ochiai 1978; Egami 1975). It is also associated with some of the iron-sulfur enzymes found in all organisms. In aerobic organisms, it is needed for the multielectron reduction of dinitrogen, nitrate and sulfate (Williams 1981). Molybdenum is an unusual transition metal in that it exists in quite high concentrations (10^{-7} M) in seawater as a conservative salt with no surface depletion (Bruland 1983). As a result, one does not expect the algae to have any problem obtaining molybdenum if natural seawater is used in the culture media.

Cobalt is a particularly interesting nutritional trace metal. Its chief function in the cell appears to be as the metal occupying the center of the corrin ring in vitamin B12 (Williams 1981). The role of vitamin B12 in the cell is not well understood but it appears to be important in transferring methyl groups and methylating toxic elements such as arsenic, selenium, mercury, tin, thallium, platinum, gold and tellurium (Jernelov and Martin 1975; Brinckman 1985). Of the species of marine phytoplankton in culture, approximately half need vitamin B12 supplied in their culture media. The others presumably obtain ionic cobalt from seawater and synthesize their own vitamin B12. It is best to provide both ionic cobalt and vitamin B12 to the culture media. In oceanic surface waters, cobalt is present at about 2×10^{-11} M (Bruland 1983) while vitamin B 12 is estimated to be around 10^{-14} to 10^{-13} M (Swift 1980). Some complexities with vitamin B12 will be discussed in the section on vitamins.

Lewin (1966) has shown boron is needed by diatoms and other types of phytoplankton. The supply of this element in natural seawater, however, is not a problem, as borate is a major conservative salt in seawater at 4×10^{-4} M (Bruland 1983). Obviously, it should be supplied at the appropriate concentration in artificial seawater.

A number of elements have been demonstrated to be essential to various laboratory animals. Many can probably be presumed to be required by all organisms including phytoplankton, even though no studies have been conducted to demonstrate such a requirement in phytoplankton.

Nickel is needed in urease and probably in hydrogenase (Williams 1981; Bowen 1979). It has also been shown to be involved in iron assimilation in animals, although the details are not clear (Mertz 1981). Vertical profiles of nickel in the open ocean show a nutrient type profile with significant surface depletion to 2×10^{-9} M (Sclater et al. 1976; Bruland 1980; Bruland and Franks 1983). The relative roles of biological nutritional uptake and particulate scavenging in causing this surface depletion is not known. Although no absolute nickel requirement has ever been demonstrated in phytoplankton, it seems prudent to add a small amount to culture media.

Arsenic has been shown to be essential to laboratory animals, although its biochemical role is unknown (Mertz 1981). One finds arsenate at a concentration of around 2×10^{-8} M in the ocean, with only a small amount of surface depletion

(Andreae 1979). Competitive inhibition occurs between arsenate and phosphate and, as a result, arsenate is generally thought of as a toxic substance (Sanders 1979). It has been argued that arsenate/phosphate ratios in surface waters of the open ocean can become high enough to become slightly toxic (Bruland 1983). Probably the oceanic phytoplankton are better adapted to these high ratios than the coastal species that have been studied experimentally. In any case, arsenate toxicity should not be a problem in standard culture media with its much higher phosphate concentrations. One wonders, however, if the opposite problem might occur. If a small amount of arsenate is needed by phytoplankton and oceanic phytoplankton are adapted to high arsenate/phosphate ratios, they may be starved of arsenate in high phosphate culture media.

Vanadium has been demonstrated to be a required element in some organisms, but its biochemical role is not known (Mertz 1981). Vanadate exists in the ocean at around $2-3 \times 10^{-8}$ M and shows some, but not extreme surface depletion (Morris 1975; Bruland 1983). The amount already present in natural seawater is probably adequate for culturing purposes and none needs to be added. Vanadate is also known to be a competitive inhibitor of phosphate in ATPase, because of its chemical similarity to phosphate (Cantley et al. 1977). Therefore, the reverse is possible and high phosphate concentrations could inhibit the uptake and utilization of vanadate, if it is needed by phytoplankton.

Chromium has been found to be important in the nicotinic acid–glutathione complex of yeast and to be involved in the glucose metabolism of animals (Mertz 1981). Although it is generally considered toxic at higher concentrations, one should at least consider the possibility of its being needed by phytoplankton. Chromium is present in oceanic seawater at a concentration of around 3×10^{-9} M with some depletion at the surface (Bruland 1983). Its acquisition by phytoplankton at this concentration should not be a problem.

Selenium is found in a number of enzymes — glutathione peroxidase, glycine reductase, formate dehydrogenase, nicotinic acid hydroxylase, and xanthine dehydrogenase (Stadtman 1980; Williams and Da Sylva 1978). A selenium requirement has been clearly demonstrated in several freshwater phytoplankton species (Lindstrom and Rodhe 1978; Wehr and Brown 1985). Seawater contains more selenium than most freshwaters, but there is significant surface depletion in oceanic waters down to 5×10^{-11} M of Se(IV) and 5×10^{-10} M of Se(VI) (Measures et al. 1983). The existence of selenium in two redox states in seawater indicates the operation of a redox cycle, probably biologically mediated (Cutter and Bruland 1984). R. Guillard and M. Keller (pers. comm.) have observed selenium limitations in some phytoplankton species occasionally, although it has been difficult to prove an absolute requirement to date in marine phytoplankton. It appears prudent to add small amounts of selenium to culture media.

Many trace elements exist at lower concentrations in the surface waters of the open ocean than anywhere else in the biosphere. This leads to the question of whether or not there may be a selective advantage for organisms living in such an environment to substitute one element for another with similar chemical characteristics. If this occurs, the nutritional requirements of coastal phytoplankton may not be an adequate guide to the nutritional needs of oceanic phytoplankton. That such a process is plausible is illustrated by the example of freshwater cyanobacterial mutants that have been obtained that utilize tungsten or chromium in place of molybdenum (Singh et al. 1978). Many enzymes are known to function with many different metals in vitro (Coleman and Vallee 1961; Frausto da Silva 1978). Although there is at present no evidence for this occurring in vivo except for the mutants of cyanobacteria that have been developed, we have to consider the possibility that novel elements may be required by oceanic phytoplankton species that have yet to be cultured. The best guidance as to which elements might be used would be similarity in chemical characteristics to the elements known to be needed.

218

While it has long been thought that trace metals must be taken up by specific uptake enzymes or carriers, Simkiss (1983) has recently presented evidence that many metals (nutritional and non-nutritional) diffuse across cell membranes as uncharged complexes. Cadmium and mercury tend to complex with chloride, and zinc and copper tend to complex with carbonate and hydroxide to form uncharged complexes that apparently diffuse easily across lipid membranes. The cell membrane is therefore probably not a very good barrier to excessively high concentrations of metals, and internal sites and enzymes may be exposed to a trace metal regime that reflects the external environment more than previously thought. The data of Fisher et al. (1984) and Fisher (1985) show that internal concentrations reflect external concentrations quite well. The lack of an effective membrane barrier may be the reason methylation, precipitation and strong complexation (such as metallothionen) are such important mechanisms in algae for the detoxification of various metals (Wood and Wang 1983; Jernelov and Martin 1975; Bowen 1966; May et al. 1978). The ease with which metals apparently reach internal sites suggests that trace metals should be kept as low as possible in culture media. Because of the competitive inhibition among trace metals, the ratios provided may be extremely important.

Organic Nutrients

While many phytoplankton have been demonstrated to be able to take up and use organic compounds (Mahoney and McLaughlin 1977; Antia et al. 1975; Antia et al. 1980; Fisher and Cowdell 1982; Ukeles and Rose 1976; Morrill and Loeblich 1979), it has generally been thought that they are unable to compete with bacteria for the low concentrations of organic compounds found in natural waters (Droop 1974; Bonin and Maestrini 1981). To a large extent this has been based on the fact that phytoplankton in culture can grow only on high concentrations of organic compounds and that bacteria are generally much smaller than algae. Thus, bacteria have a higher surface to volume ratio, an inherent advantage over algae in taking up low concentrations of a limiting substrate. The small size and large surface to volume ratio of autotrophic picoplankton, however, forces a reevaluation of the argument. Indeed, some field data do support the idea that phytoplankton can compete effectively with bacteria for organic compounds (Wheeler et al. 1977). We should consider the possibility that many picoplankton are mixotrophic (need both light and organic compounds), photoheterotrophic (need light to take up and use organic compounds) or amphitrophic (can use either light or organic compounds as an energy source). Such biochemical diversity is generally more prevalent in the more primitive algal lineages. Many species of cyanobacteria and dinoflagellates are heterotrophic (Wolk 1973; Smith 1982). Some phytoplankton are even known to be phagotrophic (Bird and Kalff 1986; Droop 1974). The combination of small size and primitive phylogeny makes mixotrophy, photoheterotrophy and amphitrophy seem rather likely in picoplanktonic cyanobacteria. That there are few examples to point to may be a result of the fact that most cyanobacteria in culture were obtained with inorganic media designed for autotrophic growth. It remains to be seen how many mixotrophic, photoheterotrophic and amphitrophic species can be obtained when media designed specifically for such nutritional modes are used.

Another possibility is that only specific organic compounds are needed to alleviate particular biochemical deficiencies rather than an entire suite of compounds. Because oceanic phytoplankton experience a more predictable and constant abiotic and biotic environment than the coastal phytoplankton, they may be exposed to a more constant source of organic compounds and be able to depend upon certain organic nutrients. In studies with protozoa and animals, the specific deficiencies in diets are usually lipids rather than carbohydrates and proteins, which have more versatile synthetic pathways (Kanazawa et al. 1979; Stryer 1981; Conklin and Provasoli 1977, 1978; Provasoli 1977). The work of Droop (1970) and Droop and Pennock (1971) on the nutritional requirements of protozoa may be good models to follow in elucidating the organic nutritional requirements of picoplankton and ultraplankton.

One group of organic compounds known to be required by many phytoplankton is the vitamins. To date, only three vitamins have been demonstrated to be needed by phytoplankton (Swift 1980; McLachlan 1973; Provasoli and Carlucci 1974; Bonin et al. 1981). Approximately half of all phytoplankton species tested in culture need vitamin B12. About 20% need thiamine and less than 5% have been demonstrated to need biotin (Provasoli and Carlucci 1974). These three vitamins are routinely added to most culture media, whether the particular species needs them or not. The provision of vitamin B12 is probably not as simple as is generally assumed. There exists not one compound, but a whole family of cobalamins with various side groups (Swift 1980). Guillard (1968) has shown that some diatoms are highly specific, being able to use only one of the compounds, whereas other diatoms can use a variety of related compounds. Ohki and Fujita (1982) have suggested that *Trichodesmium* is stimulated only by hydroxycobalamin. Complicating the situation is the observation that many of the analogues of vitamin B12 seem to competitively inhibit one another (Rahat and Reich 1963; Epstein and Timmis 1963; Bonin et al. 1981). Thus the addition of one cobalamin in the culture media may not only not provide the correct analogue but could also inhibit the use of the analogue the species can use if it is present in the seawater. A study of cobalamin specificity similar to the one of Guillard (1968) needs to be conducted on a wide range of phytoplankton. Another complicating factor is the glycoprotein B12 binding factor that appears to be produced by most eukaryotes, but not by prokaryotes (Pintner and Altmeyer 1979; Messina and Baker 1982; Droop 1968; Bonin et al. 1981). This compound can sequester large amounts of vitamin B12 in a form that algae cannot obtain. The advantage to algae of producing this glycoprotein compound is not clear, but it may have a distinct disadvantage in culture. While not a problem in sparse natural populations, at high biomass levels enough may be produced to bind up all the vitamin B12 in the culture media and starve the phytoplankton of the vitamin. The addition of much higher concentrations of vitamin B12 would be needed to alleviate this problem in cultures of eukaryotic picoplankton. Thiamine is heat sensitive at moderate or high pH, being destroyed if autoclaved in seawater (Gold et al. 1966). Therefore, it and other vitamins should be sterilized by filtration or autoclaved in acidified distilled water and later added aseptically to sterilized culture media. Even at 20°C, however, thiamine is slowly destroyed in seawater (Gold et al. 1966), so culture media should be used soon after preparation.

One must always wonder if other vitamins besides the standard three may be needed by phytoplankton. Unknown vitamin requirements may be the reason some species can only be grown in bacterized culture and others are stimulated by bacteria. These bacteria may be secreting vitamins needed by the phytoplankton but not provided in the culture media. Consider, for example, riboflavin which has never been demonstrated to be a requirement for any alga. It is destroyed by light within minutes (Oster et al. 1962). Addition of riboflavin to culture media in the standard way would be useless, as it would be destroyed in the first photoperiod. Standard techniques would never allow a riboflavin requiring species to be cultured or tested for this requirement. In nature, riboflavin would be supplied continuously by bacteria. Recent data (S. Vastano, pers. comm.) show riboflavin concentrations to increase steadily at night in the photic zone (up to 10^{-10} M) and then drop precipitously at dawn and remain extremely low during the day (undetectable to 5×10^{-12} M). Presumably phytoplankton take up riboflavin at night if they need it. Clearly the culture strategy would be to add riboflavin each day during the dark period so that the algae can take it up before it is destroyed by light the next day. Even vitamin B12, thiamine and biotin are destroyed by light (Carlucci et al. 1969), only slower (weeks), so it may be beneficial to add all vitamins repeatedly to cultures.

Other possibilities to consider are the various hormones known to be important in higher plants such as the gibberillins, indole acetic acid, cytokinen and kinetin.

Numerous studies (reviewed by Provasoli and Carlucci 1974 and Bonin et al. 1981) have examined the effects of hormones on phytoplankton, with basically equivocal results. Some studies show stimulation, others show inhibition, while most show essentially no effect. Those studies that do show effects use environmentally unrealistic concentrations. The lack of any dramatic effects of hormones on phytoplankton is perhaps not surprising, given that the role of these hormones in higher plants is intracellular communication for coordinated multicellular growth, not nutritional per se.

Water Quality

A factor occasionally encountered by most phytoplankton culturists is the elusive "water quality". This is one of the problems of using natural seawater for the preparation of culture media. At certain times of the year, especially wintertime in temperate areas, coastal waters apparently contain substances inhibitory to the growth of phytoplankton in culture (Smayda 1974). It is generally thought the inhibition results from the excretion of various organic compounds by some phytoplankton species, but these compounds have not been identified. A number of methods are used by different culturists to reduce or eliminate these inhibitory effects. One can use charcoal to remove organic matter (including the inhibitory substances) from the seawater (McLachlin 1973), but one must consider that it may be removing beneficial organic compounds such as organics that complex toxic metals or keep iron in solution as well (Huntsman and Barber 1975; Wood et al. 1983; Giesy 1976). In addition, the charcoal must be cleaned extremely well because the commercially available charcoal has rather high concentrations of toxic metals that would contaminate the seawater. A more routine method is simply to let the seawater age for several months (McLachlan 1973). Presumably bacteria decompose the noxious organic compounds during this period of time, resulting in much less inhibition. Similarly, autoclaving seems often to destroy or reduce the inhibitory factor (but not always). Another approach is simply to use oceanic seawater, which has little biomass to begin with and no reported inhibitory properties. The use of oceanic rather than coastal seawater is also advantageous in that it is more uniform and lower in nutrients, metals and pollutants. The use of oceanic seawater seems to be the most reliable method, but unfortunately not everyone has access to this source.

The method of sterilizing the culture media must be considered in evaluating laboratory effects on phytoplankton. The most widely used technique is autoclaving — a process that has many effects on seawater and its constituents. As mentioned earlier, it can alter or destroy inhibitory organic compounds, but one must consider that it can do the same to beneficial organic molecules as well. Even the most oligotrophic waters contain around one milligram of organic matter per litre, making organic matter an important constituent of seawater. Although there is still much research to be conducted on the diverse organic substances in seawater, it seems likely that they have an influence on phytoplankton growth (Huntsman and Barber 1975; Fisher and Frood 1980). It is known that they complex a substantial fraction of some trace metals in seawater (Mackey 1983; Fisher and Fabris 1982; Wood et al. 1983; Batley and Florence 1976). Autoclaving would be expected to alter the organic molecules in seawater and alter their effects on phytoplankton, but the details of those effects are not yet known. Autoclaving also introduces trace metal contamination to the culture media (Brand et al. 1983), apparently transported by the steam. Because of the steam atmosphere in the autoclave, carbon dioxide is driven out of the seawater and the pH is raised to about 10. This high pH can cause the precipitation of ferric phosphates and hydroxides and calcium carbonate. If the seawater is autoclaved in glass vessels, the high pH will dissolve the glass, releasing not only high concentrations of silicate to the seawater, but also any contaminants in the glass. For this reason, it is best to autoclave seawater in teflon bottles. One way to avoid some of the effects of autoclaving seawater is to pasteurize it in teflon bottles. Pasturizing avoids trace metal contamination and probably alters organic molecules to a lesser extent than does autoclaving, but it does not completely sterilize the seawater. It kills all eukaryotes and most bacteria, but some bacterial spores probably survive. One must weigh the relative advantages and disadvantages of altering the seawater less and not maintaining completely axenic cultures. Ultraviolet radiation can be used to sterilize seawater, but rather high intensities are needed to kill everything in the seawater (Hamilton and Carlucci 1966). Such light necessarily alters and destroys the organic molecules in seawater and generates many long lived free radicals and other toxic reactive chemical species. Seawater exposed to intense ultraviolet light must be stored for many days to allow the level of highly reactive chemical species to decline before it can be used to culture phytoplankton. Sterile filtration is probably the best method of sterilizing seawater in that it does not alter the chemistry of the water as long as care is taken not to contaminate the seawater with a dirty filtration apparatus. The main disadvantage is the time necessary to filter large volumes of seawater with pore sizes small enough to retain the smallest bacteria and cyanobacteria found in the ocean.

Autotoxicity

One must consider the possibility that phytoplankton species may produce and excrete compounds that are rather innocuous at the low concentrations generated by the natural population abundances found in the ocean, but are inhibitory or toxic at concentrations generated by biomass levels a thousand to ten thousand times higher in culture. This may explain why many oceanic phytoplankton species do not become very dense in culture despite the availability of excess nutrients. The buildup of byproducts may inhibit the continued growth of the phytoplankton in culture. In addition, bacteria may decompose these waste products in natural communities, preventing a buildup of toxic substances, whereas the waste products would simply accumulate in axenic cultures. This could be another reason why some species of phytoplankton will grow in bacterized culture but not in axenic culture. Several lines of reasoning lead to the possibility that autotoxicity may be more of a problem in cultures of oceanic phytoplankton than coastal phytoplankton. First, one might expect oceanic phytoplankton to produce and excrete more organic compounds as a result of their being more nutrient limited and having a relative excess of light and carbon. If, as is thought, competition is more intense in the nutrient poor oceanic waters, oceanic phytoplankton may use some of these excreted organic compounds for interference competition with other species to augment their exploitive competitive abilities. Secondly, it is also possible that oceanic phytoplankton are generally more sensitive to toxic organic substances. Fisher et al. (1973) and Fisher (1977) have shown this to be true for exotic and anthropogenic organic compounds, indicating a general trend. Thus it is possible that oceanic phytoplankton species excrete more organic compounds and are more sensitive to their own inhibitory compounds in dense culture.

To prove this autotoxicity hypothesis for the difficulty in culturing some species is not easy because one must have the species in culture in order to demonstrate the mechanism by which it inhibits or kills itself in culture. Nevertheless, a number of particular types of compounds known to be excreted by phytoplankton can be envisioned as causing autotoxicity in high biomass cultures of some species.

A number of phytoplankton are known to produce compounds toxic to other organisms (Pinter and Altmeyer 1979; Chan et al. 1980; Maestrini and Bonin 1981). This ability seems especially prevalent among the most primitive groups of algae, the cyanobacteria and the dinoflagellates (Wolk 1973; Collins 1978; Schmidt and Loeblich 1979; Steidinger and Baden 1984). The ecological function (if there is one) of these toxins is not known, but some might result in autotoxicity. Elbrachter (1976) and Kayser (1979) have shown that many dinoflagellate species inhibit each other in culture. Many terrestrial plants are known to produce toxins for interference competitive purposes, resulting in autotoxicity (Smith 1979; Whittaker and Feeny 1971). The production of such autotoxins is advantageous because the compounds are more toxic to competing species than to themselves. If the same situation is particularly prevalent among cyanobacteria and dinoflagellates, one might expect many of them to be especially difficult to culture at high biomass levels.

Is it possible that many cyanobacteria and dinoflagellates produce enough iron chelating siderophores strong enough to starve themselves of iron in dense cultures? They are able to starve other species of algae of iron (Murphy et al. 1976). Do high enough concentrations of the glycoprotein vitamin B12 binding factor accumulate in culture to starve the cells of vitamin B12?

Some phytoplankton are known to produce pheromones (Kochert 1978; Derenbach 1985). They are known to act as gamete attractants, but may also alter cellular physiology in shifting the balance between sexual and asexual reproduction and life cycle stages. Is it possible that the accumulation of high concentrations of pheromones in culture alter the cellular physiology and life cycle to such an extent that vegetative growth cannot be sustained?

Although the reason is not completely clear, many phytoplankton produce large amounts of dimethyl sulfide. This results in concentrations in the photic zone of around 10^{-9} M (Andreae and Raemdonck 1983). In cultures with high biomass levels one finds concentrations of around 10^{-6} M (Vairavamurthy et al. 1985). One wonders if there are species that produce even more dimethylsulfide but cannot be obtained in culture because the dimethyl sulfide (and its oxidation product dimethylsulfoxide) levels accumulate to toxic concentrations.

Many organisms are known to produce highly reactive compounds for oxidative attack against competitors, predators, parasites and diseases. The compounds most often used are various oxygen radicals and compounds with highly polar bonds with halogens (Williams and Da Silva 1978). Not only macroalgae, but also cyanobacteria have been demonstrated to produce a number of these compounds for chemical defenses (Fenical 1981). Atmospheric chemists have recently discovered an incredible range of volatile, highly reactive organic compounds in the lower troposphere over the ocean (G. Harvey, per.comm.). It is suspected that they are produced by phytoplankton. No one has measured how much accumulates in dense cultures, but their highly reactive nature could cause chemical damage to organisms in high concentrations.

Another possibility worth considering is the production of highly reactive compounds such as superoxide anions, hydroxyl radicals, hydrogen peroxide and singlet oxygen. These chemical species are the necessary byproducts of many enzymatic reactions that evolved under the anaerobic conditions of the early earth and must now operate in the presence of molecular oxygen (Elstner 1982; Chapman and Schopf 1983; Halliwell 1974; Fridovich 1977; 1978a, b; Pryor 1978; Fee 1980; Krinsky 1979; Cerutti 1985). Because of their reactive nature, they are rather ephemeral and hard to measure directly and the concentrations in seawater of most are not yet known. It is known, however, that hydrogen peroxide is present in the photic zone at a concentration of about 10^{-7} M (Van Baalen and Marler 1966; Zafiriou 1983; Zafiriou et al. 1984; Zika et al. 1985), higher than almost all the nutrients in surface oceanic waters. It has been estimated that superoxide anions could be present at around 10^{-8} M in seawater (Zafiriou et al. 1984). Various nitrogen oxide and halogen radicals are also thought to be present in the photic zone (Zafiriou 1983; Zafiriou et al. 1984), but their concentrations have not yet been quantified. These highly reactive species peroxydize membranes and damage many organic molecules and are implicated in a wide variety of metabolic disorders and diseases (Chapman and Schopf 1983; Fridovich 1978b; Cerutti 1985). As a result of the destructive nature of these reactive chemicals, organisms have a variety of defense mechanisms to protect themselves (Halliwell 1974; Fridovich 1977, 1978a, b; Asada et al. 1977; Meister 1983; Stadtman 1980; Elstner 1982; Fee 1980). It seems plausible that organisms that evolved when the earth was still anaerobic may be more sensitive to superoxide anions, hydroxyl radicals, hydrogen peroxide and singlet oxygen. Indeed, Van Baalen (1965) and Marler and Van Baalen (1965) have demonstrated that some cyanobacteria inhibit their own growth and catalase alleviates this inhibition. If these various reactive oxygen species accumulate in dense cultures, the addition to culture media of substances that reduce them, such as catalase and superoxide dismutase, may be beneficial.

Light

The quantity, quality, and timing of light is as important to the successful culture of picoplankton and ultraplankton as with any other phytoplankton. Although there is considerable overlap, cyanobacteria generally show both light saturation and photo-inhibition at somewhat lower light levels than most eukaryotic algae (Barlow and Alberte 1985; Glover et al. 1985; Glover 1985). The difference is not great but slightly dimmer lights might be advisable for the culture of cyanobacteria.

The quality of light is probably a much more important factor to consider. It is quite difficult to simulate the quality of light phytoplankton experience deep within the photic zone of the open ocean. While most algae possess a number of accessory pigments to allow them to utilize a wide range of wavelengths (Prezelin 1981; Jeffrey 1980), studies have shown that light quality can have considerable effect on physiology (Vesk and Jeffrey 1977). Although the reason for this is unclear, Hauschild et al. (1962a, b), Wallen and Geen (1971), and Jones and Galloway (1979) have demonstrated that more protein is synthesized in blue light and relatively more carbohydrate is synthesized in white light. Glover et al. (1985) have suggested that picoplankton grow better in blue or green light than white light. Most laboratories use fluorescent lights for the culture of phytoplankton. Ohki and Fujita (1982) have suggested that fluorescent lighting is detrimental to *Trichodesmium*. The spectral characteristics of fluorescent lights are quite different from the light in the photic zone, especially at greater depths. It provides relatively less blue and more red light than what is found in the photic zone. Blue light with wavelengths less than 450 nm is important for photoreactivation of DNA damage repair (Teramura 1982; Halldal 1977) and photoinduction of many biochemical reactions (Ninnemann 1980; Schmidt 1984) in many organisms, including algae. This short wavelength light travels to great depths in the photic zone (Kirk 1983), but fluorescent lighting provides very little (Jones and Galloway 1979). Is it possible that some phytoplankton have not been successfully cultured because DNA damage accumulates as a result of inadequate light for the photoreactivation of repair mechanisms or photoinduction of other processes? At the same time, fluorescent lights provide considerable red light with wavelengths longer than 600 nm, yet such red light is absorbed within the top few meters of seawater (Kirk 1983). It is this long wavelength red light that controls the phytochrome system of plants. The phytochrome system is known to exert strong control over the metabolism of terrestrial plants (Marme 1977; Danks et al. 1983). Although much less is known about them, phytochrome systems have been detected in microalgae as well (Zucker 1972; Lipps 1973; Haupt 1968; Taylor and Bonner 1967). The phytochrome system is based on a red-light induced structural change in the phytochrome molecule (Kendrick and Spruit 1977). Proper metabolic timing is dependent on the ratio and abundance of the two forms (Pr and Pfr) of phytochrome. It has been demonstrated that excess red light causes a detrimental shift in the ratio of the two forms and can destroy the phytochrome molecule (Marme 1977; Pratt 1978). Lipps (1973) has demonstrated the inhibitory effects of far red light on phytoplankton. If phytoplankton are adapted to the extremely low levels of red light found in the photic zone, one must ask if the excess red light given off by fluorescent lights alters or destroys the phytochrome system of algae, resulting in the mistiming of metabolic processes. Many enzymes are known to be photoactivated (Anderson et al. 1982; Teramura 1982; Montagnoli 1977; Zucker 1972). Racusen and Galston

(1980) have demonstrated that red light can interfere with the operation of blue light receptors. The ratio of blue to red light may be extremely important for metabolic coordination, as different biochemical events are photoactivated by different wavelengths of light. Vastly different ratios of wavelengths could cause major metabolic disruption. Because light is used not only as an energy source, but also as an information source for environmental cuing, light quality should be considered at least as important as light quantity. At any given depth, the ratio of blue to red light is much more constant and predictable than is light intensity. It seems likely that efforts to better simulate the light quality of the photic zone may be rewarded.

The timing of light might also be expected to be rather important to phytoplankton, especially the smaller cells with less storage capacity. Many oceanic phytoplankton species are known to be inhibited or killed by continuous light (Brand and Guillard 1981). Because this inhibition is independent of light intensity, it is thought to be a result of the disruption of metabolic timing. Clearly, a light–dark cycle should be provided for phytoplankton cultures. Other aspects of light timing to be considered are that the effective photoperiod is shorter deeper in the photic zone because of reflection at the surface and longer path length at low sun angles; that most laboratories provide an on–off light regime rather than a sinusoidal light intensity time course with dawn and dusk; and that light quality changes with light intensity throughout the day (Gibson and Jewson 1984; Kirk 1983). The provision of a more realistic light regime may be important in culturing many of the species presently not in culture.

Biotic Interactions

If a species is susceptible to a disease, the best way of ensuring an epidemic is to crowd the organisms together. This is precisely what we do in culturing phytoplankton at cell densities thousands of times higher than are found in natural communities. The proximity of phytoplankton cells makes it quite easy for the progeny of disease organisms to find new host cells and proliferate. Phytoplankton are known to be susceptible to viral and bacterial attack. It has been well demonstrated that viruses lyse many cyanobacteria and can have considerable effect on their population dynamics (Wolk 1973; Padan and Shilo 1973). While less is known about viruses in eukaryotic algae, viruses or virus-like particles have been found in many eukaryotic phytoplankton as well (Melkonian 1982; Misra et al. 1982). Mayer and Taylor (1979) has demonstrated that a virus can lyse the picoplankter *Micromonas pusilla.*

Bacteria are also known to attack phytoplankton cells (Cole 1982). It appears that bacteria are more prone to attack unhealthy cells and cells without walls (Jones 1982; Kogure et al. 1982). Cyanobacteria appear to be especially susceptible to bacterial attack (Jones 1982). Studies have also found more bacteria attached to cells in culture than to cells in the natural environment (Jones 1982; Kogure et al. 1982). This suggests that even the phytoplankton we do have in culture may be less healthy in the laboratory than in the ocean and that our culture techniques do not simulate the natural environment very well. Phytoplankton species that naturally live at relatively high biomass levels might be expected to be more exposed to epidemics than the sparse oceanic species. Therefore, they may have better mechanisms for resisting viral or bacterial attack and surviving in dense cultures. If oceanic species have poorer mechanisms for resisting attack, they may be destroyed more easily in culture by viruses or bacteria brought in with the isolated cell. This may be why some oceanic species can be cultured for a few weeks with active growth, but eventually die.

The reverse situation must also be considered. Several studies (Provasoli and Pintner 1972; Machlis 1973; Jones 1982) have shown that the presence of particular bacteria is necessary for proper morphology or sexual reproduction in algae. It seems likely that bacteria are necessary for the survival and growth of some species as well. This is obviously difficult to demonstrate because one must have the species in culture to demonstrate that it will not live without the bacteria. Some phytoplankton species have been brought into unialgal culture but cannot be maintained axenically (L. Provasoli and R. Guillard, pers.comm.). Most likely, this is the result of an obligate symbiotic or consortial relationship with bacteria. One generally finds more symbiotic relationships in oligotrophic habitats with less seasonality, where population dynamics are more stable and relationships can be coordinated over evolutionary time (Valiela 1984), so one might expect a much higher percentage of oceanic phytoplankton to need symbiotic or consortial bacteria in culture. Absence of such biological relationships in the laboratory environment may be one reason why oceanic phytoplankton species are so much more difficult to culture than coastal species. The same probably applies for symbiotic or consortial relationships between species of phytoplankton. Intimate symbiotic relationships such as *Richelia* in *Rhizosolenia* (Taylor 1980) are quite obvious. Much less obvious and more difficult to detect and demonstrate are obligate consortial relationships between species with no direct physical contact. Undoubtedly, loose consortial relationships exist in the ocean between species that excrete vitamins and those that need particular vitamins (Swift 1980; Bonin et al. 1981; Ukeles and Bishop 1975; Jones 1982). This particular consortial relationship can be circumvented in the laboratory by adding the known vitamins to the culture media. One wonders, however, how many other specific compounds are excreted by bacteria and phytoplankton and are necessary for the successful growth of various phytoplankton species, particularly oceanic species. Until such compounds can be identified, a useful shortcut may be to try coculturing several species together.

Cocultured consortial species can provide a continual supply of compounds at low concentrations rather than one large pulse, as is common in standard laboratory culturing techniques. The use of consortial species may avoid problems that could arise from excessively high concentrations of trace nutrients. Obviously, an impossibly large number of species combinations exists, but the attempt might prove fruitful. The attempt will be difficult, as very few combinations will probably provide the proper consortial relationship, many will have little effect and some interactions will be detrimental. Numerous studies have shown that various combinations of phytoplankton and bacteria can inhibit each other as well (Cole 1982; Pintner and Altmeyer 1979; Chan et la. 1980; Kayser 1979; Berland et al. 1972; Kogure et al. 1979; Jones 1982; Maestrini and Bonin 1981). They could produce substances toxic to the other species, outcompete the other species for nutrients, or produce compounds that sequester iron, vitamin B12 and perhaps other nutrients from the other species. Another consideration is the possibility that a second species may provide vital compounds to the species of interest in the ocean or in an extremely dilute culture, but may start competing severely for nutrients or producing detrimental compounds at high cell densities. Thus, the coculturing of species could be successful in getting some species into culture, but it also seems fraught with many potential difficulties.

Conclusions

Marine phytoplankton have now been cultured for close to a century. Improvements continue to be made in culturing techniques and more species are still being brought into culture. Nevertheless, the majority of species in the ocean have not yet been successfully cultured and it appears that there is a strong habitat and phylogenetic bias in our present culture collections. Our current techniques are quite good for culturing species such as coastal diatoms adapted for living in eutrophic waters, but apparently quite poor at culturing species such as dinoflagellates from the oligotrophic subtropical central gyres. A general evaluation of the types of species that can and cannot be cultured by current techniques leads to the suspicion that most picoplankton and ultraplankton species, especially from oceanic waters, have not been successfully cultured. The species that dominate our culture collections are not the species that dominate the ocean. As a result, our studies on the biochemistry and physiology of marine phytoplankton have not been conducted on representative species and our perception of how marine phytoplankton interact with their environment is probably strongly biased. The successful culture of the entire range of phytoplankton species should provide us with a much better perspective. Obviously, the solution to the problem of culturing the more K-selected species in the ocean is not yet known, but some of the potential sources of the problem have been suggested and discussed. It is suspected that some of the problems may result from the high nutrient concentrations used, the high biomass levels generated and the lack of interaction with other species in laboratory cultures. The species we cannot culture at present tend to be among the first groups of algae to have evolved and today live in stratified, oligotrophic waters with a low biomass but high diversity plankton community. Future work on the development of better culture techniques should take into consideration the ways in which oceanic picoplankton and ultraplankton differ from the species we presently have in culture.

References

ABELIOVICH, A., AND Y. AZOV. 1976. Toxicity of ammonia to algae in sewage oxidation ponds. Appl. Environ. Microbiol. 31: 801–806.

AHRLAND, S. 1985. Inorganic chemistry of the ocean, p. 65–88. *In* [ed.] K. J. Irgolic and A. E. Martell Environmental inorganic chemistry, VCH Publishers.

ANDERSON, L. E., A. R. ASHTON, A. H. MOHAMED, AND R. SCHEIBE. 1982. Light/dark modulation of enzyme activity in photosynthesis. Bioscience 32: 103–107.

ANDERSON, M. A., AND F. M. M. MOREL. 1982. The influence of aqueous iron chemistry on the uptake of iron by the coastal diatom *Thalassiosira weissflogii*. Limnol. Oceanogr. 27: 789–813.

ANDERSON, M. A., AND F. M. M. MOREL. 1980. Uptake of Fe (II) by a diatom in oxic culture medium. Mar. Biol. Lett. 1: 263–268.

ANDERSON, M. A., F. M. M. MOREL, AND R. R. L. GUILLARD. 1978. Growth limitation of a coastal diatom by low zinc ion activity. Nature 276: 70–71.

ANDREAE, M. O. 1979. Arsenic speciation in seawater and interstitial waters: The influence of biological–chemical interactions on the chemistry of a trace element. Limnol. Oceanogr. 24: 440–452.

ANDREAE, M. O., AND H. RAEMDONCK. 1983. Dimethylsulfide in the surface ocean and the marine atmosphere: a global view. Science 221: 744–747.

ANTIA, N. J., B. R. BERLAND, AND D. J. BONIN. 1980. Proposal for an abridged nitrogen turnover cycle in certain marine planktonic systems including hypoxanthine-guanine excretion by ciliates and their reutilization by phytoplankton. Mar. Ecol. Prog. Ser. 2: 97–103.

ANTIA, N. J., B. R. BERLAND, D. J. BONIN, AND S. Y. MAESTRINI. 1975. Comparative evaluation of certain organic and inorganic sources of nitrogen for phototrophic growth of marine microalgae. J. Mar. Biol. Assoc. U.K. 55: 519–539.

ANTIA, N. J., AND V. CHORNEY. 1968. Nature of the nitrogen compounds supporting phototrophic growth of the marine cryptomonad *Hemiselmis virescens*. J. Protozool. 15: 198–201.

ARMSTRONG, J. E., AND C. VAN BAALEN. 1979. Iron transport in microalgae: the isolation and biological activity of a hydroxymate siderophore from the blue-green alga *Agmenellum quadruplicatum*. J. Gen. Microbiol. 111: 253–262.

ASADA, K., S. KANEMATSU, AND K. UCHIDA. 1977. Superoxide dismutases in photosynthetic organisms: Absence of the cuprozinc enzyme in eukaryotic algae. Arch. Biochem. Biophys. 179: 243–256.

BANNISTER, J. V., AND G. ROTILIO. 1984. A decade of superoxide dismutase activity, p. 146–189. *In* J. V. Bannister and W. H. Bannister [ed.] The biology and chemistry of active oxygen.

BARLOW, R. G., AND R. S. ALBERTE. 1985. Photosynthetic characteristics of phycoerythrin-containing marine *Synechococcus* spp. Mar. Biol. 86: 63–74.

BATLEY, G. E., AND T. M. FLORENCE. 1976. Determination of the chemical forms of dissolved cadmium, lead and copper in seawater. Mar. Chem. 4: 347–363.

BENDER, M. L., G. P. KLINKHAMMER, AND D. W. SPENCER. 1977. Manganese in seawater and the marine manganese balance. Deep Sea Res. 24: 799–812.

BERLAND, B. R., D. J. BONIN, S. Y. MAESTRINI. 1972. Are some bacteria toxic for marine algae? Mar. Biol. 12: 189–193.

BIRD, D. F., AND J. KLAFF. 1986. Bacterial grazing by planktonic lake algae. Science 231: 493–495.

BONIN, D. J., S. Y. MAESTRINI, AND J. W. LEFTLEY. 1981. The role of phytohormones and vitamins in species succession of phytoplankton, p. 310–322. *In* T. Platt [ed.] Physiological bases of phytoplankton ecology. Can. Bull. Fish. Aquat. Sci. 210.

BONIN, D. J., AND S. Y. MAESTRINI. 1981. Importance of organic nutrients for phytoplankton growth in natural environments: Implications for algal species succession, p. 279–291. *In* T. Platt [ed.] Physiological bases of phytoplankton ecology. Can. Bull. Fish. Aquat. Sci. 210.

BOTHE, H. 1982. Nitrogen fixation, p. 87–104. *In* N. G. Carr and B. A. Whitton [ed.] The biology of cyanobacteria.

BOWEN, H. J. M. 1966. Trace elements in biochemistry. Academic Press, New York, NY.
1979. Environmental chemistry of the elements. Academic Press, New York, NY. 333 p.

BOYLE, E. A., AND J. M. EDMOND. 1975. Copper in surface waters south of New Zealand. Nature 253: 107–109.

BRAND, L. E., AND R. R. L. GUILLARD. 1981. The effects of continuous light and light intensity on the reproduction rates of twenty-two species of marine phytoplankton. J. Exp. Mar. Biol. Ecol. 50: 119–132.

BRAND, L. E., R. R. L. GUILLARD, AND L. S. MURPHY. 1981. A method for the rapid and precise determination of acclimated phytoplankton reproduction rates. J. Plankton Res. 3: 193–201.

BRAND, L. E., W. G. SUNDA, AND R. R. L. GUILLARD. 1983. Limitation of marine phytoplankton reproductive rates by zinc, manganese, and iron. Limnol. Oceanogr. 28: 1182–1198.
1986. Reduction of marine phytoplankton reproduction rates by copper and cadmium. J. Exp. Mar. Biol. Ecol. 96: 225–250.

BRINCKMAN, F. E., 1985. Environmental inorganic chemistry of main group elements with special emphasis on their occurrence as methyl derivatives, p. 195–238. *In* K.J. Irgolic and A. E. Martell [ed.] Environmental inorganic chemistry. VCH Publishers.

BRULAND, K. W. 1980. Oceanographic distribution of cadmium, zinc, nickel, and copper in the North Pacific. Earth Planet. Sci. Lett. 47: 176–198.
1983. Trace elements in seawater. Chem. Oceanogr. 8: 157–20.

BRULAND, K. W., AND R. P. FRANKS. 1983. Mn, Ni, Cu, Zn, and Cd in the western North Atlantic, p. 395–414. *In* C. S. Wong et al. [ed.] Trace metals in sea water. NATO Conf. Ser. 4, Mar. Sci., Vol. 9, Plenum, New York, NY.

BRULAND, K. W., R. P. FRANKS, G. A. KNAUER, AND J. H. MARTIN. 1979. Sampling and analytical methods for the determination of copper, cadmium, zinc, and nickel at the nanogram per liter level in seawater. Anal. Chem. Acta 105: 233–245.

BRULAND, K. W., G. A. KNAUER, AND J. H. MARTIN. 1978. Zinc in northeast Pacific water. Nature 271: 741–743.

BURTON, J. D., W. A. MAHER, AND P. J. STATHAM. 1983. Some recent measurements of trace metals in Atlantic Ocean waters, p. 415–426. *In* C. S. Wong et al. [ed.] Trace metals in sea water. NATO Conf. Ser. 4, Mar. Sci. Vol. 9, Plenum Press, New York, NY.

BUTLER, E. I., S. KNOX, AND M. I. LIDDICOAT. 1979. The relationship between inorganic and organic nutrients in sea water. J. Mar. Biol. Assoc. U.K. 59: 239–250.

BYRNE, R. H., AND D. R. KESTER. 1976. Solubility of hydrous ferric oxide and iron speciation in seawater. Mar. Chem. 4: 255–274.

CANTLEY, L. C., L. JOSEPHSON, R. WARNER, M. YANAGISAWA, C. LECHENE, AND G. GUIDOTTI. 1977. Vanadate is a potent (Na, K)-ATPase inhibitor found in ATP derived from muscle. J. Biol. Chem. 252: 7421–7423.

CARLUCCI, A. F., S. B. SILBERNAGEL, AND P. M. McNALLY. 1969. Influence of temperature and solar radiation on persistence of vitamin B12, thiamine, and biotin in seawater. J. Phycol. 5: 302–305.

CARPENTER, E. J., AND J. S. LIVELY. 1980. Review of estimates of algal growth using ^{14}C tracer techniques, p. 161–178. In Primary Productivity in the Sea Ed. P.G. Falkowski.

CERUTTI, P. A., 1985. Prooxidant states and tumor promotion. Science 227: 379–381.

CHAN, A. T., R. J. ANDERSEN, M. J. LE BLANC, AND P. J. HARRISON. 1980. Algal plating as a tool for investigating allelopathy among marine microalgae. Mar. Biol. 59: 7–13.

CHAPMAN, D. J., AND J. W. SCHOPF. 1983. Biological and biochemical effects of the development of an aerobic environment, p. 302–320. *In* J. W. Schopf [ed.] Earth's earliest biosphere: its origin and evolution. Princeton University Press, Princeton.

COLE, J. V. 1982. Interactions between bacteria and algae in aquatic ecosystems. Ann. Rev. Ecol. Syst. 13: 291–314.

COLEMAN, J. E., AND B. E. VALLEE. 1961. Metallo-carboxypeptidases: Stability constants and enzymatic characteristics. J. Biol. Chem. 236: 2244–2249.

COLLIER, R, AND J. EDMOND. 1984. The trace element geochemistry of marine biogenic particulate matter. Prog. Oceanogr. 13: 113–199.

COLLINS, M. 1978. Algal toxins. Microbiol. Rev. 42: 725–746.

CONKLIN, D. E., AND L. PROVASOLI. 1978. Biphasic particulate media for the culture of filter-feeders. Biol. Bull. 154: 47–54.

CONKLIN, D. E., AND L. PROVASOLI. 1977. Nutritional requirements of the water flea *Moina macrocopa*. Biol. Bull. 152: 337–350.

CUTTER, G. A., AND K. W. BRULAND. 1984. The marine biogeochemistry of selenium: A reevaluation. Limnol. Oceanogr. 29: 1179–1192.

DANKS, S. M., H. EVANS, AND P. A. WHITTAKER. 1983. Photosynthetic systems: structure, function and assembly. Wiley, NY. 162 p.

DERENBACH, J. B. 1985. Investigations into a small fraction of volatile hydrocarbons II. Indication for possible pheromones in open sea phytoplankton. Mar. Chem. 15: 305–309.

DROOP, M. R., AND J. F. PENNOCK. 1971. Terpenoid quinones and steroids in the nutrition of *Oxyrrhis marina*. J. Mar. Biol. Assoc. U.K. 51: 455–470.

DROOP, M. R., 1968. Vitamin B12 and marine ecology. IV. The kinetics of uptake, growth and inhibition in *Monochrysis lutheri*. J. Mar. Biol. Assoc. U.K. 48: 689–733.

DROOP, M. R. 1970. Nutritional investigation of phagotrophic protozoa under axenic conditions. Helgol. wiss. Meeresunters. 20: 272–277.

1974. Heterotrophy of carbon, p. 530–559. W. D. P. Stewart [ed.] *In* Algal physiology and biochemistry.

EGAMI, F. 1975. Origin and early evolution of transition element enzymes. J. Biochem. 77: 1165–1169.

ELBRACHTER, M. 1976. Population dynamic studies on phytoplankton cultures. Mar. Biol. 35: 201–207.

ELSTNER, E. F. 1982. Oxygen activation and oxygen toxicity. Ann. Rev. Pl. Physiol. 33: 73–96.

EMERICK, R. J. 1974. Consequence of high nitrate levels in feed and water supplies. Fed. Proc. 33: 1183–1187.

EPPLEY, R. W., R. W. HOLMES, AND J. D. H. STRICKLAND. 1967. Sinking rates of marine phytoplankton measured with a fluorometer. J. Exp. Mar. Biol. Ecol. 1: 191–208.

EPSTEIN, S. S., AND G. M. TIMMIS. 1963. Simple antimetabolites of vitamin B 12. J. Protozool. 10: 63–73.

FEE, J. A. 1980. Superoxide, superoxide dismutases, and oxygen toxicity, p. 209–237. *In* T. G. Spiro [ed.] Metal activation of dioxygen.

FENICAL, W., 1981. Natural halogenated organics, p. 375–393. *In* E. K. Duursma and R. Dawson [ed.] Marine organic chemistry, Elsevier.

FINDEN, D. A. S., E. TIPPING, G. H. M. JAWORSKI, AND C. S. REYNOLDS. 1984. Light induced reduction of natural iron (III) oxide and its relevance to phytoplankton. Nature 309: 783–784.

FISHER, N. S., 1977. On the differential sensitivity of estuarine and open-ocean diatoms to exotic chemical stress. Am. Nat. 111: 871–895.

1985. Accumulation of metals by marine picoplankton. Mar. Biol. 87: 137–142.

FISHER, N. S., AND R. A. COWDELL. 1982. Growth of marine planktonic diatoms on inorganic and organic nitrogen. Mar. Biol. 72: 145–155.

FISHER, N. S., AND J. G. FABRIS. 1982. Complexation of Cu, Zn and Cd metabolites excreted from marine diatoms. Mar. Chem. 11: 245–255.

FISHER, N. S., AND F. FROOD. 1980. Heavy metals and marine diatoms: Influence of dissolved organic compounds on toxicity and selection for metal tolerance among four species. Mar. Biol. 59: 85–93.

FISHER, N. S., M. BOHE, AND J. L. TEYSSIE. 1984. Accumulation and toxicity of Cd, Zn, Ag, and Hg in four marine phytoplankters. Mar. Ecol. Prog. Ser. 18: 201–213.

FISHER, N. S., L. B. GRAHAM, E. J. CARPENTER, AND C. F. WURSTER. 1973. Geographic differences in phytoplankton sensitivity to PCBs. Nature 241: 548–549.

FITZWATER, S. E., G. A. KNAUER, AND J. H. MARTIN. 1982. Metal contamination and its effect on primary production measurements. Limnol. Oceanogr. 27: 544–551.

FOGG, G. E. 1974. Nitrogen fixation, p. 560–582. *In* W. D. P. Stewart [ed.] Algal physiology and biochemistry.

FRAUSTO DA SILVA, J. J. R. 1978. Interaction of the chemical elements with biological systems, p. 449–484. *In* New trends in bio-inorganic chemistry. Academic Press, New York, NY.

1978a. The biology of oxygen radicals. Science 201: 875–880.

1978b. Superoxide radicals, superoxide dismutases and the aerobic lifestyle. Photochem. Photobiol. 28: 733–741.

FRIDOVICH, I. 1977. Oxygen is toxic! Bioscience 27: 462–466.

GARSIDE, C. 1985. The vertical distribution of nitrate in open ocean surface water. Deep Sea Res. 32: 723–732.

GIBSON, C. E., AND D. H. JEWSON. 1984. The utilization of light by microorganisms, p. 97–127. *In* G. A. Codd [ed.] Aspects of microbial metabolism and ecology. Academic Press, New York, NY.

GIESY, J. P. 1976. Stimulation of growth in *Scenedesmus obliquus* (Chlorophyceae) by humic acids under iron limited conditions. J. Phycol. 12: 172–179.

GIVAN, C. V. 1979. Metabolic detoxification of ammonia in tissues of higher plants. Phytochemistry 18: 375–382.

GLOVER, H. E. 1985. The physiology and ecology of the marine cyanobacterial genus *Synechococcus* Adv. Aquat. Microbiol. 3: 49–107.

GLOVER, H. E., D. A. PHINNEY, AND C. S. YENTSCH. 1985. Photosynthetic characteristics of picoplankton compared with those of larger phytoplankton populations in various water masses in the Gulf of Maine. Biol. Oceanogr. 3: 223–248.

GOLD, K., O. A. ROELS, AND H. BANK. 1966. Temperature dependent destruction of thiamine in seawater. Limnol. Oceanogr. 11: 410–413.

GORDON, R. M., J. H. MARTIN, AND G. A. KNAUER. 1982. Iron in Northeast Pacific waters. Nature 299: 611–612.

GOULD, S. J., AND R. C. LEWONTIN. 1979. The spandrels of San Marco and the Panglossian paradigm: A critique of the adaptationist programme. Proc. R. Soc. Lond. 205B: 581–598.

GUILLARD, R. R. L., AND M. D. KELLER. 1984. Culturing dinoflagellates, p. 391–442. *In* D. L. Spector [ed.] Dinoflagellates. Academic Press, New York, NY.

GUILLARD, R. R. L. 1968. B12 specificity of marine centric diatoms. J. Phycol. 4: 59–64.

1975. Culture of phytoplankton for feeding marine invertebrates, p. 29–60. *In* W. L. Smith and M. H. Chanley [ed.] Culture of marine invertebrate animals.

HALLDAL, P. 1977. Effects of changing levels of ultraviolet radiation on phytoplankton, p. 21–34. *In* A. K. Biswas [ed.] The ozone layer.

HALLIWELL, B. 1974. Superoxide dismutase, catalase and glutathione: Solutions to the problems of living with oxygen. New Phytol. 73: 1075–1086.

HAMILTON, R. D., AND A. F. CARLUCCI. 1966. Use of ultraviolet irradiated sea water in the preparation of culture media. Nature 211: 483–484.

HAUPT, W. 1968. Der phytochrosystem der *Mougeatiazella*: Weitere untersuchunger uber dem P. 730-gradienten. Z. Planzenphysiol. 59:

45-55.

HARRISON, P. J., R. E. WATERS, AND F. J. R. TAYLOR. 1980. A broad spectrum artificial seawater medium for coastal and open ocean phytoplankton. J. Phycol. 16: 28-35.

HAUSCHILD, A. H. W., C. D. NELSON, AND G. KROTKOV. 1962a. The effect of light quality on the products of photosynthesis in *Chlorella vulgaris*. Can. J. Bot. 40: 179-189.

———. 1962b. The effect of light quality on the products of photosynthesis in green and blue-green algae and in photosynthetic bacteria. Can. J. Bot. 40: 1619-1630.

HO, K. K., AND D. W. KROGMANN. 1982. Photosynthesis, p. 191-214. *In* N. G. Carr and B. A. Whitton. [ed.] The biology of cyanobacteria.

HOLLAND, H. D. 1972. The geologic history of sea water — an attempt to solve the problem. Geochim. Cosmochim. Acta. 36: 637-651.

HOLLAND, H. D., AND M. SCHIDLOWSKI. 1982. Mineral deposits and the evolution of the biosphere. Springer-Verlag, New York, NY. 332 p.

HULBURT, E. M. 1964. Succession and diversity in the plankton flora of the western North Atlantic. Bull. Mar. Sci. 14: 33-44.

———. 1966. The distribution of phytoplankton, and its relationship to hydrography, between southern New England and Venezuela. J. Mar. Res. 24: 67-81.

———. 1967. A note on regional differences in phytoplankton during a crossing of the southern North Atlantic Ocean in January, 1967. Deep Sea Res. 14: 685-690.

———. 1976. Limitation of phytoplankton species in the ocean off western Africa. Limnol. Oceanogr. 21: 193-211.

———. 1985. Adaptation and niche breadth of phytoplankton species along a nutrient gradient in the ocean. J. Plankt. Res. 7: 581-594.

HUNTSMAN, S. A., AND W. G. SUNDA. 1980. The role of trace metals in regulating phytoplankton growth, p. 285-328. *In* I. Morris [ed.] The physiological ecology of phytoplankton. University of California, Berkeley, CA.

HUNTSMAN, S. A., AND R. T. BARBER. 1975. Modification of phytoplankton growth by excreted compounds in low-density populations. J. Phycol. 11: 10-13.

JACKSON, G. A., AND P. M. WILLIAMS. 1985. Importance of dissolved organic nitrogen and phosphorus to biological nutrient cycling. Deep Sea Res. 32: 223-235.

JACKSON, G. A. 1980. Phytoplankton growth and zooplankton grazing in oligotrophic oceans. Nature 284: 439-441.

JEFFREY, S. W. 1980. Algal pigment systems, p. 33-58. *In* P. G. Falkowski [ed.] Primary productivity in the sea.

JENSEN, A., B. RYSTAD, AND S. MELSON. 1974. Heavy metal tolerance of marine phytoplankton I. The tolerance of the algal species to zinc in coastal sea waters. J. Exp. Mar. Biol. Ecol. 15: 145-157.

JERNELOV, A., AND A. L. MARTIN. 1975. Ecological implications of metal metabolism by microorganisms. Ann. Rev. Microbiol. 29: 61-77.

JONES, A. L. 1982. The interaction of algae and bacteria, p. 189-247. *In* A. T. Bull and J. H. Slater [ed.] *Microbial* interactions and communities. Academic Press, New York, NY.

JONES, T. W., AND R. A. GALLOWAY. 1979. Effect of light quality and intensity on glycerol content in *Dunaliella tertiolecta* (Chlorophyceae) and the relationship to cell growth/osmoregulation. J. Phycol. 15: 101-106.

JONES, C. W. 1977. Aerobic respiratory systems in bacteria, p. 23-59. *In* B. A. Haddock and W. A. Hamilton [ed.] 27th Symp. Soc. Gen. Microbiol.

KANAZAWA, A., S. LASHIMA, AND K. OMO. 1979. Relationship between the essential fatty acid requirements of aquatic animals and their capacity for bioconversion of linolenic acid to highly unsaturated fatty acids. Comp. Biochem. Physiol. 633: 295-298.

KAYSER, H. 1979. Growth interactions between marine dinoflagellates in multispecies culture experiments. Mar. Biol. 52: 357-363.

KENDRICK. R. E., AND C. K. P. SPRUIT. 1977. Phototransformations of phytochrome. Photochem. Photobiol. 26: 201-214.

KIRK, J. T. O. 1983. Light and photosynthesis in aquatic ecosystems. Cambridge University Press, New York, NY. 388 p.

KLAVENESS, D., AND R. R. L. GUILLARD. 1975. The requirement for silicon in *Synura petersenii* (Chrysophyceae). J. Phycol. 11: 349-355.

KLINKHAMMER, G. P., AND M. L. BENDER. 1980. The distribution of manganese in the Pacific Ocean. Earth Planet. Sci. Lett. 46: 361-384.

KOCHERT, G. 1978. Sexual pheromones in algae and fungi. Ann. Rev. Pl. Physiol. 29: 461-486.

KOGURE, K., U. SIMIDU, AND N. TAGA. 1979. Effect of *Skeletonema costatum* (Grev.) Cleve on the growth of marine bacteria. J. Exp. Mar. Biol. Ecol. 36: 201-215.

———. 1982. Bacterial attachment to phytoplankton in seawater. J. Exp. Mar. Biol. Ecol. 56: 197-204.

KREMER, B. P. 1981. Dark reactions of photosynthesis, p. 44-54. *In* T. Platt [ed.] Physiological bases of phytoplankton ecology. Can. Bull. Fish. Aquat. Sci. 210.

KRINSKY, N. I. 1979. Carotenoid protection against oxidation. Pure Appl. Chem. 51: 649-660.

LEWIN, R. 1984. How microorganisms transport iron. Science 225: 401-402.

LEWIN, J., AND C.-H. CHEN. 1971. Available iron: a limiting factor for marine phytoplankton. Limnol. Oceanogr. 16: 670-675.

LEWIN, J. 1966. Boron as a growth requirement for diatoms. J. Phycol. 2: 160-163.

LEWONTIN, R. C. 1978. Adaptation. Sci. Am. 239: 213-220.

LINDSTROM, K., AND W. RODHE. 1978. Selenium as a micronutrient for *Peridinium cinctum* fa. *westii* Mitt. Internat. Verein. Limnol. 21: 168-173.

LIPPS, M. J. 1973. The determination of the far red effect in marine phytoplankton. J. Phycol. 9: 207-240.

LIPPS, J. H. 1970. Plankton evolution. Evolution

24: 1-22.

LOEBLICH, A. R. 1975. A seawater medium for dinoflagellates and the nutrition of *Cachonina niei*. J. Phycol. 11: 80-86.

LUMSDEN, J. AND D. O. HALL. 1975. Superoxide dismutase in photosynthetic organisms provides an evolutionary hypothesis. Nature 257: 670-672.

MACHLIS, L. 1973. The effects of bacteria on the growth and reproduction of *Oedogonium cardiacum*. J. Phycol. 9: 342-344.

MACKEY, D. J. 1983. Metal-organic complexes in seawater — an investigation of naturally occurring complexes of Cu, Zn, Fe, Mg, Ni, Cr, Mn and Cd using high-performance liquid chromatography with atomic fluorescence detection. Mar. Chem. 13: 169-180.

MAESTRINI, S. Y., AND D. J. BONIN. 1981. Allelopathic relationships between phytoplankton species, p. 322-338. *In* T. Platt [ed.] Physiological bases of phytoplankton ecology. Can. Bull. Fish Aquat. Sci. 210.

MAHONEY, J. B., AND J. J. A. MCLAUGHLIN. 1977. The association of phytoflagellate blooms in lower New York Bay with hypertrophication. J. Exp. Mar. Biol. Ecol. 28: 53-65.

MARLER, J. E., AND C. VAN BAALEN. 1965. Role of H_2O_2 in single-cell growth of the blue-green alga, *Anacystis nidulans*. J. Phycol. 1: 180-185.

MARME, D. 1977. Phytochrome: Membranes as possible sites of primary action. Ann. Rev. Pl. Physiol. 28: 173-198.

MAY, P. M., D. R. WILLIAMS, AND P. W. LINDER. 1978. Biological significance of low molecular weight iron (III) complexes, p. 29-76. *In* H. Sigel. [ed.] Metal ions in biological systems. Vol.7.

MAYER, J. A., AND F. J. R. TAYLOR. 1979. A virus which lyses the marine nanoflagellate *Micromonas pusilla*. Nature 281: 299-301.

MCCARTHY, J. J. 1980. Nitrogen, p. 191-233. *In* I. Morris [ed.] The Physiological ecology of phytoplankton.

MCLACHLAN, J. 1973. Growth media-marine, p. 25-51. *In* J. R. Stein [ed.] *Handbook of phycological methods: culture methods and growth measurements*. Cambridge University Press, London.

MEASURES, C. I., B. C. GRANT, B. J. MANGUM, AND J. M. EDMOND. 1983. The relationship of the distribution of dissolved selenium IV and VI in three oceans to physical and biological processes, p. 73-83. *In* C. S. Wong et al. [ed.] Trace metals in sea water. NATO Conf. Ser 4. Mar. Sci. Vol. 9, Plenum Press, New York, NY.

MEISTER, A. 1983. Selective modification of glutathione metabolism. Science 220: 472-477.

MELKONIAN, M. 1982. Virus-like particles in the scaly green flagellate *Mesostigma viride*. Br. Phyciol. J. 17: 63-68.

MERTZ, W. 1981. The essential trace elements. Science 213: 1332-1338.

MESSINA, D. S., AND A. L. BAKER. 1982. Interspecific growth regulation in species succession through vitamin B12 competitive inhibition. J. Plankton Res. 4: 41-46.

MISRA, A., R. SINKA, S. JHA, V. JHA, B. SINGH, AND B. P. SHARMA. 1982. Virus infection of marine algae, p. 289-296. *In* Marine algae in pharmaceutical science. Walter de Gruyter and co.

MONTAGNOLI, G. 1977. Biological effects of light on protein: enzyme activity modulation. Photochem. Photobiol. 26: 679-683.

MOPPER, K. AND E.T. DEGENS. 1977. Organic carbon in the ocean: Nature and cycling, p. 293-316. *In* Bolin et al. [ed.] The global carbon cycle.

MOREL, F. M. M., AND N. M. L. MOREL-LAURENS. 1983. Trace metals and plankton in the oceans, p. 841-869. *In* C. S. Wong et al. [ed.] Trace metals in sea water. Plenum Press, New York, NY.

MOREL, F. M. M., J. G. RUETER, D. M. ANDERSON, AND R. R. L. GUILLARD. 1979. Aquil: A chemically defined phytoplankton culture medium for trace metal studies. J. Phycol. 15: 135-141.

MORRILL, L. C., AND A. R. LOEBLICH. 1979. An investigation of heterotrophic and photoheterotrophic capabilities in marine Pyrrhophyta. Phycologia 18: 394-404.

MORRIS, A. W. 1975. Dissolved molybdenum and vanadium in the northeast Atlantic Ocean. Deep Sea Res. 22: 49-54.

MURPHY, L. S., AND E. M. HAUGEN. 1985. The distribution and abundance of phototrophic ultraplankton in the North Atlantic. Limnol. Oceanogr. 30: 47-58.

MURPHY, T. P., D. R. S. LEAN, AND C. NALEWAJKO. 1976. Blue-green algae: Their excretion of iron-selective chelators enables them to dominate other algae. Science 192: 900-902.

NINNEMANN, H. 1980. Blue light photoreceptors. Bioscience 30: 166-170.

OCHIAI, E.-I. 1978. The evolution of the environment and its influence on the evolution of life. Origins of Life 9: 81-91.

OHKI, K., AND Y. FUJITA. 1982. Laboratory culture of the pelagic blue-green alga *Trichodesmium thiebautii*: Conditions for unialgal growth. Mar. Ecol. Prog. Ser. 7: 185-190.

OSTER, G. J. S. BELLIN, AND B. HOLMSTROM. 1962. Photochemistry of riboflavin. Experientia 18: 249-296.

OSTERBERG, P. 1974. Origins of metal ions in biology. Nature 249: 382-383.

PADAN, E., AND M. SHILO. 1973. Cyanophages-viruses attacking blue-green algae. Bact. Rev. 37: 343-370.

PINTNER, I. J., AND V. L. ALTMEYER. 1979. Vitamin B12 binder and other algal inhibitors. J. Phycol. 15: 391-398.

POSTGATE, J. R. 1982. The fundamentals of nitrogen fixation. Cambridge University Press, New York, NY. 252 p.

PRATT, L. H. 1978. Molecular properties of phytochrome. Photochem. Photobiol. 27: 81-105.

PREZELIN, B. B. 1981. Light reactions in photosynthesis, p. 1-43. *In* T. Platt [ed.] Physiological bases of phytoplankton ecology. Can. Bull. Fish. Aquat. Sci. 210.

PROVASOLI, L. 1977. Cultivation of animals: Axenic

cultivation, pp. 1295-1319. *In* O. Kinne [ed.] Marine ecology. Volume 3, Part 3, Wiley and Sons, New York, NY.

PROVASOLI, L, AND A. F. CARLUCCI. 1974. Vitamins and growth regulators, p. 741-787. *In* W. D. P. Stewart [ed.] Algal physiology and biochemistry.

PROVASOLI, L., AND I. J. PINTER. 1972. Effects of bacteria on seaweed morphology. J. Phycol. 8: 10.

PROVASOLI, L., J. J. A. MCLAUGHLIN, AND M. R. DROOP. 1957. The development of artificial media for marine algae. Arch. Mikrobiol. 25: 392-428.

PRYOR, W. A. 1978. The formation of free radicals and the consequences of their reactions *in vivo*. Photochem. Photobiol. 28: 787-801.

RACUSEN, R. H., AND A. W. GALSTON. 1980. Phytochrome modifies blue light induced electrical changes in corn coleoptiles. Pl. Physiol. 66: 534-535.

RAHAT, M., AND K. REICH. 1963. The B12 vitamins and growth of the flagellate *Prymnesium parvum*. J. Gen. Microbiol. 31: 195-202.

RAVEN, J. A., AND J. BEARDALL. 1981. Respiration and photorespiration, p. 52-82. *In* T. Platt [ed.] Physiological bases of phytoplankton ecology. Can Bull. Fish. Aquat. Sci. 210.

REDFIELD, A. C. 1958. The biological control of chemical factors in the environment. Am. Sci. 46: 205-221.

REDFIELD, A. C., B. H. KETCHUM, AND F. A. RICHARDS. 1963. The influence of organisms on the composition of sea water, p. 26-77. *In* Hill [ed.] The sea. Vol. 2.

SANDERS, J. G. 1979. Effects of arsenic speciation and phosphate concentration on arsenic inhibition of *Skeletonema costatum* (Bacillariophyceae). J. Phycol. 15: 424-428.

SCHOPF, J. W. 1970. Precambrian microorganisms and evolutionary events prior to the origin of vascular plants. Biol. Rev. 45: 319-352.

1978. The evolution of the earliest cells. Sci. Am. 239(3): 111-129.

SCLATER, F. R., E. BOYLE, AND J. M. EDMOND. 1976. On the marine geochemistry of nickel. Earth Planet. Sci. Lett. 31: 119-128.

SCHMIDT, W. 1984. Bluelight physiology. Bioscience 34: 698-704.

SCHMIDT, R. J., AND A. R. LOEBLICH. 1979. Distribution of paralytic shellfish poison among pyrrophyta. J. Mar. Biol. Assoc. U.K. 59: 479-487.

SIMKISS, K. 1983. Lipid solubility of heavy metals in saline solutions. J. Mar. Biol. Assoc. UK. 63: 1-7.

SIMPSON, F. B., AND J. B. NEILANDS. 1976. Siderochromes in Cyanophyceae: Isolation and characterization of schizakinen from *Anabaena* sp. J. Phycol. 12: 44-48.

SINGH, H. N., A. VAISHAMPAYAN, AND R. K. SINGH. 1978. Evidence for the involvement of a genetic determinant controlling functional specificity of group VI B elements in the metabolism of N_2 and NO_3^- in the blue-green alga *Nostoc muscorum*. Biochem. Biosphys. Res. Comm. 81: 67-74.

SMAYDA, T. J. 1974. Bioassay of the growth potential of the surface water of lower Narragansett Bay over an annual cycle using the diatom *Thalassiosira pseudonana* (oceanic clone, 13-1). Limnol. Oceanogr. 19: 889-901.

1980. Phytoplankton species succession, p. 493-570. *In* I. Morris [ed.] The physiological ecology of phytoplankton. University of California, Berkeley, CA.

SMITH, A. J. 1982. Modes of cyanobacterial carbon metabolism, p. 47-85. *In* N. G. Carr and B. A. Whitton [ed.] The biology of cyanobacteria.

SMITH, A. P. 1979. The paradox of autotoxicity in plants. Evol. Theory. 4: 173-180.

STADTMAN, T. C. 1980. Selenium dependent enzymes. Ann. Rev. Biochem. 49: 93-110.

STEIDINGER, K. A, AND D. G. BADEN. 1984. Toxic marine dinoflagellates, p. 201-261. *In* D. L. Spector [ed.] *Dinoflagellates*.

STRYER, L. 1981. Biochemistry. Freeman and Co., San Francisco, CA.

SUGIMURA, Y., Y. SUZUKI, AND Y. MIYAKE. 1978. The dissolved organic iron in seawater. Deep Sea Res. 25: 309-314.

SUNDA, W. G., S. A. HUNTSMAN, AND G. R. HARVEY. 1983. Photoreduction of manganese oxides in seawater and its geochemical and biological implications. Nature 301: 234-236.

SUNDA, W. G., AND R. R. L. GUILLARD. 1976. The relationship between cupric ion activity and the toxicity of copper to phytoplankton. J. Mar. Res. 34: 511-529.

SWIFT, D. G. 1980. Vitamins and phytoplankton growth, p. 329-368. *In* I. Morris [ed.] The physiological ecology of phytoplankton. University of California Press, Berkeley, CA. 625 p.

SYMES, J. L. AND D. R. KESTER. 1985. The distribution of iron in the northwest Atlantic. Mar. Chem. 17: 57-74.

TAPPAN, H. 1980. The paleobiology of plant protists, Freeman and Co., CA. 1028 p.

TAPPAN, H., AND A. R. LOEBLICH. 1973. Evolution of the oceanic plankton. Earth Sci. Rev. 9: 207-240.

TAYLOR, F. J. R. 1980. Basic biological features of phytoplankton cells, p. 3-55. *In* I. Morris [ed.] The physiological ecology of phytoplankton. University of California Press, Berkeley, CA. 625 p.

TAYLOR, D. L. 1973. Algal symbionts of invertebrates. Ann. Rev. Microbiol. 27: 171-187.

TAYLOR, A. D., AND B. A. BONNER. 1967. Isolation of phytochrome from the algae *Mesotaenium* and liverwort *Sphaerocarpos*. Pl. Physiol. 42: 762-766.

TERAMURA, A. G. 1982. The amelioration of UV-B effects on productivity by visible radiation, p. 367-382. *In* J. Calkins [ed.] The Role of Solar Ultraviolet Radiation in Marine Ecosystems. Plenum Press, New York, NY. 724 p.

THEIS, T. L., AND P. C. SINGER. 1974. Complexation of iron (II) by organic matter and its effect on iron (II) oxygenation. Environ. Sci. Technol. 8: 569-573.

TRICK, C. G., R. J. ANDERSEN, A. GILLAM, AND P. J. HARRISON. 1983a. Prorocentrin: an extracellular siderophore produced by the marine dinoflagellate *Prorocentrum minimum*. Science 219: 306–308.

TRICK, C. G., R. J. ANDERSEN, N. M. PRICE, A. GILLAM, AND P. J. HARRISON. 1983b. Examination of hydroxamate siderophore production by neritic eukaryotic marine phytoplankton. Mar. Biol. 75: 9–17.

UKELES, R., AND W. E. ROSE. 1976. Observations on organic carbon utilization by photosynthetic marine microalgae. Mar. Biol. 37: 11–28.

UKELES, R., AND J. BISHOP. 1975. Enhancement of phytoplankton growth by marine bacteria. J. Phycol. 11: 142–149.

VAIRAVAMURTHY, A., M. O. ANDREA, AND R. L. IVERSON. 1985. Biosynthesis of dimethylsulfide and dimethylpropiothetin by *Hymenomonas carterae* in relation to sulfur source and salinity variations. Limnol. Oceanogr. 30: 59–70.

VAN BAALEN, C. 1965. Quantitative surface plating of coccoid blue-green algae. J. Phycol. 1: 19–22.

VAN BAALEN, C., AND J. E. MARLER. 1966. Occurrence of hydrogen peroxide in sea water. Nature 211: 951.

VALIELA, I. 1984. Marine ecological processes. Springer-Verlag, New York, NY. 546 p.

VESK, M., AND S. W. JEFFREY. 1977. Effect of blue-green light on photosynthetic pigments and chloroplast structure in unicellular marine algae from six classes. J. Phycol. 13: 280–288.

WAITE, T. D., AND F. M. M. MOREL. 1984. Photoreductive dissolution of colloidal iron oxides in natural waters. Environ. Sci. Technol. 18: 860–868.

WALLEN, D. G., AND G. H. GEEN. 1971. Light quality in relation to growth, photosynthetic rates and carbon metabolism in two species of marine phytoplankton algae. Mar. Biol. 10: 34–43.

WEHR, J. D., AND L. M. BROWN. 1985. Selenium requirement of a bloom-forming planktonic alga from softwater and acidified lakes. Can. J. Fish. Aquat. Sci. 42: 1783–1788.

WELLS, M. L., N. Z. ZORKIN, AND A. G. LEWIS. 1983. The role of colloid chemistry in providing a source of iron to phytoplankton. J. Mar. Res. 41: 731–746.

WHEELER, P., B. NORTH, M. LITTLER, AND G. STEPHENS. 1977. Uptake of glycine by natural phytoplankton communities. Limnol. Oceanogr. 22: 900–910.

WHITTAKER, R. H., AND P. P. FEENY. K 1971. Allelochemicals: Chemical interactions between species. Science 171: 757–770.

WILLIAMS, R. J. P. 1981. Natural selection of the chemical elements. Proc. R. Soc. Lond. 213B: 361–397.

WILLIAMS, R. J. P., AND J. J. R. F. DA SYLVA. 1978. High redox potential chemicals in biological systems, p. 121–171. *In* R. J. P. Williams and J. R. R. F. Da Sylva [ed.] New trends in bio-inorganic chemistry. Academic Press, New York, NY.

WILLIAMS, P. J. LeB. 1975. Biological and chemical aspects of dissolved organic material in sea water. Chem. Oceanogr. 2: 301–363.

WOLK, C. P. 1973. Physiology and cytology chemistry of blue-green algae. Bact. Rev. 37: 32–101.

WOOD, A. M., D. W. EVANS, AND J. J. ALBERTS. 1983. Use of an ion exchange technique to measure copper complexing capacity on the continental shelf of the southeastern United States and in the Sargasso Sea. Mar. Chem. 13: 305–326.

WOOD, J. M., AND H. K. WANG. 1983. Microbial resistance to heavy metals. Environ. Sci. Technol. 17: 582–590.

WOOD, P. M. 1978. Interchangeable copper and iron proteins in algal photosynthesis. Studies on plastocyanin and cytochrome c 522 *Chlamydomonas*. Eur. J. Biochem. 87: 9–19.

YOCH, D. C., AND R. P. CARITHERS. 1979. Bacterial iron-sulfur proteins. Microbiol. Rev. 43: 384–421.

ZAFIRIOU, D. C. 1983. Natural water photochemistry. Chem. Oceanogr. 8: 339–379.

ZAFIRIOU, O. C., AND M. B TRUE. 1980. Interconversion of iron (III) hydroxy complexes in seawater. Mar. Chem. 8: 281–288.

ZAFIRIOU, O. C., J. GOUSSET-DUBIEN, R. G. ZEPP, AND R. G. ZIKA. 1984. Photochemistry of natural waters. Environ. Sci. Technol. 18: 358–371.

ZIKA, R. G., J. W. MOFFETT, R. G. PETASNE, W. J. COOPER, AND E. S. SALTZMAN. 1985. Spatial and temporal variations of hydrogen peroxide in Gulf of Mexico waters. Geochim. Cosmochim. Acta 49: 1173–1184.

ZUCKER, M. 1972. Light and enzymes. Ann. Rev. Pl. Physiol. 23: 133–156.

Photosynthetic Response of Marine Picoplankton at Low Photon Flux

MARLON R. LEWIS

*Department of Oceanography, Dalhousie University,
Halifax, Nova Scotia B3H 4H9*

RODERICK E. WARNOCK

Department of Biology, Dalhousie University, Halifax, Nova Scotia B3H 4H9

AND TREVOR PLATT

*Department of Fisheries and Oceans, Marine Ecology Laboratory,
Dartmouth, Nova Scotia B2Y 4A2*

Introduction

Recent discoveries establishing the existence (e.g. Waterbury et al. 1979; Johnson and Sieburth 1979) and physiological competence (e.g. Li et al. 1983; Platt et al. 1983) of the oceanic, autotrophic picoplankton, have profound implications for estimation of global fluxes of carbon. These organisms may account for the major part of photoautotrophic carbon reduction over large parts of the world's oceans, particularly at low levels of local irradiance. This we conclude from field work (Platt et al. 1983) and from laboratory observations on physiological performance of unialgal cultures (e.g. Morris and Glover 1981; Glover and Morris 1981; Cuhel and Waterbury 1984; Barlow and Alberte 1985; Campbell 1985; Wood 1985; Glover et al. 1986). Further evidence comes from theoretical computations concerning the efficiency of irradiance absorption by cells of small size (Kirk 1975; Morel and Bricaud 1981; Bricaud et al. 1983): for a particular intracellular concentration of photosynthetic pigment, the pigment-specific absorption coefficient varies strongly, and inversely, with the size of organism. The smallest cells absorb light with the greatest efficiency.

What are the implications for calculation of global primary production? Here, we argue that the photosynthetic response at the intensities and spectral composition of irradiance in the deeper water of the ocean is paramount in determination of rates of depth-integrated primary production, particularly in the open ocean, "blue-water" regions. We consider the photosynthetic performance of the autotrophic picoplankton at low light and how that performance may affect their success in different oceanic conditions. We also consider briefly an inverse problem: how the unique characteristics of the picoplankton might affect remote observation of photosynthetic pigment and primary production.

Background

We take as a point of departure the photosynthesis–irradiance curve, $P^B = f(I)$, where the photosynthetic rate P^B has been normalized to chlorophyll a biomass (B) and $f(I)$ is some saturating function of irradiance, I. Two parameters are sufficient to characterize $f(I)$: they are the initial slope α^B,

$$(1) \qquad \alpha^B = \left. \frac{\partial P^B}{\partial I} \right|_{I \to 0}$$

and the maximum rate of photosynthesis, P^B_m. We neglect photoinhibition effects. Many explicit forms have been used for $f(I)$ (Platt et al. 1977); one useful form (Platt et al. 1980) is

(2) $\qquad P^B = P_m^B (1\text{-}\exp(-\alpha^B I/P_m^B))$

Assuming uniform attenuance with depth (K_s is the attenuation coefficient for diffuse, downwelling irradiance) and chlorophyll (B), an expression for the distribution of irradiance can be substituted into Eq. 2 and the equation integrated from the sea-surface to infinity to yield P', the areal photosynthetic rate (Lewis et al. 1985),

$$(3) \qquad P' = \int_0^\infty P^B B \, dz = \frac{-P_m^B B}{K_s} \sum_{N=1}^{\infty} \frac{(-I_o/I_k)^n}{n \cdot n!}$$

whose units are Mass C $(\text{L}^2\text{T})^{-1}$ and where I_k is the derived parameter $I_k = P_m^B/\alpha^B$. Note that all saturating photosynthesis–irradiance curves can be reduced to a form similar to Eq. 3 such that,

$$(4) \qquad P' = \frac{P_m^B B}{K_s} \, f(I_o/I_k)$$

where the functions of I_o/I_k will differ slightly in form. Therefore, our conclusions will not be biased through choice of photosynthesis model.

A problem of some interest here is the error introduced into estimates of P' from Eq. 3 because of error in either P_m^B or α^B. Lewis et al. (1985a) compared the relative magnitudes of such errors (see Table 1); they are equal at $I_o/I_k = 4$. With I_o/I_k less than 4, an error in α^B produces larger effect in P' than the same relative error in P_m^B. With I_o/I_k greater than 4, the opposite is true. As will be seen below, I_o/I_k is usually less than 4 for most oceanic waters on a daily-average basis; consequently, accurate determinations or estimates of *light-limited* photosynthesis are of utmost importance for estimating integral photosynthesis in the sea.

TABLE 1. Partial derivatives with respect to parameters of equation (3), for use in sensitivity analysis.

$$\frac{\partial P'}{\partial P_m} = \frac{-1}{K_s} \sum_{n=1}^{\infty} \frac{(-I_o/I_k)^n (1-n)}{n \cdot n!}$$

$$\frac{\partial P'}{\partial \alpha} = \frac{I_o}{K_s} \sum_{n=1}^{\infty} \frac{(-I_o/I_k)^{n-1}}{n \cdot n!}$$

$$\frac{\partial P'}{\partial K_s} = \frac{P_m}{K_s^2} \sum_{n=1}^{\infty} \frac{(-I_o/I_k)^n}{n \cdot n!}$$

We also note that the errors discussed here are all inverse functions of the attenuation coefficient, and furthermore, that the error introduced into estimates of P' because of error in determining K_s is proportional to K_s^{-2} (Lewis et al. 1985a). Therefore, the region of the worlds' oceans where the estimation of integral photosynthesis is most sensitive to error in parameter estimates is the blue-water, optically-clear regime: it is also the region where we have the least confidence in the parameters because of sparse spatial and temporal coverage (Platt 1984). Moreover, it is the zone which has been shown to have a large component of the autotrophic biomass associated with cells less than 1 μm in size whose physiological parameters are poorly characterized.

Remote Sensing Considerations

A central problem in biological oceanography is characterization of the mean and variance of biomass and production in the sea. For phytoplankton, remote sensing,

236

from air or space borne instruments of light leaving the ocean's surface has been invoked as an aid to this end. At present, estimates of sea-surface chlorophyll concentration can be made from observations using the Coastal Zone Color Scanner (Hovis et al. 1980; Gordon et al. 1980; Gordon and Morel 1983). The problem remains to estimate primary production from such measurements of the biomass (Sathyendranath, this volume).

Eppley et al. (1985) compiled a considerable body of shipboard measurements on sea-surface chlorophyll concentrations and simultaneous measurements of integral production. The stated goal was to estimate P' empirically from B as determined from satellite observation (in practice defined as the average B over a depth equal to one attenuation length, $1/K_s$). The proportionality factor F ($= P'/B$) was determined from stations scattered around the world, and an analysis was made of the variability in F. From Eqs. 3 and 4 it is immediately obvious that

$$(5) \qquad F = \frac{P_m^B}{K_s} \; f(I_o/I_k),$$

and it is clear that F will vary directly as a function of incident photon flux density (PFD) (as demonstrated statistically in their paper); with temperature (as has been demonstrated for P_m^B); and inversely with the optical attenuation coefficient (as demonstrated statistically in their paper). The qualitative agreement between empirical analysis and the theoretical treatment is therefore good.

The quantitative agreement is also encouraging. Under the assumption (see below) that $I_o < I_k$, only the first term on the right hand side of Eq. 3 is important and Eq. 3 reduces to,

$$(6) \qquad P' = \frac{P_m^B \, B \, I_o}{K_s \, I_k}$$

or

$$(7) \qquad P' = \frac{B \, \alpha^B}{K_s} \; I_o$$

where the factor F (Eppley et al. 1985) can be identified as $\alpha^B I_o / K_s$. The variation in F with respect to variation in I_o within a given cruise is shown to be linear with slope $= 14.3$ mg C/(mg Chl m^{-3}). The intercept does not appear to be significantly different from zero. If a value of 0.01 mg C/(mg Chl a^{-1} h^{-1} (μE m^{-2} s^{-1})$^{-1}$) for α^B and a value of 0.1 m^{-1} for K_s can be considered typical, then one can calculate from Eq. 8 that the value of the slope should be 28 mg C/(mg Chl m^{-3}). The order of magnitude is correct and the overestimate is in the direction and magnitude one would expect given a linear approximation to saturating photosynthesis.

A method based in first principles for computing primary production of the water column as a function of surface light intensity has recently been published (Platt 1986). It differs from the above analysis in that the production is normalized to the integral taken through the entire euphotic zone rather than the surface concentration. The two analyses are easily inter-converted. In Platt (1986), an error analysis shows the bias in integral production introduced by the assumption that photosynthesis is linear in available light. The relevant dimensionless number that determines the magnitude of the error is I_o/I_k. When it takes the numerical value of 4, P' is overestimated, using the linear model, by approximately 100%.

But what is the magnitude of I_o/I_k in nature? In an unpublished compilation of results from over 800 photosynthesis–irradiance curves determined on populations from the high Arctic to the tropics we find a mean I_k of 54 Wm^{-2} (or ca. 250 μE m^{-2} s^{-1}). On a daily average basis, Reed (1977) shows I_o to be maximal at latitude 30° and roughly 130 W m^{-2} (or ca. 620 μE m^{-2} s^{-1}). Of course, the unweighted daily average may not be the most useful average, but it does serve to set the appropriate scale for questions addressed here. We conclude that for much of the ocean, much of the time, and for the problem of estimating the integral photosynthetic rate from remote observations, it is important to increase precision in the determination of rates of photosynthesis at low irradiance levels, particularly with respect to the smallest of the photoautotrophic organisms.

Variability in Photosynthetic Rate at Low Irradiance

At low irradiances, photosynthesis (P) is a linear function of irradiance (I), with slope $= \alpha^B$. It can be shown that α^B is proportional to the product of the light absorbed by the light-harvesting antenna and the quantum efficiency with which absorbed quanta are used to generate reduced carbon. Because α^B depends upon the amount of light absorbed, it is wavelength dependent such that

$$(8) \qquad \alpha^B(\lambda) = a_c (\lambda) \, \phi_A (\lambda)$$

where $(a_c(\lambda))$ is the biomass-normalized (diffuse) absorption cross-section [m^2 biomass^{-1}] and $\phi_A(\lambda)$ is the dimensionless apparent quantum yield [mol C(mol photons)$^{-1}$]. Chlorophyll a is the biomass index most frequently employed. Significant variations in α^B have been found in both cultured algae and natural assemblages of phytoplankton; we now consider why this may be so.

Variations in α^B may come about through variability in either $\phi_A(\lambda)$ or $a_c(\lambda)$. Current mechanistic models of the photosynthetic process require a minimum of 8 quanta absorbed per molecule of oxygen evolved and a quantum requirement ($1/\phi$) of 10 is favored on thermodynamic grounds (Raven and Beardall 1982). This places an upper bound to the quantum yield for oxygen evolution (ϕ_m) of 0.1. The quantum yield actually realised may be considerably less. In addition to the possibility of light-absorbing photosynthetic pigments that are not connected to reaction centres (and therefore photosynthetically incompetent), some pigments may be preferentially associated with one or other of the photosystems thereby compromising the balanced flow of electrons envisaged by the serial photosystem model. This is particularly pertinent with respect to cyanobacteria where the phycobiliprotein pigments are known to be associated exclusively with Photosystem II, while chlorophyll a is predominantly associated with Photosystem I (Öquist 1974; Mimuro and Fujita 1977; Tel-Or and Malkin 1977; Wang et al. 1977). The stoichiometry of the two photosystems has also been shown to vary in cyanobacteria (Myers et al. 1978, 1980; Kawamura et al. 1979).

Secondly, a slightly greater number of quanta/CO_2 molecule reduced may be expected as a result of competition for photoreductant by pathways other than CO_2 assimilation (for example the photoreduction of nitrate or nitrite to ammonia). These considerations suggest a maximum quantum yield for carbon fixation (ϕ_A) <0.10 (Radmer and Kok 1977a,b). Variations in $\phi_A(\lambda)$ may thus contribute to variations in α^B. Estimates of the magnitude and variability of $\phi_A(\lambda)$ for natural phytoplankton assemblages have only recently been determined (Lewis et al. 1985).

There are two primary sources of variability in $a_c(\lambda)$. First, the localisation of the photosynthetic pigments within discrete units implies a significant effect due to size of the organisms if the intracellular pigment concentration remains constant. Second, the presence of accessory pigments will alter the magnitude and spectral variability in $a_c(\lambda)$.

Estimates of the *in vivo* chlorophyll-specific absorption cross section are complicated by the fact that the pigment molecules of the light-harvesting antenna are not uniformly distributed but occur in discrete packages within chloroplasts or cells. This "package effect" results in a lower specific absorption per unit pigment and is proportionately greatest in the spectral bands where absorption is strongest (Duysens 1956). As a result, the *in vivo* chlorophyll-specific absorption cross-section is a variable quantity which

238

depends not only upon the particular kind and quantity of pigment proteins involved in light harvesting relative to the concentration of chl *a*, but also is heavily dependent upon the physical and geometrical properties of the absorbing cells. The implications of this discrete packaging of pigments with respect to light absorption by phytoplankton cells have been investigated theoretically (Kirk 1975a,b, 1976; Morel and Bricaud 1981; Bricaud and Morel 1983). A major conclusion is that for a given intracellular concentration of pigment, the *in vivo* chlorophyll-specific absorption cross-section of a suspension of cells increases with diminishing cell size and approaches that of a true solution. Picoplankton cells, by virtue of their small size, are likely to exhibit considerably higher values of $a_c(\lambda)$ than larger phytoplankton cells. Higher values of $a_c(\lambda)$ would be reflected in higher rates of photosynthesis under low irradiance.

The amount of light absorbed is a function of both the spectral distribution of the incident irradiance and the spectral response of the absorption cross-section. In the absence of non-linear effects, an experimentally derived estimate of the average initial slope $\bar{\alpha}$ may be expressed:

$$(9) \qquad \bar{\alpha} = \frac{\int_{400}^{700} \alpha(\lambda)\, I(\lambda)\, d\lambda}{\int_{400}^{700} I(\lambda)\, d\lambda}$$

where the integration is carried out over the photosynthetically active waveband, 400–700 nm.

Because of this spectral dependence of $\bar{\alpha}$, use of a particular $\bar{\alpha}$ determined in one irradiance field $I(\lambda)$ to estimate the photosynthetic efficiency in a second irradiance field of similar intensity but different spectral distribution $I'(\lambda)$ will result in error. The relative error introduced ($\epsilon = P - \hat{P})/P$ where $\hat{P}$ is the predicted photosynthetic rate and P is the true photosynthetic rate) will be:

$$(10) \qquad \epsilon = \frac{\int_{400}^{700} \alpha(\lambda)\, I(\lambda)\, d\lambda}{\int_{400}^{700} \alpha(\lambda)\, I'(\lambda)\, d\lambda} - 1$$

Similarly, estimates of $\bar{\alpha}$ for two cell populations of differing spectral responses $\alpha(\lambda)$ and $\alpha'(\lambda)$ will be in error in the absence of explicit information on the spectral distribution of the irradiance field $I(\lambda)$.

Variations in accessory pigments in addition to chlorophyll *a* are responsible for such spectral variations in the "chlorophyll-specific" absorption coefficient. Cyanobacteria, which may constitute a significant fraction of the picoplankton, possess phycobiliproteins as their major light-harvesting antenna pigments. These pigments absorb maximally at wavelengths between 490 and 620 nm, in the spectral region where absorption by chlorophyll, and to a lesser extent carotenoids, is weak. It has been suggested that these differences in spectral absorption of cellular pigments may confer some advantage upon phycobiliprotein-containing cells in the irradiance fields characteristic of natural waters (Ong et al. 1984; Wood 1985). The degree and efficiency with which these pigments participate in photosynthetic carbon reduction remains to be demonstrated and some preliminary evidence (Wyman et al. 1985) suggests that under certain conditions the quantum yield of photons absorbed by phycobiliproteins may be very small.

Absorption and Action Spectra of Picoplankton

Data are starting to accumulate on rates of light absorption and photosynthesis for the many picoplankton strains now held in culture. With respect to marine cyanobacteria, three broad groups have been distinguished, spectrally, among clonal isolates of marine *Synechococcus* spp. (Wood 1985). The first to be identified were clones lacking phycoerythrin (PE) and containing phycocyanin (PC) and allophycocyanin (APC) with strong absorption at ca. 630–650 nm (i.e. WH5701, WH8007). Clonal isolates containing PE as the major light-harvesting phycobiliprotein could be further divided into two groups: those whose PE contained both phycoerythrobilin (PEB) and phycourobilin (PUB) chromophores (designated Type 1, i.e. WH7803, WH6501) and those whose PE contained only the PEB chromophore (designated Type 2, i.e. WH8010, WH8018). The PUB chromophore absorbs maximally at ca. 490–500 nm compared to the PEB chromophore which absorbs maximally at ca. 550 nm. Within the clonal group whose PE possesses both PEB and PUB chromophores, different clones exhibit considerable variation in the ratio of the two chromophores (Ong et al. 1984; Wood 1985, Wood (these proceedings)). These pigments, together with various carotenoids (Guillard et al. 1985) result in strong absorption in the 500–550 nm range, a waveband contributing significantly to submarine irradiance fields in both the open ocean and coastal waters.

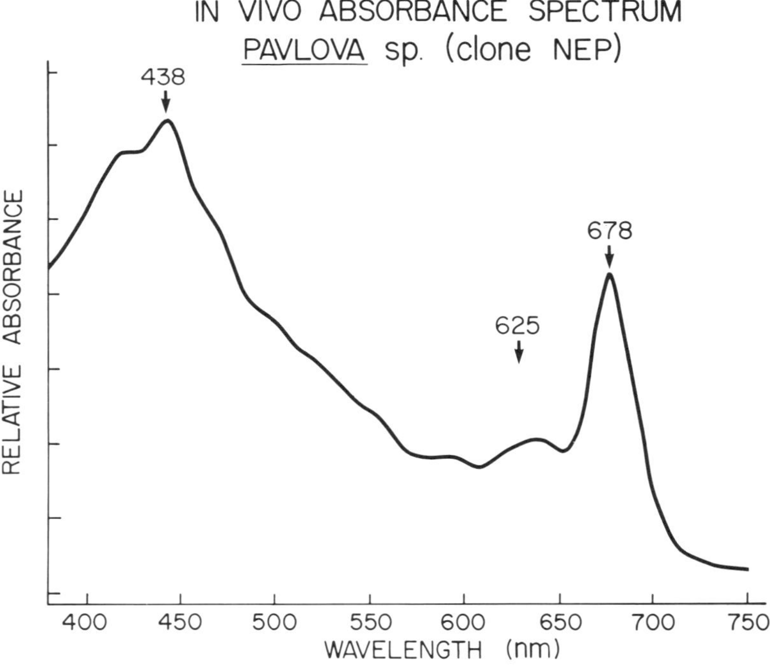

FIG. 1. *In vivo* absorbance spectrum of the picoplankton eukaryote *Pavlova* sp. (clone NEP).

Less information is available concerning the spectral absorption characteristics of the eukaryotic picoplankton component. Figure 1 shows the *in vivo* absorption spectra of the prymnesiophyte *Pavlova* sp. The peaks at 438 and 678 nm may be attributed

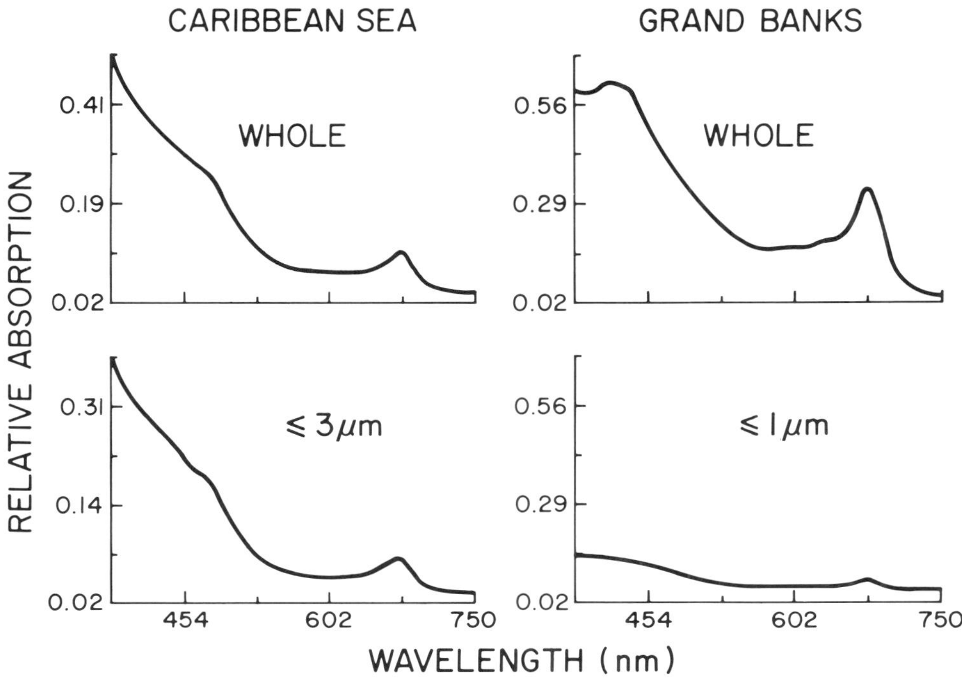

FIG. 2. *In vivo* absorbance spectra of natural phytoplankton assemblages from the Caribbean Sea and the Grand Banks off Newfoundland.

to chlorophyll *a*, the shoulder at ca. 470–500 is attributable to fucoxanthin and chlorophyll *c*, and the minor peak at 625 nm is due to chlorophyll *c*. In all respects the absorption spectrum of *Pavlova* sp. resembles those absorption spectra determined for natural phytoplankton populations (Fig. 2).

The first determinations of the photosynthetic action spectra of the marine cyanobacteria and the picoplanktonic eukaryote *Pavlova* sp. are given in Fig. 4–6. These curves result from the measurement of the initial sope of the photosynthesis–irradiance curve at 12 wavebands spanning the range 400–675 nm (25 nm half-maximum bandpass, every 25 nm, 8 intensities per waveband, Lewis et al. 1985a,b). These curves permit a very precise estimate of $\alpha(\lambda)$ at each of the 12 wavebands; a set of 12 photosynthesis–irradiance curves derived for the eukaryote *Pavlova* sp. are shown in Fig. 3. Due to increases in initial slope at wavebands corresponding to phycoerythrin (WH7803) and phycocyanin (WH5701) absorption, the curves for the prokaryotes are clearly distinguishable from the curve for *Pavlova* sp. which results primarily from chlorophyll *a* and carotenoid absorption. The phycobilipigment peaks are the most characteristic feature of both absorption and photosynthetic action spectra; one is led to expect similar features when these measurements are carried out on natural populations containing large numbers of cyanobacteria.

But what do the field data show? Extensive measurements have been made of the optical properties of the constituents of the upper ocean other than water (e.g. Jerlov 1976; Morel and Prieur 1977; Morel 1978; Smith and Baker 1978; Prieur and Sathyendranath 1981) as well as optical properties of the particulate matter concentrated on filters (Yentsch 1957, 1962, 1980; Kiefer and SooHoo 1982; Lewis et al. 1985a). None of these studies, to our knowledge, have demonstrated clear absorption peaks which can be identified with phycobilipigments.

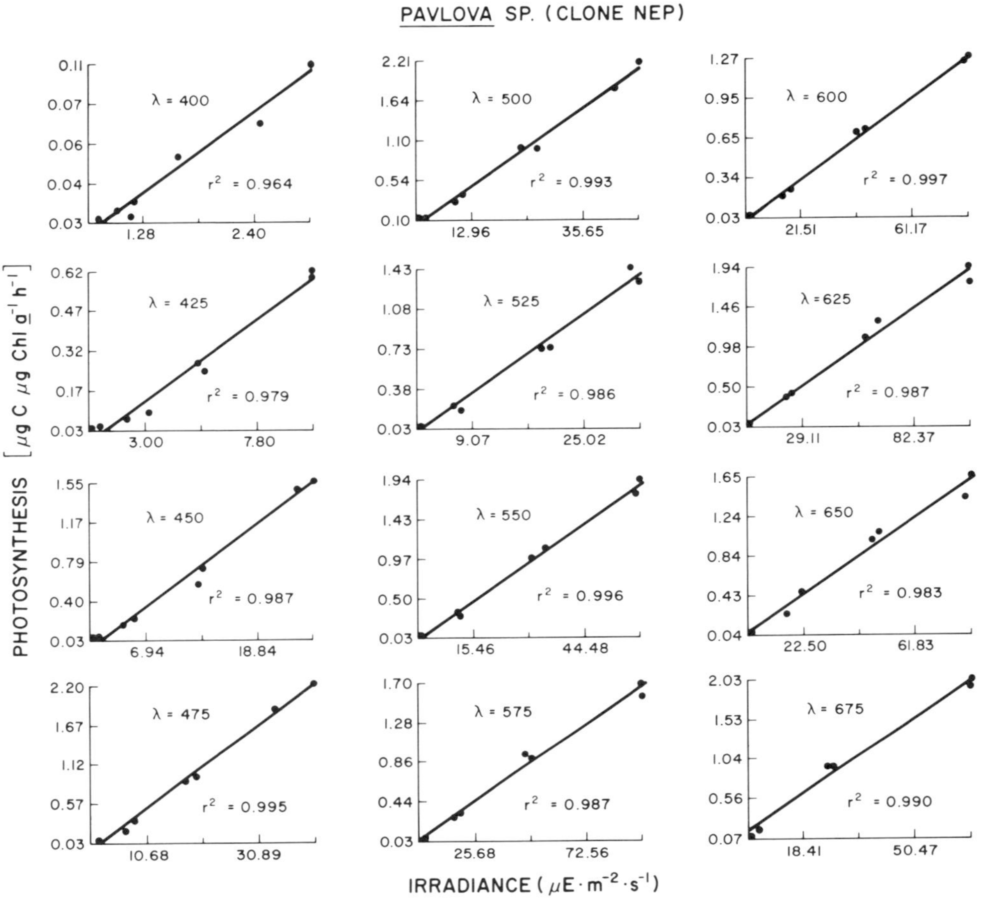

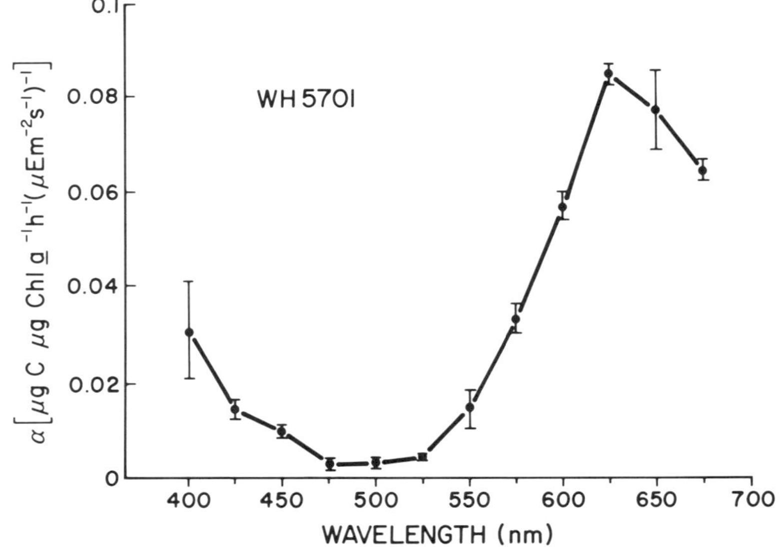

FIG. 3. P vs I curves for a set of 12 wavelengths from a single experiment with *Pavlova* sp. The slopes are used to construct the action spectrum ($\alpha^B(\lambda)$) shown in Fig. 6.

FIG. 4. Photosynthetic action spectrum ($\alpha^B(\lambda)$) for the phycocyanin-dominant cyanobacterium WH5701.

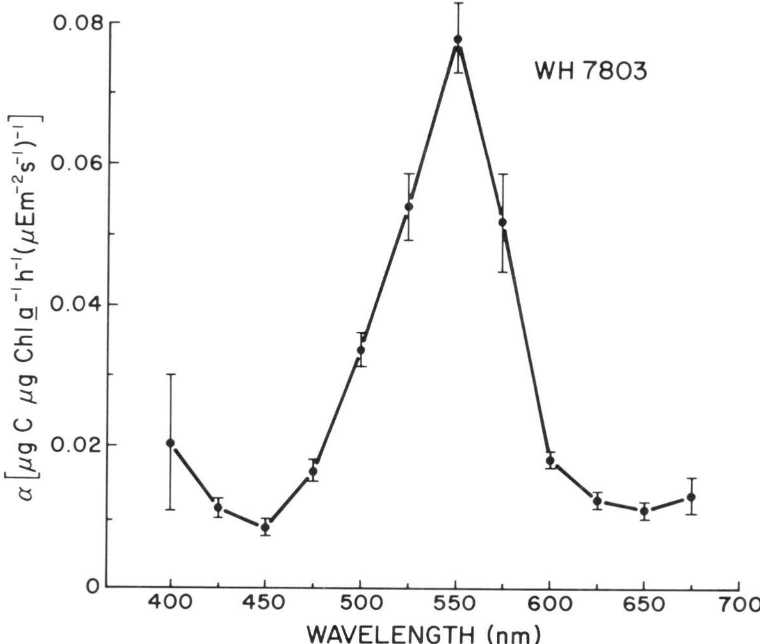

FIG. 5. Photosynthetic action spectrum ($\alpha^B(\lambda)$) for the phycoerythrin-dominant cyanobacterium WH7803.

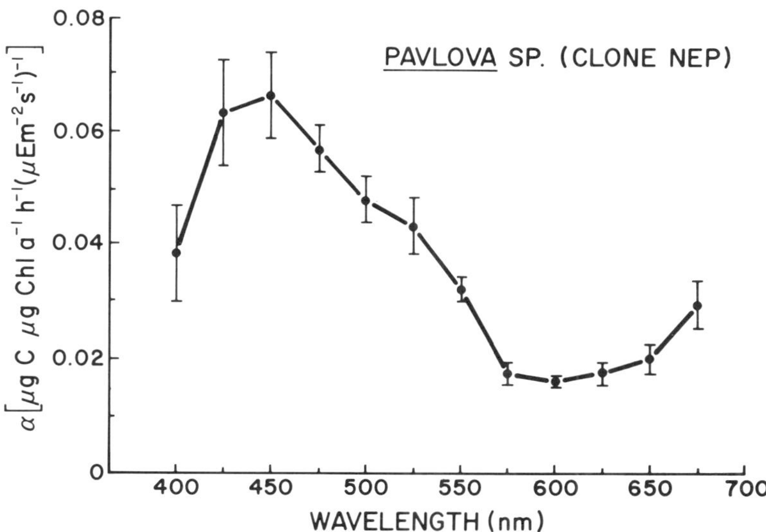

FIG. 6. Photosynthetic action spectrum ($\alpha^B(\lambda)$) for the eukaryote *Pavlova* sp. (clone NEP).

Nor do those photosynthetic action spectra which have been determined for natural phytoplankton populations show any phycobilipigment involvement (Lewis et al. 1985a,b; Harrison et al. 1985; see Fig. 7). These spectra have been measured on populations from the high Arctic to the Sargasso and Caribbean Seas; all of them appear to result solely from absorption by chlorophylls and carotenoids (which also are present in the marine cyanobacteria; see Guillard et al. 1985). There was no evidence of significant peaks in wavebands corresponding to phycobilipigment absorption. Significant absorption by size fractions corresponding to the picoplankton was demonstrated for Sargasso Sea and Caribbean Sea samples, but there was little significant absorption in the <1–3 μm fractions in the Arctic or Grand Banks (Fig. 2). For the blue water stations, the absorption spectra appear to be due to particulate material with a chlorophyll-like spectral composition; again, even with size-fractionated material no evidence was found for significant phycobilipigment absorption.

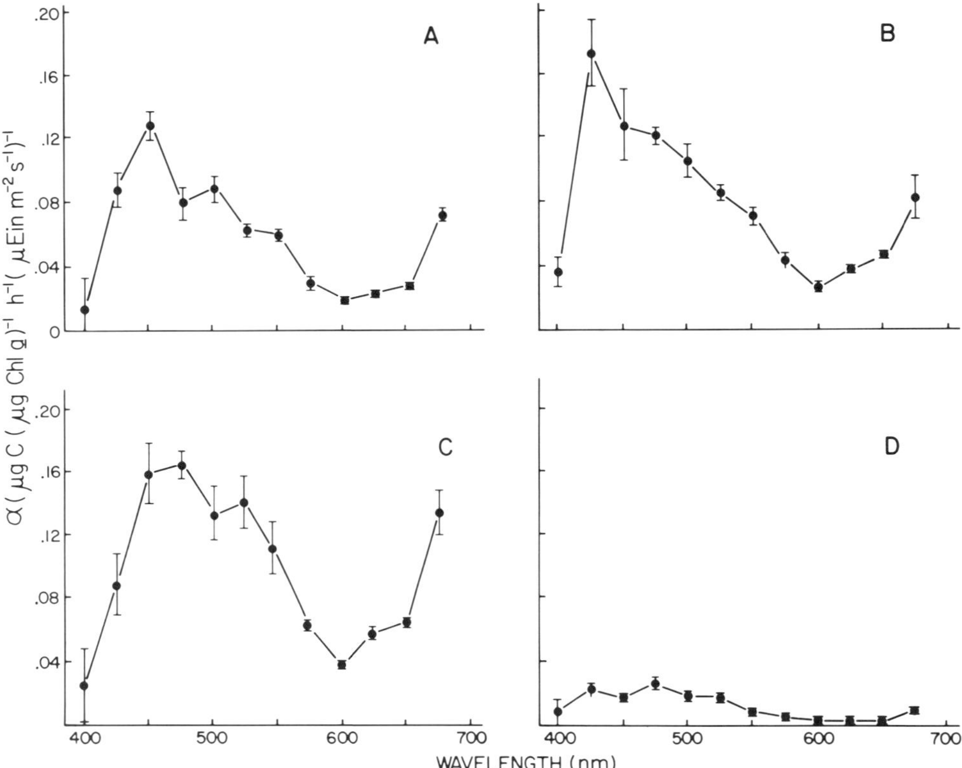

FIG. 7. Photosynthetic action spectra ($\alpha^B(\lambda)$) for natural phytoplankton assemblages from (A) Arctic, (B) Ice Algae, (C) Grand Banks, and (D) Sargasso Sea.

Cyanobacterial populations containing these pigments were present, however, at least for the Sargasso Sea experiment. Li and Dickie (1985) demonstrated carbon uptake by photoautotrophic cells less than 1 μm, which could not be inhibited by cycloheximide, an inhibitor of eukaryote protein synthesis. Campbell (1985) showed that water samples collected on this cruise contained concentrations of cells of order $10^6 \cdot L^{-1}$ and which had proteins with a high titer for immunofluorescent antibodies produced against phycoerythrin. Evidently, even though the cyanobacterial cells were present, their concentration of pigment was insufficient to be observed in the absorption spectra, and was not sufficient to contribute significantly to the photosynthetic action spectra in contrast to that observed in the laboratory cultures.

Even if the phycobilipigments were present in concentrations high enough to influence the absorption of irradiance, one could not be certain of their photosynthetic competence. The intense *in vivo* fluorescence observed at 580 nm in natural assemblages of phytoplankton both detected by submersible fluorometers (Glover et al. 1985) and by remote sensing (Hoge and Swift 1983; Exton et al. 1983) has been ascribed to phycoerythrin (PE) (Yentsch and Yentsch 1979) found to be associated with the picoplankton fraction containing chroococcoid cyanobacteria (Yentsch 1983). The high yield of PE fluorescence in both natural assemblages and isolated strains of PE-containing cyanobacteria (Kursar et al. 1981; Barlow and Alberte 1985) implies

244

that these pigments are not effectively transferring absorbed energy to the photo-chemical reaction centres and that a substantial fraction of the cellular phycoerythrin is not coupled to the photosynthetic energy transduction chain of the phycobilisome. In contrast to isolated physobilisomes which exhibit coupled energy transfer with little or no PE fluorescence, whole cells of a phycoerythrin-containing strains (WH7803 type + WH8018 type) both show high PE fluorescence (Kursar et al. 1981). The magnitude of PE fluorescence appears to be related to the irradiance and possibly also the nutrient history of the cells. Glover et al. (1985) found that in natural assemblages of picoplankton in the Gulf of Maine, PE fluorescence intensity per cyanobacterial cell increased with depth; cyanobacterial cells from depth showed 5–6 times the cellular fluorescence of those near the surface. In addition Barlow and Alberte (1985) suggest that a significant inverse correlation exists between PE fluorescence and both photosynthesis and growth rate for two PE-containing clones of *Synechococcus* (WH8018 and WH7803). They report a poor correlation between PE fluorescence and pigment content implying that the proportion of photosyntheti-cally competent PE is variable. Finally, the recent work of Wyman et al. (1985) elegantly demonstrates that at least under certain growth conditions a proportion of the phycobilipigment pool is not photosynthetically competent and functions primarily for nitrogen storage in the marine cyanobacteria WH7803.

Whole cell absorption measurements made on three cyanobacterial clones clearly show the selective loss of phycobiliproteins after nutrient starvation (Fig. 8). Here, for all three clones, nutrient starvation virtually eliminates the accessory phycobili-protein pigment; the difference spectra between the nutrient sufficient and deficient

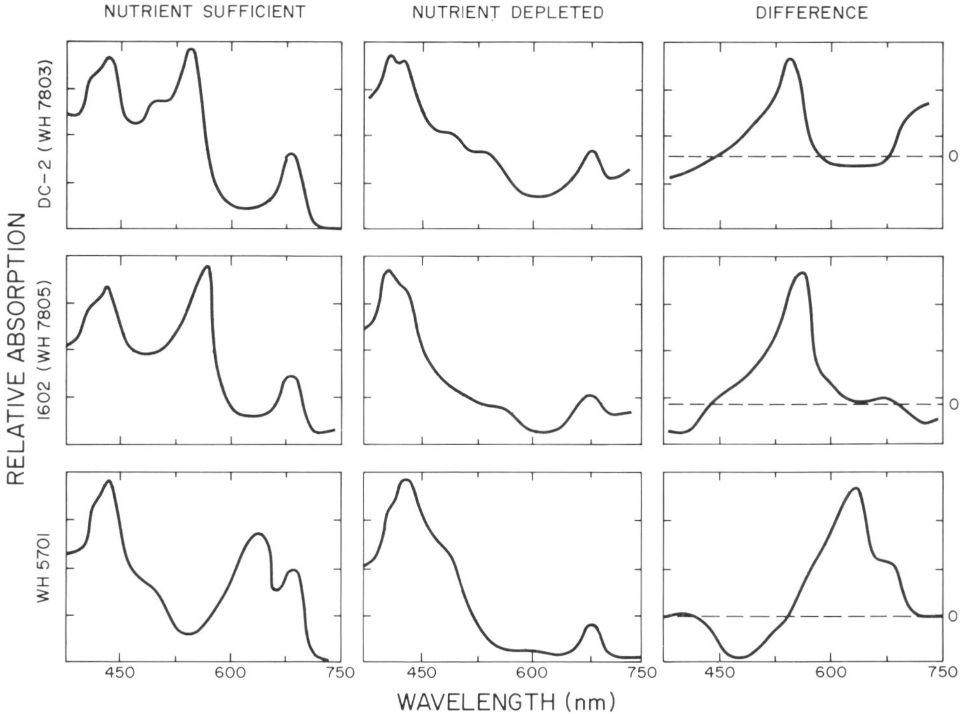

FIG. 8. *In vivo* absorption spectra of three cyanobacterial clones (WH7803, WH7805, and WH5701) grown under nutrient sufficient and nutrient depleted conditions. The difference spectra indicate the *in vivo* absorbance spectra of the pigments degraded following nutrient depletion, and corresponds to either phycoerythrin (WH7803, WH7805) or phycocyanin (WH5701).

cells is that of phycoerythrin (for WH7803 and WH7805) and phycocyanin (for WH5701).

In contrast to the cyanobacteria examined, the action spectrum determined for *Pavlova* sp. is similar in virtually every respect to that determined for natural assemblages (see Fig. 6 and 7). Both exhibit peaks at 425–450 and 675 nm due to chlorophyll *a* and a shoulder at 475–525 nm due to carotenoid accessory pigments. This similarity suggests a closer look at the contribution eukaryotic picoplankton species make to the primary productivity attributed to the picoplankton size fraction.

Photosynthetic Response of Picoplankton in Water of Different Optical Types

We have observed distinctive spectral signatures for $\alpha^B(\lambda)$ between three laboratory clones of oceanic picoplankton (Fig. 4–6). It is of some interest to determine how these differences might influence the photosynthetic performance in the spectral light field at depth in the sea. Wood (1985) compared photosynthetic performance of various picoplankton clones suspended at various depths in the optically clear blue water regions where the wavelength of maximum transmission is ca. 480 nm (Morel and Prieur 1977). In these experiments, the phycoerythrin-containing cyanobacterial clones which had *both* PEB and PUB chromophores had higher rates of photosynthesis at low light deep in the water column (on a per cell basis) than clones with PEB only. Phycocyanin dominant clones lacking phycoerythrin performed the worst and on two occasions, showed no net photosynthesis at all. The result is in the direction that one would expect based on the spectral absorption characteristics of the pigment. Even more interesting, eukaryotic cells of similar size (*Pavlova* sp. and *Thalassiosira oceanica*) outcompeted all cyanobacterial clones at low light.

A similar experiment can be conducted using laboratory derived determinations of photosynthetic action spectra convoluted with measured spectral distributions of downwelling irradiance (Lewis et al. 1985b; Harrison et al. 1985). For "blue water" we take the experimentally determined values of the initial slope of the spectral photosynthesis–irradiance curves (Fig. 4–6), and weight them by the *in situ* irradiance spectrum (e.g. Morel 1978). These spectrally-weighted average α^B values, when normalized such that the value for *Pavlova* sp. is 1.0, are 0.32 for WH7803 and 0.14 for WH5701. For a given chlorophyll *a* concentration, the eukaryote is 3 times more efficient at absorbing and using light than the PE containing strain and 7 times more efficient than the PE lacking strain. It would be difficult to imagine an action spectrum less suitable for photosynthesis in blue water than that of WH5701. Indeed, no strain of PE-lacking *Synechococcus* isolated to date has come from the open sea (J. Waterbury cited in Wood 1985). Of course, normalization to chlorophyll *a* may not be the most appropriate way to evaluate relative photosynthetic performance; it would be preferable to normalize to carbon and compare rate constants for photosynthesis. This option is invalidated by the presence of detrital carbon in natural populations. One can note, however, that the carbon to chlorophyll ratio for the cyanobacterial strains currently in culture (e.g. Barlow and Alberte 1985) is high relative to those more typical of eukaryotes; this would tend to accentuate the differences between physiological performance of cyanobacterial and eukaryotic picoplankton calculated above.

For green water, such as the Gulf of Maine and Grand Banks of Newfoundland, the situation is somewhat different. Using the same convention as above, strain WH7803 outperforms *Pavlova* by a factor of 1:1.31, while the PE lacking strain is still 3 times less efficient than the PE containing strain. It would appear that if all else is equal, eukaryotic picoplankton would dominate the deep water flora of the open sea while cyanobacteria containing phycoerythrin would be favored inshore. This situation agrees with at least some of the recent distributional evidence (see Glover et al. 1985; Murphy and Haugen 1985).

The analysis is complicated by the fact that our experimentally determined action spectra are single-waveband measurements and as such represent unenhanced action spectra. The presence of significant non-linear effects (such as enhancement) may cause some error when we convolute the action and incident irradiance spectra, since we assume strict additivity. Enhancement spectra of cyanobacteria show that all phycobiliprotein absorption is by PSII, while light absorbed by chlorophyll is delivered largely to PSI (Jones and Myers 1964; Mimuro and Fujita 1977; Wang et al. 1977). Therefore, light absorbed by PSI (<525 nm, >625 nm) shows a distinct reduction in quantum yield if it is not supplemented by a background green light. This is apparent in the action spectra presented in Fig. 4 and 5 for the cyanobacteria WH7803 and WH5701 which show little or no photosynthetic activity attributable to chlorophyll *a* despite its presence in the absorption spectra. The extent to which such non-linear effects bias the results of the convolution exercise are currently being investigated. In blue water, with maximum transmission around 475 nm combined with strong spectral narrowing, the error may be small.

Conclusions

The estimation of the mean and variance of marine autotrophic biomass and production continues to be of highest priority in biological oceanography, particularly in the open ocean (see Platt 1984). Here, we have presented strong evidence that the prediction of rates of photosynthesis at low photon flux is of paramount importance in achieving this goal. The presence of high concentrations of autotrophic organisms in the picoplankton size range deep in the euphotic zone, coupled with theoretical calculations on the size-dependence of photosynthetic efficiency, leads to the conclusion that estimates of rates of photosynthesis of these organisms at low light would add greatly to our predictive ability. In particular, insufficient attention has been paid to examination of the wavelength-dependence of rates of light absorption and photosynthesis at the low irradiances by the organisms found at these depths.

Cyanobacteria appear to be the most abundant picoplankton in nature, however, we can find no evidence that the phycobilipigments characterizing the group significantly influence either the underwater light field, the absorption of light by particulate matter, or the photosynthetic action spectra of natural phytoplankton populations. Of the thousands of measurements that have been made of the spectral characteristics of the submarine light, none that we have seen to date show any evidence of the presence of phycobilipigments. This is also true for the relatively fewer measurements of particulate matter absorption (on filtered material) and for the even fewer direct measurements of photosynthetic action spectra. The conclusion we are forced to draw is that these pigments are not important light-harvesting pigments in the ocean relative to the chlorophylls and carotenoids. This does not necessarily reduce the importance of the marine cyanobacteria, but it does focus some deserved attention on the less well-known eukaryotic picoplankton.

Current algorithms in use for estimating chlorophyll *a* from remote observations of ocean colour rely on the ratio of water-leaving radiance at two measured spectral bands centered at 440 and 550 nm. Fortunately for the purpose of remotely measuring chlorophyll in the open ocean, the lack of a significant phycobilin signal (which could potentially interfere by absorbing strongly at 550 nm) means that the algorithms can be used there with some confidence. Furthermore, such measurements, coupled with remote measurements of sea-surface irradiance (e.g. Gautier 1982) can be used to estimate synoptically the rate of primary production in the ocean to within acceptable accuracy (Platt 1986); this accuracy will be enhanced significantly as better information on the mean and variance of the photosynthetic action spectra of natural phytoplankton populations becomes available.

References

ALBERTE, R. S., A. M. WOOD, T. A. KURSAR, AND R. R. L. GUILLARD. 1984. Novel phycoerythrins in marine *Synechococcus* spp. Characterization and evolutionary and ecological implications. Plant Physiol. 75: 732–739.

BARLOW, R. G., AND R. S. ALBERTE. 1985. Photosynthetic characteristics of phycoerythrin-containing marine *Synechococcus* spp. I. Responses to growth photon flux density. Mar. Biol. 86: 63–74.

BRICAUD, A., A. MOREL, AND L. PRIEUR. 1983. Optical efficiency factors of some phytoplankters. Limnol Oceanogr. 28: 816–832.

BROWN, O. B., R. H. EVANS, J. W. BROWN, H. R. GORDON, R. C. SMITH, AND K. S. BAKER. 1985. Phytoplankton blooming off the U.S. East Coast: A satellite description. Science 229: 163–167.

CAMPBELL, L. 1985. Investigations of marine, phycoerythrin-containing *Synechococcus* spp. (Cyanobacteria): Distribution of serogroups and growth rate measurements. Ph.D. thesis, SUNY at Stony Brook, NY.

CUHEL, R. L., AND J. B. WATERBURY. 1984. Biochemical composition and short term nutrient incorporation patterns in a unicellular marine cyanobacterium, *Synechococcus* (WH 7803). Limnol. Oceanogr. 29: 370–374.

DUYSENS, L. N. M. 1956. The flattening of the absorption spectrum of suspensions, as compared to that of solutions. Biochim. Biophys. Acta 19: 1–12.

EPPLEY, R. W., E. STEWART, M. R. ABBOTT, AND U. HEYMAN. 1985. Estimating ocean primary production from satellite chlorophyll. Introduction to regional differences and statistics for the Southern California Bight. J. Plankton Res. 7: 57–70.

EXTON, R. J., W. M. HOUGHTON, W. ESAIAS, L. W. HAAS, AND D. HAYWARD. 1983. Spectral differences and temporal stability of phycoerythrin fluorescence in estuarine and coastal waters due to the domination of labile cryptophytes and stabile cyanobacteria. Limnol. Oceanogr. 28: 1225–1231.

GAUTIER, C. 1982. Mesoscale insolation variability derived from satellite data. J. Appl. Meteorol. 21: 51–58.

GLOVER, H. E., AND I. MORRIS. 1981. Photosynthetic characteristics of coccoid marine cyanobacteria. Arch. Microbiol. 129: 42–46.

GLOVER, H. E., D. A. PHINNEY, AND C. S. YENTSCH. 1985. Photosynthetic characteristics of picoplankton compared with those of larger phytoplankton populations, in various water masses in the Gulf of Maine. Biol. Oceanogr. 3: 223–248.

GLOVER, H. E., M. D. KELLER, AND R. R. L. GUILLARD. 1986. Light quality and oceanic ultraphytoplankters. Nature 319: 142–143.

GORDON, H. R., AND A. Y. MOREL. 1983. Remote assessment of ocean color for interpretation of satellite visible imagery: A review. Springer-Verlag, Berlin and New York.

GORDON, H. R., D. K. CLARK, J. L. MUELLER, AND W. A. HOVIS. 1980. Phytoplankton pigments from the Nimbus-7 Coastal Zone Color Scanner: comparisons with surface measurements. Science 210: 63–66.

GORDON, H. R., D. K. CLARK, W. A. HOVIS, R. W. AUSTIN, AND C. S. YENTSCH. 1985. Ocean color measurements. Adv. Geophys. 27: 297–333.

GUILLARD, R. R. L., L. S. MURPHY, P. FOSS AND S. LIAAEN-JENSEN. 1985. *Synechococcus* spp. as likely zeaxanthin-dominant ultraphytoplankton in the North Atlantic. Limnol Oceanogr. 30: 412–414.

HARRISON, W. G., T. PLATT, AND M. R. LEWIS. 1985. The utility of light-saturation models for estimating marine primary productivity in the field: A comparison with conventional "simulated" in situ methods. Can J. Fish. Aquatic Sci. 42: 864–872.

HOGE, F. E., AND R. N. SWIFT. 1983. Airborne dual laser excitation and mapping of phytoplankton photopigments in a Gulf Stream Warm Core Ring. Appl. Optics 22: 2272–2281.

HOVIS, W. A., D. K. CLARK, F. ANDERSON, R. W. AUSTIN, W. H. WILSON, E. T. BAKER, D. BALL, H. R. GORDON, J. L. MUELLER, S. Z. EL-SAYED, B. STURM, R. C. WRIGLEY, AND C. S. YENTSCH. 1980. Nimbus-7 Coastal Zone Color Scanner: System description and initial imagery. Science 210: 60–63.

JERLOV, N. G. 1976. Marine Optics. Elsevier, New York, NY. 231 p.

JOHNSON, P. W., AND J. MCN. SIEBURTH. 1979. Chroococcoid cyanobacteria in the sea: A ubiquitous and diverse phototrophic biomass. Limnol. Oceanogr. 24: 928–935.

JONES, L. W., AND J. MYERS. 1964. Enhancement in the blue-green alga, *Anacystis nidulans*. Plant. Physiol. 39: 938–946.

KAWAMURA, M., M. MIMURO, AND Y. FUJITA. 1979. Quantitative relationship between two reaction centers in the photosynthetic system of blue-green algae. Plant Cell Physiol. 20: 697–705.

KIEFER, D. A., AND J. B. SOOHOO. 1982. Spectral absorption by marine particles of coastal waters of Baja California. Limnol. Oceanogr. 27: 492–499.

KIRK, J. T. O. 1975a. A theoretical analysis of the contribution of algal cells to the attenuation of light within natural waters. I. General treatment of suspensions of pigmented cells. New Phytol. 75: 11–20.

——— 1975b. A theoretical analysis of the contribution of algal cells to the attenuation of light within natural waters. II. Spherical cells. New Phytol. 75: 21–36.

——— 1976. A theoretical analysis of the contribution of algal cells to the attenuation of light within natural waters. III. Cylindrical and spheroidal cells. New Phytol. 77: 341–358.

KURSAR, T. A., H. SWIFT, AND R. S. ALBERTE. 1981. Morphology of a novel cyanobacterium and characterization of light-harvesting complexes from it: implications for phycobiliprotein evolution. Proc. Natl. Acad. Sci. 78: 6888–6892.

LEWIS, M. R., R. E. WARNOCK, AND T. PLATT. 1985a. Absorption and photosynthetic action spectra for natural phytoplankton populations: Implications for production in the open ocean. Limnol. Oceanogr. 30: 794– 806.

LEWIS, M. R., R. E. WARNOCK, B. IRWIN, AND T. PLATT. 1985b. Measuring photosynthetic action spectra of natural phytoplankton populations. J. Phycol. 21: 310–315.

LI, W. K. W., AND P. M. DICKIE. 1985. Metabolic inhibition of size-fractionated marine plankton radiolabelled with amino acids, glucose, bicarbonate and phosphate in the light and dark. Microb. Ecol. 11: 11–24.

LI, W. K. W., D. V. SUBBA RAO, W. G. HARRISON, J. C. SMITH, J. J. CULLEN, B. IRWIN, AND T. PLATT. 1983. Autotrophic picoplankton in the tropical ocean. Science 219: 292–295.

MIMURO, M., AND Y. FUJITA. 1977. Estimation of chlorophyll *a* distribution in the photosynthetic pigment systems I and II of the blue-green alga *Anabaena variabilis*. Biochim. Biophys. Acta 459: 376–389.

MOREL, A. 1978. Available, usable, and stored radiant energy in relation to marine photosynthesis. Deep-Sea Res. 25: 673–688.

MOREL, A., AND A. BRICAUD. 1981. Theoretical results concerning light absorption in a discrete medium and application to specific absorption of phytoplankton. Deep-Sea Res. 28: 1375–1393.

MOREL, A, AND L. PRIEUR. 1977. Analysis of variations in ocean color. Limnol Oceanogr. 22: 709–722.

MORRIS, I., AND H. GLOVER. 1981. Physiology of photosynthesis by marine coccoid cyanobacteria — some ecological implications. Limnol. Oceanogr. 26: 957–961.

MURPHY, L. S., AND E. M. HAUGEN. 1985. The distribution and abundance of phototrophic ultraplankton in the North Atlantic. Limnol. Oceanogr. 30: 47–58.

MYERS, J., J.-R. GRAHAM, AND R. T. WANG. 1978. On spectral control of pigmentation in *Anacystis nidulans* (Cyanophyceae). J. Phycol. 14: 513–518.

——— 1980. Light harvesting in *Anacystis nidulans* studied in pigment mutants. Plant Physiol. 66: 1144–1149.

ÖQUIST, G. 1974. Distribution of chlorophyll between the two photoreactions in photosynthesis of the blue-green alga *Anacystis nidulans* grown at two different light intensities. Physiol. Plant. 30: 38–44.

ONG, L. J., A. N. GLAZER, AND J. B. WATERBURY. 1984. An unusual phycoerythrin from a marine cyanobacterium. Science 224: 80–83.

PLATT, T. 1984. Primary productivity in the central North Pacific: Comparison of oxygen and carbon fluxes. Deep-Sea Res. 31: 1311–1319.

——— 1986. Primary production of the ocean water column as a function of surface light intensity. Deep-Sea Res. 33: 149–163.

PLATT, T., D. V. SUBBA RAO, AND B. IRWIN. 1983. Photosynthesis of picoplankton in the oligotrophic ocean. Nature 300: 702–704.

PLATT, T., C. L. GALLEGOS, AND W. G. HARRISON. 1980. Photoinhibition of photosynthesis in natural assemblages of marine phytoplankton. J. Mar. Res. 38: 687–701.

PLATT, T., K. L. DENMAN, AND A. D. JASSBY. 1977. Modelling the productivity of phytoplankton, p. 807–856, *In* E. D. Goldberg [ed.], The sea: ideas and observations on progress in the study of the seas. John Wiley, New York, NY.

PRIEUR, L., AND S. SATHYENDRANATH. 1981. An optical classification of coastal and oceanic waters based on the specific spectral absorption curves of phytoplankton pigments, dissolved organic matter, and other particulate materials. Limnol. Oceanogr. 26: 671–689.

RADMER, R. J., AND B. KOK. 1977a. Light conversion efficiency in photosynthesis. *In* A. Trebst and M. Avron [ed.] Photosynthesis I: Photosynthetic electron transport and photophosphorylation. Springer-Verlag, New York, 1977.

——— 1977b. Photosynthesis: limited yields, unlimited dreams. Bioscience 27: 599–605.

RAVEN, J. A., AND J. BEARDALL. 1982. The lower limit of photon fluence rate for photoautotrophic growth — the significance of slippage reactions. Plant Cell 5: 117–124.

REED, R. K. 1977. On estimating insolation over the ocean. J. Phys. Oceanogr. 7: 482–485.

SMITH, R. C., AND K. S. BAKER. 1978. The bio-optical state of ocean waters and remote sensing. Limnol. Oceanogr. 23: 247–259.

TEL-OR, E., AND S. MALKIN. 1977. The photochemical and fluorescence properties of whole cells, spheroplasts and spheroplast particles from the blue-green alga *Phormidium luridum*. Biochim. Biophys. Acta 459: 157–174.

WANG, R. T., C. L. R. STEPHENS, AND J. MYERS. 1977. Action spectra for photoreactions I and II of photosynthesis in the blue-green alga *Anacystis nidulans*. Photochem. Photobiol. 25: 103–108.

WATERBURY, J. B., S. W. WATSON, R. R. L. GUILLARD, AND L. E. BRAND. 1979. Widespread occurrence of a unicellular, marine, planktonic, cyanobacterium. Nature 277: 293–294.

WOOD, A. M. 1985. Adaptation of photosynthetic apparatus of marine ultraphytoplankton to natural light fields. Nature 316: 253–255.

WOOD, A. M., P. K. HORAN, K. MUIRHEAD, D. A. PHINNEY, C. M. YENTSCH, AND J. B. WATERBURY. 1985. Discrimination between types of pigments in marine *Synechococcus* spp. by scanning spectroscopy, epifluorescence microscopy and flow cytometry. Limnol. Oceanogr. 30: 1303–1315.

WYMAN, M., R. P. F. GREGORY, AND N. G. CARR. 1985. Novel role for phycoerthyrin in a marine cyanobacterium, *Synechococcus* strain DC2. Science 230: 818–820.

YENTSCH, C. S. 1957. A non-extractive method for the quantitative estimation of chlorophyll in algal cultures. Nature 179: 1302–1304.

——— 1962. Measurement of visible light absorption by particulate matter in the ocean. Limnol. Oceanogr. 7: 207–217.

——— 1980. Light attenuation and phytoplankton photosynthesis. p. 95–127. *In* I. Morris [ed.], The physiological ecology of phytoplankton. Univ. California Press, CA.

——— 1983. Remote sensing of biological substances, p. 263–297. *In* A. P. Cracknell [ed.], Remote sensing applications in marine science and technology. Reidel.

YENTSCH, C. S., AND C. M. YENTSCH. 1979. Fluorescence spectral signatures: the characterization of phytoplankton populations by the use of excitation and emission spectra. J. Mar. Res. 37: 471–483.

Experimental Approaches to Field Measurements: Methods and Interpretation

W. K. W. Li

Marine Ecology Laboratory, Bedford Institute of Oceanography,
Department of Fisheries and Oceans,
Dartmouth, N.S. B2Y 4A2

1. Introduction

The organisms that comprise the picoplankton community are trophically diverse (Sieburth 1979): in addition to the photoautotrophs and chemoorganotrophs usually considered (Sieburth et al. 1978), there are also chemoautotrophs (Ward 1982; Ward et al. 1982), methanotrophs (Sieburth et al. 1983; Sieburth 1984; Strand and Lidstrom 1984), and bacterivores (= "picoplankton biotrophs" of Sieburth 1983, 1984) (Cynar et al. 1985; Fuhrman and McManus 1984; Wright and Coffin 1984). Given this diversity, the task of studying the photosynthetic picoplankton in multi-trophic assemblages is fraught with potential difficulties. The nature of these difficulties is in assuring that what is being measured is in fact what is intended to be measured.

The ecophysiological study of photosynthetic picoplankton, like that of phytoplankton in general, begins with the recognition, characterization and enumeration of these cells; and it continues with the estimation of their metabolic rates *in situ*. The first of these aspects is relatively straightforward: there are several useful features that distinguish the phototrophs from the rest of the picoplankton community. However, the measurements of photosynthesis and growth, as commonly conducted in field experiments, are subject to some degree of uncertainty because potentially interfering processes are difficult to eliminate or to account. Many of the methodological concerns confronting ecological studies of phytoplankton as a whole also have a bearing on studies of photosynthetic picoplankton. Since these topics are regularly reviewed (Berman and Eppley 1974; Morris 1974, 1982; Carpenter and Lively 1980; Peterson 1980; Sakshaug 1980; Leftley et al. 1983), it is the intent here to discuss only selected problems which relate specifically to picoplankton.

I will first survey some of the methods used to determine the occurrence and abundance of photosynthetic picoplankton. This survey serves as background to the subsequent discussions since the interpretation of many physioecological methods depends on the way that organisms (or biomass) are measured. In methods that do not rely on a visual or electronic recognition of cell size, the most practical way to ensure that the measurements pertain to picoplankton is to size fractionate the plankton sample. Size fractionation is made possible by using nucleation-track perforated pore membranes which act as sieves. Some of the problems that arise when plankton are passed through such membranes will also be addressed. Finally, there will be a discussion of some factors affecting the interpretation of common field techniques to assess the primary productivity of photosynthetic picoplankton.

2. Methods for Determining Occurrence and Abundance

Glover (1985) has recently reviewed this subject for marine *Synechococcus* spp. so that the following discussion emphasizes only particular aspects of each method. Those interested in fuller descriptions, as well as advantages and disadvantages of many of these methods should consult Leftley et al. (1983) and Glover (1985). The characteristic most useful for distinguishing photosynthetic from non-photosynthetic forms in the picoplankton is the presence of autofluorescent pigments in the former.

This is the basis for methods such as fluorometry, spectrofluometry, epifluorescence microscopy, and flow cytometry. There are other methods that do not principally rely on the presence of pigments: these include electron microscopy, microautoradiography, immunofluorescence, and measurement of photosynthetic enzyme activity.

2.1 *IN VITRO* FLUOROMETRY

In the present context, *in vitro* fluorometry is taken to mean the standard oceanographic practice of determining plant pigment content by measuring the intensity of fluorescence emission from a solvent extract of the plankton (Strickland and Parsons 1972). The use of this fluorometric approach to estimate picoplankton biomass must necessarily be in conjunction with a size fractionation of the water sample. This method has been used extensively for descriptions of how chlorophyll *a* is distributed among various size classes of the plankton (e.g. Saijo and Takesue 1965; Gieskes et al. 1979; Malone 1980; Takahashi and Bienfang 1983; Bienfang 1984, 1985; Takahashi and Hori 1984; Herbland et al. 1985). An implicit assumption in this approach is that the origin of fluorescence in each size fraction be the same, i.e. from chlorophyll *a*. However, in comparing estimates of chlorophyll *a* measured by conventional fluorometry and high-performance liquid chromatography. Gieskes and Kraay (1983) found overestimates by the former method that could largely be explained by an unidentified chlorophyll *a* derivative. Moreover, this derivative was associated with particles that passed 1 μm filters. These particles contained only traces of "normal" chlorophyll *a* but had zeaxanthin as the major carotenoid. Guillard et al. (1985) have since provided evidence that *Synechococcus* spp. were the likely zeaxanthin-dominant particles observed by Gieskes and Kraay (1983). These findings indicate that there should now be a re-evaluation of the basis and intent of bulk chlorophyll *a* fluorometry as applied to size-fractionated plankton, especially the picoplankton.

Recently, Stewart and Farmer (1984) reported on a fluorometric method to quantitate phycobiliprotein pigments from plankton. To overcome the problem of effective pigment extraction from cyanobacteria, such as that to which Glover et al. (1985) alluded, Stewart and Farmer (1984) disrupted cells in a tissue grinder and then digested the homogenate with lysozyme. In principle, the proposed technique could allow phycoerythrin of picoplankton to be distinguished from that of larger plankton. This is because the phycoerythrin in picoplankton (i.e. cyanobacteria) fluoresces maximally at a wavelength shorter than that of the phycoerythrin in larger phytoplankton (i.e. cryptomonads) (Stewart and Farmer 1984).

2.2 SPECTROFLUOROMETRY

Spectrofluorometry, or fluorescence spectrophotometry, of plankton *in vivo* is a useful relative measure of different phytoplankton "colour groups" (Yentsch and Yentsch 1979; Yentsch 1983; Yentsch and Phinney 1984, 1985a, b). Although similar to bulk fluorometry in that plankton must first be concentrated on filters, this method differs in that the pigments are not extracted into solvent and that both exciting and emitted light are scanned across the wavelength spectrum. Results arising from *in vivo* spectrofluorometry are uncompromised by the presence of chlorophyll degradation products as these are not highly fluorescent *in vivo* (Yentsch 1974). The ability to vary both the excitation and emission wavelengths allows a gross characterization of the relative amounts of pigments. The major problem of relating *in vivo* fluorescence to pigment biomass is that the fluorescence per unit pigment varies with environmental conditions (Cullen 1982). Furthermoere, when phytoplankton are size fractionated, the different size classes may have different *in vivo* fluorescence yields (Alpine and Cloern 1985). However, much useful information can be gained by considering ratios of fluorescence excited at wavelengths selected to maximize the differences between major pigments (Yentsch and Yentsch 1979; Yentsch and Phinney 1984, 1985a, b). An example of this approach is the study of picoplankton by Glover et al. (1985) in which variations in the presumptive ratio of phycoerythrin to chlorophyll *a* were confirmed by cell counts of cyanobacteria and other phytoplankton.

2.3 EPIFLUORESCENCE MICROSCOPY

Undoubtedly, the examination of plankton samples collected on membrane filters by epifluorescence microscopy is the standard method for recognizing and enumerating photosynthetic picoplankton (Waterbury et al. 1979; Johnson and Sieburth 1979; Li et al. 1983; Glover et al. 1985a, b; Murphy and Haugen 1985). Depending on the excitation and emission filter combinations, phycoerythrin-containing cells appear yellow or orange and other phytoplankton appear red. Estimates of abundance can therefore be obtained of phycoerythrin-containing and other picoplankton from the same preparation. Glover (1985) has tabulated the optical filter combinations used by various workers and the resulting colours of the cells. The attractiveness of epifluorescence microscopy is that it gives quantitative estimates of abundance by direct visual means. Moreover, samples can be stained (Burney et al. 1981, 1982; Davis and Sieburth 1982; Caron 1983; Davis et al. 1985) or heated (Caron et al. 1985) to visualize the heterotrophic componenets. Furuya (1982) described an image analyzer system based on epifluorescent detection and subsequently reported on its use in a study of the size distribution of phytoplankton in the Western Pacific Ocean (Furuya and Marumo 1983).

Unfortunately, if values of biomass are desired, it is necessary to adopt some arithmetical factor to convert cell numbers to (carbon) weight. Such factors cannot be expected to be the same for different cells, nor for cells of different physiological states. For *Synechococcus* sp. WH7803 doubling once every 14.3 h in laboratory culture, each cell contains 294 fg carbon (Cuhel and Waterbury 1984). For cells in natural samples, this factor cannot be measured directly as in the *Synechococcus* laboratory culture; instead, it must be estimated indirectly. First, an estimate for the volume of the cell is necessary; in exercises of this sort, values of 0.2 μm^3 (Johnson and Sieburth 1979; Sieburth 1984), 0.46 μm^3 (Schmaljohann 1984), and 0.524 μm^3 (Krempin and Sullivan 1981; Caron et al. 1985) have been used. Second, volume must be converted to carbon weight: the factor for this is itself further decomposed into three experimentally determinable subfactors, namely the wet weight to volume ratio, the dry weight to wet weight ratio, and the carbon weight to dry weight ratio. Factors ranging from 84.7 to 165 fg C $\cdot$ μm^{-3} have been published (Bowden 1977; Hagstrom et al. 1979). The value apparently used most often is 121 fg C $\cdot$ μm^{-3} but Takahashi et al. (1985) recently measured a value of 400 fg C $\cdot$ μm^{-3} for synechococcoid picocyanobacteria isolated from Japanese coastal waters. The latter value is supported by recent re-evaluations of this factor which suggest that perhaps values twice (Bratbak and Dundas 1984) or five times (Bratbak 1985) as large as 121 fg C $\cdot$ μm^{-3} may be more appropriate. Uncertainties as large as these underline the difficulties of expressing cell numbers in terms of biomass.

2.4 FLOW CYTOMETRY

Another method that relies on the autofluorescence of native pigments for detection and enumeration of phototrophs in the picoplankton is flow cytometry (Yentsch et al. 1983; Olson et al. 1985; Wood et al. 1985). In microscopy, human eyes are used to detect fluorescence emissions; in flow cytometry, photomultipliers tubes are used (Shapiro 1985). Cell enumeration by flow cytometry is therefore more rapid and less tedious than microscopy. There is an added advantage in the former technique in that cells are enumerated according to their degree of autofluorescence; in other words, a frequency distribution of cells in generated and the variable over which the cells are distributed is relative fluorescence emission. However, as in epifluorescence

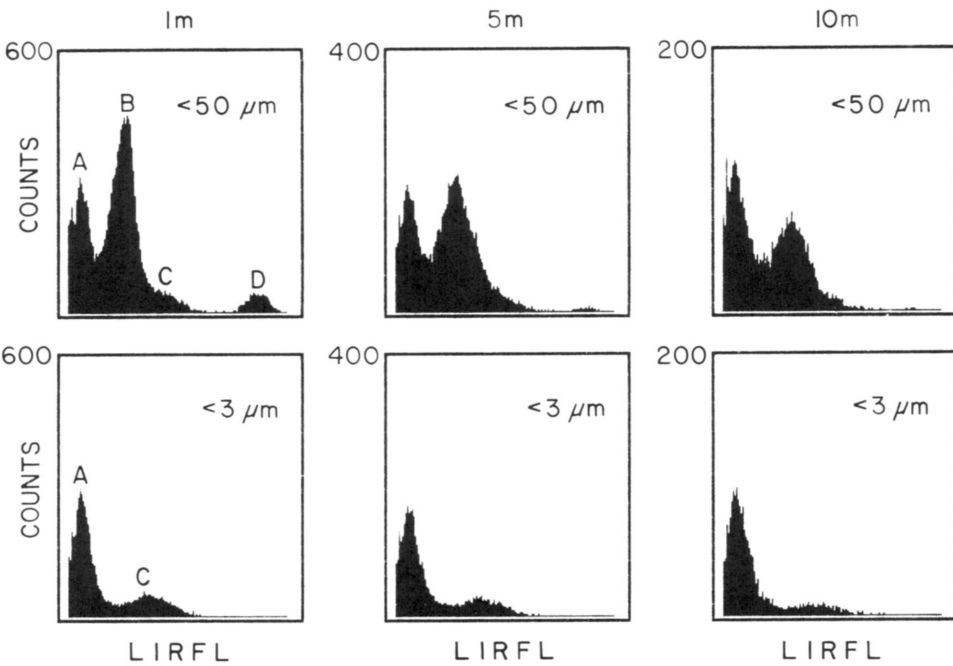

FIG. 1. Flow cytometric analyses of natural plankton samples (1, 5, and 10 m, Bedford Basin, Canada; Sept 5, 1985) filtered through 50 μm mesh (upper panels) or 3 μm Nuclepore membrane (lower panel). Frequency distribution of events counted with respect to LIRFL which is distributed over 256 channels in a three decade logarithmic scale. LIRFL represents fluorescence >630 nm excited by 514 nm light. Note differences in ordinate scale. Labelled peaks are discussed in Section 2.4

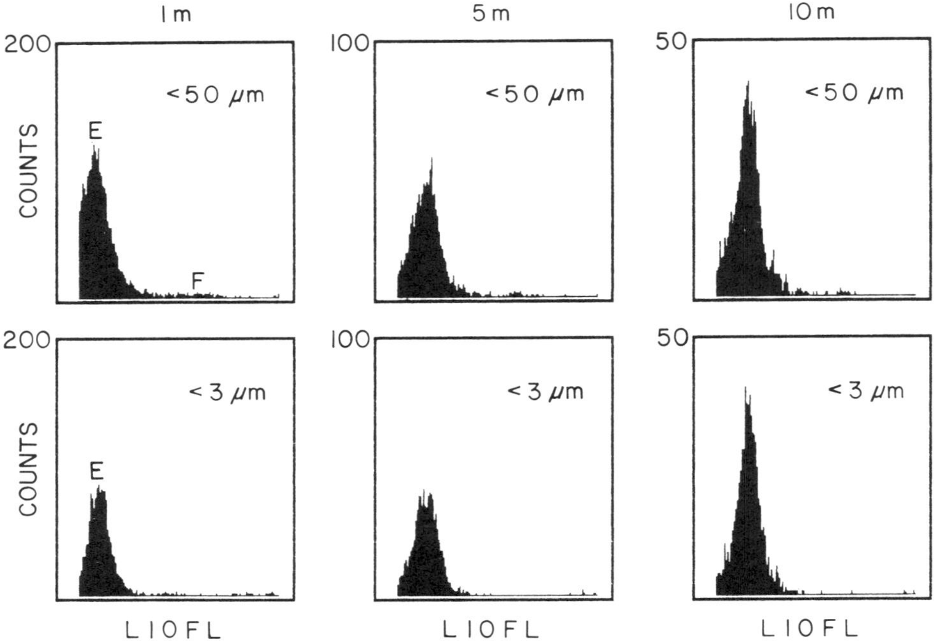

FIG. 2. Same as in Fig. 1 except showing LIOFL instead of LIRFL. LIOFL represents fluorescence between 530 and 590 nm excited by 514 nm light.

254

microscopy, if an estimate of biomass is desired, the same conversion (see Section 2.3) must necessarily be applied.

Examples of such frequency distributions are shown in Fig. 1 and 2. Analyses were made of plankton from 1, 5, and 10 m depths in Bedford Basin, Nova Scotia (Sept 5, 1985) filtered through 50 μm mesh (to prevent clogging the flow orifice) or 3 μm Nuclepore filter. In Fig. 1, LIRFL (log integrated red fluorescence) represents fluorescence >630 nm excited by 514 nm laser light: this is taken to represent chlorophyll a fluorescence (although fluorescence from phycocyanin would also be included). In Fig. 2 LIOFL (log integrated orange fluorescence) represents fluorescence between 530 and 590 nm excited by the same 514 nm light: this is taken to represent phycoerythrin fluorescence. The results indicate that picoplankton (<3 μm) comprised a significant percentage of the <50 μm cells: they accounted for 38, 40, and 47% of the LIRFL at 1, 5, and 10 m respectively; and 62, 75, and 85% of the LIOFL at 1, 5, and 10 m respectively. The picoplankton at all three depths were characterized by low and intermediate LIRFL (peaks A and C) and low LIOFL (peak E) signals. LIRFL peaks B and D and LIOFL F belonged solely to 3–50 μm cells.

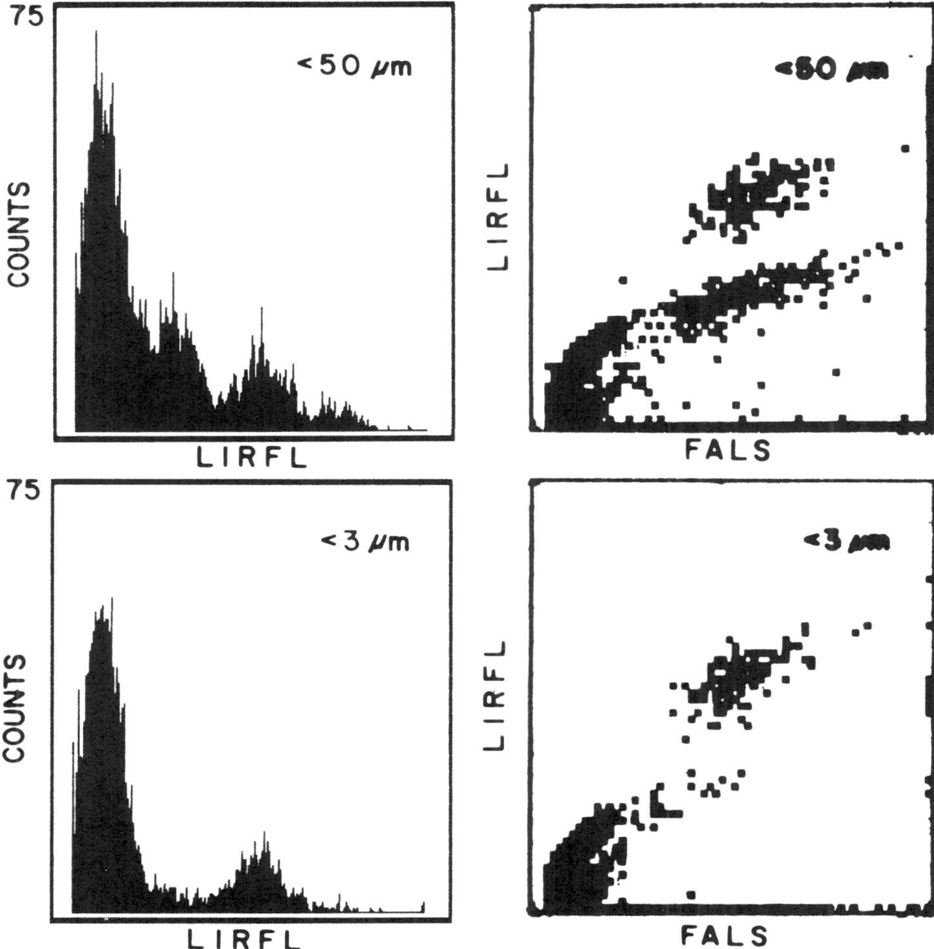

FIG. 3. Flow cytometric analyses of natural plankton sample (10 m, Bedford Basin, Sept. 23, 1985) filtered through 50 μm mesh (upper panels) or 3 μm Nuclepore membrane (lower panels). One-parameter histograms (left panels) show the frequency distribution of events counted with respect to LIRFL. Two-parameter histograms (right panels) show the relationship between LIRFL and FALS. LIRFL as defined in Fig. 1; FALS represents light scatter in the forward angles between 2° and 19°.

A more detailed study of plankton can be made by simultaneously collecting several variables for single cells. By this method, it is possible to construct two-dimensional histograms which show the relationship between two variables for all cells. The analysis of another sample from Bedford Basin (10 m, Sept. 23, 1985) illustrates this approach (Fig. 3–5). FALS stands for Forward Angle (2° to 19°) Light Scatter: it is governed to a large extent by cell size, to some extent by the refractive index, and to a small extent by cell shape (Yentsch and Yentsch 1984). Figure 3 shows that by passing plankton through a 3 μm filter, (i) virtually all the cells with high FALS (regardless of their LIRFL) were removed; and (ii) a distinct group of cells characterized by intermediate LIRFL and FALS intensities was also removed. Figure 4 shows that some of the high FALS cells removed by the 3 μm filter also possessed high LIOFL. The picoplankton ($<3\mu$m) in this sample (i) included cells that had no phycoerythrin (LIOFL) but had varying amounts of chlorophyll a or phycocyanin (LIRFL); (ii) included cells that had phycoerythrin but very low intensity of LIRFL; (iii) did not include any cryptomonads (high LIOFL and high LIRFL) (Fig. 5).

The successful use of flow cytometry to determine cell concentration depends on knowing the exact volume of suspension (seawater plankton sample) analyzed. As there is no such instrument indication on the popular Coulter® EPICS flow cytometer,

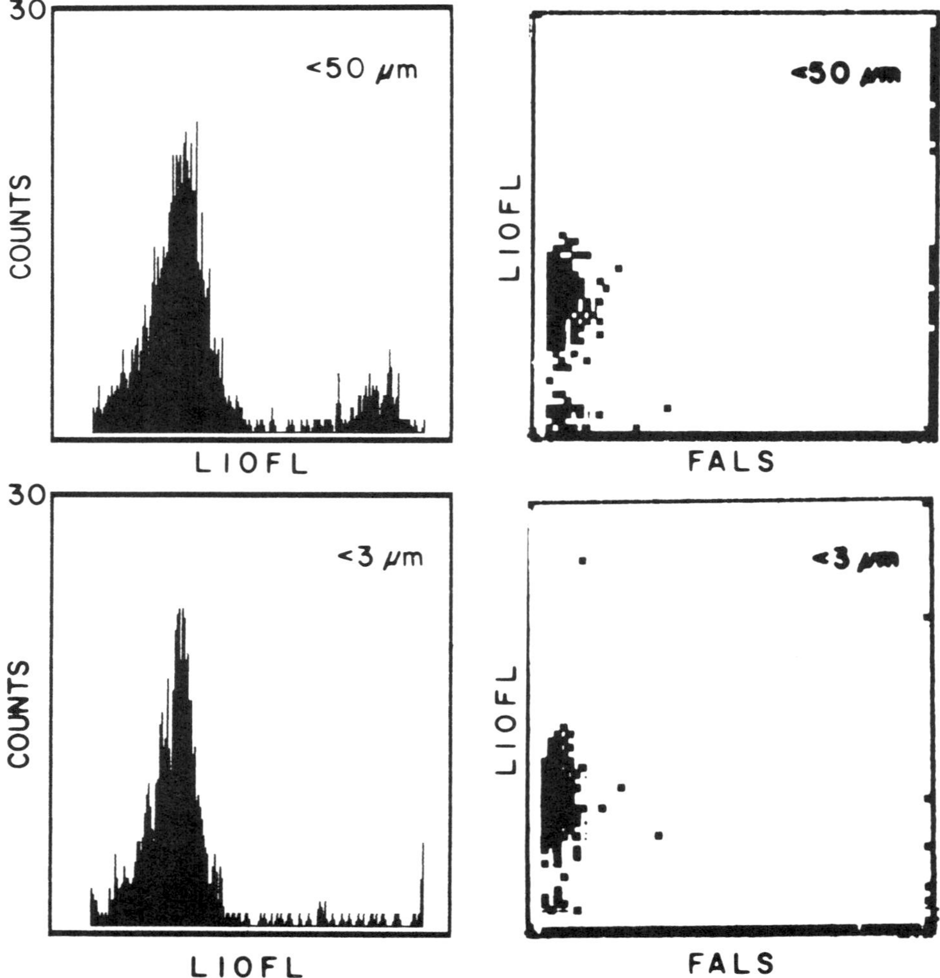

FIG. 4. Same as in Fig. 3 except showing LIOFL instead of LIRFL. LIOFL as defined in Fig. 2; FALS as defined in Fig. 3.

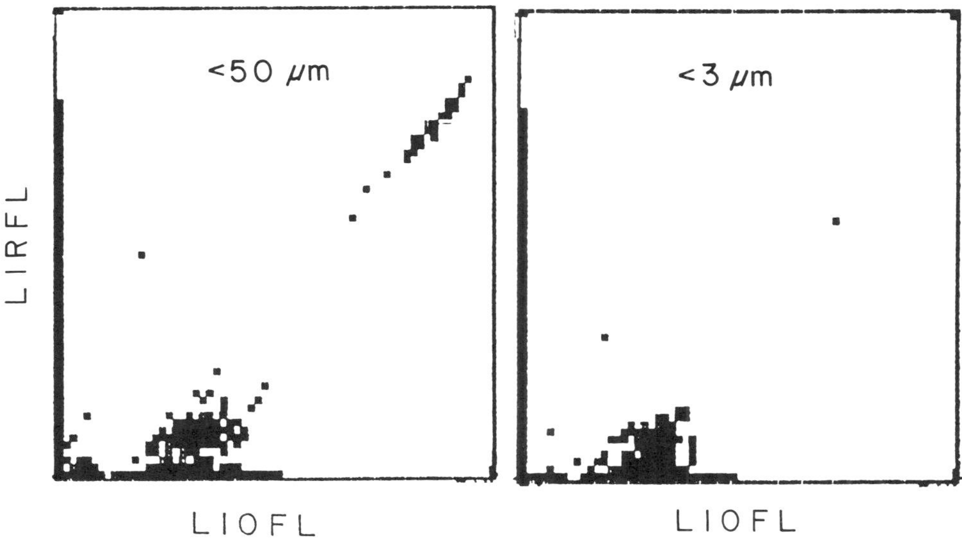

FIG. 5. Flow cytometric analyses of natural plankton sample (10 m, Bedford Basin; Sept. 23, 1985) filtered through 50 μm mesh or 3 μm Nuclepore membrane. Two-dimensional histograms showing the relationship between LIRFL and LIOFL. LIRFL as defined in Fig. 1; LIOFL as defined in FIG. 2.

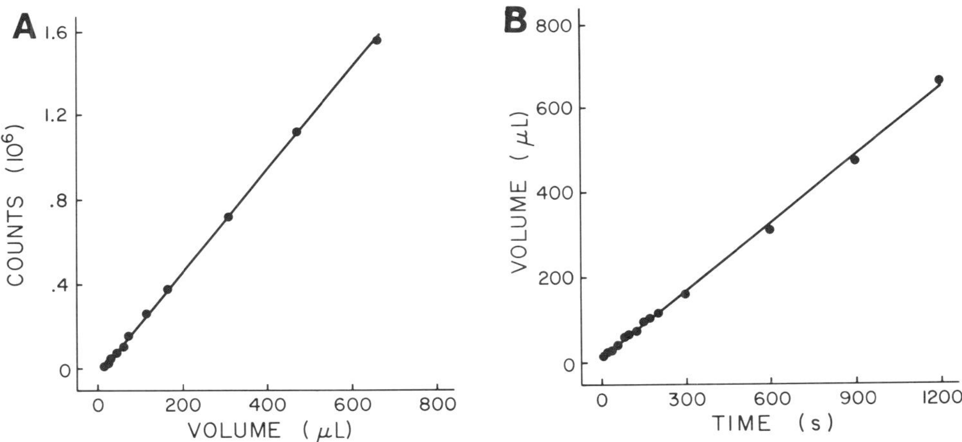

FIG. 6. Flow calibration of Coulter® EPICS V flow cytometer by gravimetric method using *Synechococcus* WH7805. (A) relationship between number of LIOFL events counted and volume of sample analyzed; (B) relationship between volume of sample analyzed and analysis time. LIOFL as defined in Fig. 1.

indirect methods have to be used. One such method is to include a known concentration of fluorescent calibration beads as internal standards in the sample (Yentsch et al. 1984; Olson et al. 1985). A simpler method is to remove the sample cuvette and measure the change in sample volume after analysis (Olson, pers. comm.). When the flow cytometer is used in a shore-based installation, the latter method may be modified by measuring the change in sample weight. The estimate of volume analyzed is simply the change in weight divided by the specific gravity of seawater. This method can be verified by a linear relationship between the number of cells counted and the volume analyzed (Fig. 6); the slope of this line is the cell concentration. If the pressure differential between sample and sheath fluid is unchanged, the sample flow rate is constant and a linear relationship is obtained between volume analyzed and the duration of analysis (Fig. 6): in this case, the counting duration (whick is an operator-determined quantity in the Coulter® EPICS) may be used as a measure of volume.

The use of flow cytometers to detect and enumerate photosynthetic cells according to their autofluorescence characteristics represents only a most rudimentary application of this instrument to phytoplankton research (Yentsch and Yentsch 1984; Yentsch et al. 1984). The ability to distinguish, in natural plankton samples, between cells containing phycoerythrin and those that do not has already been demonstrated (Yentsch et al. 1983; Olson et al. 1985; Wood et al. 1985). Wood et al. (1985) have taken this approach one step further: they demonstrated the ability of flow cytometry to distinguish between cyanobacteria with phycoerythrin containing phycoerythrobilin and phycourobilin chromophores (Type I) and those containing only phycoerythrobilin (Type II). Because the fluorescence excitation and emission maxima occur at shorter wavelengths for Type I cells, discrimination between the two cell types was possible simply by judicious choices of the common excitation wavelength and the cutoff of the dichroic mirror used to accept the emitted fluorescence.

2.5 Methods that do not Principally Rely on Pigments

Electron microscopy, both scanning and transmission, has been used to study picoplankton cells (Johnson and Sieburth 1979, 1982; Sieburth 1979; Wilhelm et al. 1982; Perkins et al. 1981; Sarokin and Carpenter 1981; Easterbrook and Subba Rao 1984; Joint and Pipe 1984; Thinh and Griffiths 1985). This is the only method by which ultrastructural and morphological details can be studied and reliable identifications made.

Microautoradiography, when combined with epifluorescent microscopy, is a powerful method for detecting and enumerating cells that are metabolically active. Following methods such as those used to study heterotrophic activities of individual bacteria (Meyer-Reil 1978; Tabor and Neihof 1982a, b), Douglas (1984) and Craig (unpublished data) developed methods to investigate light dependent $H^{14}CO_3^-$ uptake (photosynthesis) by individual picoplankters. The same microscope slide preparation is viewed to enumerate both autofluorescent particles and particles that have silver grains associated with them. This method is the only one that has the potential of discriminating not only between cells that take up ^{14}C in the light (presumably autotrophs, but see later discussion) and those that do not, but also between those pigmented cells that do (metabolically active) and those pigmented ones that do not (metabolically inactive). However, when used as a quantitative technique, great care must be taken as the combined method suffers from all the technical and standardization difficulties of conventional microautoradiography (Smith and Kalff 1983).

Immunofluorescence, or fluorescent antibody techniques are well established in microbial ecology (Stanley et al. 1979; Bohlool and Schmidt 1980) but they have not been extensively used for the detection and enumeration of unicellular cyanobacteria in nature (Fliermans and Schmidt 1977; Koz'yakov et al. 1979; Campbell et al. 1983; Glover et al. 1986). These methods are the only ones that have the potential of yielding information about individual or closely related strains. In indirect immunofluorescence, antibody (e.g. from rabbits) is first prepared against a particular microbial strain (antigen); the antiserum is then applied to the plankton sample of interest. If the strain in question is present, the highly specific antigen-antibody reaction will occur. Detection is made possible by applying a second antibody (e.g. from goats or sheep) which has been made fluorescent (e.g. with fluorescein isothiocyanate) and which was prepared in response to rabbit globulin. The stained samples can then be examined by epifluorescent microscopy or flow cytometry (e.g. Tyndall et al. 1985). We should perhaps expect a quickening of the gradual pace at which immunofluorescence techniques are applied in contemporary oceanography (Ward and Perry 1980; Ward 1982; Ward et al. 1982; Campbell et al 1983; Karl et al. 1984; Glover et al. 1986) now that epifluorescence microscopy is routine and flow cytometry gradually becoming so. Moreover, the demonstration that immunofluorescence can be combined with autoradiography and fluorescence microscopy to study chemoautotrophs (Fliermans and Schmidt 1975; Ward 1984a) invites a similar study for photoautotrophs.

D-ribulose-1,5-bisphosphate carboxylase (E.C.4.1.1.39) is an enzyme associated with autotrophic CO_2 fixation so that its activity can be used as a marker for photosynthetic cells in a way comparable to chlorophyll *a*. This method relies on the number of photoautotrophic cells being very much greater than the number of all other autotrophic cells. Smith et al. (1985) compared the activity of this enzyme with chlorophyll *a* content in <1 μm arctic plankton samples. In general, the percentage of total enzyme activity <1 μm was slightly less than two times the percentage of total chlorophyll *a* <1 μm. The hypothesis that this discrepancy was due to non-chlorophyll bearing autotrophs was unsupported by very low rates of dark $H^{14}CO_3{}^-$ uptake. These results remain a puzzle.

3. Size Fractionation

3.1 STATEMENT OF THE PROBLEM

The ideal size fractionation scheme should consist of a large pore size filter through which only those organisms of interest and smaller ones pass, together with a small pore size filter on which all organisms of interest are retained but through which all smaller organisms pass. With this in mind, we need first to define our organisms of interest. Sieburth et al. (1978) defined the picoplankton in such a way that they might be distinguished from other plankton by size fractionation using perforated membrane filters. Although picoplankton are those organisms between 0.2 and 2.0 μm in width, "from a practical viewpoint a 0.1 to 1.0 μm fraction is preferred, since it will include the narrowest minibacteria and most of the larger bacterioplankton while excluding the smallest flagellates" (Sieburth 1979: p. 82). "(Bacterioplankton) can be selectively filtered with 0.1- and 1.0 μm porosity (sic) Nuclepore membranes that yield particles in the 0.2–2.0 μm size range" (Sieburth et al. 1978: p. 1260). ["Porosity is the degree of openness of the membrane, and is expressed quantitatively by its void volume, whereas pore size is the size of the individual pores." (Brock 1983)]. Sicko-Goad and Stoermer (1984) proposed that picoplankton refer to cells between 0.1 and 1.0 μm but did not make it clear if these dimensions refer to cell widths or nominal filter pore sizes. The following sections discuss three aspects concerning the use of filters in relation to the above definition: upper and lower nominal pore sizes that bound our organisms of interest, and the influence of vacuum pressure during filtration.

3.2 THE UPPER BOUND IN NOMINAL FILTER PORE SIZE

The dimensions reported for small photosynthetic cells suggest that many of them would pass through 1.0 μm Nuclepore filters. It appears that the width of a cell, and not its length, determines whether the cell is retained by these filters: when water flows through the pores, streamlines are established and the small cells are aligned longitudinally (Sieburth et al. 1978; Sieburth 1979). Johnson and Sieburth (1979, 1982) reported cyanobacteria (Type I cells) of about 0.8×1.0 μm, *Micromonas pusilla* of 1×1.5 μm, and a ubiquitous scaled, non-flagellated prasinophyte of 0.5×1.0 μm. [Type II and III cells which were even smaller turned out to be nitrifying and methanotrophic bacteria (Sieburth 1983)]. Schmaljohann (1984) found many cyanobacteria in the size range $0.6–0.8 \times 0.9–1.5$ μm. One of the largest cyanobacterium observed by Joint and Pipe (1984) was only 0.94×1.07 μm, leading the authors to suspect that "all of the cyanobacteria were probably small enough to pass through a 1 μm pore size sieve". Furthermore, "many of the small eukaryotic algae would also pass through a 1 μm pore-size sieve" as they were <1 μm wide (Joint and Pipe 1984). Many workers have therefore relied on 1 μm Nucleore filters to separate picoplankton from nanoplankton (Morris and Glover 1981; Joint and Pomroy 1983; Li et al. 1983; Platt et al. 1983; Douglas 1984; Ward 1984b; Herbland et al. 1985; Probyn 1985; Smith et al. 1985).

However, when 1 μm Nuclepore filters are examined by epifluorescence microscopy, small phycoerythrin-containing autofluorescent cells are frequently found on them (Krempin and Sullivan, 1981; Caron et al. 1985; Craig 1985). Realizing this, many other workers have used 3 μm Nuclepore filters to effect the separation of picoplankton (Larsson and Hagstrom 1982; Bienfang and Takahashi 1983; Takahashi and Bienfang 1983; Bienfang et al. 1984a, b; Craig 1984, 1986; Laws et al. 1984; Takahashi and Hori 1984; Takahashi et al. 1985; Glover et al. 1985a, b; Murphy and Haugen 1985).

To verify the effectiveness of 3 μm filters and the inadequacy of 1 μm filters, I filtered several clones of cultured marine *Synechococcus* through various Nuclepore membranes and enumerated, by flow cytometry, the number of cells found in the filtrates. In the case of the two phycoerythrin-containing clones WH7803 and WH7805 (Fig. 7), as many cells were found in the 3 μm and 2 μm filtrates as were in the unfiltered culture; however, only 24% and 33% respectively of these cultures were found in the 1 μm filtrates. In the case of the phycocyanin-containing clone WH5701 (Fig. 8), two separate cultures (of different ages) were tested. In one, about 92% of the cells were in the 3 and 2 μm filtrates; but only 31% of them were in the 1 μm filtrate. In the other, 74% of the cells were in the 3 and 2 μm filtrates; but only 7% of them were in the 1 μm filtrate.

An interesting observation was that the frequency distribution of fluorescence could be changed by filtration even when the total number of cells remained about the same. This phenomenon was very obvious in the case of WH5701 (Fig. 8): a LIRFL peak (centred at channel 85) absent in the unfiltered culture was apparent in the 3 and 2 μm filtrates. Similarly, in the case of WH7805 (Fig. 7), a LIOFL shoulder (centred at channel 70) absent in the unfiltered culture was apparent in the 3 and 2 μm filtrates. The significance of this redistribution of cells into higher fluorescence channels is addressed in a following section.

These observations indicate that marine coccoid cyanobacteria have a size distribution spanning the 1 μm nominal pore size of Nuclepore filters. In view of the importance of this group of small phototrophs, the practical definition of picoplankton should be made to fully encompass it: this means that filters of 3 μm nominal pore size should be used (Craig 1986). Useful information can be obtained when both a small pore size filter (such as 0.6, 0.8, or 1 μm) and a large pore size filter (such as 3 or 5 μm) are systematically used: Glover et al. (1985b, 1986), Herbland et al. (1985), Craig (1986), and Prezelin et al. (1986) have all demonstrated depth-related variations in the ratio of smaller to larger cyanobacteria.

The general recognition that photosynthetic organisms pass easily into 3 μm or 1 μm filtrates means that we should carefully re-evaluate those studies in which heterotrophic bacteria are assumed to have been the only significant organisms in the picoplankton. For example, Derenbach and Williams (1974) assumed, on the basis of the literature then available, that no phytoplankton would be found in 3 μm filtrates. They then interpreted the radiocarbon labelling of <3 μm cells as due to heterotrophic uptake of ^{14}C-labelled organic exudates from >3 μm algae. In fact, Joint and Pomroy (1983) and Joint and Pipe (1984) showed that in waters nearby to the working sites of Derenbach and Williams (1974), there are abundant populations of cyanobacteria and eukaryotic picoplankton all of which could pass through 3 μm filters. Furthermore, Joint and Pomroy (1983) concluded that ^{14}C in the picoplankton could not have been due to heterotrophic uptake of exudates because the bacterial production (estimated by independent means) was <10% of the carbon fixed by the picoplankton.

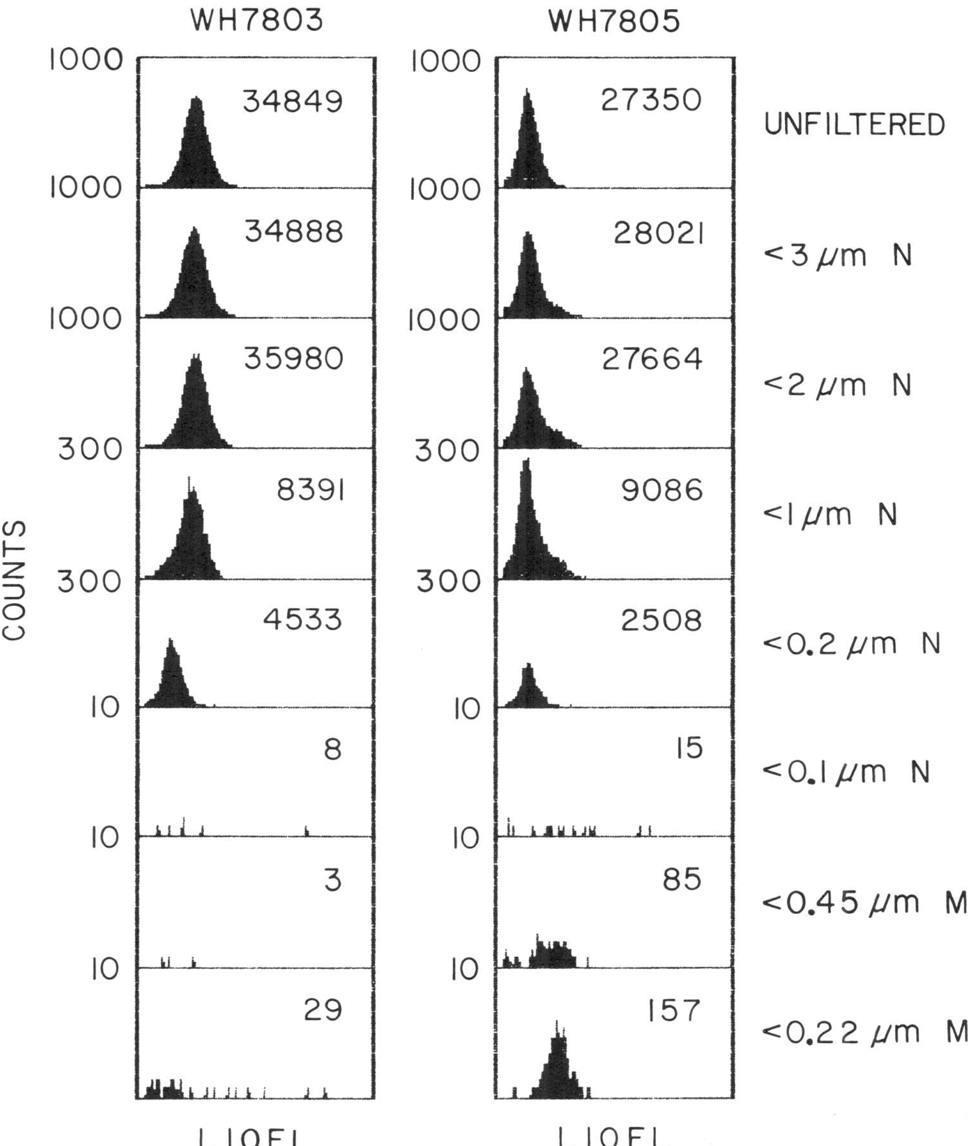

FIG. 7. Flow cytometric analyses of *Synechococcus* WH7803 (left panels) and WH7805 (right panels) in various filtrates. Frequency distribution of events counted with respect to LIOFL. The total number of events in each histogram is indicated within each panel. Note the differences in ordinate scale N = Nuclepore polycarbonate filters; M = Millipore filters (mixed esters of cellulose acetate and nitrate). Vacuum pressure = 100 mm Hg. LIOFL as defined in Fig. 2.

3.3 THE LOWER BOUND IN NOMINAL FILTER PORE SIZE

Filters most commonly used to retain picoplankton include 0.2 μm Nuclepore, Whatman GF/F and 0.45 μm Millipore (or equivalents from other manufacturers. In contrast to the attention given towards determining the practical upper bound for picoplankton, there has been little consideration of what constitutes an effective practical lower bound. An exception is the recent study by Phinney and Yentsch (1985) who showed the inadequacy of GF/F filters for retaining chlorophyll in oligotrophic

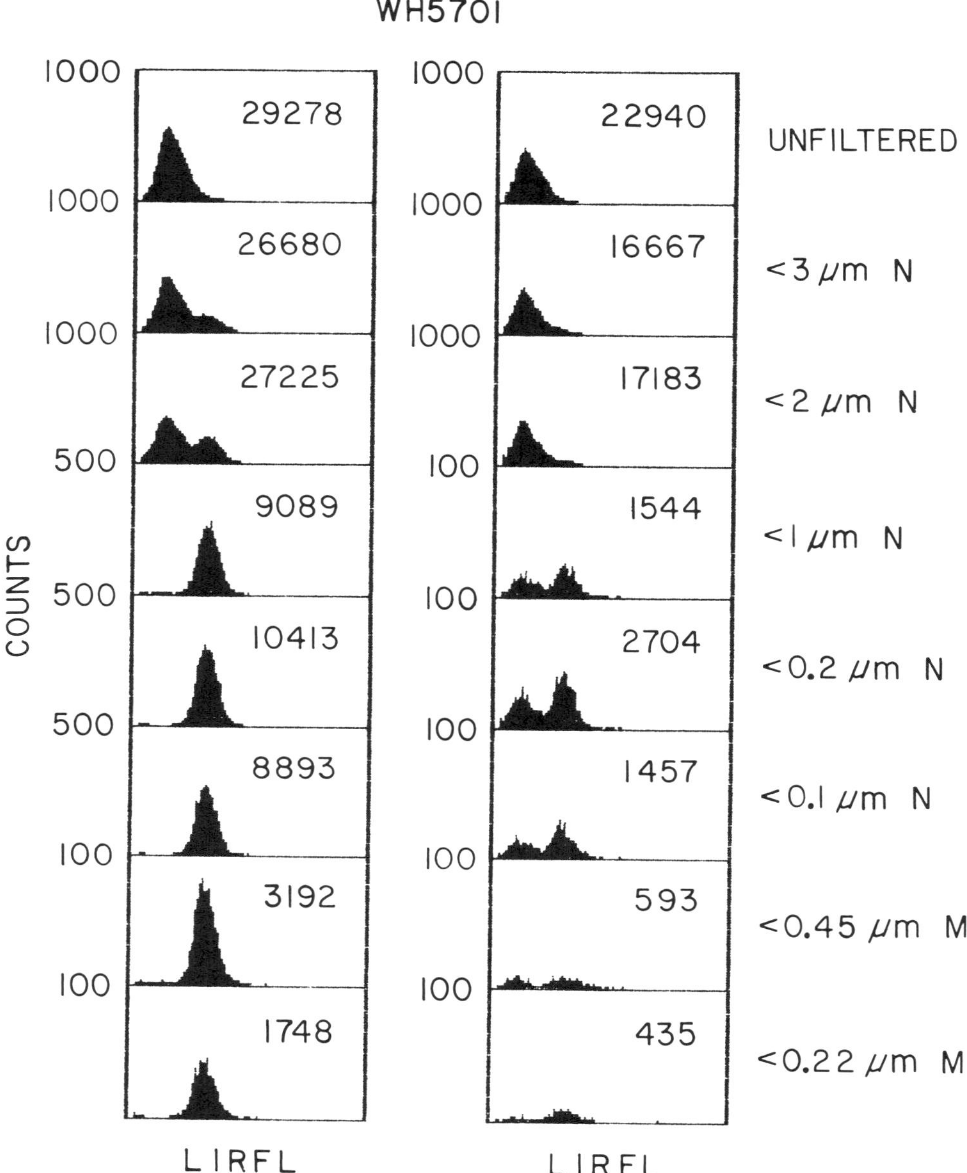

FIG. 8. Same as in Fig. 7 except *Synechococcus* WH5701 was counted with respect to LIRFL. Results from two different cultures as shown (left and right panels). Note differences in ordinate scale. LIRFL as defined in Fig. 1.

waters. Some of their measurements indicated that as much as 60% of the chlorophyll present in low concentration samples may pass through GF/F filters (to be retained on 0.45 μm Millipore filters).

As noted previously, Sieburth et al. (1978) intended the picoplankton to include the minibacteria (about 0.1 μm in diameter). In the literature on heterotrophic bacteria, the existence of minibacteria, or ultramicrobacteria (Morita 1982) is well-recognized. That these cells can be found in 0.2 μm Nuclepore filtrates was recently re-emphasized by Li and Dickie (1985b) who further showed that once isolated from the rest of the plankton community, these cells could grow rapidly ($\mu = 0.18$ h^{-1}) in 0.2 μm filtrates. Sheldon et al. (1967) had earlier demonstrated (by Coulter counting) that particles with diameters between 1.58 and 5.0 μm could increase in concentration in seawater filtered through 0.45 μm and 0.22 μm Millipore membranes. It was suggested that some of these particles were probably aggregates of bacteria. Smith et al. (1985) reported substantial activity of ribulose-1,5-bisphosphate carboxylase in 0.2 μm Nuclepore filtrates from some arctic plankton samples. This is an indication that autotrophic activity (though not necessarily photoautotrophic) is incompletely retained by such filters. There are few systematic investigations that evaluate the effectiveness of various filters used to retain picoplankton (but see Munawar et al. 1982). It is for this reason that I made the following measurements of cells passing into filtrates of 0.2 and 0.1 μm Nuclepore, 0.45 and 0.22 μm Millipore, and Whatman GF/F filters.

The phycoerythrin-containing clones WH7803 and WH7805 (Fig. 7) were incompletely retained by 0.2 μm Nuclepore filters (87% and 91% retention, respectively); on the other hand, 0.1 μm Nuclepore, 0.45 and 0.22 μm Millipore filters performed much better (>99.4% retention). The phycocyanin-containing clone WH5701 (Fig. 8) provided startling results: the number of cells found in 0.2 μm and 0.1 μm filtrates was about the same as that found previously in the 1.0 μm filtrate. Furthermore, substantial numbers were found even in the filtrates of the 0.45 μm and 0.22 μm Millipore filters. In a separate experiment, the effectiveness of Whatman GF/F filters was tested. These filters retained 98.2% of WH7803, 99.4% of WH7805 and 98.1% of WH5701 (Fig. 9). As assurance that the filters used were not faulty, measurements were also made of larger cells, *Pavlova* sp. (clone NEP): Fig. 10 shows that none of these cells were detected in any of the filtrates below 1 μm.

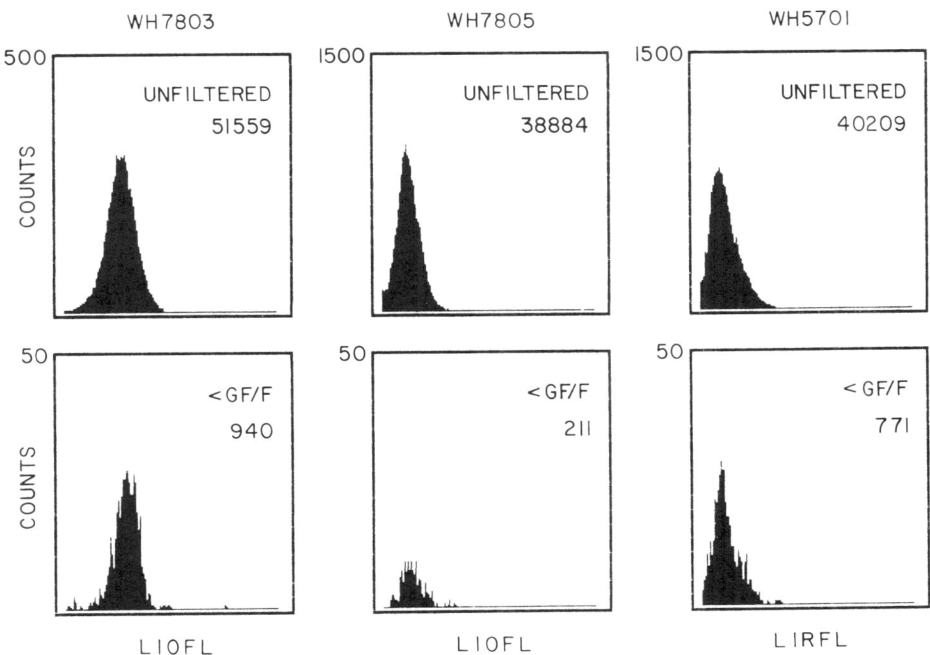

FIG. 9. Flow cytometric analyses of *Synechococcus* WH7803, WH7805 and WH5701 in filtrates of Whatman GF/F filters. Frequency distribution of events counted with respect to LIOFL (WH7803, WH7805) and LIRFL (WH5701). The total number of events in each histogram is indicated within each panel. Note the differences in ordinate scale. Vacuum pressure = 100 mm Hg. LIRFL as defined in Fig. 1; LIOFL as defined in Fig. 2.

263

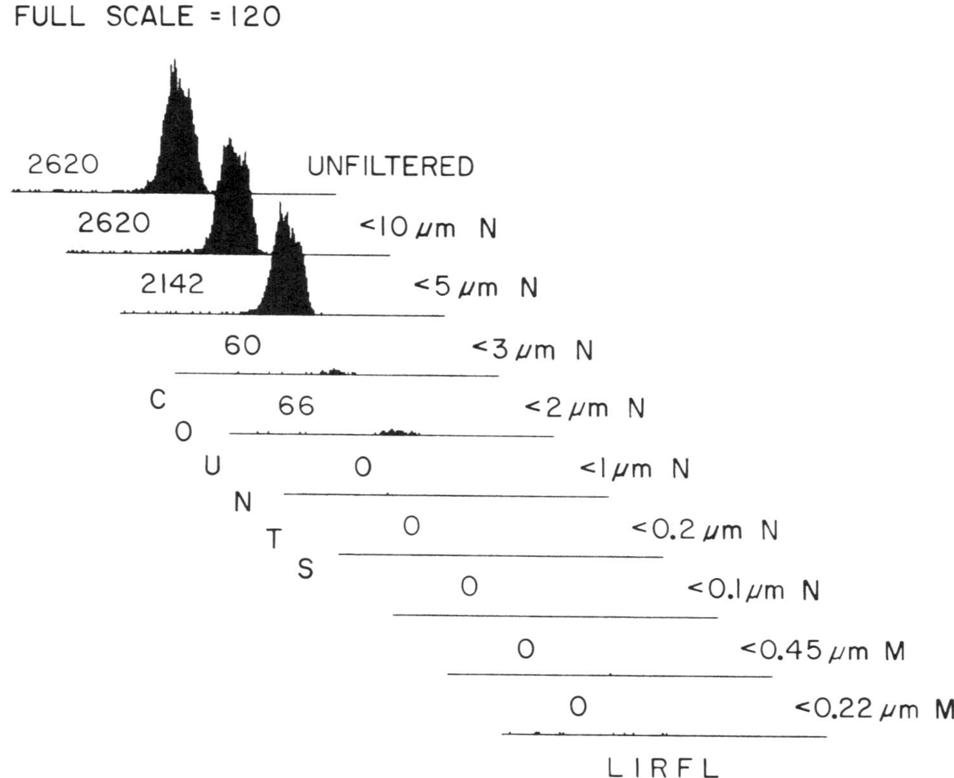

FIG. 10. Same as in Fig. 8 except for *Pavlova* sp. (Nep). Note that the ordinate scale remains the same in all panels (full scale = 120 events).

As a final illustration of the retention characteristics of filters, Fig. 11 shows flow cytometric analyses of a plankton sample (Bedford Basin, Nova Scotia) passed through filters of various pore sizes. FALS/LIRFL (Fig. 11) stands for Forward Angle Light Scatter gated on LIRFL. FALS/LIRFL means that cells are enumerated only if their fluorescence in the >630 nm region exceeds a minimum value (i.e. the LIRFL gate). In this case, the gate was set to exclude particles whose LIRFL was less than that of a laboratory culture of the phycocyanin-containing *Synechococcus* WH5701. In effect, this excluded the phycoerythrin-containing cyanobacteria which do not fluoresce measurably in the red region: in other words, FALS/LIRFL is a measure pertaining to all phytoplankton except the phycoerythrin-containing cyanobacteria. Conversely, FALS/LIOFL (Fig. 11) means that cells are enumerated only if their fluorescence from 530 to 590 nm exceeds a minimum value (i.e. the LIOFL gate). In this case, the gate was set to exclude particles whose LIOFL was less than that of a laboratory culture of the phycoerythrin-containing *Synechococcus* WH7803. In effect, this excluded all *but* the phycoerythrin-containing cyanobacteria and crypto-monads: in other words, FALS/LIOFL is a measure pertaining to all phycoerythrin-containing cells.

Cells that did not contain phycoerythrin (Fig. 11: FALS/LIRFL) passed through 10 μm and 5 μm Nuclepore filters undiminished in numbers. About 30% of these cells was retained on both 3 and 2 μm Nuclepore filters while 77% was retained on

1 μm Nuclepore. The filter commonly used to retain picoplankton, 0.2 μm Nuclepore, in fact allowed 4% of the cells (with low FALS intensity) to pass through. The number of events detected in the 0.22 μm Millipore filtrate was not significantly different from that in the instrument sheath fluid (i.e. blank measurement).

Relatively more of the phycoerythrin-containing cells were found in the smaller size fractions (Fig. 11: FALS/LIOFL). These cells also passed through 10 and 5 μm

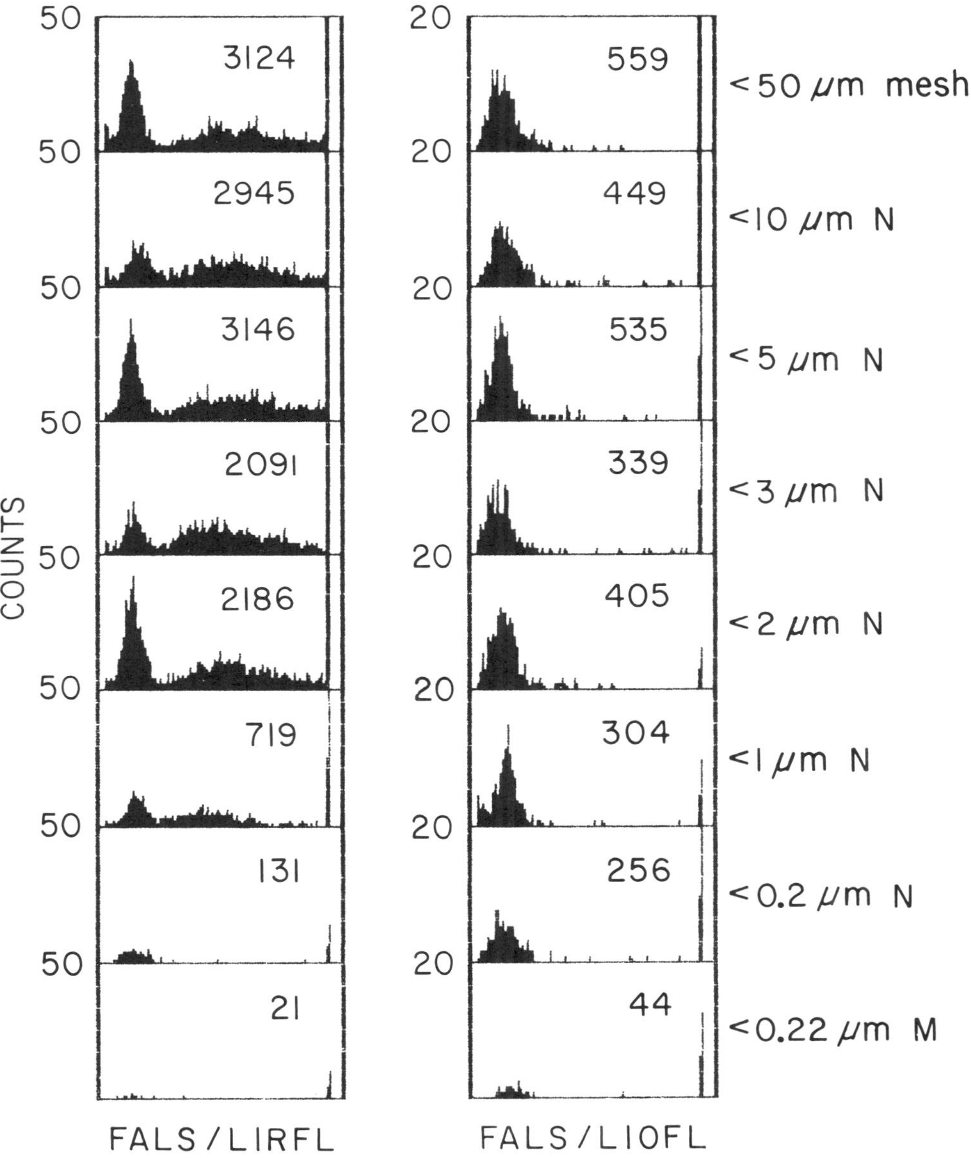

FIG. 11. Flow cytometric analyses of a natural plankton sample (5 m, Bedford Basin; Aug. 19, 1985) in various filtrates. Frequency distribution of events counted with respect to FALS gated on LIRFL (left panels) and FALS gated on LIOFL (right panels). The total number of events in each histogram is indicated within each panel. Note differences in ordinate scale. N = Nuclepore polycarbonate filters; M = Millipore filters (mixed esters of cellulose acetate and nitrate). Vacuum pressure = 100 mm Hg. FALS as defined in Fig. 3; LIRFL as defined in Fig. 1; LIOFL as defined in Fig. 2.

Nuclepore filters largely undiminished in numbers. About 34% of these cells was retained on both 3 and 2 μm Nuclepore filters while 46% was retained on 1 μm Nuclepore. The 0.2 μm Nuclepore filter allowed a surprising 46% of the phycoerythrin-containing cells to pass through. The number of events detected in the 0.22 μm Millipore filtrate, although small, was perhaps just slightly greater than that detected in the sheath fluid (but this conclusion has not been tested statistically). Sheldon et al. (1967) demonstrated by Coulter counting that 0.2% of natural marine particles between 1.58 and 5.0 μm could pass into 0.45 μm Millipore filtrates.

The difference between nucleation-track perforated pore membranes (e.g. Nuclepore polycarbonate), colloidal polymer film membranes (e.g. Millipore cellulose ester) and depth filters (e.g. Whatman glass fibre) have been discussed by Brock (1983). Both the Millipore and Whatman filters retain particles much smaller than their stated pore sizes (Sheldon 1972; Zierdt 1979) because of numerous reasons related to filter structure, capillary action and absorption phenomena (Brock 1983; Johnson and Wangersky 1985). Although Nuclepore filters act much like sieves because they have cylindrical pores of precisely defined size, retention on these filters depends on cell concentration, cell shape and rigidity, the volume filtered and filtration pressure (Brock 1983). For these reasons, it should be emphatically stressed that the results shown in Fig. 7–11 pertain only to the specific conditions under which the experiments were performed. If, for example, cultures of different ages were tested, or if different sample volumes were filtered, or if different cell concentrations were used, the percentage retention by each filter would almost certainly be different.

Figure 12 is an example of how the retention of cultured cyanobacteria by various filters depends on the density of the cell suspension. Nominal dilutions of 1:10, 1:100, and 1:1000 were made of a WH5701 culture (1×10^7 cells $\cdot$ mL^{-1}): the diluted cell suspensions were then passed through various filters as indicated (Fig. 12) and counts were made by flow cytometry of the number of fluorescent events in each filtrate. The most relevant feature of the results was the greater passage of cells (in both absolute numbers and percentage of the unfiltered suspension) into 0.45 μm Millipore, 0.22 μm Millipore and Whatman GF/F filtrates when the cells were suspended at the intermediate dilution of 1:100 (1×10^5 cells $\cdot$ mL^{-1}).

In light of the results presented in this section, the question of which filter is best at retaining photosynthetic picoplankton remains difficult to answer. Considering the extreme case of WH5701 (Fig. 8), we could say that none of the filters performed well. On the other hand, 0.1 μm Nuclepore, 0.45 and 0.22 μm Millipore filters retained almost all the other cells. Whatman GF/F allowed through a small but significant percentage of cyanobacteria. In almost all cases tested, a substantial percentage of photosynthetic picoplankters passed through 0.2 μm Nuclepore filters. The suggestion by Johnson and Wangersky (1985) that a pore size of 2.0 μm might provide a "reasonable delineation between dissolved and particulate material" cannot be upheld.

If these observations are generally valid, they should force a careful scrutiny of data pertaining to phytoplankton excretion of "dissolved" ^{14}C-labelled organic matter. Normally, the fraction that passes many of these filters is considered "dissolved". If small picoplankters are actually found in these filtrates, values ascribed to excretion would be overestimated. As an example, consider the widely-cited work of Mague et al. (1980). In this study which included an investigation of the phytoplankton in the Gulf of Maine, Mague et al., like Shifrin (1980) used Whatman GF/C filters to separate phytoplankton from their exudates. According to Sheldon (1972), these filters have a median effective retention of 0.7 μm. More recently, Hickel (1984) showed empirically that GF/C filters have a retention characteristic like that of 0.4 μm Uni-pore filters for seston in the German Bight and Wadden Sea, and like that of 1.0 μm Uni-pore filters for seston in the Elbe estuary. In waters of Lake Ontario, Munawar et al. (1982) found that 42% (in biomass) of the <3 μm cells could be filtered through GF/C filters. Assume then that the retention of cyanobacteria

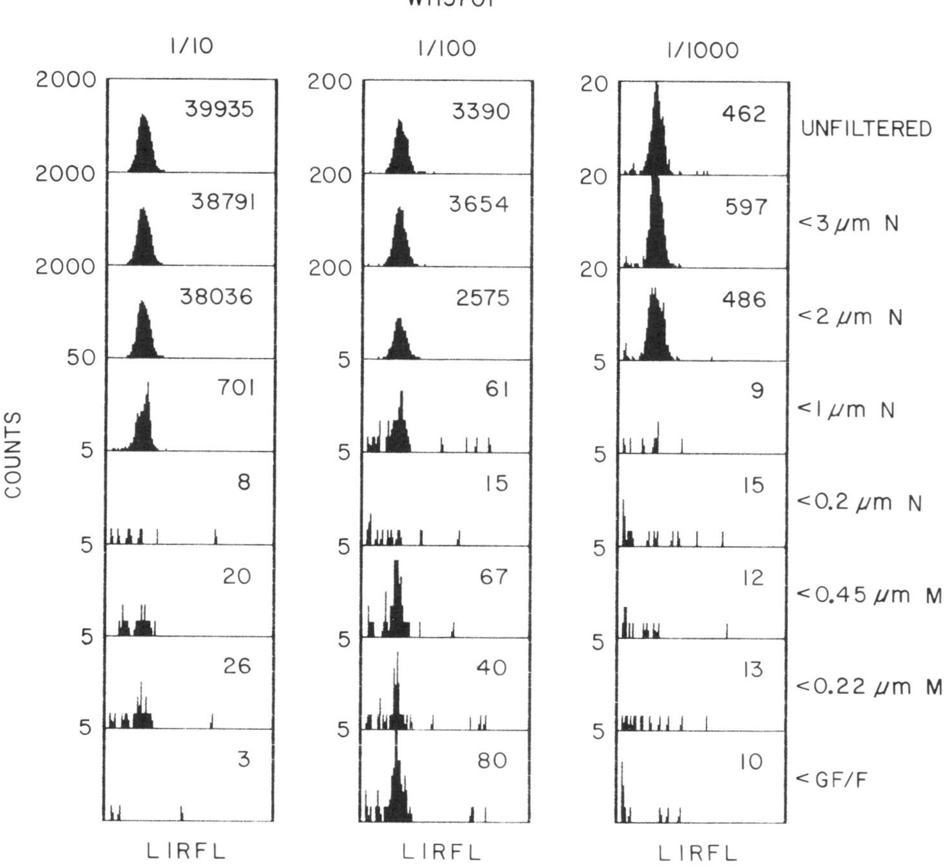

FIG. 12. Flow cytometric analyses of a culture of *Synechococcus* WH5701 (1×10^7 cells • mL^{-1}) diluted 1:10 (left panels), 1:100 (centre panels) and 1:1000 (right panels) in various filtrates. Frequency distribution of events counted with respect to LIRFL. The total number of events (40 µL sample) in each histogram is indicated within each panel. Note differences in ordinate scale. A count of 9 LIRFL events was recorded for the sheath fluid. N = Nuclepore polycarbonate filters; M = Millipore filters (mixed esters of cellulose acetate and nitrate). Vacuum pressure = 100 mm Hg. LIRFL as defined in Fig. 1.

by GF/C filters is intermediate between 1.0 µm Nuclepore and Whatman GF/F filters: the results from flow cytometry presented above indicate mean values of 29.4 and 1.4%, respectively, for the percentage of cells passing these filters. Glover et al. (1985a,b) indicated that in the Gulf of Maine, ultraphytoplankton (< 3 µm) contribute about 60% to the total primary production. Thus, from 1% ($0.014 \times 60\%$) to 18% ($0.294 \times 60\%$) of the total ^{14}C activity might have passed through GF/C filters. The values reported by Mague et al. (1980) for percent extracellular release were almost all within this range. In fact, the calculation probably yields a conservative (lower) estimate because Glover et al. (1985b) showed that in near-surface waters of the Gulf of Maine (in which the experiments of Mague et al. were conducted), the cyanobacteria were small (passed 0.8 µm Nuclepore filters). This example should not of course be taken to mean that phytoplankton do not excrete: Mague et al. (1980) clearly showed that *Skeletonema costatum* does; rather, it means that if filters of too large a pore size are used (including 0.2 µm Nuclepore as suggested above), we cannot be assured that what is in the filtrate is truly dissolved.

3.4 VACUUM PRESSURE

The pressure differential across membrane filters affects the passage (Brock 1983) and structural integrity (Sharp 1977) of cells. The latter has been of great concern to those who use membrane filters to separate phytoplankton from their excretion products since cell breakage during filtration artifactually increases the amount of intracellular compounds in the filtrate (Arthur and Rigler 1967; Nalewajko and Lean 1972; Berman 1973; Herbland 1974; Sharp 1977; Lean and Burnison 1979; Mague et al. 1980; Shifrin 1980; Vegter and De Visscher 1984; Goldman and Dennett 1985). The two main concerns in the present context are: what is the optimum pressure that ensures complete passage of picoplankton through the upper bound filter; and what is the optimum pressure that ensures minimal detectable alteration of the filtered cells? Irwin et al. (1982) demonstrated that in the pressure range 70 to 640 mm Hg, the passage of oceanic chlorophyll a through 1 μm Nuclepore filters was a monotonically increasing function of pressure. Craig (1986) reported that at one site in the Atlantic Ocean, whereas 71% of autofluorescent picoplankton was retained on 1 μm Nuclepore membrane after gravity filtration, only 38% was retained after vacuum filtration at pressures of 100 and 250 mm Hg. To study the effect of vacuum pressure in more detail, I filtered 3 cyanobacterial cultures through 3 and 1 μm Nuclepore filters at various pressures and then counted the number of cells in the filtrates by flow cytometry. The detailed effects of pressure appeared to be clone-specific but some useful generalizations were forthcoming.

For WH7803 (Table 1), increasing the pressure from 25 to 600 mm Hg resulted in only relatively small changes to cell recovery in 3 μm filtrates (from 99.6 to 92.4%) and 1 μm filtrates (from 59.4 to 47.5%). Significantly, the worst recovery was associated with gravity filtration for which only 78.8% of cells passed 3 μm Nuclepore

TABLE 1. The effect of vacuum pressure on filtration of *Synechococcus* WH7803 through 3 and 1 μm Nuclepore membranes. Flow cytometric analyses of LIOFL counts, mean and coefficient of variation (CV) of the frequency distributions.

Pressure	<3 μm counts	<3 μm % unfiltered	<3 μm Mean	<3 μm CV
unfiltered	59899	100.0%	82.8	21.9
gravity	47192	78.8%	86.8	18.2
25	59165	98.8%	83.4	19.6
50	59677	99.6%	83.3	20.4
100	57610	96.2%	84.4	19.3
200	54721	91.4%	85.1	19.1
300	57967	96.8%	84.9	19.9
400	58095	97.0%	84.6	20.1
500	57360	95.8%	83.6	19.6
600	55372	92.4%	84.9	20.6

Pressure	<1 μm counts	<1 μm % unfiltered	<1 μm Mean	<1 μm CV
gravity	14916	24.9%	81.6	20.4
25	30023	50.1%	78.9	19.4
50	35564	59.4%	80.1	18.9
100	33783	56.4%	80.9	18.7
200	30013	50.1%	81.2	19.1
300	28471	47.5%	81.7	19.4
400	30027	50.1%	80.9	19.1
500	32571	54.4%	81.2	19.2
600	33280	55.6%	81.3	18.9

and 24.9% passed 1 μm Nucleopre filters. The mean and coefficient of variation of the LIOFL frequently distributions were not substantially different for any of the treatments, indicating that cellular fluorescence characteristics were unchanged by filtration.

For WH7805 filtered through 3 μm Nuclepore (Table 2), gravity filtration gave a poor recovery (79%). Recoveries were quantitative (100%) at low pressures, poorest (85%) at intermediate pressures, and variable but high (88 to 97%) at high pressures. The mean of the LIOFL frequency distribution was shifted to higher values in those filtrates with poor recoveries, indicating that there were relatively more cells with high fluorescence yields. The effect of pressure was generally the same for filtration through 1 μm Nuclepore (Table 2) except that recoveries in the filtrate were of course much less.

For WH5701 (Table 3), gravity filtration gave the highest, though not quantitative recovery (96%). Cell recoveries in both the 3 and 1 μm filtrates were largely decreasing functions of pressure. The LIRFL distribution means, although largely invariant with pressure, were significantly higher in 3 μm than in 1 μm filtrates (indicating the retention of cells with higher pigment content on the 1 μm filter).

To assess these results, it is useful to recall that particles can be retained by filters in a number of ways (Brock 1983). For particles considerably larger than the pore size, interception by the filter (sieving) is obvious. For particles just slightly smaller than the pore size, part of the particles may touch the filter matrix and adhere to it (surface adsorption). For particles considerably smaller than the pore size, electrostatic and van der Waals forces between the matrix and the particles may be important. Conceivably, the force of gravity alone may not be strong enough to overcome some of these forces so that recovery of cells in the filtrate may be poor when no vacuum is applied. Consistent with this idea was the observation that in many

TABLE 2. The effect of vacuum pressure on filtration of *Synechococcus* WH7805 through 3 and 1 μm Nuclepore membranes. Flow cytometric analyses of LIOFL counts, mean and coefficient of variation (CV) of the frequency distributions.

Pressure	<3 μm counts	<3 μm % unfiltered	<3 μm Mean	<3 μm CV
unfiltered	56584	100.0%	42.5	32.1
gravity	44481	78.6%	51.4	23.1
25	56720	100.2%	43.2	28.8
50	56292	99.5%	43.4	28.8
100	47826	84.5%	49.9	27.9
200	48010	84.8%	50.4	28.2
300	50549	89.3%	50.4	29.7
400	54938	97.1%	47.3	31.8
500	51906	91.7%	47.3	31.8
600	49886	88.2%	48.1	29.1

Pressure	<1 μm counts	<1 μm % unfiltered	<1 μm Mean	<1 μm CV
gravity	42168	74.5%	51.2	27.1
25	38781	68.5%	40.8	27.2
50	39139	69.2%	41.4	28.1
100	24678	43.6%	47.8	25.5
200	21692	38.3%	50.9	25.5
300	35966	63.6%	49.6	28.5
400	29857	52.8%	50.2	25.2
500	36196	64.0%	46.8	25.5
600	30480	53.9%	47.3	25.9

TABLE 3. The effect of vacuum pressure on filtration of *Synechococcus* WH5701 through 3 and 1 μm Nuclepore membranes. Flow cytometric analyses of LIRFL counts, mean and coefficient of variation (CV) of the frequency distributions.

Pressure	<3 μm counts	<3 μm % unfiltered	<3 μm Mean	<3 μm CV
unfiltered	36006	100.0%	45.4	47.3
gravity	34721	96.4%	44.1	43.1
25	31938	88.7%	44.6	43.4
50	30021	83.4%	43.9	43.9
100	26066	72.4%	44.8	44.1
200	25243	70.1%	44.8	44.2
300	23481	65.2%	45.1	44.5
400	21252	59.0%	44.9	44.4
500	19574	54.4%	43.9	44.1
600	22848	63.5%	45.8	44.3

Pressure	<1 μm counts	<1 μm % unfiltered	<1 μm Mean	<1 μm CV
gravity	lost	lost	lost	lost
25	1684	4.7%	31.5	47.5
50	1245	3.5%	33.6	47.3
100	1344	3.7%	33.4	43.1
200	1654	4.6%	35.1	57.5
300	971	2.7%	32.6	41.3
400	477	1.3%	33.9	54.5
500	230	.6%	34.6	51.1
600	228	.6%	33.3	36.8

cases, the highest recoveries were obtained when slight pressures were applied. At higher pressures, there may be impaction of particles on the filter surface (Brock 1983) or cells may be ruptured (Goldman and Dennett 1985): both of these would lower cell recovery in the filtrate. Brock (1983) also noted that "the rate at which the pressure is applied to the filter will influence the retention characteristics. Rapid opening of the pressure valve causes a hydraulic shock, which may push more organisms through the filter": this factor has hardly been considered in any of the phytoplankton size fractionation studies.

An interesting observation in this and previous experiments (Section 3.1: Fig. 7 and 8; vacuum pressure in those experiments was 100 mm Hg) was that filtration could result in a change to the frequency distribution of cellular fluorescence in some cultures even when few cells were retained on the filters. In other words, it appeared that fluorescence characteristics could be altered in cells which have simply gone through the pores of Nuclepore filters. The mechanism that accounts for this apparent alteration in fluorescence yield cannot be inferred from the data. Nevertheless, if this change in fluorescence is an indication that photosynthetic intensity is also changed, these results will have important implications for measurements of photosynthesis in cyanobacteria that have been size fractionated prior to incubation. It is worthwhile noting that both Ferguson et al. (1984) and Fuhrman and Bell (1985) measured increases in dissolved amines after plankton was passed through Nuclepore filters at low vacuum pressure: these increases were ascribed to filtration-induced cell damage.

It is apparent from the results presented in this and previous sections that cell retention on filters depends not only on pore size and vacuum pressure, but also on clonal and culture differences (reflected in factors such as population density, cell size, flexibility, susceptibility to rupture). It is unlikely that there is a particular vacuum pressure that optimizes the recovery of picoplankton in a physiologically unaltered state under all circumstances. However, it does appear that in some cases, filtration at a low vacuum pressure (25–50 mm Hg) is preferable to gravity filtration.

4. ^{14}C Incorporation

4.1 STATEMENT OF THE PROBLEM

The rate of phytoplankton photosynthesis is most often measured by the extent to which particulate matter is radiolabelled after a period of incubation during which the plankton is exposed to $H^{14}CO_3^-$ under illumination. Many methodological and interpretative difficulties plague this apparently simple technique (Carpenter and Lively 1980; Peterson 1980; Morris 1982; Leftley et al. 1983). When the technique is used in conjunction with size fractionation for picoplankton, some of these difficulties are brought into even sharper focus.

In essence, the major difficulty is knowing how much of the particulate-^{14}C in small plankton (< 1 or 3 μm) is derived by direct photosynthetic uptake. The trophic structure of microbial plankton is complex (Sieburth 1983, 1984) and this leads to a requirement that, even at its minimal level, consideration be given to several pools of carbon and pathways of its transfer among these pools (Peterson 1980; Smith 1982; Gieskes and Kraay 1984; Smith and Platt 1984; Smith et al. 1984). When $H^{14}CO_3^-$ is introduced to a plankton sample, the photoautotrophs will take it up by photosynthesis; the chemoautotrophs will take it up by chemosynthesis; the chemoorganotrophs will take up ^{14}C-labelled exudates from the autotrophs; the biotrophs will take up ^{14}C-labelled cells; and all organisms will take up $H^{14}CO_3^-$ in anaplerotic reactions. The problem is not one of ascertaining whether these many modes of ^{14}C uptake occur in the ocean; rather it is to assess their importance in relation to photosynthetic uptake. In this latter regard, the conclusion will almost certainly depend on the community structure of the plankton in question. In general, two approaches to the problem have been adopted. In the first, a difference is sought between all uptake processes and those that occur without the requirement of light: this difference is presumed to represent the light-dependent processes — largely but not exclusively photosynthesis. In the second, methods are employed to isolate (physically or chemically) the photosynthetic process: e.g. by selective metabolic inhibition, microautoradiography or flow cytometry. These two approaches are frequently used in conjunction with each other as well as with size fractionation (Ward 1984b; Li and Dickie 1985a). In fact, there is substantial literature on the use of size fractionation, light–dark incubation and metabolic inhibition (not necessarily all together) in studies of phytoplankton exudates and their assimilation by heterotrophic bacteria (Derenbach and Williams 1974; Larsson and Hagstrom 1979, 1982; Berman and Gerber 1980; Iturriaga and Zsolnay 1981; Cole et al. 1982; Coveney 1982; Wolter 1982; Jensen 1983; Jones et al. 1983; Brock and Clyne 1984).

4.2 DARK ^{14}CO$_2$ UPTAKE

The validity of subtracting ^{14}C fixed in the dark from ^{14}C fixed in the light to estimate photosynthesis is questionable. There are concerns that dark uptake is not simply a background process which is invariant over the light–dark cycle and on which is superimposed the process of photosynthesis (Morris et al. 1971; Legendre et al. 1983). It is important to remember that we are concerned here with the dark uptake by all organisms, not solely the phototrophs. Many instances have been recorded where the magnitude of dark ^{14}C values constitutes a large percentage of the light ^{14}C values (Taguchi and Platt 1977; Gieskes et al. 1979; Peterson 1979). This appears especially so in lower latitude open ocean waters and towards the base of the euphotic zone (Saijo and Takesue 1965; Morris et al. 1971; Taguchi 1983): these are also the regions where photosynthetic picoplankton appear to be most important (Murphy and Haugen

1985). Furthermore, the highest rates of dark fixation are generally found in the
< 1 μm fraction (Saijo and Takesue 1965; Ward 1984b). Similarly, in tropical and
subtropical oceanic plankton < 1 μm, rates of ^{14}C uptake at night can often
approach or equal the rates during the lighted hours (Li et al. 1983; Herbland et al.
1985; Fig. 13).

Nonphotosynthetic CO_2 fixation (chemosynthetic or anaplerotic) is a feature
common to all cells (Wood and Utter 1965; Kornberg 1966; Sorokin 1965, 1966;
Overbeck 1976, 1979). It is my contention that in some cases, a large proportion of
the dark (i.e. nonphotosynthetic or light-independent) ^{14}C uptake in the
picoplankton can be accounted for by chemoorganotrophic (heterotrophic) bacteria.
Furthermore, there are reasons to believe that bacterial processes do not necessarily
proceed at rates invariant over the light–dark cycle. Therefore there is no direct reason
to suppose that ^{14}C uptake in the dark (or at night) somehow serves as an
"experimental blank" against which light uptake can be judged. My contention is
based on estimates of the magnitude of heterotrophic bacterial CO_2 uptake in
relation to picoplankton photosynthesis (Table 4). These estimates proceed from
Sieburth's (1984) compilation of biovolumes for heterotrophic picoplankton (H-pico)
and photosynthetic picoplankton (P-pico), the latter of which includes prokaryotic
and eukaryotic members. Biovolumes were converted to carbon standing stocks using

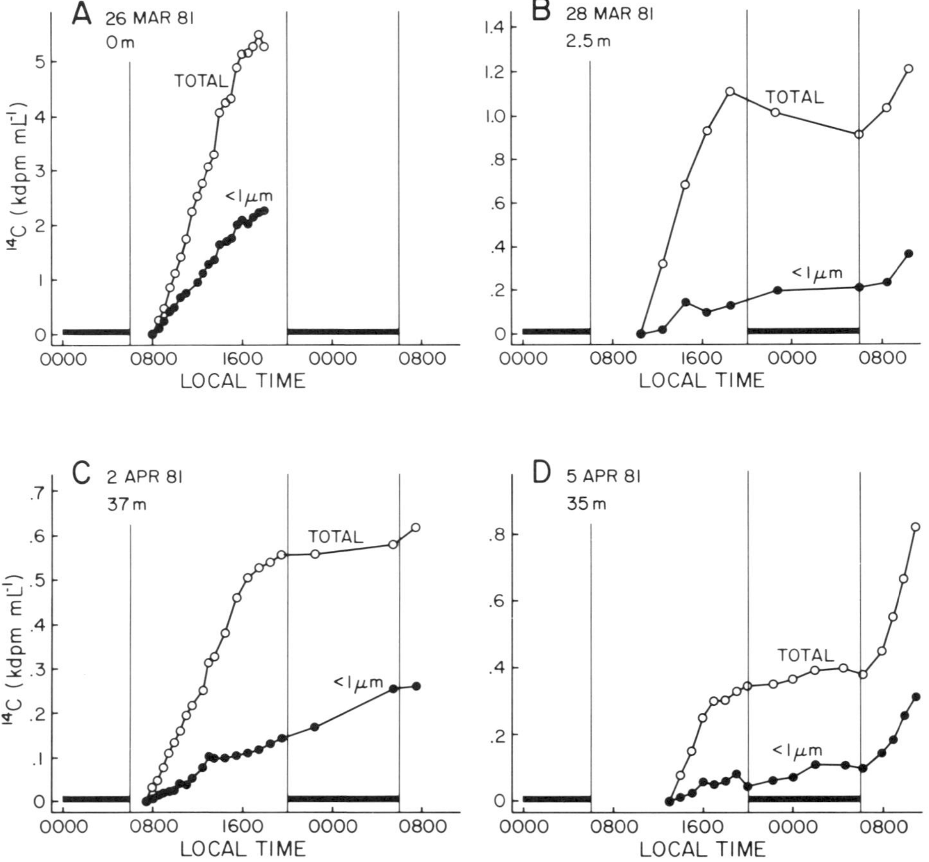

FIG. 13. Time course of ^{14}C-labelling of plankton (post-incubation filtration protocol). Total
= all plankton collected on 0.2 μm Nuclepore filters. 1 μm = plankton passing < 1 μm
Nuclepore filters retained on 0.2 μm Nuclepore filters. Station locations: 9°25′N, 89°30′W
(Mar. 26 and 28, 1981); 9°45′N, 93°45′W (Apr. 2 and 5, 1981).

TABLE 4. Calculation of the CO_2 uptake by picoplankton in three oceanic provinces. Heterotrophic picoplankton (H-pico) are assumed to grow at 1/day, and require CO_2 at 6% of their production rate. Photosynthetic picoplankton (P-pico) are assumed to grow at 1.5/day (prokaryotes) and 3.5/day (eukaryotes), and carbon requirement is from CO_2 exclusively (100%). All cells are assumed to contain 220 fg $C/\mu m^3$. See text for details. Based on biovolume compilation by Sieburth (1984).

	H-Pico	P-Pico	
		Prokaryote	Eukaryote
Estuary			
Biovolume ($\mu m^3 \cdot L^{-1}$)	3×10^8	1×10^6	4×10^6
Carbon stock ($\mu g\ C \cdot L^{-1}$)	66	0.22	0.88
Production ($\mu g\ C \cdot L^{-1} \cdot d^{-1}$)	66	0.33	3.1
CO_2 uptake ($\mu g\ C \cdot L^{-1} \cdot d^{-1}$)	4.0	0.33	3.1
Relative CO_2 uptake	1.0	0.08	0.78
Shelf			
Biovolume ($\mu m^3 \cdot L^{-1}$)	1×10^8	1×10^7	2×10^6
Carbon stock ($\mu g\ C \cdot L^{-1}$)	22	2.2	0.44
Production ($\mu g\ C \cdot L^{-1} \cdot d^{-1}$)	22	3.3	1.5
CO_2 uptake ($\mu g\ C \cdot L^{-1} \cdot d^{-1}$)	1.3	3.3	1.5
Relative CO_2 uptake	1.0	2.5	1.2
Open sea			
Biovolume ($\mu m^3 \cdot L^{-1}$)	2×10^7	1×10^6	4×10^4
Carbon stock ($\mu g\ C \cdot L^{-1}$)	4.4	0.22	0.009
Production ($\mu g\ C \cdot L^{-1} \cdot d^{-1}$)	4.4	0.33	0.03
CO_2 uptake ($\mu g\ C \cdot L^{-1} \cdot d^{-1}$)	0.26	0.33	0.03
Relative CO_2 uptake	1.0	1.2	0.12

the factor suggested by Bratbak and Dundas (1984): 220 fg $C \cdot \mu m^{-3}$. Carbon production values were calculated as the product of standing stocks and specific growth rates (μ). For heterotrophic bacteria, μ was taken to be 1 d^{-1}: considering the range of estimates compiled in the literature (Van Es and Meyer-Reil 1982; Ducklow and Hill 1985a, b), I felt this value fairly represents the mode. For prokaryotic photosynthetic marine picoplankton, μ was taken to be 1.5 d^{-1}. This is a generous estimate so that uncertainties in the final analysis will favour the importance of photosynthesis over heterotrophic uptake. This value (1.5 d^{-1}) represents an upper limit to μ_{max} reported by various workers for different clones of *Synechococcus* in laboratory cultures (Brand and Guillard 1981; Morris and Glover 1981; Brand 1984; Cuhel and Waterbury 1984; Barlow and Alberte 1985). For eukaryotic photosynthetic marine picoplankton, μ was taken to be 3.5 d^{-1}. Once again, this is a generous estimate. It is the maximum value for *Micromonas pusilla* reported by Throndsen (1976). Wilhelm and Wild (1982) and Thinh and Griffiths (1985) reported much lower values for *Nanochlorum eucaryotum* (0.3 d^{-1}) and *Chlorella nana* (1.4 d^{-1}), respectively. The exercise (Table 4) was continued by calculating CO_2 uptake from carbon production rates. For P-pico, all of the production was assumed to have come from CO_2. For H-pico, CO_2 uptake was assumed to be 6% of carbon production (Romanenko et al. 1972). It is recognized that the value of this last conversion factor can be variable (see compilation of Taguchi 1983). The results of this calculation indicate that CO_2 uptake by H-pico may at times approach or even exceed CO_2 uptake by P-pico. If dark ^{14}C values are subtracted from light ^{14}C values, photosynthetic rates would be low or even assume negative values: clearly this would be unrealistic in view of the growth rates that have been estimated for P-pico by independent means (see later section). A simple solution to this apparent enigma is to postulate that activities of H-pico are reduced in the light.

There is some evidence to support such a postulate. Sieburth (1983, 1984) proposed a scheme in which the light–dark cycle regulates a diel succession of events in oceanic plankton communities. He cited direct and indirect evidence that pointed to an apparent photoinhibition of heterotrophic and chemosynthetic processes. Other diel studies, taken together, do not offer such a clearcut impression of enhanced night-time heterotrophic activity although such a conclusion could be reached from particular experiments (Riemann et al. 1982, 1984; Carlucci et al. 1984; Riemann and Sondergaard 1984). Similarly, direct comparisons of heterotrophic activity in light and dark bottle incubations are equivocal: rates in the dark may be higher or lower than those in the light (Azam and Holm-Hansen 1973; Williams and Yentsch 1976; Bailey et al. 1983; Carlucci et al. 1985; Li and Dickie 1985a). The study of Riemann and Sondergaard (1984) provides the most direct test to the $^{14}CO_2$ uptake question in consideration. Here, the authors measured <1 μm dark $^{14}CO_2$ uptake in short incubation assays conducted on plankton sampled over a diel cycle. In spite of the stated conclusion (based on an ANOVA for results from 4 methods) that no significant diel changes occurred, the results actually show that within any particular diel experiment, the highest values for < 1 μm dark $^{14}CO_2$ uptake were almost always measured during the night.

In Fig. 14, results are shown of some $H^{14}CO_3^-$ labelling experiments conducted in the Caribbean Sea which agree with the above interpretation of Riemann and Sondergaard's (1984) results. In these experiments, plankton were placed into clear

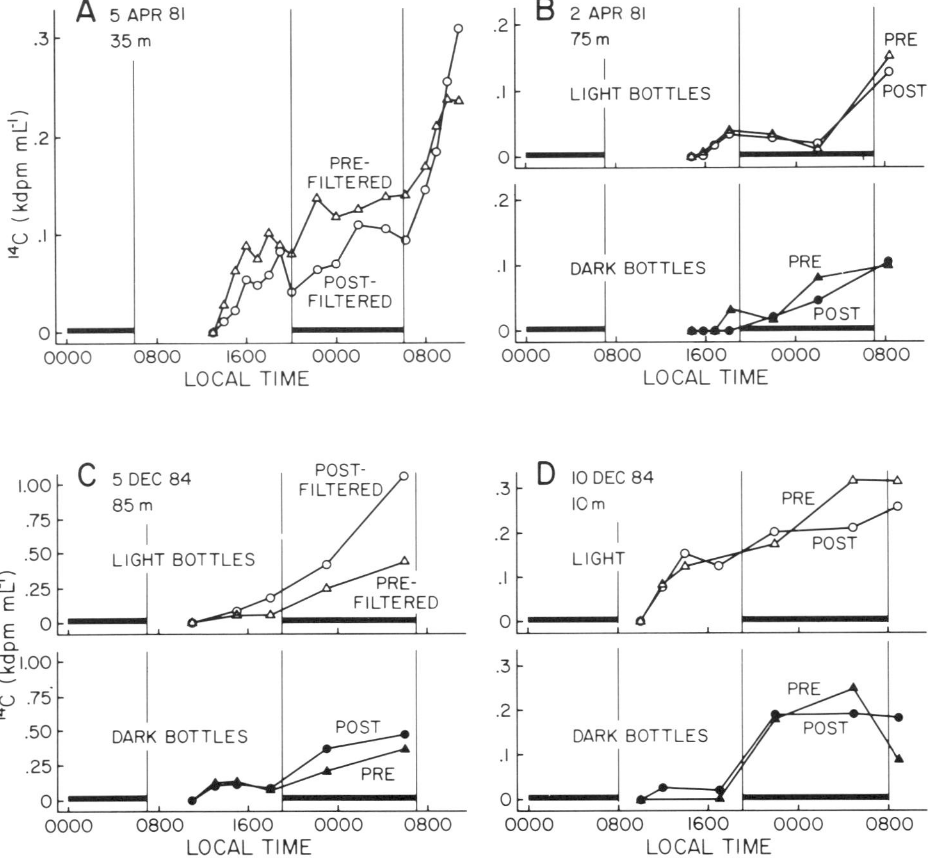

FIG. 14. Time course of ^{14}C-labelling of <1 μm plankton: comparison between post-incubation and pre-incubation filtration protocols; comparison between incubations in clear ("light") and opaque ("dark") bottles. Station locations: 9°45′N, 93°45′W (Apr. 5, 1981); 14°41′N, 64°51′W (Dec. 2, 1984); 14°38′N, 64°55′W (Dec. 5, 1984); 26°42′N, 73°39′W (Dec. 10, 1984).

or opaque bottles and incubated on deck through a light–dark period. Plankton were either fractionated (<1 μm) before radiolabelling and incubation ("pre-incubation filtration" = pre-IF) or after ("post-incubation filtration" = post-IF). "Time zero" blanks (Morris et al. 1971) have been subtracted from the results shown. The important features of the results are: (1) uptake at night was substantial in both clear and opaque bottles, (2) in dark bottles, uptake during the day was sometimes undetectable but at night, uptake was always very high. These results indeed indicate that what goes on in the picoplankton at night is different from what goes on in an opaque bottle during the day. The results further indicate that organisms collected after sunrise do not necessarily realize their nighttime potential simply by being enclosed in an opaque bottle; only the passage of time into real night brings about the high rates of uptake. These results however do not indicate the nature of the dark labelling process: whether it is anaplerotic uptake, chemosynthesis, heterotrophic uptake of radiolabelled organic exudates or a combination of these. The interpretation of differences between post-IF and pre-IF values is difficult. They may reflect heterotrophic uptake of ^{14}C organic exudates from >1 μm organisms, release of grazing pressure exerted by >1 μm organisms, curtailment of nutrient supply usually regenerated by >1 μm organisms, physiological alteration to <1 μm cells as a result of pre-IF, loss of intracellular radiolabel by cell rupture as a result of post-IF, or a great number of other reasons — all of which can be invoked, either singly or in combination to explain any differences between pre-IF and post-IF values.

Despite the discussions in this section and in other reviews of dark ^{14}C values (Morris 1974, 1982) we are still generally unsure of how to deal with them. When more information becomes available on the abundance and distribution of chemoautotrophs and methylotrophs, it will be important to assess their contributions to CO_2 uptake. In the meantime, if much of the dark uptake in the picoplankton is due to heterotrophic bacteria, and if these are most active at night, it may be easiest to ignore dark ^{14}C values when trying to evaluate light ^{14}C values of the picoplankton in terms of photosynthesis. Morris et al. (1971) arrived at the same recommendation. Alternatively, it may be feasible to regard treatment with the noncyclic photophosphorylation inhibitor DCMU as an appropriate control (Legendre et al. 1983). In a field test of this DCMU method, Li and Dickie (1985a) demonstrated the expected inhibitory effect on light $H^{14}CO_3{}^-$ uptake; at the same time, they also reported on the unexpected inhibition by DCMU of heterotrophic activites. Clearly, a more extensive investigation is required before the DCMU method can be used with confidence.

4.3 Metabolic Inhibitors

The rationale for using metabolic inhibitors in studies of photosynthesis by picoplankton depends on the aim of the experiments. For example, an inhibition of all oxygenic photosynthesis may be attempted by the use of DCMU as already described. A more ambitious undertaking would be to selectively inhibit prokaryotic or eukaryotic processes of ^{14}C uptake. In principle, much information could be gained by considering the effects of prokaryotic and eukaryotic inhibitors on the uptake of ^{14}C into separate size fractions of the plankton in clear and opaque bottles incubated through a light–dark cycle (Li, unpublished). Li and Dickie (1985a) attempted a less detailed version of this experiment. The most significant feature of their results was that cycloheximide (an inhibitor of eukaryotic protein synthesis) apparently had no effect on light dependent ^{14}C-labelling of the <1 μm fraction (post-IF) in a Sargasso Sea community. In contrast, ^{14}C uptake by the >1 μm fraction was fully inhibited 2 hours after cycloheximide was added. The results suggested that ^{14}C-labelling of the picoplankton in this sample was very likely due mainly to cyanobacterial photosynthesis. Neither photosynthesis due to eukaryotic picoplankton nor heterotrophic uptake of ^{14}C-exudates from eukaryotic phytoplankton seemed to be important.

The successful use of inhibitors in microbial ecology depends on the extent to which the primary metabolic effect (e.g. uncoupling of phosphorylation or inhibition of protein biosynthesis) is manifested at the level of responses that ecologists measure. In principle, because cellular metabolism is coordinated, an injury to some part will possibly lead to the manifestation of that injury at all levels of the cell's biology, including ecology. In practice, the usefulness of an inhibitor depends on a consistent correlation between primary response and measured response within the duration of the experiment (Li and Dickie 1985a; Sanders et al. 1985).

4.4 MICROAUTORADIOGRAPHY

This method was discussed in section 2.5. In principle, microautoradiographic examination of autofluorescent cells labelled with $H^{14}CO_3^-$ gives information about photosynthetic picoplankton that is uncompromised by activities of all other picoplankters. It is the most direct technique of verifying what type of cells (autofluorescent vs. non-fluorescent or weakly fluorescent) are responsible for taking up ^{14}C. Unfortunately, it cannot distinguish autofluorescent prokaryotes from autofluorescent eukaryotes (Douglas 1984). Furthermore, although dark uptake of ^{14}C can be ascribed to non-fluorescent cells, great care has to be taken to account for background levels of silver grains which can often be quite high (Craig, unpublished). The method also suffers from low sensitivity and the need for extensive experience and patience on the part of the experimentor.

4.5 FLOW CYTOMETRY

General aspects of flow cytometric sorting of small marine particles has been discussed by Yentsch and Yentsch (1984). In principle, if a cell (or subpopulation of cells) can be distinguished from all others, then the flow cytometer/sorter will be able to physically isolate it from all others. A simple scheme that separates photosynthetic picoplankton by "colour groups" (Yentsch and Yentsch 1979, 1984; Yentsch and Phinney 1985a,b) would rely on filtering plankton through 3 μm Nuclepore membranes followed by flow cytometric detection of phycoerythrin-containing cyanobacteria on the one hand, and chlorophyll a fluorescing algae and phycocyanin-containing cyanobacteria on the other. By analogy to pre-IF and post-IF ^{14}C experiments, pre-incubation sort (pre-IS) or post-incubation sort (post-IS) experiments can be proposed for flow cytometers. In pre-IS, it would be necessary to contend with the possibility that cells may be physiologically altered as a result of the sorting procedure.

An example of post-IS is shown in Fig. 15. In this experiment, <50 μm plankton from Bedford Basin (Nova Scotia, Canada) was labelled with $H^{14}CO_3^-$ in the light for 2 hours and then sorted into 6 groups as shown. Each group represented a different region on the LIRFL–LIOFL map. Previously, these regions had been mapped as follows using laboratory cultures: I and II correspond to where phycoerythrin-containing cyanaobacteria (*Synechococcus* WH7803, WH7805) were found; III corresponded to where a cryptomonad (*Choomonas salina* 3C) was found; IV and V corresponded to where non-phycoerythrin containing phytoplankton (*Pavlova, Dunaliella, Phaeodactylum*) were found; and VI corresponded to where a phycocyanin-containing cyanobacterium (*Synechococcus* WH5701) was found. It should be noted that fluorescence emission spectra for the pigments do not have sharp wavelength cutoffs that match the dichroic mirror and absorbance filter wavelength characteristics in the flow cytometer. This means that there is some overlap of the phycoerythrin signal into the LIRFL region, and vice-versa, some overlap of the chlorophyll a signal into the LIOFL region. This accounts for the extension of the phycoerythrin-containing cyanobacterial region (I and II) up along the LIRFL axis; and vice-versa, the extension of the chlorophyll a region (IV and V) along the LIOFL

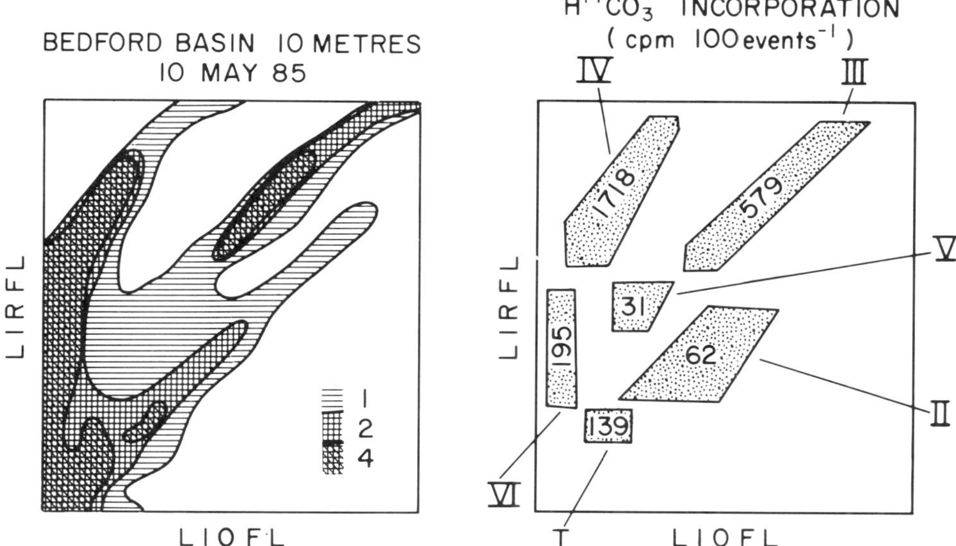

FIG. 15. Flow cytometric analysis of a <50 μm natural plankton sample (10 m, Bedford Basin; May 10, 1985). Left panel: Two dimensional contour plot of LIRFL-LIOFL signature. LIRFL represents fluorescence >590 nm excited by 488 nm light; LIOFL represents fluorescence between 530 and 590 nm excited by 488 nm light. Pulse subtractor not invoked. Right panel: Display of radioactivity per 100 events sorted into 6 polygons as indicated.

axis. The fluorescent signatures of plankton obtained by others (Yentsch et al. 1983; Wood et al. 1985) also show these features. It is possible to correct for these emission overlaps by invoking "pulse subtraction" on the Coulter® EPICS V instrument (e.g. Fig. 5). In the experiment shown (Fig. 15), the amount of radioactivity associated with 100 cells was different in all 6 groups, indicating that even under the same environmental conditions, individual components of the phytoplankton community do not necessarily photosynthesize at the same rate.

5. Growth

In the simplest sense, growth means an increase in biomass. Thus to measure growth, one merely has to monitor changes in biomass over time. Of course there are many methodological concerns that have to be addressed in practice; nevertheless, this simple idea remains as the essence of the task.

5.1 POPULATION CHANGES IN NATURE

To study net growth of populations in nature, allowing for losses to grazing, advection and other factors, it is necessary only to adopt an observational approach: biomass is monitored in samples collected from nature at intervals suitable to the time scale of interest. Examples of diel and seasonal changes in the abundance of photosynthetic picoplankton are given in Krempin and Sullivan (1981), Burney et al. (1982) and Caron et al. (1985). From June to July in Lake Ontario, the apparent doubling time of picoplanktonic cyanobacteria was 8 days (Caron et al. 1985).

5.2 H¹⁴CO₃⁻ UPTAKE

To estimate the inherent growth of photosynthetic picoplankton, it is necessary to adopt experimental approaches which eliminate or account for loss processes. Almost all such methods rely on confining plankton samples in an incubation vessel and monitoring changes in biomass with time. The most common of such confinement methods is the conventional $H^{14}CO_3^-$ labelling experiment which is usually presumed to yield the synthetic rate of new carbon (photosynthesis). If the amount of phytoplankton carbon is estimated, a calculation can be made of the presumed growth rate (Eppley and Strickland 1968). Apart from the uncertainties in assuring that net photosynthesis is being measured and from the difficulties of estimating phytoplankton carbon, there are numerous other problems in translating these results into growth estimates (Eppley 1981; Morris 1981, 1982). Nevertheless, this simple calculation has been made for size-fractionated ^{14}C-labelled picoplankton from various study sites: some published estimates for specific growth rate (d^{-1}) are 0.15 (Platt et al. 1983) 0.9 to 8.9 (Douglas 1984) and 0.5 to 0.8 (Putt and Prezelin 1985). Refinements to this method which depend on determining the specific activity of ^{14}C-labelled chlorophyll a have been proposed by Redalje and Laws (1981), Redalje (1983) and Welschmeyer and Lorenzen (1984). To my knowledge, these techniques have yet to be applied to the picoplankton.

5.3 REMOVAL OF GRAZERS: SELECTIVE FILTRATION

Another confinement method depends on the measurement of phytoplankton biomass over time in water from which grazers are presumably removed by filtration (pre-IF method). It is assumed that although photosynthetic picoplankton are actively growing in nature, population size does not increase much because of a close check by their grazers; when such grazers are removed by retention on filters, the population of photosynthetic picoplankton should increase. This method yielded specific growth rate estimates (d^{-1}) of 0.9 to 1.7 (Bienfang and Takahashi 1983), 0.7 to 2.5 (Bienfang et al. 1984) and 1.0 to 1.3 (Laws et al. 1984) when changes in chlorophyll a were measured in 3 μm filtrates. Furnas (1982a,b) considered the growth of "non-motile ultraplankton" (enumerated by bright-field microscopy) in < 10 μm filtrates incubated within *in situ* diffusion chambers: the maximum growth rate was 1.4 d^{-1} but in over half of all experiments, the rate was less than 0.5 d^{-1}. Landry et al (1984) performed a similar *in situ* diffusion chamber experiment except that the growth of cyanobacteria (enumerated by autofluorescence microscopy) was monitored in 1 μm filtrates: growth rate was estimated to be 1.42 d^{-1}. A potential problem with this method is that some grazers may not be removed by filtration. Protozooplankton are the major predators on bacteria (Sieburth 1983, 1874; Clarholm 1984) and they have been shown to consume and utilize chroococcoid cyanobacteria for growth (Johnson et al. 1982). Many of these animals pass into 3 or 1 μm filtrates (Wright and Coffin 1984); there are even indications that 0.6 and 0.4 μm filters may not effectively retain them all (Fuhrman and McManus 1984; Cynar et al. 1985). Bienfang and Takahashi (1983) were concerned with this problem: they could not detect grazers in their samples but admitted that "difficulty in the detection/identification of small phagoflagellates means that their potential presence cannot be discounted altogether". In the case of Furnas' (1982a,b) experiments, "some smaller tintinnids and ciliates passed through the 10 μm fractionating screen; however, these small populations of protozoans disappeared or declined . . .".

5.4 INACTIVATION OF GRAZERS: SELECTIVE INHIBITION

A possible alternative to removing the grazers is to inactivate them by eukaryotic inhibitors (Newell et al. 1983; Fuhrman and McManus 1984; Campbell 1985; Sanders et al. 1985). The major difficulty in this appears to be that of inhibitor specificity: in other words, the measured (ecological) response may not necessarily correspond to that which is expected on the basis of the presumed primary effect of the inhibitor (see comments in section 4.3). To my knowledge, this method has not yet been used for studying the cyanobacteria (the photosynthetic eukaryotes would presumably also be sensitive to the inhibitors). Cycloheximide, an inhibitor of eukaryotic protein synthesis, may prove useful in such experiments as Campbell (1985) has demonstrated that this inhibitor (100 mg $\cdot$ L^{-1}) had no significant effect on the growth of *Synechococcus* WH7803 over 24 hours.

5.5 REDUCTION OF GRAZERS: DILUTION EXPERIMENTS

Instead of trying to remove or inhibit the grazers, it may be possible to simply account for their activity. This is the rationale for "dilution experiments" (Kirchman et al. 1982; Landry and Hassett 1982; Landry et al. 1984; Campbell 1985; Ducklow and Hill 1985a,b; Li and Dickie 1985b). In this type of experiment, a small volume of the intact plankton sample is diluted into the same seawater from which all organisms have presumably been removed by filtration. The number of bacteria in these diluted samples often increases substantially over several hours and their specific growth rate is in inverse proportion to the ratio of unfiltered to filtered water (Landry et al. 1984). These results suggest a decreasing rate of encounter between grazers and prey as dilution increases (Kirchman et al. 1982; Landry and Hassett 1982; Landry et al. 1984). In independent applications of this method using diluent prepared by passing seawater through a 0.2 μm filters, Landry et al. (1984) estimated the specific growth rate of cyanobacteria to be 1.98 d^{-1} and Campbell (1985) estimated values between 0.55 and 0.84 d^{-1}

The assumptions underlying dilution experiments have been detailed by Landry and Hassett (1982). In this connection, I wish to note one further possible complication. As discussed in section 3.3, some cyanobacteria appear to be able to pass through small pore size filters. Even if the number of cells in freshly prepared diluent is checked and found to be very low, this cannot be taken as assurance that they remain at the low level throughout the duration of an experiment. In the almost assured absence of predators, these 0.2 μm filtrate cells may grow in numbers and size at a high rate. If this is so, the interpretation of these experiments becomes very difficult. In light of this possibility, it becomes important to monitor the diluent for cell growth throughout the experiment. Li and Dickie (1985b) did this in dilution experiments which considered the growth of heterotrophic bacteria and reported a high rate of growth (0.18 h^{-1}) for cells found in 0.2 μm Nuclepore filtrates. Similarly, Sheldon et al. (1967) had earlier demonstrated (by Coulter counting) that particles with diameters between 1.58 and 5.0 μm could increase in concentration in seawater filtered through 0.45 or 0.22 μm Milipore membranes.

5.6 FREQUENCY OF DIVIDING CELLS

The technique of estimating population growth rate from measurements of the frequency of dividing cells (FDC) is based upon the relationship between the number of cells in a given state of division and the growth rate. This method is attractive because problems incurred as a result of sample incubation and experimental manipulation are eliminated. The requirements of this method are that the cells of interest are recognizable, that cells undergoing division are also recognizable, and that the duration of division (which should be constant with respect to growth rate and temperature) is known. Given these, it is possible to calculate specific growth rate from measurements of FDC (McDuff and Chisholm 1982).

Campbell (1985) estimated *in situ* specific growth rates for *Synechococcus* at various oceanic sites using the FDC method and a 2-h division duration (derived from laboratory experiments using *Synechococcus* WH7803 and WH8012): values between 0.47 and 0.75 d^{-1} were obtained.

6. Concluding Remarks

The existence of small phytoplankton in the sea has been recognized for a long time (Lohmann 1911). Speculations that they may be of far greater importance than the diatoms and dinoflagellates (Wood and Davis 1956) were generally supported by results of size-fractionated ^{14}C uptake and chlorophyll a (see references in Malone 1980). An early suggestion was made that some phytoplankton might be less than about 1 μm in size (Holmes 1958; Holmes and Anderson 1963), but many workers were inclined to believe that evidence of photosynthetic cells in 1 μm filtrates could be better explained as fragments of algal cells broken by filtration (Anderson 1965; Berman 1975; Herbland and Le Bouteiller 1981). Recent work showing the ubiquity of intact, photosynthetically active picoplankton cells (Johnson and Sieburth 1979; Waterbury et al. 1979; Li et al. 1983; Platt et al. 1983) has confirmed the early suggestions. Research in this area has accelerated tremendously in the last 5 years. The prospects of new and significant insights into the physioecology of photosynthetic picoplankton appear very promising.

In the next few years, many methods currently used for phytoplankton will no doubt be applied to the picoplankton. More exciting will be new techniques developed to take advantage of the peculiarities of the photosynthetic picoplankton (e.g. small size, prokaryotic nature, phycobiliproteins). Rapid advances should be expected from immunofluorescence and flow cytometry. There is no single method that is "best" for studying photosynthetic picoplankton: the most fruitful research will result from applying the technique, or combination of techniques most appropriate to the question posed.

Acknowledgements

I thank E. P. W. Horne for help in making the flow cytometer operational. C. M. Yentsch and D. A. Phinney provided much useful information on flow cytometry. The cultures were obtained from the Culture Collection of Marine Phytoplankton through the courtesy of R. R. L. Guillard.

References

ALPINE, A. E., AND J. E. CLOERN. 1985. Differences in *in vivo* fluorescence yield between three phytoplankton size classes. J. Plankton Res. 7: 381–390.

ARTHUR, C. R., AND F. H. RIGLER. 1967. A possible error in the ^{14}C method of measuring primary productivity. Limnol. Oceanogr. 12: 121–124.

AZAM, F., AND O. HOLM-HANSEN. 1973. Use of tritiated substrates in the study of heterotrophy in seawater. Mar. Biol. 23: 191–196.

BAILEY, C. A., R. A. NEIHOF, AND P. S. TABOR. 1983. Inhibitory effect of solar radiation on amino acid uptake in Chesapeake Bay bacteria. Appl. Environ. Microbiol. 46: 44–49.

BARLOW, R. G., AND R. S. ALBERTE. 1985. Photosynthetic characteristics of phycoerythrin-containing marine *Synechococcus* spp. I. Responses to growth photon flux density. Mar. Biol. 86: 63–74.

BERMAN, T. 1973. Modifications in filtration methods for the measurement of inorganic ^{14}C uptake by photosynthesizing algae. J. Phycol. 9: 327–330.

——— 1975. Size fractionation of natural aquatic populations associated with autotrophic and heterotrophic carbon uptake. Mar. Biol. 33: 215–220.

BERMAN, T., AND R. W. EPPLEY. 1974. The measurement of phytoplankton parameters in nature. Sci. Prog., Oxf. 61: 219–239.

BERMAN, T., AND C. GERBER. 1980. Differential filtration studies of carbon flux from living algae to microheterotrophs, microplankton size distribution and respiration in Lake Kinneret. Microbial. Ecol. 6: 189–198.

BIENFANG, P. K. 1984. Size structure and sedimentation of biogenic microparticulates in a subarctic ecosystem. J. Plankton Res. 6: 985–995.

——— 1985. Size structure and sinking rates of various microparticulate constituents in oligotrophic Hawaiian waters. Mar. Ecol. Prog. Ser. 23: 143–151.

BIENFANG, P. K., AND M. TAKAHASHI. 1983. Ultraplankton growth rates in a subtropical ecosystem. Mar. Biol. 76: 213–218.

BIENFANG, P. K., L. MORALES, K. KLEIN, AND M. TAKAHASHI. 1984a. Picoplankton growth rates in subtropical Hawaiian embayments. Pac. Sci. 38: 134–140.

BIENFANG, P. K., J. P. SZYPER, M. Y. OKAMOTO, AND E. K. NODA. 1984b. Temporal and spatial variability of phytoplankton in a subtropical ecosystem. Limnol. Oceanogr. 29: 527–539.

BOHLOOL, B. B., AND E.L. SCHMIDT. 1980. The immunofluorescence approach in microbial ecology. Adv. Microbial. Ecol. 4: 203–241.

BOWDEN, W. B. 1977. Comparison of two direct-count techniques for enumerating aquatic bacteria. Appl. Environ. Microbiol. 33: 1229–1232.

BRAND, L. E. 1984. The salinity tolerance of forty-six marine phytoplankton isolates. Estuarine Coastal Shelf Sci. 18: 543–556.

BRAND, L. E., AND R. R. L. GUILLARD. 1981. The effects of continuous light and light intensity on the reproduction rates of twenty-two species of marine phytoplankton. J. Exp. Mar. Biol. Ecol. 50: 119–132.

BRATBAK, G. 1985. Bacterial biovolume and biomass estimations. Appl. Environ. Microbiol. 49: 1488–1493.

BRATBAK, G., AND I. DUNDAS. 1984. Bacterial dry matter content and biomass estimations. Appl. Environ. Microbiol. 48: 755–757.

BROCK, T. D. 1983. Membrane filtration: a user's guide and reference manual. Science Tech., Madison, WI. 381 p.

BROCK, T. D., AND J. CLYNE. 1984. Significance of algal excretory products for growth of epilimnetic bacteria. Appl. Environ. Microbiol. 47: 731–734.

BURNEY, C. M., P. G. DAVIS, K. M. JONHSON, AND J. MCN. SIEBURTH. 1981. Dependence of dissolved carbohydrate concentrations upon small scale nanoplankton and bacterioplankton distributions in the western Sargasso Sea. Mar. Biol. 65: 289–296.

——— 1982. Diel relationship of microbial trophic groups and *in situ* dissolved carbohydrate dynamics in the Caribbean Sea. Mar. Biol. 67: 311–322.

CAMPBELL, L. 1985. Investigations of marine, phycoerythrin-containing *Synechococcus* spp. (Cyanobacteria): Distribution of serogroups and growth rate measurements. Ph.D. thesis, State University of New York at Stony Brook, NY. 186 p.

CAMPBELL, L., E. J. CARPENTER, AND V. J. IACONO. 1983. Identification and enumeration of marine chroococcoid cyanobacteria by immuno-fluorescence. Appl. Environ. Microbiol. 46: 553–559.

CARLUCCI, A. F., D. B. CRAVEN, AND S. M. HENRICHS. 1984. Diel production and microheterotrophic utilization of dissolved free amino acids in waters off Southern California. Appl. Environ. Microbiol. 48: 165–170.

——— 1985. Surface-film microheterotrophs: amino acid metabolism and solar radiation effects on their activities. Mar. Biol. 85: 13–22.

CARON, D. A. 1983 Technique for enumeration of heterotrophic and phototrophic nanoplankton, using epifluorescence microscopy, and comparison with other procedures. Appl. Environ. Microbiol. 46: 491–498.

CARON, D. A., F. R. PICK, AND D. R. S. LEAN. 1985. Chroococcoid cyanobacteria in Lake Ontario: vertical and seasonal distributions during 1982. J. Phycol. 21: 171–175.

CARPENTER, E. J., AND J. S. LIVELY. 1980. Review of estimates of algal growth using ^{14}C tracer techniques, p. 161–178. *In* P. G. Falkowski [ed.] Primary productivity in the sea. Plenum Press, New York, NY.

CLARHOLM, M. 1984. Heterotrophic, free-living protozoa: neglected microorganisms with an important task in regulating bacterial populations, p. 321–326. *In* M. J. Klug and C. A. Reddy [ed.] Current perspectives in microbioal ecology. American Society for Microbiology, Washington, DC.

COLE, J. J., G. E. LIKENS, AND D. L. STRAYER. 1982. Photosynthetically produced dissolved organic carbon: an important carbon source for planktonic bacteria. Limnol. Oceanogr. 27: 1080–1090.

COVENEY, M. F. 1982. Bacterial uptake of photosynthetic carbon from freshwater phytoplankton. Oikos 38: 8–20.

CRAIG, S. R. 1984 Productivity of algal picoplankton in a small meromictic lake. Verh. Internat. Verein. Limnol. 22: 351–354.

——— 1986. Picoplankton size distributions in marine and fresh waters: problems with filter fractionation studies. FEMS Microbiol. Ecol. 38: 171–177.

CUHEL, R. L., AND J. B. WATERBURY. 1984. Biochemical composition and short term nutrient incorporation patterns in a unicellular marine cyanobacterium (WH7803). Limnol. Oceanogr. 29: 370–374.

CULLEN, J. J. 1982. The deep chlorophyll maximum: comparing vertical profiles of chlorophyll *a*. Can. J. Fish. Aquat. Sci. 39: 791–803.

CYNAR, S. J., K. W. ESTEP, AND J. MCN. SIEBURTH. 1985. The detection and characterization of bacteria-sized protists in "protist-free" filtrates and their potential impact on experimental marine ecology. Microbial. Ecol. 11: 281–288.

DAVIS, P. G., AND J. MCN. SIEBURTH. 1982. Differentiation of phototrophic and heterotrophic nanoplankton populations in marine waters by epifluorescence microscopy. Ann. Inst. Oceanogr., Paris, 58(S): 249–260.

DAVIS, P. G., D. A. CARON, P. W. JOHNSON, AND J. MCN. SIEBURTH. 1985. Phototrophic and apochlorotic components of picoplankton and nanoplankton in the North Atlantic: geographic, vertical, seasonal and diel distributions. Mar. Ecol. Prog. Ser. 21: 15–26.

DERENBACH, J. B., AND P. J. LEB. WILLIAMS. 1974. Autotrophic and bacterial production: fractionation of plankton populations by differential filtration of samples from the English Channel. Mar. Biol. 25: 263–269.

DOUGLAS, D. J. 1984. Microautoradiography-based enumeration of photosynthetic picoplankton with estimates of carbon-specific growth rates. Mar. Ecol. Progr. Ser. 14: 223–228.

DUCKLOW, H. W., AND S. M. HILL. 1985a. The growth of heterotrophic bacteria in the surface waters of warn core rings. Limnol. Oceanogr. 30: 239–259.

——— 1985b. Tritiated thymidine incorporation and the growth of heterotrophic bacteria in warm core rings. Limnol. Oceanogr. 30: 260–272.

EASTERBROOK, K. B., AND D. V. SUBBA RAO. 1984. Conical spinae associated with a picoplanktonic procaryote. Can. J. Microbiol. 30: 716–718.

EPPLEY, R. W. 1981. Relations between nutrient assimilation and growth in phytoplankton with a brief review of estimates of growth rate in the ocean, p. 251–263. *In* T. Platt [ed.]

Physiological bases for phytoplankton ecology. Can. Bull. Fish. Aquat. Sci. 210.

EPPLEY, R. W., AND J. D. H. STRICKLAND. 1968. Kinetics of marine phytoplankton growth, p. 23–62. *In* M. R. Droop and E. J. Ferguson Wood [ed.] Advances in microbiology of the sea 1. Academic Press, New York, NY.

FERGUSON, R. L., BUCKLEY, E. N., AND A. V. PALUMBO. 1984. Response of marine bacterioplankton to differential filtration and confinement. Appl. Environ. Microbiol. 47: 49–55.

FLIERMANS, C. B., AND E. L. SCHMIDT. 1975. Autoradiography and immunofluorescence combined for autoecological study of single cell activity with *Nitrobacter* as a model system. Appl. Microbiol. 30: 676–684.

1977. Immunofluorescence for autecological study of a unicellular bluegreen alga. J. Phycol. 13: 364–368.

FUHRMAN, J. A., AND T. M. BELL. 1985. Biological considerations in the measurement of dissolved free amino acids in seawater and implications for chemical and microbiological studies. Mar. Ecol. Prog. Ser. 25: 13–21.

FUHRMAN, J. A., AND G. B. MCMANUS. 1984. Do bacteria-sized marine eukaryotes consume significant bacterial production? Science 224: 1257–1260.

FURNAS, M. J. 1982a. An evaluation of two diffusion culture techniques for estimating phytoplankton growth rates in situ. Mar. Biol. 70: 63–72.

1982b. Growth rates of summer nanoplankton ($<10\ \mu$m) populations in Lower Narragansett Bay, Rhode Island, USA. Mar. Biol. 70: 105–115.

FURUYA, K. 1982. Measurement of phytoplankton standing stock using an image analyser system. Bull. Plank. Soc. Japan 29: 131–132.

FURUYA, K., AND R. MARUMO. 1983. The structure of the phytoplankton community in the subsurface chlorophyll maxima in the western North Pacific ocean. J. Plankton Res. 5: 393–406.

GIESKES, W. W., AND G. W. KRAAY. 1983. Unknown chlorophyll *a* derivations in the North Sea and the tropical Atlantic ocean revealed by HPLC analysis. Limnol. Oceanogr. 28: 757–766.

1984. State-of-the-art in the measurement of primary production, p. 171–190. *In* M. J. R. Fasham [ed.] Flows of energy and materials in marine ecosystems. Plenum Press, New York, NY.

GIESKES, W. W. C., G. W. KRAAY, AND M. A. BAARS. 1979. Current ^{14}C methods for measuring primary production: gross underestimates in oceanic waters. Neth J. Sea Res. 13: 58–78.

GLOVER, H. E. 1985. The physiology and ecology of marine cyanobacteria, *Synechococcus* sp. Adv. Aquat. Microbiol. 3: 49–107.

GLOVER, H. E., L. CAMPBELL, AND B. B. PREZELIN. 1986. Contribution of *Synechococcus* spp. to size-fractionated primary productivity in three water masses in the Northwest Atlantic. Mar.

Biol. 91: 193–203.

GLOVER, H. E., D. A. PHINNEY, AND C. S. YENTSCH. 1985a. Photosynthetic characteristics of picoplankton compared with those of larger phytoplankton populations, in various water masses in the Gulf of Maine. Biol. Oceanogr. 3: 223–248.

GLOVER, H. E., A. E. SMITH, AND L. SHAPIRO. 1985b. Diurnal variations in photosynthetic rates: comparisons of ultraphytoplankton with a larger phytoplankton size fraction. J. Plankton Res. 7: 519–535.

GOLDMAN, J. C., AND M. R. DENNETT. 1985. Susceptibility of some marine phytoplankton species to cell breakage during filtration and post-filtration rinsing. J. Exp. Mar. Biol. Ecol. 86: 47–58.

GUILLARD, R. R. L., L. S. MURPHY, P. FOSS, AND S. LIAAEN-JENSEN. 1985. *Synechococcus* spp. as likely zeaxanthin-dominant ultraphytoplankton in the North Atlantic. Limnol. Oceanogr. 30: 412–414.

HAGSTROM, A., U. LARSSON, P. HORSETEDT, AND S. NORMARK. 1979. Frequency of dividing cells, a new approach to the determination of bacterial growth rates in aquatic environments. Appl. Environ. Microbiol. 37: 805–812.

HERBLAND, A. 1974. Influence de la dépression de filtration sur la mesure simultanée de l'assimilation et de l'excrétion organique du phytoplancton. Cah. O.R.S.T.O.M. Ser. Oceanogr. 12: 173–177.

HERBLAND, A., AND A. LE BOUTEILLER. 1981. The size distribution of phytoplankton and particulate organic matter in the equatorial Atlantic ocean: importance of ultraseston and consequences. J. Plankton Res. 3: 659–673.

HERBLAND, A., A. LE BOUTEILLER, AND P. RAIMBAULT. 1985. Size structure of phytoplankton biomass in the equatorial Atlantic ocean. Deep-Sea Res. 32: 819–836.

HICKEL, W. 1984. Seston retention by Whatman GF/C glass fiber filters. Mar. Ecol. Prog. Ser. 16: 185–191.

HOLMES, R. W. 1958. Size fractionation of photosynthesizing phytoplankton, p. 69–71. *In* R. W. Holmes et al. [ed.] Scripps cooperative oceanic productivity expedition, 1956. U.S. Fish and Wildlife Service, Special scientific report: Fisheries No. 279.

HOLMES, R. W., AND G. C. ANDERSON. 1963. Size fractionation of ^{14}C-labeled natural phytoplankton communities, p. 241–250. *In* C. H. Oppenheimer [ed.] Marine microbiology, C. C. Thomas Publishers, Springfield, IL.

IRWIN, B. D., J. J. CULLEN, AND T. PLATT. 1982. Size fractionation by differential filtration of particulate matter in the sea: effect of vacuum pressure (Unpublished)

ITURRIAGA, R., AND A. ZSOLNAY. 1981. Differentiation between auto- and heterotrophic activity: problems in the use of size fractionation and antibiotics. Bot. Mar. 24: 399–404.

JENSEN, L. M. 1983. Phytoplankton release of extracellular organic carbon, molecular weight composition, and bacterial assimilation. Mar. Ecol. Prog. Ser. 11: 39–48.

JOHNSON, B. D., AND P. J. WANGERSKY. 1985. Seawater filtration: particle flow and impaction considerations. Limnol. Oceanogr. 30: 966–971.

JOHNSON, P. W., AND J. McN. SIEBURTH. 1979. Chroococcoid cyanobacteria in the sea: a ubiquitous and diverse phototrophic biomass. Limnol. Oceanogr. 24: 928–935.

——— 1982. In-situ morphology and occurrence of eucaryotic phototrophs of bacterial size in the picoplankton of estuarine and oceanic waters. J. Phycol. 18: 318–327.

JOHNSON, P. W., H.-S. XU, AND J. McN. SIEBURTH. 1982. The utilization of chroococcoid cyanobacteria by marine protozooplankters but not by calanoid copepods. Ann. Inst. Oceanogr., Paris 58(S): 297–308.

JOINT, I. R., AND R. K. PIPE. 1984. An electron microscope study of a natural population of picoplankton from the Celtic Sea. Mar. Ecol. Prog. Ser. 20: 113–118.

JOINT, I. R., AND A. J. POMROY. 1983. Production of picoplankton and small nanoplankton in the Celtic Sea. Mar. Biol. 77: 19–27.

JONES, J. G., B. M. SIMON, AND C. R. CUNNINGHAM. 1983. Bacterial uptake of algal extracellular products: an experimental approach. J. Appl. Bacteriol. 54: 355–365.

KARL, D. M., G. A. KNAUER, J. H. MARTIN, AND B. B. WARD. 1984. Bacterial chemolithotrophy in the ocean is associated with sinking particles. Nature 309: 54–56.

KIRCHMAN, D., H. DUCKLOW, AND R. MITCHELL. 1982. Estimates of bacterial growth from changes in uptake rates and biomass. Appl. Environ. Microbiol. 44: 1296–1307.

KORNBERG, H. L. 1966. Anaplerotic sequences and their role in metabolism, p. 1–31. In P. N. Campbell and G.D. Greville [ed.] Essays in biochemistry 2. Academic Press, London and New York.

KOZ'YAKOV, S. Y., B. V. GROMOV, AND E. B. KOZ'YAKOVA. 1979. Application of the immunofluorescence technique to the identification of unicellular cyanobacteria in natural water bodies. Microbiology 48: 919–921.

KREMPIN, D. W., AND C. W. SULLIVAN. 1981. The seasonal abundance, vertical distribution, and relative microbial biomass of chroococcoid cyanobacteria at a station in southern California coastal waters. Can J. Microbiol. 27: 1341–1344.

LANDRY, M. R., AND R. P. HASSETT. 1982. Estimating the grazing impact of marine microzooplankton. Mar. Biol. 67: 283–288.

LANDRY, M. R., L. W. HASS, AND V. L. FAGERNESS. 1984. Dynamics of microbial plankton communities: experiments in Kaneohe Bay, Hawaii. Mar. Ecol. Prog. Ser. 16: 127–133.

LARSSON, U., AND A. HAGSTROM. 1979. Phytoplankton exudate release as an energy source for the growth of pelagic bacteria. Mar. Biol. 52: 199–206.

——— 1982. Fractionated phytoplankton primary production, exudate release and bacterial production in a Baltic eutrophication gradient. Mar. Biol. 67: 57–70.

LAWS, E. A., D. G. REDALJE, L. W. HAAS, P. K.

BIENFANG, R. W. EPPLEY, W. G. HARRISON, D. M. KARL, AND J. MARRA. 1984. High phytoplankton growth and production rates in oligotrophic Hawaiian coastal waters. Limnol. Oceanogr. 29: 1161–1169.

LEAN, D. R. S., AND B. K. BURNISON. 1979. An evaluation of errors in the ^{14}C method of primary production measurement. Limnol. Oceanogr. 24: 917–928.

LEFTLEY, J. W., D. J. BONIN, AND S. Y. MAESTRINI. 1984. Problems in estimating marine phytoplankton growth, productivity and metabolic activity in nature: an overview of methodology. Oceanogr. Mar. Biol. Ann. Rev. 21: 23–66.

LEGENDRE, L., DEMERS, C. M. YENTSCH, AND C. S. YENTSCH. 1983. The ^{14}C method: patterns of dark CO_2 fixation and DCMU correction to replace the dark bottle. Limnol. Oceanogr. 28: 996–1003.

LI, W. K. W., AND P. M. DICKIE. 1985a. Metabolic inhibition of size-fractionated marine plankton radiolabeled with amino acids, glucose, bicarbonate, and phosphate in the light and dark. Microbial. Ecol. 11: 11–24.

——— 1985b. Growth of bacteria in seawater filtered through 0.2 μm Nuclepore membranes: implications for dilution experiments. Mar. Ecol. Prog. Ser. 26: 242–252.

LI, W. K. W., D. V. SUBBA RAO, W. G. HARRISON, J. C. SMITH, J. J. CULLEN, B. IRWIN, AND T. PLATT. 1983. Autotrophic picoplankton in the tropical ocean. Science 219: 292–295.

LOHMANN, H. 1911. Uber das nannoplankton und die zentrifugierung kleinster wasserproben zur gewinnung desselben in lebendem zustande. Int. Rev. Hydrobiol. Hydrogr. 4: 1–38.

MAGUE, T. H., E. FRIBERG, D. J. HUGUES, AND I. MORRIS. 1980. Extracellular release of carbon by marine phytoplankton; a physiological approach. Limnol. Oceanogr. 25: 262–279.

MALONE, T. C. 1980. Algal size, p. 433–463. In I. Morris [ed.] The physiological ecology of phytoplankton. University of California Press, Berkeley and Los Angeles, CA.

McDUFF, R. E., AND S. W. CHISHOLM. 1982. The calculation of in situ growth rates of phytoplankton populations from fractions of cells undergoing mitosis: a clarification. Limnol. Oceanogr. 27: 783–788.

MEYER-REIL, L.-A. 1978. Autoradiography and epifluorescence microscopy combined for the determination of number and spectrum of actively metabolizing bacteria in natural waters. Appl. Environ. Microbiol. 36: 506–512.

MORITA, R. Y. 1982. Starvation-survival of heterotrophs in the marine environment. Adv. Microbial. Ecol. 6: 171–198.

MORRIS, I. 1974. The limits to the productivity of the sea. Sci. Prog., Oxf. 61: 99–122.

——— 1981. Photosynthetic products, physiological state, and phytoplankton growth, p. 83–102. In T. Platt [ed.] Physiological bases of phytoplankton ecology, Can. Bull. Fish. Aquat. Sci. 210.

——— 1982. Primary production of the oceans, p. 239–252. In R. G. Burns and J. H. Slater [ed.]. Experimental microbial ecology.

Blackwell Scientific Publications, Oxford.

MORRIS I., AND H. E. GLOVER. 1981. Physiology of photosynthesis by marine coccoid cyanobacteria — some ecological implications. Limnol. Oceanogr. 26: 957–961.

MORRIS, I., C. M. YENTSCH, AND C. S. YENTSCH. 1971. Relationship between light carbon dioxide fixation and dark carbon dioxide fixation by marine algae. Limnol. Oceanogr. 16: 854–858.

MUNAWAR, M., I. F. MUNAWAR, P. E. ROSS, AND A. DAGENAIS. 1982. Microscopic evidence of phytoplankton passing through glass-fibre filters and its implications for chlorophyll analysis. Arch. Hydrobiol. 94: 520–528.

MURPHY, L. S., AND E. M. HAUGEN. 1985. The distribution and abundance of phototrophic ultraplankton in the North Atlantic. Limnol. Oceanogr. 30: 47–58.

NALEWAJKO, C., AND D. R. S. LEAN. 1972. Retention of dissolved compounds by membrane filters as an error in the ^{14}C method of primary production measurement. J. Phycol. 8: 37–43.

NEWELL, S. Y., B. F. SHERR, E. B. SHERR, AND R. D. FALLON. 1983. Bacterial response to presence of eukaryote inhibitors in waters from a coastal marine environment. Mar. Environ. Res. 10: 147–157.

OLSON, R. J., D. VAULOT, AND S. W. CHISHOLM. 1985. Marine phytoplankton distributions measured using shipboard flow cytometry. Deep-Sea Res. 32: 1273—1280.

OVERBECK, J. 1976. Some remarks on the ecology of the CO_2- metabolism of heterotrophic and methylotrophic bacteria, p. 263–266. In H. G. Schlegel, G. Gottschalk, and N. Pfenning [ed.] Symposium on microbial production and utilization of gases (H_2, CH_4, CO). E. Goltze KG, Gottingen.

——— 1979. Dark CO_2 uptake — biochemical background and its relevance to in situ bacterial production. Arch. Hydrobiol. Beih. Ergebn. Limnol. 12: 38–47.

PERKINS, F.O., L. W. HAAS, D. E. PHILLIPS, AND K. L. WEBB. 1981. Ultrastructure of a marine *Synechococcus* possessing spinae. Can. J. Microbiol. 27: 318–329.

PETERSON, G. H. 1979. On the analysis of dark fixation in primary production computations. J. Cons. Int. Explor. Mer. 38: 326–330.

PETERSON, B. J. 1980. Aquatic primary productivity and the ^{14}C-CO_2 method: a history of the productivity problem. Ann. Rev. Ecol. Syst. 11: 359–385.

PHINNEY, D. A., AND C. S. YENTSCH. 1985. A novel phytoplankton chlorophyll technique: toward automated analysis. J. Plankton Res. 7: 633–642.

PLATT, T., D. V. SUBBA RAO, AND B. IRWIN. 1983. Photosynthesis of picoplankton in the oligotrophic ocean. Nature 300: 702–704.

PREZELIN, B. B., M. PUTT, AND H. E. GLOVER. 1986. Diurnal patterns in photosynthetic capacity and depth-dependent photosynthesis-irradiance relationship in *Synechococcus* spp. and larger phytoplankters in three water masses in the Northwest Atlantic. Mar. Biol. 91: 205–217.

PROBYN, T. A. 1985. Nitrogen uptake by size-fractionated phytoplankton populations in the southern Benguela upwelling system. Mar. Ecol. Prog. Ser. 22: 249–258.

PUTT, M., AND B. B. PREZELIN. 1985. Observations of diel patterns of photosynthesis in cyanobacteria and nanoplankton in the Santa Barbara Channel during 'el Nino'. J. Plankton Res. 7: 779–790.

REDALJE, D. G. 1983. Phytoplankton carbon biomass and specific growth rates determined with the labeled chlorophyll *a* technique. Mar. Ecol. Prog. Ser. 11: 217–225.

REDALJE, D. G., AND E. A. LAWS. 1981. A new method for estimating phytoplankton growth rates and carbon biomass. Mar. Biol. 62: 73–79.

RIEMANN, B., J. FUHRMAN, AND F. AZAM. 1982. Bacterial secondary production measured by ^{3}H-thymidine incorporation method. Microb. Ecol. 8: 101–114.

RIEMANN, B., P. NIELSEN, M. JEPPESEN, B. MARCUSSEN, AND J. A. FUHRMAN. 1984. Diel changes in bacterial biomass and growth rates in coastal environments, determined by means of thymidine incorporation into DNA, frequency of dividing cells (FDC), and microautoradiography. Mar. Ecol. Prog. Ser. 17: 227–235.

RIEMAN, B., AND M. SONDERGARD. 1984. Measurements of diel rates of bacterial secondary production in aquatic environments. Appl. Environ. Microbiol. 47: 632–638.

ROMANENKO, V. I., J. OVERBECK, AND Y. I. SOROKIN. 1972. Estimation of production of heterotrophic bacteria using ^{14}C, p. 82–85. In Y. I. Sorokin and H. Kadota [ed.] Techniques for the assessment of microbial production and decomposition in fresh waters. IBP Handbook No. 23. Blackwell, Oxford.

SAIJO, Y., AND K. TAKESUE. 1965. Further studies on the size distribution of phytosynthesizing phytoplankton in the Indian Ocean. J. Oceanogr. Soc. Japan 20: 10–17.

SAKSHAUG, E. 1980. Problems in the methodology of studying phytoplankton, p. 57–91. In I. Morris [ed.] The physiological ecology of phytoplankton. University of California Press, Berkeley and Los Angeles, CA.

SANDERS, R. W., R. J. McDONOUGH, AND K. G. PORTER. 1985. Variable effects of eukaryotic and procaryotic inhibitors on bacterial production and grazing by bacteriovores. Am. Soc. Limnol. Oceanogr. Ann. Meet. Abstracts 48: 93.

SAROKIN, D. J., AND E. J. CARPENTER. 1981. Cyanobacterial spinae. Bot. Mar. 27: 425–436.

SCHMALJOHANN, R. 1984. Morphological investigations on bacterioplankton of the Baltic Sea, Kattegat and Skagerrak. Bot. Mar. 27: 425–436.

SHAPIRO, H. M. 1985. Practical flow cytometry. Alan Liss Inc., New York, NY, 295 p.

SHARP, J. H. 1977. Excretion of organic matter by marine phytoplankton: do healthy cells do it? Limnol. Oceanogr. 22: 381–399.

SHELDON, R. W. 1972. Size separation of marine sestion by membrane and glass-fiber filters. Limnol. Oceanogr. 17: 494–498.

SHELDON, R. W., T. P. T. EVELYN, AND T. R. PARSONS. 1967. On the occurrence and formation of small particles in seawater. Limnol. Oceanogr. 12: 367–375.

SHIFRIN, N. S. 1980. The measurement of dissolved organic carbon released by phytoplankton. Estuaries 3: 230–233.

SICKO-GOAD, L., AND E. F. STOERMER. 1984. The need for uniform terminology concerning phytoplankton size fractions and examples of picoplankton from the Laurentian Great Lakes. J. Great Lakes Res. 10: 90–93.

SIEBURTH, J. McN. 1979. Sea microbes. Oxford University Press, New York, NY, 491 p.

———— 1983. Microbiological and organic-chemical processes in the surface and mixed layers, p. 121–172. *In* P.S. Liss and W. G. N. Slinn [ed.] Air – sea exchange of gases and particles. D. Reidel Publishing, Dordrecht, Boston, Lancaster.

———— 1984. Protozoan bacterivory in pelagic marine waters, p. 405–444. *In* J. E. Hobbie and P. J. LeB. Williams [ed.] Heterotrophic activity in the sea. Plenum Press, New York and London.

SIEBURTH, J. McN., V. SMETACEK, AND J. LENZ. 1978. Pelagic ecosystem structure: heterotrophic compartments of the plankton and their relationship to plankton size fractions. Limnol. Oceanogr. 23: 1256–1263.

SIEBURTH, J. McN., P. W. JOHNSON, M. A. EBERHARDT, AND M. E. SIERACKI. 1983. Methane-oxidizing bacteria from the mixing layer of the Sargasso Sea and their photosensitivity. Eos 64: 1054.

SMITH, J. C., T. PLATT, W. K. W. LI, E. P. W. HORNE, W. G. HARRISON, D. V. SUBBA RAO, AND B. D. IRWIN. 1985. Arctic marine photoautotrophic picoplankton. Mar. Ecol. Prog. Ser. 20: 207–220.

SMITH, R. E. H. 1982. The estimation of phytoplankton production and excretion by ^{14}C. Mar. Biol. Lett. 3: 325–334.

SMITH, R. E. H., AND J. KALFF. 1983. Sample preparation for quantitative autoradiography of phytoplankton. Limnol. Oceanogr. 28: 383–389.

SMITH, R. E. H., R. J. GEIDER, AND T. PLATT. 1984. Microplankton productivity in the oligotrophic ocean. Nature 311: 252–254.

SMITH, R. E. H., AND T. PLATT. 1984. Carbon exchange and ^{14}C tracer methods in a nitrogen-limited diatom, *Thalassiosira pseudonana*. Mar. Ecol. Prog. Ser. 16: 75–87.

SOROKIN, J. I. 1965. On the trophic role of chemosynthesis and bacterial biosynthesis in water bodies. Mem. Ist. Ital. Idrobiol. 18 (Suppl.): 187–205.

———— 1966. On the carbon dioxide uptake during the cell synthesis by microorganisms. Z. Allg. Mikrobiol. 6: 69–73.

STANLEY, P. M., M. A. GAGE, AND E. L. SCHMIDT. 1979. Enumeration of specific populations by immunofluorescence p. 46–55. *In* J. W. Costerton and R. R. Colwell [ed.] Native aquatic bacteria: enumeration, activity, and ecology. ASTM STP 695.

STEWART, D. E., AND F. H. FARMER. 1984. Extraction, identification, and quantitation of phycobiliprotein pigments from phototrophic plankton. Limnol. Oceanogr. 29: 392–396.

STRAND, S. E., AND M. E. LIDSTROM. 1984. Characterization of a new marine methylotroph. FEMS Microbiol. Lett. 21: 247–251.

STRICKLAND, J. D. H., AND T. R. PARSONS. 1972. A practical handbook of seawater analysis. Bull. Fish. Res. Board Canada 167.

TABOR, P. S., AND R. A. NEIHOF. 1982a. Improved method for determination of respiring individual microorganisms in natural waters. Appl. Environ. Microbiol. 43: 1249–1255.

———— 1982 b. Improved microautoradiographic method to determine individual microorganisms active in substrate uptake in natural waters. Appl. Environ. Microbiol. 44: 945–953.

TAGUCHI, S. 1983. Dark fixation of CO_2 in the subtropical north Pacific Ocean and the Weddell Sea. Bull. Plankt. Soc. Japan 30: 115–124.

TAGUCHI, S., AND T. PLATT. 1977. Assimilation of $^{14}CO_2$ in the dark compared to phytoplankton production in a small coastal inlet. Est. Coast. Mar. Sci. 5: 679–684.

TAKAHASHI, M., AND P. K. BIENFANG. 1983. Size structure of phytoplankton biomass and photosynthesis in subtropical Hawaiian waters. Mar. Biol. 76: 203–211.

TAKAHASHI, M., AND T. HORI. 1984. Abundance of picophytoplankton in the subsurface chlorophyll maximum layer in subtropical and tropical waters. Mar. Biol. 79: 177–186.

TAKAHASHI, M., K. KIKUCHI, AND Y. HARA. 1985. Importance of picocyanobacteria biomass (unicellular, blue-green algae) in the phytoplankton population of the coastal waters off Japan. Mar. Biol. 89: 63–69.

THINH, L.-V., AND D. J. GRIFFITHS. 1985. A small marine *Chlorella* from the waters of a coral reef. Bot. Mar. 28: 41–46.

THRONDSEN, J. 1976. Occurrence and productivity of small marine flagellates. Norw. J. Bot. 23: 269–293.

TYNDALL, R. L., R. E. HAND Jr., R. C. MANN, C. EVANS, AND R. JERNIGAN. 1985. Application of flow cytometry to detection and characterization of *Legionella* spp. Appl. Environ. Microbiol. 49: 852–857.

VAN ES, F. B., AND L.-A. MEYER-REIL. 1982. Biomass and metabolic activity of heterotrophic marine bacteria. Adv. Microb. Ecol. 6: 111–170.

VEGTER, F., AND P. R. M. DE VISSCHER. 1984. Extracellular release by phytoplankton during photosynthesis in Lake Grevelingen (SW Netherlands). Neth. J. Sea Res. 18: 260–272.

WARD, B. B. 1982. Oceanic distribution of ammonium-oxidizing bacteria determined by immunofluorescent assay. J. Mar. Res. 40: 1155–1172.

———— 1984a. Combined autoradiography and immunofluorescence for estimation of single cell activity by ammonium-oxidizing bacteria. Limnol. Oceanogr. 29: 402–409.

———— 1984b. Photosynthesis and bacterial utilization of phytoplankton exudates: results from pre- and post-incubation size fractionation. Oceanol. Acta 7: 337–343.

WARD, B. B., AND M. J. PERRY. 1980. Immunofluorescent assay for the marine ammonium-oxidizing bacterium *Nitrosococcus oceanus*. Appl. Environ. Microbiol. 39: 913–918.

WARD, B. B., R. J. OLSON, AND M. J. PERRY. 1982. Microbial nitrification rates in the primary nitrite maximum off southern California. Deep-Sea Res. 29: 247–255.

WATERBURY, J. B., S. W. WATSON, R. R. L. GUILLARD, AND L. E. BRAND. 1979. Widespread occurrence of a unicellular, marine, planktonic, cyanobacterium. Nature 277: 293–294.

WELSCHMEYER, N. A., AND C. J. LORENZEN. 1984. Carbon-14 labeling of phytoplankton carbon and chlorophyll *a* carbon: determination of specific growth rates. Limnol. Oceanogr. 29: 135–145.

WILHELM, C., G. EISENBEIS, A. WILD, AND R. ZAHN. 1982. *Nanochlorum eucaryotum*: a very reduced coccoid species of marine chlorophyceae. Z. Naturforsch. 37c: 107–114.

WILHELM, C., AND A. WILD. 1982. Growth and photosynthesis of *Nanochlorum eucaryotum*, a new and extremely small eucaryotic green alga. Z. Naturforsch. 37c: 115–119.

WILLIAMS, P. J. LEB., AND C. S. YENTSCH. 1976. An examination of photosynthetic production, excretion of photosynthetic products, and heterotrophic utilization of dissolved organic compounds with reference to results from a coastal subtropical sea. Mar. Biol. 35: 31–40.

WOLTER, K. 1982. Bacterial incorporation of organic substances released by natural phytoplankton populations. Mar. Ecol. Prog. Ser. 7: 287–295.

WOOD, A. M., P. K. HORAN, K. MUIRHEAD, D. A. PHINNEY, C. M. YENTSCH, AND J. B. WATERBURY. 1985. Discrimination between pigment types of marine *Synechococcus spp.* by scanning spectroscopy, epifluorescence microscopy, and flow cytometry. Limnol. Oceanogr. 30: 1303–1315.

WOOD, E. J. F., AND P. S. DAVIS. 1956. Importance of smaller phytoplankton elements. Nature 177: 438.

WOOD, H. G., AND M. F. UTTER. 1965. The role of CO_2 fixation in metabolism, p. 1–27. *In* P. N. Campbell and G. D. Greville [ed.] Essays in biochemistry 1. Academic Press, London and New York.

WRIGHT, R. T., AND R. B. COFFIN. 1984. Measur-

ing microzooplankton grazing on planktonic marine bacteria by its impact on bacteria production. Microb. Ecol. 10: 137–149.

YENTSCH, C. M., P. K. HORAN, K. MUIRHEAD, Q. DORTCH, E. HAUGEN, L. LEGENDRE, L. S. MURPHY, M. J. PERRY, D. A. PHINNEY, S. A. POMPONI, R. W. SPINARD, M. WOOD, C. S. YENTSCH, AND B. J. ZAHURANEC. 1983. Flow cytometry and cell sorting: a technique for analysis and sorting of aquatic particles. Limnol. Oceanogr. 28: 1275–1280.

YENTSCH, C. M., L. CUCCI, AND D. A. PHINNEY. 1984. Flow cytometry and cell sorting: problems and promises for biological ocean science research, p. 141–155. *In* O. Holm-Hansen, L. Bolis, and R. Gilles [ed.] Marine phytoplankton and productivity. Springer-Verlag, Berlin.

YENTSCH, C. M., AND C. S. YENTSCH. 1984. Emergence of optical instrumentation for measuring biological properties. Oceanogr. Mar. Biol. Ann. Rev. 22: 55–98.

YENTSCH, C. S. 1974. Some aspects of the environmental physiology of marine phytoplankton: a second look. Oceanogr. Mar. Biol. Ann. Rev. 12: 41–75.

——— 1983. A note on the fluorescence characteristics of particles that pass through glass-fiber filters. Limnol. Oceanogr. 28: 597–599.

YENTSCH, C. S., AND D. A. PHINNEY. 1984. Observed changes in spectral signatures of natural phytoplankton populations: the influence of nutrient availability, p. 129–140. *In* O. Holm-Hansen, L. Bolis, and R. Gilles [ed.] Marine phytoplankton and productivity, Springer-Verlag, Berlin.

——— 1985a. Fluorescence spectral signatures for studies of marine phytoplankton, p. 259–274. *In* A. Zirino [ed.] Mapping strategies in chemical oceanography. American Chemical Society, Washington, DC.

——— 1985b. Spectral fluorescence: an ataxonomic tool for studying the structure of phytoplankton populations. J. Plankton Res. 7: 617–632.

YENTSCH, C. S., AND C. M. YENTSCH. 1979. Fluorescence spectral signatures: the characterization of phytoplankton populations by the use of excitation and emission spectra. J. Mar. Res. 37: 471–483.

ZIERDT, C. H. 1979. Adherence of bacteria, yeast, blood cells, and latex spheres to large-porosity membrane filters. Appl. Envir. Microbiol. 38: 1166–1172.

Physiological Ecology of Picoplankton in Various Oceanographic Provinces

I. R. JOINT

*NERC Institute for Marine Environmental Research,
Prospect Place, The Hoe, Plymouth, PL1 3DH, U.K.*

Introduction

The reports by Waterbury et al. (1979) and Johnson and Sieburth (1979) of the widespread distribution of small marine cyanobacteria must be considred as the beginning of the current upsurge of interest and research on small photosynthetic microbes in the sea. Although these reports were the first to show that cyanobacteria were widely distributed, the presence of small autotrophic algae has been known for many years. Lohmann (1911) described unicellular blue-green algae in marine plankton and Gross (1937) commented on the difficulties of establishing cultures of diatoms because flagellates, some as small as 2 μm, would grow in vast numbers in the enrichment cultures. However, it is only in the last 5 years that data have been presented which show the quantitative importance of small phytoplankton cells to pelagic ecosystems.

The aim of this paper is to examine data obtained on picoplankton from different regions of the worlds oceans and to see if the measurements made in different environments can be used to draw any generalisations about the physiology and ecology of natural populations of picoplankton. The term picoplankton was proposed by Sieburth et al. (1978) for organisms that are smaller than 2 μm but larger than 0.2 μm; however, most workers have adopted slightly different size categories. Perhaps the most common functional definition has arisen as a result of doing size fractionation experiments with Nuclepore® membrane filters and the most commonly used pore-size is 1 μm. Sheldon (1972) showed only 30% retention of 1 μm diameter particles by 1 μm pore-size Nuclepore membrane but less than 30% of particles >2 μm passed through a 1 μm pore-size Nuclepore. So a reasonable functional definition of picoplankton, and one which has been adopted by many experimentalists, would appear to be those cells which pass through a 1 μm pore-size sieve.

This paper will consider the evidence available on the geographical distribution of picoplankton and the quantitative significance of picoplankton primary production in regions from the tropics to the Arctic. The seasonal cycle of picoplankton abundance and production in temperate waters will then be discussed. Picoplankton is usually considered to be adapted to living at very low light levels. However, how much is this a response to the environment in which the cells are found and how much does it reflect the basic physiology of the cells? Data on photosynthesis–light curves from different hydrographic regimes will be examined to see whether picoplankton cells from the deep chlorophyll maximum have the same photosynthetic characteristics as cells found at high light in the surface mixed layer. Details of ultrastructure, as they influence our perception of cellular function, will then be considered for natural populations from different regimes. Finally, I will briefly consider what is known about the abundance and distribution of those protozoa (ciliates and microflagellates) which are the presumed grazers of picoplankton in the oceans.

Geographical Distribution

It is now clear that picoplankton cells are extremely widespread; indeed, such appears to be the ubiquitous nature of picoplankton that it would almost be more interesting to discover if there are any environments where picoplankton is not found and to determine why it is not present.

The early reports by Waterbury et al. (1979) and Johnson and Sieburth (1979) were concerned with cyanobacteria which were readily identified with epifluorescence microscopy by the orange fluorescence due to phycoerythrin. However, it is now known that very small eukaryotic algae (<2 μm) are also present in many regions and are a true component of the picoplankton. Much less data are available on the abundance and distribution of this eukaryotic picoplankton; it is more difficult to distinguish it from nanoplankton using epifluorescence microscopy because eukaryotic cells do not possess a unique photosynthetic pigment with a characteristic fluorescence. Size is the only criterion that can be used to distinguish the red fluorescence of eukaryotic picoplankton from that of larger chlorophyll containing nanoplankton. Data are, however, becoming available to suggest that eukaryotic picoplankton can be a significant component of certain picophytoplankton populations.

Table 1 shows data on abundance of picoplankton from the tropics, subtropics, temperate oceanic and shelf seas, and the Arctic. Most data are available for cyanobacteria because of the relative ease with which they can be distinguished. The

TABLE 1. Abundance of cyanobacteria and eukaryotic picoplankton from various geographical locations.

| Region | Depth of sample (m) | Picoplankton number $\cdot$ mL^{-1} | | Reference |
		Cyanobacteria	Eukaryotes	
Off Peru	0–20	0.5 to 8.8 $\times$ 10^4	n.d.	Waterbury et al. (1979)
Arabian Sea	0–100	3 to 12 $\times$ 10^4	n.d.	Waterbury et al. (1979)
Caribbean	50 and 100	0.6 to 6.6 $\times$ 10^3	n.d.	Johnson and Sieburth (1979)
Sargasso Sea	50 and 100	1.3 to 2.2 $\times$ 10^3	n.d.	Johnson and Sieburth (1979)
S. California coastal	1	0.1 to 7 $\times$ 10^4	n.d.	Krempin and Sullivan (1981)
Tropical Pacific	0–50	0.5 to 1.5 $\times$ 10^6	n.d.	Li et al. (1983)
Azores	65–89	0.4 to 1.7 $\times$ 10^4	n.d.	Platt et al.(1983)
Great Barrier Reef	0–10	0.1 to 20 $\times$ 10^4	n.d.	Moriarty et al. (1985)
North Atlantic	1–25	0.2 to 18 $\times$ 10^4	3.5 to 28 $\times$ 10^3 [a]	Murphy and Haugen (1985)
Sargasso Sea	25–100	0.46 to 22 $\times$ 10^4	0.05 to 6.5 $\times$ 10^3 [a]	Murphy and Haugen (1985)
Gulf of Maine	1 to 40	0.1 to 34 $\times$ 10^4	2 to 42 $\times$ 10^3 [a]	Murphy and Haugen (1985)
Canadian Arctic	5–55	0.9 to 2.1 $\times$ 10^3	n.d.	Smith et al. (1985)
Gulf of Maine	1–38	1.5 to 11.7 $\times$ 10^4	0.6 to 44 $\times$ 10^3	Glover et al. (1985)

[a] Numbers are total eukaryotic cells but population dominated (up to 96%) by cell <3 μm.
n.d. No data reported.

data in Table 1 demonstrate that in most oceanic areas cyanobacteria are present at between 10^3 and 10^4 cells $\cdot$ mL^{-1}. One of the most comprehensive data sets on eukaryotic picoplankton abundance is that of Murphy and Haugen (1985) who counted numbers of eukaryotic algae as well as cyanobacteria; although these data include all eukaryotic algae, most of the populations encountered were dominated by cells <3 μm which frequently accounted for more than 90% of the total number of chlorophyll-containing organisms; so these data contain a component of the larger phytoplankton but can be taken as a reasonable approximation of the numbers of eukaryotic picoplankton present. Murphy and Haugen (1985) found that the numbers of eukaryotic cells in the north Atlantic were generally an order of magnitude less than the numbers of cyanobacteria.

There have been other reports that have demonstrated that eukaryotic picoplankton can be very abundant. Throndsen (1978) reported that *Micromonas pusilla* (cell length 1.5 to 2.5 μm) was present at numbers in excess of 10^4 $\cdot$ mL^{-1} in March 1974 in Oslofjord. *Micromonas pusilla* was also found by Johnson and Sieburth (1982) in an electron microscope study of natural populations from estuarine and oceanic waters of the western North Atlantic. *Micromonas pusilla* was confined to the estuarine waters of Narragansett Bay, R.I. but another unidentified microalga, which had the same cell dimensions as bacteria, was found to be abundant in oceanic waters of the western North Atlantic; this alga was a scaled, non-flagellated prasinophyte which had cell dimensions of approximately 0.5 × 1.0 μm. Johnson and Sieburth estimated the abundance of this alga from transmission electron micrographs to be between 10^3 and 10^4 cells • mL^{-1}. Other electron microscope studies of natural populations have also demonstrated the presence of very small eukaryotic algae. Takahashi and Hori (1984) found cyanobacteria, with the distinctive peripheral arrangement of the thylakoid within the cell, in water samples from the western North Pacific and South China Sea; the cyanobacteria had cell dimensions of 0.5–2 μm and were more abundant than a similar size eukaryotic alga. This "*Chlorella*-like" cell had a single cup-shaped chloroplast, starch storage grain and had cell dimensions of 1.2–1.5 μm. As in other studies, picoplankton organisms were the major component of the phytoplankton population and >70% of the chlorophyll was found in cells that passed through a 3 μm pore-size Nuclepore membrane.

Small "*Chlorella*-like" cells, possessing neither scales more flagella, were also found in an electron microscope study of a natural population by Joint and Pipe (1984). In a sample from the temperate, shelf waters of the Celtic Sea, we found cyanobacteria with cell dimensions of 0.5–1.0 μm and small eukaryotic algae with diameters of 0.85–2 μm; these algae had a single cup-shaped chloroplast and starch grain and were very similar to the cells found by Takahashi and Hori (1984) in the western North Pacific.

Murphy and Haugen (1985) sampled a total of 50 stations in the North Atlantic and were able to draw some conclusions about geographical distribution of the cyanobacteria component of the picoplankton. In particular, a plot of latitude and cyanobacterial abundance (averaged over the surface 50 m) showed a clear trend with

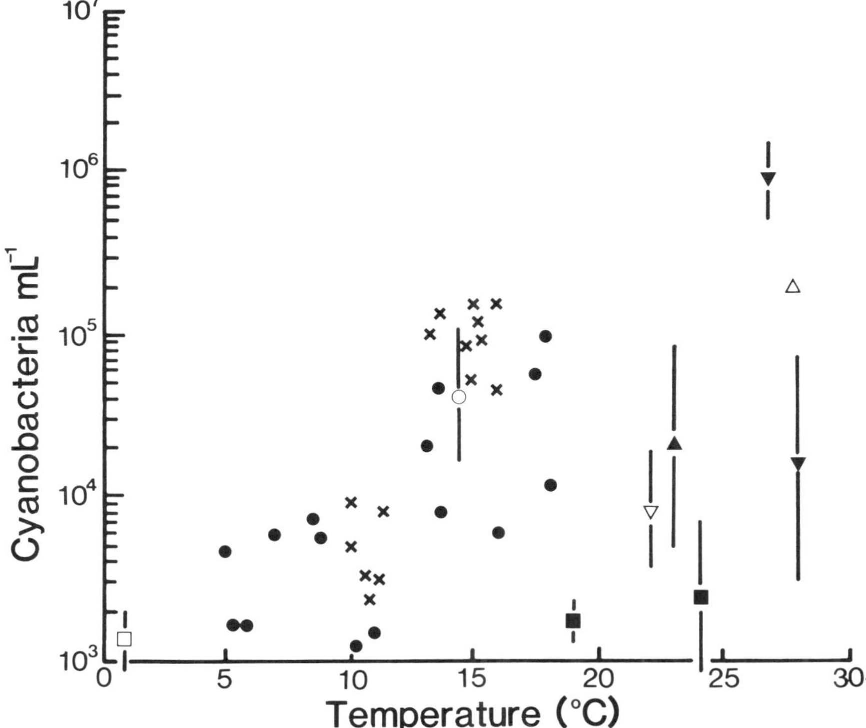

FIG 1. Ranges in number of cyanobacteria from Table 1 plotted against temperature; where temperature data were not reported, typical temperatures for the region have been taken from standard oceanographic atlas. (●) Murphy and Haugen 1985; (×) Joint, unpublished data; (■) Johnson and Sieburth 1979; (▲) Waterbury et al. 1979; (▼) Li et al. 1983; (▽) Platt et al. 1983; (□) Smith et al. 1985; (△) Moriarty et al. 1985.

decreasing numbers at higher latitudes. These authors suggest that cyanobacterial abundance is related in part to temperature but other factors are also likely to be important. Figure 1 shows the data on cyanobacterial abundance taken from Table 1 and plotted against temperature. Clearly the relationship is not quite as robust as with the data of Murphy and Haugen (1985) but nevertheless there seems to be a strong correspondence between increasing temperature and the upper limit of cyanobacteria number. It is unlikely that temperature per se is the controlling factor since cyanobacterial abundance is a balance of two dynamic processes, growth and grazing, neither of which are likely to have a different temperature response. It is much more likely that temperature covaries with other factors, such as light or nutrient advection into the euphotic zone, which have a greater effect on cyanobacterial ecology. A reasonable generalisation at this stage in research on marine picoplankton is that cyanobacteria are most abundant in the tropics and least abundant in polar seas; however, as Fig. 1 demonstrates, these are relative terms and there are still 10^3 cyanobacteria $\cdot$ mL^{-1} in Arctic waters which are only slighlty warmer than 0°C (Smith et al. 1985). Interestingly, a recent paper by Caron et al. (1985) also found a similar relationship between temperature and the abundance of picoplanktonic chroococcoid cyanobacteria (0.7–1.3 μm) in Lake Ontario. These data illustrate the ubiquitous distribution of picoplankton in the euphotic zone of the worlds oceans.

VERTICAL DISTRIBUTION

It is generally accepted that picoplankton is adapted to growth at low light. If this generalisation is true, the vertical distribution of picoplankton in different sea areas should reflect this adaptation and picoplankton might be expected to be restricted to regions such as the pycnocline, where dispersion is reduced and the cells are less likely to be advected into surface waters where irradiance is high. However, this does not appear to be the case; in their initial report of cyanobacterial picoplankton, Waterbury et al. 1979 found highest numbers in the surface waters. Johnson and Sieburth (1979) reported very similar cells densities at 50 and 100 m in the Sargasso Sea and Caribbean, but they did comment that a morphologically different cyanobacterium, with more closely space thylakoids, was only found in samples taken at 100 m in oceanic waters. However, neither of these studies specifically sampled the chlorophyll maximum which is often associated with a well developed pycnocline.

In warm oligotrophic regions of the oceans, seasonal variations in temperature are small and there is a permanent, strong pycnocline; a distinct chlorophyll maximum is usually associated with this pycnocline and chlorophyll concentrations are often an order of magnitude greater in this layer than in the surface mixed layer. Bienfang and Szyper (1981) found that small phytoplankton (<5 μm) accounted for more than 80% of the biomass in the chlorophyll maximum in waters off Hawaii. In a subsequent study, Takahashi and Bienfang (1983) used 3 μm pore-size filters and again found ca. 80% of the chlorophyll associated with cells smaller than 3 μm. Therefore, the subsurface chlorophyll maximum in these waters appears to be dominated by picoplankton. However, picoplankton was also present in surface waters (Bienfang and Takahashi, 1983) and no information is available to indicate whether there were differences in species composition of the picoplankton from the subsurface chlorophyll maximum and from the surface mixed layer. Given the difficulties in identification of these small cells, it may not be possible to obtain such data; Furuya and Marumo (1983) were able to show differences in the species composition of larger phytoplankton (microplankton) between the surface waters and the subsurface chlorophyll maximum in the western North Pacific off Japan, but they presented no data on picoplankton per se. However, *Micromonas* (1–3 μm) was the most abundant genus and was much more abundant in the subsurface chlorophyll maximum than in the surface waters; again this suggests that picoplankton, in general, may be more abundant in the subsurface maximum. Takahashi and Hori (1984) certainly believe so and, as a result of an electron microscope study of samples from the subsurface chlorophyll maximum in the western North Pacific, they concluded that picoplankton dominated that layer. However, their electron microscope study also showed the presence of prokaryotic and eukaryotic picoplankton in the surface waters and there were no morphological differences between cells from these two depths.

Two questions arise from these reports; is picoplankton always more abundant in the subsurface chlorophyll maximum and is picoplankton always responsible for that chlorophyll maximum? In oligotrophic waters with a permanent pycnocline, the evidence is fairly strong in favour of the first point, namely that picoplankton is most abundant in the subsurface chlorophyll maximum (Bienfang and Szyper 1981; Furuya and Marumo 1983; Takahashi and Hori 1984) but there are too few data available to say whether or not picoplankton is usually responsible for the chlorophyll maxima.

However, the situation must be different in temperate shelf waters where the thermocline that develops is seasonal. Chlorophyll maxima that develop in association with these hydrographic features may only be present for a few months and, because of the influence of tidal variation over a spring–neap cycle, will not experience the stable conditions that exist in the permanent pycnocline of warmer waters. In addition to chlorophyll maxima associated with the seasonal thermocline, large concentrations of chlorophyll can be found associated with tidal fronts (Pingree et al. 1975, 1978, 1982). However, in all cases the chlorophyll maxima appear to be the result of increased numbers of nano- and microplankton; picoplankton has not been implicated. Joint et al. (1986) found a chlorophyll maximum associated with the developing thermocline in the Celtic Sea in May 1984, but this was the result of increased numbers of the diatom *Ceratulina pelagica* and an unidentified nanoflagellate; numbers of cyanobacteria were no higher in this chlorophyll maximum than in the rest of the euphotic zone.

The numbers of cyanobacteria usually show homogeneous distribution with depth in the surface mixed layer of the Celtic Sea (Fig. 2). However, in periods of low wind, the degree of surface mixing is greatly reduced and a secondary thermocline can

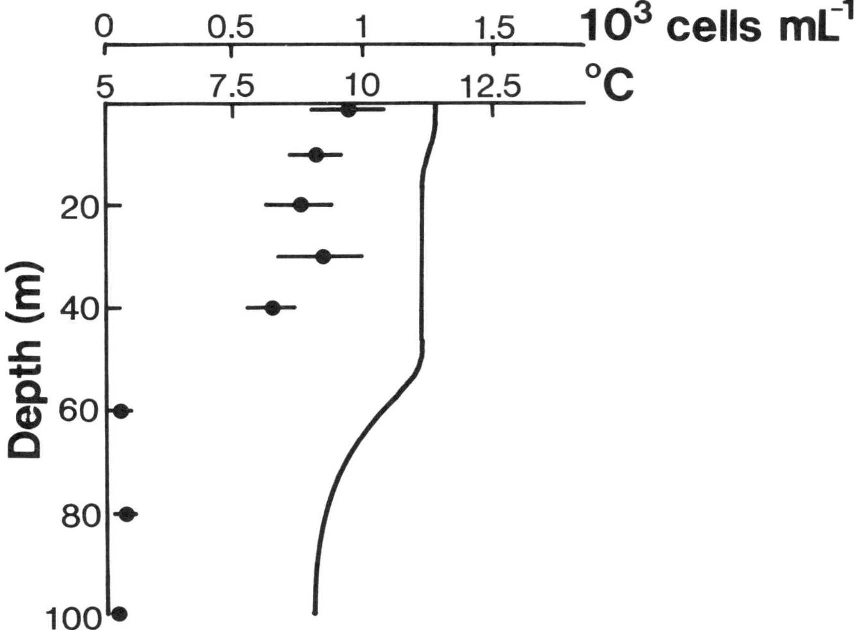

FIG. 2. Depth distribution of cyanobacteria (mean and 95% confidence interval) and temperature at a station in the Porcupine Seabight on 28 May 1984.

develop; Fig 3. shows such a situation in early July 1985. The numbers of cyanobacteria increased with depth and the highest numbers were found in the thermocline. This situation is short lived and increasing wind speed results in total mixing of the surface mixed layer and uniform vertical distribution of cyanobacteria. There is no evidence that picoplankton is responsible for chlorophyll maxima associated with the thermocline in temperate waters.

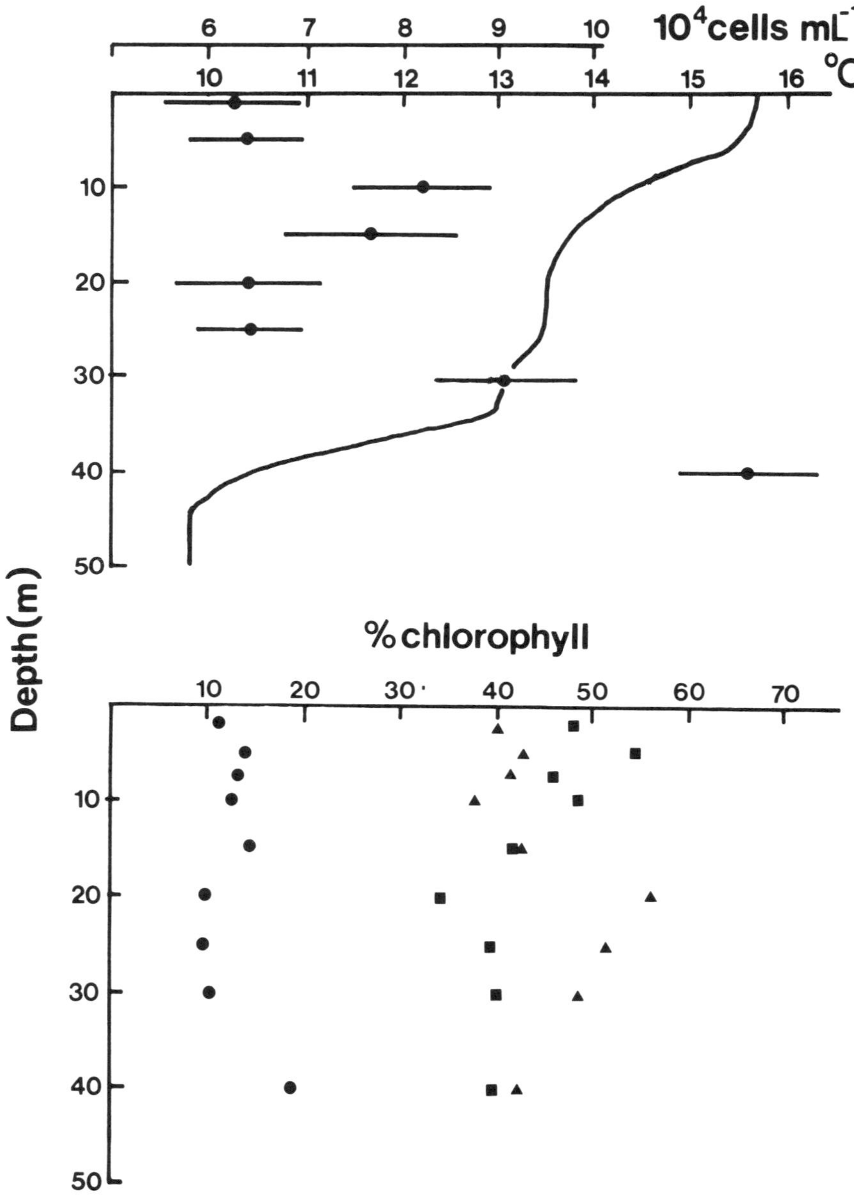

FIG. 3. Vertical distribution of temperature and numbers of cyanobacteria at station CS2 in the Celtic Sea on 3 July 1985; numbers of cyanobacteria are expressed as mean and 95% confidence limits. The distribution of chlorophyll in the different size fractions is shown in the lower figure; (▲) retained by 5 μm pore-size Nuclepore filter, (■) passing a 5 μm but retained by 1 μm Nuclepore, (●) picoplankton fraction which passed a 1 μm but was retained by 0.2 μm Nuclepore.

CONTRIBUTION OF PICOPLANKTON TO TOTAL PHYTOPLANKTON PRODUCTION

The preceding data suggest that picoplankton is abundant; but is photosynthesis by picoplankton a significant input of carbon into the pelagic ecosystem? Measurements of picoplankton production have been made less frequently than abundance has been assessed, but the answer to the question must be that in certain environments, particularly oligotrophic, oceanic regions, picoplankton is responsible for the majority of photosynthetic production. The paper by Gieskes et al. (1979) was one of the first to report significant carbon fixation by cells which passed through a 1 μm pore-size sieve; in their study of primary production in the North Equatorial Current in the tropical Atlantic, they found 20–30% of ^{14}C and 43–53% of the chlorophyll a passed through a 1 μm Unipore® filter. These measurements were done in 1977 and 1978, before the first publications of cyanobacterial picoplankton abundance and Gieskes et al. (1979) did not comment on the ecological significance of this fixation; however, it is probable that the ^{14}C fixation which passed through the 1 μm pore filter was indeed the result of picoplankton activity. Therefore, this early work by Gieskes et al. suggests that 20–30% of the primary production in the tropical Atlantic was by cells <1 μm, 16% was by cells <3 μm and 16–31% was by cells 3–8 μm.

More recent studies in the tropical oceans have had the specific objective of measuring picoplankton production. Li et al. (1983) studied two sites in the tropical Pacific and found that the percentage of ^{14}C fixation which passed a 1 μm Nuclepore filter increased with depth. In experiments with light-attenuated deck incubators, 20% of ^{14}C fixation in the surface waters was by picoplankton <1 μm but this proportion increased with depth and was up to 80% of total fixation at the equivalent of 70 m depth. Carbon fixation by picoplankton at one station at the Costa Rica Dome was about 3 mg C • m^{-3} • h^{-1} at 30% of surface illumination and maximum production was 4.5 mg C • m^{-3} • h^{-1}. Li et al. (1983) calculated that carbon fixation at this station was equal to the maximum cell-specific rate reported by Morris and Glover (1981) for *Synechococcus* strain DC-2 in culture.

In the subtropical Atlantic, Platt et al. (1983) also found significant picoplankton production and calculated that picoplankton <1 μm was capable of supplying about 60% of the total primary production at this open-ocean ecosystem west of the Azores. These estimates were again based on experiments using ^{14}C and were made on water samples taken from the subsurface chlorophyll maximum. Takahashi and Bienfang (1983) found 77–82% of the total ^{14}C fixation in waters off Hawaii was by phytoplankton <3 μm. Using a different approach, Bienfang and Takahashi (1983) estimated specific growth rates by following the change in chlorophyll concentration after size fractionation of water samples through 3 μm sieves; the prefiltration was designed to remove grazing organisms and, hence, any change in chlorophyll concentration should be the result of growth of the population on the absence of grazing pressure. The assumptions made in this type of experiment are open to criticism because it is difficult to demonstrate the total absence of grazing organisms; nevertheless, these experiments suggested that picoplankton <3 μm had growth rates equivalent to 1.3 to 2.5 doublings • d^{-1}. In contrast the calculated growth rate of picoplankton from the chlorophyll maximum studied by Platt et al. was 0.22 doublings • d^{-1} for the picoplankton <1 μm and 0.07 doublings • d^{-1} for phytoplankton >1 μm. There is a large discrepancy between these two estimates of picoplankton production in subtropical waters. The experiments of Bienfang and Takahashi (1983) were carried out on water samples taken from near shore and incubated at high light (250 μE • m^{-2} • s^{-1}) and it is difficult to draw any comparison with the estimate of Platt et al. which indicate a growth rate only 10% that suggested by Bienfang and Takahashi; it is important to known which estimate is more typical of oligotrophic waters and more experiments are urgently needed. In temperate waters in early summer, Joint and Pomroy (1986) estimated the doubling time of the cyanobacterial component of the shelf and oceanic picoplankton to be between 6.5 and 8.5 h at saturating irradiance. However, such growth rates could only be achieved by natural populations in the surface 10 m under cloud-free conditions and we considered that the actual doubling time was likely to be of the order of 1 to 2 days.

Picoplankton production is also significant in temperate waters. In waters of the European continental shelf, Joint and Pomroy (1983) found that picophytoplankton (<1 μm) accounted for 20–30% of the primary production in the summer months. In a more recent paper (Joint et al. 1986) we have extended our measurements of size fractionated photosynthesis to all seasons and estimated total annual primary production in the Celtic Sea to be 102 g C $\cdot$ m^{-2} $\cdot$ yr^{-1}. Of this production, 22.4% was by picoplankton (<1 μm), and 40.7% was by cells <5 to >1 μm and large phytoplankton (<5 μm) only dominated production for two months during the spring diatom bloom. In the upwelling region of the Benguela current off southern Africa, Probyn (1985) found picoplankton in the oceanic stations to be responsible for 27% of the production, as assessed by the uptake of ^{15}N: in the coastal stations, this proportion had declined to 10%. In the eastern Canadian Arctic, Smith et al. (1985) found picoplankton responsible for 10–25% of the primary production in late summer.

These data suggest that picoplankton can be responsible for significant proportion of primary production; generally picoplankton accounts for 20–30% of the summer production in temperate regions, >50% in the tropics and subtropics and between 10 and 25% in the Arctic in summer. However, in polar and temperate seas, this production is unlikely to be sustained because of the strong influence of season on production cycles and the next section considers what is known about the seasonal production of picoplankton.

SEASON PICOPHYTOPLANKTON PRODUCTION

Krempin and Sullivan (1981) reported the seasonal abundance of cyanobacteria in coastal waters off southern California; maximum numbers were found in August when the population reached 0.1×10^4 $\cdot$ mL^{-1} and the population minimum was in February and March when 7×10^4 cells $\cdot$ mL^{-1} were present. The seasonal pattern of picoplankton abundance varied in the same way as phytoplankton and heterotrophic bacteria but no data were given on how picoplankton production varied with season. Recently, we have completed a study of seasonal primary production in the Celtic Sea by different size fractions of phytoplankton (Joint et al. 1986). In a total of nine cruises over a 2-yr period, phytoplankton production was measured in three size fractions, picoplankton (<1 μm), small nanoplankton (<5 to >1 μm) and other phytoplankton (>5 μm). Figure 4 shows the estimated monthly carbon fixation by the three size fractions. The greatest measured monthly production was in the <5 μm fraction in April when the spring diatom bloom occurred in the Celtic Sea (Fasham et al. 1983); the production of 18.5 g C $\cdot$ m^{-2} $\cdot$ yr^{-1} in April accounts for almost half of the annual production by phytoplankton >5 μm and occurred before the development of maximum copepod biomass in the area. Therefore, a large proportion of the phytoplankton which is traditionally considered as the food of zooplankton will not be directly available (Joint et al. 1986) and probably sinks out of the water column (Malone and Chervin 1979; Davies and Payne 1984).

An estimate of total primary production (but excluding dissolved organic carbon production which was about 10–15% of the total carbon fixation) shows that small nanoplankton is most significant in the Celtic Sea and fixed about 42 g C $\cdot$ m^{-2} $\cdot$ yr^{-1} (40.7% of the total fixation). Picoplankton (<1 μm) fixed 23.1 g C $\cdot$ m^{-2} $\cdot$ yr^{-1} (22.4% of the total) and phytoplankton >5 μm fixed 37.9 g C $\cdot$ m^{-2} $\cdot$ yr^{-1} (36.8% of the total). Picoplankton production was highest in spring, immediately after the spring diatom bloom and again in August (Fig. 4). At both times, nutrient levels were low (nitrate <0.1 μmol $\cdot$ L^{-1}) and chlorophyll a concentrations in the surface mixed layer were <0.5 mg $\cdot$ m^{-3}. Picoplankton pro-

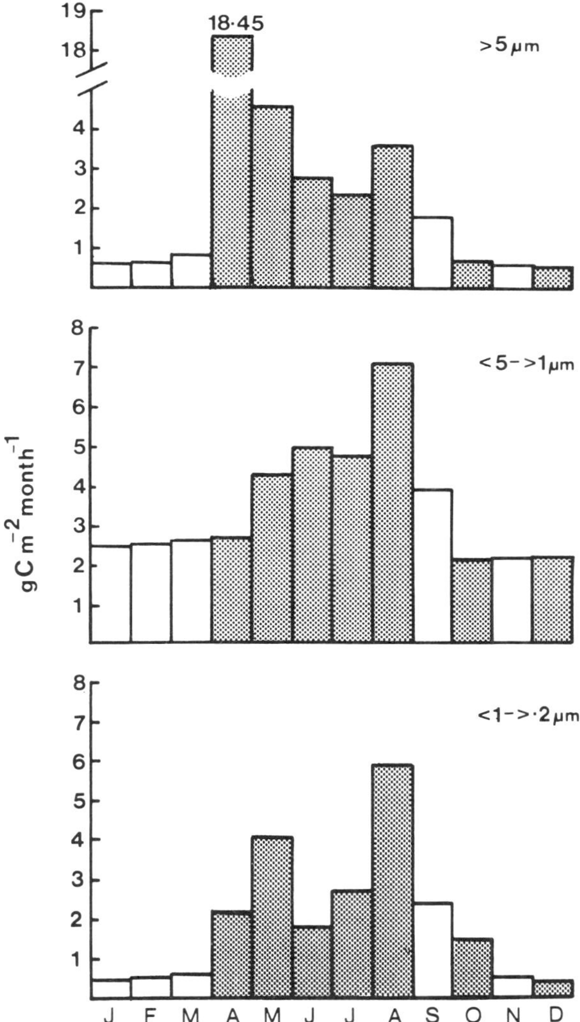

FIG. 4. Estimated monthly production for three size fractions of phytoplankton in the Celtic Sea taken from Joint et al. (1986); shaded parts of the figure indicate the months when measurements were made and other values were estimated by linear interpolation.

duction was low in winter (13 mg C fixed $\cdot$ m^{-2} $\cdot$ d^{-1} by cells <1 μm, out of a total daily production of 99 mg C $\cdot$ m^{-2} $\cdot$ d^{-1}); small nanoplankton (<5 to >1 μm) were the most productive fraction in winter and fixed 67.4 mg C $\cdot$ m^{-2} $\cdot$ d^{-1}. Hannah and Boney (1983) working in the nearshore waters of the Firth of Clyde, Scotland, also found that small nanoplankton was dominant in winter and commented that all the carbon fixation in the winter could be attributed to the nanoplankton. Therefore, picoplankton production in the temperate waters of the European continental shelf is most significant in the summer months in those regions where a seasonal thermocline results in long periods of stable surface stratification and when ambient nutrient concentrations are low.

Low nutrient concentrations are, by definition, characteristic of oligotrophic waters and, here again, picoplankton appears to be abundant. Although tropical and subtropical oligotrophic waters have relatively constant water temperatures, there can still be large changes in primary production rate. Bienfang et al. (1984) reported that primary production in the oligotrophic waters off Hawaii was 11 times greater from January to May than from May to November. Most of this production appears to have been by picoplankton <3 μm. The increased production rate was due to temporal variations in the physical system, associated with changes in trade winds; current speeds at 65 m were noticably higher at the times of higher primary production and Bienfang et al. (1984) suggest that the sharper velocity gradient generated greater shear forces which increased vertical mixing of deep waters containing elevated levels of nutrients. Photosynthesis was slightly enhanced at 60–80 m which these authors took as evidence of increased supply of nutrients from deeper water. A consequence of this enhanced production at 60–80 m is that temporal changes in production rate are the result of changes occurring at the base of the euphotic zone. This contrasts with the situation in temperate waters where production is highest in the surface mixed layer (Joint and Pomroy 1983).

These data from different oceanic regimes show that picoplankton is successful in a variety of different environments. In summary, picoplankton biomass appears to be greatest at times of nutrient limitation in temperate, subtropical and tropical waters. There is some evidence that picoplankton concentrations in the tropics and subtropics are highest in the subsurface maximum and, indeed, picoplankton may be responsible for this chlorophyll maximum. In temperate waters, picoplankton production is greatest in the summer months when nutrient concentrations are extremely low, but there is no evidence of accumulation of picoplankton cells in the chlorophyll maximum that develops on the seasonal pycnocline. The information that is available on picoplankton physiology will now be discussed to see if there is some intrinsic feature of picoplankton biology that would explain its geographical and seasonal distribution.

Aspects of Picoplankton Physiology

ARE NATURAL POPULATIONS OF PICOPLANKTON ADAPTED TO PHOTOSYNTHESIZE AT LOW LIGHT?

It appears that picoplankton is adapted to photosynthesis at low light, but is not clear if this a physiological (phenotypic) or a genotypic adaptation. Data from cultures suggest that cyanobacterial picoplankton is a low light organism. Morris and Glover (1981) working with a laboratory culture of *Synechococcus* sp., strain DC-2, found optimal growth at low light (45 μE $\cdot$ m^{-2} $\cdot$ s^{-1}) and most marine isolates of cyanobacteria appear to grow best at low light. However, it is difficult to draw any conclusions on whether or not this is genotypic adaptation of cells from cultures which are effectively the product of intense artificial selection in the laboratory.

It should be possible to determine the degree to which phenotypic adaptation occurs by comparing the photosynthetic characteristics of picoplankton from different oceanic regimes. Although data on natural populations are often difficult to interprete because water movements and mixing mean that it is impossible to establish the previous light history of the cells, certain environments are more predictable than others. For example, phytoplankton in the subsurface chlorophyll maximum of subtropical oceans is unlikely to be mixed into the surface layers and there is a high probability that this population has experienced low light conditions for a considerable period of time. How would the photosynthetic characteristics of such a population compare with those of a population from the surface mixed layer of temperate waters where phytoplankton cells would experience continuously variable light conditions as a result of vertical mixing in the euphotic zone?

Picoplankton cells from the subsurface chlorophyll maximum appear to be more adapted to low light conditions than larger phytoplankton. Platt et al. (1983) demonstrated significant differences in the photosynthesis–light curve of picoplankton and other phytoplankton sampled from the subsurface chlorophyll maximum in the subtropical Atlantic. Specifically, assimilation number (P_m^B) and initial slope α^B were higher for the picoplankton fraction than for the rest of the population and optimal irradiance (I_m), adaptation parameter (I_k, equivalent to P_m^B/α^B) and photoinhibition parameter (I_b) were all lower. That is, picoplankton photosynthesis saturated at lower light than other phytoplankton and picoplankton was more susceptible to photoinhibition. The optimal irradiance for picoplankton was about 30 W $\cdot$ m^{-2} (approximately 125 μE $\cdot$ m^{-2} $\cdot$ s^{-1}, assuming a conversion factor for photosynthetically active radiation of 2.5 $\times$ 10^{18} quanta $\cdot$ s^{-1} $\cdot$ W^{-1}; Morel and Smith 1974) and 46 W $\cdot$ m^{-2} (ca. 190 μE $\cdot$ m^{-2} $\cdot$ s^{-1}) for other phytoplankton. Platt et al. (1983) reported that the average light available at the depth of the subsurface chlorophyll maximum was 6 W $\cdot$ m^{-2} (ca. 25 μE $\cdot$ m^{-2} $\cdot$ s^{-1}) so photosynthesis was well below maximum at this depth and the picoplankton was severely light limited. Nevertheless, the higher values of a α^B suggest that picoplankton should have a competitive advantage over large phytoplankton at this depth.

Do these photosynthetic characteristics imply that picoplankton abundance and production should be greatest in low light environments and should we expect to find picoplankton restricted to low-light regions like the subsurface chlorophyll maximum of subtropical waters? Although picoplankton does occur in such low light environments, it is by no means restricted to them. The data of Li et al. (1983) demonstrate that although maximum picoplankton carbon fixation in a simulated in situ incubator occurred at 25% of surface irradiance, photosynthesis at 100% surface light was still significant (>50% of maximum measured rate). Therefore, although photosynthesis is not maximal at surface irradiances, it is still significant; indeed one might argue

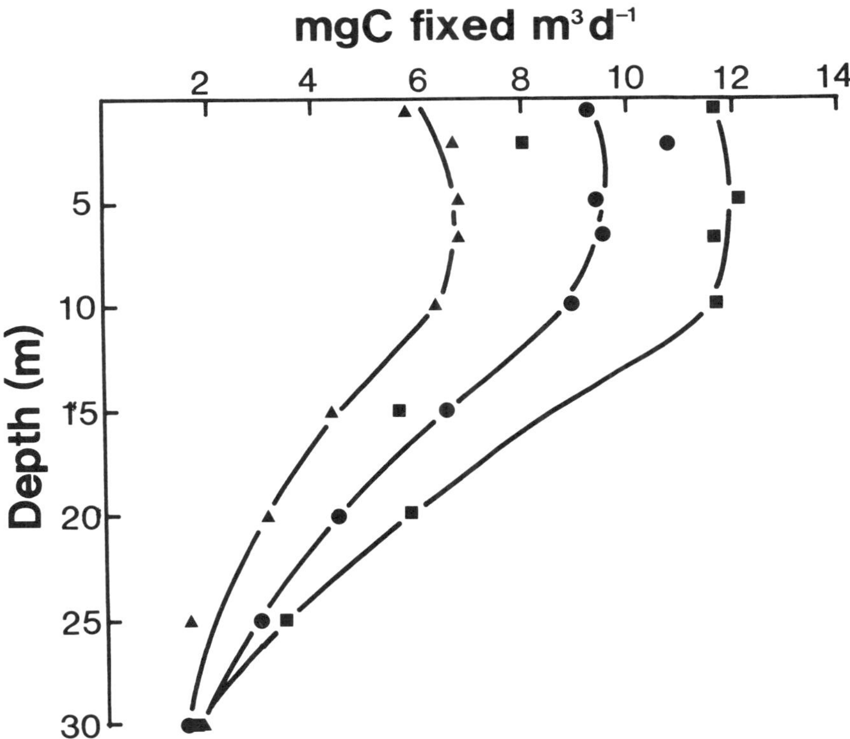

FIG. 5. ^{14}C fixation during a 24 h in situ experiment on 23 August 1982 at station CS2 in the Celtic Sea (from Joint and Pomroy 1983); ($\blacktriangle$) >5 μm fraction; ($\blacksquare$) <5 to >1 μm fraction; ($\bullet$) <1 to >0.2 μm fraction.

that the amount of carbon fixed in surface waters by picoplankton, although photo-inhibited to half the maximum, could be greater than that occurring under the severe light limitation at the subsurface chlorophyll maximum. The data of Platt et al. (1983) for the picoplankton population of the subsurface chlorophyll maximum show that photosynthetic carbon fixation at the irradiance at that depth would also be about 50% of that at optimum light. So photosynthetic characteristics alone do not dictate the development of picoplankton populations at low light in the subsurface chlorophyll maximum and population growth must be influenced by other factors such as the availability of nutrients from below the pycnocline.

The population studied by Platt et al. (1983) would not experience large variations in irradiance and should be fully adapted to low light conditions. However, picoplankton in the Celtic Sea (Joint and Pomroy, 1983) would experience a very different light regime. Maximum carbon fixation occurred in the surface 10 m and

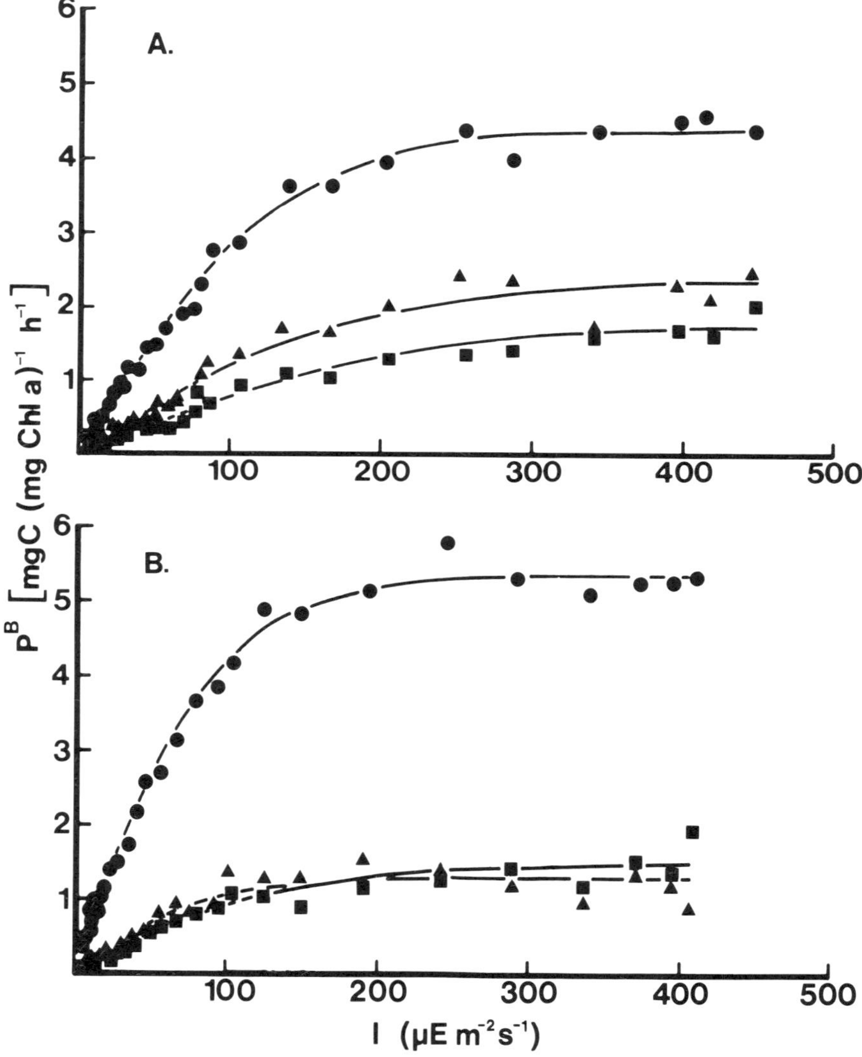

FIG. 6. Photosynthesis/irradiance curves obtained with samples taken at station CS2 in the Celtic Sea on 3 July 1985, (A) samples taken from 10 m and (B) taken from 25 m. (▲) >5 μm fraction; (■) <5 to >1 μm fraction; (●) picoplankton fraction, <1 to >0.2 μm.

there was little difference in the depth profiles of photosynthesis for any of the size fractions studied (Fig. 5). This does not appear to be compatible with the suggestion that picoplankton is adapted to low light. Being in the surface mixed layer, the picoplankton cells would experience large variations in light, both as a result of vertical mixing and because of the diurnal variations in irradiance (Marra 1978, 1980; Falkowski and Wirick 1981; Gallegos and Platt 1982; Lewis et al. 1984). Would such a picoplankton population show the same adaptation to low light as the population studied by Platt et al. (1983)? Cells at the base of the euphotic zone might do so if the time scale of mixing was long compared with the time scale of photoadaptation.

Joint and Pomroy (1986) have recently determined photosynthetic parameters for three-size-fractions of phytoplankton; Fig. 6 shows results of photosynthesis/ irradiance curves obtained for samples taken from two depths at a shelf station (CS2) in the Celtic Sea in July 1985. Other data were obtained at two oceanic stations in the Porcupine Seabight. At all stations, very calm conditions persisted during the cruise and reduced wind-driven mixing resulted in the development of a secondary thermocline within the surface mixed layer (Fig. 3). Therefore, populations at 20–30 m must have been at low light conditions for at least 1 wk before the experiments were done and phytoplankton from the surface 10 m experienced diurnal variations in irradiance, up to full sunlight.

TABLE 2. Photosynthetic parameters compared for natural phytoplankton populations from the Celtic Sea and subtropical Atlantic.

Station and depth	α			P_m^B		
	$(\text{mg C} \cdot [\text{mg chl a}]^{-1} \cdot h^{-1})/(\mu E \cdot m^{-2} \cdot s^{-1})$			$(\text{mg C} \cdot [\text{mg chl a}]^{-1} \cdot h^{-1})$		
	>5 μm	<5->1 μm	<1 μm	>5 μm	<5->1 μm	<1 μm
Celtic Sea 10 m	0.013	0.008	0.033	2.407	1.851	4.436
" 25 m	0.014	0.011	0.056	1.287	1.509	5.378
Porcupine Seabight 10 m	0.012	0.022	0.0234	5.509	4.849	3.615
" 20 m	0.008	0.019	0.038	1.855	2.152	3.368
Azores Chlorophyll Max.[a]	—	0.009	0.018	—	0.49	0.68

[a] Data taken from Platt et al. (1983), who had two size fractions, >1 μm and <1 μm; mean values are used and data for α are converted from units of W m^{-2} to $\mu E \cdot m^{-2} \cdot s^{-1}$ using a conversion factor of 2.5 × 10^{18} quanta s$^{-1} \cdot$ W^{-1}.

Table 2 shows values of α^B and P_m^B obtained for >5 μm, <5 to >1 μm and picoplankton taken from above and below the secondary thermocline which developed at 10 to 15 m. Values of α^B were not significantly different for >5 μm and <5 to >1 μm phytoplankton taken from the surface or from 20 to 30 m; however, there were significant differences in α^B for picoplankton and at both stations there was 60% increase in the value of α^B for the "deep" population. In contrast, picoplankton did not show any variation in the value of P_m^B at either depth but the larger phytoplankton did show significant decreases with depth. The light source used in these experiments was not sufficiently powerful to induce photoinhibition in these samples so it is not possible to compare any photoinhibition parameters. However, for comparative purposes, Table 2 shows the values of α^B and P_m^B obtained by Platt et al. (1983) for phytoplankton from the subsurface chlorophyll maximum in the subtropical Atlantic; values of both α^B and P_m^B are numerically much smaller than those obtained for the Celtic Sea populations.

However, not all measurements of photosynthetic parameters made by Joint and Pomroy (1986) conform to this pattern and there was considerable day-to-day variation in photosynthetic parameters. P_m^B of >5 μm phytoplankton was always greater for the surface than the "deeper" populations. Small nanoplankton (<5 to >1 μm) from the oceanic stations showed changes similar to the >5 μm phytoplankton. However, the small nanoplankton from the shelf station showed little change in P^B_m but an increase in α^B with depth. Values of α^B were sometimes greater for the picoplankton fraction than for the larger phytoplankton; however, the adaptation parameter, I_k (P^B_m/α^B), which is often used as a measure of the degree to which phytoplankton is adapted to low irradiance, was not always lower for picoplankton than for the larger phytoplankton. Indeed, the greatest reductions in I_k were the result of low values of P^B_m, rather than increases in α; such decreases in I_k do not indicate any enhancement in the ability of the cell to use low irradiance and there was little evidence from the data of Joint and Pomroy (1986) to support the hypothesis that these populations of picoplankton were more adapted to low light conditions than the larger phytoplankton cells.

The decreases in P^B_m values observed for phytoplankton >5 μm at 20–30 m are consistent with data obtained with phytoplankton cultures; Beardall and Morris (1976), working with *Phaeodactylum tricornutum*, and Senger and Fleischhacker (1978), with *Scenedesmus obliquus*, showed higher values of P^B_m when cells were grown at high rather than low irradiance. Such changes in P^B_m have been suggested to indicate that the size of the photosynthetic unit can alter in response to changes in irradiance (Gallegos and Platt 1981); the data from the Celtic Sea (Joint and Pomroy 1986) suggest that sometimes picoplankton may compensate for low light by increasing the size of the phytosynthetic unit, but this does not always happen. The initial slope α^B is related to maximum quantum yield (Rabinowitch and Govindjee 1969); phytoplankton >1 μm from the Celtic Sea again shows considerable day-to-day variability in α^B and there did not appear to be any predictable changes in the values of α^B. Changes in α^B value did not appear to be related to the irradiance received by the population, either on the day of the experiment or on the previous day. Changes in α^B might be related to changes in accessory pigment concentration or to changes in cell morphology, which would affect package effect. However, we have no data from these populations which might explain the observed variability in photosynthetic parameters.

The time-scale of photoadaptation may be an important factor in determining the values of photosynthetic parameters but it is likely to be relatively long; data are not available for picoplankton but Marra (1980) found that the characteristic changes in photosynthesis–light curves of the diatom *Lauderia borealis* took 10 h to develop after a change from high to low light. Similarly, Gallegos et al. (1983) found that a phytoplankton population in the Arctic took 2–6 wk to develop susceptibility to photoinhibition after the cells were isolated below the pycnocline; in contrast, a much shorter time (4–6 h) was required to remove this susceptibility to photoinhibition once the population was returned to surface light. Although it is difficult to extrapolate from these data, which were obtained with a culture and with a population growing at very low temperatures, nevertheless it seems likely that the time-scale of photoadaptation of picoplankton will be several hours, if not days, i.e., of the same time-scale as vertical mixing in the surface mixed layer in the Celtic Sea.

RESPONSE TO LIGHT QUALITY

The quality as well as quantity of light varies with depth and the spectral composition of light at the base of the euphotic zone is very different from that at the surface. It has been suggested that cyanobacteria are better adapted to utilise blue and green light, which are the only wavelengths that will penetrate deep into the euphotic zone. Major light harvesting components of cyanobacteria are biliproteins (phycoerythrin, phycocyanin and allophycocyanin) which absorb strongly in the blue-green region of the visible spectrum. Phycoerythrin is a major pigment in marine cyanobacterial picoplankton and, indeed, the orange-yellow fluorescence of phycoerythrin is the means of distinguishing cyanobacteria from other phytoplankton using the fluorescence microscope; several novel phycoerythrins have now been characterised from marine cyanobacteria (Alberte et al. 1984; Ong et al. 1984). Alberte et al. report that the presence of phycoerythrin resulted in greater photosynthetic efficiency (increased values of α in P/I curves when normalised to chlorophyll a concentration); they suggested that photosynthetic performance of *Synechococcus* grown at low light (30–50 μE $\cdot$ m^{-2} $\cdot$ s^{-1}) was equivalent to that of eukaryotic algae grown at high light (300 μE $\cdot$ m^{-2} $\cdot$ s^{-1}); this they ascribed to several factors, including a higher concentration of light harvesting pigment-proteins. Therefore, the presence of these pigments might suggest another mechanism by which cyanobacteria will photosynthesize efficiently in low light conditions.

Experiments done in simulated in situ incubators and the type of artificial light incubators used for derivation of photosynthetic parameters, are deficient because they take no account of the varying spectral properties of light with increasing depth. In a recent attempt to gain information on the influence of spectral changes on photosynthetic activity, Lewis et al. (1985) measured photosynthesis at 12 wavelengths and at 8 irradiance levels at each wavelength. There were significant differences in action spectra for populations taken from the Arctic, North Atlantic and the Sargasso Sea; the response of picoplankton was not specifically measured. However, Glover et al. (1985) did study the effect of spectral composition on picoplankton activity. Using coloured filters in an on-deck incubator, these authors found that picoplankton did utilise green light more efficiently for photosynthesis and that the lower the light level, the greater was the contribution of picoplankton to total production. Glover et al. suggested that it was the ability of the cyanobacteria to utilise blue-green light that was a major factor in the higher photosynthetic efficiency of picoplankton. Clearly, these results emphasis how important it is to control the spectral characteristics of the light used in measuring primary production, especially if an estimate of depth integrated production is to be made. However, the problem of controlling light quality does not arise if the incubations are carried out in situ, as we have done in the Celtic Sea (Joint and Pomroy 1983; Joint et al. 1986).

PHOTORESPIRATION

In their experiments with laboratory cultures, Morris and Glover (1981) and Glover and Morris (1981) demonstrated high potential photorespiration in two strains of *Synechococcus*. This conclusion was mainly based on observations of oxygen inhibition of photosynthetic CO_2 fixation which was light dependant and occurred at atmospheric O_2 concentrations; at 100% oxygen concentration there was a post-illumination enhancement of oxygen uptake which decreased when the bicarbonate concentration was increased, and there was a significant increase in the proportion of label found in phosphoglycolate and glycolate when ambient oxygen concentrations were increased to 100%. Oxygenase activity of the ribulose-1,5-biscarboxylase-oxygenase (RUBISCO, EC 4.1.1.39) was three times that of carboxylase activity in *Synechococcus* but in *Phaeodactylum tricornutum* oxygenase activity was only half that of the carboxylase and in *Dunaliella tertiolecta* less than a third. All these data suggest that *Synechococcus* spp. should exhibit considerable photorespiration, particularly at high light and high O_2 concentrations. Morris and Glover (1981) argued that these characteristics, coupled with the low saturating irradiances of photosynthesis, might suggest that picoplankton production would be of greatest significance at the base of the euphotic zone. As we have seen, this is true for some environments, particularly in the tropics and subtropics, but is not the case in temperate waters with a relatively shallow surface mixed layer. Would cyanobacteria in the surface mixed layer exhibit high rates of photorespiration? There is no evidence either to confirm or refute the hypothesis.

The evidence in the literature suggests that some freshwater cyanobacteria do not appear to exhibit photorespiration. Lloyd et al. (1977) suggested that photorespiration did not occur in algae when grown at low ΣCO_2 concentrations but Birmingham et al. (1982) found oxygen inhibition of CO_2 fixation in some algae. Two cyanobacteria were susceptible to oxygen and these were *Anabaena flos-aquae* and *Anacystis nidulans* but two others, *Coccochloris peniocystis* and *Phormidium molle* did not show evidence of photorespiration. There is some controversy over algal photorespiration since there is no evidence that glycolate is involved and the phenomenon found in algae and cyanobacteria may not be photorespiration in the higher plant sense. Although this conclusion is not universally accepted, there is some reason to question whether cyanobacteria have high potential photorespiration. The evidence concerns the role of carboxysomes in natural populations and whether they may protected RUBISCO from oxygen and hence not exhibit high oxygenase activity.

A conspicuous feature in transmission electron micrographs of marine *Synechococcus* is the presence of polyhedral bodies (Johnson and Sieburth 1979, 1982; Joint and Pipe 1984; Takahashi and Hori 1984) which are assumed to be carboxysomes; RUBISCO is present in a soluble and particulate form in cyanobacteria and the particulate form of the enzyme is located in carboxysomes (Shively 1974; Stewart and Codd 1975; Codd and Stewart; 1976; Codd and Marsden 1984). Although carboxysomes are readily observed in electron micrographs, the function of these inclusion bodies is poorly understood; however, one suggested role is that carboxysomes protect the carboxylase function of RUBISCO from the competitive inhibition by oxygen (Codd and Marsden 1984). Evidence for this role was largely indirect; it was argued that since carboxysomes are present in cyanobacteria, which have oxygenic photosynthesis, and are also present in aerobic chemolithotrophic bacteria which also fix CO_2 *via* RUBISCO, but are absent from purple-sulphur and purple-nonsulphur photosynthetic bacteria which have anoxygenic photosynthesis, then a possible explanation would be if carboxysomes had an oxygen protective role. Recently, Coleman et al. (1982; cited by Turpin et al. 1984) have presented evidence that RUBISCO is, indeed, insensitive to oxygen when sequestered in carboxysomes. The carboxysome content of the freshwater cyanobacterium *Synechococcus leopoliensis (Anacystic nidulans)* varies according to growth conditions; Turpin et al. (1984) found low carboxysome content in cells grown with excess nutrients and excess ΣCO_2. However, under ΣCO_2 limitation, carboxysome content increased approximately six times. Nitrogen or phosphorus limited cultures did not result in any variation in carboxysome content, an observation which is inconsistent with another proposed function of carboxysomes, that of acting as a nitrogen storage body (Codd and Marsden 1984). However, the effects of ΣCO_2 concentration suggests that carboxysome concentration might be very variable in response to variations in culture conditions.

Given that ΣCO_2 concentrations are much higher in seawater than in freshwater, perhaps it would be surprising if there were ever sufficient variations in concentration to affect carboxysome content in cultures of marine *Synechococcus*. However, we have recently begun to examine the fine structure of *Synechococcus* strain DC-2 and have found that carboxysomes are less frequently observed in exponentially growing cultures. Clearly a negative result could be due to a number of factors, including variations in sample preparation for electron microscopy; however, we have consistently found low numbers of carboxysomes in exponentially growing cultures. This evidence, albeit of a negative nature, does raise the question of the generality of the finding of high potential photorespiration in marine *Synechococcus* (Morris and Glover 1981; Glover and Morris 1981). If carboxysome abundance is low in actively growing *Synechococcus* strain DC-2, then there might be little protection for the carboxylase activity of RUBISCO and photorespiration might be high. However, natural populations appear to have a significant carboxysome content and the assumption of high photorespiration in these cells may not be justified. Given the difficulties of measuring photorespiration in natural populations, the question of photorespiration and carboxysome content needs to be examined in strains of marine *Synechococcus* other than DC-2.

Picoplankton cells seem to be most abundant in waters with very low nutrient concentrations, whether in tropical oligotrophic waters or in the surface waters of seasonally stratified, temperate waters. Data on the affinity of picoplankton for nitrate, ammonia and phosphate are not available for natural populations and it is inappropriate to speculate on how picoplankton appear to be so successful in these low nutrient conditions. The small size (and hence large surface area: volume ratio) of the picoplankton must surely play a beneficial role in the uptake of nutrients that are present at extremely low concentrations but nutrient uptake will not be considered here. However, there is one aspect of growth of natural populations at low nutrient concentrations which appears paradoxical. Cyanobacteria can be readily identified in natural populations by the presence of phycoerythrin; however, phycoerythrin is a "nitrogen-rich" molecule and it is perhaps surprising that organisms that have to divert a significant proportion of a scarce resource into a light harvesting pigment can be so successful in nutrient limiting situations. Clearly, in the light limiting conditions that exist in the subsurface chlorophyll maximum in tropical and subtropical waters, the cyanobacteria are growing close to a source of nutrient advecting from below the pycnocline. Therefore, the effects of light limitation can be mitigated by the synthesis of more light harvesting pigment and, if the population is growing close to a source of nutrients and is not nutrient limited, then the cell will benefit from the investment of nitrogen into phycoerythrin synthesis. However, this argument cannot apply to cells growing in the surface mixed layer. These cells are unlikely to be light limited but are probably nutrient limited; they will have to divert nitrogen away from growth into phycoerythrin synthesis and the population would presumably grow less than if this investment was not required. Would natural populations of cyanobacteria from the surface mixed layer sequester nitrogen into phycoerythrin?

The evidence from cultures suggests that phycoerythrin from marine cyanobacteria does act as a nitrogen storage molecule. Wyman et al. (1985) have recently suggested that phycoerythrin has two functions; it absorbs quanta for photosynthesis but, when nitrogen is available in excess, phycoerythrin acts as a dynamic pool of stored nitrogen and most of the light energy then absorbed by this molecule is not utilized in photosynthesis but is given out as fluorescence. Phycoerythrin is present in phycobilisomes; in transmission electron micrographs of cultures of marine *Synechococcus*, these structures appear as ovoid structures which are densely packed along the thylakoid membranes (Alberte et al. 1984). However, in electron micrographs of natural populations, phycobilisomes are not a prominant feature (Johnson and Sieburth 1979, 1982; Joint and Pipe 1984). Of course, this could be a result of sample preparation and phycobilosmes may indeed be present in cyanobacteria from natural populations. Such populations certainly contain phycoerythrin but it is by no means certain that these cells have committed nitrogen to extensive phycobilisome synthesis rather than to growth. It would be extremely interesting to study phycoerythrin synthesis and phycobilisome formation in cultures of *Synechococcus* to see the effect of nitrogen starvation.

Most of this section has dealt with cyanobacterial picoplankton but transmission electron micrographs of eukaryotic picoplankton also show morphological features that may be the consequence of environmental conditions. Many of the eukaryotes described by Johnson and Sieburth (1982), Takahashi and Hori (1984), and Joint and Pipe (1984) have a prominent starch storage grain. Since these cells were all sampled from nutrient limited waters, it is possible that the starch grain is larger than in cells grown under nutrient replete conditions. It is known that nitrogen limited cells fix more carbon into carbohydrate than into protein and this may be reflected in an increase in starch grain size; however, no data are yet available on the effect of nutrient limitation on the morphology of eukaryotic picoplankton.

The Presumed Grazers of Picoplankton

The preceding sections have dealt with aspects of picoplankton physiology that influence its ecology and how that influence can vary in different geographical regions. But there are factors other than physiology which influence the success of any component of the ecosystem; in particular, population abundance is a balance between growth and loss terms. In the case of picoplankton, losses from the system can occur as a result of grazing or death. Very few data are available about the natural death of phytoplankton, that is loss of vitality as a result of nutrient or light deprivation or other factors which result in the death of the organisms. One significant process that can result in the loss of phytoplankton biomass from the pelagic ecosystem is sinking of phytoplankton cells; Walsh (1983) estimated the quantity of phytoplankton sinking as detritus and suggests that up to 50% of the annual primary production of continental shelf waters may be exported to the benthos and upper continental shelf slope as phytodetritus. However, it is also clear that picoplankton, because of its small size, cannot be lost from the surface layers at any appreciable rate unless significant aggregation occurs. Takahashi and Bienfang (1983) measured sinking rates of different size fractions of phytoplankton in subtropical surface waters and found sinking rates of picoplankton <3 μm to be so low as to be unmeasurable. There can, therefore, be little if any loss of picoplankton cells by sinking and even if picoplankton cells die, they must enter the detritus pool in surface waters and be available for consumption by other components of the pelagic ecosystem.

If picoplankton does not sink it must be grazed, but which organisms are responsible for controlling the biomass of picoplankton in the surface layers of the oceans? Crustacean zooplankton is usually considered as the primary group of grazing organisms in the sea but it seems probable that copepods or euphausiids are unable to graze particles as small as picoplankton; Conover (1978) found that neritic copepods did not ingest particles smaller than 2 to 3 μm and Bartram (1980) showed that particle capture efficiency by *Paracalanus parvus* decreased with particles <12 μm and was less than 30% with particles smaller than 5 μm. However, if is also clear that picoplankton cells can be captured by copepods; Johnson et al. (1982) found cyanobacteria in the faecal pellets of *Calanus finmarchicus*, but since there was no degradation of the cells, they concluded that cyanobacteria could not contribute to the nutrition of the copepod. Therefore, macrozooplankton is unlikely to be a significant grazer of photosynthetic picoplankton.

The problem of which organisms graze heterotrophic bacteria in the sea has recently been considered by many people; because of the similarily in size, photosynthetic picoplankton must be considered in the same category as the heterotrophic bacteria. The prime candidates for the role of dominant grazers on bacteria are the protozoan zooplankton. Indeed, Azam et al. (1983) have suggested that the interactions occurring between bacteria, microflagellates and ciliates might form the basis of a previously unrecognised food web which they termed the "microbial loop". Evidence is accumulating to suggest that the components of such a food web are present in marine waters and are capable of consuming bacteria at significant rates; much of the evidence comes from cultures and studies in estuaries (Hass and Webb 1979; Fenchel 1982; Sieburth and Davis 1982; Davis and Sieburth 1984; Sherr et al. 1983; Fuhrman and McManus 1984; Wright and Coffin 1984; Davis et al. 1985; Andersen and Fenchel 1985). Sieburth and Davis (1982) and Davis et al. (1985) did quantify heterotrophic microflagellates in oceanic waters in the Sargasso Sea, North Atlantic and Caribbean Sea; numbers varied from 10^4 cells mL^{-1} in estuarine waters to 10^2 mL^{-1} in oceanic waters. It has usually been assumed that heterotrophic microflagellates are bactivorous and their role is usually considered to be one of controlling the abundance of heterotrophic bacteria in the sea; however, they would presumably also be capable of grazing on picoplankton. Indeed, there is evidence that the assumed role of microflagellates as grazers of bacteria may be much too narrow; Goldman and Caron

(1985) have recently shown that a microflagellate could be successfully cultured with a variety of phytoplankton species as well as bacteria as food organisms. If microflagellates in natural populations are also capable of feeding on nanophytoplankton, and indeed on other microflagellates, then Goldman and Caron suggest that the conceptual "microbial loop" model of Azam et al. (1983) must be modified to include interactions between microflagellates and other nanoplankton as well as picoplankton.

Data on the geographical distribution of heterotrophic nanoplankton in the oceans is sparse; Davis et al. (1985) showed a general decrease in heterotrophic nanoplankton biomass with increasing depth of the water column and there were approximately 700 cells $\cdot$ mL^{-1} in waters off the continental shelf; for all stations sampled there appeared to be a reasonably constant ratio between the numbers of heterotrophic microflagellate and the numbers of total picoplankton (which Davis et al. 1985, define as heterotrophic bacteria, by far the largest numerical component, and photosynthetic picoplankton). Other protozoa that might graze picoplankton are phagotrophic dinoflagellates (Kimor 1981) and ciliates (Johnson et al. 1982; Capriulo and Ninivaggi 1982; Stoecker et al. 1983). However, these organisms are unlikely to be significant grazers of picoplankton. Fenchel (1980) pointed out that the numbers of bacteria in the sea are too small for ciliates to survive and the same arguments that apply to bacteria must also apply to picoplankton. Also, Jørgensen (1983) has shown that ciliates adapted to feed on small particles, of the size of bacteria and picoplankton, have a low capacity for water processing because of the high resistance of the filtering mechanism; this explains why ciliates that are capable of filtering bacteria require a very high density of cells (of the order of 10^7 to 10^8 bacteria mL^{-1}) before growth is possible. Ciliates from oceanic regimes appear to have more porous filters which can process larger volumes of water but which cannot filter efficiently particles of bacterial size; these oceanic ciliates probably graze on both photosynthetic and heterotrophic nanoplankton.

Protozoan microflagellates, therefore, appear to be the strongest candidates for the role of grazer of picoplankton but the data which would confirm their quantitative importance does not yet exist. Joint and Williams (1985) have recently taken literature data on microflagellate grazing rates that were derived mostly from culture work and have applied these rates to the biomass of microflagellates found in the Celtic Sea; our conclusion was that there was insufficient production by heterotrophic bacteria to meet microflagellate carbon requirement but this shortfall could be made up if microflagellates were grazing on photosynthetic picoplankton. In contrast, Andersen and Fenchel (1985) found that heterotrophic microflagellates in Aarhus Bay could clear 10^{-5} mL h^{-1}and, by extrapolating to other sea areas, they suggested that between 5 and 250% of the water column could be cleared of bacteria per day; Andersen and Fenchel also suggested that microflagellate growth could be sustained by bacterial populations greater than 10^6 mL^{-1}, a value which is commonly reported for many oceanic regimes. However, Joint and Williams (1985) argued that data obtained from culture experiments and with populations obtained from estuaries and coastal waters need not be directly comparable with populations in the open ocean. Oceanic heterotrophic bacteria are much smaller than bacteria found in estuaries and nearshore and are typically 0.2–0.4 μm diameter (Sieburth 1979); estuarine bacteria (typically 1 μm diameter), therefore, have a biomass 15–120 times than of oceanic bacteria and it is questionable whether microflagellate populations could be maintained by the biomass of oceanic heterotrophic bacteria. However, the presence of picophytoplankton (that is, a smaller number of larger cells) in addition to the large number of small heterotrophic bacteria may present the microflagellates with a food source that can sustain the population in oceanic waters. Experimental work on microflagellates grazing under oceanic conditions is desperately needed before any conclusions can be drawn on the significance of the "microbial loop" in pelagic food webs.

It appears that the traditional herbivore community of crutacean zooplankton are unlikely to graze picoplankton and the interactions with the protozoa in the "microbial loop" are unclear. However, other components of the pelagic ecosystem could make significant demands on picoplankton production. In particular, the mucous-net feeders have the potential to graze picoplankton efficiently; Harbison and Gilmer (1976) found that salps could ingest particles <3 μm with high efficiency and could retain particles as small as 0.7 μm with a lesser, unknown efficiency. Harbison and McAlister (1979) conducted feeding experiments at sea on the slope waters of the northeastern U.S. and suggested that salps could make a significant grazing impact on microbes of 1 μm diameter. Salps are clearly capable of capturing picoplankton cells but could this be quantitatively significant? Deibel (1982), on the basis of laboratory measurements, calculated that the biomass of salps normally found in coastal waters in the Georgia Bight would take one month to clear the volume of water in which they were found; however, when concentrated swarms of salps occurred, they could clear their resident water volume in less than a day. This would imply that salps are only likely to be significant grazers of picoplankton in oceanic waters when swarms are present. However, Pace et al. (1984) in their recent simulation model of a continental shelf ecosystem suggested that significant flows could occur through the mucous-net feeders. Like the "microbial loop" hypothesis, more data are required on the grazing of mucous-net feeder in different environments before it will be possible to quantify their significance as grazers of picoplankton.

It is still by no means certain which organisms are the major consumers of picoplankton. Protozoa and mucous-net feeders are clearly capable of grazing cells of picoplankton size but the quantitative significance of these processes and the way in which picoplankton are incorporated into the pelagic food web is still far from clear.

Acknowldegements

This work forms part of the Shelf-Sea Ecology Programme of the Institute for Marine Environmental Research, a component of the UK Natural Environment Research Council.

References

ALBERTE, R. S., A. M. WOOD, T. A. KURSAR, AND R. R. L. GUILLARD. 1984. Novel phyco-erythrins in marine *Synechococcus* spp. Plant Physiol. 75: 732–739.

ANDERSEN, P., AND T. FENCHEL. 1985. Bactivory by microheterotrophic flagellates in seawater samples. Limnol. Oceanogr. 30: 198–202.

AZAM, F., T. FENCHEL, J. C. FIELD, J. S. GRAY, L. A. MEYER-REIL, AND F. THINGSTAD. 1983. The ecological role of water-column microbes in the sea. Mar. Ecol. Prog. Ser. 10: 257–263.

BARTRAM, W.C. 1980. Experimental development of a model for the feeding of neritic copepods on phytoplankton. J. Plankton Res. 3: 25–51.

BIENFANG, P. K., AND J. P. SZYPER. 1981. Phytoplankton dynamics in the subtropical Pacific Ocean off Hawaii. Deep-Sea Res. 28A: 981–1000.

BIENFANG, P. K., J. P. SZYPER, M. Y. OKAMOTO, AND E.K. NODA. 1984. Temporal and spatial variability of phytoplankton in a subtropical ecosystem. Limnol. Oceanogr. 29: 527–539.

BIENFANG, P. K., AND M. TAKAHASHI. 1983. Ultraplankton growth rates in a subtropical ecosystem. Mar. Biol. 76: 213–218.

BIRMINGHAM, B. C., J. R. COLEMAN, AND B. COLEMAN. 1982. Measurement of photo-respiration in algae. Plant Physiol. 69: 259–262.

CAPRIULO, G. M., AND D. V. NINIVAGGI. 1982. A comparison of the feeding activities of field col-lected tintinnids and copepods fed identical natural particle assemblages. Ann. Inst. Oceanogr., Paris 58(S): 325–334.

CARON, D. A., F. R. PICK, AND D. R. S. LEAN. 1985. Chroococcoid cyanobacteria in Lake Ontario: vertical and seasonal distributions during 1982. J. Phycol. 21: 171–175.

BEARDALL, J., AND I. MORRIS. 1976. The concept of light adaptation in marine phytoplankton: some experiments with *Phaeodactylum tricor-nutum*. Mar. Biol. 37: 377–387.

CODD, G. A., AND W. J. N. MARSDEN. 1984. The carboxysomes (polyhedral bodies) of auto-trophic prokaryotes. Biol. Rev. 59: 389–422.

CODD, G. A., AND W. D. P. STEWART. 1976. Polyhedral bodies and ribulose 1,5-diphosphate carboxylase of the blue-green alga *Anabaena cylindrica*. Planta (Berl.) 130: 323–326.

COLEMAN, J. R., J. R. SEEMAN, AND J.A. BERRY. 1982. RuBP carboxylase in carboxysomes of blue-green algae. Carnegie Inst. Wash. Year Book. 81: 83–87.

CONOVER, R. J. 1978. Feeding interactions in the pelagic zone. Rapp. P.-v. Réun. Cons. Perm. Int. Explor. Mer. 173: 66–76.

DAVIES, J. M. AND R. PAYNE. 1984. Supply of organic matter to the sediment in the northern North Sea during a spring phytoplankton bloom. Mar. Biol. 78: 315–324.

DAVIS, P. G., AND J. McN. SIEBURTH. 1984. Estuarine and oceanic microflagellate predation of actively growing bacteria: estimation by frequency of dividing-divided bacteria. Mar. Ecol. Prog. Ser. 19: 237–246.

DAVIS, P. G., D. A. CARON, P. W. JOHNSON, AND J. McN. SIEBURTH. 1985. Phototrophic and apochlorotic components of picoplankton and nanoplankton in the North Atlantic: geographic, vertical, seasonal and diel distributions. Mar. Ecol. Prog. Ser. 21: 15–26.

DIEBEL, D. 1982. Laboratory-measured grazing and ingestion rates of the salp, *Thalia democratica* Forskal, and the doliolid, *Dolioletta gegenbauri* Uljanin (Tunicata, Thaliacea). J. Plankton Res. 4: 189–201.

FALKOWSKI, P. G., AND C. D. WIRICK. 1981. A simulation model of the effects of vertical mixing on primary productivity. Mar. Biol. 65: 69–75.

FASHAM, M. J. R., P. M. HOLLIGAN, AND P.R. PUGH. 1983. The spatial and temporal development of the spring phytoplankton bloom in the Celtic Sea, April 1979. Prog. Oceanogr. 12: 87–145.

FENCHEL, T. 1980. Suspension feeding in ciliated Protozoa: feeding rates and their ecological significance. Microb. Ecol. 6: 13–25.

——— 1982. Ecology of heterotrophic microflagellates. IV. Quantitative occurrence and importance as bacterial consumers. Mar. Ecol. Prog. Ser. 9: 35–42.

FUHRMAN, J. A., AND G. B. McMANUS. 1984. Do bacteria-sized marine eukaryotes consume significant bacterial production? Science (N.Y.) 224: 1257–1260.

FURUYA, K., AND R. MARUMO. 1983. The structure of the phytoplankton community in the subsurface chlorophyll maxima in the western North Pacific Ocean. J. Plankton Res. 5: 393–406.

GALLEGOS, C. L., AND T. PLATT. 1981. Photosynthesis measurements on natural populations of phytoplankton: numerical analysis. Can. Bull. Fish. Aquat. Sci. 210: 103–112.

——— 1982. Phytoplankton production and water motion in surface mixed layers. Deep-Sea Res. 29: 65–76.

GALLEGOS, C. L., T. PLATT, W. G. HARRISON, AND B. IRWIN. 1983. Photosynthetic parameters of arctic marine phytoplankton. Vertical variations and time scales of adaptation. Limnol. Oceanogr. 28: 698–708.

GIESKES, W. W. C., G. W. KRAAY, AND M.A. BAARS. 1979. Current ^{14}C methods for measuring primary production: gross underestimates in oceanic waters. Neth. J. Sea Res. 13: 58–78.

GLOVER, H. E., AND I. MORRIS. 1981. Photosynthetic characteristics of coccoid marine cyanobacteria. Arch. Microbiol. 129: 42–46.

GLOVER, H. E., D. A. PHINNEY, AND C. S. YENTSCH. 1985. Photosynthetic characteristics of picoplankton compared with those of larger populations, in various water masses in the Gulf of Maine. Biol. Oceanogr. 3: 223–248.

GOLDMAN, J. C., AND D. A. CARON. 1985. Experimental studies on an omnivorous microflagellate: implications for grazing and nutrient regeneration in the marine microbial food chain. Deep-Sea Res. 32: 899–915.

GROS, P., AND M. RYCKAERT. 1983. Étude de la production primaire phytoplanctonique dans les eaux littorales de la côte normande (Manche orientale). Oceanol. Acta 6: 435–450.

GROSS, F. 1937. Notes on the culture of some marine plankton organisms. J. Mar. Biol. Assoc. U.K. 21: 753–768.

HAAS, L. W., AND K. L. WEBB. 1979. Nutritional mode of several non-pigmented microflagellates from the York River estuary, Virginia. J. Exp. Mar. Biol. Ecol. 39: 125–134.

HARBISON, G. R., AND R. W. GILMER. 1976. The feeding rates of the pelagic tunicate *Pegea confederata* and two other salps. Limnol. Oceanogr. 21: 517–528.

HARBISON, G. R., AND V. L. McALISTER. 1979. The filter-feeding rates and particle retention efficiencies of three species of *Cyclosalpa* (Tunicata, Thaliacea). Limnol. Oceanogr. 24: 875–892.

JOHNSON, P. W., AND J. McN. SIEBURTH. 1979. Chroococcoid cyanobacteria in the sea: a ubiquitous and diverse biomass. Limnol. Oceanogr. 24: 928–935.

——— 1982. In-situ morphology and occurrence of eucaryotic phototrophs of bacterial size in the picoplankton of estuarine and oceanic waters. J. Phycol. 18: 318–327.

JOHNSON, P. W., H.-S. XU, AND J. McN. SIEBURTH. 1982. The utilization of chroococcoid cyanobacteria by marine protozooplankters but not by calanoid copepods. Ann. Inst. Oceanogr., Paris, 58(S): 297–308.

JOINT, I. R., AND R. K. PIPE. 1984. An electron microscope study of a natural population of picoplankton from the Celtic Sea. Mar. Ecol. Prog. Ser. 20: 113:118.

JOINT, I. R., AND A. J. POMROY. 1983. Production of picoplankton and small nanoplankton in the Celtic Sea. Mar. Biol. 77: 19–27.

——— 1986. Photosynthetic characteristics of nanoplankton and picoplankton from the surface mixed layer. Mar. Biol. 92: 465–474.

JOINT, I. R., N. J. P. OWENS, AND A. J. POMROY. 1986. The seasonal production of picoplankton and nanoplankton in the Celtic Sea. Mar. Ecol. Prog. Ser. 28: 251–258.

JOINT, I.R., AND R. WILLIAMS. 1985. Demands of the herbivore community on phytoplankton production in the Celtic Sea. Mar. Biol. 87: 297–306.

JØRGENSEN, C. B. 1983. Fluid mechanical aspects of suspension feeding. Mar. Ecol. Prog. Ser. 11: 89–103.

KIMOR, B. 1981. The role of phagotrophic dinoflagellates in marine ecosystems. Kieler Meeresforsch. 5: 164–173.

KREMPIN, D. W., AND C. W. SULLIVAN. 1981. The seasonal abundance, vertical distribution, and relative biomass of chroococcoid cyanobacteria at a station in southern California coastal waters. Can J. Microbiol. 27: 1341–1344.

LEWIS, M. R., J. J. CULLEN, AND T. PLATT. 1984. Relationships between vertical mixing and photoadaptation of phytoplankton: similarity criteria. Mar. Ecol. Prog. Ser. 15: 141–149.

LEWIS, M. R., R. E. WARNOCK, B. IRWIN, AND T. PLATT. 1985. Measuring photosynthetic action spectra of natural phytoplankton populations. J. Phycol. 21: 310–315.

LI, W. K. W., D. V. SUBBA RAO, W. G. HARRISON, J. C. SMITH, J. J. CULLEN, B. IRWIN, AND T. PLATT. 1983. Autotrophic picoplankton in the tropical ocean. Science (N.Y.) 219: 292–295.

LLOYD, N. D. H., D. T. CANVIN, AND D. A. CULVER. 1977. Photosynthesis and photorespiration in algae. Plant Physiol. 59: 936–940.

LOHMANN, H. 1911. Uber das Nannoplankton und die Zentrifugierung kleinster Wasserproben zur Gewinnung desselben in lebendem Zustand. Int. Rev. Gesamten. Hydrobiol. Hydrog. 4: 1–38.

MALONE, T. C., AND M. B. CHERVIN. 1979. The production and fate of phytoplankton size fractions in the plume of the Hudson River, New York Bight. Limnol. Oceanogr. 24: 683–696.

MARRA, J. 1978. Phytoplankton photosynthetic response to vertical movement in a mixed layer. Mar. Biol. 46: 203–208.

———— 1980. Time course of light intensity adaptation in a marine diatom. Mar. Biol. Lett. 1: 175–183.

MOREL, A., AND R. C. SMITH. 1974. Relation between total quanta and total energy for aquatic photosynthesis. Limnol. Oceanogr. 19: 591–600.

MORIARTY, D. J. W., P. C. POLLARD, AND W. G. HUNT. 1985. Temporal and spatial variation in bacterial production in the water column over a coral reef. Mar. Biol. 85: 285–292.

MORRIS, I., AND GLOVER. 1981. Physiology of photosynthesis by marine coccoid cyanobacteria-some ecological implications. Limnol. Oceanogr. 26: 957–961.

MURPHY, L. S., AND E. M. HAUGEN. 1985. The distribution and abundance of phototrophic ultraplankton in the North Atlantic. Limnol. Oceanogr. 30: 47–58.

ONG, L.J., A. N. GLAZER, AND J.B. WATERBURY. 1984. An unusual phycoerythrin from a marine cyanobacterium. Science (N.Y.) 224: 80–83.

PACE, M. L., J. E. GLASSER, AND L. R. POMEROY. 1984. A simulation analysis of continental shelf food webs. Mar. Biol. 82: 47–63.

PINGREE, R. D., P. R. PUGH, P. M. HOLLIGAN, AND G. R. FORSTER. 1975. Summer phytoplankton blooms and red tides along tidal fronts in the approaches to the English Channel. Nature (Lond.) 258: 672–677.

PINGREE, R. D., P. M. HOLLIGAN, AND G. T. MARDELL. 1978. The effects of vertical stability on phytoplankton distributions in the summer on the northwest European Shelf. Deep-Sea Res. 25: 1011–1028.

PINGREE, R. D., P. M. HOLLIGAN, G. T. MARDELL, AND R. P. HARRIS. 1982. Vertical distribution of plankton in the Skagerrak in relation to doming of the seasonal thermocline. Continental Shelf Res. 1: 209–219.

PLATT, T., D. V. SUBBA RAO, AND B. IRWIN. 1983. Photosynthesis of picoplankton in the oligotrophic ocean. Nature (Lond.) 301: 702–704.

PROBYN, T.A. 1985. Nitrogen uptake by size-fractionated phytoplankton populations in the southern Benguela upwelling system. Mar. Ecol. Prog. Ser. 22: 249–258.

RABINOWITCH, E., AND GOVINDJEE. 1969. Photosynthesis. Wiley, New York, NY. 273 p.

SENGER, H., AND P. FLEISCHHACKER. 1978. Adaptation of the photosynthetic apparatus of *Scenedesmus obliquus* to strong and weak light conditions. Physiol. Plant. 43: 35–42.

SHELDON, R.W. 1972. Size separation of marine seston by membrane and glass-fibre filters. Limnol. Oceanogr. 17: 494–498.

SHERR, B. F., E. V. SHERR, AND T. BERMAN. 1983. Grazing, growth and ammonium excretion rates of a heterotrophic microflagellate fed with four species of bacteria. Appl. Environ. Microbiol. 45: 1196–1201.

SHIVELY, J. M. 1974. Inclusion bodies of prokaryotes. Annu. Rev. Microbiol. 28: 167–187.

SIEBURTH, J. MCN. 1979. Sea microbes. Oxford University Press, New York, NY. 491 p.

SIEBURTH, J. MCN., V. SMETACEK, AND J. LENZ. 1978. Pelagic ecosystem structure: heterotrophic compartments of the plankton and their relationship to plankton size. Limnol. Oceanogr. 23: 1256– 1263.

SIEBURTH, J. MCN., AND P.G. DAVIS. 1982. The role of heterotrophic nanoplankton in the grazing and nurturing of planktonic bacteria in the Sargasso and Caribbean Seas. Ann. Inst. Oceanogr., Paris 58(S): 285–296.

SMITH, J. C., T. PLATT, W. K. W. LI, E. P. W. HORNE, W. G. HARRISON, D. V. SUBBA RAO, AND B. D. IRWIN. 1985. Arctic marine photoautotrophic picoplankton. Mar. Ecol. Prog. Ser. 20: 207–220.

STEWART, W. D. P., AND G. A. CODD. 1975. Polyhedral bodies (carboxysomes) of nitrogen-fixing blue-green algae. Br. Phycol. J. 10: 273–278.

STOECKER, D. L., L. H. DAVIS, AND A. PROVAN. 1983. Growth of *Flavella* sp. (Ciliata: Tintinnina) and other microzooplankters in cages incubated in situ and comparison to growth in vitro. Mar. Biol. 75: 293–302.

TAKAHASHI, M., AND P. K. BIENFANG. 1983. Size structure of phytoplankton biomass and photosynthesis in subtropical Hawaiian waters. Mar. Biol. 76: 203–211.

TAKAHASHI, M., AND T. HORI. 1984. Abundance of picoplankton in the subsurface chlorophyll maximum layer in subtropical and tropical waters. Mar. Biol. 79: 177–186.

THRONDSEN, J. 1978. Productivity and abundance

of ultra- and nanoplankton in Oslofjorden. Sarsia 63: 273–284.

WALSH, J. J. 1983. Death in the sea: enigmatic phytoplankton losses. Prog. Oceanogr. 12: 1–86.

WATERBURY, J. B., S. W. WATSON, R. R. L. GUILLARD, AND L. E. BRAND. 1979. Widespread occurrence of a unicellular, marine, planktonic, cyanobacterium. Nature (Lond.) 277: 293–294.

WRIGHT, R.T., AND R. B. COFFIN. 1984. Measuring microzooplankton grazing on planktonic marine bacteria by its impact on bacterial production. Microb. Ecol. 10: 137–149.

WYMAN, M., R. P. F. GREGORY, AND N. G. CARR. 1985. Novel role for phycoerythrin in a marine cyanobacterium, *Synechococcus* strain DC2. Science (N.Y.) 230: 818–820.

The Association of Photosynthetic Picoplankton and Ultraplankton with Pelagic Detritus through the Water Column (0–2000 m)

MARY WILCOX SILVER AND MARCIA M. GOWING

Institute of Marine Sciences, University of California, Santa Cruz, CA 95064, USA

AND PETER J. DAVOLL

Harbor Branch Foundation, Inc., Fort Pierce, FL 33450, USA

Introduction

Photosynthetic plankton occur not only as individual cells in the open sea, but also in combination with other organisms and with pelagic detritus. The association of photosynthetic ultraplankton (cells ≤ 5 μm) with non-living, particulate organic matter, or detritus, places them in microhabitats substantially different from those of their free-living counterparts suspended in the sea. The dispersal potential of these smallest of photosynthetic organisms is markedly altered by the larger particles. The purpose of this review is to discuss the association of the picoplankton-sized (≤ 2 μm) procaryote, *Synechococcus*, and ultraplankton-sized, photosynthetic eucaryotes, which are mostly slightly larger than *Synechococcus* (see Murphy and Haugen 1985), with detritus in surface and deep waters. We consider the mechanisms leading to the association, the ecological consequences of life on detritus, and the reasons for the presence of extensive populations of small, pigmented cells in the meso- and bathypelagic regions of the ocean.

Size is an important feature for predicting the behavior of particles, including both cells and detritus. Studies of field samples indicate that most oceanic particles are very small, but that roughly equal concentrations of material occur throughout the 1 to 100 μm range; larger sizes are less frequent and harder to sample but their masses may be about the same as those in the smaller classes (Sheldon et al. 1972; McCave 1984). Recently, special attention has been paid to the larger particles (100s of μm and larger), even though these may be numerically rare, because of their disproportionate importance in vertically transferring materials through the water (Menzel 1974; McCave 1975). Ultraplankton become associated with the large particles by the same general physical and biological processes that add small particles to large ones. The fates of the particles and their contents are then strongly influenced by the aggregate sizes and thus sinking rates.

Studies of associations between microorganisms and natural detritus have been hindered by the lack of collecting devices that maintain the integrity of fragile detrital aggregates. Some large particles, especially those classified as "marine snow" (detrital aggregates >0.5 mm), are readily disrupted by standard particle collecting devices and processing methods (Nishizawa et al. 1954; Jannasch 1973; Hamner et al. 1975; Trent et al. 1978), and the extent to which this also occurs with the smaller, more abundant particles is presently unknown. Larger detrital aggregates are especially poorly known because they are comparatively rare, and thus substantial volumes of water are needed for quantitative collections. Furthermore, large aggregates may break apart when collected, and the resulting detrital fragments and dislodged cells can not be distinguished from free particles originally present in the sample. Because of these difficulties, the most unequivocal data showing associations of organisms with large detrital particles come from alternative collecting methods. These approaches include the *in situ* collection of large detrital aggregates by divers, obviously a difficult and inefficient method for surveying large numbers of particles, and collections of particles that settle into sediment traps, which will contain the large, rapidly sinking materials.

The paucity of data on associations between small autotrophs and non-living particulates also results from difficulty in identifying the cells within detrital aggregates. Photosynthetic organisms are particularly difficult to recognize when present in mixtures that include the decomposing plant and animal remains commonly found in detrital aggregates. Fluorescence microscopy and even electron microscopy (especially transmission electron microscopy [TEM]) may be required to distinguish intact cells from other particles.

In this review of the detrital microcosms in euphotic and aphotic zones, we consider the associations of phytoplankton of all size classes as well as those of the ultraplankton component. The full photoautotrophic community has received more attention than has the ultraplankton fraction, because serious, systematic study of these smallest photosynthetic organisms has only recently begun. As we will show, many of the same processes that bring about the associations of ultraplankton with non-living particulates also cause associations between larger phytoplankton and detritus. The more extensive data base from the entire phytoplankton community is reviewed to give a better prediction of the variations likely to be expected of the smallest size component. For that, we draw on a relatively small number of papers published specifically on ultraplankton in detritus and on some of our own previously unpublished studies on detrital–picoplankton associations, and make predictions based on the larger literature of the entire phytoplankton community on detritus.

I. Euphotic Zone Associations of Ultraplankton with Detritus

Although detritus has long been known to carry microorganisms, principal areas of research on detritus have dealt mainly with broader topics related to its role as a carbon pool in the sea. In the earliest literature, non-living particulate material was usually viewed as the decomposed remains of various organisms. By the 1950s and especially the 1960s, the non-living fraction was recognized as one of the several interactive pools of organic material — along with dissolved carbon reservoirs and living organisms (Riley 1970). Of primary interest was the quantitative relationship between organic particles and dissolved organic matter, the processes forming the particles, the general morphological and chemical characteristics of the particles as they revealed origins of the detritus, and the possible use of that detritus as food by marine organisms.

Associations between intact cells and non-living particulates were noted in some of the early studies. Bacteria were most commonly investigated, possibly because particulates were viewed as decomposition products. It was thought that bacteria could not flourish on the low ambient concentrations of dissolved organics in sea water and therefore required richer particulate substrates (Wiebe and Pomeroy 1972; Riley 1970; Jannasch 1973). Although many studies showed the presence of phytoplankton on detritus, the abundances were not particularly noteworthy, likely because of the small size of most of the particles examined and because fluorescence microscopy usually was not used to detect minute cells. After the early microscopic work, chemical analyses or automated particle counting and sizing devices were used increasingly, and little further was discovered about associated microorganisms. Even in relatively early reports, however, investigators recognized that some classes of detritus, especially the large detrital aggregates of marine snow, were conspicuously colonized, but the data were qualitative (see below). Quantitative studies have been made only in the last decade and almost exclusively have involved larger sized particles. These are the ones that can be readily recognized and collected intact from standard sampling devices (e.g. some intact fecal pellets and some larvacean mucus houses), by divers, from submersibles, or with pelagic sediment traps.

A. Associations of the Entire Phytoplankton Community with Detritus

Small particles — The pioneering studies of Gordon Riley and his associates were concerned with the quantitative distribution and origin of non-living particles in the sea. Their microscopic analyses of particles filtered from water collected by buckets or bottles showed that seawater contained large numbers (10^3 to $10^5 \cdot L^{-1}$) of small (usually <50 μm) organic particles. The numbers sometimes correlated with phytoplankton cycles and hydrographic conditions, and particulate abundances showed characteristic seasonal patterns in surface waters (e.g. Riley 1963; Riley et al. 1964, 1965; Johannes 1967; Kane 1967; Gordon 1970). They also investigated possible correlations between algal biomass or production and the non-living particles, because the latter were thought to originate from phytoplankton-leached dissolved matter via a bubble mechanism (Baylor and Sutcliffe 1963; see discussion by Biddanda 1985).

In 1963, Riley presented one of the first quantitative studies on "organic aggregates", a name he gave to the naturally occurring particles of non-living material that often appeared to consist of accumulations of smaller particles. (The name was originally applied to only one type of particulate, but Riley and others soon used it as a general term for organic detritus, as we also do in this review.) In a 2-year study of Long Island waters, Riley (1963) noted that up to 50% of the phytoplankton occurred in association with particles, that these aggregates showed a seasonal cycle, and that the cycle only partially coincided with phytoplankton abundance patterns. Later studies on aggregates showed that phytoplankton occasionally occurred on particles, but no other quantitative associations appear to have been recorded, excepted for the comment of Wiebe and Pomeroy (1972) that " >15% of the small aggregates appear to have at least one whole algal cell associated with them." However, even in his 1963 study, Riley noted the importance of phytoplankton association with the aggregates by commenting that the aggregates (probably larger marine snow particles) came to the sea surface on calm days because of the entrapped bubbles from photosynthesis by colonizing algae.

Subsequent studies by Riley and others showed that two dominant classes of particles existed in water obtained by water bottles and that colonization of these classes differed. The scale-like, semi-transparent "flakes", evenly distributed throughout the ocean depths, were only sparsely populated, whereas the more complex flocculent "aggregates", most abundant in the euphotic zone were variably colonized and sometimes contained large populations of microorganisms (reviewed in Wiebe and Pomeroy 1972; Riley 1970). In the coastal waters of Peru, where flocculent aggregates were particularly abundant, Pomeroy and Johannes (1968) described the phytoplankton and bacteria-rich aggregates as "centers of metabolic activities in the ocean", although this richness was later considered atypical of most oceanic environments (Wiebe and Pomeroy 1972). Microscopic studies of some of the aggregates during this period, however, suggested that, in some places or on occasion, the particles were protozoan-rich decomposer centers analogous to those found in the benthos. The particles could harbor many algae and their internal microenvironment could be substantially different from the surrounding waters (Riley 1963; Pomeroy and Johannnes 1968; Wiebe and Pomeroy 1972).

Marine snow — Unlike the small aggregates normally obtained with water-bottles, the large detrital associations of marine snow occasionally seen by direct visual observation attracted early attention as microenvironments for planktonic organisms. Possible the most dramatic observations of particles were made by scientists from Hokkaido Institute, who described large numbers of readily visible flocs from their submersible *Kuroshio* in the nearshore waters of the Sea of Japan. These particles appeared suspended and were designated "marine snow" by Suzuki and Kato (1953) after a phenomenon named by Carson (1951). Suzuki and Kato (1953) collected some of the particles and stated that they were "chiefly the aggregates of the remains of plankton, sinking in some stages of disintegration by marine bacteria." They thought that marine snow originated from nuclei supplied either as decomposition products of living organisms, particularly diatoms, or from land-derived debris. Tsujita (1952), who presented the first detailed microscopic description of the material, noted that many types of organisms, including diatoms, were richly represented on the snow and that the cells were associated with a mucus-rich matrix.

Quantitative studies on the colonization by microorganisms of the larger marine snow aggregates began in the late 1970s. Like previous studies, these considered the entire phytoplankton community. In contrast to the earlier investigations, though, there was a focus on the role of detritus as a biological center in the water. This work thus contrasted populations free in the water column with those on the non-living particles and determined the relative contribution of populations on detritus to the entire pelagic community.

Detritus can be a center for photosynthetic organisms, depending on the hydrographic regime and the successional state of the aggregate. In a study in the neritic waters of Monterey Bay, California, marine snow contained 0.1–8% of the phytoplankton, as determined by chlorophyll a content (Trent et al. 1978). In another case, marine snow collected from the same area but at a different time contained 11–58% of the total primary production (Knauer et al. 1982). The authors of both studies noted that differences in production associated with snow likely resulted from changes in the abundance and physical characteristics of the snow. Aggregates in shelf waters off Santa Barbara, California contained an average of 7% of the chlorophyll a, with a maximum of 27%. During this time primary production on the aggregates averaged about 2% (range 0.1–9%) of the total (Alldredge and Cox 1982). The low contribution to primary production was thought to reflect the relatively small numbers of snow particles during the study period.

The association of microorganisms with marine snow detritus has also been presented as an "enrichment factor." That is, the ratio of material found in a given volume of detritus to that found in the same volume of water surrounding the detritus can be used to express concentrations of cells, photosynthetic pigments, or microbial rates of activity. This ratio is particularly useful because it gives a relative measure of the density of organisms in the detrital microhabitat and allows direct comparisons of community richness between particles from different collections. Enrichments of chlorophyll a were 4–750× for the neritic Monterey Bay study (Trent et al. 1978), and averaged 252× for the neritic Santa Barbara one (Alldredge and Cox 1982). Production was enriched by an average of 53× in the Santa Barbara study.

Succession on pelagic detritus — Changes occur in the populations of microorganisms accompanying flocculent aggregates. Jannasch (1973) collected specimens of marine snow from surface waters and, finding them bacteria-poor, suggested that the particles were so newly formed that they lacked the expected microbial complement. Wiebe and Pomeroy (1972) and Johannes (1967) described flocculent particles in coral reef waters and suggested that changes both in the particles and in their colonizer populations — primarily changes in bacterial number — were due to aging of the particles.

The most extensive account of successional changes in pelagic detritus, however, was provided by Pomeroy and his colleagues (Pomeroy and Deibel 1980; Pomeroy et al. 1984), who studied flocculent fecal aggregates produced by some pelagic tunicates (species of salps and doliolids). All feces showed successional changes of microbes that varied to some extent depending upon the organisms originally present inside the pellets and those colonizing from outside. Bacterial growth in the aggregates began after a lag phase and, after a period of rapid growth for a day or so, populations declined rapidly due to cropping by small invading protozoans. Some of the protozoans also ate the associated phytoplankton, both cyanobacteria and eucaryotic cells, that had survived passage through the tunicate gut. Local patches of bacteria,

314

and presumably algae, remained only in areas of the aggregate not yet invaded by the protozoa. Several day old feces resembled the organism-poor flocculent aggregates found abundantly in water samples, and suggested to the authors that feces were an important source of particulates in seawater.

Quantitative studies of marine snow support the earlier descriptions of succession on detritus. Prezelin and Alldredge (1983) studied aggregates during an upwelling event and several weeks later. During the initial sampling, aggregates were small, very abundant, and physically variable. At this time, they were populated by the same algal groups found in the water and contained 20% of the chlorophyll a pigments and about 20% of the primary production. Enrichment factors for the pigments and productivity were both 74×. On the second sampling date, the aggregates were larger, flocculent, contained <0.3% of the chl a, and contributed <0.4% of the production. Enrichment factors for chl a were only 3× and many of the cells appeared moribund. In the first sample series, the photosynthesis-irradiance relationships for the photoautotrophs on snow and in the water were the same, indicating similar populations in both environments. On the later date, however, the relationship was depressed for the snow populations, suggesting senescent cells.

Davoll (1984) studied community succession in one common neritic form of marine snow, the abandoned mucus feeding structure, or house, of the larvacean tunicate. He hypothesized that the aggregates went through several distinct phases. The first stage was the period when the larvacean occupied the house and used it to concentrate microbial populations from the surrounding water: bacteria and algae <5 μm became enriched by factors of 10^2 on the mucus houses. After abandonment of the house by the larvacean, or the second stage, growth of the introduced microflagellates and further immigration and growth of larger protozoa (mostly ciliates) held the growing algal and bacterial population levels nearly stationary. Now, in the third stage, the numbers of bacteria remained low because they had exhausted readily utilizible organic matter in the house. Numbers of photosynthetic ultraplankton stayed high and populations of small dinoflagellates appeared to increase with the lessening of grazing pressure. The successional events on the house were often out of phase with successional events in the free-living populations in the water surrounding the detritus. Thus the relative contribution of the houses to populations in the water changed over time.

B. Association of Photosynthetic Ultraplankton with Detritus.

Marine snow — The findings presented above show that the entire phytoplankton community associated with detritus will vary under different hydrographic regimes and with different successional phases of detrital particles. The limited research on ultraplankton associates of marine snow, the only detrital particle examined in this context (aside from fecal pellets, discussed below), show some variability in the abundance of the cells within and between hydrographic regimes. Caron (1984) found photosynthetic picoplankton, predominantly chroococcoidal cyanobacteria, to be enriched 19–24× on marine snow in various environments in the North Atlantic. The picoplankton producers were, on the average, less concentrated than algae larger than 20 μm, and more enriched than bacteria. Furthermore, the photosynthetic picoplankton were least enriched in the shelf environments and most enriched in oceanic waters. Caron emphasized the importance of the detrital habitats to microflagellate grazers, which must exert considerable grazing pressure on their photoautotrophic picoplankton and bacterial prey.

A similar study was carried out in neritic waters of Monterey Bay, an oceanographic environment particularly rich in marine snow. In this study, photosynthethic ultraplankton (excluding cyanobacteria) were examined on aggregates that were derived from larvacean houses (Davoll and Silver, unpubl. data). The algae were enriched about a hundred-fold on the houses, with relatively low variability between different sampling dates. Protozoans and larger phytoplankton were even more enriched on the houses, but their concentrations were considerably more variable on the snow. In both this neritic study and the oceanic one (Caron et al. 1986), the bacterial and ultraplankton–picoplankton sized photoautotrophs were more predictable in their enrichments, and the enrichments were lower than those of the protozoans that accompanied them. Possibly grazing by protozoans accounts for the relative stability of the smaller organisms.

The fraction of the total ultraplankton on marine snow has not yet been directly assessed. However, from the two studies just discussed, an estimate can be made of the role of these large particles as population centers for ultraplankton, using the enrichment factors measured in them. The estimate is based on observations made in a number of studies on the volumes of marine snow present in neritic waters, including some from Monterey Bay. By multiplying the volumes of snow (i.e. proportion of the water occupied by snow) by the enrichments of photosynthetic cells on snow, the proportions of the total population on snow can be predicted. Values for this calculation are shown in Table 1. The importance of snow in neritic waters is variable, but the predictions indicate that, on the average, a few percent of the total ultraplankton populations likely occurs on marine snow in the euphotic zone.

TABLE 1. Predicted abundance of photosynthetic ultraplankton on marine snow in neritic waters. Estimates are based on average enrichment factors, EF (conc. on marine snow/conc. in surrounding water) measured (1) by Caron (1984) for marine snow in North Atlantic neritic waters and (2) by Davoll (1984) for larvacean house aggregates in Monterey Bay, California. Predicted contributions are EF $\times$ sea water volume occupied by aggregate, with the latter estimated from the average value in the literature source cited.

Source	Site	Depth	Ultraplankton on Snow (EF:19$\times$)	Ultraplankton on Snow (EF:142$\times$)
Trent et al. (1978)	Monterey Bay, CA	3–17 m	2.3%	17%
Alldredge (1979)	Gulf of Calif.	10 m	0.09%	0.71%
	Santa Barbara Channel	10 m	0.26%	2.0%
Alldredge and Cox (1982)	So. Calif. Bight	10 m	0.5%	3.4%
Prezelin and Alldredge (1983)	Santa Barbara (upwelling)	7m	5.1%	38%
	Santa Barbara (nonupwelling)	10 m	1.9%	14%
Honjo et al. (1984)	Monterey Bay, CA	<100 m	0.48%	3.6%

Fecal pellets — Fecal wastes of zooplankton are an important and readily recognizable form of detritus. Autofluorescent ultraplankton-sized algal cells, together with bacteria, occur abundantly in pellets from field collections and in pellets produced by freshly collected zooplankton (Pomeroy and Deibel 1980; Pomeroy et al. 1984; Caron et al. 1982; Caron 1984). The presence of small autotrophs in pellets has been recognized for some time using fluorescence microscopy (e.g. Pomeroy and Johannes 1968). Further verification of the presence of intact cells came through the study of field collected pellets with transmission electron microscopy (TEM) (Sieburth 1979; Silver and Bruland 1981; Gowing and Silver 1983, 1985).

Concentrations of cells inside certain types of fecal pellets are known. Gowing and Silver (1985) studied water bottle collections of "minipellets" (wastes <50 μm) from the Eastern Tropical Pacific with TEM and showed that about 0.5% of the volume of a pellet consists of ultrastructurally intact cells; this is an ultraplankton cell concentration of about $10^6 \cdot$ mm^{-3}. The minipellets, which numbered over $10^2 \cdot$ L^{-1}(Gowing and Silver 1985), contained about 0.03% of the total ultraplankton-sized photosynthetic cells in the euphotic zone (Silver and Gowing, unpubl. data).

Large fecal pellets can harbor sizeable populations of phytoplankton, but such pellets, because of their more rapid sinking speeds, quickly settle out of the euphotic zone. For instance, the detritivorous swimming crab *Pleuroncodes planipes* in the

Eastern Tropical Pacific produces pellets that contain an average of 1×10^5 cells $\cdot$ mm^{-3} (counted by autofluorescence); 2×10^4 cells therefore are enclosed in a single, averaged-sized (Gowing and Silver 1983) pellet of the crab. Pellets of herbivorous crustacea (euphausiids or sergestids) collected in the Gulf of Maine averaged 2×10^3 pico and nanoplankton sized eucaryotes and 7×10^4 cyanobacteria per pellet (Davoll and Youngbluth, unpubl. data). In the Gulf of Maine study, the cyanobacteria were more enriched in the pellets, compared to the water column, than were the eucaryotes, suggesting greater survivorship during gut passage (Davoll, unpubl. data). Even higher numbers of cells are likely present in larger pellets, such as those produced by some species of salps (Silver and Bruland 1981). The pellets of the swimming crab, some salps, and the euphausiids, however, are sufficiently large that they leave the euphotic zone within a few hours of their production (Bruland and Silver 1981).

C. Taxonomic Composition of Photosynthetic Ultraplankton on Detritus

The ultraplankton associated with detritus in the euphotic zone have received little taxonomic attention. The chroococcoidal cyanobacteria are probably the best known associates, because they can be recognized by their characteristic yellow or orange autofluorescence and small size. They have been found commonly in the fecal pellets of zooplankton (Pomeroy and Deibel 1980; Pomeroy et al. 1984; Caron et al. 1982; Caron 1984; Sieburth 1979). TEM examination of pellets collected from the field has revealed the presence of many cells with cytomembrane structure and general appearance of cyanobacteria (Fig. 1) (Johnson and Sieburth 1979; Sieburth 1979; Silver and Bruland 1981).

A second common associate of marine detritus belongs to a group of small, nonflagellated green algae that are *Chlorella*-like in ultrastructure, though similar cells taken in oceanic environments have the pigments expected of a prasinophyte (Foss et al. 1984). These cells contain 1 chloroplast (sometimes with a starch grain), a heavy wall that often possesses several layers, a single mitochondrion, and a nucleus. The cells are scaleless and frequently fix very poorly (Fig. 1). Cells with similar ultrastructure have been found in many different environments (Johnson and Sieburth 1982; Joint and Pipe 1984; Takashi and Hori 1984; Thinh and Griffiths 1985). The small chlorophytes occasionally can be abundant and even numerically dominant in the chlorophyll maximum layer (Takashi and Hori 1984; Takashi and Bienfang 1983). We have cultured a similar cell from a depth of 500 m in Monterey Bay (Fig. 2). Like the field-collected specimens, the cell in culture has a wall of trilaminar construction, similar to that found in *Chlorella* known to possess sporopollenin (Atkinson et al. 1972). After acetolysis (digestion with sulfuric acid and acetic anhydride), the wall residues show infrared spectra characteristic of sporopollenin (Fig. 3). This highly resistant biopolymer may account for the digestion resistance of this chlorophyte, which along with the chroococcoid cyanobacteria, is the numerically dominant picoplankter inside fecal pellets from the euphotic zone (Silver and Bruland 1981; Gowing and Silver 1983, 1985). The cells also maintain pigmentation and can grow very slowly in the dark on organic growth media (Silver, unpubl. data).

Other ultraplankton and picoplankton have been found associated with marine detritus from the euphotic zone, but few forms are described in the general literature. For example, diatoms, coccolithophores, and some dinoflagellates are within this size range and are known from marine snow and fecal pellets (Trent et al. 1978; Silver et al. 1978; Davoll 1984; Turner 1984; Silver, unpubl. data), but the species were not identified. For larger marine snow, however, the associated photosynthetic assemblages likely resemble those in the water, since much of this detrital material is produced as a food collecting device. We have encountered a variety of very small cells on marine snow and in fecal material; examples of common cell types are shown in Fig. 4 and 5.

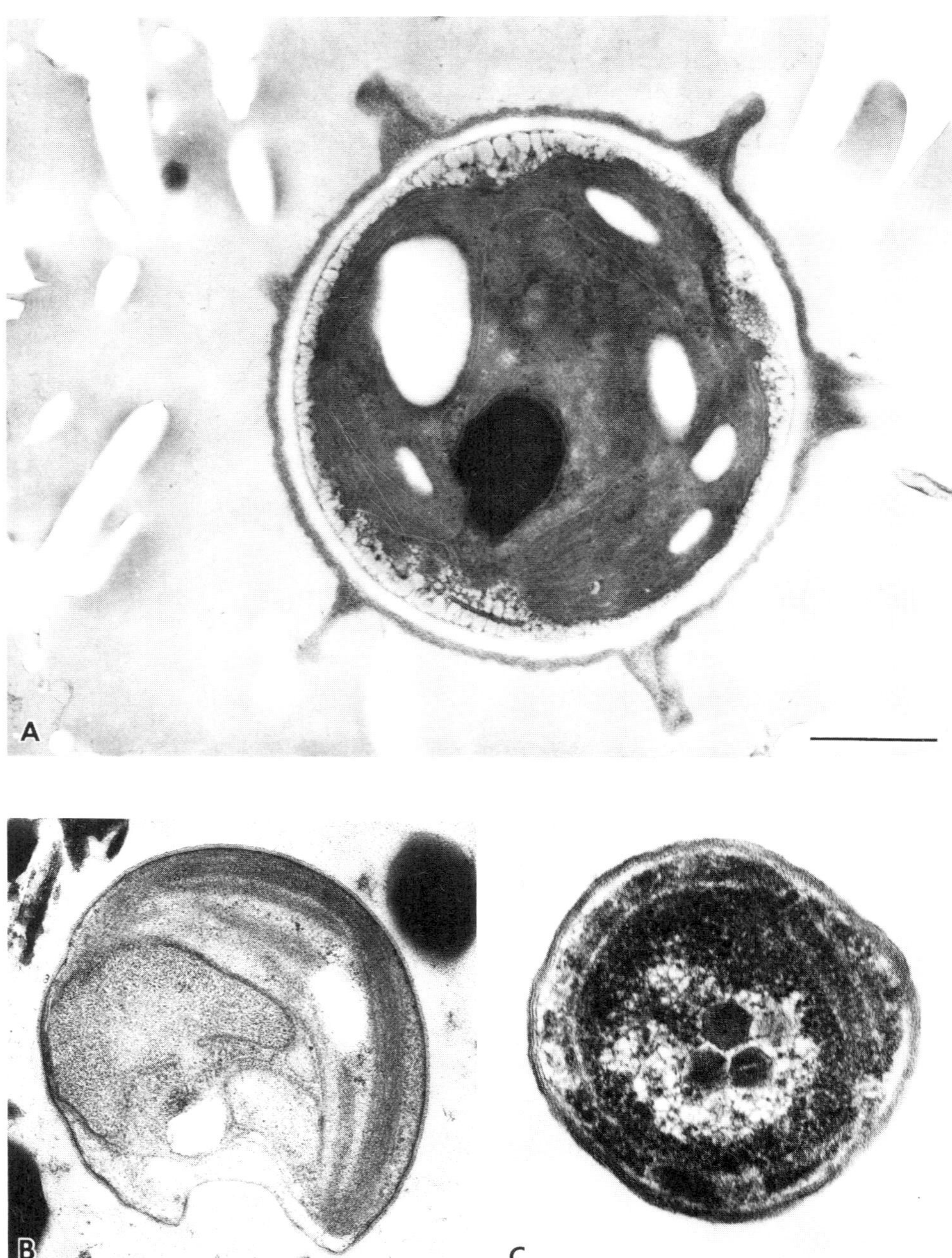

FIG. 1. Photosynthetic ultraplankton from fecal pellets in the euphotic zone. (a) Heavy-walled eucaryote from a sediment trap at 100 m, VERTEX 1, 200 km off central California. (b) *Chlorella*-like algal cell from a fecal pellet of the swimming crab *Pleuroncodes planipes*, from the 100 m sediment trap on VERTEX 2, 400 km off Mexico. (c) Chroococcoidal cyanobacterium from a salp fecal pellet collected 350 km off Pt. Sur, California. All bars = 0.5 μm.

318

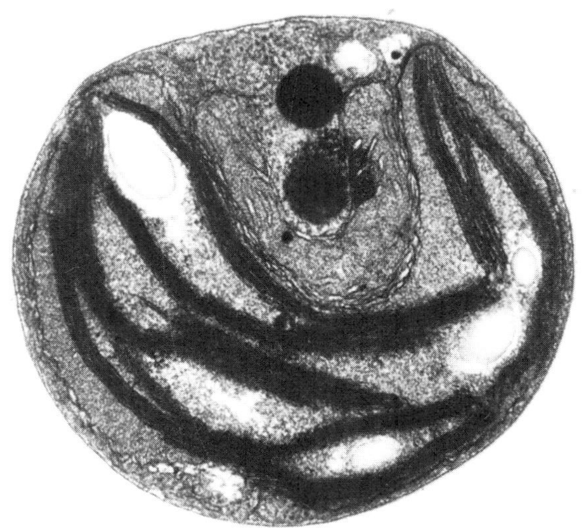

FIG. 2. Transmission electron micrograph of a *Chlorella*-like cell cultured from a water sample taken at 500 m depth in Monterey Bay. Cells possess one chloroplast and mitochondrion and a trilaminar wall. Upon division, two daughter cells are formed; they remain for a while inside the mother cell wall. Cell diameter about 2 μm.

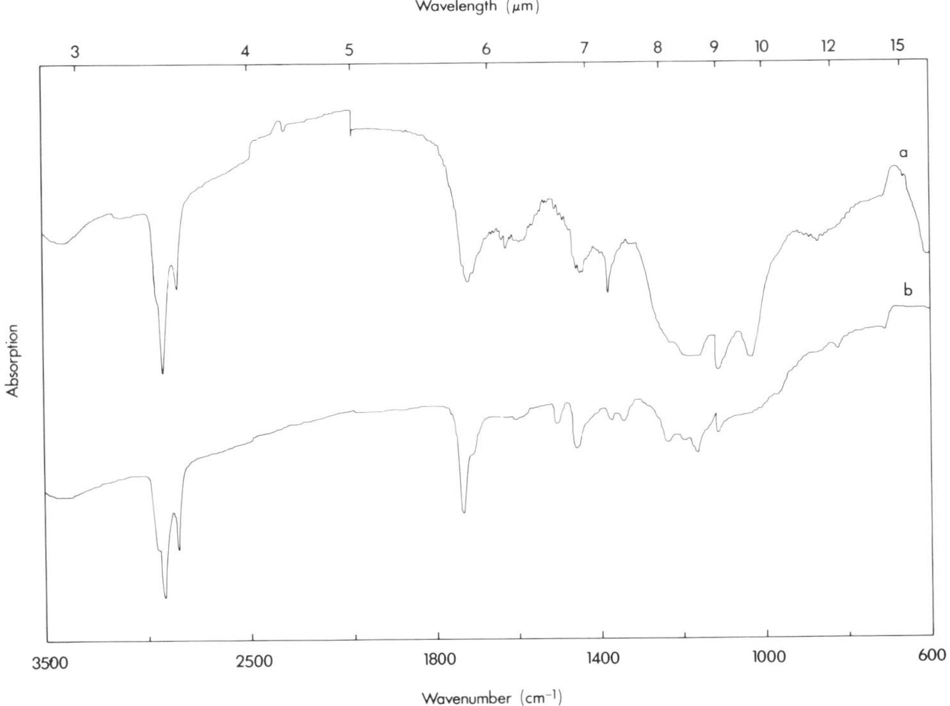

FIG. 3. Infrared spectrum from acetolysis-resistant wall residues of pine pollen standard (a) and cultured *Chlorella*-like cells (b) shown in Fig. 2, showing sporopollenin absorption features. Residues were incorporated into a potassium bromide pellet and spectra measured on a Perkin–Elmer (model 621) grating infrared spectrophotometer. (Preparation methods are outlined in Atkinson et al. 1972 and in Konig and Peveling 1984.)

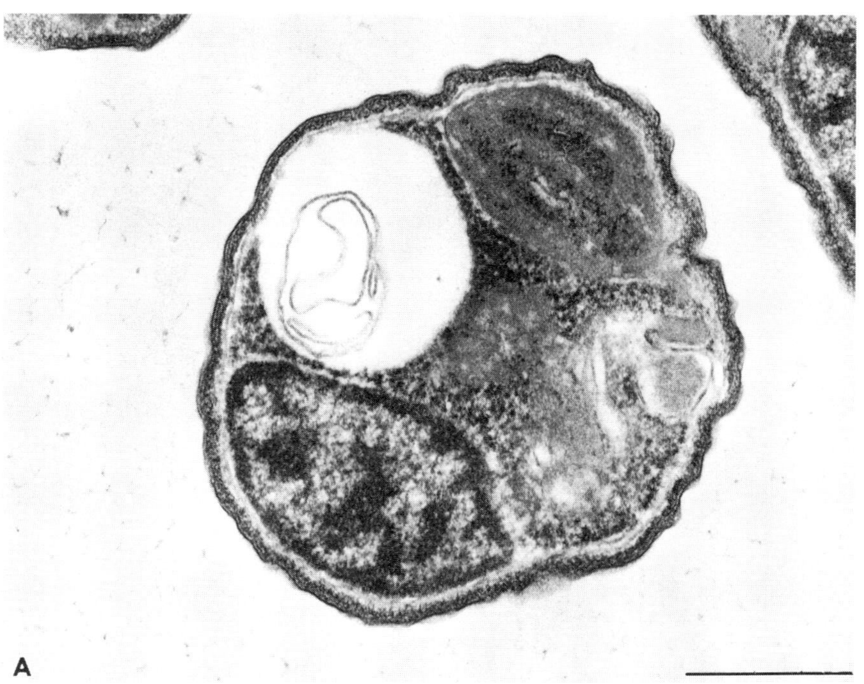

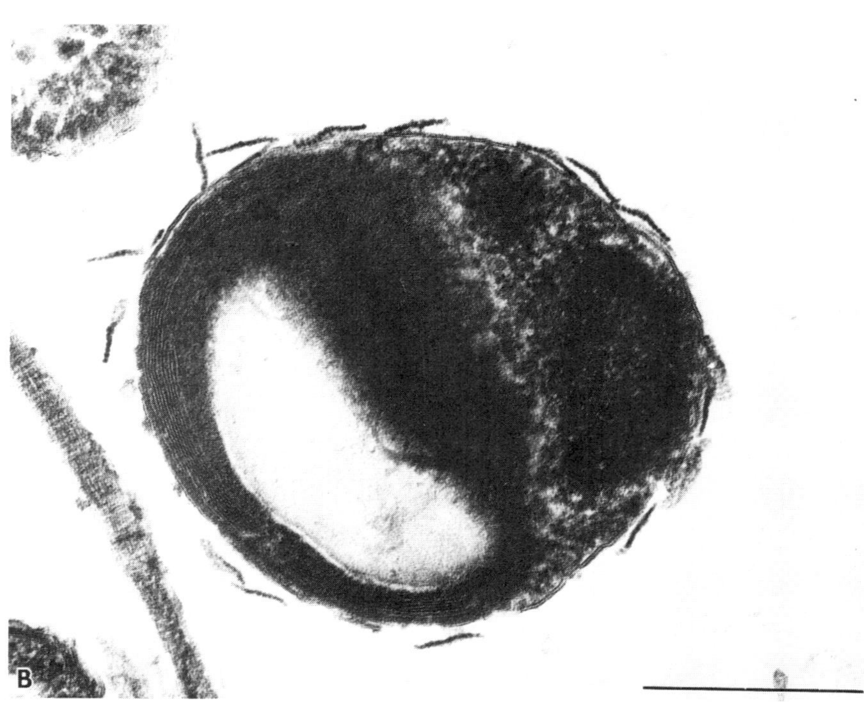

FIG. 4. Photosynthetic ultraplankton from marine snow in the euphotic zone. (a) eucaryotic alga from an aggregate hand-collected at 10 m on VERTEX 1, 200 km off central California. (b) scaly alga, possibly similar to the one shown by Johnson and Sieburth (1982), Fig 3A. From hand-collected larvacean house at 10 m, Monterey Bay, CA. Bars = 0.5 μm.

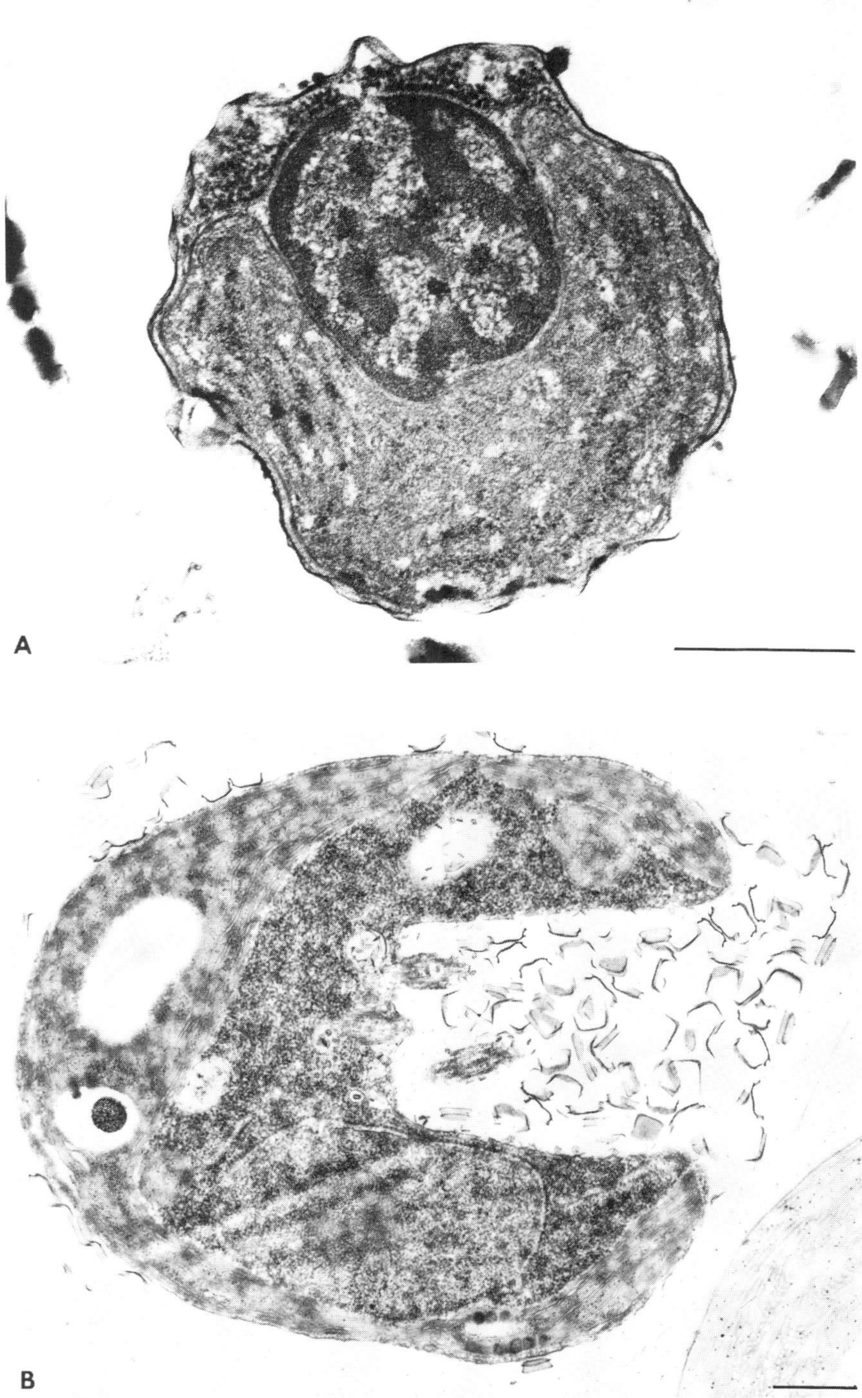

FIG. 5. Photosynthetic ultraplankton from marine snow in the euphotic zone (cont.). (a) Alga with single, large cup-shaped chloroplast, from marine snow collected in a sediment trap at 50 m depth, 200 km off central California (VERTEX 1). (b) Scaly green alga (prasinophyte) with 4 lobed chloroplast (seen in other sections). From hand-collected "comet-shaped" aggregate from Monterey Bay at 15 m depth. Bars = 0.5 μm.

D. Mechanisms Responsible for the Association of Ultraplankton with Detritus

Physical mechanisms — The literature on detritus has frequently inferred that organisms become associated through simple physical contact and adhesion. McCave (1984) has reviewed the mechanisms bringing about physical aggregation of particles in the sea. He distinguished between processes that make particles collide initially and those that make particles remain together once they have collided. He noted that "stickiness," which is enhanced by biologically produced slimes and mucus, increases the likelihood that particles will remain together, but the presence of such coatings does not increase the frequency of collisions. For non-motile particles in the size range of picoplankton, contact is governed by Brownian motion. However, the collisions between large particles such as marine snow and small cells are physically dominated by shear-controlled coagulation, and the mechanics of such collisions indicate that capture efficiencies are greatest for particles of similar sizes. The abundances of large particles, McCave notes, are much too high to indicate particle origins by such contacts. Rather, he argues, "larger particles, particularly 'snow', are not produced by aggregations of small ones; they start out with a large organic substrate" (McCave 1984, p. 340). He concludes that only biological processes can explain the rate at which aggregation of suspended material occurs in the sea. Although large particles, once formed, may scavenge cells during their residence in the water, the rates of such processes are not high. Thus we must look to biological processes to explain the initial association of ultraplankton cells with detrital particles.

Biological mechanisms — Mucus nets are one of the most effective biological mechanisms of concentrating small particles onto large ones. These are used by a number of invertebrates to collect their food, but when abandoned, remain as concentrated, particle-rich centers in the water. The best known of these devices are the mm to cm sized mucus houses of larvaceans (Alldredge 1976). In some species 6 of these may be produced over a 24-hour period (Lohmann 1909). They contain many microorganisms on their surfaces and fecal pellets of their former occupant inside. The houses are a common source of marine snow in many seas (Alldredge 1972, 1976). Pelagic pteropods produce a variety of free-floating mucus nets (Gilmer 1972, 1974; Gilmer and Harbison 1986) and these are thought to be abandoned from time to time. Other invertebrates, including larval forms of polychaete worms and gastropod snails (Hamner et al. 1975), produce external mucus structures apparently used in food collection. Youngbluth (1982) reports a bizarre form of collecting device produced by mysid shrimps, consisting of a peritrophic membrane or fecal string left out for colonization by algae and other organisms, which are subsequently harvested by their mysid producer. (Some of these strings probably detach from their producers and become marine snow.) Furthermore, many pelagic organisms, especially the gelatinous ones such as ctenophores and salps, produce copious mucus that is released into the water, possibly with inocula of microorganisms (Caron 1984).

Once the aggregates are biologically produced, their populations can be augmented both *in situ* growth and by immigration of motile microorganisms from surrounding waters. Growth of microorganisms on detritus follows a pattern of succession, with waves of producers followed by population expansion of various consumers (Fenchel and Jorgensen 1977). Population growth of photoautotrophs can be expected as long as the detrital particles remain in the euphotic zone (a function of particle sinking rate and depth of origin of the particle), nutrient resources are available, and grazing does not overrun growth. For detritus formed in the euphotic zone, population increases of both bacteria and small photoautotrophs usually can occur for several days before microflagellates arrest net population growth (Linley and Newell 1984). Nutrients, especially ammonia–nitrogen, are present in higher concentrations in detrital aggregates than in surrounding water (Shanks and Trent 1979; Prezelin and Alldredge 1983), and thus should favor growth here under nutrient limited conditions in the

322

water column. That photoautotrophic growth does occur has been suggested by studies, discussed above, showing relatively high rates of carbon fixation on marine snow.

Immigration has been discussed in the literature, but the extent of its occurrence in the field is unknown. The high concentrations of larger protozoans on marine snow, such as the highly motile ciliates, may be due to their swimming abilities (Davoll 1984; Silver et al. 1984). However, high concentrations of smaller, ultraplankton sized cells on detritus are harder to explain. Jacobsen and Azam (1984) noted that fecal pellets suspended in a mixing apparatus during experiments contained more bacteria on their surfaces than pellets resting on the bottom of a container. They suggested that motile bacteria are able to chemotactically locate the pellets and had a greater opportunity to contact the detrital particles as the pellets moved through the water. Goldman (1984a) hypothesized that the widespread occurrence of motility in very small plankters may "allow the cell to migrate from nutrient-depleted zones to sites of nutrient enrichment" (p. 468). He argued that the abundant flakes and microaggregates of the type described by Riley in 1970 (particles <50 μm and numbering over $10^4 \cdot L^{-1}$) were abundant enough that the random swimming motions of bacteria, possibly coupled with chemotaxis in some cases, allow frequent contact of even weakly motile forms. Waterbury et al. (1985) suggested the swimming ability of some strains of *Synechococcus* may allow them to locate nutrient rich microenvironments, such as aggregates. Mitchell et al. (1985) further discussed the ability of motile bacteria to locate nutrient sources in the pelagic zone. They argued that the cells can only orient to a particle once they encounter the chemical gradient which lies within the viscous boundary layer around a particle source, because turbulence removes the chemical signal outside the boundary layer. For a particle such as a diatom, the chemical boundary layer is several hundred micrometres thick and depends on the alga's exudation rate and the molecular size of the attractant. Data for the boundary layer and molecular attractants are not available for large aggregates, but would be important in predicting their ability to attract motile microorganisms from surrounding waters.

A well-known mechanism whereby algae become concentrated in detritus is by placement into fecal pellets. Some cyanophytes, green algae, diatoms and dinoflagellates (reviewed by Porter 1977) have been shown to pass through metazoan guts. Porter (1976), and Epp and Lewis (1981) hypothesized that some autotrophs may benefit from passage through the digestive tract of an herbivore by utilizing nutrients released during lysis of non-resistant cells. Viability of some of these marine algae in pellets has been shown by subsequent growth of the cells in culture (Porter 1973; Fowler and Fisher 1983) and by the ultrastructural integrity of the cells (reviewed above). Why some cells are comparatively unaffected by grazers is not well understood, but the walls of some (e.g. species of cyanobacteria and gelatinous green algae) are known to be relatively inert to lytic enzymes (Porter 1977). The sporopollenin walls of *Chlorella* are highly resistant to chemical attack and may prevent digestion of the protoplast (Brooks and Shaw 1971). Furthermore, Johnson et al. (1982) showed that digestive enzymes differ among taxa: cyanobacteria passed through guts of copepods but were digested by protozoans. The ability to digest and utilize cyanobacteria may vary among protozoans (Caron 1984; Gowing and Silver 1985).

II. Aphotic Zone Associations of Photosynthetic Picoplankton with Detritus

The occurrence of most photosynthetic cells below the euphotic zone can be explained by their settling from overlying waters. The maximum sinking rates of algal cells are positively correlated with their sizes (Smayda 1970). Because of their small size, the sinking rate of individual picoplankton and ultraplankton cells is considered negligible (Bienfang 1984, 1985). Thus one expects photosynthetic cells in these size classes to be absent from aphotic waters, but the opposite is true. In fact, pigmented cells in the size range of picoplankton and ultraplankton were among the earliest reported microorganisms in the deep sea (Schiller 1931). Such small cells must be transferred to depth via an association with large particles. Alternatively, the presence of photosynthetic cells could be explained by heterotrophic growth at depth, a phenomenon that probably is not important for the chroococcoidal cyanobacteria (Waterbury et al. 1985). Heterotrophic growth is known for phytoplankton cells (e.g. Sloan and Strickland 1966; Ukeles and Rose 1976), and we have observed it for one of the *Chlorella*-like cells (see above), but the long-term sustenance and growth of photosynthetic populations in subeuphotic zones is not yet supported by experimental evidence.

A. Rate of Supply of Photosynthetic Picoplankton to the Aphotic Zone

The rate at which biogenic materials leave the surface ocean and enter the deep sea is presently a subject of active research. The bulk characteristics and quantities of materials leaving the euphotic zone are reasonably well known. More productive systems export more organic matter than do lower productivity ones, and a larger proportion of the production is exported in eutrophic than in oligotrophic ecosystems (Eppley and Peterson 1979; Betzer et al. 1984; and references therein). Explanations for the differential export rates center around trophic structure and suggest that eutrophic systems lose primary producers, dominantly netplankton (Eppley and Peterson 1979; Walsh 1983; Smetacek 1985), or fecal pellets bearing large quantities of intact algae (Staresinic et al. 1983). The low proportional export rates of oligotrophic systems are thought to result from the nearly complete utilization of the dominantly pico- and nanoplanktonic sized producers by the small-sized consumers there; wastes large enough to settle out of the euphotic zone are only produced by the larger, higher level consumers that process a proportionally small amount of material (Eppley and Peterson 1979).

During the VERTEX expedition (Martin et al. 1983) we have made the first measurements of the export rates of picoplankton and ultraplankton from zones of different primary production. (Traps, sites and sampling, and counting methods are described in Gowing and Silver 1983; Martin et al. 1983). The flux of cyanobacteria from the euphotic zone ranges from 10^7 to 10^9 cells $\cdot$ m^{-2} $\cdot$ d^{-1} (determined from phycoerythrin autofluorescence of trap collections of ultraplankton-sized cells), as shown in Fig. 6. Such daily losses represent 0.3–0.9% of the standing crop of these cells (Table 2). For contrast, diatom fluxes range between 10^5 to 10^8

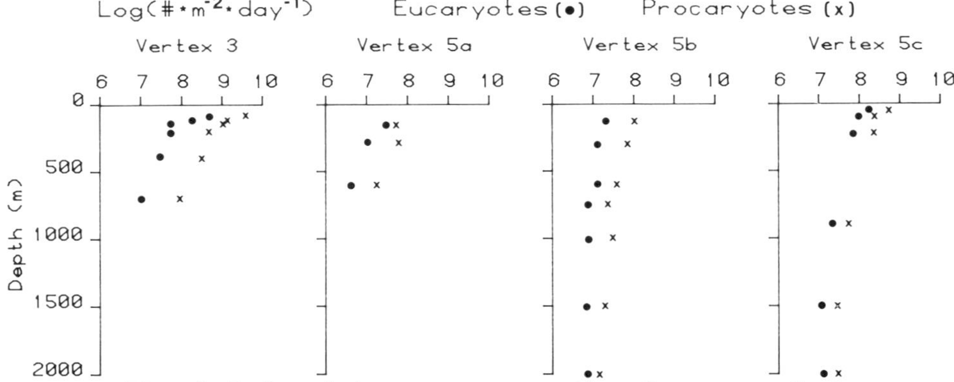

FIG. 6. Flux of cells through the water, as measured by sediment trap collections on VERTEX expeditions. Eucaryotes are mostly cells ≤ 3 μm and procaryotes are likely *Synechococcus*. VERTEX 3 at 15°55′N, 107°12′W was 330 km off central Mexico, VERTEX 5a was 1800 km off central California at 30°N, 139°W; VERTEX 5b was 600 km off central California at 35°N, 129°W, and 5c was about 90 km off central California at 36°N, 123°W.

TABLE 2. Cumulative standing stock of photosynthetic cells, measured by autofluorescence, in the euphotic zone (depth stratum above the 1% light level); number flux out of euphotic zone; % of respective stock lost in the flux; and cumulative primary production in zone (from Knauer et al. 1984).

Photosynthetic cell type	Cumulative stock $(\# \cdot m^{-2})$ in euphotic zone	Flux $(\# \cdot m^{-2} \cdot d^{-1})$ leaving euphotic zone	% Euphotic Zone stock lost daily in the flux	Cumulative Production of all photosynthetic forms in the euphotic zone $(mg\ C \cdot m^2 \cdot d^{-1}$
VERTEX 3[a]				470
Synechococcus	5.8×10^{11}	2.5×10^{9}	0.43	
1–5 μm eucaryotes	1.5×10^{11}	3.2×10^{8}	0.22	
VERTEX 5a				245
Synechococcus	7.0×10^{9}	5.2×10^{7}	0.74	
1–5 μm eucaryotes	2.0×10^{10}	2.3×10^{7}	0.11	
VERTEX 5b				325
Synechococcus	3.2×10^{10}	1.0×10^{8}	0.33	
1–5 μm eucaryotes	5.2×10^{10}	2.1×10^{7}	0.04	
VERTEX 5c				1140
Synechococcus	6.0×10^{10}	5.4×10^{8}	0.90	
1–5 μm eucaryotes	2.2×10^{10}	1.7×10^{8}	0.78	

[a]Cruise and station data given in Fig. 6 legend.

cells $\cdot$ m^{-2} $\cdot$ d^{-1}, a daily loss of a few to 75% of the euphotic zone stocks from a rich, upwelling system (Silver, unpubl. data from the California Current). Fluxes of eucaryotic cells <5 μm, most of which are ≤ 3 μm, possessing chlorophyll a fluorescence associated with chloroplasts, ranged from 10^7 to 10^8 cells $\cdot$ m^{-2} $\cdot$ d^{-1} or 0.1 to 0.8% of their standing crop (Fig. 6, Table 2). Thus the sinking of the small cells constitutes negligible losses of euphotic stocks, but delivers large numbers of cells to the aphotic zone. The highest flux was observed in the most productive environment, but there is no apparent overall correlation in these data between export rates of ultraplankton and primary production.

B. Quantitative Association of Ultraplankton with Deep Sea Detritus

Two types of cell populations are found in the deep sea. One group of cells is present on large, rapidly sinking particles. These particles are rare and thus usually are not sampled in standard sized water bottles; they require particle traps or large volume filtration systems for collection (Menzel 1974; McCave 1975; Bishop et al. 1978) or *in situ* collections from submersibles. To date, few quantitative studies of the photosynthetic picoplankton and ultraplankton have been carried out from these collections. The second group of cells can be considered "suspended"; that is, they are sinking very slowly, at most. They are rather homogeneously distributed, and are readily collected in standard water bottles. These cells are often associated with small detrital particles (our own unpublished observations), and their uniformity suggests that the detritus is relatively homogeneously distributed, small in size, and abundant. The extent of association between the cells and detritus is difficult to measure because of the likelihood that the associations are disrupted during initial sampling, as discussed above, or during the preparation of aliquots for counting.

Cells associated with rapidly sinking detritus — Large numbers of photosynthetic cells occur in association with rapidly sinking detritus throughout the water column. The "flux", or number of photosynthetic cells passing through an area during a period of time (i.e. numbers $\cdot$ m^{-2} $\cdot$ d^{-1}), has been measured using sediment traps at various sites in the North Pacific (Silver and Gowing, unpubl. data). Numbers of settling photosynthetic ultraplankton exceeded 10^7 $\cdot$ m^{-2} $\cdot$ d^{-1} to depths of 2000 m. Cyanobacteria were generally more abundant in the flux than the eucaryotes (Table 2), and most of the cells occurred on amorphous detritus (likely marine snow) rather than in fecal pellets (Table 3). However, the proportion of both cyanobacteria and the photosynthetic eucaryotes found in fecal pellets generally increased from the base of the euphotic zone to the aphotic region (Table 3).

As expected, the numbers of sinking photosynthetic cells decline with depth (Fig. 6); the greatest losses occur in the upper 500 m. This depth stratum is also considered

TABLE 3. Photosynthetic cells associated with rapidly sinking detritus. Values show percentages of autofluorescent cells inside fecal pellets from sediment trap samples. (The remainder of the autofluorescent cells were on marine snow.) "Base euphotic" means flux at about the 1% light depth and "aphotic" is the average for samples below the 1% light depth down to 2000 m (average of 6–7 values).

	Synechococcus % cells in pellets		1–5 μm Eucaryotes % cells in pellets		Ratio # Cells (*Synechococcus*/ 1–5 μm Eucaryotes) in pellets	
Cruise[a]	Base euphotic	Aphotic	Base euphotic	Aphotic	Base euphotic	Aphotic
VERTEX 3	24	31	30	33	8	10
VERTEX 5a	35	—[b]	44	—[b]	0.2	—[b]
VERTEX 5b	8	50	39	50	5	3
VERTEX 5c	15	16	3	10	3	3

[a]Cruise explanation in Fig. 6 legend.
[b]Values not calculated because only 2 depths were sampled.

the zone of regeneration for most organic matter because most organic matter is lost from the flux here (Bishop et al. 1978; Knauer and Martin 1981). Indeed, the losses of cells correspond rather closely with losses of detritus, as shown by the nearly constant concentrations of cells on the sinking detritus (Fig. 7; Silver and Gowing, unpubl. data). However, even when plotted as concentration over depth, a cell decrease still occurs in some cases. These two overlapping loss patterns suggest several removal processes co-occur on detritus as it settles.

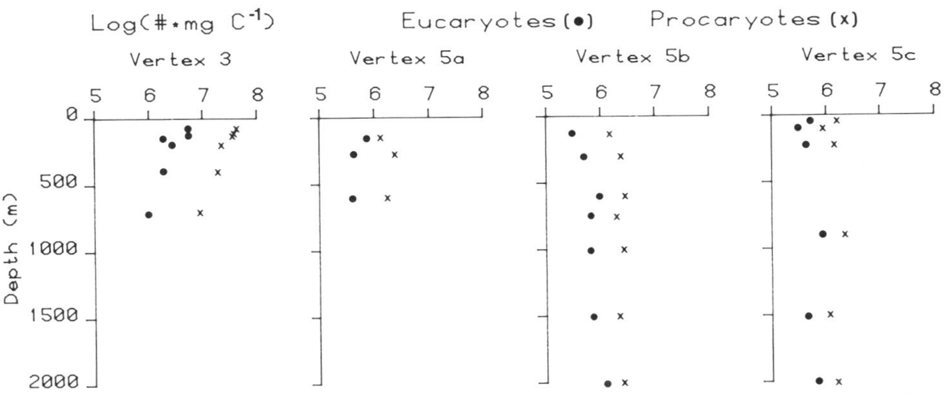

Fig. 7. Cell contents of rapidly sinking detritus in the flux, with materials derived from sediment traps on the VERTEX expeditions. Stations and cells are those described in Fig. 6.

As microorganism-rich particles settle below the euphotic zone, some processes remove large pieces of the detritus and all attendant microorganisms, and — in some cases — other processes selectively remove individual cells from the matrix. The former processes could be the disruption and fragmentation of particles or their wholesale consumption by animals. Disruption would be indicated by increases in the total amount of suspended particulates in the water at selected depths; particles would not be destroyed but rather broken apart. The disrupted fragments then would settle more slowly, resulting in local increases in the abundance of suspended materials. Mid-depth maxima in suspended particles generally have not been observed, indicating that disruption processes do not accelerate at depth (Menzel 1974). Consumption by grazers does occur, with major stocks of metazoans located at the upper mesopelagic zone where the greatest loss of particulate material occurs (Vinogradov 1968; Pearcy et al. 1977; Wiebe et al. 1985). Further evidence for detrital consumption comes from the observation that detritus is "repackaged"; new classes of fecal pellets appear for the first time at depth and contain reworked material similar to detritus found at shallower depths (Honjo 1978; Urrère and Knauer 1981; Gowing and Silver 1985), and guts of deep sea plankton contain detritus (Harding 1974; Gowing and Wishner, unpubl. data).

The processes reducing cell numbers on the detritus that remains in the flux are probably the consumption of cells by accompanying protozoans and the death of cells (or their fluorescence loss) in the inhospitable environment of the deep sea. Study of the microbial communities on sinking detritus indicates that some losses are due to local consumers. Ciliates are abundant on sinking detritus in the deep sea, numbering 10^6 to 10^7 · g C^{-1} to 2000 m (Silver et al. 1984). Microflagellates are also abundant and range from 10^8 to 10^9 · g C^{-1} over the same depths (calculated from data of G. Taylor, Hawaii Institute of Geophysics, University of Hawaii at Manoa, Honolulu, Hawaii, 96822 USA [pers. comm.] for the VERTEX 5 samples). Both the ciliates and the microflagellates probably consume the photosynthetic cells they accompany. The largest biomass of ultraplankton consumers on sinking detritus, however, appears to be sarcodine protozoans, the ubiquitous phaeodarian radiolarians (Gowing 1986). These cryptic protozoans (their phaeodium of wastes makes many of them resemble fecal pellets) contain abundant ultraplankton in their feeding vacuoles and "minipellet" waste bodies (Gowing and Silver 1985), indicating incomplete digestion of some of the photosynthetic ultraplankton and picoplankton.

"Suspended" cells — After cells have been brought to depth by rapidly sinking detritus, they can be transferred into the pool of "suspended" cells by several mechanisms. Some may be scattered by grazers manipulating the loosely associated aggregates: free cells could be released or more slowly sinking fragments could be broken away from the original particles. Alternatively, *in situ* microbial decomposition and chemical dissolution can combine to weaken the association of some detritus, resulting in fragmentation of larger particles into smaller ones and/or releasing the accompanying cells into the water. Bottle samples collect these populations and detrital particles.

Very small, pigmented particles have long been known in the deep sea. Many of these particles are called "olive green cells" (reviewed by Fournier 1970; Silver and Bruland 1981; Silver and Alldredge 1981) because of their distinctive color: otherwise they were generally featureless when viewed with the light microscope. (Distinctive algal cells that could be assigned to known taxa occur occasionally in deep water and were reviewed by Bernard 1961). Although counts of some of the olive green cells were made even on the early expeditions, their concentrations were likely underestimated because of concentration by hand-turned centrifuges (Lohmann 1920; Hentschel 1936). Their affinities have been problematic; most of the small pigmented particles are now considered waste bodies rather than living cells (Silver and Bruland 1981; Gowing and Silver 1985). However, a fraction of the olive green cells are ultrastructurally intact, photoautotrophic ultraplankton (Fournier 1970). The earlier studies indicated maxima of over $10^5 \cdot L^{-1}$ of "pigmented microorganisms" at depths between 200 and 500 m in some samples (e.g. Fournier 1970, 1971; Hentschel 1936). The proportion of these counts that consisted of intact cells, however, was not known. In all cases, filtered samples from water bottle collections in the deep sea contained pigmented "cells" exceeding $10^4 \cdot L^{-1}$.

We (MS and MG) have measured cell numbers of suspended algae by autofluorescence of freshly collected, preserved samples (samples usually <24 h old, obtained during the 4 VERTEX expeditions). We commonly find $>10^4$ autofluorescent, phycoerythrin-containing picoplankton cells per litre at depths to 2000 m (Fig. 8)

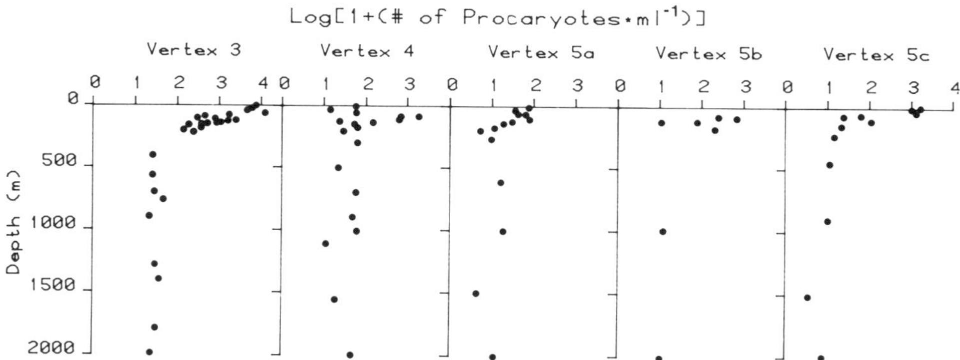

FIG. 8. Concentrations of suspended procaryotic cells (i.e. *Synechococcus*), measured by water bottle collections on the VERTEX expeditions. Station positions described in Fig. 6 legend. Station 4 was 700 km north of Hawaii at 28° 10′N, 154° 59′W.

in the North Pacific. Red fluorescing cells that appear to be eucaryotes, almost all ≤ 3 μm, are slightly less abundant in the aphotic zone ($>5 \times 10^2 \cdot L^{-1}$; Fig. 9). One of us (PD) has obtained counts within these ranges for samples collected at aphotic depths at several sites in the North Atlantic (Bermuda, the Bahamas, and the Gulf

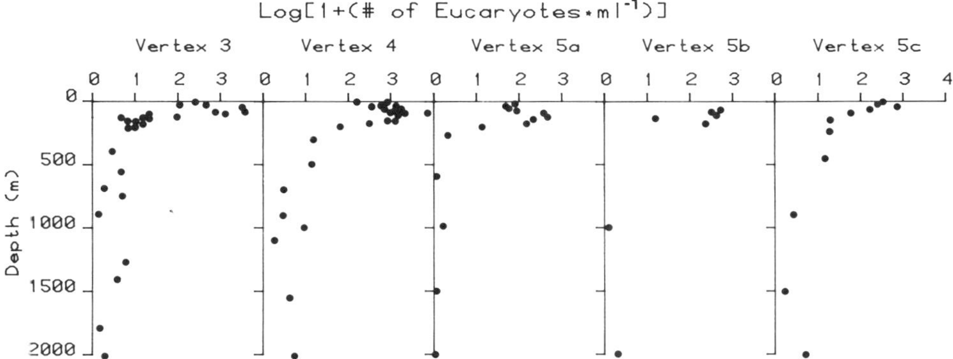

FIG. 9. Concentrations of suspended eukaryotic cells (most cells $\leq 3\ \mu$m), measured by water bottle collections on VERTEX expeditions. Station positions described in Fig. 6 and 8 legends.

of Maine). Although secondary maxima occasionally appeared immediately below the euphotic zone, we usually did not see the types of mesopelagic zone peaks observed by Fournier and Hentschel.

Cells on individual marine snow particles — A few collections of individual specimens of marine snow have been made in the deep sea, allowing the direct comparison of the quantities of associated cells on aggregates with cells in the surrounding water. (Snow can also be collected from the deep sea in sediment traps, but individual particles can not be retrieved or recognized from the mixtures in the traps.) Submersibles (e.g. Silver and Alldredge 1981) or special diving suits (Alldredge et al. 1984) have been used. The origins of some of the particles have been identified by their appearance in the water at the time of collection. Such particles are likely the source of much of the particulate matter captured in sediment traps, but the relationship between trap-collected and submersible collected particles is unknown because sinking rates of single marine snow particles in the deep sea have not yet been measured.

Silver and Alldredge (1981) found high concentrations of picoplankton on aggregates from bathypelagic water off Southern California. One of the aggregates contained a larvacean house. All the particles had concentrations of both cyanobacteria and *Chlorella*-like cells numbering about 10^4 cells $\cdot$ mL^{-1}. (Because counting by autofluorescence was delayed several months after the collection date, these counts are underestimates). If cell concentrations at the time of sampling were similar to those obtained in deep water off Southern California (VERTEX 5C data shown in Fig. 8 and 9), the enrichments of these cells in detritus over that in surrounding waters would be about 10^3. The richness and composition of associated materials on these snow samples indicated a surface origin. If these aggregates are similar to the types of particles obtained in sediment traps, their particle loads suggest that they would sink rapidly.

Marine snow is also produced by organisms at depth, including by larvaceans, pteropods, and other gelatinous zooplankton. One of us (PD) has made collections of newly produced particles of marine snow using the *Johnson Sea Link* submersible in the North Atlantic. In a May 1984 collection off the south side of Bermuda at about 640 m (water depth nearly 1100 m), samples were obtained of occupied larvacean houses and ones likely recently abandoned. Several hundred associated cells of both cyanobacteria and photosynthetic eukaryotes $<5\ \mu$m occurred on an average house; enrichment factors exceeded 2.5×10^2 and 1.7×10^2, respectively, assuming the larvaceans had concentrated water from the depth of capture. Larvacean houses from the Bahamas at about 600 m depth were enriched 1.5×10^2 times in eucaryotes and 44-fold in cyanobacteria.

C. *Taxonomic Composition of Photosynthetic Picoplankton in the Deep Sea*

Picoplankton cells are less diverse at depth than in the euphotic zone. Chroococcoidal cyanobacteria and *Chlorella*-like species dominate in the deep sea; they are abundant inside fecal pellets and common in larger detrital aggregates. Fournier (1970) made the first ultrastructural studies of such cells. He collected water samples to depths of 5000 m in various locations in the North Atlantic, taking precautions to avoid surface contamination. His TEM micrographs of "class 2" type cells — 1.3–2.5 μm in diameter and possessing a chloroplast and a nucleus — appear identical to those we describe as *Chlorella*-like. Other studies, using light microscopic analysis, showed abundant small, non-motile chlorophytes that could be cultured from deep water samples (Amos et al. 1972; Hamilton et al. 1968). Silver and Alldredge (1981) showed that *Chlorella*-like cells averaging 1.6 μm diameter occurred abundantly in detritus collected at depth. Zooplankton fecal pellets collected at depth (Gowing and Silver 1983) and the abundant "minipellets" in detritus and in radiolarians (Gowing and Silver 1985; Gowing 1986) also contain similar *Chlorella*-like cells.

Electron micrographs of cell types we have encountered on deep sea detritus are shown in Fig. 10–13.

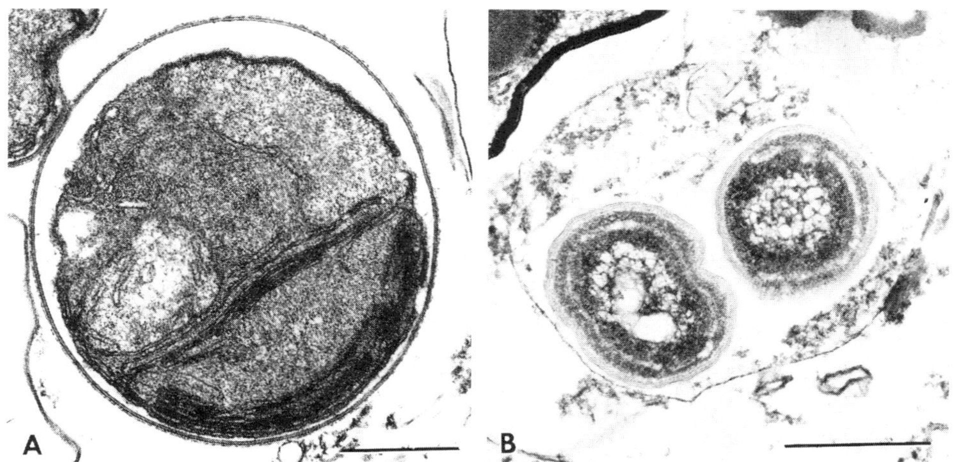

FIG. 10. Photosynthetic ultraplankton from fecal pellets in the aphotic zone. (a) *Chlorella*-like alga from marine snow sample collected by *Alvin* at 1650 m in the San Clemente Basin, off San Diego, CA. (b) chroococcoidal cyanobacteria from a fecal pellet collected in a 900 m sediment trap during VERTEX 2, off central Mexico. Bars = 0.5 μm.

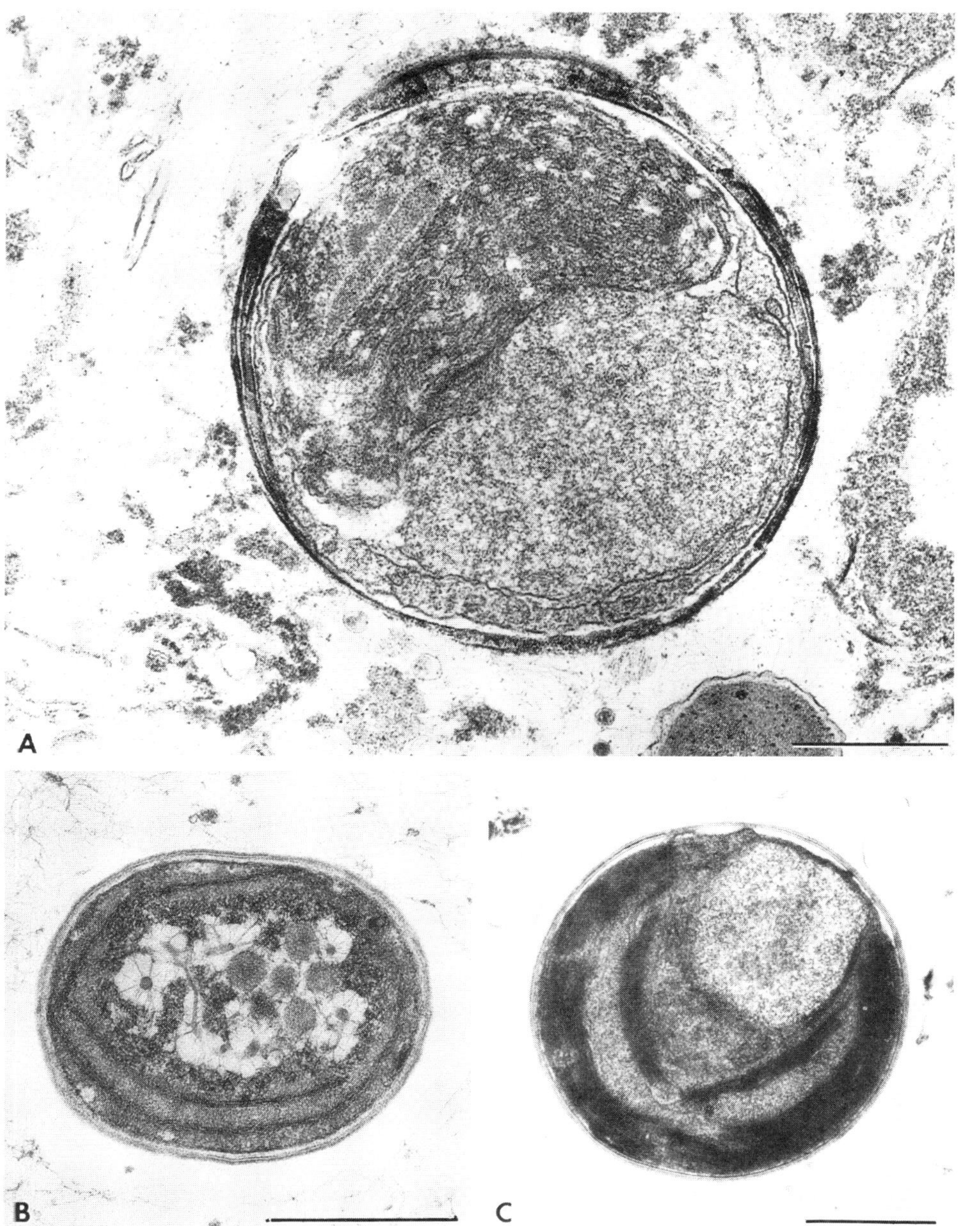

FIG. 11. Photosynthetic ultraplankton from marine snow in the aphotic zone, cont. (a) alga from *Alvin* collected snow sample at 1650 m in the San Clemente Basin, off San Diego, CA. (b) *Chlorella*-like alga from *Alvin* collected snow sample at 1000 m in the East Cortez Basin off San Diego, CA. Bars = 0.5 μm.

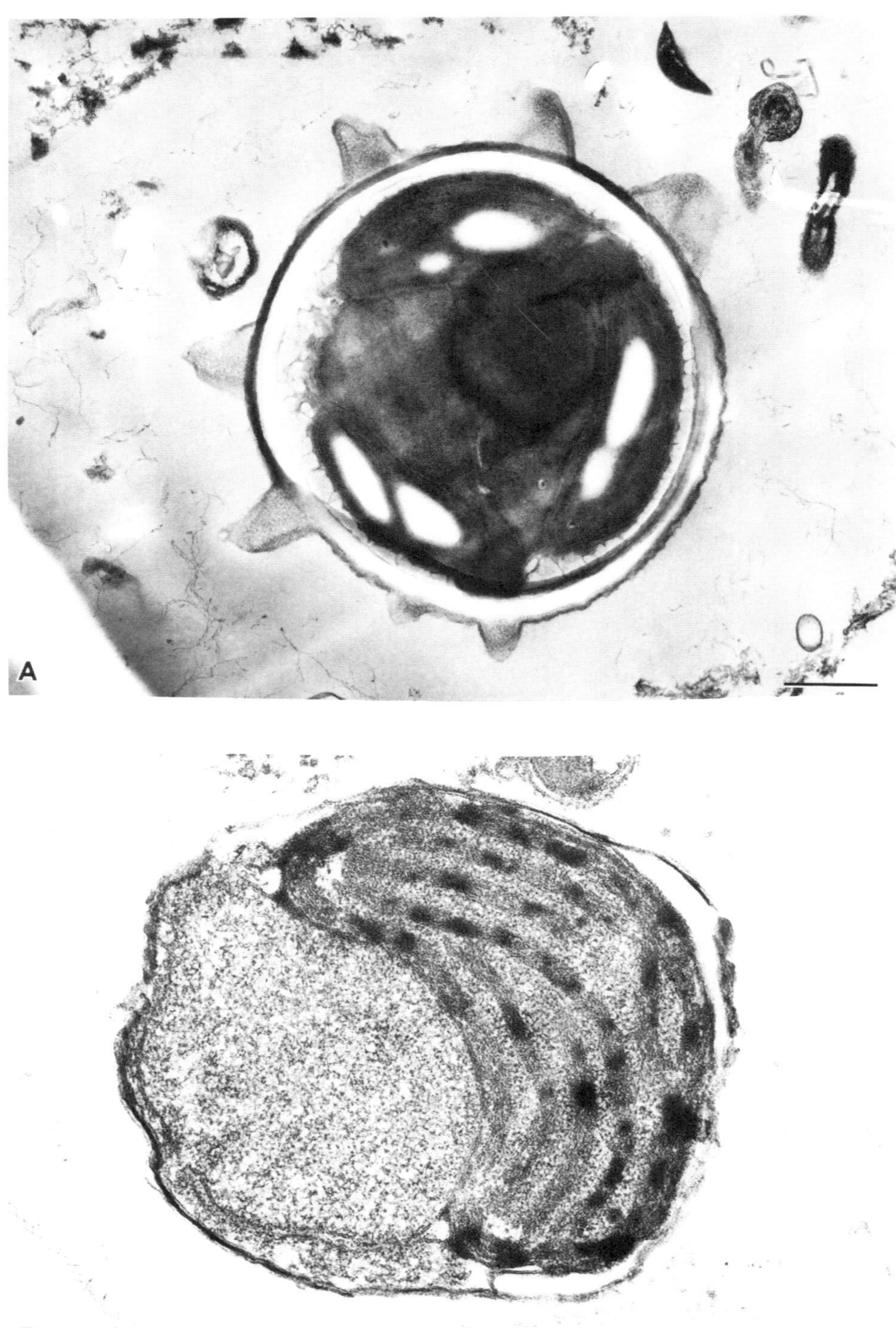

FIG. 12. Photosynthetic ultraplankton from marine snow in the aphotic zone. (a) Heavy-walled alga (cf. Fig. 1a) from sediment trap at 600 m on VERTEX 2, 400 km off central Mexico. (b) Eukaryotic cell from *Alvin* collected sample of marine snow at 1650 m in the San Clemente Basin, off San Diego, CA. Bars = 0.5 μm.

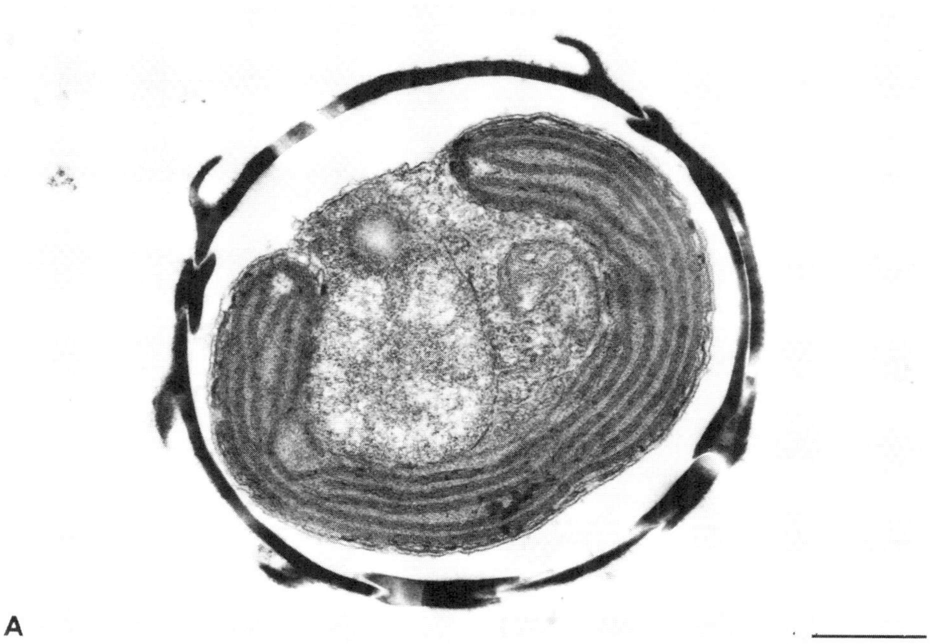

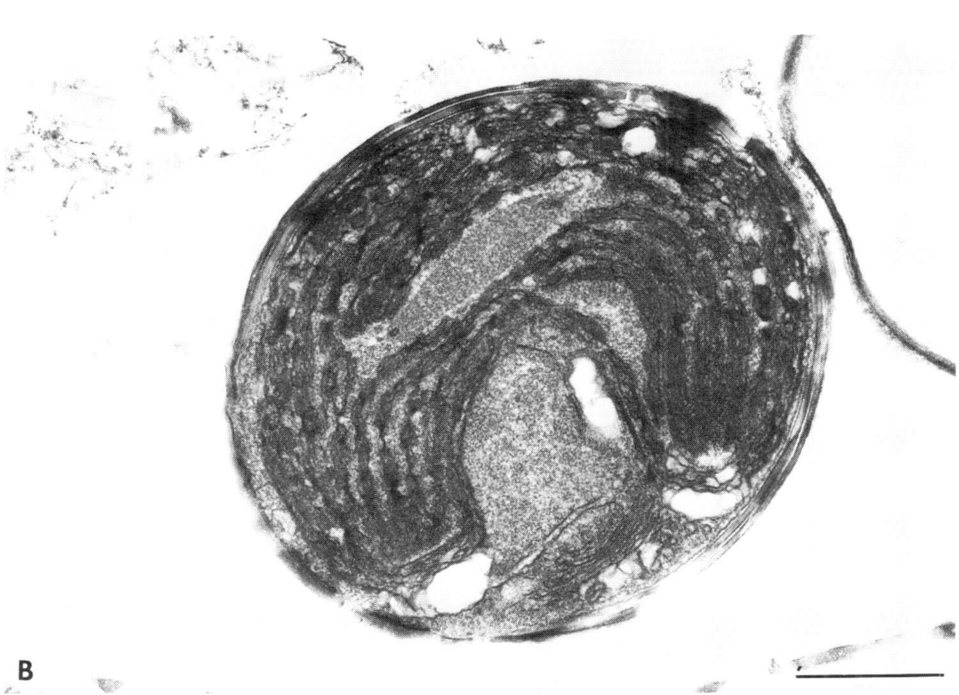

FIG. 13. Photosynthetic ultraplankton from marine snow in the aphotic zone, cont. (a) Walled alga with large, cup shaped chloroplast. Marine snow collected from *Alvin*, 1650 m in the San Clemente Basin, off San Diego, CA. (b) Another alga from the same depth and sample as (a). Bars = 0.5 μm.

A. *Detritus as a Habitat*

Cells associated with detritus are in physically and chemically different environments than cells living freely in the water column. Detrital habitats can be oases for organisms in the water column (Seiburth 1986; Goldman 1984b; Caron et al. 1986). Growth conditions as well as predation rates may differ. We noted above that very high ammonia concentrations occur in marine snow. Compared to background oceanic values for ammonia in the range of $<.2$ μM or coastal values of about 1 μM (Goldman and Glibert 1983), the concentrations found in marine snow can be high: values up to about 0.5 mM have been measured, and these may even be underestimates because of dilution from surrounding water (Shanks and Trent 1979). Such values may approach inhibitory levels, which can occur at 1 mM, for some phytoplankton (Syrett 1962). The presence of chemoautotrophs, possibly nitrifiers, has been hypothesized in snow, based on high RUBPCase activity found in photosynthetically depressed populations (Prezelin and Alldredge 1983). Substantial oxygen depletion has been found in some aggregates of marine snow and inside fecal pellets (A. Alldredge, Marine Science Institute, University of Calif., Santa Barbara, CA. 93106, USA, pers. comm.), suggesting these habitats could be suboxic or even anoxic under certain circumstances. Sharp redox gradients also occur in detrital communities (Paerl 1984), allowing microzone-oriented nutrient processes not possible in fully aerobic environments around the detritus.

These chemical alterations of the environment may favor growth of particular organisms and limit others: the microenvironment of detritus likely is a highly selective one. For photosynthetic organisms, these physiological differences are apparent from the few studies that have simultaneously measured carbon fixation rates and plant pigment concentrations. In two of the three measurements, production rates per unit chlorophyll *a* on marine snow were considerably less than in surrounding water; in the third case they were about the same (Prezelin and Alldredge 1983; Alldredge and Cox 1982).

The detrital habitat harbors dense communities with multiple trophic levels, especially grazers. Various unicells, as discussed above, are usually enriched over populations free in the water. Bacteria are enriched to a moderate degree — 10^1 to 10^2 times in surface marine snow samples — even though their populations are cropped by large numbers of consumers. Concentrations can be even higher for other organisms: photosynthetic picoplankton are enriched 10^3 times on large snow particles in the deep sea (Silver and Alldredge 1981), and ciliates are concentrated 10^4 times on marine snow from surface waters (Silver et al. 1978). The detrital habitat is densely populated compared to the pelagic habitat and the abundant prey (bacteria and microalgae) attract large numbers of protozoan consumers. The survival of many microflagellates in oligotrophic waters may depend on the presence of these enriched microhabitats. Caron et al. (1986) hypothesize that the sparse populations free in the water, in some cases, may be either transients between particles or swarmers in search of new particles. The importance of detritus for protozoans is further suggested by data indicating that minimum numbers of prey must be present for particular types of protozoans to survive. Bactivorous ciliates and microflagellates can not be sustained by the normally dilute prey concentrations in oceanic environments (Caron et al. 1986; Fenchel 1980).

Detritus is a transient habitat. In some cases, such as fecal pellets with large inocula of bacteria, the major growth phase may last only a few days (Pomeroy and Deibel 1980; Pomeroy et al. 1984) because of both grazing pressure and exhaustion of readily available substrates. Large predator–prey oscillations occur in detrital populations (Davoll 1984; Pomeroy et al. 1984), making detritus a site of complex biological interactions. Early occupants of detritus may escape predation and undergo rapid growth for a period, whereas later-developing populations likely experience controls by better developed communities of predators. Finally, the microenvironment itself is destroyed or radically altered by intervening chemincal and physical processes, such as dissolution or mechanical disruption of the matrix, or settlement to an inhospitable new environment.

B. Euphotic Zone Characteristics revealed by Deep Populations

Digestion resistance — Unlike the diatoms, which may export modest to sizable fractions of their standing crop, the ultraplankton stocks in the euphotic zone are not heavily depleted by sinking losses, even in productive environments. If ultra-plankton populations are to remain approximately constant over time and not export major population fractions, then low growth rates or consumption or death are indicated. The growth rates of small photosynthetic cells do not appear particularly low, as discussed elsewhere in this volume. Microflagellates are hypothesized to be major local consumers (e.g. by Azam et al. 1983; Sherr and Sherr 1984; Goldman and Caron 1985). Laboratory studies have shown that some picoplankton can sustain microflagellates (Landry et al. 1984; Johnson and Sieburth 1982; Caron 1984). In contrast, metazoans may not use some of these cells effectively, as evidenced both from laboratory grazing studies (Johnson et al. 1982) and from the sizeable populations of living cyanobacteria and *Chlorella*-like cells seen in field collected pellets (see references above).

Although only an insignificant fraction of the total photosynthetic ultraplankton stock leaves the euphotic zone daily, this constitutes a large number of cells. The loss occurs through the association of cells with sinking fecal pellets and marine snow rather than via individual cell settlement. Picoplankton populations inside intact pellets account for 3–44% of the euphotic zone exports (Table 3); the chroococcoidal cyanobacteria and the *Chlorella*-like chlorophytes are particularly common here. Digestion resistance may provide a competitive advantage under certain circumstances: when eaten by consumers that produce wastes remaining in the euphotic zone (either loosely packaged ones or small ones that sink slowly), ultraplankton may benefit from nutrient inputs associated with gut passage (Porter 1976; Silver and Alldredge 1981). The minimal losses, then, indicate that the major grazers on the digestion-resistant cells produce wastes that remain in surface waters, as would be expected for protozoan consumers.

Dark survival — The total numbers of suspended cells in the deep sea argue for rapid delivery rates, ability to survive in the dark, or some combination of both. Platt et al. (1983) showed the photosynthetic competence of some algae at depth, and the ultrastructural integrity and autofluorescence, discussed above, all indicate the cells are viable. If the flux of cells associated with large particles represents the source of the deep populations, then the life expectancy of the cells in the water at depth can be calculated.

To predict the life expectancy of the photosynthetic ultraplankton in a depth interval within the aphotic zone, we used the observed delivery rate of cells into our sediment traps (data shown in Fig. 6). We assumed that all cells lost in this interval entered the suspended pool (data shown in Fig. 8 and 9), i.e. that grazers did not use the cells, though they might repackage them, and that cell death for settling cells was not important. (If either death is important or substantive losses occur though assimilation by grazers [as we suggest, above] then the life expectancies calculated below are longer by the inverse of the fraction removed.) We then determine the average residency for cells entering the suspended pool. This residence time is the average period that the cells remain in the suspended pool, with their "death" including cell demise, loss of autofluorescence (i.e. the cell can no longer be detected), or consumption and assimilation by a grazer.

For fluxes in one dimension (i.e. simple cell losses with depth, excluding lateral advection),

$$\frac{dN}{dt} = \Gamma_{x_1} - \Gamma_{x_2}$$

where N is the number of cells in the depth interval V, a volume of unit cross-sectional area (1 m^2) and depth Z(m) and Γ_{x_1} and Γ_{x_2} are, respectively, the numbers of cells $\cdot$ m^{-2} $\cdot$ d^{-1} (measured by traps) entering and leaving the depth interval V. In steady state, $\Gamma_{x_1} - \Gamma_{x_2}$ is equal to the rate at which the cells in the pool are "dying". This rate for exponentially dying populations is simply given by

$$\frac{N_p}{\tau} = \Gamma_{x_1} - \Gamma_{x_2}$$

where N_p is the number of suspended cells (concentrations measured by bottle casts) in the depth interval V, and τ is the cell's average life expectancy. One easily calculates τ knowing the entry and exit fluxes and the pool size. In Table 4 we show the results of the calculation for cyanobacteria and eucaryotic cells obtained during the VERTEX expeditions.

Results from this calculation (Table 4) suggest that cells in the suspended pool in the aphotic zone have life expectancies of weeks to years. Errors in our assumptions mostly tend to underestimate residence time. The results for the cyanobacteria are reasonably similar over the various expeditions and not completely unexpected given the hardiness of some forms (J. Waterbury, Woods Hole Oceanographic Institution, Woods Hole, MA. 02543, USA, pers. commun.). The results for the eucaryotes are more variable and, especially for those from the California Current transect (VERTEX 5 samples), indicate life expectancies for the cells of many months. This calculated residence time is due to the relatively low delivery rate for the cells compared to the large suspended pool.

TABLE 4. Residence time, τ, for intact photosynthetic cells in the aphotic zone. See text for method of calculating τ.

Cruise[a]	Depth stratum	Synechococcus τ (days)	1-5 μm Eucaryotes τ (days)
VERTEX 3	120–200 m	64	10.5
	200–700 m	92	58
VERTEX 5a	275–600 m	91	306
5b	150–600 m	106	740
5c	130–750 m	79	139

[a]Cruise explanations given in Fig. 6 legend.

The longer residence time of the eucaryotic cells could result from population fluctuations on the surface, causing variable delivery to depth, and thus violating our steady state assumptions. Alternatively, the results could be the consequence of the slow division of cells at depth. For some of the chlorophytes, as discussed above, heterotrophic growth and division in the dark may occur, especially when the cells are in contact with organic substrates on detritus. (Growth in the dark appears less likely for the cyanobacteria, as discussed above.) However, indefinite maintenance would seem unlikely for individual aphotic zone cells in the probably inhospitable environment of the deep sea.

Traits allowing survival in the deep sea must have been selected in the euphotic zone or waters just below, where survivors could return to favorable growth environments. Thus the adaptive value of the traits should be sought by considering how these features would contribute to the success of populations nearer to the surface. An explanation might be found by considering that life in the surface ocean, especially in oligotrophic environments, may be feast-or-famine for some photosynthetic species. Other evidence for such conditions is provided by the existence of rapid uptake kinetics in phytoplankton from oligotrophic habitats, a feature that allows cells to take advantage of pulsed nutrient availability in largely unfavorable growth environments (McCarthy and Goldman 1979; Goldman 1984b). The photosynthetic ultraplankton that survive in the deep sea may reflect selection for characteristics that allow cell maintenance between these episodic, favorable growth conditions.

Summary and Conclusions

Detritus serves as an alternative habitat to open water for photosynthetic ultraplankton. Detritus is enriched in some growth limiting nutrients, such as ammonia–nitrogen, and likely experiences greater diel oscillations in its biotic characteristics (e.g. pH, O_2 conc.) than does open water, due to the activities of concentrated communities of microorganisms here. The habitat can serve as a temporary refuge from predators, but soon after its formation, protozoan populations increase and exert grazing pressure. For photosynthetic organisms, the detrital habitat in the euphotic zone may favor opportunists or particularly digestion-resistant species. However, the limited date suggest relatively modest ultraplankton populations with limited taxonomic diversity.

The photosynthetic ultraplankton, because of their small size and limited association with rapidly sinking detritus, mostly remain within the euphotic zone. Data from sediment traps indicate <1 % of the euphotic populations, even in highly productive waters, leave this zone. The sinking populations exit both on marine snow and inside fecal pellets and, though they constitute minor losses for the surface stocks, they constitute numerically high fluxes of cells: typically $>10^8$ cells • m^{-2} • d^{-1}. Such rates account for the presence, long known in the literature, of abundant, small, pigmented cells in the deep sea. As these cells settle through the aphotic zone, the highest losses occur within the upper 500 m, but fluxes of photosynthetic ultraplankton still average $>5 \times 10^6$ • m^{-2} • d^{-1} at 2000 m.

The rapidly sinking particles moving through the aphotic zone supply cells to the "suspended" (stationary or more slowly settling) pool of cells. Calculations based on flux and pool size at depth suggest life expectancy in the mesopelagic zone of a minimum of a few months for the cyanobacteria and even longer for some eucaryotic algae. (The latter result may indicate a non-steady state condition during the studies or may reflect a slow heterotrophic growth of some cells, a possibility particularly likely for one of the common coccoidal chlorophytes on detritus.)

The widespread occurrence of digestion resistance in photosynthetic ultraplankton on detritus is emphasized by their presence within fecal pellets of all sizes. Large pellets may contain up to 10^3 or 10^4 intact photosynthetic ultraplankton cells or more; ultraplankton cells are even more concentrated in the abundant "minipellets" (pellets <50 μm) or wastes of larger protozoa and small invertebrates. Between 10–50 % of the photosynthetic sinking cells are commonly inside recognizable pellets to depths of 2000 m. Delivery to depth via animal wastes tends to be more pronounced in oligotrophic than in productive environments, particularly for the eucaryotic cells, emphasizing the close coupling of grazing and production in the low productivity environments.

The presence of large numbers of photosynthetic ultraplankton in the deep sea, in spite of their individually negligible sinking rates, emphasizes several aspects of the biology of the deep sea survivors. First, their resistance to digestion by many types of consumers is evidenced by their presence in a variety of pellet types and abundance in the pellets. Second, they can survive under conditions unfavorable for photosynthetic growth; this may be accomplished by reducing metabolic rates or utilizing organic susbtrates for heterotrophic maintenance. Since the deep-sea populations are continually recruited from overlying euphotic waters, these two traits must have some adaptive value in the euphotic zone. They may reflect strategies particularly important for survival in the consumer dominated, nutrient-poor environment from which they originate.

Acknowledgments

The work was supported by NSF grants OCE 79-19244, IDOE OCE80-03200 and OCE 83-15059. We thank R. Hinegardner for reviewing the manuscript, A. Michaels for help with graphics, R. Franks with help with the sporopollenin analysis, and the Harbor Branch Foundation for providing submersible time to P. Davoll for some of the observations reported here. Many people helped with sample collection and processing; we thank J. Mitchell, D. Hurley, S. Coale, and E. Bay-Schmith.

References

ALLDREDGE, A. L. 1972. Abandoned larvacean houses: A unique food source in the pelagic environment. Science 177: 885–887.

——— 1976. Discarded appendicularian houses as sources of food, surface habitats, and particulate organic matter in the plankton environment. Limnol. Oceanogr. 21: 14–34.

ALLDREDGE, A. L., AND J. L. COX. 1982. Primary productivity and chemical composition of marine snow in surface waters of the Southern California Bight. J. Mar. Res. 40: 517–527.

ALLDREDGE, A. L., B. H. ROBISON, A. FLEMINGER, J. J. TORRES, J. M. KING, AND W. H. HAMNER. 1984. Direct sampling and in situ observation of a persistent copepod aggregation in the mesopelagic zone of the Santa Barbara Basin. Mar. Biol. 80: 75–81.

ALLDREDGE, A. L., J. COLE, AND D. A. CARON. 1986. Production of heterotrophic bacteria inhabiting macroscopic organic aggregates (marine snow) from surface waters. Limnol. Oceanogr. 31: 68–78.

AMOS, A. F., C. GARSIDE, K. C. HAINES, AND O. A. ROELS. 1972. Effects of surface-discharged deep sea mining effluent. Mar. Technol. Soc. J. 6: 40–45.

ATKINSON, A. W., JR., B. E. S. GUNNING, AND P. C. L. JOHN. 1972. Sporopollenin in the cell wall of *Chlorella* and other algae: ultrastructure, chemistry, and incorporation of ^{14}C-acetate, studied in synchronous cultures. Planta 107: 1–32.

AZAM, F., T. FENCHEL, J. G. FIELD, J. S. GRAY, L. A. MEYER-REIL, AND F. THINGSTAD. 1983. The ecological role of water-column microbes in the sea. Mar. Ecol. Prog. Ser. 10: 257–263.

BAYLOR, E. R., AND W. H. SUTCLIFFE. 1963. Dissolved organic matter in seawater as a source of particulate food. Limnol. Oceanogr. 8: 369–371.

BERNARD, F. 1961. Density of flagellates and myxophyceae in the heterotrophic layers related to environment, p. 215–228. *In*, C. H. Openheimer [ed.] Symposium on marine microbiology. Charles C. Thomas, Springfield, IL.

BETZER, P. R., W. J. SHOWERS, E. A. LAWS, C. D. WINN, G. R. DITULLIO, AND P. M. KROOPNICK. 1984. Primary productivity and particle fluxes on a transect of the equator at 153° W in the Pacific Ocean. Deep-Sea Res. 31: 1–11.

BIDDANDA, B. A. 1985. Microbial synthesis of macroparticulate matter. Mar. Ecol. Prog. Ser. 20: 241–251.

BIENFANG, P. K. 1984. Size structure and sedimentation of biogenic microparticulates in a subarctic ecosystem. J. Plank. Res. 6: 985–995.

——— 1985. Size structure and sinking rates of various microparticulate constituents in oligotrophic Hawaiian waters. Mar. Ecol. Prog. Ser. 23: 143–151.

BISHOP, J. K. B., D. R. KETTEN, AND J. M. EDMOND. 1978. The chemistry, biology, and vertical flux of particulate matter from the upper 400 m of the Cape Basin in the southeast Atlantic Ocean. Deep-Sea Res. 24: 1121–1161.

BROOKS, J., AND G. SHAW. 1971. Recent developments in the chemistry, biochemistry, geochemistry and post-tetrad ontogeny of sporopollenins derived from pollen and spore exines, p. 99–114. *In* J. Heslop-Harrison [ed.] Pollen: development and physiology. Butterworths, London.

BRULAND, K. W., AND M. W. SILVER. 1981. Sinking rates of fecal pellets from gelatinous zooplankton (salps, pteropods, doliolids). Mar. Biol. 63: 295–300.

CARON, D. A. 1984. The role of heterotrophic microflagellates in plankton communities. Ph.D. thesis, Woods Hole Oceanogr. Instit., WHOI Rep. No. 84–35: 268 p.

CARON, D. A., P. G. DAVIS, K. M. JOHNSON, AND J. MCN. SIEBURTH. 1982. Heterotrophic bacteria and bactivorous protozoa in oceanic macroaggregates. Science 218: 795–797.

CARON, D. A., P. G. DAVIS, L. P. MADIN, AND J. MCN. SIEBURTH. 1986. Enrichment of microbial populations in macroaggregates (marine snow) from surface waters of the North Atlantic. J. Mar. Res. 44: 543–565.

CARSON, R. 1951. The sea around us. Oxford Univ. Press, Inc., New York, NY.

DAVOLL, P. J. 1984. Microcosms in the pelagic zone: a study of the microbial community on larvacean house aggregates from Monterey Bay. PhD. thesis, Univ. Calif., Santa Cruz, 113 p.

EPP, R. W., AND W. M. LEWIS. 1982. Photosynthesis in copepods. Science 214: 1349–1350.

EPPLEY, R. W., AND B. J. PETERSON. 1979. Particulate organic matter flux and planktonic new production in the deep ocean. Nature 282: 677–680.

FENCHEL, T. 1980. Suspension feeding in ciliated protozoa: feeding rates and their ecological significance. Microb. Ecol. 6: 13–25.

FENCHEL, T. M., AND B. B. JORGENSEN. 1977. Detritus food chains of aquatic ecosystems: the role of bacteria. Adv. Microb. Ecol. 1: 1–58.

FOSS, P., R. R. L. GUILLARD, AND S. LIAAEN-JENSEN. 1984. Prasinoxanthin — a chemosystematic marker for algae. Phytochem. 23: 1629–1633.

FOURNIER, R. O. 1970. Studies on pigmented microorganisms from aphotic marine environments. Limnol. Oceanogr. 15: 675–682.

——— 1971. Studies on pigmented microorganisms from aphotic marine environments. II. North Atlantic distribution. Limnol. Oceanogr. 16: 952–961.

FOWLER, S. W., AND N. S. FISHER. 1983. Viability of marine phytoplankton in zooplankton fecal pellets. Deep-Sea Res. 30: 963–969.

GILMER, R. W. 1972. Free-floating mucus webs: a novel feeding adaptation for the open ocean. Science 176: 1239–1240.

——— 1974. Some aspects of feeding in thecosomatous pteropod molluscs. J. Exp. Mar. Biol. Ecol. 15: 127–144.

GILMER, R. W., AND G. R. HARBISON. 1986. Field behavior and morphology of pteropod molluscs: feeding methods in the families Cavoliniidae and Peraclididae (Gastropoda, Thecosomata). Mar. Biol. 91: 47–57.

GOLDMAN, J. C. 1984a. Conceptual role for microaggregates in pelagic waters. Bull. Mar. Sci. 35: 462–476.

——— 1984b. Oceanic nutrient cycles, p. 137–170. In M. J. Fasham [ed.] Flows of energy and materials in marine ecosystems: theory and practice. Plenum, London.

GOLDMAN, J. C., AND P. M. GLIBERT. 1983. Kinetics of inorganic nitrogen uptake by phytoplankton, p. 233–274. In E. J. Carpenter and D. G. Capone, Nitrogen in the marine environment. Academic Press, New York, NY.

GOLDMAN, J. C., AND D. A. CARON. 1985. Experimental studies on a omnivorous microflagellate: implications for grazing and nutrient regeneration in the marine microbial food chain. Deep-Sea Res. 32: 899–915.

GORDON, D. C. 1970. A microscopic study of organic particles in the North Atlantic Ocean. Deep-Sea Res. 17: 175–185.

GOWING, M. M., AND M. W. SILVER. 1983. Origins and microenvironments of bacteria mediating fecal pellet decomposition in the sea. Mar. Biol. 73: 7–16.

GOWING, M. M., AND M. W. SILVER. 1985. Minipellets: a new and abundant size class of marine fecal pellets. J. Mar. Res. 43: 395–418.

GOWING, M. M., 1986. Trophic biology of phaeodarian radiolarians and flux of living radiolarians in the upper 2000 m of the north Pacific central gyre. Deep-Sea Res. 33: 655–674.

HAMILTON, R. D., O. HOLM-HANSEN, AND J. D. H. STRICKLAND. 1968. Notes on the occurrence of living microscopic organisms in deep water. Deep-Sea Res. 15: 651–656.

HAMNER, W. M., L. P. MADIN, A. L. ALLDREDGE, R. W. GILMER, AND P. P. HAMNER. 1975. Underwater observations of gelatinous zooplankton: sampling problems, feeding biology, and behavior. Limnol. Oceanogr. 20: 907–917.

HARDING, G. C. H. 1974. The food of deep sea copepods. J. Mar. Biol. Assoc. U.K. 54: 141–155.

HENTSCHEL, E. 1936. Allgemeine Biologie des Sudatlantischen Ozeans. Deutsch. Atl. Exped. *Meteor* 1925–1927. Wiss. Ergeb. 11: 1–344.

HONJO, S. 1978. Sedimentation of materials in the Sargasso Sea at a 6,367 m deep station. J. Mar. Res. 36: 469–492.

JACOBSEN, T. R., AND F. AZAM. 1984. Role of bacteria in copepod fecal pellet decomposition: colonization, growth rates and mineralization. Bull. Mar. Sci. 35: 495–502.

JANNASCH, H. W. 1973. Bacterial content of particulate matter in offshore surface waters. Limnol. Oceanogr. 18: 340–342.

JOHANNES, R. E. 1967. Ecology of organic aggregates in the vicinity of coral reefs. Limnol. Oceanogr. 12: 189–195.

JOHNSON, P. W., AND J. McN. SIEBURTH. 1979. Chroococcoid cyanobacteria in the sea: a ubiquitous and diverse phototrophic biomass. Limnol. Oceanogr. 24: 928–935.

——— 1982. In situ morphology and occurrence of eucaryotic phototrophs of bacterial size in the picoplankton of estuarine and oceanic waters. J. Phycol 18: 318–327.

JOHNSON, P. W., H. S. XU, J. McN SIEBURTH. 1982. The utilization of chroococcoid cyanobacteria by marine protozooplankters but not by calanoid copepods. Ann. Inst. Oceanogr., Paris 58S: 297–308.

JOINT, I. R., AND R. K. PIPE. 1984. An electron microscope study of a natural population of picoplankton from the Celtic Sea. Mar. Ecol. Prog. Ser. 20: 113–118.

KANE, J. E. 1967. Organic aggregates in surface waters of the Ligurian Sea. Limnol. Oceanogr. 12: 287–294.

KNAUER, G. W., AND J. H. MARTIN. 1981. Primary production and carbon-nitrogen fluxes in the upper 1500 m of the northeast Pacific. Limnol. Oceanogr. 26: 181–182.

KNAUER, G. W., D. HEBEL, AND F. CIPRIANO. 1982. Marine snow: major site of primary production in coastal waters. Nature 300: 630–631.

KNAUER, G. W., J. H. MARTIN, AND D. M. KARL. 1984. The flux of particulate organic matter out of the euphotic zone, p. 136–150. In Global Ocean Flux Study, Proc. Workshop, 1984, Woods Hole, Mass. National Academy Press, Wash. DC.

Konig, J., and E. Peveling. 1984. Cell walls of the phycobionts *Trebouxia* and *Pseudotrebouxia*: constituents and their localization. Lichenologist 16: 129–144.

Landry, M. R., L. W. Haas, and V. L. Fagerness. 1984. Dynamics of microbial plankton communities: experiments in Kaneohe Bay, Hawaii. Mar. Ecol. Prog. Ser. 16: 127–133.

Linley, E. A. S., and R. C. Newell. 1984. Estimates of bacterial growth yields based on plant detritus. Bull. Mar. Sci. 35: 409–425.

Lohmann, H. 1909. Die Gehause and Gallertblasen der Appendicularian und ihre Bedeutung fur die Erforschung des Lebens im Meer. Verh. Dtsch. Zool. Ges. 19: 200–239.

———. 1920. Die Bevolkerung des Ozean mit Plankton wahrend des Ausreise des Deutschland 1911. Arch. f. Biontologie 4: 1–600.

Martin, J., G. A. Knauer, W. W. Broenkow, K. W. Bruland, D. M. Karl, L. F. Small, M. W. Silver, and M. M. Gowing. 1983. Vertical transport and exchange of materials in the upper waters of the oceans (VERTEX): introduction to the program, hydrographic conditions and major component fluxes during VERTEX I. Moss Landing Marine Lab. Tech. Publ. 83-2: 40 p.

McCave, I. N. 1975. Vertical flux of particles in the ocean. Deep-Sea Res. 22: 491–502.

———. 1984. Size spectra and aggregation of suspended particles in the deep ocean. Deep-Sea Res. 31: 329–352.

McCarthy, J. J., and J. C. Goldman. 1979. Nitrogenous nutrition of marine phytoplankton in nutrient-depleted waters. Science 203: 670–672.

Menzel, D. W. 1974. Primary productivity, dissolved and particulate organic matter, and the sites of oxidation of organic matter, p. 715–798. *In* E. D. Goldberg [ed.] The sea, Vol. 5. John Wiley & Sons, New York, NY.

Mitchell, J. G., A. Okubo, and J. A. Fuhrman. 1985. Microzones surrounding phytoplankton from the basis for a stratified marine microbial ecosystem. Nature 316: 58–59.

Murphy, L. S., and E. M. Haugen. 1985. The distribution and abundance of phototrophic ultraplankton in the North Atlantic. Limnol. Oceanogr. 30: 47 -58.

Nishizawa, S., M. Fukuda, and N. Inoue. 1954. Photographic study of suspended matter and plankton in the sea. Bull. Fac. Fish. Hokkaido Univ. 5: 36–40.

Paerl, H. W. 1984. Alteration of microbial metabolic activities in association with detritus. Bull. Mar. Sci. 35: 393–408.

Pearcy, W. G., E. E. Krygier, R. Mesecar, and F. Ramsey. 1977. Vertical distribution and migration of oceanic micronekton off Oregon. Deep-Sea Res. 24: 223–245.

Platt, T., D. V. Subba Rao, J. C. Smith, W. K. Li, B. Irwin, E. P. W. Horne, and D. D. Sameoto. 1983. Photosynthetically-competent phytoplankton from the aphotic zone of the deep ocean. Mar. Ecol. Prog. Ser. 10: 105–110.

Pomeroy, L. R., and R. E. Johannes. 1968. Occurrence and respiration of ultraplankton in the upper 500 m of the ocean. Deep-Sea Res. 15: 381–391.

Pomeroy, L. R., and D. Deibel. 1980. Aggregation of organic matter by pelagic tunicates. Limnol. Oceanogr. 25: 643–652.

Pomeroy, L. R., R. B. Hanson, P. A. McGillivary, B. F. Sherr, D. Kirchman, and D. Deibel. 1984. Microbiology and chemistry of fecal products of pelagic tunicates: rates and fates. Bull. Mar. Sci. 35: 426–439.

Porter, K. G. 1973. Selective grazing and differential digestion of algae by zooplankton. Nature 244: 179–180.

———. 1976. Enhancement of algal growth and productivity by grazing zooplankton. Science 192: 1332–1334.

———. 1977. The plant–animal interface in freshwater ecosystems. Am. Sci. 65: 159–170.

Prezelin, B. B., and A. L. Alldredge. 1983. Primary production of marine snow during and after an upwelling event. Limnol. Oceanogr. 28: 1156–1167.

Riley, G. A. 1963. Organic aggregates in seawater and the dynamics of their formation and utilization. Limnol. Oceanogr. 8: 372–381.

———. 1970. Particulate organic matter in sea water. Adv. Mar. Biol. 8: 1–118.

Riley, G. A., P. J. Wangersky, and D. VanHemert. 1964. Organic aggregates in tropical and subtropical surface waters of the North Atlantic Ocean. Limnol. Oceanogr. 9: 546–550.

Riley, G. A., D. VanHemert, and P. J. Wangersky. 1965. Organic aggregates in surface and deep waters of the Sargasso Sea. Limnol. Oceanogr. 10: 354–363.

Schiller, J. 1931. Uber autochthone pflanzliche Organismen in der Tiefsee. Biol. Zbl. 51: 329–344.

Shanks, A. L., and J. D. Trent. 1979. Marine snow: microscale nutrient patches. Limnol. Oceanogr. 24: 850–854.

Sheldon, R. W., A. Prakash, and W. H. Sutcliffe, Jr. 1972. The size distribution of particles in the ocean. Limnol. Oceanogr. 17: 327–340.

Sherr, B. F., and E. B. Sherr. 1984. Role of heterotrophic protozoa in carbon and energy flow in aquatic ecosystems, p. 412–423. *In* M. J. Klug and C. A. Reddy [ed.] Current perspectives in microbial ecology. Amer. Soc. Microbiol. Proc. 3rd Intl. Symp. Microb. Ecol.

Sieburth, J. McN. 1979. Sea microbes. Oxford Univ. Press, NY, 491 p.

———. 1986. Dominant microorganisms of the upper ocean: form and function, spatial distribution and photoregulation of biochemical processes. *In* J. D. Burton, P. G. Brewer and R. P. Chesselet [ed.] NATO Advanced Research Institute on Dynamic Processes in the Chemistry of the Upper Ocean. NATO Conference Series IV: Marine Sciences. Plenum Press, New York, NY. (in press).

Silver, M. W., A. W. Shanks, and J. D. Trent. 1978. Marine snow: microplankton habitat and source of small-scale patchiness in pelagic

340

populations. Science 21: 371–373.

SILVER, M. W., AND A. L. ALLDREDGE. 1981. Bathypelagic marine snow: deep sea algal and detrital community. J. Mar. Res. 39: 501–530.

SILVER, M. W., AND K. W. BRULAND. 1981. Differential feeding and fecal pellet composition of salps and pteropods, and the possible origin of the deep-water flora and olive-green "cells". Mar. Biol. 62: 263–273.

SILVER, M. W., M. M. GOWING, D. C. BROWNLEE, AND J. O. CORLISS. 1984. Ciliated protozoa associated with sinking oceanic detritus. Nature 309: 246–248.

SLOAN, P. R., AND J. D. H. STRICKLAND. 1966. Heterotrophy of four marine phytoplankters at low substrate concentrations. J. Phycol. 2: 29–32.

SMAYDA, T. J. 1970. The suspension and sinking of primary producers in the sea. Oceanogr. Mar. Biol. Ann. Rev. 8: 353–414.

SMETACEK, V. 1985. Role of sinking in diatom life-history cycles: ecological, evolutionary and geological significance. Mar. Biol. 84: 239–251.

STARESINIC, N., J. FARRINGTON, R. B. GAGOSIAN, C. H. CLIFFORD, AND E. M. HURLBURT. 1983. Downward transport of particulate matter in the Peru coastal upwelling: role of the anchoveta *E. ringens*. p. 225–240. *In* E. Seuss and J. Thiede [ed.] Coastal upwelling. Plenum Pub. Co., New York, NY.

SUZUKI, N., AND K. KATO. 1953. Studies on suspended materials marine snow in the sea. PART I. Sources of marine snow. Bull. Fac. Fish. Hokkaido Univ., 4: 132–135.

SYRETT, P. J. 1962. Nitrogen assimilation, p. 171–188. In R. A. Lewin [ed.] Physiology and biochemistry of algae. Academic Press, New York, NY.

TAKASHI, M., AND T. HORI. 1984. Abundance of picophytoplankton in the subsurface chlorophyll maximum layer in subtropical and tropical waters. Mar. Biol. 79: 177–186.

TAKASHI, M., AND P. K. BIENFANG. 1983. Size structure of phytoplankton biomass and photosynthesis in subtropical Hawaiian waters. Mar. Biol. 76: 213–218.

THINH, L.-V., AND D. J. GRIFFITHS. 1985. A small marine *Chlorella* from the waters of a coral reef. Bot. Mar. 28: 41–46.

TRENT, J. D., A. L. SHANKS, AND M. W. SILVER. 1978. In situ and laboratory measurements on macroscopic aggregates in Monterey Bay, California. Limnol. Oceanogr. 23: 626–636.

TSUJITA, T. 1952. A preliminary study on naturally occurring suspended organic matter in water adjacent to Japan. J. Oceanogr. Soc. Japan 8: 113–126.

TURNER, J. T. 1984. Zooplankton feeding ecology: contents of fecal pellets of the copepods *Acartia tonsa* and *Labidocera aestiva* from continental shelf waters near the mouth of the Mississippi River. P.S. Z. N. I: Mar. Ecol. 5: 265–282.

UKELES, R., AND W. E. ROSE. 1976. Observations on organic carbon utilization by photosynthetic marine microalgae. Mar. Biol. 37: 11–28.

URRÈRE, M. A., AND G. A. KNAUER. 1981. Zooplankton fecal pellet fluxes and vertical transport of particulate organic material in the pelagic environment. J. Plank. Res. 3: 369–387.

VINOGRADOV, M. E. 1968. Vertical distribution of the oceanic plankton. Nauka, Moscow (Translation: Israel Program for Scientific Translation, Jerusalem, 1970, 339 p.)

WALSH, J. J. 1983. Death in the sea: enigmatic phytoplankton loss. Prog. Oceanogr. 12: 1–86.

WATERBURY, J. B., J. M. WILLEY, D. G. FRANKS, F. W. VALOIS, AND S. W. WATSON. 1985. A cyanobacterium capable of swimming motility. Science 230: 74–76.

WIEBE, P. H., G. R. FLIERL, C. DAVIS, V. BARBER, AND S. H. BOYD. 1985. Macrozooplankton biomass in Gulf Stream warm-core rings: spatial distribution and temporal changes. J. Geophys. Res. 90(C5): 8885–8901.

WIEBE, W. J., AND L. R. POMEROY. 1972. Microorganisms and their association with aggregates and detritus in the sea: a microscopic study. Mem. 1ˢᵗ Ital. Idrobiol. 29 Suppl.: 325–352.

YOUNGBLUTH, M. J. 1982. Utilization of a fecal mass as food by the pelagic mysis larva of the Penaeid shrimp *Solenocera atlantidis*. Mar. Biol. 66: 47–51.

The Individual Cell in Phytoplankton Ecology: Cell Cycles and Applications of Flow Cytometry

SALLIE W. CHISHOLM

48–425 Ralph M. Parsons Laboratory,
Massachusetts Institute of Technology, Crambridge, MA. 02139, USA

E. VIRGINIA ARMBRUST AND ROBERT J. OLSON

Biology Department, Woods Hole Oceanographic Institution, Woods Hole, MA. 02543, USA

Introduction

Populations of cells are composed of unique individuals. At any instant in time in an asynchronous population, each cell has a unique character which is a function of the time that has elapsed since it was "born" … i.e. its position in its cell cycle. This heterogeneity is reflected in the steady state population structure, which can be defined in terms of the distribution of cell age, volume, DNA, and protein, for example. Even two daughter cells born at exactly the same time will not be at the same position in their cell cycle some unit of time later, and they will not complete their cell cycle simultaneously. This is due to the intrinsic variability in all biological systems, and as a result of this variability, populations of cells can never be perfectly synchronized.

In a fluctuating environment such as the sea, the population structure reflects not only the intrinsic variability between the cells, but also variability resulting from their individual past histories. The critical components of the environment of autotrophic phytoplankton are light, nutrients and predators, all of which vary in time and space. The cell experiences this variability in time, but the population as a whole experiences it in both time and space. It is the fabric of this complex matrix that dictates the probability that a given cell will see light or nutrients, or be eaten at a particular point in its cell cycle; it is the integrated past history of each cell which dictates its rate of cell cycle progression and its generation time. This, along with the intrinsic variability, defines the population structure at any instant and ultimately determines the population growth rate.

The proximate goal of this paper is to review what is known about the regulation of cell cycle progression in prokaryotic, autotrophic picoplankton, and to contrast their behavior with the more well understood eukaryotic phytoplankton. The ultimate goal is to convince the reader that we cannot understand the dynamics of marine plankton populations (and ultimately communities) unless we understand the "lives" of the individual cells; that is, unless we understand the intrinsic variability between cells and the extrinsic factors that start and stop them as they progress through their cell cycles. The relatively recent introduction of flow cytometric techniques to the study of phytoplankton populations has buttressed our ability to maintain this perspective, and these techniques have played a critical role in molding the evidence presented in this paper.

FLOW CYTOMETRY

Although flow cytometry has been used in biomedical research for over a decade, its application to the study of marine organisms is quite recent (Trask et al. 1982; Paau et al. 1978, 1979; Yentsch et al. 1983a,b; Olson et al. 1983; Olson et al. 1985; Li 1986). Detailed reviews of this technology can be found in Melamed et al (1979),

Shapiro (1983), and Visser and Van Den Engh (1982), and most extensively in Shapiro (1985); only a brief description is necessary here.

The essence of a flow cytometer is a flow cell, which consists of a capillary containing a flowing sheath fluid (Fig. 1). When a sample is injected into the center

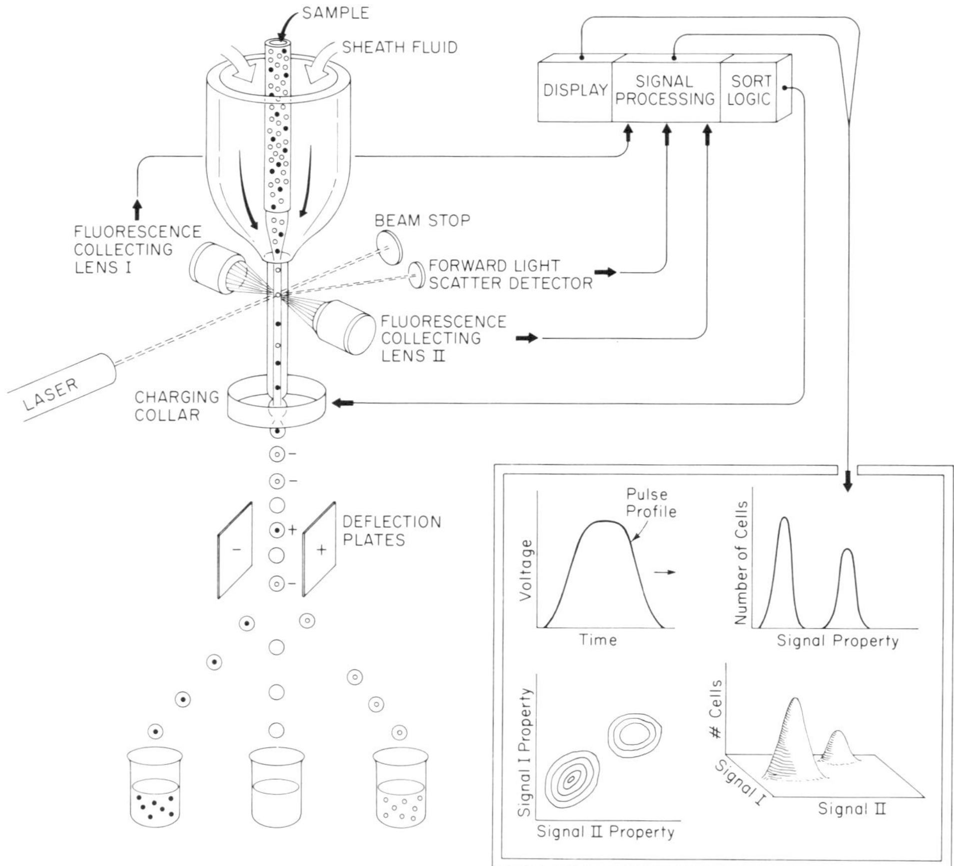

FIG. 1. Diagram of a flow cytometer/cell sorter. The sample is injected into sheath fluid at a rate that permits analysis of properties of individual cells as they pass through a laser beam. The pulse profiles of fluorescence emission and light scatter signals of each cell are processed and displayed as one-dimensional histograms, bivariate two-dimensional contour plots, or three-dimensional plots. Based on a predesignated sort logic, particles of interest can be physically sorted from the emerging droplet stream by electrostatic deflection (see text for details).

of the flowing fluid, the cells are diluted and fluid focussed such that they are carried through the capillary single file in a thin optical plane. A laser beam or focussed light source is focussed on the capillary and each particle is illuminated as it passes through the beam. Light emitted as autofluorescence (e.g. from photosynthetic pigments) or induced fluorescence from fluorescent dyes (e.g. DNA-specific stains) can be collected quantitatively from each particle. Forward angle and perpendicular light scatter, which yield information about particle size and surface properties, can also be measured. In a typical system, three signals can be measured simultaneously on each cell at a rate of about 50 000 cells per minute.

In addition to measuring properties of cells, many flow cytometers have the capability of physically sorting particles according to pre-determined optical properties. This is accomplished by imposing a high frequency vibration in the flow cell, creating

344

droplets as the sample stream emerges; the sample concentration is adjusted such that each droplet contains no more than one particle. When a particle of interest passes through the laser beam, the sample stream is charged, and the resulting charged droplet is deflected either to the right or to the left of the main sample stream, according to pre-set fluorescence or light scattering characteristics (Fig. 1). This sorting capability is essential for the interpretation and verification of flow cytometric signatures and invaluable for the isolation of cells or populations from natural plankton communities.

The introduction of flow cytometry to biological oceanography has expanded the horizons of plankton research. As will be revealed in the following sections, this instrument enables us to measure the relative concentration of cellular constituents such as DNA, in individual cells, and thus identify the positions of cells in their cycle. We can then relate the rate of progression through the cycle to the environment the cell has experienced. We can also use the unique fluorescent pigments of phytoplankton to detect specific types of cells *in situ*, and measure the distribution of properties among the cells, such as their pigment content and ability to scatter light. This gives us an entirely new perspective on the plankton community. For the first time we can measure characteristics of individuals, and then choose the appropriate boundaries for calculating average properties of relevant subsets of the community. Until now, we have been forced to look almost exclusively at average properties of entire communities, which at a minimum sacrifices information, and at a maximum distorts the true nature of the relationships between organisms and their environment.

The Cell Cycle

General Characteristics

There are certain properties of the "generic" eukaryotic and prokaryotic cell cycle which are generally accepted as typical, thus providing a framework for the discussions in the following sections. A more detailed treatment of the cell cycle can be found in John (1981), Lloyd et al. (1982), and Nurse and Streiblova (1984).

For both eukaryotic and prokaryotic cells, the cell cycle can be defined by two key events: the interval of DNA synthesis and the event of cell division. The latter marks the beginning of the cycle for the daughter cells and the end of the cycle (and the individual) for the mother cell. In eukaryotic cells, the intervals that precede and follow DNA synthesis (the S phase) are designated as "gaps" (G_1 and G_2 respectively), and mitosis (M) occurs between G_2 and cell division. The DNA content of the cell doubles during S phase, and at no time does it exceed twice the content of a G_1 cell. In general, the lengths of the cell cycle stages S, G_2, and M are independent of the cell generation time (i.e. of environmental conditions); it is the G_1 phase of the cell cycle that expands as the generation time of the cell is lengthened by growth limiting conditions (Fig. 2). This expansion is hypothesized to be a reflection of restriction or transition points located in G_1 where a cell is arrested in its cell cycle until it is supplied with a critical amount of a required substance, or has reached a critical size (Pardee 1974; Prescott 1976; Fantes 1977).

Prokaryotic cells, here modelled after *E. coli*, have a single circular chromosome and the chromosome replication period (C) is the time required for a single DNA replication fork to travel around the chromosome (Fig. 3). Cell division is initiated when the cell reaches a critical volume (twice the minimum cell volume), and requires a constant interval (D) to complete. In cells which are growing rapidly, such that their generation time is less than the sum of the intervals C + D, multiple replication forks are found on the chromosome. Under these circumstances DNA synthesis occurs throughout the cell cycle, and the DNA content per cell can be much greater than twice that of the single chromosome. In other words, the DNA replication time is

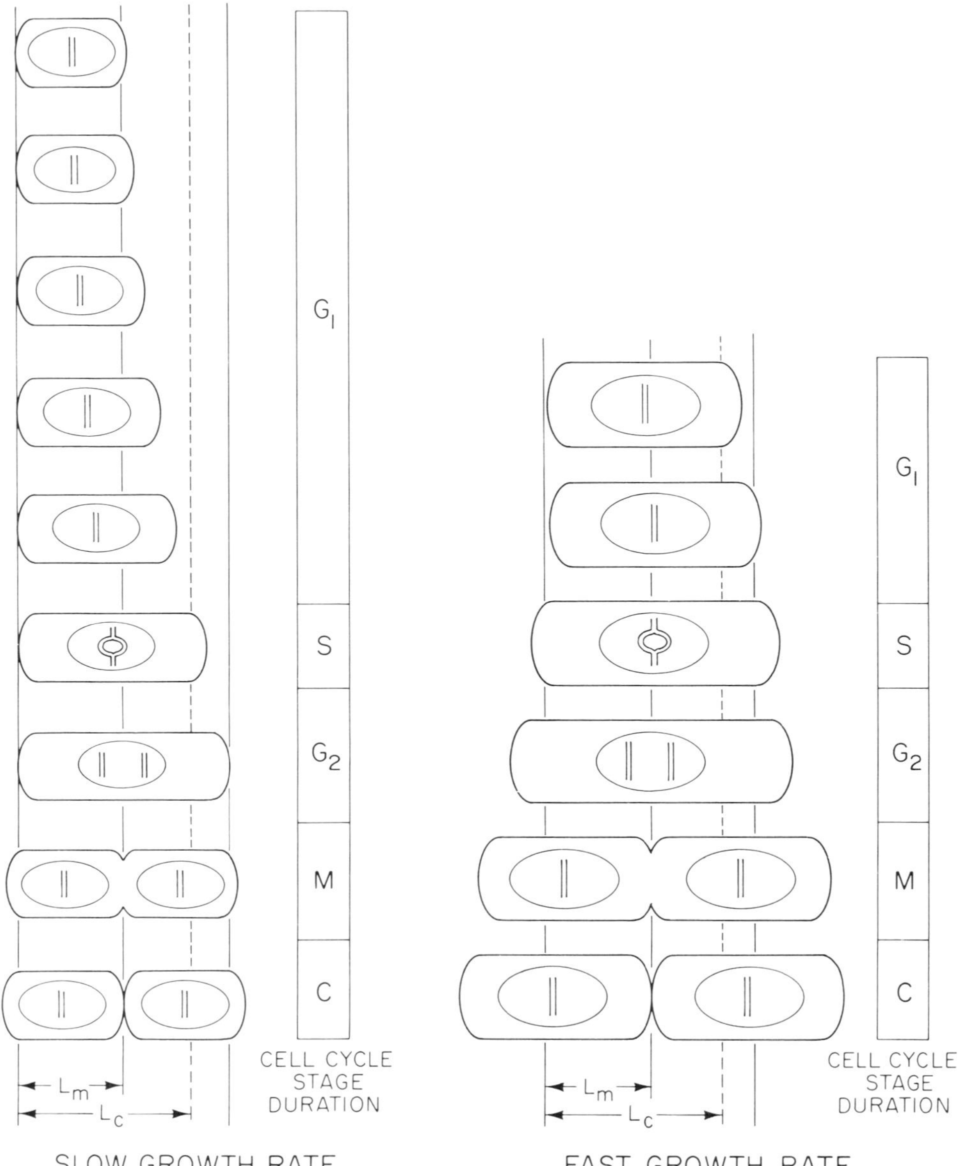

FIG. 2. The cell cycle of a "typical" eukaryotic cell grown under resource limited (slow growth) and unlimited (fast growth) conditions. A single chromosome is depicted in the cells in the G_1 phase of the cycle. The DNA is replicated during the S phase, which is followed by a second "gap" in the cycle (G_2). At this stage the cell has twice the DNA of a G_1 cell, and it can then progress through mitosis (M) and cytokinesis (C). L_m is the minimum possible size of a cell at birth, and L_c is the hypothesized (Fantes 1977) minimum cell size for the initiation of DNA synthesis; other models for the coupling between cell growth and division could also be considered (Tyson 1985). Note that only the G_1 phase of the cycle lengthens with increasing generation time, and cells are never smaller than L_c at the fastest growth rate, implying a mandatory G_1 interval which is size independent.

346

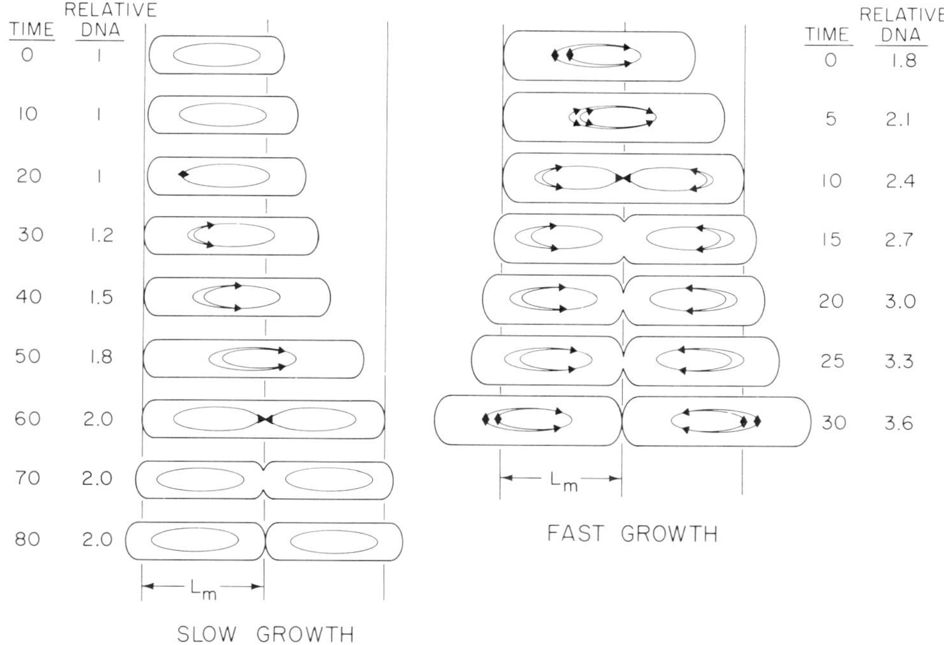

FIG. 3. The cell cycle of a "typical" prokaryotic cell under conditions promoting slow and fast growth (adapted from Donachie 1981). The cell has a single circular chromosome, and the chromosome replication period (C) is the time required for a single round of replication, which is 40 min in this example. Septum formation begins when the cell reaches twice the minimum cell length (L_m), and the process of division (D) takes 20 min to complete. When the generation time of the cell is longer than the duration of C + D (slow growth), the cell cycle resembles that of a eukaryotic cell; there is a gap before and after DNA replication, and the cell never has more than two complements of DNA. When the generation time is shorter than C + D (fast growth), the cell must begin new rounds of DNA replication before the previous round is completed, and DNA synthesis is continuous throughout the lifetime of the cell.

much faster than that of the chromosome. When cells are growing slowly, the prokaryotic cell cycle resembles that of the eukaryotic cell cycle in that it contains gaps before and after the interval of DNA synthesis (Fig. 3).

CELL CYCLE REGULATION IN EUKARYOTIC PICOPLANKTON

Because we know nothing about the cell cycles of eukaryotic picoplankton, we must try to understand them by analogy with their larger (roughly 10 μm in diameter) cousins. Our work on the cell cycles of a marine diatom (*Thalassiosira weissflogii*), coccolithophore, (*Hymenomonas carterae*), and dinoflagellate (*Amphidinium carteri*) has shown that the cell cycle and its regulation in these species shares many general characteristics with the "model" eukaryote (Vaulot 1985; Olson et al. 1986; Olson and Chisholm 1986), but some significant differences can be identified, as will be discussed below.

All of these species have a restricted interval of DNA synthesis (S) bounded by G_1 and G_2 (Fig. 4). The duration of all of these stages is a function of temperature (Fig. 5); each stage expands as the generation time lengthens at sub-optimal temperatures, although not to an equal degree within and between species (Olson et al. 1986). In contrast, when the growth rate of the cells is retarded by nitrogen limitation, only the G_1 phase of the cell cycle is extended; the length of S and

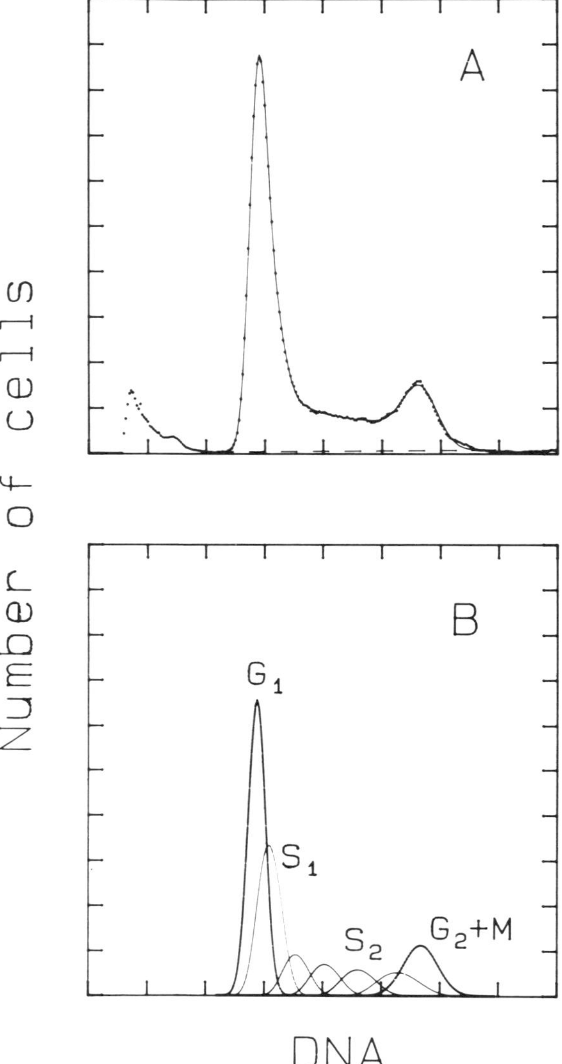

FIG. 4. Distribution of relative DNA per cell in a steady state culture of the marine coccolithophorid, *Hymenomonas carterae* (from Vaulot et al. 1986). A cell population was grown in continuous light (100 μE $\cdot$ m^{-2} $\cdot$ s^{-1}) at 20°C, harvested in log-phase growth, stained with propidium iodide in the presence of RNase, and analyzed by flow cytometry (Vaulot et al. 1986). (A) Raw data. (B) The DNA distribution was modelled as a sum of Gaussian peaks fitted using a non-linear algorithm. The continuous line in panel A is the sum of the Gaussian peaks shown in panel B. If the mean generation time is known, the duration of each cell cycle stage can be calculated from the proportion of cells in each stage according to the equations of Slater et al. (1977).

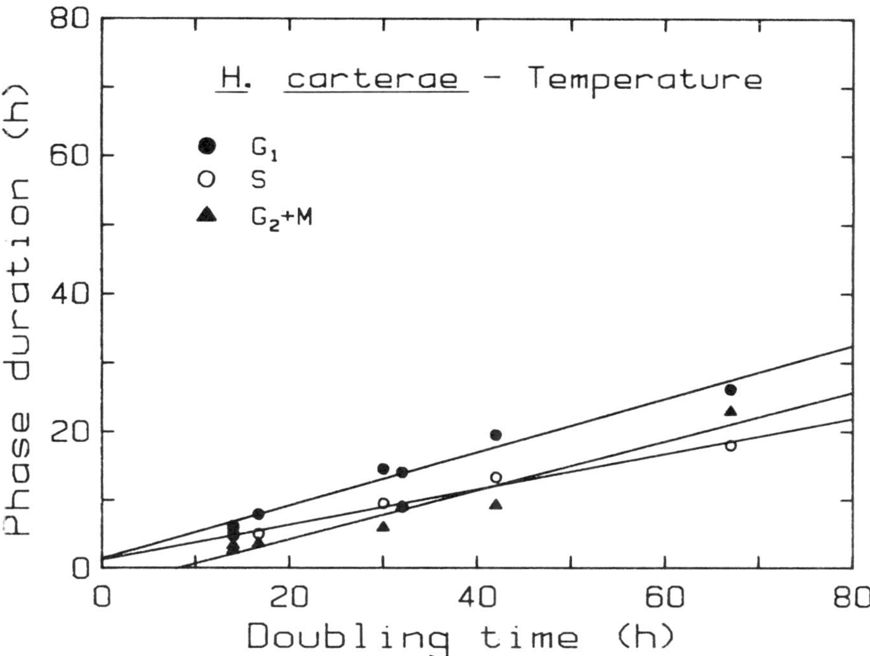

FIG. 5. Durations of cell cycle stages of *H. carterae* as a function of temperature controlled growth rate. Growth rates were regulated by increasing the temperature in increments between 13°C and 23°C. When the cultures reached steady state, samples were analyzed as shown in Fig. 4. Note that all of the cell cycle stages expand with increasing generation times (after Olson et al. 1986).

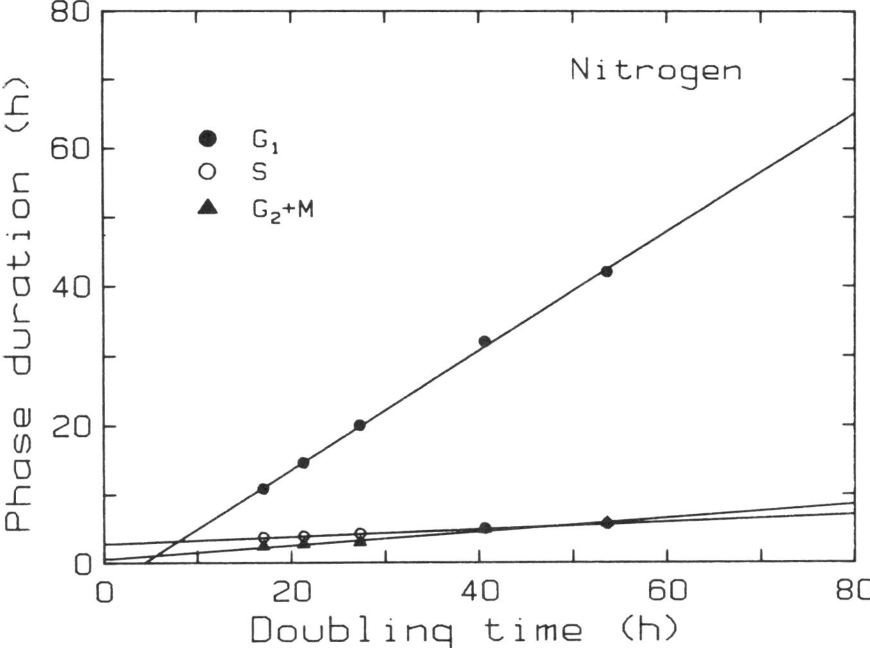

FIG. 6. Durations of cell cycle stages in *H. carterae* as a function of nitrogen-limited growth rate. Cells were grown in N-limited chemostat cultures maintained over a range of steady state growth rates. Samples were analyzed as shown in Fig. 4. Note that only the G_1 stage of the cell cycle expands with increasing generation time. The duration of S and G_2 + M remains constant (after Olson et al. 1986).

G_2 + M remains constant regardless of the growth rate of the population (Fig. 6). Furthermore, when population growth stops completely due to nitrogen limitation, essentially all of the cells are arrested in G_1 (Fig. 7). These results are consistent with the concept of a nitrogen dependent (undoubtedly involving a protein(s)) transition point in G_1, similar to those described for other types of eukaryotic cells (e.g. Gould et al. 1981; Johnston and Singer 1978). They also explain why cell division in nitrogen limited populations of these species can be entrained by a regular pulse of nitrogen (Olson and Chisholm 1984).

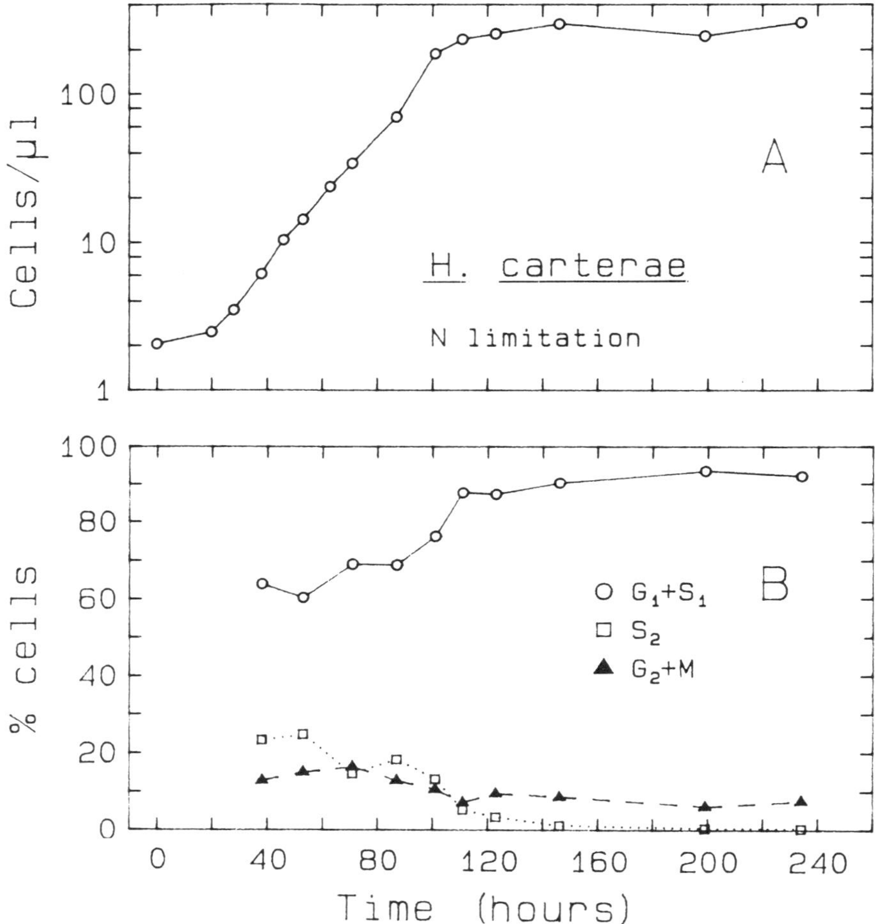

FIG. 7. Dynamics of cell cycle progression in nitrogen depleted cells of *H. carterae*. A batch culture was monitored as it progressed from exponential growth into N-limited stationary phase. Cell number (A) and percent of cells in the G_1, S, or G_2 + M phase of the cell cycle (B) were measured using flow cytometric analysis of DNA per cell. Note that when growth stops due to nitrogen depletion at 120 hours, essentially all of the cells are arrested in the G_1 stage of the cell cycle (after Vaulot et al. 1986).

When the growth of these same three species is limited by light instead of nitrogen, the coccolithophore and the dinoflagellate behave essentially as they did under nitrogen limited growth; only the G_1 phase expands with increasing light limitation, and all of the cells are arrested in G_1 in the dark (Olson et al. 1986; Vaulot et al. 1986). Again, this is indicative of a light dependent transition point in the G_1 phase of the

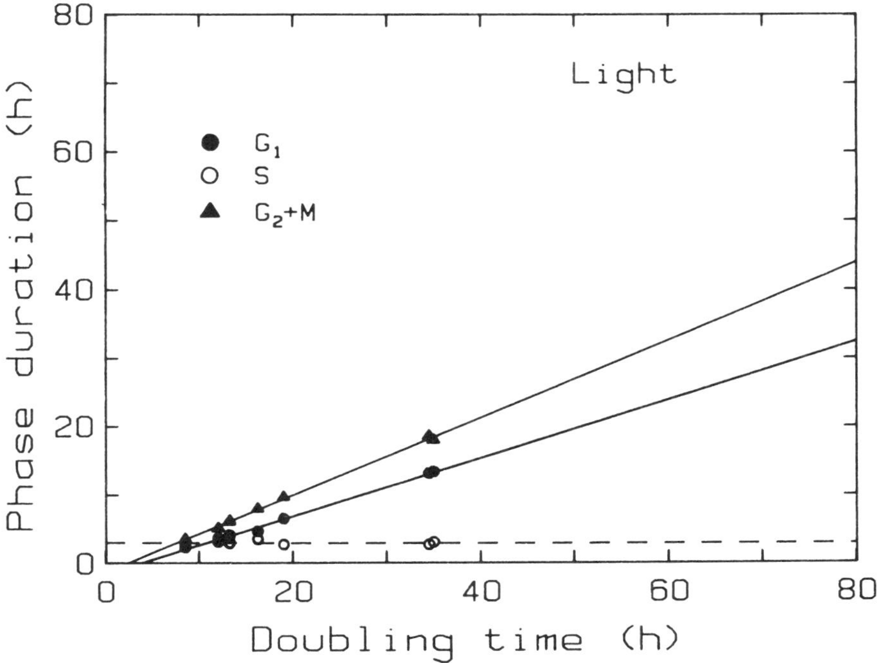

FIG. 8. Durations of cell cycle stages in *T. weissflogii* as a function of light-limited growth rate. Growth rates were regulated by increasing the light intensity in increments between 10–70 μE $\cdot$ m^{-2} $\cdot$ s^{-1}. When the cultures reached steady state, samples were analyzed as shown in Fig. 4. Note that both the G_1 and $G_2 + M$ stages of the cell cycle increase with increasing generation times of the population, suggesting a light requirement for cell cycle progression in both of these phases (Olson et al. 1986).

cell cycle in these species (Chisholm et al. 1984). The diatom behaves quite differently, however. In *T. weissflogii* both the G_1 and G_2 phases of the cell cycle expand during light-limited growth (Fig. 8), and cells arrest in both G_1 and G_2 in the dark (Fig. 9). We have hypothesized, based on Blank and Sullivan's (1979) demonstration of a light-dependent component of the silicon transport system, that the light dependent block point in G_2 in the diatom is due to the silicon requirement for frustule formation (Vaulot et al. 1986). It also seems likely that the unique division patterns displayed by diatoms grown on light/dark cycles (Chisholm et al. 1980) have their origin in this unusual light requirement for cell cycle progression in diatoms (Vaulot 1985).

The concept of a light-dependent transition point in the cell cycle of phytoplankton is not new. It was first alluded to by Bernstein (1964) in an attempt to explain the synchronous division patterns displayed by *Chlamydomonas* when grown on light/dark cycles. More recently, Spudich and Sager (1980) have demonstrated the existence of a transition point in the G_1 phase of the cell cycle in this species, which cannot be traversed until the cell has been exposed to a fixed amount of light energy. The subsequent studies of Donnan et al. (1985) and McAteer et al. (1985) have demonstrated convincingly that control of the *Chlamydomonas* cell cycle by light is mediated by a growth-dependent "hour glass timer" which controls the time of first commitment to divide. Once the commitment to division is made, a second timer period is initiated which runs from that point to the formation of daughter cells. This second interval is light independent and temperature compensated. A similar type of analysis has been carried out by Heath and Spencer (1985) for the marine diatom *Thalassiosira pseudonana*.

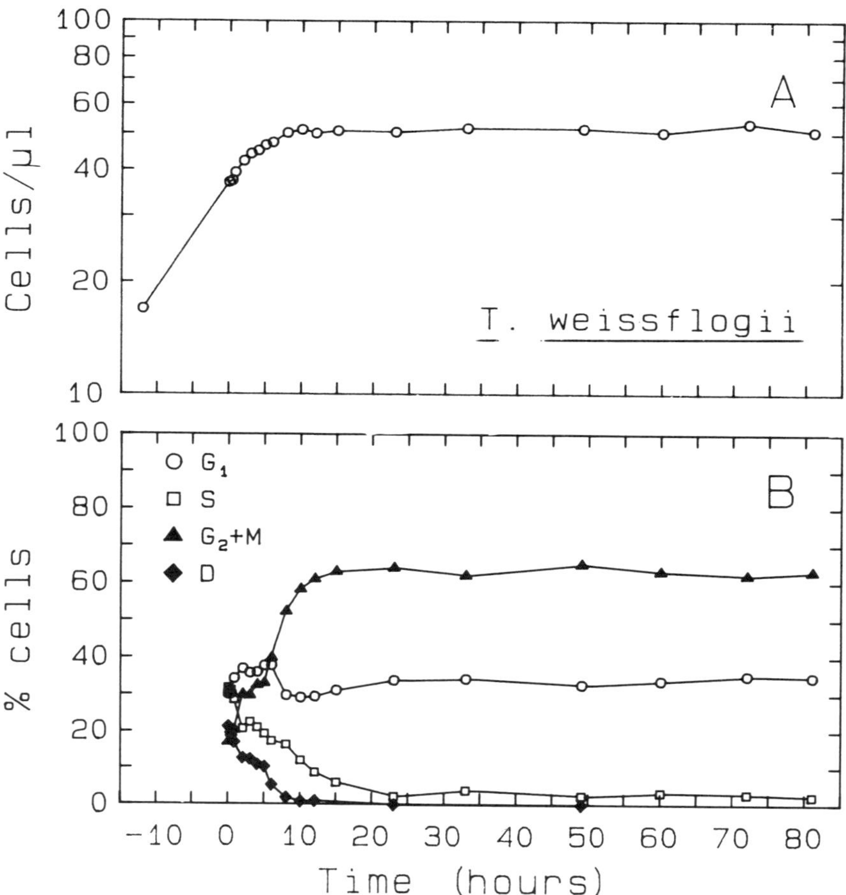

FIG. 9. Dynamics of cell cycle progression in a population of *T. weissflogii* transferred from steady state growth in continuous light to darkness (at $t = 0$). Cell number (A) and percent of cells in the G_1, S, or $G_2 + M$ phase of the cell cycle (B) were measured using flow cytometric analysis of DNA per cell. D is a post-mitotic doublet stage found in this species which was monitored by microscopic counts because it cannot be discriminated using the flow cytometer. Note that cells are arrested in the dark in both the G_1 and $G_2 + M$ stage of the cell cycle, suggesting a light-dependent block point in both of these segments (Vaulot et al. 1986).

Although the study of cell cycle regulation in eukaryotic autotrophic phytoplankton is in its infancy, this brief review should serve to demonstrate that we cannot assume that these cells will act like the model eukaryotic cell. In particular, it cannot be assumed that the G_1 phase of the cell cycle is the only phase that lengthens under suboptimal growth conditions. Environmentally controlled block points in the G_2 phase of the cell cycle, as we have demonstrated for the diatom *T. weissflogii*, are rarely observed in other cell types (Pardee et al. 1978) and certainly do not appear to be a general characteristic of eukaryotic photoautotrophs. The likely proposition that the G_2 block point in *T. weissflogii* underlies its unique division patterns (Vaulot et al. 1986) speaks to the power of cell cycle analysis for interpreting population phenomena.

We have not yet mentioned the relationship between cell volume, cell cycle, and cell generation time, as is implied in Fig. 2 for our model eukaryote. Although we have not investigated this relationship *per se*, we do know that the average cell volume

352

of our three experimental species decreases with increasing light limitation (Olson et al. 1986; Olson and Chisholm 1986), which is at least consistent with the model in Fig. 2. In nitrogen limited populations, average cell volume is much smaller than in nitrogen replete populations, but the cells do not continue to decrease in volume with increasing degrees of nutrient limitation. Temperature, on the other hand, has the opposite effect on average cell size; the cells get larger as they become more and more temperature limited, as has been observed by others (see e.g. Goldman 1977; Eppley 1972). We have no mechanistic interpretation for these results, other than the obvious observation that the coupling between the cell growth cycle and the DNA cycle is not rigid. It seems clear that a systematic study of volume regulation in *individual cells* is the key to understanding the relationships between the average volume of a population and growth conditions. Phytoplankton cells could prove to be an excellent model system for unravelling the mysteries surrounding the coupling between cell growth and division (Tyson 1985). In particular, because of the "non-scalability" of cellular DNA content (Raven 1986), eukaryotic picoplankton represent a very unique system for cell cycle studies. More work is needed in this area.

CELL CYCLE REGULATION IN PROKARYOTIC PICOPLANKTON

Anacystis nidulans

Just as our description of the generic eukaryotic cell cycle was based primarily upon results from studies of yeast and mammalian cell populations, most of what is known about prokaryotes comes from studies of *E. coli*. The autotrophic analogue to *E. coli* in this area of research is the freshwater cyanobacterium *Anacystis nidulans*. This is the only species of cyanobacteria whose cell cycle has been studied to any extent, and since it is a close relative of the marine cyanobacterium *Synechococcus*, it serves as a good model for comparison.

To begin, we can learn something about cell cycle regulation in this species by examining the conditions required to synchronize cell division in a population. Populations have been successfully synchronized with repeated temperature cycles (Lorenzen and Vankataraman 1969) and repeated cycles of [bright light]:[dim light]:[darkness]: (Lorenzen and Kaushik 1976); there are no reports of success (or of the details of failures) at inducing synchrony using simple repeated light:dark cycles. This is in direct contrast to most eukaryotic species (diatoms are the exception; cf. Chisholm and Costello 1980) in which the cell cycle is tightly coupled to light and population division patterns are easily entrained to photocycles (Chisholm 1981).

The most commonly used method of inducing synchrony in *A. nidulans* has been prolonged dark exposure followed by release into continuous light (Herdman et al. 1970; Csatorday and Horvath 1977; Asato and Folsome 1970 Asato 1979). Although this treatment cannot induce sustained synchrony (and can be criticized in this regard), it results in at least one synchronous burst of division in the population after release from darkness. This implies, but does not demonstrate (see below), that all of the cells are arrested at a common cell cycle stage in the dark. It further suggests that synchrony should be achievable by repeated photocycles if the right frequency were identified. Why then, does this seem impossible to achieve?

Herdman et al. (1970) and Herdman and Carr (1977) examined carefully the behavior of a culture subjected to 24 h of darkness and released into continuous light. Through experiments and modeling, they deduced that a cell's behavior in the dark depended on where it was in its cell cycle when it was put into the dark. Those cells which were about to divide finished their cell cycles in the dark; those which were in the midst of DNA replication or were about to initiate it, completed DNA replication but did not divide, and cells which were very early in their cell cycle at the onset of darkness did not replicate their DNA and were arrested at that early stage of the cycle (Fig. 10A). Thus, at the end of the dark induction period, two populations of cells

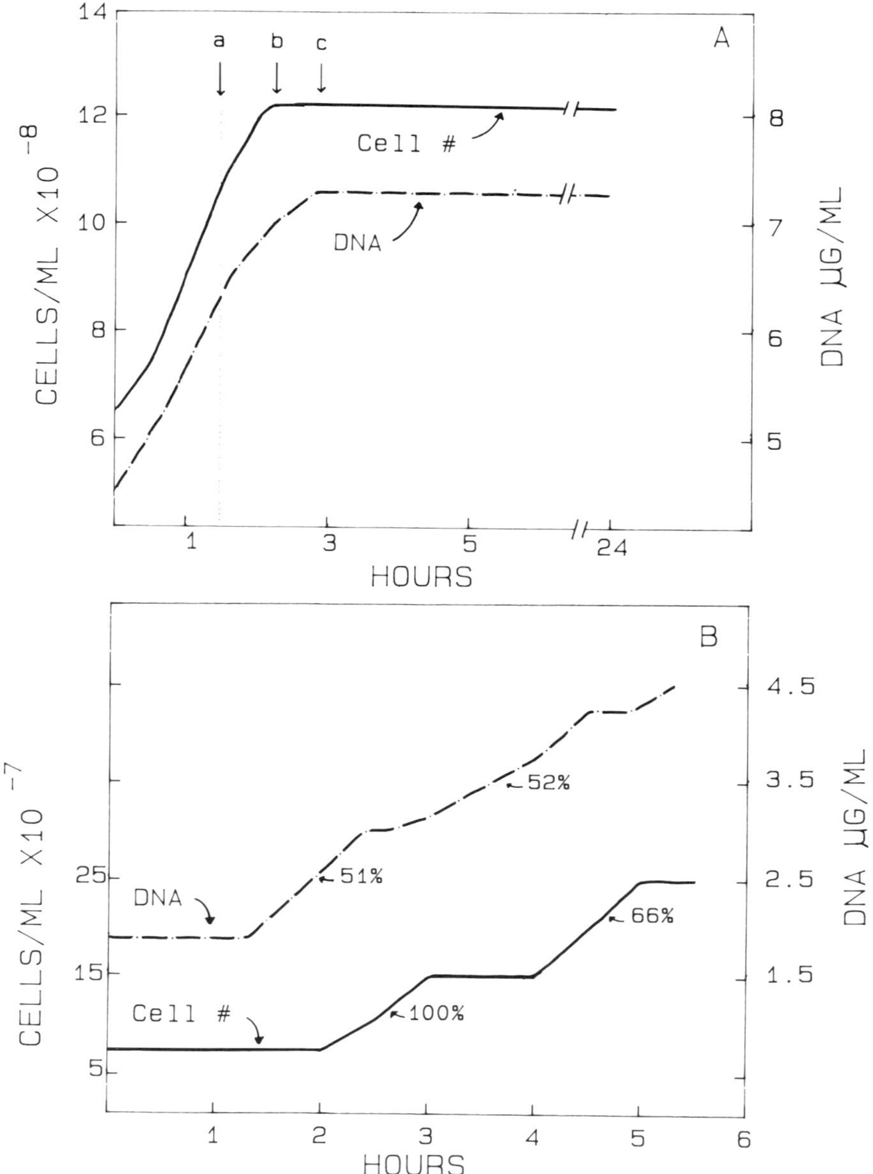

FIG. 10. Patterns of DNA systhesis and cell division in dark-arrested *A. nidulans* (after Herdman et al. 1970; see also Herdman and Carr 1971). (A) Cells which had been growing in continuous light with a generation time of 2 h were placed in the dark at point a. Cell division ceased after 30 min (point b), cell number having increased by 15%. Cells which were undergoing or about to initiate DNA replication at the time the cells were placed in the dark, completed replication at point c. New rounds of replication were not initiated in the dark, and after point c 68% of the cells contained one genome and 32% contained 2 genomes. (B) After 24 h in the dark, normal growth conditions were reestablished at time zero. DNA synthesis began 75 min after light was resupplied, and according to their model the 51% increase was attributable to the cells which were arrested in the dark with a single genome. Cell division was initiated in the entire population after 2 h of light exposure, and all of the cells divided in this round of division. Only 66% of the cells divided in the second round of division. Those that did not divide or replicate their DNA were deduced to be the non-viable offspring of the cells (32%) which had completed DNA synthesis during the dark arrest period.

existed; some had one copy of the genome, and others had two. This is analogous to diatom populations in which cells arrest in both G_1 and G_2 in the dark. It is noteworthy in this context that diatom populations do not phase tightly to light/dark photocycles.

When the dark arrested population described above was released into continuous light, the population did not commence division until 2 hours after exposure to light (which was the mean generation time in continuous light), indicating that the cells which had replicated their DNA in the dark had regressed in their growth cycle in the dark (Fig. 10B). Although these cells did eventually divide, their daughter cells were not viable and thus did not go through a second round of division. (Note that this behavior has serious consequences for trying to estimate *in situ* growth rates from the frequency of dividing cells . . . see below.) The cells that had a single genome at the time of light exposure commenced DNA synthesis after one hour and executed their cell cycles synchronously for two rounds of division.

The results of this experiment, and those of Asato (1983), imply that the synchronous burst of division observed in *A. nidulans* after release from prolonged darkness does not reflect synchrony in the actual cell cycles of the cells. To put it another way, the DNA cycle and the light-regulated growth cycle in these cells appear to be loosely coupled, at best. This is consistent with the general lack of regulation at the transcriptional level in these obligate phototrophs (Carr 1973) as is discussed in detail in Carr and Wyman (1986). It is also in direct contrast to the tight coupling between light, the DNA cycle, and the growth cycle characteristic of eukaryotic phytoplankton.

Our model of the generic prokaryotic cell cycle (Fig. 3) allows for more than two genome copies to exist during the cell cycle in rapidly growing cells. Mann and Carr (1974) observed that in asynchronous, light limited populations the DNA content per cell increased exponentially with increasing growth rate, reaching 16 times the level of non-growing cells at the fastest growth rate. The generation time of the cells having 16 genome equivalents, however, was longer than the combined durations of the C and D phases in this species (180 min; Herdman et al. 1970) and thus could not be explained in the context of the multiple replication fork model used for heterotrophic bacteria (Mann and Carr 1974). These observations become even more confusing in the light of more recent data (Parrott and Slater 1980) in which DNA/cell was found to decrease with decreasing generation times in light and CO_2-limited chemostats of *A. nidulans*. Clearly the control of the cell cycle in *A. nidulans* is not well understood, and bears little resemblance to that of phototrophic eukaryotes, or heterotrophic prokaryotes.

Synechococcus sp.

Our understanding of the cell cycle of marine *Synechococcus* species has just begun to nucleate. Studies of growth patterns on 24-h photocycles all indicate that cell division occurs primarily in the light period of the photocycle (Chisholm 1981; Campbell and Carpenter 1986; Waterbury et al. 1986). Our early suspicion that this pattern results from a cessation of cell division in the dark (Chisholm 1981) . . . i.e. an arresting of the late cell cycle stages by darkness . . . is confirmed by more recent studies. Both Campbell and Carpenter (1986) and Waterbury et al. (1986) have observed that the percentage of cells in the doublet stage is fixed at a relatively low but constant value during the second half of the dark period (Fig. 11). Since the dark period in these studies is much longer than the average duration of the paired cell stage (2-4 h) we can conclude that the cells which enter the doublet stage in the dark are arrested at that point. Whether or not these cells complete their cell cycle upon reexposure to light cannot be determined from existing data sets.

We can examine more closely the regulation of the cell cycle in *Synechococcus* through flow cytometric analysis of the DNA cycle, as we have done for eukaryotic

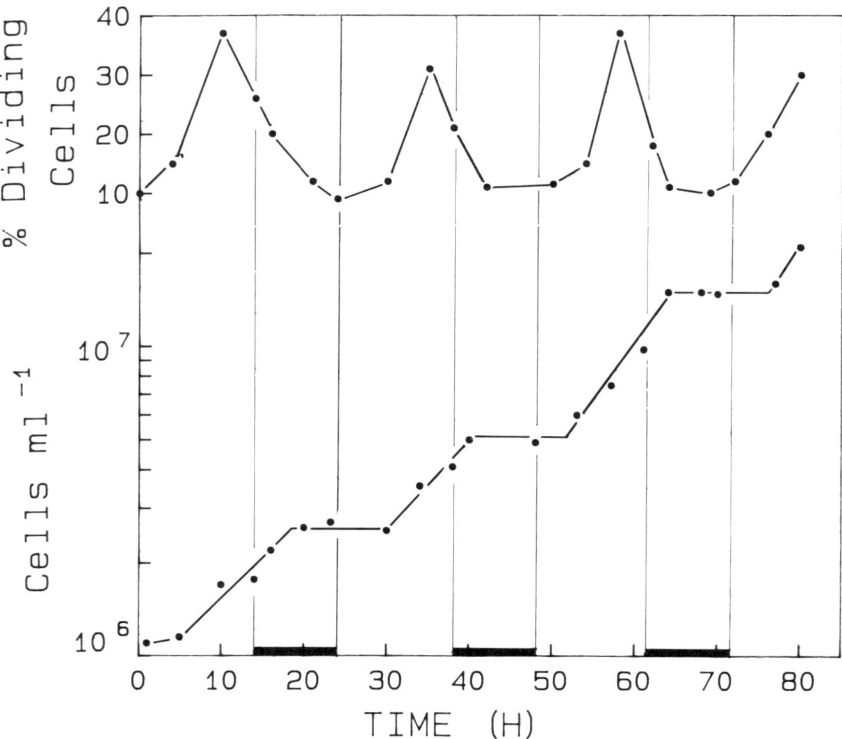

FIG. 11. Cell division patterns of *Synechococcus* sp. grown on diel photocycles. Change in cell number and percent doublet cells as a function of time in strain 7803 grown at 25°C (after Waterbury et al. 1986).

species. Over a range of doubling times from 14 to 55 h, we found that the DNA cycle of these prokaryotes was very similar to that of the eukaryotic species. Regardless of the growth rate, the cells had a distinct period of DNA synthesis, and the average increase in the DNA content per cell was never more than one doubling (Fig. 12A). For convenience of comparison, we have adopted the eukaryotic terminology to describe the various stages of the DNA cycle in *Synechococcus*; the unequivocal differences in mechanisms at the molecular level, however, must not be forgotten. Moreover, we have not tried to culture these populations under conditions which might promote faster doubling times. It is possible that if higher growth rates could be achieved, we might observe continuous DNA synthesis, and/or more than two genomes per cell; the fastest growth rate we have studied in *Synechococcus* is equal to the slowest growth rate in the Mann and Carr (1974) study of *Anacystis* described above.

FIG. 12. Analysis of cell cycle characteristics of *Synechococcus* sp. (strain WH 8101) grown at different light intensities supporting doubling times between 14.4 and 55.2 h. Cells were harvested from exponential growth, and stained with propidium iodide for the DNA analysis (as described in Olson et al. 1986). The relative amount of DNA per cell in the populations grown at the highest (doubling time = 14.4 h) and lowest (doubling time = 55.2 h) light intensities are shown in (A). For illustrative purposes these histograms were analyzed as described in Fig. 4, with the recognition that the analysis rests on assumptions about the DNA cycle that may not be correct for this prokaryotic species. The durations of the various cell cycle stages as a function of population doubling time are shown in (B). Note that both the "G_1" and "G_2" phases of the cell cycle expand with increasing generation time.

FIG. 12.
NUMBER OF CELLS
DOUBLING TIME = 14.4 H
DOUBLING TIME = 55.2 H
A
RELATIVE DNA
B
Phase Duration (h)
"G₁"
S
"G₂"
Doubling Time (h)

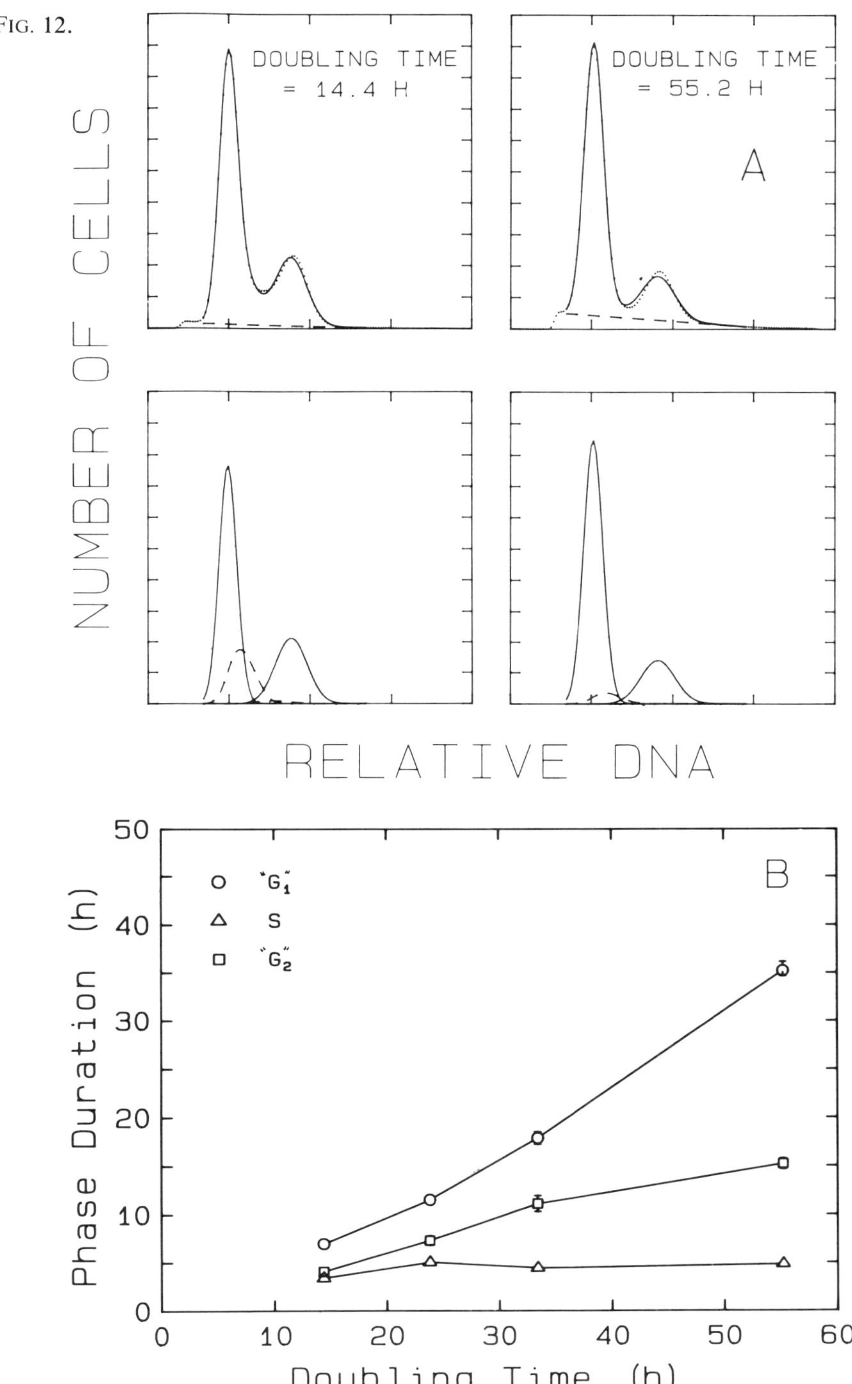

How do the lengths of the various cell cycle stages in *Synechococcus* change with increasing generation times? Does only the "G_1" period expand as in our eukaryotic species, or are all the stages flexible in duration? To address this question we grew *Synechococcus* over a range of light intensities supporting doubling times from 14 to 55 h, and analyzed the relative DNA content per cell in each population using flow cytometry. Analysis of the DNA histograms showed that under light-limited growth, the "S" phase has a constant duration regardless of the mean generation time, but both "G_1" and "G_2" expand as the generation time increases (Fig. 12B). This response is analogous to the response of the diatom *T. weissflogii* described above (Olson et al. 1986), and is consistent with the observation that both of these species divide primarily during the light period of the diel photocycle. In other words, species which have light-dependent processes toward the end of the cell cycle divide primarily during the day.

Based on the results shown in Fig. 12, we would predict that, like the diatom, *Synechococcus* should arrest in both "G_1" and "G_2" when placed in the dark for extended periods. This prediction is not only based on our observation of G_2 expansion in response to light limitation, but also on the persistent doublet population observed in the dark of populations grown on diel photocycles (Fig. 11). Indeed, in preliminary experiments designed to test this hypothesis, we have found that dark-arrested populations of *Synechococcus* do contain cells blocked in both "G_1" and "G_2". Our data is not complete enough to resolve whether or not cells also arrest in S, but the *Anacystis* model would suggest that they do not, as would the constancy of the duration of the S phase regardless of the energy supply (Fig. 12B).

We conclude that cell cycle regulation in marine *Synechococcus* is similar in some regards to that of its close cousin, *Anacystis*. In particular, both species appear to have light-dependent segments at the end of their cell cycle; i.e. when placed in the dark they continue synthesizing DNA, but cannot proceed to division unless reexposed to light. Our limited experience working on the *Synechococcus* cell cycle, however, suggests that there are also some fundamental differences between the DNA cycles of these two cyanobacteria. Under all growth conditions we have examined, asynchronous populations of *Synechococcus* (strain WH 8101) exhibit a biomodal DNA distribution similar to those shown in Fig. 12A. This reflects a discrete interval of DNA synthesis in the cell cycle similar to that found in eukaryotes. We know, however, that flow cytometric analysis of the DNA per cell in asynchronous *E. coli* cultures yields a continuous distribution (Hutter and Eipel 1979) which is what one would expect if DNA synthesis were continuous throughout the cell cycle. Although similar data are not yet available for *Anacystis*, we suspect that this species will resemble more the *E. coli* model than the *Synechococcus* model, especially at near maximal growth rates. If true, this could represent an interesting paradox.

Estimating in situ Growth Rates from Cell Cycle Analysis

In recent years, biological oceanographers have attempted to use recognizable cell cycle markers as indicators of how fast populations of cells are growing in the sea. Just as one can measure the rate of execution of cell cycle events by examining the frequency of labelled mitosis in a heterogeneous cell population (Quastler and Sherman 1959), one should be able to gain some information regarding the growth rate of an *in situ* population from the frequency of cells in a given cell cycle stage. The history, applications, and limitations of this technique as applied to eukaryotic phytoplankton are reviewed more extensively in McDuff and Chisholm (1982).

In an asynchronous culture of cells with a steady state growth rate of μ, the frequency of cells undergoing division at any point in time can be related to μ as:

$$(1) \qquad \mu = 1/t_D \, [\ln (1 + f)]$$

where f is the fraction of the cells identified as undergoing division (usually as paired cells or cells with two nuclei), and t_D is the duration of the division stage of an individual cell (Cook and James 1964). The relationship between these three variables is shown in Fig. 13. It is critical to recognize that *this formulation assumes that the value of t_D is the same for all cells in the population*, and that all cells are actively growing.

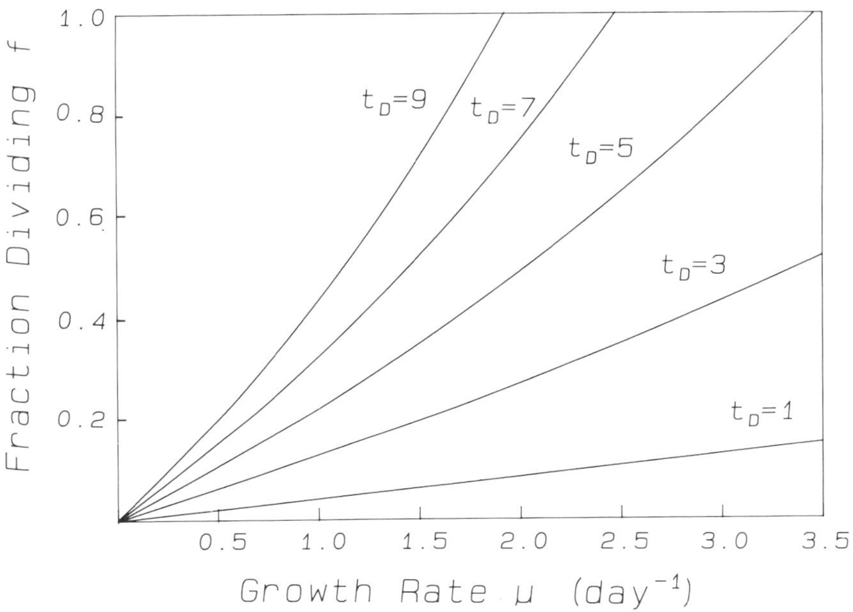

FIG. 13. Relationship between population growth rate and fraction of post-mitotic or paired cells for various values of t_D, the duration of the dividing cell stage (see eq. 1).

In a non-steady state culture, such as one in which division is phased to a light/dark cycle, or in samples collected in the field, the average value of μ can be approximated with sequential observations of f according to the equation:

$$(2) \qquad \mu = (1/nt_D) \sum_{i=1}^{i=n} \ln (1 + f_i)$$

where μ is the average population growth rate over an interval during which n samples were taken. If the interval between samples is equal to t_D, then the estimate of μ is exact; if the sampling interval is greatly different from t_D uncertainty is introduced into the estimate, however, this is rarely significant relative to other errors inherent in the method (McDuff and Chisholm 1982).

In order to apply equation (2) to data from field populations, t_D must be known for the species in question *under the environmental conditions present in the field at the time the samples were collected*. This can be done by growing the species in the laboratory, measuring μ and f over the required intervals, and solving equation (2) for t_D. If t_D does not remain constant over a variety of environmental conditions (e.g. a range of temperatures), and thus varies with growth rate, the utility of the technique is severely limited. One can use different t_D values as the environmental temperature changes, but if growth rate is regulated by temperature alone, such that temperature, μ and t_D covary, knowledge of f is peripheral; one could just as easily estimate the growth rate of the cells *in situ* from a laboratory calibration curve between growth rate and temperature (e.g. see Rubin 1981).

There are several data sets in the literature describing the relationship between cell number increase and frequency of doublets in rapidly growing, synchronized populations of *A. nidulans* (Asato and Folsome 1970; Bagi et al. 1979; Csatorday and Horvath 1977). Applying equation (2) to .these data, we discover that t_D ranges between 1 and 2 h in these rapidly growing cultures. Similar data are now available for marine *Synechococcus* grown under a variety of conditions. Campbell and Carpenter (1986) calculated t_D for a variety of clones grown on light/dark photocycles at different temperatures and light intensities. They found that t_D ranged from 2 to 4 h (depending on the clone) except in populations with growth rates less than 0.5 day^{-1}, in which it ranged from 3 to 9 h. In a similar set of experiments with *Synechococcus* strain 7803 Waterbury et al. (1986) calculated values of 2 h for the t_D of cells grown in continuous light, and 4 h for cells grown on a 24 h photocycle. He has also demonstrated that t_D increases with decreasing temperature in cells grown in continuous light.

Although upon first impression the values of t_D calculated from this diversity of studies may appear surprisingly similar, reexamination of equation (2) and Fig. 13 forces us to remember that a one or two hour error in the estimate of t_D can cause large errors in the estimate of μ for small values of t_D. It is optimistic to think that we can know the value of t_D in field populations with any greater accuracy than this. Moreover, if our understanding of the cell cycle of *A. nidulans* is transferrable to that of *Synechococcus* . . . i.e. if cells arrested after DNA replication in the dark do not produce viable progeny . . . the assumptions of equations (1) and (2) are violated. Finally, recent observations of nitrogen limited populations of *Synechococcus* indicate that when population growth stops completely due to nitrogen depletion some of their cells arrest in the doublet stage as they do in darkness (Waterbury pers. comm.; Glibert et al. 1986). Clearly this behavior would result in a completely erroneous estimate of growth rate.

The inescapable conclusion from this analysis is that the frequency of dividing cells technique is not refining our estimates of growth rates of *Synechococcus* in the sea. It has yielded corroborative numbers, i.e. values that are not in conflict with rates measured by other methods (see Campbell 1985), which is comforting. Until cell cycle regulation in *Synechococcus* is better understood, however, this method should be used with caution.

Single Cell Analyses at Sea

The previous section should serve to expose the power of single cell analysis for the study of phytoplankton growth dynamics in clonal cultures. What promise does this perspective have for analyses at sea? Other than the mitotic index assay discussed above and traditional microscopic analysis of phytoplankton community composition, there have been very few studies of the characteristics and behavior of individual cells *in situ*. Historically, the only way to do this kind of work has been through microscopic analysis and/or manual isolation of cells using a micropipette (e.g. Rivkin and Seliger 1981; Rivkin 1985; Boulding and Platt 1986), which is too tedious for routine or widespread use, and is restricted to large-celled species. The introduction of flow cytometry into biological oceanography has changed this situation.

Although it is no substitute for the microscope in resolving the impressive diversity of form in the plankton, the flow cytometer provides multiparameter data (up to 6) on each particle at extraordinary rates, and with great precision. The resulting flow cytometric "signature" provides a snapshot of the assemblage of organisms being examined. Although we cannot yet interpret all aspects of these signatures, with experience and improved data analysis techniques they will soon become as familiar to us as are microscope images. The sorting capability of the flow cytometer enables us to physically sort the particles responsible for a particular component of the signature, so we can confirm or establish their identity using a microscope.

We have taken a commercially available flow cytometer (Coulter Electronics, EPICS V) to sea on three occasions, and have found that the instrument functions normally on board ship with only minor modifications (Olson et al. 1985). The picoplankton population with a fluorescence emission and light scatter signature which has become very familiar to us is a group of *Synechococcus* which is distinguished by its unique fluorescence emission peak between 550 and 590 nm (due to phycoerythrin). In natural water samples, the phycoerythrin containing *Synechococcus* have relatively low red fluorescence (from chlorophyll) and relatively high orange fluorescence (from phycoerythrin), and a very low light scatter signal due to their small size. Another group of organisms that is readily discernable are members of the *Cryptophyta*, which also contain phycoerythrin, but have relatively high chlorophyll and thus can be distinguished from the *Synechococcus* (Fig. 14). The light scatter signal from the flow cytometer not only gives us information regarding cell size, but also cell surface properties, which in some cases provides a diagnostic signature for certain groups of organisms such as the coccolithophorids (Olson and Chisholm 1987).

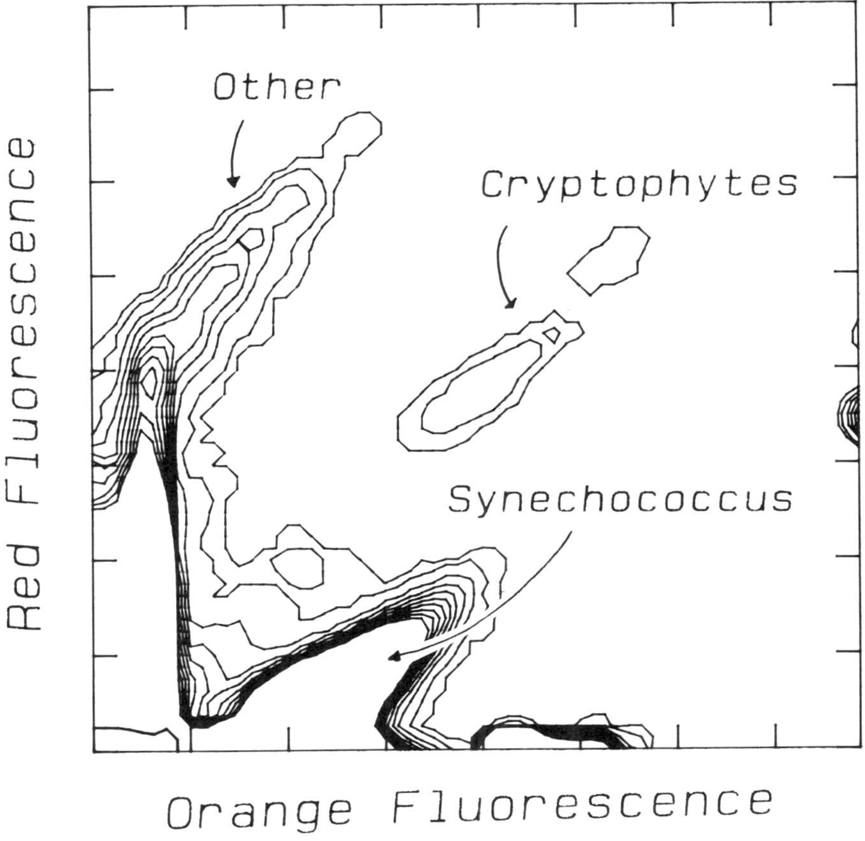

FIG. 14. Flow cytometric signature of a surface sea water sample from near shore off Woods Hole, MA. The sample was excited with 500 mW of 488 nm wavelength light, and fluorescence emission was measured between 550–590 nm (orange) and 660–700 nm (red) and recorded on a three decade log scale. Note that *Synechococcus* has relatively low chlorophyll (red) and high phycoerythrin (orange), cryptophytes have high values of both, and other cells are distinguished by their high chlorophyll and low orange fluorescence. Identification of the components in the signature was confirmed by sorting and microscopic examination.

361

Another application of flow cytometric analysis at sea is for fine scale resolution and diagnostic purposes in the analysis of experiments done with natural populations. This instrument gives us the capability to examine what is going on inside incubation bottles, in terms of changes in community composition, by comparing signatures over time. Even if we cannot interpret fully the signatures, the change in the signals has information content in and of itself. For demonstrative purposes,

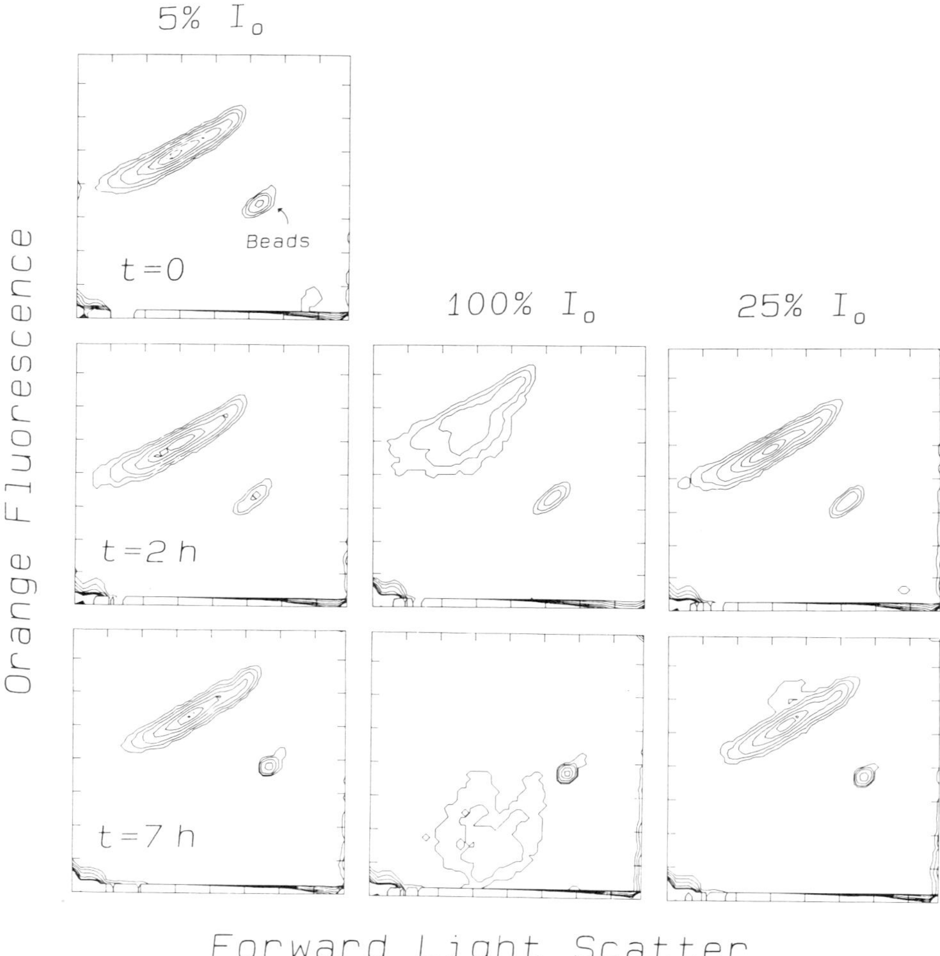

FIG. 15. Flow cytometric analysis of the impact of light exposure on deep water samples containing *Synechococcus* sp. during bottle incubations done at a station at 14°N, 65°W. The 5% I_o sample was collected from 85 m and incubated on deck at various light intensities. The tight cluster of cells ($t = 0$) to the left of the standard beads (0.9 μm in diameter) is *Synechococcus*, as identified by sorting and microscopic verification. Each frame is a 3 decade log–log plot of orange fluorescence (550–590 nm) vs. light scatter. Note the dramatic change in the signature after a two hour incubation at surface light intensity, and the significant reduction in fluorescence intensity after 7 h.

consider an experiment in which a sample was collected from a depth at which the light level was 5% that of surface intensity (I_o) and incubated at 5% I_o, 25% I_o and 100% I_o (Fig. 15). When incubated at the light intensity from which it was collected (5% I_o), the "signature" of the population remained unchanged over a 7 h incubation period. Similarly, when incubated at 25% I_o there was very little change in the signature over the incubation period, although a slight deviation is apparent at the seven hour time point. In contrast, when incubated at surface light intensities, the fluorescence signature of the population underwent dramatic changes during the first 2 h of incubation, and fluorescence intensity had deteriorated to very low values by the end of the experiment. We have also observed similar dramatic changes in signatures in nutrient enrichment experiments done at sea.

Perhaps the most exciting aspect of flow cytometric analysis at sea, is the insights it gives us into the dynamic behavior of phytoplankton populations *in situ*. On our first cruise with the instrument (Olson et al. 1985), we were able to demonstrate that changes in the *average* phycoerythrin fluorescence in a typical depth profile did not

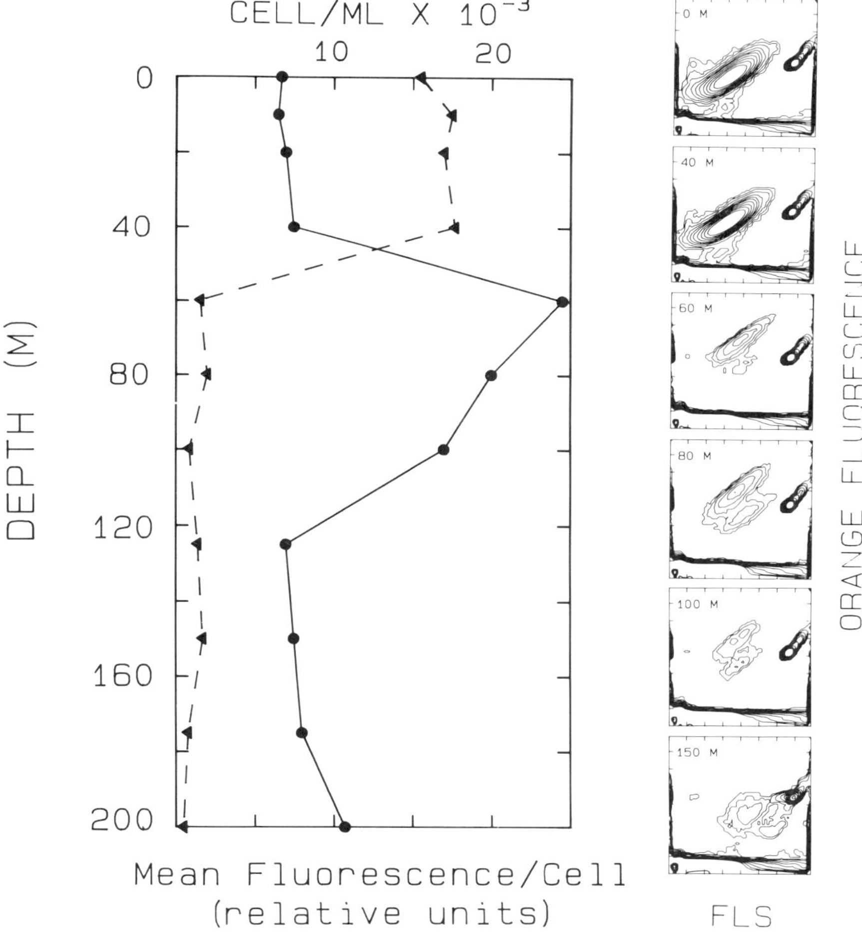

FIG. 16. Depth profile of the distribution of *Synechococcus* at a station in the Gulf Stream (after Olson et al. 1985). The cell concentrations (dashed line/triangles) were determined with the flow cytometer, and the mean orange fluorescence (solid line/circles) was calculated by averaging the fluorescence per cell from the flow cytometric signatures from each depth. Orange fluorescence (550–590 nm) vs. light scatter is shown on the right for 6 selected depths on a 3 decade log-log plot. The bead internal standard can be seen at the far right in each frame. Note that two populations of *Synechococcus* appear at 60, 80, and 100 m, the relative proportions of which dictate the shape of the average fluorescence profile on the left.

363

reflect changes in the properties of the cells in a given *Synechococcus* population; on the contrary, they reflected changes in the composition of the *Synechococcus* community, i.e. changes in the relative proportions of the "bright" and "dim" populations with depth (Fig. 16). We have also seen this same appearance and disappearance of "bright" and "dim" *Synechococcus* populations in surface water samples collected from an onshore to offshore transect crossing a frontal region off the coast of North Carolina (Fig. 17), and in a surface transect running from Puerto Rico to Woods Hole, MA, and have observed three or four different populations appearing in nutrient enriched, surface water cultures (unpubl. data).

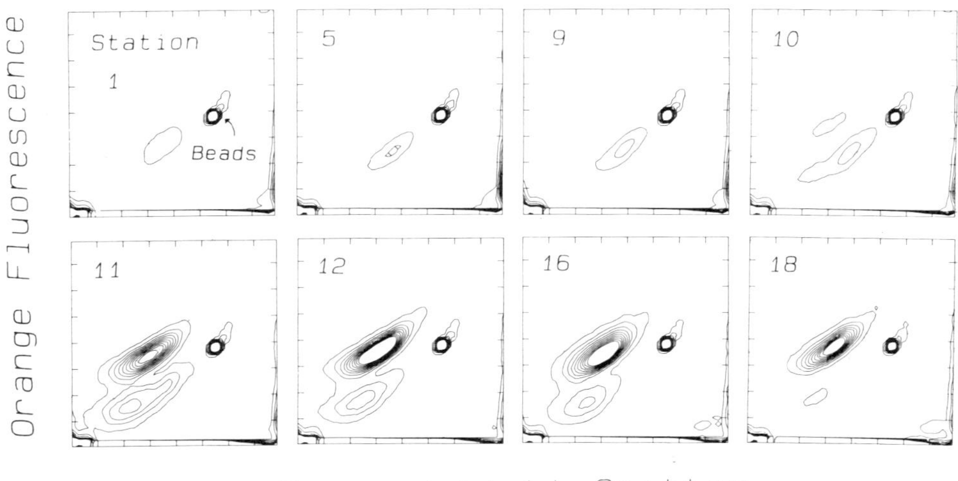

FIG. 17. Flow cytometric analysis of a surface water from an onshore to offshore transect near Cape Hatteras, North Carolina. Samples were taken at 1.5 mile intervals (and numbered sequentially), but only selected stations are shown here to highlight the changes. Stations that are not shown had signatures similar to those adjacent to them. Orange fluorescence (550–590 nm) vs. forward light scatter signatures are shown which identify the *Synechococcus* fraction of the community.

To determine whether or not these two types of *Synechococcus* populations were different strains with different pigment compositions or two populations of the same strain with different past histories, we sorted the two populations using the cell sorting capability of the EPICS V and grew them in culture medium. We found that the populations maintained their fluorescence differences after many weeks in culture under identical conditions, indicating that they are distinct strains. Fluorescence excitation and emission spectra of the strains indicate that they differ in their phycoerythrin composition. In addition to the major absorption peak at 550 nm (caused by the chromophore phycoerythrobilin) the "bright" strains have an additional peak around 500 nm caused by the chromophore phycourobilin (Wood 1985; Ong et al. 1984). Since this peak is missing in the "dim" strain, it has lower fluorescence intensity because the 488 nm laser line of the flow cytometer is relatively less effective (as much as 8-fold) at exciting fluorescence via the erythorobilin than the urobilin-containing phycoerythrin.

As a final example of the power of single cell analyses at sea, we examine a depth profile of *Synechococcus* sp. in the Caribbean, which was analyzed using the flow cytometer (Fig. 18). Samples were collected at dawn and dusk at the same station on two consecutive days, and the *Synechococcus* fraction was analyzed for cell concentration (Fig. 18A), forward light scatter, or mode relative size per cell

364

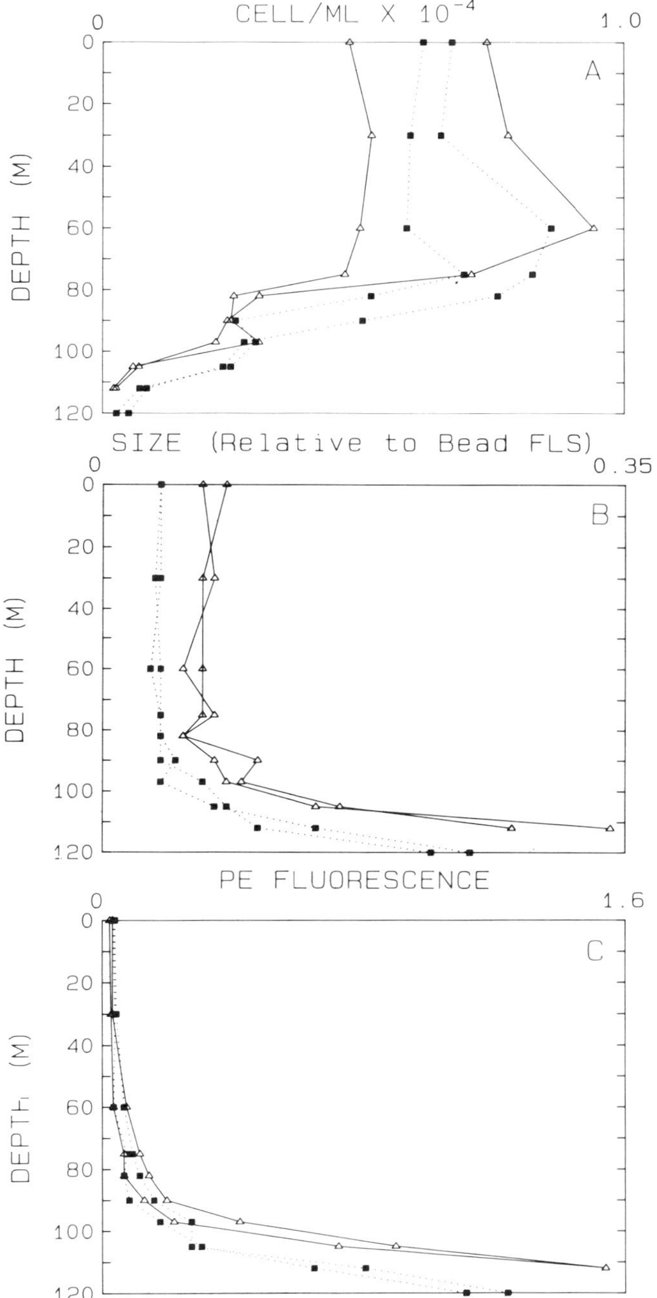

FIG. 18. Changes in depth profiles of *Synechococcus* populations as a function of time of day over two diel cycles at a station in the Caribbean using flow cytometric analysis. All data points were calculated from the flow cytometric signatures and are the modes of the distributions; squares represent samples collected at dawn, and triangles represent samples collected at dusk on 2 consecutive days. (A) Cell concentration as a function of depth; (B) relative forward light scatter per cell, a reflection of cell size, as a function of depth; (C) relative phycoerythrin per cell fluorescence as a function of depth. The depth of the mixed layer was at 60 m, the 1% light level was at 97 m, and the nitracline was at 105 m.

(Fig. 18B), and mode phycoerythrin fluorescence (Fig. 18C). Without interpreting the general shapes of these profiles for the moment, let us focus on the reproducibility of the signals on successive days. The cell concentration data shows no consistent diel pattern from one day to the next, which could be a result of real variability, or variability due to changing water masses. The data on cell size and pigment fluorescence, on the other hand, is surprisingly consistent from one day to the next. The light scatter data suggests that the cells get larger during the day, and "shrink" at night, and the fluorescence intensity data also increases during the day, at least in the deeper waters. Thus, by allowing us to look at cellular properties, the flow cytometer reveals patterns which would have been difficult to discern using traditional methods of analysis . . . methods which most often treat the cells as though they are an intrinsic property of the water. These patterns reflect the physiological ecology of the group in question, which must be understood for a mechanistic understanding of the system.

One aspect of flow cytometric analyses at sea that has been imagined but not explored to date is the use of molecular probes . . . i.e. specific fluorescent stains . . . to aid in the characterization of natural plankton communities. For example, we should be able to use protein-specific, DNA-specific, and vital fluorescent stains to distinguish between living and dead organisms, and organic detritus (Yentsch et al. 1983a). With the help of these probes, it should be possible to study the heterotrophic plankton community, which has thus far received limited attention by users of flow cytometry. We also should be able to use fluorescent antibodies as markers for specific types of organisms (e.g. Ward and Perry 1980; Campbell 1985), or for the detection of cells with specific biochemical properties. There are some barriers to the development of these techniques such as non-specific staining and sensitivity problems, but none of these barriers is insurmountable. Lack of progress in this area is due solely to the limited number of instruments that are now available for oceanographic applications.

We have only scratched the surface with regard to the overall potential of flow cytometric analysis of plankton in the sea. As this potential is realized, we will better appreciate the predictable variability between individual cells in populations and how this variability is woven into the fabric of the community structure. The variability itself contains information that is important for an accurate description and understanding of the dynamics of these organisms. In concert with cell kinetics studies in the laboratory, we are optimistic that the analysis of this variability will lead to new insights into the ecology of the plankton.

Acknowledgements

This work was supported in part by NSF grants OCE 8121270, 8118475, 8316616, CEE 8211525 and ONR Grant NR 104-192 (N00014–83-K-0661). We thank Jim Bowen for assistance with the graphics and John Waterbury for useful discussions. We also thank Trevor Platt, Holgar Jannasch, and John Waterbury for inviting us to participate in their cruises. E. V. Armbrust is a student in the MIT/Woods Hole Joint Program and has been supported in part by the Education Department at Woods Hole. Travel funds from the Ippen Award in the Parsons Lab are also appreciated.

References

ASATO, Y. 1983. Dark incubation causes reinitiation of cell cycle events in *Anacystis nidulans*. J. Bacter. 153(3): 1315–1321.

——— 1979. Macromolecular synthesis in synchronized cultures of *Anacystis nidulans*. J. Bacter. 140(1): 65–72.

ASATO, Y., AND C. F. FOLSOME. 1970. Temporal genetic mapping of the blue-green alga, *Anacystis nidulans*. Genetics 65: 407–419.

BAGI, G., K. CSATORDAY, AND G. L. FARKAS. 1979. Drifts in DNA level in the Cyanobacterium *Anacystis nidulans* during synchrony induction and in synchronous culture. Arch. Microbiol. 123: 109-111.

BERNSTIEN, E. 1964. Physiology of an obligate photoautotroph (*Chlamydomonas moewusii*) I. Characteristics of synchronously and randomly reproducing cells and an hypothesis to explain their population curves. J. Protozool. 11: 56-74.

BLANK, G., AND C. W. SULLIVAN. 1979. Diatom mineralization of silicic acid. III. Silicic acid binding and transport regulation in the marine diatom *Nitzschia angularis*. Arch. Microbiol. 122: 157-164.

BOULDING, E. B., AND T. PLATT. 1986. Variation in photosynthetic rates among individual cells of a marine dinoflagellate. Mar. Ecol. Prog. Ser. 29: 199-203.

CAMPBELL, L. 1985. Investigation of marine phycoerythrin containing *Synechococcus* sp. (Cyanobacteria): Distribution of serogroups and growth rate measurements. Ph.D. dissertation, S.U.N.Y., Stonybrook, NY.

CAMPBELL, L., AND E. J. CARPENTER. 1986. Diel patterns of cell division in marine *Synechococcus* (Cyanobacteria): The use of the frequency of dividing cells technique to measure growth rate. Mar. Ecol. Prog. Ser. 32: 139-148.

CARR, N. 1973. Metabolic control and autotrophic physiology, p. 39-68. *In* N. G. Carr and B. A. Whitton [ed.] The biology of blue green algae. U. Calif. Press., Los Angeles, CA.

CARR, N. G., AND M. WYMAN 1986. Cyanobacteria: their biology in relation to the oceanic picoplankton, p. 159-204. *In* T. Platt and W. K. W. Li [ed.] Photosynthetic picoplankton. Can. Bull. Fish. Aquat. Sci.: 214.

CHISHOLM, S. W. 1981. Temporal patterns of cell division in unicellular algae, p. 150-181. *In* T. Platt [ed.] Physiological bases of phytoplankton ecology. Can. Bull. Fish Aquat. Sci. 210.

CHISHOLM, S. W., AND J. C. COSTELLO. 1980. Influence of environmental factors and population composition on the timing of cell division in *Thlassiosira fluviatilis* (Bacillariophyceae) grown on light/dark cycles. J. Phycol. 16: 375-383.

CHISHOLM, S. W., F. M. M. MOREL, AND W. S. SLOCUM. 1980. The phasing and distribution of cell division in marine diatoms, p. 281-299. *In* P. Falkowski [ed.] Primary productivity in the sea. Plenum, New York, NY.

CHISHOLM, S. W., D. VAULOT, AND R. J. OLSON. 1984. Cell cycle controls in phytoplankton: Comparative physiology and ecology, p. 365-394. *In* L. N. Edmonds [ed.] Cell cycle clocks, M. Dekker.

COOK, J. R., AND T. W. JAMES. 1964. Age distribution of cells in logarithmically growing cell populations. *In* E. Zeuthen [ed.] Synchrony in cell division and growth. Interscience, NY.

CSATORDAY, K., AND G. HORVATH. 1977. Synchronization of *Anacystis nidulans*. Oxygen evolution during the cell cycle. Arch. Microbiol. 111: 245-246.

DONACHIE, W. D. 1981. The cell cycle of *Escherichia coli*, p. 63-83. *In* P. C. L. John [ed.] The cell cycle. Society for Experimental Biology, Seminar Series 10. Cambridge University Press, Cambridge.

DONNAN, L., E. P. CARVILL, T. J. GILLILAND, AND P. C. L. JOHN. 1985. The cell cycles of *Chlamydomonas* and *Chlorella*. 99: 1-40.

EPPLEY, R. W. 1972. Temperature and phytoplankton growth in the sea. Fish. Bull. 70: 1063-1085.

FANTES, P. A. 1977. Control of cell size and cycle time in *Schizosaccharomyces pombe*. J. Cell. Sci. 24: 51-67.

GLIBERT, P. M., T. M. KANA, R. J. OLSON, R. S. ALBERTE, AND D. L. KIRCHMAN. 1986. The response of *Synechococcus* clones WH7803 and WH8018 to nitrogen depletion: Growth and Photosynthesis. J. Exp. Mar. Biol. Ecol. 101: 199-208.

GOLDMAN, J. C. 1977. Temperature effects on phytoplankton growth in continuous culture. Limnol. Oceanogr. 22: 932-935.

GOULD, A. R., N. P. EVERETT, T. L. WANG, AND H. E. STREET. 1981. Studies on the control of the cell cycle in cultured plant cells. I. Effects of nutrient limitation and nutrient starvation. Protoplasma. 106: 1-13.

HEATH, M. R., AND C. P. SPENCER. 1985. A model of the cell cycle and cell division phasing in a marine diatom. J. Gen. Microbiol. 131(3): 411-425.

HERDMAN, M., AND N. G. CARR. 1971. Observations on replication and cell division in synchronous cultures of the blue-green alga *Anacystis nidulans*. J. Bact. 107: 583-584.

HERDMAN, M., B. M. FAULKNER, AND N. G. CARR. 1970. Synchronous growth and genome replication in the blue-green alga *Anacystis nidulans*. Arch. Microbiol. 73: 238-249.

HUTTER, K. J., AND H. E. EIPEL. 1979. Microbial determinations by flow cytometry. J. Gen. Microbiol. 113: 369-375.

JOHN, P. C. L. 1981. The cell cycle. Society for Experimental Biology, Seminar Series 10. Cambridge University Press, Cambridge.

JOHNSTON, G. C., AND R. A. SINGER. 1978. RNA synthesis and control of cell division in the yeast *S. cerevisiae*. Cell 14: 951-958.

LI, W. K. W. 1986. Experimental approaches to field measurements: methods and interpretation, p. 251-286. *In* T. Platt and W. K. W. Li [ed.] Photosynthetic picoplankton. Can. Bull. Fish. Aquat. Sci.: 214.

LORENZEN, H., AND B. D. KAUSHIK. 1976. Experiments with synchronous *Anacystis nidulans*. Ber. Deutsch. Bot. Ges. Bd. 89: 491-498.

LORENZEN, H., AND G. S. VENKATARAMAN. 1969. Synchronous cell divisions in *Anacystis nidulans* Richter. 67: 251-255.

LLOYD, D., R. K. POOLE, AND S. W. EDWARDS. 1982. The Cell Division Cycle. Temporal Organization and Control of Cellular Growth and Reproduction. Academic Press, New York, NY. 523 p.

MANN, N., AND N. G. CARR. 1974. Control of macromolecular composition and cell division in the blue-green alga *Anacystis nidulans*. J. Gen. Microbiol. 83: 399-405.

MCATEER, M., L. DONNAN, AND P. C. L. JOHN. 1985. The timing of division in *Chlamydomonas*. New Phytol. 99: 41-56.

MCDUFF, R. E., AND S. W. CHISHOLM. 1982. The calculations of *in situ* growth rates of phytoplankton populations from fractions of cells undergoing mitosis: A clarification. Limnol. Oceanogr. 27: 783-788.

MELAMED, M. R., P. F. MULANEY, AND M. L. MENDELSON. 1979. Flow cytometry and sorting, Wiley, New York, NY.

NURSE, P., AND E. STREIBLOVA. 1984. The microbial cell cycle. CRC Press, Inc., Boca Raton, Florida. 278 p.

OLSON, R. J., AND S. W. CHISHOLM. 1984. Effects of photocycles and periodic ammonium supply of three marine phytoplankton species. I. Cell division patterns. J. Phycol. 19: 522-528.

——— 1986. Effects of light and nitrogen limitation on the cell cycle of the dinoflagellate *Amphidinium carteri*. J. Plankton Res. 8: 785-793.

——— 1987. Flow cytometry: applications and prospects. *In* R. Alberte and R. Barber [ed.] Photosynthesis in the sea. (in press)

OLSON, R. J., S. L. FRANKEL, S. W. CHISHOLM, AND H. M. SHAPIRO. 1983. An inexpensive flow cytometer for the analysis of fluorescence signals in phytoplankton: Chlorophyll and DNA distributions. J. Exp. Mar. Biol. Ecol. 68: 129-144.

OLSON, R. J., D. VAULOT, AND S. W. CHISHOLM. 1985. Marine phytoplankton measured using shipboard flow cytometry. Deep Sea Res. 10: 1273-1280.

——— 1986. Effects of environmental stresses on the cell cycle of two marine phytoplankton species. Plant Physiol. 80: 918-925.

ONG, L. G., A. N. GLAZER, AND J. B. WATERBURY. 1984. An unusual phycoerythrin from a marine cyanobacterium. Science. 224: 80-83.

PARDEE, A. B. 1974. A restriction point for control of normal animal cell proliferation. Proc. Natl. Acad. Sci. U.S.A. 71: 1286-1290.

PARDEE, A. B., R. DUBRON, J. L. HAMLIN, AND R. F. KLEVTZIEN. 1978. Animal cell cycles. Ann. Rev. Biochem. 47: 715-750.

PARROTT, L. M., AND J. H. SLATER. 1980. The DNA, RNA and protein composition of the Cyanobacterium *Anacystis nidulans* grown in light- and carbon dioxide-limited chemostats. Ann. Microbiol. 127: 53-58.

PAUU, A. S., J. R. COWLES, J. ORO, A. BARTEL, AND E. HUNGERFORD. 1979. Separation of algal mixtures and bacterial mixtures using chlorophyll and ethidium bromide fluorescence. Arch. Microbiol. 23: 271-273.

PAUU, A. S., J. ORO, AND J. R. COWLES. 1978. Application of flow cytometry to the study of algal cells and isolated chloroplasts. J. Exp. Bot. 29: 1011-1020.

PRESCOTT, D. 1976. Reproduction in eucaryotic cells. Academic Press, New York, NY.

QUASTLER, H., AND F. G. SHERMAN. 1959. Cell population kinetics in the intestinal epithelium of the mouse. Exp. Cell. Res. 17: 420-438.

RAVEN, J. 1986. Physiological consequences of extremely small size for autotrophic organisms in the sea, p. 1-70. *In* T. Platt and W. K. W. Li [ed.) Photosynthetic picoplankton. Can. Bull. Fish. Aquat. Sci. 214.

RIVKIN, R. 1985. Carbon-14 labelling patterns of individual marine phytoplankton from natural populations. Mar. Biol. 89: 135-142.

RIVKIN, R., AND H. H. SELIGER. 1981. Liquid scintillation counting for ^{14}C uptake of single algal cells isolated from natural populations. Limnol. Oceanogr. 26: 780-781.

RUBIN, C. G. 1981. Measurements of *in situ* growth rates of *Gonyaulax tamarensis*: The New England red tide organism. M.S. thesis, Massachusetts Institute of Technology, MA. 54 p.

SHAPIRO, H. M. 1985. Practical flow cytometry. Alan R. Liss, Inc., New York, NY. 295 p.

——— 1983. Multistation, multiparameter flow cytometry: A critical review and rationale. Cytometry. 3: 227-243.

SLATER, M. L., S. O. SHARROW, AND J. J. GART. 1977. Cell cycle of *Saccharomyces cerevisiae* in populations growing at different rates. Proc. Nat'l. Acad. Sci., USA. 74: 3850-3854.

SPUDICH, J. L., AND R. SAGER. 1980. Regulation of the *Chlamydomonas* cell cycle by light and dark. J. Cell Biol. 85: 136-145.

TRASK, B. J., G. J. VAN DEN ENGH, AND J. H. B. W. ELGERSHUIZEN. 1982. Analysis of phytoplankton by flow cytometry. Cytometry 2: 258-264.

TYSON, J. J. 1985. The coordination of cell growth and division — Intentional or incidental? BioEssays 2(2): 72-77.

VAULOT, D. 1985. Cell cycle controls in marine phytoplankton. Ph.D. thesis, Massachusetts Institute of Technology, Cambridge, MA, 268 p.

VAULOT, D., R. J. OLSON, AND S. W. CHISHOLM. 1986. Light and dark control of the cell cycle in two marine phytoplankton species. Exp. Cell Res. 167: 38-52.

VISSER, J. W. M., AND G. R. VAN DEN ENGH. 1982. Immunofluorescence measurements by flow cytometry, p. 95-128. *In* Wick et al. [ed.] Immunofluorescence technology: selected theoretical and clinical aspects. Elsevier Biomedical Press, New York, NY.

WARD, B. B., AND M. J. PERRY. 1980. Immunofluorescent assay for the marine ammonium-oxidizing bacterium *Nitrosococcus oceanus*. Appl. Environ. Microbiol. 39: 253-255.

WATERBURY, J. B., S. W. WATSON, F. W. VALOIS, AND D. G. FRANKS. 1986. Biological and ecological characterization of the marine unicellular cyanobacterium *Synechococcus*, p. 71-120. *In* T. Platt and W. K. W. Li [ed.] Photosynthetic picoplankton. Can. Bull. Fish. Aquat. Sci. 214.

WOOD, A. M. 1985. Adaptation of the photosynthetic apparatus of marine ultraplankton to natural light fields. Nature 316: 253-255.

YENTSCH, C. M., F. C. MAGUE, P. K. HORAN, AND K. MUIRHEAD. 1983a. Flow cytometric DNA determinations on individual cells of the dinoflagellate *Gonyaulax Tamarensis* var. *excavata*. J. Exp. Biol. Ecol. 67: 175–183.

YENTSCH, C. M., P. K. HORAN, K. MUIRHEAD, Q. DORTCH, E. M. HAUGEN, L. LEGENDRE, L. S. MURPHY, M. J. PERRY, D. PHINNEY, S. A. POMPONI, R. W. SPINRAD, A. M. WOOD, C. S. YENTSCH, AND B. J. ZAHURENEC. 1983b. Flow cytometry and sorting: A powerful technique with potential applications in aquatic sciences. Limnol. Oceanogr. 28: 1275–1280.

Physiology and Ecology of the Marine Eukaryotic Ultraplankton

Lynda P. Shapiro and Robert R. L. Guillard

Bigelow Laboratory for Ocean Sciences
West Boothbay Harbor, Maine 04575, USA

Introduction

Although the existence of minute eukaryotic phytoplankters has been known for at least 50 years, their enormous abundance in the marine biota has only recently been recognized. This paper summarizes what is known of the distribution of this flora, and of its ecological relationships and basic physiological responses. We begin with some definitions and limitations.

The Eukaryotes

We use the term "eukaryotic" in the simplest sense: cells possessing a nucleus. Thus, we include the dinoflagellates, relatively primitive plankters, lacking a mitotic spindle, which are sometimes referred to as mesokaryotic. The group contrasted to the eukaryotes, the prokaryotes, is discussed at length by others (Waterbury, Carr, Bryant) in this volume, and will be mentioned here only for purposes of comparison with their eukaryotic counterparts.

We define "ultraplankton" as plankters <5 μm in diameter. Since cell flexibility permits passage through a pore slightly smaller than the cell's smallest diameter (Murphy and Haugen 1985), and since the cell's smallest diameter is the significant one in this regard (Thomsen, this volume), we can define ultraplankton operationally as those plankters that pass a 3 μm Nuclepore filter.

Sieburth et al. (1978) established a clear and logical system of size classification, with boundaries corresponding to changes in volume of three orders of magnitude. They placed the photosynthetic phytoplankton in the size category of microplankton (20–200 μm) and nanoplankton (2.0–20 μm), reserving the next smallest size category, the picoplankton (0.2–2.0 μm), for the heterotrophic bacterioplankton.

Since 1978, two facts have become clear; first, that the phytoplankton includes many forms smaller than the nanoplankton, and secondly that these picophytoplankton may at times dominate the plant biomass as well as the numerical abundance. In the open ocean, most prokaryotic phytoplankters are picoplankton. The eukaryotes are often as small, but members of the eukaryotic phytoplankton range upward three orders of magnitude in volume. Based on examination of live natural assemblages and clones in culture, we call attention to the existence and importance of a group of eukaryotic phytoplankters in the size range of 1 μm to perhaps 5 or 6 μm. Thus, an important functional group lies across the boundary between two size categories, picoplankton and nanoplankton. The older term "ultraplankton" has been used historically to describe plankters <5-10 μm (Sverdrup et al. 1942); Jorgensen 1966; Throndsen 1976), and seems appropriate here.

We limit this paper to a consideration of the oxygen-evolving, photosynthetic eukaryotic ultraplankton. Although the heterotrophic eukaryotic ultraplankton may sometimes equal its photosynthetic counterparts in biomass (Davis et al. 1985), these heterotrophs are not the subject of this paper, and, in any case, little is known about their ecology and physiology.

History and Systematics

In the 1930's and 1940's a number of papers described minute flagellates in British and adjacent coastal waters (Cole 1937; Parke 1949; Knight-Jones 1951; others). Many of these papers concerned nutrition of larval forms of commercially valuable bivalve species. However, an overall appreciation of the contribution of these flagellates to biomass and productivity did not follow for several decades.

Hulburt et al. (1960) wrote of the importance of minute flagellates and coccoid forms in the Sargasso Sea and of the difficulties involved in enumerating them. Wood (1963) enumerated ultraplankton in the Pacific and Indian Oceans. The category "unidentified flagellate" frequently appeared in the literature as the most abundant plankter in an otherwise detailed species list (Hulburt and Jones 1977, for example). Then, in the past 10 years, a series of papers (Throndsen 1976; Lewin et al. 1977; Moestrup 1979; Johnson and Sieburth 1982) made obvious the importance of this group. They often outnumber prokaryotes in polluted waters (Ryther 1954; Murphy et al. 1982) and in the open ocean at depth (Murphy and Haugen 1985; Glover et al. 1985a).

The taxonomic affinities of these plankters are described in detail by Thomsen (this volume). Here we give only a brief classification scheme from our own perspective, that is, based more on pigment characteristics (Guillard et al. 1985) than on the elegant ultrastructural details described by Thomsen. Table 1 lists the Classes of marine oxygen evolving eukaryotes and estimates the contribution of each to the ultraplankton.

TABLE 1. Marine oxygen-evolving eukaryotes found in the plankton.

Class	Ultraplankton contribution
Prasinophyceae (*sensu latu*)	widespread, abundant
Chlorophyceae (*sensu strictu*)	mainly in estuaries
Charophyceae	no marine forms of any size
Euglenophyceae	none <5 μm
Arachnidophyceae	none <3 μm
Eustigmatophyceae	estuarine and eutrophic
Rhodophyceae	gametes and 2+ genera
Prymnesiophyceae	important component
Chrysophyceae	important component
Bacillariophyceae	many are <5 μm
Phaeophyceae	gametes only
Rhaphidophyceae	none <5 μm
Xanthophyceae	no certain marine forms
Dinophyceae	a few <3 μm
Cryptophyceae	important component

The Prasinophyceae (*sensu latu*) are a major component of the ultraplankton, especially in open-ocean waters. We include here Mattox and Stewart's (1984) Micromonadophyceae and Christensen's (1962) Loxophyceae. Important genera include *Micromonas*, which at slightly less than 1 μm may be the smallest eukaryote, and a newly described oceanic genus to be named in honor of L. Provasoli (Guillard, Flyod, and O'Kelley, unpubl. data) which we believe to be very important in oligotrophic waters. Also important is the unidentified, but ubiquitous, "scaled prasinophyte" described by several authors including Johnson and Sieburth (1982) and Joint and Pipe (1984). This oceanic species either resists culture, or loses its scales in culture and remains unrecognized. If the latter, then we may have cultured it and described it as the new oceanic genus mentioned above.

Most prasinophytes contain chlorophylls *a* and *b*, with higher *b* to *a* ratios than found in the higher plants or in other chlorophyll *a* and *b*-containing phytoplankton. This is due to the presence of a larger photosynthetic unit containing more chlorophyll *b*, not to the presence of any less chlorophyll *a* (Wood 1979). The major carotenoid in at least three genera, including *Micromonas* and the new oceanic genus, is prasinoxanthin (Foss et al. 1984), which may well be a good systematic marker for one portion of the group (Guillard et al. 1985). Prasinophytes not grouped with *Micromonas* generally have lutein rather than prasinoxanthin (Ricketts 1970). The absence of lutein in oceanic water samples, and the major contribution of prasinoxanthin to the carotenoid complement of those samples (Bidigare, pers. comm.) suggests that the *Micromonas* group is the major component of oceanic prasinophytes. Also important as a marker may be the chlorophyll *c*-like pigment Mg 2,4-divinylphaeoporphyrin a_5 monomethyl ester (Ricketts 1970).

The Chlorophyceae (*sensu strictu*) contain many ultraplanktonic genera (*Chlorella*, *Nannochloris*, some *Chlamydomonas*), but distribution is largely restricted to bays and estuaries. "*Chlorella*-like" cells have been described from the open ocean (Johnson and Sieburth 1982; Joint and Pipe 1984; Takahashi and Hori 1984). These have no scales, but based on pigment content of similar cultured forms, they are likely to be more closely allied with the prasinophytes than with chlorophytes. (In fact, they appear to be unscaled prasinophytes.) The true Chlorophyceae contain chlorophylls *a* and *b*, in ratios similar to terrestrial plants; that is, chlorophyll *a* content exceeds content of the antenna pigment chlorophyll *b*.

The ultraplanktonic marine Eustigmatophyceae are contained in the single genus *Nannochloropsis*. They are abundant in eutrophic, estuarine waters, but are not known from open seas. The cells contain chlorophyll *a* only, no accessory chlorophylls.

Of the Classes containing chlorophylls *a* and *c*, the Prymnesiophyceae (which includes the coccolithophores) and the Bacillariophyceae are important components of the ultraplankton, and many representative clones are in culture. The Chrysophyceae may be fewer and overall less abundant than the prymnesiophytes and bacillariophytes. This has been observed in Norwegian waters (Throndsen 1976), and in Australian waters (Jeffrey and Hallegraeff 1986). We note, though, that relative dominance of a group in a preserved sample may differ from the relative dominance of that group in the natural community prior to preservation. We have found that chrysophytes are lost in preserved samples far more rapidly than are diatoms and coccolithophores. *Pelagococcus subviridis* is widespread (Lewin et al. 1977), and the colonial *Dinobryon*, though seldom reported, often accompanies *Phaeocystis* in spring blooms (unpublished obs.).

Recently, we have identified and cultured phycoerythrin-containing Cryptophycean species <5 μm, and based on our enumerations they can be a major ultraplankton component, both in coastal waters and in the open ocean (Booth et al. 1982; Chang 1983; Haugen and Shapiro, unpubl. data). Like the chrysophytes, they preserve poorly and are probably under-represented in preserved samples. Their presence can be inferred from the presence of the marker carotenoid alloxanthin even when they cannot be visually identified in preserved samples (Gieskes and Kraay 1985). We have also cultured a few Dinophyceaen species in this size range. A few such small species, mostly unarmored, have been described (Hasle 1960).

Yentsch's (1983) observation of chlorophylls *a* and *c* in the material passing a glass fiber filter could be explained by the presence of members of these several classes. Or, the chlorophyll *c* measured may actually have been the chlorophyll *c*-like Mg 2,4-divinylphaeoporphyrin a_5 monomethyl ester, characteristic of the *Micromonas*-group of prasinophytes.

The Rhodophyceae and Phaeophyceae contribute gametes in coastal waters. The remaining phytoplankton classes either do not contain marine forms, or do not contain species known to be small enough to be considered ultraplankton. Thus, the important classes containing ultraplankton in the open ocean are Prasinophyceae, Prymnesiophyceae, Chrysophyceae, Bacillariophyceae, and Cryptophyceae. In estuarine waters, the Chlorophyceae and Eustigmatophyceae are also important, and often most important.

Life Histories

Just as most algal classes are represented in the marine ultraplankton, so also are most kinds of life cycles (though a survey of the literature reveals that for most groups the number of species that have been investigated is very small, and in the case of some groups, it is zero). Chapters in Cox (1980) summarize current knowledge concerning most groups.

The diatoms are diploid. In the larger forms, sexual reproduction appears to occur regularly in nature. There is a progressive diminution in size resulting from asexual reproduction, and obligative sexual restoration of maximal species dimensions. Based on observations in culture, this progressive diminution may not lead to obligate sexual reproduction in the most minute diatoms, and they may be capable of indefinite asexual reproduction. Drebes (1977) provides a detailed review. Coastal dinoflagellates, which are haploid in the vegetative stage, usually have long-lived diploid cysts. Oceanic dinoflagellates may also have life cycles involving non-motile (although not benthic) stages (Tangen et al. 1982). Their ploidy has not been established, though the motile cells are presumably haploid.

What little information exists for prasinophytes and chlorophytes suggests that the planktonic stages are haploid. Sexuality is known to exist in these groups, though it has only been documented in a few species. Probably they can reproduce asexually indefinitely. In the coccolithophores (Prymnesiophyceae) the hapioid and diploid stages of a few species were first described as different genera. Both stages bear coccoliths (though of different kinds) and are planktonic. This situation may be true of many species. Sexuality is not known in the cryptomonads or eustigmatophytes.

Distribution

Table 2 lists marine locations exclusive of estuaries, ponds and tide pools, from which eukaryotic ultraplankters have been reported. Locations included are oceanic and coastal, oligotrophic and eutrophic, polar, temperate and equatorial. We are not aware of reports from Antarctic waters although nanoplankters are important components (El-Sayed and Turner 1977). However, neither are we aware of any systematic attempt to find ultraplankters in the Antarctic. In fact, eukaryotic ultraplankters have been found in all marine ecosystems, oceans, seas and bays that have been examined purposely for them. The group is indeed ubiquitous.

In coastal and estuarine environments, ultraplankton dominance is often indicative of a stressed environment. Parson et al. (1976) found an increase in microflagellates following the addition of low levels of mineral hydrocarbons to enclosed natural assemblages. Thomas and Seibert (1977) further observed that the addition of copper to the same enclosures led to dominance of microflagellates and certain pennate diatoms and disappearance of larger centric diatoms and dinoflagellates. O'Connors et al. (1978) noted changes in the components of natural communities subjected to organic pollutants; these changes favored smaller sized species. The most obvious examples of this have occurred in overly eutrophic estuarine waters, such as Great South Bay, Long Island (USA), where up to 10^9 cells • L^{-1} of a population consisting of the eustigmatophyte *Nannochloropsis salina (then mis-identified* as a *Stichococcus* species) and the chlorophyte *Nannochloris atomus* destroyed a shellfish industry (Ryther 1954). A similar outbreak, with similar consequences, occurred in 1985 in Narragansett Bay, R.I. and Long Island Sound (H. D. Freudenthal, pers. comm.).

TABLE 2. Reported occurrences of marine eukaryotic ultraplankton.

Location	Source
Central North Atlantic	Ryther et al. (1960); Throndsen (1976); Johnson and Sieburth (1982); Estep et al. (1984); Murphy and Haugen (1985); Glover et al. (1986)
Equatorial Atlantic	Gieskes and Kraay (1986)
British Coastal Waters	Cole (1937); Parke (1949); Knight-Jones (1951); Knight-Jones and Walne (1951)
Norwegian Coastal Waters	Throndsen (1969, 1979)
Celtic Sea	Joint and Pomroy (1983); Joint and Pipe (1984)
Gulf of Maine	Murphy and Haugen (1985); Glover et al. (1885a, b); Shapiro and Haugen, unpubl. data
Chesapeake Bay	Van Valkenburg et al. (1978)
Arctic Seas	Throndsen (1970)
Central Pacific	Takahashi and Bienfang (1983); Takahashi and Hori (1984)
Western North Pacific	Furuya and Marumo (1983)
Eastern North Pacific	Lewin et al. (1977)
Bering Sea	Shapiro and Haugen, unpubl. data
Gulf of Alaska	Shapiro and Haugen, unpubl. data; Booth et al. (1982)
Saanich Inlet, Vancouver	Parsons et al. (1976); Thomas and Siebert (1977)
Australian Waters	Wood (1983); Hallegraeff (1981, 1983); Hallegraeff and Jeffrey (1984); Jeffrey and Hallegraeff (1986)
New Zealand Waters	Moestrup (1979); Chang (1983)
South China Sea	Takahashi and Hori (1984)
Japanese Waters	Takahashi et al. (1985)
Indian Ocean	Wood (1963)

It is not clear why ultraplankters are more successful in stressed environments than their larger counterparts. It does seem likely that their resistance to pollution is not a species property, but rather, a property acquired by specific cells. When resistant isolates of a given species have been found, they have always come from environments known or suspected to be polluted. Ultraplankton isolates from near-pristine estuarine and coastal environments have been far more sensitive to chemical stress than closely related isolates from polluted environments (Murphy et al. 1982). The apparent increased survival of certain microflagellates (probably chlorophytes) may be due to their haploid genome which is more directly exposed to selection.

In stratified water, the chlorophyll maximum occurs below the surface layer and is the combined result of an increased number of cells and an increase in the chlorophyll content of each cell. The eukaryotic ultraplankton shows a subsurface maximum below that of the larger eukaryotes and that of the cyanobacteria as well. This peak is not only a chlorophyll maximum, but also a cell maximum. It occurs just below the mixed layer, and therefore in the vicinity of the nutricline (Murphy and Haugen 1985; Glover et al. 1985a). Cells in this zone are largely naked plankters; minute diatoms (and possibly coccolithophores), important ultraplankton components in the mixed layer, are relatively unimportant here (pers. obs.). Altered community dominance is reflected in a changed pigment content of the entire community; there is an increase at depth of chlorophyll b, 19'-hexanoyloxy-fucoxanthin and an unknown fucoxanthin derivative. These changes are indicative of dominance of prasinophytes, prymnesiophytes and chrysophytes (Gieskes and Kraay 1986).

Figure 1 shows the vertical distribution, in stratified waters, of both prokaryotic and eukaryotic ultraplankton in examples of three marine environments. In coastal water, prokaryote numbers dominate at all depths; in oceanic waters and sometimes in slope water, prokaryotes dominate only to the depth of the thermocline. In the low light, high nutrient zone just below the mixed layer, eukaryotic ultraplankton abundance equals and exceeds that of the prokaryotes.

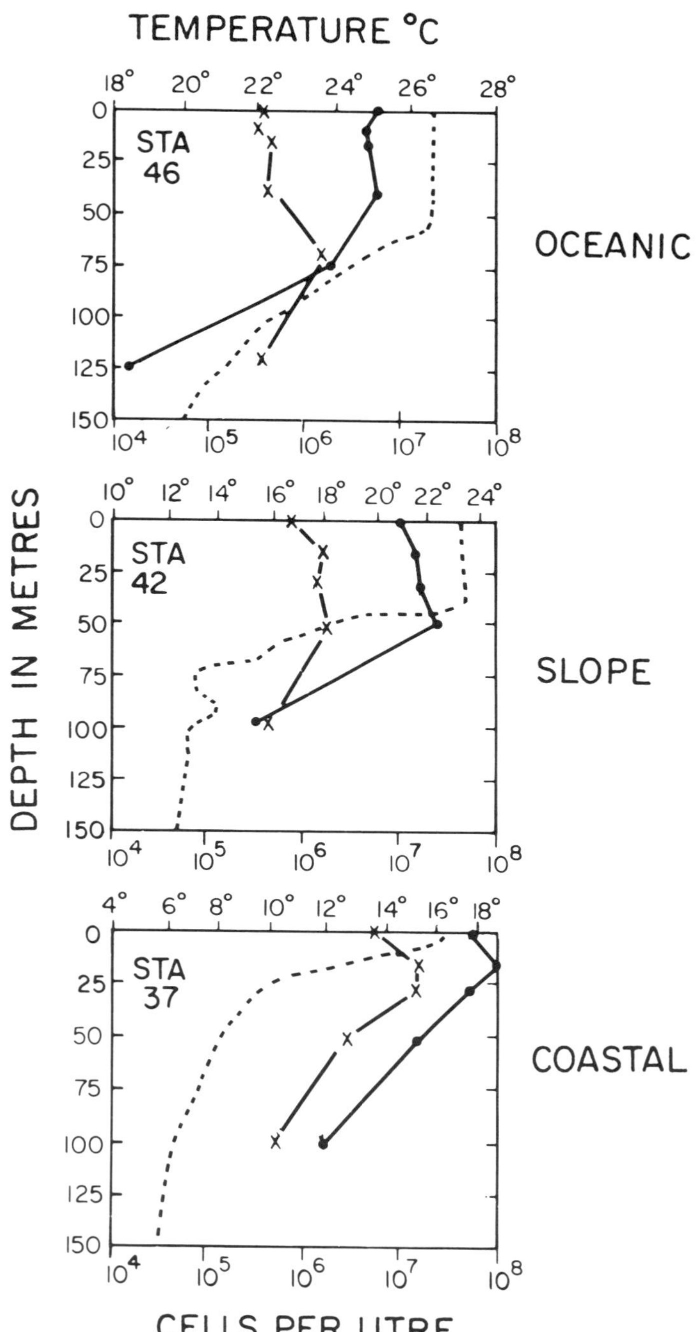

FIG. 1. Vertical distribution of prokaryotic (•———•) and eukaryotic (×———×) ultraplankton in examples of three marine environments (from Murphy and Haugen 1985).

Some General Characteristics

Since both the systematic and geographic extents of the ultraplankton are as wide as those of the larger phytoplankton, it is not surprising to find that the ultraplankton possesses as wide a range of physiological responses as is found in the nano- and microplankton. Further, as do their larger counterparts, ultraplankton species show both genetic variability within populations and genetic differentiation between populations, at least as judged by the meager evidence available. This evidence does include purposeful genetic studies of two diatoms and three coccolithophores. These studies are listed in Table 3, which also refers to certain other experiments in which more than one clone of the same species was examined, and differences revealed that appear to be genotypic. All of the species listed have at least one cell dimension of approximately 3 μm at some point in the life cycle.

Many observations leave little doubt that clones of the same ultraplankton species isolated from shallow bays and from oligotrophic open ocean waters are physiologically quite different, and different in ways that parallel the distinctions between

TABLE 3. Ultraplankton species showing intraspecific genetic variability.

Organism	Study	Source
Thalassiosira pseudonana	Enzyme variability	Murphy and Guillard (1976)
	Temperature effect on growth rate	Brand et. al. (1981)
	Sensitivity to an industrial waste	Murphy and Belastock (1980)
	Effect of free copper ion activity on growth	Gavis et al. (1981)
	Wastes & copper sensitivity	Murphy et al. (1982)
	Interaction of Fe, Mn, Cu on growth	Murphy et al. (1984)
Skeletonema costatum	Enzyme and physiological variability	Gallagher (1982)
	Photoadaptive variability	Gallagher et al. (1984)
	Effect of free copper ion activity on growth	Gavis et al. (1981)
	Wastes & copper sensitivity	Murphy and Belastock (1980); Murphy et al. (1982)
	Sensitivity to Zn toxicity	Jensen et al. (1974)
	Lipid fractions related to Cu tolerance	Shifrin and Chisholm (1981)
Minutocellulus (ex Bellerochea)	Responses to vitamins, salinity, temperature	Hargraves and Guillard (1974)
polymorphus	Differences in nitrage uptake capability	Carpenter and Guillard (1971)
	Influence of *p*Cu on growth	Gavis et al. (1981)
Extubocellulus (ex *Bellerochea*) *spinifer*	Responses to vitamins, salinity, temperature	Hargraves and Guillard (1974)
Fragilaria pinnata	Responses to vitamins, salinity, temperature	Hargraves and Guillard (1974)
	Differences in nitrate uptake capability	Carpenter and Guillard (1971)
Leptocylindrus danicus	Alternate life cycles in culture studies	French and Hargraves (1985)
Navicula pelliculosa	Responses to vitamins, salinity, temperature	Hargraves and Guillard (1979); Lewin (1955)
Phaeodactylum tricornutum	Growth responses to light, salinity, temperature, organic substances (one strain differed from 4 others)	Hayward (1968)
Emiliania huxleyi	Intraspecific differences in growth rates	Brand (1981, 1982)
	Responses to salinity	Brand (1984)
	Responses to artificial seawater formulations	Guillard, unpubl. data
Gephyrocapsa oceanica	Responses to artificial seawater formulations	Brand (1981, 1982)
	Cell division periodicity	Nelson and Brand (1979)
Cyclococcolithina leptopora	Cell division periodicity	Brand (1981)
Phaeocystis pouchetii	Responses to temperature Excretion of metabolites	Guillard and Hellebust (1971)
Nannochloropsis (ex *Monallantus*) *salina*	Growth responses to free cupric ion activity	Gavis et al. (1981)
	Lipid fractions related to Cu tolerance	Shifrin and Chisholm (1981)

neritic and oceanic species of the larger phytoplankton. A coastal clone of *Micromonas* is more like the neritic *Skeletonema* in its culture requirements than it is like an oceanic clone of *Micromonas*. A clone of *Emiliania huxleyi* from the Oslofjord grows easily in artificial seawater, but a conspecific clone from the Sargasso Sea resisted such culture for decades. Within clones of coccoid prasinophytes there have been differences in culture behavior, even though all are of oceanic origins.

The ultraplankton's only common denominator is size, and there are certainly consequences of being small (Raven, this volume). They are more buoyant than larger plankton (Smayda and Bienfang 1983), and approach neutral density (Takahashi and Bienfang 1983) unless they occur in aggregates (Silver, this volume). The large surface-to-volume ratio can increase uptake rates, growth rates, respiration rates, and photosynthetic capacity (Malone 1980). Nutrient reserves may be less than those of larger forms since the limited volume may limit storage capacity. After packing the required amounts of DNA, RNA, essential enzymes, membranes, and at least one chloroplast and mitochondrion into a sphere of ~ 1 μm, there may not be much room left for particles of solid food reserves or lipid droplets. And, as cell size approaches the wavelengths of visible light, cells become increasingly efficient photon absorbers (Morel, this volume).

Beyond these features, all attributable to size, the species comprising the ultraplankton share little in common; their physiological responses, like those of the larger plankton, are as varied as their origins and systematic affinities. The coastal forms are far better studied than their oceanic counterparts; some of the earliest studies of phytoplankton physiology utilized members of the ultraplankton. The oceanic ultraplankton is of much more recent attention, and the balance of this paper will emphasize recent findings on the eukaryotic ultraplankton from the oligotrophic open ocean.

Primary Productivity

It has become obvious that cells <3 μm in diameter make a major contribution to marine productivity and that in oligotrophic waters they can account for as much as 80% of that productivity (Li et al. 1983; Bienfang and Takahashi 1983; Berman et al. 1985; Glover et al. 1985a). Even these numbers may be underestimates. Based on the notorious difficulty of maintaining oceanic ultraplankters in culture, it seems reasonable to suspect that container effects inhibit the smaller oceanic cells more than larger ones.

Far less obvious at this time is the relative contribution of eukaryotes vs. prokaryotes to the <3 μm productivity total. Since these populations overlap in size, and since in this size range small differences in dimension can be reflected in relatively large differences in volume, filter fractions dominated by one or the other type cannot give us unequivocal answers. Consider, for example, a 1 μm filtrate sample containing 80% cyanobacteria. Assume the cyanobacteria average 0.8 μm in diameter, and thus 0.085 μm^3 in volume. Since the eukaryotes' much greater cell wall flexibility allows cells >1 μm to pass a 1 μm filter, assume the 20% eukaryotes average 1.3 μm in diameter. Their average volume would be 0.366 μm^3 and their total contribution to the biomass would be similar to that of the prokaryotes. Further, if the photosynthetic rates being measured experimentally were carried out in blue light, as should be since that is the light that penetrates to the region of their maximum abundance, then the greater spectral efficiency of the eukaryotes in blue light means that they are responsible for a disproportionately larger amount of the measured photosynthesis (see below). Thus, productivity attributed to the numerically dominant prokaryotes may, in fact, be due largely to eukaryotic cells.

Research is now in progress attempting to evaluate the productivity of these two moieties separately. One promising approach uses inhibitors specific either to prokaryotes or to eukaryotes. Another approach physically separates the two types based on their different fluorescence properties. When the separate productivities are determined (as surely they will be) a second problem must be addressed: the effect of separation of community components on the rate processes of each component. We cannot assume productivity rates will remain unaltered in the face of the drastic alteration caused by the separation. Mutualism, allelopathy, and grazing may operate significantly.

Until the results of such studies are available, most of our understanding of the pro- and eukaryotic contributions will be based on relative photosynthetic responses of clones in culture. Some recent studies illustrate the complexities of the problem.

Light Responses

Glover (1985) compared the saturating light fluxes for growth and photosynthesis of eukaryotic ultraplankters with those of prokaryotic ultraplankters and those of larger eukaryotes, using as light source unfiltered "cool-white" fluorescent bulbs ("white light"). Table 4 shows that the saturating fluxes for most of the smaller

TABLE 4. Light requirements (white light, $\mu E \cdot m^{-2} \cdot s^{-1}$).[a]

Photoautotroph	Approximate cell volume (μm^3)	Light Intensity for Saturation	
		Growth	Photosynthesis
Prokaryote ultraplankton			
various *Synechococcus*	1	45–55	85–130
Eukaryote ultraplankton			
Chlorophytes			
7–15	~6	92	433
Prasinophytes			
IB4	4	100	216
Ω48–23	20	181	357
25–2E	34	50	249
Chrysophytes			
1935	14	130	473
Prymnesiophytes			
IIB3	35	141	390
Iso	48	—	178
IG7	65	101	448
IIB6	65	161	282
Larger eukaryotes			
4 clones	120–524	79–118	188–227

[a]Modified from Glover (1985).

eukaryotes were higher than those of the prokaryotes, and essentially similar to those of the larger eukaryotes. It should be borne in mind that all the clones used, both pro- and eukaryotic, were isolated from near-surface waters except for the eukaryotic ultraplankter 25-2E, which was collected from 88 m at an oceanic station just north of the Azores. Its saturating flux in white light more nearly resembled that of the prokaryotes, a tantalizing hint of habitat-related adaptation.

Provocative though this experiment was, implying better adaptation of the prokaryote *Synechococcus* to lower light fluxes and hence to life deeper in the euphotic zone, it adopted a primarily eukaryotic model of the role of accessory pigments in photosynthesis. That is, accessory light harvesting pigments were viewed as increasing both the number of photons which could be absorbed by the photosynthetic apparatus, and as adding to the chromatic window of photosynthetically usable light.

If, however, for different phytoplankters, the chlorophyll fluorescence emission at 680 nm is measured over an excitation scan of 300–700 nm (Fig. 2), it can be seen that the excitation spectrum maximum is characteristically different for eukaryotes and prokaryotes. The *Synechococcus* clone DC-2 shows maximal chlorophyll emission

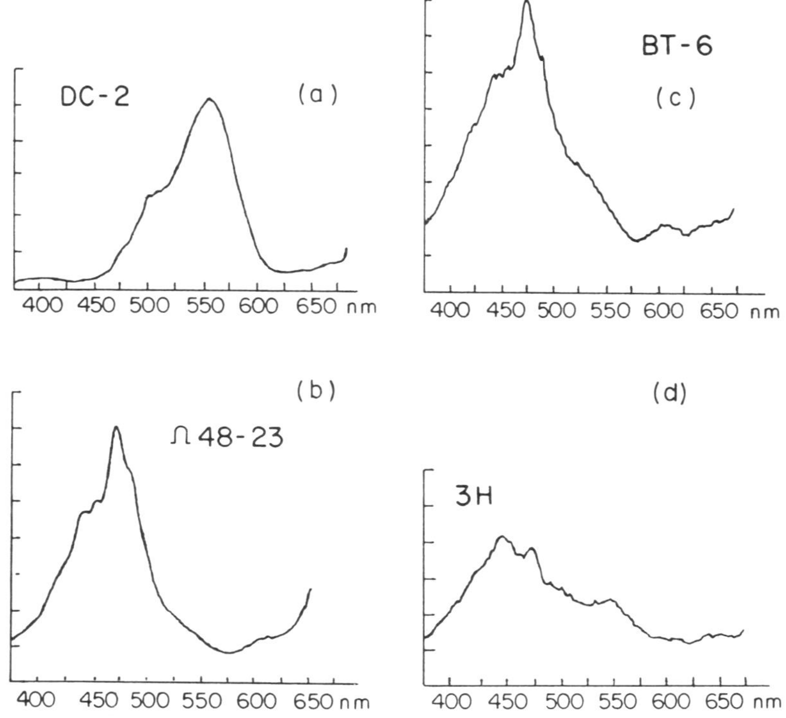

FIG. 2. Fluorescence excitation spectra of ultraplankton-sized clones. (a) *Synechococcus* clone DC-2; (b) prasinophyte clone Ω48–23; (c) coccolithophore clone BT-6; (d) diatom clone 3H.

following excitation at 540 nm, which means that chlorophyll excitation is by way of the phycobiliprotein. There is no indication of direct excitation of chlorophyll emission (Yentsch and Phinney 1985; Wood et al. 1985). The prasinophyte Ω48–23, the coccolithopore BT-6, and the diatom 3H, show clear excitation maxima at 465 nm; the coccolithophore shows a minor excitation bulge at ~530 nm and the diatom a more pronounced one, due to chlorophyll excitation by way of accessory pigments (19′ hexanoyloxyfucoxanthin, Haxo (1985), in the case of the coccolithophore, and fucoxanthin in the case of the diatom). This pattern was first observed by Wood (1985) who noted that it results from fundamental differences in the organization of the photosynthetic apparatus. In eukaryotic organisms, the accessory light harvesting pigments are intrinsic to the thylakoid membrane and include an abundance of chlorophyll pigment proteins which absorb light between 400–450 nm; in prokaryotes, the principal accessory light harvesting pigments are found in phycobilisomes which are extrinsic to the thylakoid membrane and which absorb longer wavelengths. Thus it would be expected that eukaryotic ultraphytoplankton would be better adapted for photosynthesis in the blue light which predominates at the bottom of the euphotic zone in oligotrophic waters than prokaryotic ultraphytoplankters. Two recent studies have shown just that.

Glover et al. (1986b) compared growth and photosynthetic response of prokaryotic and eukaryotic ultraplankton exposed to variable light quality. Figure 3 shows the characteristics of the colored filters passing blue-violet, blue and green light. Growth

380

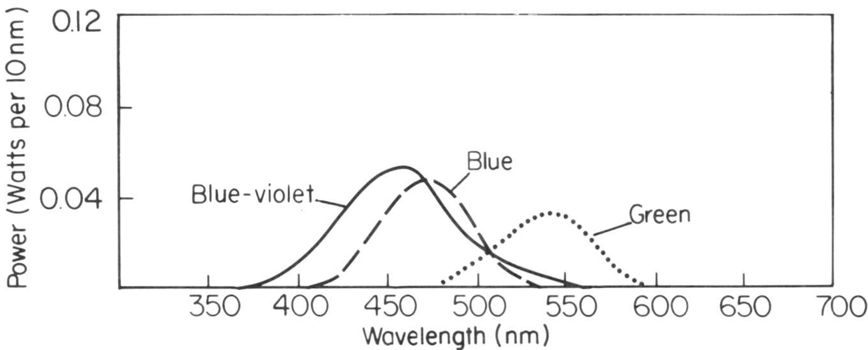

FIG. 3. Transmission characteristics of colored filters used by Glover et al. (1986b) (Courtesy of H. E. Glover).

efficiencies of a prokaryote (the *Synechococcus* clone DC-2) and a minute unidentified prymnesiophyte (clone IIE6) were compared in these three light fields and in white light of comparable photon fluxes (Fig. 4a). While growth efficiencies of both clones were similar in white light, that of the eukaryote was greater in blue-violet and blue light and that of the prokaryote was greated in green light. Similarly, the prokaryotes (clones DC-2 and WH7805) and a minute prasinophyte (clone Ω48–23) showed qualitatively the same pattern as the first experiment (Fig. 4b).

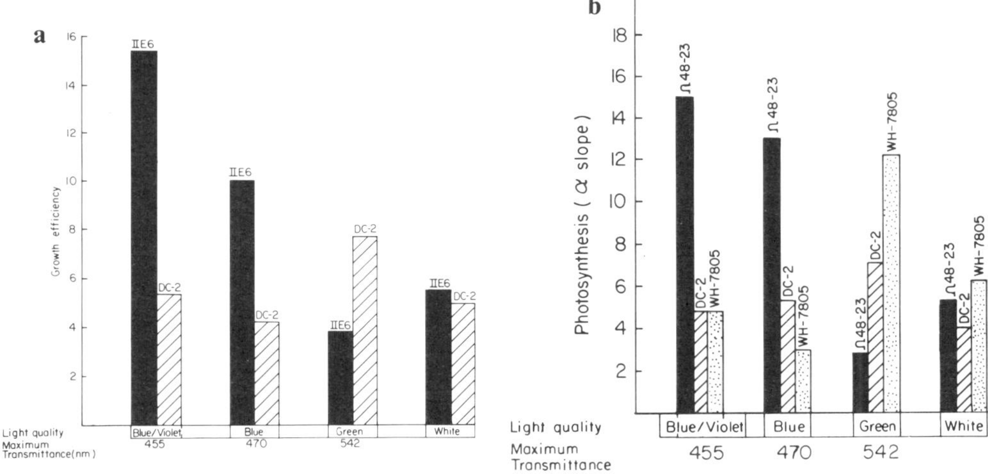

FIG. 4. Effects of light qualities on prokaryotic and eukaryotic ultraplankton clones: (a) growth efficiencies, (b) photosynthetic efficiencies (α slope) where

$$\alpha \text{ slope} = \frac{\text{photosynthesis (mg C} \cdot \text{mg Chl } a^{-1} \cdot \text{h}^{-1})}{\text{irradiance (quanta} \times 10^{-16} \cdot \text{cm}^{-2} \cdot \text{s}^{-1})}$$

(from Glover et al. 1986b).

The second study, by Wood (1985), compared, in various light fields, photosynthesis of clones of *Synechococcus* having different kinds of phycoerythrins, with photosynthesis of eukaryotes clones. The clones were all acclimated to low fluxes of unfiltered "cool white" fluorescent light, then placed in the different light fields (natural light at 20, 40, and 60 m depths in clear ocean water and artificial white light of the same fluxes). Not surprisingly, the eukaryotes showed greater ability to adapt to the changing spectral composition that occurs with increasing depth (Table 5). They photosynthesized more rapidly than the prokaryotes in natural light, and

TABLE 5. Comparison of photosynthetic rates of different clones.[a]

Environmental conditions	Total PAR (μE $\cdot$ m^{-2} $\cdot$ s^{-1})	% Surface irradiance	Rate of photosynthesis (fg C $\cdot$ cell^{-1} $\cdot$ h^{-1})		
			WH 7803	13-1	NEP
AWL	260	43.3	29.6	3 451.3	1 181.9
AWL	70	11.7	14.4	2 115.3	576.2
AWL	36	6.0	5.0	725.4	190.8
20 m	240	40.0	21.4	3 688.5	910.7
40 m	100	16.7	10.5	2 242.7	561.9
60 m	40	6.7	4.3	1 049.6	323.0

[a]Modified from Wood (1985).

more rapidly in natural light than in artificial white light. Wood concludes that simulating the natural light field with artificial white light of equal flux results in an overestimate of the prokaryotic contribution and an underestimate of the eukaryotic contribution. Thus, whenever prokaryotic activity is minimal (see Lewis et al., this volume), determination of photosynthetic rates in artificial white light will lead to an underestimate of the total ultraplankton contribution, even at low light intensities.

It is clear, therefore, that the eukaryotic ultraplankton is better adapted to the light regime at the depth below that of the maximum abundance of *Synechococcus*, which is the depth at which they are most abundant in nature.

Nutrition

Brand (this volume) discusses in detail the difficulties involved in culturing ultraplankton clones. We include here only some comments based on our own experiences.

Eukaryotic ultraplankton in culture show enormous individuality in their responses to different media. All respond well to relatively high levels of nutrients (i.e., they grow rapidly), which may relate to their dominance at a depth in the vicinity of the nutricline (Murphy and Haugen 1985). All are favored by, or at least tolerate, high levels of chelation, implying that they do best at low ambient levels of trace metals, including those supplied inadvertently, as contaminants. They can be flexible in meeting their nitrogen requirement: of some 30 clones tested, all but one (which required ammonium) grew well on nitrate, most did well on ammonium or urea at the levels tested. All used inorganic or organic phosphate. All required thiamine, most required B_{12}, and 2 non-motile green clones (probably prasinophytes) required biotin (L. Provasoli and I. Pintner, pers. comm.), which is most unusual.

Alternate modes of nutrition — heterotrophy and phagotrophy — are known in these chlorophyll-containing cells (Wood 1965; Beers et al. 1975; Hallegraeff 1983; Joint and Pomroy 1983; Fuhrman and McManus 1984; Davis et al. 1985; Bird and Kalff 1986; Estep et al. 1986). The ability of pigmented cells to ingest solid food particles may be limited to the chrysophytes and dinoflagellates (Wood 1965). We note that many ultraplankton clones in culture do well as long as they are accompanied by bacteria, and much less when axenic. (This relationship is also known for some larger phytoplankters and seaweeds.) Many of these clones, and many cells in live samples, have a low cellular chlorophyll content, again supporting the idea that photosynthesis is not their only means of meeting energy requirements.

Several trace elements, in addition to those customarily added to marine phytoplankton media, frequently enhanced growth of certain eukaryotic ultraplankton clones; absolute requirements have not been demonstrated. Nickel and vanadium have augmented growth in at least some clones. Selenium enhanced growth in an oceanic

Micromonas clone (Fig. 5a) but not an estuarine clone (Fig. 5b) (Keller and Guillard, unpubl. data). These effects are not (yet) reliably repeatable.

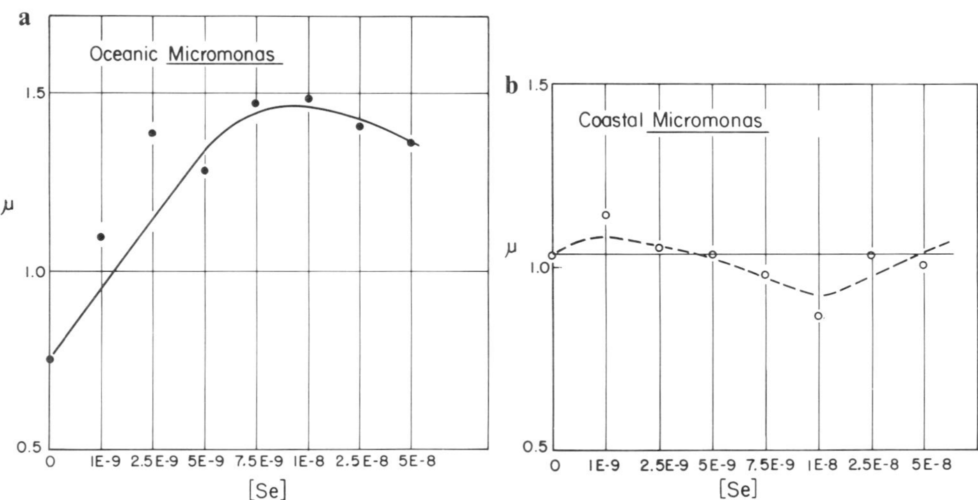

FIG. 5. Effects of selenium addition on growth rates of *Micromonas*: (a) oceanic clone, (b) coastal clone.

Oceanic phytoplankton species of all sizes can grow at levels of iron, manganese, and zinc too low for coastal species (Brand et al. 1983; Murphy et al 1984). This habitat-related difference is an evolutionary consequence of the characteristically different trace metal regimes of the two environments and appears not to be size-related.

Trace Metal Toxicity

The ratio of availabilities of trace metals, including the micronutrients iron and manganese, is an important factor in evaluating the toxicity of any one trace metal. Table 6 lists some important interactions evaluated to date. These studies utilized two ultraplankton-sized species of the diatom *Thalassiosira* (*T. pseudonana* and *T. oceanica*) and a related, somewhat larger species (*T. weissflogii*). To summarize briefly, the responses of these and other plankters to the Mn/Cu ratio, and the effects of iron availability on that ratio, differ significantly in the various species studied and are habitat-related. There is good reason to think that the Fe–Mn–Cu interaction is particularly significant in the case of the eukaryotic ultraplankton because oceanic and coastal species (or races) are subjected to trace metal regimes as different as any that exist in the marine environment outside of polluted estuaries. Since the size range

TABLE 6. Trace metal interactions on growth.

Species	Metal interaction	Source
Thalassiosira		
pseudonana	Cu/Zn	Rueter and Morel (1981)
	Cu/Mn	Sunda and Huntsman (1983)
	Cu/Mn/Fe	Murphy et al. (1984)
Thalassiosira oceanica	Cu/Mn	Sunda and Huntsman (1983)
	Cu/Mn/Fe	Murphy et al. (1984)
Thalassiosira weissflogii	Cd/Fe	Harrison and Morel (1983)
	Cu/Mn/Fe	Murphy et al. (1984)

of plankters studied to date is so small, the presence or absence of a size-related response pattern cannot yet be determined.

Polluted estuaries are usually dominated by small "green" flagellates and coccoid cells; a similar flora arose following copper addition to an enclosed community in a relatively clean estuary (Thomas and Siebert 1977).

Outside of estuaries, little is known about the trace metal tolerances of the eukaryotic ultraplankton as a distinct flora. The broadest study to date of the relation between tolerance on one hand, and systematic position or habitant on the other, was that of Brand et al. (1986). They found that diatoms were less sensitive to copper and cadmium than were coccolithophores and dinoflagellates, and that eukaryotes in general were less sensitive than were the chroococcoid prokaryotes. No habitat-related differences were observed. The study did include several minute diatoms and coccolithophores, but representatives of the other groups, the prasinophytes, chlorophytes, eustigmatophytes, chryosphytes, cryptomonads and prymnesiophytes other than coccolithophores, were not included. The responses of these groups is of special interest because of their distributional maximum, deep in the euphotic zone, at the nutricline, where the high copper levels (up to $10\times$ the surface activities), combined with low manganese levels (about $10\times$ lower than at the surface) present the worst combination for phytoplankton growth (Sunda et al. 1981; Sunda and Huntsman 1983). As far as we are aware, no isolates from these depths have ever been tested for heavy metal tolerance. As for the ultraplankton-sized diatoms and coccolithophores tested by Brand and co-workers, they responded no differently than did larger members of their respective groups.

Nutrient Supply and Ultraplankton Growth

Liebig's Law of the Minimum states that the magnitude of any crop cannot exceed that permitted by complete utilization of whichever nutrient is present in least supply relative to need. Ketchum (1954) made the important point that supply "relative to need" meant "relative to the minimum requirement," since plant cells in general, and phytoplankton cells in particular, may change their nutrient contents (cell quotas) in response to various environmental influences. Further, he linked the "optimum concentration" of a nutrient with cellular nutrient accumulation (uptake). Thus, he said "The optimum concentration is based upon the concept that too low a concentration may be limiting (to growth rate) as a result of a decreased rate of assimilation . . . " Almost simultaneously, Harvey (1955) began advancing the idea that growth rate, rather than the crop, could be limited by low nutrient levels in nature, and that the species composition of the flora could be dictated, at least in part, by nutrient levels. His reference was to selection in favor of small cells by low nutrient conditions.

The intuitively simple idea that low nutrients levels reduce growth rates and hence offer the possibility for control of both crops and relative species abundance has proven unexpectedly difficult to verify. Reviews of note include McCarthy (1982) on kinetics of nutrient adaptation, Dugdale et al. (1981) on adaptation in assimilation processes, and Eppley (1981) on the relations between assimilation and growth. The only habitat-related or size-related differences found were in the uptake kinetics of nitrate (or ammonium) (Eppley et al. 1969) using phytoplankton species of different sizes, systematic groups, and habitats, and of nitrate uptake kinetics in clones of small diatoms from oceanic or estuarine waters (Carpenter and Guillard 1971). These studies showed that overall, smaller species and those species or clones from oceanic (low nutrient) environments had lower half saturation constants.

In view of the many other environmental factors that can influence nitrate uptake, and in spite of other complications associated with cellular function, these early studies may be essentially correct in the main conclusion to be drawn from them, which is that the availability of nitrogen has been a significant ecological selection force. This remains to be demonstrated rigorously for any alga, and in particular for any of the species characteristic of the deeper portions of the euphotic zone, where light is certainly limiting, but where the effect of the relatively high nutrient regime is unknown. We can expect that constant or episodic upward diffusive processes, as well as regeneration via grazing, will be important. Whether or not the species inhabiting the region are selected for efficiency in nitrogen source uptake, or versatility in the use of chemical species, or both, remains to be determined. The only facts at present are that ultra-sized clones now in culture from oligotrophic waters are versatile, and they are able to tolerate about as high levels of nitrogen sources as can most of the larger species studied. Behavior at the low end of the nutrient scale is unknown.

Ultraplankton as Food

One of the decade's biggest surprises has been the realization that much of chlorophyll-related carbon in the sea is packaged in very small particles. Since approximately 40% of global primary production is marine (Bolin et al. 1977) and up to 80% of the chlorophyll in seawater is packaged in particles that can pass a 1 μm filter (Li et al. 1983), as much as one-third of the global primary production is ultraplanktonic. Basing an estimate simply on biomass, and ignoring possible differences in photosynthetic efficiency, much less than half of this production can be attributed to eukaryotes. Nonetheless, the eukaryotic ultraplankton represent a significant portion of global carbon.

Recognition of this extensive biomass has led to a number of recent investigations of its role in marine food as webs. The following is a synopsis of past experience and an emerging picture.

Young stages of copepods (Nival and Nival 1976) and bivalve larvae (Jørgensen 1966; Fritz et al. 1984) feed only on ultraplankton-sized cells, but adult copepods are not efficient at filtering particles less than 5 μm in diameter (Frost 1972; Boyd 1976; Nival and Nival 1976), and adult bivalves are most efficient at filtering particles greater than 3–5 μm (Jørgensen 1966). Salps are less efficient at filtering particles smaller than 4 μm, but do filter even 1 μm particles, although with reduced efficiency (Harbison and McAlister 1979; Deibel 1985). It is reasonable to assume that in waters where over 90% of the particles are less than 5 μm in diameter, these particles are difficult to avoid; animals no doubt ingest them by whatever processes they acquire food and, if the particles can be digested, they will serve as nutrition.

Copepod fecal pellets often appear to be composed entirely of cyanobacteria and small, green or olive-green coccoid cells. When the pellets are teased apart, the cells appear intact (Silver and Alldredge 1981), suggesting that although they can be ingested, they are not utilized, or at least not fully utilized. Mucus net feeders also produce fecal pellets rich in cyanobacterial and minute eukaryotic cells (Silver and Bruland 1981). While it is obvious that at least some portion of the ultraplankton is not digested by these herbivores, there is no evidence yet available establishing whether or not another portion is digested.

Intact cells sink rapidly when contained in fecal pellets and marine snow (Smayda 1969; Small et al. 1979; Bruland and Silver 1981; Silver, this volume). By this route, undigested ultraplankton cells, or certain types at least, can be transported downward. It is not known whether or not these cells can be digested by benthic feeders.

Evidence to date suggests that eukaryotic and prokaryotic ultraplankters are preyed upon by heterotrophic flagellates and ciliates (Fenchel 1982; David et al. 1985; Gifford 1985; Wood, Sherr, and Sherr, unpublished data). The major predators may be the heretotrophic flagellates and these in turn are eaten by ciliates. The protozoans then presumably are eaten by the larger zooplankton. This may be a major route by which ultraplankton enter the food web. If so, the importance of the ultraplankton as a source of carbon to larger animals would depend on the efficiency of energy transfer at these lower trophic levels.

Acknowledgements

We thank E. M. Haugen, H. E. Glover, A. M. Wood, and S. E. Shumway for helpful comments and suggestions. This work was supported by NSF grant OCE-8415963. This is Bigelow Laboratory for Ocean Sciences contribution number 86003.

References

BEERS, J. R., F. M. H. REID, AND G. L. STEWART. 1975. Microplankton of the North Pacific Central Gyre, Population structure and abundance, June 1973. Int. Rev. Gesamten Hydrobiol. 60: 607–638.

BERMAN, T., D. W. TOWNSEND, S. Z. EL SAYED, C. C. TREES, AND Y. ASOV. 1985. Optical transparency, chlorophyll and primary productivity in the Eastern Mediterranean near the Israeli coast. Oceanol. Acta 7: 367–372.

BIENFANG, P. K., AND M. TAKAHASHI. 1983. Ultraplankton growth rates in a subtropical ecosystem. Mar. Biol. 76: 213–218.

BIRD, D. F., AND J. KALFF. 1986. Bacterial grazing by planktonic lake algae. Science 231: 493–495.

BOLIN, B., E. T. DEGENS, P. DUVIGNEAUD, AND S. KEMPE. 1977. The global biogeochemical carbon cycle, p. 1–53. In B. Bolin, E. T. Degens, S. Kempe and P. Ketner [ed.] The global carbon cycle. Wiley, New York, NY.

BOOTH, B. C., J. LEWIN, AND R. E. NORRIS. 1982. Nanoplankton species predominate in the subarctic Pacific in May and June 1978. Deep-Sea Res. 29: 185–200.

BOYD, C. M. 1976. Selection of particle sizes by filter-feeding copepods: a plea for reason. Limnol. Oceanogr. 21: 175–180.

BRAND, L. E. 1984. The salinity tolerance of forty-six marine phytoplankton isolates. Estuarine Coastal Shelf. Sci. 18: 543–556.

——— 1982. Genetic variability and spatial patterns of genetic differentiation in the reproduction rates of the marine coccolithophores Emiliania huxleyi and Gephyrocapsa oceanica. Limnol. Oceanogr. 27: 236–245.

——— 1981. Genetic variability in reproduction rates in marine phytoplankton populations. Evolution 35: 1117–1127.

BRAND, L. E., R. R. L. GUILLARD, AND L. S. MURPHY. 1986. Reduction of marine phytoplankton reproduction rates by copper and cadmium. Mar. Biol. (in press)

——— 1981. A method for the rapid and precise determination of acclimated phytoplankton reproduction rates. J. Plank. Res. 3: 193–201.

BRAND, L. E., W. G. SUNDA, AND R. R. L. GUILLARD. 1983. Limitation of marine phytoplankton reproduction rates by zinc, manganese, and iron. Limnol. Oceanogr. 28: 1182–1198.

BRULAND, K. W., AND M. W. SILVER. 1981. Sinking rates of fecal pellets from gelatinous zooplankton (salps, pteropods, doliolids). Mar. Biol. 63: 295–300.

CARPENTER, E.J., AND R. R. L. GUILLARD. 1971. Intraspecific differences in nitrate half-saturation constants for three species of marine phytoplankton. Ecology 52: 183–185.

CHANG, F. H. 1983. Winter phytoplankton and microzooplankton populations off the coast of Westland, New Zealand, 1979. J. Mar. Freshw. Res. 17: 279–304.

CHRISTENSEN, T. 1962. Alger. In T. W. Böcher, M. Lange, and T. Sørensen [ed.] Botanik, II, No. 2. Munksgaard, Copenhagen. 178 p.

COLE, H. A. 1937. Experiments in the breeding of oysters in tanks with special reference to the food of the larva and spat. Fish. Invest. Ser. 2, 15, No. 14

COX, E. R. 1980. Phytoflagellates. Elsevier, North Holland. 473 p.

DAVIS, P. G., D. A. CARON, P. W. JOHNSON, AND J. McN. SIEBURTH. 1985. Phototrophic and apochlorotic components of picoplankton and nanoplankton in the North Atlantic: geographic, vertical, seasonal and diel distributions. Mar. Ecol. Prog. Ser. 21: 15–26.

DEIBEL, D. 1985. Clearance rates of the salp Thalia democratica fed naturally occurring particles. Mar. Biol. 86: 47–54.

DREBES, G. 1977. Sexuality, p. 250–283. In D. Werner [ed.] The biology of diatoms. Blackwell Sci. Pubs.

DUGDALE, R. C., B. H. JONES, JR., J. J. MACISAAC, AND J. J. GOERING. 1981. Adaptation of nutrient assimilation, p. 234–250. In T. Platt [ed.] Physiological bases of phytoplankton ecology. Can. Bull. Fish. Aquat. Sci. 210.

EL SAYED, S., AND J. T. TURNER. 1977. Productivity of the Antarctic and tropical/subtropical regions: A comparative study, p. 463–501. In Polar Oceans, Proc. Polar Oceans. Conf. Montreal.

EPPLEY, R. W. 1981. Relations between nutrient assimilation and growth in phytoplankton with a brief review of estimates of growth rate in the ocean, p. 251–263. In T. Platt [ed.] Physiological bases of phytoplankton ecology. Can. Bull. Fish. Aquat. Sci. 210.

EPPLEY, R.W., J. N. ROGERS, AND J. J.

MacCarthy. 1969. Half-saturation constants for uptake of nitrate and ammonium by marine phytoplankton. Limnol. Oceanogr. 14: 912–920.

Estep, K. W., P. G. Davis, P. E. Hargraves, and J. McN. Sieburth. 1984. Chloroplast containing microflagellates in natural populations of North Atlantic nanoplankton, their identification and distribution; including a description of five new species of *Chrysochromulina* (Prymnesiophyceae). Protistologica 20: 613–634.

Estep, K. W., P. G. Davis, M. D. Keller, and J. McN. Sieburth. 1986. How important are oceanic algal nanoflagellates in bacterivory? Limnol. Oceanogr. 31: 646–650.

Fenchel, T. 1982. Ecology of heterotrophic microflagellates. II. Bioenergetics and growth. Mar. Ecol. Prog. Ser. 8: 225–231.

Foss, P., R. R. L. Guillard, and S. Liaaen-Jensen. 1984. Prasinoxanthin — a chemosystematic marker. Phytochemistry 23: 1629–1633.

French, F. W. III, and P. Hargraves. 1985. Spore formation in the life cycles of the diatoms *Chaetoceros diadeina* and *Leptocylindrus danicus*. J. Phycol. 21: 477–483.

Fritz, L. W., R. A. Lutz, M. A. Foote, C. L. Van Dover, and J. W. Ewart. 1984. Selective feeding and grazing rates of oyster (*Crassostrea virginica*) larvae on natural phytoplankton assemblages. Estuaries 7: 513–518.

Frost, B. W. 1972. Effects of size and concentration of food particles on the feeding behavior of the marine planktonic copepod *Calanus pacificus*. Limnol. Oceanogr. 17: 805–815.

Fuhrman, J. A., and G. B. McManus. 1984. Do bacteria-sized marine eukaryotes consume significant bacterial production? Science 224: 1257–1260.

Furuya, K., and R. Marumo. 1983. The structure of the phytoplankton community in the subsurface chlorophyll maxima in the western North Pacific Ocean. J. Plank. Res. 5: 393–406.

Gallagher, J. C. 1982. Physiological variation and electrophoretic banding patterns of genetically different seasonal populations of *Skeletonema costatum*. J. Phycol. 18: 148–162.

Gallagher, J. C., A. M. Wood, and R. S. Alberte. 1984. Ecotypic differentiation in a marine diatom. I. Influence of light intensity on the photosynthetic apparatus. Mar. Biol. 82: 121–134.

Gavis, J., R. R. L. Guillard, and B. L. Woodward. 1981. Cupric ion activity and the growth of phytoplankton clones isolated from different marine environments. J. Mar. Res. 39: 315–333.

Geskes, W. W. C., and G. W. Kraay. 1986. Floristic and physiological differences between the shallow and the deep nanoplankton community in the euphoric zone of the open tropical Atlantic revealed by HPLC analysis of pigments. Mar. Biol. 91: 567–576.

———. 1983. Dominance of Cryptophyceae during the phytoplankton spring bloom in the central North Sea detected by HPLC analysis

of pigments. Mar. Biol. 75: 179–185.

Gifford, D. 1985. Grazing impact of microzooplankton field assemblages. EOS 66: 1314.

Glover, H. E. 1985. The physiology and ecology of the marine cyanobacterial genus *Synechococcus*. Adv. Microbiol. 3: 49–107.

Glover, H. E., M. D. Keller, and R. R. L. Guillard. 1986a. Light quality and oceanic ultraphytoplankton. Nature 319: 142–143.

Glover, H. E., L. Campbell, and B. B. Prézelin. 1986b. Contribution of *Synechococcus* spp. to size-fractioned primary productivity in three water masses in the Northwest Atlantic. Mar. Biol. 91: 193–203.

Glover, H. E., D. A. Phinney, and C. S. Yentsch. 1985a. Photosynthetic characteristics of picoplankton compared with those of larger phytoplankton populations in various water masses of the Gulf of Maine. Biol. Oceanogr. 3: 223–248.

Glover, H. E., A. E. Smith, and L. P. Shapiro. 1985b. Diurnal variations in photosynthetic rates: comparisons of ultraphytoplankton with a larger size fraction. J. Plank. Res. 7: 519–535.

Guillard, R. R. L., and J. A. Hellebust. 1971. Growth and production of extracellular substances by two strains of *Phaeocystis pouchetii*. J. Phycol. 7: 330–338.

Guillard, R. R. L., L. S. Murphy, P. Foss, and S. Liaaen-Jensen. 1985. *Synechococcus* spp. as likely zeaxanthin-dominant ultraphytoplankton in the North Atlantic. Limnol. Oceanogr. 30: 412–414.

Hallegraeff, G. M. 1983. Scale-bearing and loricate nanoplankton from the East Australian Current. Bot. Marina 26: 493–515.

———. 1981. Seasonal study of phytoplankton pigments and species at a coastal station off Sydney: importance of diatoms and the nanoplankton. Mar. Biol. 61: 107–118.

Hallegraeff, G. M., and S. A. Jeffrey. 1984. Tropical phytoplankton species and pigments of continental shelf waters of North and North-West Australia. Mar. Ecol. 20: 59–74.

Harbison, G. R., and V. L. McAlister. 1979. The filter-feeding rates and particle retention efficiencies of three species of *Cyclosalpa* (Tunicata, Thaliacea). Limnol. Oceanogr. 24: 875–892.

Hargraves, P. E., and R. R. L. Guillard. 1974. Structural and physiological observations on some small marine diatoms. Phycologia 13: 163–172.

Harrison, G. I., and F. M. M. Morel. 1984. Antagonism between cadmium and iron in the marine diatom *Thalassiosira weissflogii*. J. Phycol. 19: 495–507.

Harvey, H. W. 1985. The chemistry and fertility of sea water. Cambridge Univ. Press. 224 p.

Hasle, G. R. 1960. Phytoplankton and ciliate species from the tropical Pacific. Skrift. norske Vidensk-Akad. I. Mat.-Naturv. Klesse, No. 2, 50 p.

Haxo, F. T. 1985. Photosynthetic action spectrum of the coccolithophorid, *Emiliania huxleyi* (Haptophyceae): 19′ hexanoyloxyfucoxanthin as antenna pigment. J. Phycol. 21: 282–287.

HAYWARD, J. 1968. Studies on the growth of *Phaeodactylum tricornutum*. IV. Comparison of different isolates. J. Mar. Biol. Assoc. U.K. 48: 657–666.

HULBURT, E. M., AND C. M. JONES. 1977. Phytoplankton in the vicinity of Deep Water Dumpsite 106. NOAA Dumpsite Evaluation Report 77-1. Volume II: 219–231.

HULBURT, E. M., J. H. RYTHER, AND R. R. L. GUILLARD. 1960. The phytoplankton of the Sargasso Sea off Bermuda. J. Cons. Int. Explor. Mer. 25: 115–128.

JEFFREY, S. W., AND G. M. HALLEGRAEFF. 1986. Phytoplankton pigments, species and light climate in a complex warm-core eddy (Mario) of the East Australia Current. Deep-Sea Res. (In press)

JENSEN, A, B. RYSTAD, AND S. MELSOM. 1974. Heavy metal tolerance of marine phytoplankton. I. The tolerance of three algal species to zinc in coastal seawater. J. Exp. Mar. Biol. Ecol. 15: 145–157.

JOHNSON, P. W., AND J. MCN. SIEBURTH. 1982. *In situ* morphology and occurrence of eucaryotic phototrophs of bacterial size in the picoplankton of estuarine and oceanic waters. J. Phycol. 18: 318–327.

JOINT, I. R., AND R. K. PIPE. 1984. An electron microscope study of a natural population of picoplankton from the Celtic Sea. Mar. Ecol. Prog. Ser. 20: 113–118.

JOINT, I. R., AND A. J. POMROY. 1983. Production of picoplankton and small nanoplankton in the Celtic Sea. Mar. Biol. 77: 19–27.

JØRGENSEN, C. B. 1966. Biology of suspension feeding. Pergamon Press, NY, 357 p.

KETCHUM, B. H. 1954. Mineral nutrition of phytoplankton. Ann. Rev. Plant. Physiol. 5: 55–74.

KNIGHT-JONES, E. W. 1951. Preliminary studies of nanoplankton and ultraplankton systematics and abundance by a quantitative culture method. J. Cons. Int. Explor. Mer. 17: 140–155.

KNIGHT-JONES, E. W., AND P. R. WALNE. 1951. *Chromulina pusilla* Butcher, a dominant member of the ultraplankton. Nature 167: 445–446.

LEWIN, J. C. 1955. Physiological races of the diatom, *Navicula pelliculosa*. Biol. Bull. (Woods Hole) 107: 343.

LEWIN, J., R. E. NORRIS, S. W. JEFFREY, AND B. E. PEARSON. 1977. An aberrant Chrysophycean alga *Pelagococcus subviridis* gen. nov. et spec. nov. from the North Pacific Ocean. J. Phycol. 13: 259–266.

LI, W. K. W., D. V. SUBBA RAO, W. G. HARRISON, J. C. SMITH, J. J. CULLEN, B. IRWIN, AND T. PLATT. 1983. Autotrophic picoplankton in the tropical ocean. Science 219: 292–295.

MALONE, T. C. 1980. Algal size, p. 433–464. *In* I. Morris [ed.] The physiological ecology of phytoplankton. Univ. California Press, Berkeley, CA.

MATTOX, K. R., AND K. D. STEWART. 1984. Classification of the green algae: A concept based on comparative cytology, p. 29–72. *In* D. E. G. Irvine and D. M. John [ed.] Systematics Assoc. Special Vol. No. 27, "Systematics of the green algae." Academic Press, New York, NY.

MCCARTHY, J. J. 1981. The kinetics of nutrient utilization, p. 211–233. *In* T. Platt [ed.] Physiological bases of phytoplankton ecology. Can. Bull. Fish. Aquat. Sci. 210: 346 p.

MOESTRUP, Ø. 1979. Identification by electron microscopy of marine nanoplankton from New Zealand, including the description of four new species. New Zealand J. Bot. 17: 61–95.

MURPHY, L. S., AND R. R. L. GUILLARD. 1976. Biochemical taxonomy of marine phytoplankton by electrophoresis of enzymes: I. The centric diatoms *Thalassiosira pseudonana* and *T. fluviatilis*. J. Phycol. 12: 9–13.

MURPHY, L. S., AND E. M. HAUGEN. 1985. The distribution and abundance of phototrophic ultraplankton in the North Atlantic. Limnol. Oceanogr. 30: 47–58.

MURPHY, L. S., R. R. L. GUILLARD, AND J. F. BROWN. 1984. The effects of iron and manganese on copper sensitivity in diatoms: differences in the responses of closely related neritic and oceanic species. Biol. Oceanogr. 3: 187–201.

MURPHY, L. S., R. R. L. GUILLARD, AND J. GAVIS. 1982. Evolution of resistant phytoplankton strains through exposure to marine pollutants, p. 401–412. *In* G. Mayer [ed.] Ecological stress and the New York Bight. Estuarine Res. Foundation, Columbia, SC.

MURPHY, L. S., P. R. HOAR, AND R. A. BELASTOCK. 1981. The effect of industrial wastes on marine phytoplankton, p. 399–410. *In* B. H. Ketchum, D. R. Kester and P. K. Park [ed.] Ocean dumping of industrial wastes. Plenum Press, New York, NY.

NIVAL, P., AND S. NIVAL. 1976. Particle retention efficiencies of an herbivorous copepod, *Acartia clausi* (adult and copepodite stages): Effects on grazing. Limnol Oceanogr. 21: 24–38.

O'CONNORS, H. B., JR. C. F. WURSTER, C. D. POWERS, D. C. BIGGS, AND R. G. ROWLAND. 1978. Polychlorinated biphenyls may alter marine trophic pathways by reducing phytoplankton size and production. Science 201: 737–739.

OMORI, M., AND T. IKEDA. 1984. Methods in marine zooplankton ecology. Wiley-Interscience, New York, NY., 332 p.

PARKE, M. W. 1949. Studies on marine flagellates. I. J. Mar. Biol. Assoc. U.K. 28: 255–286.

PARSONS, T. R., W. K. W. LI, AND R. WATERS. 1976. Some preliminary observations on the enhancement of phytoplankton growth by low levels of mineral hydrocarbons. Hydrobiologia 51: 85–89.

RICKETTS, T. R. 1970. The pigments of the Prasinophyceae and related organisms. Phytochemistry 9: 1835–1842.

RUETER, J., AND F. M. M. MOREL. 1981. The interaction between zinc deficiency and copper deficiency as it affects the silicic acid uptake mechanisms in *Thalassiosira pseudonana*. Limnol. Oceanogr. 26: 67–73.

RYTHER, J. H. 1954. The ecology of phytoplankton blooms in Moriches Bay and Great South Bay, Long Island, New York. Biol. Bull. (Woods Hole) 106: 198–209.

SHIFRIN, N. S., AND S. W. CHISHOLM. 1981. Phytoplankton lipids: interspecific differences and effects of nitrate, silicate and light-dark cycles. J. Phycol. 17: 374–384.

SIEBURTH, J. MCN., V. SMETACEK, AND J. LENZ. 1978. Pelagic ecosystem structure: heterotrophic compartments of the plankton and their relation to plankton size fractions. Limnol. Oceanogr. 23: 1256–1263.

SILVER, M. W., AND A. L. ALLDREDGE. 1981. Bathylpelagic marine snow: Deep-sea algal and detrital community. J. Mar. Res. 39: 501–530.

SILVER, M. W., AND K. W. BRULAND. 1981. Differential feeding and fecal pellet composition of salps and pteropods, and the possible origin of deep water flora and olive-green "cells". Mar. Biol. 62: 263–273.

SMALL, L. F., S. W. FOWLER, AND M. Y. UNLU. 1979. Sinking rates of natural fecal pellets. Mar. Biol. 51: 233–241.

SMAYDA, T. J. 1969. Some measurements of the sinking rate of fecal pellets. Limnol. Oceanogr. 14: 621–625.

SMAYDA, T. J., AND P. K. BIENFANG. 1983. Suspension properties of various phyletic groups of phytoplankton and tintinnids in an oligotrophic, subtropical system. Mar. Ecol. 4: 289–300.

SUNDA, W. G., AND S. A. HUNTSMAN. 1983. Effect of competitive interactions between manganese and copper on cellular manganese and growth in estuarine and oceanic species of the diatom *Thalassiosira*. Limnol. Oceanogr. 28: 924–934.

SUNDA, W. G., R. T. BARBER, AND S. A. HUNTSMAN. 1981. Phytoplankton growth in nutrient rich seawater: importance of copper-manganese cellular interactions. J. Mar. Res. 39: 567–586.

SVERDRUP, H. U., M. W. JOHNSON, AND R. H. FLEMING. 1942. The oceans. Prentice-Hall, NJ.

TAKAHASHI, M., AND P. K. BIENFANG. 1983. Size structure of phytoplankton biomass and photosynthesis in subtropical Hawaiian waters. Mar. Biol. 76: 203–211.

TAKAHASHI, M., AND T. HORI. 1984. Abundance of picophytoplankton in the subsurface chlorophyll maximum layer in subtropical and tropical waters. Mar. Biol. 79: 177–186.

TAKAHASHI, M., K. KIKUCHI, AND Y. HARA. 1985. Importance of picocyanobacteria biomass (unicellular, blue-green algae) in the phytoplankton population of the coastal waters off Japan. Mar. Biol. 89: 63–69.

TANGEN, K., L. E. BRAND, P. L. BLACKWELDER, AND R. R. L. GUILLARD. 1982. *Thoracosphaera heimii* (Lohmann) Kamptner is a dinophyte: observations on its morphology and life cycle. Mar. Micropaleontol. 7: 193–212.

THOMAS, W. H., AND D. L. R. SEIBERT. 1977. Effects of copper on the dominance and the diversity of algae: Controlled ecosystem pollution experiment. Bull. Mar. Sci. 27: 23–33.

THRONDSEN, J. 1979. The significance of ultraplankton in marine primary production. Acta Bot. Fennica 110: 53–56.

——— 1976. Occurrence and productivity of small marine flagellates. Norw. J. Bot. 23: 269–293.

——— 1970. Flagellates from Arctic waters. Nytt Mag. Bot. 17: 49–57.

——— 1969. Flagellates from Norwegian coastal waters. Nytt Mag. Bot. 16: 161–216.

VAN VALKENBURG, S. D., J. K. JONES, AND D. R. HEINLE. 1978. A comparison by size class and volume of detritus versus phytoplankton in Chesapeake Bay. Estuarine Coastal Mar. Sci. 6: 569–582.

WOOD, A. M. 1985. Adaptation of photosynthetic apparatus of marine ultraphytoplankton to natural light fields. Nature 316: 253–255.

WOOD, A. M., P. K. HORAN, K. MUIRHEAD, D. A. PHINNEY, C. M. YENTSCH, AND J. B. WATERBURY. 1985. Discrimination between types of pigments in marine Synechococcus spp. by scanning spectroscopy, epifluorescence microscopy, and flow cytometry. Limnol. Oceanogr. 30: 1303–1315.

——— 1979. Chlorophyll a:b ratios in marine planktonic algae. J. Phycol. 15(3): 330–332.

WOOD, E. J. F. 1965. Marine microbial ecology. Reinhold Publ. Corp., New York, NY. 243 p.

——— 1963. The relative importance of groups of protozoa and algae in marine environments of the Southwest Pacific and East Indian Oceans, p. 236–241. *In* C. H. Oppenheimer [ed.] Symposium on marine microbiology. C. H. Thomas, Springfield, IL.

YENTSCH. C. S. 1983. A note on the fluorescence characteristics of particles that pass through glass-fiber filters. Limnol Oceanogr. 28: 597–599.

YENTSCH, C. S., AND D. A. PHINNEY. 1985. Fluorescence spectral signatures for studies of marine phytoplankton, p. 259–274. *In* A. Zirino [ed.] Mapping strategies in chemical oceanography. American Chemical Society, Washington, DC.

Ecophysiology of Nutrient Uptake, Photosynthesis and Growth

WANDA ZEVENBOOM[1]

*Laboratory of Microbiology, University of Amsterdam, Nieuwe Achtergracht 127,
1018 WS Amsterdam, The Netherlands*

1. Introduction

Nutrient uptake and photosynthesis can be regarded as key processes in the growth of phytoplankton. These physiological activities are influenced by both the prevailing and the former growth environments. Through adaptation to different sets of environmental factors, different phenotypes arise with distinct differences in the values of their physiological parameters. The aim of ecophysiological research is to gain insight in the physiological properties and the adaptive strategies of various species, and to use this insight to assess, explain and predict phytoplankton distribution and its physiological performance in natural environments.

The internal triggers for the adaptive responses of phytoplankton species are difficult to assess in nature where environmental conditions are often variable, and unpredictable, in space and time. One can use instead continuous culture systems in which the species is exposed to a constant, well-defined set of growth conditions. The nature of the growth-limitation and the growth rate of the population are exactly known. One can make this originally simple system more complex, step by step, by introducing other factors, e.g. non-saturating light levels during nutrient-limited growth, transitions from one steady state to another, or continuously pulsed conditions (of intermittent light- or nutrient supply). By increasing the complexity of continuous culture systems, the insight into the basic triggers for the adaptation strategies of the algal cell is not lost, and the extrapolations from chemostat results to natural environments may become more realistic.

The purpose of this review is not to go into detail in chemostat theory which has been well described by Tempest (1970) and for algal studies by Rhee (1980). This chapter will focus on some basic physiological mechanisms of phytoplankton adaptation in relation to environmental parameters. Many ecophysiological studies have been undertaken over the past few years to explain successfully a number of ecological observations. Some examples will be given of what has been learned through chemostat studies about phytoplankton behaviour, including small phytoplankters in their natural environment.

2. Boundaries and Outlines

It is clearly not possible to focus on the entire field of algal adaptation strategies, and, therefore, it is appropriate to make some boundaries around this subject. The choice of topics is certainly not based on differences in their importance; it is one out of numerous possible approaches to deal with the subject.

1) *Physiology.* The regulation in nutrient uptake will be considered in more detail than the regulation mechanisms of pigment composition/concentration and the (concomitant) photosynthetic response of an algal cell.

---

[1] Present address: Ministry of Transport and Public Works, North Sea Directorate, P.O. Box 5807, 2280 HV Ryswyk (z. h.) The Netherlands.

2) *Adaptive responses*. Less mention will be made on the effects of pulsed conditions of limiting substrates or light, known to have an impact on cell division/cycle, cell physiology and phytoplankton community structure (Chisholm 1981; Prézelin and Matlick 1980; Rhee et al. 1981; Turpin and Harrison 1979; Van Gemerden 1974). Instead, more attention is paid to the adaptive responses to different steady state growth environment and during transitions from one steady state to another. The adaptation comprises changes in many physiological functions (for reviews see Falkowski 1980; Harris 1978; Rhee 1980; 1982; Richardson et al. 1983; Tempest et al. 1983), among which changes in the nutrient uptake–photosynthesis–and growth kinetic parameters form a part that will be discussed here.

3) *Growth kinetics*. In a number of laboratory studies, attention is paid to ecologically relevant growth rate determining factors. The most important factors that may limit (determine) the algal growth rate periodically or for a longer time are the (macro)-nutrients nitrogen, phosphorus or silicon (in the case of diatoms) (Maestrini and Kossut 1981; Sakshaug et al. 1983; Tilman 1977; Van Donk 1983; Zevenboom et al. 1982). Also trace metals (De Haan et al. 1982; Patrick 1978), vitamins (Droop 1968), temperature (Eppley 1982; Goldman 1977) and light (Nalewjko et al. 1981; Scott et al. 1980; Zevenboom et al. 1982) have been mentioned as potentially growth rate limiting factors. In this review nutrient-limited growth (under-saturating conditions) and light-limited growth kinetics (with nutrients in excess) will be considered.

There are three aspects concerning light-limited growth, namely, the effects of light intensity, light periodicity and light quality on phytoplankton growth rate. All three have direct ecological relevance. In waters with dense algal populations, the amount of light available is greatly decreased and spectral quality changes with depth (Kirk 1983; Van Liere and Walsby 1982). Not only is it important to consider diurnal L/D (light/dark) cycles (Foy et al. 1976; Loogman et al. 1980), but also the more rapid L/D fluctuations (Loogman 1982; Gibson 1985; Walsh and Legendre 1983). The latter conditions arise when algae circulate through the vertical light gradient. This gradient can be very steep in waters with high phytoplankton biomass concentration and other dissolved/ or suspended matter, sometimes leading to situations in which the mixing depth exceeds the euphotic zone. Effects of light intensity and light periodicity on algal growth rate will be considered while no further mention is made of the effects of light quality.

4) *Interactions*. During nutrient-limited growth in nature, other factors such as low non-saturating light conditions, low temperatures, low concentrations of other nutrients, may be potentially growth rate limiting. Nutrient-light interaction is one of these interactions that will be considered here.

5) *Physiological indicators*. In studies on carbon flows, primary production, and growth potential of phytoplankton, and for water-management, it has become important to identify the factor that limits the growth rate of algal populations in their natural habitats. The use of physiological data and of the insights in physiological mechanisms of algal adaptation is a rather new and also more revealing approach to this subject compared with the use of bioassays. Two examples of this "physiological indicator" approach will be presented from our own work. One which deals with the assessment of factors limiting growth rate of a freshwater cyanobacterium, the other marine picoplankton.

3. Nutrient Uptake and Growth

Nutrient-limited growth rate is the end result of cell's ability to take up, transport and assimilate the limiting nutrient and/or to use a previously stored product. Through regulation of enzymes, specifically involved in the various stages of cell-environment (nutrient) interaction, the cell can achieve a certain growth rate. Before discussing the end result, nutrient-limited growth, it is appropriate to focus first on nutrient uptake.

3.1.1. Steady-State and Short-Term Nutrient Uptake

In the study of nutrient uptake, two aspects can be considered, namely, steady state and short-term (transient) nutrient uptake. The first deals with cells under steady state conditions, that exhibit a specific actual nutrient uptake, q, proportional to the growth rate (μ) (which equals dilution rate) according to:

$$(1) \qquad q = \mu \cdot \frac{S_R - \bar{s}}{\bar{x}}$$

where S_R and $\bar{s}$ are the nutrient concentrations in the inflow and chemostat, respectively, and $\bar{x}$ is the steady state biomass concentration. When it is assumed that there is not turnover and/or excretion of the nutrient, the yield (Y) can be calculated, which is the amount of biomass ($\bar{x}$) produced per nutrient consumed ($S_R - \bar{s}$) thus:

$$(2) \qquad Y = \frac{\bar{x}}{S_R - \bar{s}}$$

Since, under the same assumptions, the cellular concentration of the (limiting) nutrient (Q) is the reciprocal value of the yield:

$$(3) \qquad Q = \frac{S_R - \bar{s}}{\bar{x}}$$

equation (1) can be transformed into (Droop 1973);

$$(4) \qquad q = \mu \cdot Q$$

The second approach in the study of nutrient uptake is to examine the *initial* short term nutrient uptake rate (V) (the uptake *potential*) expressed by cells harvested from

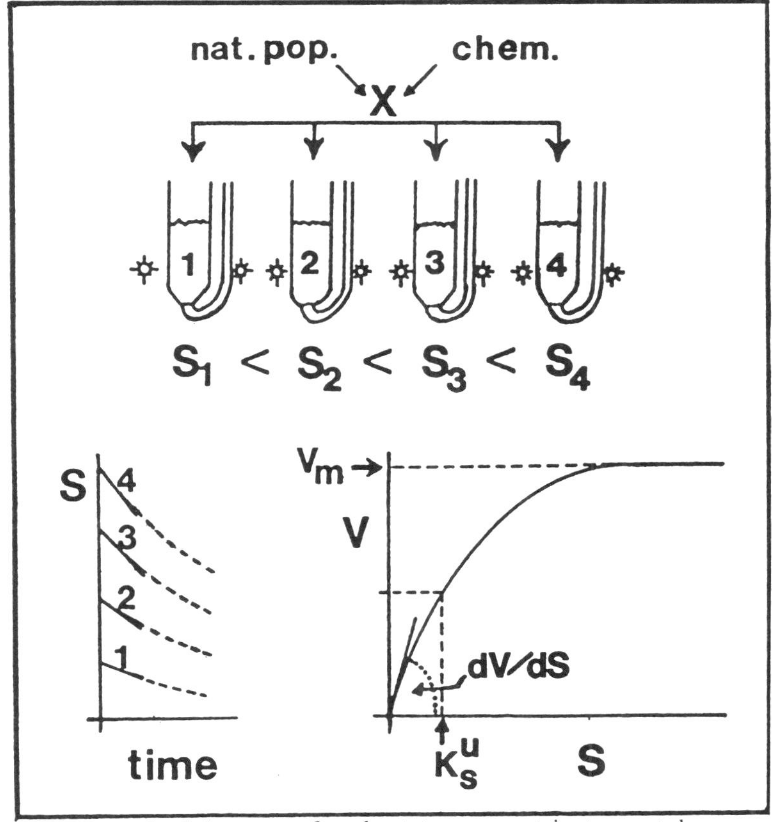

FIG. 1. Scheme of short-term nutrient uptake study. chem. = chemostat; nat. pop. = natural population.

chemostats or nature and pulsed with various concentrations of the nutrient S. The quantity V is not necessarily in balance with μ (as will be discussed later), and considered over a *short* time, which is best defined as the time much shorter than (i.e. 10% of) the generation time of the species. The procedure is schematically shown in Fig. 1. From the *early* disappearance of the nutrient concentration with time, V is determined at each initially present nutrient concentration S and calculated according to:

$$(5) \qquad V = - \frac{dS}{dt} \cdot \frac{1}{X}$$

where X is the biomass concentration used in the experiment. It is generally assumed that the $V - S$ relationship is the result of an enzyme-mediated uptake process, and can be similarly described as steady state enzyme kinetics by the Michael-Menten equation, because cell biomass concentration (X) is, like *in vitro* enzyme concentration, of constant composition. Indeed, it has been demonstrated for a number of species from different habitats and for a variety of nutrients that V is related to S according to Michaelis — Menten kinetics:

$$(6) \qquad V = V_m \cdot \frac{S}{K_s^u + S}$$

in which V_m is the potential maximum short term nutrient uptake rate, or maximum nutrient uptake capacity, and K_s^u is the half-saturation constant for nutrient uptake, or S at which $V = 1/2\ V_m$ (Fig. 1) (*see* for reviews on literature data e.g. Button 1985; Rhee 1980, 1982). There are a few reports on the involvements of diffusion-limited nutrient uptake (Mierle 1985; Riegman and Mur 1984a) and of two different enzyme systems in the uptake of the same nutrient (Zevenboom and Mur 1979), which lead to deviation from Michaelis–Menten kinetics.

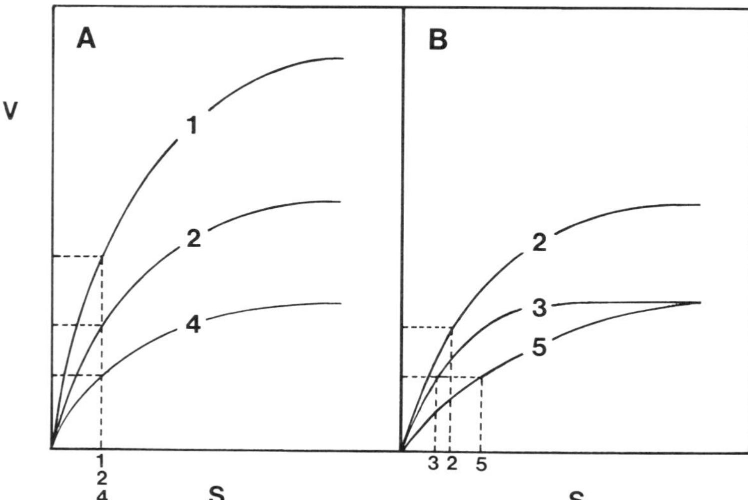

FIG. 2. Some examples of V–S curves of different shape.
A = differences in V_m values and initial slopes (dV/dS), while K_s^u is constant.
B (case 2 compared with cases 3 and 5) = differences in V_m, K_s^u and dV/dS.
Cases 3, 4 and 5 = differences in K_s^u and dV/dS, while V_m is constant.
Estimates of affinities (dV/dS) 1 > (dV/dS) 2 > (dV/dS) 3 > (dV/dS) 4 > (dV/dS) 5
(After Zevenboom 1980).

The K_s^u value has long been considered as the "affinity constant" and often used to explain succession phenomena in nature in the competitive interactions between species for a growth rate limiting substrate. The idea was that a lower K_s^u would imply a higher affinity and was typical for species occupying oligotrophic environments, while a higher K_s^u would result in a lower affinity and would be possessed by species living in eutrophic habitats (e.g. Eppley et al. 1969). However, considering K_s^u alone is not sufficient, since a high V_m can compensate for a disadvantageous high K_s^u value. In later studies it was therefore suggested to use instead the *initial* slope (dV/dS) (Fig. 1) of the V–S curve as a measure of affinity so that, when S approaches zero, the affinity approaches the ratio of V_m/K_s^u (Button 1983; Healey 1980; Zevenboom 1980; Zevenboom et al. 1980). This is illustrated in Figs. 2A and 2B. In Fig. 2A the *V–S* curves have different V_m values ($V_m1 > V_m2 > V_m4$), but an identical value of K_s^u. However, in this example, the affinities (dV/dS) are not the same. In Fig. 2B another example is shown of differences in both V_m and K_s^u. Although we have $K_s^u3 K_s^u2 < K_s^u5$, the differences in affinities are: (dV/dS) 2 > (dV/dS) 3 > (dV/dS) 5. Finally, an example is shown of identical V_m values and different K_s^u values: $K_s^u3 < K_s^u4 < K_s^u5$ (Fig. 2A and 2B). Only in this case the K_s^u value can be taken as a measure of affinity since we have: (dV/dS) 3 > (dV/dS) 4 > (dV/dS) 5 (Fig. 2 legend).

3.1.2 Transient Short-Term Nutrient Uptake

Let us consider next the *initial* short-term nutrient uptake rate V in more detail. V is not necessarily in balance with species-specific growth rate μ, but depends on the nutrient status of the algal cell, the external concentration of the nutrient and the environmental conditions (temperature, light, pH, ion strength, etc.) used in the short-term experiments, as will be shown below. When a cell is *not* limited in its growth rate by a particular nutrient and when short-term environmental conditions are identical to the *in situ* growth conditions, V of that *non-limiting nutrient*, supplied in excess, proceeds linearly with time at a rate prescribed by the steady state growth

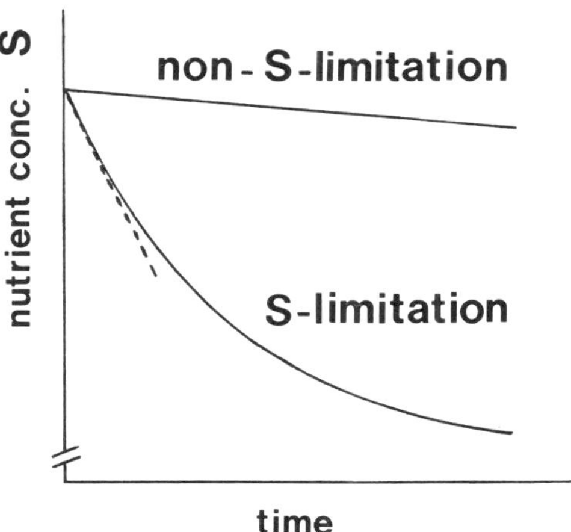

FIG. 3. Short-term nutrient uptake after a saturating pulse with nutrient S. In case of S-limitation, an exponential decrease in the initial high, early "linear", disappearance of S will be found, while the decrease in S (thus uptake rate) proceeds linear with time when S is *not* limiting growth rate. In the latter case, uptake rate of non-limiting nutrient is: $V = \mu Q$ (eq. 4). After Zevenboom 1980; Riegman 1985).

rate μ. In this case, V equals the value of $q(q = \mu \cdot Q$; eq. 4), as is illustrated in Fig. 3. However, when the cell is *limited* by that nutrient and pulsed with a high ($S >> K_s^u$) concentration of the *growth rate-limiting nutrient*, V is initially high, but will decrease when uptake is followed for longer periods of time (Fig. 3). The time course of changes in the saturated nutrient uptake rate of a nutrient-limited cell can be explained as follows. Under nutrient-limited growth, the internal pool of the limiting nutrient (Q) is low, but the affinity for nutrient uptake and/or the cellular content of the uptake system is high so that the cell exhibits a *high initial* saturated uptake rate (V_m) when pulsed with a high saturating concentration of the limiting nutrient. During short-term uptake, V_m becomes under feed back control by the increased Q or increased concentration of substrates available for assimilation (Riegman and Mur 1984c). It is clear that when a non-saturating concentration of the limiting nutrient is used, the decrease in V with time is due to the decrease in external nutrient concentration and the increase in Q of the nutrient-limited cell.

The transient (short-term) nutrient uptake phenomenon of nutrient-limited cells has been observed in both marine and freshwater species, grown under different nutrient-limitations and pulsed with the respective limiting nutrient (*see* e.g. review of Droop 1983). Caperon and Meyer (1972) calculated V from the time course of changes in the nutrient concentration of a single, saturating pulse ($S >> K_s^u$). It is clear that the V–S curve, obtained by this procedure, is difficult to interpret, especially at lower V (and correspondingly decreased S) values, since those were obtained in a later stage of the experiment, where the physiological state of the cells deviates from that present at the start.

In a study on several cyanobacteria, Riegman (1985) showed that the transient short-term uptake of phosphate, in P-starved species, followed first order kinetics. The decrease in external *saturating* phosphate concentration could be described by the equation:

$$(7) \qquad S_t = (S_o - S_t) \, e^{-kt} + S_t$$

where S_t is the external concentration (S) during the transient state at any time t, S_o and S_t are the initial and final values of S, respectively, and k is the specific adaptation rate constant of transient uptake. The relatively high k values (ranging from 0.7 to 2.2 h^{-1}, much higher than specific growth rate) found may indicate that cyanobacteria, previously subjected to P-limiting conditions, can handle quite rapidly an extremely high phosphate pulse (Riegman 1985). In the same study, a mathematical equation was developed that describes the decrease in transient (short-term) phosphate uptake rate with time in P-limited cyanobacteria. In this equation (eq. 8) the inhibition of V by short-term accumulated phosphate and by decrease in phosphate concentration was taken into account:

$$(8) \qquad V_t = V_m \left(1 - \frac{\Delta Q}{Q_m}\right) \left(\frac{S}{K_s^u + S}\right)$$

where V_t is V at time t, Q_m is the maximum amount of phosphate that can be accumulated and ΔQ is the short-term accumulated phosphate at time t. Using eq. 8 it was possible to prescribe transient phosphate uptake of phosphate pulses, of both saturating and non-saturating concentrations, by P-limited cyanobacteria (Riegman 1985).

3.2 Adaptation in Short-Term Nutrient Uptake Parameters

Over the last ten years evidence has accumulated such that one cannot speak of *one* initial potential maximum uptake rate (V_m) or *one* affinity (V_m/K_s^u) for a nutrient. These uptake parameters are found to be affected, firstly by the growth environment, thus by the nature of the growth limitation and the growth rate of the species expressed under the given limitation, and secondly by the experimental conditions used in the short term nutrient uptake assay. Let us first consider the effects of growth environment on the nutrient uptake potentials of species.

3.2.1. Effect of Growth Rate

Some adaptation patterns to low nutrient concentrations, known to occur in microorganisms, and in detail compiled by Tempest and Neijssel (1979) are relevant here. Organisms can increase their short-term nutrient uptake rate (V) at any concentration of limiting nutrient (S) in several ways (Fig. 4):

(i) By increasing their affinity for the limiting nutrient while keeping V_m constant, which implies that K_s^u is decreased. Based on enzyme kinetics, this pattern of adaptation can be described as "competitive" (Fig. 4A).

(ii) Increasing V_m (i.e. increasing the cellular content of the uptake enzymes) and increasing their affinity, keeping then K_s^u constant, which can be described as "non-competitive" (Fig. 4C).

(iii) Increasing their affinity to a higher extent than increasing V_m, so that K_s^u is decreased (Fig. 4D). We suggested calling this pattern of adaptation "non-coherent" (Zevenboom and Mur 1984).

(iv) Increasing V_m, while keeping their affinity constant (Fig. 4B).

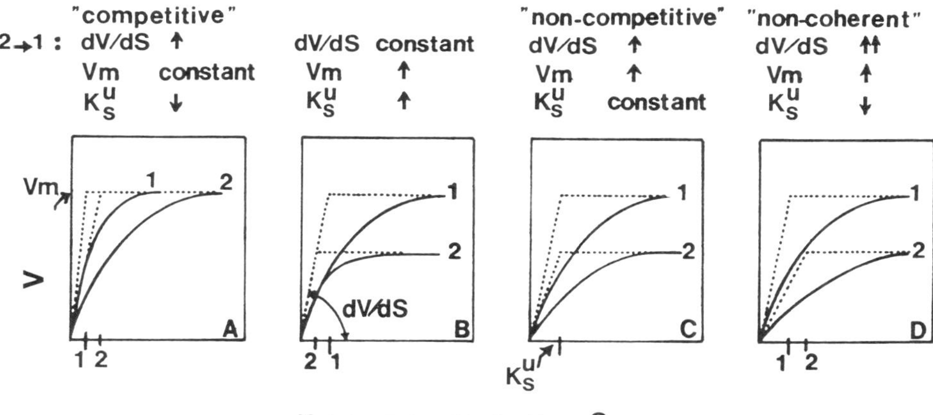

FIG. 4. Adaptation patterns in short-term nutrient uptake. 2 →1: Increase in nutrient-limitation, V_m, dV/dS and K_s^u are indicated.

In one of these ways, the net effect on the cell will be to increase its "scavenging capacity" for the limiting nutrient so that the ability to compete successfully with others is enhanced. The different adaptation patterns in nutrient uptake may be related to the different types of growth environment. One may speculate that adaptation in affinity with constant V_m (Fig. 4A) is important in environments with constant low limiting nutrient supply, while adaptation in V_m (Fig. 4B, C, D) is relevant for species exposed to discontinuous, relatively high pulses of limiting substrates. Although studies in which species are subjected to either continuous of discontinuous nutrient supply point in this direction (e.g. Turpin and Harrison 1979; Van Gemerden 1974), the hypothesis formulated above, still needs further studies.

The patterns of adaptation in V_m and affinity with changes in nutrient-limited growth rate are not the same for all nutrients and also seems to be species-specific, as can be concluded from the short review below. Some P-or N-limited green algae and diatoms showed a pattern of non-competitive adaptation (Fig. 4C): thus, V_m and affinity increased (and K_s^u remained, more or less, constant) with decreasing P- or N-limited growth rate (Rhee 1973; 1978; Gotham and Rhee 1981a; 1981b). Rhee

(1973, 1978) suggested the presence of a long-term feedback regulation of V_m by the internal concentration of an inhibitor (polyphosphate or total intracellular free amino acid pool, in case of P- or N-limited growth, respectively). When the limitation becomes less severe (i.e., at higher growth rates) the cellular concentration of the inhibitor increased and thereby its inhibiting effect on V_m. It was further suggested by Rhee (1973) that the short-term uptake rate could be described by an equation that resembles non-competitive enzyme inhibition:

$$(9) \qquad V = \frac{V_m}{(1 + K_s^u/S)(1 + i/K_i)}$$

where i is the concentration of the inhibitor and K_i is the inhibition constant. It is however, questionable whether eq. 9 describes satisfactorily all data sets provided by Rhee and Gotham. The uncertainty is in the K_s^u value which was not constant with changes in growth rate, so that the adaptation patterns may well resemble those presented in Fig. 4D or Fig. 4B.

In P-limited *Oscillatoria agardhii*, the initial short term uptake was acccording to the patterm shown in Fig. 4B: V_m increased but the affinity for P-uptake remained constant with decreasing P-limited growth rate (Riegman and Mur 1984b). In this species, the diffusion-limited P uptake that is likely to be involved in the uptake of low external phosphate concentrations (Riegman and Mur 1984a) seemed not to be affected by the P-limited growth rate. However, the enzymatic phosphate uptake (V_m) was supposed to be under a short- and long-term feed back control (Riegman and Mur 1984b).

In other P- or N-limited species the pattern shown in Fig. 4A, was found: thus, V_m, for phosphate or nitrogen, remained constant while the affinity increased (and thus K_s^u decreased) with decreasing P- or N-limited growth rate (e.g. Healey and Hendzel 1975; Zevenboom and Mur 1981a). Figure 5 shows the effect of the N-limited growth rate (μ) on V_m for nitrate and ammonium in *O. agardhii* in more detail. At

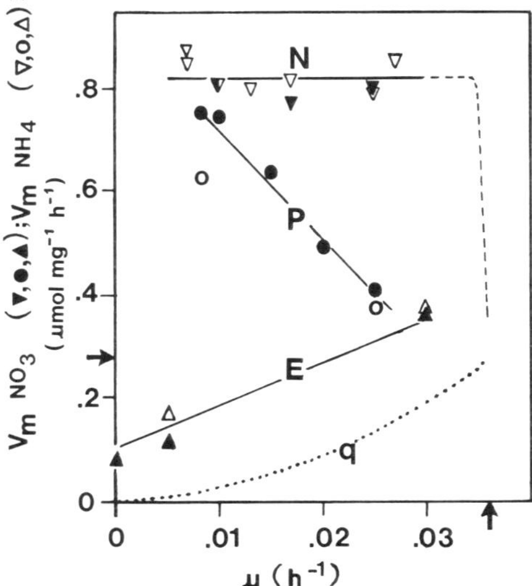

FIG. 5. Influence of specific growth rate (μ) on the uptake capacity (V_m) of *O. agardhii* for nitrate (closed symbols) and ammonium (open symbols). The species was grown in continuous cultures that were, respectively, nitrogen (nitrate or ammonium)-limited (N), phosphate-limited (P), and light-energy-limited (E). Dotted lines: specific N uptake during steady state (q) of N-limited species. Arrows: μm and q_m values, expressed by this organism under saturating conditions of nutrients and light at T = 20°C, pH = 8.0 (Data of Zevenboom 1980; Zevenboom et al. 1980; Zevenboom and Mur 1981a).

398

low growth rate, V_m can be 30 times the value of the actual steady state uptake rate q (Zevenboom and Mur 1979, 1981a). When μ approaches μ_m it is clear that the high maximum uptake potential (V_m) will decrease and approaches q_m (being the maximum steady state uptake rate of the non-limiting nutrient at μ_m) (Fig. 5). For phosphate uptake by P-limited species, the differences between V_m and q can even be larger (e.g. Riegman and Mur 1984b), while for silicon uptake by Si-limited species a smaller difference between V_m and q is found (Tilman and Kilham 1976).

3.2.2 Effect of Type of Limitation

One may expect that the enzymatic uptake potential and affinity for uptake of the non-limiting nutrient are still under metabolic control, so that their values are not only changing with growth rate but also with the nature of the growth limitation. This was, for example, shown for nitrate uptake by N-, P- or light-limited *O. agardhii* (Zevenboom and Mur 1978). Effects of the different growth limitations on V_m values for nitrate and ammonium in this species are summarized in Fig. 5. Light-limited (= nitrate-sufficient) cells showed lower V_m values than N-limited cells, as was expected and already mentioned above. Although P-limited cells were facing an excess of N (nitrate) in the chemostat, V_m for nitrate and ammonium were markedly stimulated in the short-term assay, where phosphate was present in excess. These results indicate that under a P-limitation the enzymes involved in uptake, transport and/or assimilation of nitrate are depressed, concomitantly, in this species (Zevenboom 1980; Zevenboom et al. 1982). P-limited cells are thus expected to contain a relatively low N content, which was indeed found (Zevenboom 1980; Ahlgren 1985) (Fig. 6). From results on a P-limited *Anabaena* and a P-limited *Scenedesmus* (Healey and Hendzel 1975) and reports on suppressed N levels in P-limited marine dinoflagellates (Sakshaug et al. 1984), it can be concluded that an interaction between P-limitation and N-metabolism is not necessarily restricted to the cyanobacterium *Oscillatoria*. In several marine phytoplankters a nitrate-chloride activated ATPase seems to be primarily involved in the active transport of nitrate ions across the plasmolemma (Falkowski 1975). One may speculate that the feedback control of P-limited growth on nitrate uptake is on the level of nitrate-transport.

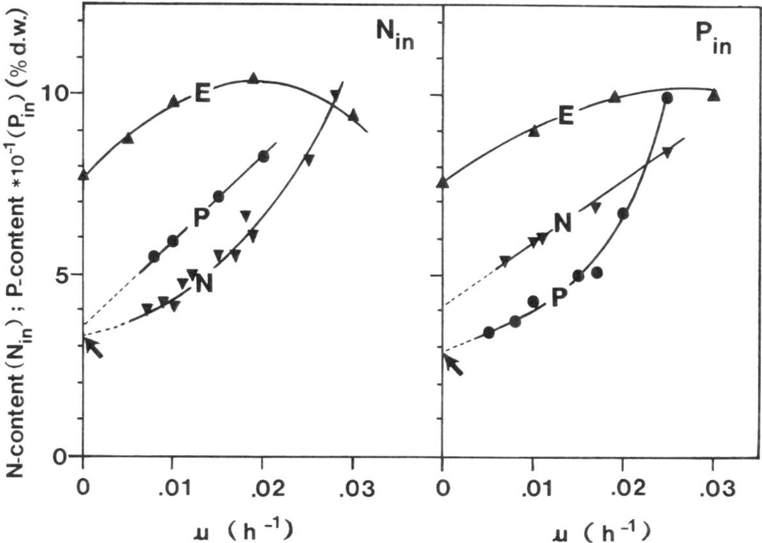

FIG. 6. Influence of specific growth rate (μ) on cellular nitrogen content (Nin) and cellular phosphorus content (Pin) of *O. agardhii* grown in continuous cultures that were, respectively, nitrogen (nitrate)-limited (N, ▼), phosphate-limited (P, ●) and light-energy-limited (E, ▲).

3.2.3 Effect of Ambient Conditions

The environmental conditions with respect to light, temperature, pH, etc. during the short-term experiment are known to influence the potential nutrient uptake rates of species, by acting directly on the activity(-ies) of the enzyme(s) involved in the nutrient uptake. For example, high experimental pH values have an inhibiting effect on the maximum ammonium uptake potential (V_m) of a number of species (Healey 1977; Zevenboom and Mur 1981a;), while low temperatures or non-saturating light conditions during the short-term experiments can lead to lower values of short-term nitrate uptake (e.g. Falkowski 1977; Zevenboom 1980). One can also mention here the interference of ammonium ions with the potential nitrate uptake rate, shown in a number of species that exhibit a preferential use of ammonium ions over nitrate (see e.g., the review by Syrett 1981). We found that the ambient ammonium ions inhibited the short-term nitrate uptake kinetics of nitrate-limited *O. agardhii* in a non-competitive manner (Zevenboom and Mur 1981b). A likely explanation for the reduced expression of the nitrate uptake potential is that the nitrate reductase is inhibited by ammonium.

The above mentioned examples may illustrate that the short-term nutrient uptake is not only a reflection of the physiological state of the cell. The expression of cell's uptake potential depends on the experimental condition used in the short-term experiments. These findings imply that one should specify and standardize the experimental conditions if one wants to use the data of nutrient uptake capacity values as indicators to assess the physiological state of natural populations. However, if one wants to assess the actual specific nutrient uptake rate then the environmental conditions during the pulse experiment should be comparable to those prevailing *in situ*. In this case one can assess the growth rate of the species in its natural environment from the short-term uptake rate (V) of the non-limiting nutrient that proceeds linaerly with time (Fig. 3) and by using eq. 4 ($V = \mu \cdot Q$).

3.3 Nutrient-Limited Growth

Before dealing with several growth models, it is relevant to define first the specific growth rate (μ). It is defined as the balanced increase in the amount of *all* cell components, at a same rate, per unit of biomass (X) per unit of time, thus:

$$(10) \qquad \mu = \frac{1}{X} \cdot \frac{dX}{dt}$$

The specific growth rate of a species may, therefore, never be confused with the yield (i.e. the amount of biomass produced per nutrient consumed, eq. 2). One should also bear in mind that μ cannot directly be assessed from the increase in dry weight concentration with time when the cell composition is changing concomitantly. This will be discussed later (Fig. 10A, B).

Among several growth models, there are two models widely used to describe nutrient-limited growth. The models are shown in Fig. 7. First, one has the model proposed by Monod (1950), which states that μ is related to the external *steady state* concentration ($\bar{s}$) of the limiting nutrient by a hyperbolic function (Fig. 7A):

$$(11) \qquad \mu = \mu m \cdot \frac{\bar{s}}{K_{\bar{s}}^g + \bar{s}}$$

in which μm is the maximum specific growth rate and $K_{\bar{s}}^g$ is the half-saturation constant for nutrient-limited growth, or $\bar{s}$ (Fig. 7) which $\mu = 1/2\,\mu m$. The assumption in this model is that cell properties do not vary with μ. In previous sections it has already been discussed that in algae, studied so far, nutrient uptake parameters (V_m, *affinity, K_s^u)* vary with the nutrient-limited growth rate. Also, the yield (Y) and thus the cell nutrient quota (Q) (the reciprocal of the yield, equations 2, 3) is not constant under nutrient-limited growth, as has been shown for a number of species (see below).

400

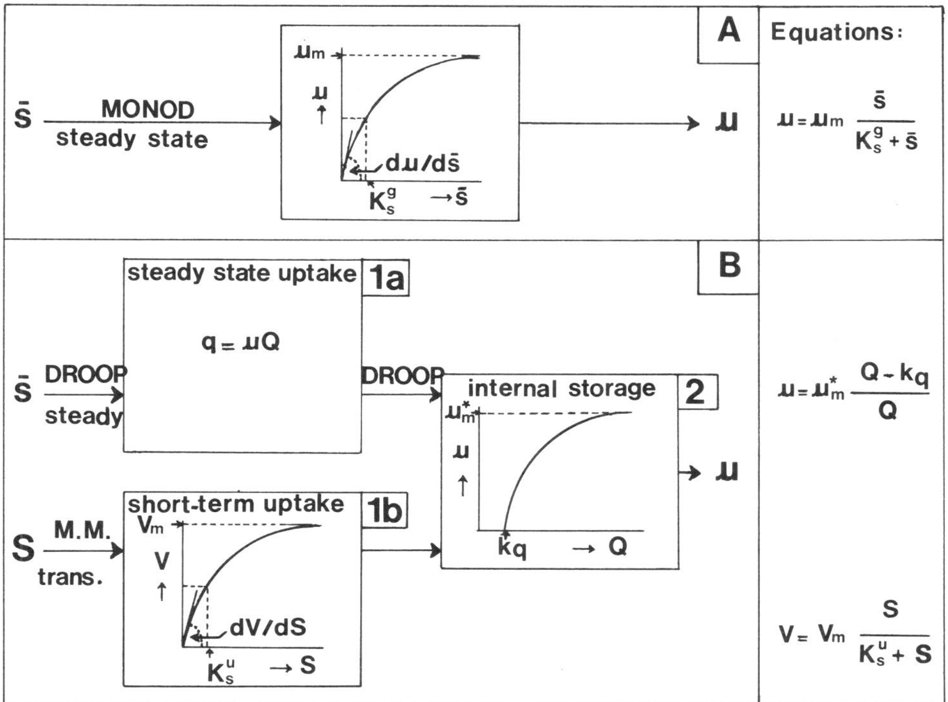

FIG. 7. Survey of nutrient-limited growth models. A = Monod model, used to describe steady state conditions; B = Droop model, a two-step model of nutrient uptake (1) and storage (2) for steady state (1a) and transient-state (1b, Michaelis–Menten = M.M.) uptake conditions. See further text.

The assumption in the Monod model seems thus unlikely to hold true, and eq. 11 can thus only give an approximation of the relationship between μ and external limiting nutrient concentration, in steady-state conditions. In this connection it is worthwhile to note the differences between the steady state growth parameter K_g^s (eq. 11) and the variable short-term nutrient uptake parameter K_u^s (eq. 6). Droop (1973) already showed that they cannot be equal; K_s^u is much larger than K_s^g (see for detailed discussion Rhee 1980).

The second, also empirical, model is the one proposed by Droop (1968), in which μ is related to the total cellular concentration of the limiting nutrient (Q) (Fig. 7B), by the equation:

$$(12) \qquad \mu = \mu\mathrm{m}^* \cdot \frac{Q - kq}{Q}$$

where $\mu\mathrm{m}^*$ is the asymptote of the hyperbole and refers to the "infinite" internal nutrient content, and k_q is the value of Q at $\mu = 0$. The theoretical $\mu\mathrm{m}^*$ value is always greater than the true $\mu\mathrm{m}$ value in eq. 11 (Droop 1973):

$$(13) \qquad \mu\mathrm{m} = \mu m^* \cdot \frac{Q_m - kq}{Q_m}$$

in which Q_m is the value of Q achieved when $\mu = \mu\mathrm{m}$. The difference between $\mu\mathrm{m}$ and $\mu\mathrm{m}^*$ is low for nutrients P (e.g. Riegman and Mur 1984a,b), Fe and Vitamin B12 (Droop 1973), large for Si and C (Goldman and McCarthy 1978) and in between for N (e.g. Zevenboom and Mur 1981a).

The model of Droop (Fig. 7B) involves steady-state (1a) or short-term (1b) uptake and "internal storage" (2). The assumption here is again, that cell properties do not change so that the μ–Q relationship is uniform for a certain species. It was suggested that the Droop model, unlike the Monod model, can be applied to transient state nutrient-limited growth (Droop 1975), but this can only be done when the assumption holds true. However, a lag period is observed between the increase in growth rate and cell quota when nutrient-starved cells are supplied with nutrients, due to rapid nutrient uptake and delayed increase in μ (Cunningham and Maas 1978; Healey 1979). Moreover, with nutrient-limited growing cells it has been repeatedly shown that during transient state the uptake characteristics are affected (see previous section, and Fig. 3) directly or indirectly, by the increased value of the internal nutrient content, Q. Loogman et al. (1979) developed a nitrogen-limited growth model in which these two aspects, growth on internal nigrogen and regulation in the nitrate uptake parameters, have been incorporated and by which the transient nitrate-limited growth response of *O. agardhii* could be successfully described and predicted. From studies dealing with light or temperature interactions on steady state nutrient-limited growth (see next section) it further appears that the yield on the limiting substrate (thus also the μ–Q relationship) is affected by non-saturating light or low temperature conditions (e.g. Rhee and Gotham 1981a, b; Zevenboom 1980; Zevenboom et al. 1980).

In conlusion, both Monod and Droop are empirical models and can only give an approximation of nutrient-limited algal growth. In a review on cell quota, Droop (1983) listed more than 30 examples of steady-state nutrient limited growth kinetics of marine and freshwater species that could be described satisfactorily by the Droop model. The number of cases is still growing like it is for the Monod relation (see review by Rhee (1982)). However, when transient conditions are present, the regulation of enzymes involved in nutrient uptake and assimilation should be taken into account. Both models describe steady-state nutrient-limited growth of freshwater species equally "well", when environmental parameters such as light and temperature are constant and comparable between the various steady-states, as was shown for example in Si- or P-limited diatoms (Tilman and Kilman 1976) and in the N-limited cyanobacterium *O. agardhii* (Zevenboom et al. 1980; Zevenboom and Mur 1981a).

For a number of steady-state nutrient-limited marine phytoplankters one has been unable to determine the K_s^g value, due to the inability to measure the low limiting external substrate concentration prevailing at each steady-state growth rate. Droop (1986) suggested that the K_s^g value could then be calculated from growth and uptake constants:

$$(14) \qquad K_s^g = \frac{\mu \mathrm{m}^* \cdot k_q \cdot K_s^u}{V_m}$$

However, since the ratio K_s^u/V_m and the values of $\mu \mathrm{m}^*$ and k_q are not true constants, as was emphasized before, it is clear that one should be cautious in following this procedure.

Although the Droop equation for nutrient-limited growth rate (μ) is expressed in terms of the total cellular content of the limiting nutrient, it is likely that μ is more directly related to a certain metabolic, active, cellular fraction. In *Scenedesmus* sp. the phosphate- and nitrate-limited growth rate could be related to polyphosphate and intracellular free amino acid pool, respectively (Rhee 1973, 1978). However, Davis et al. (1978) found in Si-limited *Skeletonema* the active pool to be associated with the cell wall silicon and not with the soluble fraction. Riegman and Mur (1984c) suggested relating μ to the concentration of substrates available for assimilation, rather than to the concentration of the assimilation products. In their paper a theoretical model is presented that gives an explanation of the empirical Droop equation.

4. Photosynthesis and Growth

In the photosynthetic process, species-specific pigments are involved in harvesting light quanta from the ambient light field to convert them into chemical energy.

Reducing equivalents and ATP, produced in the light reactions (photosystems I and II) are used in the enzymatically mediated photosynthetic CO_2 fixation into organic compounds (Calvin cycle, dark reaction). (for review see e.g. Harris 1978; Ho and Krogmann 1982; Kirk 1983). Through regulation of the photosynthetically active pigments and biochemical processes, a species exhibits a certain photosynthetic rate (measured as O_2 production rate, or CO_2-fixation rate) and growth rate under the given light conditions. Light may thus be limiting growth, as nutrients are. In this section, attention is paid to some similarities and differences in procedures of measurements and in adaptive responses between photosynthesis and nutrient uptake and between light- and nutrient-limited growth.

4.1 Steady-State and Short-Term Photosynthesis

The two methods most widely used for determining the initial short-term photosynthetic response of phytoplankton are the ^{14}C technique (e.g. Steemann Nielsen 1952) and the oxygen technique (e.g. Talling 1973). Although the ^{14}C technique is more sensitive than the oxygen technique, it has technical and theorical problems, as has been reviewed in detail by Harris (1978). Moreover, the ^{14}C measurements are often not performed on a short time scale (i.e. lasting longer than 10% of the generation time of the species) and usually only consider the initial and final values. The great advantages of oxygen electrode technique are, among others, the ability to measure the dark respiration rate and to continuously monitor oxygen concentrations. Hence, analogous to transient nutrient uptake studies (see Fig. 3) the changes in oxygen production rate can be examined with time (e.g. Harris and Piccinin 1977; Zevenboom et al. 1983).

The general procedure in the laboratory studies of the short-term photosynthetic rates is to expose subsamples to a range of irradiance values (I) under standard constant conditions (with respect to light, temperature, pH, nutrients) and to relate initial net oxygen production rate (P_n) to the corresponding I value to construct the P–I curve. The gross oxygen production rate (P) is obtained according to:

$$(15) \qquad P = P_n - R$$

in which R is the dark respiration rate (which has a negative sign) and with the assumption that R in light and dark are equal.

In the P–I curve, two parameters can be distinguished: P_m (maximum photosynthetic capacity; value of P at saturating light) and α (initial slope of P–I curve; photosynthetic efficiency). Jassby and Platt (1976) discussed 8 different equations to describe the P–I curve below the onset of photoinhibition, while Platt et al. (1980) also focussed on the latter feature, where P_m decreases at inhibiting light irradiance. When photoinhibition was excluded, the best fit for their experimental data was found to be the hyperbolic tangent equation (Jassby and Platt 1976):

$$(16) \qquad P = P_m \cdot \tanh\left(\frac{\alpha \cdot I}{P_m}\right)$$

A number of freshwater species (green algae, diatoms), examined by short-term oxygen electrode measurements, showed a rising linear relationship between P and I over a wide range of limiting light irradiances. The transition to P_m was sharp so that from the ratio of P_m/α the value of I_k can be calculated (Talling 1957) according to:

$$(17) \qquad I_k = \frac{P_m}{\alpha}$$

in which I_k is the irradiance at which the rising linear portion of the P–I curve intersects with P_m. Data for light-limited cultures of the cyanobacterium *Microcystis aeruginosa*, also obtained by the short-term oxygen method, could be fitted by the Michaelis–Menten equation (eq. 6). Analogous to nutrient uptake affinity, dV/dS, the value of α is then the initial rising slope of the *P–I* curve. One should bear in mind, however, that the latter represents the ratio of flux/flux, while the former is a ratio of flux/concentration.

It is not surprising that there is no single mathematical equation to describe the complex series of events in photosynthesis. The value of α depends on the cellular contents of chlorophyll *a* and accessory pigments (such as chlorophyll *b* and *c*, the species-specific photosynthetically active carotenoids, and the biliproteins phycocyanin, phycoerythrin and allophycocyanin), and on their absorption characteristics (see e.g. review by Harris 1978). The light-limited part of the *P*-I curve (α) is thus essentially a biophysical event, while the light-saturated part (P_m) involves the biochemical processes in photosynthesis. It has been suggested that α is proportional to the number and size (absorption cross section) of photosynthetic units (PSUs), while P_m is a function of the number of PSUs and the rate limiting step in the photosynthetic pathway (Herron and Mauzerall 1972; Falkowski and Owens 1980).

The short-term photosynthetic response is time-dependent and differs from the steady-state (long-term) photosynthetic rate. We observed for a number of species, grown under light-limitation, a decrease in the initially high oxygen production rates when these were examined over longer periods of time at incubation irradiances higher than the growth irradiance (Zevenboom and Post 1983). The initially high P_m value of a light-limited species is higher than its actual photosynthetic rate (q_{O_2}) expressed during steady-state growth at prevailing growth irradiance (Zevenboom and Mur 1984). A second, general phenomenon is that the irradiance value that saturates photosynthesis (P_m) is much higher than the one that saturates growth (μ_m) (Harris 1978). In these respects, short-term photosynthesis of light-limited cells resembles short-term nutrient uptake of nutrient-limited cells (see Fig. 3 and 5) (Zevenboom and Mur 1984).

Several papers deal with attempts to assess *in situ* growth rates from short-term phosynthetic responses of the phytoplankton community and difficulties encountered in these attempts (e.g. Li and Goldman 1981; Eppley 1981). The problems are among others, the diel pattern in short-term photosynthesis (e.g. Prézelin and Matlick 1980; Harding et al. 1981), and the initially high short-term photosynthetic rates which level off when incubation period at various fixed light irradiances is extended (see e.g. Harris 1978). Another problem is that the rates of change in photosynthetic response and growth rate do not necessarily parallel each other during transition from one light level to another (see e.g. Post et al. 1985c). In long-term incubations (i.e., incubation periods approaching the generation time of the species), photosynthesis may be in good correlation with growth rate. From the work of Morris et al. (1981) it appears that the growth rate can be "safely" assessed from the C^{14} incorporation into protein. Also continuous culture studies suggest that under intermittent light the protein synthesis rate is a good approximation of μ, as long as it remains constant over the entire L/D cycle (Cuhel et al. 1984; Loogman 1982; Post et al. 1985b) (see later Fig. 10B).

4.2 Adaptation in Short-Term Photosynthetic Parameters

Changes in P_m and α in response to changed experimental growth-environment are mediated through adaptation in the photosynthetically active pigments and (on P_m level) enzyme activities and electron transport. In studies on temperature adaptation it is not surprising to find that the biophysical part (α) is not affected, while P_m (per unit of cellular biomass) is higher at higher-temperature, due to acceleration of biochemical processes involved in photosynthesis (see reviews by Harris 1978 and e.g. recent paper by Post et al. 1985a). Some species also achieve higher P_m values (per unit of cellular biomass) when adapted to higher light conditions, probably due to lower turnover time (i.e. time required to transfer an electron through the photosynthetic electron transfer chain) and/or higher concentration of ribulose

1,5 diphosphate carboxylase (Raps et al. 1983; Senger and Fleischhacker 1978a, b). However, it seems to be more common that P_m is equal or lower under higher light condition, which may suggest that in a number of species light adaptation is mediated through regulation in pigments. It has been suggested that species respond to decreased light level by increasing the size or the number of the PSU value (Falkowski and Owens 1980; Perry et al. 1981). Either change would promote an increased photosynthetic response (per unit of cellular biomass) under low light conditions (see also review of Richardson et al. 1983). Before dealing further with light-adaptation in photosynthesis it is worthwhile to focus shortly on the adaptation patterns in pigments first.

In many species, the contents of chlorophyll *a* (Chl. *a*) and other species-specific photosynthetic pigments increase with decreasing growth irradiances. In some species, the increase in light-harvesting capacity is further enhanced through an increase in the ratio of accessory pigment/Chl. *a*. This was, for example, demonstrated for the marine dinoflagellate *Glenodinium* (increase in ratio of peridinin/Chl. *a*)(Prézelin 1976;1981) and for several cyanobacteria (increase in ratio of C-phycocyanin/Chl. *a*) (Foy and Gibson 1982; Post et al. 1985a). Conformational changes in chloroplasts, in response to lightshifts, have also been reported (Yentsch et al. 1985). Under non-saturating low light levels, combined with nutrient-limitation, the pigment content is largely increased, as has been shown for N-limited *Oscillatoria agardhii* (Van Liere et al. 1979a). In studies with this species it was found that the increased contents of Chl. *a* and C-phycocyanin resulted in a large increase in the internal N content of the N-limited species, as will be discussed later (Fig. 13) Zevenboom et al. 1980). Under saturating light conditions or nutrient-limitation, the pigment content of the cell, and also its light-harvesting capacity is, in general, low (see e.g. Konopka and Schnur 1981; Van Liere et al. 1979a). If regulation in biochemical processes can be neglected, one can expect that under light saturating, nutrient-limiting conditions the pigment content and Pm correlates with the nutrient-limited growth rate, as was found for *Oscillatoria* (Van Liere et al. 1979a; Riegman 1985) and *Scenedesmus* (Smith 1983). In conclusion literature data on the regulation in pigmentation of algae seem to suggest that algae synthesize the most "useful" pigments, in certain ratios and amounts adjusting thus their light-harvesting capacity and thereby, partly or totally, regulating their photosynthetic response to changed light conditions.

From an ecological point of view it is worthwhile to note that adaptation patterns in short-term photosynthesis of light-limited cells may show some similarities with those of short-term nutrient uptake by nutrient-limited organisms (shown in Fig. 4). Different adaptation patterns may well be related to the different ecological niches of the respective species. Adaptation in α with constant P_m (compare with Fig. 4A) seems to be crucial in low-light environments. This pattern was found in the cyanobacterium *Oscillatoria* (Foy and Gibson 1982; Post et al. 1985a), a species that often dominates waters with high phytoplankton biomass concentrations and steep light gradients (Ahlgren 1970; Foy et al. 1976; Zevenboom et al. 1982). The strategy of changing Pm with constant α seems to be important in species such as *Scenedesmus* and several marine species that grow in nature under relatively high light conditions (e.g. Senger and Fleischhacker 1978a; Falkowski and Owens 1980). One may also find changes in both P_m and α (Fig. 4C and D). *Microcystis* showed a response to low light levels that resembled the pattern shown in Fig. 4D (Zevenboom and Mur 1984). We suggested that this adaptation pattern may provide *Microcystis* with a flexability to invade and to flourish in low-light environments (α-adaptation) and to cope as well with high-light conditions (P_m-adaptation) that are encountered at the surface when buoyancy regualtion is lost.

4.3.1 Light-Limited Growth; Effect of Irradiance

To examine the relationship between growth rate (μ) and growth irradiance (I) it is of utmost importance to establish an even light distribution in the culture vessel.

Factors that influence the average irradiance are the concentration and pigmentation
of the algae, the geometry of culture vessel and light source, and the distance between
these two. In dilute suspensions in turbidostat cultures, differences in irradiance in
the culture vessel are minimized (Loogman and Van Liere 1978; Van Liere et al. 1978).
Using these systems, we found for a number of species that their light-limited growth
kinetics could be described by the Monod equation, thus:

$$(18) \qquad \mu = \bullet \; \frac{\bar{I}}{K_I + \bar{I}}$$

in which K_I is $\bar{I}$ at which $\mu = 1/2\ \mu$m (Van Liere 1979; Zevenboom et al. 1981c; Post
et al. 1985a).

A distinct feature of light-limited growth is the cell's requirement of a minimum
amount of light irradiance (or light pulse, see next session). Thus, upon extrapolation
to $I = 0$ one finds the negative growth rate, μ_e, the specific maintenance rate
constant which is required for maintenance energy purposes. The importance of μ_e
was pointed out by Gons and Mur (1975) and later by Van Liere and Mur (1979).
They described the relationship between μ and specific light-energy uptake rate of
photoautotrophs with the equation:

$$(19) \qquad \mu = c \bullet q_E - \mu_e$$

in which c is the growth efficiency factor and q_E is the specific light-energy uptake
rate (h^{-1}) (energy absorbed per unit time (J.h^{-1}) per unit of biomass (J)). Equation

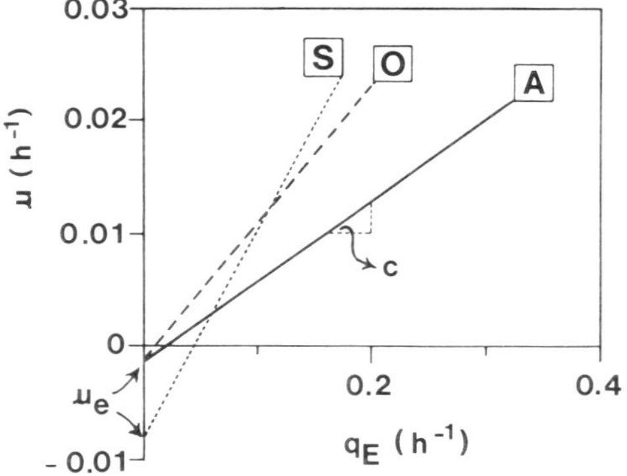

FIG. 8. Energy balances of light-limited
continuous cultures of *Aphanizomenon flos-
aquae* (A), *Oscillatoria agardhii* (O) and
Scenedesmus protuberans (S). Arrows: μ_e
values (Data after Van Liere and Mur 1979;
Gons and Mur 1980; Zevenboom and Mur
1980). Species were grown at 20°C and under
identical incident irradiance.

19 is essentially the same as the formula derived by Pirt (1965) for heterotrophic
bacteria. Thus, under light limitation, the relationship between μ and q_E is linear (eq.
19) with μ_e as the negative intercept of the μ-axis and c as the slope of the line (see
three examples, given in Fig. 8). One can argue that the maintenance energy requirement is not a fixed cost paid by the cell, but depends on the ambient growth conditions. Indeed it has been found that μ_e increased with increasing temperature (Gons
and Mur 1980) and under very short L/D cycles (Gibson 1985). Thus, when different
species are compared, one should take these effects into account.

Studies in our laboratory, on various species, demonstrated that both μ_e and the
value of c are species-specific (Gons and Mur 1980; Van Liere and Mur 1979;

406

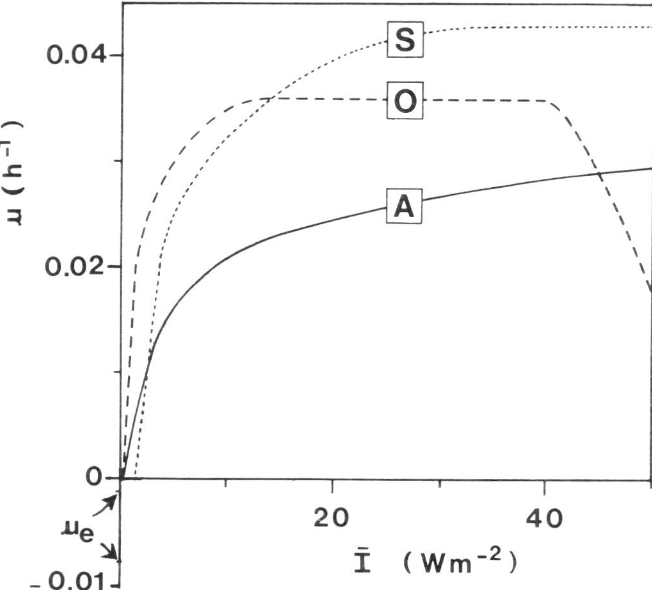

FIG. 9. Specific growth rate (μ) as function of average irradiance in, dilute, continuous cultures of three different species. Species symbols and literature as in Fig. 8. Arrows: μ_e values.

Zevenboom and Mur 1980; Zevenboom et al. 1981c). The differences in growth parameters are also reflected in the growth responses of the different species to the prevailing light conditions (μ-I curves, see examples in Fig. 9). Cyanobacteria, being prokaryotes, show a low μ_e value (of 0.001 h^{-1}; Table 1), much lower than eukaryotic species like *Scenedesmus protuberans*. This feature enables cyanobacteria to outgrow the latter group of species in low-light environments. However, by virtue of its high growth efficiency (Fig. 8, Table 1), *Scenedesmus* grows faster than cyanobacteria under conditions of higher irradiance values, which may even inhibit the growth rate of some cyanobacteria (in this case: *Oscillatoria agardhii*, see Fig. 9). Although all cyanobacteria studied sofar, show comparable low μ_e values (see also Zevenboom and Mur 1984), they may differ in the growth efficiency value. In the presented example (Fig. 8 and Table 1), the heterocystous N$_2$-fixer *Aphanizomenon flos-aquae* has a lower growth efficiency than the non-N$_2$-fixer *Oscillatoria*. This means that *Aphanizomenon* must absorb more light than *Oscillatoria* to attain the same growth rate. The higher light-energy requirement of the N$_2$-fixer may explain why this species is unable to outgrow *Oscillatoria agardhii* in waters with low ambient light, even when the latter is growing under N-limitation (Zevenboom et al. 1982; Zevenboom and Mur 1980).

TABLE 1. Steady-state data of *A. flos-aquae*, *O. agardhii* and *S. protuberans* in light-limited continuous cultures. Growth parameters μ_e and c were calculated from Fig. 8, using equation 19 *q_E expressed in J · mg^{-1} · h^{-1}, calculated for identical growth rate: $\mu = 0.01$ h^{-1}, identical incident irradiance and temperature.

Organism	N$_2$-fixation	c	μ_e	q_E*	Reference
A. flos-aquae	+	0.07	0.001	3.14	Zevenboom et al. (1981c)
					Zevenboom and Mur (1980)
O. agardhii	—	0.12	0.001	1.83	Van Liere and Mur (1979)
S. protuberans	—	0.18	0.008	2.0	Gons and Mur (1980)

In conclusion, one can readily explain species-shifts and succession phenomena in waters of high trophic state on the basis of information of growth kinetics of the species involved. On the bases of light-limited growth kinetics, results of competition experiments and a simulation model, Mur and coworkers were able to put forth a hypothesis to explain the shift from green algae to cyanobacteria in hypertrophic lakes (Mur et al. 1977; Mur and Beydorff 1978; Loogman et al. 1980). As in the case for nutrients (see e.g. Tilman 1977), species competing for light can be placed along an irradiance gradient, their position being defined by their light-energy requirements (Zevenboom and Mur 1980).

4.3.2 Light-Limited Growth; Effect of L/D Cycle

Under continuous illumination with a constant nutrient supply, algae generally show an ideal steady state growth rate μ ($\mu = \frac{1}{X} \cdot \frac{dX}{dt}$, eq. 10). However, under a L/D cycle, periodic changes in cell components and activities are expected to occur (e.g. Rhee et al. 1981), so that μ is more likely to be in steady state over a complete L/D cycle (denoted by $\bar{\mu}$). In eukaryotic cells, the periodic changes can be more complex than in prokaryotes, due to group synchronization, as was observed in *Scenedesmus* and *Chlamydomonas* grown under various L/D cycles (Post et al. 1985d; Zevenboom, unpublished results; see also Chisholm 1981). One of the cell components that vary greatly in magnitude is the carbohydrate content of the cell. An example is shown in Fig. 10A, B. With culture studies it has been shown that carbohydrate is accumulated during the light period, and subsequently used in the dark as to allow

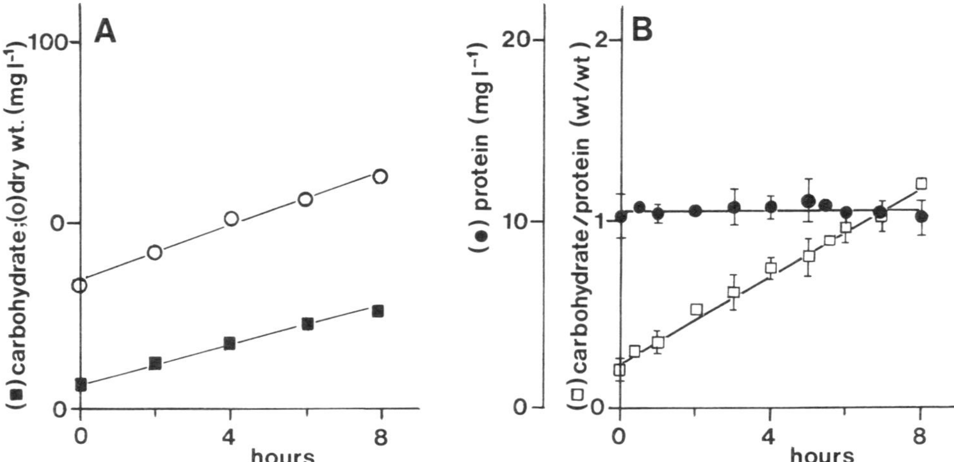

FIG. 10. (A) Patterns of changes during light period, in carbohydrate concentration (■) and dry weight concentration (○) for *O. agardhii* grown with an 8/16 h L/D cycle at 30 W m^{-2}. (After Loogman 1982). (B) As Fig. 10A but now patterns of changes in protein concentration (●) and carbohydrate content (per unit protein) (□). Bars indicate standard deviations around mean values of triplicates (After Post et al. 1985b).

continued protein synthesis (Cuhel et al. 1984; Foy and Smith 1980; Loogman 1982; Van Liere et al. 1979b; Post et al. 1985b). The specific accumulation rate of carbohydrate thus exceeds μ several fold. The diel changes in carbohydrate are parallelled by changes in culture biomass concentration (Fig. 10A). Therefore, it is important to calculate μ over a complete L/D cycle interval so that the influence of diel patterns in storage products are excluded.

In several species, a non-linear relationship was found between μ and photoperiod length (Foy and Smith 1980, Loogman 1982), while other species showed a more linear relationship (e.g. Gibson and Foy 1983) or a biphasic pattern (Zevenboom and Mur

408

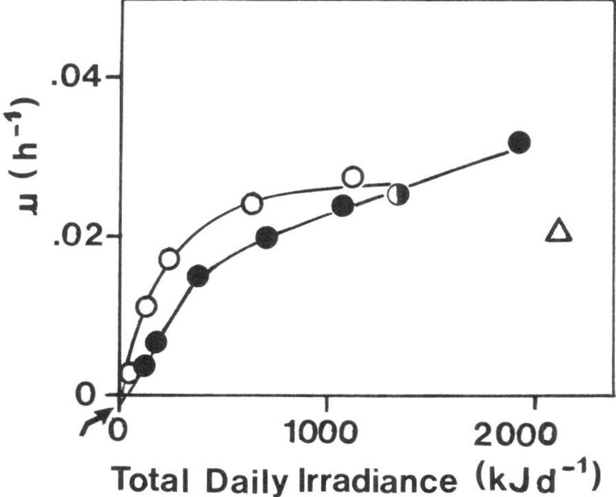

FIG. 11. Specific growth rate (μ) as function of total daily irradiance for *M. aeruginosa* grown with various L/D cycles at saturating light irradiance ($\bullet$), under various growth irradiances with a 16/8 L/D cycle ($\bigcirc$), and under continuous, saturating light ($\triangle$). Arrow: μ_e value (After Zevenboom and Mur 1984).

1984). Extrapolation of the curve yields the μ_e value (*see* previous section) and the minimal light period that keeps the cell viable. In general, cyanobacteria require much lower light levels (irradiance, photoperiod length) than eukyarotes to stay alive. At very short light hours, the dark period may determine μ (Foy 1983). Some diatoms (Paasche 1968) and *Microcystis* (Zevenboom and Mur 1984) show submaximum growth rates under continuous light (see also Fig. 11). This indicates that these species require a dark period (shorter than 2 h in the case of *Microcystis*) to attain maximum growth rate.

Some species show growth rates that are proportional to the product of irradiance and photoperiod length (Total Daily Irradiance; TDI value) (Talling 1955; Eppley and Coatsworth 1966). One can expect however, that μ is not uniformly related to TDI, since photoperiod length and irradiance value regulate growth rate in different ways. Under limiting irradiance, μ is limited by the light harvesting potential of the cells, so that the growth affinity (initial slope of $\mu - I$ curve) is the important growth rate determining parameter. Under light pulses of saturating intensity, μ is not determined by growth affinity, but rather by the maximum attained photosynthetic potential. Species are thus expected to grow faster under limiting light intensity than under short photoperiod length with saturating light, at identical TDI value, as was indeed found for *Microcystis* (Fir. 11) and *Oscillatoria* (Zevenboom and Mur 1984; Post et al. 1986). It appears that one thus needs to discriminate between photoperiod and irradiance-limitation, and this restricts the use of simple algal growth models, as was proposed by Bannister and Laws (1980).

5. Nutrient-Light Interactions

Culture studies on the interactions of nutrients, light, temperature have been started in a simple way, by examining steady state growing cells under nutrient limitation at various "non-saturating" levels of the other parameter. One should realize, however, that these steady state results yield no direct information on the kinetics of regulation. This information can be obtained when one considers the changes during transition from one level to another. Thus, the study on interactions of various growth parameters is still in the beginning of the complex road to bridge gaps between laboratory and field.

In these studies the question is emphasized whether the steady-state interactions are multiplicative or according to a threshold. If the latter is the case, they will resemble nutrient-nutrient interaction, for which it was demonstrated that a species is limited in its growth by only one nutrient at a time, namely the one which is in the shortest supply relative to its need (Droop 1974; Rhee 1978). Nyholm (1978) and Riegman (1985) proposed using a multiplicative formula to describe steady state interacting effects of light and nutrient-limited growth. However, results of other papers may indicate that steady state interactions are not multiplicative, but show a threshold pattern (Zevenboom et al. 1980; Rhee and Gotham 1981a; Droop et al. 1982; Healey 1985). Thus, growth rate is, also here, determined by one factor at a time, (either by the availability of the nutrient or that of light), while the other factor is present in subsaturating concentration, thereby affecting the yield on the limiting nutrient and other physiological properties of the species, but *not* determining its specific growth rate.

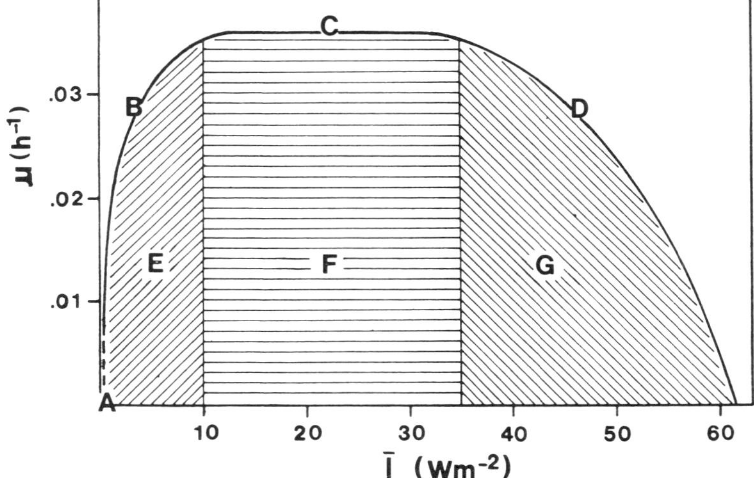

FIG. 12. Light (growth irradiance, $\bar{I}$)–nutrient interaction for *O. agardhii* grown in continuous culture. A.: Light is used for maintenance energy purposes. B.: Light-limited growth, nutrients are saturating; the μ–I curve sensu stricto. C.: $\mu = \mu_m$. D.: Growth rate is limited by light-inhibition. E.: Nutrient-limited growth under light-non-saturating conditions. F.: Nutrient-limited growth under light-saturating conditions. G.: Nutrient-limited growth under light-inhibiting conditions. (From Zevenboom et al. 1980).

To illustrate the above mentioned findings, steady-state date on nitrate-limited *Oscillatoria agardhii*, grown under a set of various irradiance values ($\bar{I}$) and various fixed steady state growth rates (= dilution rates) will be used (Zevenboom et al. 1980). (Fig. 12, and 13). By considering the μ–I curve, one can see what is meant by non-saturating and saturating-light during nutrient-limited growth (Fig. 12, area E and F, respectively). In Fig. 12, seven areas are distinguished in the steady state interaction between nutrient and light (*see* legend on Fig. 12). For any given *fixed nitrate*-limited growth rate it was found that the nitrogen content (Q) (Fig. 13) increased when the non-saturating irradiance was lowered (which is for this species; $I < \sim 10$ W m^{-2}, *see* Fig. 12). A large part of the increase in Q was associated with the quantitative

410

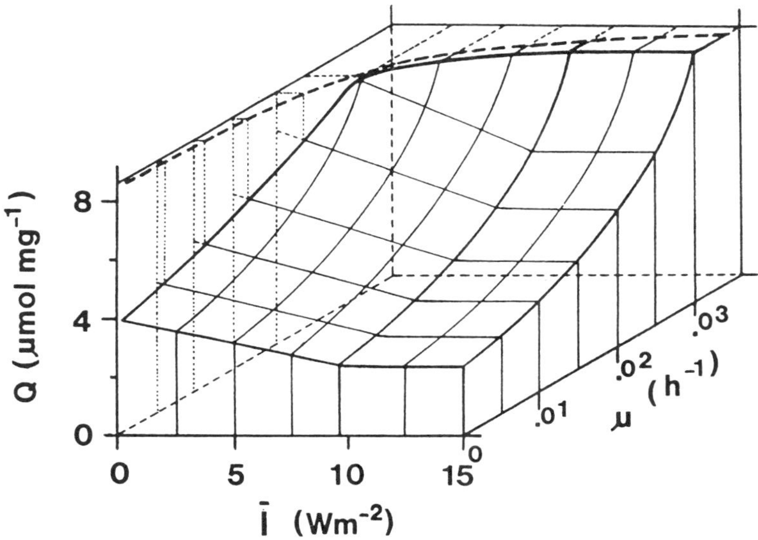

FIG. 13. Dependence of the nitrogen content (Q) of nitrate-limited *O. agardhii* on the specific growth rate (μ) and on the average light irradiance (growth irradiance $\bar{I}$), calculated according to the equations:

$$Q = \frac{-0.0066\,\bar{I} + 0.1633}{0.042 - \mu} \quad \text{for } 0 < \bar{I} < 9.4 \text{ W m}^{-2}$$

$$Q = \frac{0.101}{0.042\,\mu} \quad \text{for } \bar{I} \geq 9.4 \text{ W m}^{-2}$$

(from Zevenboom et al. 1980)

changes in the light-harvesting capacity of the cells, namely the nitrogen-containing pigments chlorophyll *a* and C-phycocyanin. The increased pigment content is necessary to keep photosynthesis at the same rate, in balance with the nitrate-limited growth rate, when less light is available. We further demonstrated that the μ–Q relationship could be described by the Droop equation (eq. 12) for any *fixed* irradiance (Fig. 13). New formulas were derived by which a three-dimensional graph was constructed (Fig. 13), defining the nitrogen content (Q) as a function of, the *nitrate*-limited growth rate (μ) and growth irradiance ($\bar{I}$).

In conclusion, the steady state studies suggest that species can maintain their growth rate under suboptimal conditions of light (or temperature, *see* Rhee and Gotham 1981b; Zevenboom 1980), by increasing the value of Q, which results in a decreased yield on the limiting substrate (since, $Q = \frac{1}{Y}$). It is tempting to speculate that under transient conditions, interactions are multiplicative, or according to a threshold with rapid alternation of growth limitation, and mediated by adaptation in pigments when sub-saturating light levels are involved.

6. Role of Continuous Cultures in Ecology

In the preceding sections it has been repeatedly illustrated that physiological properties within one single species are influenced by the growth environment. Phenotypes are established that are appropriate to and characteristic of the different types of growth-limitations, so that they can successfully compete with others under the given conditions. In mixed populations, different species may be limited in their growth rate by different factors, in which case they will coexist. This has been described, for example, for silicon-phosphorus interaction in mixed populations of diatoms (Tilman 1977)), and nitrogen-light interaction in systems with populations of N_2-fixing and non-N_2-fixing cyanobacteria (Zevenboom and Mur 1980). A patchy distribution of a limiting factor (in space and/or time) can also lead to coexistence of species, each of them exploiting the limiting resource at different moments (e.g. Van Gemerden 1974).

6.1 Physiological Indicators

Using such physiological information obtained by continuous culture studies, it becomes possible to explain shifts in population composition and species-dominance, and to identify the nature of the primary growth rate limiting factor of the dominant species in its natural habitat. Using physiological indicators to assess the type of growth limitation has been successfully applied (e.g. Healey and Hendzel 1979; Riegman 1985; Sakshaug et al. 1983; Zevenboom et al. 1982). Before giving two examples, alternative approaches will be discussed briefly.

The most widely used method for identifying the factor limiting growth rate has been the bioassay (e.g. Gerhart and Likens 1975; Goldman 1978). After nutrient enrichment in a closed system, the change in biomass is followd over several days and used as a measure of nutrient-deficiency. During such long-term incubations, the physiological properties of the enclosed population are markedly changed and one limiting factor may very well have replaced the other. Especially, when the concentrations of potentially limiting nutrients are rather low, it is difficult to obtain an unequivocal assessment of the limiting factor for the population at the start of the investigation (see for detailed discussion, Zevenboom et al. 1982). Also, *in situ* nutrient *concentrations* are by no means reliable indicators for nutrient-sufficient or limiting conditions. Nutrient *fluxes* (supply and uptake rates) should be known for valid conclusions to be drawn (see also Kilham and Kilham 1978).

Since species show specific responses, one may argue that physiological indicators are of limited use and time-consuming, when physiological characteristics of less well studied species are to be examined first. However, one can list some general features (Table 2) that seem to be sufficiently characteristic and specific of different types

TABLE 2. List of physiological indicators. * Values of cell components are not constant, *see* further text.

Parameter	Nutrient-limitation	Light-limiation
Cell components:		
Pigments	low*	high*
Cellular nutrient	low*	high*
Carbohydrates	high*	low*
Protein	low*	high*
Cell activities:		
Mode of nutrient uptake	exponential	linear
Nutrient uptake potential	high	low
Photosynthetic potential	low	high

of growth limitation. Based on literature noted in previous sections one can summarize as follows: Under severe nutrient-limitation (which is, at low growth rate), algae show low pigment contents, low photosynthetic potentials, low cellular contents of the limiting nutrients, high potentials for uptake of the limiting nutrient, high carbohydrate and low protein contents. The reversed trends are, in general, characteristic for light-limited species. Not all parameters, listed in Table 2, are of equal validity in their use as physiological indicators. The pigment content of a cell is a reflection of the "light-history" of the cell and, thus, cannot discriminate between different types of nutrient-limitation. The cellular content of the limiting nutrient is related to the nutrient-limited growth rate (Droop equation 12), so that a *high* value in a *nutrient-limited* population can be found. Furthermore, influences of temperature and ambient light on the cellular content of the limiting nutrient cannot be excluded (see Fig. 13) and synergisms between nutrients may exist, as described for N and P in *Oscillatoria agardhii* (Fig. 6). Also protein and the fluctuating carbohydrate pool (see Fig. 10) are not specific enough to discriminate between different types of limitations. The consequence of the above mentioned findings is that cell *components* (and their ratios) are poor references to assess a specific limitation in natural populations, without prior knowledge of growth rate and environmental factors. So, it is necessary to combine them with other, more characteristic parameters, namely cell *activities* (Table 2), which are nutrient uptake and photosynthetic potentials (V_m, P_m) and the mode of nutrient uptake (exponential vs. linear; Fig. 3). All phytoplankton species studied so far, show a high uptake potential for the limiting nutrients with an exponential decrease in nutrient concentration with time. The reason for this pattern has been discussed before. Since this strategy is of such fundamental importance (Tempest et al. 1983), one can expect it to be found in all phytoplankters. Thus, by determining the time course of uptake, the nutrient uptake potentials for various nutrients and the photosynthetic potentials, one can obtain an accurate and direct assessment of the nature of the primary growth limiting factor of a natural population, and its growth rate.

6.2.1 Case Study: Freshwater Cyanobacteria

To illustrate the approach outlined above the study on growth limiting factors of *Oscillatoria agardhii* will be shortly summarized. This species dominates the phytoplankton in several Dutch hypertrophic lakes, such as lake Wolderwijd (Zevenboom et al. 1982) and was studied in its natural environment over four

TABLE 3. Survey of physiological data of *O. agardhii*, grown in continuous cultures under P (phosphate), N (nitrate or ammonium), or E (light) limitation. (From Zevenboom et al. 1982).

Parameter	Limiation		
	P	N	E
P-content (% dry wt)	0.28–1.1[a]	0.53–1.1[b]	0.75–1.2[c]
N-content (% dry wt)	5.6–10.5[b]	3.4–11[a,c]	7.5–11[c]
N:P (wt/wt)	10–14[d]	7.5–13[c,e]	9–11[c]
Chl. *a* content (% dry wt)	0.3–0.8[b]	0.15–1.5[c,e]	0.5–1.7[c]
V_m P (μmol $\cdot$ mg^{-1} $\cdot$ h^{-1})	0.06–0.9[d]	0.01	0.01
V_m NH$_4$ (μmol $\cdot$ mg^{-1} $\cdot$ h^{-1})	0.4–0.7[d]	0.8[f]	0.1–0.4[e]
V_m NO$_3$ (μmol $\cdot$ mg^{-1} $\cdot$ h^{-1})	0.4–0.7[d]	0.8–1.5[c,f]	0.1–0.4[e]

[a] Increase in nutrient content (Q) with increasing growth rate (μ), according to Droop equation (eq. 12). Lowest values of nutrient contents listed here are the k$_q$ values.
[b] Linear increase in value of the parameter with increasing μ.
[c] Increase in value of the parameter with decreasing light irradiance.
[d] Increase in value of the parameter with decreasing μ.
[e] Increase in value of the parameter with increasing μ.
[f] Independent of μ.

successive years (Zevenboom 1980). Data gathered over a single year (1977) will serve the point. Table 3 gives some of the physiological data for this species under various limitations.

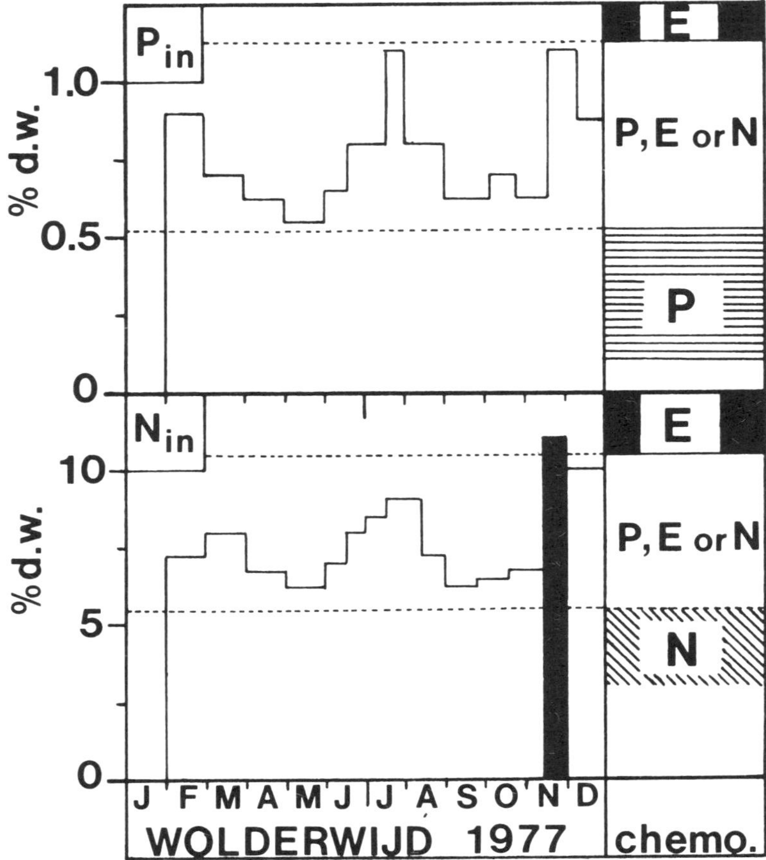

FIG. 14. Total cell phosphorus (Pin) (% dry wt) and total cell nitrogen (% dry wt) of the natural population of *O. agardhii* in Lake Wolderwijd, examined through 1977. Right-hand column: data of chemostat-cultured (chemo.) *O. agardhii*. Solid: light (E) limitation; hatched: nitrogen (N) limitation; horizontal ruled: phosphate (P) limitation. P, E or N: "overlap area" (no clear distinction between P, E or N limitation). (After Zevenboom 1980).

As expected, measurements of the cellular phosphorus, nitrogen (Fig. 14) and pigment contents and N/P ratios (Table 3) provided no firm basis to conclude whether N, P or light (E) limitation prevailed during the whole year that they were studied. However, the less frequently studied uptake potentials for nitrate, phosphate (Fig. 15) and ammonia (Table 3), revealed clearly that the natural population was growing under a P-limitation in early June and early November, under a N-limitation in early August and mid-October, while a light-limitation prevailed throughout the second half of June and all July and again in late-November.

In the above described study it is even possible to specifiy in several cases, the light conditions and growth rate when all available data are incorporated. A more direct assessment of light-limited growth and growth rate can be made by including data from measurements on photosynthetic potentials and mode of nutrient uptake, respectively, as was noted before. This was done in a study on *Microcystis* (Zevenboom and Mur 1984) and in recently undertaken study on marine species.

414

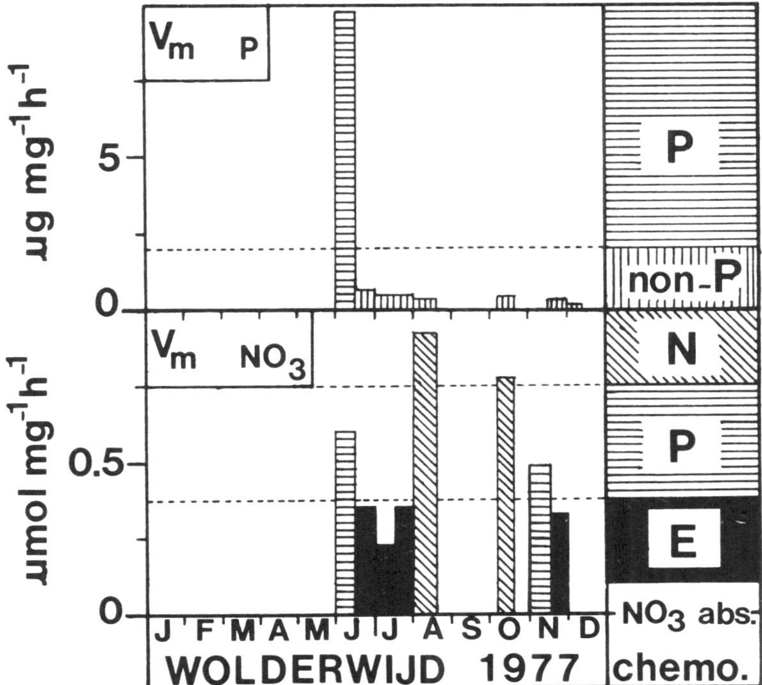

FIG. 15. Uptake capacity values for phosphate (V_mP; $\mu g \cdot mg^{-1}h^{-1}$ and for nitrate (V_mNO_3; $\mu mol \cdot mg^{-1} \cdot h^{-1}$) of the natural population in Lake Wolderwijd 1977 and of chemostat cultures (right) of *O. agardhii*. Symbols as in Fig. 14. NO_3abs. indicates that $V_m NO_3$ is zero, which occurs when cells are grown in the absence of nitrate. Vertical ruled: non-P-limitation. (After Zevenboom 1980).

6.2.2. Case Study: Marine Picoplankton

The growth limitation for (pico) phytoplankton was investigated, as part of a study on red-pigmented coccoid cyanobacteria, during the *Snellius II* Expedition in the Banda Sea, in August 1984 and February–March 1985. The small ($< 1.0 \, \mu m$) red-pigmented cyanobacteria were found in large concentrations (10^4–10^5 cells mL^{-1}) and showed a preference for deeper layers where light intensity was low (Zevenboom 1985). The latter observation suggests that these species are, like the freshwater cyanobacteria (see previous sections), well adapted to grow and photosynthesize at relatively low irradiances, which is consistent with laboratory results (Morris and Glover 1981; Zevenboom 1986).

Although not all data are available yet, a short summary for the first cruise may serve the point to be made here. Nutrient uptake data revealed that most probably light, but certainly not nutrients were limiting the phytoplankton growth, except for one Station where N-limitation prevailed (Zevenboom 1985). It has been suggested (Gieskes and Kraay 1984; Waterbury et al. 1979) and shown (Platt et al. 1983, *see* Fig. 16) that small photoautotrophs can form an important component of the primary productivity in tropical waters. Phytoplankton growth rate data obtained from the Banda Sea support this view (Zevenboom 1985). An example, presented in Table 4, shows that the deep-small-sized photoautotrophs (mainly red-pigmented cyanobacteria) are able to grow at relatively low irradiance at considerably high rates that may exceed the growth rate of larger-sized species by a factor of three.

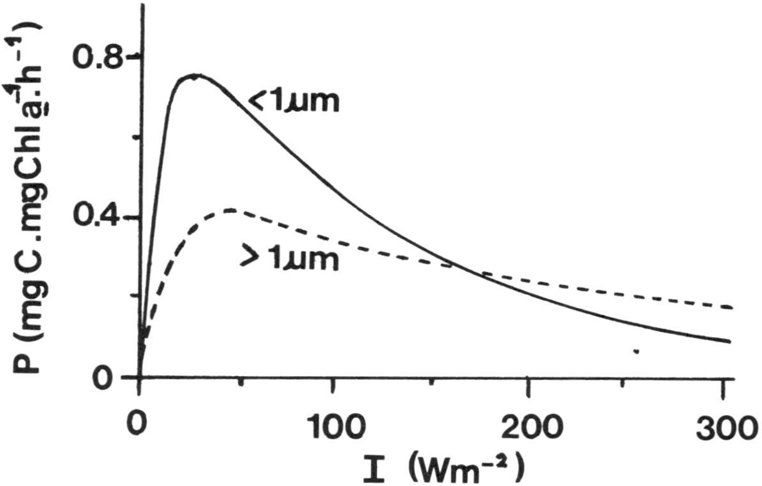

FIG. 16. P–I curves for two size-fractions of the sample from station 18 in the tropical North Atlantic, July 1982. (redrawn from Platt et al. 1983).

TABLE 4. One example of kinetic data of non-nutrient-limited Banda Sea samples. Station 29. August 6. 1984. The value of μ was calculated using eq. 4 ($V = \mu \cdot Q$) see also Fig. 3.

Fraction	q μmol N $\cdot$ mg^{-1} $\cdot$ h^{-1}	μ h^{-1}
$< 1\mu$m	0.63	0.088
$> 1\mu$m	0.19	0.027

7. Conclusions

Through ecophysiological research, insights are obtained in the various adaptation strategies of phytoplankton species in response to changed growth environment. Culture studies, including suboptimal factors and transient conditions, are important, since these studies can reveal the triggers of the adaptive responses and make extrapolations to the field more realistic. Although the level of adaptation to growth environment is species-specific, both freshwater and marine phytoplankters examined so far use at least two fundamental strategies to obtain an optimum growth response in order to compete successfully with others for growth-limiting resources. Firstly, under nutrient- or light-limitation the scavening capacity will be increased, leading to increased nutrient uptake- or photosynthetic potentials. Secondly, or concomitantly, the efficiency of assimilation of the limiting resource will increase in response to nutrient- or light-limitation. Physiological responses of phytoplankton are influenced by both the ambient and past growth environment. To understand physiological adaptation patterns in nature it is thus of importance to consider their physiological state at present in relation to their ambient and past growth environment. Their physiological state and growth rate can be assessed by determining their mode of nutrient uptake, and their nutrient uptake and photosynthetic potentials. The use of physiological indicators, characteristic of a certain type of growth limitation, is a fruitful line in testing and predicting species succession — phenomena in both freshwater and marine habitats.

8. Achnowledgements

This paper was written, for a part, during an extended expedition from Indonesia to the Netherlands. I am greatful for the fine atmosphere that was created by a number of colleagues of various disciplines during this long sea-trip. Especially thanked are my colleagues of the lab., Tineke Burger-Wiersma, Jacco Kromkamp, Luuc Mur, Anton Post and Roel Riegman, for their valuable comments on the manuscript and Jerien Braakman-de Vriend and Truus Smit for typing the manuscript. This work was supported in part by WOTRO (Netherlands Foundation for the Advancement of Tropical Research), NRZ (Netherlands Council of Oceanic Research), and ZWO (Netherlands Organization for the Advancement of Pure Research).

9. Symbols

μ	specific growth rate (h^{-1}); balanced growth: $\frac{1}{X} \cdot \frac{dX}{dt}$
$\bar{\mu}$	specific growth rate over a complete L/D cycle (d^{-1})
μ_e	specific maintenance rate constant (h^{-1})
μm	maximum specific growth rate (h^{-1})
μm*	asymptote of Droop-hyperbole (eq. 12), is greater than μ_m (h^{-1})
K_s^g	half-saturation constant for nutrient-limited growth; $\bar{s}$ at which $\mu = 0{,}5$ μm (μM)
K_I	half-saturation constant for light-limited growth; $\bar{I}$ at which $\mu = 0{,}5\ \mu_m$ (W m^{-2})
$d\mu/d\bar{s}$	initial slope of μ–$\bar{s}$ curve; affinity for nutrient-limited growth ($= \sim \mu_m/K_s^g$)
$d\mu/d\bar{I}$	initial slope of μ–$\bar{I}$ curve, affinity for light-limited growth. ($= \sim \mu_m/K_I$)
$\bar{s}$	nutrient concentration in the chemostat during steady state (μM)
S_R	nutrient concentration in the inflow (μM)
$\bar{I}$	average irradiance in culture vessel, during steady state (W m^{-2})
q_E	specific light energy uptake (h^{-1}, or J $\cdot$ mg^{-1} $\cdot$ h^{-1})
c	efficiency factor for growth (no dimensions)
$\bar{x}$	culture biomass concentration during steady state (mg $\cdot$ L^{-1})
Q	cell nutrient quota; amount of internal (limiting) nutrient per unit population (μmol $\cdot$ mg^{-1}; or % of dry weight)
Q_m	Q at μ_m, or Q at nutrient-saturated growth rate (μmol $\cdot$ mg^{-1})
k_q	minimum cell nutrient quota; Q at μ is zero (μmol $\cdot$ mg^{-1})
Y	yield value; biomass ($\bar{x}$) produced per nutrient (S_R–$\bar{s}$) consumed (mg $\cdot$ μmol^{-1})
q	specific nutrient uptake rate, related to μ (μmol $\cdot$ mg^{-1} $\cdot$ h^{-1})
q_m	maximum specific nutrient uptake rate, related to μm (μmol $\cdot$ mg^{-1} $\cdot$ h^{-1})
V	short term, initial nutrient uptake rate during "linear" decrease of nutrient pulse S (μmol $\cdot$ mg^{-1} $\cdot$ h^{-1}); V equals q when $S = \bar{s}$, during steady state
V_m	potential maximum nutrient uptake rate; maximum nutrient uptake capacity (μmol $\cdot$ mg^{-1} $\cdot$ h^{-1})
K_s^u	half-saturation constant for nutrient uptake; S at which $V = 0{,}5\ V_m$ (μM)
k	adaptation rate constant of transient nutrient uptake (h^{-1})
K_i	inhibition constant (μM)

dV/ds	initial slope of V–S curve; affinity for nutrient-limited uptake ($= \sim V_m/K_m$)
S	external nutrient concentration used in the short-term pulse experiment (μM)
i	concentration of inhibitor
P	short-term initial gross photosynthetic (oxygen production) rate (μg $O_2 \cdot mg^{-1} \cdot h^{-1}$)
Pn	short-term initial net photosynthetic rate (μg $O_2 \cdot mg^{-1} \cdot h^{-1}$)
R	dark respiration rate (μg $O_2 \cdot mg^{-1} \cdot h^{-1}$) (negative sign)
P_m	maximum light-saturated phtosynthetic rate; maximum photosynthetic capacity (μg $O_2 \cdot mg^{-1} \cdot h^{-1}$)
q_{O_2}	specific oxygen production rate during steady state μ (μg $O_2 \cdot mg^{-1} \cdot h^{-1}$); P equals qO_2 when $I = \bar{I}$ during steady state.
α	rising initial slope of P–I curve; photosynthetic affinity (μg $O_2 \cdot mg^{-1} \cdot h^{-1} \cdot W^{-1} \cdot m^2$)
I_k	I at which initial rising portion of P–I curve intersects with P_m ($W\ m^{-2}$)
I	irradiance used in short-term photosynthetic measurements ($W\ m^{-2}$)
L/D	Light/dard cycle (h)
TDI	Total daily irradiance ($kJ \cdot d^{-1}$)

References

AHLGREN, G. 1970. Limnological studies of Lake Norrviken, a eutrophicated Swedish Lake. II. Phytoplankton and its production. Schweiz. Z. Hydrol. 32: 354–396.

———— 1985. Growth of *Oscillatoria agardhii* in chemostat culture. 3. Simultaneous limitation of nitrogen and phosphorus. Br. Phycol. J. 20: 249–261.

BANNISTER, T. T., AND E. A. LAWS. 1980. Modelling phytoplankton carbon metabolism, p. 243–258. *In* P. G. Falkowski [ed.] Primary productivity in the sea. Plenum Publishing Corporation, New York, NY.

BUTTON, D. K. 1983. Differences between the kinetics of nutrient uptake by micro-organisms, growth and enzyme kinectics. Trends Biochem. Sci. 8: 121–124.

———— 1985. Kinetics of nutrient-limited transport and microbial growth. Microbiol. Rev. 49: 270–297.

CAPERON, J. A., AND J. MEYER. 1972. Nitrogen-limited growth of marine phytoplankton. II. Uptake kinetics and their role in nutrient limited growth of phytoplankton. Deep-Sea Res. 19: 619–632.

CHISHOLM, S. W. 1981. Temporal patterns in cell division in unicellular algae. Can. Bull. Fish. Aquat. Sci. 210: 150–181.

CUHEL, R. L., P. B. ORTNER, AND D. R. S. LEAN. 1984. Night synthesis of protein by algae. Limnol. Oceanogr. 29: 731–744.

CUNNINGHAM, A., AND P. MAAS. 1978. Time lag and nutrient storage effects in the transient growth response of *Chlamydomonas reinhardtii* in nitrogen-limited batch and continuous culture. J. Gen. Microbiol. 104: 227–231.

DAVIS, J. O., N. F. BREITNER, AND P. J. HARRISON. 1978. Continuous culture of marine diatoms under silicon limitation. 3. A model of silicon-limited diatom growth. Limnol. Oceanogr. 23: 41–52.

DEHAAN, H., J. B. W. WANDERS, AND J. R. MOED. 1982. Multiple addition bioassay of Tjeukemeer water. Hydrobiologia 88: 233–244.

DROOP, M. R. 1968. Vit. B12 and marine ecology: IV. The kinetics of uptake, growth and inhibition in *Monochrysis lutheri*. J. Mar. Biol. Assoc. U.K. 48: 689–733.

———— 1973. Some thoughts on nutrient limitation in algae. J. Phycol. 9: 264–272.

———— 1974. The nutrient status of algal cells in continuous culture. J. Mar. Biol. Assoc. U.K. 54: 825–855.

———— 1975. The nutrient status of algal cells in batch culture. J. Mar. Biol. Assoc. U.K. 55: 541–555.

———— 1983. 25 Years of algal growth kinetics. A personal view. Bot. Mar. Vol. XXVI: 99–112.

DROOP, M. R., M. J. MICKELSON, J. M. SCOTT, AND M. F. TURNER. 1982. Light and nutrient status of algal cells. J. Mar. Biol. Assoc. U.K. 62: 403–434.

EPPLEY, R. W. 1972. Temperature and phytoplankton growth in the sea. Fish Bull. 70: 1063–1085.

———— 1981. Relations between nutrient assimilation and growth in phytoplankton with a brief review of estimates of growth rate in the ocean. p. 251–263. *In* T. Platt [ed.] Physiological bases of phytoplankton ecology. Can. Bull. Fish. Aquat. Sci. 210: 346 p.

EPPLEY, R. W., AND J. L. COATSWORTH. 1966.

Culture of marine phytoplankter, *Dunaliella tertiolecta* with light-dark cycles. Arch. Mikrobiol. 55: 66–80.

EPPLEY, R. W., J. N. ROGERS, AND J. J. MCCARTHY. 1969. Half-saturation constants for uptake of nitrate and ammonium by marine phytoplankton. Limnol. Oceanogr. 14: 912–920.

FALKOWSKI, P. G. 1975. Nitrate uptake in marine phytoplankton: (nitrate, chloride)-activated adenosine triphosphatase from *Skeletonema costatum* (Bacillariophyceae). J. Phycol. 11: 323–326.

——— 1977. A theoretical description of nitrate uptake kinetics in marine phytoplankton based on bisubstrate kinetics. J. Theor. Biol. 64: 375–379.

——— 1980. Light-shade adaptation in marine phytoplankton. p. 99–119. *In* P. G. Falkowski [ed.] Primary Productivity in the sea. Plenum Publishing Corporation.

FALKOWSKI, P. G., AND T. G. OWENS. 1980. Light-shade adaptation — two strategies in marine phytoplankton. Plant. Physiol. 66: 592–595.

FOY, R. H. 1983. Interaction of temperature and light on the growth rates ot two planktonic *Oscillatoria* species under a short photoperiod regime. Br. Phycol. J. 18: 267–273.

FOY, R. H., AND C. E. GIBSON. 1982. Photosynthetic characteristics of planktonic blue-green algae: changes in photosynthetic capacity and pigmentation of *Oscillatoria redekei* van Goor under high and low light. Br. Phycol. J. 17: 183–193.

FOY, R. H., C. E. GIBSON, AND R. B. SMITH. 1976. The influence of daylength, light intensity and temperature on the growth rates of planktonic blue-green algae. Br. Phycol. J. 11: 151–163.

FOY, R. H., AND R. V. SMITH. 1980. The role of carbohydrate in the growth of planktonic *Oscillatoria* species. Br. Phycol. J. 7: 139–150.

GERHART, D. A., AND G. E. LIKENS. 1975. Enrichment experiments for determining nutrient limitation: Four methods compared. Limnol. Oceanogr. 20: 649–654.

GIBSON, C. E. 1985. Growth rate, maintenance energy and pigmentation of planktonic cyanophyta during one-hour light:dark cycles. Br. Phycol. J. 20: 155–161.

GIBSON, C. E., AND R. H. FOY. 1983. The photosynthesis and growth efficiency of a planktonic blue-green alga *Oscillatoria redekei*. Br. Phycol. J. 18: 39–45.

GIESKES, W. W., AND G. W. KRAAY. 1984. State-of-the Art in the measurement of primary production, p. 171–190. *In M. J. R. Fasham* [ed.] Flows of energy and materials in marine ecosystems. Plenum Publishing Corporation.

GOLDMAN, C. R. 1978. The use of natural phytoplankton populations in bioassay. Mitt. Int. Verein. Limnol. 21: 364–371.

GOLDMAN, J. C. 1977. Biomass production in mass cultures of marine phytoplankton at varying temperatures. J. Exp. Mar. Biol. Ecol. 27: 161–169.

GOLDMAN, J. C., AND J. J. MCCARTHY. 1978. Steady-state growth and ammonium uptake of a fast-growing marine diatom. Limnol. Oceanogr. 23: 695–703.

GONS, H. J., AND L. R. MUR. 1975. An energy balance for algal populations in light-limiting conditions. Verh. Int. Ver. Limnol. 19: 2719–2723.

——— 1980. Energy requirements for growth and maintenance of *Scenedesmus protuberans* Fritsch in light-limited continuous cultures. Arch. Microbiol. 125: 9–17.

GOTHAM, I. J., AND G-Y. RHEE. 1981a. Comparative kinetic studies of phosphate-limited growth and phosphate uptake in phytoplankton in continuous culture. J. Phycol. 17: 257–265.

——— 1981b. Comparative kinetic studies of nitrate-limited growth and nitrate uptake in phytoplankton in continuous culture. J. Phycol. 17: 309–314.

HARDING, L. W. JR., B. B. PRÉZELIN, B. M. SWEENEY, AND J. L. COX. 1981. Diel oscillations in the photosynthesis–irradiance relationship of a planktonic marine diatom. J. Phycol. 17: 389–394.

HARRIS, G. P. 1978. Photosynthesis, productivity and growth: The physiological ecology of phytoplankton. Ergeb. Limnol. 10: 171 p.

HARRIS, G. P., AND B. B. PICCININ. 1977. Photosynthesis by natural phytoplankton populations. Arch. Hydrobiol. 80: 405–457.

HEALEY, F. P. 1977. Ammonium and urea uptake by some freshwater algae. Can. J. Bot. 55: 61–69.

——— 1979. Short-term responses of nutrient-deficient algae to nutrient addition. J. Phycol. 15: 289–299.

——— 1980. Slope of the Monod equation as an indicator of advantage in nutrient competition. Microbiol. Ecol. 5: 281–286.

——— 1985. Interacting effects of light and nutrient limitation on the growth rate of *Synechococcus linearis* (Cyanophyceae). J. Phycol. 21: 134–146.

HEALEY, F. P., AND L. L. HENDZEL. 1975. Effect of phosphorus deficiency on two algae growing in chemostats. J. Phycol. 11: 303–309.

——— 1979. Indicators of phosphorus and nitrogen deficiency in five algae in culture. J. Fish. Res. Board Can. 36: 1364–1369.

HERRON, H. A., AND D. MAUZERALL. 1972. The development of photosynthesis in a greening mutant of *Chlorella* and an analysis of the light saturation curve. Plant. Physiol. 50: 141–148.

HO, K. K., AND D. W. KROGMANN. 1982. Photosynthesis, p. 191–214. *In* N. G. Carr and B. A. Whitton [ed.] The Biology of cyanobacteria. Botanical Monographs. Vol. 19. Blackwell Scientific Publications, Oxford.

JASSBY, A. D., AND T. PLATT. 1976. Mathematical formulation of the relationship between photosynthesis and light for phytoplankton. Limnol. Oceanogr. 21: 540–547.

KILHAM, S. S., AND KILHAM, P. 1978. Natural community bioassay: predictions of results based on nutrient physiology and competition. Ver. Int. Ver. Limnol. 20: 68–74.

KIRK, J. T. O. 1983. Light and photosynthesis in aquatic systems. Cambridge University Press.

KONOPKA, A., AND M. SCHNUR. 1981. Biochemical composition and photosynthetic carbon metabolism of nutrient-limited cultures of *Merismopedia tenuissima* (Cyanophyceae). J. Phycol. 17: 118-122.

LI, W. K. W., AND J. C. GOLDMAN. 1981. Problems in estimating growth rates of marine phytoplankton from short-term ^{14}C assays. Microb. Ecol. 7: 113-121.

LOOGMAN, J. G. 1982. Influence of photoperiodicity on algal growth kinetics. Thesis. University of Amsterdam.

LOOGMAN, J. G., AND L. VAN LIERE. 1978. An improved method for measuring irradiance in algal cultures. Verh. Int. Ver. Limnol. 20: 2322-2328.

LOOGMAN, J. G., A. F. POST, AND L. R. MUR. 1980. The influence of periodicity in light conditions, as determined by the trophic state of the water; on the growth of the green algae *Scenedesmus protuberans* and the cyanobacterium *Oscillatoria agardhii*, p. 79-82. *In* J. Barica and L. R. Mur [ed.] Hypertrophic Ecosystems. Developments in Hydrobiology. Vol. 2. Dr. W. Junk b.v. Publishers. The Hague.

LOOGMAN, J. G., J. SAMSON, AND L. R. MUR. 1979. The use of experimental ecology in simulating algae blooms: a nitrogen-limited algal growth model. State in the Art in Ecological Modelling 7: 617-646.

MAESTRINI, S. Y., AND M-G. KOSSUT. 1981. In situ cell depletion of some marine algae enclosed in dialysis sacks and their use for the determination of nutrient-limiting growth in Ligurian Coastal Waters (Mediterranean Sea). J. Exp. Mar. Biol. Ecol. 50: 1-19.

MIERLE, G. 1985. Kinetics of phosphate transport by *Synechococcus leopoliensis* (Cyanophyta): Evidence for diffusion limitation of phosphate uptake. J. Phycol. 21: 177-181.

MONOD, J. 1950. La technique de culture continue, théorie et applications. Ann. Inst. Pasteur 77: 390-410.

MORRIS I., AND H. E. GLOVER. 1981. Physiology of photosynthesis by marine coccoid cyanobacteria–Some ecological implications. Limnol. Oceanogr. 26: 957-961.

MORRIS, I., A. E. SMITH, AND H. E. GLOVER. 1981. Products of photosynthesis in phytoplankton off the Orinoco River and in the Caribbean Sea. Limnol. Oceanogr. 26: 1034-1044.

MUR, L. R., AND R. O. BEYDORFF. 1978. A model for the succession from green algae to blue-green algae based on light limitation. Verh. Int. Verh. Limnol. 20: 2314-2321.

MUR, L. R., H. J. GONS, AND L. VAN LIERE. 1977. Some experiments on the competition between green algae and blue-green bacteria in light-limited environments. FEMS Microbiol. Letters 1: 335-338.

NALEWAJKO, C., K. LEE, AND H. SHEAR. 1981. Phosphorus kinetics in Lake Superior: light intensity and phosphate uptake in algae. Can. J. Fish. Aquat. Sci. 38: 224-232.

NYHOLM, N. 1978. Mathematical model for growth of phytoplankton. Mitt. Int. Ver. Limnol. 21: 193-206.

PAASCHE, E. 1968. Marine plankton algae grown with light–dark cycles. II. *Ditylum brightwellii* and *Nitzschia turgidula*. Physiol. Plant. 21: 66-77.

PATRICK, R. 1978. Effects of trace metals in the aquatic ecosystem. Am. Sci. 66: 185-191.

PERRY, M. J., M. C. TALBOT, AND R. S. ALBERTE. 1981. Photoadaptation in marine phytoplankton: Responses of the photosynthetic unit. Mar. Biol. 62: 91-101.

PIRT, S. J. 1965. The maintenance energy of bacteria in growing cultures. Proc. R. Soc. London B. 163: 224-231.

PLATT, T., C. L. GALLEGOS, AND W. G. HARRISON. 1980. Photoinhibition of photosynthesis in natural assemblages of marine phytoplankton. J. Mar. Res. 38: 687-701.

PLATT, T., D. V. SUBBA RAO, AND B. IRWIN. 1983. Photosynthesis of picoplankton in the oligotrophic ocean. Nature 300: 702-704.

POST, A. F., R. DE WIT, AND L. R. MUR. 1985a. Interactions between temperature and light intensity on growth and photosynthesis of the cyanobacterium *Oscillatoria agardhii*. J. Plankton Res. 7: 487-495.

POST, A. F., J. G. LOOGMAN, AND L. R. MUR. 1985b. Regulation of growth and photosynthesis by *Oscillatoria agardhii* grown with a light/dark cycle. FEMS Microbiol. Ecol. 31: 97-102.

POST, A. F., Z. DUBINSKY, K. WYMAN, AND P. G. FALKOWSKI. 1985c. Physiological responses of a marine planktonic diatom to transitions in growth irradiance. Mar. Ecol. Prog. Ser. 25: 141-149.

POST, A. F., F. EIJGENRAAM, AND L. R. MUR. 1985d. Influence of light period length on photosynthesis and synchronous growth of the green alga *Scenedesmus protuberans*. Br. Phycol. J. 20: 391-397.

POST, A. F., J. G. LOOGMAN, AND L. R. MUR. 1986. Photosynthesis and growth of *Oscillatoria agardhii* Gomont in environments with a periodic supply of light energy. J. Gen. Microbiol. (in press).

PRÉZELIN, B. B. 1976. The role of peridinin-chlorophyll *a* proteins in the photosynthetic light adaptation of the marine dinoflagellate *Glenodinium* sp. Planta 130: 225-233.

——— 1981. Light reactions in photosynthesis, p. 1-43. *In* T. Platt. [ed.] Physiological bases of phytoplankton ecology. Can. Bull. Fish. Aquat. Sci. 210: 346 p.

PRÉZELIN, B. B., AND H. A. MATLICK. 1980. Time courses of photoadaptation in the photosynthesis irradiance relationship of a dinoflagellate exhibiting photosynthetic periodicity. Mar. Biol. 58: 85-96.

RAPS, S., K. WYMAN, H. W. SIEGELMAN, AND P. G. FALKOWSKI. 1983. Adaptation of the cyanobacterium *Microcystis aeruginosa* to light intensity. Plant Physiol. 72: 829-832.

RHEE, G.-Y. 1973. A continuous culture study of phosphate uptake, growth rate and polyphosphate in *Scenedesmus* sp. J. Phycol. 9: 495-506.

——— 1978. Effects of N: P atomic ratios and

420

nitrate limitation on algal growth, cell composition and nutrient uptake. Limnol. Oceanogr. 23: 10 25.

——— 1980. Continuous culture in phytoplankton ecology, p. 151–203. *In* M. R. Droop and H. W. Jannasch [ed.] Advances in aquatic microbiology. Vol. 2 Academic Press, New York, NY.

——— 1982. Effects of environmental factors and their interactions on phytoplankton growth, p. 33–74. *In* K. C. Marshall [ed.] Advances in Microbiol. ecology. Vol. 6. Plenum Publishing Corporation.

RHEE, G.-Y., AND I. J. GOTHAM. 1981a. The effects of environmental factors on phytoplankton growth: Light and the interactions of light with nitrate limitation. Limnol. Oceanogr. 26: 649–659.

——— 1981b. The effects of environmental factors on phytoplankton growth: Temperature and the interactions of temperature with nutrient limitation. Limnol. Oceanogr. 26: 635–648.

RHEE, G.-Y., I. J. GOTHAM, AND S. W. CHISHOLM. 1981. Use of cyclostat cultures to study phytoplankton ecology, p. 159–186. *In* P. C. Calcott [ed.] Continuous culture of cells. Vol. 2. C.R.C. Press, Cleveland.

RICHARDSON, K., J. BEARDALL, AND J. A. RAVEN. 1983. Adaptation of unicellular algae to irradiance: an analysis of strategies. New Phytol. 93: 157–191.

RIEGMAN, R. 1985. Phosphate–phytoplankton interactions. Thesis. University of Amsterdam.

RIEGMAN, R., AND L. R. MUR. 1984a. Phosphate uptake by P-limited *Oscillatoria agardhii*. FEMS Microbiol. Letters 21: 335–339.

——— 1984b. Regulation of phosphate uptake kinetics in *Oscillatoria agardhii*. Arch. Microbiol. 139: 38–32.

——— 1984c. Theoretical considerations on growth kinetics and physiological adaptation of nutrient-limited phytoplankton. Arch. Microbiol. 140: 96–100.

SAKSHAUG, E., ANDRESEN, S. MYKLESTAD, AND Y. OLSEN. 1983. Nutrient status of phytoplankton communities in Norwegian waters (marine, brackish and fresh) as revealed by their chemical composition. J. Plankton Res. 5: 175–196.

SAKSHAUG, E., E. GRANELI, M. ELBRACHTER, AND H. KAYSER. 1984. Chemical composition and alkaline phosphatase activity of nutrient-saturated and P-deficient cells of four marine dinoflagellates. J. Exp. Mar. Biol. Ecol. 77: 241–254.

SCOTT, W. E., P. J. ASHTON, R. D. WALMSLEY, AND M. T. SEAMAN. 1980. Hartbeespoort Dam: a case study of a hypertrophic, warm, monomictic impoundment, p. 317–322. *In* J. Barica and L. R. Mur [ed.] Hypertrophic Ecosystems. Developments in Hydrobiology. Vol. 2. W. Junk b.v. Publishers, The Hague.

SENGER, H., AND PH. FLEISCHHACKER. 1978a. Adaptation of photosynthetic apparatus of *Scenedesmus obliquus* to strong and weak light conditions. I. Differences in pigments, photosynthetic capacity, quantum yield and dark reactions. Physiol. Plant. 43: 35–42.

——— 1978b. Adaptation of photosynthetic apparatus of *Scenedesmus obliquus* to strong and weak light conditions. II. Differences in photochemical reactions, the photosynthetic electron transport and photosynthetic units. Physiol. Plant. 43: 43–51.

SMITH, V. H. 1983. Light and nutrient dependence of photosynthesis by algae. J. Phycol. 19: 306–313.

STEEMANN NIELSEN, E. 1952. The use of radioactive (C^{14}) for measuring organic production in the sea. J. Cons. Cons. Int. Explor. Mer. 18: 117–140.

SYRETT, P. J. 1981. Nitrogen metabolism of microalgae, p. 182–210. *In* T. Platt. [ed.] Physiological bases of phytoplankton ecology. Can. Bull. Fish. Aquat. Sci. 210: 346 p.

TALLING, J. F. 1955. The relative growth rates of three plankton diatoms in relation to underwater radiation and temperature. Ann. Bot. 19: 329–341.

——— 1957. Photosynthetic characteristics of some freshwater plankton diatoms in relation to underwater radiation. New. Phytol. 56: 29–50.

——— 1973. The application of some electrochemical methods to the measurement of photosynthesis and respiration in freshwaters. Freshwater Biol. 3: 335–362.

TEMPEST, D. W. 1970. The continuous cultivation of microorganisms. I. Theory of the chemostat, p. 259–277. *In* J. R. Norris and D. W. Ribbons [ed.] Methods in Microbiology. Vol. 2. Academic Press, London, New York.

TEMPEST, D. W., AND O. M. NEIJSSEL. 1978. Ecophysiological aspects of microbial growth in aerobic nutrient-limited environments. Adv. Microbial. Ecol. 2: 105–153.

TEMPEST, D. W., O. M. NEIJSSEL, AND W. ZEVENBOOM. 1983. Properties and performances in laboratory cultures; their relevance to growth in natural ecosystems, p. 119–152. *In* J. H. Slater, R. Whittenbury, and J. W. T. Wimpenny [ed.] Microbes in their natural environments. Symp. 34. Soc. Gen. Microbiol. University Press, Cambridge.

TILMAN, D. 1977. Resource competition between planktonic algae: An experimental and theoretical approach. Ecology 58: 338–348.

TILMAN, D., AND S. S. KILHAM. 1976. Phosphate and silicate growth and uptake kinetics of the diatom *Asterionella formosa* and *Cyclotella meneghiniana* in batch and semi-continuous cultures. J. Phycol. 12: 375–383.

TURPIN, D. A., AND P. J. HARRISON. 1979. Limiting nutrient patchiness and its role in phytoplankton ecology. J. Exp. Mar. Biol. Ecol. 39: 151–166.

VAN DONK, E. 1983. Factors influencing phytoplankton growth and succession in lake Maarsseveen (I). Thesis. University of Amsterdam.

VAN GEMERDEN, H. 1974. Coexistence of organisms competing for the same substrate: an example among the purple sulfur bacteria. Microbiol. Ecol. 1: 104–119.

VAN LIERE, L. 1979. On *Oscillatoria agardhii* Gomont, experimental ecology and physiology of a nuisance bloomforming cyanobacteria. Thesis. University of Amsterdam.

VAN LIERE, L. G. J. DE GROOT, AND L. R. MUR. 1979a. Pigment variation with irradiance in *Oscillatoria agardhii* Gomont in nitrogen (nitrate)-limited chemostat cultures. FEMS Microbiol. Letters 6: 337–340.

VAN LIERE, L., L. R. MUR, C. E. GIBSON, AND M. HERDMAN. 1979b. Growth and physiology of *Oscillatoria agardhii* Gomont cultivated in continuous culture with a light/dark cycle. Arch. Microbiol. 123: 315–318.

VAN LIERE, L., J. G. LOOGMAN, AND L. R. MUR. 1978. Measuring light irradiance in cultures of phototrophic micro-organisms. FEMS Microbiol. Lett. 3: 161–164.

VAN LIERE, L., AND L. R. MUR. 1979. Growth kinetics of *Oscillatoria agardhii* Gomont in continuous culture, limited in its growth by the light-energy supply. J. Gen. Microbiol. 115: 153–160.

VAN LIERE, L., AND A. E. WALSBY. 1982. Interactions of cyanobacteria with light, p. 9–45. *In* N. G. Carr and B. A. Whitton [ed.] The biology of Cyanobacteria. Blackwell. Oxford.

WALSH, P., AND L. LEGENDRE. 1983. Photosynthesis of natural phytoplankton under high frequency light fluctuations simulating those induced by sea surface waves. Limnol. Oceanogr. 28: 688–697.

WATERBURY, J. B., S. W. WATSON, R. R. GUILLARD, AND L. E. BRAND. 1979. Widespread occurrence of a unicellular, marine, planktonic cyanobacterium. Nature 277: 293–294.

YENTSCH, C. M., T. L. CUCCI, D. A. PHINNEY, R. SELVIN, AND H. E. GLOVER. 1985. Adaptation to low photon flux densities in *Protogonyaulax tamarensis* var. *excavata*, with reference to chloroplast photomorphogenesis. Mar. Biol. 89 9–20.

ZEVENBOOM, W. 1980. Growth and nutrient uptake kinetics of *Oscillatoria agardhii*. Thesis. University of Amsterdam.

——— 1985. Studies on red-pigmented coccoid marine cyanobacteria. Abstract V Internat. Symp. on Photosynthetic Prokaryotes. Grindelwald. Switzerland.

——— 1986. Tracing red-pigmented marine cynobacteria using *in vivo* absorption maxima.

FEMS Microbiol. Ecol. 38: 267–275.

ZEVENBOOM, W., A. BIJ DE VAATE, AND L. R. MUR. 1982. Assessment of factors limiting growth rate of *Oscillatoria agardhii* in hypertrophic Lake Wolderwijd. 1978, by use of physiological indicators. Limnol. Oceanogr. 27: 39–52.

ZEVENBOOM, W., G. J. DE GROOT, AND L. R. MUR. 1980. Effects of light on nitrate-limited *Oscillatoria agardhii* in chemostat cultures. Arch. Microbiol. 125: 59–65.

ZEVENBOOM, W., AND L. R. MUR. 1978. On nitrate uptake by *Oscillatoria agardhii*. Verh. Int. Ver. Limnol. 20: 2302–2307.

——— 1979. Influence of growth rate on short term and steady state nitrate uptake by nitrate-limited *Oscillatoria agardhii*. FEMS Microbiol. Lett. 6: 209–212.

——— 1980. N_2-fixing cyanobacteria: why they do not become dominant in Dutch, hypertrophic lakes, p. 123–130. *In* J. Barica and L. R. Mur [ed.] Hypertrophic ecosystems. Developments in hydrobiology. Vol. 2 Dr. W. Junk b.v. Publishers. The Hague.

——— 1981a. Ammonium-limited growth and uptake by *Oscillatoria agardhii* in chemostat culture. Arch. Microbiol. 129: 61–60.

——— 1981b. Simultaneous short-term uptake of nitrate and ammonium by *Oscillatoria agardhii* grown in nitrate or light-limited continuous culture. J. Gen. Microbiol. 126: 353–363.

——— 1984. Growth and photosynthetic response of the cyanobacterium *Microcystis aeruginosa* in relation to photoperiodicity and irradiance. Arch. Microbiol. 139: 232–239.

ZEVENBOOM, W., AND A. F. POST. 1983. Use of physiological parameters. *In* F. Colijn, W. W. C. Gieskes and W. Zevenboom [ed.] The measurement of primary production: problems and recommendations. Hydrobiol. Bull. 17: 29–51.

ZEVENBOOM, W., A. F. POST, U. M. VAN HES, AND L. R. MUR. 1983. A new incubator for measuring photosynthetic activity of phototrophic microorganisms, using the amperometric oxygen method. Limnol. Oceanogr. 28: 787–791.

ZEVENBOOM, W., J. VAN DER DOES, K. BRUNING, AND L. R. MUR. 1981c. A non-heterocystous mutant of *Aphanizomenon flos-aquae*, selected by competition in a light-limited continuous culture. FEMS Microbiol. Lett. 10: 11–16.

The Cyanobacterial Photosynthetic Apparatus: Comparisons to Those of Higher Plants and Photosynthetic Bacteria

D. A. BRYANT

Department of Molecular and Cell Biology, Pennsylvania State University, University Park, Pennsylvania 16802, USA

Introduction

The purpose of this review is to discuss the structure, function, and organization of the cyanobacterial photosynthetic apparatus. For this discussion the term "photosynthetic apparatus" is defined as those cellular components which are required for the light-driven formation of the ATP and NADPH, the energy and reducing power, for carbon dioxide fixation. Emphasis will be given to structural and genetic aspects. Since most aspects have been extensively reviewed, only selected and recent results will be discussed. Where appropriate, the reader will be directed to specialized reviews which can provide more detailed discussions. Efforts will be made to extrapolate from advances made with higher plants and the purple photosynthetic bacteria so as to give a more complete view of the current state of understanding and thinking concerning the structure and function of the cyanobacterial photosynthetic apparatus.

As has been the case in all areas of biological science, molecular genetics and recombinant DNA methods are currently having an enormous impact upon our understanding of the photosynthetic process. As a result of the recent cloning and nucleotide sequencing of several genes encoding polypeptides involved in the photosynthetic process, new insights are being achieved. X-ray crystallographers have also contributed significantly by providing structural details at the molecular level for many proteins which play roles in the photosynthetic process. The information gathered by the molecular biologists and X-ray crystallographers has in turn been employed by biochemists and biophysicists to refine our ideas about the light-harvesting, charge-separation, electron-transport, and energy-conservation processes.

Thylakoid Organization in Cyanobacteria

Cyanobacteria are the largest and most diverse group of photosynthetic procaryotes (Stanier and Cohen-Bazire 1977). Cyanobacteria share the property of performing oxygenic photosynthesis with the procaryotic *Prochloron* sp. (Lewin 1984; Burger-Wiersma et al. 1986) and with all photosynthetic eucaryotes. The cyanobacterial photosynthetic apparatus is clearly similar in structure, function, and molecular aspects to that which occurs in the eucaryotic algae and higher plants (Ho and Krogmann 1982; Hladik and Sofrova 1983). Each system is comprised of five or six multi-protein complexes which are localized in or upon the photosynthetic (thylakoid) membranes. Four of these complexes are common to both cyanobacteria and higher plants: (1) Photosystem II (PS II)/oxygen-evolving complex (OEC); (2) Plastoquinol-plastocyanin oxidoreductase; (3) Photosystem I (PS I); and (4) ATP synthase (coupling factor). The fifth type of complex, which is functionally but not structurally homologous, serves as the light-harvesting antenna for PS II. In cyanobacteria this complex is the phycobilisome. In chlorophyll *b*-containing organisms, including *Prochloron* sp. and higher plants, this light-harvesting complex is a chlorophyll-protein complex CP II, the chlorophyll *a/b* binding protein (Thornber 1975; Lewin 1984; Arntzen 1978). The sixth protein complex, which is uniquely found in higher plants and green algae, is a chlorophyll a/b protein complex which serves as the peripheral antenna for Photosystem I. (Mullet et al. 1980a,b; Ortiz et al. 1984). The interactions between these multi-protein complexes are accomplished via several small, diffusable proteins and molecules: plastoquinone, plastocyanin (cytochrome *c*-553), ferredoxin

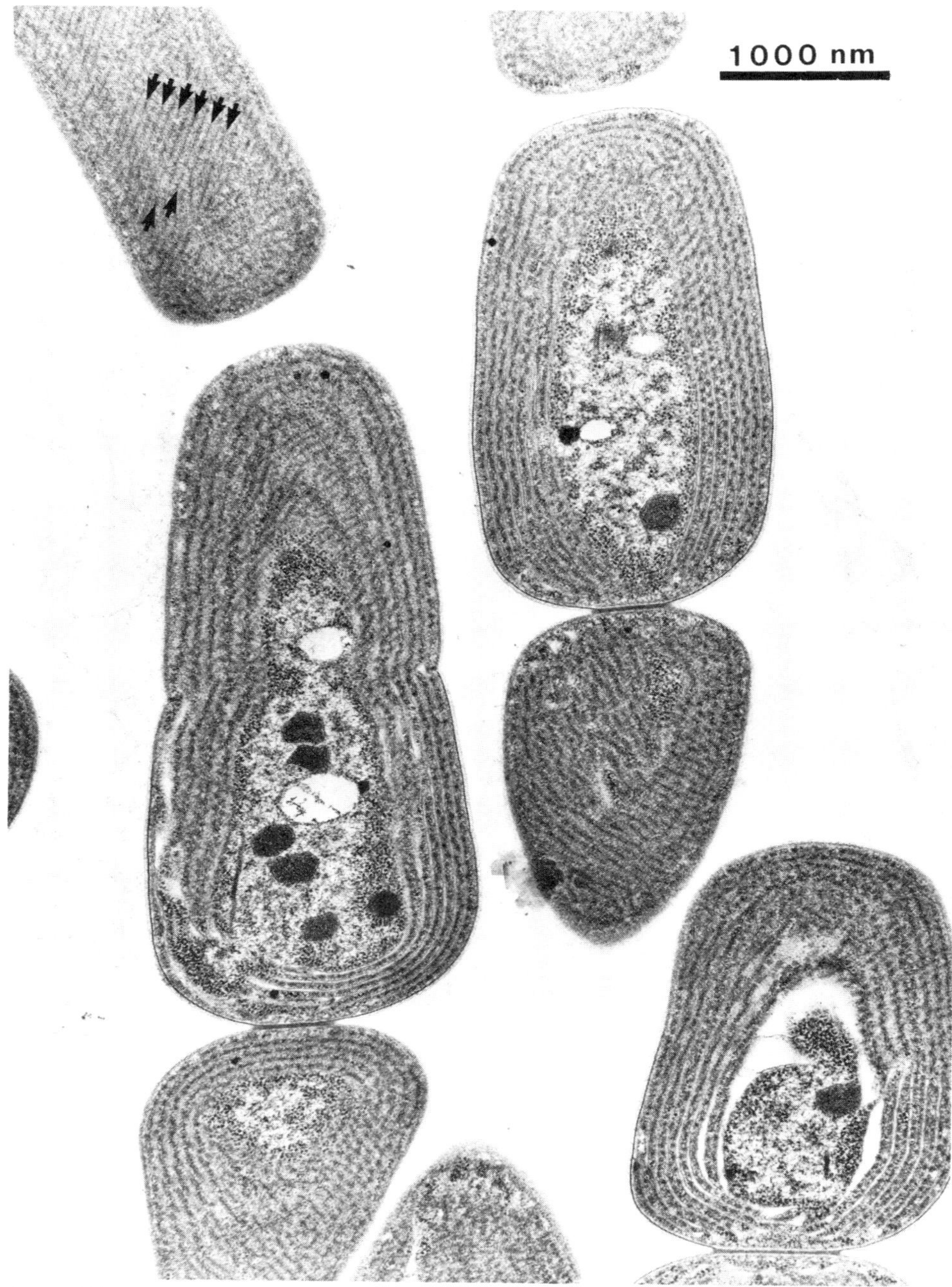

FIG. 1. Electron micrograph of a typical cyanobacterium. This section of *Pseudanabaena* sp. PCC 7408 cells showing the arrangement of thylakoids and phycobilisomes. The rows of phycobilisomes are seen in cross section, longitudinal section, and tangential section. In regions of tangential sectioning (arrows), long rows of closely packed phycobilisomes appear as parallel electron-opaque cords which run at an angle to the long axis of the cells. When the rows of phycobilisomes are cut in cross section, the structures have an irregular semi-circular outline and interdigitate between opposing stromal surfaces of the thylakoids. ($\times$32 000).

(flavodoxin), and protons. Each of the cyanobacterial components will be discussed separately and in relationship to its frequently better understood higher plant homologue in this review.

In cyanobacteria and *Prochloron* sp., the photosynthetic membranes are found dispersed throughout the cytoplasm. This property of course differentiates these procaryotic organisms from the photosynthetic eucaryotes in which the photosynthetic membranes are always localized in an organelle, the chloroplast. In the plastids of two eucaryotic groups, the red algae ("rhodoplasts") and the taxonomically enigmatic eucaryotes such as *Cyanophora paradoxa* ("cyanelles"; Trench 1982), the organization of the photosynthetic apparatus is considered to be identical to that which occurs in the cyanobacteria. For this reason results obtained with red algae and *C. paradoxa* will be included in this discussion. Although they synthesize phycobiliproteins, the organization of the photosynthetic membranes in the plastids of the cryptomonads differs strikingly from that which will be described for the cyanobacteria (for a review of cryptomonads, see Gantt 1980a). The other groups of eucaryotic algae, higher plants, and the procaryote *Prochloron* sp. synthesize membrane-bound, light-harvesting protein complexes in which the chromophores are chlorophylls *a*, *b*, and/or *c*, and carotenoids.

The cyanobacterial photosynthetic apparatus is localized in and on intracytoplasmic membranes, the thylakoids (see Fig. 1–3), whose organization is similar to that of the chloroplast membranes of higher plants and green algal chloroplasts (Staehelin et al. 1978a). Although the intracytoplasmic membranes of the purple photosynthetic bacteria are clearly derived from infoldings of the cytoplasmic membrane (Oelze and Drews 1972), a similar relationship between the cytoplasmic membrane and the thylakoids in cyanobacterial has remained a controversial issue. As stated by Stanier and Cohen-Bazire (1977), the two membrane systems generally appear to be topologically distinct, even in species of *Synechococcus* and *Pseudanabaena* in which the membrane systems appear to be fairly simple (see Fig. 1 and 2). However, contacts between the thylakoid membranes and the cytoplasmic membranes have frequently been reported in the literature, and the suggestion that the thylakoids arise from invaginations of the cytoplasmic membrane has frequently been made (for a discussion of this point, see Nierzwicki-Bauer et al. 1983). There are a few species, e.g. *Arthrospira jenneri* (Wildman and Bowen 1974), in which the thylakoids obviously intersect with the cytoplasmic membrane and for which the thylakoids can easily be interpretted as infoldings of the cytoplasmic membrane. Moreover, there is good evidence that light causes protons to be extruded from cells into the medium (Scholes et al. 1969) as observed for the purple bacteria (Crofts and Wraight 1983). This observation implies that topologically the thylakoid lumen is extracellular space and strengthens the argument that the thylakoids are infoldings of the cytoplasmic membrane. It should be noted that Kunkel (1982) has reported the detection of a structure which he refers to as the "thylakoid center". These open-ended cylindrical structures, which have obvious connections to thylakoids, are found in several cyanobacteria and have been suggested to be involved in membrane biogenesis.

Phycobilisomes are never observed on the inner surface of the cytoplasmic membrane; this strongly implies that the two membrane systems are functionally distinct (Cohen-Bazire and Bryant 1982). Omata and Murata (1983, 1984) have recently reported the isolation and initial characterization of the cytoplasmic membranes from *Synechococcus* sp. PCC 6301 and *Synechocystis* sp. PCC 6714. Although the lipid compositions of the two membranes were similar, the distribution of carotenoids in the two membranes was remarkably different. Furthermore, the polypeptide compositions of the two membranes were clearly different. This is consistent with the view that the thylakoids are differentiated infoldings of the cytoplasmic membranes.

The arrangement of the cyanobacterial thylakoids varies considerably and depends upon species, cell type (vegetative cells, heterocysts, akinetes, hormogonia) and culture

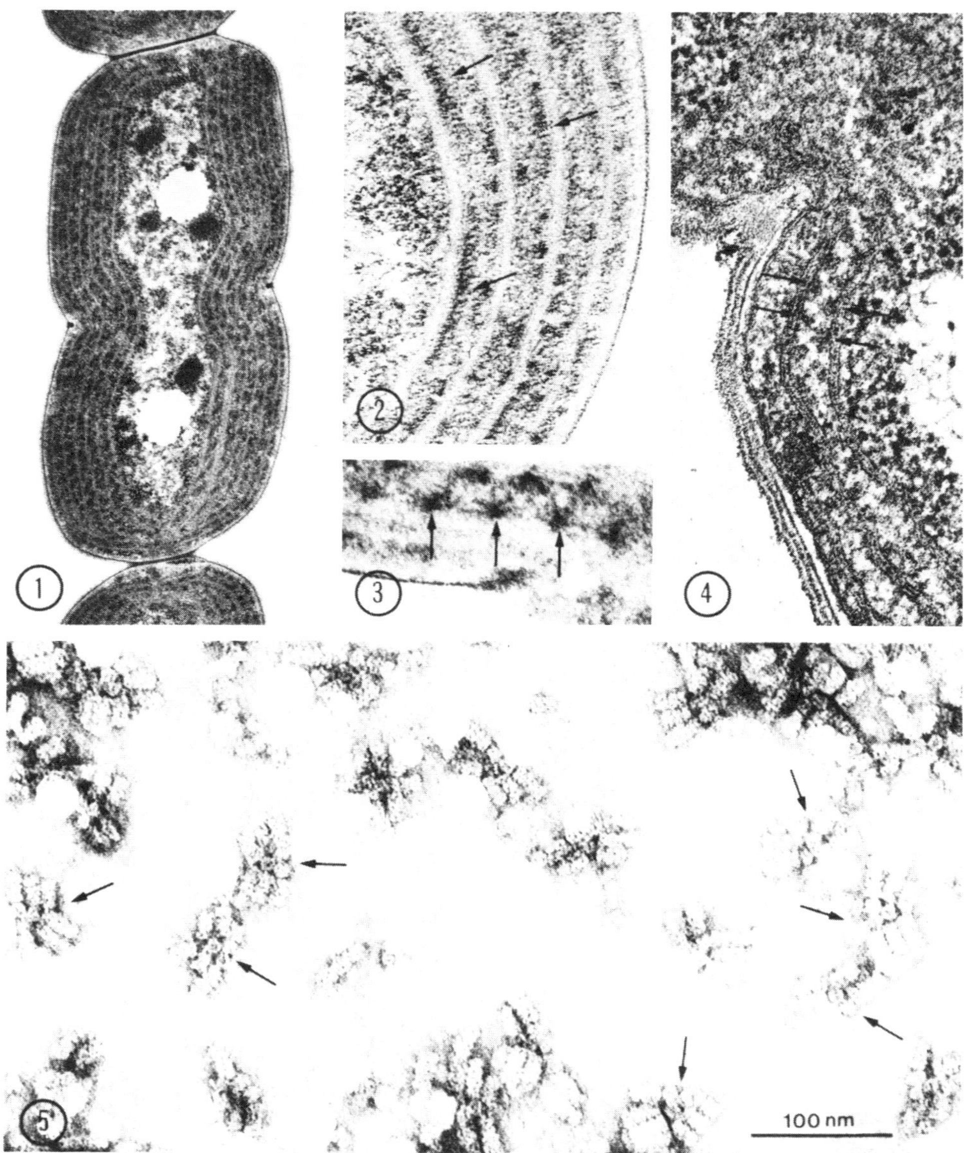

Fig. 2. Electron micrographs of *Pseudanabaena* sp. PCC 7409, *Synechococcus* sp. PCC 6312, and isolated phycobilisomes from *Pseudanabaena* sp. PCC 7409. Part 1 shows a thin section of a dividing cell of *Pseudanabaena* sp. PCC 7409 with numerous concentric thylakoids at the periphery of the cytoplasm. Phycobilisomes are visible as electron-opaque granules on the stromal surfaces of the thylakoids ($\times 27\,000$). Part 2 shows an enlargement of a portion of a *Pseudanabaena* sp. PCC 7409 cell in which the plane of sectioning was nearly perpendicular to the thylakoid surface. Rows of phycobilisomes (arrows) in longitudinal section can be seen standing on edge on the thylakoid surface ($\times 120\,000$). Part 3 shows an enlargement of a *Pseudanabaena* sp. PCC 7409 cell in which the rows of phycobilisomes have been cross-sectioned. The phycobilisomes (arrows) sometimes reveal arms radiating from a more densely stained core ($\times 150\,000$). Part 4 shows an enlargement of a thin section showing a dividing cell of *Synechococcus* sp. PCC 6312. The arrows point to phycobilisomes, visible in face view, which have an irregularly lobed but approximately semi-circular outline ($\times 120\,000$). Part 5 shows the typical appearance of isolated phycobilisomes (arrows) from *Pseudanabaena* sp. PCC 7409 cells grown in red light after fixation and negative staining ($\times 250\,000$). Reprinted with permission from Bryant et al. (1979).

426

conditions. At least one cyanobacterium, *Gloeobacter violaceus*, has no thylakoids (Rippka et al. 1974; Guglielmi et al. 1981). In this organism the photosynthetic apparatus is housed in and upon a multi-functional cytoplasmic membrane. In the most common thylakoid arrangement, three to six layers, forming concentric shells parallel to the cytoplasmic membrane, are found in the periphery of the cytoplasm (Stanier and Cohen-Bazire 1977; Golecki and Drews 1982). Organisms are known in which the tylakoids radiate from the periphery of the cell into the central cytoplasmic or in which the thylakoids are perpendicular to the longitudinal cell wall (*Oscillatoria rubescens*, Jost (1965); *Arthrospira jenneri*, Wildman and Bowen (1974); and *Starria* sp. Lang 1977). There are numerous species in which the thylakoids meander randomly throughout the cell. Although thylakoids do occasionally fuse giving rise to interconnecting networks (Nierzwicki-Bauer et al. 1983; Kunkel 1982), stacking of thylakoids as occurs in chloroplasts is rarely observed.

The appearance of the thylakoid membrane in ultrathin sections observed in the electron microscope is dependent upon the fixation procedure employed. After osmium tetroxide fixation, the thylakoid membrane has the appearance of two parallel electron dense lines 3.0 nm in diameter separated by an electron-transparent zone about 3–5 nm wide (Golecki and Drews 1982). Cyanobacterial thylakoids are generally closely appressed to one another, but the two membranes can be separated by an electron-transparent zone referred to as the thylakoid lumen or the intrathylakoidal space. Edwards et al. (1968) suggested that the appearance of the lumen was dependent upon the physiological state of the cell and the osmotic conditions prevalent during the fixation procedure. The outer (stromal or plasmic) surface of the thylakoid bears regularly spaced rows of electron-dense structures referred to as phycobilisomes (see Fig. 1–3; Gantt and Conti 1965, 1966, 1969; Edwards and Gantt 1971). These light-harvesting, pigment-protein complexes will be described in detail later in this discussion.

Cyanobacterial thylakoids have also been studied by the freeze-etching technique (Branton 1966). The appearance of the plasmic and exoplasmic fracture faces of the thylakoid membrane have characteristic and remarkably different appearances when examined by this method (see Golecki and Drews 1982, for a review). The plasmic fracture face, which results from the membrane leaflet adjacent to the stromal surface of the thylakoid, is densely covered with particles (3000–5000 particles μm^{-2}). These particles, considered to represent intrinsic membrane proteins, range in size from about 2.5 nm — 12 nm in diameter (mean diameter 7–9 nm). The exoplasmic fracture face, which results from the membrane leaflet adjacent to the thylakoid lumen, carries regularly spaced, parallel rows of 10 nm particles. In the cyanelles of *Cyanophora paradoxa*, these particles frequently appear to be composed of two subunits and have a row periodicity of 10 nm (Giddings et al, 1983). Rows of two-subunit particles can occasionally be observed on the plasmic membrane surface in freeze-etch micrographs of membrane regions where the phycobilisomes have become detached from the thylakoid (Giddings et al. 1983). The rows of particles on the exoplasmic fracture face are generally separated by 30–50 nm; it is believed that the rows of particles on this face fit into the grooves observed on the plasmic fracture face. The parallel rows of particles and the complementary grooves have the same spacing as the rows of phycobilisomes on the stromal (plasmic) surface of the thylakoids (see Fig. 1–3). Since it is known that the primary function of the phycobilisome is to deliver photons to the PS II reaction center (Manodori et al. 1984), it has been suggested that the 10 nm particles observed on the thylakoid membrane in fact represent the PS II reaction centers. Melis and co-workers (Manodori et al. 1984; Manodori and Melis 1985) have recently suggested that there is a strict structural-functional relationship between the phycobilisomes and the PS II complexes in *Synechococcus* sp. PCC 6301 (also see Khanna et al. 1983). In this organism, each phycobilisome is associated with two PS II reaction centers which compete for the harvested excitation energy. It is satisfying to see that similar conclusions are reached

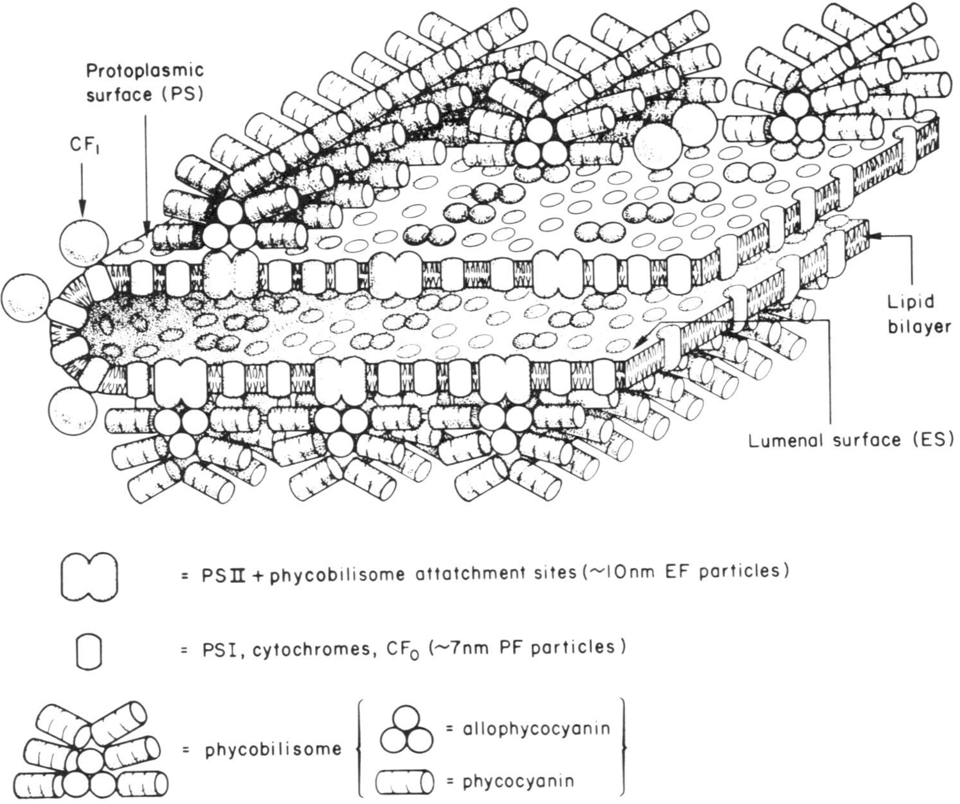

FIG. 3. Model of the cyanobacterial photosynthetic apparatus. Rows of hemidiscoidal phycobilisomes are shown in the stromal (protoplasmic) surfaces of the thylakoids. Each phycobilisome is associated with two PS II reaction centers. The arrangement of the other components is speculative. Reprinted with permission from Giddings et al. (1983).

from biophysical measurements and electron microscopic examination of membranes. Figure 3 shows a model for the thylakoid membranes of the *C. paradoxa* cyanelle as proposed by Giddings et al. (1983). This model is similar to those proposed for cyanobacteria (Gantt 1980b) and red algae (Morschel et al. 1977), and accurately depicts the phycobilisome/PS II ratio and interaction.

Because chemical and physical factors greatly affect the synthesis of components of the photosynthetic apparatus, it is not surprising that these factors influence thylakoid number and arrangement. Thylakoid number and/or area is higher when cells are grown at low light intensities than when cells are grown at high light intensities (Allen 1968; Golecki and Drews 1982). This phenomenon is similar to the variation in intracytoplasmic membrane content observed in purple photosynthetic bacteria grown under different light intensities (Drews 1978). Carbon dioxide deprivation of cyanobacteria can cause the nearly total degradation of photosynthetic pigments, proteins, and thylakoids (Eley 1971; Miller and Holt 1977). This process is reported to be reversible, but this system has not been exploited effectively to study thylakoid biogenesis. Some studies have shown that temperature can affect pigmentation, and hence temperature could also affect thylakoid production and structure (Halldal 1958, 1970).

The membrane lipids of cyanobacteria have been analyzed but most studies have been performed on unpurified membranes or whole cells (The work of Omata and Murata (1983) is an exception.). The major lipids of cyanobacterial membranes are

428

monogalactosyl diglyceride, digalactosyl diglyceride, sulfoquinovosyl diglyceride and phosphatidyl glycerol. The three glycolipids are invariantly and specifically associated with the thylakoids of eucaryotic chloroplasts, but none of these lipid classes are found in the membrane of the purple bacteria (Stanier and Cohen-Bazire 1977). The fatty acid composition of most cyanobacteria is little affected by growth conditions (Kenyon 1972) and this trait is a stable and useful phenotypic property for taxonomic studies (Kenyon 1972; Kenyon et al. 1972). Considerable diversity is observed when the distribution of fatty acids among cyanobacteria is considered, but no obvious correlation of thylakoid morphology and fatty acid composition is discernable. Many cyanobacteria, particularly filamentous forms, contain high proportions of polyunsaturated C18 fatty acids as observed in chloroplasts (Stanier and Cohen-Bazire 1977). Other cyanobacteria resemble typical procaryotes in synthesizing primarily saturated and mono-unsaturated fatty acids. Others resemble neither group; many Pleurocapsalean cyanobacteria contain a high proportion of polyunsaturated C16 fatty acids.

The pigments which comprise the cyanobacterial photosynthetic apparatus are the phycobiliproteins, chlorophyll *a*, and carotenoids. Phycobiliproteins are water-soluble pigment proteins, and the properties of these light-harvesting antenna proteins will be detailed in the following section. Chlorophyll *a* is the only chlorophyll synthesized by cyanobacteria (Stanier and Cohen-Bazire 1977). Most of the chlorophyll (85–90%) is associated with Photosystem I, which is a structurally distinct entity and which probably receives little excitation from other pigments (Manodori et al. 1984). The properties of the chlorophyll-protein complexes of PS I and PS II will be discussed in detail in later sections. The carotenoid compositions of many cyanobacteria have been summarized by Hertzberg et al. 1971), and the role of carotenoids in various photosynthetic organisms including the cyanobacteria has recently been reviewed by Siefermann-Harms (1985). β-carotene is universally found and is usually accompanied by either or both of two oxycarotenoids, zeaxanthin and echinone. Most cyanobacteria also synthesize group-specific carotenoids. Holt and Krogmann (1981) reported the isolation and characterization of a water-soluble carotenoprotein from several cyanobacteria (The carotenoid component was 3-hydroxy echinone?). However, the role of this carotenoprotein, if any, in photosynthesis is unknown. In *Synechococcus* sp. PCC 6301 β-carotene was greatly enriched in the thylakoids, while zeaxanthin was enriched in the cytoplasmic membrane (Omata and Murata 1983). The carotenoid distribution between cytoplasmic and thylakoid membranes was also highly assymmetric in *Synechocystis* sp. PCC 6714 (Omata and Murata 1984). There is little information concerning the association of specific carotenoid species with specific proteins in PS I and PS II complexes. If such specific associations do occur, then carotenoid compositions could vary depending upon the growth conditions employed. Environmental factors such as light intensity, light wavelength, nutrient availability, and temperature are known to affect cyanobacterial pigmentation and the ratio of PS II/PS I centers (Halldal 1970; Stanier and Cohen-Bazire 1977; Kawamura et al. 1979; Myers et al. 1980; Manodori and Melis 1984). Although chromatic illumination has been reported to alter the carotenoid content and composition of *Calothrix* sp. PCC 7601 and 7101 (Fiksdahl et al. 1983), it is not clear from the experimental protocal whether this was truly a light wavelength effect or a light intensity effect.

Phycobilisomes: The Light-Harvesting Complex For Photosystem II

The anoxygenic photosynthetic bacteria, i.e., the purple bacteria and the green bacteria have antenna complexes which are comprised of a variety of bacteriochlorophylls, carotenoids, and proteins. The purple bacteria are characterized by either one, two, or three discreet protein complexes which are spectroscopically distinct (Drews 1985; Zuber 1985). For example, *Rhodospirillum rubrum* and *Rhodo-*

pseudomonas viridis have only one antenna complex which absorbs at either 870–890 nm or at 1015 nm, respectively. *Rps. sphaeroides, Rps. capsulata,* and *Rps. gelatinosa* each have two types of antenna complexes which absorb at 800 and 850 nm and at 870–890 nm. Finally, some purple bacteria (e.g. *Chromatium vinosum*) have three antenna complexes which absorb at 800–820 nm, 800–850 nm, and 870–890 nm. Each of these complexes are composed of equimolar amounts of two small, non-identical polypeptides composed of 50–60 amino acids. The amino acid sequences of several of these hydrophobic polypeptides have been determined; they form a related family and can be divided into α- and β-type sequences (see Zuber 1985). Zuber and coworkers have proposed and tested models for the interaction of these polypeptides with bacteriochlorophyll and for the organization of these complexes in the membranes as they surround the reaction centers. The genes for the four subunits of the *Rps. capsulata* antenna complexes have been cloned, sequenced, and the validity of the structural models are being tested by site-directed mutagenesis (Youvan et al. 1984, 1985; Youvan and Ismail 1985; Bylina et al. 1986). The genes encoding the subunits of the antenna complex of *R. rubrum* have also been cloned recently (Berard et al. 1986).

The green photosynthetic bacteria are characterized by the presence of a unique light-harvesting antenna structure, the chlorosome (Staehelin et al. 1978b; Staehelin et al. 1980; see Olson 1980 for a review). These antenna structures are composed of large amounts of protein-associated bacteriochlorophyll *c* and smaller amounts of protein-associated bacteriochlorophyll *a*. The reaction centers and electron transport reactions of green bacteria have recently been reviewed by Blankenship (1985). The ultrastructure of the cigar-shaped chlorosomes, found attached to the inner surface of the cytoplasmic membrane has been described in detail (Staehelin et al. 1980). Feick and Fuller (1984) recently proposed a detailed model for the chlorosomes of *Chloroflexus aurantiacus*. The complete amino acid sequences of several green bacterial chlorophyll-binding proteins, including both bacteriochlorophyll *a* and *c* binding proteins, have recently been described (Weschler et al. 1985 a, b; Daurat-Larroque et al. 1985). The structure of one of these proteins, a water-soluble bacteriochlorophyll *a*-binding protein, has been solved at 1.9 Å resolution (Matthews et al. 1979; Tronrud et al. 1986). This structure, the first to be described for a chlorophyll–protein complex, can be described as a "protein coccoon" encapsulating seven bacteriochlorophyll *a* molecules.

Oxygenic photoautotrophs vary widely with respect to the light-harvesting antenna systems which are associated with PS II (Glazer 1983). In higher plants and green algal chloroplasts (Thornber 1975; 1986) and in the procaryotic *Prochloron* sp. (Lewin 1984), the light-harvesting complex for PS II is a complex of intrinsic proteins which are complexed with chl *a*, chl *b*, and carotenoids. The apoproteins of this complex are encoded in the nucleus by a small family of genes, the *cab* genes, and are synthesized on cytoplasmic ribosomes as soluble precursors which are processed proteolytically during uptake into the chloroplast (Schmidt et al. 1982). A model for the structure of the chl *a/b* protein has recently been proposed, and experiments to test this model are already in progress (Karlin-Neumann et al. 1985; Kohorn et al. 1986; Chitnis et al. 1986). With the exception of the red algae (see below), other eucaryotic algal families have major light-harvesting antenna systems which are generally much less well characterized (see Glazer 1983, for a review). These organisms have antenna systems composed of intrinsic protein complexes carrying specific carotenoid and chlorophylls *b* or c_1 and/or c_2.

The light harvesting complexes for PS II in cyanobacteria, red algae, and the chloroplast-like cyanelles of certain dinoflagellates such as *Cyanophora paradoxa*, are supramolecular protein complexes known as phycobilisomes (PBS; for reviews, see Gantt 1980b, 1981; Glazer 1982, 1983, 1984, 1985; Cohen-Bazire and Bryant 1982; Tandeau de Marsac 1983; Wehrmeyer 1983b; Zuber 1985). PBS deliver as much as 95% of the light which they absorb to the reaction centers of PS II (Manodori et al. 1984, Manodori and Melis 1985). PBS are primarily composed of phycobiliproteins, a brilliantly colored family of water-soluble proteins bearing covalently attached open-chain tetrapyrrole chromophores or phycobilins. In addition, PBS also contain smaller amounts of proteins, most of which do not bear chromophores, which are referred to as "linker polypeptides". These 8–120 kDa components are absolutely required for proper assembly and functional organization of the structure. PBS are the best-studied of all photosynthetic antenna systems, and the following discussion will consider various aspects of the structure, composition, and function of these organelles.

Location, Morphology and Isolation

PBS are typically found on the stromal (protoplasmic) surfaces of the thylakoid membrane pair. Remarkably regular arrays, in the form of parallel rows, are frequently observed (see Lichtle and Thomas 1976; Gantt 1980b, for examples). The regularity of the PBS arrays suggests that the chlorophyll-protein complexes of PS II, to which the PBS are presumably attached, must likewise be regularly displaced in the plane of the membrane. Freeze-fracture and freeze-etching studies, performed with PBS-containing organisms (Staehelin et al. 1978a; Golecki and Drews 1982; Giddings et al. 1983) provide strong evidence for this contention. These studies suggest that each PBS is associated with a pair of membrane complexes approximately 10 nm in diameter. Giddings et al. 1983). The freeze-fracture/freeze-etch data are strongly supported by physical measurements indicating that two PS II reaction centers compete for the excitation energy of one PBS in *Synechococcus* sp. PPC 6301 (Manodori and Melis 1985).

PBS morphology can vary significantly and is dependent upon the source organism (1980b). Three PBS types have been shown to occur in cyanobacteria: 1. bundle-shaped; 2. hemidiscoidal; and 3. hemi-ellipsoidal. A fourth morphology class, "block-shaped," has also been reported, but thusfar these structures have only been reported for the red alga *Griffithsia pacifica* (Gantt and Lipshcultz 1980). The only organism known to have bundle-shaped PBS is *Gloeobacter violaceus*, a cyanobacterium which has no thylakoid membranes (Guglielmi et al. 1981). The PBS of *G. violaceus* consist of bundles of six rods; each rod is 50–70 nm in length, 10–12 nm in diameter, and is composed of eight to twelve disc-shaped subunits about 6 nm in thickness. The rods are apparently hexagonally packed and attached to a poorly defined basal substructure which is presumed to attach the entire assembly to the PS II units embedded in the cytoplasmic membrane (Guglielmi et al. 1981). These PBS are attached to the inner surface of the cytoplasmic membrane and stand perpendicular to the plane of that membrane. In thin section electron micrographs, these PBS characteristically appear as an electron-opaque cortical layer at the inner surface of the cytoplasmic membrane (Rippka et al. 1974; Guglielmi et al. 1981).

Hemidiscoidal PBS are the most common forms and have been observed for most cyanobacteria (Bryant et al. 1979; Glazer et al. 1979; Nies and Wehrmeyer 1980; Rosinski et al. 1981; Ohki and Gantt 1983; Ohki et al. 1985; and Raps et al. 1985), for the red algae *Rhodella violacea* (Morschel et al. 1977) and *Porphyridium aerugineum* (Gantt et al. 1968), and for the cyanelles of the dinoflagellate *C. paradoxa* (Giddings et al. 1983; see Fig. 1–5). The appearance of these PBS in electron micrographs is dependent upon the plane of sectioning (Gantt 1980b). When the rows of PBS are viewed in cross-section, the PBS have a semi-circular outline. When the rows of PBS are viewed in longitudinal or tangential (relative to the membrane surface) section, these PBS appear as regularly spaced (~ 10–15 nm, center-to-center) electron-opaque rods. Hemidiscoidal PBS from different sources are 45–75 nm in width at their base, are 30 to 40 nm in height, but are rather uniform in thickness (~ 14–17 nm).

Hemidiscoidal PBS are constructed from either eight or nine cylindrical building blocks which comprise two discreet substructures. The PBS "core" is composed of

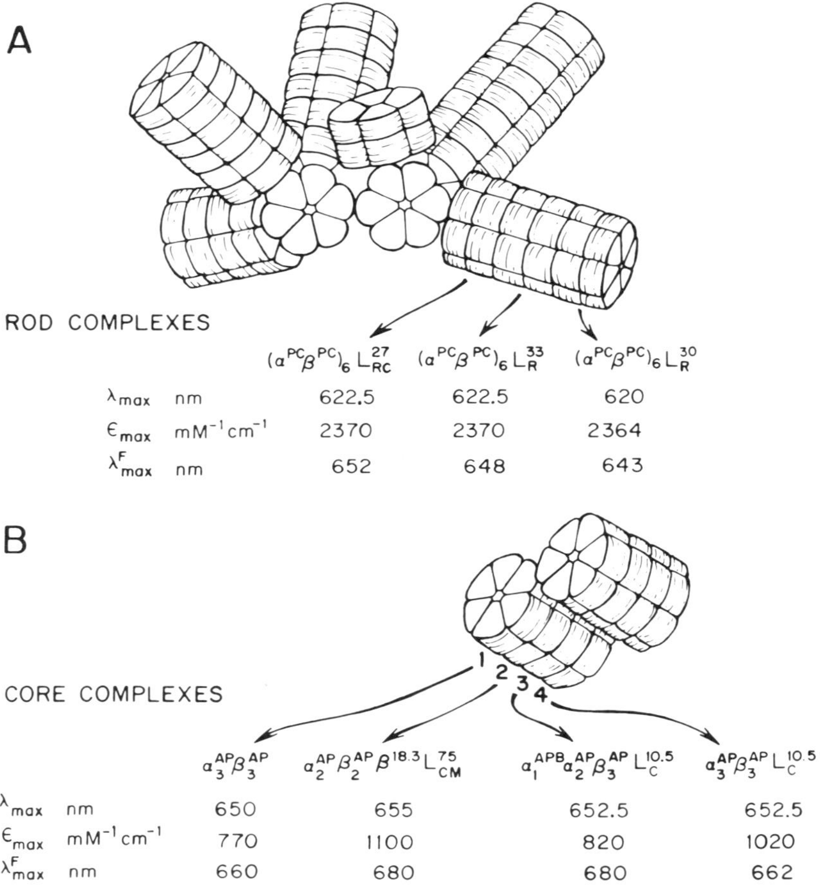

FIG. 4. Diagrammatic representation of the hemidiscoidal phycobilisomes of *Synechococcus* sp. PCC 6301. The composition and some properties of the subcomplexes comprising the peripheral rods and the core are shown in parts A and B, respectively. The superscripts AP, APB, and PC denote the biliproteins allophycocyanin, allophycocyanin B, and phycocyanin, respectively. The symbols L_R, L_{RC}, L_C, and L_{CM} denote linker polypeptides associated with the peripheral rods, the rod-core junction, the core and the core-thylakoid membrane junction, respectively; the superscript numbers indicte the apparent molecular masses of these polypeptides in kilodaltons. Reprinted with permission from Glazer (1985), which provides additional details of the model.

either two (Glazer et al. 1979) or more commonly three (Bryant et al. 1979) cylinders which have a diameter of about 11 nm and a length of 14–17 nm. These core cylinders are composed of a stack of four disc about 3.5 nm in thickness. In PBS with tri-cylindrical cores, these cylinders are stacked along their long axes to produce a structure which approximates a regular triangular prism (see Fig. 4 and 5). Radiating from each of two sides of this core substructure are six "peripheral rod" substructures. Each peripheral rod is composed of a stack of discs which have a diameter of 11 to 12 nm and a thickness of about 6 nm. The number of discs per rod can range from two to about six and is dependent upon both the source organism and the growth conditions employed (Bryant et al. 1979; Glazer et al. 1979). Electron microscopy of these rods indicates that each subunit disc is comprised of two discs approximately 11 to 12 nm in diameter and 3 nm thick. A more complete description of the hemidiscoidal PBS is given below.

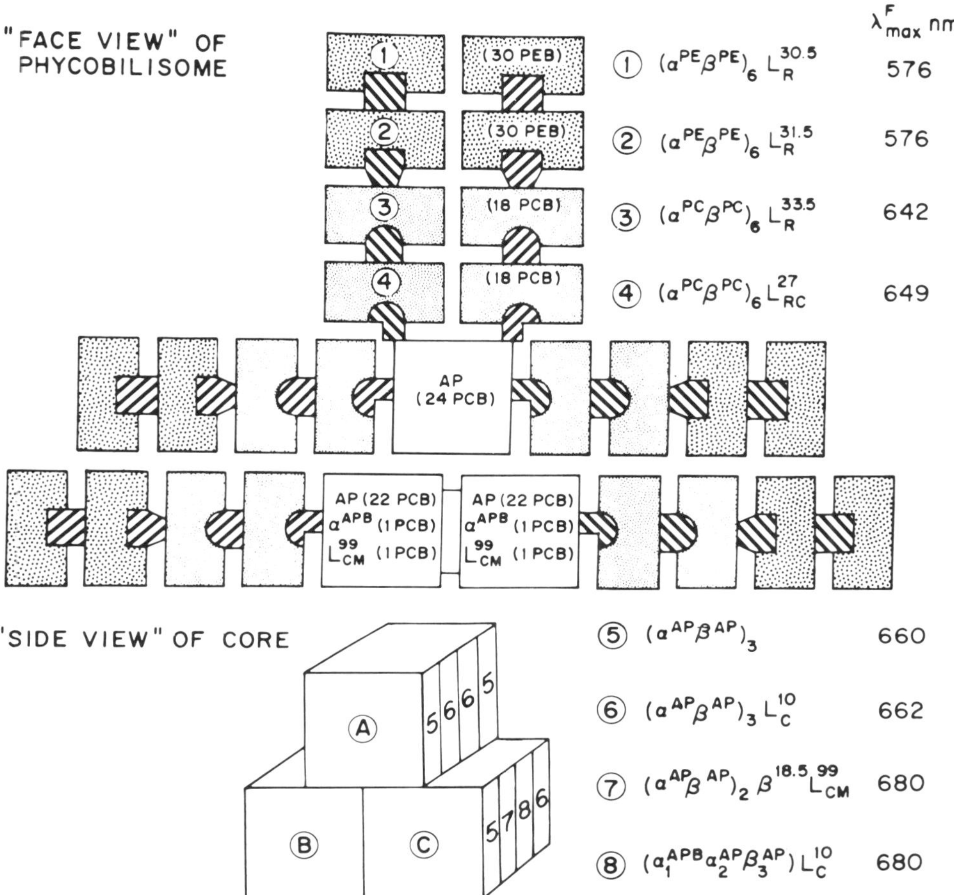

FIG. 5. Diagrammatic representation of the hemidiscoidal phycobilisome of *Synechocystis* sp. PCC 6701. The composition and fluorescence emission properties of the subcomplexes are shown. The superscripts AP, APB, PC, and PE denote the biliproteins allophycocyanin, allophycocyanin B, phycocyanin, and phycoerythrin, respectively. The symbols L_R, L_{RC}, L_C, and L_{CM} denote linker polypeptides associated with the peripheral rods, the rod-core junctions, the core and the core-thylakoid membrane junction, respectively; the superscript numbers indicate the apparent molecular masses of these polypeptides in kilodaltons. The numbers of phycocyanobilin (PCB) and phycoerythrobilin (PEB) chromosomes of the subcomplexes are also shown. Reprinted with permission from Glazer and Clark (1986).

The hemi-ellipsoidal morphology type was the first structural form to be isolated and examined (Gantt and Lipschultz 1972). Until recently these PBS had only been reported to occur in red algae such as *Porphyridium cruentum* (Gantt and Lipschultz 1972) and *Gastroclonium coulteri* (Glazer et al. 1983). However, Guglielmi and Cohen-Bazire (1984) have recently shown that the PBS of the cyanobacterium LPP Group sp. PCC 7376 are hemi-ellipsoidal (63 nm × 40 nm × 25 nm). No accurate model for these PBS has been developed because their large size produces a superpositioning of stain layers that generates electron microscopic images that are difficult to interpret. However, since some views of the hemi-ellipsoidal PBS closely resemble those of the hemidiscoidal class, it is possible that the hemi-ellipsoidal structures represent a "double-thickness" version of that structure (see fig. 7 in Gantt et al. 1976).

Rapid advances in our understanding of PBS structure, assembly and function have occurred because these antenna structures can be easily isolated in high yield and purity. The techniques for isolation employed today are only slight variations on the procedures originally introduced by Gantt and Lipschultz (1972). PBS are very large structures (5–20 000 000 Da) and are easily isolated in high purity from whole cell extracts by sucrose gradient centrifugation. PBS are not stable at low ionic strength, but intact structures can readily be isolated in high ionic strength buffers composed of polyvalent anions such as phosphate or sulfate (Gantt and Lipschultz, 1972; Gantt et al. 1979; Yamanaka et al. 1978). Phosphate buffers (pH 7–8, 0.75 M) are most commonly employed. A non-ionic detergent such as Triton X-100 (Gantt and Lipschultz 1972) or a zwitterionic detergent (Glazer et al. 1979) is employed to release the PBS from the thylakoid membranes. Factors affecting PBS integrity have been systematically assayed by Gantt et al. (1979). The functional assay for PBS integrity is the fluorescence emission spectrum of the purified product. Intact PBS exhibit a strong emission at about 680 nm when excited with light at wavelengths shorter than 650 nm. If dissociation has occurred, then the characteristic fluorescence emissions of the component biliproteins (phycocyanin, ~646 nm; phycoerythrin, ~575 nm; see Table 1) will be observed. Other criteria employed for routine characterization of PBS include absorption spectroscopy, polyacrylamide gel electrophoretic analysis, uniformity of sedimentation properties, and uniformity of appearance by electron microscopy.

Phycobilisome Composition: Phycobiliproteins and Linker Polypeptides

PBS are believed to entirely be composed of two types of proteins: phycobiliproteins and linker polypeptides. Phycobiliproteins typically account for 85–90% of the total PBS protein, with the linker polypeptides accounting for the difference. PBS do not contain chlorophyll as routinely isolated. Numerous recent reviews have detailed the properties of the phycobiliproteins (Scheer 1981, 1982; Zuber 1983, 1985; Wehrmeyer 1983b; Cohen-Bazire and Bryant 1982; Tandeau de Marsac 1983; Glazer 1981, 1982, 1984, 1985), and their properties will only briefly be discussed here.

Table 1 summarizes some of the spectroscopic, chemical, and physical properties of the major phycobiliproteins which have been characterized from various cyanobacteria. The phycobiliproteins can be classified into three major groups on the basis of their visible absorbance properties: phycoerythrins (λ max = 490–570 nm); phycocyanins (λ max = 610–630 nm); and allophycocyanins (λ max = 650–670 nm; see Table 1). The fundamental structure of all phycobiliproteins is a monomer (protomer), $\alpha\beta$, composed of two dissimilar subunits with molecular masses in the range 17 000–21 000 daltons. Complete primary structures of a variety of phycobiliproteins, representing all spectroscopic classes and derived from phylogenetically diverse sources have been determined (Wehrmeyer 1983a). All primary sequences can be assigned to one or two classes: α-type and β-type. Within any spectroscopic class, a very high degree of homology exist when comparing α or β subunits no matter how distantly related the source organisms may be. Finally, the α and β subunit families are sufficiently homologous to one another to establish that the phycobiliproteins constitute a family of proteins descended from a common ancestral gene.

All phycobiliprotein subunits carry at least one linear tetrapyrrole chromophore; the chromophores are covalently attached to the polypeptide chain by one or two thioether linkages. (Glazer 1985). It is these chromophores which impart to the proteins their characteristic, intense visible absorption properties. Four chemically distinct chromophores are known to occur among cyanobacterial biliproteins: phyco-cyanobilin, phycoerythrobilin, phycourobilin, and a biliviolin-type chromophore of as yet undetermined structure (Absorption spectra of the various chromophores can be found in Cohen-Bazire and Bryant 1982; also see Glazer 1985). Much of the spectroscopic diversity of the phycobiliproteins is the direct result of these chemi-

TABLE 1. Properties of cyanobacterial phycobiliproteins[a].

Protein	Aggregation State[b]	Bilin content per subunit[c]	Molecular mass $\times 10^{-3}$ Da	Visible absorption maximum (nm)	Fluorescence emission maximum (nm)
Allophycocyanin B	$(\alpha^{APB}\ \beta^{AP})_3$	α : 1 PCB β : 1 PCB	89	670 > 618	675
Allophycocyanin	$(\alpha^{AP}\ \beta^{AP})_3$	α : 1 PCB β : 1 PCB	100	650	660
Phycocyanin	$(\alpha^{PC}\ \beta^{PC})_n$ (n = 1–6)	α : 1 PCB β : 2 PCB	36.5–220	620	630–645
Phycoerythrocyanin	$(\alpha^{PEC}\ \beta^{PEC})_3$	α : 1 PXB β : 2 PCB	120	568, 590 (sh)[d]	625
Phycoerythrin[e]	$(\alpha^{PE}\ \beta^{PE})_n$ (n = 1–6)	α : 2 PEB β : 3 PEB	40–240	565	575–580

[a]Data taken from Cohen-Bazire and Bryant (1982) and Glazer (1985).
[b]The molecular masses and spectroscopic properties are those of the aggregates specified in this column.
[c]Abbreviations: PCB, phycocyanobilin; PEB, phycoerythrobilin; PXB, phycobiliviolin-type chromophore.
[d]sh, shoulder.
[e]Some cyanobacterial phycoerythrins are spectroscopically similar to those found in red algae (R-phycoerythrins). Their subunit structure is $(\alpha\beta)_6\ \gamma$ and they carry phycourobilin chromophores (Ong et al. 1984; Glazer 1985).

cally distinct chromophores with different numbers of conjugated double bonds (see Table 1; for chromophore structures see Glazer 1985).

Other elements also contribute to the spectroscopic diversity exhibited by this family of proteins. While all covalent linkages to the chromophores are thioether linkages to the vinyl substituents carried by the pyrrole rings, there are at least three distinct modes of linkage for the phycobilin chromophores (Glazer 1985). Linkages may occur through the vinyl substituents on the A pyrrole ring, on the D pyrrole ring, or through both the A and the D pyrrole rings. The other major source of spectroscopic diversity is the chromophore environment contributed by the polypeptide chain(s). Although allophycocyanin and phycocyanin each carry phycocyanobilin chromphores, the visible absorption properties of these two proteins are remarkably different (see Table 1). In summary, at least three elements are combined to generate the remarkable spectroscopic diversity exhibited by the phycobiliprotein family: (1) Chemically distinct chromophores; (2) Chemically distinct chromophore–protein linkages; (3) Distinct chromophore environments (may include chromophore-chromophore interactions).

The basic assembly form of all phycobiliproteins is a disc-shaped $(\alpha\beta)_3$ trimer which is $\sim 3 \times 12$ nm (Bryant et al. 1976). An extremely important advance in understanding PBS structure and function has been provided by the solution of the three-dimensional structures of two biliproteins: that of the *Mastigocladus laminosus* phycocyanin $(\alpha\beta)_3$ trimer at 0.3 nm resolution (Schirmer et al. 1985); and that of the *Synechococcus* sp. PCC 7002 (*Agmenellum quadruplicatum* PR-6) phycocyanin $(\alpha\beta)_6$ hexamer at 0.25 nm resolution (Schirmer et al. 1986). The phycocyanin trimer is a torus-shaped molecule with a diameter of 11.0 nm and a thickness of 3.0 nm; the central hole is about 3.5 nm in diameter. This structure agrees nicely with images of the phycocyanin and allophycocyanin trimers observed by negative electron microscopy (Morschel et al. 1980). The hexamer structure is also toriodal with a diameter of about 10.65 nm, a thickness of 6.05 nm, and a central cavity 3.5–4.0 nm in diameter. It is presumed that the linker polypeptides (see below) interact with phycocyanin by inserting into this central cavity. The level of resolution for the trimer clearly defines the polypeptide backbones of both the α and β subunits, establishes the extent and location of eight α-helices for each subunit, and defines the positions and conformations of the nine phycocyanobilin chromophores which occur in the trimer. An interesting point concerning the structure of phycocyanin is that six α-helices (A, B, E–H) of each biliprotein subunit are similar in three-dimensional arrangement to the equivalent helices of myoglobin (Schirmer et al. 1985). This observation suggests that the biliproteins and the globin heme-protein family are actually members of a superfamily of proteins descended from a common ancestral gene.

The phycocyanin hexamers of *Synechococcus* sp. PCC 7002 are composed of two trimers, aggregated head-to-head, which closely resemble the *M. laminosus* trimer structures. The two trimers fit together in a complementary fashion and are held together by polar and ionic interactions. In contrast, the interactions of the α and β subunits are predominantly hydrophobic. Schirmer et al. (1985) have analyzed in detail the functions of conserved amino acids in protein-chromophore interactions and protein–protein interactions. This analysis suggests that many aspects of the structure will be common structural features in all biliproteins. The hexameric structure found in the crystals is probably quite similar to the structure adopted by this protein in the peripheral rods of the phycobilisomes. An analysis of the chromophore positions, intrachromophore distances, and orientations should greatly advance our knowledge of how energy is harvested and transferred rapidly and efficiently through these molecules to the PS II reaction centers.

The solution of these structures will undoubtedly provide new insights into the structure and function of other phycobiliproteins as well. Crystals of phycoerythrocyanin from *M. laminosus* have also been obtained, and these are isomorphous with those of the phycocyanin trimer (Rumbeli et al. 1985). The data from the *M. laminosus* phycocyanin may also allow the solution by Patterson search techniques of the crystal structure for *P. cruentum* B-phycoerythrin (Fisher et al. 1980). One imagines that phycoerythrins have grossly similar structures to those described, since these proteins share extensive primary sequence homology (50–60%) with phycocyanins (W. Sidler and H. Zuber, personal communication). Although allophycocyanin shares some sequence homology with phycocyanin (35–40%), it would be desirable to obtain crystal structure information for this protein as well to complete, at a first approximation, the basic structure for the complete light-harvesting system.

Highly purified biliproteins do not interact with one another to form PBS-kike aggregates even when mixed at very high protein concentrations (Bryant et al. 1976). PBS assembly is now known to require a small family of related polypeptides which have been termed "linker polypeptides" to reflect their essential role in the assembly process. Tandeau de Marsac and Cohen-Bazire (1977) were the first to demonstrate that these polypeptides were integral components of the cyanobacterial PBS; they showed that the linker polypeptides varied in cells which are capable of chromatic adaptation but are constant in cells which do not undergo complementary chromatic adaptation (see Tandeau de Marsac 1983). The presence of these polypeptides in PBS was subsequently confirmed by immunoprecipitation studies performed by Glazer et al. (1978). Tandeau de Marsac and Cohen-Bazire (1977) postulated that these polypeptides might play roles in PBS assembly and attachment of the PBS to the thylakoid membrane; both roles have been confirmed by more recent studies (see below).

Typical PBS require from six to ten distinct linker polypeptides for their assembly and membrane attachment. This number is dependent upon the cyanobacterial species being considered but is mostly dependent upon the number of spectroscopically different biliproteins which comprise the PBS (see Fig. 4 and 5). Glazer (1985) has proposed a systematic nomenclature and abbreviated symbols to represent the linker polypeptides as well as the subunits of the phycobiliproteins, and this system will be employed as a convenience in the remainder of this discussion. Linker polypeptides can arbitrarily be divided into four groups: those which participate in the assembly of the core (L_C); and those which participate in the assembly of the peripheral rods (L_R); those which participate in the attachment of the core to the thylakoid membrane (L_{CM}); and those which participate in the attachment of the peripheral rods to the PBS core (L_{RC}). Although initial studies suggested that the linker polypeptides do not carry chromophores (see Cohen-Bazire and Bryant 1982), more recent studies have shown that certain linker polypeptides carry bilin chromophores and hence can participate in the light-energy capture and transfer processes. Examples of chromophore-bearing linker polypeptides include the γ subunits of certain phycoerythrins (Glazer and Hixson 1977; Ong et al. 1984) and the large core linker phycobiliprotein (L_{CM}) which plays roles in membrane attachment, core assembly, and energy transfer to the chl *a* (Lundell et al. 1981b; Redlinger and Gantt 1982).

Glazer and coworkers have determined that the linker polypeptides perform three roles in PBS structure, function and assembly: (1) They determine the aggregation state and geometry of the biliprotein with which they interact; (2) They modulate the spectroscopic properties of that biliprotein; and (3) They determine the location of the biliproteins within the PBS structure and form "bridges" between the various biliprotein complexes within that structure. Lundell et al. (1981a) were the first workers to systematically examine the structure and function of the linker polypeptides. They successfully purified four linker polypeptides from the PBS of *Synechococcus* sp. PCC 6301 and determined their amino acid compositions, peptide maps, and assembly functions in *in vitro* reconstitution experiments with purified phycocyanin. The conclusion from these studies was that each of the linker polypeptides was a distinct protein which played a unique role in the assembly process and which conferred upon phycocyanin a unique set of spectroscopic properties (see Fig. 4). An interesting observation is that the linker polypeptides have basic isoelectric points while the biliproteins have acidic isoelectric points. This suggests that electrostatic interactions could play very important roles in the assembly process. Several invariant acidic residues line the inner surface of the phycocyanin trimer (T. Schirmer, personal communication), and it is possible that these residues are responsible for strong interactions with arginine or lysine residues on the surface of the linker polypeptides.

Detailed structural information has recently been obtained for several linker polypeptides. Gantt et al. (1985) and Zilinskas and Howell (1986) have shown that the large core linker phycobiliprotein (L_{CM}) exhibits a high degree of structural similarity by immunological techniques. Although the apparent molecular masses of the two proteins are quite different, the amino terminal amino acid sequences of the core linker phycobiliproteins of *Synechococcus* sp. PCC 6301 (75 kDa) and *Synechococcus* sp. PCC 7002 (94 kDa) are 75% homologous (G. Guglielmi and D. A. Bryant, unpublished results). Complete primary structure information is now available for several other linker polypeptides. Fuglistaller et al. (1984, 1985) have determined the complete amino acid sequences of two *M. laminosus* linker polypeptides, $L_C^{8.9}$ and $L_R^{8.9}$; in addition, they have determined the amino terminal sequences of the $L_R^{34.5\ PEC}$ and $L_R^{34.5\ PC}$ linker polypeptides and the carboxyl terminal sequence of $L_R^{34.5\ PEC}$. The complete amino acid sequence of two linker polypeptides, $L_R^{8.9}$ and $L_R^{33\ PC}$, have been deduced from a translation of the genes encoding these polypeptide in *Synechococcus* sp. PPC 7002 (R. de Lorimier, D. A. Bryant, G. Guglielmi, and S. E. Stevens, Jr., manuscript in preparation). Finally, the complete amino acid sequence of the core linker $L_C^{8.9}$ of *Synechococcus* sp. PCC 6301 (J. Houmard and N. Tandeau de Marsac, manuscript in preparation) and the partial sequence of the core linker $L_C^{8.5}$ of *Synechococcus* sp. PCC 7002 (V. L. Stirewalt and D. A. Bryant, unpublished observations) have also been deduced from translations of the respective genes. Comparisons of these data reveal several new and intriguing aspects about the linker polypeptides. Firstly, these data indicate that the linker polypeptides from different cyanobacteria are homologous and share considerable sequence homology. For example, the two $L_R^{8.9}$ proteins are 60% homologous in sequence although *M. laminosus* and *Synechococcus* sp. PCC 7002 are phylogenetically very distantly related. Secondly, the sequence data indicate that the linker polypeptides $L_C^{8.9}$, $L_R^{8.9}$, $L_R^{34.5\ PEC}$, $L_R^{34.5\ PC}$ and $L_R^{33\ PC}$ share considerable amounts of sequence homology and thus form a family of related proteins. These proteins share a region of very high homology at their carboxyl termini. It can be suggested that this highly conserved domain might be required for interactions with the equally highly conserved biliproteins. This conserved domain in the linker polypeptides is rich in basic amino acids, and as noted above these amino acids could be important in electrostatic interactions of the biliproteins and linker polypeptides. The most striking observation is that sequence comparisons of peripheral rod linkers (Mr ~ 30–33 kDa) with the sequences of the biliproteins suggest that these linker polypeptides are distantly but detectably homologous to the α and β subunits of the biliproteins. Fuglistaller et al. (1985) found that the amino terminus of $L_R^{34.5\ PEC}$ was distantly related to the phycoerythrocyanin β subunit and that the carboxyl terminus of this linker was distantly related to the α subunit of phycoerythrocyanin. A similar analysis of the complete sequence of the $L_R^{33\ PC}$ linker polypeptide from *Synechococcus* sp. PCC 7002 also revealed significant homology (R. de Lorimier, D. A. Bryant, G. Guglielmi, and S. E. Stevens, Jr., manuscript in preparation); however, the domains homologous to the α and β subunits were opposite that observed by Fuglistaller et al. (1985). While the significance of the differences between these two observations is not clear, the larger significance of these observations is clear. All components of the PBS can reasonably be argued to be the descended from a single ancestral gene. If this assertion is correct then the globin-biliprotein-linker gene family is one of the largest (supergene) families identified to date.

Structure of Hemidiscoidal Phycobilisomes

Diagrammatic representations of the PBS of *Synechococcus* sp. PCC 6301 and *Synechocystis* sp. PCC 6701 are shown in Figures 4 and 5 (Glazer 1984, 1985). The PBS of *Synechococcus* sp. PCC 6301 are largely composed of phycocyanin and allophycocyanin. The peripheral rods of the PBS are composed of phycocyanin-linker polypeptide complexes as shown. Each linker polypeptide-phycocyanin complex has slightly different spectroscopic properties; these subtle differences are biased to cause a unidirectional flow of excitation energy towards the PBS core and are partly responsible for the overall rate and efficiency of energy transfer to the chl *a* of the PS II reaction centers (Lundell et al. 1981a). As can be seen in Fig. 5, the organization of the peripheral rods of *Synechocystis* sp. PCC 6701 is similar to that observed for *Synechococcus* sp. PCC 6301. The phycoerythrin-linker polypeptide complexes are localized at the distal ends of the peripheral rods as shown (Bryant et al. 1979; Gingrich et al. 1982a, 1982b).

Hemidiscoidal PBS have a complex core structure which must necessarily be asymmetric to accommodate its interaction with both the thylakoid membrane surface and the peripheral rods. Although the actual assembly pathway for the cores is not known, Glazer and coworkers have shown that the PBS cores of *Synechococcus* sp. PCC 6301 can be dissociated into several well-defined multi-protein subcomplexes (Yamanaka et al. 1982; Lundell and Glazer 1983a). Each of the two core cylinders of the *Synechococcus* sp. PCC 6301 PBS is composed of four subcomplexes: $(\alpha^{AP}\beta^{AP})_3$, $(\alpha^{AP}\beta^{AP})_2$, $\beta^{18.3}L_{CM}^{75}$, $\alpha_1^{APB}\,\alpha_2^{AP}\,\beta_3^{AP}\,L_C^{10.5}$, $(\alpha^{AP}\beta^{AP})_3\,L_C^{10.5}$ (see Fig. 4.). Two of these complexes, which are believed to occupy the central positions of the core cylinders, contain the terminal energy acceptors for the PBS which are responsible for the characteristic fluorescence emission properties of intact PBS and which are probably responsible for the transfer of excitation energy to the chl *a* molecules associated with the PS II reaction centers.

Figure 5 presents one model for the tri-cylindrical core structure of the *Synechocystis* sp. PCC 6701 PBS. Gingrich et al. (1983) succeeded in isolating a series of subcomplexes from this core that resembled the subcomplexes isolated from the *Synechococcus* sp. PCC 6301 core by chromatography of dissociated core fractions on DEAE cellulose columns. On the basis of the relative stoichiometries of the subcomplexes obtained in this manner, these workers suggested that the third cylinder was composed of two $(\alpha^{AP}\beta^{AP})_3$ and two $(\alpha^{AP}\beta^{AP})_3\,L_C^{10}$ complexes. The lower two cylinders were proposed to have the same composition and organization as had been deduced for the core of *Synechococcus* sp. PCC 6301. Anderson and Eiserling (personal communication) have recently isolated core substructures from the *Synechocystis* sp. PCC 6701 core which are inconsistent with the model proposed by Gingrich et al. (1983). These workers suggest that the subcomplexes of the lower cylinders are not arranged in the anti-parallel fashion shown in Fig. 5 but are asymmetrically distributed. Resolution of the conflicting proposals will require further testing of the two models.

Energy Transfer in Phycobilisomes

Greater than 95% of the light energy absorbed by PBS is initially transferred to PS II (Manodori and Melis 1984, 1985; Manodori et al. 1984). Since a typical hemidiscoidal PBS carries 300 to 700 phycobilin chromophores, the PBS greatly augment the limited absorption cross-section of the approximately 50 chl *a* molecules that are associated with each PS II reaction center (see below). Light energy absorbed by the phycobiliproteins can be delivered to the PS II reaction centers with an overall efficiency of greater than 90%. This implies that the energy transfer mechanisms must proceed rapidly to avoid energy losses by competing radiative or non-radiative decay processes. What are the structural and functional properties of the system which allow these process to occur with such remarkable efficiency? While some aspects, such as the obvious proximity of the various chromophores achieved by bringing the component proteins together in a multi-protein complex, are obvious, other points are more subtle. The results of X-ray crystallography, pico-second spectroscopy, and even molecular genetics are helping to unravel the mysteries of these processes and replace them with some well-founded insights.

Using pico-second fluorescence techniques, Glazer and coworkers have studied the excitation energy transfer process in the PBS of *Synechocystis* sp. PCC 6701 and in mutants of this organism which produce PBS deficient in either phycoerythrin or phycocyanin and phycoerythrin (Glazer et al. 1985a, b; Glazer and Clark 1986). Light absorbed by any of the component biliproteins of this PBS is emitted as fluorescence by either or both of the terminal energy acceptors, L_{CM}^{99} and α^{APB}, at about 676–680 nm. When excited with light at wavelengths primarily absorbed by phycoerythrin, the fluorescence emission at 680 nm occurs in 56 ± 8 ps; excitation at wavelengths primarily absorbed by phycocyanin, in either the wild-type or the phycoerythrin-less mutant, results in fluorescence emission at 680 nm in 28 ± 4 ps; excitation of the cores isolated from the mutant deficient in both peripheral rod proteins results in emission at 680 nm in 6.6 ± 3.6 ps. Measurements of the energy transfer among the chromophores within a disc suggest that these processes are very rapid and occur in less than 8 ps. Glazer et al. (1985a, b) conclude that it is the time required to transfer the excitation energy from one disc substructure to the next which is the rate-limiting step in the energy transfer process.

Model building experiments suggest that the chromophores of adjacent peripheral rods would largely be non-interacting in hemidiscoidal PBS (Bryant et al. 1979). Thus, the peripheral rod organization is probably functionally important in reducing the overall random walk possibilities available to excitation energy once it has entered the chromophore bed of the PBS (Glazer 1984). Since intradisc transfer events appear to be relatively more rapid than inter-disc transfer events, each PBS substructure can approximately be regarded as a single chomophore for the purpose of modeling energy transfer. Mimuro et al. (1986) have carefully studied the phycocyanin trimer of *M. laminosus* whose crystal structure has been solved. From their analyses they conclude that within the trimer disc, light energy is concentrated from the chromophores on the periphery of the trimer (the α subunit phycocyanobilin and β subunit phyco-cyanobilin attached to cysteine-155) to the β_f chromophores (the phycocyanobilins attached to cysteine-84) which extend into the central cavity of the trimer. The crystal structure of the *Synechococcus* sp. PCC 7002 hexamer shows that these β chromophores of the two trimers are oriented with their transition dipole moments nearly parallel to one another and are separated by approximately 5 nm (Schirmer et al. 1986). While this separation is too large to allow a strong, excitonic coupling of the chromophore pairs, the distance is within the distance over which Forster-type, dipole, induced-dipole transfers can occur. Other subtleties of the energy migration pathway in the phycocyanin hexamer must await further detailed analysis of the potential chromophores interactions within and between hexamers as well as additional spectroscopic analyses.

In *Synechocystis* sp. PCC 6701 PBS (see Fig. 5), the energy transfer process can be represented by five inter-substructure transfer events (Glazer and Clark 1986). Energy flow through the PBS is unidirectional because the component complexes of the structure are arranged in order of decreasing energy from the periphery of the PBS to the terminal acceptors at the center of the core. It is relatively easy to see that the large energy differences between phycoerythrin and phycocyanin, or between phycocyanin and allophycocyanin, will contribute to unidirectional energy flow through the PBS. In a sense this energy flow is ultimately driven by the largely irreversible photochemistry which occurs at PS II. This arrangement is also true for PBS, such as those of *Synechococcus* sp. PCC 6301 (see Fig. 4), which have peripheral rods composed of only phycocyanin and linker polypeptides. Otherwise identical phycocyanobilin chromophores are poised at slightly different energy levels in the peripheral rod complexes by the combined effects of protein-protein, protein-chromophore, and chromophore-chromophore interaction. Finally, it appears likely that light energy is concentrated from the periphery of the component proteins of the peripheral rods to their interior. Light energy thus concentrated could quickly migrate from the distal ends of the peripheral rods to the core substructure through the "excitation pipeline" of aligned chromophores at the interior of the peripheral rods (Schirmer et al. 1986).

Molecular Genetics

The genes encoding several PBS components have recently been isolated and characterized. Table 2 summarizes those genes which have been isolated and indicates the organisms from which they were obtained. In *Synechococcus* sp. PCC 7002, the genes encoding the PBS components are organized into two apparent operons: a "rod" operon and a "core" operon. The rod operon consists of the *cpcB*, *cpcA*, *cpcC*, and *cpcD* genes which are transcribed in that order (de Lorimier et al. 1984; Pilot and Fox 1984; Bryant et al. 1985, 1986). Two unidentified open reading frames, one of 38 codons upstream from the *cpcB* gene and one of greater than 160 codons downstream from the *cpcD* gene have also been found in this cluster. Mutational

TABLE 2. Genes for phycobilisome components which have been cloned and characterized from cyanobacteria and cyanelles.[a]

Locus	Encoding	7002[b]	6301[c]	7409[d]	7601[e]	*C. paradoxa*[f]
cpcA	α subunit, phycocyanin	X	X			X
cpcB	β subunit, phycocyanin	X	X			X
cpcC	L_R^{33} linker polypeptide	X			(X)[g]	
cpcD	L_R^{9} linker polypeptide	X				
apcA	α subunit, allophycocyanin	X	X		X	X
apcB	β subunit, allophycocyanin	X	X		X	X
apcC	L_C linker polypeptide	X	X		X	
cpeA	α subunit, phycoerythrin			X	X	
cpeB	β subunit, phycoerythrin			X	X	
gpcA (*pcyA*2)	α subunit, constitutive phycocyanin			X	X	
gpcB (*pcyB*2)	β subunit, constitutive phycocyanin			X	X	
rpcA (*pcyA*1)	α subunit, red-light inducible phycocyanin			X	X	
rpcB (*pcyB*1)	β subunit, red-light inducible phycocyanin			X	X	

[a]Data is taken from: Pilot and Fox (1984); de Lorimier et al. (1984); Lemaux and Grossman (1984, 1985); Bryant et al. (1985, 1986); Lambert et al. (1985); Conley et al. (1985, 1986); and Lind et al. (1985).
[b]7002, *Synechococcus* sp. PCC 7002.
[c]6301, *Synechococcus* sp. PCC 6301.
[d]7409, *Pseudanabaena* sp. PCC 7409.
[e]7601, *Calothrix* sp. PCC 7601 (*Fremyella diplosiphon*).
[f]*C. paradoxa*, *Cyanophora paradoxa*
[g]Conley et al. (1986) have reported that a linker polypeptide lies downstream from the *pcyA*1 gene of *Calothrix* sp. PCC 7601. The amino terminal sequence of that polypeptide is about 60% homologous to that of the 33 kDa linker polypeptide (cpcC gene product) of *Synechococcus* sp. PCC 7002 (T. Lomax, personal communication).

analysis suggests the latter may be a trans-acting regulatory element; the deduced amino acid sequences of the two putative genes do not correspond to those of any known structural components of the *Synechococcus* sp. PCC 7002 PBS (R. de Lorimier, G. Guglielmi, D. A. Bryant, and S. E. Stevens, Jr., unpublished results). The core operon of *Synechococcus* sp. PCC 7002 consists of the *apcA*, *apcB*, and *apcC* genes which are transcribed in that order order (Bryant et al., 1986). This arrangement has also been found in *Synechococcus* sp. PCC 6301 (J. Houmard, D. Mazel, C. Moguet, D. A. Bryant, and N. Tandeau de Marsac, manuscript in preperation), and *Calothrix* sp. PCC 7601 (Conley et al. 1986).

An analysis of PBS structure of *Synechococcus* sp. PCC 7002 is being pursued through the construction of defined mutations. Thusfar, three structural mutations have been characterized: 1. A deletion of the *cpcA* and *cpcB* genes (see Table 2); 2. A deletion of the *cpcD* gene; and 3. An insertion mutation of the *cpcC* gene (Bryant et al. 1986; R. de Lorimier, D. A. Bryant, G. Guglielmi, and S. E. Stevens, Jr., manuscripts in preparation). A fourth mutation, a deletion of the *apcA* and *apcB* genes is presently being constructed (V. L. Stirewalt and D. A. Bryant, unpublished observations). In general these studies have confirmed the existing ideas concerning PBS assembly which were developed by Glazer and coworkers through *in vitro* experiments (Glazer 1982; 1984; 1985). The cloned genes are also being employed to investigate the control of biliprotein gene expression by light wavelength (e.g. see Conley et al. 1985, 1986; Bryant et al. 1986), and by light intensity and nitrogen availability (Bryant et al. 1986). These aspects are discussed in a separate section below (see DYNAMICS ASPECTS, below).

Photosystem II and the Oxygen-Evolving Complex

The PS II Reaction Center

Photosystem II (PS II) refers to the chlorophyll–protein complex (and associated antenna chlorophylls) which is functionally defined by its ability to catalyze the light-driven transfer of electrons from water, thereby evolving oxygen, to plastoquinone ("water-plastoquinone photo-oxidoreductase"). Recent reviews have focused on various aspects of this photosystem: oxygen evolution (Renger and Govindjee, 1985; Murata and Miyao 1985; Zimmerman and Rutherford 1985); the role of chloride in the oxygen evolution process (Critchley 1985; Izawa 1986); absorption changes (Lavergne 1985); polypeptide composition and structure (Ghanotakis and Yocum 1985; Satoh 1985); the role of manganese and metal clusters in the oxygen evolution complex (Amesz 1984; Dismukes 1985); the 32 kDa polypeptide *psbA* gene product (Kyle 1985); primary photochemistry, charge stabilization, and electron transport processes (Hoff 1982; Parson and Ke 1982; Okamura et al., 1982; Cogdell 1983; Crofts and Wraight 1983; Haehnel 1984; van Gorkom 1985; Mathis 1985); and herbicide sensitivity (Trebst 1980). A detailed discussion of PS II is beyond the scope of this article, and the reader is urged to examine the review articles listed for specific details concerning this photosystem. This discussion will briefly consider electron transport properties, will examine the structure and composition of the PS II reaction center core, and will then consider the oxygen-evolution complex (OEC).

Figure 6 outlines the flow of electrons believed to occur in PS II in both higher plants, green algae, and cyanobacteria. The carriers between Z and PQ_b are believed to be associated with the PS II core complex (described below). The structure of the P680 chlorophyll center is not known with certainty. Although arguments in support of a monomeric chl *a* molecule have been presented (see Hoff 1982), the overall similarity of PS II to the reaction centers of photosynthetic bacteria strongly suggests that the P680 center consists of a "special pair" of chl *a* molecules (Haehnel 1984). The redox potential for the P680 center is believed to be approximately $+1120$ mV and clearly must be higher than the H_2O/O_2 couple ($+810$ mV)

442

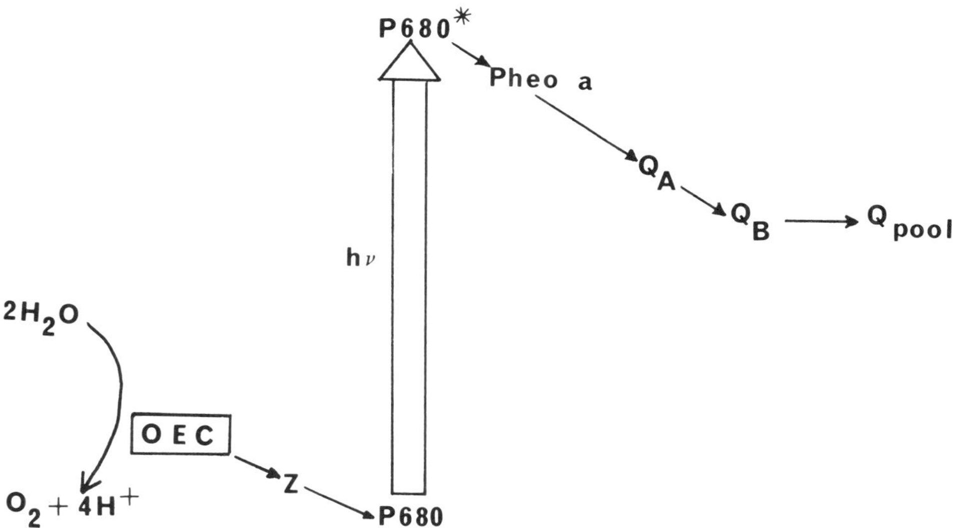

FIG. 6. The path of electron flow around Photosystem II. Q_A, Q_B, and Q_{pool} indicate plasto-quinone molecules and OEC indicates the oxygen evolution complex. Oxidized "Z" has been proposed to be a plastosemiquinol cation. For additional details, see text.

(Blankenship and Prince 1985; Renger and Govindjee 1985). Under the influence of light, the PS II primary donor P680 is brought to the excited state and an electron is transferred to the intermediate acceptor pheophytin a (Pheo a; Em = -610 mV; Klimov et al. 1977; Klimov et al. 1980; Nuijs et al. 1986). This reaction occurs rapidly (probably < 1 psec) and the electron transfer could pass via an intermediary chl a molecule. In about 200 psec the electron is transferred from P680 via a pheo a to PQ_a, a plastoquinone — iron center; this is believed to be the electrogenic step in the separation of charge in PS II from the inside to the outside of the thylakoid.

Further electron transfers from PQ_a^- to PQ_b occur via a two-step, electron gate mechanism (for a review, see Crofts and Wraight 1983). The secondary acceptor PQ_b can only exchange with the PQ pool when it is fully reduced. The scheme below indicates how this portion of the electron transport chain in PS II functions:

1st flash or photon: $PQ_a\ PQ_b\ h\nu\ PQ_a^-\ PQ_b \rightarrow PQ_a PQ_b^-$

2nd flash or photon:

$$PQ_a\ PQ_b^- \xrightarrow{h\nu} PQ_a^-\ PQ_b^- \xrightarrow[\text{2H}^+]{} \quad PQ_a\ PQ_bH_2 \xrightarrow[\text{PQpool}]{} \quad PQ_a\ PQ_b + PQH_2pool$$

The halftimes for PQ_a-oxidation are about 200 μs for the first flash, 400–800 μs for the second flash, and about 600 μs under steady state conditions (Haehnel 1984). Diuron (DCMU) and the atrazine family of herbicides are competitive inhibitors of PQ binding at the PQ_b site (see Crofts and Wraight 1983).

The halftime for reduction of P680$^+$ is about 25–45 ns for dark-adapted chloroplasts but is considerably slower (400 ns) under steady-state conditions (Haehnel 1984). The oxidation of Z, the electron donor to P680$^+$, can be monitored by EPR spectroscopy. Oxidation of Z gives rise to the rapidly reversible EPR transient known as Signal IIvf (vf = very fast). The kinetics of appearance of Signal IIvf are essentially equal to the disappearance of the EPR signal from P680$^+$; this provides strong evidence that Z is the electron donor to P680$^+$ (Boska et al. 1983). The oxidized donor Z$^+$ is believed to be plastosemiquinol cation on the basis of EPR and absorbance difference spectroscopy measurements (see Renger and Govindjee 1985). The redox potential of the PQH_2/PQH_2^+ has been estimated to be >900 mV, a value which would be well suited for a role between the water-splitting reactions and the P680 center (Rich 1985).

TABLE 3. Stoichiometry of components of the photosynthetic oxygen-evolving complex of spinach PS II particles prepared according to Kuwabara and Murata (1982)[a].

Component	
Photochemical reaction center II[b]	
P680	1
Q_A	1
Pheophytin *a*	2
Cytochrome b-559	1–2
Plastoquinone-9	4
Chlorophyll *a*	50
Extrinsic proteins	
33 kDa	1
24 kDa	1
18 kDa	1
Mn	4
Light-harvesting chlorophyll *a/b* protein	
Apoprotein	25
Chlorophyll a + b	150
Total chlorophyll	200

[a]Data is taken from Murata and Miyao (1985).
[b]The apoproteins of the spinach Photosystem II reaction center are polypeptides with apparent molecular masses of 47, 43, 34, 32, 9, and 4 kDa. Other small polypeptides may also be required, in addition to the three extrinsic proteins listed above, for oxygen evolution activity.

Well-characterized PS II reaction center preparations of simple polypeptide composition have now been obtained from several higher plants including spinach (Satch 1979; Satoh et al. 1983; Omata et al. 1984; Yuasa et al. 1984; Yamada et al. 1985; Gounaris and Barber 1985; Bricker et al. 1985), barley (Hinz 1985), and wheat (Ikeuchi and Inoue 1986) as well as the green alga *C. reinhardii* (Diner and Wollman 1980; de Vitry et al. 1984) While the spinach preparations have probably been most extensively characterized (see Table 3), it is clear that all of these reaction center preparations are very similar in composition, structure, and function. Gounaris and Barber (1985) have found that the PS II as prepared with Triton X-100 is a lipoprotein complex. The spinach PS II reaction center contains one P680 center, one "Z," one PQ_a, and 10 carotenoid molecules per 50 chlorophyll *a* (Satoh and Mathis 1981; Omata et al. 1984). Quantitation of plastoquinone and pheophytin in this reaction center showed that two molecules of plastoquinone and two molecules of pheophytin are present per 50 chlorophyll *a*. Since there are 50 chlorophyll *a* per P680 center, there are two pheophytins per P680 center (Omata et al. 1984; Murata et al. 1986). The relative insensitivity of these reaction center preparations to DCMU and atrazine suggests that PQ_b is not present (Satoh et al. 1983; Omata et al. 1984). Tabata et al. (1985) have obtained similar results to those of Omata et al. (1984) with hexane-extracted PS II particles. The hexane extraction left two plastoquinones per reaction center and these reaction centers did not contain PQ_b but did retain the donor Z. Thus, the presence of two PQ per P680 center in these preparations supports the notion that the secondary donor Z may be plastoquinone in spinach. High concentrations of DCMU/atrazine have been shown to inhibit a reaction on the oxidizing side of PS II (Carpenter et al. 1985), and azido-derivates of phenolic inhibitors label polypeptides other than the 32 kDa PQ_b-binding protein, principally the 51 kDa polypeptide although the 43 kDa polypeptide was also labeled (Oettmeier and Trebst 1983; Neumann et al. 1984). These results are also consistent with the notion that Z is a plastoquinone species and suggest that Z may be bound to one or more of the 43–50 kDa chlorophyll a-binding proteins. In spite of these arguments, the actual polypeptides responsible for the binding of P680, Z and PQ_a are not unambiguously identified.

The spinach PS II reaction center cores are most likely composed of six polypeptides with molecular masses of 47, 43, 34, 30, 9.4, and 4.4 kDa (Satoh et al. 1983; W. Widger et al. 1985; Satoh 1985). The stoichiometry of these polypeptides, generally considered to be 1:1:1:1:2:2, can only be considered approximate at this time, since a precise determination has not yet been performed. A study employing uniform 14-C labeling would be useful in this regard. Similar results have been obtained with pea PS II preparations (Satoh et al. 1983), with barley (Hinz 1985), and with wheat (Ikeuchi and Inoue 1986). The spinach PS II complex has been studied by chemical modification and protease sensitivity methods in an effort to learn more about the topography of the complex in the thylakoid membrane (Lam and Malkin 1985). Ljungberg et al. (1986) have recently suggested that two additional polypeptides with molecular masses of 5.5 and 5 kDa also occur in spinach PS II core particles incapable of oxygen evolution. The largest two polypeptides have been shown to bind chl a (Delepelaire and Chua 1979; Machold et al. 1979; Satoh et al. 1983; Satoh 1985). Considerable efforts have been expended in attempting to correlate the P680 center with either the 47 or 43 kDa chl a-binding polypeptide (Nakatani et al. 1984). The tentative conclusion from these studies was that the P680 reaction center was associated with the 47 kDa chlorophyll-a binding polypeptide. Nakatani et al. (1984) suggested that the 43 kDa chl a-binding polypeptide is an "internal antenna" for the PS II reaction center. Tang and Satoh (1984) have recently purified the 47 kDa polypeptide-chl a complex in a relatively native form by octyl-β-D-glucopyranoside extraction of PS II reaction center cores. The purified 47 kDa complex carried 6–7 chl a molecules, β-carotene, and perhaps pheophytin a but no plastoquinone. The complex was photochemically inactive; it was incapable of photoaccumulating reduced pheophytin or of DCIP reduction with diphenylcarbazide.

None of the remaining polypeptides of the spinach PS II core preparations have yet been conclusively shown to bind chlorophyll. The 30 kDa polypeptide is the DCMU/atrazine binding protein (Satoh et al. 1983; Satoh 1985). The identity of the 34 kDa polypeptide was in doubt until recently; however, it now appears likely that the 34 kDa polypeptide is the "D2" polypeptide originally identified by Chua and Gillman (1977). A postulated role for this polypeptide in PS II will be discussed below. The 9.4 kDa and 4.4 kDa polypeptides are the apoproteins of cytochrome b-559 (Metz et al. 1983; Widger et al. 1985; Herrman et al. 1984). Cytochrome b-559 is currently thought to be a heterodimer ($\alpha\beta$); each dimer would bind one heme via the unique histidine that occurs on each polypeptide (Babcock et al. 1985; W. Widger et al. 1985). Although some reports indicate that there are two cytochrome b-559 hemes per P680 center (Murata et al. 1984; Yamamoto et al. 1984b), other reports indicate that there is only a single cytochrome b-559 per reaction center (see Ghanotakis and Yocum 1985). This cytochrome variously occurs in two potential forms: a high (Em = 340–380 mV) and a low (Em = 115 mV) potential form. The function of cytochrome b-559 in PS II is not known. Butler (1978) and Hervas et al. (1985) have discussed potential roles for cytochrome b559 function in PS II. It has been shown, however, that the high potential form of this cytochrome plays no essential role in water oxidation/ O_2-evolution (Briantais et al. 1985; also see below). Falkowski et al. (1986) have obtained evidence suggesting cyclic electron flow can occur around PS II. They have proposed that cytochrome b-559 could participate, possibly conserving energy in the process, in that cyclic flow. It would be interesting to compare wild-type and the $psbE/psbF$ deletion mutant of *Synechocystis* sp. PCC 6803 by the methods employed by Falkowski et al. (1986; see below).

Diner and Wollman (1980) have isolated a very highly purified PS II reaction center preparation from *C. reinhardii*. de Vitry et al. (1984) recently reported that this reaction center contained one photoreducible pheophytin and one photoreducible quinone per 45 chl a molecules. This complex was composed of five intrinsic popypeptides with molecular masses of 50, 47, 32, 31, and 10 kDa. The stoichiometry of the first four polypeptides has been reported to be 1:1:1:1 by uniform 14-C labeling (Satoh 1985). As in the spinach complex, the largest two polypeptides have been shown to bind chl a (Delepelaire and Chua 1979) and de Vitry et al. (1984) concluded that these polypeptides carried approximately 23 molecules of chl a per polypeptide. These workers also suggested that the P680 center was associated with the 50 kDa polypeptide. However, it should be noted that all fractions which contained both the 50 kDa polypeptide and which exhibited P680 activity, also contained the polypeptides of 31 and 32 kDa. These latter polypeptides correspond to polypeptides D1 (DCMU/atrazine binding protein) and D2 (Chua and Gillham 1977), respectively. The similarity of the *C. reinhardii* and spinach PS II reaction center preparations is obvious, although one potential difference has been reported (de Vitry et al. 1986). Quantitative analysis of the plastoquinone-9 content of this reaction center, which exhibited light-driven electron transport from Z to PQ_a. indicated only 1.16 ± 0.14 plastoquinone molecules per reaction center. Since PQa is present and is known to be plastoquinone-9 from spectroscopic analysis, this result indicates that Z is not plastoquinone-9 in *C. reinhardii* PS II reaction centers. These results do not exclude the possibility that Z is a different quinone, a more polar form of plastoquinone, or a covalently bound plastoquinone.

Yamagishi and Katoh (1983, 1984, 1985) have prepared highly purified PS II reaction center preparations from the thermophilic cyanobacterium *Synechococcus* sp. Two forms of the reaction center have been obtained. The first is composed of five polypeptides (47, 40, 31, 28, and 9 kDa) in the apparent molar ratio 1:1:1:1:2. These reaction centers were reported to contain photo-reducible pheophytin (1/32 chl a), plastoquinone A (PQ_a; 1/46 chl a), cytochrome b-559 (1/33 chl a), and virtually no manganese. The second form, although devoid of the 40 kDa chl a-binding polypeptide was nonetheless capable of pheophytin and PQ_a photoreduction (at reduced efficiency; Yamagishi and Katoh, 1985). The finding that the reaction centers depleted of the 40 kDa polypeptide retain photochemical activity supports the notion that the 40 kDa chl a-binding protein functions as an "internal antenna" for PS II. Takahashi and Katoh (1986) have recently reported that the plastoquinone contents of several different PS II complexes from this *Synechococcus* sp. Their data suggests that PQ_a, PQ_b, and Z are plastoquinone molecules in *Synechococcus* sp., as was described above for spinach. Characterization of PS II-enriched particles from *Phormidium laminosum* gave similar results (Ke et al. 1982), although the polypeptide composition of these preparations was not as well defined as those described by Yamagishi and Katoh. The *P. laminosum* particles, which retain Q_2-evolution capacity, contain 1 P680 center: 1 photoreducible pheophytin a: 4 PQ: 4 Mn: 44 chlorophyll. Murata et al. (1986) have recently determined that the ratio of pheophytin/ PS II reaction centers in *Synechococcus* sp. PCC 6301 and *Synechocystis* sp. PCC 6714 is two, as found for higher plants. Immunological relatedness of cyanobacterial PS II components to those occurring in eucaryotic chloroplasts has been reported (Pakrasi et al. 1985; Bullerjahn et al. 1985; Liveanu et al. 1986). From the results cited above and from other studies, it is clear that the PS II reaction center of cyanobacteria is fundamentally identical to that which occurs in higher plants and green algae.

446

In addition to the 47 and 40 kDa chl-protein complexes usually associated with PS II reaction center core preparations, Sherman and coworkers report the occurrence of at least three additional chl-protein complexes associated with PS II preparations of *Synechococcus* sp. PCC 7942 and *Synechocystis* sp. PCC 6714 (Pakrasi et al. 1985; Bullerjahn et al. 1985). The apoproteins of these complexes are 64, 42, and 36 kDa. The chl-protein complex associated with the 36 kDa polypeptide represented a substantial amount of the protein-bound chl associated with fractions enriched in PS II; the amount of chl associated with the 64 kDa and 42 kDa polypeptides was much smaller. None of these three polypeptides cross-reacted with antisera produced against the 50 or 47 kDa polypeptides of the *C. reinhardii* PS II core complex, although these antisera did cross-react with their true homologues in the cyanobacteria. The specific roles of these complexes in light-harvesting and/or photochemistry is unknown.

Molecular details concerning the PS II reaction center are accumulating at a rapid pace, and it is clear that cyanobacterial PS II will be very important in the structural and functional analysis of this photosystem. Much of the stimulus for the progress in analyzing PS II has been provided by molecular biologists and X-ray crystallographers. After years of effort no primary structural information existed for any PS II component. Then, in a period of about two years, the genes (*psbA*, *psbB*, *psbC*, *psbD*, *psbE*, *psbF*; see Table 4) encoding each of the six polypeptides (30, 47, 43, 34, 9.4, and 4.4 kDa), respectively) of the PS II reaction center core of spinach were cloned, sequenced, and the deduced amino acid sequences determined (Zurawski et al. 1982a, Morris and Herrmann 1984; Alt et al. 1984; Holschuh et al. 1984; Westhoff et al. 1983a, 1985b; Herrmann et al. 1984). These results, in combination

TABLE 4. Gene locus designations for components of the photosynthetic apparatus.

LOCUS	ENCODING
Photosystem II.	
psbA	Q_B-binding (herbicide-binding) protein, D1.
psbB	47 kDa chl *a*-binding polypeptide.
psbC	43 kDa chl *a*-binding polypeptide.
psbD	D2 polypeptide.
psbE	larger (9 kDa) subunit of cytochrome b-559.
psbF	smaller (4 kDa) subunit of cytochrome b-559.
Plastoquinol-plastocyanin oxidoreductase (Cytochrome b_6/f complex).	
petA	cytochrome *f* apoprotein.
petB	cytochrome b_6 apoprotein.
petC	Rieske Fe–S apoprotein.
petD	"Subunit IV".
Photosystem I.	
psaA, *psaB*	apoproteins of the P700 complex (chl *a*-binding proteins).
ATP Synthase.	
atpA	α subunit of CF_1 ATP synthase.
atpB	β subunit of CF_1 ATP synthase.
atpC	γ subunit of CF_1 ATP synthase.
atpD	δ subunit of CF_1 ATP synthase.
atpE	ϵ subunit of CF_1 ATP synthase.
atpF	b subunit of CF_0 ATP synthase (= Subunit I).
atpG	b' subunit (?) of CF_0 ATP synthase (= Subunit II).
atpH	c subunit of CF_0 ATP synthase (= Subunit III).
atpI	a subunit of CF_0 ATP synthase (= Subunit IV).

with sequences obtained for genes encoding the polypeptides of the purple bacterial reaction centers (Youvan et al. 1984); Williams et al. 1983, 1984; Michel et al. 1985) have established several new findings and have led to further refinements of models for the PS II reaction center.

As might be expected, comparisons of the sequences deduced for the L and M subunits of the reaction centers of *Rhodopseudomonas capsulata* (Youvan et al. 1984) and *Rps. sphaeroides* (Williams et al. 1983, 1984), and *Rps. viridis* (Diesenhofer et al. 1985a) indicate that these polypeptides are homologous to one another. Moreover, the L and M subunit sequences are approximately 30% homologous; this observation suggests that these proteins are the products of genes which likely arose via gene duplication of a single ancestral gene. Comparison of *psbA* gene sequences from higher plants (e.g., Zurawski et al. 1982a; Oishi et al. 1984; Hirschberg and McIntosh 1983), green algae (Erickson et al. 1984; Keller and Stutz 1984) and cyanobacteria (Curtis and Haselkorn 1984; Mulligan et al. 1984; Golden and Haselkorn 1985) indicates that these genes are very highly conserved (85–100% homologous) and related, albeit distantly, to the genes encoding subunits L and M (overall, ~25% homology). Likewise, the *psbD* genes of higher plants (Alt et al. 1984; Rasmussen et al 1984; Holschuh et al. 1984;) and green algae (Rochaix et al. 1984; Erickson et al. 1985) are highly conserved members of this gene family. The 30 kDa herbicide-binding polypeptide product of the *psbA* gene has been known to play a role in PQ_b binding for some time (see Kyle 1985). Hearst and Sauer (1983) postulated that a particular, highly conserved amino acid sequence in this gene family, is important in PQ_b binding. Their argument was supported by the fact that azido-atrazine specifically labels the tryptic peptide which includes the highly conserved sequence (Wolber and Steinbeck 1984). However, their argument was not supported by DNA sequence analysis of *psbA* genes from DCMU/atrazine-resistant biotypes of *Amaranthus hybridis* (Hirschberg and McIntosh 1983), *Solanum nigrum* (Hirschberg et al. 1984), and *C. reinhardii* (Erickson et al. 1984a, 1985a). Each of these genes carried single base changes in the same amino acid codon (effecting a SER→GLY/ ALA conversion) which is in a region some linear distance (~50 amino acids) from the conserved sequence. While it is possible that the sequences are brought close to one another by the folding of the protein, it is also possible that the conserved sequence is highly conserved for other reasons. The polypeptides L and M, in addition to being quinone-binding proteins, are chlorophyll-binding proteins. In fact, a portion of the sequence targeted by Hearst and Sauer (1983) has been shown to provide the histidine ligand to the special pair of chlorophylls in the bacterial reaction center (Diesenhofer et al. 1985a). There is no evidence to date indicating that the polypeptide products (D1 and D2) of the genes *psbA* and *psbD* bind chlorophyll. However, the striking similarity of the *psbA* and *psbD* gene products in oxygenic photoautotrophs to the reaction center L and M polypeptides of the purple photosynthetic bacteria, strongly suggests that the *psbA* and *psbD* products might, in fact, represent the PS II reaction center (Diesenhofer et al. 1985a). The argument that conservation of structure implies conservation of function seems compelling. The counter argument, that the functionally important residues should be conserved, can also be made. The regions of the L and M polypeptides which are involved in liganding the special pair of bacteriochlorophylls, or which are involved in liganding the Fe^{++} and forming the quinone binding sites, are precisely those regions which are the most highly conserved in the *psbA* and *psbD* gene products. Trebst (1986) has recently summarized these comparisons and has presented models for the topologies of the *psbA* and *psbD* gene products based upon their respective homologies to the L and M polypeptides. Finally, it should be noted that Bricker et al. (1985) identified a chl-protein complex derived from a PS II reaction center core preparation of spinach which carried an efficient fluorescence quencher. This complex, which they named CP*, might be an excellent candidate for the "missing" link in the argument above: a demonstration that the *psbA* and *psbD* gene products can in fact bind chl *a*.

It should be noted that not all of the evidence supports the argument. Several authors have presented arguments that the P680 center resides on the 47 kDa polypeptide product of the psbB gene (see Satoh 1985). Perhaps the most problematic data are those of Rutherford (1985a; Rutherford and Acker 1986). Using partially dried films of PS II reaction center complexes, these workers measured the orientation dependence of the EPR signal from the triplet state of the primary donor chlorophyll of PS II (^{3}P680). From these studies they concluded that the chlorophyll macrocycle of the ^{3}P680 was oriented parallel to the plane of the natural membrane. This result is clearly different from the situation observed in the reaction center of photosynthetic bacteria, where the triplet resides on the special pair of chlorophylls which are oriented almost exactly perpendicular to the plane of the membrane. However, Rutherford (1985b) has pointed out that it has not been conclusively shown that the reaction center triplet of PS II actually resides on the primary donor chlorophyll. If the triplet were to reside on the equivalent of the "voyeur" chlorophylls of the bacterial reaction center, the plane of the macrocycle would lie only 30° out of the plant of the membrane (Rutherford 1985b; Diesenhofer et al. 1984). A final problem of considerable magnitude is the fact that the *psbA* polypeptide is known to undergo very rapid turnover *in vivo* (Kyle 1985). It is somewhat difficult to imagine that this crucial complex could be constantly in a state of assembly and disassembly.

Nonetheless, if the much studied *psbA* gene product and the less studied psbD gene product do represent the PS II reaction center core, then there is even greater reason for excitement. The crystal structure of the reaction center of the purple photosynthetic bacterium *Rhodopseudomonas viridias* has recently been solved at 0.3 nm resolution (Diesenhofer et al. 1984, 1985a, b). The complete amino acid sequence of the H subunit of this reaction center has recently been deduced from the nucleotide sequence of the gene which encodes it (Michel et al. 1985). The sequences of the genes encoding the L and M subunits are complete and that of a cytochrome associated with this reaction center is nearly complete (H. Michel, personal communication). This information will allow still further refinement of the crystal structure and produce additional molecular details. X-ray structure analysis of the better characterized reaction centers of *Rps. sphaeroides* are progressing rapidly because of the similarity of the structures (Allen and Feher 1984; Chang et al. 1985). The compositional, structural, and functional homology of the PS II and photosynthetic bacterial reaction centers suggest that the crystal structure of the bacterial reaction centers may provide an excellent approximation to the organization and structure of the PS II reaction center in cyanobacteria and higher plants.

The other components of the PS II reaction center are also highly conserved among oxygen-evolving photosynthetic organisms, as revealed by heterologous hybridizations carried out using the spinach genes as probes. Photosystem II components have been mapped and/or cloned in a variety of plant species including tomato (Phillips 1985), *Oenothera hookeri* and *Nicotiana tabacum* (Carrillo et al. 1986), pea (Courtice et al. 1985; Berends et al. 1986; Bookjans et al. 1986) and wheat (Courtice et al. 1985; Hird et al. 1986). The genes *psbB* and *psbC* encode the known chlorophyll-binding proteins of 50 and 43 kDa (The deduced molecular masses are somewhat larger: 56 and 52 kDa). Hinz (1985) has confirmed the reading frame assignment for *psbB* by determining the partial amino acid sequence of a peptide derived from the barley protein. This peptide was 90% homologous over 20 amino acids to a portion of the sequence deduced for the spinach *psbB* gene (Morris and Herrmann 1983). The deduced amino acid sequences of the *psbC* genes of spinach, pea, and maize are 95% homologous to one another (Bookjans et al. 1986). The *psbB* and *psbC* polypeptides do not have much if any homology to the bacterial reaction center subunits L and M. However, the sequences of the *psbB* and *psbC* genes do show distant relatedness at the amino acid sequence level to one another (Morris and Herrman 1983; Alt et al. 1984; Holschuh et al. 1984). Moreover, there is pronounced similarity in the hydrophobicity profiles for the two polypeptides (Alt et al. 1984; Cramer et al. 1985). If the arguments concerning the *psbA* and *psbD* gene products above is correct, then the presumptive major roles of the *psbB* and *psbC* gene products would be to bind the chlorophyll *a* antenna for PS II. There is already good evidence that the *psbC* product functions as an antenna component (Yamagishi and Satoh 1984, 1985). The *psbE* and *psbF* genes, which encode the two apoprotein subunits of the spinach cytochrome *b*-559, are also very highly conserved and have been mapped and/or cloned from a variety of organisms (Courtice et al. 1985; Hird et al. 1986; Carrillo et al. 1986;; also see below).

The similarities between the cyanobacterial and higher plant PS II reaction centers, as noted above, are further re-inforced by studies employing molecular genetics/recombinant DNA. Hybridization analyses indicate that the *Calothrix* sp. PCC 7601 and *Nostoc* sp. strain MAC genomes contain 5–6 copies of *psbA*-homologous sequences and that the *Synechococcus* sp. PCC 7942 (*Anacystis nidulans* R2) genome carries three copies (Mulligan et al. 1984). These workers have recently cloned two of the *Calothrix* sp. PCC 7601 sequences and determined the nucleotide sequence of one of these. Curtis and Haselkorn (1984) have isolated and sequenced two of five *psbA*-homologous sequences that occur in *Anabaena* sp. PCC 7120. Their data indicates that only one of these genes is actually expressed, however. The deduced amino acid sequences of the cyanobacterial *psbA* genes are 90% homologous to one another and to higher plant *psbA* sequences. This confirms the similarities of the two PS II reaction centers. Golden et al. (1985) have confirmed that *Synechococcus* sp. PCC 7942 has three *psbA*-homologous sequences; a DCMU-resistant mutant of *Synechococcus* sp. PCC 7942 had a point mutation in one of the three *psbA* genes (Golden and Haselkorn 1985). Cloned DNA fragments carrying the altered *psbA* sequence were successfully used to transform wild-type cells to DCMU resistance, unequivocally demonstrating that the point mutation in the *psbA* gene confers the resistance phenotype. Hybridization analysis of *Synechococcus* sp. PCC 7002 suggest that this cyanobacterium also carries three copies of the *psbA* gene (Bryant et al. 1986). The significance of the multiple *psbA* genes is not clear; however, Golden et al. (1985) have recently reported that more than one copy of the genes in *Synechococcus* sp. PCC 7942 can be expressed.

Studies with genes for other structural components of cyanobacterial and cyanellar PS II reaction centers are in progress. Lambert et al. (1985) have mapped and cloned the six genes encoding PS II reaction center components from the cyanelle genome of *C. paradoxa*. The nucleotide sequence of the *pbsE* gene and most of the *psbF* gene of *C. paradoxa* has been determined (A. Cantrell and D. Bryant, unpublished results). Interestingly, these results indicate that the *C. paradoxa* polypeptides are intermediate in their properties between those of the cyanobacterium *Synechocystis* sp. PCC 6803 (see below) and those of the spinach chloroplast (Herrmann et al. 1984). The cloned *C. paradoxa* genes and probes derived from higher plant sequences have in turn been used to clone the cyanobacterial homologues in *Synechocystis* sp. PCC 6803 (see below) and *Synechococcus* sp. PCC 7002 (Bryant et al. 1986). *Synechococcus* sp. PCC 7002 has two copies of the *psbD* gene, and but has single copies of the genes *psbB*, *psbC*, *psbE*, and *psbF*. Nucleotide sequencing of the *psbB* gene is in progress and confirms the expected high degree of homology (~80%) to the deduced amino acid sequence of the spinach gene (J. Gingrich and D. Bryant, unpublished results).

Both *Synechocystis* sp. PCC 6803 and *Synechococcus* sp. PCC 7002 are transformable (for a review, see Porter 1986), and both are good photoheterotrophs (Rippka et al. 1979; Lambert and Stevens 1986). These two organisms are ideally suited for a complete genetic dissection of the PS II reaction center. Remarkable progress has already been achieved with *Synechocystis* sp. PCC 6803 (J. G. K. Williams, W. Vermaas, H. Pakrasi, and C. Arntzen, personal communication). *Synechocystis* sp. PCC 6803 has two copies of both the *psbA* and *psbD* genes, and single copies of the *psbB*, *psbC*, *psbE*, and *psbF* genes. As occurs in the spinach chloroplast genome (Alt et al. 1984; Holschuh et al. 1984) and in the cyanelle genome of *C. paradoxa* (Lambert et al. 1985), the *psbC* sequence is immediately downstream from one of the *psbD* genes. As found in spinach, the *Synechocystis* sp. PCC 6803 *psbC* and *psbD* genes overlap, and the overlapping sequence is nearly identical to that in spinach. A mutant created by deletion of the *psbC* and *psbD* genes is photosynthetically incompetent. The *psbB* gene of *Synechocystis* sp. PCC 6803 has been completely sequenced and is highly homologous ($\sim 80\%$) to the higher plant gene. A mutant constructed by insertion of a drug-resistance gene into the coding sequence of the gene is photosynthetically incompetent. No $P680^+$ was detectable by EPR in this mutant. The *pbsE* and *psbF* genes are adjacent to one another, as observed in the spinach chloroplast genome; the deduced amino acid sequences of the cyanobacterial genes are about 70% homologous to the spinach chloroplast genes (H. Pakrasi, personal communication). A deletion of the cytochrome *b*-559 genes results in a photosynthetically incompetent cell. Hence, cytochrome *b*-559 is functionally or structurally required for PS II activity in *Synechocystis* sp. PCC 6803. The preliminary results described above shows the analytical power of the cyanobacterial systems for unravelling the mysteries of PS II function. Williams and Arntzen are planning a more critical test of the *psbA/psbD* reaction center hypothesis through site-directed mutagenes studies of these genes in *Synechocystis* sp. PCC 6803. This approach should be extendable to analysis of the oxygen-evolution complex as well (see below).

The Oxygen-Evolution Complex

Until recently, the oxygen-evolving complex (OEC) associated with PS II was the "inner sanctum" of the photosynthesis process. However, the OEC black box is beginning to open and researchers are beginning to peek into both the higher plant and cyanobacterial complexes. Much progress has recently been achieved in understanding this portion of photosynthesis because of improved isolation techniques that have allowed oxygen-evolving PS II preparations to be made from cyanobacteria (Stewart and Bendall 1979, 1981) and spinach chloroplasts (Berthold et al. 1981; Yamamoto et al. 1981; and Kuwabara and Murata 1982). Another isolation procedure that has greatly facilitated studies on the OEC is the technique developed by Albertsson and coworkers which allows chloroplast thylakoids to be fractionated into right-side out and inside-out vesicles (Akerlund et al. 1976).

The oxidation of two water molecules, resulting in the evolution of one oxygen molecule, is a four-electron oxidation. Since the P680 center of PS II undergoes a one-electron photooxidation/reduction cycle, it is clear that a coupling system is required to allow these two processes to occur. Joliot et al. (1969) reported oscillations with a period of four in the yield of oxygen observed by illuminating dark-adapted *Chlorella vulgaris* cells with short saturating light flashes. Shortly thereafter, Kok et al. (1970) postulated the "S-state scheme" for oxygen evolution. Kok proposed that the OEC undergoes a cycle of reactions in which the complex increases in oxidation state from S_0 to S_4 in one-electron increments. After the complex reaches the state S_4, oxygen is released and the complex returns to the S_0 state. The S_2 and S_3 states are unstable and collapse to the stable S_1 state in the dark. A model for the S-state cycle, also known as "Kok's oxygen clock," is shown in Fig. 7. The kinetics of the various electron transfer events in the cycle were determined by measuring the reduction of Z^+ by EPR (via Signal IIvf) when short light flashes are employed (Babcock et al. 1976). As shown in Fig. 7, protons are released as the oxidation of the OEC complex procedes from S_0 to S_4.

The requirement Lent for manganese in the OEC is well established (for reviews, see Amesz 1983; Critchley 1985; Ghanotakis and Yocum 1985; and Renger and Govindjee

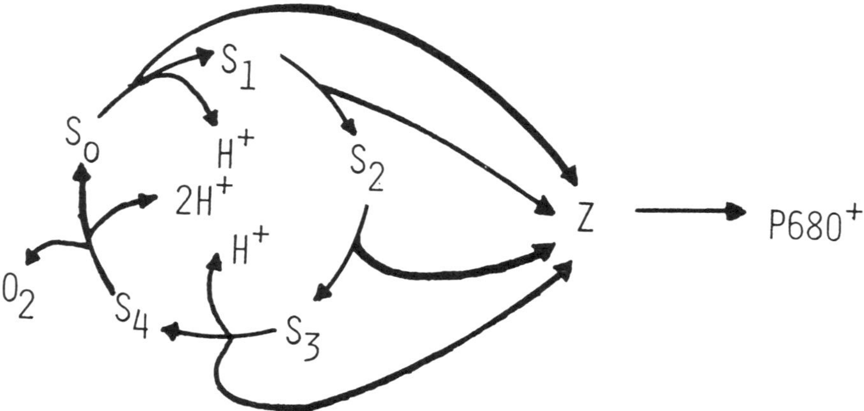

FIG. 7. Scheme of the reactions at the oxidizing side of Photosystem II. Electrons are delivered to oxidized P680 via Z. The oxidized form of Z, Z^+, gives rise to EPR Signal II_{vf} and is believed to be a plastosemiquinol cation. S_0, S_1, S_2, S_3, and S_4 represent the various states of the oxygen-evolution complex.

1985), but a precise determination of the role of this component remains elusive. At least four Mn ions are functionally associated with O_2 evolution activity in intact chloroplast thylakoids (Yocum et al. 1981) in chloroplast PS II particles (Murata et al. 1984; Yamamoto et al. 1984b; Ikeuchi et al. 1985), and in cyanobacterial PS II particles (Stewart and Bendall 1981; Ke et al. 1982; Bowes et al. 1983). The Mn is probably bound to protein(s) at the lumenal surface of the thylakoid membrane. EPR studies suggest that the Mn exists as a binuclear or tetranuclear cluster (Dismukes and Siderer 1981; see Dismukes (1985) and Renger and Govindjee (1985) for reviews). Extraction studies suggest that only two Mn ions may actually be required for OEC catalytic activity. Kambara and Govindjee (1985) have recently proposed a detailed reaction scheme for the water oxidation process that suggests how the manganese centers could function. Two other inorganic cofactors, chloride and calcium ions, also play important roles in the oxygen evolution process (for reviews, see Critchley (1985); Ghanotakis and Yocum (1985); Renger and Govindjee (1985); and Izawa (1986)).

In higher plants the thylakoids are functionally and compositionally differentiated into two domains (for reviews, see Anderson and Andersson 1982; Anderson 1982; Haehnel 1984). The grana-stack regions are highly enriched for the chlorophyll *a/b*-binding protein and the other components of PS II and can be isolated as inside-out vesicles by a phase partition separation or as oxygen-evolving PS II particles. Both types of materials have been useful in defining OEC function, but the PS II particles can be obtained with less contamination by other membrane complexes. PS II particles which are highly active in oxygen evolution have been prepared from spinach chloroplasts by two similar strategies by Berthold et al. (1981) and Kuwabara and Murata (1982). The former authors employed 5 mM Mg^{++} ions and the latter 150 mM NaCl to cause tight appression of the grana membranes via the chlorophyll *a/b* binding protein prior to detergent disruption of the intact thylakoids with Triton X-100. The structural and functional properties of such O_2-evolving PS II particles, as well as those prepared by other procedures, have been studied by Dunahay et al. (1984). These "particles" appear to be unsealed, appressed double-membranes in which the lumenal surfaces are exposed to the solvent ("inside-out"); they have O_2-evolving abilities similar to those of intact thylakoids. The particles are almost totally depleted of PS I components, the cytochrome b_6/f complex, and ATP synthase, and hence provide an excellent, enriched starting material for purification of PS II reaction centers and

OEC components. The composition of the PSII particles prepared from spinach by the procedure of Kuwabara and Murata (1982) is shown in Table 3.

Starting with these membranes enriched in PS II and their OEC complexes, highly purified, oxygen-evolving PS II core preparations have been prepared from spinach (Ikeuchi et al. 1985; Ghanotakis and Yocum 1986; Ljungberg et al. 1986) and wheat (Ikeuchi and Inoue 1986). These complexes are similar in polypeptide composition to the PS II reaction center core preparations described in the previous section, but all retain an extrinsic polypeptide of 33 kDa and the complex retains about 4 Mn per P680 center. Some minor differences result from the different procedures employed. The preparation of Ghanotakis and Yocum (1986) retains most of the sensitivity of PS II complexes to DCMU, while the preparation of Ikeuchi et al. (1985) does not. Ljungberg et al. (1986) provide evidence that the active spinach complex also contains polypeptides of 7.0 and 6.5 kDa; although not discussed, the data presented by Ikeuchi et al. (1985) support the proposal of Ljungsberg et al. (1986) that some small polypeptides may be integral components of the complex. The ability to prepare quickly and easily large amounts of such well-defined, active complexes indicates that this research area will undergo additional rapid advances in the immediate future.

Although the preparations described above define a minimum oxygen-evolving PS II complex, in higher plants it is now well established that at least three extrinsic proteins, associated with the inner surface of the thylakoids (Andersson et al. 1984) and primarily in the grana stacks (Goodchild et al. 1985; Vallon et al. 1985), are associated with the OEC (for reviews, see Murata and Miyao 1985; Ghanotakis and Yocum 1985; Renger and Govindjee 1985; Andersson et al. 1985; and Zimmerman and Rutherford 1985). In spinach these hydrophilic polypeptides have molecular masses of approximately 33, 23–24, and 16–18 kDa; are translated in the cytoplasm and presumably are nuclear-encoded (Westhoff et al. 1985c; Minami et al. 1986); do not carry bound chlorophyll; have isoelectric points of 5.2, 6.4, and 9.5, respectively; and generally have been isolated free of bound Mn or other metals (see Murata and Miyao 1985). These polypeptides probably occur in a 1:1:1 molar ratio; although some workers report that one copy of each is associated with each P680 reaction center (Murata et al. 1984; Yamamoto et al. 1984b; Murata and Miyao 1985), this stoichiometry is still not clearly established (Ghanotakis and Yocum 1985). While most studies have concentrated on these three polypeptides, there is evidence that other polypeptides may be involved in the OEC as well. Analyses of PS II mutants of *C. reinhardii* also suggest a fourth extrinsic polypeptide may be involved in the OEC/oxidizing side of PS II (Bennoun et al. 1983). Ljungberg et al. (1984a) reported that, in addition to the three extrinsic proteins described above, a 10 kDa extrinsic polypeptide could also be released from inside-out thylakoid vesicles and PS II particles by alkaline Tris treatment. Further analysis of the OEC system by immunological methods suggested that the 10 kDa polypeptide, along with polypeptides of 33, 24, 23, 22, and 16 kDa are all subunits of an OEC located at the lumenal surface of the grana thylakoids in higher plants (Ljungberg et al. 1984b; Andersson et al. 1985). Finally, Ljungberg et al. (1986) have shown that, in addition to the 33 kDa polypeptide, intrinsic 6.5 and 7.0 kDa polypeptides are missing from PS II core preparations which are incapable of oxygen evolution.

Selective extraction conditions can be employed to extract either all three polypeptides (0.8–1.0 M $CaCl_2$ or $MgCl_2$; Ono and Inouye 1984a; Kuwabara et al. 1985) or only the 24 and 18 kDa polypeptides (0.8–2.0 M NaCl; Akerlund et al. 1982; Kuwabara and Murata 1983). Stoichiometric rebinding of the three extrinsic polypeptides to depleted PS II complexes can restore O_2-evolution (Akerlund et al. 1982; Lindberg-Moller and Hoj 1983; Miyao and Murata 1984a, 1984b; Ono and Inouye 1984b; Kuwabara et al. 1985). The 33 kDa and 24 kDa proteins apparently bind directly to the intrinsic polypeptides of the PS II core. Protease sensitivity experiments employing PS II particles depleted of the 24 and 18 kDa proteins alone or all three extrinsic proteins, suggest that the 33 kDa protein is associated with the 43 kDa polypeptide of the PS II core and can shield it from trypsin attack (Isogai et al. 1985). However, these workers also report that trypsin digestion of the 47 kDa polypeptide also depends upon the amount of 33 kDa polypeptide removed from the PS II complex. Bowlby and Frasch (1986) have recently reported some interesting results employing chemically modified forms of the 33 kDa polypeptide. Radiolabelled 33 kDa protein could induce a 63% recovery of oxygen evolution activity when 70% of the high-affinity binding sites on PS II particles were filled. Using a cleavable photo-activatable cross-linking agent, N-succinimidyl ((4-azido-phenyl) dithio) proprionate, it was shown that the 33 kDa protein could be cross-linked to 22, 24, 26, 28, 29, and 31 kDa polypeptides. The cross-linked complexes retained protein-bound Mn. Cross-linking to the 43 and 47 kDa polypeptides was not reported. The inconsistency between the data of Isogai et al. (1985) and Bowlby and Frasch (1986) will have to be resolved by additional studies. In any event, rebinding of the 33 kDa polypeptide enhances the rebinding of the 24 kDa polypeptide (Andersson et al. 1984). The 18 kDa polypeptide binds to the OEC only when the 24 kDa protein is previously bound (Murata and Miyao 1985).

The 33 kDa polypeptide seems to preserve Mn atoms in the OEC, although a high Cl^- ion concentration (100–150 mM) can partially sustitute for this polypeptide (Miyao and Murata 1984a, c; Kuwabara et al. 1985). Although this protein can be extracted under conditions where little if any bound Mn is released (Ono and Inouye 1984a; Miyao and Murata 1984a; Andersson et al. 1985; Ghanotakis and Yocum 1985; Kuwabara et al. 1985; Imaoka et al. 1986), subsequent low ionic strength incubations release two Mn/OEC. Observations from a variety of laboratories suggest that the 33 kDa polypeptide influences and/or directly interacts with two Mn (e.g., Miyao and Murata 1984a,c; Kuwabara et al. 1985; Miller 1985). Abramowicz and Dismukes (1985a, b) have recently reported that under appropriate oxidizing conditions the 33 kDa polypeptide can be isolated as a Mn-protein complex (2 Mn/polypeptide) from spinach grana membranes. Yamamoto et al. (1984a) reported that 0.1–0.25 g-atom Mn per mole 33 kDa polypeptide could be isolated in a protein-bound state by a butanol, phase-partioning technique. Although these results have not yet been confirmed by other laboratories (see Murata and Miyao 1985), it is tempting to speculate that the 33 kDa polypeptide plays a direct role in the binding of at least two of the Mn. The complete amino acid sequence of the 33 kDa polypeptide of spinach has recently been determined (Yamamoto et al. 1986; Oh-oka et al. 1986). The sequence of this protein exhibits considerable local homology to a region known to be involved in liganding manganese in the Mn-superoxide dismutases of *E. coli* and *Bacillus sterothermophilus*. Although several polypeptides associated with the PS II core complex have been suggested to be the site of Mn binding, at present an intrinsic polypeptide of 34 kDa appears to be the most likely candidate (Andersson et al. 1985). Ono and Inoue (1984a) have proposed that the Mn is bound to either or both of the 47 and 43 kDa polypeptides, and that this binding is stabilized by the 33 kDa extrinsic polypeptide. In any event the extrinsic polypeptide appears to form a structural shield around the manganese center (Ghanotakis and Yocum 1985).

The function of the 24 kDa protein can be studied by its readdition to PS II preparations depleted of the 24 and 18 kDa polypeptides. The 24 kDa protein restores oxygen-evolution capacity although >5 mM Ca^{++} can substitute for this function (Ghanotakis et al. 1984a; Miyao and Murata 1984b; Andersson et al. 1984). This polypeptide appears to be required for high-affinity binding of Ca^{++} at low physiological Ca^{++} concentrations (see Ghanotakis et al. 1984b, 1985). This protein also reduces the Cl^- requirement for oxygen-evolution (Ghanotakis et al. 1985; Miyao and Murata 1985; Imaoka et al. (1986). Miller (1985) has studied the effects of zinc treatments on the OEC. Her studies show that the 23 kDa polypeptide may function as a "gate" to an aqueous compartment into which Mn^{++} can be released by Zn^{++} ions. The 18 kDa protein is necessary for maximum oxygen evolution at

454

Cl$^-$ concentrations lower than 10 mM (Akabori et al. 1984; Imaoka et al. 1986). The role of chloride in the oxygen evolution process is not unambiguously established (for reviews, see Critchley 1985; Izawa 1986). Sandusky and Yocum (1986) suggest that Cl$^-$ ions may serve as bridging ligands to Mn centers of the OEC; Homann (1985) has discussed the possible role of Cl$^-$ ions in protonation-deprotonation events associated with water oxidation. Recent studies indicate that the extraction of the 18 and 24 kDa proteins allow only partial advancement through the S-state cycle; a block prevents the formation of the S$_2$ state (Franzen et al. 1985; Boska et al. 1985). When the 33 kDa polypeptide was also extracted, coupling of the S-state cycle reactions to the secondary donor Z was greatly decreased or blocked. It is interesting to note that calmodulin-type inhibitors and inhibitors known to influence Ca^{++} and/or Cl$^-$ channels in mammalian tissues also inhibit the oxidizing side of PS II preparations (Carpentier and Nakatani 1985).

Oxygen-evolving PS II particles have also been prepared from several cyanobacteria (Stewart and Bendall 1979; England and Evans 1981; Miyairi and Schatz 1983; Pakrasi and Sherman 1984; Schatz and Witt 1984; Koike and Inouye 1985; Satoh et al. 1985) and the red alga *Porphyridium cruentum* (Clement-Metral and Gantt 1983). The ability to isolate such active complexes from *Phormidium laminosum* (Stewart and Bendall 1979) created considerable interest in this area nearly ten years ago. However, the polypeptide composition of such preparations is considerably more complex than those which have subsequently been obtained from higher plants (see above), and consequently these preparations are not as well defined as their higher plant counterparts. These preparations nonetheless were crucial in demonstrating the remarkable similarity of the cyanobacterial and higher plant PS II reaction centers. These particles and others like them are now being used to demonstrate the overall similarity of the OEC as well.

As described above for higher plants, Ca^{++} is known to be important for oxygen evolution activity in cyanobacteria (Brand 1979; Astier et al. 1986). England and Evans (1983) have reported that Ca^{++} is crucial to the retention of OEC activity for isolation of PS II particles from *Synechococcus* sp. PCC 6301. The presence of Ca^{++} in the isolation buffers caused three polypeptides (30, 33, and 36 kDa) to remain bound to the PS II particles. The presence of a Ca^{++} binding protein in these preparations was suggested by the sensitivity of the preparations to the calmodulin antagonist chlorpromazine. In further studies by England and Evans (1985), three polypeptides (36, 27, and 15 kDa) were identified as being either Tris-extractable, trypsin-sensitive, or both. These authors suggest that these polypeptides are homologous of the three proteins of similar molecular masses and properties from the spinach OEC. England and Evans (1985) also identified a 12 kDa extrinsic polypeptide in their studies which is enriched in a lead-tolerant mutant.

Koike and Inouye (1985) have recently isolated PS II particles which retained a high rate of O$_2$-evolution from the thermophilic cyanobacterium *Synechococcus vulcanus* Copeland. The particles contained an extrinsic polypeptide of 34 kDa but did not contain polypeptides corresponding to the 24 and 16 kDa proteins of higher plants. The 34 kDa polypeptide had a pI of 5.2, as does its spinach homologue, and could be extracted from the PS II particles with 1.0 M CaCl$_2$. This polypeptide restored O$_2$-evolution to 75% of the original activity when this polypeptide was rebound by the PS II particles. Moreover, this polypeptide rebound to and restored O$_2$-evolution capacity (28%) to CaCl$_2$-washed, spinach PS II particles. In the reciprocal experiment, the spinach 33 kDa polypeptide rebound to and stimulated O$_2$-evolution ($\sim$60%) with the *S. vulcanus* PS II particles. These results indicate that cyanobacteria and higher plants have highly homologous 33 kDa proteins. The absence of the 24 kDa and 16 kDa polypeptides from the *S. vulcanus* particles does not necessarily imply that cyanobacteria do not have such proteins, since these could have been released from the PS II preparations during their preparation. The possibility that the 34 kDa protein from *S. vulcanus* might carry bound Mn was not examined in this study.

Stewart et al. (1985a) and Astier et al. (1986) have shown that cyanobacteria indeed have a polypeptide homologous to the 33 kDa polypeptide of spinach. An antiserum to the spinach 33 kDa polypeptide cross-reacted strongly with similar sized polypeptides from the thylakoids of *P. laminosum*, *Anabaena variabilis*, *Synechococcus leopoliensis*, and *Synechocystis* sp. PCC 6714. In contrast antisera to the 23 and 16 kDa polypeptides of spinach did not cross-react with any cyanobacterial proteins (Stewart et al. 1985). These results confirm those of Koike and Inouye (1985) and suggest that cyanobacteria may not contain polypeptides corresponding to the two smaller chloroplast proteins. The apparent absence of these polypeptides from the cyanobacterial OEC constitutes a major difference between the photosynthetic apparatus of cyanobacteria and higher plants. This difference might indicate that cyanobacteria may exhibit differential sensitivity to some herbicides which act on the oxidizing side of PS II. Trypsin digestion of the cyanobacterial thylakoids caused the loss of the 33 kDa polypeptide and caused a parallel decline in oxygen evolution (Stewart et al. 1985a). A variety of salt washes also inhibited the oxygen evolution and concomitantly caused the quantitative release of a 9 kDa polypeptide from the PS II particles of *P. laminosum*. These treatments, except for a Tris wash, did not release manganese from the membrane and did not quantitatively release the 33 kDa polypeptide. Astier et al. (1986) have reported, however, that Tris washing of *Synechocystis* sp. PCC 6714 membranes does release a 32 kDa polypeptide, which is immunologically related to the 33 kDa polypeptide of spinach. The 9 kDa polypeptide is lost from PS II preparations of *P. laminosum* prepared in the absence of glycerol (Stewart et al. 1985b). This polypeptide does not appear to be associated with manganese or with the Ca^{++} or Cl^- requirements of oxygen evolution. Additionally, the relationship (if any) between this polypeptide and the small polypeptides associated with the oxygen evolution complex in higher plants has not yet been determined.

The most highly resolved, cyanobacterial PS II preparation which still exhibits significant oxygen evolution capacity is that described by Satoh et al. (1985). This preparation consisted of five major polypeptides: essentially a core complex plus a 35 kDa polypeptide which was shown to be the equivalent of the spinach 33 kDa protein. The active preparations contained about one PQ_a per 51 chl *a* and about 3.2 Mn per PQ_a. These results clearly establish the similarity of the cyanobacterial and higher plants PS II reaction centers and OEC.

The results of other studies of the OEC complex of cyanobacteria are more difficult to intrepret. Okada and Asada (1983) have reported the extraction of an instrinsic 13 kDa Mn-protein from the thylakoids of *Plectonema boryanum* with a cholate-deoxycholate mixture. The purified protein had a catalase-like activity, and antibodies to this protein inhibited oxygen evolution from *P. boryanum* thylakoids when dichlorophenol indophenol (DCIP) was employed as electron acceptor; reduction of DCIP by diphenylcarbazide was not inhibited by the antibody. Further studies with this protein will be required to establish the role of this protein (if any) in the OEC.

Gantt and coworkers have isolated O_2-evolving photosynthetic vesicles which still carry intact phycobilisomes from both the cyanobacterium *Anabaena variabilis* and the red alga *Porphyridium cruentum* (Katoh and Gantt 1979; Dilworth and Gantt 1983). Clement-Metral and Gantt (1983) subsequently reported the isolation of PBS-PS II-OEC particles from *P. cruentum* which were greatly depleted in PS I/P700 but which retained high rates of O_2-evolution when silicomolybdate was added as electron acceptor. DCIP reduction by these particles is sensitive to DCMU and atrazine, suggesting that the PQ_b-binding protein is a component of the complex (Chereskin et al. 1985). The activity of the complex was greatest with light wavelengths shorter than 665 nm; this suggests that the PBS are serving as functional antennae for the PS II-OEC. When examined in the electron microscope, clusters of the two to three PBS are seen to surround a particle which may arise from the thylakoid

(Clement-Metral et al. 1985). Single PBS attached to small membrane fragments and small vesicles with attached PBS were also observed. These preparations should allow further characterization of the interaction of the PBS with PS II.

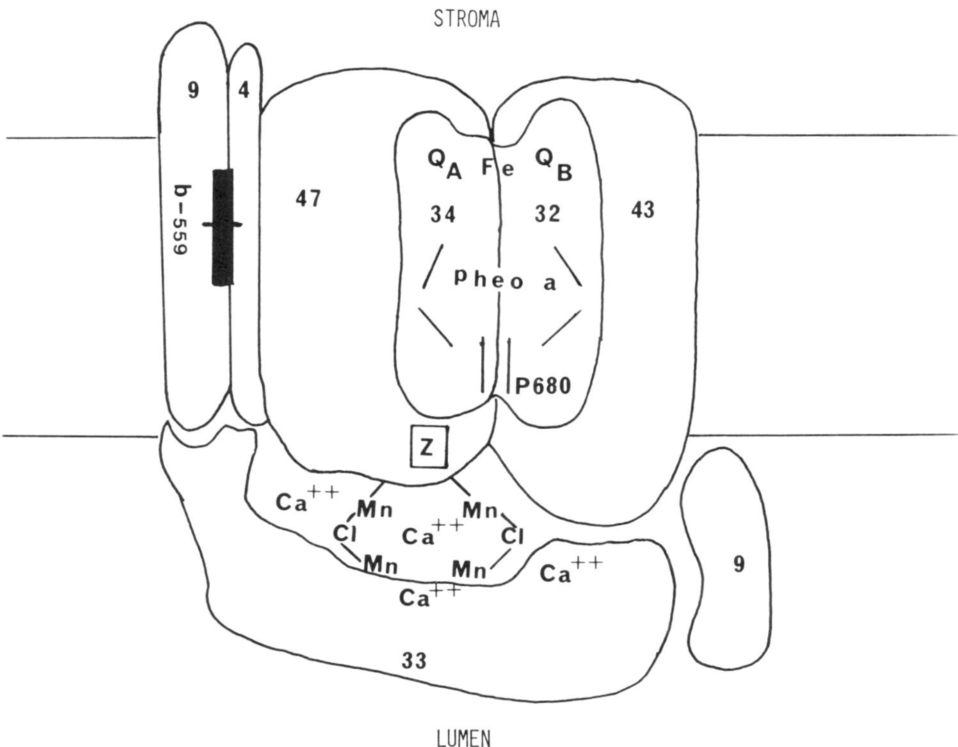

FIG. 8. Model for the cyanobacterial Photosystem II complex adapted from Fig. 1 of Ghanotakis and Yocum (1985). The model presumes the argument of Michel and coworkers (see Deisenhofer et al. 1985a), i.e., the conservation of structure of the photosynthetic bacterial and higher plant Photosystem II reaction centers, is correct. The numbers indicate the apparent molecular masses of the constituent polypeptides. The sites and types of Mn ligands are speculative. The heme group of cytochrome b-559 is indicated by the bar.

Figure 8 shows a model for PS II which incorporates the information obtained from both cyanobacteria and higher plants. Although some elements of this model are still extremely speculative, other elements of the model now seem well established.

Plastoquinol-plastocyanin Oxidoreductase (Cytochrome B_6/F Complex)

During respiratory electron transport in the mitochondria of animals, plants, and fungi, electrons are transferred from ubiquinol to oxidized, soluble cytochrome c. This transfer is accomplished by a membrane-bound redox cluster comprised of a high-potential Fe–S protein ("Rieske protein"), cytochrome b, and cytochrome c_1 (Hauska et al. 1983). Similar complexes occur in other electron-transport chains which are known to oxidize quinols in related fashion. For example, in the cyclic electron transport process that occurs in photosynthetic purple bacteria, electrons from the reducing side of the photochemical reaction centers are shuttled back to the primary donor P865 on the oxidizing side via a series of reactions involving ubiquinol, a membrane-bound redox complex, and soluble cytochrome c_2 (Dutton and Prince 1978; Prince 1985; Robertson et al. 1986). In the oxygenic photosynthetic process of algal and higher plant chloroplasts and of cyanobacteria, plastoquinol oxidation is accomplished by electron transfer via an homologous complex to the copper protein plastocyanin or to a soluble c-type cytochrome (Bendall 1982). It is this electron transfer reaction which couples the two photosystems and which is responsible for the generation of much of the protonmotive force which drives ATP synthesis (noncyclic photophosphorylation; Bendall 1982). This same complex is also believed to function in cyclic electron transport around PS I (Bouges-Bouquet 1980).

Through the efforts of Hauska and coworkers, who have isolated the quinol-cytochrome c (plastocyanin) oxidoreductases from *Rhodopseudomonas spaeroides*, *Anabaena variabilis*, and spinach, some general properties of these complexes are now established (for reviews, see Hauska et al. 1983; Hauska 1985). The redox center composition of each complex is identical: one cytochrome c_1 or cytochrome f: one 2Fe–2S "Rieske" center: two cytochromes b. The two cytochrome b hemes have different redox potentials, but are associated with a single polypeptide. The polypeptide compositions of the photosynthetic complexes are more simple than those observed for the homologous mitochondrial complexes. It is not known whether the "functional complex" is a monomer or a dimer; however, topologically the arrangement of the oxidoreductase subunits is similar from all sources. Mechanistically, all of the complexes are similar, and all of the complexes function as electrogenic proton translocators when the purified complexes are inserted into lipid vesicles. Each of the complexes appears to have two quinone binding sites. Finally, the complex from *Rps. sphaeroides* appears to be more similar to that in mitochondria than do the complexes isolated from cyanobacteria or chloroplasts. This is evident from differences in the cytochrome b polypeptide, the redox properties of the b hemes, as well as differential sensitivity to inhibitors (Hauska et al. 1983).

The ubiquinol-cytochrome c_2 oxidoreductase isolated from *Rps. sphaeroides* is composed of five polypeptides: cytochrome b (40 kDa; high potential and low potential forms present); cytochrome c_1 (33 kDa); Rieske 2Fe–2S protein (24 kDa); and 8 and 10 kDa polypeptides of unknown function (Gabellini et al. 1982; Prince 1985; Hauska 1985). The genes encoding the apoproteins of the first three components (*fbcB*, *fbcC*, and *fbcF*, respectively) have recently been cloned and characterized (Gabellini et al. 1985; Gabellini and Sebald 1986). These three genes are organized into an operon (*fbc*) which is transcribed into a polycistronic mRNA 2.9 kB in length. The *Rps. sphaeroides* components are highly homologous to their mitochondrial homologues. It should be noted that the *Rps. sphaeroides* cytochrome b apoprotein is a 40 kDa polypeptide. This is similar in size to that observed for all mitochondrial cytochromes b which have been characterized (~ 42 kDa; see Widger et al. 1984).

The spinach chloroplast plastoquinol-plastocyanin oxidoreductase (cytochrome b_6/f complex) was originally reported by Hurt and Hauska (1981) to be comprised of five polypeptides (34, 33, 23.5, 20, and 17.5 kDa). These workers subsequently showed that the 34 and 33 kDa polypeptides were derived from cytochrome f, that the 23 kDa polypeptide was the apoprotein of cytochrome b_6, and that the 20 kDa polypeptide carried the Rieske 2Fe–2S center (Hurt et al. 1981; Hurt and Hauska 1982). The complex may also contain small, hydrophobic polypeptides of less than 8 kDa (Hauska 1985). Using a different procedure, a complex composed of five polypeptides has been isolated from spinach chloroplasts by Hind and coworkers (Clark and Hind 1983; Clark et al. 1984). As isolated by Hind and coworkers, who took precautions to avoid proteolytic degradation and who did not employ a NaBr wash to remove peripheral membrane proteins, the complex was composed of polypeptides of molecular masses 37, 33.5, 22, 19, and 16.5 kDa. The redox centers associated with this preparation were the same ones associated with the complex as described by Hurt and Hauska (1982). The 33.5 and 22 kDa polypeptides exhibit heme-associated peroxidase activity and are presumed to represent cytochrome f and cytochrome b_6, respectively. The 37 kDa polypeptide has recently been shown to be ferredoxin-NADP$^+$ oxidoreductase (Clark et al. 1984); the functional significance of this association is discussed separately (see below).

The topology of the spinach chloroplast plastoquinol-plastocyanin oxidoreductase complex has been extensively studied. Each of the constituent polypeptides has been reported to have both stroma- and lumen-exposed regions (Ortiz and Malkin 1985; Mansfield and Anderson 1985). Mansfield and Anderson (1985) also concluded from carboxypeptidase digestion experiments that the carboxyl termini of the four major polypeptides of the complex are exposed on the stromal side of the thylakoid. Lam (1986) has recently reported that Subunit IV (the 16.5 kDa polypeptide) and the Rieske Fe–S protein are apparently cross-linked by low concentrations of glutaraldehyde. This result suggests that Subunit IV and the Rieske Fe–S protein are closely interacting in the cytochrome b_6/f complex, although the functional siginificance, if any, of such an interaction is not yet known. Hauska et al. (1984) have shown that plastocyanin, the electron acceptor for the complex, binds directly to cytochrome f through chemical cross-linking studies. Finally, Morschel and Staehelin (1983) have suggested that the functional complex is a dimer as has been concluded from studies of the equivalent mitochondrial complex. When the purified cytochrome b_6/f complex was reconstituted into liposomes and studied by freeze-fracture electron microscopy, the observed particle size was too large to be the protomeric complex.

The genes (see Table 4) for cytochrome f (*petA*), cytochrome b_6 (*petB*), and the Subunit IV (16.5–17.5 kDa; *petD*) polypeptide of the spinach complex are encoded by the chloroplast genome (Alt et al. 1983). The Rieske 2Fe–2S protein (*petC* product) is translated in the cytoplasm as a larger molecular mass precursor and is presumably encoded in the nucleus (Alt et al. 1983a). Hybrid-selection mapping and cell-free transcription/translation of plastid genome restriction fragments have been employed to identify and clone the coding sequences of the chloroplast-encoded components of the spinach complex (Alt et al. 1983) as well as the equivalent genes for other chloroplasts (Willey et al. 1983; Willey et al. 1984a; Phillips and Gray 1984). Several *petA* genes, including those from spinach (Alt and Herrmann 1984), pea (Willey et al. 1984a), wheat (Willey et al. 1984b), and *Oenothera hookeri* (Tyagi and Herrmann 1986) have been cloned and sequenced. The spinach *petB* and *petD* genes (Heinemeyer et al. 1984), as well as the wheat *petD* gene (Phillips and Gray 1984), have also been cloned and sequenced.

The molecular biologists have contributed some significant insights into the structure and function of the cytochrome b_6/f complex. Unlike most chloroplast-encoded polypeptides, the apoprotein of cytochrome f is apparently synthesized as a precursor some 35 amino acids larger than the mature protein (Willey et al. 1984a; Alt and Herrmann 1984). The 35 amino acid segment which is missing from the mature protein resembles a leader sequence and may help direct and orient the polypeptide in the thylakoid membrane. Consistent with this proposal, Willey et al. (1984a) have shown that the carboxyl terminus of cytochrome f is exposed on the stromal side of the thylakoid while the majority of the protein and the amino terminus is exposed at the intrathylakoidal surface. More recent studies of *petA-lacZ* fusion proteins show that the cytochrome f signal sequence is recognized by the secretory apparatus of *E. coli* and can direct the fusion proteins to the inner membrane (Rothstein et al. 1985). This finding suggests that protein secretion in bacteria and chloroplasts may be highly conserved. The heme binding site of cytochrome f is identical to that observed for other cytochromes *c* and is located near the amino terminus. Hydropathy plots suggest that the polypeptide is anchored in the membrane by a single membrane spanning segment which occurs near the carboxyl terminus of the molecule (Willey et al. 1984a, b; Alt and Herrmann 1984). These observations are topologically consistent with results for cytochrome f homologues in other complexes (Hauska et al. 1983) and are consistent with the location of plastocyanin, the electron acceptor reduced by cytochrome f. Plastocyanin is localized in the thylakoid lumen where it shuttles electrons between cytochrome f and PS I/P700 (Anderson and Andersson 1982).

When the deduced amino acid sequences for the spinach chloroplast *petB* and *petD* products were compared to several mitochondrial cytochromes *b* (Widger et al. 1984) some surprising features emerged. The heme-binding cytochrome b_6 polypeptide (*petB* product) was 30–40% homologous to the amino terminal two-thirds of the mitochondrial cytochromes b. The Subunit IV polypeptide (*petD* product) was about 30% homologous to the carboxyl terminal one-third of the mitochrondrial cytochromes b. Furthermore, four histidines in the *petB* gene product were found to lie in completely conserved sequences, which probably represent hydrophobic membrane-spanning regions, when the primary sequences of all cytochromes b were aligned. This observation is significant since there are two b hemes per c heme in the complex (Hauska et al. 1983) and since each heme is probably coordinated by two histidine residues (Widger et al. 1984). This data, in combination with hydropathy information, has been used to formulate a detailed structural model for the chloroplast cytochrome b_6 (Widger et al. 1984; Cramer et al. 1985). In this model the planes of the two hemes are perpendicular to the plane of the membrane. The two hemes are separated by 12 Å edge-to-edge, can be rigidly held in place by ionics bonds between the side chain proprionates and four arginyl residues, but are completely buried within the hydrophobic, membrane-spanning regions of the polypeptide. Figure 9 presents a model for the higher plant plastiquinol-plastocyanin oxidereductase.

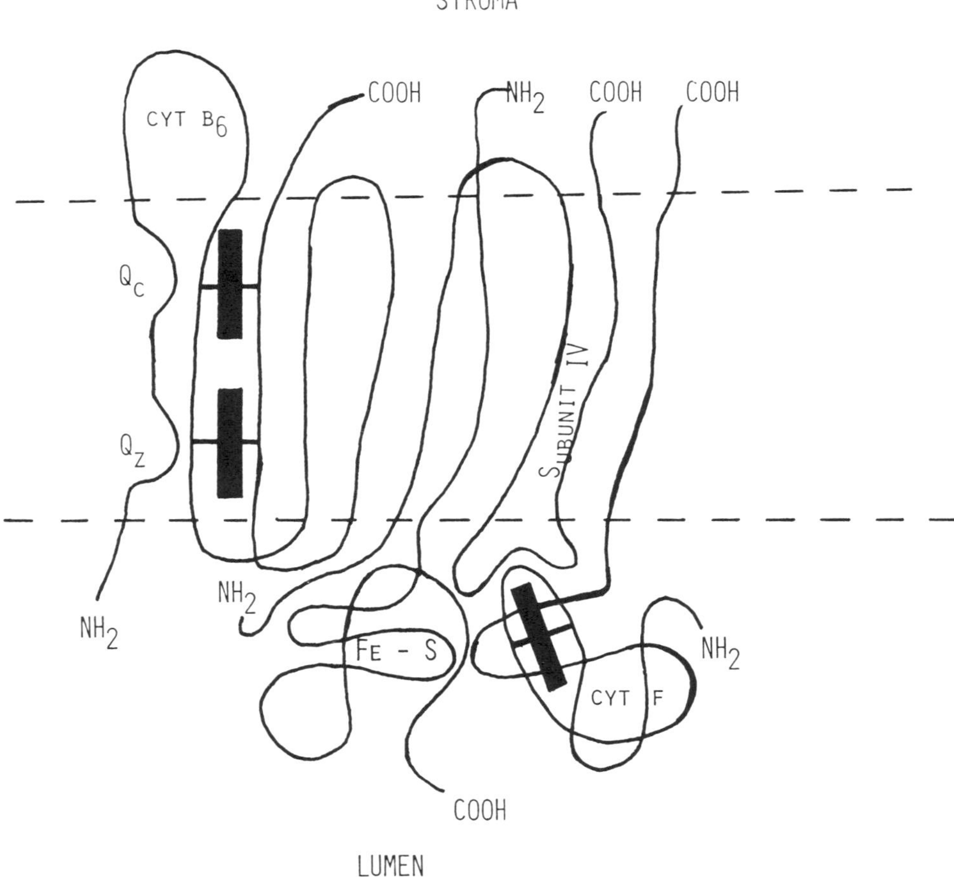

FIG. 9. Topographical model for the plastoquinol-plastocyanin oxidoreductase (cytochrome b_6/f complex). The dashed lines represent the membrane surface. The two quinone binding sites (Q_c and Q_z) are labeled and the heme groups (bars) of the cytochromes are shown. Redrawn after fig. 3 of Hauska (1985).

460

Considerably less is known about cyanobacterial cytochrome b_6/f complexes than is known about their purple bacterial and chloroplast homologues. Nonetheless, a complex similar to that isolated from spinach chloroplasts (see Fig. 9) has been isolated from *Anabaena variabilis* (Krinner et al. 1982; Hauska et al. 1983; Hauska 1985). This complex consists of four major polypeptides: cytochrome f (31 kDa); cytochrome b_6 (22.5 kDa); Rieske 2Fe–2S protein (22 kDa); and a polypeptide of 16 kDa. The *A. variabilis* complex was equally active in reducing *A. variabilis* cytochrome *c*-553 and plastocyanin (Krinner et al. 1982). The *A. variabilis* complex probably functions in both respiratory and photosynthetic electron transport, since no evidence for any other cytochrome complex was detected during isolation of the complex described above (Krinner et al. 1982). It should be noted that the *A. variabilis* cytochrome b_6 appears to be more similar to that observed in chloroplasts than to that observed in *Rps. sphaeroides* which is in turn more closely related to that observed in mitochondria (Hauska 1985).

Ho et al. (1979) have developed methods for isolating cytochrome f in large amounts from cyanobacteria. Amino acid sequence analyses have shown the amino termini of the cytochromes f from *Spirulina maxima* and *Aphanizomenon flos-aquae* are identical (Ho and Krogmann 1980). Moreover, these sequences are similar to that determined for spinach cytochrome f (Ho and Krogmann 1980). These similarities may indicate that cyanobacterial cytochrome f is also synthesized with a leader sequence which can direct the polypeptide across the thylakoid membrane to the thylakoid lumen. Little structural information is available for the other components of the cytochrome b_6/f complex in cyanobacteria. The chloroplast genes which have been cloned will allow the cloning of the genes encoding the cyanobacterial cytochrome b_6/f components. Preliminary results with *Nostoc* sp. PCC 7906 have recently been reported by Kallas and Malkin (1985). Heterologous hybridization analyses have been successfully employed in mapping and cloning the *petA*, *petB*, and *petD* genes from the cyanelle genome of *C. paradoxa* (Lambert et al. 1985; Bohnert et al. 1985; Ko et al. 1985). For the cyanobacterial genomes it will be interesting to see whether these genes are clustered as in *Rps. sphaerodies* or dispersed as in higher plants.

Photosystem I and the P700 Reaction Center

Photosystem I can be defined functionally as the membrane protein complex, with associated antenna chlorophylls, which catalyzes the light-driven transport of electrons from reduced plastocyanin (or cytochrome c-553) to oxidized ferredoxin ("plastocyanin-ferredoxin photo-oxidoreductase"). This complex contains a reaction center chlorophyll species (Em = +480 mV) referred to as "P700." This designation reflects the approximate wavelength maximum of bleaching observed when the primary donor chlorophyll is photo-oxidized (Kok 1956). The photobleaching that occurs at 700 nm provides a convenient assay for the isolation of chlorophyll–protein complexes which contain this reaction center chlorophyll. In addition to P700, PS I preparations typically exhibit three low-temperature electron paramagnetic resonance signals which are believed to arise from bound iron–sulfur centers. These centers, denoted A, B, and X, are believed to function as electron acceptors in the PS I complex. PS I complexes have been purified from several cyanobacteria, algae and higher plants. Their properties are detailed in a number of recent reviews (Thornber 1975; Okamura et al. 1982; Malkin 1982; Cogdell 1983; Glazer 1983; Haehnel 1984; Malkin et al. 1985; Rutherford and Heathcote 1985).

The properties of PS I preparations from higher plants (Mullet et al. 1980a; Lam et al. 1984a) or green algal chloroplasts (Anderson 1985) depend upon the type of detergent employed in their isolation and the severity of the detergent treatment. PS I particles prepared by extraction of pea thylakoids with a low concentration of Triton X-100 contain 110 chl a + b/P700 (chl a/chl b ratio = 18) and are composed of at least eleven polypeptides (68, 66, 24, 22.5, 21, 17, 16.5, 11.5, 11, and 10.5 kDa); Mullet et al., 1980a). Similar extraction of spinach chloroplast membranes results in the isolation of a particle which contains ten polypeptides and about 200 chl a + chl b per P700 center (Lam et al. 1984a), an antenna size similar to that estimated for the unfractionated thylakoids (Malkin et al. 1985). The difference in the chl : P700 ratios for the two preparations might be due to differences in the methods for determining the concentration of P700 (Malkin et al. 1985). On the basis of chemical modification and protease sensitivity studies, a model of the PS I reaction center complex of spinach has been proposed (Ortiz et al. 1985).

Further treatment of the complexes described above with detergents results in a particle with 65–100 chl /P700. This complex contained no chl b, and three polypeptides (24.5, 24, and 22.5 kDa for the pea complex; 23, 22, and 20 kDa for the spinach complex) were missing from this complex (Mullet et al. 1980a; Malkin et al. 1985). From these results and from PS I compositions observed in chl b-less mutants or developing chloroplasts, it was suggested that higher plant PS I consists of a chl a/chl b antenna and a core complex which contains the P700 reaction center (Mullet et al. 1980a, 1980b). This antenna complex has been isolated and characterized from *Chlamydomonas reinhardi* (Wollman and Bennoun 1982), *Codium* sp. (Chu and Anderson 1985), spinach (Anderson et al. 1983; Lam et al. 1984a; Lam et al. 1984b; Ortiz et al. 1984), pea (Haworth et al. 1983), and maize (Bassi et al. 1985). Recent immunological studies indicate that the chl a + chl b proteins associated with PS I in spinach are similar to the major chl a + chl b binding proteins associated with PS II (Evans and Anderson 1986). Core complexes, similar to that described above and similar to cyanobacterial PS I preparations (see below), have been isolated and characterized from a wide variety of higher plants and algae (e.g. see Bengis and Nelson 1975, 1977). It should be noted that the characteristic EPR signals of Fe–S centers X, A, and B are present in both types of complexes described above.

When thylakoids of higher plants are treated with the detergents sodium dodecylsulfate or lithium dodecylsulfate, the PS I chlorophyll–protein complex which can be isolated contains (per P700) 40–60 chl a, 1–2 molecules of β-carotene, and no Fe–S centers (Malkin 1982). The apoproteins of such complexes generally consist of one or two polypeptides with molecular masses of 60–70 kDa (Malkin 1982; Okamura et al. 1982). The actual number of 60–70 kDa polypeptides per P700 center does not appear to be well established. Nelson (cited in Okamura et al. 1982) concluded that PS I contained two 70 kDa polypeptides per P700 center from cross-linking studies. However, Vierling and Alberte (1983) suggested that there are four polypeptides associated with a single P700 reaction center.

The roles of the small molecular mass subunits of the PS I complex are largely unknown at present. There is some evidence to suggest that some of these may participate in binding some of the chlorophyll of the core complex (see Lundell et al. 1985). Efforts to assign the three Fe–S centers of Photosystem I to specific polypeptides have been attempted by several workers (see Discussion in Bonnerjea et al. 1985). Recent results with the Photosystem I complex of spinach suggest that only the 19 kDa and the 60 and 58 kDa polypeptides could play roles in binding or stabilizing the Fe–S centers A and B (Bonnerjea et al. 1985; Malkin et al. 1985). Golbeck and coworkers have reported that the Fe–S centers A and B are rapidly lost from the Photosystem I complex when the complex is treated with lithium dodecylsulfate, but that the large polypeptides of the Photosystem I complex retains an Fe–S center which they identified as the redox center X (Golbeck and Cornelius 1986; Warden and Golbeck 1986).

Vierling and Alberte (1983) have recently shown that antibodies prepared against the 58–62 kDa polypeptides of the barley P700 complex crossreact with similar size polypeptides from soybean, tobacco, petunia, tomato, corn, and *C. reinhardii*. Similar results were obtained by Nechustai et al. (1983). These workers showed that antibodies prepared against the 70 kDa polypeptide(s) of Swiss chard PS I crossreact with the analogous polypeptides from the PS I complexes of the green alga *C. reinhardii* and the cyanobacterium *Mastigocladus laminosus*. Antibodies prepared against the 24 kDa polypeptide of the Swiss chard complex also cross-reacted with polypeptides from the other two complexes. The results re-inforce the fundamental equivalence of the higher plant and cyanobacterial PS I complexes.

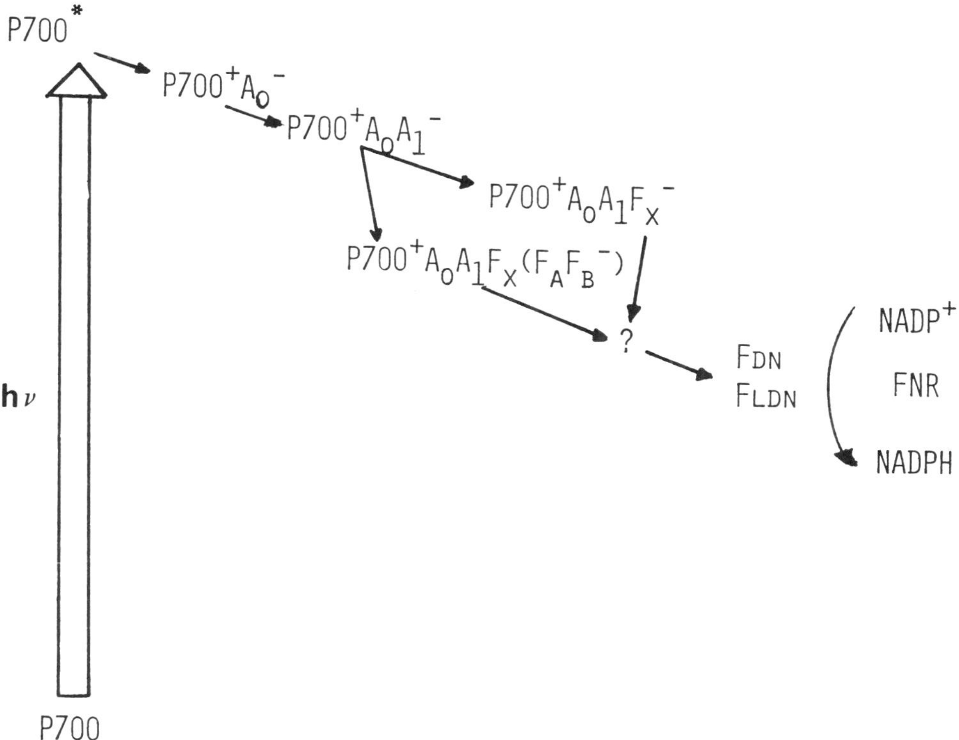

Fig. 10. A scheme for photochemistry and forward electron transfer of the Photosystem I/P700 reaction center. The complete pathway is not certain, which is why there are branches and question marks. A_0, the primary electron acceptor which is believed to be a monomeric chlorophyll a molecule; A_1, a secondary electron acceptor which is now believed to be a phylloquinone (vitamin K_1) molecule; F_X, F_A, and F_B denote the Fe–S centers X, A, and B, respectively; Fdn, soluble ferredoxin; Fldn, soluble flavodoxin; FNR, ferredoxin (flavodoxin)-$NADP^+$ oxidoreductase. Figure is redrawn from Fig. 1 of Rutherford and Heathcote (1985).

Figure 10 shows the postulated sequence of electron transfer events which are believed to occur when Photosystem I preparations are illuminated (Rutherford and Heathcote 1985). The identity of the P700 center is unknown. Although the EPR spectrum of oxidized P700 has been interpreted by some workers to arise from a special pair of chl *a* molecules, other investigators have suggested that a chemically modified monomeric chlorophyll *a* could better describe the properties of the P700 center (Rutherford and Heathcote 1985). Two such modified chlorophyll species have been reported. Dornemann and Senger (1984) have reported the occurrence of a modified chlorophyll *a*, 20-chloro, 13-hydroxy, chlorophyll *a*, in P700/Photosystem I preparations from the green alga *Scenedesmus obliquus*. This modified chlorophyll (Chl RCI) species was also present in stoichiometric amounts (Chl RCI : P700 :: 1:1) in spinach chloroplasts and the cyanobacteria *Synechococcus* sp. PCC 6301 and *Anabaena gracilis* (Katoh et al. 1985). Scheer et al. (1984) have also reported this chlorophyll species to occur in PS I/P700 particles from the cyanobacterium *Spirulina geitleri*. Another modified chlorophyll *a* species, chlorophyll *a'* (the C10 epimer of chlorophyll *a*), has been reported to occur in spinach chloroplasts and in PS I fractions derived from them (Watanabe et al. 1985). These workers found a stoichiometry of chl *a'* /P700 = 2, found similarities between the spectroscopic properties of oxidized chl *a'* and P700, and suggested that a chl *a'* dimer could account for the observed properties of the P700 center. Further complicating matters, Watanabe et al. (1985) found no evidence for the species detected by Dornemann and Senger (1984) and Scheer et al (1984). While the roles (if any) of these chlorophyll species in the electron transfer events of Photosystem I have not been precisely established, the discovery of these species is certain to lead to further characterization of the P700 center. It should be noted that PS I does not contain pheophytin *a* (Murata et al. 1986).

The primary reactions of Photosystem I exhibit a number of similarities to those which occur in the reaction centers of the purple bacteria and photosystem II (Rutherford and Heathcoat 1985). The primary charge separation apparently occurs between two chlorophyll-like molecules ($P700^+$ A_o^-) in less than 10 picoseconds. The optical difference spectrum of the acceptor A_o is similar to that of a chlorophyll *a* anion monomer, and the species is present in reaction centers in the ratio 0.94 A_o per 1.0 P700 (Mansfield and Evans 1985). Further charge separation occurs within a few hundred picoseconds by transfer of the electron from A_o^- to a non-chlorophyllous acceptor, A_1, which may be a quinone. Quinones have long been known to be components of Photosystem I (Thornber et al. 1976), but until recently have received little attention. However, Thurnauer and Gast (1985) and Mansfield and Evans (1985) have recently suggested that the species A_1 may be a quinone. Takahashi et al. (1985) and Schoeder and Lockau (1986) have recently reported that Photosystem I preparations from a thermophilic *Synechococcus* sp. and from *Anabaena variabilis*, respectively, each contain two moles of phylloquinone (vitamin K_1) per P700 center. If the A_1 acceptor is previously reduced, then the primary radical pair ($P700^+$ A_o^-) decays to the ground state within a few nanoseconds but can form a relatively long-lived triplet as well (Fig. 10). These events are clearly similar to those known to occur in purple bacterial reaction centers and Photosystem II (see Fig. 6 and above).

Other features of the photochemistry of Photosystem I clearly distinguish it from Photosystem II and the reaction centers of the purple bacteria. Unlike the latter there are iron-sulfur centers (A, B, and X) which are believed to act as tertiary electron acceptors as shown in Fig. 10. Although other interpretations of the properties of X have been made, it now appears likely that this electron carrier is an Fe–S center with a mid-point potential of — 705 mV (Rutherford and Heathcoat 1985; Golbeck and Cornelius 1986; Warden and Golbeck 1986). The reduction of this center is probably responsible for the optical signal "A_2" (Rutherford and Heathcote 1985); Warden and Golbeck 1986). The redox centers A (-550 mV) and B (-590 mV) have been characterized in detail (see Malkin 1982). Both are 4 Fe-4 S centers, and their reduction is detectable by EPR spectroscopy or by absorbance changes which occur at 430 nm. The electron transfer sequence among the centers X, A, and B is still not clear. The redox potentials of the three centers suggest that these centers could act in series: $X \rightarrow B \rightarrow A$. However, there is little or no kinetic evidence that centers A and B act in series or that center X is reduced prior to reduction of centers A or B. Centers A and B may act in parallel under physiological conditions (Rutherford and Heathcote 1985). In contrast to the situation in Photosystem II and the reaction centers of the purple bacteria, the stabilizing energy losses in these forward electron transfer steps is small and is spread over several steps. For a more detailed description of the electron transfer events for Photosystem I, see Malkin (1982), Haehnel (1984), or Rutherford and Heathcote (1985).

In cyanobacterial cells, the antenna size and the composition of cyanobacterial PS I do not vary with growth conditions (Mimuro and Fujita 1977; Myers et al. 1980; Manodori and Melis 1984). Myers et al. (1980) found that, under all growth conditions tested, the ratio of PS I chl *a*/P700 was 123 in either whole cells or thylakoid membranes of *Synechococcus* 6301. Mimuro and Fujita (1977) reported a value of 130 chl *a*/P700 for PS I in *Anabaena variabilis*. PS I/P700 preparations have been generated from an extensive list of cyanobacteria (for a comprehensive listing, see Hladik and Sofrova 1983). As pointed out by Lundell et al. (1985) it is nearly impossible to compare the results that have been obtained by various workers because of experimental variations in detergent employed for membrane solubilization, purification techniques, and methods employed for the measurement of P700. Nonetheless, it is clear that all cyanobacterial PS I preparations are likely to be similar to that described by Lundell et al. (1985) for *Synechococcus* 6301. The purified PS I complex contained 130 $\pm$ 5 chl *a*/P700, displayed the EPR signals characteristic of Fe–S centers A, B, and X, and had a protein/chl ratio of 2.9–3.1. The PS I complex was composed of polypeptides of 70, 18, 17.7, 16, and 10 kDa; the most likely molar ratio of the polypeptides, determined by uniform 14-C labeling, was 4:1:1:1:2. No evidence for a peripheral chlorophyll antenna complex was obtained, and the similarity of the chl *a* : P700 ratio detected *in vivo* and in vitro suggests that such complexes do not exist in the cyanobacteria. The 70 kDa polypeptide was sometimes resolved into two discreet polypeptides which were very similar in mobility on SDS-PAGE. In addition the complex contained 16 moles of carotenoids, 13.7 $\pm$ 1.0 g-atoms of Fe, and 12.2 $\pm$ 1.1 g-atoms of labile sulfide per 130 moles of chl *a*. The PS I particles, when negatively stained, had the appearance of prolate ellipsoids 18 x 8 nm (Williams et al. 1983). When this PS I complex was treated with SDS, approximately 40 chl and all of the small polypeptides and Fe were released. Although the Fe–S centers A, B, and X were absent, the SDS-treated complex was still capable of P700 photooxidation; however, the rate of oxidation was 100-fold slower. The complex described by Lundell et al. (1985) is clearly similar to the core complexes derived from higher plant chloroplasts (see discussion above).

Katoh and coworkers have obtained results very similar to those described above for *Synechococcus* 6301 with PS I preparations from a thermophilic *Synechococcus* sp. (Nakayama et al. 1979; Takahashi et al. 1982; Takahashi and Katoh 1982; Takahashi et al. 1985). Their PS I preparations contain 70 chl/P700 and consisted of four or five polypeptides (62, 60, 14, (13), and 10 kDa). The complexes also contained 10–14 g-atoms Fe, 10 moles of β-carotene, 2 moles of phylloquinone (vitamin K_1), and Fe–S centers A, B, and X. The possible identity of phylloquinone with the electron acceptor A_1 was discussed above. Sonoike and Katoh (1986) have recently reported the isolation of the 60 kDa polypeptide of this complex which retained about 60% of the chl *a* of the original PS I complex but which was no longer photochemically active. Although these authors suggest that this protein–chl *a* complex functions only as an intrinsic antenna complex in PS I, there is little evidence to support such a contention. Preparations similar to those derived from the two *Synechococcus* sp. described above have been obtained from many cyanobacteria including *Synechococcus cedrorum* (Newman and Sherman 1978), *Mastigocladus laminosus* (Nechustai et al. 1983), *Phormidium luridum* (Reinman and Thornber 1979; Huang et al. 1984), *Phormidium laminosum* (Stewart 1980); *Spirulina platensis* (Hoarau et al. 1977), *Synechococcus lividus* (Oquist et al. 1981), Nostoc sp. (Ruskowski and Zilinskas 1980), and *Synechocystis* sp. PCC 6714 (Bullerjahn et al. 1985). Interestingly, the PS I complex from *Prochloron* sp. is similar in polypeptide composition to those isolated from cyanobacteria; no evidence for a chl *a* + chl *b* antenna complex has yet been obtained for this organism (Schuster et al. 1985).

Several reports indicate that in higher plants the 60–70 kDa apoprotein(s) of PS I are synthetized on chloroplast ribosomes (Green 1980; Zielinski and Price 1980; Ortiz and Stutz 1980; Nechustai et al. 1981). Some of the smaller polypeptides of PS I are also synthetized in the chloroplast but others are translated in the cytosol and presumably are encoded in the nucleus (Nechustai et al. 1981; Mullet et al. 1982; Hoyer-Hansen et al. 1985). The apoproteins of the peripheral chl *a*/*b* antenna are among those believed to be encoded in the nucleus (Mullet et al. 1982). Stable mRNAs for the P700 chlorophyll *a* apoprotein (67 kDa) and the 22 kDa subunit of the spinach PS I core complex have been detected by *in vitro* translation of chloroplast and cytosolic RNA fractions, respectively (Westhoff et al. 1983b). Hybrid selection mapping and DNA programmed, *in vitro* transcription-translation has been used to map the coding sequences for the P700 chlorophyll *a* apoprotein(s) on the spinach (Westhoff et al. 1983b) and pea (Smith and Gray 1984) chloroplast genomes.

A major contribution to the understanding of PS I/P700 structure-function has been made by Fish et al. (1985a, 1985b). These workers have recently reported the nucleotide sequences for the two presumed, chlorophyll-binding apoproteins of the PS I/P700 complex in maize. The maize chloroplast genome contains two light-inducible genes, denoted *psaA* and *psaB*, that encode polypeptides of 83.2 and 82.5 kDa that are 45% homologous to one another. Both genes appear to be expressed, as a 4.9 kb mRNA is produced from this region of the chloroplast genome. The light-stimulated expression of this mRNA in fact allowed the isolation of these "photogenes". Two lines of evidence indicate that *psaA* codes for a PS I apoprotein. Firstly, antibodies to the barley CP I complex (PS I minus small polypeptides) immunoprecipitate polypeptides produced from an *in vitro* transcription translation reaction in which the cloned *psaA* gene served as template. Secondly, antibodies prepared against a synthetic peptide, with a sequence deduced from the DNA sequence of an unconserved, hydrophilic segment of the *psaA* gene, cross-reacted with PS I polypeptides of maize and pea. Fish et al. (1985b) have presented a model for the two gene products and suggest that both polypeptides are incorporated into the PS I reaction center. Although some workers have suggested that PS I/P700 centers were composed of two molecules of a single 60–70 kDa protein species (see Okamura et al. 1982), reports that there are two apoproteins of this size have consistently appeared (e.g., Mullet et al. 1980a). The deduced amino acid sequences of the spinach *psaA* and *psaB* genes was used by Nelson (1985) to predict that, if the *psaB* gene product were a subunit of the PS I complex, then a unique 30.7 kDa cyanogen bromide fragment should be produced from cleavage of the intact PS I complex. A fragment of this size was indeed produced confirming that the *psaB* product is an integral component of the PS I/P700 reaction center complex. An interesting feature of the deduced amino acid sequences is that the product of *psaB* would have only two cysteine residues. These two cysteines are conserved in the deduced amino acid sequence of the *psaA* gene, although two additional cysteines appear in hydrophobic (possibly membrane-spanning) regions of this polypeptide. It is tempting to speculate that the conserved cysteine residues of the two polypeptides could together participate in the formation of the Fe–S centers which function as electron acceptors in PS I/P700. The results of Golbeck and Cornelius (1986) suggest that these cysteines could participate in the formation of Fe–S center X.

Similar genes have been demonstrated on the spinach (Herrmann et al. 1985), pea (Berends et al. 1986; Lehmbeck et al. 1985), tomato (Phillips 1985) and *C. reinhardii* chloroplast genomes (Kuck 1985). The complete nucleotide sequence of the pea *psaA* and *psaB* genes has been determined; the translation products of these genes are 89% and 95% homologous to the maize polypeptides (Lehmbeck et al. 1985). The exceptional degree of sequence conservation for the two genes is additional proof that both genes are expressed. Interestingly, the two genes in *C. reinhardii* are not adjacent to one another but are separated by 45 Kbp on the chloroplast genome (Kuck 1985). In the cyanobacteria-like cyanelles of *C. paradoxa* the *psaA* and *psaB* genes are adjacent to one another (Lambert et al. 1985; Bohnert et al. 1985).

Heterologous hybridization experiments, using cloned DNA fragments from maize or spinach *psa* genes, indicate that cyanobacterial *psa* genes are highly homologous to those which occur in higher plants (Bryant et al. 1986). In the cyanobacterium *Synechococcus* sp. PCC 7002, the *psaA* and *psaB* genes are adjacent to one another although the intragenic region is larger than found in chloroplasts. The *psaA* and *psaB* polypeptides, determined by translation of the partially sequenced genes, are greater than 80% homologous to the equivalent polypeptides of maize chloroplasts (D. A. Bryant and A. Cantrell, unpublished results). The predicted sizes for the *psaA* and *psaB* polypeptides are nearly identical to those established for maize and pea. Since the cyanobacterial PS I particle contains four large polypeptides per P700 center, the PS I particle described by Lundell et al. (1985) would be predicted to have the subunit structure α_2, β_2, γ, δ, ϵ, ζ_2. The calculated molecular mass of the entire complex would be about 520 000–530 000 Da. Although this value is somewhat higher than that estimated from electron microscopy of the isolated PS I complex (Williams et al. 1983), it is nonetheless consistent with the protein/chl *a* (3.0) and chl *a*/P700 (130 : 1) ratios calculated by Lundell et al. (1985).

ATP Synthase

Mitchell's chemiosmotic theory (Mitchell 1968) predicted that the essential element of energy-transducing membranes is the electrochemical proton gradient across the membrane. The proton gradient (or proton-motive force) for ATP synthesis can be generated by light-driven electron transport (photophosphorylation) or respiratory electron transport (oxidative phosphorylation). Cyanobacteria are apparently capable of coupling either type of process to ATP synthesis (Binder 1982). Falkner et al. (1976) showed that a light-driven movement of protons occurs from the cytoplasm to the intrathylakoidal space in *Synechococcus* 6301. The gradient formed is comparable in size to that observed for chloroplasts, is sensitive to the uncoupling agent CCCP, and consists primarily of a pH gradient with nearly no electrical potential component. Similar results have been obtained with other cyanobacteria (see Binder 1982). As occurs in other procaryotic cells (Crofts and Wraight 1983), light or oxygen energization of spheroplasts or whole cells causes protons to be extruded into the medium (Scholes et al. 1969; Erber et al. 1986; Peschek et al. 1986). This indicates that topologically the intrathylakoidal space must be equivalent to extracellular space.

The chloroplast ATP synthase (coupling factor) is an essential component of the photosynthetic apparatus and is one of the major protein complexes of the thylakoid membrane (for reviews, see Strotmann and Bickel-Sandkotter 1984; McCarty 1985; Merchant and Selman 1985). The higher plant enzyme is a complex, multi-subunit protein which is remarkably similar to the enzyme which occurs in *E. coli* (Futai and Kanazawa 1983; Walker et al. 1984), the photosynthetic bacterium *R. rubrum* (Falk et al. 1985), the cyanobacteria *Synechococcus* sp. PCC 6301 (Walker and Tybulewicz 1985; Cozens et al. 1986 and *Anabaena* sp. PCC 7120 (S. Curtis, personnal communication), or even beef heart mitochondria (Racker 1981). Chloroplast ATP synthase is composed of a well-characterized extrinsic complex (CF1) of five polypeptides: α, β, γ, δ, and ϵ; the stoichiometry of these five polypeptides is generally accepted to be $\alpha_3:\beta_3:\gamma:\delta:\epsilon$ (Moroney et al. 1983; Tiedge et al. 1985). Although no chloroplast enzyme has yet been crystallized, a low resolution structure of the beef heart mitochondrial F_1-ATPase has been reported (Amzel et al. 1982). A model for the cyanobacterial and higher plant coupling factor is shown in Fig. 11. The intrinsic, membrane-bound portion of the complex (CF_o) is much less well characterized and until recently was generally believed to be composed of three subunits (Nelson et al. 1980). However, other studies suggesting that the CF_o complex consists of four subunits (Pick and Racker 1979) have recently been confirmed (Westhoff et al. 1985a; Cozens et al. 1986; Hennig and Herrmann 1986). The presently accepted subunit structure for chloroplast CF_o is as follows: Subunit I = *E. coli* subunit b; Subunit II, no. *E. coli* homologue (Although no *E. coli* homologue exists for this component, cyanobacteria possess two genes encoding *b* subunits: *b* and *b'*; see below); Subunit III = *E. coli* subunit c; and Subunit IV = *E. coli* subunit a (Cozens et al. 1986). The stoichiometry of the subunits for the CF_o is not known with certainty, but the ratio, Subunit I: Subunit II: Subunit III: Subunit a = 1:1:6–10:1, is a reasonable approximation.

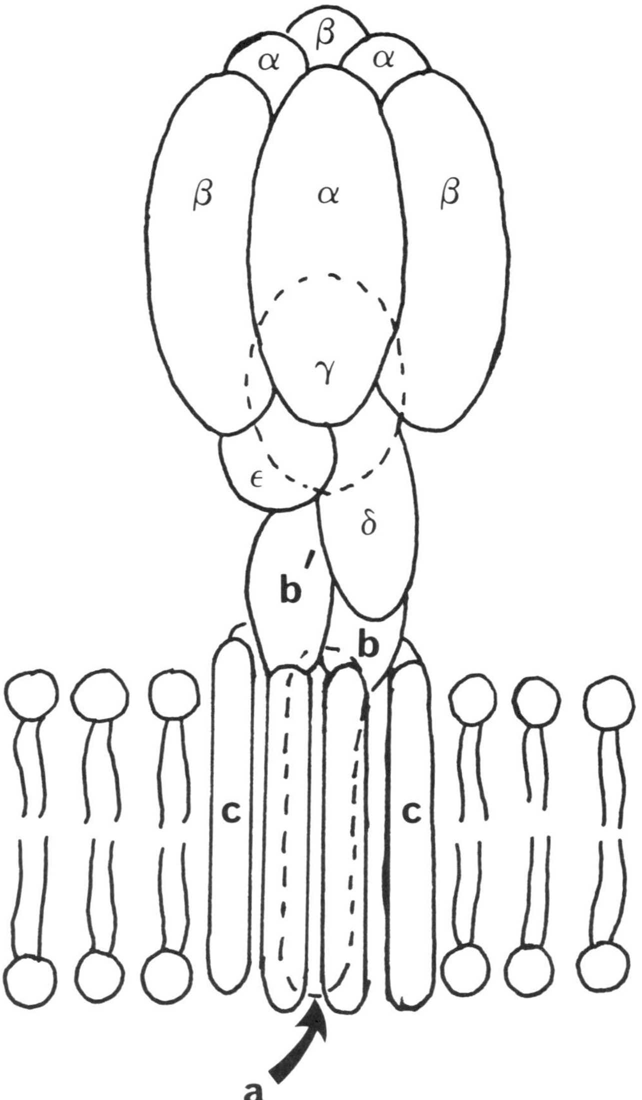

FIG. 11. Topographical model for the cyanobacterial CF_0CF_1 ATP synthase based upon the similarity of the cyanobacterial and higher plant complexes to that of *E. coli*. The CF_0 subunits a, b, and c are the products of the *atpI*, *atpF*, and *atpH* genes, respectively and correspond to higher plant subunits IV, I, and III, respectively. The subunit b' could be equivalent to higher plant subunit II, the product of the nuclear-encoded *atpG* gene. The model is redrawn after Futai and Kanazawa (1983) and is not drawn to scale.

In *E. coli* the genes encoding all eight ATP synthase subunits form a single operon, the *unc* or *atp* operon (Walker et al. 1984). In the photosynthetic bacteria *Rhodopseudomonas blastica* (Tybulewicz et al. 1984) and *Rhodospirillum rubrum* (Falk et al. 1985), the genes encoding the F_1 portion of the complex are clustered and arranged in the same order as in *E. coli* (δ, α, γ, β, ϵ). However, the components of the F_0 complex are not associated with this locus. In higher plants the subunits γ, δ, and II are synthesized in the cytoplasm as larger precursors and are presumably encoded by nuclear DNA (Westhoff et al. 1981, 1985a). The genes encoding these subunits have not yet been isolated and characterized. Subunits α, β, ϵ, I, III and

IV are synthetized on chloroplast ribosomes and are encoded by chloroplast DNA. Examples of all chloroplast-encoded genes (for locus designations, see Table 4) have been cloned and nucleotide sequences have been determined for all (for examples, see Westhoff et al. 1981; Krebbers et al. 1982; Zurawski et al. 1982b; Howe et al. 1982; Alt et al. 1983b; Huttly and Gray 1984; Howe et al. 1985; Bird et al. 1985; Walker and Tybulewicz 1985; Passavant and Hallick 1985; Westhoff et al. 1985a; Cozens et al. 1986; Hennig and Herrmann 1986). The chloroplast-encoded genes are arranged in two clusters: the genes encoding the β and ϵ subunits (*atpB* and *atpE*) are adjacent and those encoding the α, I, III, and IV subunits (*atpA*, *atpF*, *atpH* and *atpI*) are adjacent (Krebbers et al. 1982; Howe et al. 1985; Bird et al. 1985; Westhoff et al. 1985a; Cozens et al. 1986; Hennig and Herrmann 1986; Zurawski et al. 1986).

The α and β subunits have partially homologous amino acid sequences suggesting that they arose by a gene duplication event (Walker et al. 1984; Howe et al. 1985). Each subunit has sequences which are characteristic of adenine nucleotide binding sites, and each subunit has been shown to bind such nucleotides (Selman and Merchant 1985). The β subunits from many sources are highly conserved ($\sim 70\%$ homology) and are believed to carry the catalytic site(s). The α subunits from various sources are less highly conserved ($\sim 55\%$ homology) but may also form a portion of the catalytic site. The γ subunit is believed to function as a proton gate; only bacterial and cyanobacterial (see below) amino acid sequence data is presently available, but these sequences are homologous ($\sim 35\%$). The δ and ϵ subunits appear to play a role in attachment of CF_1 and CF_0. The δ subunit stoichiometry is uncertain (Selman and Merchant, 1985) since this subunit is rather easily lost from the CF_1 complex. The amino acid sequence homology among the δ subunits and the ϵ subunits is rather limited (25–30%; Walker and Tybulewicz 1985). A more detailed description of the properties and functions of the CF_1 subunits can be found in Merchant and Selman (1985).

Although studies of the cyanobacterial coupling factor have been limited, the cyanobacterial enzyme is clearly similar in structure and function to the much more highly characterized chloroplast enzymes. Binder and Bachofen (1979) have isolated the coupling factor from the thermophile *Mastigocladus laminosus* and found four of the five subunit types seen in the higher plant enzyme (δ subunit missing). Binder et al. (1980) were also successful in isolating a coupling factor complex (CF_1–CF_0) with ATP–P_i exchange activities from *M. laminosus*. The ATP synthase of *Synechococcus* sp. PCC 6716 has also been isolated and described (Lubberding et al. 1983; van Walraven et al. 1983, 1984). Hicks and Yocum (1986a, 1986b) recently reported the purification of the *Spirulina maxima* enzyme. The highly purified enzyme was comprised of subunits with apparent molecular masses of 53.4, 51.6, 36, 21.1, and 14.7 kDa; these values are quite similar to those observed for the spinach CF1. ATP synthase has also been isolated from the cyanelles of *C. paradoxa*; as found for the enzyme from *M. laminosus*, the δ subunit had been lost. The molecular masses of the remaining four subunits of the CF_1 portions of the complex were similar to those observed for higher plant enzymes (Klein et al. 1983). The *M. laminosus* DCCD-binding protein, which forms the proton channel of the CF_0 portion of the complex, has been sequenced and shown to be similar to its spinach chloroplast homologue (Binder 1982).

The genes encoding the ATP synthase subunits of *Synechococcus* sp. PCC 6301 (J. Walker, personnal communication; see also Walker and Tybulewicz 1985 and Cozens et al. 1986) and *Anabaena* sp. PCC 7120 (S. Curtis, personnal communication) have recently been cloned and sequenced. In both organisms the genes are arranged into two unlinked clusters which are similar to those observed in chloroplast genomes (see above). The *atpB* and *atpE* genes are adjacent and form one cluster; the genes encoding subunits a, c, b, b', δ, α and γ are adjacent and occur in the same order as they do in *E. coli*. Both cyanobacteria are unusual in having a duplicated and divergent copy of the gene encoding subunit b (= subunit b'). The cyanobacterial and pea chloroplast subunits a are 70% homologous in amino acid sequence and are much more similar to one another than either is to the equivalent polypeptides of *E. coli* or bovine mitochondria (Cozens et al. 1986). Finally, the genes encoding the cyanelle-encoded *atpA, atpB, atpE*, and *atpH* genes of *C. paradoxa* have been mapped and cloned using heterologous hybridization probes (Lambert et al. 1985; Bohnert et al. 1985; Ko et al. 1985). The arrangement of these genes into two clusters is similar to that observed in both cyanobacterial and chloroplast genomes. A determination of whether the γ and/or δ subunits of *C. paradoxa* are encoded within the cyanellar or the nuclear genomes should prove most interesting and could provide additional insight into the origins of chloroplasts.

Other Electron Transport Components

Plastoquinone

Synechococcus sp. thylakoids contain approximately seven plastoquinone molecules per PS II reaction center (Takahashi and Katoh 1986). The function of plastoquinone in cyanobacterial and higher plant thylakoids is apparently identical. Plastoquinone is the lipid-soluble molecule believed to couple electron transport between PS II and the cytochrome b_6/f complex. Millner and Barber (1984) have recently reviewed the role of plastoquinone as a mobile electron carrier in the photosynthetic membranes of chloroplasts. They suggest that plastoquinone might be located primarily within the fluid bilayer-midplane region and that "tunnelling" along the midplane could allow for extremely rapid diffusion of this electron carrier. Since plastoquinone participates in vectorial transport of reducing equivalents across the membrane, it should be noted that either the quinol head group does occasionally move to the membrane surface or that electron/proton exchanges occur at quinone-binding sites on PS II and the cytochrome b_6/f complexes buried within the hydrophobic lipid matrix (Millner and Barber 1984).

A second role for plastoquinone in photosynthesis has recently been postulated. It has been suggested (Nugent et al. 1982; O'Malley and Babcock 1983; see Ghanotakis and Yocum 1985) that Z, the primary donor to P680, may be a plastoquinol cation radical. Consistent with this suggestion is the finding that 3–3.5 molecules of plastoquinone are found tightly associated with chloroplast PS II particles which are highly active in oxygen evolution (Murata et al. 1984; Yamamoto et al. 1984b). *Phormidium laminosum* PS II particles exhibit Signal IIvf EPR signals from Z^+ which are similar to those observed in chloroplast PS II preparations. If Z^+ is due to a plastoquinol cation, then cyanobacteria probably also employ this quinone species as the primary donor to P680.

Plastocyanin/Cytochrome C-553

In cyanobacteria and algae, the source of electrons for the reduction of oxidized P700 is reduced plastocyanin or cytochrome *c*-553. In many, perhaps most, cyanobacteria, these small proteins with similar redox potentials are alternative choices for electron transport. The conditions which determine whether plastocyanin or cytochrome *c*-553 is synthesized will be discussed separately (see DYNAMIC ASPECTS, below). The functioning of these carriers presumably mimics the situation observed in higher plants, in which plastocyanin is localized in the intrathylakoid space and serves as a mobile carrier of electrons from the cytochrome b_6/f complex to the P700 centers of PS I (Haehnel 1984; Bottin and Mathis 1985). Cross-linking studies have shown that spinach plastocyanin binds to the cytochrome *f* portion of the plastoquinol-plastocyanin oxidoreductase complex (Hauska et al. 1984). Two recent studies have examined the electron transfer reactions mediated by plastocyanin.

Matsuura and Itoh (1985) found that the flash-induced oxidation of cytochrome f was highly salt dependent. These workers suggest that plastocyanin reacts collisionally with the PS I reaction center and with cytochrome f in a manner that is controlled by the surface electrostatic potential. Bottin and Mathis (1985) studied the reduction kinetics of flash photo-oxidized P700 in pea chloroplasts and purified spinach PS I particles. These authors report that plastocyanin is indeed the immediate electron donor to oxidized P700, and that the plastocyanin which is responsible for the donation is bound directly to the PS I reaction center. Chemical cross-linking studies would be desirable to prove this assertion and to indicate to which of the many polypeptide subunits of PS I plastocyanin binds. The differences obtained could either be due to differences in reaction conditions or inherent differences in the reactions of plastocyanin with cytochrome f and PS I. Further investigations will be required to clarify this difference. Nanba and Katoh (1985) tested the ability of cytochrome c-553 to function as a mobile electron carrier. These authors concluded that this cytochrome could shuttle electrons between the two complexes in *Synechococcus* sp. but that the cytochrome probably does not diffuse over very large distances to couple electron transport.

Plastocyanin is a blue copper-containing protein whose redox potential is approximately $+370$–380 mV (Katoh et al. 1962; Takabe et al. 1983). The three-dimensional structure of a higher plant plastocyanin has been determined at 1.6 Å resolution (Guss and Freeman 1983; Colman et al. 1978). The copper atom is co-ordinated by a cysteine sulfur atom, a methionine sulfur atom, and two imidazole (histidine) nitrogen atoms. X-ray crystallographic analysis of apoplastocyanin indicate that the irregular geometry of the copper site is apparently imposed upon the metal atom by the polypeptide moiety (Garrett et al. 1984). The complete amino acid sequences of sixteen plastocyanins, mostly from eucaryotic algae and higher plants, have been determined; amino-terminal sequences have been determined for more than 80 proteins (Ramshaw 1982; Guss and Freeman 1983). Aitken (1975, 1978)) has reported the complete amino acid sequence of plastocyanin from *Anabaena variabilis* and a partial sequence for the *Plectonema boryanum* protein. These sequences are highly homologous to both higher plants ($\sim 40\%$) and green algae ($\sim 50\%$). Homology in the vicinity of the copper-binding site is much higher ($\sim 90\%$). In higher plants plastocyanin is encoded in the nucleus (*petE*) and is translated on cytoplasmic ribosomes as a higher molecular mass precursor which is processed to mature plastocyanin upon import into the chloroplast (Grossman et al. 1982). Although no cyanobacterial plastocyanin gene has yet been cloned, Smeekens et al. (1985b) have isolated and sequenced a cDNA clone from *Silene pratensis*.

Amino acid sequences for a variety of cyanobacterial cytochromes c-553 have been determined (Ambler and Bartsch 1975; Aitken 1976, 1977, 1979; see Ho and Krogmann 1984). However, the three-dimensional structure of this cytochrome class has not yet been determined. The gene encoding the cytochrome c-553 apoprotein has not yet been cloned. However, an oligonucleotide probe, synthesized from the amino acid sequence of a conserved region of the protein, gave strong, specific hybridization signals when used to probe genomic Southern blots of several cyanobacterial DNAs including *Synechococcus* strains 7002 and 6301 (D. A. Bryant and V. Stirewalt, unpublished results). This cytochrome is widely distributed among cyanobacteria and may function uniquely in P700 reduction in red algae (Evans and Krogmann 1983) and cyanelles (Jaynes and Vernon 1982). Although Aitken (1978) has reported the immunological detection of plastocyanin in the red alga *Porphyridium cruentum*, Evans and Krogmann (1983) found no plastocyanin and only cytochrome c-553 in this alga. A curious feature of cyanobacterial electron donors to the PS I P700 center is the considerable variation in isoelectric points (pI 3.8–9.3) which occurs for these proteins. Ho and Krogmann (1982, 1984) have discussed the consequences of this variability with respect to the function of these proteins.

Ferredoxin, Flavodoxin, and the Reducing Side of PS I

The powerful reducing equivalents generated by PS I are delivered to a variety of enzymes in cyanobacterial cells via the interchangeable, soluble electron carriers ferredoxin and flavodoxin. The conditions which cause one or the other of these carriers to be synthesized will be discussed separately (see DYNAMIC ASPECTS, below). Primary destinations of reducing equivalents include $NADP^+$ via ferredoxin $^-NADP^+$ oxidoreductase; nitrate via nitrate reductase; nitrate via nitrite reductase; dinitrogenase gas via nitrogenase; and possibly sulfate and organic molecules by as yet undocumented reactions.

Ferredoxin is a small (11 kDa) non-heme iron–sulfur (2 Fe–2S) protein with a redox potential which can vary from -350 to -450 mV. Ferredoxin was the first photosynthetic electron carrier to be purified and has been characterized in considerable detail. The amino acid sequences of ferredoxins from a wide variety of cyanobacteria, eucaryotic algae, and higher plants have been determined (see Matsubara and Hase 1983). The three-dimensional structure of *Spirulina platensis* ferredoxin was recently determined at 2.5 Å resolution (Tsukihara et al. 1981). Some cyanobacteria synthesize two distinct ferredoxins which have been shown to differ in mid-point potential and amino acid sequence (Hase et al. 1982; Cohn et al. 1985). While the significance of these observations has remained obscure, Schrantmeir and Bohme (1985) have recently reported that the heterocysts of *Anabaena variabilis* contain a ferredoxin whose properties differ significantly from those of the vegetative cell ferredoxin. The heterocyst ferredoxin appears to be specifically adapted for direct electron donation to nitrogenase. In higher plants ferredoxin is encoded in the nucleus, translated in the cytoplasm as a precursor, and imported and processed during movement into the chloroplast (Smeekens et al. 1985a). These workers have also reported the cDNA cloning and nucleotide sequence of the ferredoxin of *Silene pratensis*. Cloning of a cyanobacterial ferredoxin gene has not yet been reported, but the higher plant gene should easily allow the cloning of the closely related cyanobacterial gene.

Cyanobacterial flavodoxins have molecular masses of 18–20 kDa and carry a single FMN prosthetic group (Fitzgerald et al. 1977). The amino acid sequence of a portion of the *Synechococcus* 6301 protein, as well as the crystal structure of this protein at 2.5 Å resolution, has been reported by Smith et al. (1983) and Tarr (1983). The redox potential (-220 mV) is somewhat more positive than that observed for typical ferredoxins. In some cyanobacteria, two ferredoxins and flavodoxin can be synthesized simultaneously, although the ferredoxins predominate in Fe-replete conditions (Hutber et al. 1977). Although no gene encoding the apoprotein of a cyanobacterial flavodoxin has been cloned, the *nifF* gene of *Klebsiella pneumoniae*, which encodes the flavodoxin which is the electron donor to nitrogenase, has recently been cloned and sequenced (Drummond 1985). The protein product of this *nifF* gene is clearly homologous (~33%) to the flavodoxin of *Synechococcus* sp. PCC 6301 (Smith et al. 1983).

One of the primary products of photosynthetic electron transport is the NADPH required for carbon dioxide reduction. The enzyme responsible for the delivery of electrons to $NADP^+$ is the flavoprotein ferredoxin-$NADP^+$ oxidoreductase (FNR). This protein has a molecular mass of approximately 35 kDa and has been isolated from a variety of cyanobacteria (Susor and Krogmann 1966; Wada et al. 1983) and higher plants (e.g. Shin et al. 1963). The amino acid sequence of the enzymes from *Spirulina* sp. (Yao et al. 1984) and spinach (Karplus et al. 1984) are homologous (55%) throughout their entire length. The crystal structure of the spinach enzyme has been determined at 3.7 Å resolution (Sheriff and Herriott 1981). The enzyme is kidney-shaped and has two lobes, one of which contains the $NADP^+$ binding site. Affinity labeling studies of the spinach enzyme indicate that the $NADP^+$ binding domain occurs in the carboxyl terminal domain of the protein (Chan et al. 1985). The second lobe is presumed to bind the FAD prosthetic group. In higher plants FNR is translated on cytoplasmic ribosomes as a larger molecular mass precursor and is presumably encoded by nuclear DNA (Grossman et al. 1982). At this time neither a higher plant nor a cyanobacterial FNR gene has been cloned. However, a comparison of the spinach and *Spirulina* sp. amino acid sequences suggests that it should be quite straight-forward to produce oligonucleotide hybridization probes to facilitate the cloning of the gene encoding the FNR apoprotein.

The chloroplast ferredoxin-NADP$^+$ oxidoreductase is rather tightly bound to the stromal surface of the thylakoid membrane; the protein apparently binds to specific sites possibly via an electrostatic mechanism (Carrillo and Vallejos 1982). The interaction of the enzyme with the thylakoid had marked effects on the activity of the enzyme (Carrillo and Vallejos 1983). The enzyme can be isolated as two distinct complexes with membrane proteins. The spinach enzyme can be isolated as a complex with a 17.5 kDa intrinsic membrane protein (Vallejos et al. 1984). The stoichiometry of the peptides in this complex is approximately 1:3, which suggests that the membrane protein may occur as a trimer (Ceccarelli et al. 1985). Nozaki et al. (1985) have reported that a second small protein, "connectein," is also required for attachment of the ferredoxin-NADP$^+$ oxidoreductase molecule to the thylakoid membrane.

The primary role of the ferredoxin-NADP$^+$ oxidoreductase has been considered to be the transfer of reducing equivalents from the one-electron carrier ferredoxin to the two-electron coenzyme acceptor NADP$^+$. However, some studies have suggested that the enzyme might also participate in cyclic electron transport, although this point is controversial (Shahak et al., 1981). Hind and coworkers recently reported that the ferredoxin-NADP$^+$ oxidoreductase can be isolated as a peripheral component of the plastoquinol-plastocyanin oxidoreductase (cytochrome b_6/f; see above) complex (Clark and Hind 1983; Clark et al. 1984). Although this complex contains a polypeptide (Subunit IV) of ~ 17 kDa molecular mass, this polypeptide is not the intrinsic protein which was described by Vallejos et al. (1984) (Coughlan et al. 1985; Ceccarelli et al. 1985). While one could imagine that positioning the enzyme on the plastoquinol-plastocyanin oxidoreductase could permit the enzyme to regulate electron flow between the non-cyclic and cyclic pathways, the significance of the association with plastoquinol-plastocyanin oxidoreductase is not clear at present. This problem requires further study for resolution.

Dynamic Aspects of the Cyanobacterial Photosynthetic Apparatus

One of the most remarkable properties of the cyanobacteria is their ability to alter the composition of their photosynthetic apparatus in response to environmental stresses. Although most detailed studies have been performed with laboratory cultures, it is clear that these adaptations occur in naturally occurring cyanobacterial populations. Krogmann and coworkers (Ho et al. 1979; Ho and Krogmann 1982) have noted changes in electron transport components in naturally occurring cyanobacterial blooms which suggest responses to either copper or iron limitation. Castenholtz (personal communication) has found that naturally occurring mats of thermophilic *Synechococcus* sp. can adapt to changes in light intensity in a matter of hours. It is presumed that such adaptability confers special selective advantages to those organisms which can make the appropriate response(s). The range and nature of these responses indicate that this aspect of cyanobacterial physiology should provide a fertile research area for those that are interested in the control of gene expression in photoautotrophic procaryotes.

The composition of the photosynthetic apparatus of cyanobacteria is known to be affected by numerous and varied environmental factors. These include temperature (Halldal 1958; Goedheer 1976; Anderson et al. 1983); carbon dioxide concentration (Eley 1971; Goedheer 1976; Miller and Holt 1977; Yamanaka and Glazer 1981; Manodori and Melis 1984); phosphate availability (Ihlenfelt and Gibson 1975; Stevens et al. 1981); sulfate availability (Schmidt et al. 1982); nitrogen availability (Allen and

Smith 1969; Lau et al. 1977; Yamanaka and Glazer 1980); copper availability (Sandmann and Boger 1980; Sandmann and Boger 1983); iron availability (Oquist 1971, 1974a, 1974b; Guikema and Sherman 1983; Hardie et al. 1983; Sandmann and Malkin 1983; Sandmann 1985); light intensity (Myers and Kratz 1955; Halldal 1958; Allen 1968; Oquist 1974c; Bryant 1982; Raps et al. 1983; Lonneborg et al. 1985; Wyman and Fay 1986a); and light wavelength (Jones and Myers 1965; Bogorad 1975; Myers et al. 1978, 1980; Tandeau de Marsac 1983; Fujita et al. 1985; Wyman and Fay 1986b). Although the effects of many of these factors on phycobiliproteins and phycobilisomes are known in some detail, far less is known of how these factors alter the relative amounts of the other components of the photosynthetic apparatus: the PS I and PS II reaction centers, electron transport components, and ATP synthase. While total chlorophyll levels may not change significantly in stressed cells, the relative amounts of PS I and PS II centers can change dramatically (see Eley 1971; Manodori and Melis 1984).

Effects of Nutrient Limitation on the Cyanobacterial Photosynthetic Apparatus

The effects of nitrogen limitation on the phycobiliprotein content of cyanobacterial cells are well documented (see Cohen-Bazire and Bryant 1982 and Allen 1984, for discussions). Allen and Smith (1969) were the first workers to provide evidence that the phycobiliproteins might serve as nitrogen storage compounds in cyanobacteria. Their work clearly demonstrated that the cellular levels of phycocyanin were markedly lower in nitrogen-limited cells of *Synechococcus* sp. PCC 6301 than in cells growing under nitrogen-replete conditions. In *Spirulina platensis* grown under nitrogen limitation conditions, 30–50% of the phycocyanin was lost with no change in the maximal growth rate (Boussiba and Richmond, 1980). Either nitrogen starvation or the addition of the glutamine synthetase inhibitor methionine sulfoximine triggered the loss of phycocyanin from the cells. When cells grown under nitrogen-replete conditions were shifted to nitrogen-deficient conditions, the cells grew normally for a period during which the phycocyanin content of the cells declined; growth continued at normal rates until the phycocyanin level fell below 50% of the normal value (Boussiba and Richmond, 1980). These results strongly suggest that phycocyanin can act as a nitrogen storage protein in *Spirulina platensis*. Wyman et al. (1985) have suggested that marine *Synechococcus* sp. use phycoerythrin as a storage compound in much the same fashion.

In some strains as much as 95% of the phycobiliproteins are degraded in a 24-h starvation period (Paone and Stevens 1981). Lau et al. (1977) showed that the loss of phycocyanin was due to the co-ordinated degradation of both the α and β subunits. Although the synthesis of other proteins continues, the synthesis of phycocyanin appears to be specifically repressed under these conditions. After four hours of nitrogen starvation, phycocyanin mRNA was not detectable in cells of *Synechococcus* sp. PCC 7002 (D. A. Bryant, unpublished results). A similar absence of detectable phycocyanin mRNA has also been made in nitrogen-starved cells of *Synechococcus* sp. PCC 6301 (N. Tandeau de Marsac and D. A. Bryant, unpublished results. These results establish that the repression of phycocyanin synthesis during nitrogen starvation results from a failure to accumulate phycocyanin mRNA (probably the result of transcriptional control).

The mobilization of the normally stable phycobilisome components during nitrogen starvation apparently is caused by the action of a specific proteolytic activity which is inducible (Foulds and Carr 1977; Wood and Haselkorn 1977, 1980). The enzyme(s) responsible for the turnover have not been isolated. In *Synechococcus* sp. PCC 6301 the protease(s) degrade phycocyanin more rapidly than allophycocyanin and also degrade linker polypeptides (Yamanaka and Glazer 1980). When protein synthesis is inhibited by chloramphenicol or methionine deprivation of methionine auxotrophic mutants, the phycobiliproteins are not degraded (Wood and Haselkorn 1980;

Yamanaka and Glazer 1980). Since the addition of chloramphenicol after the onset of PBS degradation stops the process, the proteolytic activity itself must turn over rapidly (Yamanaka and Glazer 1980).

In *Synechococcus* sp. PCC 6301, the first stage of nitrogen starvation is characterized by the preferential degradation of phycocyanin and the linker polypeptide $L_R^{30, PC}$ (Yamanaka and Glazer 1980). During this stage the pool of unassembled phycobiliproteins disappears, phycobilisomes become smaller, the ratio of phycocyanin to allophycocyanin decreases, but the residual PBS is functionally intact for energy transfer. More severe nitrogen starvation leads to the degradation of the L_R^{33} and L_{CM}^{75} linker polypeptides; at this stage the PBS presumably detach from the thylakoid and are completely degraded. This pattern may not be typical of all cyanobacteria. In *Pseudanabaena* sp. PCC 7409 moderate nitrogen starvation seems to cause a decrease in PBS number without a significant shift in their size (Tandeau de Marsac 1978).

The effect of nitrogen starvation on the expression of the *cpc* operon of *Synechococcus* sp. PCC 7002 is being studied with a translational fusion of the *cpcB* gene, with upstream regulatory sequences intact, to the *E. coli lacZ* gene (Bryant et al. 1986; G. Gasparich and D. A. Bryant, unpublished results). When *Synechococcus* sp. PCC 7002 cell harboring this fusion (on shuttle vector pAQE19) are subjected to nitrogen-starvation, β-galactosidase activities continue to increase for 2 h. After this lag, the β-galactosidase activities decline sharply (about ten-fold) over a 24-h period. In control cells in nitrogen-replete media, β-galactosidase levels increase five-fold to ten-fold during the same time period. β-galactosidase activities in control cells, which carry the wild-type *E. coli lacZ* gene, do not change whether the cells are grown under nitrogen-replete or nitrogen-deficient conditions. If nitrate is added to the cells after two hours of starvation, β-galactosidase levels continue to increase as though no starvation had occurred. Nitrate readdition at later times leads to a rapid recovery of the β-galactosidase levels in the cells. These results are consistent with a regulatory mechanism by which the *cpc* genes are not transcribed during nitrogen starvation.

Although Allen and Smith (1969) reported that carotenoid and chl *a* concentrations do not change significantly in *Synechococcus* sp. PCC 6301, Stevens et al. (1981) have reported that chl *a* levels decline and thylakoid membranes disappear in severely starved *Synechococcus* sp. PCC 7002 cells. Hence, it seems likely that the relative amounts of other components of the photosynthetic apparatus change in response to nitrogen starvation. At present there have been no systematic measurements of changes in either turnover or synthesis of such components.

The phycobiliproteins may also serve as a fixed carbon reserve which can be mobilized under conditions of carbon dioxide limitation. Although carbon dioxide limitation causes major differences in the phycobiliprotein content of cyanobacterial cells, the relative proportions of other components of the photosynthetic apparatus are also altered. Eley (1971) reported that *Synechococcus* sp. PCC 6301 cells grown in 1% carbon dioxide contain about twice as much phycocyanin as cells grown on air alone. Goedheer (1976) confirmed these results and found some differences in the fluorescence characteristics of cells grown under different carbon dioxide concentrations. Miller and Holt (1977) found that chlorophyll and phycobiliproteins were not detectable in severely carbon-starved *Synechococcus lividus* cells. The cells retained viability, however, and resynthesized their photosynthetic apparatus when carbon dioxide was reintroduced. Eley (1971) suggested that the loss of phycocyanin might provide a mechanism for balancing the light reactions to the dark reactions without altering the balance between the two photoreactions. Although the chlorophyll contents of cells grown under different carbon dioxide concentrations are similar, there are significant differences in the relative ratio of PS I : PS II reaction centers. Manodori and Melis (1984) found the ratio PS I : PS II to be higher (4.17) in cells grown at low carbon dioxide concentration than in cells grown at high carbon dioxide concentration (2.5). The greater proportion of PS I centers at low carbon dioxide concentrations may be required to provide greater ATP synthesis via cyclic photophosphorylation for active carbon dioxide uptake.

Phosphate starvation causes severe and parallel reductions in the phycobiliprotein and chlorophyll contents of the cells of *Synechococcus* sp. PCC 7002 (Stevens et al. 1981). The cellular nitrogen content of the cells also declined but less dramatically than did the photopigment concentrations. As would be expected, there are substantial reductions in the thylakoid content of the starved cells. Schmidt et al. (1982) have studied the effects of sulfur starvation on *Synechococcus* sp. PCC 6301. Sulfur deficiency causes changes which are similar to those caused by nitrogen deficiency. The phycobiliprotein content of the cells declines shortly after the onset of starvation, and phycocyanin degradation appears to precede the degradation of allophycocyanin. Chlorophyll synthesis continues during sulfur deficiency at a reduced rate. Readdition of a sulfur source to starved cells results in the preferential resynthesis of the phycobiliproteins (relative to chl *a*). After 16-20 h, normal growth resumes. Since there are numerous Fe–S centers associated with PS I function, it would be of interest to study the relative effects of sulfur starvation on other components of the photosynthetic apparatus, as has been done for iron starvation (see below).

In higher plants the only known electron donor to oxidized P700 is the copper protein plastocyanin. Most, but certainly not all, cyanobacteria and eucaryotic algae are genetically competent to synthesize plastocyanin (see Sandmann and Boger 1983, and Sandmann et al. 1983 for listings). Those organisms which cannot synthesize plastocyanin instead produce a small, water-soluble *c*-type cytochrome which is usually designated cytochrome *c*-552 or cytochrome *c*-553. In a variety of cyanobacteria and eucaryotic algae the electron donor to oxidized P700 is determined by the availability of copper in the growth medium. Wood (1978) first showed that in some green algae grown under copper-deficient conditions, plastocyanin is replaced by cytochrome *c*-552. Similar results have now been obtained for numerous eucaryotic algae and cyanobacteria, although some species have been shown to synthesize either plastocyanin or cytochrome *c*-553 uniquely (Sandmann and Boger 1980; Sandmann et al. 1983; Sandmann and Boger 1983). Plastocyanin is generally not present in the cells of dense natural blooms of cyanobacteria and is only found in laboratory cultures when care is take to make sufficient copper available (Ho et al. 1979).

At present nothing is known about the regulatory processes which control the synthesis and accumulation of plastocyanin and cytochrome *c*-553 in cyanobacteria. However, Merchant and Bogorad (1986) have recently reported some experiments performed with the green alga *C. reinhardii* that suggest in part how such regulation may occur. When cells are grown in copper-deficient conditions, the plastocyanin apoprotein is not detectable although the level of preapoplastocyanin mRNA is similar to that found in cells under copper-replete conditions. The mRNA for preapocytochrome *c*-552, however, is only detected in cells grown under conditions where cytochrome *c*-552 accumulates (i.e. copper-deficiency). Hence, in *C. reinhardii* the copper-mediated regulation of plastocyanin and cytochrome *c*-552 occurs at different levels: the former at the level of stable protein accumulation and the latter at the level of stable mRNA. Merchant and Bogorad (1986) present arguments suggesting that *C. reinhardii* possesses a metal-sensing system which would measure and respond to copper concentration.

The effects of iron limitation on cyanobacterial cells have been extensively studied (Oquist 1971; 1974 a, b; Sherman and Sherman 1983; Guikema and Sherman 1983, 1984; Sandmann and Malkin 1983; Sandmann 1985). Substantial alterations in the photosynthetic apparatus have been observed. The phycocyanin and chlorophyll contents of iron-deficient cells are depressed (Oquist 1971, 1974a,b; Hardie et al. 1983), and the red absorption maximum of the membrane-bound chlorophyll is blue-shifted 5–8 nm (Guikema and Sherman 1983). Iron-starved cells have fewer photosynthetic membranes, phycobilisomes, and carboxysomes, but show increased glycogen storage granules (Sherman and Sherman 1983). Chlorophyll fluorescence kinetics and the fluorescence emission spectra of thylakoids are considerably altered (Oquist 1974b; Guikema and Sherman 1983). Significant changes in the polypeptide profiles of membranes from iron-replete and iron-starved cells have also been reported (Guikema and Sherman 1984).

Moderate iron depletion causes little effect on the photosynthetic electron transport reactions and growth of *Synechocystis* sp. PCC 6714 (Sandmann and Malkin 1983). Nonetheless, membrane-bound non-heme iron decreased sharply and ferredoxin was replaced by the flavin-containing protein, flavodoxin. Rogers and coworkers (Fitzgerald et al. 1977; Hutber et al. 1977) were the first to report the replacement of ferredoxin by flavodoxin in cyanobacteria. Cyanobacterial cells from natural blooms frequently do not contain ferredoxin (Ho et al. 1979), an observation which suggests that these cells are often moderately iron-starved. More severe iron limitation causes growth inhibition and decreased rates of photosynthetic electron flow in *Synechocystis* sp. PCC 6714 (Sandmann and Malkin 1983; Sandmann 1985). The PS I reaction center was most affected by severe iron depletion but even redox components which do not contain iron are affected by severe iron starvation. At present, nothing is known about the regulatory processes which cause the replacement of ferredoxin by flavodoxin.

Adaptation to Light Wavelength: Complementary Chromatic Adaptation

Complementary chromatic adaptation is probably the most dramatic example of the ability of cyanobacterial cells to alter their photosynthetic apparatus (for reviews, see Bogorad 1975 and Tandeau de Marsac 1983). This adaptive response to light wavelength is unique to cyanobacteria and among them only occurs in certain of those species which are capable of phycoerythrin synthesis (Tandeau de Marsac 1977). Reports that two phycoerythrocyanin-producing strains, *Oscillatoria splendida* (Thomas et al. 1984) and *Mastigocladus laminosus* (Fuglistaller et al. 1981) are chromatic adapters have appeared. However, red-light/green-light reversibility has not yet been demonstrated for either of these strains. Other data suggests that phycoerythrocyanin synthesis is regulated by light intensity and not light wavelength (Bryant 1982). For the moment this question remains unresolved.

In strains which exhibit chromatic adaptation, growth in green light promotes the synthesis of the red-colored, green-absorbing protein phycoerythrin. During growth in red light, the blue-colored, red-absorbing biliprotein phycocyanin is produced in abundance. These changes in phycobiliprotein synthesis can cause rapid and remarkable changes in the coloration of cultures subjected to a shift in light wavelength. The chromatic adaptation response is mediated by a photoreceptor pigment whose action spectral properties have been clearly defined but which has not yet been isolated (Cohen-Bazire and Bryant 1982; Tandeau de Marsac 1983). It should be mentioned that this photoreceptor bears remarkable similarities to the higher plant photomorphogenic/photoregulatory pigment phytochrome. The photoreceptor of the complementary chromatic adaptation phenomenon could, in fact, be the evolutionary precursor of phytochrome, although this speculation remains to be established.

Cyanobacteria which synthesize phycoerythrin can be assigned to three groups (Tandeau de Marsac 1977; Bryant 1982). The first group does not chromatically adapt: i.e. the ratio of phycoerythrin to phycocyanin remains constant whether cells are grown in green, red, or white light. In the second group, only phycoerythrin synthesis is under light-wavelength control. Phycoerythrin synthesis is promoted by green light which can of course be absorbed by this biliprotein. Phycoerythrin synthesis is repressed in red light. Phycocyanin synthesis in strains of Group II is constant whether cells are grown in green, red, or white light. In strains of Group III, the synthesis of both phycocyanin and phycoerythrin are modulated by light wavelength. In green light the differential rate of phycoerythrin synthesis is high, and that of phycocyanin is low but never zero. Growth in red light severely represses the synthesis of phycoerythrin and concommitantly causes the synthesis rate of phycocyanin to be maximal. It should be noted that these changes in phycobiliprotein synthesis are also accompanied by changes in the synthesis of the linker polypeptides required to assemble either phycoerythrin or phycocyanin, respectively (Tandeau de Marsac and Cohen-Bazire 1977; Bryant and Cohen-Bazire 1981).

The rate of synthesis of allophycocyanin, the composition of the PBS core substructure, and the appearance of the core substructures in the electron microscope are unaltered in cells undergoing chromatic adaptation (Tandeau de Marsac 1977; Bryant et al. 1979). In members of Group II, the length of the peripheral rod substructures decreases in direct proportion to the phycoerythrin content of the PBS (Bryant et al. 1979; Williams et al. 1980). In these organisms, phycoerythrin can be added or removed from the distal ends of the peripheral rod substructures of the PBS. In *Synechocystis* sp. PCC 6701, two phycoerythrin-specific linker polypeptides are required to assemble phycoerythrin onto the PBS (Bryant , 1981; Gingrich et al. 1982a,b). In members of Group III an additional pair of phycocyanin subunits is expressed specifically in cells grown in red light (Bryant and Cohen-Bazire 1981; Bryant 1981). These phycocyanin subunits compensate for the phycoerythrin lost from the peripheral rods during growth in red light such that there is little change in the length of the peripheral rods when cells growing in green light are shifted to red light (Bryant et al. 1979; Bryant and Cohen-Bazire 1981). In *Nostoc* sp. strain MAC the six phycocyanin subunits expressed in red light are mixed in phycocyanin hexamers in proportion to their relative abundances (i.e. the red-light-specific phycocyanin subunits apparently do not preferentially assemble into discreet trimers or hexamers in this organism; G. Gugleilmi and D. A. Bryant, unpublished observations).

Several laboratories are now studying the chromatic adaptation phenomenon using recombinant DNA methods. The genes encoding phycoerythrin (*cpeA* and *cpeB*), the phycocyanin subunits (*gpcA* and *gpcB*) expressed "constitutively," and the phycocyanin subunits (*rpcA* and *rpcB*) only expressed in cells grown in red light have been cloned from the Group III chromatic adapter *Pseudanabaena* sp. PCC 7409 (Bryant et al. 1986). Sequence analysis of the phycoerythrin genes is nearly complete (J. Dubbs and D. A. Bryant, unpublished results). These genes are similar in arrangement to those encoding phycocyanin; the *cpeB* gene, which encodes the β subunit is located 5' to the *cpeA* gene which encodes the α subunit. As deduced from the nucleotide sequence, the amino acid sequences of the α subunit of *Pseudanabaena* sp. PCC 7409 phycoerythrin is nearly identical ($>90\%$ homologous) to that of the phycoerythrin of *Calothrix* sp. PCC 7601 (W. Sidler and H. Zuber, personal communication).

Grossman and coworkers have reported the cloning and characterization of four genes encoding phycocyanin subunits of the Group III chromatic adapter *Calothrix* sp. PCC 7601 (Conley et al. 1985, 1986). In *Calothrix* sp. PCC 7601 the genes encoding the α and β subunits of allophycocyanin, the α and β subunits of the "inducible" phycocyanin, and the α and β subunits of the "red-light-inducible" phycocyanin are clustered within a 13 Kbp region of the chromosome (Conley et al. 1986). At least two linker polypeptides, L_c and L_R^{33} are also encoded within this region of the genome (Conley et al. 1986). The two phycocyanin gene sets have distinctly different transcriptional properties. The constitutively expressed phycocyanin genes are transcribed as a single 1600 bp mRNA which is present in cultures grown in either red or green light. This mRNA is more abundant (three-fold) in cultures grown in red light, however. The second phycocyanin gene pair is transcribed as two different mRNA species (1600 bp and 3800 bp) which are only detectable in RNA isolated from cultures grown in red light (Conley et al. 1985). The 3800 bp mRNA contains 3' sequences which encode at least one phycocyanin-associated linker polypeptide. Three transcripts are produced from the *apc* locus, and the levels of these transcripts do not change in cells grown in red or green light. The *apcA* and *apcB* genes hybridize to transcripts of 1750 and 1400 bp; the *apcC* gene, which encodes the L_c linker of *Calothrix* sp. PCC 7601, hybridizes to the 1750 bp transcript and to an abundant

370 bp mRNA species (Conley et al. 1986). The 1750 bp transcript may be a precursor to the 1400 bp and 370 bp species.

Tandeau de Marsac and coworkers (personal communication) have recently cloned the *cpeA* and *cpeB* genes of *Calothrix* sp. PCC 7601. These genes are transcribed as a unique 1600 bp mRNA which is detectable only in cells grown in green light. Although differential mRNA stability cannot be rigourously excluded, the combined data for *Calothrix* sp. PCC 7601 clearly suggest that the chromatic adaptation phenomenon is the result of light-wavelength control of transcription initiation. The complete description of the factors which mediate this novel gene-regulatory phenomenon should provide a most interesting story.

Adaptation to Light Intensity

When higher plants are grown at high irradiances, they have higher chl *a*/chl *b* ratios and less chlorophyll per chloroplast than when grown at lower irradiances (Anderson 1986). Similarly, "shade plants" have more chlorophyll per chloroplast and lower chl *a*/chl *b* ratios ($\sim$2.0–2.4) than "sun plants" ($\sim$2.8–3.6; Anderson, 1986). Among photosynthetic procaryotes there is frequently an inverse correlation between light intensity and photopigment content. Most cyanobacteria generally appear to fit this pattern (Myers and Kratz 1955; Allen 1968; Vierling and Alberte 1980; Barlow and Alberte 1985; Wyman and Fay 1986a). In *Synechococcus* sp. PCC 6301, it has been shown that pigment concentration and thylakoid content vary inversely with light intensity (Allen 1968; Vierling and Alberte 1980). In *Microcystis aeruginosa* the chl *a* and phycocyanin contents decreased while carotenoid, cellular carbon, and cellular nitrogen contents did not decrease as the light intensity for growth increased (Raps et al. 1983). Although the number of photosynthetic units per cell is greatly reduced at high light intensities, these conditions generally result in PS I : PS II ratios near unity (Kawamura et al. 1979). Similar total chl *a* reductions and PS I : PS II ratios have been obtained with *Synechococcus* sp. PCC 6301 by growing cells in far-red light (the so-called inverse chromatic adaptation; Jones and Myers 1965; Myers et al. 1978, 1980; Fujita et al. 1985). Growth of cells under low light intensities or under conditions which preferentially excite PS II results in a larger PS I : PS II ratio; values approaching 4.0 have been reported (Fujita et al. 1985). Fujita et al. (1985) suggest that these changes in the PS I : PS II ratio are solely due to changes in the PS I population in the cell. However, other workers find that the size of the PS II population also varies as a function of either light intensity of light wavelength (Khanna et al. 1983; Barlow and Alberte 1985). In any event these light intensity/light wavelength effects also affect the cytochrome content of the thylakoids (Fujita et al. 1985) and possibly alter the relative amounts of other components of the photosynthetic apparatus as well. Although considerable variation in the PS I : PS II ratio occurs, the size of the PS I and PS II units do not vary significantly (Myers et al. 1980; Fujita et al. 1985). In some species there is both size and compositional variation in the PBS antenna complexes, however. Since the phycobiliprotein contents of cyanobacterial cells vary as a function of a large number of other variables (temperature and virtually all nutrients; see Goedheer 1976, for examples), it is nearly impossible to compare the data of different workers concerning the direct effects of light intensity on phycobiliprotein synthesis.

The effect of light intensity on PBS size has been investigated in *Microcystis aeruginosa* by Raps et al. (1985). PBS isolated from cells grown at high or at low light intensities were identical in size. These workers conclude that adaptation of *M. aeruginosa* to high light intensity results in a decrease in the number of PBS per cell but that no alteration in PBS composition or structure occurs. Since two PS II reaction centers appear to be associated with each hemidiscoidal PBS (Manodori and Melis 1985), a change in the number of PBS per cell would require a change in the number of PS II reaction centers per cell as well.

In some cyanobacteria the size and/or composition of the PBS may vary as a function of light intensity. Yamanaka and Glazer (1981) have shown that the phycocyanin/allophycocyanin ratio of *Synechococcus* sp. PCC 6301 can be varied over a two-fold range by changes in culture conditions which included light intensity and wavelength differences. The length of the peripheral rods of the PBS changed as did the relative amounts of the $L_R^{30, PC}$ and $L_R^{33, PC}$ kDa linker polypeptides. Lonneborg et al. (1985) have shown that, in combined intensity/wavelength shift experiments, *Synechococcus* sp. PCC 6301 cells rapidly adjust their phycocyanin content (and concomitantly their linker polypeptide content) when shifted from a high, white-light intensity to a low, red-light intensity. Most cyanobacteria which are capable of synthesizing phycoerythrocyanin preferentially synthesize this biliprotein at low light intensity (Bryant 1982). Since this biliprotein is assembled at the distal ends of the peripheral rods of the PBS (Glazer 1982), the size of the PBS must change in these cells at low light intensity.

The mechanism(s) by which changes in light intensity effect changes in PBS size and number is not known. Indeed, one may reasonably question whether light intensity has a direct effect on phycobiliprotein synthesis (see Goedheer 1976). To examine this phenomenon, a translational fusion of the *cpcB* gene, complete with upstream regulatory sequences, has been made to the *E. coli lacZ* gene (Bryant et al. 1986). This construct, which was cloned on shuttle vector pAQE19, has been introduced into *Synechococcus* sp. PCC 7002. In cells grown at low light intensity, β-galactosidase levels are 1.8-fold higher than β-galactosidase levels in cells grown at high light intensity. In a control strain which harbors a plasmid carrying the wild-type *E. coli lacZ* gene and expression sequences, β-galactosidase levels are equal in cells grown in high- and low-intensity light. These results strongly imply that the increased phycobiliprotein content of cells grown at low light intensity is mediated at the transcriptional level. Similar experiments are planned with the *psaA* gene of *Synechococcus* sp. PCC 7002. A convenient restriction site (*Bam*HI) for fusion formation occurs near 5' end of the *psaA* gene (codon 8). Since PS I levels change in response to low intensity chromatic illumination, studies with a *psaA-lacZ* fusion could help ellucidate the real effects of light intensity on the composition of the PS apparatus in cyanobacteria.

Acknowledgments

The author would like to thank the organizers of the NATO Advanced Study Institute on the "Physiological Ecology of Marine Picoplankton," especially Drs. Noel G. Carr and Trevor Platt, for the opportunity to meet other scientists and discuss with them the physiology of marine cyanobacteria. Original research from the authors lab which is described in this review was supported by grants U.S.P.H.S. GM-31625, NSF DMB-8504294, and NSF INT-8514249.

References

ABRAMOWICZ, D. A., AND G. C. DISMUKES. 1984a. Manganese proteins isolated from spinach thylakoid membranes and their role in O_2 evolution. I. A 56 kilodaltion manganese-containing protein, a probable component of the coupling factor enzyme. Biochim. Biophys. Acta 765: 309–317.

—— 1984b. Manganese proteins isolated from spinach thylakoid membranes and their role in O_2 evolution. II. A binuclear manganese-containing 34 kilodalton protein, a probably component of the water dehydrogenase enzyme. Biochim. Biophys. Acta 765: 318–328.

AITKEN, A. 1975. Prokaryote–eukaryote relationships and the amino acid sequence of plastocyanin from *Anabaena variabilis*. Biochem. J. 149: 675–683.

—— 1976. Protein evolution in cyanobacteria. Nature 263: 793–796.

—— 1977. Purification and primary structure of cytochrome *f* from the cyanobacterium, *Plectonema boryanum*. Eur. J. Biochem. 78:

273-279.

——— 1978. Plastocyanin and the evolutionary relationship between chloroplasts and cyanobacteria, p. 251-263. *In* H. Matsubara and T. Yamanaka [ed.] Evolution of Protein Molecules. Japanese Scientific Society Press, Tokyo.

——— 1979. Purification and primary structure of cytochrome *c*-552 from the cyanobacterium *Synechococcus* PCC 6312. Eur. J. Biochem. 110: 297-308.

AKABORI, K., A. IMAOKA, AND Y. TOYOSHIMA. 1984. The role of lipids and 17-kDa protein in enhancing the recovery of O_2-evolution in cholate-treated thylakoid membranes. FEBS lett. 173: 36-40.

AKERLUND, H.-E., B. ANDERSSON, AND P.-A. ALBERTSSON. 1976. Isolation of photosystem II enriched membrane vesicles from spinach chloroplasts by phase partition. Biochim. Biophys. Acta 449: 525-535.

AKERLUND, H.-E., C. JANSSON, AND B. ANDERSSON, 1982. Reconstitution of photosynthetic water-splitting in inside-out thylakoid vesicles and identification of a participating polypeptide. Biochim. Biophys. Acta 681: 1-10.

ALLEN, J. P., AND G. FEHER. 1984. Crystallization of reaction center from *Rhodopseudomonas sphaeroides*: Preliminary characterization. Proc. Nat. Acad. Sci. USA 81: 4795-4799.

ALLEN, M. M. 1968. Photosynthetic membrane system in *Anacystis nidulans*. J. Bacteriol. 96: 836-841.

——— 1984. Cyanobacterial cell inclusions. Ann. Rev. Microbiol. 38: 1-25.

ALLEN, M. M., AND A. J. SMITH. 1969. Nitrogen chlorosis in blue-green algae. Arch. Mikrobiol. 69: 114-120.

ALT, J., AND R. G. HERRMANN. 1984. Nucleotide sequence of the gene for pre-apocytochrome *f* in the spinach plastid chromosome. Curr. Genet. 8: 551-557.

ALT, J., J. MORRIS, P. WESTHOFF, AND R. G. HERRMANN. 1984. Nucleotide sequence of the clustered genes for the 44 kd chlorophyll *a* apoprotein and the "32 kd"-like protein of the photosystem II reaction center in the spinach plastid chromosome. Curr. Genet. 8: 597-606.

ALT, J., P. WESTHOFF, B. B. SEARS, N. NELSON, E. HURT, G. HAUSKA, AND R. G. HERRMANN. 1983a. Genes and transcripts for the polypeptides of the cytochrome b_6/f complex from the spinach thylakoid membranes.

ALT, J., P. WINTER, W. SEBALD, J. G. MOSER, R. SCHEDEL, P. WESTHOFF, AND R. G. HERRMANN. 1983b. Localization and nucleotide sequence of the gene for the ATP synthase proteolipid subunit on the spinach plastid chromosome. Curr. Genet. 7: 129-138.

AMBLER, R., AND R. G. BARTSCH. Amino acid sequence similarity between cytochrome *f* from a blue-green bacterium and algal chloroplasts. Nature 253: 285-288.

AMESZ, J. 1983. The role of manganese in photosynthetic oxygen evolution. Biochim. Biophys. Acta 726: 1-12.

AMZEL, L. M., M. MCKINNEY, P. NARAZANAN, AND P. L. PEDERSEN. 1982. Structure of the mitochondrial F_1ATPase at 9Å resolution. Proc. Nat. Acad. Sci. USA 79: 5852-5856.

ANDERSON, L. K., M. C. RAYNER, R. M. SWEET, AND F. A. EISERLING. 1983. Regulation of *Nostoc* sp. phycobilisome structure by light and temperature. J. Bacteriol. 155: 1407-1416.

ANDERSON, J. M. 1982. The role of chlorophyll-protein complexes in the function and structure of chloroplast thylakoids. Mol. Cell. Biochem. 46: 161-172.

——— 1984. A chlorophyll *a/b*-protein complex of photosystem I. Photobiochem. Photobiophys. 8: 221-228.

——— 1985. Chlorophyll-protein complexes of a marine green alga, *Codium* species (Siphonales). Biochim. Biophys. Acta 806: 145-153.

——— 1986. Photoregulation of the composition, function and structure of thylakoid membranes. Ann. Rev. Plant Physiol. 37: 93-136.

ANDERSON, J. M., AND B. ANDERSSON. 1982. The architecture of photosynthetic membranes: lateral and transverse organization. Trends Biochem. Sci. 7: 288-292.

ANDERSON, J. M., J. S. BROWN, E. LAM, AND R. MALKIN. 1983. Chlorophyll *b*: an integral component of photosystem I of higher plant chloroplasts. Photochem. Photobiol. 38: 205-210.

ANDERSSON, B., C. JANSSON, U. LJUNGBERG, AND H.-E. AKERLUND. 1985. Polypeptides on the oxidizing side of photosystem II, p. 21-31. 1985. *In* K. E. Steinbeck, S. Bonitz, C. J. Arntzen, and L. Bogorad [ed.] Molecular biology of the photosynthetic apparatus. Cold Spring Harbor Laboratory, Cold Spring Harbor.

ANDERSSON, B., C. LARSSON, C. JANSSON, U. LJUNGBERG, AND H.-E. AKERLUND. 1984. Immunological studies on the organization of proteins in photosynthetic oxygen evolution. Biochim. Biophys. Acta 766: 21-28.

ARNTZEN, C. J. 1978. Dynamic structural features of chloroplast lamellae. Current Topics Bioenerget. 8: 111-160.

ASTIER, C., S. STYRING, B. MAISON-PETERI, AND A.-L. ETIENNE. 1986. Preparation and characterization of thylakoid membranes and Photosystem II particles from the facultative phototrophic cyanobacterium *Synechocystis* 6714. Photobiochem. Photobiophys. 11: 37-47.

BABCOCK, G. T., R. E. BLANKENSHIP, AND K. SAUER. 1976. Reaction kinetics for positive charge accumulation on the water side of chloroplast photosystem II. FEBS Lett. 61: 286-289.

BABCOCK, G. T., W. R. WIDGER, W. A. CRAMER, W. A. OERTLING, AND J. G. METZ. 1985. Axial ligands of chloroplast cytochrome *b*-559: Identification and requirement for a heme-cross-linked polypeptide structure. Biochemistry 24: 3638-3645.

BARLOW, R. G., AND R. S. ALBERTE. 1985. Photosynthetic characteristics of phycoerythrin-containing marine *Synechococcus* spp. I. Responses to growth photon flux density. Mar. Biol. 86: 63–74.

BASSI, R., O. MACHOLD, AND D. SIMPSON. 1985. Chlorophyll–proteins of two photosystem I preparations from maize. Carlsberg Res. Commun. 50: 145–162.

BENDALL, D. S. 1982. Photosynthetic cytochromes of oxygenic organisms. Biochim. Biophys. Acta 683: 119–151.

BENGIS, C., AND N. NELSON. 1975. Purification and properties of the photosystem I reaction center from chloroplasts. J. Biol. Chem. 250: 2783–2788.

BENGIS, C., AND N. NELSON. 1977. Subunit structure of chloroplast photosystem I reaction center. J. Biol. Chem. 252: 4564–4569.

BENNOUN, P., B. A. DINER, F.-A. WOLLMAN, G. SCHMIDT, AND N.-H. CHUA. 1981. Thylakoid polypeptides associated with Photosystem II in *Chlamydomonas reinhardtii*: comparison of system II mutants and particles, p. 839–849. *In* G. Akoyunoglou [ed.] Photosynthesis III. Structure and Molecular Organization of the Photosynthetic Apparatus. Balaban, Philadelphia.

BERARD, J., G. BELANGER, P. CORRIVEAU, AND G. GINGRAS. 1986. Molecular cloning and sequence of the B880 holochrome gene from *Rhodospirillum rubrum*. J. Biol. Chem. 261: 82–87.

BERENDS, T., Q. KUBICEK, AND J. E. MULLET. 1986. Localization of the genes coding for the 51 kDa PS II chlorophyll apoprotein, apocytochrome b_6, the 65–70 kDa PS I chlorophyll apoproteins and the 44 kDa PS II chlorophyll apoprotein in pea chloroplast DNA. Plant Mol. Biol. 6: 125–134.

BERTHOLD, D. A., G. T. BABCOCK, AND C. F. YOCUM. 1981. A highly resolved, oxygen-evolving photosystem II preparation from spinach thylakoid membranes. FEBS Lett. 134: 231–234.

BINDER, A. 1982. Respiration and photosynthesis in energy-transducing membranes of cyanobacteria. J. Bioenerg. Biomem. 14: 271–286.

BINDER, A., AND R. BACHOFEN. 1979. Isolation and characterization of a coupling factor I ATPase of the thermophilic blue-green alga (cyanobacterium) *Mastigocladus laminosus*. FEBS Lett. 104: 66–70.

BINDER, A., M. C. BOHLER, M. WOLF, AND R. BACHOFEN. 1980. The proton APTase of the thermophilic cyanobacterium *Mastigocladus laminosus*, p. 771–778. *In* G. A. Akoyunoglou [ed.] Proceedings of the Vth Internat. Congress of Photosynthesis, Vol. 2. Balaban, Philadelphia.

BIRD, C. R., B. KOLLER, A. D. AUFFRET, A. K. HUTTLY, C. J. HOWE, T. A. DYER, AND J. C. GRAY. 1985. The wheat chloroplast gene for CF_0 subunit I of ATP synthase contains a large intron. EMBO J. 4: 1381–1388.

BLANKENSHIP, R. E. 1985. Electron transport in green photosynthetic bacteria. Photosynthesis Res. 6: 317–333.

BLANKENSHIP, R. E., AND R. C. PRINCE. 1985. Excited-state redox potentials and the Z scheme of photosynthesis. Trends Biochem. Sci. 10: 382–383.

BOGORAD, L. 1975. Phycobiliproteins and complementary chromatic adaptation. Ann. Rev. Plant Physiol. 26: 369–401.

BOHNERT, H. J., C. MICHALOWSKI, S. BEVACQUA, H. MUCKE, AND W. LOFFELHARDT. 1985. Cyanelle DNA from *Cyanophora paradoxa*. Physical mapping and location of protein coding regions. Mol. Gen. Genet. 201: 565–574.

BONNERJEA, J., W. ORTIZ, AND R. MALKIN. 1985. Identification of a 19-kDa polypeptide as an Fe–S center apoprotein in the photosystem I primary electron acceptor complex. Arch. Biochem. Biophys. 240: 15–20.

BOOKJANS, G., B. M. STUMMANN, O. F. RASMUSSEN, AND K. W. HENNINGSEN. 1986. Structure of a 3.2 Kb region of pea chloroplast DNA containing the gene for the 44 kD photosystem II polypeptide. Plant Mol. Biol. 6: 359–366.

BOSKA, M., N. V. BLOUGH, AND K. SAUER. 1985. The effects of mono- and divalent salts on the rise and decay kinetics of EPR signal II in photosystem II preparations from spinach. Biochim. Biophys. Acta 808: 132–139.

BOSKA, M., K. SAUER, W. BUTTNER, AND G. T. BABCOCK. 1983. Similarity of EPR signal II, rise and P-680$^+$ decay kinetics in tris-washed chloroplast photosystem II preparations as a function of pH. Biochim. Biophys. Acta 722: 327–330.

BOTTIN, H., AND P. MATHIS. 1985. Interaction of plastocyanin with the photosystem I reaction center: a kinetic study by flash absorption spectroscopy. Biochemistry 24: 6453–6460.

BOUGES-BOCQUET, B. 1980. Electron and proton transfer from P-430 to ferredoxin-NADP-reductase in *Chlorella* cells. Biochim. Biophys. Acta 590: 223–233.

BOUSSIBA, S., AND A. E. RICHMOND. 1980. C-phycocyanin as a storage protein in the blue-green alga *Spirulina platensis*. Arch. Microbiol. 125: 143–147.

BOWES, J. M., A. C. STEWART, AND D. S. BENDALL. 1983. Purification of photosystem II particles from *Phormidium laminosus* using the detergent dodecyl-β-maltoside. Biochim. Biophys. Acta 725: 210–219.

BOWLBY, N. R., AND W. D. FRASCH. 1985. Isolation of a manganese-containing protein complex from photosystem II preparations of spinach. Biochemistry 25: 1402–1407.

BRAND, J. J. 1979. The effect of Ca^{2+} on oxygen evolution in membrane preparations from *Anacystis nidulans*. FEBS Lett. 103: 114–117.

BRANTON, D. 1966. Fracture faces of frozen membranes. Proc. Nat. Acad. Sci. USA 55: 1048–1056.

BRIANTAIS, J.-M., C. VERNOTTE, M. MIYAO, N. MURATA, AND M. PICAUD. 1985. Relationship between O_2 evolution capacity and cytochrome b-559 high-potential form in Photosystem II particles. Biochim. Biophys. Acta 808: 348–351.

BRICKER, T. M., H. B. PAKRASI, AND L. A.

SHERMAN. 1985. Characterization of a spinach Photosystem II core preparation isolated by a simplified method. Arch. Biochem. Biophys. 237: 170–176.

BRYANT, D. A. 1981. The photoregulated expression of multiple phycocyanin species. A general mechanism for the control of phycocyanin synthesis in chromatically adapting cyanobacteria. Eur. J. Biochem. 119: 425–429.

BRYANT, D. A. 1982. Phycoerythrocyanin and phycoerythrin: properties and occurrence in cyanobacteria. J. Gen. Microbiol. 128: 835–844.

BRYANT, D. A., AND G. COHEN-BAZIRE. 1981. Effects of chromatic illumination on cyanobacterial phycobilisomes. Evidence for the specific induction of a second pair of phycocyanin subunits in *Pseudanabaena* 7409 grown in red light. Eur. J. Biochem. 119: 415–424.

BRYANT, D. A., J. M. DUBBS, P. I. FIELDS, R. D. PORTER, AND R. DE LORIMIER. 1985. Expression of phycobiliprotein genes in *Escherichia coli*. FEMS Lett. 29: 343–349.

BRYANT, D. A., A. N. GLAZER, AND F. A. EISERLING. 1976. Characterization and structural properties of the major biliproteins of *Anabaena* sp. Arch. Microbiol. 110: 61–75.

BRYANT, D. A., G. GUGLIELMI, N. TANDEAU DE MARSAC, A.-M. CASTETS, AND G. COHEN-BAZIRE. 1979. The structure of cyanobacterial phycobilisomes: a model. Arch. Microbiol. 123: 113–127.

BRYANT, D. A., R. DE LORIMIER, D. H. LAMBERT, J. M. DUBBS, V. L. STIREWALT, S. E. STEVENS, JR., R. D. PORTER, J. TAM, AND E. JAY. 1985. Molecular cloning and nucleotide sequence of the α and β subunits of allophycocyanin from the cyanelle genome of *Cyanophora paradoxa*. Proc. Nat. Acad. Sci. USA 82: 3242–3246.

BRYANT, D. A., R. DE LORIMIER, G. GUGLIELMI, V. L. STIREWALT, J. M. DUBBS, B. ILLMAN, G. GASPARICH, J. S. BUZBY, A CANTRELL, R. C. MURPHY, J. GINGRICH, R. D. PORTER, AND S. E. STEVENS, JR. 1986. The cyanobacterial photosynthetic apparatus: a molecular genetic analysis. *In* D. C. Youvan and F. Daldal [ed.] Microbial Energy Transduction: Genetics, Structure and Function. Cold Spring Harbor Laboratory, Cold Spring Harbor (in press)

BRYANT, D. A., R. DE LORIMIER, R. D. PORTER, D. H. LAMBERT, J. M. DUBBS, V. L. STIREWALT, P. I. FIELDS, S. E. STEVENS, JR., W.-Y. LIU, J. TAM, AND E. JAY. 1985. Phycobiliprotein genes in cyanobacteria and cyanelles. *In* K. Steinbeck, S. Bonitz, C. Arntzen, and L. Bogorad [ed.] pp. 249–258. Molecular Biology of the Photosynthetic Apparatus. Cold Spring Harbor Laboratory, Cold Spring Harbor.

BULLERJAHN, G. S., H. C. RIETHMAN, AND L. A. SHERMAN. 1985. Organization of the thylakoid membrane from the heterotrophic cyanobacterium, *Aphanocapsa* 6714. Biochim. Biophys. Acta 810: 148–157.

BURGER-WIERSMA, T., M. VEENHUIS, H. J. KORTHALS, C. C. M. VAN DE WIEL, AND L. R. MUR. 1986. A new procaryote containing chlorophylls *a* and *b*. Nature 320, 262–264.

BUTLER, W. L. 1978. On the role of cytochrome *b*-559 in oxygen evolution in photosynthesis. FEBS Lett. 95: 19–25.

BYLINA, E. J., S. ISMAIL, AND D. C. YOUVAN. 1986. Site-specific mutagenesis of bacteriochlorophyll binding sites affects biogenesis of the photosynthetic apparatus. Nature, submitted.

CARPENTIER, R., E. P. FUERST, H. Y. NAKATANI, AND C. J. ARNTZEN. 1985. A second site for herbicide action in photosystem II. Biochim. Biophys. Acta 808: 293–299.

CARPENTIER, R., AND H. Y. NAKATANI. 1985. Inhibitors affecting the oxidizing side of photosystem II at the Ca^{2+} — and Cl^{--} sensitive sites. Biochim. Biophys. Acta 808: 288–292.

CARRILLO, N., P. SEYER, A. TYAGI, AND R. G. HERRMANN. 1986. Cytochrome b-559 genes from *Oenothera hookeri* and *Nicotiana tabacum* show a remarkably high degree of conservation as compared to spinach. Curr. Genet. 10: 619–624.

CARRILLO, N., AND R. R. VALLEJOS. 1982. Interaction of ferredoxin-NADP oxidoreductase with the thylakoid membrane. Plant Physiol. 69: 210–213.

1983. The light-dependent modulation of photosynthetic electron transport. Trends Biochem. Sci. 8: 52–56.

CECCARELLI, E. A., R. L. CHAN, AND R. H. VALLEJOS. 1985. Trimeric structure and other properties of the chloroplast reductase binding protein. FEBS Lett. 190: 165–168.

CHAN, R. L., N. CARRILLO, AND R. H. VALLEJOS. 1985. Isolation and sequencing of an active-site peptide from spinach ferredoxin-NADP + oxidoreductase after affinity labelling with periodate-oxidized NADP + . Arch. Biochem. Biophys. 240: 172–177.

CHANG, C.-H., M. SCHIFFER, D. TIEDE, U. SMITH, AND J. NORRIS. 1985. Characterization of bacterial photosynthetic reaction center crystals from *Rhodopseudomonas sphaeroides* R-26 by X-ray diffraction. J. Mol. Biol. 186: 201 -203.

CHERESKIN, B. M., J. D. CLEMENT-METRAL, AND E. GANTT. 1985. Characterization of a purified photosystem II-phycobilisome particle preparation from *Porphyridium cruentum*.

CHITNIS, P. R., E. HAREL, B. D. KOHRON, E. M. TOBIN, AND J. P. THORNBER. 1986. Assembly of the precursor and processed light-harvesting chlorophyll *a/b* protein of *Lemna* into the light-harvesting complex II of barley etiochloroplasts. J. Cell Biol. 102: 982–988.

CHU, Z.-X., AND J. M. ANDERSON. 1985. Isolation and characterization of a siponaxanthin-chlorophyll *a/b*-protein complex of photosystem I from I *Codium* species (Siphonales). Biochim. Biophys. Acta 806: 154–160.

CHUA, N.-H., AND N. W. GILLHAM. 1977. The sites of synthesis of the principal thylakoid membrane polypeptides in *Chlamydomonas reinhardii*. J. Cell. Biol. 74: 441–452.

CLARK, R. D., M. J. HAWKESFORD, S. J. COUGHLAN, AND G. HIND. 1984. Association of ferredoxin-NADP+ oxidoreductase with chloroplast cytochrome b-f complex. FEBS Lett. 174: 137–142.

CLARK, R. D., AND G. HIND. 1983. Isolation of a five-polypeptide cytochrome b-f complex from spinach chloroplasts. J. Biol. Chem. 258: 10348–10354.

CLEMENT-METRAL, J. D., AND E. GANTT. 1983. Isolation of oxygen-evolving phycobilisome II particles from *Porphyridium cruentum*. FEBS Lett. 156: 185–188.

CLEMENT-METRAL, J. D., E. GANTT, AND T. REDLINGER. 1985. A photosystem II-phycobilisome preparation from the red alga *Porphyridium cruentum*: Oxygen evolution, ultrastructure, and polypeptide resolution. Arch. Biochem. Biophys. 238: 10–17.

COGDELL, R. J. 1983. Photosynthetic reaction centers. Ann. Rev. Plant Physiol. 34: 21–45.

COHEN-BAZIRE, G., AND D. A. BRYANT. 1982. Phycobilisomes: composition and structure, p. 143–190. *In* N. G. Carr and B. A. Whitton [ed.] The Biology of the Cyanobacteria. Blackwell Scientific, Oxford.

COHN, C. L., J. ALAM, AND D. W. KROGMANN. 1985. Multiple ferredoxins from cyanobacteria. Physiol. Veg. 23: 659–667.

COLMAN, P. M., H. C. FREEMAN, J. M. GUSS, M. MURATA, V. A. NORRIS, J. A. M. RAMSHAW, AND M. P. VENKATAPPA. 1978. X-ray crystal structure analysis of plastocyanin at 2.7 Å resolution. Nature 272: 319–324.

CONLEY, P. B., P. G. LEMAUX, AND A. R. GROSSMAN. 1985. Cyanobacterial light-harvesting complex subunits encoded in two red light-induced transcripts. Science 230: 550–553.

CONLEY, P. B., P. G. LEMAUX, T. L. LOMAX, AND A. R. GROSSMAN. 1986. Genes encoding major light-harvesting polypeptides are clustered on the genome of the cyanobacterium *Fremyella diplosiphon*. Proc. Nat. Acad. Sci. USA 83: 3924–3928.

COUGHLAN, S., H. C. P. HATTHIJS, AND G. HIND. 1985. The ferredoxin-NADP+ oxidoreductase-binding protein is not the 17-kDa component of the cytochrome b/f complex. J. Biol. Chem. 260: 14891–14893.

COURTICE, G. R. M., C. M. BOWMAN, T. A. DYER, AND J. C. GRAY. 1985. Localization of genes for components of photosystem II in chloroplast DNA from pea and wheat. Curr. Genet. 10: 329–333.

COZENS, A. L., J. E. WALKER, A. L. PHILLIPS, A. K. HUTTLY, AND J. C. GRAY. 1986. A sixth subunit of ATP synthase, an F_0 component, is encoded in the pea chloroplast genome. EMBO J. 5: 217–222.

CROFTS, A. R., AND C. WRAIGHT. 1983. The electrochemical domain of photosynthesis. Biochim. Biophys. Acta 726: 149–185.

CRAMER, W. A., W. R. WIDGER, R. G. HERRMANN, AND A. TREBST. 1985. Topography and function of thylakoid membrane proteins. Trends Biochem. Sci. 10: 125–129.

CRITCHLEY, C. 1985. The role of chloride in Photosystem II. Biochim. Biophys. Acta 811: 33–46.

CURTIS, S. E., AND R. HASELKORN. 1984. Isolation, sequence and expression of two members of the 32 kd thylakoid membrane protein gene family from the cyanobacterium *Anabaena* 7120. Plant Mol. Biol. 3: 249–258.

DAURAT-LARROQUE, K. BREW, AND R. E. FENNA. 1986. The complete amino acid sequence of a bacteriochlorophyll a-protein from *Prosthecochloris aestuarii*. J. Biol. Chem. 261: 3607–3615.

DEISENHOFER, J., O. EPP, K. MIKI, R. HUBER, AND H. MICHEL. 1984. X-ray structure analysis of a membrane protein complex. Electron density map at 3Å resolution and a model of the chromophores of the photosynthetic reaction center from *Rhodopseudomonas viridis*. J. Mol. Biol. 180: 385–398.

———. 1985a. Structure of the protein subunits in the photosynthetic reaction center of *Rhodopseudomonas viridis* at 3Å resolution. Nature 318: 618–624.

DEISENHOFER, J., O. H. MICHEL, AND R. HUBER. 1985b. The structural basis of photosynthetic light reactions in bacteria. Trends Biochem. Sci. 10: 243–248.

DELEPELAIRE, P., AND N.-H. CHUA. Lithium dodecyl sulfate/polyacrylamide gel electrophoresis of thylakoid membranes at 4°C: Characterization of two additional chlorophyll a-protein complexes. Proc. Nat. Acad. Sci. USA 76: 111–115.

DE VITRY, C., C. CARLES, AND B. A. DINER. 1986. Quantitation of plastoquinone-9 in photosystem II reaction center particles. FEBS Lett 196: 203–206.

DE VITRY, C., F.-A. WOLLMAN, AND P. DELEPELAIRE. 1984. Function of the polypeptides of the photosystem II reaction center in *Chlamydomonas reinhardii*. Biochim. Biophys. Acta 767: 415–422.

DILWORTH, M. F., AND E. GANTT. 1981. Phycobilisome-thylakoid topography on photosynthetically active vesicles of *Porphyridium cruentum*. Plant Physiol. 67: 608–612.

DINER, B. A., AND F.-A. WOLLMAN. 1980. Isolation of highly active photosystem II particles from a mutant of *Chlamydomonas reinhardii*. Eur. J. Biochem. 110: 521–526.

DISMUKES, G. C. 1986. The metal centers of the photosynthetic oxygen-evolving complex. Photochem. Photobiol. 43: 99–115.

DISMUKES, G. C., AND Y. SIDERER. 1981. Intermediates of a polynuclear manganese center involved in photosynthetic oxidation of water. Proc. Nat. Acad. Sci. USA 78: 274–278.

DORNEMANN, D., AND H. SENGER. 1984. The novel chlorophyll RC I associated with PS I, p. 77–80. *In* C. Sybesma [ed.] Proceedings of the VIth International Congress on Photosynthesis, Vol. II. Martinus Nijhoff/Dr. W. Junk Publishers, The Hague.

DREWS, G. 1978. Structure and Development of the membrane system of photosynthetic bacteria. Curr. Topics Bioenerget. 8: 161–207.

———. 1985. Structure and functional organization of light-harvesting complexes and

484

photochemical reaction centers in membranes of phototrophic bacteria. Microbiol. Rev. 49: 59–70.

DRUMMOND, M. H. 1985. The base sequence of the *nifF* gene of *Klebsiella pneumoniae* and homology of the predicted amino acid sequence of its protein product to other flavodoxins. Biochem. J. 232: 891–896.

DUNAHAY, T. G., L. A. STAEHELIN, M. SEIBERT, P. D. OGILVIE, AND S. P. BERG. 1984. Structural, biochemical, and biophysical characterization of four oxygen-evolving photosystem II preparation from spinach. Biochim. Biophys. Acta 764: 179–193.

DUTTON, P. L., AND R. C. PRINCE. 1978. Reaction center driven cytochrome interactions in electron and proton translocation and energy coupling, p. 525–610. *In* R. K. Clayton and W. R. Sistrom [ed.] The Photosynthetic Bacteria. Academic, New York.

EDWARDS, M. R., D. S. BERNS, W. C. GHIORSE, AND S. C. HOLT. 1968. Ultrastructure of the thermophilic blue-green alga, *Synechococcus lividus* Copeland. J. Phycol. 4: 283–289.

EDWARDS, M. R., AND E. GANTT. 1971. Phycobilisomes of the thermophilic blue-green alga *Synechococcus lividus*. J. Cell Biol. 50: 896–900.

ELEY, J. H. 1971. The effect of carbon dioxide concentration on pigmentation of the blue-green alga *Anacystis nidulans*. Plant Cell Physiol. 12: 311–316.

ENGLAND, R. R., AND E. H. EVANS. 1981. A rapid method for extraction of oxygen–evolving photosystem 2 preparations from cyanobacteria. FEBS Lett. 134: 175–177.

1983. A requirement for Ca^{2+} in the extraction of O_2-evolving Photosystem 2 preparations from the cyanobacterium *Anacystis nidulans*. Biochem. J. 210: 473–476.

1985. Characterization of the Photosystem II polypeptides of *Anacystis nidulans* by trypsin digestion, Tris washing, and lead incubation. Biochim. Biophys. Acta 808: 323–327.

ERBER, W. W. A., W. H. NITSCHMANN, R. MUCHL, AND G. A. PESCHEK. 1986. Endogenous energy supply to the plasma membrane of dark aerobic cyanobacterium *Anacystis nidulans*: ATPase-independent efflux of H + and Na + from respiring cells. Arch. Biochem. Biophys. 247: 28–39.

ERICKSON, J. M., M. RAHIRE, P. BENNOUN, P. DELEPELAIRE, B. DINER, AND J.-D. ROCHAIX. 1984a. Herbicide resistance in *Chlamydomonas reinhardii* results from a mutation in the chloroplast gene for the 32-kilodalton protein of photosystem II. Proc. Nat. Acad. Sci. USA 81: 3617–3621.

ERICKSON, J. M., M. RAHIRE, AND J.-D. ROCHAIX. 1984b. *Chlamydomonas reinhardii* gene for the 32000 mol. wt. protein of photosystem II contains four large introns and is located entirely within the chloroplast inverted repeat. EMBO J. 3: 2753–2762.

ERICKSON, J. M., M. RAHIRE, J.-D. ROCHAIX, AND L. METS. 1985a. Herbicide resistance and cross-resistance: Changes at three distinct sites in the herbicide-binding protein. Science 228: 204–207.

ERICKSON, J. M., J.-D. ROCHAIX, AND P. DELEPELAIRE. 1985a. Analysis of genes encoding two photosystem II proteins of the 30–34 kD size class, p. 53 -65. *In* K. E. Steinbeck, S. Bonitz, C. J. Arntzen, and L. Bogorad [ed.] Molecular Biology of the Photosynthetic Apparatus. Cold Spring Harbor Laboratory, Cold Spring Harbor.

EVANS, P. K., AND J. M. ANDERSON. 1986. The chlorophyll *a/b*-proteins of PS I and PS II are immunologically related.

EVANS, P. K., AND D. W. KROGMANN. 1983. Three c-type cytochromes from the red alga *Porphyridium cruentum*. Arch. Biochem. Biophys. 227: 494–510.

FALK, G., A. HAMPE, AND J. E. WALKER. 1985. Nucleotide sequence of the *Rhodospirillum rubrum atp* operon. Biochem. J. 228: 391–407.

FALKNER, G., F. HORNER, K. WERDAN, AND H. W. HELDT. 1976. pH changes in the cytoplasm of the blue-green alga *Anacystis nidulans* caused by light dependent proton flux into the thylakoid space. Plant Physiol. 58: 717–718.

FALKOWSKI, P. G., Y. FUJITA, A. LEY, AND D. MAUZERALL. 1986. Evidence for cyclic electron flow around photosystem II in *Chlorella pyrenoidosa*. Plant Physiol. 81: 310–312.

FEICK, R. G., AND R. C. FULLER. 1984. Topography of the photosynthetic apparatus of *Chloroflexus aurantiacus*. Biochemistry 23: 3693–3700.

FIKSDAHL, A. P. FOSS, AND S. LIAAEN-JENSEN. 1983. Carotenoids of blue-green algae. 11. Carotenoids of chromatically-adapted cyanobacteria. Comp. Biochem. Physiol. 76B: 599–601.

FISH, L. E., U. KUCK, AND L. BOGORAD. 1985a. Two partially homologous adjacent light-inducible maize chloroplast genes encoding polypeptide of the P700 chlorophyll *a*-protein complex of photosystem I. J. Biol. Chem. 260: 1413–1421.

1985b. Analysis of the two partially homologous P700 chlorophyll *a* proteins of maize photosystem I: Predictions based on the primary sequences and features shared by other chlorophyll proteins, p. 111–120. *In* K. E. Steinbeck, S. Bonitz, C. J. Arntzen, and L. Bogorad [ed.] Molecular Biology of the Photosynthetic Apparatus. Cold Spring Harbor Laboratory, Cold Spring Harbor.

FISHER, R. G., N. E. WOODS, H. E. FUCHS, AND R. M. SWEET. 1980. Three-dimensional structure of the phycobiliproteins I: C-phycocyanin and B-phycoerythrin at 5 Å resolution. J. Biol. Chem. 255: 5082–5089.

FITZGERALD, M. P., A. HUSAIN, G. N. HUTBER, AND L. J. ROGERS. 1977. Studies on the flavodoxins from a cyanobacterium and a red alga. Biochem. Soc. Trans. 5: 1505–1506.

FOULDS, I. J., AND N. G. CARR. 1977. A proteolytic enzyme degrading phycocyanin in the cyanobacterium *Anabaena cylindrica*. FEMS Microbiol. Lett. 2: 117–119.

FRANZEN, L.-G., O. HANSSON, AND L.-E. ANDREASSON. 1985. The roles of the extrinsic subunits in Photosystem II as revealed by EPR.

Biochim. Biophys. Acta 808: 171–179.

FUGLISTALLER, P., R. RUMBELI, F. SUTER, AND H. ZUBER. 1984. Minor polypeptides from the phycobilisome of the cyanobacterium *Mastigocladus laminosus*. Isolation, characterization and amino-acid sequences of a colour-less 8.9-kDa polypeptide and of a 16.2-kDa phycobiliprotein. Hoppe-Seyler's Z. Physiol. Chem. 365: 1085–1096.

FUGLISTALLER, P., F. SUTER, AND H. ZUBER. 1985. Linker polypeptides of the phycobilisome from the cyanobacterium *Mastigocladus laminosus*: amino-acid sequences and relationships. Biol. Chem. Hoppe-Seyler 366: 993–1001.

FUGLISTALLER, P. F., H. WIDMER, W. SIDLER, G. FRANK, AND H. ZUBER. 1981. Isolation and characterisation of phycoerythrocyanin and chromatic adaptation of the thermophilic cyanobacterium *Mastigocladus laminosus*. Arch. Microbiol. 129: 268–274.

FUJITA, Y., K. OHKI, AND A. MURAKAMI. 1985. Chromatic regulation of photosystem composition in the photosynthetic system of red and blue-green algae. Plant Cell Physiol. 26: 1541–1548.

FUTAI, M., AND H. KANAZAWA. 1983. Structure and function of proton translocating ATPase (F_oF_1): biochemical and molecular biological approaches. Microbiol. Rev. 47: 285–312.

GABELLINI, N., J. R. BOWYER, E. HURT, B. A. MELANDRI, AND G. HAUSKA. 1982. A cytochrome b/c_1 complex with ubiquinol-cytochrome c_2 oxidoreductase activity from *Rhodopseudomonas sphaeroides* GA. Eur. J. Biochem. 126: 105–111.

GABELLINI, N., U. HARNISCH, J. E. G. MCCARTHY, G. HAUSKA, AND W. SEBALD. 1985. Cloning and expression of the *fbc* operon encoding the FeS protein, cytochrome *b* and cytochrome c_1 from the Rhodopseudomonas sphaeroides b/c_1 complex. EMBO J. 4: 549–553.

GABELLINI, N., AND W. SEBALD. 1986. Nucleotide sequence and transcription of the *fbc* operon from *Rhodopseudomonas sphaeroides*. Evaluation of the deduced amino acid sequences of the FeS protein, cytochrome *b*, and cytochrome c_1. Eur. J. Biochem. 154: 569–579.

GANTT, E. 1980a. Photosynthetic cryptophytes, p. 381–405. *In* E. Cox [ed.] Phytoflagellates: Form and Function. Elsevier North Holland, Amsterdam.

1980b. Structure and function of phycobilisomes: light harvesting pigment complexes in red and blue-green algae. Int. Rev. Cytol. 66: 45–80.

1981. Phycobilisomes. Ann. Rev. Plant Physiol. 32: 327–347.

GANTT, E., AND S. F. CONTI. 1965. The ultrastructure of *Porphyridium cruentum*. J. Cell Biol. 26: 365–381.

1966. Granules associated with the chloroplast lamellae of *Porphyridium cruentum*. J. Cell Biol. 29: 423–434.

1969. Ultrastructure of blue-green algae. J. Bacteriol. 97: 1486–1493.

GANTT, E., M. R. EDWARDS, AND S. F. CONTI. 1968. Ultrastructure of *Porphyridium aerugineum*, a blue-green colored rhodophytan. J. Phycol. 4: 65–71.

GANTT, E., AND C. A. LIPSCHULTZ. 1972. Phycobilisomes of *Porphyridium cruentum*. I. Isolation. J. Cell Biol. 54: 313–324.

1980. Structure and phycobiliprotein composition of phycobilisomes from *Griffithsia pacifica* (Rhodophyceae). J. Phycol. 16: 394–398.

GANTT, E., C. A. LIPSCHULTZ, AND B. ZILINSKAS. 1976. Further evidence for a phycobilisome model from selective dissociation, fluorescence emission, immunoprecipitation, and electron microscopy. Biochim. Biophys. Acta 430: 375–388.

GANTT, E., C. A. LIPSCHULTZ, J. GRABOWSKI, AND B. K. ZIMMERMAN. 1979. Phycobilisomes from blue-green and red algae. Isolation criteria and dissociation characteristics. Plant Physiol. 63: 615–620.

GANTT, E., C. A. LIPSCHULTZ, AND T. REDLINGER. 1985. Phycobilisomes: a terminal acceptor pigment in cyanobacteria and red algae, p. 223–229. *In* K. E. Steinbeck, S. Bonitz, C. J. Arntzen, and L. Bogorad [ed.] Molecular Biology of the Photosynthetic Apparatus. Cold Spring Harbor Laboratory, Cold Spring Harbor.

GARRETT, T. R. J., D. J. CLINGELEFFER, J. M. GUSS, S. J. ROGERS, AND H. C. FREEMAN. 1984. The crystal structure of poplar apoplastocyanin at 1.8 Å resolution. J. Biol. Chem. 259: 2822–2825.

GHANOTAKIS, D. F., G. T. BABCOCK, AND C. F. YOCUM. 1984a. Calcium reconstitutes high rates of oxygen evolution in polypeptide depleted photosystem II preparations. FEBS Lett. 167: 127–130.

1984b. Structural and catalytic properties of the oxygen-evolving complex. Correlation of polypeptide and manganese release with the behavior of Z^+ in chloroplasts and a highly resolved preparation of the PS II complex. Biochim. Biophys. Acta 765: 388–398.

1985. On the role of water-soluble polypeptides (17, 23 kDa), calcium and chloride in photosynthetic oxygen evolution. FEBS Lett. 192: 1–3.

GHANOTAKIS, D. F., AND C. F. YOCUM. 1985. Polypeptides of photosystem II and their role in oxygen evolution. Photosyn. Res. 7: 97–114.

1986. Purification and properties of an oxygen-evolving reaction center complex from photosystem II membranes. FEBS Lett. 197: 244–248.

GIDDINGS, T. H. JR., C. WASSMAN, AND L. A. STAEHELIN. 1983. Structure of the thylakoids and envelope membranes of the organelles of *Cyanophora paradoxa*. Plant Physiol. 71: 409–419.

GINGRICH, J. C., L. K. BLAHA, AND A. N. GLAZER. 1982a. Rod substructure in cyanobacterial phycobilisomes: analysis of *Synechocystis* 6701 mutants low in phycoerythrin. J. Cell Biol. 92: 261–268.

486

GINGRICH, J. C., R. C. WILLIAMS, AND A. N. GLAZER. 1982b. Rod substructure in cyanobacterial phocobilisomes: phycoerythrin assembly in *Synechocystis* 6701 phycobilisomes. J. Cell Biol. 95: 170–178.

GINGRICH, J. C., D. J. LUNDELL, AND A. N. GLAZER. 1983. Core substructure in cyanobacterial phycobilisomes. J. Cell. Biochem. 22: 1–14.

GLAZER, A. N. 1981. Photosynthetic accessory proteins with bilin prosthetic groups, p. 51–96. *In* M. D. Hatch and N. K. Boardman [ed.] The Biochemistry of Plants, Vol. 8. Academic, New York.

——— 1982. Phycobilisomes: structure and dynamics. Ann. Rev. Microbiol. 36: 173–198.

——— 1983. Comparative biochemistry of photosynthetic light-harvesing systems. Ann. Rev. Biochem. 52: 125–157.

——— 1984. Phycobilisome. A macromolecular complex optimized for light energy transfer. Biochim. Biophys. Acta 768: 29–51.

——— 1985. Light harvesting by phycobilisomes. Ann. Rev. Biophys. Biophys. Chem. 14: 47–77.

GLAZER, A. N., C. CHAN, R. C. WILLIAMS, S. W. YEH, AND J. H. CLARK. 1985a. Kinetics of energy flow in the phycobilisome core. Science 230: 1051–1053.

GLAZER, A. N., AND J. H. CLARK. 1986. Phycobilisomes. Macromolecular structure and energy flow dynamics. Biophys. J. 49: 115–116.

GLAZER, A. N., AND C. S. HIXSON. 1977. Subunit structure and chromophore composition of rhodophytan phycoerythrins. *Porphyridium cruentum* B-phycoerythrin and b-phycoerythrin. J. Biol. Chem. 252: 32–42.

GLAZER, A. N., D. J. LUNDELL, G. YAMANAKA, AND R. C. WILLIAMS. 1983. The structure of a "simple" phycobilisome. Ann. Microbiol. (Inst. Pasteur) 134B: 159–180.

GLAZER, A. N., R. C. WILLIAMS, G. YAMANAKA, AND H. K. SCHACHMAN. 1979. Characterization of cyanobacterial phycobilisomes in zwitterionic detergents. Proc. Nat. Acad. Sci. USA 76: 6162–6166.

GLAZER, A. N., S. W. YEH, S. P. WEBB, AND J. H. CLARK. 1985b. Disk-to-disk transfer as the rate-limiting step for energy flow in phycobilisomes. Science 227: 419–423.

GOEDHEER, J. C. 1976. Spectral properties of the blue-green alga *Anacystis nidulans* grown under different environmental conditions. Photosynthetica 10: 411–422.

GOLBECK, J. H., AND J. M. CORNELIUS. 1986. Photosystem I charge separation in the absence of centers A and B. I. Optical characterization of center "A_2" and evidence for its association with a 64-kDa peptide. Biochim Biophys. Acta 849, 16–24.

GOLDEN, S. S., AND R. HASELKORN. 1985. Mutation to herbicide resistance maps within the *psbA* gene of *Anacystis nidulans* R2. Science 229: 1104–1107.

GOLDEN, S. S., R. HASELKORN, AND L. A. SHERMAN. 1985. Analysis of a cyanobacterial gene encoding herbicide resistance, p. 73–77. *In* K. E. Steinbeck, S. Bonitz, C. J. Arntzen, and L. Bogorad [ed.] Molecular biology of the photosynthetic apparatus. Cold Spring Harbor Laboratory, Cold Spring Harbor.

GOLECKI, J. R., AND G. DREWS. 1982. Supramolecular organization and composition of membranes, p. 125–142. *In* N. G. Carr and B. A. Whitton [ed.] The biology of the cyanobacteria. Blackwell Scientific, Oxford.

GOODCHILD, D. J., B. ANDERSON, AND J. M. ANDERSON. 1985. Immunocytochemical localization of polypeptides associated with the oxygen evolving system of photosynthesis. Eur. J. Cell Biol. 36: 294–298.

GOUNARIS, K., AND J. BARBER. 1985. Isolation and characterisation of a photosystem II reaction center lipoprotein complex. FEBS Lett. 188: 68–72.

GREEN, B. R. Protein synthesis by isolated *Acetabularia* chloroplasts. *In vivro* synthesis of the apoprotein of the P-700-chlorophyll *a*-protein complex. Biochim. Biophys. Acta. 609: 107–120.

GROSSMAN, A. R., S. G. BARTLETT, G. W. SCHMIDT, J. E. MULLET, AND N.-H. CHUA. 1982. Optimal conditions for post-translational uptake of proteins by isolated chloroplasts. J. Biol. Chem. 257: 1558–1563.

GUGLIELMI, G., AND G. COHEN-BAZIRE. 1984. Étude taxonomique d'un genre de cyanobactérie Oscillatoriacee: le genre *Pseudanabaena* Lauterborn. II. Analyse de la composition moléculaire et de la structure des phycobilisomes. Protistologica 20: 393–413.

GUGLIELMI, G., G. COHEN-BAZIRE, AND D. A. BRYANT. 1981. The structure of *Gloeobacter violaceus* and its phycobilisomes. Arch. Microbiol. 129: 181–189.

GUIKEMA, J. A., AND L. A. SHERMAN. 1983. Organization and function of chlorophyll in membranes of cyanobacteria during iron starvation. Plant Physiol. 73: 250–256.

——— 1984. Influence of iron deprivation on the membrane composition of *Anacystis nidulans*. Plant Physiol. 74: 90–95.

GUSS, J. M., AND H. C. FREEMAN. 1983. Structure of oxidized poplar plastocyanin at 1.6 Å resolution. J. Mol. Biol. 169: 521–563.

HAEHNEL, W. 1984. Photosynthetic electron transport in higher plants. Ann. Rev. Plant Physiol. 35: 659–693.

HALLDAL, P. 1958. Pigment formation and growth in the blue-green algae in crossed gradients of light intensity and temperature. Physiol. Plant. 11: 401–420.

——— 1970. The photosynthetic apparatus of microalgae and its adaptation to environmental factors, p. 17–55. *In* P. Halldal [ed.] Photobiology of microorganisms. Wiley, London.

HARDIE, L. P., D. L. BALKWILL, AND S. E. STEVENS, JR. 1983. Effects of iron starvation on the physiology of the cyanobacterium *Agmenellum quadruplicatum*. Appl. Environ. Microbiol. 45: 999–1006.

HASE, T., H. MATSUBARA, G. N. HUTBER, AND L.

J. ROGERS. 1982. Amino acid sequences of *Nostoc* Strain MAC ferredoxins I and II. J. Biochem. 92: 1347–1355.

HAUSKA, G. 1985. Organization and function of cytochrome $b_6 f/bc_1$ complexes, p. 79–87. *In* K. E. Steinbeck, S. Bonitz, C. J. Arntzen, and L. Bogorad [ed.] Molecular biology of the photosynthetic apparatus. Cold Spring Harbor Laboratory, Cold Spring Harbor.

HAUSKA, G., E. HURT, N. GABELLINI, J. W. DAVENPORT, AND W. O. LOCKAU. 1984. Organization and function of photosynthetic and respiratory cytochrome *b/c*-FeS complexes, p. 243–249. *In* C. Sybesma [ed.] Advances in Photosynthesis Research, Vol. III. Martinus Nijhoff/Dr. W. Junk Publishers, The Hague.

HAUSKA, G., E. HURT, N. GABELLINI, AND W. LOCKAU. 1983. Comparative aspects of quinol-cytochrome *c*/plastocyanin oxidoreductases. Biochim. Biophys. Acta 726: 97–133.

HAWORTH, P., J. L. WATSON, AND C. J. ARNTZEN. 1983. The detection, isolation and characterization of a light-harvesting complex which is specifically associated with photosystem I. Biochim. Biophys. Acta 724: 151–158.

HEARST, J. E., AND K. SAUER. 1983. Protein sequence homologies between portions of the L and M subunits of reaction centers of *Rhodopseudomonas capsulata* and the Q_b-protein of chloroplast thylakoid membranes: a proposed relation to quinone-binding sites. Z. Naturforsch. 39c: 421–424.

HEINEMEYER, W., J. ALT, AND R. G. HERRMANN. 1984. Nucleotide sequence of the clustered genes for apocytochrome *b6* and subunit 4 of the cytochrome *b/f* complex in the spinach plastid chromosome. Curr. Gent. 8: 543–549.

HENNIG, J., AND R. G. HERRMANN. 1986. Chloroplast ATP synthase of spinach contains nine nonidentical subunit species, six of which are encoded by plastid chromosomes in two operons in a phylogenetically conserved arrangement. Mol. Gen. Genet. 203: 117–128.

HERRMANN, R. G., J. ALT, B. SCHILLER, W. R. WIDGER, AND W. A. CRAMER. 1984. Nucleotide sequence of the gene for apocytochrome *b*-559 on the spinach plastid chromosome: implications for the structure of the membrane protein. FEBS Lett. 176: 239–244.

HERRMANN, R. G., P. WESTHOFF, J. ALT, J. TITTGEN, AND N. NELSON. 1985. Thylakoid membrane proteins and their genes. *In* L. van Vloten-Doting, G. S. P. Groot, and T. C. Hall [ed.] Molecular form and function of the plant genome. NATO ASI Series A: Life Sciences Vol. 83. Plenum, New York, NY.

HERTZBERG, S., S. LIAAEN-JENSEN, AND H. W. SIEGELMAN. 1971. The carotenoids of the blue-green algae. Phytochem. 10: 3121–3127.

HERVAS, M., J. M. ORTEGA, M. A. DE LA ROSA, R. R. DE LA ROSA, AND M. LOSADA. Location and function of cytochrome *b*-559 in the chloroplast noncyclic electron transport chain. Physiol. Veg. 23: 593–604.

HICKS, D. B., AND C. F. YOCUM. 1986a. Properties of the cyanobacterial coupling factor ATPase from *Spirulina platensis*. I. Electrophoretic characterization and reconstitution of photophosphorylation. Arch. Biochem. Biophys. 245: 220–229.

1986b. Properties of the cyanobacterial coupling factor ATPase from *Spirulina platensis*. II. Activity of the purified and membrane-bound enzymes. Arch. Biochem. Biophys. 245: 230–237.

HINZ, U. G. 1985. Isolation of the photosystem II reaction center complex from barley. Characterization by circular dichroism spectroscopy and amino acid sequencing. Carlsberg. Res. Commun. 50: 285–298.

HIRD, S. M., D. L. WILLEY, T. A. DYER, AND J. C. GRAY. 1986. Location and nucleotide sequence of the gene for cytochrome *b*-559 in wheat chloroplast DNA. Mol. Gen. Genet. 203: 95–100.

HIRSCHBERG, J., A. BLEECKER, D. J. KYLE, L. MCINTOSH, AND C. J. ARNTZEN. 1984. The Molecular basis of triazine-herbicide resistance in higher-plant chloroplasts. Z. Naturforsch. 39c: 412–420.

HIRSCHBERG, J., AND L. MCINTOSH. 1983. Molecular basis of herbicide resistance in *Amaranthus hybridus*. Science 222: 1346–1349.

HLADIK, J., AND D. SOFROVA. 1983. The structure of cyanobacterial thylakoid membranes. Photosynthetica 17: 267–288.

HO, K. K., AND D. W. KROGMANN. 1980. Cytochrome *f* from spinach and cyanobacteria: purification and characterization. J. Biol. Chem. 255: 3855–3861.

1982. Photosynthesis, p. 191–214. *In* N. G. Carr and B. A. Whitton [ed.] The biology of the cyanobacteria. Blackwell Scientific, Oxford.

1984. Electron donors to P700 in cyanobacteria and algae. An instance of unusual genetic variability. Biochim. Biophys. Acta 766: 310–316.

HO, K. K., E. L. ULRICH, D. W. KROGMANN, AND G. GOMEZ-LOJERO. 1979. Isolation of photosynthetic catalysts from cyanobacteria. Biochim. Biophys. Acta 545: 236–248.

HOARAU, J., R. REMY, AND J. C. LECLERC. 1977. Hétérogénité des variations spectrales photoinduites vers 700 nm observées sur les membranes chlorophylliennes et les complexes chlorophyll-protéines isolés de divers organismes photosynthétiques. Biochi. Biophys. Acta 462: 659–670.

HOFF, A. J. 1982. Photooxidation of the reaction center chlorophylls and structural properties of photosynthetic reaction centers. *In* F. K. Fong [ed.] Light reaction path of photosynthesis. pp. 80–151. Mol. Biol., Biochem., Biophys. Vol. 35, Springer-Verlag, Berlin.

HOLSCHUH, K., W. BOTTOMLEY, AND P. R. WHITFIELD. 1984. Structure of the spinach chloroplast genes for the D2 and 44 kd reaction-center proteins of photosystem II and for tRNA-Ser (UGA). Nucleic Acids Res. 12: 8819–8834.

HOLT, T., AND D. W. KROGMANN. 1981. A carotenoid protein complex from cyanobacteria. Biochim. Biophys. Acta 637: 408–414.

HOMANN, P. H. 1985. The association of functional anions with the oxygen-evolving center of chloroplasts. Biochim. Biophys. Acta 809: 311–319.

HOWE, C. J., A. D. AUFFRET, A. DOHERTY, C. M. BOWMAN, T. A. DYER, AND J. C. GRAY. 1982. Location and nucleotide sequence of the gene for the proton-translocating subunit of wheat chloroplast ATP synthase. Proc. Nat. Acad. Sci. USA 79: 6903–6907.

HOWE, C. J., I. M. FEARNLEY, J. E. WALKER, T. A. DYER, AND J. C. GRAY. 1985. Nucleotide sequences of the genes for the alpha, beta and epsilon subunits of wheat chloroplast ATP synthase. Plant Mol. Biol. 4: 333–345.

HOYER-HANSEN, G., L. S. HONBERG, AND P. B. HOJ. 1985. Probing *in vitro* translation products with monoclonal antibodies to a 15.2 kD polypeptide subunit of photosystem. I. Carlsberg Res. Commun. 50: 211–221.

HUANG, C., D. S. BERNS, AND D. U. GUARINO. Characterization of components of P-700-chlorophyll *a*-protein complex from a blue-green alga, *Phormidium luridum*. Biochim. Biophys. Acta 765: 21–29.

HURT, E., AND G. HAUSKA. 1981. A cytochrome f/b_6 complex of five polypeptides with plastoquinol-plastocyanin-oxidoreductase activity from spinach chloroplasts. Eur. J. Biochem. 117: 591–599.

1982. Identification of the polypeptides in the cytochrome b_6/f complex from spinach chloroplasts with redox-center-carrying subunits. J. Bioenerget. Biomembr. 14: 405–424.

HURT, E., G. HAUSKA, AND R. MALKIN. 1981. Isolation of the Rieske iron-sulfur protein from the cytochrome b_6/f complex of spinach chloroplasts. FEBS Lett. 134: 1–5.

HUTBER, G. N., K. G. HUTSON, AND L. J. ROGERS. 1977. Effect of iron deficiency on levels of two ferredoxins and flavodoxin in a cyanobacterium. FEMS Microbiol. Lett. 1: 193–196.

HUTBER, G. N., A. J. SMITH, AND L. J. ROGERS. 1978. Isolation of ferredoxin-adenine dinucleotide phosphate (oxidized) reductase from a prokaryote. Biochem. Soc. Trans. 6: 1212–1216.

HUTTLY, A. K., AND J. C. GRAY. 1984. Localisation of genes for four ATP synthase subunits in pea chloroplast DNA. Mol. Gen. Genet. 194: 402–409.

IHLENFELDT, M. J. A., AND J. GIBSON. 1975. Phosphate utilization and alkaline phosphatase activity in *Anacystis nidulans* (*Synechococcus*). Arch. Microbiol. 102: 23–28.

IKEUCHI, M., AND Y. INOUE. 1986. Characterization of O_2 evolution by a wheat photosystem II reaction center complex isolated by a simplified method: disjunction of secondary acceptor quinone and enhanced Ca^{++} demand. Arch. Biochem. Biophys. 247: 97–107.

IKEUCHI, M., M. YUASA, AND Y. INOUE. 1985. Simple and discrete isolation of an O_2-evolving PS II reaction center complex retaining Mn and the extrinsic 33 kDa protein. FEBS Lett. 185: 316–322.

IMAOKA, A., K. AKABORI, M. YANAGI, K. IZUMI, Y. TOYOSHIMA, A. KAWAMORI, H. NAKAYAMA, AND J. SATO. 1986. Roles of three lumen-surface proteins in the formation of S_2 state and O_2 evolution in Photosystem II particles from spinach thylakoid membranes. Biochim. Biophys. Acta 848: 210–211.

ISOGAI, Y., Y. YAMAMOTO, AND M. NISHIMURA. 1985. Association of the 33-kDa polypeptide with the 43 kDa component in photosystem II particles. FEBS Lett. 187: 240–244.

IZAWA, S. 1986. The effect of chloride on oxygen evolution. Photosyn. Res. (in press)

JAYNES, J. M., AND L. P. VERNON. 1982. The cyanelle of *Cyanophora paradoxa*: almost a cyanobacterial chloroplast. Trends Biochem. Sci. 7: 22–24.

JOLIOT, P., G. BARBIERI, AND R. CHABAUD. 1969. Un nouveau modèle des centres photochimiques du système II. Photochem. Photobiol. 10: 309–329.

JONES, L. W., AND J. MYERS. 1965. Pigment variation in *Anacystis nidulans* induced by light of selected wavelengths. J. Phycol. 1: 7–13.

JOST, M. 1985. Die ultrastruktur von *Oscillatoria rubescens* D. C. Arch. Mikrobiol. 50: 211–245.

KALLAS, T., AND R. MALKIN. 1985. Genomic organization and isolation of genes for the cytocrome b_6/f complex from the cyanobacterium *Nostoc* PCC 7906. 1st International Congress of Plant Molecular Biology, Abstracts, p. 123.

KAMBARA, T., AND GOVINDJEE. 1985. Molecular mechanism of water oxidation in photosynthesis based on the functioning of manganese in two different environments. Proc. Nat. Acad. Sci. USA 82: 6119–6123.

KARLIN-NEUMANN, G. A., B. D. KOHORN, J. P. THORNBER, AND E. M. TOBIN. 1985. A chlorophyll *a/b* protein encoded by a gene containing an intron with characteristics of a transposable element. J. Mol. Applied Genet. 3: 45–61.

KARPLUS, P. A., K. A. WALSH, AND J. R. HERRIOTT. 1984. Amino acid sequence of spinach ferredoxin:NADP+ oxidoreductase. Biochemistry 23: 6576–6583.

KATOH, S., I. SHIRATORI, AND A. TAKAMIYA. 1962. Purification and properties of spinach plastocyanin. J. Biochem. 51: 32–40.

KATOH, T., D. DORNEMANN, AND H. SENGER. 1985. Evidence for chlorophyll RC I in cyanobacteria. Plant Cell Physiol. 26. 1583–1586.

KATOH, T., AND E. GANTT. 1979. Photosynthetic vesicles with bound phycobilisomes from *Anabaena variabilis*. Biochim. Biophys. Acta 546: 383–393.

KAWAMURA, M., M. MIMURO, AND Y. FUJITA. 1979. Quantitative relationship between two reaction centers in the photosynthetic system of blue-green algae. Plant Cell Physiol. 20: 697–705.

KE, B., H. INOUE, G. T. BABCOCK, Z.-X. FANG, AND E. DOLAN. 1982. Optical and EPR characterizatins of oxygen-evolving photosystem II subchloroplast fragments isolated from the thermophilic blue-green alga *Phormidium laminosum*. Biochim. Biophys. Acta 682: 297–306.

KELLER, M., AND E. STUTZ. 1984. Structure of the *Euglena gracilis* chloroplast gene (*psbA*) coding for the 32-kDa protein of photosystem II. FEBS Lett. 175: 173–177.

KENYON, C. N. 1972. Fatty acid composition of unicellular strains of blue-green algae. J. Bacteriol. 109: 827–834.

KENYON, C. N., R. RIPPKA, AND R. Y. STANIER. 1972. Fatty acid composition and physiological properties of some filamentous blue-green algae. Arch. Mikrobiol. 83: 216–236.

KHANNA, R., J.-R. GRAHAM, J. MYERS, AND E. GANTT. Phycobilisome composition and possible relationship to reaction centers. 1983. Arch. Biochem. Biophys. 224: 534–542.

KLEIN, S. M., L. P. VERNON, AND S. S. KENT. 1983. Photosynthetic properties and components of the endosymbiont *Cyanophora paradoxa*, p. 175–183. *In* G. Akoyunoglou [ed.] Photosynthesis, Vol. III. Proceedings of the Vth International Congress on Photosynthesis.

KLIMOV, V. V., A. V. KLEVANIK, V. A. SHUVALOV, AND A. V. KRASNOVSKY. 1977. Reduction of pheophytin in the primary light reaction of photosystem II. FEBS Lett. 82: 183–186.

KLIMOV, V. V., E. DOLAN, AND B. KE. 1980. EPR properties of an intermediary electron acceptor (phaeophytin) in photosystem II reaction centers at cryogenic temperatures. FEBS Lett. 112: 97–100.

KO, K., J. M. JAYNES, AND N. STRAUSS. 1985. Homology between the cyanelle DNA of *Cyanophora paradoxa* and the chloroplast DNA of *Vicia faba*. Plant Science 42: 115–123.

KOKORN, B. D., E. HAREL, P. R. CHITNIS, J. P. THORNBER, AND E. M. TOBIN. 1986. Functional and mutational analysis of the light-harvesting chlorophyll *a/b* protein of thylakoid membranes. J. Cell Biol. 102: 972–981.

KOIKE, H., AND Y. INOUE. 1985. Properties of a peripheral 34 kDa protein in *Synechococcus vulcanus* photosystem II particles. Its exchangeability with spinach 33 kDa protein in reconstitution of 02 evolution. Biochim. Biophys. Acta 807: 64–73.

KOK, B. 1956. On the reversible absorption change at 705 mμ in photosynthetic organisms. Biochim. Biophys. Acta 22: 399–401.

KOK, B., B. FORBUSH, AND M. MCGLOIN. 1970. Cooperation of charges in photosynthetic O_2 evolution-I. A linear four step mechanism. Photochem. Photobiol. 11: 457–475.

KREBBERS, E. T., I. M. LARRINUA, L. MCINTOSH, AND L. BOGORAD. 1982. The maize chloroplast genes for the β and ϵ subunits of the photosynthetic coupling factor CF1 are fused. Nucleic Acids Res. 10: 4985–5002.

KRINNER, M., G. HAUSKA, E. HURT, AND W. LOCKAU. 1982. A cytochrome *f-b*$_6$ complex with plastoquinol-cytochrome *c* oxidoreductase activity from *Anabaena variabilis*. Biochim. Biophys. Acta 681: 110–117.

KUCK, U. 1985. DNA sequence analysis of the gene encoding the P700 chlorophyll a-protein of photosystem I from the green alga *Chlamydomonas reinhardii*. Ist Internat. Congress Plant Mol. Biol. Abstracts, p. 125.

KUNKEL, D. D. 1982. Thylakoid centers: Structures associated with the cyanobacterial photosynthetic membrane system. Arch. Microbiol. 133: 97–99.

KUWABARA, T., M. MIYAO, T. MURATA, AND N. MURATA. 1985. The function of the 33-kDa protein in the photosynthetic oxygen-evolution system studied by reconstitution experiments. Biochim. Biophys. Acta 806: 283–289.

KUWABARA, T., AND J. MURATA. 1982. Inactivation of photosynthetic oxygen evolution and concomitant release of three polypeptides in the photosystem II particles of spinach chloroplasts. Plant Cell Physiol. 23: 533–539.

——— 1983. Quantitative analysis of the inactivation of photosynthetic oxygen evolution and the release of polypeptides and maganese in the photosystem II particles of spinach chloroplasts. Plant Cell Physiol. 24: 741–747.

KYLE, D. J. 1985. The 32000 dalton Q_b protein of photosystem II. Photochem. Photobiol. 41: 107–116.

LAM, E. 1986. Nearest-neighbor relationships of the constituent polypeptides in plastoquinol-plastocyanin oxidoreductase. Biochim. Biophys. Acta 848: 324–332.

LAM, E., AND R. MALKIN. 1985. Topography of the protein complexes of the chloroplast thylakoid membrane. Plant Physiol. 79: 1118–1124.

LAM, E., W. ORTIZ, S. MAYFIELD, AND R. MALKIN. 1984a. Isolation and characterization of a light-harvesting chlorophyll *a/b* protein complex associated with photosystem I. Plant Physiol. 74: 650–655.

LAM, E., W. ORTIZ, AND R. MALKIN. 1984b. Chlorophyll *a/b* proteins of Photosystem I. FEBS Lett. 168: 10–14.

LAMBERT, D. H., D. A. BRYANT, V. L. STIREWALT, J. M. DUBBS, S. E. STEVENS, JR., AND R. D. PORTER. 1985. Gene map for the *Cyanophora paradoxa* cyanelle genome. J. Bacteriol. 164: 659–664.

LAMBERT, D. H., AND S. E. STEVENS, JR. 1986. Photoheterotrophic growth of *Agmenellum quadruplicatum* PR-6. J. Bacteriol. 165: 654–656.

LANG, N. J. 1977. *Starria zimbabweensis* (Cyanophyceae) gen. nov. et sp. nov.: a filament triradiate in transverse sections. J. Phycol. 13: 288–296.

LAU, R. H., M. M. MACKENZIE, AND W. F. DOOLITTLE. 1977. Phycocyanin synthesis and degradation in the blue-green bacterium *Anacystis nidulans*. J. Bacteriol. 132: 771–778.

LAVERGNE, J. 1985. Absorption changes in photosystem II: recent developments. Physiol. Veg. 23: 411–423.

LEHMBECK, J., O. F. RASSMUSSEN, G. B. BOOKJANS, B. JEPSON, B. M. STUMMANN, AND K. W. HENNINGSEN. 1985. Sequence of the genes in pea chloroplast DNA coding for the 84 and 83 kD polypeptides of the PS I complex: the 83 kD gene contains a putative origin or replication. I[st] Internat. Congress Plant Mol. Biol., Abstracts, p. 123.

LEMAUX, P. G., AND A. GROSSMAN. 1984. Isolation and characterization of a gene for a major light-harvesting polypeptide from *Cyanophora paradoxa*. Proc. Nat. Acad. Sci. USA 81: 4100–4104.

——— 1985. Major light-harvesting polypeptides encoded in polycistronic transcripts in a eukaryotic alga. EMBO J. 4: 1911–1919.

LEWIN, R. A. 1984. *Prochloron* — A status report. Phycologia 23: 203–208.

LICHTLE, C., AND J. C. THOMAS. 1976. Étude ultrastructurale des thylacoides des algues à phycobiliprotéines, comparaison des résultats obtenus par fixation classique et cryodécapage. Phycologia 15: 393–404.

LIND, L. K., S. R. KALLA, A. LONNEBORG, G. OQUIST, AND P. GUSTAFSSON. 1985. Cloning of the β-phycocyanin gene from *Anacystis nidulans*. FEBS Lett. 188: 27–32.

LINDBERG-MOLLER, B., AND P. B. HOJ. 1983. A thylakoid polypeptide involved in the reconstitution of photosynthetic oxygen evolution. Carlsberg Res. Commun. 48: 161–185.

LIVEANU, V., C. F. YOCUM, AND N. NELSON. 1986. Polypeptides of the oxygen-evolving photosystem II complex. Immunological detection and biogenesis. J. Biol. Chem. 261: 5296–5300.

LJUNGBERG, U., H.-E. AKERLUND, AND B. ANDERSSON. 1984a. The release of a 10-kDa polypeptide from everted photosystem II thylakoid membranes by alkaline Tris. FEBS. Lett. 175: 255–258.

LJUNGBERG, U., H.-E. AKERLUND, C. LARSSON, AND B. ANDERSSON. 1984b. Identification of polypeptides associated with the 23 and 33 kDa proteins of photosynthetic oxygen evolution. Biochim. Biophys. Acta 767: 145–152.

LJUNGBERG, J., T. HENRYSSON, C. P. ROCHESTER, H.-E. AKERLUND, AND B. ANDERSSON. 1986. The presence of low-molecular-weight polypeptides in spinach Photosystem II core preparations. Isolation of a 5 kDa hydrophilic polypeptide. Biochim. Biophys. Acta 949: 112–120.

LONNEBORG, A., L. K. LIND, S. R. KALLA, P. GUSTAFSSON, AND G. OQUIST. Acclimation processes in the light-harvesting system of the cyanobacterium *Anacystis nidulans* following a light shift from white to red light. Plant Physiol. 78: 110–114.

DE LORIMIER, R., D. A. BRYANT, R. D. PORTER, W.-Y. LIU, E. JAY, AND S. E. STEVENS, JR. 1984. Genes for the α and β subunits of phycocyanin. Proc. Nat. Acad. Sci. USA 81: 7946–7950.

LUBBERDING, H. J., G. ZIMMER, H. S. VAN WALRAVEN, J. SCHRICKX, AND R. KRAAYENHOF. 1983. Isolation, purification, and characterization of the ATPase complex from the thermophilic cyanobacterium *Synechococcus* 6716. Eur. J. Biochem. 137: 95–99.

LUNDELL, D. J., AND A. N. GLAZER. 1981.

Allophycocyanin B. A common β subunit in *Synechococcus* allophycocyanin B (λ max 670 nm) and allophycocyanin (λ max 650 nm). J. Biol. Chem. 256: 12600–12606.

——— 1983a. Molecular architecture of a light-harvesting antenna. Structure of the 18 S core-rod subassembly of the *Synechococcus* 6301 phycobilisome. J. Biol. Chem. 258: 894–901.

——— 1983b. Molecular architecture of a light-harvesting antenna. Core substructure in *Synechococcus* 6301 phycobilisomes: two new allophycocyanin and allophycocyanin B complexes. J. Biol. Chem. 258: 902–908.

——— 1983c. Molecular architecture of a light-harvesting antenna. Quaternary interactions in the *Synechococcus* 6301 phycobilisome core as revealed by partial tryptic digestion and circular dichroism studies. J. Biol. Chem. 258: 8708–8713.

LUNDELL, D. J., R. C. WILLIAMS, AND A. N. GLAZER. 1981a. Molecular architecture of a light-harvesting antenna. *In vitro* assembly of the rod substructures of *Synechococcus* 6301 phycobilisomes. J. Biol. Chem. 256: 3580–3592.

LUNDELL, D. J., A. N. GLAZER, A. MELIS, AND R. MALKIN. 1985. Characterization of a cyanobacterial photosystem I complex. J. Biol. Chem. 260: 646–654.

LUNDELL, D. J., G. YAMANAKA, AND A. N. GLAZER. 1981b. A terminal energy acceptor of the phycobilisome: the 75,000-dalton polypeptide of *Synechococcus* 6301 phycobilisomes — a new biliprotein. J. Cell Biol. 91: 315–319.

MACHOLD, O., D. J. SIMPSON, AND B. LINDBERG-MOLLER. 1979. Chlorophyll-proteins of thylakoids from wild-type and mutants of barley (*Hordeum vulgare* L.). Carlsberg Res. Commun. 44: 235–254.

MALKIN, R. 1982. Photosystem I. Ann. Rev. Plant Physiol. 33: 455–479.

MALKIN, R., W. ORTIZ, E. LAM, AND J. BONNERJEA. 1985. Structural organization of thylakoid membrane electron transfer complexes: photosystem I. Physiol. Veg. 23: 619–625.

MANODORI, A., M. ALHADEFF, A. N. GLAZER, AND A. MELIS. 1984. Photochemical apparatus organization in *Synechococcus* 6301 (*Anacystis nidulans*). Effect of phycobilisome mutation. Arch. Microbiol. 139: 117–123.

——— 1985. Phycobilisome-photosystem II association in *Synechococcus* 6301 (Cyanophyceae). FEBS Lett. 181: 79–82.

MANODORI, A., AND A. MELIS. 1984. Photochemical apparatus organization in *Anacystis nidulans* (Cyanophyceae). Effect of CO_2 concentration during cell growth. Plant Physiol. 74: 67–71.

MANSFIELD, R. W., AND J. M. ANDERSON. 1985. Transverse organization of components within the chloroplast cytochrome *b-563/f* complex. Biochim. Biophys. Acta 809: 435–444.

MANSFIELD, R. W., AND M. C. W. EVANS. 1985. Optical difference spectrum of the electron acceptor A_o in photosystem I. FEBS Lett. 190: 237–241.

MATHIS, P. 1986. Structural aspects of vectorial electron transfer in photosynthetic reaction centers. Photosynthesis Res. 8: 97–111.

MATSUBARA, H., AND T. HASE. 1983. Phylogenetic consideration of ferredoxin sequences in plants, particularly algae, p. 168–181. *In* U. Jensen and D. E. Fairbrothers [ed.] Proteins and nucleic acids in plant systematics. Springer-Verlag, Berlin.

MATSUURA, K., AND S. ITOH. 1985. Comparison between electron transfers through plastocyanin in spinach chloroplasts and cytochrome c_2 in *Rhodopseudomonas sphaeroides*. Plant Cell Physiol. 26: 1057–1065.

MATTHEWS, B. W., R. E. FENNA, M. C. BOLOGNESI, M. F. SCHMID, AND J. M. OLSON. 1979. Structure of bacteriochlorophyll *a*-protein from the green photosynthetic bacterium *Prosthecochloris aestuarii*. J. Mol. Biol. 131: 259–285.

MCCARTY, R. E. 1985. H^+-ATPases in oxidative and photosynthetic phosphorylation. BioScience 35: 27–30.

MERCHANT, S., AND L. BOGORAD. 1986. Regulation by copper of the expression of plastocyanin and cytochrome *c*-552 in *Chlamydomonas reinhardi*. Mol. Cell. Biol. 6: 462–469.

MERCHANT, S., AND B. SELMAN. 1985. Photosynthetic ATPase: purification, propreties, subunit isolation and function. Photosyn. Res. 6: 3–31.

METZ, J. G., G. ULMER, T. M. BRICKER, AND D. MILES. 1983. Purification of cytochrome *b*-559 from oxygen-evolving photosystem II preparations of spinach and maize. Biochim. Biophys. Acta 725: 203–209.

MICHEL, H., K. A. WEYER, H. GRUENBERG, AND F. LOTTSPEICH. 1985. The "heavy" subunit of the photosynthetic reaction centre from *Rhodopseudomonas viridis*: isolation of the gene, nucleotide and amino acid sequence. EMBO J. 4: 1667–1672.

MILLER, L. S., AND S. C. HOLT. 1977. Effect of carbon dioxide on pigment and membrane content in *Synechococcus lividus*. Arch. Microbiol. 115: 185–198.

MILLER, M. 1985. The release of polypeptides and manganese from oxygen-evolving photosystem II preparations following zinc-treatment. FEBS Lett. 189: 355–360.

MILLNER, P. A., AND J. BARBER. 1984. Plastoquinone as a mobile redox carrier in the photosynthetic membrane. FEBS Lett. 169: 1–6.

MIMURO, M., P. FUGLISTALLER, R. RUMBELI, AND H. ZUBER. 1986. Functional assignment of chromophores and energy transfer in C-phycocyanin isolated from the thermophilic cyanobacterium *Mastigocladus laminosus*. Biochim. Biophys. Acta 848: 155–166.

MIMURO, M., AND Y. FUJITA. 1977. Estimation of chlorophyll *a* distribution in the photosynthetic pigment systems I and II of the blue-green alga *Anabaena variabilis*. Biochim. Biophys. Acta 459: 376–389.

MINAMI, E.-I., K. SHINOHARA, T. KUWABARA, AND A. WATANABE. 1986. *In vitro* synthesis and assembly of photosystem II proteins of spinach chloroplasts. Arch. Biochem. Biophys. 244: 517–527.

MITCHELL, P. 1968. Chemiosmotic Coupling and Energy Transduction. Glynn Research Ltd., Bodmin, Cornwall.

MIYAIRI, S., AND G. H. SCHATZ. 1983. Oxygen-evolving extracts from a thermophilic cyanobacterium *Synechococcus* sp. Z. Naturforsch. 38c: 44–48.

MIYAO, M., AND N. MURATA. 1984a. Effect of urea on photosystem II particles. Evidence for an essential role of the 33 kilodalton polypeptide in photosynthetic oxygen evolution. Biochim. Biophys. Acta 765: 253–257.

——— 1984b. Calcium ions can be substituted for the 24-kDa polypeptide in photosynthetic oxygen evolution. FEBS Lett. 168: 118–120.

——— 1984c. Role of the 33-kDa polypeptide in preserving Mn in the photosynthetic oxygen-evolution system and its replacement by chloride ions. FEBS Lett. 170: 350–354.

MORONEY, J. V., L. LOPRESTI, B. F. MCEWEN, R. E. MCCARTY, AND G. G. HAMMES. 1983. The M_r-value of chloroplast coupling factor 1. FEBS Lett. 158: 58–62.

MORRIS, J., AND R. G. HERRMANN. 1984. Nucleotide sequence of the gene for the P680 chlorophyll *a* apoprotein of the photosystem II reaction center from spinach. Nucleic Acids Res. 12: 2837–2850.

MORSCHEL, E., K.-P. KOLLER, AND W. WEHRMEYER. 1980. Biliprotein assembly in the disc-shaped phycobilisomes of *Rhodella violacea*. Electron microscopial and biochemical analyses of C-phycocyanin and allophycocyanin aggregates. Arch. Microbiol. 125: 43–51.

MORSCHEL, E., K.-P. KOLLER, W. WEHRMEYER, AND H. SCHNEIDER. 1977. Biliprotein assembly in the disc-shaped phycobilisomes of *Rhodella violacea*. I. Electron microscopy of phycobilisomes in situ and analysis of their architecture after isolation and negative staining. Cytobiologie 16: 118–129.

MORSCHEL, E., AND L. A. STAEHELIN. 1983. Reconstitution of cytochrome f/b_6 and CF_o-CF_1 ATP synthase complexes into phospholipid and galactolipid liposomes. J. Cell Biol. 97: 301–310.

MULLET, J. E., J. J. BURKE, AND C. J. ARNTZEN. 1980a. Chlorophyll proteins of photosystem I. Plant Physiol. 65: 814–822.

——— 1980b. A developmental study of photosystem I peripheral chlorophyll proteins. Plant Physiol. 65: 823–827.

MULLET, J. E., A. R. GROSSMAN, AND N.-H. CHUA. 1982. Synthesis and assembly of the polypeptide subunits of photosystem I. Cold Spring Harbor Symp. Quant. Biol. 46: 979–984.

MULLIGAN, B., N. SCHULTES, L. CHEN, AND L. BOGORAD. 1984. Nucleotide sequence of a multiple-copy gene for the B protein of photosystem II of a cyanobacterium. Proc. Nat. Acad. Sci. USA 81: 2693–2697.

MURATA, N., S. ARAKI, Y. FUJITA, K. SUZUKI, T. KUWABARA, AND P. MATHIS. 1986. Stoichiometric determination of pheophytin in photosystem II of oxygenic photosynthesis. Photosyn. Res. 9: 63–70.

MURATA, N., AND M. MIYAO. 1985. Extrinsic membrane proteins in the photosynthetic oxygen-evolving complex. Trends Biochem. Sci. 10: 122–124.

MURATA, N., M. MIYAO, T. OMATA, H. MATSUNAMI, AND T. KUWABARA. 1984. Stoichiometry of components in the photosynthetic oxygen evolution system of photosystem II particles prepared with triton X-100 from spinach chloroplasts. Biochim. Biophys. Acta 765: 363–369.

MYERS, J., J.-R. GRAHAM, AND R. T. WANG. 1978. On spectral control of pigmentation in *Anacystis nidulans* (Cyanophyceae). J. Phycol. 14: 513–518.

——— 1980. Light harvesting in *Anacystis nidulans* studied in pigment mutants. Plant Physiol. 66: 1144–1149.

MYERS, J., AND W. A. KRANTZ. 1955. Relations between pigment content and photosynthetic characteristics in a blue-green alga. J. Gen. Physiol. 39: 11–22.

NAKATANI, H. Y., B. KE, E. DOLAN, AND C. J. ARNTZEN. 1984. Identity of the photosystem II reaction center polypeptide. Biochim. Biophys. Acta 765: 347–352.

NAKAYAMA, K., T. YAMAOKA, AND S. KATOH. 1979. Chromatographic separation of photosystems I and II from the thylakoid membrane isolated from a thermophilic blue-green alga. Plant Cell Physiol. 20: 1565–1576.

NANBA, M., AND S. KATOH. 1985. Electron transport from cytochrome b_6-f complexes to photosystem I reaction center complexes in *Synechococcus* sp. Is cytochrome c-553 a mobile electron carrier? Biochim. Biophys. Acta 808: 39–45.

NECHUSHTAI, R., P. MUSTER, A. BINDER, V. LIVEANU, AND N. NELSON. 1983. Photosystem I reaction center from the thermophilic cyanobacterium *Mastigocladus laminosus*. Proc. Nat. Acad. Sci. USA 80: 1179–1183.

NECHUSHTAI, R., N. NELSON, A. K. MATTOO, AND M. EDELMAN. 1981. Site of synthesis of subunits to photosystem I reaction center and the proton-ATPase in *Spirodela*. FEBS Lett. 125: 115–119.

NELSON, N. 1985. Functional assembly of the chloroplast H^+-ATPase and photosynthetic reaction centres. Biochem. Soc. Trans. 14: 5–7.

NELSON, N., H. NELSON, AND G. SCHATZ. 1980. Biosynthesis and assembly of the proton-translocating adenosine triphosphatase complex from chloroplasts. Proc. Nat. Acad. Sci. USA 77: 1361–1364.

NEUMANN, E., B. DEPKA, AND W. OETTMEIER. 1984. Plastoquinone and inhibitor binding to photosystem II particles from *Chlamydomonas reinhardii*, p. 473–476. *In* C. Sybesma [ed.] Advances in Photosynthesis Research, Vol. I. Martinus Nijhoff/Dr. W. Junk Publishers, The Hague.

NEWMAN, P. J., AND L. A. SHERMAN. 1978. Isolation and characterization of photosystem I and II membrane particles from the blue-green alga, *Synechococcus cedrorum*. Biochim. Biophys. Acta 503: 343–361.

NIERZWICKI-BAUER, S. A., D. L. BALKWILL, S. E. STEVENS, JR. 1983. Three-dimensional ultrastructure of a unicellular cyanobacterium. J. Cell Biol. 97: 713–722.

NIES, M., AND W. WEHRMEYER. 1980. Isolation and biliprotein characterization of phycobilisomes from the thermophilic cyanobacterium *Mastigocladus laminosus* Cohn. Planta 150: 330–337.

NOZAKI, Y., M. TAMAKI, AND M. SHIN. 1985. The reconstituted NADP+ photoreducing system by recombination of ferredoxin-NADP+ reductase and connectein with thylakoids. Physiol. Veg. 23: 627–633.

NUGENT, J. H. A., M. C. W. EVANS, AND B. D. DINER. 1982. Characteristics of the photosystem II reaction center. II. Electron donors. Biochim. Biophys. Acta 682: 106–114.

NUIJS, A. M., H. J. VAN GORKHOM, J. J. PLIJTER, AND L. N. M. DUYSENS. 1986. Primary-charge separation and excitation of chlorophyll a in photosystem II particles from spinach as studied by picosecond absorbance-difference spectroscopy. Biochim. Biophys. Act 848: 167–175.

OELZE, J., AND G. DREWS. 1972. Membranes of photosynthetic bacteria. Biochim. Biophys. Acta 265: 209-239.

OETTMEIER, W., AND A. TREBST. 1983. Inhibitor and plastoquinone binding to photosystem II. *In* Y. Inoue [ed.] Oxygen-evolving System of Plant Photosynthesis. Academic Press, New York. NY.

OHKI, K., AND E. GANTT. 1983. Functional phycobilisomes from *Tolypothrix tenuis* (Cyanophyta) grown heterotrophically in the dark. J. Phycol. 19: 359–364.

OHKI, K., E. GANTT, C. A. LIPSCHULTZ, AND M. C. ERNST. 1985. Constant phycobilisome size in chromatically adapted cells of the cyanobacterium *Tolypothrix tenuis*, and Variation in *Nostoc* sp. Plant Physiol. 79: 943–948.

OH-OKA, H., S. TANAKA, K. WADA, T. KUWABARA, AND N. MURATA. 1986. Complete amino acid sequence of 33 kDa protein isolated from spinach photosystem II particles. FEBS Lett. 197: 63–66.

OISHI, K. K., D. R. SHAPIRO, AND K. A. TEWARI. 1984. Sequence organization of a pea chloroplast DNA gene coding for a 34,500-dalton protein. Mol. Cell. Biol. 4: 2556–2563.

OKADA, S., AND K. ASADA. 1983. A manganese protein from thylakoids of a cyanobacterium, *Plectonema boryanum*; inhibition of oxygen evolution with its antibody. Plant Cell Physiol. 24: 163–172.

OKAMURA, M. Y., G. FEHER, AND N. NELSON. 1982. Reaction centers, p. 195–272. *In* Govindjee [ed.] Photosynthesis: Energy Conversion by Plants and Bacteria, Part I. Academic Press, New York, NY.

OLSON, J. M. Chlorophyll organization in green photosynthetic bacteria. 1980. Biochim. Biophys. Acta 594: 33–51.

O'MALLEY, P. J., AND G. T. BABCOCK. 1983. Origin of signal II in spinach chloroplasts and its role in the water splitting process. Biophys. J. 41: 315.

OMATA, T., AND N. MURATA. 1983. Isolation and characterization of the cytoplasmic membranes from the blue-green alga (cyanobacterium) *Anacystis nidulans*. Plant Cell Physiol. 24: 1101–1112.

———. 1984. Isolation and characterization of three types of membranes from the cyanobacterium (blue-green alga) *Synechocystis* PCC 6714. Arch. Microbiol. 139: 113–116.

OMATA, T., N. MURATA, AND K. SATOH. 1984. Quinone and pheophytin in the photosynthetic reaction center II from spinach chloroplasts. Biochim. Biophys. Acta 765: 403–405.

ONG, L. J., A. N. GLAZER, AND J. B. WATERBURY. 1984. An unusual phycoerythrin from a marine cyanobacterium. Science 224: 80–83.

ONO, T. A., AND Y. INOUE. 1984a. Mn-preserving extraction of 33, 24, and 16 kDa proteins from O_2-evolving PS II particles by divalent salt-washing. FEBS Lett. 164: 255–260.

———. 1984b. Ca2 + -dependent restoration of O_2-evolving activity in $CaCl_2$-washed PS II particles depleted of 33, 24, and 16 kDa proteins. FEBS Lett. 168: 281–286.

ÖQUIST, G. 1971. Changes in pigment composition and photosynthesis induced by iron deficiency in the blue-green alga, *Anacystis nidulans*. Physiol. Plant. 25: 188–191.

———. 1974a. Iron deficiency in the blue-green alga, *Anacystis nidulans*: changes in pigmentation and photosynthesis. Physiol. Plant. 30: 30–37.

———. 1974b. Iron deficiency in the blue-green alga, *Anacystis nidulans*: fluorescence and absorption spectra recorded at 77 K. Physiol. Plant. 33: 55–58.

———. 1974c. Distribution of chlorophyll between the two photoreactions in photosynthesis of the blue-green alga *Anacystis nidulans* grown at two different light intensities. Physiol. Plant. 30: 38–44.

ÖQUIST, G., D. C. FORK, S. SCHOCH, AND G. MALMBERG. 1981. Solubilization and spectral characterization of chlorophyll-protein complexes isolated from the thermophilic blue-green alga *Synechococcus lividus*. Biochim. Biophys. Acta 638: 192–200.

ORTIZ, W., E. LAM, M. GHIRARDI, AND R. MALKIN. 1984. Antenna function of a chlorophyll *a/b* protein complex of photosystem I. Biochim. Biophys. Acta 766: 505–509.

ORTIZ, W., E. LAM, S. CHOLLAR, D. MUNT, AND R. MALKIN. 1985. Topography of the protein complexes of the chloroplast thylakoid membrane. Studies of photosystem I using a chemical probe and proteolytic digestion. Plant Physiol. 77: 389–397.

ORTIZ, W., AND R. MALKIN. 1985. Topographical studies of the polypeptide subunits of the thylakoid cytochrome b_6-f complex. Biochim. Biophys. Acta 808: 164–170.

ORTIZ, W., AND E. STUTZ. 1980. Synthesis of polypeptides of the chlorophyll-protein complexes in isolated chloroplasts of *Euglena gracilis*. FEBS Lett. 116: 298–302.

PAKRASI, H. B., H. C. RIETHMAN, AND L. A. SHER-MAN. 1985. Organization of pigment proteins in the photosystem II complex of the cyanobacterium *Anacystis nidulans* R2. Proc. Nat. Acad. Sci. USA 82: 6903–6907.

PAKRASI, H. B., AND L. A. SHERMAN. 1984. A highly active oxygen-evolving photosystem II preparation from the cyanobacterium *Anacystis nidulans*. Plant Physiol. 74: 742–745.

PAONE, D. A. M., AND S. E. STEVENS, JR. 1981. Nitrogen starvation and the regulation of glutamine synthetase in *Agmenellum quadruplicatum*. Plant Physiol. 67: 1097–1100.

PARSON, W. W., AND B. KE. 1982. Primary photochemical reactions, p. 331–385. *In* Govindjee [ed.] Photosynthesis. Academic Press, New York, NY.

PASSAVANT, C. W., AND R. B. HALLICK. 1985. Location, nucleotide sequence and expression of the proton-translocating subunit gene of the *E. gracilis* chloroplast ATP synthase. Plant Mol. Biol. 4: 347–354.

PESCHEK, G. A., B. HINERSTOISSER, M. RIEDLER, R. MUCHL, AND W. H. NITSCHMANN. 1986. Exogenous energy supply to the plasma membrane of dark anaerobic cyanobacterium *Anacystis nidulans*: thermodynamic and kinetic characterization of the ATP synthesis effected by an artificial proton motive force. Arch. Biochem. Biophys. 247: 40–48.

PHILLIPS, A. L., AND J. C. GRAY. 1984. Location and nucleotide sequence of the gene for the 15.2 kDa polypeptide of the cytochrome *b-f* complex from pea chloroplasts. Mol. Gen. Genet. 194: 477–484.

PHILLIPS, A. L. 1985. Localisation of genes for chloroplast components in tomato plastid DNA. Curr. Genet. 10: 153–161.

PICK, U., AND E. RACKER. 1979. Purification and reconstitution of the N, N' dicyclohexyl-carbodiimide sensitive ATPase complex from spinach chloroplasts. J. Biol. Chem. 254: 2793–2799.

PILOT, T. J., AND J. L. FOX. 1984. Cloning and sequencing of the gene encoding the α and β subunits of C-phycocyanin from the cyanobacterium *Agmenellum quadruplicatum*. Proc. Nat. Acad. Sci. USA 81: 6983–6987.

PORTER, R. D. 1986. Transformation in cyanobacteria. Critical Reviews in Microbiology (in press)

PRINCE, R. C. 1985. Redox-driven proton gradients. BioScience 35: 22–26.

RACKER, E. 1981. Subunit functions of ATP-driven pumps, p. 337 355. *In* C. P. Lee, G. Schatz, and G. Dallner [ed.] Mitochondria and microsomes. Addison-Wesley, Reading.

RAMSHAW, J. A. M. 1982. Structures of plant proteins, p. 229–240. *In* D. Boulter and B. Parthier [ed.] Nucleic acids and proteins in plants. I. Encyclopedia of Plant Physiology, Vol. 14A. Springer-Verlag, Berlin.

RAPS, S., J. H. KYCIA, M. C. LEDBETTER, AND H. W. SIEGELMAN. 1985. Light intensity adaptation and phycobilisome composition of *Microcystis aeruginosa*. Plant Physiol. 79: 983–987.

494

RAPS, S., K. WYMAN, H. W. SIEGELMAN, AND P. G. FALKOWSKI. 1983. Adaptation of the cyanobacterium *Microcystis aeruginosa* to light intensity. Plant Physiol. 72: 829–832.

RASMUSSEN, O. F., G. BOOKJANS, B. M. STUMMANN, AND K. W. HENNINGSEN. 1984. Localization and nucleotide sequence of the gene for the membrane polypeptide D2 from pea chloroplast DNA. Plant Mol. Biol. 3: 191–199.

REDLINGER, T., AND E. GANTT. 1982. A Mr 95,000 polypeptide in *Porphyridium cruentum* phycobilisomes and thylakoids: possible function in linkage of phycobilisomes to thylakoids and in energy transfer. Proc. Nat. Acad. Sci. USA 79: 5532–5546.

REINMAN, S., AND J. P. THORNBER. 1979. The electrophoretic isolation and partial characterization of three chlorophyll-protein complexes from blue-green algae. Biochim. Biophys. Acta 547: 188–197.

RENGER, G., AND GOVINDJEE. 1985. The mechanism of photosynthetic water oxidation. Photosyn. Res. 6: 33–55.

RICH, P. R. 1985. Mechanisms of quinol oxidation in photosynthesis. Photosyn. Res. 6: 335–348.

RIPPKA, R., J. DERUELLES, J. B. WATERBURY, M. HERDMAN, AND R. Y. STANIER. 1979. Generic assignments, strain histories, and properties of pure cultures of cyanobacteria. J. Gen. Microbiol. 111: 1–61.

RIPPKA, R., J. B. WATERBURY, AND G. COHEN-BAZIRE. 1974. A cyanobacterium which lacks thylakoids. Arch. Microbiol. 100: 419–436.

ROBERTSON, D. E., E. DAVIDSON, R. C. PRINCE, W. H. VAN DEN BERG, B. L. MARRS, AND P. L. DUTTON. 1986. Discrete catalytic sites for quinone in the Ubiquinol-cytochrome c_2 oxidoreductase of *Rhodopseudomonas capsulata*. J. Biol. Chem. 261: 584–591.

ROCHAIX, J.-D., M. DRON, M. RAHIRE, AND P. MALNOE. 1984. Sequence homology between the 32 K dalton and the D2 chloroplast membrane polypeptides of *Chlamydomonas reinhardii*. Plant. Mol. Biol. 3: 363–370.

ROSINSKI, J., J. F. HAINFELD, M. RIGBI, AND H. W. SIEGELMAN. 1981. Phycobilisome ultrastructure and chromatic adaptation in *Fremyella diplosiphon*. Ann. Bot. 47: 1–12.

ROTHSTEIN, S. J., A. A. GATENBY, D. L. WILLEY, AND J. C. GRAY. 1985. Binding of pea cytochrome *f* to the inner membrane of *Escherichia* coli requires the bacterial *secA* gene product. Proc. Nat. Acad. Sci. USA 82: 7955–7959.

RUMBELI, R., T. SCHIRMER, W. BODE, W. SIDLER, AND H. ZUBER. 1985. Crystallization of phycoerythrocyanin from the cyanobacterium *Mastigocladus laminosus* and preliminary characterization of two crystal forms. J. Mol. Biol. 186: 197–200.

RUSCKOWSKI, M., AND B. A. ZILINSKAS. 1980. Chlorophyll-protein complexes of the cyanophyte, *Nostoc* sp. Plant Physiol. 65: 392–396.

RUTHERFORD, A. W. 1985a. Orientation of EPR signals arising from components in Photosystem II membranes. Biochim. Biophys. Acta 807: 189–201.

——— 1985b. How close is the analogy between the reaction centre of Photosystem II and that of purple bacteria? Biochem Soc. Trans. 14: 15–17.

RUTHERFORD, A. W., AND S. ACKER. 1986. Orientation of the primary donor in isolated photosystem II reaction centers studied by electron paramagnetic resonance. Biophys. J. 49: 101–102.

RUTHERFORD, A. W., AND P. HEATHCOTE. 1985. Primary photochemistry in photosystem-I. Photosyn. Res. 6: 295–316.

SANDMANN, G. 1985. Consequences of iron deficiency on photosynthetic and respiratory electron transport in blue-green algae. Photosyn. Res. 6: 261–271.

SANDMANN, G., AND P. BOGER. 1980. Copper-induced exchange of plastocyanin and cytochrome *c*-553 in cultures of *Anabaena variabilis* and *Plectonema boryanum*. Plant Sci. Lett. 17: 417–424.

——— 1983. The enzymological function of heavy metals and their role in electron transfer processes of plants, p. 563–596. *In* A. Lauchli and R. L. Bieleski [ed.] Encyclopedia of Plant Physiology, New Series, Vol. 15B. Inorganic Plant Nutrition. Springer-Verlag, Berlin.

SANDMANN, G., H. RECK, E. KESSLER, AND P. BOGER. 1983. Distribution of plastocyanin and soluble plastidic cytochrome *c* in various classes of algae. Arch. Microbiol. 134: 23–27.

SANDUSKY, P. O., AND C. F. YOCUM. 1985. The chlorimde requirement for photosynthetic oxygen evolution: factors affecting nucleophilic displacement of chloride from the oxygen-evolving complex. Biochim. Biophys. Acta 849: 85–93.

SATOH, K. 1979. Polypeptide composition of the purified photosystem II pigment-protein complex from spinach. Biochim. Biophys. Acta 546: 84–92.

——— 1985. Protein-pigments and photosystem II reaction center. Photochem. Photobiol. 42: 845–853.

SATOH, K., AND P. MATHIS. 1981. Photosystem II chlorophyll *a*-protein complex: A study by flash absorption spectroscopy. Photobiochem. Photobiophys. 2: 189–198.

SATOH, K., H. Y. NAKATANI, K. E. STEINBECK, J. WATSON, AND C. J. ARNTZEN. 1983. Polypeptide composition of a photosystem II core complex. Presence of a herbicide-binding protein. Biochim. Biophys. Acta 724: 142–150.

SATOH, K., T. OHNO, AND S. KATOH. 1985. An oxygen-evolving complex with a simple subunit structure — a water-plastoquinone oxidoreductase — from the thermophilic cyanobacterium *Synechococcus* sp. FEBS Lett. 180: 326–330.

SCHATZ, G. H., AND H. T. WITT. 1984. Extraction and characterization of oxygen-evolving photosystem II complexes from a thermophilic cyanobacterium *Synechococcus* spec. Photobiochem. Photobiophys. 7: 1–14.

SCHEER, H. 1981. Biliproteins. Ange. Chemie Int.

Ed. Eng. 20: 241–261.

———. 1982. Phycobiliproteins: molecular aspects of photosynthetic antenna system, p. 7–45. *In* F. K. Fong [ed.] Light reaction path of photosynthesis. Mol. Biol. Biochem., and Biophys., Vol. 35. Springer-Verlag, Berlin.

SCHEER, H., H. WIESCHOFF, W. SCHAEFER, E. CMIEL, B. NITSCHE, AND H.-R. SCHULTEN. 1984. Structure of a chlorophyll RC I, p. 81 -84. *In* Sybesma [ed.] Proceedings of the VIth International Congress on Photosynthesis, Vol. II. Martinus Nijhoff/Dr. W. Junk Publishers, The Hague.

SCHIRMER, T., W. BODE, R. HUBER, W. SIDLER, AND H. ZUBER. 1985. X-ray crystallographic structure of the light-harvesting biliprotein C-phycocyanin from the thermophilic cyanobacterium *Mastigocladus laminosus* and its resemblance to globin structures. J. Mol. Biol. 184: 257–277.

SCHIRMER, T., R. HUBER, M. SCHNEIDER, W. BODE, M. MILLER, AND M. L. HACKERT. Crystal structure analysis and refinement at 2.5 Å of hexameric C-phycocyanin from the cyanobacterium *Agmenellum quadruplicatum*. The molecular model and its implications for light-harvesting. J. Mol. Biol. 188: 651–676.

SCHMIDT, A., I. ERDLE, AND H.-P. KOST. 1982. Changes of C-phycocyanin in *Synechococcus* 6301 in relation to growth on various sulfur compounds. Z. Naturforsch. 37c: 870–876.

SCHMIDT, G. W., S. G. BARTLETT, A. R. GROSSMAN, A. R. CASHMORE, AND N.-H. CHUA. 1982. Biosynthetic pathways of two polypeptide subunits of the light-harvesting chlorophyll *a/b* protein complex. J. Cell Biol. 91: 468–478.

SCHOEDER, H.-U., AND W. LOCKAU. 1986. Phylloquinone copurifies with the large subunit of photosystem I. FEBS Lett. 199: 23–27.

SCHOLES, P., P. MITCHELL, AND J. MOYLE. 1969. The polarity of proton translocation in some photosynthetic microorganisms. Eur. J. Biochem. 8: 450–454.

SCHRAUTEMEIER. B., AND H. BOHME. 1985. A distinct ferredoxin for nitrogen fixation isolated from heterocysts of the cyanobacterium *Anabaena variabilis*. FEBS Lett. 184: 304–308.

SCHUSTER, G., R. NECHUSHTAI, N. NELSON, AND I. OHAD. 1985. Purification and composition of photosystem I reaction center of *Prochloron* sp., and oxygen-evolving prokaryote containing chlorophyll *b*. FEBS Lett. 191: 29–33.

SHAHAK, Y., D. CROWTHER, AND G. HIND. 1981. The involvement of ferredoxin-NADP + reductase in cyclic electron transport in chloroplasts. Biochim. Biophys. Acta 636: 234–243.

SHERIFF, S., AND J. R. HERRIOTT. 1981. Structure of ferredoxin-NADP + oxidoreductase and the location of the NADP binding site. J. Mol. Biol. 145: 441–451.

SHERMAN, D. M., AND L. A. SHERMAN. 1983. Effect of iron deficiency and iron restoration on ultrastructure of *Anacystis nidulans*. J. Bacteriol. 156: 393–401.

SHIN, M., K. TAGAWA, AND D. I. ARNON. 1963. Crystallization of ferredoxin-TPN reductase

and its role in the photosynthetic apparatus of chloroplasts. Biochem. Z. 338: 84–96.

SIEFERMANN-HARMS, D. 1985. Carotenoids in photosynthesis. I. Location in photosynthetic membranes and light-harvesting function. Biochim. Biophys. Acta 811: 325–355.

SMEEKENS, S., J. VAN BINSBERGEN, AND P. WEISBEEK. 1985a. The plant ferredoxin precursor: nucleotide sequence of a full length cDNA clone. Nucleic Acids Res. 13: 3179–3194.

SMEEKENS, S., M. DE GROOT, J. VAN BINSBERGEN, AND P. WEISBEEK. 1985b. Sequence of the precursor of the chloroplast thylakoid lumen protein plastocyanin. Nature 317: 456–458.

SMITH, A. G., AND J. C. GRAY. 1984. Localization of the gene for P700 chlorophyll a protein in pea chloroplast DNA. Mol. Gen. Genet. 194: 471–476.

SMITH, W. W., K. A. PATTRIDGE, M. L. LUGWIG, G. A. PETSKO, D. TSERNOGLOU, M. TANAKA, AND K. T. YASUNOBU. 1983. Structure of oxidized flavodoxin from *Anacystis nidulans*. J. Mol. Biol. 165: 737–753.

SONOIKE, K., AND S. KATOH. 1986. Isolation of an intrinsic antenna chlorophyll *a*-protein from the photosystem I reaction center complex of the thermophilic cyanobacterium *Synechococcus* sp. Arch. Biochem. Biophys. 244: 254–260.

STAEHELIN, L. A., T. H. GIDDINGS, P. BADAMI, AND W. W. KRZYMOWSKI. 1978a. A comparison of the supramolecular architecture of photosynthetic membranes of blue-green, red and green algae and of higher plants, p. 335–355. *In* D. W. Deamer [ed.] Light transducing membranes: structure, function and evolution. Academic Press, New York, NY.

STAEHELIN, L. A., J. R. GOLECKI, R. C. FULLER, AND G. DREWS. 1978b. Visualization of the supramolecular architecture of chlorosomes (*Chlorobium* type vesicles) in freeze-fractured cells of *Chloroflexus aurantiacus*. Arch. Microbiol. 119: 269–277.

STAEHELIN, L. A., J. R. GOLECKI, AND G. DREWS. 1980. Supramolecular organization of chlorosomes (chlorobium vesicles) and of their membrane attachment sites in *Chlorobium limicola*. Biochim. Biophys. Acta 589: 30–45.

STANIER, R. Y., AND G. COHEN-BAZIRE. 1977. Phototrophic procaryotes: The cyanobacteria. Ann. Rev. Microbiol. 31: 225–274.

STEVENS, S. E. JR., D. L. BALKWILL, AND D. A. M. PAONE. 1981. The effects of nitrogen limitation on the ultrastructure of the cyanobacterium *Agmenellum quadruplicatum*. Arch. Microbiol. 130: 204–212.

STEVENS, S. E. JR., D. A. M. PAONE, AND D. L. BALKWILL. 1981. Accumulation of cyanophycin granules as a result of phosphate limitation in *Agmenellum quadruplicatum*. Plant Physiol. 67: 716–719.

STEWART, A. C. 1980. The chlorophyll-protein complexes of a thermophilic blue-green alga. FEBS Lett. 114: 67–72.

STEWART, A. C., AND D. S. BENDALL. 1979. Preparation of an active, oxygen-evolving photosystem 2 particles from a blue-green alga.

FEBS Lett. 107: 308–312.

——— 1981. Properties of oxygen-evolving Photosystem-II particles from *Phormidium laminosum*, a thermophilic blue-green alga. Biochem. J. 194: 877–887.

STEWART, A. C., U. LJUNGBERG, H.-E. AKERLUND, AND B. ANDERSSON. 1985a. Studies on the polypeptide composition of the cyanobacterial oxygen-evolving complex. Biochim. Biophys. Acta 808: 353–362.

STEWART, A. C., M. SICZKOWSKI, AND U. LJUNGBERG. 1985b. Glycerol stabilizes oxygen evolution and maintains binding of a 9 kDa polypeptide in photosystem II particles from the cyanobacterium, *Phormidium laminosum*. FEBS Lett. 193: 175–179.

STROTMANN, H., AND S. BICKEL-SANDKOTTER. 1984. Structure, function and regulation of chloroplast ATPase. Ann. Rev. Plant Physiol. 35: 97 -120.

SUSOR, W., AND D. W. KROGMANN. 1966. TPN photoreduction with cell-free preparations of *Anabaena variabilis*. Biochim. Biophys. Acta 120: 65–72.

TABATA, K., S. ITOH, Y. YAMAMOTO, S. OKAYAMA, AND M. NISHIMURA. 1985. Two plastoquinone A molecules are required for photosystem II activity: analysis in hexane-extracted photosystem II particles. Plant Cell Physiol. 26: 855–863.

TAKABE, T., H. ISHIKAWA, S. NIWA, AND Y. TANAKA. 1984. Electron transfer reactions of chemically modified plastocyanin with P700 and cytochrome *f*. Importance of local charges. J. Biochem. 96: 385–393.

TAKAHASHI, Y., K. HIROTA, AND S. KATOH. 1985. Multiple forms of P700-chlorophyll *a*-protein complexes from *Synechococcus* sp.: The iron, quinone and carotenoid contents. Photosyn. Res. 6: 183–192.

TAKAHASHI, Y., AND S. KATOH. 1982. Functional subunit structure of photosystem I reaction center in *Synechococcus* sp. Arch. Biochem. Biophys. 219: 219–227.

——— 1986. Numbers and functions of plastoquinone molecules associated with photosystem II preparations from *Synechococcus* sp. Biochim. Biophys. Acta 848: 183–192.

TAKAHASHI, Y., H. KOIKE, AND S. KATOH. 1982. Multiple forms of chlorophyll-protein complexes from a thermophilic cyanobacterium *Synechococcus* sp. Arch. Biochem. Biophys. 219: 209–218.

TANDEAU DE MARSAC, N. 1977. Occurrence and nature of chromatic adaptation in cyanobacteria. J. Bacteriol. 130: 82–91.

——— 1978. Étude sur la Biosynthèse des Phycobiliprotéines chez les Cyanobactéries. 166 p. Thèse Doctorat d'État, Université Pierre et Marie Cure, Paris VI.

——— 1983. Phycobilisomes and complementary chromatic adaptation in cyanobacteria. Bull. Inst. Pasteur 81: 201–254.

TANDEAU DE MARSAC, N., AND G. COHEN-BAZIRE. 1977. Molecular composition of cyanobacterial phycobilisomes. Proc. Nat. Acad. Sci. USA 74: 1635–1639.

TANG, X.-S., AND K. SATOH. 1984. Characterization of a 47-kilodalton chlorophyll-binding polypeptide (CP-47) isolated from a photosystem II core complex. Plant Cell Physiol. 25: 935–945.

TARR, G. E. 1983. Determination of the sequence which spans the beginning of the insertion region in *Anacystis nidulans* flavodoxin. J. Mol. Biol. 165: 754–755.

THOMAS, J. C., A. MOUSSEAU, AND N. HAUSWIRTH. 1984. Photocontrol of phycoerythrocyanin synthesis in an *Oscillatoria* strain (Cyanobacteria), p. 691–694. *In* C. Sybesma [ed.] Advances in Photosynthesis Research, Vol. II. Martinus Nijhoff/Dr. W. Junk Publishers, The Hague.

THORNBER, J. P. 1975. Chlorophyll-proteins: light-harvesting and reaction center components of plants. Ann. Rev. Plant Physiol. 26: 127–158.

——— 1986. Biochemical characterization and structure of pigment-proteins of photosynthetic organisms, p. 98–142. *In* L. A. Staehelin and C. J. Arntzen [ed.] Encyclopedia of plant physiology. New Series. Volume 19. Photosynthesis III. Photosynthetic membranes. Springer-Verlag, Berlin.

THORNBER, J. P., R. S. ALBERTE, F. A. HUNTER, J. A. SHIOZAWA AND K.-S. KAN. 1976. The organization of chlorophyll in the plant photosynthetic unit. Brookhaven Symp. Biol. 28: 132–148.

THURNAUER, M. C., AND P. GAST. 1985. Q-band (35 GHz) EPR results on the nature of A_1 and the electron spin polarization in photosystem I particles. Photobiochem. Photobiophys. 9: 29–38.

TIEDGE, H., H. LUNSDORF, G. SCHAFER, AND H. U. SCHAIRER. 1985. Subunit stoichiometry and juxtaposition of the photosynthetic coupling factor 1: Immunoelectron microscopy using monoclonal antibodies. Proc. Nat. Acad. Sci. USA 82: 7874–7878.

TREBST, A. 1980. Inhibitors in electron flow: Tools for the functional and structural localization of carriers and energy conservation sites. Methods Enzymol. 69: 675–715.

——— 1986. The topology of the plastoquinone and herbicide binding peptides of photosystem II in the thylakoid membrane. Z. Naturforsch 41c: 240–245.

TRENCH, R. K. 1982. Physiology, biochemistry, and ultrastructure of cyanellae, p. 257–288. *In* F. E. Round and D. J. Chapman [ed.] Progress in phycological research, Vol. I. Elsevier Biomedical Press, Amsterdam.

TRONRUD, D. E., M. F. SCHMID, AND B. W. MATTHEWS. 1986. Structure and X-ray amio acid sequence of a bacteriochlorophyll *a* protein from *Prosthecochloris aestuarii* refined at 1.9 Å resolution. J. Mol. Biol. 188: 443–454.

TSUKIHARA, T., K. FUKUYAMA, M. NAKAMURA, Y. KATSUBE, N. TANAKA, M. KAKUDO, K. WADA, T. HASE, AND H. MATSUBARA. 1981. X-ray analysis of a (2Fe–2S) ferredoxin from Spirulina platensis. Main chain fold and location of side chains at 2.5 Å resolution. J. Biochem. 90: 1763–1773.

TYAGI, A. K., AND R. G. HERRMANN. 1986. Location and nucleotide sequence of the pre-apocytochrome *f* gene on the *Oenothera hookeri* chromosome (*Euoenothera* plastome I). Curr. Genet. 10: 481–486.

TYBULEWICZ, V. L. J., G. FALK, AND J. E. WALKER, 1984. *Rhodopseudomonas blastica atp* operon: Nucleotide sequence and transcription. J. Mol. Biol. 179: 185–214.

VALLEJOS, R. H., E. A. CECCARELLI, AND R. L. CHAN. 1984. Evidence for the existence of a thylakoid intrinsic protein that binds ferredoxin-NADP+ oxidoreductase. J. Biol. Chem. 259: 8048–8051.

VALLON, O., F. A. WOLLMAN, AND J. OLIVE. 1985. Distribution of intrinsic and extrinsic subunits of the PS II protein complex between appressed and non-appressed regions of the thylakoid membrane: an immunocytochemical study. FEBS Lett. 183: 245–250.

VAN GORKHAM, H. J. 1985. Electron transfer in photosystem II. Photosyn. Res. 6: 97–112.

VAN WALRAVEN, H. W., E. KOPPENAAL, H. J. P. MARVIN, M. J. M. HAGENDOORN, AND R. KRAAYENHOF. Lipid specificity for the reconstitution of well-coupled ATPase proteoliposomes and a new method for lipid isolation from photosynthetic membranes. Eur. J. Biochem. 144: 563–569.

VAN WALRAVEN, H. S., H. J. LUBBERDING, H. J. P. MARVIN, AND R. KRAAYENHOF. 1983. Characterization of reconstituted ATPase complex proteoliposomes prepared from the thermophilic cyanobacterium *Synechococcus* 6716. Eur. J. Biochem. 137: 101–106.

VIERLING, E., AND R. S. ALBERTE. 1980. Functional organization and plasticity of the photosynthetic unit of the cyanobacterium *Anacystis nidulans*. Physiol. Plant. 50: 93–98.

——— 1983. P700 chlorophyll *a*-protein. Purification, characterization, and antibody preparation. Plant Physiol. 72: 625–633.

WADA, K., T. TAMURA, H. MATSUBARA, AND K. KODO. *Spirulina* ferredoxin-NADP+ reductase. Further characterization with an improved preparation. J. Biochem. 94: 387–393.

WALKER, J. E., M. SARASTE, AND J. J. GAY. 1984. The unc operon. Nucleotic sequence, regulation, and structure of ATP-synthase. Biochim. Biophys. Acta 768: 164–200.

WALKER, J. E., AND V. L. J. TYBULEWICZ. 1985. Comparative genetics and biochemistry of light-driven ATP synthases, p. 141–153. *In* K. E. Steinbeck, S. Bonitz, C. J. Arntzen, and L. Bogorad [eds.] Molecular Biology of the Photosynthetic Apparatus. Cold Spring Harbor Laboratory, Cold Spring Harbor.

WARDEN, J. T., AND J. H. GOLBECK. 1986. Photosystem I charge separation in the absence of centers A and B. II. ESR spectral characterization of center "X" and correlation with optical signal "A_2". Biochim. Biophys. Acta 849: 25–31.

WATANABE, T., M. KOBAYASHI, A HONGU, M. NAKAZATO, T. HIYAMA, AND N. MURATA. 1985. Evidence that a chlorophyll *a'* dimer constitutes the photochemical reaction centre I (P700) in photosynthetic apparatus. FEBS Lett. 191: 252–256.

WECHSLER, T., F. SUTER, R. C. FULLER, AND H. ZUBER. 1985a. The complete amino acid sequence of the bacteriochlorophyll *c* binding polypeptide from chlorosomes of the green photosynthetic bacterium *Chloroflexus aurantiacus*. FEBS Lett. 181: 173–178.

WECHSLER, T., R. BRUNISHOLZ, F. SUTER, R. C. FULLER, AND H. ZUBER. 1985b. The complete amino acid sequence of a bacteriochlorophyll *a* binding polypeptide isolated from the cytoplasmic membrane of the green photosynthetic bacterium *Chloroflexus aurantiacus*. FEBS Lett. 191: 34–38.

WEHRMEYER, W. 1983a. Phycobiliproteins and phycobiliprotein organization in the photosynthetic apparatus of cyanobacteria, red algae, and cryptophytes, p. 143–167. *In* U. Jensen and D. E. Fairbrothers [ed.] Proteins and nucleic acids in Plant Systematics. Springer-Verlag, Berlin.

——— 1983b. Organization and composition of cyanobacterial and rhodophycean phycobilisomes, p. 1–22. *In* G. C. Papageorgiou and L. Packer [ed.] Photosynthetic prokaryotes: cell differentiation and function. Elsevier Biomedical, New York, NY.

WESTHOFF, P., J. ALT, AND R. G. HERRMANN. 1983a. Localization of the genes for the two chlorophyll *a*-conjugated polypeptides (mol. wt. 51 and 44 kd) of the photosystem II reaction center on the spinach plastid chromosome. EMBO J. 2: 2229–2237.

WESTHOFF, P., J. ALT, N. NELSON, W. BOTTOMLEY, H. BUNEMANN, AND R. G. HERRMANN. 1983b. Genes and transcripts for the P700 chlorophyll *a* apoprotein and subunit 2 of the photosystem I reaction center complex from spinach thylakoid membranes. Plant Mol. Biol. 2: 95–107.

WESTHOFF, P., J. ALT, N. NELSON, AND R. G. HERRMANN. 1985a. Genes and transcripts for the ATP synthase CF_0 subunits I and II from spinach thylakoid membranes. Mol. Gen. Genet. 199: 290–299.

WESTHOFF, P., J. ALT, W. R. WIDGER, W. A. CRAMER, AND R. G. HERRMANN. 1985b. Localization of the gene for apocytochrome *b*-559 on the plastid chromosome of spinach. Plant Mol. Biol. 4: 103–110.

WESTHOFF, P., C. JANSSON, L. KLEIN-HITPASS, R. BERZBORN, C. LARSSON, AND S. G. BARTLETT. 1985c. Intracellular coding sites of polypeptides associated with photosynthetic oxygen evolution of photosystem II. Plant Mol. Biol. 4: 137–146.

WESTHOFF, P, N. NELSON, H. BUNEMANN, AND R. G. HERRMANN. 1981. Localization of genes for coupling factor subunits on the spinach plastid chromosome. Curr. Genet. 4: 109–120.

WIDGER, W. R., W. A. CRAMER, M. HERMODSON, AND R. G. HERRMANN. 1985. Evidence for a hetero-oligomeric structure of the chloroplast cytochrome *b*-559. FEBS Lett. 191: 186–190.

WIDGER, W. R., W. A. CRAMER, R. G. HERRMANN, AND A. TREBST. 1984. Sequence homology and structural similarity between cytochrome *b* of mitochondrial complex III and the chloroplast b6/f complex: Position of the cytochrome *b* hemes in the membrane. Proc. Nat. Acad. Sci. USA 81: 674–678.

WILDMAN, R. B., AND C. C. BOWEN. 1974. Phocobilisomes in blue-green algae. J. Bacteriol. 117: 866–881.

WILLEY, D. L., A. K. HUTTLY, A. L. PHILLIPS, AND J. C. GRAY. 1983. Localization of the gene for cytochrome *f* in pea chloroplast DNA. Mol. Gen. Genet. 189: 85–89.

WILLEY, D. L., A. D. AUFFRET, AND J. C. GRAY. 1984a. Structure and topology of cytochrome *f* in pea chloroplast membranes. Cell 36: 555–562.

WILLEY, D. L., C. J. HOWE, A. D. AUFFRET, C. M. BOWMAN, T. A. DYER, AND J. C. GRAY. 1984b. Location and nucleotic sequence of the gene for cytochrome *f* in wheat chloroplast DNA. Mol. Gen. Genet. 194: 416–422.

WILLIAMS, J. C., L. A. STEINER, G. FEHER, AND M. I. SIMON. 1984. Primary structure of the L subunit of the reaction center from *Rhodopseudomonas sphaeroides*. Proc. Nat. Acad. Sci. USA 81: 7303–7307.

WILLIAMS, J. C., L. A. STEINER, R. C. ODGEN, M. I. SIMON, AND G. FEHER. 1983. Primary structure of the M subunit of the reaction center from *Rhodopseudomonas sphaeroides*. Proc. Nat. Acad. Sci. USA 80: 6505–6509.

WILLIAMS, R. C., J. C. GINGRICH, AND A. N. GLAZER. 1980. Cyanobacterial phycobilisomes. Particles from *Synechocystis* 6701 and two pigment mutants. J. Cell Biol. 85: 558–566.

WILLIAMS, R. C., A. N. GLAZER, AND D. J. LUNDELL. 1983. Cyanobacterial protosystem I: Morphology and aggregation behavior. Proc. Nat. Acad. Sci. USA 80: 5923–5926.

WOLBER, P. K., AND K. E. STEINBECK. 1984. Identification of the herbicide binding region of the Q_b protein by photoaffinity labeling with azidoatrazine. Z. Naturforsch. 39c: 424–429.

WOLLMAN, F.-A., AND P. BENNOUN. 1982. A new chlorophyll-protein complex related to photosystem I in *Chlamydomonas reinhardii*. Biochim. Biophys. Acta 680: 352–360.

WOOD, N. B., AND R. HASELKORN. 1977. Protein degradation during heterocyst differentiation in the blue-green alga *Anabaena* 7120. Fed. Proc. (Abstr.) 36: 886.

——— 1980. Control of phycobiliprotein proteolysis and heterocyst differentiation in *Anabaena*. J. Bacteriol. 141: 1375–1385.

WOOD, P. M. 1978. Interchangeable copper and iron proteins in algal photosynthesis. Studies on plastocyanin and cytochrome *c*-552 in *Chlamydomonas*. Eur. J. Biochem. 87: 9–19.

WYMAN, M., AND P. FAY. 1986a. Underwater light climate and the growth and pigmentation of planktonic blue-green algae (Cyanobacteria). I. The influence of light quantity. Proc. R. Soc. Lond. B 227: 367–380.

——— 1986b. Underwater light climate and the growth and pigmentation of planktonic blue-green algae (Cyanobacteria). II. The influence of light quality. Proc. R. Soc. Lond. B 227: 381–393.

WYMAN, M., R. P. F. GREGORY, AND N. G. CARR. 1985. Novel role for phycoerythrin in a marine cyanobacterium, *Synechococcus* strain DC2. Science 230: 818–820.

YAMADA, Y., N. ITOH, AND K. SATOH. 1985. A versatile chromatographic procedure for purifying PS II reaction center complex from digitonin extracts of spinach thylakoids. Plant Cell Physiol. 26: 1263–1271.

YAMAGISHI, A., AND S. KATOH. 1983. Two chlorophyll-binding subunits of the photosystem 2 reaction center complex isolated from the thermophilic cyanobacterium *Synechococcus* sp. Arch. Biochem. Biophys. 225: 836–846.

——— 1984. A photoactive photosystem II reaction-center complex lacking a chlorophyll-binding 40 kilodalton subunit from the thermophilic cyanobacterium *Synechococcus* sp. Biochim. Biophys. Acta 765: 118–124.

——— 1985. Further characterization of the two Photosystem II reaction center complex preparations from the thermophilic cyanobacterium *Synechococcus* sp. Biochim. Biophys. Acta 807: 74–80.

YAMAMOTO, Y, M. DOI, N. TAMURA, AND M. NISHIMURA. 1981. Release fo polypeptides from highly active Q_2-evolving photosystem-2 preparation by Tris treatment. FEBS Lett. 133: 265 –268.

YAMANOTO, Y., M. A. HERMODSON, AND D. W. KROGMANN. 1986. Improved purification and N-terminal sequence of the 33-kDa protein in spinach. FEBS Lett. 195: 155–158.

YAMAMOTO, Y., H. SHINKAI, Y. ISOGAI, K. MATSUURA, AND M. NISHIMURA. 1984a. Isolation of an Mn-carrying 33-kDa protein from an oxygen-evolving photosystem-II preparation by phase partioning with butanol. FEBS Lett. 175: 429–432.

YAMAMOTO, Y, K. TABATA, Y. ISOGAI, M. NISHIMURA, S. OKAYAMA, K. MATSUURA, AND S. ITOH. 1984b. Quantitative analysis of membrane components in a highly active O_2-evolving photosystem II preparation from spinach chloroplasts. Biochim. Biophys. Acta 767: 493–500.

YAMANAKA, G., AND A. N. GLAZER. 1980. Dynamic aspects of phycobilisome structure. Phycobilisome turnover during nitrogen starvation in *Synechococcus* sp. Arch. Microbiol. 124: 39–47.

——— 1981. Dynamic aspects of phycobilisome structure: modulation of phycocyanin content of *Synechococcus* phycobilisomes. Arch. Microbiol. 130: 23–30.

YAMANAKA, G., A. N. GLAZER, AND R. C. WILLIAMS. 1978. Cyanobacterial phycobiliomes. Characterization of the phycobilisomes of *Synechococcus* sp. 6301. J. Biol. Chem. 253: 8303–8310.

YAMANAKA, G., D. J. LUNDELL, AND A. N. GLAZER. 1982. Molecular architecture of a light-harvesting antenna. Isolation and characterization of phycobilisome subassembly particles. J. Biol. Chem. 257: 4077–4086.

YAO, Y., T. TAMURA., K. WADA, H. MATSUBARA, AND K. KODO. 1984. Spirulina ferredoxin-NADP + reductase. The complete amino acid sequence. J. Biochem. 95: 1513–1516.

YOCUM, C. F., C. T. YERKES, R. E. BLANKENSHIP, R. R. SHARP, AND G. T. BABCOCK. 1981. Stoichiometry, inhibitor sensitivity, and organization of manganese associated with photosynthetic oxygen evolution. Proc. Nat. Acad. Sci. USA 78: 7507–7511.

YOUVAN, D. C., E. J. BYLINA, M. ALBERTI, H. BEGUSCH, AND J. E. HEARST. 1984. Nucleotide and deduced polypeptide sequences of the photosynthetic reaction-center, B870 antenna, and flanking polypeptides from *R. capsulata*. Cell 37: 949–957.

YOUVAN, D. C., AND S. ISMAIL. 1985. Light-harvesting II (B800–B850 complex) structural genes from *Rhodopseudomonas capsulata*. Proc. Nat. Acad. Sci. USA 82: 58–62.

YOUVAN, D. C., S. ISMAIL, AND E. J. BYLINA. 1985. Chromosomal deletion and plasmid complementation of the photosynthetic reaction center and light-harvesting genes from *Rhodopseudomonas capsulata*. Gene 38: 19–30.

YUASA, M., T.-A. ONO, AND Y. INOUE. 1984. Isolation of photosystem II reaction center complex retaining 33 kDa protein and Mn, a possible structural minimum of photosynthetic O_2-evolving system. Photobiochem. Photobiophys. 7: 257–266.

ZIELINSKI, R. E., AND C. A. PRICE. 1980. Synthesis of thylakoid membrane proteins by chloroplasts isolated from spinach. Cytochrome *b*-559 and P700-chlorophyll *a*-protein. J. Cell Biol. 85: 435–445.

ZILINSKAS, B. A., AND D. A. HOWELL. 1986. The immunologically conserved phycobilisome-thylakoid linker polypeptide. Plant Physiol. 80: 829–833.

ZIMMERMAN, J.-L., AND A. W. RUTHERFORD. 1985. The O_2-evolving of photosystem II. Recent advances. Physiol. Veg. 23: 425–434.

ZUBER, H. 1985. Structure and function of light-harvesting complexes and their polypeptides. Photochem. Photobiol. 42: 821–844.

——— 1983. Structure and function of the light-harvesting phycobiliproteins from the cyanobacterium *Mastigocladus laminosus*, p. 23–42. *In* G. C. Papageorgiou and L. Packer [ed.] Photosynthetic Prokariotes. Cell differentiation and function. Elsevier Biomedical, New York, NY.

ZUBER, H., W. SIDLER, P. FUGLISTALLER, R. BRUNISHOLZ, AND R. THEILER. 1985. Structural studies on the light-harvesting polypeptides from cyanobacteria and bacteria, p. 183–196. *In* K. Steinbeck, S. Bonitz, C. J. Arntzen, and L. Bogorad [ed.] Molecular biology of the photosynthetic apparatus. Cold Spring Harbor Laboratory. Cold Spring Harbor.

ZURAWSKI, G., H. J. BOHNERT, P. R. WHITFIELD, AND W. BOTTOMLEY. 1982a. Nucleotide sequence of the gene for the M_r 32,000 thylakoid membrane protein from *Spinacia oleracea* and *Nicotiana debneyi* predicts a totally conserved primary translation product of Mr 38,950. Proc. Nat. Acad. Sci. USA 79: 7699–7703.

ZURAWSKI, G., W. BOTTOMLEY, AND P. R. WHITFIELD. 1982b. Structures of the genes for the β and ϵ subunits of spinach chloroplast ATPase indicate a dicistronic mRNA and an overlapping translation stop/start signal. Proc. Nat. Acad. Sci. USA 79: 6260–6264.

——— 1986. Sequence of the gens for the β and ϵ subunits of ATP synthase from pea chloroplasts. Nucl. Acids. Res. 14: 3974.

Optical Properties of Picoplankton Suspensions

JOHN T.O. KIRK

CSIRO, Division of Plant Industry, Canberra, Australia

Introduction

The optical properties of the picoplankton can be considered from two quite different, but complementary, aspects. There are the optical properties seen from the point of view of the algal cells, in particular the extent to which they succeed in absorbing and scattering the light which impinges upon them. And then there are the optical properties seen from outside the algal cells, namely, the extent to which, as a result of these absorption and scattering processes, the picoplankton population modifies the underwater light field, thus changing the physical environment for all the inhabitants of the ecosystem, including, of course, itself. I shall in this paper explore both aspects.

Before doing so, in order that we shall all be quite clear in our minds about the meaning of the terms being used, I propose briefly to define some of the optical properties relevant to our discussion. We shall first consider that set of optical properties known as the *inherent* optical properties. These are properties which belong to any medium through which light passes, and which determine the way light behaves in that medium. The medium in question might be the ocean itself, or a solution of pigments in the sample cell of a spectrophotometer, or a stack of photosynthetic membranes within an algal cell. The inherent optical properties are those which belong to the medium itself, which are determined entirely by its composition, regardless of the prevailing light field.

The ones we need to know about are the absorption coefficient, the scattering coefficient, and the normalized volume scattering function. These are defined with the help of an imaginary, infinitesimally thin, layer of medium, illuminated at right angles by a parallel beam of monochromatic light. Some of the light is absorbed within the thin layer. Some is scattered — that is, caused to diverge from its original path. The fraction of the incident flux which is absorbed divided by the thickness of the layer, is the absorption coefficient — a. The fraction of the incident flux which is scattered, divided by the thickness of the layer is the scattering coefficient — b, and this can be partitioned in accordance with

$$b = b_f + b_b$$

where b_f relates to forward scattering and b_b to backward scattering. Absorption and scattering coefficients have the units, m^{-1}.

The remaining inherent optical property, the normalized volume scattering function, $\tilde{\beta}(\theta)$, I shall not discuss in detail. It is sufficient for our present purposes to say that it describes the angular distribution of scattering from that point in the thin layer on which the light beam is incident. In all natural waters which have been examined, the volume scattering function is such that most of the scattering is in the forward direction, within a narrow angle (about 50% within 5 degrees) relative to the original direction of the beam. Note that the volume scattering function describes only single-event scattering: once you start getting multiple scattering, as of course you do in any substantial depth of water, then much broader angular distributions can be generated.

These inherent optical properties are defined, as you have just seen, in terms of an incident beam which is parallel. Without at this stage expanding on it, I wish to draw to your attention the possibility of defining analogous sets of optical properties in terms of the absorption and scattering of incident light streams which are not parallel. I shall explain later what I mean, and why it is important.

The methods for measuring the inherent optical properties I shall indicate only briefly. So far as absorption coefficients are concerned, spectrophotometry with long pathlength cells can often be used for determining soluble colour, particularly when as in inland and some coastal waters, this is comparatively high. Absorption due to particulate material must be assessed by special techniques, such as concentration by filtration and resuspension, followed by spectrophotometric procedures suitable for scattering samples. From absorbance values measured in the laboratory the absorption coefficient due to a given component, soluble or particulate, in the water body itself can readily be determined by converting the values to a 1-m pathlength, and from a base-10 to a natural logarithm basis. The total absorption coefficient at any wavelength is obtained simply by adding the absorption coefficients of the different components present: soluble colour, particulate colour, and water itself.

In most oceanic, and many coastal waters it must however be said that the amount of colour present is so low as to be impossible to measure with acceptable accuracy with the sort of pathlength that can be accommodated in a normal spectrophotometer. In such cases the absorption coefficients may have to be determined indirectly by calculation from *in situ* measurements of attenuation with depth of the natural light flux in the appropriate wavebands.

In principle the scattering coefficient should be determined by measuring scattering from a parallel beam over all angles, and then summing to give total scattering. In practice, advantage may be taken of the fact that the volume scattering function (or angular distribution of scattering) has approximately the same shape for most natural waters, and so from a scattering measurement at a single fixed angle the total scattering coefficient can be obtained with reasonable accuracy. Like the absorption coefficient, the scattering coefficient can also be estimated indirectly by calculation from *in situ* measurements on the underwater light field.

The picoplankton both absorb and scatter light. They therefore make their own additive contribution to the inherent optical properties of the whole aquatic medium. The nature of their contribution we shall return to later.

The second set of optical properties that we need to know about are those for which Preisendorfer (1961) coined the term "*apparent* optical properties". These are properties, strictly speaking, not of the aquatic medium but of the light field which, under the incident stream of solar radiation, is established within it. The one which concerns us most is the vertical attenuation coefficient, K_d, for downward irradiance. It is a fact of observation that within any well mixed region of a natural water body the downward irradiance, E_d, i.e. the downwelling radiant flux per unit area, diminishes with depth, as a result of absorption and scattering processes, in an approximately exponential manner, in accordance with

$$(1) \qquad E_d(z) = E_d(0)e^{-K_d z}$$

where $E_d(z)$ and $E_d(0)$ are the values of downward irradiance at z metres and just below the surface, respectively. This equation is obeyed very satisfactorily for monochromatic light. It is less satisfactory for a broad waveband, such as that of photosynthetically available radiation (PAR, 400–700 nm), due to the changing spectral character of the light flux with depth, but even in this case it is still often a useful approximation. The value of K_d most commonly used is the average value, $K_d(av)$, for that layer (the euphotic zone) in which the downward irradiance falls to 1% of the subsurface value.

Another apparent optical property of some interest in the phytoplankton field is the irradiance reflectance, $R(z)$, which is the ratio at any depth, z, of the upward irradiance, E_u, to the downward irradiance.

$$(2) \qquad R(z) = E_u(z)/E_d(z)$$

In waters sufficiently deep for reflection from the bottom to be ignored, the upward irradiance is generated mainly by upward scattering of the downwelling light flux (together sometimes with a small contribution from phytoplankton fluorescence). It is the reflectance, $R(0)$, just below the surface which is of most relevance because of its central role in the remote sensing of aquatic ecosystems.

R and K_d are by definition properties of the underwater light field, not of the aquatic medium. Nevertheless their values for any given water body are largely determined by the inherent optical properties (a and b) of that particular water, and are not very sensitive to, although they are influenced by, changes in the solar flux incident on the surface. These parameters can therefore as a rough but useful approximation be attributed to the water body itself and can be used to compare one water body at a given time with another, or as a guide to changes in the optical character of a given water body with time. We can say, for example, that a particular oceanic station has a particular $K_d(av)$, or $R(0)$, value for a certain waveband, and this statement will remain approximately valid, independently of the time of day or the weather conditions, for as long as the composition (and therefore the inherent optical properties) of the water remain unchanged. It was in recognition of this useful aspect of these properties of the field that Preisendorfer suggested they be referred to as *apparent* optical properties of the water. The picoplankton, through their contribution to the inherent optical properties, affect also the apparent optical properties and to this we shall return later.

I present a general account of the inherent and apparent optical properties of natural waters — their definition, measurement, and range of values observed — elsewhere (Kirk 1983).

INHERENT OPTICAL PROPERTIES AND THE PICOPLANKTON

The optical properties of individual phytoplankton cells or colonies depend, of course, on their size, shape, refractive index, pigment composition and pigment concentration. In the ecological context our concern is more with the optical properties of the phytoplankton suspension. We shall here explore the relationships between the optical properties of the cells or colonies, and the resultant optical properties of the population. Except where otherwise stated we shall be considering monochromatic light. For simplicity we shall from here on refer only to "cells", but it must be remembered that in some cases, such as chain-forming diatoms, the optically significant free-floating unit is a colony.

Absorption

A quantity of central importance for an understanding of the photosynthetic behaviour of a phytoplankton population is the fraction of the incident flux which it absorbs. In deep waters such as the ocean, all those solar photons which enter the water but do not escape again through the surface are extinguished by absorption within the aquatic medium. The proportion of these photons which are captured by any given component of the medium is equal to a_i/a_{TOT}, where a_i is the absorption coefficient due to the ith component and a_{TOT} is the total absorption coefficient. The absorption coefficient due to phytoplankton is given by

$$(3) \qquad a_p = n\,\overline{sA}$$

where n is the number of phytoplankton cells per cubic metre, and $\overline{sA}$ is the mean value, for all the randomly oriented cells in the population, of the product of the projected area (s) of a cell (in the direction of the light beam) and the absorptance (A) of the cell (the fraction of the light incident on the cell which it absorbs). It should be noted here that the cell absorptance, A, is identical to the efficiency factor for absorption, Q_a, which André Morel will be referring to in the next paper.

The averaged product, $\overline{sA}$, has the dimensions of area and is the average absorption cross-section of the phytoplankton cells. Thus in any water body the absorption coefficient due to a particular phytoplankton population is equal to the number of cells per cubic metre multiplied by the average absorption cross-section per cell, and so if we want to know how the light absorption properties of the total suspension vary with the size, shape, pigment characteristics etc. of the individual cells, then we must examine the dependence of the absorption cross-section on these cell properties.

The first point to be made is that it can be shown (Kirk 1975 a,b, 1983) that the absorption coefficient (a_{SUS}) of a suspension of pigmented particles is always less than that of the same amount of pigment uniformly dispersed. If the average volume per particle is $\bar{v}$ m³, the concentration of pigment within the particle is C mg • m⁻³, and specific absorption coefficient of dispersed pigment is γ m² • mg⁻¹, then the absorption coefficient of the system if the pigment were to be uniformly dispersed in solution (a_{SOL}) would be $nC\bar{v}\gamma$ • m⁻¹. We can now write

$$(4) \qquad \frac{a_{SUS}}{a_{SOL}} = \frac{\overline{sA}}{C\bar{v}\gamma} < 1.0$$

i.e. when pigment is confined to discrete packages instead of being uniformly dispersed, its effectiveness in the absorption of light is inavoidably diminished. A convenient name for this phenomenon — the diminution of a_{SUS}/a_{SOL} below 1.0 — is the "package effect". The greater the package effect, the less light the cells collect per mg photosynthetic pigment.

What determines the extent of the package effect? Since detailed discussions of this topic may be found elsewhere (Kirk 1975a,b, 1976, 1983; Morel and Bricaud 1981) I shall here only briefly refer to some of the relationships which apply. One important rule is that for a given cell size and shape, when cell absorptance, A, is increased by raising the intracellular pigment concentration or altering the wavelength to one more strongly absorbed, then the package effect intensifies (a_{SUS} does not increase in proportion to a_{SOL}). The basis for this can be seen by considering equation (4). We can in principle increase C or γ in the denominator as much as we please, but A in the numerator can never increase beyond 1.0, and as we raise C or γ to high values, all A can do is increase asymptotically towards 1.0, and so a_{SUS}/a_{SOL} must decrease. A corollary of this rule is that the more weakly pigmented the cells are, the less of a package effect they exhibit, i.e. pale cells harvest light more efficiently per unit pigment than do strongly coloured cells.

Since the package effect increases with intensity of absorption, it follows that in an absorption spectrum the peaks are affected more than the troughs, i.e. the spectrum is flattened. This is evident in Fig. 1 which compares the absorption spectrum of a suspension of intact cells of *Euglena gracilis* with that of the same concentration of cellular material finely dispersed by ultrasonic disruption. The dispersed material absorbs one and a half times as strongly as the intact cells in the trough at 550 nm, but two and a half times as strongly at the peak in the blue at 440 nm.

A second important rule is that for a given pigment concentration within the cells, as cell size increases at constant cell shape, so the package effect intensifies, i.e. the efficiency of light collection per unit pigment decreases. An illustration of this is given in Fig. 2 which shows that the specific absorption coefficient per mg chlorophyll *a* at the red peak, calculated for a suspension of spherical cells, decreases several-fold as diameter increases over the range 0–100 μm. One of the reasons the package effect is so marked in the spectra in Fig. 1 is that the cells of *Euglena gracilis* are large — about 50 μm long and 15 μm wide.

Another rule is that the more extended the shape, the less the package effect. Calculation shows, for example, that long thin cylinders harvest light almost as well

504

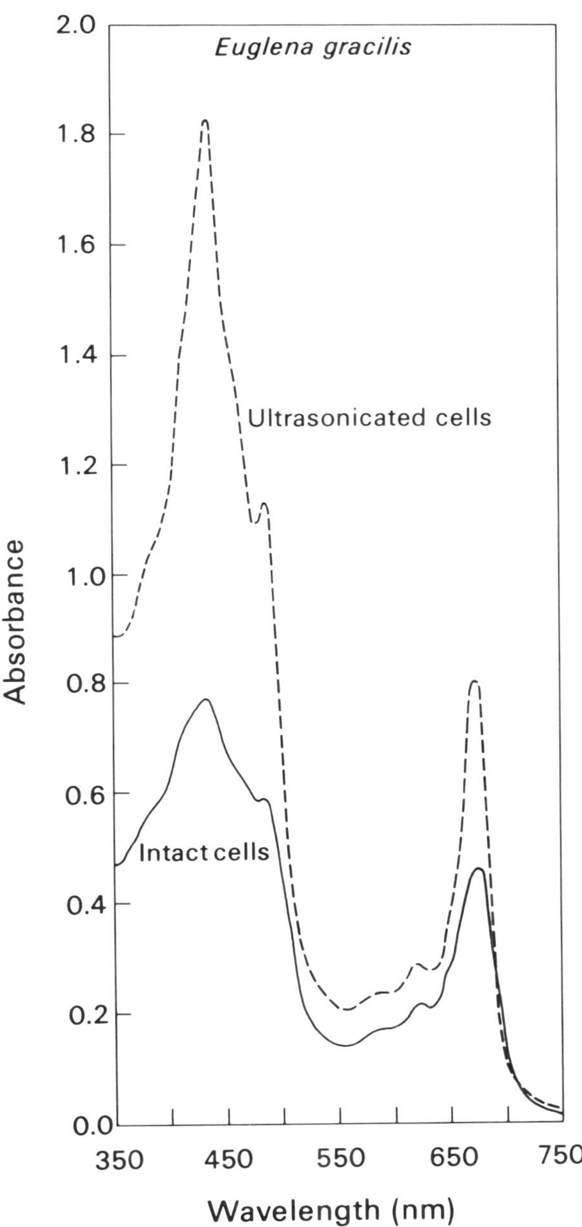

FIG. 1. Absorbance spectrum of whole cells of *Euglena gracilis* compared with that of disrupted cells in which absorption is essentially due to thylakoid fragments. The spectra of intact cells (———) and of cells fragmented by ultrasonication (------) have been corrected for scattering and in both cases correspond to 12 μg chlorophyll a • mL^{-1} and 1 cm pathlength.

as spheres of the same diameter, despite their much greater volume (Kirk 1976).

Now where does all this leave the picoplankton? The answer is that it leaves them sitting pretty since in their very small size range, 0.2–2.0 μm, there is not very much package effect. This is clearly evident, for example, from Fig. 2, where picoplankton would occupy the extreme left-hand end of the curve and even at 2 μm diameter would suffer a diminution in specific absorption coefficient at the chlorophyll red peak of only about 5%. For the nanoplankton, 2–20 μm, the package effect begins to represent a significant source of inefficiency in light harvesting. From the curve in Fig. 2 we can, for example, see that for 10 μm cells the specific absorption coefficient at the chlorophyll peak is lowered by about 25% as a consequence of the package effect: a comparable diminution would be found throughout the greater part of the broad absorption band in the blue.

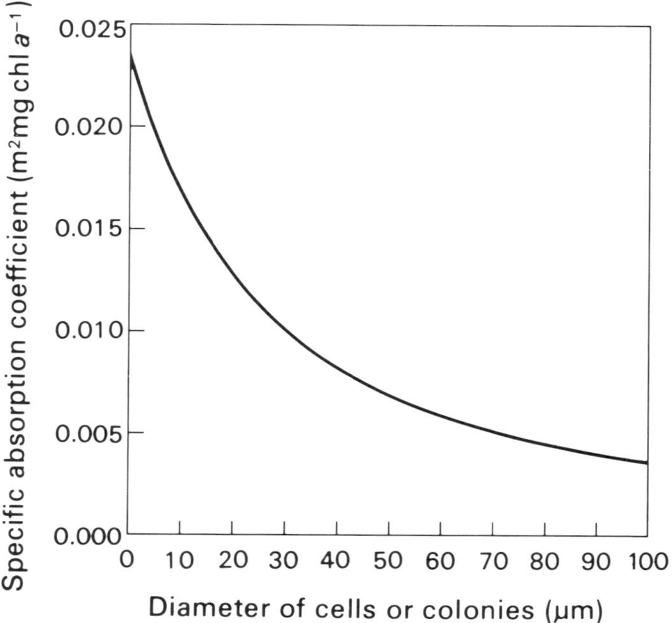

FIG. 2. Specific absorption coefficient of phytoplankton chlorophyll at the red maximum (670–680 nm) as a function of cell or colony size. The particles are assumed to be spherical and the absorptance value at each diameter was calculated using the equation of Duysens (1956), assuming the cells to contain 2%, dry weight, chlorophyll, a (4 g $\cdot$ L^{-1} cell volume) and that the specific absorption coefficient of chlorophyll a in a solution at the red peak is 0.0233 m^2 $\cdot$ mg^{-1}. The specific absorption coefficient of the phytoplankton suspension was calculated from eq. (3).

Photosynthetic pigment synthesis is biochemically expensive, requiring as it does the concomitant synthesis of large amounts of protein. It is therefore to the advantage of planktonic algae to harvest as much light per mg of pigment as they can. It is plausible to suppose that the increased efficiency of light harvesting conferred by their small size is part of the reason for the ecological success of the picoplankton. It might reasonably be speculated that minimization of the package effect becomes increasingly important as light availability diminishes. In this context it is of interest that in vertically stable waters in the North Atlantic, picoplankton and the smaller nanoplankton are most abundant at some distance below the surface (Murphy and Haugen 1985).

To round off our examination of light absorption by phytoplankton, it is of interest to consider how the instantaneous rate of light absorption by the population is related to the cell and population optical properties discussed above. The rate of light absorption (in W $\cdot$ m^{-2}, or quanta $\cdot$ s^{-1} $\cdot$ m^{-2}) by the ith component of the

506

medium within a thin horizontal layer of thickness Δz m, at depth z m is given by

$$(5) \qquad E_i(z) = E_o(z)a_i\Delta z$$

where $E_o(z)$ is the scalar irradiance (the radiant intensity per square metre from all directions at a given point), and a_i is the absorption coefficient due to the ith component. The rate of absorption of light by the phytoplankton within the thin layer can therefore be expressed either as

$$(6) \qquad E_p(z) = E_o(z)\,\overline{nsA}\Delta z$$

or

$$(7) \qquad E_p(z) = E_o(z)B_c a_c\Delta z$$

where B_c *is the concentration of phytoplankton expressed in terms of mg chlorophyll a* $\cdot$ m^{-3}, and a_c is the specific absorption coefficient per mg chlorophyll a (units — m^2 $\cdot$ mg chl a^{-1}) of the phytoplankton population at the wavelength in question. The rate of absorption of total PAR by the phytoplankton in the thin layer would correspond to the integral of equations (6) and (7) over the whole waveband.

Scattering

In totally pure water without any particles, scattering is due to localized microscopic fluctuations of density resulting from the random motion of the molecules. These density fluctuations are small relative to the wavelength of light, and so, as in the case of Rayleigh scattering by gas molecules, scattering by pure water increases inversely with the fourth power of the wavelength. Scattering in natural waters is, however, almost entirely due to suspended particles and the contribution of the water itself is in most cases impossible to detect. Most of the scattering in the sea is due to particles of diameter greater than 2 μm, which is not small relative to the wavelength of visible light, and this scattering is relatively insensitive to wavelength, although there is evidence for some increase in scattering as wavelength diminishes (Morel 1973).

Density fluctuation scattering, weak though it is, occurs to the same extent in the forward and backward directions, whereas scattering by particles in natural waters is 96–98% in the forward direction. As a consequence, in oceanic waters density fluctuation scattering does make a significant contribution to the upwelling light flux, which arises mainly from backscattering.

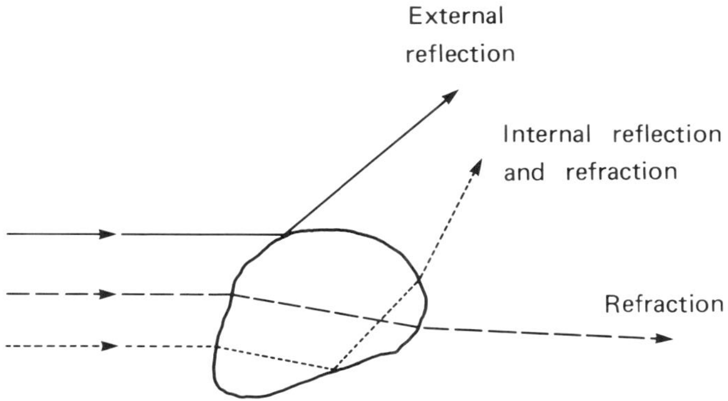

FIG. 3. Scattering of light by a particle: reflection and refraction processes.

Particles in the sea scatter light because their refractive index is greater than that of the water in which they are suspended. This can lead (Fig. 3) to reflection of photons at the particle surface, or refraction within the particle, with a consequent deviation from the original direction, i.e. scattering. Most of the scattering at large angles, $>10°$, can be attributed to reflection and refraction. In addition to these processes which can be understood in terms of geometrical optics, there is scattering resulting from the wave nature of light. The particles remove a part of the wavefront — this is the phenomenon of diffraction. In the case of a transparent particle there is interference between the light which passes though the particle and the light diffracted from its edge. The nature (constructive or destructive) and extent of this interference on the far side of the particle varies in a characteristic manner with increasing angular distance from the axis, and this perturbation of the angular distribution of the light field beyond the particle also constitutes light scattering.

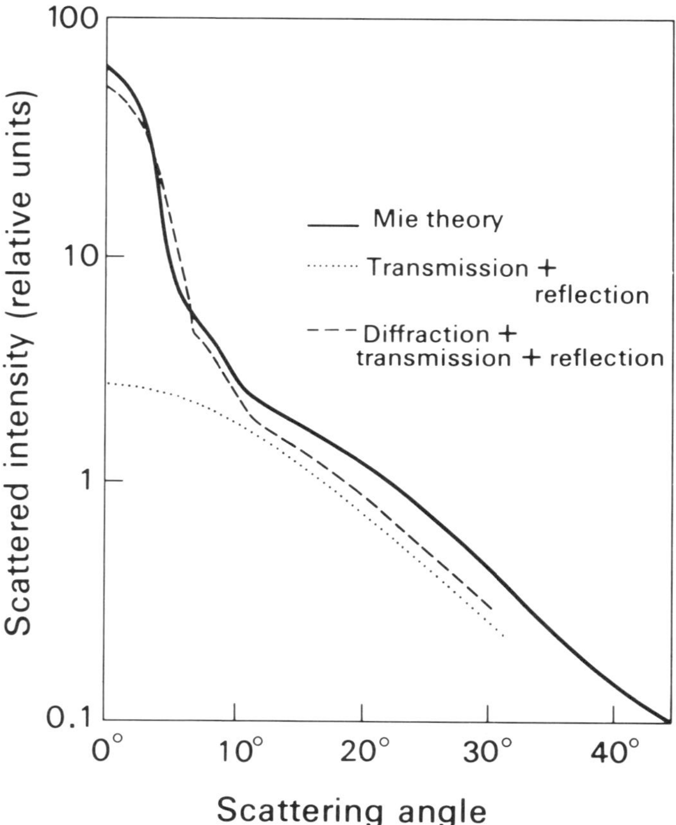

FIG. 4. Angular distribution of scattered intensity from transparent spheres calculated from electromagnetic (Mie) theory or on the basis of transmission and reflection, or diffraction, transmission and reflection. The particles have a refractive index of 1.20 relative to the surrounding medium, and diameters 5–12 times the wavelength of the light. After Hodkinson and Greenleaves (1963).

A more fundamental treatment of light scattering by particles is provided by the theory of Mie (1908) who approached the problem from the point of view of electromagnetism. He considered a polarizable spherical particle in an electromagnetic field. Oscillations are set up within it which then in turn lead to light being re-radiated (i.e. scattered) from it. Although the Mie theory is all-embracing — it can be used for particles of any size — it unfortunately does not lend itself to easy numerical calculations. For particles larger than the wavelength of light, both Mie and diffraction/reflection/refraction calculations predict that most of the scattering is in the forward direction within small angles of the axis of the incident light beam (Fig. 4). The scattering at small angles, up to $10°$–$15°$, can be attributed to diffraction, whereas most of the scattering at greater angles is due to external reflection and transmission with refraction (Hodkinson and Greenleaves 1963).

Picoplankton, small though they are, are in the size range to which these considerations apply, and so will scatter light predominantly at small angles. A feature of their angular distribution of scattering, of particular relevance to the remote sensing of these small plankters, is the extent to which they backscatter light. André Morel will discuss this aspect in the following paper.

Any particle in a beam of light will scatter a certain fraction of the beam, and the radiant flux scattered will be equivalent to that in a certain cross-sectional area of the incident beam. This area is the scattering cross-section of the particle. It is equal to the product of the projected area (s) of the particle (in the direction of the light beam) and the efficiency factor, Q_b, for scattering. The scattering coefficient due to a particular phytoplankton population in a natural water is given by

$$(8) \qquad b_p = n s \overline{Q_b}$$

where n is the number of phytoplankton cells per cubic meter and $s\overline{Q_b}$ is the mean value of scattering cross-section (the product of projected area and scattering efficiency factor) for all the randomly oriented cells in the population.

Efficiency of scattering by particles in the size range of picoplankton is highly dependent on particle size. Furthermore in the case of pigmented particles, such as the photosynthetic cells we are considering, there are marked changes in the scattering properties in the region of the absorption bands. These aspects too will be discussed in the following paper.

APPARENT OPTICAL PROPERTIES AND THE PICOPLANKTON

The phytoplankton are not only subject to the influence of the underwater light field, but they in turn influence it. We shall here consider the way in which the phytoplankton, especially its picoplankton component, affects the submarine light climate, and we shall do so specifically by examining the effects on one of the apparent optical properties of the aquatic medium, namely K_d, the vertical attenuation coefficient for downward irradiance.

Partial Vertical Attenuation Coefficients

The phytoplankton, as we have seen, through their own absorption and scattering properties contribute to the inherent optical properties of the aquatic medium in an additive manner. This means that we can in principle attribute a certain proportion of the scattering coefficient, or the absorption coefficient, of the medium to the phytoplankton in accordance with, for example

$$(9) \qquad a_{\text{TOT}} = a_W + a_G + a_{TR} + a_{PH}$$

where the subscripts TOT, W, G, TR and PH indicate total, water, gilvin (soluble yellow colour, gelbstoff), tripton (non-living particulate matter) and phytoplankton respectively. The question I now wish to address is — can we validly write analogous equations for the apparent optical properties of the medium? i.e. can K_d be partitioned in a simple additive manner amongst the different components of the aquatic medium in accordance with, for example

$$(10) \qquad K_{\text{TOT}} = K_W + K_G + K_{TR} + K_{\text{PH}}$$

Although this is already common practice it is in no way obvious that it is correct, nor is it even clear what a partial vertical attenuation coefficient for a single component actually means. K_d is by definition a property of the underwater light field. Although it is frequently convenient to treat it as though it is an optical property of the aquatic medium, it is important to remember that it may not have all the attributes of the true optical properties such as, for example, additivity of the contributions of the different components, or linear proportionality to concentration of a given component.

It is quite easy to give a physical meaning to the partitioning of the absorption coefficient in accordance with eq. (9). We can say that each individual absorption coefficient is the absorption coefficient that the component would have if it were the only absorbing constituent present. It is readily shown that the absorption coefficient of a mixture of more than one component is obtained by combining the separate absorption coefficients additively. To try in a similar way to give a physical meaning to the notion of separate attenuation coefficients we could say, for example, that K_i is the vertical attenuation coefficient that the ith component would give rise to if it were the only absorbing and scattering component present. Partial vertical attenuation coefficients defined in this way could not, however, be combined additively to give the total attenuation coefficient as the following considerations show. Say component i absorbs strongly but scatters weakly, and component j scatters strongly but has little absorption. When both are present together the vertical attenuation coefficient will be substantially greater than the sum of the two partial attenuation coefficients as defined above, since the greatly increased photon pathlength resulting from scattering by component j will lead to a marked increase in absorption by component i. Thus

$$K_{i+j} > K_i + K_j$$

Clearly we must find another definition for partial vertical attenuation coefficients if we are going to be able to combine them additively in the manner of eq. (10). To assist us in this quest let us return once again to the absorption coefficient. An alternative definition to the one given above for the absorption coefficient, a_i, due to component i, is that it represents that part of the total absorption coefficient which is due to absorption carried out by component i. As it happens this gives rise to the same numerical value for a_i as the previous definition. Let us now apply the same reasoning to K and say that K_i is that part of the total vertical attenuation coefficient which is due to attenuation of the downwelling light stream actually carried out by component i. Accepting for the moment this definition at its face value, it necessarily follows from it that the total vertical attenuation coefficient is equal to the sum of the partial attenuation coefficients due to each of the components present, i.e. with this definition of K we may indeed use eq. (10).

It might be thought that we have now solved the problem. But what in fact do we mean when we refer to "attenuation of the downwelling light stream actually carried out by component i"? Given the complexity of what happens to the light, the absorption of photons, the scattering in all directions including upwards, the intensification of absorption resulting from scattering-induced changes in the angular distribution of the light field, it is far from obvious what attenuation by any one component considered on its own could mean.

It can, however, be given a meaning. To do so we will examine the mechanisms by which attenuation of the downwelling light stream is taking place at any given depth within a water body, and define a new set of optical properties with the help of which the concept of a partial vertical attenuation coefficient can be given both a precise meaning and mathematical expression.

510

Although there is no simple conceptual basis for saying what proportion of the vertical attenuation is carried out by any given component of the medium, it is quite easy, and legitimate, to think in terms of the proportions of the absorbing and scattering events going on within the medium which are actually carried out by that particular component. Thus if we knew the way in which vertical attenuation depends on rates of absorption and scattering, we might be in a position to apportion responsibility for attenuation amongst the different components of the medium.

A theoretical relationship suitable for our purpose does in fact exist. Before examining it in detail it will be helpful first to look more closely at the mechanisms responsible for attenuation of downward irradiance at a given depth, z metres, within the water. The change in irradiance with increasing depth is the net result of three different processes. Firstly, within any small increment of depth some of the downwelling light stream is removed by absorption. Secondly, some is removed by upward scattering, i.e. it is transferred from the downwelling to the upwelling stream. Thirdly, some of the photons in the existing upwelling light stream are scattered downwards again so that they once again become part of the downwelling light stream: this third mechanism of course, by increasing downward irradiance, partly counteracts the effect of the other two.

To give these three processes quantitative expression we need a new set of optical properties. You will recall that the definition of the inherent optical properties is based on what happens to a parallel beam of light traversing a thin layer of medium, and I pointed out that in principle one could define analogous optical properties in terms of the absorption and scattering of incident light streams with any specified angular distributions. In particular, absorption and scattering coefficients can be defined for incident light streams corresponding to the downwelling and upwelling streams that exist at particular depths in real water bodies. We shall refer to these as the *diffuse* absorption and scattering coefficients for the downwelling or upwelling light streams at a given depth. Since the angular distributions of the light streams vary with depth the diffuse coefficients are themselves functions of depth.

For our present purposes we need to define three diffuse coefficients, one for absorption and two for backscattering. The diffuse absorption coefficient for the downwelling light stream at depth z, $a_d(z)$, is the proportion of the incident radiant flux which would be absorbed from the downwelling stream by an infinitesimally thin layer of medium at that depth, divided by the thickness of the layer. The back-scattering coefficient for the downwelling light stream at depth z, $b_{bd}(z)$, is the proportion of the incident radiant flux which would be scattered backwards (i.e. upwards) by the thin layer of medium divided by the thickness of the layer. The backscattering coefficient for the upwelling light stream at depth z, $b_{bu}(z)$, is defined in the same way in terms of the light scattered downwards again from that stream. The diffuse absorption and scattering coefficients can usefully be referred to as "quasi-inherent" optical properties to indicate the close relationship between them and the inherent optical properties, differing as they do only in the angular distribution of the light flux that is imagined to be incident upon the hypothetical thin layer of medium. Any given quasi-inherent optical property is always greater in value than the corresponding inherent property for the same medium since any departure from a normally incident, parallel, light field must increase the pathlength of the light across the thin layer.

We are now in a position to write an explicit equation for the vertical attenuation coefficient for downward irradiance at depth z metres in terms of the quasi-inherent optical properties

$$(11) \qquad K_d(z) = a_d(z) + b_{bd}(z) - b_{bu}(z)R(z)$$

This equation was derived by Preisendorfer (1961) from fundamental radiation transfer theory. The three terms on the right-hand side correspond to the three different attenuation processes I referred to earlier. Thus the overall attenuation of downward irradiance can be seen to be made up of a term for absorption of the downwelling stream, $a_d(z)$, a term for upward scattering of the downwelling stream, $b_{bd}(z)$, and a negative term, $-b_{bu}(z)R(z)$, corresponding to reinforcement of the downwelling stream by light scattered downwards from the upwelling stream $(R(z)$ being the irradiance reflectance at depth z).

Now, after that detour, we are in a position to give a meaning to the concept of partial vertical attenuation coefficients for the different components of the aquatic medium. To do this we take advantage of a particularly useful feature of the quasi-inherent optical properties, namely that they can, quite legitimately, be partitioned amongst the different components of the medium. Any given component of the medium can be considered to carry out a certain proportion of the total absorption from the downwelling stream, a certain proportion of the total upward scattering from that stream, and a certain proportion of the total downward scattering from the upwelling stream. The value of any given quasi-inherent optical property attributable to a particular component of the medium is the proportion of the radiative process in question (e.g. absorption from the downwelling stream) going on at depth z which is carried out by that component, multiplied by the total value of the given quasi-inherent property due to all components of the medium. Thus partial quasi-inherent optical properties can be unambiguously defined in the same way as partial inherent optical properties.

An important difference, however, is that any given partial quasi-inherent property does not (unlike the corresponding inherent properties) increase linearly with value with the concentration of the component in question. This is because the value of any quasi-inherent optical property is a function of the angular distribution of the light field at the depth in question, as well as of the concentration of the given component. Any change in the concentration of the component will by its effects on the ratio, and total amounts, of absorption and scattering, change the angular distribution of the field at depth z m, and this will have effects on the quasi-inherent optical properties over and above those due to concentration alone.

Nevertheless, given that we can partition any quasi-inherent optical property amongst the various constituents of the medium, it can be shown (Kirk 1983) that for a medium containing n absorbing/scattering components the vertical attenuation coefficient for downward irradiance at depth z can be partitioned amongst the different components in accordance with

$$(12) \qquad K_d(z) = K_d(z)_1 + K_d(z)_2 + \ldots + K_d(z)_i + \ldots + K_d(z)_n$$

the partial vertical attenuation coefficient for the ith component of the medium being given by

$$(13) \qquad K_d(z)_i = a_d(z)_i + b_{bd}(z)_i - b_{bu}(z)_i R(z)$$

where the quasi-inherent optical properties on the right-hand side of the equation are the diffuse absorption, and the two diffuse backscattering, coefficients attributable to that specific component of the medium.

It will be noted that since the angular distribution of the light field varies with depth, the values of $K_d(z)$ and the three quasi-inherent optical properties are all functions of depth. In fact, however, the average value of K_d through the euphotic zone is, for a very wide range of optical water types, within a few percent of the value of K_d at the mid-point of the euphotic zone (Kirk 1981), and so these concepts can legitimately be applied in the case of depth-averaged K_d values also.

Our theoretical treatment so far has been for monochromatic light, but the conclusions can, with certain limitations, be extended to broadband radiation. For

512

example, eqs. (11), (12) and (13) can be considered to apply to the photosynthetic waveband, but in this case the diffuse absorption and backscattering coefficients in eqs. (11) and (13) apply to absorption and backscattering of light fluxes not only with the angular distributions existing at the specified depth, z, but also with the spectral distributions existing at that depth. The problem is that the spectral distribution of the light field varies with depth more than the angular distribution does, and so the agreement between $K_d(z)$ at any given depth within the euphotic zone, and the depth-averaged value for the whole euphotic zone may not be very good. Nevertheless, the principle that K_d for PAR can be partitioned amongst the different components of the medium is an important and useful one. When using it we should simply remember that it strictly applies only to the localized value of K_d at any given depth, and its application to the depth-averaged K_d is an approximation. A convenient standard depth is the mid-point of the euphotic zone, z_m, and so for PAR as for monochromatic light we can regard the depth-averaged K_d value as being approximately equal to $K_d(z_m)$, to which the equations can be applied.

It is important also to remember that although the contributions of the different components of the medium to total attenuation can be simply added together in accordance with eq. (12), the nature of their contribution to attenuation can be quite different for different components. Consider, for example, a water in which component i is dissolved yellow colour, and component $i + 1$ consists of highly scattering mineral particles with little colour. For the ith component, $K_d(z)$ will consist mainly of the absorption term, $a_d(z)$, in eq. (13), whereas for the $(i + 1)$th component, $K_d(z)$ will consist mainly of the upward scattering term, $b_{bd}(z)$.

Quantitative Analysis of the Attenuation Process

It might be objected that although eqs. (11) and (13) do seem, on the face of it, to apportion responsibility for overall attenuation to the various absorption and scattering processes, in an explicit manner, in reality the values of the three quasi-inherent optical properties in the equations will not be known, and it is not easy to see how they could be measured. It might therefore seem that these equations, although of some conceptual value in helping us to understand the attenuation process, are not of practical use. As we shall now see, however, this gloomy conclusion is not in fact warranted.

If we know the inherent optical properties of the water then for any given incident flux on the surface it is possible, using a Monte Carlo computer simulation procedure, to arrive at a complete description of the light field at any depth, including the radiance distribution — the angular distribution of intensity. From the radiance distribution existing at a given depth and the inherent optical properties we can calculate the quasi-inherent optical properties of the medium at that depth (Kirk 1981). This means that not only can we bring the hitherto purely hypothetical quasi-inherent optical properties to life, but we can also calculate for any depth in any given optical water type the relative significance of the three different processes, discussed earlier, which determine the rate of attenuation. This we shall now do since it can help us understand the influence of picoplankton on the underwater light field.

Figure 5 shows the way in which the three different attenuation processes change in value as the optical character of the water — specified in terms of the ratio of scattering to absorption (b/a) — is progressively altered. We consider here a water body in which the absorption coefficient is, for simplicity, made equal to 1.0, but the scattering coefficient is varied from zero to 30, i.e. the ratio of scattering to absorption also changes from zero to 30. Light is vertically incident on the surface, so that at zero scattering the vertical attenuation coefficient for downward irradiance — that is K_d — is equal to the absorption coefficient, 1.0. As scattering increases at constant absorption, K_d also increases, as shown by the upper, continuous curve. The question we wish to address is — what contribution do each of the three attenuation processes make to K_d in different parts of this range of scattering/ absorption ratios?

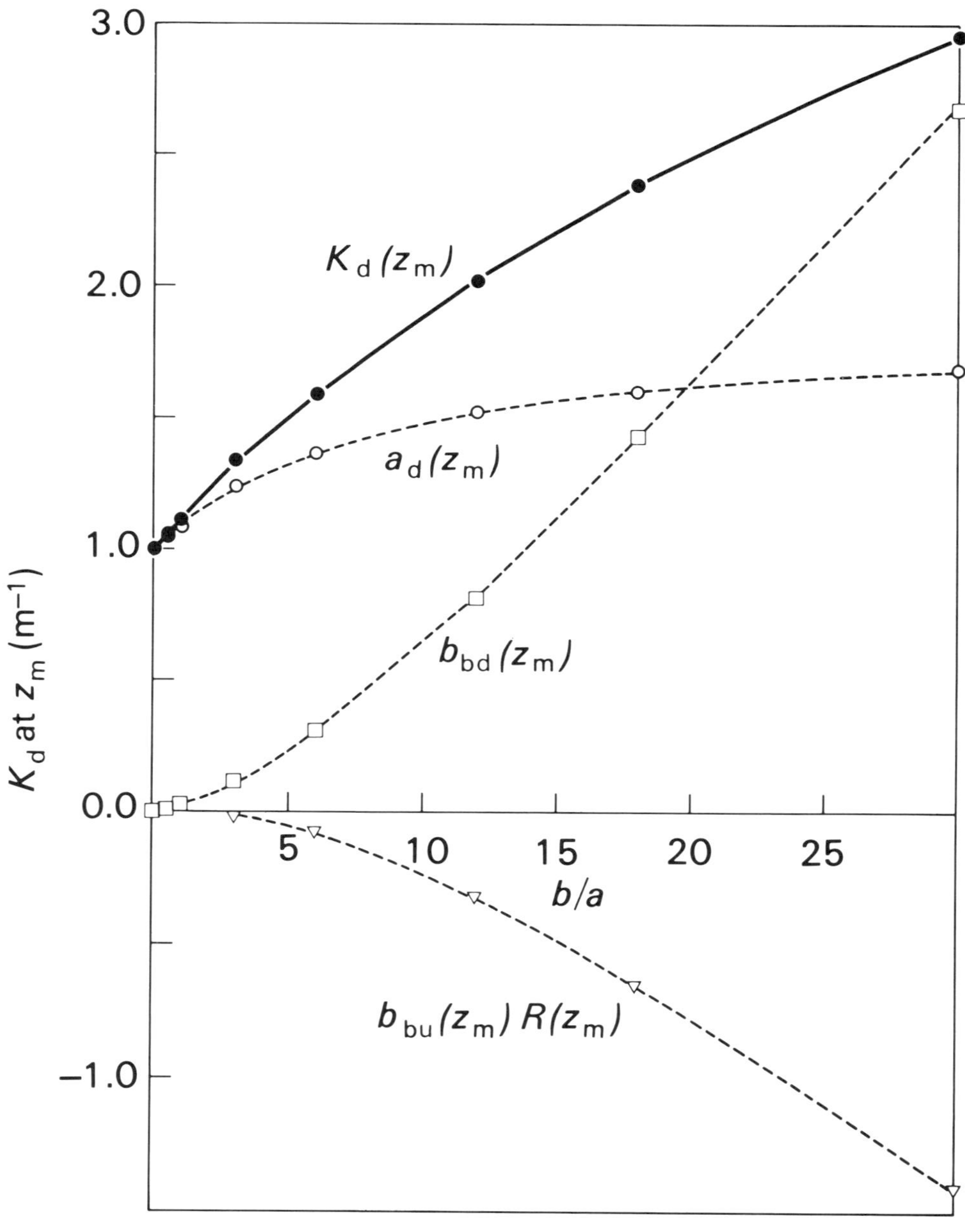

Fig. 5. Changes in the three different components of the irradiance attenuation process as a function of the ratio of scattering to absorption (b/a). The values of the three right-hand terms in eq. (11), calculated for the depth (z_m) at which irradiance is reduced to 10% of the subsurface value, are plotted alongside the corresponding values of K_d at z_m (data from Kirk 1981).

The answers are to be found in the three broken curves each of which corresponds to one of the terms on the right-hand side of eq. (11). The uppermost one is the absorption of light from the downwelling stream, accentuated as scattering increases, by the resultant change in angular distribution of this downwelling flux. With increasing scattering, absorption rises steeply at first but then slows down and levels off. The curve rising from the origin represents the backscattering, i.e. upward scattering, of the downwelling light. After a slow start it then begins to increase rapidly as b/a increases. The bottom curve is for the backscattering, i.e. downward scattering, of the upwelling stream, and this of course, since it opposes the other processes heads off in the negative direction, having very little effect at first but becoming of substantial magnitude at high scattering/absorption ratios.

So how is K_d made up? At low values of scattering, up to b/a ratios of about 3, attenuation is due almost entirely to absorption, and furthermore, the scattering-induced increase in attenuation is mainly a consequence of the increase in absorption resulting from the changed angular distribution of the downwelling flux. Upward scattering of the downwelling stream contributes only slightly to attenuation, and downward scattering of the upwelling stream is negligible.

When b/a has risen to about 7, upward scattering of the downwelling stream accounts for about 25% of all the attenuation and contributes as much to the increase in attenuation as does the increase in absorption of the downwelling light. When b/a is 20, upward scattering of the downwelling radiation accounts for as much attenuation as does absorption, and at higher b/a values it becomes the major mechanism for attenuation.

The downward scattering of the upwelling stream, which acts to diminish attenuation, has little effect over the lower part of the range of b/a values but becomes significant from about $b/a = 6$ onwards, and counteracts a large part of the attenuation due to the other two processes at b/a values in the range 12 to 30.

Phytoplankton and Vertical Attenuation

Where do the phytoplankton fit in the range of scenarios covered in Fig. 5? The answer is — depending on wavelength and species, just about everywhere. Bricaud, Morel and Prieur (1983) measured the absorption and scattering coefficients of suspensions of four marine phytoplankton species at wavelengths throughout the photosynthetic range. In the case of *Platymonas*, for example, b/a ranged from a minimum of about 3 at the blue absorption peak to a maximum of about 10 in the yellow–green, where absorption was relatively low. The corresponding ratios were about 3 and 12 for *Tetraselmis maculata* and 2 and 10 for *Hymenomonas elongata*. In the case of the coccolithophorid, *Coccolithus huxleyi*, however, with its highly scattering calcareous scales, b/a ranged from about 7 at the blue peak to as high as about 60 in the yellow–green region.

What this means is that phytoplankton contribute to attenuation of solar radiation through their scattering as well as their absorption effects, the relative importance varying as we have seen with wavelength throughout the photosynthetic spectrum.

Now that we have, in the foregoing sections, established a firm theoretical basis for partitioning the total vertical attenuation coefficient amongst the different components of the medium, it becomes meaningful to ask how much the phytoplankton contribute to the attenuation of photosynthetically available radiation in the ocean. Our consideration of this matter is simplified if we make the assumption that the contribution of phytoplankton to K_d increases approximately linearly with the concentration of the algae, as specified for example in terms of the concentration of phytoplankton chlorophyll a in the water (B_c mg • m^{-3}). Since, as we noted earlier, the values of the quasi-inherent optical properties themselves do not increase strictly linearly with the concentrations of the components responsible, the value of the partial vertical attenuation coefficient will also not be simply proportional to concentration, but field data for real phytoplankton populations indicate that the assumption of linearity is obeyed well enough to be useful, even in the case of the total photosynthetic waveband. We shall therefore assume that the partial vertical attenuation coefficient due to phytoplankton is given by

$$(14) \qquad K_{PH} = B_c k_c$$

where k_c is the specific vertical attenuation coefficient (units m^2 • mg chl a^{-1}) per unit phytoplankton concentration.

Thus, to characterize the contribution of a given phytoplankton population to the vertical attenuation of light we need to determine k_c for that population. This can be done by field measurement of $K_d(PAR)$ or $K_d(\lambda)$ at a series of phytoplankton concentrations, e.g. at different times during a period of natural increase of the population. Most literature values were in fact obtained in this way. Such data do not, however, make it possible for us to systematically explore the effects on k_c of variation in cell size, shape and pigment concentration and composition. For this we must carry out calculations for idealized model populations which have the desired characteristics. Such calculations show (Kirk 1975b, 1976) that, as might be expected from our previous consideration of the relationship between cell size and efficiency of light absorption, the smaller the cells (for a given pigment composition and concentration), the higher the value of k_c for the population, i.e. the greater the contribution of the phytoplankton to attenuation of solar radiation. For example, for spherical blue-green algal cells or colonies of diameter 58, 8 and 0.8 μm, the calculated values of k_c were 0.0063, 0.015 and 0.018 m^2 • mg chl a^{-1}, respectively. The direction and order of magnitude of these size-associated changes in k_c, predicted on theoretical grounds, have been confirmed in the field by Robarts and Zohary (1984) from measurements in natural populations of *Microcystis aeruginosa* with variable colony size.

Given the very small size of picoplankton cells, the package effect, as we saw earlier, would only slightly influence their light absorption properties, and so within any given pigment category of phytoplankton the picoplankton should have values of specific vertical attenuation coefficients at the top end of the range. Indeed picoplankton should have k_c values about the same as those which might be calculated for corresponding concentrations of uniformly dispersed photosynthetic pigment-proteins.

Since k_c is expressed per unit chlorophyll a, then its value for broadband PAR will very much depend on what other photosynthetic pigments are present. Calculations for model cells in water with a fairly high level of yellow substances indicated that k_c for diatoms would be about 70% higher than that for green algae because of the increased absorption in the 500–560 nm region due to fucoxanthin; k_c for blue-green algae with substantial levels of the biliprotein phycocyanin, absorbing in the 550–650 nm region was calculated to be about twice that for diatoms (Kirk 1976).

The background colour of the water in which the cells are suspended can also have a marked influence on the value of the specific vertical attenuation coefficient of the phytoplankton, since by its effects on the spectral distribution of the underwater light field it determines what proportion of the light is at wavelengths that the algae absorb effectively. In yellow inland waters for example the contribution of the blue spectral region to the underwater light field is greatly diminished, and so in such waters, green algae, which have their major absorption band in the blue region, have a low value of k_c (Kirk 1975b). The localized k_c values may also change with depth as the spectral quality of the light changes. For a typical inland water, Atlas and Bannister (1980) calculated that k_c for green algae would diminish from about 0.012 m^2 • mg^{-1} near the surface to about 0.005 m^2 • mg^{-1} at the bottom of the euphotic zone. In clear, colourless waters of the oceanic type, such as we for our present purposes are interested in, there is little diminution of k_c by background colour. Furthermore, in such waters there is comparatively little change of k_c with depth (Atlas and Bannister 1980).

From measurements in a range of oceanic and coastal waters Smith and Baker (1978) obtained an average value of k_c for PAR of 0.016 m² • mg chl a^{-1}. These were of course for mixed phytoplankton populations. We do not to my knowledge yet have any field measurements of k_c for populations known to consist predominantly of picoplankton. On the grounds of the high efficiency of light absorption by small cells we might anticipate that k_c for picoplankton would be somewhat higher that the value for mixed phytoplankton. As mentioned earlier, calculations for model blue-green algal cells indicated an increase in k_c from 0.015 to 0.018 when cell diameter decreased from 8 to 0.8 μm.

In oceanic waters, because of the low level of other absorbing materials, phytoplankton can account for a surprisingly high proportion of the total attenuation despite their own low concentration. Making the assumption that the fraction of total attenuation caused by phytoplankton is approximately equal to $B_c k_c / K_{TOT}$, the data of Smith and Baker (1978) indicate that phytoplankton present at 1.0 mg chlorophyll a m^{-3} account for 10–20% of the total attenuation.

IMPLICATIONS FOR PHOTOSYNTHETIC PERFORMANCE

The rate of photosynthesis of a phytoplankton population increases at first linearly with irradiance. The graph then begins to curve over and eventually levels off when the photosynthetic system becomes light-saturated (Fig. 6). At higher intensities still, photoinhibition sets in, but that is an aspect I do not here wish to address. In the linear, light-limited region the rate of photosynthesis is proportional to the product of the rate of capture of light energy by the population and the quantum yield (the number of CO_2 molecules fixed in biomass per quantum of light absorbed by the cells). The photosynthetic light response curve of any phytoplankton population, expressed as is normal practice in terms of the photosynthetic rate per mg chlorophyll a ($P(CO_2)$), as a function of irradiance, is therefore markedly affected by the rate of light capture per mg chlorophyll a.

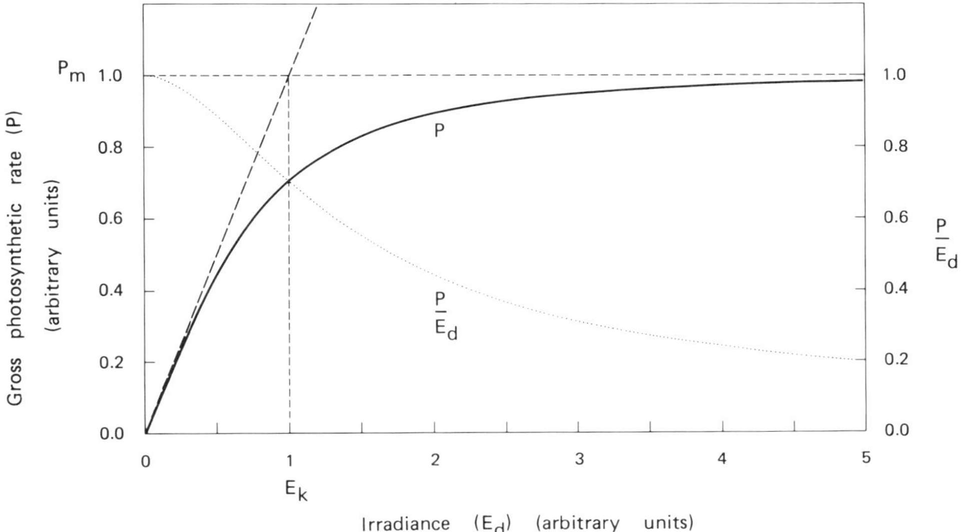

FIG. 6. Idealized curve of specific photosynthetic rate (P) as a function of irradiance (E_d). The maximum photosynthetic rate (P_m), the saturation onset parameter (E_k), and the efficiency of utilization of incident light (P/E_d) are also indicated.

517

We saw earlier (eq. (7)) that the instantaneous rate of absorption of light by the phytoplankton within any given thin layer of medium at a certain depth is equal to the product of the scalar irradiance at that depth, the concentration of phytoplankton (B_c, mg chl a • m^{-3}), the specific absorption coefficient per mg chlorophyll a, and the thickness of the layer. For PAR the relevant absorption coefficient must be the coefficient for the whole photosynthetic waveband, with whatever spectral distribution actually exists at the depth in question. Such absorption data are not readily available. We do, however, have data, admittedly approximate, for the specific vertical attenuation coefficient for phytoplankton, and so it seems worthwhile to consider the possibility of seeking to understand the rate of absorption of light by the phytoplankton population in terms of k_c rather than a_c.

It can be shown that the total rate of energy absorption per unit volume at depth z is given by

$$(15) \qquad \frac{d\Phi}{dv}(z) = \vec{E}(z)K_E(z)$$

where $\vec{E}(z)$ is the net downward irradiance ($E_d - E_u$, downward minus upward irradiance), and $K_E(z)$ is the vertical attenuation coefficient for $\vec{E}(z)$. In oceanic waters studied by Tyler and Smith (1970) the upward irradiance of total PAR at 5 m depth varied from 2 to 5% of downward irradiance. Furthermore the vertical attenuation coefficient for net downward irradiance in natural waters is normally very close in value to the attenuation coefficient for downward irradiance. We can therefore replace the exact relationship expressed in eq. (15) by the approximate, but more useful relationship

$$(16) \qquad \frac{d\Phi}{dv}(z) = E_d(z)K_d(z)$$

Making the assumption that the fraction of the total absorbed PAR captured by the phytoplankton is aproximately $B_c k_c / K_d(\mathrm{PAR})$ then the rate of absorption of light energy by phytoplankton per unit volume at depth z m may be approximately represented by

$$(17) \qquad \frac{d\Phi_{\mathrm{PH}}(z)}{dv} = B_c k_c E_d(z)$$

We have seen above that the vertical attenuation coefficient may validly be partitioned amongst the different components of the medium. What this means in relation to eqs. (15) and (16) is that the responsibility for causing energy to be absorbed at depth z may correctly be attributed to any given component of the medium in accordance with its fractional contribution to the total vertical attenuation coefficient. It should, however, be realized that the fraction of the energy absorption which a given component causes in terms of these equations is not necessarily the same as the fraction of the absorbed energy which is actually absorbed by that component. We have been led to this dilemma, to this potential trap, by our wish, in the interest of convenience, to arrive at an expression for phytoplankton light absorption making use of k_c rather than of a_c (which, pertaining as it does strictly to light absorption, would have avoided the problem).

The example given earlier of two hypothetical components of the medium, one contributing mainly to absorption, the other to scattering, can be used to make the essential point. The two components could have the same partial vertical attenuation coefficients, and thus be equally responsible in accordance with eqs. (15) and (16) for causing absorption at depth z m, but the first component would nevertheless be actually carrying out much more of the absorption within its own molecules than

518

the second component. The second component, by its effect on the angular distribution of the light field, would be acting mainly to promote absorption by the other components of the medium.

The important conclusion from the foregoing is that eq. (17), useful though it is, may in some cases be seriously in error. This is likely to arise where the ratio of scattering to absorption of PAR by a phytoplankton population is grossly different from the ratio for the rest of the medium. For example, if a highly scattering coccolithophorid population was found experimentally to have a high k_c value and this value was used in eq. (17), then its rate of energy absorption might be substantially overestimated.

Despite this qualification, eq. (17) is nevertheless of assistance in understanding some aspects of the phtosynthetic behaviour of natural phytoplankton populations. The quantum yield of photosynthesis, ϕ is given by the rate of CO_2 fixation by the population $(B_c P(CO_2))$ divided by the rate of absorption of light quanta by the cells $(B_c k_c E_d)$. Thus

$$(18) \qquad \phi = \frac{P(CO_2)}{k_c E_d(z)}$$

The quantum yield of course varies with light intensity since at higher intensities the photosynthetic apparatus cannot fix CO_2 fast enough to cope with the rate at which quanta are being absorbed. In the initial, linear part of the curve, however, each increment in light absorbed is accompanied by a proportionate increase in photosynthesis, and it is here, at low irradiance, that the quantum yield has its maximum value $-\phi_m$. The initial slope of the curve $(P(CO_2)/E_d)$ in the linear region is commonly given the symbol, α, and from eq. (18) it follows that

$$(19) \qquad \alpha = \phi_m k_c$$

Any plant species could in principle achieve a quantum yield approaching the theoretical maximum of about 0.1. This, combined with the fact that the initial slope of the light response curve should be proportional to maximum quantum yield, ϕ_m, has led sometimes to the expectation that α should not vary markedly from one species to another, or for a given species from one environment to another. Such expectations, however, ignore the extent to which k_c (which equally determines α) can vary. We have seen earlier that k_c can vary markedly with size, shape, pigment composition and content of cells, and this would be accompanied by corresponding changes in α. Part of the variation in α observed for phytoplankton populations in nature we may reasonably attribute to variation in k_c. For example the values for coastal phytoplankton were observed to be lower in deeper water, and in the winter (Platt and Jassby 1976). A possible explanation would be that lowered light intensities led to an increased pigment content leading to a greater package effect and therefore a decrease in k_c.

The inevitable uncertainty in the true value of k_c applicable to a given situation means that a corresponding uncertainty attaches to estimates of *in situ* quantum yields of natural phytoplankton populations. If for any reason a particularly accurate determination of ϕ at a particular depth at a certain oceanic station is required, then the rate at which the phytoplankton absorb light should be estimated using eq. (7), the effective absorption coefficient for PAR at the depth of interest being calculated (Morel 1978) from the absorption spectrum of the cells and the spectral distribution of the light field, literature data being used if first-hand information is not available.

As we have remarked earlier, very small cells are, for good optical reasons, particularly efficient at harvesting light. It may turn out, when the necessary measurements are done, that natural picoplankton populations have relatively high α values. If so this will be because of their high k_c values since there is no reason to suppose that decrease in size should lead to an increase in quantum yield, i.e. we expect that picoplankton will show a higher efficiency than larger phytoplankton for the conversion of available light energy, but not of absorbed light energy.

References

ATLAS, D., AND T. T. BANNISTER. 1980. Dependence of mean spectral extinction coefficient of phytoplankton on depth, water colour and species. Limnol. Oceanogr. 25: 157–159.

BRICAUD, A., A. MOREL, AND L. PRIEUR. 1983. Optical efficiency factors of some phytoplankters. Limnol. Oceanogr. 28: 816–832.

DUYSENS, L. N. M. 1956. The flattening of the absorption spectrum of suspensions compared to that of solutions. Biochim. Biophys. Acta. 19: 1–12.

HODKINSON, J. R., AND J. I. GREENLEAVES. 1963. Computations of light-scattering and extinction by spheres according to diffraction and geometrical optics and some comparisons with the Mie theory. J. Opt. Soc. Am. 53: 577–588.

KIRK, J. T. O. 1975a. A theoretical analysis of the contribution of algal cells to the attenuation of light within natural waters. I. General treatment of suspensions of pigmented cells. New Phytol. 75: 11–20.

——— 1975b. A theoretical analysis of the contribution of algal cells to the attenuation of light within natural waters. II. Spherical cells. New Phytol. 75: 21–36.

——— 1976. A theoretical analysis of the contribution of algal cells to the attenuation of light within natural waters. III. Cylindrical and spheroidal cells. New Phytol. 77: 341–358.

——— 1981. A Monte Carlo study of the nature of the underwater light field in, and the relationship between optical properties of, turbid yellow waters. Aust. J. mar. freshwater Res. 32: 517–532.

——— 1983. Light and photosynthesis in aquatic ecosystems. Cambridge University Press, Cambridge.

MIE, G. 1908. Beiträge zur Optik trüber Medien, speziell kolloidalen Metall-lösungen. Ann. Phys. 25: 377–445.

MOREL, A. 1973. Diffusion de la lumière par les eaux de mer. Résultats expérimentaux et approche théorique, p. 3.1-1 to 3.1-76. In Optics of the Sea. AGARD Lect. Ser. 61. NATO, Neuilly-sur-Seine.

——— 1978. Available, usable and stored radiant energy in relation to marine photosynthesis. Deep-Sea Res. 25: 673–688.

MOREL, A., AND A. BRICAUD. 1981. Theoretical results concerning light absorption in a discrete medium, and applications to specific absorption of phytoplankton. Deep-Sea Res. 28: 1375–1393.

MURPHY, L. S., AND E. M. HAUGEN. 1985. The distribution and abundance of phototrophic ultraplankton in the North Atlantic. Limnol. Oceanogr. 30: 47–58.

PLATT, T., AND A. D. JASSBY. 1976. The relationship between photosynthesis and light for natural assemblages of coastal marine phytoplankton. J. Mar. Res. 38: 687–701.

PREISENDORFER, R. W. 1961. Application of radiative transfer theory to light measurements in the sea. Union Geod. Geophys. Inst. Monogr. 10: 11–30.

ROBARTS, R. D., AND T. ZOHARY. 1984. *Microcystis aeruginosa* and underwater light attenuation in a hypertrophic lake (Hartbeespoort Dam, South Africa). J. Ecol 72: 1001–1017.

SMITH, R. C., AND K. S. BAKER. 1978. The bio-optical state of ocean waters and remote sensing. Limnol. Oceanogr. 23: 247–259.

TYLER, J. E., AND R. C. SMITH. 1970. Measurements of spectral irradiance underwater. Gordon and Breach, New York. NY.

Inherent Optical Properties of Algal Cells
Including Picoplankton:
Theoretical and Experimental Results

ANDRÉ MOREL AND ANNICK BRICAUD

Laboratoire de Physique et Chimie Marines
Université Pierre et Marie Curie — CNRS, UA 353
BP 8, F 06230 Villefranche-sur-Mer, France

1. Introduction

Even though they are not always accurately known, there exist direct relationships between the concentrations of various substances, dissolved or particulate, present within a water body and the "inherent" optical properties of the body. In the open ocean, that is to say in more than 95% of the world's ocean, the influence of terrigeneous materials, like sediments and dissolved yellow substance, is negligible from an optical view point. Phytoplankton, with their detrital retinue, become the dominant, practically the unique factor which causes the optical properties to depart from those of pure water. Even in oligotrophic regions, the low concentrations of algae, including picoplankton, are easily revealed by their optical "signatures". How algal cells influence the inherent properties (light absorption and scattering) of natural waters is a question to be addressed.

The penetration of the solar radiation into water, the decay of the radiative energy with increasing depth, the changes in the spatial distribution of the submarine light field, the existence of an upwelling light stream with photons crossing the interface towards the atmosphere, all these phenomenon are depicted by appropriate coefficients which form the class of the "apparent" optical properties. (This useful and logical distinction between the two classes of optical properties was introduced by Preisendorfer in 1961). The apparent optical properties obviously depend on the inherent properties. The former, however, depend also on the geometrical structure of the light field: the light field incident upon the surface or the light field within the water, which is progressively modified by the scattering process.

Conceptually a "cause-effect chain" can be imagined as being

This chain can be explored in both directions. For instance S. Sathyendranath, in a paper in this volume, evaluates the possibility of inferring algal biomass from ocean colour, i.e. from its spectral reflectance (an apparent property defined as the ratio of upwelling to downwelling irradiances).

Also J.T.O. Kirk, in this volume, examines in particular how the "K_d coefficients" (the vertical attenuation coefficient for irradiance) can be related to the phytoplankton concentration.

The present paper deals only with the first link in the above chain (and only with phytoplankton). It aims at providing a theoretical framework inside which the "optics" of an algal cell can be understood and possibly predicted (especially for the case of picoplankton). When developing a theory, it is conceivable that the most suitable level is that of a single particle, seen as an isolated body able to absorb and to scatter radiation. Optical properties at this level are not pure abstractions. They are of direct interest, in physiological studies for instance, to the extent that they govern the energy capture capabilities of the photosynthesizing organisms. Also with the recent development of optical techniques associated with flow-cytometry and cell sorting, the optical behaviour of individual cells is directly involved. This is another question to be addressed. With these ideas in mind, the organization of the present paper is as follows.

Precise definitions of the optical properties of an individual cell and subsequently of an algal suspension are recalled (Part 2). The symbols and terminology are those recommended by IAPSO (see in Morel and Smith 1982). The theoretical background is progressively developed in Part 3, until a general formulation is provided through Mie theory. Useless mathematical developments have been avoided, to the profit of the physics behind the equations and of the results, which have been approached in some detail. Picoplankton, from an optical point of view, do not constitute a special case deserving a separate treatment, although particular properties can be anticipated. They are pointed out as predictive tools since existing data are rather scarce.

Thereafter (Part 4) the relationships between the optical characteristics of a single cell and the optical coefficients of a suspension are established. Close attention is paid to the "chlorophyll-specific" optical coefficients. The angular distribution of the scattered light (the volume scattering function) is also studied in relation to the cell characteristics.

Finally (Part 5), a comparison is attempted between the theoretical predictions and some experimental results obtained with algae grown in culture. The rather high variability of the optical cross sections of algae grown in culture and also the variability of their chlorophyll-specific coefficients as evidenced by experiments, are examined in the light of the theory. A model is presented which accounts for the spectral variations of these parameters.

2. Optical Properties of Algal Cells and Algal Suspensions (Definitions)

2.1 INDIVIDUAL CELLS

The first category of optical properties we shall deal with concerns the algal cell when considered as an isolated particle. The interaction between radiation and such an "optical object" is conveniently represented by dimensionless numbers, termed as efficiency factors and defined below.

We assume that the particle (any shape and nature) is illuminated by electromagnetic radiation propagating as a plane wave. In general, two processes happen simultaneously when the wave reaches the particle: one part of the energy carried by the wave is removed by absorption inside the particle and another part is scattered, that is diverted from the direction of propagation of the incident light (the scattered waves propagate as spherical waves originating from the particle). A scattered ray is characterized geometrically by an azimuth angle ϕ and by a polar angle θ which it forms with the initial direction of propagation. Both processes remove energy from the incident plane wave, and the global effect is called attenuation (see Fig. 1a).

The efficiency factors for absorption, Q_a, and for scattering, Q_b, are respectively defined as the ratios of the radiative energy absorbed within, or scattered by, the particle to the energy impinging on its geometrical cross section (measured perpendicularly to the direction of propagation).

The efficiency factor for attenuation, Q_c, is therefore

(1) $$Q_c = Q_a + Q_b$$

If s denotes the area of the geometrical cross section, the products

(2) $$s \, Q_i = S_i$$

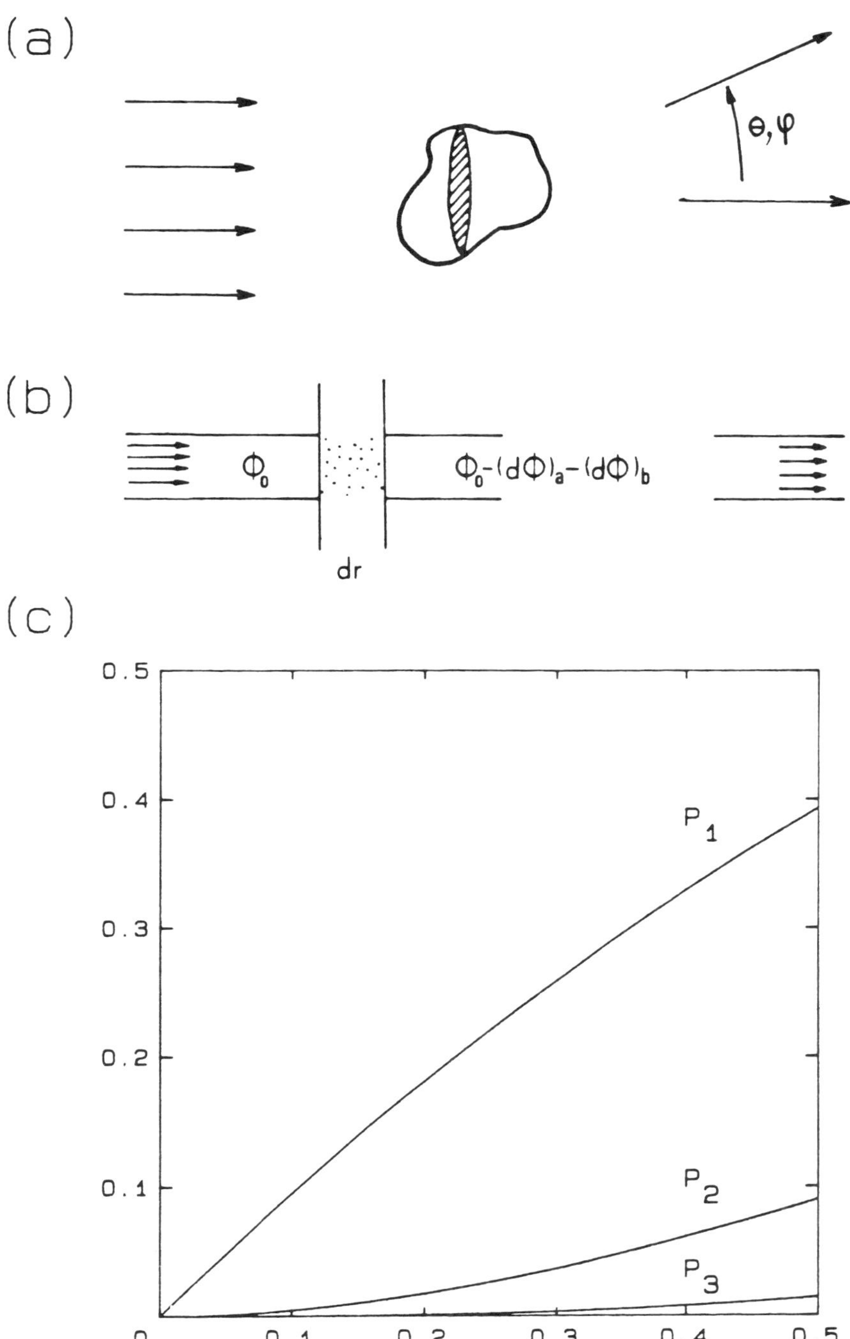

FIG. 1. (a) Schematic diagram corresponding to the definition of the scattering angle and the optical cross sections of a particle in reference to its geometrical cross section (hatched area); (b) Schematic diagram corresponding to the definition of the inherent optical properties of a suspension; (c) Probability of single ($P = 1$), double ($P = 2$) or triple ($P = 3$) scattering as a function of the optical thickness τ in a scattering (non-absorbing) medium. These results have been established by Bugnolo (1960) in the case of scattering which should be strongly peaked in the forward direction (such a pattern is that exhibited by algal cells).

(with $i = a, b, c$) define S_i, the cross sections of the particle for absorption, scattering and attenuation (they have the dimensions of L^2).

The energy scattered out by the particle propagates in all directions of space with respect to the initial direction ($\theta = 0$). The scattering process can be split into a forward ($0 < \theta < \pi/2$) and backward scattering ($\pi/2 < \theta < \pi$) processes; accordingly Q_b can also be split into Q_{bf} and Q_{bb}, respectively (with $Q_b = Q_{bf} + Q_{bb}$).

The second class of optical properties we shall consider concerns a medium in which the particles are in suspension (a water body with algal cells). These properties, namely the absorption coefficient a, the scattering coefficient b, their sum c, the attenuation coefficient, and finally the volume scattering function $\beta(\theta)$ have been referred to by Preisendorfer (1961) as "inherent" optical properties in the sense that "their magnitudes (for each wavelength) depend only on the substances comprising the hydrosol and not on the geometrical structure of the various light fields that may pervade it." (page 119, vol. 1, Preisendorfer 1976).

Their definition rests on a somewhat ideal experiment which should respect rigorous conditions (physically unattainable). This experiment would consist in illuminating an infinitely thin layer (dr) of a homogenous medium with a perpendicular beam of monochromatic light (Fig. 1b). The beam is assumed to be non divergent (containing only perfectly parallel rays) and the receiver beyond the layer is assumed to have a nul acceptance angle (receiving only the parallel rays). The diffraction limitation prevents both these conditions from being perfectly fulfilled, thus the experimental arrangements may only approach them.

If ϕ_o is the incident flux on the slab, beyond the slab it becomes:

$$\phi = \phi_o - (d\phi)_a - (d\phi)_b$$

$(d\phi)_a$ and $(d\phi)_b$ are those parts of the flux which have disappeared from the beam respectively by absorption and by scattering. The losses are proportional to the incident flux and to the thickness of the slab.

$$(3) \qquad - (d\phi)_a = a\phi_o dr$$

$$(3') \qquad - (d\phi)_b = b\phi_o dr$$

and the coefficients of proportionality are precisely the absorption coefficient a, and the scattering coefficient, b. They have the dimension of L^{-1} and the unit is in general m^{-1}.

The beam is attenuated by both the scattering and absorption processes and the corresponding coefficient c, with $c = a + b$, is the attenuation coefficient. If the medium is assumed to be homogenous along a finite path r (a and b constant), the integration of

$$- (d\phi_a + d\phi_b) = (a+b)\,\phi_o dr$$

leads to

$$(4) \qquad \phi = \phi_o e^{-cr}$$

which expresses the exponential decay of the flux along the beam and is known as Beer's law. The dimensionless product $\tau = cr$ is the optical thickness. The transmittance, ϕ/ϕ_o, is given by $e^{-\tau}$, whereas the attenuance, $(\phi_o - \phi)/\phi_o$, is given by $1-e^{-\tau}$. The rationales for the above rigorous conditions are:

524

1) No scattered photons enter the detector even if the scattering direction (θ) is close to that of the incident beam direction ($\theta = 0$); no divergent rays ($\theta \neq 0$) exist inside the beam which could generate a scattered flux into the direction $\theta = 0$.

2) Multiple interactions must not occur; in other words, incident photons must not encounter more than one particle when travelling through the slab, and scattered diverging photons must freely escape from the slab without experiencing another collision (leading to absorption or scattering).

As a matter of fact the meaningful condition is that $d\tau = c \cdot dr$ must be very small. If dr is small with a high value for c, the layer is not optically thin and the probability of multiple collision will not be negligible. For instance, if $d\tau = 0.1$, photons have an approximately 90% chance of going freely through the slab and 10% of experiencing a collision (the ratios a/c and b/c govern the fate of these photons, i.e. whether they are absorbed or scattered). The surviving photons have, at the most, again a 10% chance of another event, so perhaps 1% of photons have suffered two collisions. Such a reasoning is oversimplifying and pessimistic, and a more precise estimate of this effect is provided by Fig. 1c, established from the formulae given by Bugnolo (1960). This condition ($d\tau$ small) must be kept in mind when performing measurements on actual samples, unless the equation of radiative transfer in its one-dimensional form is inverted to compute a and c (Privoznik et al. 1978).

The flux $(d\phi)_b$ lost by the beam reappears as light scattered in all directions from the illuminated small volume dv (with $dv = dr \cdot ds$, if ds is the sectional area of the beam).

The volume scattering function $\beta(\theta,\phi)$, also an inherent property, describes the angular distribution of these scattered radiations. Since we do not examine the case of spatially organized material (such as crystal, for instance) but the case of a medium with randomly oriented particles (algal cells), we can assume without error that a rotational symmetry exists. Therefore β will not depend on the azimuth angle, and will only be a function of the polar angle θ. The infinitely small volume dv seen from a distant point, acts as a source emitting in the direction θ with an intensity $dI(\theta)$. This intensity, again, is proportional to ϕ_o and dr; and the $\beta(\theta)$ function is defined through:

$$(5) \qquad dI(\theta) = \beta(\theta)\phi_o dr$$

The flux scattered around the beam in the direction θ, and within an elementary solid angle $d\Omega$ ($d\Omega = 2\pi \sin\theta \, d\theta$) is

$$d^2\phi_b = d\Omega dI(\theta)$$

The flux $(d\phi)_b$ lost by the beam is obviously equal to the flux scattered once it is integrated over all space (θ from 0 to π), so,

$$d\phi_b = \int_{4\pi} d^2\phi_b = 2\pi\phi_0 dr \int_0^\pi \beta(\theta) \sin\theta d\theta$$

By comparing it with the definition of b it becomes straightforwardly

$$(6) \qquad b = 2\pi \int_0^\pi \beta(\theta) \sin\theta d\theta$$

Often a normalized dimensionless volume scattering function is conveniently used; its definition is

$$(5') \qquad \overline{\beta}(\theta) = \beta(\theta)/b$$

Its integral over all space leads to unity by definition. The integral (Equation 6) can be separately computed over the forward and the backward half-spaces, and thus the following are obtained

$$b_f = 2\pi \int_0^{\pi/2} \beta(\theta) \sin\theta d\theta$$

$$b_b = 2\pi \int_{\pi/2}^{\pi} \beta(\theta) \sin\theta \, d\theta$$

respectively, the forward and the backward scattering coefficients. These coefficients are advantageously replaced by the dimensionless ratios:

$$(6') \qquad \bar{b}_f = b_f/b \quad \text{and} \quad \bar{b}_b = b_b/b$$

called forward and backward scattering ratios.

The additivity principle strictly applies to the inherent coefficients. This principle implies that if several substances are present in the medium, the global phenomena of absorption and scattering result from several partial contributions. In the case of a water body constant terms, a_w, b_w, β_w (θ), are always present. The additivity property leads to:

$$a = a_w + \Sigma a_i \text{ or } b = b_w + \Sigma b_i$$

where a_i and b_i are the partial contributions of the ith substance to absorption and scattering. Among these substances, we shall deal with those parts, $a\phi$ and $b\phi$, which result from the presence of phytoplankton in suspension.

In many applications, it is interesting to relate a given partial contribution to the concentration of the substance responsible for the effect, and hence to write a_i as a_i^* $[x_i]$ where a_i^* is a specific coefficient defined for a unit concentration and $[x_i]$ is the actual concentration of the substance i in the medium.

In the aquatic medium, the most widely used (or routinely measured) indicator for quantifying the phytoplankton biomass is the chlorophyll a concentration. In reference to this index, the specific absorption coefficient a_ϕ^* and the specific scattering coefficient, b_ϕ^* are defined as the values of a and b when the chlorophyll concentration in the medium is 1 mg m^{-3}. The units for the specific coefficients are m^2 (mg Chl$a$)$^{-1}$.

Since no confusion can exist, the subscript ϕ will hereafter be abandoned. It could be imagined that a normalisation to the "living" algal carbon should be better. Some drawbacks in effect result from using a single pigment. For instance, with a considerably variable carotenoid-to-chlorophyll ratio the absorption in the blue part of the spectrum when normalized vis-à-vis Chl$a$, appears widely varying. The situation is perhaps still more dramatic for cyanobacteria which often contain a noticeable amount of phycobilins. Because of the difficulty in determining the living algal carbon in a natural environment, where detritus often predominate, this normalisation problem has still not been satisfactorily solved.

3. Theoretical Considerations Concerning the Optical Properties of a Single Cell

3.1 EMPIRICAL APPROACH

Let us consider a particle which would be "totally" absorbing. All light incident upon, and entering into the particle is absorbed; it is a perfect black body, with Q_a = 1. As a matter of fact, if the particle does not have an index of refraction which perfectly matches that of the surrounding medium, some reflection occurs at the interface and not all the light can penetrate the interior, so Q_a will be slightly less than 1, and the particle will not be a perfect black body.

526

The presence of such a particle will also perturb the electromagnetic field outside its own physical boundary. The photons travelling in the vicinity of the particle are deviated by its presence, that is scattered. If we consider a large particle (large compared to the wavelength of the incident radiation), this perturbation leads to the phenomenon known as diffraction. According to the principle of the "complementary screens", Babinet's theorem predicts that the diffraction by an obstacle (such as a particle) is identical to the diffraction through a hole of the same geometrical cross section. According to this equivalence, the number of photons forming the scattered flux is exactly equal to the number of photons impinging on the geometrical cross section of the obstacle. Thus $Q_b = 1$ and therefore $Q_c = 2$, a result which means that a "large" particle is able to remove from the incident beam twice the amount of energy that it can geometrically intercept. This somewhat surprising conclusion is often referred to as the "extinction paradox".

If some reflection takes place (leading to $Q_a < 1$), the reflected radiation is added to the diffracted radiation, hence Q_b is greater than 1, whereas Q_c remains equal to 2.

Conversely, let us consider now a totally non-absorbing large particle. All light entering into it undergoes refraction, internal reflections, but finally escapes from the particle in various directions and without any loss of energy. This light scattered out from the particle is added to the scattered light due to diffraction which remains unchanged with respect to the previous example. In terms of efficiency factors, the situation is depicted by:

$$Q_a = 0$$

$$Q_b = Q_c = 2$$

If the material forming the particle is moderately absorbing (as algal cells), always under the proviso of a larger size with respect to the wavelength, the general situation will be:

$$0 < Q_a < 1$$

$$1 < Q_b < 2$$

$$\text{in such a way that: } Q_c = 2$$

The spectrum of phytoplankton size is quite large and the wavelengths of the electromagnetic radiation we deal with are those of the visible (or photosynthetic) domain, say between 0.4 and 0.7 μm. With dimensions of algal cells ranging from less than 1 μm to more than 100 μm (more than 6 orders of magnitude in volume or biomass), the above proviso, which ensures that $Q_c = 2$, is in general not fulfilled. The general theory able to account for scattering by spherical, variously sized, and variously absorbing particles is the Mie theory. This theory is a rigorous solution of Maxwell's equations, with the appropriate boundary conditions. It provides the way of computing the efficiency factors and the intensity of scattered light, including its polarization state, for any scattering angle. It is an all embracing theory applicable to spheres of arbitrary size. For very small (compared to wavelengths) spheres it reduces to the Rayleigh theory of the radiating dipole and, for large spheres, it provides a rigorous sum of the geometrical-optics pattern (reflection and refraction) and the Fraunhofer diffraction pattern. The Mie theory initially developed for homogeneous spheres, has been extended for otherwise shaped and/or non homogeneous particles. The computations are generally cumbersome, especially for large particles.

Fortunately, a considerable simplification occurs if the index of refraction of the particle differs only slightly from that of the surrounding medium. In this case, the "anomalous diffraction approximation", developed by Van de Hulst (1957), may apply and provides an estimate of Q_c, Q_b, and Q_a. If information concerning the spatial distribution of the scattered energy is needed, it is nevertheless necessary to come back to the rigorous Mie theory.

At this point, it is necessary to precise the position of a phytoplankton cell, in the context of the theories, particularly with respect to its size and its index of refraction. In what follows, it is assumed that the cell can be seen as a homogeneous sphere. Several important parameters, needed for the theoretical treatment, will be introduced under this assumption in the following section. The actual values taken by these parameters in the case of algal cells are examined thereafter.

3.2 PARAMETERS INTERVENING IN THE THEORY

The relative size α

It is conceivable that the geometrical size (diameter d) is not an adequate parameter for examining the interaction with an electromagnetic wave. The appropriate parameter is a dimensionless number which compares the actual size to the wavelength. The relative size α is defined as:

$$(7) \qquad \alpha = \pi d / \lambda_w$$

where λ_w is the wavelength in the surrounding medium (water in our case), or $\lambda_w = \lambda / n_w$, if λ is the wavelength *in vacuo* and n_w is the index of refraction of the water.

The relative index of refraction m

The propagation of electromagnetic waves in a homogeneous substance, is characterized by the complex refractive index m_s with

$$m_s = n_s - i\, n'_s$$

The real part governs the phase velocity, and the imaginary part describes the decrease in electric field strength or the decay of the energy flux (amplitude squared of the electric field), as the wave propagates through an absorbing medium. n_s (linked to the dielectric constant of the substance) is the ratio:

$$n_s = v_0 / v_s$$

where v_0 and v_s are the speed of the light *in vacuo* and in the substance. The imaginary part n'_s is expressed as:

$$n'_s = a_s \lambda / 4\pi$$

where a_s is the absorption coefficient of the substance.

The optical behaviour of a particle suspended in a given medium, is not dependent on the absolute values of its index of refraction, but on its relative value, with respect to that of the surrounding medium. In the visible domain the refractive index of the water is a real number. Even in the red part of the spectrum, where the absorption increases, the imaginary part remains less than 2.10^{-8}. Thus the relative index of a particle is:

$$(8) \qquad m = \frac{n_s - i\, n'_s}{n_w}$$

528

simply written thereafter as $m = n - i\,n'$
where $n = n_s/n_w$

(9) and $n' = n'_s/n_w = a_s\lambda/4\pi n_w$ or $a_s\lambda_w/4\pi$

The phase lag ϱ and the optical thickness due to absorption ϱ'

It is obvious that if $m = 1$, the particle is undistinguishable from the surrounding medium. There is no scattering neither absorption. If m differs from 1 solely by the way of n' (non zero at a given wavelength) the particle will be perceptible through its coloration, without giving rise to scattering. The scattering process happens only where the real part of the index differs from that of the medium, i.e. where $(n-1)\neq0$. It can easily be imagined that the intensity of the scattering process depends not only on the size of the obstacle but also on the increment $(n-1)$. It is therefore natural to introduce in the theoretical treatment another dimensionless parameter, ϱ, which combines both these aspects

(10) $\varrho = 2\alpha(n-1)$

The physical meaning of this parameter is the phase lag suffered by the ray which crosses the sphere along its diameter. The optical thickness corresponding to absorption along the same diameter ray, i.e. the product $d \cdot a_s$, forms a similar dimensionless parameter, ϱ'

(11) $\varrho' = a_s d$

which, according to equations 7 and 9, can be written:

(12) $\varrho' = 4\alpha n'$

Between ϱ and ϱ' there exists the relationship

(12') $\varrho' = 2\varrho \tan\zeta$

where $\tan\zeta$ is defined as the ratio $n'/(n-1)$. As it will be shown later, ϱ and ϱ' govern the scattering and absorption efficiencies (Q_b and Q_a) of the particle. Their definitions deserve comment. The parameters α and $(n-1)$ play equivalent roles in fixing ϱ and hence the global scattering properties. The same proposition can be repeated for the roles of α and n' in governing the absorption properties. Thus a larger and weakly refringent particle will scatter as does a more refringent but smaller particle, provided that ϱ is maintained constant. Similarly, with ϱ' remaining constant, weakly pigmented large cells will absorb the same amount of energy as smaller but more pigmented ones.

It is important to note that this reasoning cannot be extended to the angular distribution of the scattered radiation. The volume scattering function separately depends on α, n and n' (α being predominant in determining the "shape" of this function, as it will be shown later on).

3.3 Actual Values of the Parameters α, m, ϱ and ϱ' in the Case of Algal Cells

The relative size α

The geometrical characteristics of phytoplankters can be determined by microscopic examination. Their actual size in terms of the diameter of a sphere equivalent in volume can be computed or is directly measured by electronic counters. Even if some practical difficulties remain, the size parameter α, once the wavelength λ is fixed, is straightforwardly obtained. To get an idea, if the equivalent diameter, d, varies from about 1 to 100 μm, α will vary (equation 7) from

$$
\begin{array}{rllll}
\alpha = & 10.50 & \text{to} & 1050 & \text{if } \lambda = 400 \text{ nm} \\
& 7.65 & \text{to} & 765 & \phantom{\text{if } \lambda = } 550 \\
& 6.01 & \text{to} & 601 & \phantom{\text{if } \lambda = } 700
\end{array}
$$

The estimate of the parameters ϱ and ϱ' requires that plausible values for n and n' are selected. The values needed in a simple theoretical framework are "bulk" values, representative of the whole cell seen as a homogeneous optical body.

The real part of the relative index of refraction, n

Direct determinations of n, while being difficult, are possible by refractometry (Bryant et al. 1969), and also by immersion or phase contrast methods (Mc Crone et al. 1967; Hodgson and Newkirk 1975). The two last methods provide results representative of the external shell index rather than of the bulk or "mean" index. The scarce data in literature tend to indicate that algal cells are weakly refringent with respect to the surrounding water (n varying from 1.02 to 1.08).

To get a plausible bulk value, the results of a thorough study by Aas (1981) can be briefly recalled. This approach consisted of inferring this bulk value from the chemical composition. The main points are as follows.

The most important feature to be emphasized is the consistently high water content of algal material. The water volume would represent 70 to more than 80% of the cellular volume. The refringence of the cells suspended in water can only originate from the other constituents (the "dry mass" which amounts to about 20–30% of the total mass). These constituents are mainly (or exclusively) organic and possibly mineral for those species which possess outer opal (diatom) or calcite (coccolithophorid) shells. For the organic fraction, the relative index ranges approximately from 1.10 for lipids, 1.15 for carbohydrates, to 1.20 for proteins. Opal has a rather low refractive index of about 1.07 ($\pm$ 0.02) whereas calcite is more refringent with $n = 1.19$ ($\pm$ 0.01). By assuming that a mixing rule may apply to the contribution of the diverse constituents, including water, in forming the bulk index, a central value of 1.035 can be obtained. Mainly in response to the water content, rather than to the chemical composition, the expected range of variations of n could be 1.015 to 1.070.

The imaginary part n'

Several methods have been developed to measure absorption by intact living cells and to obtain values exempt from, or corrected for, the scattering interference (Duntley 1942; Shibata et al. 1954; Yentsch 1957; Latimer and Rabinowitch 1959; Amesz et al. 1960; Doucha and Kubin 1976; Bricaud et al. 1983). From the spectral a (λ) values, $n'(\lambda)$, the imaginary part, is obtained as a function of the wavelength through Equation 9, once the absorption coefficient of the cellular substance, a_s has been computed (see 4.2). Imagining a "typical" algal cell allows the estimation of at least the magnitude of n'.

It is generally admitted that the ratio of the mass of chlorophyll a to the dry mass does not exceed a few percent. If we adopt a value of 1.5% for this ratio and by remembering that water forms say 80% of the total volume, the intracellular concentration c_i of chlorophyll a will be of the order of 3 kg m^{-3} (c_i = mass Chl$a$ / biovolume), a value compatible with the "conversion factors" given by Strickland (1960).

The specific absorption coefficient for Chl$a$ dissolved in acetone, when derived from SCOR-UNESCO equations, is 0.0207 m^2 mg^{-1} at the wavelength of the red peak (663 nm). If we assume that the Chl*a* molecules, when embedded in the watery cellular material, maintain approximately the same specific absorption (the same molar "extinction") despite the shift of the absorption peak, (from 663 for acetone solution to 675 nm for intact cells), the absorption coefficient computed for the substance forming the cell is

$$a_s = 0.0207 \ (m^2 \ mg^{-1}) \times 3 \ 10^6 \ (mg \ m^{-3}) = 6.2 \ 10^4 \ m^{-1}$$

and leads to (Equation 9)

$$n' = \frac{6.2 \ 10^4 \times 0.675 \ 10^{-6}}{4 \times 1.334} = 0.0025$$

This value is similar to that of black glass. A 10 μm thick layer of such a material will absorb 56% of the incident (red) light. In the blue part of the spectrum, n' is

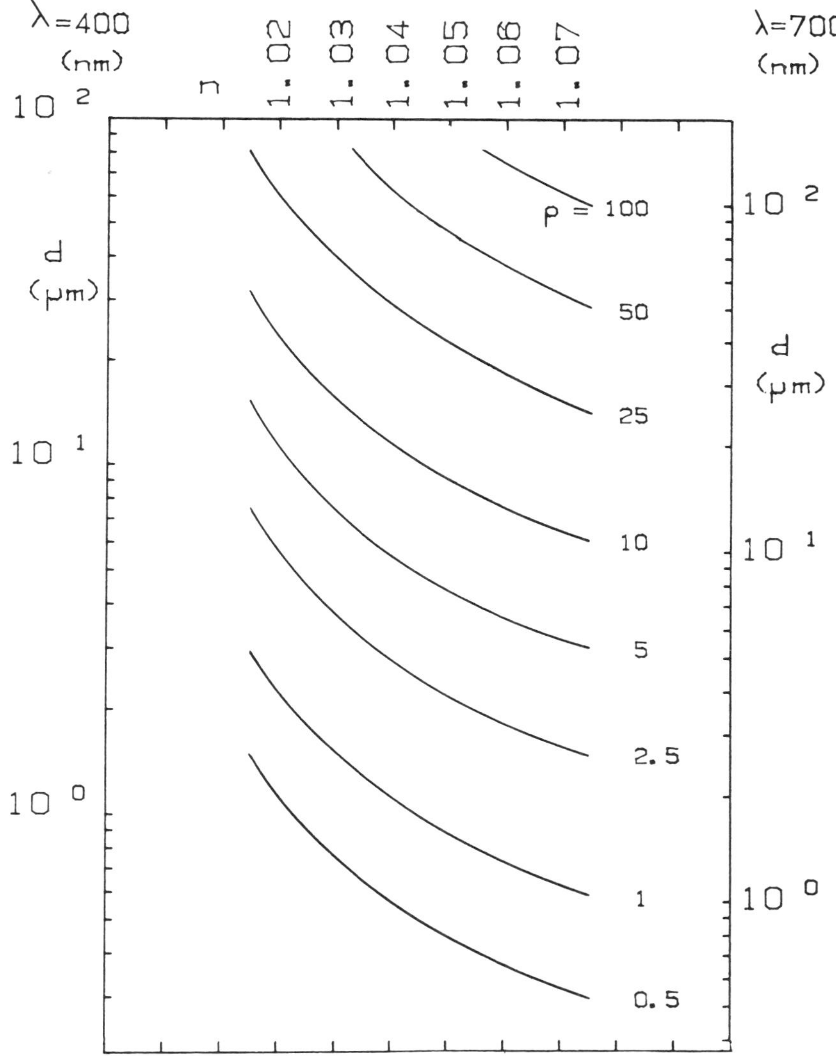

FIG. 2. Nomogram providing the value of the parameter ϱ as a function of the index of refraction (real part) ranging from 1.02 to 1.07 (horizontal scale) and as a function of the diameter of the cell in μm (logarithmic vertical scales), left side if the wavelength is λ = 400 nm or, right side, if λ = 700 nm.

in general increased by a factor 2 or 3 due to presence, beside chlorophyll a, of accessory pigments. In addition the c_i value adopted does not represent an upper limit and n' could be higher than the "typical" value given above. In any case, however, it will remain low, less than 1 or perhaps $2\ 10^{-2}$. It may become negligible and the cells can be seen as practically transparent particles, acting only as scatterers, in the yellow part of the spectrum (560–600 nm), where ordinary pigments are very weakly absorbing. The presence of phycobilin pigments in some species obviously modifies this behaviour.

In conclusion, with n close to 1 and n' very small, the aforementioned condition for making use of the anomalous diffraction approximation is fulfilled. It will be therefore possible to derive the efficiency factors from this approximation.

The values of ϱ and ϱ'

As these efficiency factors depend on ϱ, it is useful at this stage to specify the possible values of ϱ according to the size of the phytoplankters, the real part of their relative index and the wavelength considered. For this purpose, the nomogram shown on Fig. 2 has been prepared. With the values previously envisaged for d (1 to 100 μm) it can be seen that ϱ could roughly vary between 0.5 and 100.

The parameter ϱ', linked to ϱ through Equation 12', is usually much less than ϱ, since $\tan\zeta$, the ratio $n'/(n-1)$, remains low. For instance, if we keep the above n' value as being a representative value for $\lambda = 675$ nm and if again d is allowed to vary from 1 to 100 μm, ϱ' according to the Equation 11 will vary from $6.2\ 10^{-2}$ to 6.2 at this wavelength.

3.4 General Formulation in the Frame of the Anomalous Diffraction Approximation

Under the proviso that $m\ (= n - in')$ is close to 1, the efficiency factors for a homogeneous spherical particle are expressed as (see Van de Hulst 1957)

$$(13) \qquad Q_c(\varrho) = 2 - 4\exp(-\varrho\tan\zeta)\left[\frac{\cos\zeta}{\varrho}\sin(\varrho-\zeta) + \left(\frac{\cos\zeta}{\varrho}\right)^2\cos(\varrho-2\zeta)\right] + 4\left(\frac{\cos\zeta}{\varrho}\right)^2\cos 2\zeta$$

$$(14) \qquad Q_a(\varrho) = 1 + [\exp(-2\varrho\tan\zeta)(2\varrho\tan\zeta + 1) - 1]\ /\ 2\ \varrho^2\tan^2\zeta$$

$$(15) \qquad Q_b(\varrho) = Q_c(\varrho) - Q_a(\varrho)$$

With respect to the parameter ϱ' Equation 14 can be transformed into

$$(14') \qquad Q_a(\varrho') = 1 + 2\frac{\exp(-\varrho')}{\varrho'} + 2\frac{\exp(-\varrho')-1}{\varrho'^2}$$

The upper part of Fig. 3 shows, as function of ϱ, the variations of Q_c and Q_a computed according to Equations 13 and 14, and the lower panel shows $Q_b\ (\varrho)$ derived from Equation 15.

When $n' = 0$ (curves "1"), beyond a first maximum, which occurs for $\varrho = 4.09$, $Q_c\ (= Q_b)$ undergoes a series of regularly decreasing oscillations around the limiting value of 2, which is asymptotically reached for large ϱ values (the case envisaged in 3.1). These oscillations originate from favorable (constructive) or unfavorable (destructive) interferences between the rays having passed through the particle, and those diffracted around the particle. Q_c at its first maximum exceeds 3.

When absorption intervenes (curves "2" and "3"), Q_a departs from zero and increases for increasing ϱ. In any case, for sufficiently high ϱ values, Q_a tends

532

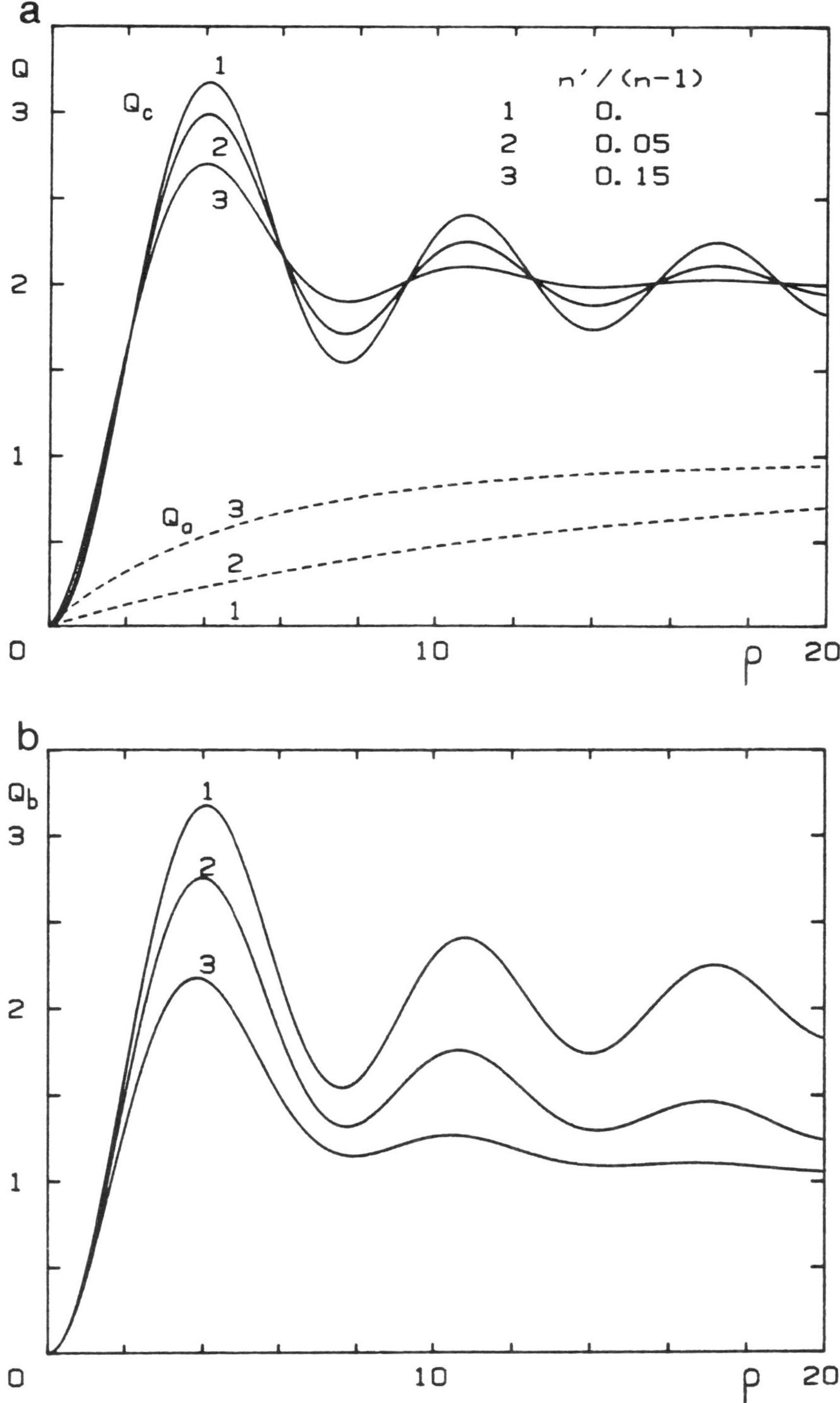

FIG. 3. Variations of the efficiency factors for attenuation, Q_c, *for absorption,* Q_a (a), and for scattering, Q_b (b) vs. the parameter $\varrho = 2\,\alpha(n-1)$, for increasing values of the ratio $n'/(n-1)$ where n and n' are the real and imaginary parts of the relative refractive index of the particles.

towards its limiting value 1, which will also be the limiting value for Q_b (cf 3.1). For lower ϱ values, it can be seen that the effect of absorption is to reduce the amplitudes of the Q_c oscillations. Since the rays passing through the particle are partly absorbed before escaping, the effect of interference with the refracted rays is accordingly diminished. By virtue of Equation 15, the smoothing effect on the Q_b curve is enhanced, and as soon as absorption arises, the scattering efficiency is lower than that of an equivalent transparent particle. This fact, obviously, has to be taken into account to explain the change in scattering by an algal cell inside an absorption band.

There is a family of Q_a (ϱ) curves according to the value given to $\tan \zeta$ (Fig. 3), whereas there is a unique absorption curve (equation 14') for all absorbing spheres if Q_a is plotted as a function of ϱ' (Morel and Bricaud 1981a).

This monotonic function, shown on Fig. 9, is important in that it represents in a condensed form the "package effect" (Kirk 1975a,b) which is discussed in detail by J. T. O. Kirk in this volume. Briefly, the properties of the Q_a (ϱ') function can be analyzed by considering:

i) the first quasi-linear part of the curve, when ϱ' is sufficiently small

ii) the last "saturating" part of the curve where Q_a approaches 1 asymptotically (black body)

iii) the intermediate curved part which represents the more general situation.

In this later situation, an increase in ϱ' by a factor k induces an increase in Q_a by a factor less than k. This factor could be k in the quasi-linear domain, and conversely is near-zero in the plateau domain. An increase in ϱ' is, in effect, an increase of the absorbing material per cell, either by increasing the size with constant cellular material, or by increasing a_s (i.e. c_i) with constant size. The above statement means that the increase in absorbing material per cell does not induce an equal increase in absorption capability. The equality only occurs in the quasi-linear domain, and for those cells which belong to this domain the effectiveness in catching the light is maximum (exactly as if the pigments were in solution). Such an optimization is likely to be the privilege of picoplankton, provided that the inner pigment concentration is moderate (we have seen before that with $c_i = 3$ kg m^{-3}, ϱ' is as low as 0.015 for $d = 1$ μm). This privilege is perhaps extended to species of medium size provided c_i remains sufficiently low.

Influence of polydispersion

The foregoing concerns one particle and can be extended to a monosized collection of particles. An algal population (a mixed natural population or a monospecific population grown in culture) is generally polydispersed with respect to the size. The polydispersion is described by a size distribution function

$$F(d) = \frac{1}{V} \frac{d(N)}{d(d)}$$

$F(d) \, d(d)$ is the number of cells per unit volume in the size range $d \pm 1/2 \, d(d)$, and the integral of this function, with appropriate limits, provides N/V, the total number of cells N in the volume V of suspension. With λ and n fixed, d can be transformed into ϱ and the distribution function can be written as $F(\varrho)$. Mean efficiency factors can be defined and computed for a "mean" particle, representative of the actual polydispersed population according to:

$$(17) \qquad Q_i(\varrho_M) = \int_o^\infty F(\varrho) \, Q_i(\varrho) \, \varrho^2 \, d\varrho \, / \int_o^\infty F(\varrho) \, \varrho^2 \, d\varrho$$

(with $i = a, b, c$), where ϱ_M is the ϱ value which corresponds to the maximum of the size distribution function.

534

Disregarding the factor introduced by the replacement of d by ϱ, the numerator represents the cumulated cross section for absorption, scattering, attenuation (Equation 2) and the denominator represents the geometrical cross section of the entire population. If Q_a is expressed with respect to the parameter ϱ' instead of ϱ, a similar relation can be written:

$$(17') \qquad Q_a(\varrho_M') = \int_0^\infty F(\varrho')\, Q_a(\varrho')\, \varrho'^2\, d\varrho' \; / \; \int_0^\infty F(\varrho')\, \varrho'^2\, d\varrho'$$

Figure 4 shows Q_c plotted as a function of ϱ_M. The functions used when preparing this graph are log-normal distributions with increasing width. As it can be easily imagined, the polydispersion results in a smoothing effect, which is more accentuated

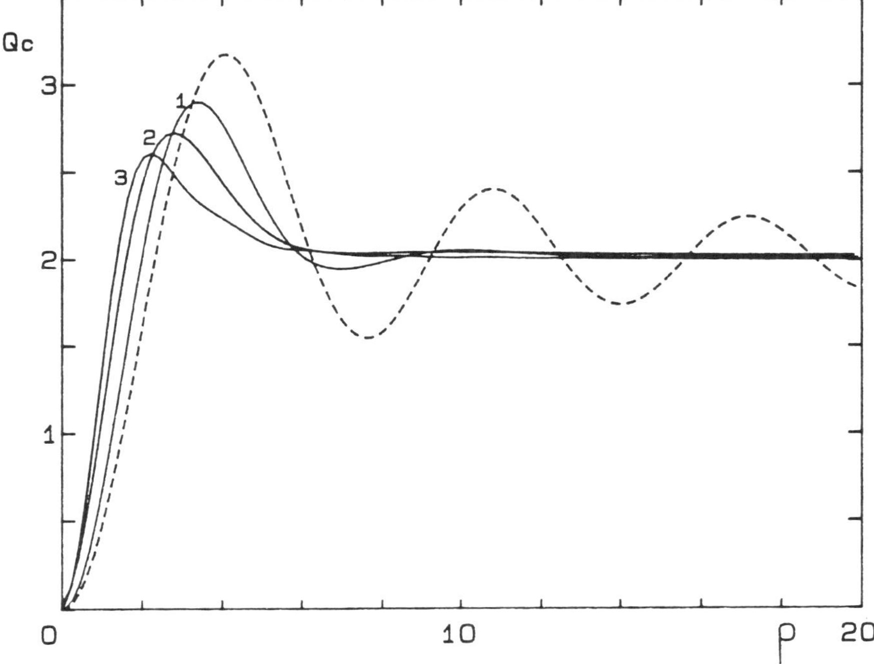

FIG. 4. Mean efficiency factor for attenuation Q_c of a "mean" particle representative of a polydispersed population, plotted as a function of ϱ_m, *the* ϱ value which corresponds to the maximum of the size distribution function $F(\varrho)$ (see Equation 17). The index of refraction is real (no absorption) and the curves 1 and 3 correspond to log-normal distributions such as $F(\varrho_M/2) = F(2\varrho_M) = $ respectively 0.01, 0.1, 0.3 $F(\varrho_M)$. The dashed curve, redrawn from Fig. 3 for $n' = 0$, represents the limiting case of a population of monosized particles.

as the distribution becomes wider. With the vanishing minor oscillations, only the first maximum remains significant. This maximum, which occurs at $\varrho = 4.09$ for monodispersion, is slightly shifted towards lower values for polydispersions. The striking feature worth recalling is the existence of the two domains delimited by this maximum. For small ϱ values ($\varrho < 3$ or 4), Q_c (and also Q_b) experience a more extended range of variations, from near-zero to 2.5 or 3. According to the nomogram (Fig. 2), this situation is that of phytoplankters smaller than say 2 (high index of refraction) or 5 μm (low n). Small phytoplankters, with $\varrho < 2$, and definitely picoplankters, are the sole plankton capable of exhibiting Q_c and Q_b values of the order of 1 or less.

Beyond this first maximum ($\varrho > 4$), Q_c is slightly more than 2 and for $\varrho > 8$ or 10, Q_c never significantly differs from 2. Consequently for cells with equivalent diameter greater than 10 μm (if n is high) or 20 μm (n small), Q_c becomes practically independent of the size. Q_b is also weakly size dependent and the scattering (the product sQ_b) by such cells appears to be simply proportional to their geometrical cross sections. This point will be re-examined later on when discussing the actual spectral values of the scattering by algae.

3.5 FORMULATION IN THE FRAME OF MIE THEORY

It is out of the scope of the present paper to enter into the details of this theory. The purposes of this part are restricted to answer the question, for what kind of problem is it necessary to use this theory? and also to provide some information about the predictions which can be made by means of this theory in the case of phytoplankton, including picoplankton species.

As soon as the angular distribution of the scattered radiation (the volume scattering function and related quantities such as Q_{bb} or partial integrals over a given angular domain) are investigated, Mie computations are needed. It is also true if the state of polarization of scattered radiation is to be studied.

The scattering pattern is described by the dimensionless angular intensity parameters $i_1 (\theta)$ and $i_2 (\theta)$ which are expressed as

$$i_1(\theta, \alpha, m) = S_1 S^*_1$$

$$i_2 (\theta, \alpha, m) = S_2 S^*_2$$

where $(\lambda/2\pi) S_1$ and $(\lambda/2\pi) S_2$ are the complex amplitudes of the scalar components of the electric field vector, respectively perpendicular and parallel to the scattering plane (in which θ is defined). The Mie solution expresses these amplitudes as converging series

$$S_1 (\theta, \alpha, m) = \sum_{n=1}^{\infty} \frac{2n + 1}{n (n + 1)} \left[a_n \pi_n (\cos\theta) + b_n \tau_n (\cos\theta) \right]$$

$$S_2 (\theta, \alpha, m) = \sum_{n=1}^{\infty} \frac{2n + 1}{n (n + 1)} \left[a_n \tau_n (\cos\theta) + b_n \pi_n (\cos\theta) \right]$$

where n is an integer. (The notations are those of Van de Hulst 1957). The number of terms to be summed in the above series, before the convergence is reached, increases with increasing size α. Roughly speaking n must be of the order of $1.1\alpha + 10$ and is weakly dependent on m. For spherical particles of the size of the algal cells (α varying from 10 to 1000), the computations are somewhat heavy, albeit feasible with fast computers.

In the above series the three variables are partially separated. The angular dependence comes from π_n and τ_n which are only functions of $\cos\theta$ (these functions are defined in terms of Legendre polynomials and their derivatives of integer order n). The so-called "Mie coefficients" a_n and b_n govern the magnitude of the phenomenon. These complex coefficients, functions only of α and m, are expressed through spherical Bessel functions of order $(n + 1/2)$.

They are also the only coefficients to appear in the exact formulation of the efficiency factors

$$Q_c (\alpha, m) = 2 \alpha^{-2} \sum_{n=1}^{\infty} (2n + 1) \, Re \, \{a_n + b_n\}$$

$$Q_b (\alpha, m) = 2 \alpha^{-2} \sum_{n=1}^{\infty} (2n + 1) \, (|a_n|^2 + |b_n|^2)$$

536

These expressions become strictly equivalent if the index of refraction m is a pure real number.

If the light incident upon the particle is not polarized and the polarization of the scattered light is disregarded, the intensity parameter $i(\alpha, \theta, m)$, simply written $i(\theta)$, is:

$$i(\theta) = (1/2) \, [i_1(\theta) + i_2(\theta)]$$

This parameter can be related to the dimensionless volume scattering function $\bar{\beta}(\theta)$ which, according to its definition (Equation 5'), can be seen as being related to a single particle, as well as to a suspension. This relationship is

$$(18) \qquad \bar{\beta}(\theta) = i(\theta) / \pi Q_b \alpha^2$$

Unlike Q_c and Q_b, the efficiency factor for backscattering Q_{bb} cannot be explicitly formulated and must be computed using the integral

$$(19) \qquad Q_{bb}(\alpha, m) = \alpha^{-2} \int_{\pi/2}^{\pi} i(\theta, \alpha, m) \, \sin\theta \, d\theta$$

Influence of polydispersion

In practical applications of such theoretical computations to a monospecific algal population, the actual size distribution has to be taken into account. There is no difficulty in principle, but it involves more calculations. The size distribution function experimentally obtained (or assumed) as a function of the geometrical size d must be written as a function of α. The $\beta(\theta)$ functions or Q_{bb} values, "pre-computed" for discrete sizes, have to be weighted by the distribution function and then summed. Equation (18) becomes

$$(20) \qquad \beta(\theta, \alpha_M, m) = (1/\pi) \frac{\int i(\theta, \alpha, m) \, F(\alpha) \, d\alpha}{\int Q_b(\alpha, m) \, F(\alpha) \, \alpha^2 \, d\alpha}$$

α_M is a modal value corresponding to the maximum frequency in the distribution and the limits for the integrals must encompass the entire population (the accuracy of the computation obviously depends on the increment $d\alpha$, i.e. on the number of precomputed i functions). Similarly $Q_{bb}(\alpha_M)$ is obtained from

$$Q_{bb}(\alpha_M) = \frac{\int Q_{bb}(\alpha) \, F(\alpha) \, \alpha^2 \, d\alpha}{\int F(\alpha) \, \alpha^2 \, d\alpha}$$

3.6 THE ANGULAR DISTRIBUTION OF THE SCATTERED RADIATION AS PREDICTED BY THE MIE THEORY

A detailed review of the numerical results obtained from Mie theory is irrelevant here; such discussions can be found elsewhere (Deirmendjian 1969; Morel 1973). The aim of this part is restricted to bring out the main features concerning the volume scattering function (VSF) and, in reference to phytoplanktonic cells, to describe the general trends allowing the variations of the VSF with size, absorption properties and the refractive index to be predicted. Algal cells, including the smallest species, are "large" particles from an optical point of view. This leads to a scattering pattern where the forward scattering predominates over the backscattering. This dissymmetry in the VSF, is in general enhanced with increasing size. For very large particles, the forward peak (central spot of diffraction) continues to be increased and simultaneously becomes narrower (Fig. 6a).

It is well known that for a given size (a specified α value), the scattering pattern exhibits a succession of maxima and minima due to interferences (see e.g. Fig. 5). The number of oscillations increases with increasing α. In practice, as an algal population exhibits a continuous size spectrum over a given interval, the oscillations

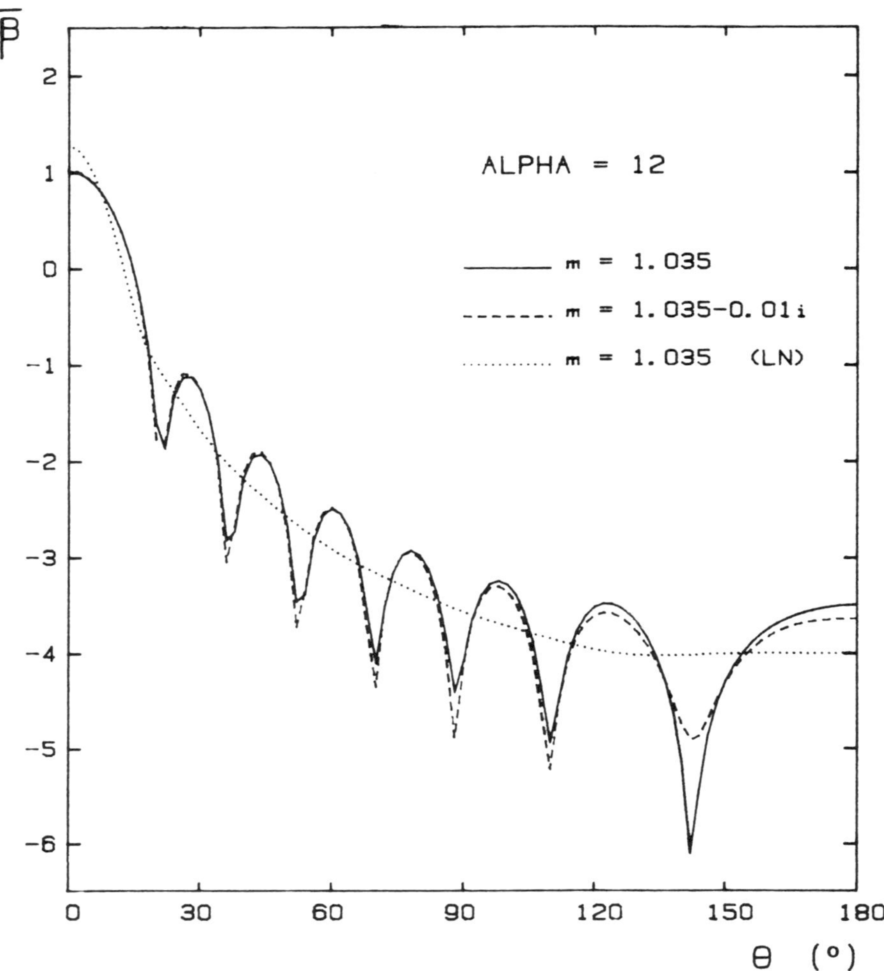

FIG. 5. Normalized volume scattering functions, $\bar{\beta}(\theta)$ (Equations 5' and 18), for a particle of relative size $\alpha = 12$, when the refractive index is 1.035 and 1.035-0.01 i. The dotted curve represents the same $\bar{\beta}(\theta)$ function for a polydispersed population of particles with $n = 1.035$, computed according to Equation 20. The size distribution function $F(\alpha)$ is a log-normal law such that the modal relative size $\bar{\alpha}_M$ is also 12, and $F(\alpha_M/2) = F(2\alpha_M) = 0.01\ F(\alpha_M)$.

are annihilated (Fig. 5 and 6a,b) and they cannot be used as a "fingerprint" for identification.

The absorption inside the particle does not modify the geometrical pathways of the light entering the particle before it re-emerges. This emergent radiation which interferes with the diffracted radiation (unchanged) has been slightly reduced by the absorption. The resulting scattered radiation is accordingly reduced, mainly in the backward directions but the position of the oscillations remains unmodified (Fig. 5). When considering the low values of the imaginary part of the refractive index for algae, the effect of absorption becomes significant only for large particles, and for backscattering.

The dependence of the VSF upon the real part of the index is much more marked everywhere in the θ range (apart from 0°) (see Fig. 7). The normalized scattering

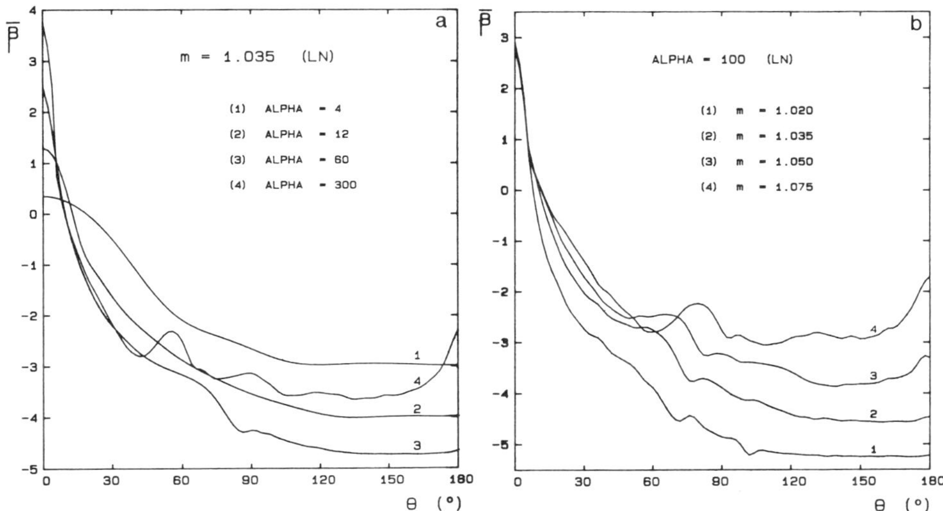

FIG. 6. (a) Normalized volume scattering function $\bar{\beta}(\theta)$ for increasing α_M values (increasing size) and for $m = 1.035$. (b) Normalized volume scattering function $\bar{\beta}(\theta)$ for increasing (real) index of refraction and for $\alpha_M = 100$. For Fig. 6a and b the log normal size distribution used is as in Fig. 5. The "bump" which occurs at about 75° for $m = 1.075$ and at smaller angles when the refractive index decreases (see also Fig. 6a) is the first "rainbow", at 138° for water droplets ($n = 1.33$). It appears for sufficiently large and perfect spheres. Thus it is unlikely that it can be observed for algal cells.

functions $\beta(\theta)$, for selected θ angles, are plotted versus α_M (the modal value of the log-normal distribution used to get a more representative gliding mean value). Disregarding the special behaviour of $\bar{\beta}(0°)$ and $\bar{\beta}(180°)$, the span of the curves from $\bar{\beta}(2°)$ to $\bar{\beta}(140°)$ is an indicator of the "shape" of the VSF and of its dissymmetry. As a rule, the smaller the index is, the more pronounced is the general dissymmetry of the VSF (see Fig. 6b).

For the sake of completeness, it must be added that this rule, in effect, progressively vanishes when the size becomes very small, as shown in Fig. 7. In the Rayleigh domain ($\alpha \cong 0.2$), the VSF is completely independent from the index, and in the extended Rayleigh domain, also called the Rayleigh–Gans domain, a progressive dependence on n appears. For picoplankton, with possible α values less than about 10, it can be conjectured that the VSF will be almost insensitive to the index and solely determined by the size.

The $\bar{\beta}(0°)$ curve is characterized by an ascending slope the value of which is equal to 2 (in the log–log coordinates). In the Rayleigh–Gans domain $i(0°)$ varies like α^6 (as in the Rayleigh domain) and Q_b like α^2. According to Equation 18, $\bar{\beta}(0°)$ is proportional to α^2. The progressive change between the Rayleigh–Gans domain and the domain where diffraction theory applies, occurs in a domain including the value of α such that $\varrho = 4.09$, the value which corresponds to the first maximum of Q_b (see Fig. 3). In the diffraction domain Q_b oscillates around 2 and becomes constant whereas $i(0°)$ varies like α^4; accordingly $\bar{\beta}(0°)$ is again proportional to α^2.

The interpretation of the complicated variations exhibited by $\bar{\beta}(180°)$ is neither straightforward nor of practical interest. The backscattering ratio $\bar{b}_b$ is more interesting particularly because the reflectance of a water body depends on the backscattering coefficient (see in S. Sathyendranath, this volume). This ratio ($= Q_{bb}/Q_b$) is simply obtained by integrating $\bar{\beta}(\theta)$ according to:

$$\bar{b}_b = 2\pi \int_{\pi/2}^{\pi} \bar{\beta}(\theta) \sin\theta \, d\theta$$

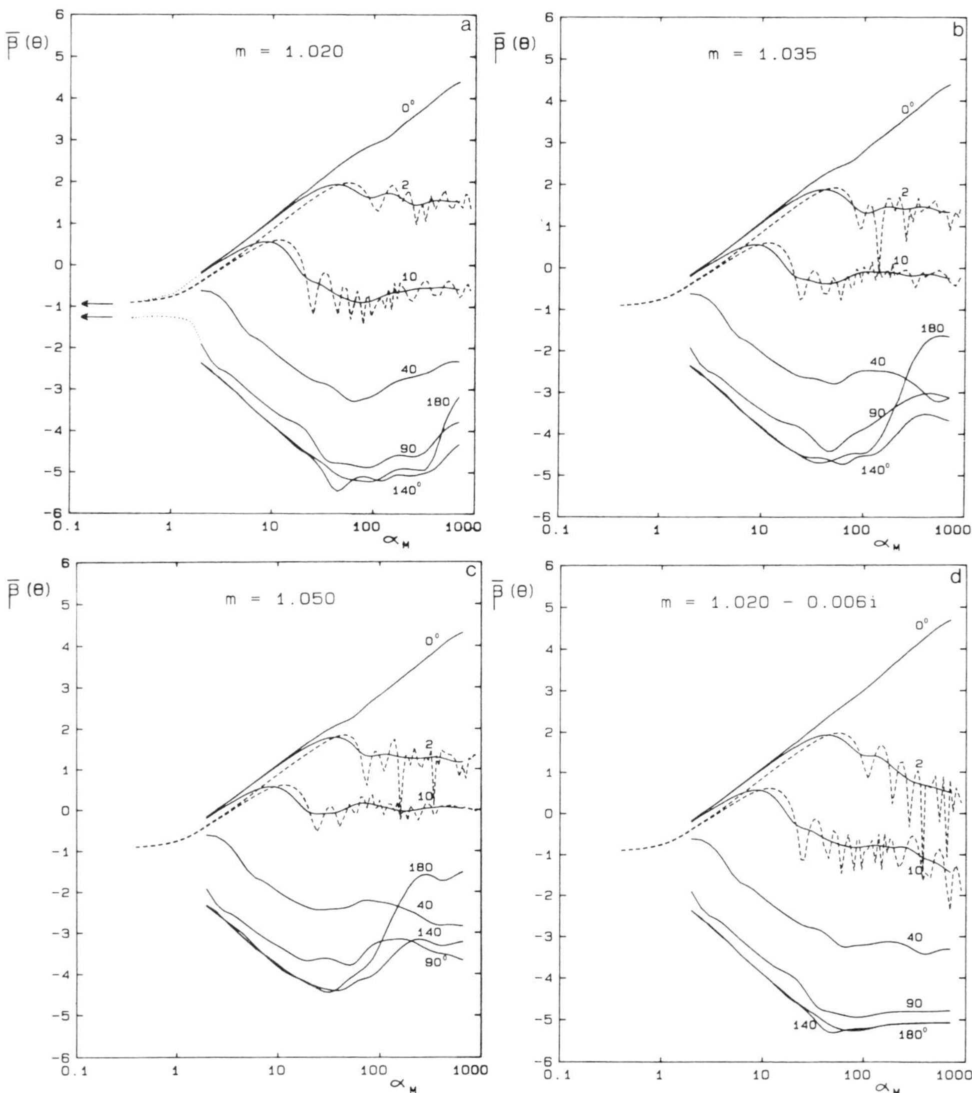

FIG. 7. Variations of $\beta(\theta)$, vs. the modal relative size α_M, for selected values indicated in degrees above the curves. The size distribution function used to smooth the curves is as for Fig. 5 and 6. The not-smoothed curves for 2° and 10° (dashed lines) correspond to monodispersed population. They are shown for comparison. Figure 7a to d corresponds to different values of the refractive index m as indicated. For all these figures, the limiting value for $\bar{\beta}(0°)$ and $\bar{\beta}(180°)$ in the Rayleigh domain (small α) is independent from the index and equal to 0.119, and half of this value if $\theta = 90°$ (arrows on Fig. 7a).

A detailed discussion concerning $\bar{b}_b$ *and* Q_{bb} has been given elsewhere (Morel and Bricaud 1981). The main points in reference to Fig. 8 are summarized below.

The value of $\bar{b}_b$ for Rayleigh scattering is exactly 1/2. When α increases $\bar{b}_b$ decreases from this value to a minimum occurring for that value of α which again leads to $\varrho = 4.09$. Therefore this minimum is retrogressing with respect to α as n increases. It is noticeable that for picoplankton with ϱ close to, or less than 10, $\bar{b}_b$ should be almost independent of the refractive index, as it was for $\beta(\theta)$.

540

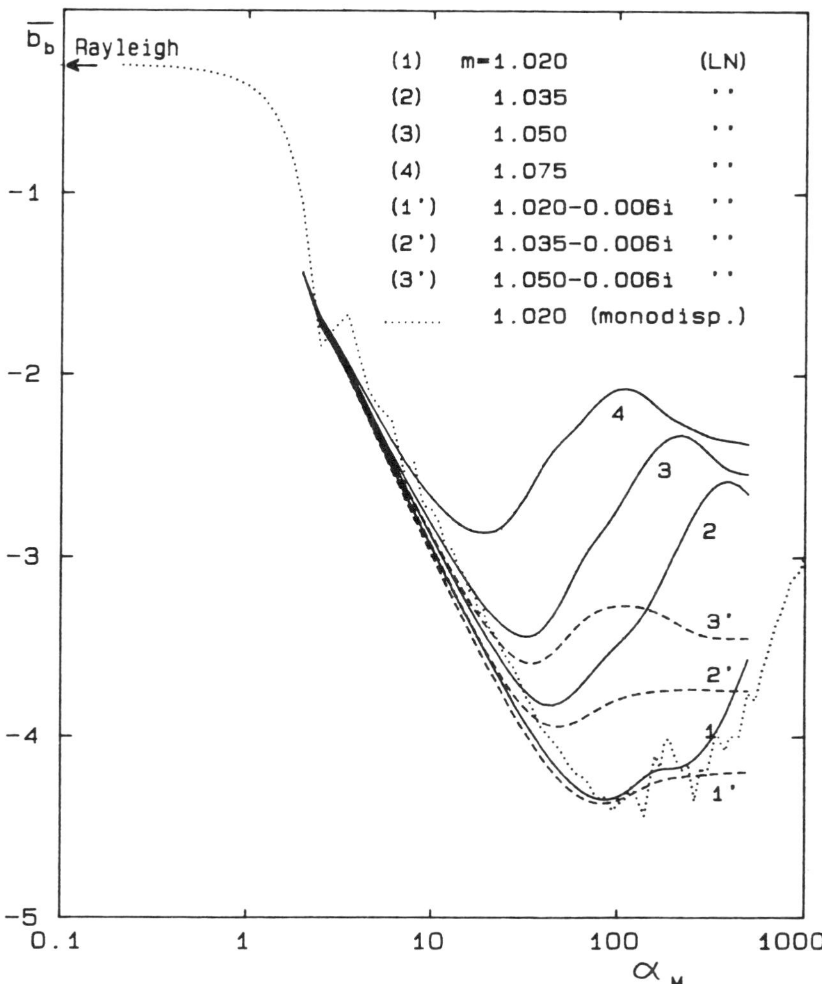

FIG. 8. Variations of the backscattering ratio $\bar{b}_b$ ($= b_b/b$) vs. the modal relative size α_M (same log-normal law as before in Fig. 5). The different curves correspond to various values of the refractive index given in inset. The curve for a monodispersed population (with $m = 1.02$) is also shown (dotted line). The arrow indicates the limiting value of b_b/b ($= 0.5$) when α tends toward 0 (Rayleigh domain).

The slope of the $\bar{b}_b$ curve is -2 before the minimum and $+2$ beyond the maximum (if the refractive index is real). The values of $\bar{b}_b$ at its minimum and before it are arranged in ascending order when n is increasing. If absorption intervenes, the ascending slope is considerably reduced and b_b remains low over an extended size range (Fig. 8 and also, Morel and Bricaud 1981b).

As a concluding remark it must be emphasized that algal cells, due to their low refractive index, are poor backscatterers with $\bar{b}_b$ of the order of 10^{-4} or 10^{-3}. The picoplankters with α below 10 could perhaps more efficient in comparison to larger species having the same index. A recent indirect evaluation (Carder et al. 1986) tends to support this hypothesis of a theoretical enhancement of the backscattering efficiency.

Other information can be drawn from the theoretical $\bar{\beta}(\theta)$ or $\bar{b}_b$ curves. They are of practical interest when interpreting the volume scattering function, $\beta(\theta)$, which is, unlike $\bar{\beta}(\theta)$, a parameter directly measurable. Such measurements are performed, for instance, in conjunction with flow-cytometry techniques. This question will be examined later on (4.4).

4. Relationships Between Inherent Optical Properties of the Suspension and the Optical Characteristics of Individual Particles

4.1 GENERAL RELATIONSHIPS

For the sake of simplicity, we assume, at first, that the cells are of uniform size. Due consideration of the polydispersion induces some formal modifications which will be given later.

If sQ_a and sQ_b (see Equation 2) are the cross sections for absorption and scattering by an individual cell, the presence of N such cells in a volume V will give rise to an absorption and a scattering coefficient for the medium according to

$$(22a) \qquad a = (N/V)\, sQ_a$$

$$(22b) \qquad b = (N/V)\, sQ_b$$

The volume scattering function of the medium $\beta(\theta)$ being (Equation 5′)

$$\beta(\theta) = \bar{\beta}(\theta) \cdot b$$

its link with the parameters intervening in the theory (Equation 18) can be expressed as:

$$\beta(\theta) = \frac{i(\theta)}{\pi Q\, \alpha^2}\, b$$

By combining the above with Equation 22b, $\beta(\theta)$ can also be related to the cell number concentration:

$$(23a) \qquad \beta(\theta) = (N/V)\, i(\theta)\, (\lambda^2/4\pi^2)$$

or

$$(23b) \qquad \beta(\theta) = (N/V)\, \bar{\beta}(\theta)\, sQ_b$$

4.2 CHLOROPHYLL-SPECIFIC OPTICAL COEFFICIENTS

With reference to the most common indicator used, we introduce the chlorophyll-specific absorption and scattering coefficients, a^* and b^*, respectively, which are defined as

$$(24a) \qquad a^* = a/C$$

$$(24b) \qquad b^* = b/C$$

where C is the chlorophyll a concentration within the medium. If, as before, c_i denotes the intracellular concentration of Chl$a$, the concentration in a volume V of suspension resulting from the presence of N cells is

$$(25) \qquad C = (N/V)\, c_i v$$

where v is the volume of one cell. If we keep the hypothesis of spherical particles, the ratio s to v of the geometrical cross section to the volume is simply $(3/2d)$. By combining Equations 22 and 24 it follows that

542

(26a) $a^* = (3/2)\, Q_a\, (dc_i)^{-1}$

(26b) $b^* = (3/2)\, Q_b\, (dc_i)^{-1}$

By recalling that Q_a depends on ϱ' (defined by Equation 11 or 12) and Q_b on ϱ (defined by Equation 10), the above equations can also be written

$$(27)\qquad a^* = (3/2)\,\frac{a_s}{c_i}\,\frac{Q_a(\varrho')}{\varrho'}$$

$$(28)\qquad b^* = 3\pi \cdot \frac{(n-1)\,n_w}{c_i\,\lambda}\,\frac{Q_b(\varrho)}{\varrho}$$

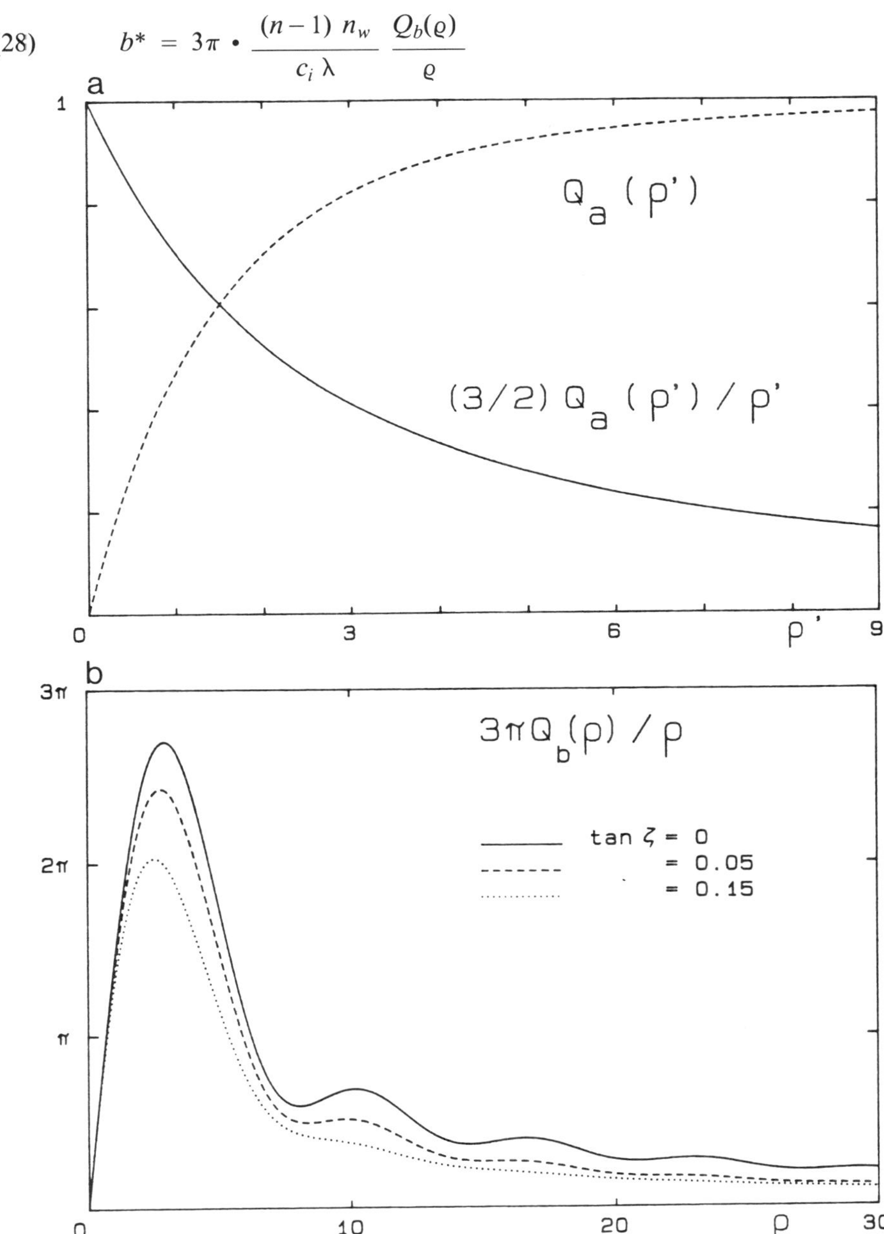

FIG. 9. (a) Variations of the function $Q_a^*(\varrho') = 3/2\,Q_a(\varrho')/\varrho'$ vs. the parameter $\varrho' = 4\alpha n'$ (in continuous line). The function $Q_a(\varrho')$ is shown as a dashed line (Equation 14′). (b) Variations of the function $Q_b^*(\varrho) = 3\pi Q_b(\varrho)/\varrho$ vs. the parameter ϱ, for increasing values of the ratio $n'/(n-1) = \tan\zeta$.

At this stage, these expressions, although similar, differ in the physics involved. In Equation 27, the ratio (a_s/c_i) appears, i.e. the specific absorption coefficient of the cellular material (noted "γ" in John Kirk's paper), whereas in Equation 28 the dependence on the refractive index is made clear. Both equations are ruled by the symmetrical dimensionless functions $Q_a(\varrho')/\varrho'$ and $Q_b(\varrho)/\varrho$. These functions, shown on Fig. 9, deserve some comment.

Because of the package effect, the cells collect a variable amount of radiation per unit of pigment. In other words, the specific absorption coefficient (here defined with respect to Chl$a$) is not a constant. Its variation is depicted by the function (see in Morel and Bricaud 1981a).

$$Q^*_a = (3/2)\, Q_a(\varrho')/\varrho'$$

which regularly decreases from 1 to 0 when ϱ' goes from 0 to ∞. The meaning is easy to understand. The specific coefficient a^* is maximal $(= a_s/c_i)$ when Q^*_a is 1, i.e. for $\varrho' = 0$ (or $d = 0$), that is as though the cells were disrupted and the pigments liberated or in solution $(a_{sol}$ in John Kirk's paper). In general, with Q^*_a less than 1, the specific coefficient for a suspension a^*_{sus} is decreased in the ratio

$$(29) \qquad a^*_{sus}/a^*_{sol} = Q^*_a$$

Another aspect of the package or discreteness effect is the flattening of the absorption spectrum by intact cells (Duysens 1956). This is due to the fact that Q^*_a decreases with increasing values of n' (at constant α), i.e. with intensity of absorption.

In so far as experimental values of d, c_i and a^*_{sus} (λ) have been obtained, Q_a can be derived by inverting equation 26a. As Q_a is a monotonous function of ϱ', ϱ' is univocally determined and leads to Q^*_a. Then a^*_{sol} (λ) can be computed by using equation 29. These spectral values are those of a hypothetical "aqueous solution" of the cells and their pigments. Such spectra obtained for different species can be directly compared, since the disturbing effect of discreteness and flattening is annihilated. Also, from a_{sol} and c_i, a_s can be inferred (allowing n' to be computed; see 3.3).

Unlike Q^*_a the other function

$$Q^*_b = 3\pi\, Q_b(\varrho)/\varrho$$

is not unique, as Q_b also depends on absorption (see Fig. 3). A limiting case worth being examined is that of $Q_b = Q_c$, when the absorption can be neglected. The introduction of a "reasonable" absorption does not drastically change the general behaviour of this function, and particularly its main feature i.e. the occurrence of a maximum which separates two domains. At a given λ, and for a given cellular material, n, n' and c_i are fixed, and b^* (Equation 28) will mimic the variations of Q^*_b when the cell size is varying. In general, with ϱ greater than about 3, b^* decreases with increasing size. The converse is true for small sized cells, particularly for picoplankton $(\varrho < 3)$ for which b^* decreases as size decreases. From one species to another, with c_i and $(n-1)$ which may suffer large variations, the b^* value is somewhat unpredictable. This point will be discussed later with reference to some real examples.

4.3 MODIFICATIONS FOR A POLYDISPERSED POPULATION

When a size distribution function is introduced, there is obviously no basic change in the above expressions and the modified expressions below are needed only when computations on actual populations are envisaged. Similar treatment applies for absorption and scattering. For instance, the absorption coefficient previously given by equation 22a becomes

$$a = (\pi/4) \int Q_a(d) \, F(d) \, d^2 \, d(d)$$

with the appropriate limits covering the whole size range for the integral. If a mean value of $\bar{Q}_a$ has been computed (from Equation 17), a can also be written

$$(30) \qquad a = (\pi/4) \, \bar{Q}_a \int F(d) \, d^2 \, d(d)$$

The computation of the specific coefficient (Equation 24a) involves the concentration within the suspension, expressed as

$$(31) \qquad C = (\pi/6) \, c_i \int F(d) \, d^3 \, d(d)$$

In fact, these expressions are used in the inverse problem, which consists in estimating Q_a, Q_b, c_i from experimental data concerning a, b, c and $F(d)$ with a view to understanding the optical properties at the level of a "mean" single cell (see e. g. Bricaud and Morel 1986).

The effect of polydispersion upon the function Q^*_a is weak, only attenuating the curvature. The functions Q^*_b are more sensitive, with a disappearance of the minor oscillations, only the first maximum remains slightly smoothed and shifted towards lower ϱ values.

4.4 Volume Scattering Function of an Algal Suspension

By remembering the description of the $\bar{\beta}(\theta)$ and $\bar{b}_b$ curves (Fig. 7 and 8) given in part 3.6, some predictions can be made concerning the VSF of an algal suspension. The links between these properties (see Equation 23b) are:

$$\beta(\theta) = \bar{\beta}(\theta) \, Q_b s \, (N/V)$$

$$\text{and } b_b = \bar{b}_b \, Q_b \, s \, (N/V)$$

For sufficiently large particles, when Q_b becomes roughly constant (and close to 2), $\beta(\theta)$ should be proportional to $(N/V)s$ the total cross sectional area, under the proviso that $\bar{\beta}(\theta)$ can be regarded as constant. An inspection of Fig. 7 demonstrates that this situation rarely occurs, except perhaps for some angles and for very restricted size ranges. If an extended size range is involved and if, in addition, the refractive index also varies (as in the case of natural populations or mixed cultures), $\bar{\beta}(\theta)$ can experience large variations. Therefore any steady linear relationship between β and the geometrical cross sectional area cannot be ensured. This does not mean that the $\beta(\theta)$ "signal" cannot be interpreted but the interpretation requires computation and has to be made carefully.

Under the same circumstances (large particles, Q_b constant), with $\bar{\beta}(0°)$ varying like α^2, $\beta(0°)$ varies like s^2, the square of the cross section. This result rejoins the Fraunhofer diffraction theory, according to which the diffracted intensity inside the central spot is proportional to the square of the area of the diffracting obstacle (or aperture). This property, valid for θ equal to, or near zero, cannot be significantly extended to other angles. In effect, the Fig. 7 shows that the $\bar{\beta}(2°)$ curve for instance is separated from the $\bar{\beta}(0°)$ curve when α is greater than about 30.

The situation for smaller α values is quite different. In the Rayleigh–Gans domain, the $\bar{\beta}(0°)$ slope is still 2. For other forward angles, significantly differing from zero, the slope is also 2. In this regime Q_b varies like α^2. Consequently $\beta(\theta)$ for all these angles will vary like s^3 (or d^6). This is a law which is derived from the Rayleigh scattering and is maintained in the Rayleigh-Gans scattering only for the forward angles. This theoretical result deserves attention. In the case of picoplankton, a forward scattering meter (1 to 10°) would measure a coefficient which is related to the cube of the cross sectional area.

Without entering into the details, the same kind of reasoning can be made for b_b and for $\beta(\theta)$ (when θ is greater than about 40°). These coefficients are proportional to the cross section when b_b (or $\beta(\theta)$) exhibits a slope -2, that is to say when α is less than that value which leads to $\varrho = 4.09$; picoplankton fall into this case. When the slope of $\bar{b}_b$ becomes positive, b_b (or $\beta(\theta)$) increases more rapidly than the cross sectional area. The exponent in this relationship, between 1 and 2, cannot be more precisely anticipated.

5. Comparison Between Experimental and Theoretical Results

The guidelines in making this comparison can be portrayed as follows:

i) As in oceanographic studies, the algal biomass is most often assessed through the chlorophyll a concentration, the specific absorption coefficient $a^*(\lambda)$ is the parameter which describes the ability of algal cells to harvest radiative energy. On the other hand, it is of major interest to know how phytoplankton (including picoplankton) influence the optical properties of a water body. This influence is again quantified by the value of the coefficients $a^*(\lambda)$, and also $b^*(\lambda)$. For both these concerns, and also because these coefficients have proved to be highly variable (Kiefer et al. 1979; Bricaud et al. 1983; Davies-Colley et al. 1986), the question arises: is this range of variation explicable and predictable in the light of theory?

(ii) From experimental data concerning $a^*(\lambda)$ and $b^*(\lambda)$, the optical properties of a single (average) cell can be extracted, for instance the efficiency factors. Are these values compatible with what we have learnt from the theory? and in turn, is the theoretical approach developed for homogeneous spherical cells an acceptable approximation?

iii) To the extent that the above questions are properly answered, the optical properties of a picoplanktonic cell could be anticipated since such a tiny cell is amenable to the same theoretical treatment.

5.1 EXPERIMENTAL DATA

Measurements were performed on monospecific healthy cultures, in active growth. The phaeopigment concentration and the amount of detritus present were insignificant. It is thus believed that the optical properties as determined are those of the only living cells and are not influenced by other detrital particulates. The methods have been presented elsewhere in detail (Bricaud et al. 1983).

The measured parameters are:

i) the chlorophyll a concentration within the suspension. Subsidiary information is also obtained concerning accessory pigments (chlorophylls b and c, carotenoids and phycobilins)

ii) the cell number concentration and the size distribution

iii) the spectral values of the absorption, attenuation and scattering coefficients $a(\lambda)$, $c(\lambda)$ and $b(\lambda)$, the latter by subtracting: $c(\lambda) - a(\lambda)$.

From (i) and (ii), the chlorophyll concentration within a mean cell, c_i, can be deduced (Equation 31). From (i) and (iii) the spectra of the specific coefficients $a^*(\lambda)$, $b^*(\lambda)$ and $c^*(\lambda)$ can be derived (Equations 24). From (ii) and (iii) the efficiency factors Q_a, Q_b, Q_c, can be computed as functions of λ (Equation 30 or similar). The real part of the index of refraction n is not measured. The absorption coefficient of the cellular substance a_s and hence the imaginary part of the index n', are not directly accessible. Their computation implies the use of theory (part 4.2).

546

TABLE 1. Data concerning the algae grown in culture and used in Fig. 11 and 12: d is the diameter of the sphere equivalent (equivalent in volume); c_i is the intracellular chlorophyll a concentration; see Equation 31; a^* is the chlorophyll-specific absorption coefficient determined for the intact cells at the wavelength 675 nm; $a^* = a/C$ where a is the absorption coefficient of the suspension and C the chlorophyll concentration within this suspension; b^* is the chlorophyll-specific scattering coefficient for the intact cells at 590 nm, except for the cyanobacteria (likely *Synechocystis*) for which the minimum of absorption occurs at 540 nm and hence b^* is given for this wavelength. $b^* = b/C$ as for a^*. Q_a and Q_b, respectively the efficiency factors for absorption and scattering, are derived from the experimental data above according to paragraph 4.3. The method for growing the algae, for determining their pigment content and size distribution, and for measuring the optical coefficients of the living cells in suspension were described in Bricaud et al. (1983) and Bricaud and Morel (1986). The data for the seven first species come from these both papers. The other measurements have been performed by A. Bricaud, A. L. Bedhomme and J. Gostan, who are duely acknowledged here. The cultures were incubated under constant irradiance (300–400 μE • m^{-2} • s^{-1}) at 18°C. The first and second cultures of *Isochrysis galbana* were grown in a turbidostat respectively at 400 and 25 μE • m^{-2} • s^{-1}, and the first and second cultures of the cyanobacteria respectively at 200 and 16 μE • m^{-2} • s^{-1}. (The strain of this species was isolated by J. Neveux from water sampled off Banyuls-sur-Mer).

	d	c_i	a^*_{675}	b^*_{590}	Q_a	Q_b
	10^{-6}m	kg • m^{-3}	m^2 • mg^{-1}			
Tetraselmis maculata	9.50	1.62	0.0200	0.178	0.205	1.824
Platymonas sp.	7.40	1.87	0.0195	0.190	0.180	1.751
Hymenomonas elongata	11.50	2.94	0.0140	0.078	0.316	1.756
Emiliania huxleyi	3.40	1.14	0.0260	0.580	0.067	1.499
Emiliania huxleyi	3.60	1.34	0.0260	0.595	0.084	1.913
Platymonas suecica	3.36	6.38	0.0155	0.185	0.220	2.643
Skeletonema costatum	5.50	0.91	0.0230	0.535	0.077	1.782
Dunaliella salina	10.20	6.16	0.0105	0.0473	0.440	1.973
Pavlova lutheri	4.50	2.27	0.0210	0.378	0.165	2.574
Pavlova pinguis	3.65	4.34	0.0170	0.185	0.179	1.947
Chroomonas fragarioides	4.60	4.34	0.0125	0.153	0.166	2.036
Prymnesium parvum	5.70	2.56	0.0150	0.220	0.145	2.139
Porphyridium cruentum	4.90	4.25	0.0153	0.175	0.212	2.425
Prasinocladus marinus	11.50	1.35	0.0160	0.138	0.166	1.426
Chaetoceros curvisetum	7.50	1.35	0.0230	0.262	0.155	1.770
Chaetoceros lauderi	25.50	0.32	0.0220	0.220	0.120	1.207
Chaetoceros protuberans	16.50	1.28	0.0150	0.080	0.211	1.122
Chaetoceros protuberans	26.50	0.12	0.0230		0.049	
Isochrysis galbana	4.20	12.60	0.0119	0.073	0.419	2.575
Isochrysis galbana	4.10	15.17	0.0110	0.059	0.459	2.453
Cyanobacteria	1.54	1.15	0.0270	0.274	0.032	0.323
Cyanobacteria	1.54	1.78	0.0384	0.186	0.070	0.340

Table 1 provides the list of species studied with values for a^* and b^* at selected wavelengths, the choice of which is discussed below. Spectral values for some species are shown on Fig. 10.

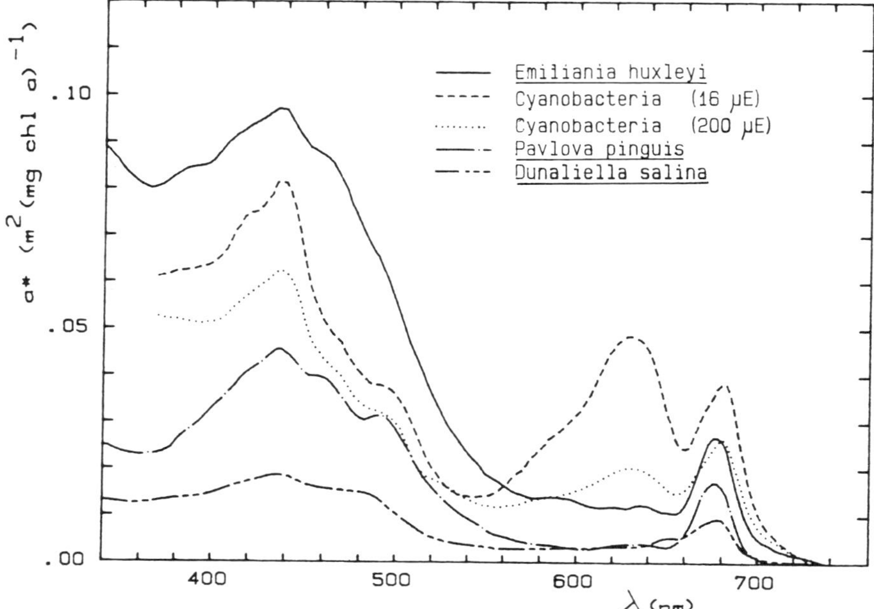

FIG. 10. Spectral variations of the chlorophyll-specific absorption coefficient, a^*, for different phytoplanktonic species. For each curve the residual absorption at 740 nm has been subtracted from the spectral values. The two spectra for the cyanobacteria have been obtained from the same strain grown under two different light intensities (200 and 16 μE $\cdot$ m^{-2} $\cdot$ s^{-1}).

As evidenced in parts 3 and 4, the influence of different parameters which govern the optical properties of algae are strongly intricated. In addition, all these parameters are wavelength dependent. Therefore it is useful to start the comparison between theory and experiment on the basis of simplified cases and by fixing the wavelength. Two kinds of simplification can be imagined when studying absorption and scattering respectively (parts 5.2 and 5.3, below). Thereafter the comparison in the general case will be presented (part 5.4).

5.2 SPECIFIC ABSORPTION AND PACKAGE EFFECT

The intracellular concentration c_i is defined with reference only to chlorophyll a, though other pigments are also present. As the package effect is related to c_i (and d), it is advisable when studying its importance in nature, to select a spectral domain where the influence of the accessory pigments is (at least) reduced. The wavelength of the "red peak", i.e. 675 nm for intact cells, appears as the best.

By looking into Table 1, $a^*(675)$ appears highly variable, from 0.01 to 0.026 m^2 $\cdot$ mg^{-1} (the values for the cyanobacteria, outside this range, form a special case because of the influence of phycocyanin). Even if a residual influence of chlorophyll b and c may subsist, the main origin of these variations in a^* is the package effect. A straightforward way to demonstrate its occurrence consists in computing Q_a (by transforming Equation 26[a])

548

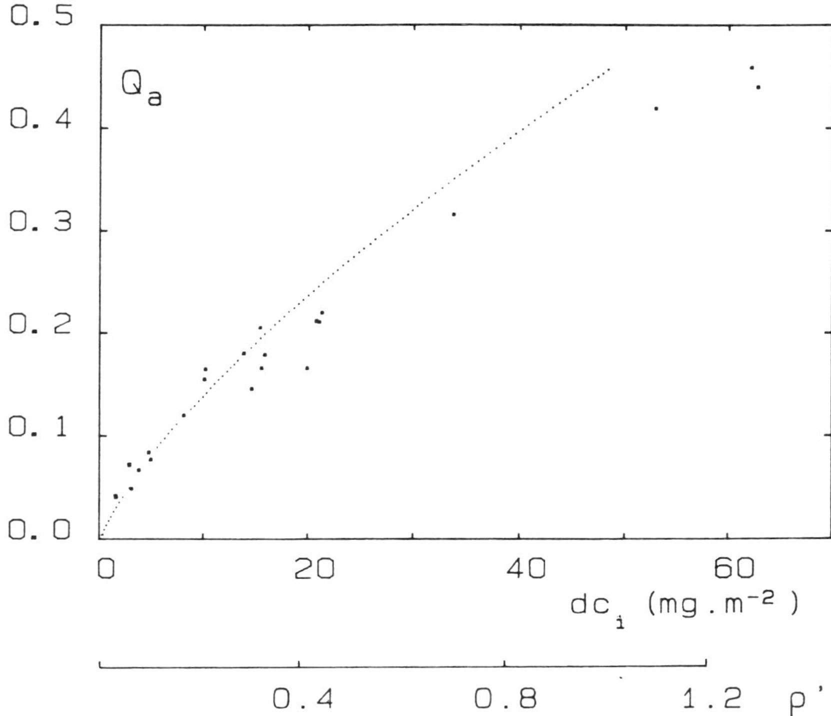

FIG. 11. Q_a values computed from experimental values of the chlorophyll-specific absorption coefficient a^* at $\lambda = 675$ nm (Table 1) plotted as a function of the product $c_i d$, expressed as mg • m^{-2} (c_i = intracellular chlorophyll concentration, d = modal diameter). The dotted curve represents the variations of Q_a as a function of ϱ' (lower scale, see text).

$$Q_a = (2/3)\, a^* \, d \, c_i$$

and plotting Q_a as a function of $d\,c_i$. Obviously a straight line would be obtained if a^* was a constant. The theory of the package effect predicts that Q_a must depart from linearity when $a_s d$ (proportional to $c_i d$) increases. The distribution of the points conforms to the expected curvature (Fig. 11). If, as previously, we assume that the value of specific absorption coefficient for Chl$a$ in acetone is maintained in the cellular material, the abscissa can be converted into a ϱ' scale ($\varrho' = a_s d = 0.0207\, c_i d$) and the theoretical Q_a curve (equation 14$'$) can be drawn. The agreement with the experimental points is satisfactory.

It could be added that the highest a^* values (or equivalently the lowest Q_a values), which correspond to an optimized effectiveness in harvesting energy, are not systematically the privilege of small species. The intracellular concentration actually seems to be a more important source of variability than the size in governing the package effect (at least for the species studied).

In the blue region, the absorption by algae, which peaks around 435 nm, is greater than in the red part of the spectrum. The package effect and the associated flattening effects are in general more pronounced in this part of the spectrum. Figure 10 provides some examples of $a^*(\lambda)$ spectra selected to point out the amplitude of actual variations. At 435 nm, a^* can be less than 0.02 m^2 • mg^{-1} with $Q_a = 0.9$, or as high as 0.085 m^{-2} • mg^{-1} with $Q_a = 0.22$ (the cyanobacteria exhibits an enhanced absorption in the green part of the spectrum due to the presence of phycocyanin).

The package effect is partially responsible for the wide variations of a^* in the blue region. The presence of variable amounts of accessory pigments also induces variations. For that reason c_i is no longer the appropriate parameter to construct a graph for 435 nm as done before on Fig. 11 (for 675 nm), or to quantitatively assess the package effect.

5.3 SPECIFIC SCATTERING AND SIZE DEPENDENCE

The scattering properties of algal cells are influenced by their absorptive properties in a complicated way. When attempting a comparison between theoretical predictions and experimental data, a simplification can be found by choosing a wavelength where absorption can (reasonably) be neglected and where only the size and the refractive index determine the scattering process (see also Morel 1987). For the most common species, an adequate spectral domain is 580–600 nm where absorption in general is at its minimum (540 nm for the cyanobacteria).

By solving Equation 26 Q_b can be obtained from experimental data. The method of judging if Q_b agrees with theory cannot be that pursued for Q_a. Q_b does not depend on the product $d\,c_i$, but on ϱ which includes an unknown, the refractive index (not experimentally determined). So an indirect verification consists in using Equation 28 and in forming the product $c_i\,b^*$. This quantity which must conform to the variations of Q_b^* (Fig. 9b), can be computed as a function of d. Several curves can be generated according to the values given to the floating parameter n, when λ is given a fixed value, 600 nm (Fig. 12). The values of $c_i b^*$ for the different species computed from the experimental data are also plotted versus d on the same figure. There exists a rather good agreement.

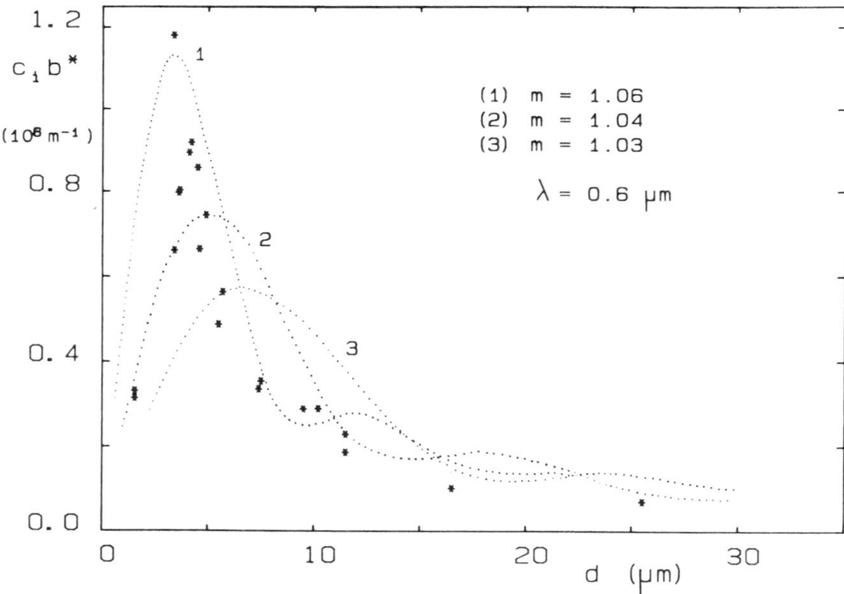

FIG. 12. Values of the product $c_i b^*$ (Table 1) expressed as m^{-1}, plotted as a function of the modal diameter d (c_i = intracellular chlorophyll concentration and b^* = chlorophyll-specific scattering coefficient at λ = 590 nm, or 540 nm for the cyanobacteria.) The dotted curves represent the variations of Q_b^* (as on Fig. 9b) as a function of d, for a given wavelength (0.6 µm) and different values given to the refractive index (real) as indicated.

550

From this Figure, it can be concluded that the quantity $c_i b^*$ is weakly variable with size and the refractive index as long as the size is greater than about 10 μm. Conversely, algal cells with an equivalent diameter around 5 μm exhibit $c_i b^*$ values sensitive to, and increasing with n. In addition, $c_i b^*$ is theoretically and actually maximum in this size range. For the small species, with d smaller than 3 μm, it can be expected that $c_i b^*$ decreases as the size decreases.

If the quantity $c_i b^*$ obeys a size-dependent law (with some flexibility due to the influence of n), there is no constraint on b^* which experiences large variations not correlated with the size (from approximately 0.06 to 0.6 m^2 $\cdot$ mg^{-1}, see Table 1). The inner concentration c_i appears to be an important, likely the dominant, source of variability in b^*, at it was for a^*.

5.4 Comparison in the General Case: Spectral Values of Absorption and Scattering

The so-called "general case" must be understood as describing the situation where absorption, no longer neglected, influences the scattering properties and where this absorption due to all the pigments present is partially discoupled from the chlorophyll a content. Such a situation is that encountered when the whole visible spectral domain is under consideration. The inner Chl$a$ concentration c_i loses its usefulness, at least for computation. The meaningful comparison has necessarily to be established on the basis of the optical cross sections, that is to say at the level of a single cell. It is to this level that the theory gives access. For experimental data, the actual optical coefficients, once determined for a suspension (see 5.1), must be transformed into efficiency factors computed for a single cell representative, on average, of the entire population. Before entering into this comparison, another phenomenon, so far left out, has to be considered. This phenomenon is known as the "anomalous dispersion" of the refractive index.

The apparition of absorption, i.e. the deviation of n' from zero, reacts on the real part of the index of refraction, n. The Ketteler–Helmholtz theory accounts for this effect. When the refractive index $m = n - in'$ remains close to unity, an approximation to this theory can be found (see e.g. Van de Hulst 1957) and will be used below. An absorption band corresponds to an oscillator which absorbs energy at its proper frequency, and also in its vicinity but with a lesser efficiency. If ν_0 is the wave number of the resonance peak and ν that of the incident radiation, the absorption in this spectral region is ruled by the number $v = 2(\nu - \nu_0)/\gamma$. The band is symmetrical with regards to ν_0 and its width is depicted by the damping constant γ. If n was equal to $1 + \epsilon$ (with ϵ small) far from the absorption region, it becomes in the neighbourhood of the absorption band

$$(32) \qquad n = 1 + \epsilon - \epsilon K \left(v/(1 + v^2) \right)$$

in response to the n' variations, expressed as

$$n' = \epsilon K \left(1/(1 + v^2) \right)$$

The oscillator strength or the absorption intensity are described by K. A schematical graph of the concomitant variations in n and n' throughout an absorption band is presented on Fig. 13 (left side). This well-known pattern is characterized by a depression of n followed by an enhancement respectively in the direction of the short and of the long wavelengths with respect to that of the absorption peak. The amplitude of the variation in n is exactly equal to that in n'.

When dealing with phytoplankton absorption, the spectrum exhibits several features (maxima, minima, shoulders . . .). Such a spectrum can be resolved by considering

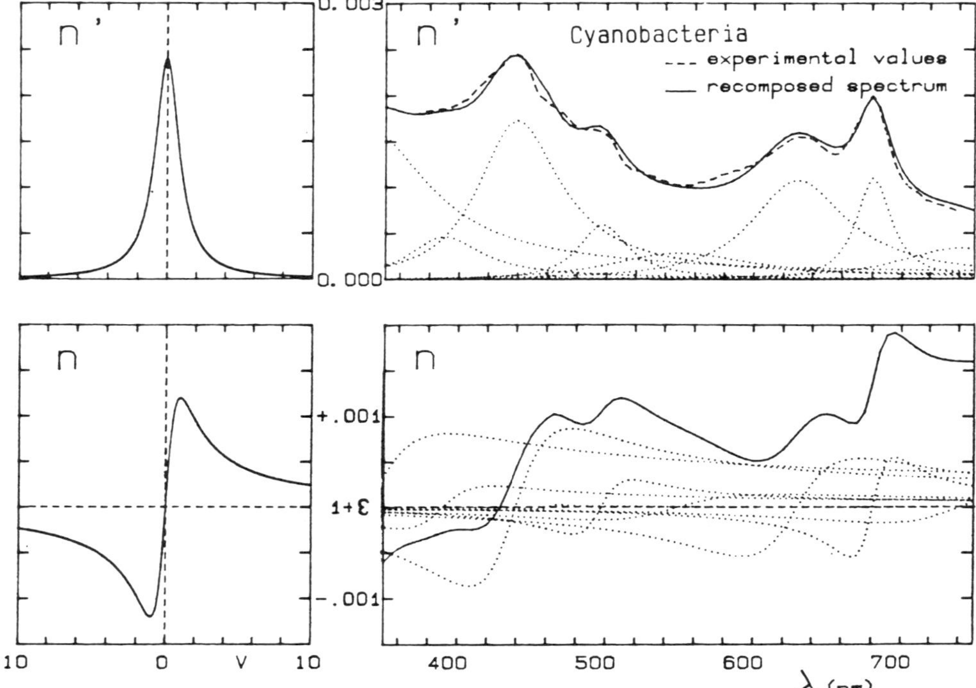

FIG. 13. On the left: variations of the real and imaginary parts of the refractive index, n (lower part) and n' (upper part), inside an absorption band. These variations are ruled by the theory of "anomalous dispersion" (Equations 32 and 33). On the right: spectral variations of n (lower Figure) and n' (upper Figure) for the cyanobacteria (grown under 200 μE • m^{-2} • s^{-1}). The "experimental" values of n', deduced from those of absorption, are shown as a dashed line. This spectrum is decomposed (by trial and error) into nine oscillators, of variable amplitude and width, which are shown as dotted lines. Corresponding values of n are computed for each oscillator by using Equation 33. By summing their effects (Equations 32' and 33') the reconstructed n and n' spectra are shown as continuous lines.

several distinct overlapping oscillators which are centered on the absorption peaks of the main pigments and which cumulate their effects. An actual $n'(\lambda)$ spectrum can be "reconstructed" in this way and the corresponding $n(\lambda)$ spectrum computed according to

$$(32') \qquad n'(\lambda) = \epsilon \sum_{j=\lambda}^{z} K_j / (1 + v_j^2)$$

$$(33') \qquad m(\lambda) = 1 + \epsilon - \epsilon \sum_{d=1}^{z} K_j v_j / (1 + v_j^2)$$

where Z is the total number of oscillators j used for the reconstruction. An example is provided on Fig. 13 (right). Obviously the variations in n are confined within the range of that in n'.

We remember that Q_b depends on n' (Fig. 3), n being assumed to be a constant. However, because of the anomalous dispersion, n is no longer steady when n' departs from zero and varies. In some sense, it can be imagined that scattering is doubly influenced by absorption. The set of Equations 13, 14 and 15 allows computations to be performed throughout the spectrum, in as much as the variations of $n(\lambda)$ and $n'(\lambda)$ have been established.

All the tools are now in hand to proceed to the comparison. By remembering (section 5.1) that the measured quantities are $F(d)$, $a(\lambda)$, $b(\lambda)$ and $c(\lambda)$, the rationale for this

552

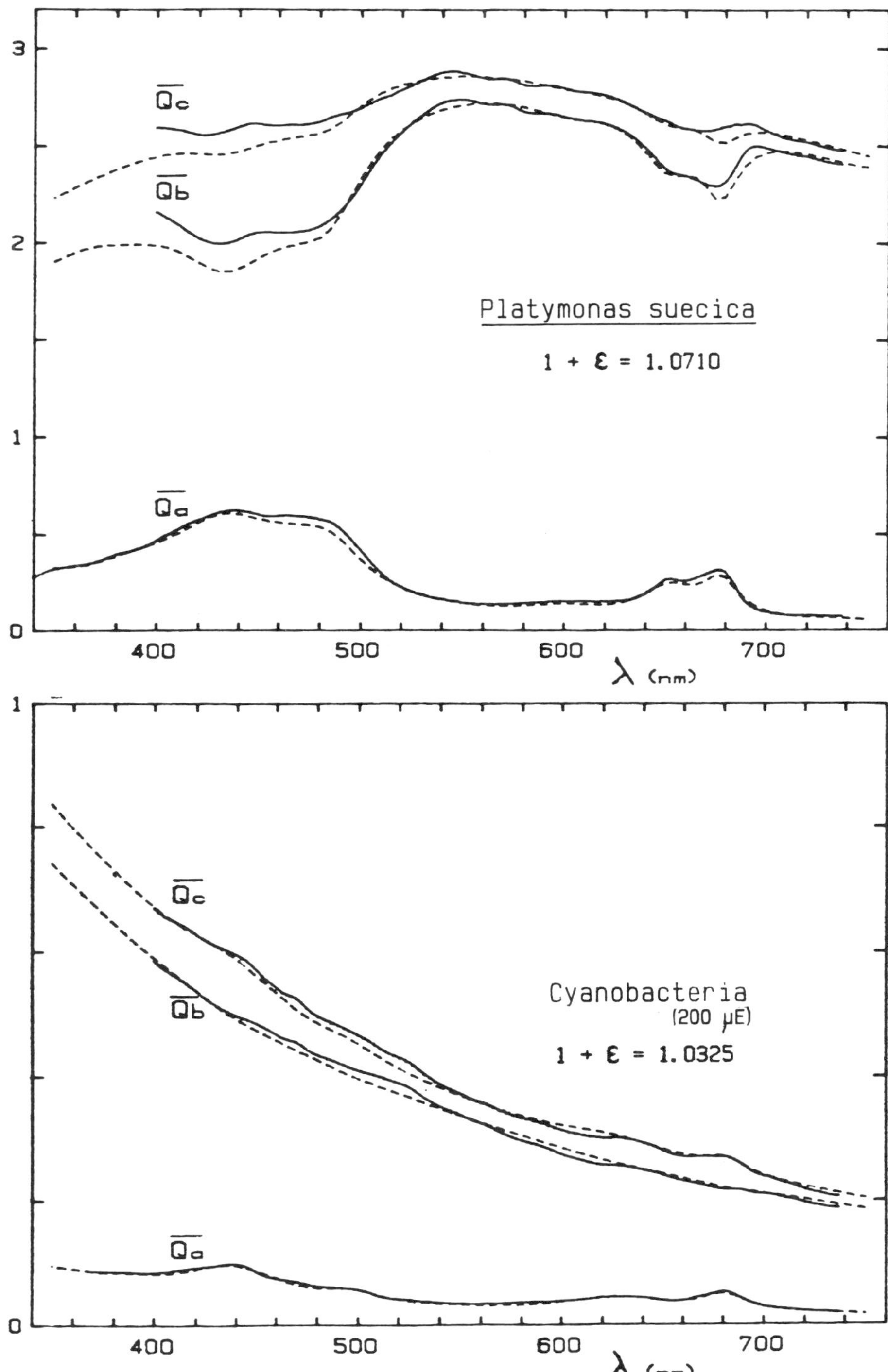

FIG. 14. Spectral variations of the mean efficiency factors for attenuation ($\bar{Q}_c$), scattering ($\bar{Q}_b$) and absorption ($\bar{Q}_a$), deduced from the attenuation and absorption coefficients experimentally determined (continuous lines), for two phytoplanktonic species. The variations of $\bar{Q}_c$, $\bar{Q}_b$ and $\bar{Q}_a$ obtained from a theoretical model (see text) are shown as dashed lines. The central value of the real part of the refractive index, $1 + \epsilon$, leading to the best theory/experiment agreement is indicated on the Figures.

comparison is as follows. $F(d)$ and $a(\lambda)$ are used as input parameters in a modeling. A floating parameter is also needed, the real part of the refractive index, introduced as $1 + \epsilon$. The products of the model are $Q_b(\lambda)$ and $Q_c(\lambda)$ which can confront the actual values. If a fit can be found, ϵ becomes determined and other properties (such as b_b, VSF) can be deduced from theory. The computations have to be performed step by step (5 nm) within the spectrum as all parameters, including the relative size α, are wavelength dependent. The details of such a modeling have been given elsewhere (Bricaud and Morel 1986). They can be summarized as follows.

By using Equation 30 the $a(\lambda)$ values are transformed into $Q_a(\lambda)$ values and from those, the parameter $\varrho'(\lambda)$ is univocally determined, because Q_a is a monotonic function of ϱ' (Equation 14'). According to the definitions (Equation 11 and 9), $\varrho'(\lambda)$ allows $a_s(\lambda)$ and then $n'(\lambda)$ to be computed. The $n'(\lambda)$ spectrum is decomposed under the form given by Equation 32'.

The second part of computations is initiated when n is given a value $1 + \epsilon$. The spectral values $n(\lambda)$, (oscillating around $1 + \epsilon$) are obtained through Equation 33'. With $n(\lambda)$, $n'(\lambda)$ and $\alpha(\lambda)$, the efficiency factors $\bar{Q}_c(\lambda)$ and $\bar{Q}_b(\lambda)$ are computed (Equations 13 and 15). They form "theoretical" values to be compared to experimental values deduced from $c(\lambda)$ and $b(\lambda)$ by using expressions similar to Equation 30. The best coincidence between theoretical and experimental values is sought by adjusting the floating parameter ϵ inside an iterative loop covering the second part of the computations. Some guidelines for this adjustment can help, as shown later on.

Figure 14 presents two examples of theoretical/experimental comparison for the case of small sized phytoplankters: *Platymonas suecica* and the cyanobacteria with $\bar{d}$ (the diameter of an equivalent average sphere) respectively equal to 3.36 and 1.54 μm. The satisfactorily agreement demonstrates that the model is able to reproduce the bulk optical properties and hence provides a clue to their understanding.

The volume scattering function for a suspension of cyanobacteria has been measured. These data are plotted (Fig. 15) together with the theoretical VSF computed by using the actual size distribution function. This function is centered on $\alpha = 12$ (corresponding approximately to $d = 1.54\ \mu$m and $\lambda = 0,546\ \mu$m). The refractive index used for the computation is $1.035-0.001\ i$, a value close to that derived from the model. As mentioned before (3.6), this value is not critical, since for small sized cells, the size (i.e. α) is the dominant factor governing the shape of the VSF. The agreement between experimental data and theoretical predictions is satisfactory. The observed ascending slope for the small θ angles and the low β values in the backward directions ($\theta > \pi/2$) are correctly reproduced by calculations. This experimental VSF is very similar to that determined by Privoznik et al. (1978) for another small phytoplankter (*Chlorella pyrenoidosa*), but differs from those obtained by Sugihara et al. (1982) also for *C. pyrenoidosa* and for *Ankistrodesmus falcatus* (these authors reported high and increasing values in the backward directions which were contradictory to all expectations).

It is worth noting that the two selected species form clearly distinct cases. Beside the obvious difference in pigment compositions and absorption capabilities, their size and refractive index (the real part as inferred from the model) lead to ϱ values which are sufficiently different to dictate different optical behaviours.

When exploring the visible part of the spectrum, say from 375 to 750 nm, the parameter α (reciprocal of λ) is exactly divided by 2 and ϱ is approximately divided by 2 (or exactly 2 if n can be regarded as constant). The $Q_c(\lambda)$ spectrum is a certain portion of the $Q_c(\varrho)$ curve (Fig. 4) delimited in such a way that ϱ may only vary in a ratio 2 to 1. In a first approximation, the effect of absorption upon Q_c can be neglected (or the reasoning can be made by using Q_c values for a hypothetical non-absorbing equivalent population, for which all the characteristics are those of the actual population, but absorption, assumed to be zero for all wavelengths). In reference to Fig. 4 and discussions in Section 3.4, several situations can be distinguished:

554

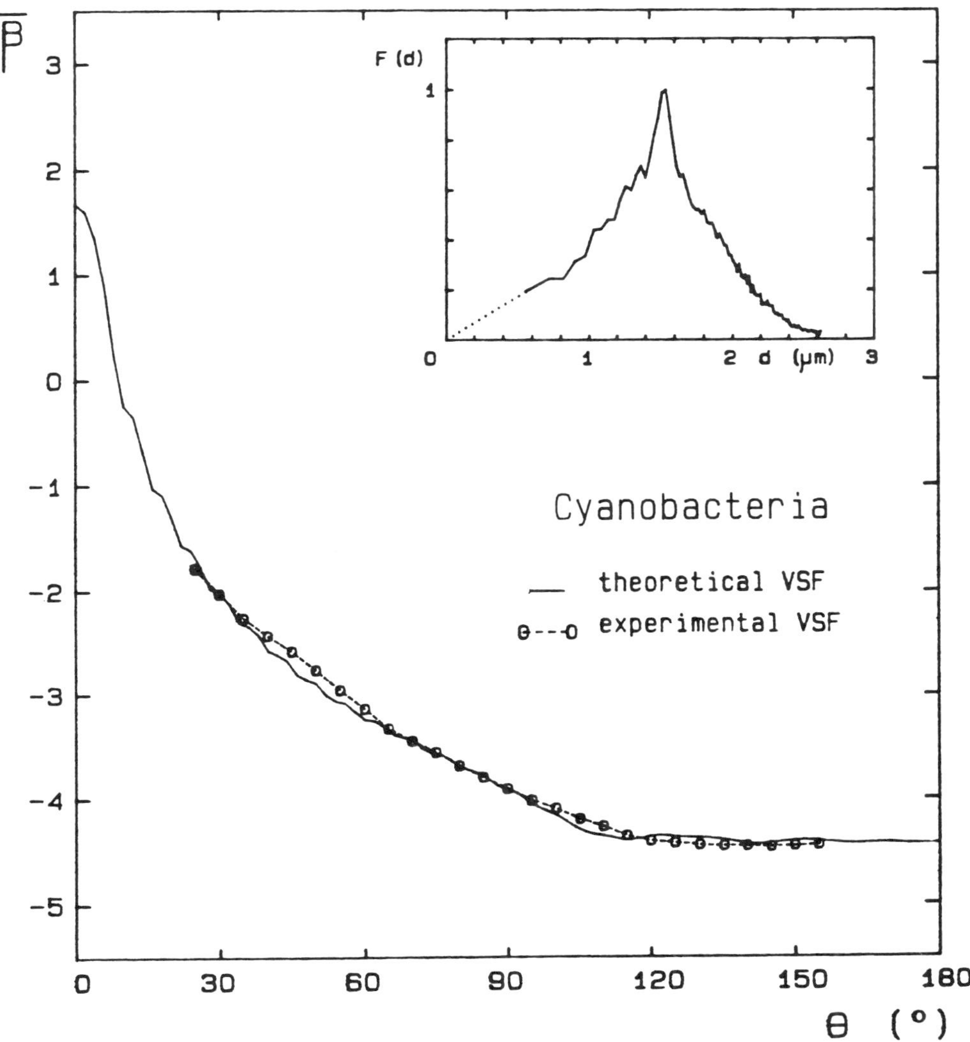

FIG. 15. Experimental volume scattering function determined in relative units for the cyanobacteria between 25 and 155° and at λ = 546 nm. The normalized volume scattering function $\bar\beta(\theta)$ is computed with m = 1.035–0.001 i and the actual size distribution function determined for this species by using a Coulter Counter ZBI (shown in inset, upper right corner). The two curves are put in coincidence at θ = 90°.

i) If ϱ, even though doubled, remains lower than about 4 (first maximum of $Q_c(\varrho)$), the $Q_c(\lambda)$ spectrum has an ascending slope towards the short wavelengths. Due to their size (see nomogram of Fig. 2), all picoplankters will exhibit this pattern, associated with low Q_c values (as for the cyanobacteria). With higher ϱ and Q_c values, the same pattern subsists for larger species (see results for *Emiliania huxleyi* and *Skeletonema costatum* in Bricaud and Morel 1986).

ii) With still larger ϱ values, overlapping the value 4 when doubled, the portion of the $Q_c(\varrho)$ curve which is involved is that which contains the first maximum. Thus the $Q_c(\lambda)$ spectrum will exhibit a maximum. This situation is exemplified by *Platymonas suecica*. Its small size is counterbalanced by a comparatively high refractive index (inferred from the model).

iii) An inverted slope is, in principle, to be expected when ϱ varies from 4 to 8 within the visible spectrum. The $Q_c(\lambda)$ curve will increase towards long wavelengths. For most of the algal species, of moderate large size, ϱ is still greater and Q_c weakly oscillates around 2. The effect of polydispersion restrains the Q_c variations and therefore the $Q_c(\lambda)$ spectrum becomes flat. In such a situation $Q_b(\lambda)$ is roughly an inverted image of $Q_a(\lambda)$ and $Q_c(\lambda)$ is equal to 2 (or must be; in practice this unavoidable value is a sensitive criterion for appreciating the quality of an experiment, especially of the size and number determination).

It has been mentioned that the floating parameter ϵ is adjusted by iterations. It is conceivable that the above remarks are helpful in orienting the choice of an initial value for ϱ. The general shape of the actual $Q_c(\lambda)$ curve provides indications about the acceptable values for ϱ. Nevertheless a flat spectrum leaves ϱ and ϵ undetermined. In this case, the measurement of the VSF could prove to be useful in the index determination.

Summary and Conclusions

The optical properties of phytoplankton, expressed on a per-cell basis or as chlorophyll-specific coefficients, appear to be varying in a rather wide range from one species to another. For a given species they are also significantly varying within the visible (or photosynthetic) spectrum. These variabilities can be assessed and reasonably well predicted from theoretical considerations based on the Mie theory and on the anomalous diffraction approximation. The same theories apply in the case of picoplankton, even if particular properties result from their small size which is within the same range as the wavelength of the radiation.

The theoretical approach makes use of two assumptions, the sphericity and the homogeneity of the algal cells. In this sense, it is a first approach which could be refined. As soon as non-spherical and inhomogeneous bodies are randomly oriented in the light field, they behave, at least in a first approximation, as sphere equivalents. Expressions have been developed (Aas 1984) for randomly oriented long cylinders and flat disks which demonstrate that the efficiency factors are not fundamentally modified. In addition, when the effect of the size polydispersion is taken into consideration, the remnant features of the Q_c curve (a unique maximum and a limiting value of 2) are similar to that for the polydispersion of spheres. It is thus believed that the bulk optical properties of algal cells are properly accounted for by the theory as presented, even if some departures presumably may occur in the case of insufficiently compact cells.

Beside the size (and the size distribution), the intracellular pigment concentration c_i is the other parameter which enters into the equations and is essential in governing the optical properties. A distinction must be made between the properties at the level of a single cell and the chlorophyll-specific coefficients which intervene in the predictions concerning a water body and its optical coefficients. By respecting this distinction, several major points can be summarized as follows, with a special emphasis put on the peculiarities of picoplankton, when appropriate.

On a per-cell basis, the efficiency factor for absorption Q_a directly related to, whilst being not linearly linked to, c_i experiences wide variations. According to Fig. 11 they encompass more than one order of magnitude. Moreover, it is not claimed that all possible situations have been explored. Low Q_a values mean a more efficient absorption for a given amount of pigment, and could be seen as the result of an optimization process. It does not seem that it occurs in nature.

It is sometimes stated or expected that picoplankton, given its small size, should exhibit a reduced "package" effect or, in other words, should benefit from the above mentioned efficiency in harvesting light. Such an expectation has to be regarded with caution. The package effect depends equally on the size and on the intracellular pigment concentration c_i, therefore the advantage provided by the smallness exists only if not counterbalanced by an increase in c_i. The data gathered by Malone (1980) tend to demonstrate that c_i would be inversely related to cell size. If such a trend was confirmed and was to be extended down to the size range of picoplankton, the advantage would be cancelled out. The results quoted in Table 1 indeed do not confirm this trend, since high and also low c_i values are observed for small species. The possible range of variation of this important parameter deserves additional studies.

The efficiency factor for scattering Q_b for individual cells is less variable than Q_a and, unlike Q_a, decreases with increasing c_i (since Q_b is lowered as soon as absorption occurs). For most of the algae of moderate and large size, with Q_c close to 2 and $Q_b = Q_c - Q_a$, Q_b may only vary within a restricted range. In this case the general rules are: the radiant energy scattered by these algae is proportional to the cross sectional area of the cells and Q_b, which acts as a proportionality coefficient, is never far from 2 (and necessarily greater than 1); it is diminished in the absorption bands.

Smaller species exhibit Q_c values higher than 2. This situation occurs when the equivalent diameter is of the order of 3 to 10 μm, inversely related to the increment of the relative index of refraction (m-1). For still smaller cells (below 3 μm) Q_c and Q_b decrease and picoplankton is the sole plankton for which Q_c and Q_b can become less than 1. Such a tiny cell scatters (and absorbs) an amount of energy less than it can geometrically intercept (the concept of the geometrical cross section is no longer adequate when describing the interaction between an electromagnetic radiation and a particle of size comparable with the wavelength).

Algal cells, even those which are armoured with opal or calcite external plates, are "soft" scatterers. The forward lobe is strongly peaked and the backscattering efficiency, which increases with the refractive index, nevertheless remains low. In the size range of picoplankton, the backscattering efficiency is enhanced with decreasing size and the dependence vis-à-vis the index disappears. It is a general rule in this size domain, that the volume scattering function becomes only size-dependent, whereas for other larger phytoplankters the VSF is determined by both the size and the refractive index.

The problem of relating a measurement of the VSF (at a given angle or within a given angular range) to a geometrical parameter, such as the mean cross sectional area of the particle, has to be tackled with care. A precise knowledge of what the theory imposes is necessary to avoid erroneous interpretation. Picoplankton, in this respect, constitute a more reliable case. It can be ascertained (under appropriate limitations concerning the upper size) that the scattering coefficient for the backward directions is simply proportional to the cross sectional area and that the forward scattering, limited to small angles, is proportional to the cube of this area. Some practical applications may be imagined from these theoretical results.

The chlorophyll-specific optical coefficients are proportional to Q_a and Q_b and, in addition, include in their definition a normalization with respect to c_i (Equations 26a and b). Therefore the wide span of variations of Q_a is reduced by virtue of such a normalization. The specific absorption coefficient, a^*, however, remains variable at the wavelength (675 nm) where chlorophyll a is still the dominant absorber (in a ratio of approximately 1 to 2 for the species studied). Elsewhere in the spectrum, in the blue region especially, a^* is still more varying because of the presence of accessory pigments in variable concentration and composition. The a^* value averaged over the whole photosynthetic spectrum is one of the parameters involved in the photosynthesis models (see also the discussion about "k_c" in Kirk's paper). Unfortunately it cannot be regarded as a constant either in the natural environment or in culture experiments.

Contrary to Q_a, Q_b is lowered when the intracellular pigment concentration increases. By the effect of dividing by c_i, this trend is thus reinforced for b^*, the specific scattering coefficient, which exhibits a high variability (in a ratio 1 to 10 for the species studied). At least in the size range 2 to 10 μm approximately, the refractive index also influences b^*, through $(m-1)$. Nevertheless $(m-l)$ cannot change as widely as c_i does. By taking the example of *Emiliania huxleyi*, it is clear that the main cause of its high b^* value lies in the low Chl$a$ content per-cell (and not in its refractive index, which is not particularly high, see in Bricaud and Morel 1986). A greater number of cells, i.e. a greater number of scattering bodies, is required to obtain in the medium a mass unit of chlorophyll a. The question of predicting the b^* values for picoplankton, is still open as for a^* until more information can be gathered about their inner pigment concentration.

Acknowledgements

This work is a contribution to the research encouraged by the IAPSO Working Group on Optical Oceanography. It was supported by Centre National de la Recherche Scientifique (UA 353 and GRECO 034) and by Centre National d'Études Spatiales under contract 83/250. The authors wish to thank E. G. Mitchelson for improving the English of the manuscript.

References

AAS, E. 1981. The refractive index of phytoplankton. Inst. Rep. Ser., Univ. Oslo, 46: 61 p.
——— 1984. Influence of shape and structure on light scattering by marine particles. Inst. Rep. Ser., Univ. Oslo, 53: 112 p.

AMESZ, J., L. N. DUYSENS, AND D. C. BRANDT. 1960. Methods for measuring and correcting the absorption spectrum of scattering suspensions. J. Theor. Biol. 1: 59–74.

BRICAUD, A., A. MOREL, AND L. PRIEUR. 1983. Optical efficiency factors of some phytoplankters. Limnol. Oceanogr. 28(5): 816–832.

BRICAUD, A., AND A. MOREL. 1986. Light attenuation and scattering by phytoplanktonic cells: a theoretical modeling. Appl. Opt. 25, 4: 571–580.

BRYANT, F. D., B. A. SEIBER, AND P. LATIMER. 1969. Absolute optical cross sections of cells and chloroplasts. Arch. Biochem. Biophys., 135: 79–108.

BUGNOLO, D. S. 1960. On the question of multiple scattering in the troposphere. J. Geophys. Res. 65, 3: 879–884.

CARDER, K. L., R. G. STEWARD, J. H. PAUL, AND G. A. VARGO. 1986. Relationships between chlorophyll and ocean color constituents as they affect remote-sensing reflectance models. Limnol. Oceanogr. 31(2): 403–413.

DAVIES-COLLEY, R. J., R. D. PRIDMORE, AND J. E. HEWITT. 1986. Optical properties of several diverse freshwater phytoplankton. Hydrobiologia. 133: 165–178.

DEIRMENDJIAN, D. 1969. Electromagnetic scattering on spherical polydispersions. American Elsevier, New York, NY. 290 p.

DOUCHA, J., AND S. KUBIN. 1976. Measurement of in vivo absorption spectra of microscopic algae using bleached cells as a reference sample. Arch. Hydrobiol. Suppl. 49: 199–213.

DUNTLEY, S. Q. 1942. The optical properties of diffusing materials. J. Opt. Soc. Am. 32 (2): 61–70.

DUYSENS, L. M. 1956. The flattening of the absorption spectra of suspensions as compared to that of solutions. Biochim. Biophys. Acta 19: 1–12.

HODGSON, R. T., AND D. D. NEWKIRK. 1975. Pyridine immersion: a technique for measuring the refractive index of marine particles. Proc. Soc. Phot. Opt. Instr. Eng. 64: 62–64.

KIEFER, D. A., R. J. OLSON, AND W. H. WILSON. 1979. Reflectance spectroscopy of marine phytoplankton. Part. 1. Optical properties as related to age and growth rate. Limnol. Oceanogr. 24: 664–672.

KIRK, J. T. O. 1975a. A theoretical analysis of the contribution of algal cells to the attenuation of light within natural waters. I. General treatment of suspensions of living cells. New Phytol. 75: 11–20.
——— 1975b. A theoretical analysis of the contribution of algal cells to the attenuation of light within natural waters. II. Spherical cells. New Phytol. 75: 21–36.

LATIMER, P., AND E. RABINOWITCH. 1959. Selective scattering of light by pigments in vivo. Arch. Biochem. Biophys. 84: 428–441.

MALONE, T. C. 1980. Algal size, p. 433–463. *In* I. Morris [ed.] The physiological ecology of phytoplankton. Studies in Ecology, Vol. 7., University of California Press, CA.

MCCRONE, W. C., R. G. DRAFTZ, AND J. G. DELLY. 1967. The particle atlas. Ann Arbor Science Publ., Ann Arbor, MI. 406 p.

MOREL, A. 1973. Diffusion de la lumière par les eaux de mer. Résultats expérimentaux et approche théorique. *In* Optics of the Sea, AGARD Lect. Ser. 61: 3.1.1–3.1.76.
——— 1987. Chlorophyll-specific scattering coef-

ficient of phytoplankton; a simplified theoretical approach. Deep-Sea Res. (in press).

MOREL, A., AND A. BRICAUD. 1981a. Theoretical results concerning light absorption in a discrete medium, and application to specific absorption of phytoplankton. Deep-Sea Res. 28A(11): 1375–1393.

——— 1981b. Theoretical results concerning the optics of phytoplankton, with special reference to remote sensing applications, p. 313–327. *In* Oceanography from space. Plenum Press, New York, NY.

MOREL, A., AND R. C. SMITH. 1982. Terminology and units in optical oceanography. Mar. Geod. 5: 335–349.

PREISENDORFER, R. W. 1961. Application of radiative transfer theory to light measurements in the sea. Int. Geophys. Geod. Monogr. 10: 11–29.

——— 1976. Hydrologics optics, Vol. 1 (Introduction), NOAA, 218 p.

PRIVOZNIK, K. G., K. J. DANIEL, AND F. P. INCROPERA. 1978. Absorption, extinction and phase function measurements for algal suspensions of *Chlorella pyrenoidosa*. J. Quant. Spectrosc. Radiat. Transfer 20: 345–352.

SHIBATA, K., A. A. BENSON, AND M. CALVIN. 1954. The absorption spectra of suspensions of living microorganisms. Biochim. Biophys. Acta 15: 461–470.

STRICKLAND, J. D. H. 1960. Definitions and conversion factors, p. 5–17. *In* Measuring the productivity of marine phytoplankton. Bull. Fish. Res. Board. Can. 122: 172 p.

SUGIHARA, S., M. KISHINO, AND N. OKAMI. 1982. Backscattering of light by particles suspended in water. 76, 1; 1–8. Sci. Papers I.P.C.R.

VAN DE HULST, H. C. 1957. Light scattering by small particles. Wiley, New York, NY.

YENTSCH, C. S. 1957. A non-extractive method for the quantitative estimation of chlorophyll in algal cultures. Nature 179: 1302–1304.

Remote Sensing of Phytoplankton:
A Review, with Special Reference to Picoplankton

SHUBHA SATHYENDRANATH

National Institute of Oceanography, Dona Paula 403 404, Goa, India

Introduction

Until recently, picoplankton was considered to be an unimportant component of marine phytoplankton, and consequently, it remained largely uninvestigated. Of late, many papers show that the contribution of picoplankton to the biomass as well as to production of phytoplankton is not negligible (Herbland et al. 1985; Johnson and Sieburth 1979; Joint and Pomroy 1983; Li et al. 1983; Platt et al. 1983; Takahashi and Beinfang 1983; Waterbury et al. 1979). It, therefore, seems an appropriate moment to re-examine procedures for phytoplankton remote sensing in the light of what we may call the "picoplankton discovery". I address the following questions:

How does picoplankton affect the remotely sensed signal?

Can remote sensing tell us anything about picoplankton? In the absence of remote sensing data on the effects of picoplankton, efforts to evaluate their implications for remote sensing will have to be based at this stage entirely on information available from theoretical, laboratory and *in situ* studies.

I have approached this problem by reviewing the capabilities and limitations of the present-day techniques for remote sensing of marine phytoplankton. This includes discussions on the various techniques for deriving phytoplankton biomass, as well as some emerging techniques for retrieving additional information on phytoplankton type and primary productivity. A persistent problem, in obtaining a phytoplankton biomass index through remote sensing, has been the variability of the phytoplankton absorption, scattering and fluorescence signatures. At each stage, the consequences of this variability in phytoplankton signatures (and whenever possible, that of picoplankton in particular) are examined. The gaps in knowledge of picoplankton optical signatures limit the conclusions that can be drawn, and these will be indicated.

The paper is organised as follows: We first derive the signal leaving the water surface from the signal received by a sensor on a remote platform. The diverse techniques of phytoplankton remote sensing (passive radiometry, passive fluorescence and active lidar techniques) are then examined. Finally, the limitations of the depths accessible by remote sensing are studied.

The Coastal Zone Color Scanner (CZCS), launched in 1978 on board the satellite NIMBUS-7, has remained the only satellite sensor ever launched that was specifically designed for ocean phytoplankton remote sensing. The CZCS orbits sun-synchronously at an altitude of about 955 km, and has a ground resolution of 825 m by 825 m. It picks up radiance at six channels, centred at 443, 520, 550, 670, 750 and 11,500 nm (the 750 nm channel has not been operational since launch). The sensor scans in a direction perpendicular to the orbit track. The scan plane can be tilted up to 20° north or south of nadir, to avoid sun glint. The swath width varies from 1300 to 2300 km, depending on the tilt. Each swath slightly overlaps the previous one. Thus the data from the satellite can be used to create contiguous mosaics of radiances or derived parameters that cover millions of square kilometres. The satellite track repeats every 6 days. All our practical experience on satellite remote sensing of phytoplankton is confined to this one sensor, and consequently, extra emphasis is given in this paper to interpretation of data from CZCS type of sensors.

A glossary of the radiometric terms and notations used in the text is given in Appendix I. All the optical quantities discussed here are wavelength-dependent. When wavelength is not specified, it implies that the statement holds for all wavelengths.

The Remotely Sensed Signal

The basic remote sensing problem posed here is to measure spectral variations in the light leaving the water surface, and to derive information on phytoplankton from them. In the case of passive remote sensing, the primary light source for the water-leaving radiance is the sunlight itself, whereas in active remote sensing, the source is a laser. Before the water-leaving signal reaches the remote sensor, it is distorted by the intervening atmospheric column. The first step in the study, therefore, is to evolve suitable schemes for correcting the total signal for atmospheric noise. This aspect is examined briefly next.

PASSIVE REMOTE SENSING

Consider a sensor at height Z above the sea surface, viewing the sea at an angle (θ', ϕ), under cloud free conditions. Let the radiance received by the sensor be $L(Z, \theta', \phi)$. The following factors would influence L (see Fig. 1 also):
   a) Attenuation of light emerging from the sea surface, by the atmospheric column between the sea surface and the sensor.

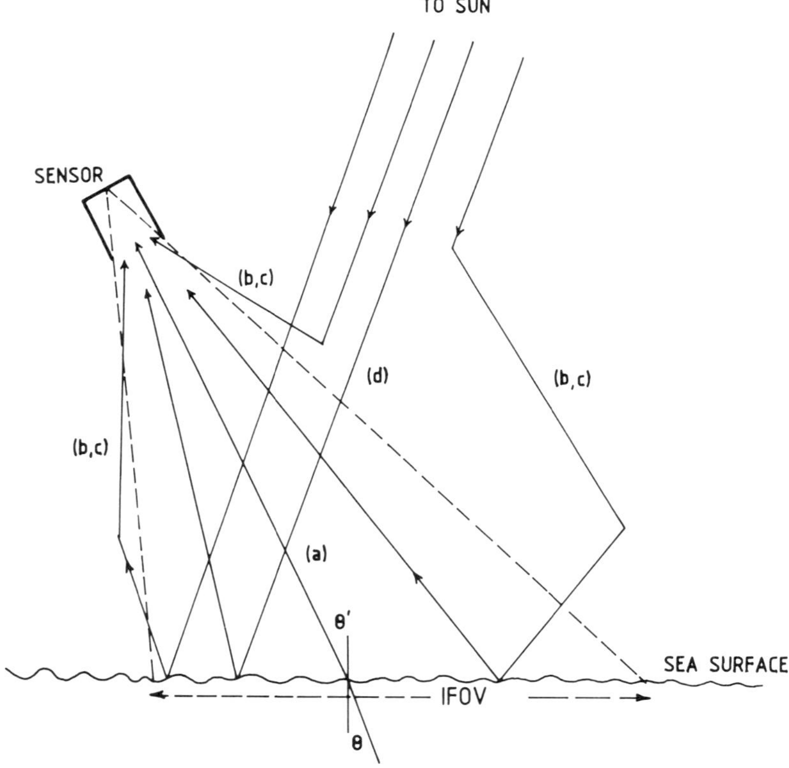

FIG. 1. Pathways of radiation reaching the remote sensor. (a) is upwelling radiation from within the water, and may consist of scattered light or fluorescence emission. This is the useful signal for the biological oceanographer. (b,c) consist of scattered light, which may or may not have been reflected at the sea surface. The atmospheric gases (b) or aerosols (c) may be responsible for the scattering process. It is convenient to treat the two types of scattering separately, because of the great differences in their volume scattering function. (d) reprensents specular reflection of sunlight into the sensor. IFOV is the Instantaneous Field of View.

b) Rayleigh scattering of diffuse sky light into the sensor by atmospheric gases. The scattered light may include that reflected at the sea surface.

c) Aerosol scattering of diffuse sky light into the sensor (Mie scattering). The scattered light may include that reflected at the sea surface.

d) Reflection of direct sunlight into the sensor.

The specular reflection term is avoided by the CZCS by pointing the sensor away from the region of sunglint. Gordon (1978) expresses the dependence of L on the remaining processes as:

$$(1) \qquad L(z, \theta', \phi) = L_w(\theta', \phi)\, t + L_R + L_A$$

The water-leaving radiance, L_w, consists of: (i) sunlight which penetrated the water, and was retransmitted into the atmosphere after scattering, and (ii) fluorescence emission from the water column, if any. This therefore, is the useful signal for the biological oceanographer.

The quantity t is the diffuse transmittance of the atmosphere defined by Gordon and Morel (1983) as:

$$(2) \qquad t = \exp\left[-(T_R/2 + T_{Oz} + (1 - \omega_A F)T_A)/\cos\theta\right]$$

where, T_R, T_{Oz} and T_A are the optical thicknesses of the intervening atmosphere due to atmospheric gases, ozone and aerosols, respectively.

ω_A is the aerosol scattering albedo.

F is the probability of forward scattering in the case of aerosol scattering.

The diffuse transmittance term allows for the fact that radiance from nearby pixels can be scattered into the sensor. If the direct transmittance were used, the exponent in Eq. 2 would have to be replaced by the sum of the three optical thicknesses. (When neighbouring pixels are sharply contrasted, contamination from the neighbours can become unacceptably high, and such areas would have to be rejected from analysis.)

L_R is the contribution to L from scattering by atmospheric gases.

L_A accounts for the contribution due to Mie scattering by aerosols.

Equation 1 assumes that the Rayleigh scattering and aerosol scattering events can be decoupled. Gordon has shown this to be a realistic and useful simplification, but it is strictly true only if there is no multiple scattering.

Gordon and co-workers, in a series of articles, have outlined practical ways of estimating L_w from L in the case of the CZCS (see review of Gordon and Morel 1983). Their method is based on the following assumptions: Aerosol scattering phase function and single scattering albedo are wavelength independent; Rayleigh scattering component does not depend on sea-surface roughness; and aerosol contribution is directly proportional to the aerosol optical thickness.

When L_w is zero at λ_o, Eq. 1 reduces to:

$$(3) \qquad L(\lambda_o) = L_R(\lambda_o) + L_A(\lambda_o)$$

In the case of the CZCS, λ_o is taken to be 670 nm. The quantity L_R is wavelength-dependent, but its value for any wavelength can be calculated theoretically from a knowledge of the atmospheric pressure, solar elevation and the look angle of the sensor. Therefore, $L_A(\lambda_o)$ can be estimated from $L(\lambda_o)$ and L_R. We now define the ratio

$$(4) \qquad S(\lambda, \lambda_o) = L_A(\lambda)/L_A(\lambda_o).$$

Here L_A is a function of T_A, the extra-terrestrial solar irradiance after correction for ozone absorption (F_o), and other terms which are assumed wavelength-independent in the model. We therefore have:

(5) $$S(\lambda, \lambda_o) \cong T_A(\lambda)\, F_o(\lambda) / [T_A(\lambda_o)\, F_o(\lambda_o)].$$

The quantity F_o can be calculated. To be able to evaluate $L_A(\lambda)$, we therefore have to estimate $T(\lambda, \lambda_o)$. Initially, this was done through ground measurements. If a direct measurement of T_A is available at ground level for any one pixel in the satellite imagery, for the time of the satellite overpass, $S(\lambda, \lambda_o)$ can be determined for that pixel from Eq. 5. The rest of the scene is processed on the assumption that this ratio remains constant over the entire scene. The assumption is only valid if the aerosol type remains constant over the area, though the total aerosol concentration may change. The quantity S can also be determined from *in situ* measurement of L_w and evaluation of tL_w, through Eqs. 1 and 4.

To calculate the term t from Eq. 2, it is assumed that the term involving the aerosol optical thickness T_A is negligible. This assumption is not considered to be critical, since T_A is typically of the order of 0.1 and $(1 - \omega_A F)$ is generally less than $1/6$. The major effect of aerosols on L is through the L_A term. The optical thicknesses due to atmospheric gases and ozone can be calculated theoretically.

An alternate method, proposed by Gordon and Clark (1981) does not require ground measurements for estimating S. In this method, atmospheric correction over the entire scene is first carried out assuming

$$S(\lambda, \lambda_o) = F_o(\lambda) / F_o(\lambda_o).$$

This permits identification of pixels with very clear waters. For these pixels, L_w values are assumed to be known, which permits recalculation of S. The atmospheric effects are now recalculated for the whole scene using the S values derived for clear water pixels. Bricaud (personal communication) uses an interation scheme which re-identifies clear water pixels in the case of the new S values (these new clear water pixels may turn out to be different from the original ones). The process is repeated until it becomes stable.

The assumption that $L_w(670)$ is zero is not valid for all waters. Smith and Wilson (1981) have proposed an iterative scheme for correcting errors arising from this assumption. Their method relies on statistical relationships between L_w values for different wavelengths taken in pairs. Guan et al. (1985) have proposed a scheme based on error analysis, for reducing the calculation time.

The atmospheric correction described here is critical, when we consider that only 20% or less of the signal received by the satellite is useful. Even though the present methods make several assumptions, they have nevertheless yielded good results.

The atmospheric correction scheme outlined here is valid for other sensors on aircraft or satellite. In the case of aircraft, the F_o terms in Eq. 4 will have to be replaced by the values of downwelling irradiance at the aircraft altitude, and the ozone optical thickness would drop to zero. In the case of low flying aircraft, naturally, atmospheric correction will be less important.

Active Remote Sensing

In the case of active remote sensing, the quantity we are interested in measuring is the laser-induced scattering and fluorescence signals leaving the water surface. All ambient light, therefore, constitutes noise. By operating the laser in pulses, it is possible to monitor the background and apply suitable corrections. Though this appears easy in principle, in practice, difficulties are introduced by the high level of background noise in the case of daytime operations, which calls for stringent instrumental requirements on sensitivity and range. (It is interesting to note that, when the active system is monitoring the background, it is in fact acting as a passive sensor.) It is also necessary to account for the atmospheric attenuation in both directions. Suitable methods can be devised for this correction, but at the present state of art, this is not a critical problem as laser techniques have so far only been tried from aircraft flying at very low altitudes (about 150 m).

Ocean Colour

I now examine how the satellite signal, after correction for atmospheric effects, can be related to the phytoplankton biomass and type.

RELATIONSHIP BETWEEN COLOUR AND UPWELLING RADIANCE AT THE WATER SURFACE

The water leaving radiance $L_w(\theta,\phi)$ is related to the upwelling radiance $L_u(\theta, \phi)$ just below the water surface through the equation:

$$(6) \qquad L_w(\theta,\phi) = L_u(\theta, \phi) [1 - \varrho(\theta, \phi)]/m^2$$

where m is the refractive index of water, $m \sin\theta = \sin\theta'$, and ϱ is the Fresnel reflectivity. The $[1 - \varrho(\theta, \phi)]$ term accounts for reflection at a calm sea surface (the reflectivity does not change much in the case of a rough sea, for small nadir viewing angles), and the $1/m^2$ term arises due to the spreading of the beam because of refraction. For nadir viewing angles of about 40° or less, the latter term is dominant, and the radiance distribution is generally found to be insensitive to changes in θ and ϕ. Therefore, for small nadir viewing angles, it is possible to simplify Eq. 6 to:

$$L_w(\theta', \phi) \cong 0.544 \, L_u$$

where L_u is the radiance in the upward direction (see Austin 1974, 1980; Kirk 1983).

Let Q be the ratio of E_u (the upwelling irradiance at depth zero) to L_u.

$$L_u = E_u(0)/Q.$$

For a perfectly diffuse medium, $Q = \pi$. Experimental and theoretical results (Austin 1980; Kirk 1983) suggest that for oceanic waters, the value of Q is close to 5 for nadir and near-nadir angles.

Now, the "intrinsic colour" of seawater is determined by the spectral form of reflectance R at depth zero, defined by:

$$R(0) = E_u(0)/E_d(0)$$

where E_d is the downwelling irradiance. The reflectance ratio at 2 wavelengths can now be expressed in terms of the water leaving radiance at those wavelengths:

$$(7) \qquad \frac{R(\lambda_1)}{R(\lambda_2)} = \frac{L_w(\lambda_1) \, Q(\lambda_1) \, E_d(\lambda_2)}{L_w(\lambda_2) \, Q(\lambda_2) \, E_d(\lambda_1)}$$

It is clear from Eq. 7 that the ratio of reflectances is proportional to the ratio of water leaving radiances if $E_d(\lambda_2)/E_d(\lambda_1)$ and $Q(\lambda_1)/Q(\lambda_2)$ are constants. Neglecting the slight variability of these parameters, the radiance ratios may be assumed equal to the reflectance ratios. Theoretically, it is possible to estimate E_d from satellite data, thereby eliminating one source of uncertainty, though generally this is not done. As we will see later, most remote sensing algorithms use radiance or reflectance ratios for phytoplankton estimations. When trying to interpret and understand these algorithms, it is convenient to think in terms of reflectance, because: (i) fluctuations in incoming solar irradiance do not affect it, as it is normalized to this variable, and (ii) reflectance can be expressed in terms of the inherent optical properties of sea water, as shown in the next section. Inherent optical properties have the advantage of being independent of variations in the angular distribution of ambient light.

Gordon and Morel (1983) have reviewed the recent theoretical results relating R to the inherent properties b_b (backscattering coefficient) and a (absorption coefficient). For oceanic waters they reduce to a simple relationship of the form:

$$(8) \qquad R \cong 0.33 \, b_b/a.$$

Equation 8 assumes that there is no fluorescence emission at the wavelength considered. It is valid for waters where b_b/a is less than 0.3. The coefficients a and b_b obey the principle of superposition.

Morel (1980) proposed that for the simplest case of oceanic waters with little land influence (no land drainage, river runoff or bottom sediments brought into resuspension by wave action), it may be reasonably assumed that the variations in a and b_b are completely determined by the optical properties of the phytoplankton present in it (this is the case 1 waters). Then a and b_b can be rewritten as:

$$(9) \qquad a = a_w + a_c$$

$$(10) \qquad b_b = b_{bw} + b_{bc}$$

where subscripts w and c stand for contributions due to pure seawater and phytoplankton respectively. In turn, a_c and b_{bc} can be expressed in terms of the specific coefficients (coefficients per unit concentration) a^*_c and b^*_{bc}, and the concentration of phytoplankton in water, C. i.e.,

$$(11) \qquad a_c = C \, a^*_c$$

$$(12) \qquad b_{bc} = C \, b^*_{bc}.$$

By common practice, the concentration of chlorophyll-a is taken as the index of phytoplankton biomass, and the specific values are expressed per unit concentration of chlorophyll-a.

Equations 9 and 10 assume simple, ideal conditions where variations by other substances like dissolved organic matter and suspended material other than phytoplankton can be accommodated provided they covary with phytoplankton. The contribution from pure seawater is invariant, and the spectral values of a_w and b_{bw} are well known (Morel 1974; Smith and Baker 1981). Therefore, if the spectral values of a^*_c and b^*_{bc} were invariant with species composition and the physiological state of the population, variations in ocean colour would be solely determined by C. It would then be a relatively simple matter to retrieve C from the magnitude and spectral form of R. This however, is not the case, and the variability in phytoplankton optical signatures merits careful examination to evaluate the consequences in remote sensing.

Backscattering by Phytoplankton

Theoretical and experimental results of Morel and Bricaud (1981a) and Bricaud et al. (1983) indicate that backscattering and absorption by phytoplankton have approximately complementary spectral forms. Phytoplankton backscattering spectra are generally characterised by two minima at around 440 and 675 nm, corresponding to the chlorophyll-a absorption at these wavelengths. Superimposed on this general form there is considerable variability. The magnitude of specific backscattering is also highly variable, but generally extremely low. Theoretically (Morel and Bricaud 1981a), the specific backscattering by picoplankton could be up to an order of magnitude or so higher than that of larger cells with the same refractive indices. Thus, the remotely sensed signal would have a higher component due to backscatter in picoplankton-rich waters. In spite of this, variations in b_{bc} would remain small compared to b_{bw} for low phytoplankton concentrations, and this effect would only be of secondary importance in contributing to variations in R, when compared to the effect of the variability in a^*_c.

SPECTRAL SPECIFIC ABSORPTION OF PHYTOPLANKTON

In vivo spectral specific absorption of phytoplankton has been the subject of several laboratory studies. These spectra are generally characterised by a maximum at 440 nm and a secondary maximum at 675 nm, both of which are associated with chlorophyll-*a* absorption bands. Additional peaks due to the auxiliary pigments chlorophylls b or c and carotenoids are superimposed on the chlorophyll-*a* absorption bands. Generally, there is a broad minimum in the 550–620 nm range. But in the case of phytoplankton with biliproteins (cyanobacteria, cryptophytes and rhodophytes), this zone is occupied by the absorption bands of phycoerythrin (500–565 nm) and phycocyanin (610–645 nm). The phytoplankton specific absorption at 440 nm per unit chlorophyll-*a* concentration is seen to vary by a factor of three. Besides the variations in the pigment composition, the "flattening effect" is another factor causing variations in the spectral form and magnitude of specific absorption. Duysens (1956), Kirk (1983) and Morel and Bricaud (1981b) have examined the effect of particle size on absorption, and noted that when the absorbing material is concentrated within suspended particles in the medium, there is a decrease in absorption, and a flattening of the spectrum, as compared to absorption by the same material in solution. Sathyendranath et al. (1987) studied the specific absorption spectra of selected monospecific cultures, and concluded that pigment composition and flattening effect account for practically all the variability in the spectra. They assumed spherical, homogeneous particles for calculating the flattening effect.

Reverting to Eq. 11, the variable a^*_c can therefore be calculated if the concentration of all the auxiliary pigments and the size distribution of the phytoplankton cells are known, using the following equations:

If n pigments with concentration C_i ($i = 1$ to n) are present in the phytoplankton population in addition to chlorophyll-*a* at concentration C, then a parameter a^*_{sol} can be calculated, which represents the absorption that the medium would have in the absence of flattening effect (absorption by a hypothetical equivalent solution):

$$(13) \qquad a^*_{sol} = \sum_{i=i}^{n} \frac{C_i}{C} a^*_{soli} + a^*_{solc}$$

where a^*_{solc} and a^*_{soli} represent specific absorption by the chlorophyll-*a* and the auxiliary pigments, respectively, in the absence of all other pigments, and if there is no package effect. The first term on the right hand side of Eq. 13 accounts for changes in specific absorption due to changes in the relative concentrations of the auxiliary pigments, with respect to chlorophyll-*a*.

The flattening effect can now be calculated as follows. We first calculate the specific absorption coefficient of the cellular material, a_{cm}, given by:

$$(14) \qquad a_{cm} = (6a^*_{sol})/(\pi N d^3)$$

where N is the number of particles per unit chlorophyll-*a* concentration, and d is the diameter of the equivalent sphere with the same volume as an individual particle. Then, with $\varrho' = d\, a_{cm}$, the efficiency factor for absorption, Q_a, is calculated:

$$(15) \qquad Q_a = 1 + (2e^{-\varrho'})/\varrho' + 2(e^{-\varrho'} - 1)/\varrho'^2.$$

Then,

$$(16) \qquad a^*_c = \pi N d^2 Q_a/4.$$

The flattening effect is given by a^*_c/a^*_{sol}, which is always less than one, but tends to one when ϱ' tends to zero. This ratio decreases non-linearly, but monotonically, with increase in ϱ'. Since ϱ' is greater at absorption maxima than at the minima, a^*_c/a^*_{sol} will be less at absorption maxima. Consequently, the spectrum will appear flattened when compared to the corresponding a^*_{sol} spectrum.

Clearly, the size of the particle influences the form and magnitude of the specific absorption spectrum. But it is to be noted that an increment in a_{cm} has an identical effect on Q_a as a similar increment in d. Obviously, picoplankton, with their small size, will tend to have higher specific absorption than larger cells with the same pigment characteristics.

We do not have much information on whether picoplankton have intracellular pigment concentrations similar to larger phytoplankton. Li et al. (1983) report chlorophyll-a concentrations of the order of 0.5 fg/cell for cyanobacteria from a tropical upwelling ecosystem. This yields a value of about 10^6 mg of chlorophyll-a per cubic metre of cell material, if we assume the cell size to be of the order of $1\mu m$. This is comparable to values obtained for larger cells in cultures (Sathyendranath et al. 1987) and is favourable for high specific absorptions in picoplankton. However, the data from Li et al. (1983) for another tropical station, from Smith et al. (1985) and from Platt et al. (1983) yield cell chlorophyll concentrations of about one or two orders of magnitude higher. Such cells, obviously, would tend to have very low specific absorption values! (One possibility, of course, is that, as suggested by Li et al. for their data, the cell numbers are underestimated. This is supported by the fact that the reported photosynthetic parameters of these picoplankton suggest high specific absorptions.) Prezelin et al. (1985) report concentrations ranging from 2 to 100 fg/cell, with the lower values at the surface, and the higher values at the bottom of the euphotic zone. (The data from Platt et al. (1983) also refer to data from the deep chlorophyll maximum.) There is a marked dearth of data on the pigment composition of picoplankton also. So the question of what a_{cm} values to attribute to picoplankton remains unsolved at present. However, the available limited data suggest that, at least for remote sensing purposes, where we are mostly concerned with surface waters, low intra-cellular pigment concentrations may be a reasonable assumption.

CASE 2 WATERS

Thus far only case 1 waters have been considered. In case 2 waters (see Gordon and Morel 1983), which are mostly encountered in coastal regions, yellow substances (dissolved organic matter) and suspended sediments may vary independently of chlorophyll-a concentrations, thereby rendering the system more complex.

The absorption coefficient in this case will contain additional terms a_y and a_x representing contributions from yellow substances and suspended matter (other than phytoplankton) respectively, while the backscattering coefficient will contain the extra term b_{bx} which accounts for backscattering by suspended sediments.

Bricaud et al. (1981) have expressed the absorption by yellow substances by an equation of the form:

$$a_y(\lambda) = a_y(440)\exp\left[-0.014(\lambda - 440)\right]$$

(see Fig. 2). The constant 0.014 is a mean value, as it has a slight variability.

Yentsch (1962) has observed a neutral absorption spectrum for suspended sediments from Woods Hole waters, and a spectrum similar to that of yellow substances for samples from waters below 100 m. Prieur and Sathyendranath (1981) have proposed an "U-shaped" absorption curve for suspended particles based on *in situ* results. Further studies are needed to determine the variability of absorption and back-scattering characteristics for different types of suspended sediments.

568

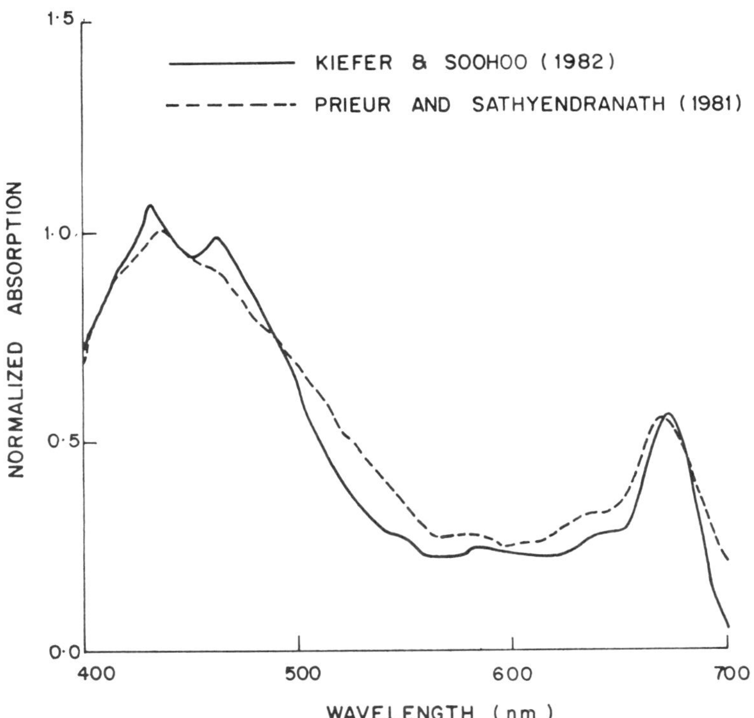

FIG. 2. Absorption spectra associated with phytoplankton, for *in situ* conditions, from Prieur and Sathyendranath (1981) and Kiefer and SooHoo (1982). Both spectra are normalized at 440 nm, in order to compare the spectral form.

In situ absorption and reflectance spectra

We examine now the *in situ* observations in the light of the theoretical and laboratory results described above. Prieur and Sathyendranath (1981) have shown that the absolute values of *in situ* specific absorption of phytoplankton in fact vary with region (at 440 nm, the value was found to vary by more than a factor of 3). The highest specific absorptions were noted in oligotrophic open ocean waters with a possibly high fraction of picoplankton. However, these specific absorptions are probably overestimates, since no special precaution was taken to retain the picoplankton fraction during filtration for chlorophyll-*a* estimation. Their study, on the other hand, did not reveal any variations in the spectral form of specific absorption and they pointed out that this would require more exhaustive and accurate data than was available, since spectral variability apparently introduced only minor changes around the average spectral form. Kiefer and SooHoo (1982) identified the *in situ* specific absorption spectrum associated with chlorophyll-*a* after correction for the effects of pheopigments. Their spectral form compares well with that proposed earlier by Prieur and Sathyendranath (1981) (see Fig. 3). Under non-bloom conditions when several types of phytoplankton would be present in the population, the mixing of types would tend to wash out the marked variations in spectral signatures which are observed in the case of single species cultures in the laboratory.

Sathyendranath (1981) showed that a model of absorption with a single spectral form for phytoplankton specific absorption, with variable efficiencies, can explain

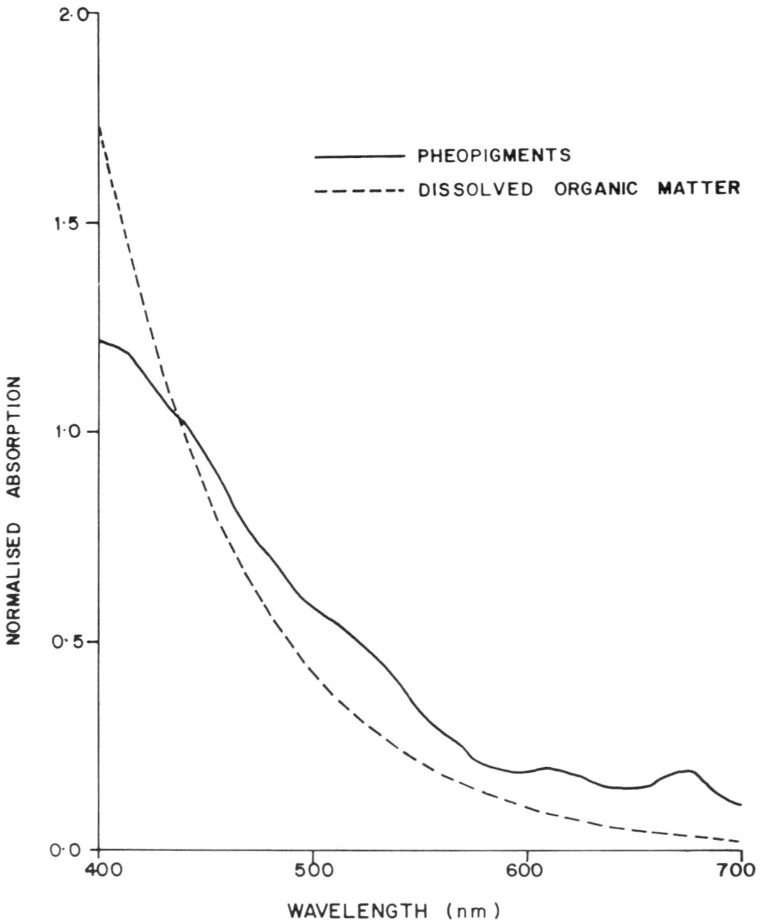

FIG. 3. Absorption spectra of dissolved organic matter, according to the Eq. of Bricaud et al. (1981). The absorption spectrum associated with pheopigments, proposed by Kiefer and SooHoo (1982) is also presented. The two spectra show considerable similarity. When spectral signatures of different substances have similar characteristics, it becomes difficult to distinguish them by remote sensing.

observed reflectance spectra, with certain additional hypotheses regarding b_b (Fig. 4). As a first approximation, the b^*_{bc} spectrum proposed had a form inversely proportional to the a^*_c curve, and a constant backscattering ratio of 0.5%. Comparison with observed values was good, which once again implies that b^*_{bc} variability is of secondary importance in determining the reflectance spectra. For case 2 waters, this model uses the "U" shaped absorption spectrum for suspended sediments and a b_{bx} spectrum which follows a λ^{-n} law. The parameter n was found to vary from station to station over a range of 0 to 2. The backscattering ratio was also found to vary over the range 0.3 to 2%.

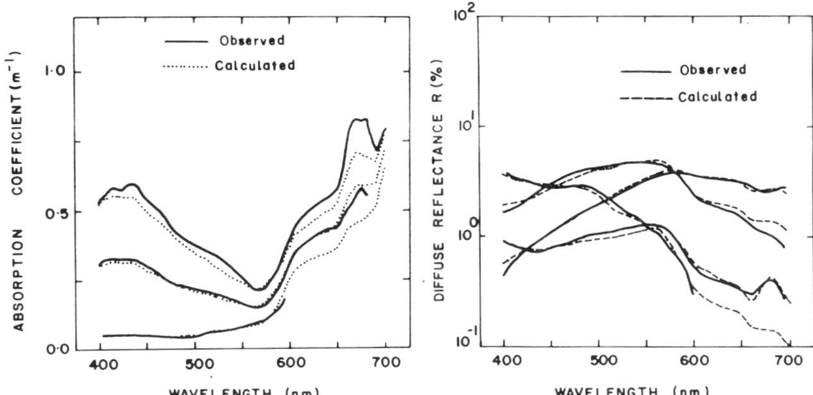

FIG. 4. *In situ* absorption and reflectance spectra compared with the spectra calculated from absorption and reflectance models (from Prieur and Sathyendranath 1981 and Sathyendranath 1981).

ALGORITHMS FOR CHLOROPHYLL RETRIEVAL FROM OCEAN COLOUR

When the phytoplankton characteristics affecting the R spectra are understood, the various techniques to solve the inverse problem of recovering phytoplankton information from the R spectra can be examined against this background. Some of the algorithms are developed directly from models of R. But the most commonly used algorithm, especially for CZCS data, is an empirical relationship of the form:

$$(17) \qquad C = A\ r_{ij}^{-B}$$

where r_{ij} stands for the ratio of radiances or reflectances at wavelengths i and j, and the coefficients A and B are determined by least square fit on log transformed data. In the case of the CZCS data, the wavelength pairs taken are (443, 550) and (520, 550). The 443 channel on CZCS corresponds to the absorption maximum of chlorophyll-*a*, while 550 corresponds to the minimum. For high phytoplankton concentrations, the signal at 443 becomes very low, and then the 520 signal is used instead. The method has been found to yield reasonable results, particularly for case 1 waters. However, the coefficients A and B appear to be subject to certain variability. Suitable coefficients are normally selected on the basis of sea truth data. Chlorophyll retrieval accuracies of $\pm$ 40% have been reported for this method. Algorithms of different forms, but essentially based on the same principle (i.e., the change in the blue-green ratio of reflectance with change in chlorophyll concentration) have also been proposed by various authors (see review of Sathyendranth and Morel 1983).

The influence of picoplankton on this type of algorithm is evaluated next, in the light of the nature of variability of phytoplankton spectral signatures:

1) Picoplankton, with their small size, would be least subjected to the flattening effect (pigment composition and intracellular concentration remaining the same) and would have maximum specific absorption values. If this factor is not taken into account in picoplankton-rich waters, it would result in an overestimation of chlorophyll-*a* concentration. It may be noted that the coefficient B in Eq. 17 takes higher values in the open ocean case 1 waters than in the coastal regions. It is probable that this higher rate of change in the blue-green ratio with change in chlorophyll concentration is related to a higher absorption efficiency associated with picoplankton.

2) Since R and a are inversely related, the rate of change of R will be higher for lower concentrations of phytoplankton. Given the relative importance of picoplankton

in oligotrophic waters, ocean colour changes provide a sensitive method for mapping small gradients in their concentration.

3) The algorithm type presented in Eq. 17, which relies on the change in the blue-green ratio of reflectance with change in chlorophyll-a concentration, can be invalidated in an area rich in blue-green algae with possible prominent absorption peak in the green due to phycoerythrin. The four CZCS channels are inadequate to cope with such problems, and the validity of the derived results in such cases will depend on the availability of sea truth in the study area at the time of the satellite overpass.

More complex algorithms have been devised for case 2 waters, in order to separate the chlorophyll signal from those of non-phytoplanktonic material. Morel and Prieur (1978) and Pelevin (1978) have proposed analytical algorithms which make use of three wavelengths. Jain and Miller (1976) and Miller et al. (1977) have used a model of R with optimization techniques, to derive C. The approach has been extended by Bukata et al. (1981) for case 2 type of waters. They also established "duo-isopleths" of chlorophyll-a and suspended mineral concentrations. The concentrations could be read off from the graphs, if radiance values of 2 wavelengths were known. (However, this method uses only 2 wavelengths, and was found to be less accurate than the multispectral optimization technique.) Carder and Steward (1985) have also used a case 2 reflectance model of R. Chlorophyll concentration is taken to be that value which yields the best comparison between theoretical and observed spectra. The problem of passive radiometric remote sensing in case 2 waters has also been examined in detail by Sathyendranath et al. (1982, 1983). Their results indicate that the role of suspended sediments is basically to enhance the magnitude of the water-leaving signal in a more or less neutral manner, while changes in the concentration of phytoplankton and yellow substances result in alteration in the spectral form of reflectance. The presence of yellow substances tends to lower the level of outgoing signal monotonically towards the blue part of the spectrum, while the phytoplankton signal can be distinguished from it primarily by the presence of the minimum at 440 nm due to the chlorophyll absorption maximum at this wavelength. The non linear interactions between the signals of these substances make it difficult to arrive at a single algorithm that will be equally satisfactory in the case of all possible combinations of these substances. Best results will probably be achieved in such waters by adjusting the algorithms to suit local conditions. Many of these multispectral algorithms are not suitable for CZCS, since they make use of different or additional wavelengths from those available on this sensor. Sathyendranth et al. (1982, 1983), for example, point out that the use of 400 nm in addition to the CZCS channels considerably enhances the possibility of separating out chlorophyll signal from that of yellow substances.

ADDITIONAL INFORMATION ON PHYTOPLANKTON FROM OCEAN COLOUR

Since we now have a working hypothesis on the causes of variation in phytoplankton absorption efficiency with respect to chlorophyll-a, and its effect on ocean colour, we can examine the inverse problem of using this variability to extract additional information on phytoplankton.

An important, if not dominant, fraction of picoplankton is cyanobacteria or the blue-green algae (Johnson and Sieburth 1979; Waterbury et al. 1979; Li et al. 1983; Platt et al. 1983). In laboratory measurements of the absorption spectra of blue-green algae, the additional absorption bands due to the biliproteins show up clearly (Yentsch and Yentsch 1982; Lazzara 1983; Jeffrey 1980; examples of blue-green algal absorption spectra from Yentsch and Yentsch (1982) are shown in Fig. 5). We may surmise from this, that it should be easy to identify the blue-greens from other phytoplankton through these distinctive absorption bands, if complete spectral information on R

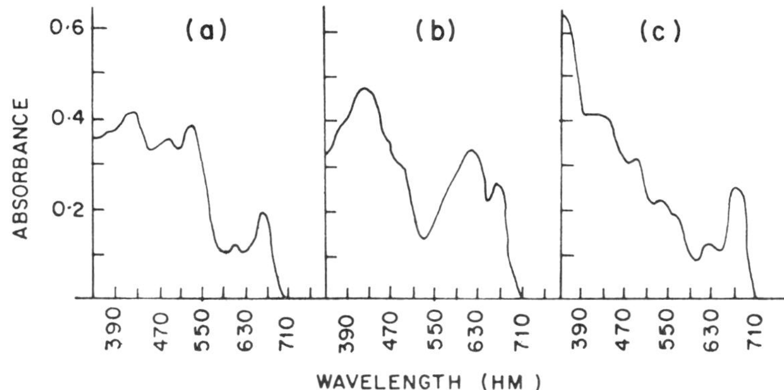

FIG. 5. The absorbance spectra of some blue-green algae, from Yentsch and Yentsch (1982). (a) Phycoerythrin containing picoplankton *Synechococcus* sp. "DC-2". (b) Phycocyanin containing picoplankton *Synechococcus* sp. "Syn". (c) Large filamentous algae *Oscillatoria erythraea*. Spectra (a) and (b) are based on laboratory cultures, while (c) is from a natural seawater sample.

is available. But there are certain problems that need to be solved before we can ascertain whether this is possible or not:

(i) The absorption spectra of blue-greens show great diversity, depending on the type of biliproteins present, and their relative importance. This variability may be attributed to some genetic differences, as well as to chromatic adaptation of cyanobacteria to the light and nutrient environment (Lazzara 1983; Prezelin and Boczar 1986). Spectral signatures for remote sensing purposes will therefore have to be carefully selected, based on the forms these spectra are likely to take *in situ*.

(ii) The *in situ* absorption spectra from surface waters of various marine regions studied by Prieur and Sathyendranath (1981) did not show clear traces of biliprotein absorption. Considering the ubiquitous nature of cyanobacteria, this can probably be explained by the fact that, in mixed populations, the differences in spectral signatures would tend to get smoothed out. Also, as noted by Prezelin and Boczar (1986), the biliproteins in surface waters may be 5 to 6 times less than in subsurface chlorophyll maxima.

(iii) Blooms of non-picoplanktonic blue-green algae are known to occur in certain areas of the world oceans (see Carpenter 1983; Devassy et al. 1978; Ulbricht 1981). Distinguishing the picoplanktonic blue-greens from their larger counterparts or the cryptophytes and rhodophytes may require some prior knowledge on the type of phytoplankton commonly occurring in the study area during different seasons.

In fact, a systematic study needs to be undertaken, to evaluate the possibilities of extracting pigment composition from ocean colour, if complete spectral information is available. Some possibilities for doing this should exist, at least for case 1 waters. A difficulty in undertaking such a study has been the lack of a data bank of spectral reflectance, together with complementary data on concentrations of all pigments, necessary for testing various retrieval techniques. Even a theoretical approach would require reliable values of the coefficient a^*_{soli} in Eq. 13. Very little information is available on this parameter, beyond the preliminary results presented by Sathyendranath et al. (1987). Certain difficulties in such a study could arise because the mixing of different classes of phytoplankton in the natural environment would tend to smooth out the differences in spectral form, necessitating working with minor changes around a mean value. This problem would be less acute in single-species bloom conditions. Furthermore, the nonlinearity of the system, coupled with the fact that the spectral signatures of the intervening variables often show similar characteristics, would make resolution of the signal into each of their components difficult. In case 2 waters, the problem is further complicated by noise from suspended sediments and dissolved organic matter, and at a first glance, the problem appears to be intractable for such waters.

Some recent results indicate the possibility of estimating primary productivity, through remote sensing. Smith (1981) and Platt and Herman (1983) showed that satellite-derived chlorophyll concentration is correlated with the total column chlorophyll concentration and the total column productivity. However, the constant of proportionality is site-specific, and for some ecosystems (eg. Peruvian shelf waters), the proportionality may not even be a reasonable assumption (Platt and Herman 1983). Platt (1986) has shown analytically that the column production, normalized by the column pigment concentration, should be a linear function of the surface irradiance $E(0)$. i.e.:

$$(18) \qquad \int P \, dz \ / \int C \, dz \ = \ \psi E(0)$$

where P is the primary production. $E(0)$ is the total photosynthetically available radiation (integral over 400 to 700 nm). He has also shown that, the slope ψ can be proportional to α, the initial slope of the photosynthesis-light curve for phytoplankton. Empirical data from various sources show ψ to be consistent over a factor of two $(0.31–0.66$ g Carbon (g Chl$a$)$^{-1}$ m^2/Einstein). The quantity $E(O)$ can be estimated from satellite data (see Gauthier et al. 1980; Gauthier 1982; Gauthier and Katsaros 1984). Initial evidence (Platt et al. 1983; Smith et al. 1985) indicates that the picoplankton α values are significantly different from those of larger phytoplankton. Therefore, the relative importance of picoplankton in a given area could have a definite role in determining the slope ψ in Eq. 18.

Passive Fluorescence Technique

Determination of chlorophyll-a via its *in vivo* fluorescence peak in the red (at around 685 nm) from passive, remotely sensed signals was initiated by Neville and Gower (1977) and Gower (1980). This may be considered an extension of the *in vivo* technique for *in situ* monitoring of chlorophyll-a introduced by Lorenzen (1966), with the following main differences:

1) Excitation for the *in situ* experiments is provided by a source emitting known flux at a given wavelength, while in passive remote sensing, sun light is the source.

2) *In situ* fluorescence is measured at a fixed wavelength, while in remote sensing, it is necessary to measure the radiation at 3 or 4 wavelengths at least, to provide correction for background noise.

3) *In situ* fluorescence is measured from a small, known volume of water, while in passive remote sensing, the total water-leaving fluorescence signal is the measured quantity.

Good correlation has been observed between the height of the fluorescence peak and chlorophyll concentration (Neville and Gower 1977; Gower 1980; Doerffer 1981; Gower and Borstad 1981). But, as in the *in situ* experiments, in remote sensing too, the fluorescence efficiency per unit chlorophyll-a, F_c, has been found to vary from region to region. Gower and Borstad (1981) report variations by a factor of 2 or 3 in remote sensing experiments from aircraft.

Various studies which have thrown light on the causes of variation of *in vivo* fluorescence efficiency of phytoplankton have been reviewed by Sathyendranath et al. (1982). The main factors which are relevant to passive remote sensing experiments may be summarized as follows:

1) *In vivo* fluorescence is a complementary function of photosynthesis, and so F_c will decrease as photosynthetic rate increases.

2) F_c can be subject to a circadian rhythm, reaching a maximum at night and minimum at around mid-day (which is in concordance with the theory of *in vivo* fluorescence as a complementary function of photosynthesis). Some results indicte that the circadian rhythm is controlled by a biological clock rather than by light availability.

3) High light intensities may inhibit fluorescence efficiency.

4) Nutrient stress can introduce changes in fluorescence efficiency.

5) F_c may also vary with phytoplankton species (the species effect may in turn be related to changes in absorption efficiency).

6) The wavelength of the peak value of fluorescence may shift a few nanometers around a mean value of 685 nm. This can result in an apparent change in F_c if it is measured at a fixed wavelength. Phycoerythrin fluorescence from cyanobacteria has a maximum at around 610 nm, and phycocyanin at about 660 nm (Yentsch and Yentsch 1984).

7) F_c varies with change in wavelength of excitation, and the excitation spectrum differs with different algal types.

8) Background fluorescence from other substances may be significant, and failure to apply suitable corrections can introduce apparent changes in F_c.

9) At high chlorophyll-*a* concentrations, re-absorption of fluoresced light by the other cells can introduce a quenching effect and nonlinearity in the fluorescence height and chlorophyll-*a* relationship.

This variability implies that, as in the case of *in situ* continuous monitoring of phytoplankton fluorescence, in remote sensing too, discrete calibration points would be necessary to convert the fluorescence signal into chlorophyll-*a* concentration. In the absence of these calibration points, passive fluorescence technique can still be used as a very useful search tool for locating phytoplankton fronts, gradients and patchiness.

A priori, it would appear difficult to see how the passive fluorescence technique can yield information on the phytoplankton type. However, laboratory results indicate that fluorescence from phycocyanin-containing cyanobacteria peaks at 660 nm (Yentsch and Yentsch 1984), while chlorophyll-*a* fluorescence peak is normally observed at 685 nm. Consequently, in natural populations with phycocyanin-containing cells, one can expect the red fluorescence peak to occur somewhere in between 660 and 685 nm. The position of the red fluorescence peak could therefore be an indicator of the presence of these blue-greens. But this has yet to be confirmed by observations.

Methods are now emerging for primary productivity estimates from these fluorescence signals. From the theory postulating fluorescence as a complementary function of photosynthesis, it follows that fluorescence efficiency would decrease with increasing photosynthetic rate (other factors being negligible). Topliss and Platt (1986) tested this hypothesis in 2 different types of marine environments and found it to hold true generally. To calculate fluorescence efficiency from fluorescence emission, we also have to know the light absorbed by phytoplankton. This can be obtained through surface irradiance, pigment concentration and attenuation coefficient derived from passive radiometric techniques. (Total attenuation coefficient at different wavelengths can be determined from changes in ocean colour. See Austin and Petzold, 1981. The algorithm has the same form as Eq. 17. In case 1 waters, the residual attenuation after substracting the contribution of pure sea water, may be attributed to phytoplankton, as a first approximation. In case 2 waters, the attenuation due to suspended sediments and dissolved organic matter would also have to be substracted from the total.) Topliss and Platt (1986) have shown the fluorescence efficiency to be inversely related to α, the initial slope of the photosynthesis-light curve. It may be noted that α derived in this manner may be used to fix the coefficient ψ in Eq. 18. This holds out the promises of a new, non-destructive technique for measuring primary production rates either *in situ* or by remote sensing.

So far we have only considered passive fluorescence from a low flying aircraft. In the case of a high flying aircraft or spacecraft, certain additional factors would have to be taken into account. Gower and Borstad (1981) have demonstrated that, with increasing altitude of aircraft, the fluorescence maximum shifts from 685 nm towards 675 nm, due to the effect of the oxygen absorption band at 688 nm. It therefore follows that from a high aircraft, chlorophyll-*a* fluorescence is best measured at 675 nm. They have also pointed out that the increase in reflectance at 685 nm is of the order of 0.02% with an increase of one mg/m^3 in chlorophyll-*a* concentration. It would be still smaller at 675 nm. As this signal will be superimposed on a background noise of the order of 5% (Sathyendranath et al. 1982), atmospheric correction becomes a very critical factor. However, atmospheric correction is easier in the red than at shorter wavelengths. Still, this demands a highly sensitive detector which would probably limit applications of this technique in waters of low chlorophyll-*a* concentrations.

Laser Fluorometry

Airborne lasers for ocean remote sensing have been described by Hoge and Swift (1981), Yentsch (1983) and Yentsch and Yentsch (1984). In the version described by Yentsch and Yentsch (1984), excitation is provided by short pulses at 532 nm with an Nd:Yag (garnet crystal) laser. The return signals are measured by 40 photomultiplier tubes. In the LIDAR mode, the 40 channels are tuned to the same wavelength, but are time gated so that the signal at each channel represents a different depth. In the fluorosensor mode, the 40 channels are set to different wavelengths, to measure the upwelling spectrum. A variation of this intrument allows excitation at 427 nm also, alternating with the pulses at 532 nm. In between pulses, the 40 receivers can act as a passive spectral radiometer.

Exton et al. (1983a) describe the return signal in terms of three components: (1) The elastically backscattered component at the excitation wavelength, (2) the spectrally shifted Raman scatter signal due to non-elastic scattering by water molecules, and (3) the fluorescence signal. For excitation at 532 nm, the predominant return signals, in addition to the backscattered light, are at 580 nm (due to phycoerythrin fluorescence), at 645 nm (Raman scattering) and at 685 nm (chlorophyll-*a* fluorescence).

The differential power dP received by the receptor due to fluorescence from a small volume of water at depth z can be expressed as follows (expression modified from Exton et al. 1983a):

$$(19) \qquad dP = \frac{\mathrm{ACF}_c(L) \exp\left[-(K(L) + K(U))z\right]dz}{(z + mZ)^2}$$

where $K(L)$ and $K(U)$ are the effective attenuation coefficients at laser and upwelling wavelengths, respectively, z is the depth within the water column, Z is height of the sensor above the water column, m is the refractive index of water, C is the concentration of fluorescing pigment and $F_c(L)$ is the fluorescence efficiency at the laser excitation wavelength. The instrumental parameters and atmospheric transmission terms are contained in A.

In the case of Raman scattered light, the equation remains effectively the same, except that the CF_c term has to be replaced by the product of N_R and σ_R representing the number of water molecules accessed by the laser, and the Raman scattering cross sections, respectively. The elastically scattered light will consist of components from molecular scattering by pure seawater and the Mie scattering component from particles. For the polidisperse suspension common to seawater, the Mie scattering term will be an integral of the form $N_M \sigma_M \, dM$ where N_M is the number of particles with scattering cross section σ_M.

The water Raman signal is a measure of the loss of signal due to attenuation, and thus provides an internal calibration. The fluorescence and scattering signals are usually normalized by the Raman signal to correct for changes in water transmissivity.

The backscattered signal at the excitation wavelength, similarly normalized by the Raman signal, can be a rough measure of the seston concentration.

Laser techniques provide certain advantages over the passive techniques: (1) They can pick up phycoerythrin fluorescence in addition to chlorophyll-*a* fluorescence. (2) They can monitor depth-wise variations in concentrations (subject to limitations due to loss of energy while penetrating and escaping the water column. This aspect is discussd further in the next section).

Laboratory results indicate possibilities of further improvements in detection capabilities. Brown et al. (1977) studied variation in fluorescence intensity at 685 nm with variation in excitation wavelength, and found that the excitation spectra were significantly different for the four main phytoplankton taxonomic groups. They used excitation at 4 wavelengths (454, 539, 598 and 617 nm) using a tuned dye laser, and the fluorescence response at 685 nm in each case. In the case of a phytoplankton population comprising a mixture of all four groups, the term $CF_c(L)$ in Eq. 19 has to be replaced by a sum $C_1F_{c1}(L_i) + \ldots\ldots + C_4F_{c4}(L_i)$ where C_j represents the concentration of chlorophyll-*a* belonging to phytoplankton of group j, and L_i is the excitation wavelength. By solving the four simultaneous equations resulting from excitation at four wavelengths, they found that they could isolate the concentration of each group of phytoplankton in a mixture. Obviously, this method ignores all the causes of variation in fluorescence efficiency listed in the previous section, other than the species effect. Further studies in this direction are needed to determine the extent of the validity of these results for *in situ* conditions.

Active remote sensing offers the best possibility at present to monitor blue-green algae by remote sensing. Biliproteins of different types of phytoplankton have characteristic differences in their absorption and fluorescence signals (see Prezelin and Boczar 1986). For example, Exton et al. (1983b) report that cyanobacterial phycoerythrin fluorescence peaks at 576 nm, while that of cryptophytes peaks at 586 nm. A Lidar with high spectral resolution could use such differences to distinguish different types of phytoplankton. Though there are some problems in measuring the fluorescence height accurately due to overlap of neighbouring peaks in the return signal (Hoge and Swift 1981) and due to background noise from fluorescence by dissolved organic matter (Exton et al. 1983a), these problems are easily overcome. Once the fluorescence height is correctly determined, the algorithm for conversion to concentration would be a simple linear relationship. But the problem lies elsewhere. Lidar techniques have only been successfully carried out from rather large airplanes flying at very low altitudes. The limiting factors that preclude higher altitude flights and use of smaller aircraft, are the prohibitive power requirements for the laser and the bulky instrumentation (Hoge and Swift 1981 used laser with 3 MW peak power from an altitude of 150 m). From Eq. 19, it is evident that a 10 fold increase in aircraft altitude would call for a 100 fold increase in laser power (neglecting increased atmospheric attenuation), to obtain the same performance, if the instrumental parameters remain the same. So it appears that until some technological break-through lowers the operational costs, it may not be possible to use laser fluorescence remote sensing techniques extensively and routinely.

Limitations to the Depths Accessible by Remote Sensing

Gordon and McCluney (1975) have shown theoretically that approximately 90% of the upwelling light leaving the water surface emanates from a water column of depth z_{90} (called penetration depth) at which the incident light reduces to $1/e$ of its surface value. If we denote the diffuse attenuation coefficient for downwelling irradiance by K, then,

$$z_{90} = 1/K.$$

As this is about 1/4.6 of the euphotic depth, this implies that the top 22% of the total euphotic zone accounts for 90% of the signal for the remote sensor. As K is wavelength dependent, the depth of water column sampled will be different for different wavelengths. Also, due to attenuation of light during penetration and escaping, the "effective concentration" as seen by the remote sensor would be a weighted average over the water column of depth $1/K$. Gordon and Clark (1980) defined the effective concentration X_i as:

$$X_i = \int_o^{1/K} X_i(z)f(z)dz \Big/ \int_o^{1/K} f(z)dz$$

where $\qquad f(z) = \exp - \int_o^z 2K(z')dz'$

and $X_i(z)$ is the concentration of the substance at depth z.

Sathyendranath (1981) and Sathyendranath et al. (1983) studied theoretically the case of remote sensing in waters where the chlorophyll-a concentration is variable with depth. Four non-uniform depth distributions of chlorophyll, each with an average chlorophyll-a concentration of 1 mg/m³ were considered. The results indicated that the maximum z_{90} values were around 8 m in the 500 to 550 nm range, while in the red (560 to 700 nm range), the penetration depth was only around 2 m (see Fig. 6).

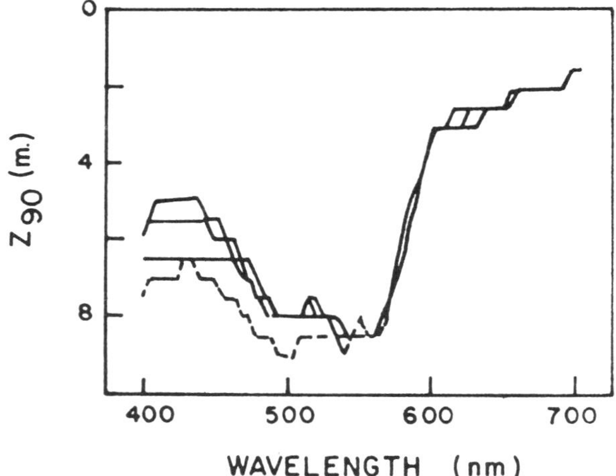

FIG. 6. Theoretically calculated wavelength dependence of penetration depth for 4 vertical distributions of chlorophyll-a (from Sathyendranath 1981). The curve in dashed line represents a uniform distribution of 1 mg/m³ of chlorophyll-a in the top 10 m. For the other 3 curves, mean chlorophyll concentration over the top 10 m remains the same, but vertical distributions are non-uniform.

The maximum observed efficiency factors for chlorophyll absorption were used in the study; using lower efficiency factors would have the effect of increasing z_{90} for the same chlorophyll-a concentrations. In this case too, the penetration depth is smaller for the red than the blue. Results of Smith (1981), based on similar calculations show that, when C varies from 0.01 to 10 mg · m⁻³, z_{90} varies from about 58 to 2 m at 443 nm, and from about 15 to 4 m at 550 nm. At 670 nm, z_{90} is insensitive to changes in chlorophyll concentration (the dominant factor here is the absorption due to pure seawater itself).

Consequently, the fluorescence technique samples a much smaller volume of water than the ocean-colour based technique which uses the blue and green part of the spectrum. Comparison of the chlorophyll values yielded by the two methods could therefore give an indication of the vertical distribution of chlorophyll. Though laser techniques permit in principle the depth profiling of chlorophyll fluorescence, in practice, the maximum sampling depths have been limited to about 4 m (Yentsch and Yentsch 1984). Cyanobacteria are known to adapt well to very low light levels, and survive well in deep waters. At the present state of the art, these picoplanktonic forms will go undetected by remote sensing techniques.

In case 2 waters, the presence of suspended sediments and dissolved organic matter would tend to increase the attenuation coefficient of the water, thereby reducing the penetration depth still further.

However, statistical results of the sort derived by Lorenzen (1970), Smith (1981) and Platt and Herman (1983), which link surface chlorophyll and satellite-derived chlorophyll to total column chlorophyll and column primary production, can be used to map these additional variables using satellite data.

Summary and Conclusion

In this paper I have evaluated the implications of picoplankton in remote sensing as it stands today, with the hope that it will help us to be suitably cautious when interpreting remotely sensed data, as well as encourage further research in this area to improve existing techniques.

Where passive radiometric techniques based on changes in ocean colour are concerned, there seems to be no doubt that the picoplankton will leave their imprint on the water-leaving radiance, which serves as signal for remote sensing. Their small size favours higher specific absorption, and so their effect on the signal may be more important than warranted by considerations of the concentrations alone. The special pigment composition of the blue-green algae which makes their absorption spectra very distinct from the other types of phytoplankton is also an important factor to be considered while interpreting spectral radiometric data from waters rich in this type of picoplankton. Since the relative abundance of picoplankton appears to be higher in oligotrophic waters, a suitable method is needed that can monitor small changes in their concentration on a synoptic scale. Passive radiometric methods seem to be suitable for such applications. The possibility of employing variations in spectral signatures of different types of phytoplankton to separate the signals from various taxonomic groups of phytoplankton remains to be explored. This would require working with much higher spectral and radiometric resolution than is currently available.

Active fluorescence techniques have demonstrated the capability of picking up the phycoerythrin fluorescence signal, thereby making it possible to monitor a blue-green fraction of picoplankton at least, if not all the picoplankton. In this case, however, investigators will have reliance on alternate supporting data to ensure that the phycoerythrin being monitored is in fact contained in picoplankton, and not the larger blue-greens such as *Oscillatoria*, which appear in bloom conditions in the tropical areas, or even cryptomonads or rhodophytes.

It is clear that the ocean-colour technique, and the emerging active and passive fluorescence techniques do not bring redundant information. By combining the methods suitably, it is becoming possible to obtain more accurate estimates, particularly in the case of primary productivity. Fluorescence technique seems to be best suited for monitoring area of high phytoplankton concentration, while the ocean colour method is more sensitive at lower pigment concentrations. Lastly, the depth of the water column sampled by the two techniques differ significantly, and this fact could be used for deriving some information on the vertical distribution of pigments.

While it appears easier to understand the role of picoplankton in affecting the remote signal, the inverse problem of inferring direct and specific information on picoplankton from the remote signal poses greater problems. There appears to be no optical method at present which would allow us to derive information on the size distribution of suspended material by remote sensing. So information pertaining to picoplankton has to be obtained indirectly, such as through the knowledge that a large fraction of picoplankton belongs to the group of cyanobacteria. As mentioned earlier, this leaves room for doubt, which can only be clarified by discrete *in situ* observations in addition to the remotely obtained data.

The power of remote sensing lies in its capability to obtain synoptic scale pictures of the ocean, and at time scales comparable to the rate of processes related to phytoplankton. The trade off is a decrease in the accuracy of estimates as compared to laboratory or *in situ* techniques. However, a point worth noting, which is often ignored, is that, when ship data are used for studying large areas, errors are introduced due to spatial variabilities in the fields, comparable to those inherent in satellite derived data (Platt and Herman 1983). At any rate, the accuracies can be increased considerably if *in situ* data points are used to validate and calibrate the output from remote sensing techniques.

Acknowledgements

The author wishes to thank Dr. H. N. Siddiquie, Director, National Institute of Oceanography, and Dr. J. S. Sastry, Head, Physical Oceanography Division, for making available the facilities, and Dr. André Morel and Dr. John T. O. Kirk for their comments on the manuscript.

Appendix 1. Radiometric Quantities.

Symbol	Quantity	Definition, Remarks	Unit
Φ	Radiant flux	Energy per unit time	Watt.
I	Radiant intensity	Flux per unit angle. $I = d\Phi/d\omega$.	Watt/steradian.
E	Irradiance	Flux per unit surface. $E = d\Phi/dA$. E_d and E_u are irradiances with respect to a surface facing upward, or downward, respectively.	Watt/m^2.
L	Radiance	Flux per unit angle per unit projected area of a surface. $L = d\Phi/dA \cos d\omega$. $E_d = \int_{\phi=0}^{2\pi} \int_{\theta=0}^{\pi/2} L(\theta, \phi) \cos\theta \sin\theta \, d\theta \, d\phi$. $E_u = \int_{\phi=0}^{2\pi} \int_{\theta=\pi}^{\pi/2} L(\theta, \phi) \cos\theta \sin\theta \, d\theta \, d\phi$.	Watt/m^2/sr.
a	Absoprtion coefficient	The energy $d\Phi$ absorbed by a medium when a parallel, monochromatic flux Φ passes through distance dr in the medium, is given by $d\Phi = a\Phi dr$. Integration over distance r yields $\Phi = \Phi_o \exp(-ar)$, where Φ_o is the incident flux.	m^{-1}
$\beta(\theta)$	Volume scattering function	Radiant intensity scattered in the direction θ with respect to the incident direction, per unit volume and unit irradiance. $\beta(\theta) = dI(\theta)/Edv$.	m^{-1}
b	Scattering coefficient	Total scattered energy per unit volume and unit irradiance. $b = \int_0^{2\pi} \beta(\theta) \, d\omega = 2\pi \int_0^{\pi} \beta(\theta) \sin\theta \, d\theta$.	m^{-1}
b_b	Backscattering coefficient	Total scattered energy within the angles $\pi/2$ to π. $b_b = 2\pi \int_{\pi/2}^{\pi} \beta(\theta) \sin\theta \, d\theta$.	m^{-1}
$\tilde{b}_b$	Backscattering ratio	$\tilde{b}_b = b_b/b$.	1
F	Forward scattering ratio	$[2\pi \int_0^{\pi/2} \beta(\theta) \sin\theta \, d\theta]/b$.	1
c	Attenuation coefficient	$c = a + b$	m^{-1}

m	Refractive index	Ratio of phase velocity of radiant energy in free space to that of the same energy in the given medium. If i is the angle of incidence, and j the angle of refraction, then $m = \sin i / \sin j$.	1
K	Irradiance attenuation coefficient	Rate of decrease of irradiance, defined by: $K = -d \ln E/dz$. K_d and K_u define attenuation for downwelling and upwelling irradiances respectively.	m^{-1}
R	Reflectance	Ratio of upwelling to downwelling irradiance at a given depth. $R(z) = E_u(z)/E_d(z)$.	1
Q_a	Efficiency factor for absoprtion by a particle	Ratio of flux absorbed by a particle to flux incident on it.	1
T	Optical thickness	Product of attenuation coefficient and distance.	1
ω	Single scattering albedo	Ratio of scattering coefficient to attenuation coefficient.	$\omega = b/c$

Here, dr, dA, $d\omega$, and dv refer to infinitesimally small distance, area, solid angle, and volume, respectively.

References

AUSTIN, R. W. 1974. The remote sensing of spectral radiance from below the ocean surface, p. 317–344. *In* N. G. Jerlov and E. Steemann Nielson [ed.] Optical aspects of oceanography. Academic, London.

———— 1980. Gulf of Mexico, ocean color surface-truth measurements. Boundary-Layer Meteorol. 18: 269–285.

AUSTIN, R. W., AND T. J. PETZOLD. 1981. Determination of diffuse attenuation coefficient of seawater using the Coastal Zone Color Scanner, p. 239–256. *In* J. F. R. Gower [ed.] Oceanography from Space. Plenum Press, New York and London.

BRICAUD, A., A. MOREL, AND L. PRIEUR. 1981. Absorption by dissolved organic matter of the sea (yellow substance) in the UV and visible domains. Limnol. Oceanogr. 26: 43–53.

———— 1983. Optical efficiency factors of some phytoplankters. Limnol. Oceanogr. 28: 816–832.

BROWN, JR., C. A., F. H. FARMER, O. JARRETT JR., AND W. L. STATON. 1977. Laboratory studies of in vivo fluorescence of phytoplankton. Proc. Fourth Joint Conference on Sensing of Environmental Polutants, New Orleans, Louisiana.

BUKATA, R. P., J. E. BURTON, J. H. JEROME, S. C. JAIN, AND H. H. ZWICK. 1981. Optical water quality model of Lake Ontario. 2: Determination of chlorophyll-a and suspended mineral concentrations of natural waters from submersible and low altitude optical sensors. Appl. Opt. 20: 1704–1714.

CARDER, K. L., AND R. G. STEWARD. 1985. A remote-sensing reflectance model of a red-tide dinoflagellate off east Florida. Limnol. Oceanogr. 30: 286–298.

CARPENTER, E. J. 1983. Physiology and ecology of marine planktonic *Oscillatoria* (*Trichodes-mium*). Mar. Biol. Lett. 4: 69–85.

DEVASSY, V. P., P. M. A. BHATTATHIRI, AND S. Z. QASIM. 1978. *Trichodesmium* phenomenon. Ind. J. Mar. Sci. 7: 168–186.

DOERFFER, R. 1981. Factor analysis in ocean colour interpretation, p. 339–345. *In* J. F. R. Gower [ed.] Oceanography from space. Plenum, New York, NY.

DUYSENS, L. M. N. 1956. The flattening of the absorption spectra of suspensions as compared to that of solutions. Biochem. Biophys. Acta 19: 1–12.

EXTON, R. J., W. M. HOUGHTON, W. ESAIAS, R. C. HARRISS, F. H. FARMER, AND H. H. WHITE. 1983a. Laboratory analysis of techniques for remote sensing of estuarine parameters using laser excitation. Appl. Opt. 22: 54–64.

EXTON, R. J., W. M. HOUGHTON, AND W. ESAIAS. 1983b. Spectral differences and temporal stability of phycoerythrin fluorescence in estuarine and coastal waters due to the domination of labile cryptophytes and stabile cyanobacteria. Limnol. Oceanogr. 28: 1225–1231.

GAUTHIER, C. 1982. Mesoscale insolation variability derived from satellite data. J. Appl. Meteorol. 21: 51–58.

GAUTHIER, C., G. DIAK, AND S. MASSE. 1980. A simple physical model to estimate incident solar radiation at the surface from GOES satellite data. J. Appl. Meteorol. 19: 1005–1012.

GAUTHIER, C., AND K. B. KATSAROS. 1984. Insolation during STREX: Comparisons between surface measurements and satellite estimates. J. Geophys. Res. 89: 11,779–11,788.

GORDON, H. R. 1978. Removal of atmospheric effects from satellite imagery of the oceans. Appl. Opt. 17: 1631–1636.

GORDON, H. R., AND D. K. CLARK. 1980. Remote sensing optical properties of a stratified ocean:

an improved interpretation. Appl. Opt. 20: 3428–3430.

———. 1981. Clear water radiances for atmospheric correction of Coastal Zone Color Scanner imagery. Appl. Opt. 20: 4175–4180.

GORDON, H. R., AND W. R. MCCLUNEY. 1975. Estimation of the depth of sunlight penetration in the sea for remote sensing. Appl. Opt. 14: 413–416.

GORDON, H. R., AND A. MOREL. 1983. Remote assessment of ocean colour for interpretation of satellite visible imagery. A review. Lecture notes on Coastal and Estuarine Studies, 4. Springer-Verlag, NY. 114 p.

GOWER, J. F. R. 1980. Observations of *in situ* fluorescence of chlorophyll-*a* in Saanich Inlet. Boundary-Layer Meteorol. 18: 235–245.

GOWER, J. F. R., AND G. BORSTAD. 1981. Use of the *in vivo* fluorescence line at 685 nm for remote sensing surveys of surface chlorophyll-*a*, p. 329–338. *In* J. F. R. Gower [ed.] Oceanography from Space. Plenum, New York, NY.

GUAN, F., J. PELAEZ, AND R. H. STEWART. 1985. The atmospheric correction and measurement of chlorophyll concentration using the coastal zone color scanner. Limnol. Oceanogr. 30: 273–285.

HERBLAND, A., A. LE BOUTEILLER, AND P. RAIMBAULT. 1985. Size structure of phytoplankton biomass in the equatorial Atlantic Ocean. Deep–Sea Res. 32A: 819–836.

HOGE, F. E., AND R. N. SWIFT. 1981. Airborne simultaneous spectroscopic detection of laser-induced water Raman backscatter and fluorescence from chlorophyll-*a* and other naturally occurring pigments. *Appl. Opt.* 20: 3197–3205.

JAIN, S. C., AND J. R. MILLER. 1976. Subsurface water parameters: optimization approach to their determination from remotely sensed water colour data. Appl. Opt. 15: 886–890.

JEFFREY, S. W. 1980. Algal pigment systems, p. 33–58. *In* P. G. Falkowski [ed.] Primary productivity in the sea. Plenum, New York, NY.

JOHNSON, P. W., AND J. MCN. SIEBURTH. 1979. Chroococcoid cyanobacteria in the sea: a ubiquitous and diverse phototrophic biomass. Limnol. Oceanogr. 24: 928–935.

JOINT, I. R., AND A. J. POMROY. 1983. Production of picoplankton and small nanoplankton in the Celtic Sea. Mar. Biol. 77: 19–27.

KIEFER, D. A., AND J. B. SOOHOO. 1982. Spectral absorption by marine particles of coastal waters of Baja California. Limnol. Oceanogr. 27(3): 492–499.

KIRK, J. T. O. 1983. Light and photosynthesis in aquatic ecosystems. Cambridge Univ. Press, Cambridge, 401 p.

LAZZARA, L. 1983. Réponse du phytoplancton (absorption spécifique et rendement quantique de la photosynthèse) aux modifications spectrales et énergétiques du flux photonique incident; mesures sur une diatomée en culture et sur des populations naturelles. Thèse 3ᵉ Cycle, Université Pierre et Marie Curie (Paris VI), 136 + 66 p.

LI, W. K. W., D. V. SUBBA RAO, W. G. HARRISON, J. C. SMITH, J. J. CULLEN, B. IRWIN, AND T. PLATT. 1983. Autotrophic picoplankton in the tropical ocean. *Science* NY 219: 292–295.

LORENZEN, C. J. 1966. A method for continuous measurement of *in vivo* chlorophyll concentration. Deep-Sea Res. 13: 223–227.

———. 1970. Surface chlorophyll as an index of the depth, chlorophyll content, and primary productivity of the euphotic layer. Limnol. Oceanogr. 15: 479–480.

MILLER, I. R., S. C. JAIN, N. T. O'NEILL, W. R., MCNEILL, AND K. P. B. THOMSON. 1977. Interpretation of airborne spectral reflectance measurements over Georgian Bay. Remote Sens. Environ. 6: 183–200.

MOREL, A. 1974. Optical properties of pure seawater, p. 1–24. *In* N. G. Jerlov and E. Steemann Nielsen [ed.] Optical aspects of oceanography. Academic, London.

———. 1980. In water and remote measurements of ocean color. Boundary-Layer Meteorol. 18: 177–201.

MOREL, A., AND A. BRICAUD. 1981a. Theoretical results concerning optics of phytoplankton with a special reference to remote sensing applications, p. 313–327. *In* J. F. R. Gower [ed.] Oceanography from space. Plenum, New York, NY.

———. 1981b. Theoretical results concerning light absorption in a discrete medium, and application to specific absorption of phytoplankton. Deep-Sea Res. 28A(11): 1375–1393.

MOREL, A., AND L. PRIEUR. 1978. Mesures par télédétection de la teneur de la mer en chlorophylle. Possibilités et limitations des méthodes. Publ. Sci. Techn. CNEXO, Actes Colloq., N° 5: 67–92.

NEVILLE, R. A., AND J. R. F. GOWER. 1977. Passive remote sensing of phytoplankton via chlorophyll *a* fluorescence. J. Geophys. Res. 82: 3487–3493.

PELEVIN, V. N. 1978. Estimation of the concentration of suspension and chlorophyll in the sea from the spectrum of outgoing radiation measured from a helicopter. Oceanology 18: 278–282.

PLATT, T. 1986. Primary production of the ocean water column as a function of surface light intensity: algorithms for remote sensing. Deep-Sea Res. 33(2): 149–163.

PLATT, T., AND A. W. HERMAN. 1983. Remote sensing of phytoplankton in the sea: surface-layer chlorophyll as an estimate of water-column chlorophyll and primary production. Int. J. Remote Sens. 4: 343–351.

PLATT, T., D. V. SUBBA RAO, AND B. IRWIN. 1983. Photosynthesis of picoplankton in the oligotrophic ocean. Nature 301: 702–704.

PREZELIN, B. B., AND B. A. BOCZAR. 1986. Molecular bases of cell absorption and fluorescence in phytoplankton: potential applications to studies in optical oceanography. *In* Progress in phycological research, 4, F. Round and D. Chapman [ed.], Elsevier Science Publishers, B. V.

PREZELIN, B. B., M. PUTT, AND H. E. GLOVER.

1986. Diurnal patterns in photosynthetic capacity and depth-dependent photosynthesis-irradiance relationships in *Synechococcus* spp. and larger phytoplankton in three water masses in the Northwest Atlantic Ocean. Mar. Biol. 91: 205–217.

PRIEUR, L., AND S. SATHYENDRANATH. 1981. An optical classification of coastal and oceanic waters based on the specific spectral absorption of phytoplankton pigments, dissolved organic matter and other particulate materials. Limnol. Oceanogr. 26: 671–689.

SATHYENDRANATH, S. 1981. Influence des substances en solution et en suspension dans les eaux de mer sur l'absorption et la réflectance. Modélisation et application à la télédétection. Doctorat 3ᵉ Cycle, Université Paris VI, 123 p.

SATHYENDRANATH, S., L. LAZZARA, AND L. PRIEUR. 1987. Variations in the spectral values of the specific absorption of phytoplankton. Limnol. Oceanogr. (In press)

SATHYENDRANATH, S., AND A. MOREL. 1983. Light emerging from the sea — interpretation and uses in remote sensing, p. 323–357. *In* A. P. Cracknell [ed.] Remote Sensing Applications in Marine Science and Technology. D. Reidel Publishing Company, Dordrecht, Holland.

SATHYENDRANATH, S., L. PRIEUR, AND A. MOREL. 1982. Interpretation of ocean colour data with special reference to OCM. Mid term report Contract ESA 4726-81-F-DD-SC, 61 p.

1983. Rapport complémentaire. Contrat ESA 4726-81-F.DD.SC. 29 p.

SMITH, J. C., T. PLATT, W. K. W. LI, E. P. HORNE, W. G. HARRISON, D. V. SUBBA RAO, AND B. D. IRWIN. 1985. Arctic marine photoautotrophic picoplankton. Marine Ecology — Progress Series 20: 207–220.

SMITH, R. C. 1981. Remote sensing and depth distribution of ocean chlorophyll. Mar. Ecol. Prog. Ser. 5: 359–361.

SMITH, R. C., AND K. S. BAKER. 1981. Optical properties of the clearest natural waters (200–800 nm). Appl. Opt. 20: 177–184.

SMITH, R. C., AND W. H. WILSON. 1981. Ship and satellite bio-optical research in the California Bight, p. 281–294. *In* J. F. R. Gower [ed.] Oceanography from Space. Plenum, New York, NY.

TAKAHASHI, M., AND P. K. BIENFANG. 1983. Size structure of phytoplankton biomass and photosynthesis in subtropical Hawaiian waters. Mar. Biol. 76: 203–211.

TOPLISS, B. J., AND T. PLATT. 1986. Passive fluorescence and photosynthesis in the ocean: implications for remote sensing. Deep-Sea Res. 33(7): 849–864.

ULBRICHT, K. A. 1981. Examples of application of digital image processing of remotely sensed phenomena, p. 285–294. *In* A. P. Cracknell [ed.] Remote sensing in meteorology, oceanography and hydrology. Ellis Horwood, Chichester.

WATERBURY, J. B., S. W. WATSON, R. R. L. GUILLARD, AND L. E. BRAND. 1979. Widespread occurrence of a unicellular marine planktonic cyanobacterium. Nature 277: 293–294.

YENTSCH, C. S. 1962. Measurement of visible light absorption by particulate matter in the ocean. Limnol. Oceanogr. 7: 207–217.

1983. Remote sensing of biological substances, p. 263–297. *In* A. P. Cracknell [ed.] Remote sensing applications in marine science and technology, D. Reidel Publishing Company, Dordrecht, Holland.

YENTSCH, C. S., AND C. M. YENTSCH. 1982. The attenuation of light by marine phytoplankton with specific reference to the absorption of near-UV radiation, p. 691–700. *In* J. Calkins [ed.] The role of solar ultraviolet radiation in marine ecosystems. Plenum, New York, NY.

1984. Emergence of optical instrumentation for measuring biological properties. Oceanogr. Mar. Biol. Ann. Rev. 22: 55–98.